OMEGA FATTY ACIDS IN BRAIN AND NEUROLOGICAL HEALTH

OMEGA FATTY ACIDS IN BRAIN AND NEUROLOGICAL HEALTH

SECOND EDITION

Edited by

Ronald Ross Watson

Victor R. Preedy

ELSEVIER

ACADEMIC PRESS

An imprint of Elsevier

Academic Press is an imprint of Elsevier
125 London Wall, London EC2Y 5AS, United Kingdom
525 B Street, Suite 1650, San Diego, CA 92101, United States
50 Hampshire Street, 5th Floor, Cambridge, MA 02139, United States
The Boulevard, Langford Lane, Kidlington, Oxford OX5 1GB, United Kingdom

Library of Congress Cataloging-in-Publication Data
A catalog record for this book is available from the Library of Congress

British Library Cataloguing-in-Publication Data
A catalogue record for this book is available from the British Library

ISBN 978-0-12-815238-6

For information on all Academic Press publications
visit our website at https://www.elsevier.com/books-and-journals

Publisher: Nikki Levy
Acquisition Editor: Natalie Farra
Editorial Project Manager: Michelle W. Fisher
Production Project Manager: Bharatwaj Varatharajan
Cover Designer: Matthew Limbert

Typeset by SPi Global, India

Contents

B

OMEGA FATTY ACIDS: BRAIN AND NEUROLOGICAL DEVELOPMENT

11. Neuroprotective Effects of Melatonin and Omega-3 on the Central Nervous System Exposed to Electromagnetic Fields in the Pre- and Postnatal Periods

KIYMET KÜBRA YURT, ELFIDE GIZEM KIVRAK, GAMZE ALTUN, ABIT AKTAS, ARIFE AHSEN KAPLAN, SÜLEYMAN KAPLAN

12. Maternal Omega-3 Status: Effects of Cross-Fostering on Offspring Brain Lipids and Hippocampal Neurogenesis

ELISE MAXIMIN, CHRISTINE HEBERDEN

13. Linking Omega-3 Fatty Acids and Depression

JUSTYNA GODOS, SABRINA CASTELLANO, FABIO GALVANO, GIUSEPPE GROSSO

14. Omega-3 Fatty Acids and the Cardiovascular System

ANDREW ELAGIZI, CARL J. LAVIE

C
DIET AND BRAIN DISEASE TREATMENT AND PREVENTION

15. Dietary Strategies and Supplements for the Prevention of Cognitive Decline and Alzheimer's Disease

LIGIA J. DOMINGUEZ, MARIO BARBAGALLO

21. Link Between Sugar Intake and Fatty Acid Metabolism in Brain

TOMISLAV MAŠEK

22. n-3 and n-6 Fatty Acids in Fish: A Focus on Non-Marine Species

SEMRA KAÇAR

23. Aspartame-Induced Neurodegeneration Versus Omega 3 Alleviation

EYAD M.T. ALI

24. DHA-Enriched Formulations as a Promising Strategy for the Treatment of Hypoxic/Ischemic Brain Injury

HUMBERTO MILANI, DANIELA VELASQUEZ OLIVEIRA

25. Alpha-Linolenic Acid and Dementia in Epidemiological Studies

KAZUMASA YAMAGISHI, UTAKO MURAI, RIE USUKI, HIROYASU ISO

26. DHA A Key Supplement for Neurologic Health—An Insight

VASANTI SUVARNA, VIKAS SINGH

27. Obesity, Cognitive Functioning, and Dementia: A Lifespan Prospective

FAYEZA S. AHMED, OLIVIA E. BOGUCKI, PETER J. DEARBORN, MERRILL F. ELIAS

Contributors

Fayeza S. Ahmed Department of Psychology, University of Maine, Orono, ME, United States

Abit Aktas Department of Histology and Embryology, Faculty of Veterinary, Istanbul University Cerrahpasa, Istanbul, Turkey

Eyad M.T. Ali Faculty of Medicine, Taibah University, Madinah, Saudi Arabia; Faculty of Medicine, Mansoura University, Mansoura, Egypt

Gamze Altun Department of Histology and Embryology, Faculty of Medicine, Ondokuz Mayis University, Samsun, Turkey

Belmira Lara da Silveira Andrade-da-Costa Laboratory of Neurophysiology, Department of Physiology and Pharmacology, Federal University of Pernambuco, Recife, Brazil

Ruth H. Asch Department of Psychiatry and Behavioral Neuroscience, University of Cincinnati College of Medicine, Cincinnati, OH, United States

Ricielle Lopes Augusto Laboratory of Neurophysiology, Department of Physiology and Pharmacology, Federal University of Pernambuco, Recife, Brazil

Graciela Avitia Universidad Autónoma de Ciudad Juárez, Cd. Juárez, Chih, Mexico

Mario Barbagallo Geriatric Unit, Department of Internal Medicine and Geriatrics, University of Palermo, Palermo, Italy

Olivia E. Bogucki Department of Psychology, University of Maine, Orono, ME, United States

John William Capps The University of Texas at El Paso, El Paso, TX, United States

Sabrina Castellano Department of Educational Sciences, University of Catania, Catania, Italy

Carol L. Cheatham University of North Carolina at Chapel Hill, Nutrition Research Institute, Kannapolis; University of North Carolina at Chapel Hill, Department of Psychology and Neuroscience, Chapel Hill, NC, United States

Andrea J. Connor School of Molecular Biosciences, Washington State University, Pullman, WA, United States

Lisette CPGM de Groot Division of Human Nutrition and Health, Wageningen University and Research, Wageningen, The Netherlands

Ricardo Augusto de Melo Reis Laboratory of Neurochemistry, Institute of Biophysics Carlos Chagas Filho, Federal University of Rio de Janeiro, Rio de Janeiro, Brazil

Raphael Fabricio de Souza Laboratory of Neurophysiology, Department of Physiology and Pharmacology, Federal University of Pernambuco, Recife; Department of Physical Education, Federal University of Sergipe, São Cristovam, Brazil

Peter J. Dearborn Department of Psychology; Graduate School of Biomedical Sciences and Engineering, University of Maine, Orono, ME, United States

Ligia J. Dominguez Geriatric Unit, Department of Internal Medicine and Geriatrics, University of Palermo, Palermo, Italy

Andrew Elagizi Department of Cardiovascular Diseases, John Ochsner Heart and Vascular Institute, Ochsner Clinical School-the University of Queensland School of Medicine, New Orleans, LA, United States

Merrill F. Elias Department of Psychology; Graduate School of Biomedical Sciences and Engineering, University of Maine, Orono, ME, United States

H.M. Francis Department of Psychology, Macquarie University, Sydney, NSW, Australia

Hércules Rezende Freitas Laboratory of Neurochemistry, Institute of Biophysics Carlos Chagas Filho, Federal University of Rio de Janeiro, Rio de Janeiro, Brazil

Fabio Galvano Department of Biomedical and Biotechnological Sciences, University of Catania, Catania, Italy

Grace E. Giles Cognitive Science Team, US Army Natick Soldier Research, Development, and Engineering Center, Natick; Center for Applied Brain and Cognitive Sciences, Medford, MA, United States

Justyna Godos Department of Biomedical and Biotechnological Sciences, University of Catania, Catania, Italy

Jacqueline F. Gould Healthy Mothers, Babies and Children, South Australian Health and Medical Research Institute, North Adelaide; School of Psychology, University of Adelaide, Adelaide, SA, Australia

Giuseppe Grosso Department of Biomedical and Biotechnological Sciences, University of Catania, Catania, Italy

Christine Heberden Micalis Institute INRA, Jouy-en-Josas, France

Alinny Rosendo Isaac Laboratory of Neurophysiology, Department of Physiology and Pharmacology, Federal University of Pernambuco, Recife; Laboratory of Neurochemistry, Institute of Biophysics Carlos Chagas Filho, Federal University of Rio de Janeiro, Rio de Janeiro, Brazil

Hiroyasu Iso Public Health, Department of Social Medicine, Osaka University Graduate School of Medicine, Suita, Japan

Luigi Janiri Institute of Psychiatry and Psychology, Department of Geriatrics, Neuroscience and Orthopedics, Fondazione Policlinico Universitario A. Gemelli IRCCS, Università Cattolica del Sacro Cuore, Rome, Italy

Corinne Joffre INRA, Nutrition et Neurobiologie Intégrée; Université de Bordeaux 2, Bordeaux, France

Semra Kaçar Mardin Artuklu University, Department of Nutrition and Dietetics, School of Health, Mardin, Turkey

Robin B. Kanarek Department of Psychology, Tufts University, Medford, MA, United States

Arife Ahsen Kaplan Department of Histology and Embryology, Faculty of Medicine, I·stanbul Medipol University, Istanbul, Turkey

Süleyman Kaplan Department of Histology and Embryology, Faculty of Medicine, Ondokuz Mayis University, Samsun, Turkey

Junhwan Kim Donald and Barbara Zucker School of Medicine at Hofstra/Northwell, Hempstead; Department of Emergency Medicine, The Feinstein Institute for Medical Research, Northwell Health System, Manhasset, NY, United States

Elfide Gizem Kıvrak Department of Histology and Embryology, Faculty of Medicine, Ondokuz Mayis University, Samsun, Turkey

M Yulianto Kurniawan Division of Human Nutrition and Health, Wageningen University and Research, Wageningen, The Netherlands

Cyrus E. Kuschner Donald and Barbara Zucker School of Medicine at Hofstra/Northwell, Hempstead, NY, United States

Carl J. Lavie Department of Cardiovascular Diseases, John Ochsner Heart and Vascular Institute, Ochsner Clinical School-the University of Queensland School of Medicine, New Orleans, LA, United States

Sophie Layé INRA, Nutrition et Neurobiologie Intégrée; Université de Bordeaux 2, Bordeaux, France

Rafael Longhi Department of Basic Health Sciences, Federal University of Health Sciences of Porto Alegre (UFCSPA), Porto Alegre, Brazil

Yolanda Loya Méndez Universidad Autónoma de Ciudad Juárez, Cd. Juárez, Chih, Mexico

Caroline R. Mahoney Cognitive Science Team, US Army Natick Soldier Research, Development, and Engineering Center, Natick; Center for Applied Brain and Cognitive Sciences, Medford, MA, United States

Gisele Gus Manfro Postgraduate Program in Neuroscience; Anxiety Disorders Program for Child and Adolescent Psychiatry (PROTAIA), Universidade Federal do Rio Grande do Sul (UFRGS), Porto Alegre, Brazil

Giuseppe Marano Institute of Psychiatry and Psychology, Department of Geriatrics, Neuroscience and Orthopedics, Fondazione Policlinico Universitario A. Gemelli IRCCS, Università Cattolica del Sacro Cuore, Rome, Italy

Tomislav Mašek Department of Animal Nutrition and Dietetics, Faculty of Veterinary Medicine, University of Zagreb, Zagreb, Croatia

Elise Maximin Micalis Institute INRA, Jouy-en-Josas, France

Marianna Mazza Institute of Psychiatry and Psychology, Department of Geriatrics, Neuroscience and Orthopedics, Fondazione Policlinico Universitario A. Gemelli IRCCS, Università Cattolica del Sacro Cuore, Rome, Italy

Salvatore Mazza Institute of Psychiatry and Psychology, Department of Geriatrics, Neuroscience and Orthopedics, Fondazione Policlinico Universitario A. Gemelli IRCCS, Università Cattolica del Sacro Cuore, Rome, Italy

Robert K. McNamara Department of Psychiatry and Behavioral Neuroscience, University of Cincinnati College of Medicine, Cincinnati, OH, United States

Humberto Milani Department of Pharmacology and Therapeutics, Heath Sciences Center, State University of Maringá, Maringá, Brazil

Robert R. Miller, Jr Hillsdale College, Biology Dept., Hillsdale, MI, United States

Utako Murai Department of Public Health Medicine, Faculty of Medicine, and Health Services Research and Development Center, University of Tsukuba, Tsukuba, Japan

Agnès Nadjar INRA, Nutrition et Neurobiologie Intégrée; Université de Bordeaux 2, Bordeaux, France

Daniela Velasquez Oliveira Department of Pharmacology and Therapeutics, Heath Sciences Center, State University of Maringá, Maringá, Brazil

Marissa Owen Michigan State University College of Veterinary Science, East Lansing, MI, United States

Georg Reiser Institute for Inflammation and Neurodegeneration (Neurobiochemistry), Medical Faculty, Otto-von-Guericke-University Magdeburg, Magdeburg, Germany

Charlotte Rey INRA, Nutrition et Neurobiologie Intégrée; Université de Bordeaux 2, Bordeaux; ITERG, Institut des corps gras, Pessac, France

Veronica Portillo Reyes Universidad Autónoma de Ciudad Juárez, Cd. Juárez, Chih, Mexico

Gilberto Reyes Leal Universidad Autónoma de Ciudad Juárez, Cd. Juárez, Chih, Mexico

Danitsa Marcos Rodrigues Postgraduate Program in Neuroscience; Anxiety Disorders Program for Child and Adolescent Psychiatry (PROTAIA), Universidade Federal do Rio Grande do Sul (UFRGS), Porto Alegre, Brazil

Erika Rodriguez-Guzman The University of Arizona, Tucson, AZ, United States

Christine Scanlan Hillsdale College, Biology Dept., Hillsdale, MI, United States

Peter Schönfeld Institute for Biochemistry and Zell Biology, Medical Faculty, Otto-von-Guericke-University Magdeburg, Magdeburg, Germany

Kelly W. Sheppard Nationwide Children's Hospital, Center for Biobehavioral Health; The Ohio State University, Department of Pediatrics, Columbus, OH, United States

Patrícia Pelufo Silveira Postgraduate Program in Neuroscience, Universidade Federal do Rio Grande do Sul (UFRGS), Porto Alegre, Brazil; Department of Psychiatry, Faculty of Medicine; Ludmer Centre for Neuroinformatics and Mental Health, Douglas Mental Health University Institute, McGill University; Sackler Program for Epigenetics & Psychobiology at McGill University, Montreal, QC, Canada

Vikas Singh Department of Pharmaceutical Chemistry and Quality Assurance, SVKM's Dr. Bhanuben Nanavati College of Pharmacy, Mithibai Campus, Mumbai, India

Lisa G. Smithers School of Public Health, University of Adelaide, North Terrace; Robinson Research Institute, School of Medicine, University of Adelaide, SA, Australia

R.J. Stevenson Department of Psychology, Macquarie University, Sydney, NSW, Australia

Vasanti Suvarna Department of Pharmaceutical Chemistry and Quality Assurance, SVKM's Dr. Bhanuben Nanavati College of Pharmacy, Mithibai Campus, Mumbai, India

Gianandrea Traversi Department of Science, University of Rome "Roma Tre"; Department of Research, Advanced Diagnostics and Technological Innovation, Regina Elena National Cancer Institute, Rome, Italy

Rie Usuki Public Health, Department of Social Medicine, Osaka University Graduate School of Medicine, Suita, Japan

Ondine van de Rest Division of Human Nutrition and Health, Wageningen University and Research, Wageningen, The Netherlands

Kathryn VanEgmond Penn State Health, Milton S. Hershey Medical Center, Hershey, PA, United States

Jennifer L. Watts School of Molecular Biosciences, Washington State University, Pullman, WA, United States

Taylor K. Wessel Pfizer Incorporated, Kalamazoo, MI, United States

Tsukasa Yagi Department of Emergency Medicine, The Feinstein Institute for Medical Research, Northwell Health System, Manhasset, NY, United States

Kazumasa Yamagishi Department of Public Health Medicine, Faculty of Medicine, and Health Services Research and Development Center, University of Tsukuba, Tsukuba, Japan

Kıymet Kübra Yurt Department of Histology and Embryology, Faculty of Medicine, Ondokuz Mayis University, Samsun, Turkey

Charles N. Zawatsky Hillsdale College, Biology Dept., Hillsdale, MI; National Institute on Alcohol Abuse and Alcoholism, Bethesda, MD, United States

PART A

OMEGA FATTY ACIDS AND BRAIN: AN OVERVIEW

CHAPTER

1

Essentiality of Omega-3 Polyunsaturated Fatty Acids for Mammalian Brain Development: A Translational Perspective

Robert K. McNamara, Ruth H. Asch

Department of Psychiatry and Behavioral Neuroscience, University of Cincinnati College of Medicine, Cincinnati, OH, United States

INTRODUCTION

Over the last 30 years there has been a substantial research devoted to understanding the role of omega-3 (*n*-3) fatty acids in normal brain development and function. The short-chain *n*-3 fatty acid α-linolenic acid (ALA) is regarded as nutritionally 'essential' because mammals cannot synthesize it endogenously and therefore require procurement through the diet. The long-chain *n*-3 (LC*n*-3) fatty acids products of ALA, including eicosapentaenoic acid (EPA, 20:5*n*-3) and docosahexaenoic acid (DHA, 22:6*n*-3), are not currently regarded as nutritionally 'essential' despite evidence that DHA biosynthesis from ALA and EPA is remarkably inefficient in humans.[1] Indeed, preformed DHA has been shown to be significantly more effective than ALA for increasing peripheral and central DHA levels in human mothers and infants[2–6] and nonhuman primates.[7–10] Therefore, developing a more comprehensive understanding of the requirement for LC*n*-3 fatty acids including DHA during human brain development has important implications for informing optimal maternal and postnatal DHA intake levels.

As DHA is the most abundant *n*-3 fatty acid in mammalian brain gray matter, and the brain DHA levels increase sharply during critical periods of perinatal mammalian brain development, the question of whether preformed DHA should be regarded as an essential nutrient for optimal human brain development requires investigation. Indeed, DHA-fortified

Omega Fatty Acids in Brain and Neurological Health
https://doi.org/10.1016/B978-0-12-815238-6.00001-8

maternal vitamins and infant formula are now widely available. The primary objective of this chapter is to provide an overview of the current translational evidence from rodent, nonhuman primate, and human developmental studies investigating the role of LC*n*-3 fatty acids and more specifically DHA in mammalian brain development.

RODENT NEURODEVELOPMENT

Unlike humans, rodents are very efficient at converting ALA to DHA, and ALA is the sole source of *n*-3 fatty acids in standard rodent laboratory chow. In the developing rat brain, frontal cortex DHA levels represent ~10% of total fatty acids at birth and increase to comprise ~18% of total fatty acids between birth and young adulthood (Fig. 1A). The rodent developmental studies have commonly utilized a 'perinatal' feeding paradigm in which female rats receive either ALA-fortified diet or ALA-free diet for a brief period

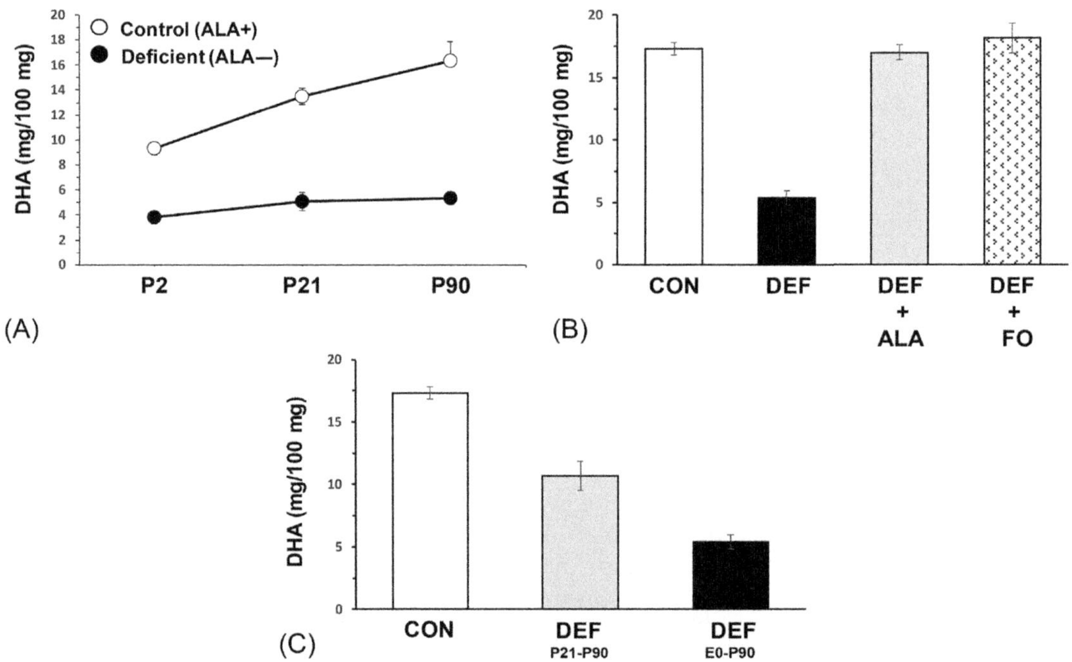

FIG. 1 (A) Maternal and postweaning dietary *n*-3 fatty acid deficiency (α-linolenic acid, ALA-) leads to significantly lower DHA composition in the neonatal (postnatal day 2, P2), prepubescent (P21) and adult (P90) rat frontal cortex compared with maternal and postweaning dietary *n*-3 fatty acid fortification (ALA+). (B) Low DHA levels observed in adult (P90) rat frontal cortex resulting from maternal and postnatal dietary *n*-3 fatty acid deficiency (DEF) are normalized when ALA-fortified or fish oil (FO)-fortified diet replaces ALA-free diet from P21 to P90. (C) Smaller frontal cortex DHA deficits are observed in adult rats subjected to dietary *n*-3 fatty acid deficiency during peri-adolescent development (P21-P90) compared with rats subjected to maternal and postnatal dietary *n*-3 fatty acid deficiency (E0-P90). Values are group mean composition (mg fatty acid/100 mg fatty acids) ± S.E.M.

(e.g., 1 month) prior to mating and during pregnancy and lactation and offspring are maintained on the maternal diet postweaning to adulthood. This paradigm leads to large (~70%) DHA deficits in adult offspring frontal cortex (Fig. 1A and B). The low DHA levels observed in adult rat frontal cortex resulting from perinatal dietary *n*-3 fatty acid deficiency can be normalized with ALA-fortified diet or fish oil-fortified diet initiated at weaning (Fig. 1B). The 'postweaning' or 'peri-adolescent' deficiency model involves placing offspring of mothers receiving an ALA-fortified diet on an ALA-free diet immediately following weaning through to adulthood. This paradigm leads to a more moderate cortical DHA deficit (~25%) compared with the perinatal model (Fig. 1C). In contrast, dietary fish oil supplementation during perinatal development is associated with greater increases (~10%) in the brain DHA levels compared with rats maintained on ALA-fortified diet (Fig. 2C). Therefore, maternal and postnatal dietary *n*-3 fatty acid intake is an important determinant of brain DHA accrual.

Rodent developmental studies have demonstrated that uncorrected deficits in DHA accrual during the perinatal period are associated with reductions in cortical neurogenesis,[11–13] neuroblast migration,[14] neuronal differentiation and arborization,[15] neurotrophic factor (NGF, BDNF) expression,[16,17] neurite outgrowth and synaptogenesis,[18–20] as well as synaptic pruning.[21] Although perinatal *n*-3 fatty acid deficiency is not associated with gross cortical lamination abnormalities or neuronal loss in the adult rat cortex, an enduring reduction in hippocampal neuronal size has been observed.[22] Perinatal deficits in DHA accrual are also associated with impaired brain glucose uptake and metabolism.[23–25] These and other findings provide proof-of-concept evidence that a robust reduction in cortical DHA accrual during development can adversely impact normal cortical maturation trajectories.

As elevated immune-inflammatory signaling during development has a long-term impact on brain functional integrity,[26–28] it is also relevant that LC*n*-3 fatty acids including EPA and DHA have anti-inflammatory actions, which in general oppose the pro-inflammatory signaling actions of omega-6 fatty acids including arachidonic acid.[29–33] Accordingly, *n*-3 fatty acid insufficiency augments peripheral and central immuno-inflammatory signaling.[34–38] In contrast, increasing dietary LC*n*-3 fatty acid status is protective against inflammation-induced white matter pathology,[39] neural apoptosis,[40,41] dendritic spine regression,[42] and histopathological features in the experimental autoimmune encephalomyelinitis model of multiple sclerosis.[43] Microglia are important regulators of central immune-inflammatory signaling, and increased DHA incorporation into microglia membranes during the development[44] can convert microglia from a pro-inflammatory phenotype associated with increased production and secretion of pro-inflammatory cytokines to a neurotrophic phenotype associated with increased BDNF biosynthesis.[34,35] In addition, DHA promotes microglia-induced phagocytosis,[34] which plays a role in synaptic pruning.[45,46] This evidence suggests that DHA-mediated regulation immune-inflammatory homeostasis may contribute to its effects on the brain development.

During development the maturation of cortical circuits requires progressive axon myelination. A role of DHA in myelination is supported by the observations that DHA accumulates in myelin during perinatal rat development[47,48] and promotes oligodendrocyte maturation.[49] Moreover, the central administration of either EPA or DHA in 2-day-old rat brain increased the expression of different myelin-related genes.[50] Using diffusion tensor imaging, we recently reported that deficits in DHA accrual during peri-adolescent development reduce indices of white matter microstructural integrity of the adult rat forebrain.[51] In adult rodents, increasing dietary LC*n*-3 fatty acid status has been shown to be protective against inflammation- and trauma-induced myelin pathology.[39,52,53] However, other studies have found that maternal fish oil supplementation, as well as maternal *n*-3 fatty acid deficiency during pregnancy and lactation transiently impair auditory brainstem responses in neonates which is consistent with dysmyelination.[47,54,55] Together these findings suggest that there are optimal DHA levels required for axonal white matter integrity and resilience over the course of perinatal development.

Synaptic Signaling Dynamics

In the developing rat brain DHA preferentially accumulates in neuronal synaptic membranes,[56] which play a critical role in modulating synaptic second messenger signaling.[57] For example, neuronal membrane concentrations of phosphatidylserine are selectively reduced in response to perinatal deficits in DHA accrual,[58,59] whereas perinatal fish oil supplementation increases membrane phosphatidylserine concentrations.[60] Membrane phosphatidylserine phospholipids modulate the activity of signal transduction proteins including protein kinase C (PKC),[61] which in turn regulate synaptic structural plasticity and stability.[62] In addition, perinatal *n*-3 fatty acid deficiency is associated with alterations in the subcellular distribution of PKC isozymes,[57] and several studies have shown that DHA directly modulates PKC activity.[63–66] These and other findings suggest that synaptic membrane DHA deficits may contribute to a dysregulation in the second messenger signal transduction processes involved in synaptic structural plasticity.

The membrane phosphatidylserine also anchors proteins to the membrane which in turn regulate neuronal phosphoinositide signaling cascades. For example, electrostatic binding anchors the myristoylated alanine-rich C kinase substrate (MARCKS) protein with the inner surface of the membrane,[67] and we previously reported that perinatal deficits in DHA accrual were associated with significant reductions in membrane-bound MARCKS expression.[57] Importantly, reductions in electrostatic binding of MARCKS with membrane phosphatidylserine upregulate the phospholipase Cβ_1-mediated phosphoinositide signaling cascades.[68] Indeed, we found that *n*-3 fatty acid deprivation during peri-adolescent development significantly increased inositol concentrations in the rat frontal cortex using proton magnetic resonance spectroscopy (^{1}H MRS) (Fig. 2). Together, these findings suggest that cortical DHA accrual has an important impact on cellular signal transduction processes which may be mediated in part through alterations in membrane phospholipid-protein interactions.

Synaptic phosphoinositide- and PKC-mediated signal transduction processes also regulate neurotransmitter vesicle trafficking and release dynamics.[69] Rodent developmental studies have demonstrated that deficits in DHA accrual are associated with abnormalities in the distribution neurotransmitter vesicles within presynaptic terminals[70,71] and the release dynamics of different neurotransmitter systems, including glutamate, acetylcholine, dopamine, and serotonin, which are reviewed in greater detail below.

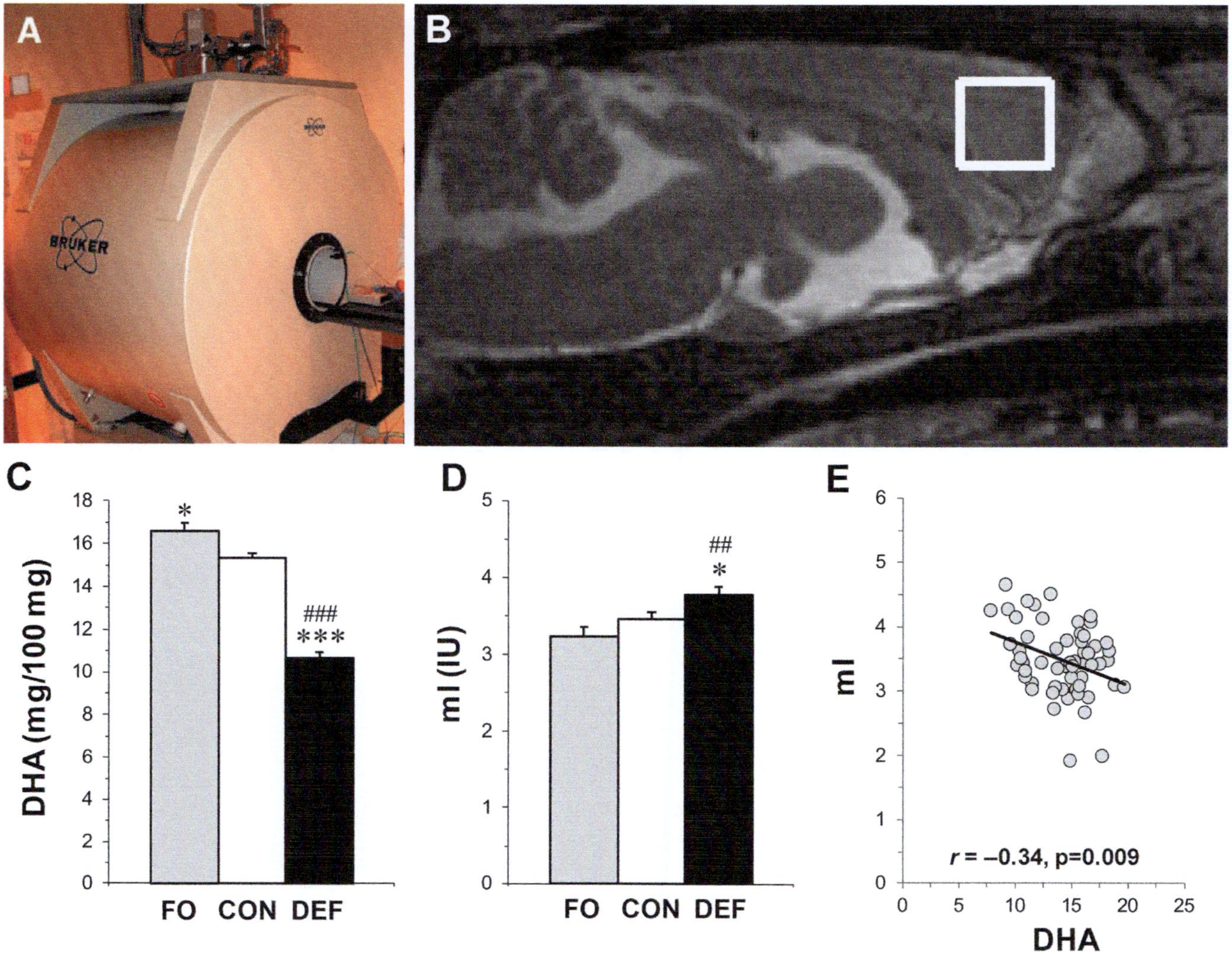

FIG. 2 Increasing DHA levels are associated with a decrease in the concentration of *myo*-inositol (mI), a product of the phosphoinositide signal transduction, in the rat frontal cortex in vivo. (A) The 7T Bruker Biospec Imaging System, (B) localization of voxel placement in the rat mPFC, (C) DHA composition in the postmortem PFC, and (D) mI concentrations in the PFC of adult rats maintained on the control diet (CON), *n*-3-free diet (DEF), and fish oil-fortified diet (FO)(n=20/group) during peri-adolescent development. (E) Linear correlations between DHA and mI levels in the PFC. Values are group mean composition (DHA, mg fatty acid/100mg fatty acids) or concentration (mI, institutional units, IU) ± S.E.M. *$P \leq 0.05$, ***$P \leq 0.0001$ vs CON, ##$P \leq 0.001$, ###$P \leq 0.0001$ vs FO.

Glutamate

Glutamate is the primary excitatory neurotransmitter in mammalian brain. A seminal study found that *n*-3 fatty acid deprivation during perinatal development was associated with marked reductions in glutamatergic synapses and impaired glutamate-mediates synaptic plasticity in the rat hippocampus.[18] Consistent with evidence for a role of glutamatergic synaptic plasticity in the formation of new synaptic connections, deficits in brain DHA accrual are also associated with learning and memory impairments across a variety of different sensory modalities.[72–75] It is also relevant that astrocytes play a key role in maintaining glutamate homeostasis,[76] and DHA is required for optimal astrocyte maturation[77–79] and also regulates astrocyte glutamate transporter activity and expression.[23,80–82] We recently reported that cortical DHA accrual during peri-adolescent development impacts glutamate concentrations in the rat frontal cortex.[83] Furthermore, DHA and its bioactive metabolites are protective against neuronal injury in response to glutamate-mediated excitoxicity,[84,85] and in animal models of ischemia[86,87] and traumatic brain injury.[52,88] Together, these findings suggest that DHA is an important modulator of glutamate homeostasis, glutamate synaptic plasticity and associated learning processes, as well as resilience against glutamate-mediated excitoxicity.

Acetylcholine

Acetylcholine plays a critical role in rodent neurocognitive processes including attention and associative learning. Rodent developmental studies have demonstrated that perinatal deficits in cortical DHA accrual are associated with increased spontaneous acetylcholine release and blunted stimulated acetylcholine release in the hippocampus.[89] In the later study, DHA deficiency was also associated with a reduction in cholinergic muscarinic receptor binding, whereas acetylcholinesterase activity and the vesicular acetylcholine transporter were not altered. In contrast, rodent maternal dietary fish oil supplementation throughout gestation and lactation significantly decreased spontaneous acetylcholine release and enhanced stimulated acetylcholine release.[90] Maternal fish oil supplementation was also shown to protect developing cholinergic neurons against excitotoxicity in the infant rat brain.[84] Moreover, hippocampal cell loss in response to acute treatment with an epileptic dose of the cholinergic muscarinic receptor agonist pilocarpine was attenuated in fish oil-supplemented rats.[91] Deficits in spatial learning resulting from treatment with the cholinergic muscarinic receptor antagonist scopolamine were attenuated in the DHA-supplemented rats.[92] Collectively, these data suggest that cortical DHA levels play an important role in acetylcholine synaptic function as well as resilience of cholinergic neurons to excitotoxic degeneration.

Dopamine

The mesocorticolimbic dopamine system is thought to regulate cognitive processes including attention and executive function as well as motivation and reward processes. Rat developmental studies have demonstrated that the perinatal deficits in brain DHA accrual are associated with a significant loss of dopamine neurons in the ventral tegmental area, the source of mesolimbic and mesocortical dopamine projections.[93] Perinatal deficits in brain DHA accrual are also associated with increased expression tyrosine hydroxylase, the rate limiting enzyme in dopamine biosynthesis, in the dorsal striatum of adolescent but not adult rats,[94] as well as significant deficits in basal and tyramine-stimulated extracellular dopamine concentrations in the frontal cortex.[95,96] Importantly, deficits in tyramine-stimulated dopamine release are reversible with early (P0-P14), but not later (P21), postnatal *n*-3 fatty acid supplementation.[95] Rats subjected to perinatal deficits in brain DHA accrual also exhibit deficits in mesolimbic dopamine neurotransmission[97,98] and alterations in dopamine D_1 and D_2 receptor expression and binding.[98–100] However, the density or function of dopamine transporters are not significantly altered in the DHA-deficit rat brain.[101] Maternal dietary fish oil supplementation throughout gestation and lactation significantly increase dopamine concentrations in the frontal cortex of adult offspring.[60] Pharmacological studies have additionally found that deficits in perinatal brain DHA accrual are associated with altered responses to drugs that target dopamine neurotransmission.[102–104] Together, these findings suggest that early perinatal brain DHA accrual is required for the functional maturation of mesocortical and mesolimbic dopamine pathways.

Serotonin

Serotonin (5-hydroxytryptamine, 5-HT) is a monoamine neurotransmitter that has been implicated in the regulation of mood, appetite, and sleep. Perinatal deficits in cortical DHA accrual are associated with impaired fenfluramine-induced elevations in extracellular serotonin concentrations,[105] reductions in midbrain expression of tryptophan hydroxylase-2, the rate-limiting enzyme in serotonin biosynthesis,[106] and elevations in 5-HT_{2A} receptor binding density in the rat frontal cortex.[99] The 5-HIAA/5-HT ratio, an index of serotonin turnover, is significantly increased in the frontal cortex, hypothalamus, and ventral striatum of adult perinatal *n*-3-deficient rats, and is associated with pro-inflammatory signalling.[37] Conversely, maternal dietary fish oil fortification significantly increases serotonin concentrations in the frontal cortex[60] and attenuates reductions in frontal cortex serotonin content in response to chronic stress.[107] Consistent with a role in central serotonin neurotransmission, perinatal or postweaning deficits in cortical DHA accrual are associated with elevated behavioral indices of aggression, anxiety, and depression,[108–110] whereas fish oil supplementation during perinatal and postweaning development or in adulthood significantly decreases depression-like behavior.[41,111–114] Together, these findings suggest that brain DHA accrual impacts the functional maturation and resilience of central serotonin neurotransmission.

NONHUMAN PRIMATE NEURODEVELOPMENT

In the developing monkey brain, frontal cortex DHA concentrations represent ~15% of total fatty acids at birth and between birth and 22 months of age increase to comprise ~22% of total fatty acids.[7,115] Feeding studies have demonstrated that a maternal diet low in ALA beginning 2 months prior to conception is associated with reduced neonatal monkey prefrontal cortex DHA concentrations at birth compared with neonates whose mothers were maintained on a standard diet containing higher levels of ALA.[7,115] Postnatal dietary FO fortification leads to a progressive increase in frontal cortex DHA concentrations in DHA-deficient neonatal monkeys, reaching control levels after approximately 10–15 weeks.[8] Baboons born preterm exhibit cortical DHA deficits compared with term births, and 4 week feeding with DHA-fortified human infant formula is more efficient for normalizing cortical DHA levels compared with human infant formula not fortified with DHA.[116,117] These results suggest that gestational length and maternal DHA status, as well as postnatal neonate dietary DHA intake, are important determinants of cortical DHA status of young adult nonhuman primates.

Primate perinatal *n*-3 fatty acid deficiency is associated with deficits in visual attention,[118] polydipsia (excessive thirst),[119] and deficits in visual acuity and electroretinogram abnormalities.[7,115] Electroretinogram abnormalities have also been observed in neonatal baboons born preterm.[120] Consistent with dysregulated mesolimbic dopamine activity, perinatal *n*-3 fatty acid deficiency is associated with increased home cage stereotypy and locomotion bouts.[121] A neuroimaging study found that resting-state functional connectivity among cortical networks was significantly reduced in monkeys raised on an *n*-3 fatty acid deficient diet compared with monkeys raised on a diet fortified with fish oil.[122] These findings provide functional evidence that reduced cortical DHA accrual is associated with a range of enduring neurodevelopmental abnormalities in the adult nonhuman primate.

HUMAN NEURODEVELOPMENT

During pregnancy the developing human fetus is reliant on maternal stores for the acquisition of DHA which is preferentially transported across the human placenta by fatty acid transport proteins.[123,124] During fetal development DHA accumulates in brain tissue at a high rate during the third trimester,[6,125] and infants born preterm exhibit lower erythrocyte and postmortem cortical DHA levels compared with term infants fed the same ALA-fortified formula postpartum.[3,4,6,125,126] It is also relevant that higher maternal DHA and fetal status during pregnancy may be an important determinant of gestational length and intrauterine growth.[127–131] Postpartum neonates are reliant on maternal breast milk (or formula) as the sole source of DHA. Human breast milk DHA concentrations are highly correlated with maternal dietary DHA but not ALA intake,[5,132–134] and lower erythrocyte or postmortem brain cortex DHA concentrations are observed in infants fed formula that does not contain preformed DHA.[4,135–141]

Observational studies indicate that higher DHA levels in maternal blood or milk or fetal cord blood during pregnancy are associated with an enduring beneficial effect on infant and child neurodevelopmental outcomes, including more mature sleep patterns, higher visual acuity, faster processing speed, reduced inattention and distractibility, and fewer emotional and behavioral problems in childhood.[142–157] Some controlled longitudinal prospective studies,[158–163] but not others,[164–167] have found that maternal DHA supplementation during pregnancy or breastfeeding has an enduring benefit on visual acuity or neurocognitive outcomes in term born infants or in early childhood. As reviewed in detail previously,[168–172] some studies, but not all, have found that infants fed formulas without DHA exhibit several neurological and neurocognitive 'soft signs' in infancy and childhood relative to infants fed formulas with DHA or maternal breast milk. However, meta-analyses of controlled trials indicate that DHA supplementation of mothers during pregnancy[173] or breastfeeding,[174] or feeding term-born infants DHA-supplemented formula,[175,176] does not confer long-term visual or neurocognitive benefits for infants and children. It is important to note, however, that many studies evaluated developmental outcomes using global standardized tests that may not be sensitive to cognitive subdomains, and in general longitudinal studies cannot account for variability in dietary DHA intake during the intervening period prior to the follow-up assessment.

Some controlled studies,[177–180] but not others,[181–184] have observed a beneficial effect of DHA supplementation on visual acuity and visual attention processes in preterm born infants. However, a meta-analysis of controlled studies found that feeding preterm infants DHA-supplemented formula did not confer any long-term benefit for visual and neurocognitive development.[185] Regarding enduring white matter pathology and associated deficits in neuronal connectivity observed in subjects born preterm,[186–192] controlled neuroimaging studies did not observe a beneficial effect of postnatal DHA supplementation on white matter volume in preterm born infants[193] or white matter microstructural integrity in preterm born children.[194] However, a cross-sectional imaging study found that preterm erythrocyte DHA levels were associated with indices of white matter microstructural integrity at 33 months of age.[195] The weight of evidence therefore suggests that the enduring brain structural or functional abnormalities associated with preterm birth are either not related to *in utero* deficits fetal brain DHA accrual or they cannot be reversed with postnatal DHA-supplementation.

During the early childhood development DHA levels continue to increase in the frontal cortex[196] in association with robust increases in frontal cortex gray matter expansion.[197] In general, the results of acute LC*n*-3 fatty acid supplementation trials in healthy developing children have not observed significant neurocognitive benefits.[198] The latter suggests that in the absence of LC*n*-3 fatty acid supplementation, the DHA status of the healthy developing youth is sufficient to prevent robust perturbations in neurodevelopment. It is notable however that emerging neuroimaging evidence suggests that there is a relationship between DHA intake/biostatus and more subtle aspects of cortical functional maturation in healthy developing children.[199–201] Furthermore, children with attention deficit hyperactivity disorder (ADHD) exhibit significant blood DHA deficits[202] and pervasive brain structural and functional impairments[203] compared with healthy developing children. While the contribution of perinatal DHA insufficiency to progressive neuropathological processes associated with ADHD is not known, some controlled studies, but not all, have observed modest improvements in neurocognitive functioning in ADHD youth following LC*n*-3 fatty acid supplementation.[204]

SUMMARY AND CONCLUSIONS

Over the past 30 years a substantial body of evidence from animal developmental studies supports the general assertion that DHA is required for optimal brain maturation. Specifically, the rodent studies have demonstrated that uncorrected deficits in perinatal brain DHA accrual are associated with transient as well as enduring structural and functional neurodevelopmental abnormalities including deficits in synaptic structure and function and a dysregulation in multiple neurotransmitter systems. Nonhuman primate studies similarly suggest that reducing perinatal brain DHA accrual is associated with long-standing deficits in functional connectivity within cortical networks, hyperactivity, and impairments in visual attention. While this research provides important proof-of-concept evidence for a vital role of DHA in normal brain development, the magnitude of the reduction in brain DHA levels is likely greater than that observed in human subjects. Nevertheless, observational findings do suggest that higher maternal DHA biostatus during pregnancy is associated with enduring neurocognitive benefits in infants and children. However, the weight of evidence from controlled human trials suggests that perinatal DHA supplementation does not confer long-term neurocognitive benefits for term-born or preterm infants or in healthy developing children. Extant translational evidence therefore suggests that although robust reductions in DHA accrual can negatively impact neurodevelopment, the DHA status of human mothers, term-born infants, and children appears to be sufficient to prevent pervasive neurodevelopmental impairment. Therefore, although DHA biosynthesis from ALA in humans is inefficient, it appears to be sufficient to sustain normal brain developmental trajectories. Nevertheless, additional research is warranted to determine whether the significant blood DHA deficits observed in children with ADHD contribute to the progressive neuropathological processes and whether DHA supplementation can prevent and/or reverse neuropathology once established.

References

1. Brenna JT, Salem Jr N, Sinclair AJ, Cunnane SC. International Society for the Study of Fatty Acids and Lipids, ISSFAL. alpha-Linolenic acid supplementation and conversion to n-3 long chain polyunsaturated fatty acids in humans. *Prostaglandins Leukot Essent Fatty Acids*. 2009;80:85–91.
2. Carlson SE, Rhodes PG, Ferguson MG. Docosahexaenoic acid status of preterm infants at birth and following feeding with human milk or formula. *Am J Clin Nutr*. 1986;44:798–804.
3. Farquharson J, Cockburn F, Patrick WA, Jamieson EC, Logan RW. Infant cerebral cortex phospholipid fatty-acid composition and diet. *Lancet*. 1992;340:810–813.
4. Farquharson J, Jamieson EC, Abbasi KA, Patrick WJ, Logan RW, Cockburn F. Effect of diet on the fatty acid composition of the major phospholipids of infant cerebral cortex. *Arch Dis Child*. 1995;72:198–203.
5. Francois CA, Connor SL, Bolewicz LC, Connor WE. Supplementing lactating women with flaxseed oil does not increase docosahexaenoic acid in their milk. *Am J Clin Nutr*. 2003;77:226–233.
6. Martinez M. Tissue levels of polyunsaturated fatty acids during early human development. *J Pediatr*. 1992;120:129–138.
7. Anderson GJ, Neuringer M, Lin DS, Connor WE. Can prenatal N-3 fatty acid deficiency be completely reversed after birth? Effects on retinal and brain biochemistry and visual function in rhesus monkeys. *Pediatr Res*. 2005;58:865–872.
8. Connor WE, Neuringer M, Lin DS. Dietary effects on brain fatty acid composition: the reversibility of n-3 fatty acid deficiency and turnover of docosahexaenoic acid in the brain, erythrocytes, and plasma of rhesus monkeys. *J Lipid Res*. 1990;31:237–247.
9. Hsieh AT, Anthony JC, Diersen-Schade DA, et al. The influence of moderate and high dietary long chain polyunsaturated fatty acids (LCPUFA) on baboon neonate tissue fatty acids. *Pediatr Res*. 2007;61:537–545.
10. Su HM, Bernardo L, Mirmiran M, Ma XH, Nathanielsz PW, Brenna JT. Dietary 18:3n-3 and 22:6n-3 as sources of 22:6n-3 accretion in neonatal baboon brain and associated organs. *Lipids*. 1999;34:347–350.
11. Beltz BS, Tlusty MF, Benton JL, Sandeman DC. Omega-3 fatty acids upregulate adult neurogenesis. *Neurosci Lett*. 2007;415:154–158.
12. Coti Bertrand P, O'Kusky JR, Innis SM. Maternal dietary (n-3) fatty acid deficiency alters neurogenesis in the embryonic rat brain. *J Nutr*. 2006;136:1570–1575.
13. Kawakita E, Hashimoto M, Shido O. Docosahexaenoic acid promotes neurogenesis in vitro and in vivo. *Neuroscience*. 2006;139:991–997.
14. Yavin E, Himovichi E, Eilam R. Delayed cell migration in the developing rat brain following maternal omega 3 alpha linolenic acid dietary deficiency. *Neuroscience*. 2009;162:1011–1122.
15. Calderon F, Kim HY. Docosahexaenoic acid promotes neurite growth in hippocampal neurons. *J Neurochem*. 2004;90:979–988.
16. Ikemoto A, Nitta A, Furukawa S, et al. Dietary n-3 fatty acid deficiency decreases nerve growth factor content in rat hippocampus. *Neurosci Lett*. 2000;285:99–102.
17. Rao JS, Ertley RN, Lee HJ, et al. n-3 polyunsaturated fatty acid deprivation in rats decreases frontal cortex BDNF via a p38 MAPK-dependent mechanism. *Mol Psychiatry*. 2007;12:36–46.
18. Cao D, Kevala K, Kim J, et al. Docosahexaenoic acid promotes hippocampal neuronal development and synaptic function. *J Neurochem*. 2009;111:510–521.

19. Innis SM, de La Presa Owens S. Dietary fatty acid composition in pregnancy alters neurite membrane fatty acids and dopamine in newborn rat brain. *J Nutr*. 2001;131:118–122.
20. Martin RE, Bazan NG. Changing fatty acid content of growth cone lipids prior to synaptogenesis. *J Neurochem*. 1992;59:318–325.
21. de Velasco PC, Mendonça HR, Borba JM, et al. Nutritional restriction of omega-3 fatty acids alters topographical fine tuning and leads to a delay in the critical period in the rodent visual system. *Exp Neurol*. 2012;234:220–229.
22. Ahmad A, Murthy M, Greiner RS, Moriguchi T, Salem Jr N. A decrease in cell size accompanies a loss of docosahexaenoate in the rat hippocampus. *Nutr Neurosci*. 2002;5:103–113.
23. Harbeby E, Jouin M, Alessandri JM, et al. n-3 PUFA status affects expression of genes involved in neuroenergetics differently in the fronto-parietal cortex compared to the CA1 area of the hippocampus: effect of rest and neuronal activation in the rat. *Prostaglandins Leukot Essent Fatty Acids*. 2012;86:211–220.
24. Pifferi F, Jouin M, Alessandri JM, et al. n-3 Fatty acids modulate brain glucose transport in endothelial cells of the blood-brain barrier. *Prostaglandins Leukot Essent Fatty Acids*. 2007;77:279–286.
25. Ximenes da Silva A, Lavialle F, Gendrot G, Guesnet P, Alessandri JM, Lavialle M. Glucose transport and utilization are altered in the brain of rats deficient in n-3 polyunsaturated fatty acids. *J Neurochem*. 2002;81:1328–1337.
26. Bauman MD, Iosif AM, Smith SE, Bregere C, Amaral DG, Patterson PH. Activation of the maternal immune system during pregnancy alters behavioral development of rhesus monkey offspring. *Biol Psychiatry*. 2014;75:332–341.
27. Careaga M, Murai T, Bauman MD. Maternal immune activation and autism spectrum disorder: From rodents to nonhuman and human primates. *Biol Psychiatry*. 2017;81:391–401.
28. Rose DR, Careaga M, Van de Water J, McAllister K, Bauman MD, Ashwood P. Long-term altered immune responses following fetal priming in a non-human primate model of maternal immune activation. *Brain Behav Immun*. 2017;63:60–70.
29. Bagga D, Wang L, Farias-Eisner R, Glaspy JA, Reddy ST. Differential effects of prostaglandin derived from omega-6 and omega-3 polyunsaturated fatty acids on COX-2 expression and IL-6 secretion. *Proc Natl Acad Sci U S A*. 2003;100:1751–1756.
30. Bazan NG, Calandria JM, Serhan CN. Rescue and repair during photoreceptor cell renewal mediated by docosahexaenoic acid-derived neuroprotectin D1. *J Lipid Res*. 2010;51:2018–2031.
31. Groeger AL, Cipollina C, Cole MP, et al. Cyclooxygenase-2 generates anti-inflammatory mediators from omega-3 fatty acids. *Nat Chem Biol*. 2010;6:433–441.
32. Serhan CN. Novel lipid mediators and resolution mechanisms in acute inflammation: to resolve or not? *Am J Pathol*. 2010;177:1576–1591.
33. Hong S, Gronert K, Devchand PR, Moussignac RL, Serhan CN. Novel docosatrienes and 17S-resolvins generated from docosahexaenoic acid in murine brain, human blood, and glial cells. Autacoids in anti-inflammation. *J Biol Chem*. 2003;278:14677–14687.
34. Hjorth E, Zhu M, Toro VC, et al. Omega-3 fatty acids enhance phagocytosis of Alzheimer's disease-related amyloid-β42 by human microglia and decrease inflammatory markers. *J Alzheimers Dis*. 2013;35:697–713.
35. Madore C, Nadjar A, Delpech JC, et al. Nutritional n-3 PUFAs deficiency during perinatal periods alters brain innate immune system and neuronal plasticity-associated genes. *Brain Behav Immun*. 2014;41:22–31.
36. Rao JS, Ertley RN, DeMar Jr JC, Rapoport SI, Bazinet RP, Lee HJ. Dietary n-3 PUFA deprivation alters expression of enzymes of the arachidonic and docosahexaenoic acid cascades in rat frontal cortex. *Mol Psychiatry*. 2007;12:151–157.
37 McNamara RK, Rider T, Jandacek RJ, Tso P, Cole-Strauss A. Lipton JW. Omega-3 fatty acid deficiency increases constitutive pro-inflammatory cytokine production in rats: Relationship with central serotonin turnover. Prostaglandins Leukot Essent Fatty Acids 2010;83:185-191.
38. Mingam R, Moranis A, Bluthé RM, et al. Uncoupling of interleukin-6 from its signalling pathway by dietary n-3-polyunsaturated fatty acid deprivation alters sickness behaviour in mice. *Eur J Neurosci*. 2008;28:1877–1886.
39. Tuzun F, Kumral A, Dilek M, et al. Maternal omega-3 fatty acid supplementation protects against lipopolysaccharide-induced white matter injury in the neonatal rat brain. *J Matern Fetal Neonatal Med*. 2012;25:849–854.
40. Orr SK, Palumbo S, Bosetti F, et al. Unesterified docosahexaenoic acid is protective in neuroinflammation. *J Neurochem*. 2013;127:378–393.
41. Dang R, Zhou X, Tang M, et al. Fish oil supplementation attenuates neuroinflammation and alleviates depressive-like behavior in rats submitted to repeated lipopolysaccharide. *Eur J Nutr*. 2018;57:893–906.

42. Chang PK, Khatchadourian A, McKinney RA, Maysinger D. Docosahexaenoic acid (DHA): a modulator of microglia activity and dendritic spine morphology. *J Neuroinflammation*. 2015;12:34.
43. Kong W, Yen JH, Ganea D. Docosahexaenoic acid prevents dendritic cell maturation, inhibits antigen-specific Th1/Th17 differentiation and suppresses experimental autoimmune encephalomyelitis. *Brain Behav Immun*. 2011;25:872–882.
44. Rey C, Nadjar A, Joffre F, et al. Maternal n-3 polyunsaturated fatty acid dietary supply modulates microglia lipid content in the offspring. *Prostaglandins Leukot Essent Fatty Acids*. 2018;133:1–7.
45. Paolicelli RC, Bolasco G, Pagani F, et al. Synaptic pruning by microglia is necessary for normal brain development. *Science*. 2011;333:1456–1458.
46. Schafer DP, Stevens B. Phagocytic glial cells: sculpting synaptic circuits in the developing nervous system. *Curr Opin Neurobiol*. 2013;23:1034–1040.
47. Haubner L, Sullivan J, Ashmeade T, Saste M, Wiener D, Carver J. The effects of maternal dietary docosahexaenoic acid intake on rat pup myelin and the auditory startle response. *Dev Neurosci*. 2007;29:460–467.
48. Yeh YY, Gehman MF, Yeh SM. Maternal dietary fish oil enriches docosahexaenoate levels in brain subcellular fractions of offspring. *J Neurosci Res*. 1993;35:218–226.
49. Bernardo A, Giammarco ML, De Nuccio C, et al. Docosahexaenoic acid promotes oligodendrocyte differentiation via PPAR-γ signalling and prevents tumor necrosis factor-α-dependent maturational arrest. *Biochim Biophys Acta*. 1862;2017:1013–1023.
50. Salvati S, Natali F, Attorri L, et al. Eicosapentaenoic acid stimulates the expression of myelin proteins in rat brain. *J Neurosci Res*. 2008;86:776–784.
51. McNamara RK, Schurdak JD, Asch RH, Peters BD, Lindquist DM. Deficits in docosahexaenoic acid accrual during adolescence reduce rat forebrain white matter microstructural integrity: An in vivo diffusion tensor imaging study. *Dev Neurosci*. 2018;40:84–92.
52. Bailes JE, Mills JD. Docosahexaenoic acid reduces traumatic axonal injury in a rodent head injury model. *J Neurotrauma*. 2010;27:1617–1624.
53. Ward RE, Huang W, Curran OE, Priestley JV, Michael-Titus AT. Docosahexaenoic acid prevents white matter damage after spinal cord injury. *J Neurotrauma*. 2010;27:1769–1780.
54. Church MW, Jen KL, Dowhan LM, Adams BR, Hotra JW. Excess and deficient omega-3 fatty acid during pregnancy and lactation cause impaired neural transmission in rat pups. *Neurotoxicol Teratol*. 2008;30:107–117.
55. Church MW, Jen KL, Jackson DA, Adams BR, Hotra JW. Abnormal neurological responses in young adult offspring caused by excess omega-3 fatty acid (fish oil) consumption by the mother during pregnancy and lactation. *Neurotoxicol Teratol*. 2009;31:26–33.
56. Suzuki H, Manabe S, Wada O, Crawford MA. Rapid incorporation of docosahexaenoic acid from dietary sources into brain microsomal, synaptosomal and mitochondrial membranes in adult mice. *Int J Vitam Nutr Res*. 1997;67:272–278.
57. McNamara RK, Ostrander M, Abplanalp W, Richtand NM, Benoit SC, Clegg DJ. Modulation of phosphoinositide-protein kinase C signal transduction by omega-3 fatty acids: implications for the pathophysiology and treatment of recurrent neuropsychiatric illness. *Prostaglandins Leukot Essent Fatty Acids*. 2006;75:237–257.
58. Hamilton L, Greiner R, Salem Jr N, Kim HY. n-3 fatty acid deficiency decreases phosphatidylserine accumulation selectively in neuronal tissues. *Lipids*. 2000;35:863–869.
59. Murthy M, Hamilton J, Greiner RS, Moriguchi T, Salem Jr N, Kim HY. Differential effects of n-3 fatty acid deficiency on phospholipid molecular species composition in the rat hippocampus. *J Lipid Res*. 2002;43:611–617.
60. Chalon S, Delion-Vancassel S, Belzung C, et al. Dietary fish oil affects monoaminergic neurotransmission and behavior in rats. *J Nutr*. 1998;128:2512–2519.
61. Huang KP, Huang FL. How is protein kinase C activated in CNS. *Neurochem Int*. 1993;22:417–433.
62. Calabrese B, Halpain S. Essential role for the PKC target MARCKS in maintaining dendritic spine morphology. *Neuron*. 2005;48:77–90.
63. Denys A, Hichami A, Maume B, Khan NA. Docosahexaenoic acid modulates phorbol ester-induced activation of extracellular signal-regulated kinases 1 and 2 in NIH/3T3 cells. *Lipids*. 2001;36:813–818.
64. Holian O, Nelson R. Action of long-chain fatty acids on protein kinase C activity: comparison of omega-6 and omega-3 fatty acids. *Anticancer Res*. 1992;12:975–980.
65. Mirnikjoo B, Brown SE, Kim HF, Marangell LB, Sweatt JD, Weeber EJ. Protein kinase inhibition by omega-3 fatty acids. *J Biol Chem*. 2001;276:10888–10896.

66. Seung Kim HF, Weeber EJ, Sweatt JD, Stoll AL, Marangell LB. Inhibitory effects of omega-3 fatty acids on protein kinase C activity in vitro. *Mol Psychiatry*. 2001;6:246–248.
67. McLaughlin S, Aderem A. The myristoyl-electrostatic switch: a modulator of reversible protein-membrane interactions. *Trends Biochem Sci*. 1995;20:272–276.
68. Glaser M, Wanaski S, Buser CA, et al. Myristoylated alanine-rich C kinase substrate (MARCKS) produces reversible inhibition of phospholipase C by sequestering phosphatidylinositol 4,5-bisphosphate in lateral domains. *J Biol Chem*. 1996;271:26187–26193.
69. Vaughan PF, Walker JH, Peers C. The regulation of neurotransmitter secretion by protein kinase C. *Mol Neurobiol*. 1998;18:125–155.
70. Yoshida S, Yasuda A, Kawazato H, et al. Synaptic vesicle ultrastructural changes in the rat hippocampus induced by a combination of alpha-linolenate deficiency and a learning task. *J Neurochem*. 1997;68:1261–1268.
71. Zimmer L, Delpal S, Guilloteau D, Aioun J, Durand G, Chalon S. Chronic n-3 polyunsaturated fatty acid deficiency alters dopamine vesicle density in the rat frontal cortex. *Neurosci Lett*. 2000;284:25–28.
72. Fedorova I, Hussein N, Baumann MH, Di Martino C, Salem Jr N. An n-3 fatty acid deficiency impairs rat spatial learning in the Barnes maze. *Behav Neurosci*. 2009;123:196–205.
73. Greiner RS, Moriguchi T, Hutton A, Slotnick BM, Salem Jr N. Rats with low levels of brain docosahexaenoic acid show impaired performance in olfactory-based and spatial learning tasks. *Lipids*. 1999;34(Suppl):239–243.
74. Greiner RS, Moriguchi T, Slotnick BM, Hutton A, Salem N. Olfactory discrimination deficits in n-3 fatty acid-deficient rats. *Physiol Behav*. 2001;72:379–385.
75. Moriguchi T, Salem Jr N. Recovery of brain docosahexaenoate leads to recovery of spatial task performance. *J Neurochem*. 2003;87:297–309.
76. Hertz L, Zielke HR. Astrocytic control of glutamatergic activity: astrocytes as stars of the show. *Trends Neurosci*. 2004;27:735–743.
77. Champeil-Potokar G, Chaumontet C, Guesnet P, Lavialle M, Denis I. Docosahexaenoic acid (22:6n-3) enrichment of membrane phospholipids increases gap junction coupling capacity in cultured astrocytes. *Eur J Neurosci*. 2006;24:3084–3090.
78. Champeil-Potokar G, Denis I, Goustard-Langelier B, Alessandri JM, Guesnet P, Lavialle M. Astrocytes in culture require docosahexaenoic acid to restore the n-3/n-6 polyunsaturated fatty acid balance in their membrane phospholipids. *J Neurosci Res*. 2004;75:96–106.
79. Joardar A, Sen AK, Das S. Docosahexaenoic acid facilitates cell maturation and beta-adrenergic transmission in astrocytes. *J Lipid Res*. 2006;47:571–581.
80. Champeil-Potokar G, Hennebelle M, Latour A, Vancassel S, Denis I. Docosahexaenoic acid (DHA) prevents corticosterone-induced changes in astrocyte morphology and function. *J Neurochem*. 2016;135:1155–1167.
81. Berry CB, Hayes D, Murphy A, Wiessner M, Rauen T, McBean GJ. Differential modulation of the glutamate transporters GLT1, GLAST and EAAC1 by docosahexaenoic acid. *Brain Res*. 2005;1037:123–133.
82. Grintal B, Champeil-Potokar G, Lavialle M, Vancassel S, Breton S, Denis I. Inhibition of astroglial glutamate transport by polyunsaturated fatty acids: evidence for a signalling role of docosahexaenoic acid. *Neurochem Int*. 2009;54:535–543.
83. McNamara RK, Asch RH, Schurdak JD, Lindquist DM. Glutamate homeostasis in the adult rat prefrontal cortex is altered by cortical docosahexaenoic acid accrual during adolescence: An in vivo 1H MRS study. *Psychiatry Res Neuroimaging*. 2017;270:39–45.
84. Högyes E, Nyakas C, Kiliaan A, Farkas T, Penke B, Luiten PG. Neuroprotective effect of developmental docosahexaenoic acid supplement against excitotoxic brain damage in infant rats. *Neuroscience*. 2003;119:999–1012.
85. Ozyurt B, Sarsilmaz M, Akpolat N, et al. The protective effects of omega-3 fatty acids against MK-801-induced neurotoxicity in prefrontal cortex of rat. *Neurochem Int*. 2007;50:196–202.
86. Blondeau N, Widmann C, Lazdunski M, Heurteaux C. Polyunsaturated fatty acids induce ischemic and epileptic tolerance. *Neuroscience*. 2002;109:231–241.
87. Belayev L, Khoutorova L, Atkins KD, Bazan NG. Robust docosahexaenoic acid-mediated neuroprotection in a rat model of transient, focal cerebral ischemia. *Stroke*. 2009;40:3121–3126.
88. Wu A, Ying Z, Gomez-Pinilla F. Dietary omega-3 fatty acids normalize BDNF levels, reduce oxidative damage, and counteract learning disability after traumatic brain injury in rats. *J Neurotrauma*. 2004;21:1457–1467.
89. Aïd S, Vancassel S, Poumès-Ballihaut C, Chalon S, Guesnet P, Lavialle M. Effect of a diet-induced n-3 PUFA depletion on cholinergic parameters in the rat hippocampus. *J Lipid Res*. 2003;44:1545–1551.

90. Aïd S, Vancassel S, Linard A, Lavialle M, Guesnet P. Dietary docosahexaenoic acid [22: 6(n-3)] as a phospholipid or a triglyceride enhances the potassium chloride-evoked release of acetylcholine in rat hippocampus. *J Nutr*. 2005;135:1008–1013.
91. Ferrari D, Cysneiros RM, Scorza CA, et al. Neuroprotective activity of omega-3 fatty acids against epilepsy-induced hippocampal damage: Quantification with immunohistochemical for calcium-binding proteins. *Epilepsy Behav*. 2008;13:36–42.
92. Vaisman N, Pelled D. n-3 phosphatidylserine attenuated scopolamine-induced amnesia in middle-aged rats. *Prog Neuropsychopharmacol Biol Psychiatry*. 2009;33:952–959.
93. Ahmad SO, Park JH, Radel JD, Levant B. Reduced numbers of dopamine neurons in the substantia nigra pars compacta and ventral tegmental area of rats fed an n-3 polyunsaturated fatty acid-deficient diet: a stereological study. *Neurosci Lett*. 2008;438:303–307.
94. Bondi CO, Taha AY, Tock JL, et al. Adolescent behavior and dopamine availability are uniquely sensitive to dietary omega-3 fatty acid deficiency. *Biol Psychiatry*. 2014;75:38–46.
95. Kodas E, Vancassel S, Lejeune B, Guilloteau D, Chalon S. Reversibility of n-3 fatty acid deficiency-induced changes in dopaminergic neurotransmission in rats: critical role of developmental stage. *J Lipid Res*. 2002;43:1209–1219.
96. Zimmer L, Hembert S, Durand G, et al. Chronic n-3 polyunsaturated fatty acid diet-deficiency acts on dopamine metabolism in the rat frontal cortex: a microdialysis study. *Neurosci Lett*. 1998;240:177–181.
97. Zimmer L, Vancassel S, Cantagrel S, et al. The dopamine mesocorticolimbic pathway is affected by deficiency in n-3 polyunsaturated fatty acids. *Am J Clin Nutr*. 2002;75:662–667.
98. Zimmer L, Delion-Vancassel S, Durand G, et al. Modification of dopamine neurotransmission in the nucleus accumbens of rats deficient in n-3 polyunsaturated fatty acids. *J Lipid Res*. 2000;41:32–40.
99. Delion S, Chalon S, Guilloteau D, Besnard JC, Durand G. alpha-Linolenic acid dietary deficiency alters age-related changes of dopaminergic and serotoninergic neurotransmission in the rat frontal cortex. *J Neurochem*. 1996;66:1582–1591.
100. Kuperstein F, Eilam R, Yavin E. Altered expression of key dopaminergic regulatory proteins in the postnatal brain following perinatal n-3 fatty acid dietary deficiency. *J Neurochem*. 2008;106:662–671.
101. Kodas E, Page G, Zimmer L, et al. Neither the density nor function of striatal dopamine transporters were influenced by chronic n-3 polyunsaturated fatty acid deficiency in rodents. *Neurosci Lett*. 2002;321:95–99.
102. Levant B, Radel JD, Carlson SE. Decreased brain docosahexaenoic acid during development alters dopamine-related behaviors in adult rats that are differentially affected by dietary remediation. *Behav Brain Res*. 2004;152:49–57.
103. McNamara RK, Sullivan J, Richtand NM. Omega-3 fatty acid deficiency augments amphetamine-induced behavioral sensitization in adult mice: prevention by chronic lithium treatment. *J Psychiatr Res*. 2008;42:458–468.
104. McNamara RK, Sullivan J, Richtand NM, et al. Omega-3 fatty acid deficiency augments amphetamine-induced behavioral sensitization in adult DBA/2J mice: relationship with ventral striatum dopamine concentrations. *Synapse*. 2008;62:725–735.
105. Kodas E, Galineau L, Bodard S, et al. Serotoninergic neurotransmission is affected by n-3 polyunsaturated fatty acids in the rat. *J Neurochem*. 2004;89:695–702.
106. McNamara RK, Able JA, Liu Y, et al. Omega-3 fatty acid deficiency during perinatal development increases serotonin turnover in the prefrontal cortex and decreases midbrain tryptophan hydroxylase-2 expression in adult female rats: Dissociation from estrogenic effects. *J Psychiatr Res*. 2009;43:656–663.
107. Vancassel S, Leman S, Hanonick L, et al. n-3 polyunsaturated fatty acid supplementation reverses stress-induced modifications on brain monoamine levels in mice. *J Lipid Res*. 2008;49:340–348.
108. Chen HF, Su HM. Exposure to a maternal n-3 fatty acid-deficient diet during brain development provokes excessive hypothalamic-pituitary-adrenal axis responses to stress and behavioral indices of depression and anxiety in male rat offspring later in life. *J Nutr Biochem*. 2013;24:70–80.
109. DeMar Jr JC, Ma K, Bell JM, Igarashi M, Greenstein D, Rapoport SI. One generation of n-3 polyunsaturated fatty acid deprivation increases depression and aggression test scores in rats. *J Lipid Res*. 2006;47:172–180.
110. Weiser MJ, Wynalda K, Salem Jr N, Butt CM. Dietary DHA during development affects depression-like behaviors and biomarkers that emerge after puberty in adolescent rats. *J Lipid Res*. 2015;56:151–166.
111. Ferraz AC, Kiss A, Araújo RL, et al. The antidepressant role of dietary long-chain polyunsaturated n-3 fatty acids in two phases in the developing brain. *Prostaglandins Leukot Essent Fatty Acids*. 2008;78:183–188.

112. Carlezon Jr WA, Mague SD, Parow AM, Stoll AL, Cohen BM, Renshaw PF. Antidepressant-like effects of uridine and omega-3 fatty acids are potentiated by combined treatment in rats. *Biol Psychiatry*. 2005;57:343–350.
113. Huang SY, Yang HT, Chiu CC, Pariante CM, Su KP. Omega-3 fatty acids on the forced-swimming test. *J Psychiatr Res*. 2008;42:58–63.
114. Naliwaiko K, Araújo RL, da Fonseca RV, et al. Effects of fish oil on the central nervous system: a new potential antidepressant? *Nutr Neurosci*. 2004;7:91–99.
115. Neuringer M, Connor WE, Lin DS, Barstad L, Luck S. Biochemical and functional effects of prenatal and postnatal omega 3 fatty acid deficiency on retina and brain in rhesus monkeys. *Proc Natl Acad Sci U S A*. 1986;83:4021–4025.
116. Sarkadi-Nagy E, Wijendran V, Diau GY, et al. Formula feeding potentiates docosahexaenoic and arachidonic acid biosynthesis in term and preterm baboon neonates. *J Lipid Res*. 2004;45:71–80.
117. Sarkadi-Nagy E, Wijendran V, Diau GY, et al. The influence of prematurity and long chain polyunsaturate supplementation in 4-week adjusted age baboon neonate brain and related tissues. *Pediatr Res*. 2003;54:244–252.
118. Reisbick S, Neuringer M, Gohl E, Wald R, Anderson GJ. Visual attention in infant monkeys: effects of dietary fatty acids and age. *Dev Psychol*. 1997;33:387–395.
119. Reisbick S, Neuringer M, Hasnain R, Connor WE. Polydipsia in rhesus monkeys deficient in omega-3 fatty acids. *Physiol Behav*. 1990;47:315–323.
120. Diau GY, Loew ER, Wijendran V, Sarkadi-Nagy E, Nathanielsz PW, Brenna JT. Docosahexaenoic and arachidonic acid influence on preterm baboon retinal composition and function. *Invest Ophthalmol Vis Sci*. 2003;44:4559–4566.
121. Reisbick S, Neuringer M, Hasnain R, Connor WE. Home cage behavior of rhesus monkeys with long-term deficiency of omega-3 fatty acids. *Physiol Behav*. 1994;55:231–239.
122. Grayson DS, Kroenke CD, Neuringer M, Fair DA. Dietary omega-3 fatty acids modulate large-scale systems organization in the rhesus macaque brain. *J Neurosci*. 2014;34:2065–2074.
123. Dutta-Roy AK. Transport mechanisms for long-chain polyunsaturated fatty acids in the human placenta. *Am J Clin Nutr*. 2000;71:315S–322S.
124. Koletzko B, Larqué E, Demmelmair H. Placental transfer of long-chain polyunsaturated fatty acids (LC-PUFA). *J Perinat Med*. 2007;35:5–11.
125. Clandinin MT, Chappell JE, Leong S, Heim T, Swyer PR, Chance GW. Intrauterine fatty acid accretion rates in human brain: implications for fatty acid requirements. *Early Hum Dev*. 1980;4:121–129.
126. Martínez M, Mougan I. Fatty acid composition of human brain phospholipids during normal development. *J Neurochem*. 1998;71:2528–2533.
127. Carlson SE, Colombo J, Gajewski BJ, et al. DHA supplementation and pregnancy outcomes. *Am J Clin Nutr*. 2013;97:808–815.
128. Cetin I, Giovannini N, Alvino G, et al. Intrauterine growth restriction is associated with changes in polyunsaturated fatty acid fetal-maternal relationships. *Pediatr Res*. 2002;52:750–755.
129. Olsen SF, Østerdal ML, Salvig JD, Weber T, Tabor A, Secher NJ. Duration of pregnancy in relation to fish oil supplementation and habitual fish intake: a randomised clinical trial with fish oil. *Eur J Clin Nutr*. 2007;61:976–985.
130. Olsen SF, Sørensen JD, Secher NJ, et al. Randomised controlled trial of effect of fish-oil supplementation on pregnancy duration. *Lancet*. 1992;339(8800):1003–1007.
131. Salvig JD, Lamont RF. Evidence regarding an effect of marine n-3 fatty acids on preterm birth: a systematic review and meta-analysis. *Acta Obstet Gynecol Scand*. 2011;90:825–838.
132. Innis SM. Polyunsaturated fatty acids in human milk: an essential role in infant development. *Adv Exp Med Biol*. 2004;554:27–43.
133. Lauritzen L, Jørgensen MH, Hansen HS, Michaelsen KF. Fluctuations in human milk long-chain PUFA levels in relation to dietary fish intake. *Lipids*. 2002;37:237–244.
134. Jensen CL, Maude M, Anderson RE, Heird WC. Effect of docosahexaenoic acid supplementation of lactating women on the fatty acid composition of breast milk lipids and maternal and infant plasma phospholipids. *Am J Clin Nutr*. 2000;71:292S–299S.
135. Gibson RA, Neumann MA, Makrides M. Effect of dietary docosahexaenoic acid on brain composition and neural function in term infants. *Lipids*. 1996;31:S177–S181.
136. Sanders TA, Naismith DJ. A comparison of the influence of breast-feeding and bottle-feeding on the fatty acid composition of the erythrocytes. *Br J Nutr*. 1979;41:619–623.

137. Putnam JC, Carlson SE, DeVoe PW, Barness LA. The effect of variations in dietary fatty acids on the fatty acid composition of erythrocyte phosphatidylcholine and phosphatidylethanolamine in human infants. *Am J Clin Nutr*. 1982;36:106–114.
138. Auestad N, Montalto MB, Hall RT, et al. Visual acuity, erythrocyte fatty acid composition, and growth in term infants fed formulas with long chain polyunsaturated fatty acids for one year. Ross Pediatric Lipid Study. *Pediatr Res*. 1997;41:1–10.
139. Byard RW, Makrides M, Need M, Neumann MA, Gibson RA. Sudden infant death syndrome: effect of breast and formula feeding on frontal cortex and brainstem lipid composition. *J Paediatr Child Health*. 1995;31:14–16.
140. Jamieson EC, Farquharson J, Logan RW, et al. Infant cerebellar gray and white matter fatty acids in relation to age and diet. *Lipids*. 1999;34:1065–1071.
141. Makrides M, Neumann MA, Byard RW, Simmer K, Gibson RA. Fatty acid composition of brain, retina, and erythrocytes in breast- and formula-fed infants. *Am J Clin Nutr*. 1994;60:189–194.
142. Cheruku SR, Montgomery-Downs HE, Farkas SL, Thoman EB, Lammi-Keefe CJ. Higher maternal plasma docosahexaenoic acid during pregnancy is associated with more mature neonatal sleep-state patterning. *Am J Clin Nutr*. 2002;76:608–613.
143. Williams C, Birch EE, Emmett PM, Northstone K, Avon Longitudinal Study of Pregnancy and Childhood Study Team. Stereoacuity at age 3.5 y in children born full-term is associated with prenatal and postnatal dietary factors: a report from a population-based cohort study. *Am J Clin Nutr*. 2001;73:316–322.
144. Colombo J, Kannass KN, Shaddy DJ, et al. Maternal DHA and the development of attention in infancy and toddlerhood. *Child Dev*. 2004;75:1254–1267.
145. Innis SM, Gilley J, Werker J. Are human milk long-chain polyunsaturated fatty acids related to visual and neural development in breast-fed term infants? *J Pediatr*. 2001;139:532–538.
146. Innis SM, Friesen RW. Essential n-3 fatty acids in pregnant women and early visual acuity maturation in term infants. *Am J Clin Nutr*. 2008;87:548–557.
147. Jacobson JL, Jacobson SW, Muckle G, Kaplan-Estrin M, Ayotte P, Dewailly E. Beneficial effects of a polyunsaturated fatty acid on infant development: evidence from the inuit of arctic Quebec. *J Pediatr*. 2008;152:356–364.
148. Kannass KN, Colombo J, Carlson SE. Maternal DHA levels and toddler free-play attention. *Dev Neuropsychol*. 2009;34:159–174.
149. Loomans EM, Van den Bergh BR, Schelling M, Vrijkotte TG, van Eijsden M. Maternal long-chain polyunsaturated fatty acid status during early pregnancy and children's risk of problem behavior at age 5-6 years. *J Pediatr*. 2014;164:762–768.
150. Steenweg-de Graaff JC, Tiemeier H, Basten MG, et al. Maternal LC-PUFA status during pregnancy and child problem behavior: the Generation R Study. *Pediatr Res*. 2015;77:489–497.
151. Kohlboeck G, Glaser C, Tiesler C, et al. Effect of fatty acid status in cord blood serum on children's behavioral difficulties at 10 y of age: results from the LISAplus Study. *Am J Clin Nutr*. 2011;94:1592–1599.
152. Krabbendam L, Bakker E, Hornstra G, van Os J. Relationship between DHA status at birth and child problem behaviour at 7 years of age. *Prostaglandins Leukot Essent Fatty Acids*. 2007;76:29–34.
153. Boucher O, Burden MJ, Muckle G, et al. Neurophysiologic and neurobehavioral evidence of beneficial effects of prenatal omega-3 fatty acid intake on memory function at school age. *Am J Clin Nutr*. 2011;93:1025–1037.
154. Escolano-Margarit MV, Ramos R, Beyer J, et al. Prenatal DHA status and neurological outcome in children at age 5.5 years are positively associated. *J Nutr*. 2011;141:1216–1223.
155. Bakker EC, Ghys AJ, Kester AD, et al. Long-chain polyunsaturated fatty acids at birth and cognitive function at 7 y of age. *Eur J Clin Nutr*. 2003;57:89–95.
156. Bakker EC, Hornstra G, Blanco CE, Vles JS. Relationship between long-chain polyunsaturated fatty acids at birth and motor function at 7 years of age. *Eur J Clin Nutr*. 2009;63:499–504.
157. Jacques C, Levy E, Muckle G, et al. Long-term effects of prenatal omega-3 fatty acid intake on visual function in school-age children. *J Pediatr*. 2011;158:83–90.
158. Helland IB, Smith L, Blomén B, Saarem K, Saugstad OD, Drevon CA. Effect of supplementing pregnant and lactating mothers with n-3 very-long-chain fatty acids on children's IQ and body mass index at 7 years of age. *Pediatrics*. 2008;122:472–479.
159. Helland IB, Smith L, Saarem K, Saugstad OD, Drevon CA. Maternal supplementation with very-long-chain n-3 fatty acids during pregnancy and lactation augments children's IQ at 4 years of age. *Pediatrics*. 2003;111:e39–e44.
160. Colombo J, Gustafson KM, Gajewski BJ, et al. Prenatal DHA supplementation and infant attention. *Pediatr Res*. 2016;80:656–662.

161. Jensen CL, Voigt RG, Llorente AM, et al. Effects of early maternal docosahexaenoic acid intake on neuropsychological status and visual acuity at five years of age of breast-fed term infants. *J Pediatr*. 2010;157:900–905.
162. Malcolm CA, McCulloch DL, Montgomery C, Shepherd A, Weaver LT. Maternal docosahexaenoic acid supplementation during pregnancy and visual evoked potential development in term infants: a double blind, prospective, randomised trial. *Arch Dis Child Fetal Neonatal*. 2003;88:F383–F390.
163. Ramakrishnan U, Gonzalez-Casanova I, Schnaas L, et al. Prenatal supplementation with DHA improves attention at 5 y of age: a randomized controlled trial. *Am J Clin Nutr*. 2016;104:1075–1082.
164. Gould JF, Treyvaud K, Yelland LN, et al. Seven-year follow-up of children born to women in a randomized trial of prenatal DHA supplementation. *JAMA*. 2017;317:1173–1175.
165. Gould JF, Makrides M, Colombo J, Smithers LG. Randomized controlled trial of maternal omega-3 long-chain PUFA supplementation during pregnancy and early childhood development of attention, working memory, and inhibitory control. *Am J Clin Nutr*. 2014;99:851–859.
166. Makrides M, Gibson RA, McPhee AJ, et al. Effect of DHA supplementation during pregnancy on maternal depression and neurodevelopment of young children: a randomized controlled trial. *JAMA*. 2010;304:1675–1683.
167. Smithers LG, Gibson RA, Makrides M. Maternal supplementation with docosahexaenoic acid during pregnancy does not affect early visual development in the infant: a randomized controlled trial. *Am J Clin Nutr*. 2011;93:1293–1299.
168. Carlson SE. Docosahexaenoic acid supplementation in pregnancy and lactation. *Am J Clin Nutr*. 2009;89:678S–684S.
169. Carlson SE, Neuringer M. Polyunsaturated fatty acid status and neurodevelopment: a summary and critical analysis of the literature. *Lipids*. 1999;34:171–178.
170. Scholtz SA, Colombo J, Carlson SE. Clinical overview of effects of dietary long-chain polyunsaturated fatty acids during the perinatal period. *Nestle Nutr Inst Workshop Ser*. 2013;77:145–154.
171. Lauritzen L, Carlson SE. Maternal fatty acid status during pregnancy and lactation and relation to newborn and infant status. *Matern Child Nutr*. 2011;7(Suppl 2):41–58.
172. Uauy R, Hoffman DR, Peirano P, Birch DG, Birch EE. Essential fatty acids in visual and brain development. *Lipids*. 2001;36:885–895.
173. Gould JF, Smithers LG, Makrides M. The effect of maternal omega-3 (n-3) LCPUFA supplementation during pregnancy on early childhood cognitive and visual development: a systematic review and meta-analysis of randomized controlled trials. *Am J Clin Nutr*. 2013;97:531–544.
174. Delgado-Noguera MF, Calvache JA, Bonfill Cosp X, Kotanidou EP, Galli-Tsinopoulou A. Supplementation with long chain polyunsaturated fatty acids (LCPUFA) to breastfeeding mothers for improving child growth and development. *Cochrane Database Syst Rev*. 2015;7:CD007901.
175. Beyerlein A, Hadders-Algra M, Kennedy K, et al. Infant formula supplementation with long-chain polyunsaturated fatty acids has no effect on Bayley developmental scores at 18 months of age—IPD meta-analysis of 4 large clinical trials. *J Pediatr Gastroenterol Nutr*. 2010;50:79–84.
176. Jasani B, Simmer K, Patole SK, Rao SC. Long chain polyunsaturated fatty acid supplementation in infants born at term. *Cochrane Database Syst Rev*. 2017;3:CD000376.
177. Clandinin MT, Van Aerde JE, Merkel KL, et al. Growth and development of preterm infants fed infant formulas containing docosahexaenoic acid and arachidonic acid. *J Pediatr*. 2005;146:461–468.
178. Carlson SE, Werkman SH. A randomized trial of visual attention of preterm infants fed docosahexaenoic acid until two months. *Lipids*. 1996;31:85–90.
179. SanGiovanni JP, Parra-Cabrera S, Colditz GA, Berkey CS, Dwyer JT. Meta-analysis of dietary essential fatty acids and long-chain polyunsaturated fatty acids as they relate to visual resolution acuity in healthy preterm infants. *Pediatrics*. 2000;105:1292–1298.
180. Werkman SH, Carlson SE. A randomized trial of visual attention of preterm infants fed docosahexaenoic acid until nine months. *Lipids*. 1996;31:91–97.
181. Collins CT, Gibson RA, Anderson PJ, et al. Neurodevelopmental outcomes at 7 years' corrected age in preterm infants who were fed high-dose docosahexaenoic acid to term equivalent: a follow-up of a randomised controlled trial. *BMJ Open*. 2015;5:e007314.
182. Molloy CS, Stokes S, Makrides M, Collins CT, Anderson PJ, Doyle LW. Long-term effect of high-dose supplementation with DHA on visual function at school age in children born at <33 wk gestational age: results from a follow-up of a randomized controlled trial. *Am J Clin Nutr*. 2016;103:268–275.
183. Makrides M, Gibson RA, McPhee AJ, et al. Neurodevelopmental outcomes of preterm infants fed high-dose docosahexaenoic acid: a randomized controlled trial. *JAMA*. 2009;301:175–182.

184. Smithers LG, Collins CT, Simmonds LA, Gibson RA, McPhee A, Makrides M. Feeding preterm infants milk with a higher dose of docosahexaenoic acid than that used in current practice does not influence language or behavior in early childhood: a follow-up study of a randomized controlled trial. *Am J Clin Nutr*. 2010;91:628–634.
185. Moon K, Rao SC, Schulzke SM, Patole SK, Simmer K. Longchain polyunsaturated fatty acid supplementation in preterm infants. *Cochrane Database Syst Rev*. 2016;12:CD000375.
186. Constable RT, Ment LR, Vohr BR, et al. Prematurely born children demonstrate white matter microstructural differences at 12 years of age, relative to term control subjects: an investigation of group and gender effects. *Pediatrics*. 2008;121:306–316.
187. Gozzo Y, Vohr B, Lacadie C, et al. Alterations in neural connectivity in preterm children at school age. *Neuroimage*. 2009;48:458–463.
188. Lubsen J, Vohr B, Myers E, et al. Microstructural and functional connectivity in the developing preterm brain. *Semin Perinatol*. 2011;35:34–43.
189. Mullen KM, Vohr BR, Katz KH, et al. Preterm birth results in alterations in neural connectivity at age 16 years. *Neuroimage*. 2011;54:2563–2570.
190. Schafer RJ, Lacadie C, Vohr B, et al. Alterations in functional connectivity for language in prematurely born adolescents. *Brain*. 2009;132:661–670.
191. Smyser CD, Snyder AZ, Shimony JS, Blazey TM, Inder TE, Neil JJ. Effects of white matter injury on resting state fMRI measures in prematurely born infants. *PLoS One*. 2013;8:e68098.
192. White TP, Symington I, Castellanos NP, et al. Dysconnectivity of neurocognitive networks at rest in very-preterm born adults. *Neuroimage Clin*. 2014;4:352–365.
193. van Wezel-Meijler G, van der Knaap MS, Huisman J, Jonkman EJ, Valk J, Lafeber HN. Dietary supplementation of long-chain polyunsaturated fatty acids in preterm infants: effects on cerebral maturation. *Acta Paediatr*. 2002;91:942–950.
194. Almaas AN, Tamnes CK, Nakstad B, et al. Diffusion tensor imaging and behavior in premature infants at 8 years of age, a randomized controlled trial with long-chain polyunsaturated fatty acids. *Early Hum Dev*. 2016;95:41–46.
195. Tam EW, Chau V, Barkovich AJ, et al. Early postnatal docosahexaenoic acid levels and improved preterm brain development. *Pediatr Res*. 2016;79:723–730.
196. Carver JD, Benford VJ, Han B, Cantor AB. The relationship between age and the fatty acid composition of cerebral cortex and erythrocytes in human subjects. *Brain Res Bull*. 2001;56:79–85.
197. Giedd JN, Lalonde FM, Celano MJ, et al. Anatomical brain magnetic resonance imaging of typically developing children and adolescents. *J Am Acad Child Adolesc Psychiatry*. 2009;48:465–470.
198. Ryan AS, Astwood JD, Gautier S, Kuratko CN, Nelson EB, Salem Jr N. Effects of long-chain polyunsaturated fatty acid supplementation on neurodevelopment in childhood: a review of human studies. *Prostaglandins Leukot Essent Fatty Acids*. 2010;82:305–314.
199. McNamara RK, Able JA, Jandacek RJ, et al. Docosahexaenoic acid supplementation increases prefrontal cortex activation during sustained attention in healthy boys: A placebo-controlled, dose-ranging, functional magnetic resonance imaging study. *Am J Clin Nutr*. 2010;91:1060–1067.
200. Almeida DM, Jandacek RJ, Weber WA, McNamara RK. Docosahexaenoic acid biostatus is associated with event-related functional connectivity in cortical attention networks of typically developing children. *Nutr Neurosci*. 2017;20:246–254.
201. McNamara RK, Jandacek RJ, Rider T, et al. Low docosahexaenoic acid status is associated with reduced indices of cortical integrity in the anterior cingulate of healthy male children: A ^{1}H MRS study. *Nutr Neurosci*. 2013;16:183–190.
202. Chang JP, Su KP, Mondelli V, Pariante CM. Omega-3 polyunsaturated fatty acids in youths with attention deficit hyperactivity disorder: a systematic review and meta-analysis of clinical trials and biological studies. *Neuropsychopharmacology*. 2018;43:534–545.
203. Rubia K, Alegría AA, Brinson H. Brain abnormalities in attention-deficit hyperactivity disorder: a review. *Rev Neurol*. 2014;58(Suppl 1):S3–16.
204. Bloch MH, Qawasmi A. Omega-3 fatty acid supplementation for the treatment of children with attention-deficit/hyperactivity disorder symptomatology: systematic review and meta-analysis. *J Am Acad Child Adolesc Psychiatry*. 2011;50:991–1000.

CHAPTER

2

Thinking in Polyunsaturated Fatty Acids, Phospholipids, and the Brain

Cyrus E. Kuschner, Tsukasa Yagi†, Junhwan Kim*,†*

*Donald and Barbara Zucker School of Medicine at Hofstra/Northwell, Hempstead, NY, United States †Department of Emergency Medicine, The Feinstein Institute for Medical Research, Northwell Health System, Manhasset, NY, United States

INTRODUCTION

The brain is frequently described as being particularly enriched with polyunsaturated fatty acids (PUFAs).[1–5] Furthermore, this high PUFA content has been proposed to be responsible for the high vulnerability of the brain to ischemia/reperfusion injury.[6–10] This belief appears to intuitively make sense. Since the bisallylic position of PUFAs is highly prone to peroxidative modifications by reactive oxygen species (ROS),[11, 12] this high reactivity could predispose the brain to postischemic and reperfusion injury. However, evidence supporting this perspective is limited.

Most cellular fatty acids are esterified to phospholipids, which are composed of a diacylglycerol moiety attached to a phosphate group that can be attached to a variety of head groups. Fatty acids, which determine the species of phospholipids within a class of phospholipids, are used to create the two acyl chains denoted as Sn-1 and Sn-2, respectively. Typically, unsaturated fatty acids or monounsaturated fatty acids (MUFAs) are placed at the Sn-1 position, while PUFAs are placed at the Sn-2 position.[13–15] Phospholipids can contain unsaturated or MUFA at the Sn-2 position; however, this is less common. The most abundant PUFAs observed in mammalian tissues are docosahexaenoic acid (DHA, 22:6) and arachidonic acid (AA, 20:4), which have widespread effects on tissue inflammation postischemia.

The degree of phospholipid unsaturation alters the properties of the individual phospholipid as well as the membrane it comprises. Increased membrane unsaturation theoretically increases the number of the bisallylic carbons which, due to increased reactivity toward ROS, enhances membrane susceptibility to oxidative damage. This is especially relevant to mitochondrial phospholipids that border the electron transport chain, the major site of ROS generation.[16, 17] Concurrently, the degree of unsaturation alters the phospholipid's fluidity, with increased unsaturation increasing membrane fluidity and permeability.[18] Furthermore, the

Omega Fatty Acids in Brain and Neurological Health
https://doi.org/10.1016/B978-0-12-815238-6.00002-X

degree of unsaturation alters signaling potential, increasing or decreasing the concentration of various biologically relevant PUFAs which can be metabolized to biologically potent molecules called eicosanoids.[19, 20]

Therefore, control of phospholipid content is integral to maintaining a physiologically functioning membrane, which will vary based on the tissue type and physiologic conditions present. To increase PUFA concentration, desaturase enzymes can increase unsaturation of fatty acid precursors produced via fatty acid synthase or imported into the cell.[8] From this point, unique combinations of desaturation, elongation, and beta-oxidation can be performed to acquire specific PUFAs.[9]

THE BRAIN IS THE LEAST ENRICHED WITH PUFAS COMPARED TO OTHER MAJOR ORGANS

Phospholipids represent the majority of cellular fatty acids and carry the capacity to be molded by the requirements of each individual tissue type as well as by stresses on the cell. As such, there is no reference for the unique phospholipid profile found in any specific tissue type. Instead the relative degree of phospholipid unsaturation in the brain can be compared with other significant organ systems commonly exposed to ischemia. To this end, we implemented normal-phase high-performance liquid chromatography (HPLC) to analyze phospholipid profiles in brain, heart, kidney, and liver tissues.[21] Class separation by normal-phase HPLC allows an easy comparison of phospholipid profiles between tissues while significantly reducing the risk of false positive class allocation.[22]

In mammalian tissues, phosphatidylcholine (PC) and phosphatidylethanolamine (PE), composed of both diacyl PE and PE plasmalogens (PEP), are the most abundant classes of phospholipids. We measured these two classes of phospholipids which account for 31% and 55% of total phospholipid content, respectively, in the rat brain.[21] Other tissue types have more significant PC concentrations in exchange for decreased PE content. This is observed in heart, kidney, and liver tissues which respectively have greater PC content (38%, 35%, and 48%) with decreased PE content (49%, 48%, and 35%). The decreased PE is attributable to decreased PEP content.[21] Although the relative contents of PC and PE vary depending on the quantitation method used for each study, most studies reported that these two classes of phospholipids make up over 75% of the membrane phospholipids in rat tissues.[23–25] Therefore, the fatty acid composition of PC and PE determines the degree of membrane saturation of a tissue.

Interestingly, brain PC content in rats is composed predominately of the non-PUFA containing species, PC (16:0/18:1), PC (16:0/16:0), PC (18:1/18:1), and PC (18:0/18:1) (Fig. 1A). In contrast, other tissues are relatively enriched with PC species containing AA as the peaks corresponding to PC(16:0/20:4) and PC(18:0/20:4) are the two most intense peaks (Fig. 1B and C).

When spectra of brain, heart, liver, and kidney tissues are compared it is readily apparent that brain phospholipids hold a uniquely high relative concentration of phospholipid species containing MUFAs and saturated fatty acids. Fig. 2 shows that in brain tissue, non-PUFA containing PC species account for ~65% of total PC while these species represent a minor or minimal portion of total PC in heart, kidney, and liver tissues. It is notable that in the brain both DHA and AA contribute a significant portion of total PC PUFA content, whereas AA

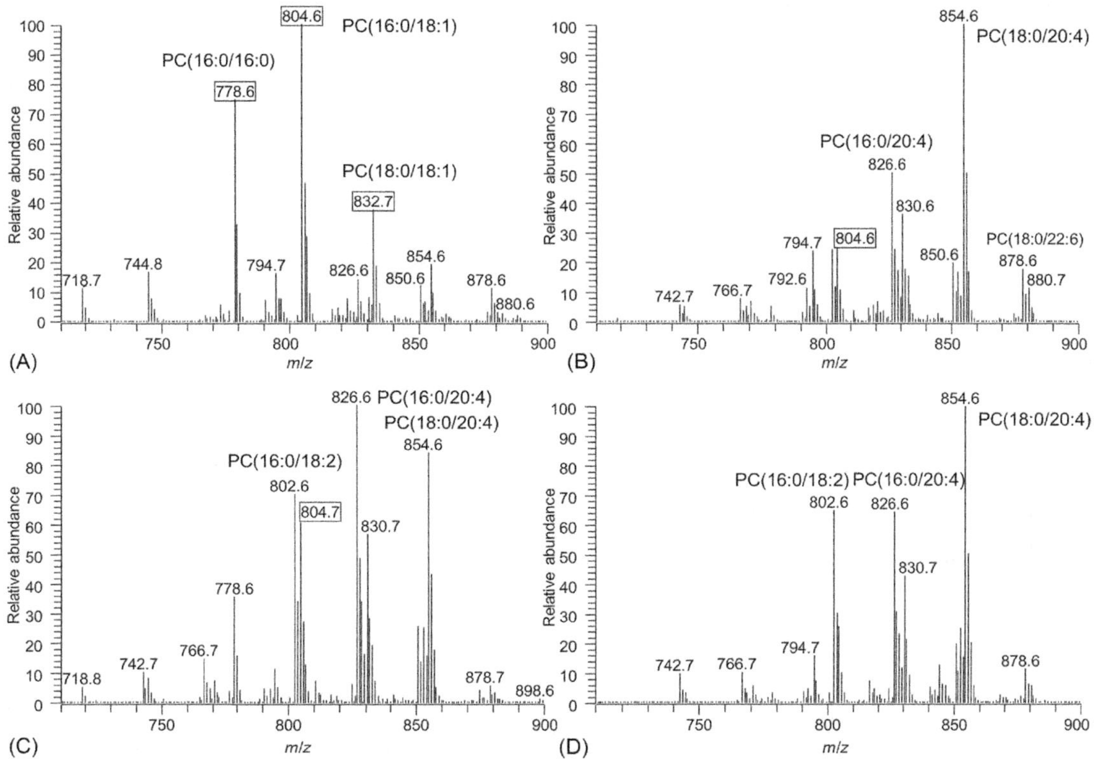

FIG. 1 MS spectra of PC from brain (A), heart (B), kidney (C), and liver (D) tissues. Labeled are the three most abundant PC species in brain tissue containing saturated fatty acid or monounsaturated fatty acids. *Reproduced with permission from Choi J, Yin T, Shinozaki K, et al. Comprehensive analysis of phospholipids in the brain, heart, kidney, and liver: brain phospholipids are least enriched with polyunsaturated fatty acids. Mol. Cell. Biochem. 2018;442(1–2), 187–201.*

is the primary PUFA in the other tissues, consistent with the perspective that omega-3 fatty acids are especially important for proper brain function.[9, 26]

Diacyl PE species in the rat brain are predominately composed of PE(18:0/22:6) and PE(18:0/20:4), as the corresponding peaks at *m/z* 790.7 and 766.8 are the two most intense peaks in MS spectrum of brain PE (Fig. 3A). These two species are also major species in heart tissue (Fig. 3B), whereas PE(18:0/20:4) is the most dominant species in kidney and liver tissues (Fig. 3C and D). Brain tissue also comprises significant quantities of non-PUFA containing species. The peaks at *m/z* 716.8 [PE(16:0/16:0)], 742.9 [PE(18:1/18:1)], and 744.9 [PE(18:0/18:1)] are readily observable in the brain (Fig. 3A), but are negligible in any of the other analyzed tissue types. The peak at *m/z* 742.8 in the heart corresponds to PE(18:0/18:2), whereas the peak with the same m/z in the brain corresponds to PE(18:1/18:1).

Direct comparison of PE profiles between tissues more clearly demonstrates that the brain is less enriched with PUFAs. In the brain, the two most abundant diacyl PE species, PE(18:0/22:6) and PE(18:0/20:4), account for 11% and 8% of total PPL content in the brain (Fig. 4A), compared to 20% and 16%, respectively, in the heart (Fig. 4B). In the kidney, arachidonic acid containing species, PE(18:0/20:4) and PE(16:0/20:4), account for 28% and 9% of total

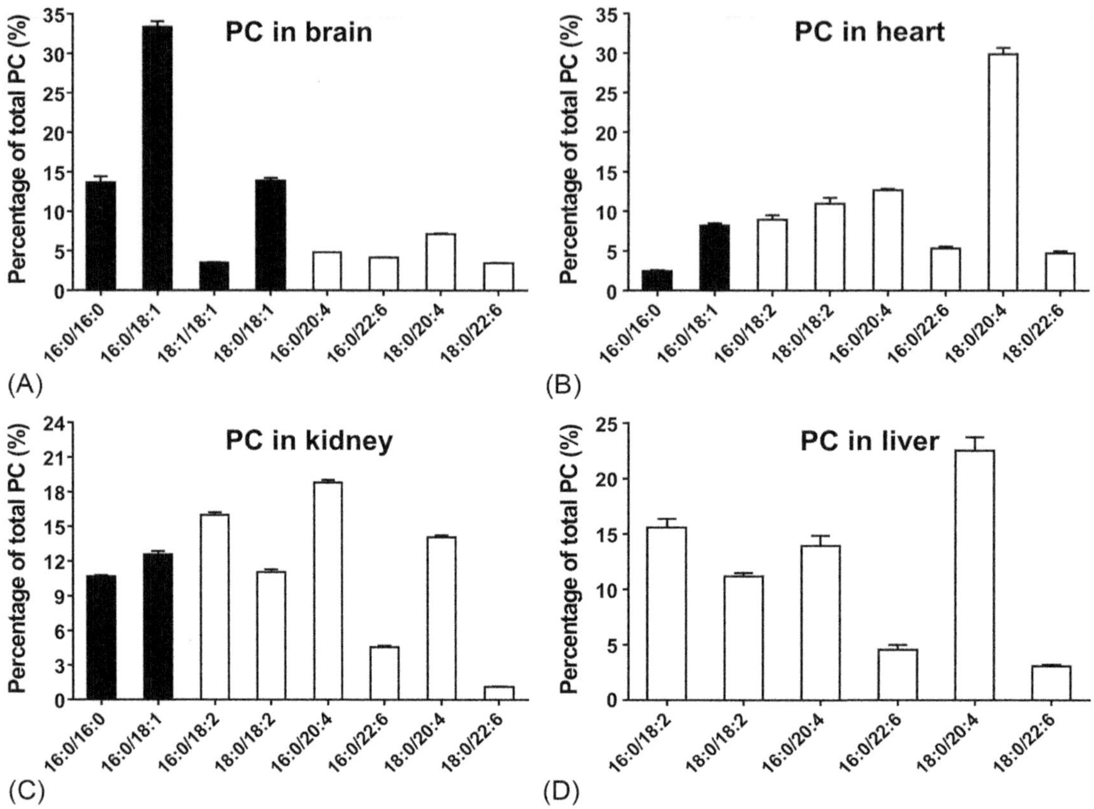

FIG. 2 Relative contents of major species of PC from the brain (A), heart (B), kidney (C), and liver (D) (filled bar = Saturated or MUFA, open bar = PUFA). *Reproduced with permission from Choi J, Yin T, Shinozaki K, et al. Comprehensive analysis of phospholipids in the brain, heart, kidney, and liver: brain phospholipids are least enriched with polyunsaturated fatty acids. Mol Cell Biochem. 2018;442(1–2), 187–201.*

PPL content, respectively (Fig. 4C). In the liver, one dominant species, PE(18:0/20:4), accounts for 30% of total PE. Other major species, PE(18:0/22:6), PE(16:0/20:4), and PE(18:0/22:6), also compose a substantial portion of PE in the liver (Fig. 4D). Although PE in brain tissue is generally more enriched with PUFAs, it also carries significant portions of PE species containing MUFAs, such as PE(16:0/18:1), PE(18:0/18:1), and PE(18:1/18:1). This is a trait found exclusively in the brain.

Rat brain tissue also contains significant quantities of PEP species enriched with MUFAs. This contrasts with heart, kidney, and liver PEP, which carry negligible quantities of MUFA containing PEP (Fig. 5). Previous analyses have found PEP to be almost exclusively enriched with PUFAs,[27] as such brain tissue's MUFA-enriched PEP profile is unique. We also observed that brain tissue carried a higher MUFA content in phosphatidylserine,[21] the third most abundant phospholipid class in the brain, compared to the other tissues. Taken as a whole, these data demonstrate that contrary to the common perception, the brain is in fact significantly less saturated with PUFA-enriched phospholipids than other major organs.

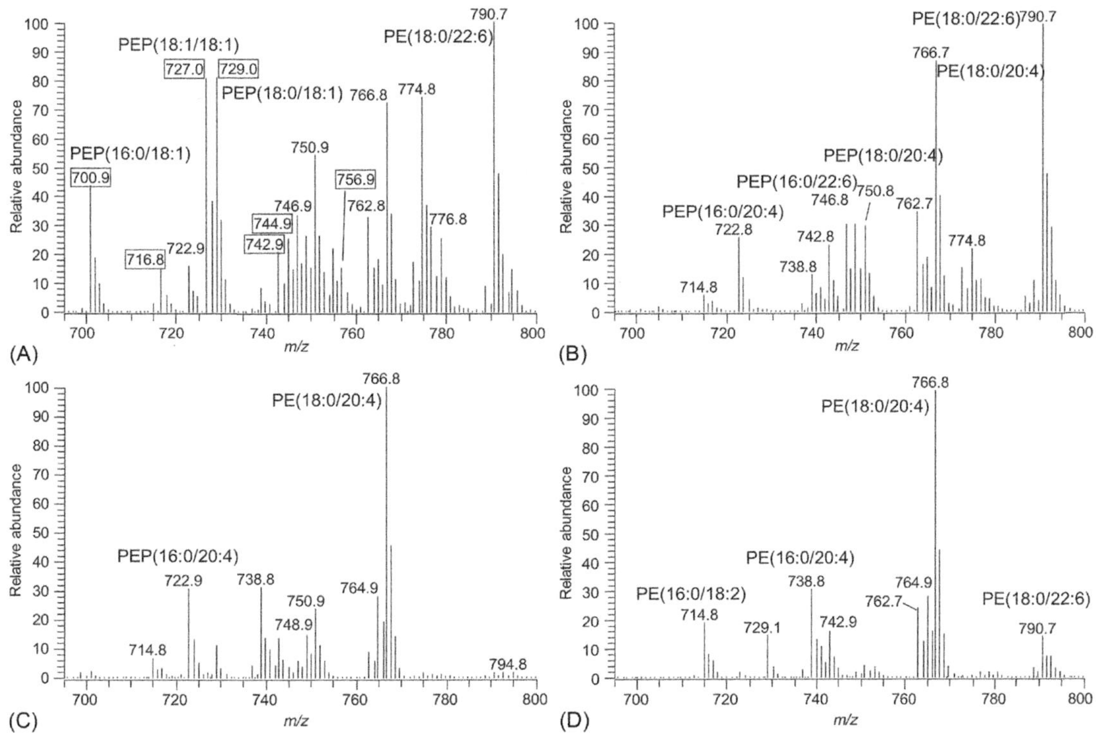

FIG. 3 MS spectra of PE from brain (A), heart (B), kidney (C), and liver (D) tissues. Major species of diacyl and plasmenyl PE in heart, kidney, and liver tissues contain PUFA (B–D) whereas PE species containing MUFA (labeled in boxes) are enriched in brain tissues (A). *Reproduced with permission from Choi J, Yin T, Shinozaki K, et al. Comprehensive analysis of phospholipids in the brain, heart, kidney, and liver: brain phospholipids are least enriched with polyunsaturated fatty acids. Mol Cell Biochem. 2018;442(1–2), 187–201.*

BRAIN MITOCHONDRIA ARE ALSO LEAST ENRICHED WITH PUFAS COMPARED TO OTHER MAJOR ORGANS

Brain tissue includes a wide variety of cell types, each with different organelles composed largely of phospholipid membranes. Among these contents, mitochondria have a significant potential to contribute to ischemic/reperfusion-induced damage due to their contribution to free-radical generation, induction of pro-inflammatory signal generation, and capacity to induce poststress cell death.[28] As such, their phospholipid contents have a unique importance in overall induction and survival of cell stressors.

Analysis of mitochondrial phospholipids in the brain, heart, kidney, and liver revealed that mitochondrial phospholipid content mimics that of their respective whole tissue profile, but with increased relative PUFA contents (Fig. 6).[29] Interestingly, brain mitochondria showed the most significant differences compared to their whole tissue phospholipid profile. This difference is clearly present both in PC and PE in brain tissue whereas it is only observed in PC in the other tissue types. Focusing on PC, brain mitochondria carried 17% more DHA and AA containing species (Fig. 6A). Kidney and liver mitochondria also held a significant increase in

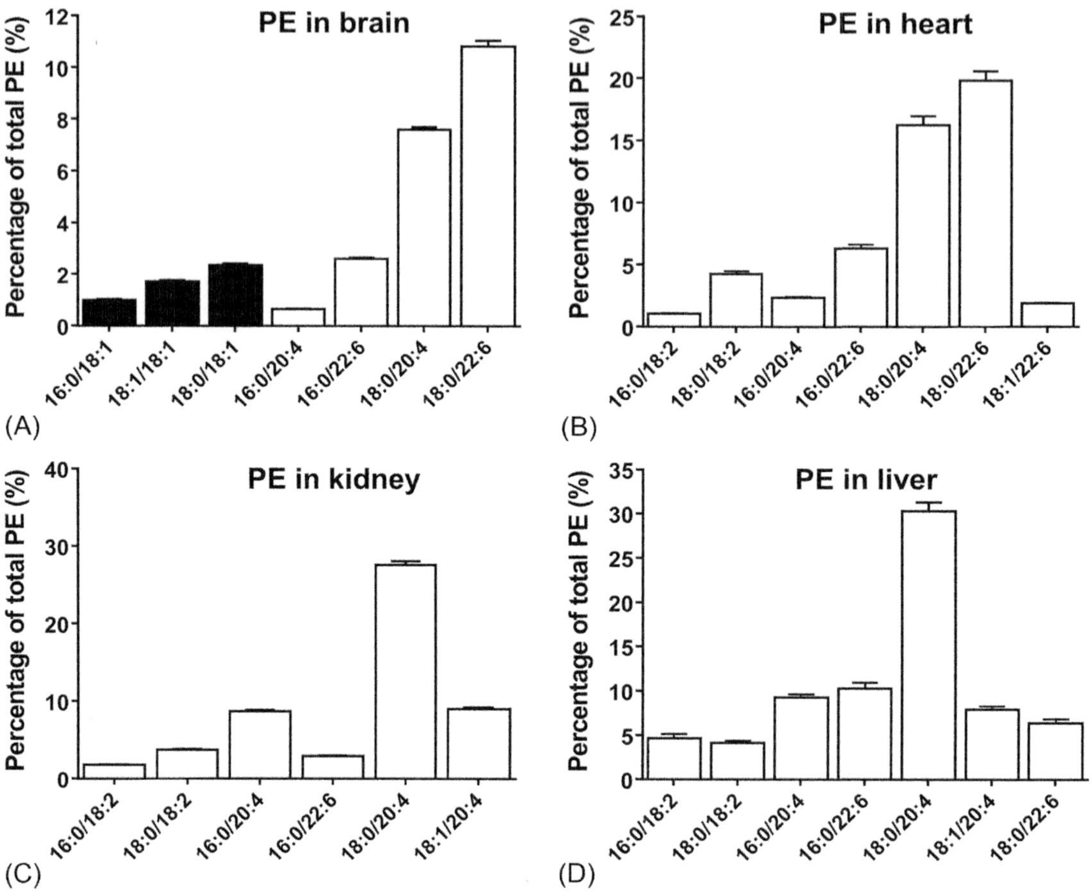

FIG. 4 Relative contents of diacyl phosphatidylethanolamine in the brain (A), heart (B), kidney (C), and liver (D) (filled bar = MUFA, open bar = PUFA). *Reproduced with permission from Choi J, Yin T, Shinozaki K, et al. Comprehensive analysis of phospholipids in the brain, heart, kidney, and liver: brain phospholipids are least enriched with polyunsaturated fatty acids. Mol Cell Biochem. 2018;442(1–2), 187–201.*

relative unsaturation (Fig. 6C and D), while this increase was comparatively minimal in heart mitochondria (Fig. 6B). We also found that DHA and AA contents in brain mitochondrial PE, including PEP, are ~20% higher than in whole brain tissue.[29] Other tissues carry only a minimal increase in mitochondrial membrane PE PUFA content. However, despite carrying the greatest increase in mitochondrial PUFA content of the four tissues, the relative mitochondrial PUFA content in the brain is still lower than the other organs.

Our findings demonstrated an increased relative content of PUFAs in mitochondria compared to whole tissue in the brain, kidney, liver, and likely in the heart. Increased mitochondrial PUFA content was most directly seen through changes in PC composition. While this phenomenon was observed in all tissue types studied, each organ system shows a distinct preference for selecting PUFAs. In the brain, increased mitochondrial unsaturation was achieved through an equivalent increase in the relative content of AA and DHA (Fig. 6A),

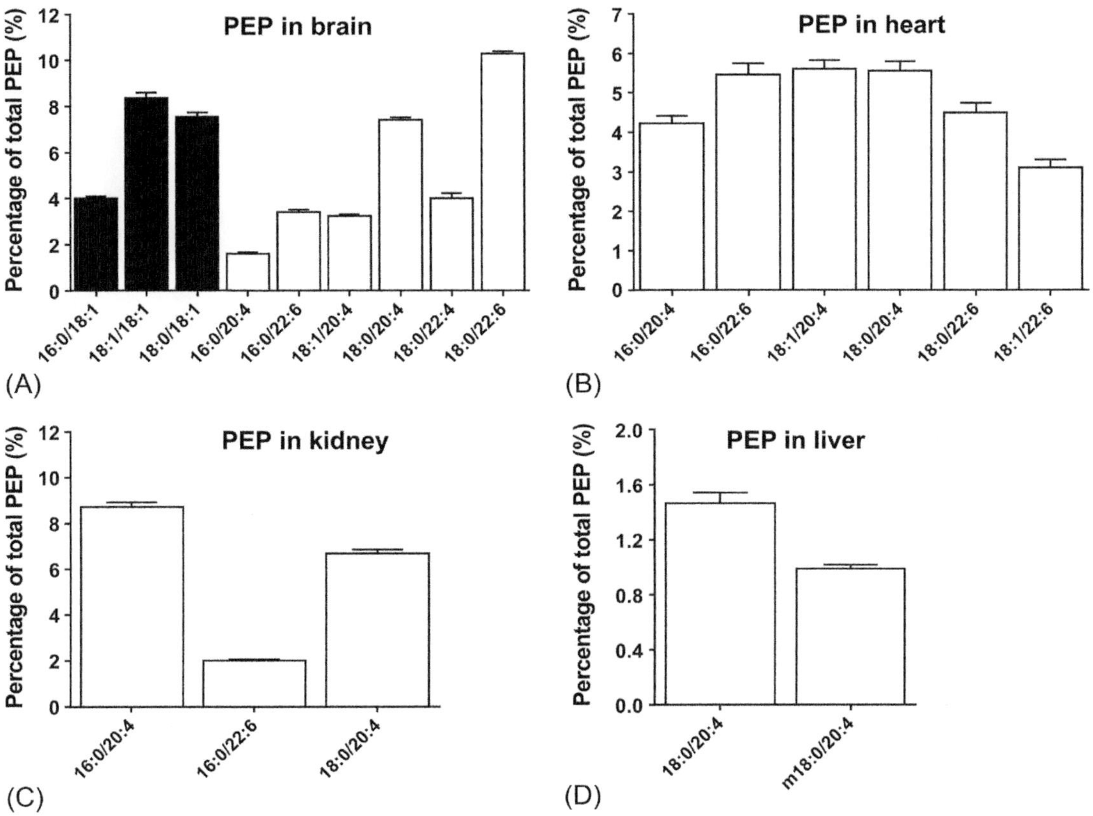

FIG. 5 Relative contents of plasmenyl phosphatidylethanolamine in the brain (A), heart (B), kidney (C), and liver (D) (filled bar = MUFA, open bar = PUFA). *Reproduced with permission from Choi J, Yin T, Shinozaki K, et al. Comprehensive analysis of phospholipids in the brain, heart, kidney, and liver: brain phospholipids are least enriched with polyunsaturated fatty acids. Mol Cell Biochem. 2018;442(1–2), 187–201.*

while kidney and liver tissues achieved greater mitochondrial PUFA content mostly through AA (Fig. 6C and D). Mitochondrial specific preference for increased concentrations of omega-3 or omega-6 may indicate a predilection for pro- or anti-inflammatory precursors.

It remains unclear why mitochondria require a higher concentration of PUFA than surrounding tissue membranes. Previously, studies have theorized that mitochondria should preferentially decrease PUFA content to protect against ROS generation.[16, 30, 31] As PUFAs are prone to peroxidation and mitochondria represent a major site of ROS generation, it is easy to assume that mitochondria would benefit from decreasing the relative content of PUFA to limit damage to the mitochondrial membrane and avoid protein degradation. However, our finding that mitochondria hold a greater relative quantity of PUFAs in multiple organ systems opposes previous perceptions regarding the function of PUFAs in health and disease. It must be recognized that factors such as membrane fluidity or downstream utility of PUFA metabolites are essential components of membrane function which need to be balanced alongside potential risk of oxidative stress.

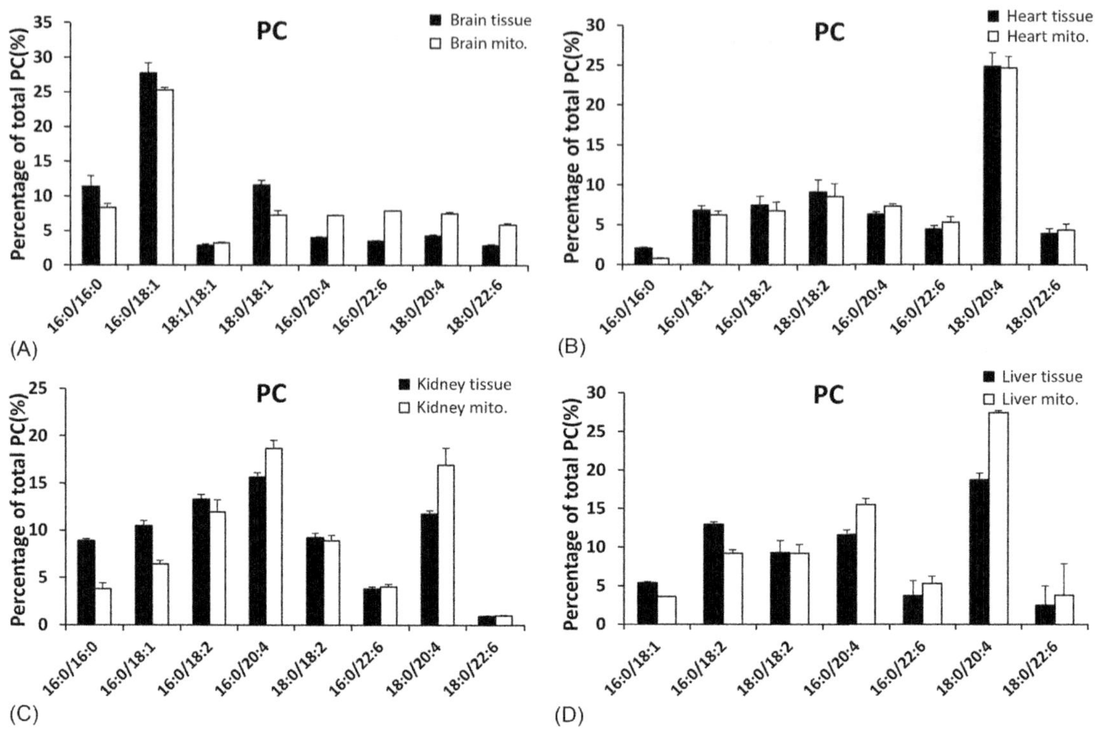

FIG. 6 Comparing relative content of PC between whole tissue and mitochondria in brain (A), heart (B), kidney (C), and liver (D) tissues.

ISCHEMIC CHANGES TO PHOSPHOLIPID PROFILES MAY EXPLAIN THE BRAIN'S UNIQUE VULNERABILITY

Any attempt to translationally apply a physiologic understanding of phospholipid composition requires analysis of how pathophysiologic conditions alter membrane composition as well as the importance of this shifting phospholipid profile on cellular function and physiological outcome. To this end, cardiac arrest and resuscitation represent a useful method for comparing how each organ, carrying unique phospholipid profiles, responds to ischemia and reperfusion. Previously, our group has used an asphyxia-induced cardiac arrest rat model that limited blood flow to induce physiologically relevant ischemia to all major organs. After a period of ischemia, rats were resuscitated using a customized cardiopulmonary bypass (CPB) system which most directly simulated extracorporeal membrane oxygenation, a developing approach to returning oxygenated blood to cardiac arrest patients with end-organ damage. This approach is unique in its capacity to both induce severe ischemia and to resuscitate animals in a highly reproducible manner.[32, 33]

Previously, it has been reported that increased lysophospholipids and free fatty acids, matched by a decrease in phospholipid content, may be an integral component of ischemic tissue damage.[33–35] Although a detailed mechanism is not clear, it has been theorized that dysfunction emerges via a variety of pathways including direct damage to membrane integrity, deleterious effects of increased lysophospholipids and free fatty acids, and depletion of phospholipid reservoirs that maintain cellular metabolic function.[8, 36–38]

Using the previously described cardiac arrest and CPB resuscitation model, we examined phospholipid alterations in brain, heart, kidney, and liver tissues following 30 min of cardiac arrest or cardiac arrest followed by 60 min of resusciation.[32] At 30 min of ischemia there were no significant changes to the phospholipid content in any of the four tissues. However, there were significant elevations in lysophosphatidylethanolamine (LPE), lysophosphatidylcholine (LPC), and lysophosphatidylinositol (LPI) concentrations in brain tissue. These elevations were unique to the postischemic brain as only an increase in LPI concentrations was found in postischemic heart, kidney, and liver (Fig. 7). Interestingly, increased LPE and LPC content in the brain was not specific to any singular fatty acid species, with species containing 18:0, 18:2. 20:4, and 22:6 all increased. This observation opposes the previously proposed theory of preferential hydrolysis of AA or DHA during ischemia.[39, 40]

Following 60 min of CPB resuscitation, heart, kidney, and liver tissues showed significant reduction in LPI levels, normalizing their lysophospholipid contents. However, brain tissue was unique with most of the lysophospholipids, which had been increased following cardiac arrest, remaining elevated following resuscitation. This suggests that while resuscitation can assist in returning key organs such as the heart, kidney, and liver to normal phospholipid status, it is unable to ameliorate phospholipid alterations induced by ischemia to the brain.

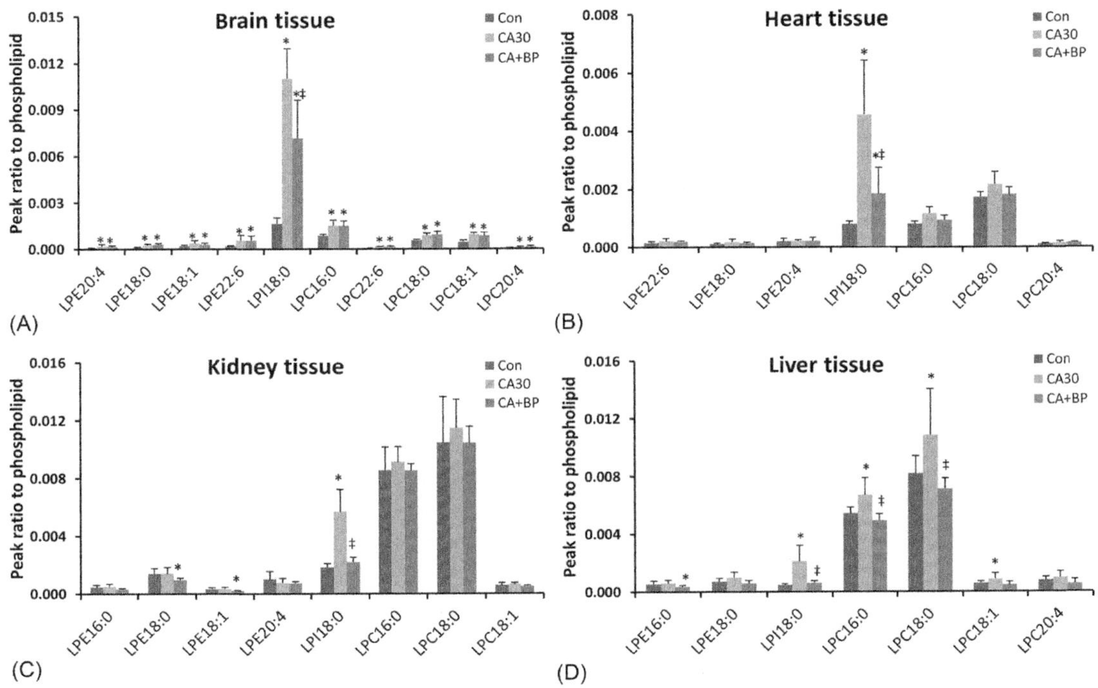

FIG. 7 Comparing LPE, LPC, and LPI content of control (Con), after 30 min of cardiac arrest (CA30), and after 30 min of cardiac arrest followed by cardiopulmonary bypass (CA+BP) in brain (A), heart (B), kidney (C), and liver (D) tissues (*<0.05, against control; ‡<0.05, against CA).

This finding is correlated with the mitochondrial respiration activities of these organs in that the respiration rates of kidney and liver mitochondria were significantly recovered by CPB resuscitation after sustaining an ischemia-mediated decrease in respiration rate. However, CPB resuscitation did not restore normal brain mitochondria respiratory function.[33] This unrecovered mitochondrial activity is further correlated with an observed lack of brain function in these animals.[41] The correlation between lysophospholipid content postischemia and reperfusion, mitochondrial function, and physiologic outcome of damaged organs, suggests that the initial production and capacity to return lysophospholipids to physiologic levels may play a role in returning organ systems to normal function postischemia and reperfusion.

Severe phospholipid alterations in brain tissue appear to stem not from inherent preischemic elevated PUFA concentrations. Instead, brain tissue is ultimately less capable of mitigating postischemic stress compared to other major organs. This unique sensitivity to ischemia and reperfusion could thus be due to a variety of causes including sensitive membrane integrity for physiologic function, decreased capacity to normalize lysophospholipid concentrations, and inability to maintain normal metabolic functions. The degree to which these interconnected factors contribute to phospholipid alteration and end-organ damage requires further research.

CONCLUSION

The brain is not especially abundant in PUFA. Therefore, excessive PUFA content does not appear to explain its unique vulnerability to ischemia and reperfusion. Instead, like all tissue types, the brain requires enough PUFA to fulfill structural, transport, signaling, and other physiologic functions of its phospholipid membranes. However, the brain may be more sensitive to external stressors due to its low stock of energy sources, making it difficult to effectively return, poststress, to physiologic conditions. Furthermore, while other organs can function effectively with mild to moderate impairment, a similar degree of damage to brain tissue has more significant impact on patient capacities. As such, the lowered PUFA content found in brain tissue may serve as a protective mechanism to decrease pro-inflammatory signaling capacity during pathologic stress. Finally, while overall PUFA content is decreased, there is an increased relative concentration of DHA which, while not directly protective, has metabolites that ameliorate inflammatory stress via resolvins and protectins. Thus, it may not be the direct quantity of PUFA that determines the vulnerability to ischemic and reperfusion damage, but instead the cellular conditions that facilitate PUFA-mediated metabolism.

References

1. Farooqui AA, Horrocks LA, Farooqui T. Modulation of inflammation in brain: a matter of fat. *J Neurochem*. 2007;101(3):577–599.
2. Miller LR, Jorgensen MJ, Kaplan JR, et al. Alterations in levels and ratios of n-3 and n-6 polyunsaturated fatty acids in the temporal cortex and liver of vervet monkeys from birth to early adulthood. *Physiol Behav*. 2016;156:71–78.
3. Shepro D. *Microvascular Research: Biology and Pathology*. Elsevier Science & Technology Books; 2005.

4. Moore SA. Polyunsaturated fatty acid synthesis and release by brain-derived cells in vitro. *J Mol Neurosci.* 2001;16(2–3):195–200 [discussion 215-121].
5. Srinivasan V, Pandi-Perumal SR, Cardinali DP, Poeggeler B, Hardeland R. Melatonin in Alzheimer's disease and other neurodegenerative disorders. *Behav Brain Funct.* 2006;2:15.
6. Noseworthy MD, Bray TM. Effect of oxidative stress on brain damage detected by MRI and in vivo 31P-NMR. *Free Radic Biol Med.* 1998;24(6):942–951.
7. Friedman J. Why is the nervous system vulnerable to oxidative stress? In: Gadoth N, Göbel HH, eds. *Oxidative Stress and Free Radical Damage in Neurology*. Totowa, NJ: Humana Press; 2011:19–27.
8. Rink C, Khanna S. Significance of brain tissue oxygenation and the arachidonic acid cascade in stroke. *Antioxid Redox Signal.* 2011;14(10):1889–1903.
9. Bazinet RP, Laye S. Polyunsaturated fatty acids and their metabolites in brain function and disease. *Nat Rev Neurosci.* 2014;15(12):771–785.
10. Esfahani BA-SM, Mirmoghtadaei M, Anaraki SB. Oxidative stress and aging. In: Massoud A, Rezaei N, eds. *Immunology of Aging*. Berlin, Heidelberg: Springer Berlin Heidelberg; 2014:323–338.
11. Kim J, Minkler PE, Salomon RG, Anderson VE, Hoppel CL. Cardiolipin: characterization of distinct oxidized molecular species. *J Lipid Res.* 2011;52(1):125–135.
12. Serhan CN, Savill J. Resolution of inflammation: the beginning programs the end. *Nat Immunol.* 2005;6(12):1191–1197.
13. Angelini R, Vitale R, Patil VA, et al. Lipidomics of intact mitochondria by MALDI-TOF/MS. *J Lipid Res.* 2012;53(7):1417–1425.
14. Ivanova PT, Cerda BA, Horn DM, Cohen JS, McLafferty FW, Brown HA. Electrospray ionization mass spectrometry analysis of changes in phospholipids in RBL-2H3 mastocytoma cells during degranulation. *Proc Natl Acad Sci U S A.* 2001;98(13):7152–7157.
15. Milne S, Ivanova P, Forrester J, Alex Brown H. Lipidomics: an analysis of cellular lipids by ESI-MS. *Methods.* 2006;39(2):92–103.
16. Pamplona R, Prat J, Cadenas S, et al. Low fatty acid unsaturation protects against lipid peroxidation in liver mitochondria from long-lived species: the pigeon and human case. *Mech Ageing Dev.* 1996;86(1):53–66.
17. Girotti AW. Lipid hydroperoxide generation, turnover, and effector action in biological systems. *J Lipid Res.* 1998;39(8):1529–1542.
18. Yang X, Sheng W, Sun GY, Lee JC. Effects of fatty acid unsaturation numbers on membrane fluidity and alpha-secretase-dependent amyloid precursor protein processing. *Neurochem Int.* 2011;58(3):321–329.
19. Michalak A, Mosinska P, Fichna J. Polyunsaturated fatty acids and their derivatives: therapeutic value for inflammatory, functional gastrointestinal disorders, and colorectal cancer. *Front Pharmacol.* 2016;7:459.
20. Calder PC. Omega-3 fatty acids and inflammatory processes: from molecules to man. *Biochem Soc Trans.* 2017;45(5):1105–1115.
21. Choi J, Yin T, Shinozaki K, et al. Comprehensive analysis of phospholipids in the brain, heart, kidney, and liver: brain phospholipids are least enriched with polyunsaturated fatty acids. *Mol Cell Biochem.* 2018;442(1–2):187–201.
22. Kim J, Hoppel CL. Comprehensive approach to the quantitative analysis of mitochondrial phospholipids by HPLC-MS. *J Chromatogr B Analyt Technol Biomed Life Sci.* 2013;912:105–114.
23. Miyazawa T, Suzuki T, Fujimoto K, Yasuda K. Chemiluminescent simultaneous determination of phosphatidylcholine hydroperoxide and phosphatidylethanolamine hydroperoxide in the liver and brain of the rat. *J Lipid Res.* 1992;33(7):1051–1059.
24. Forstner GG, Tanaka K, Isselbacher KJ. Lipid composition of the isolated rat intestinal microvillus membrane. *Biochem J.* 1968;109(1):51–59.
25. Chen CY, Lam BL, Bhattacharya SK. Mass spectrometric analyses of phospholipids in the S334ter-3 rat model of retinal degeneration. *Mol Vis.* 2014;20:1605–1611.
26. Sugasini D, Thomas R, Yalagala PCR, Tai LM, Subbaiah PV. Dietary docosahexaenoic acid (DHA) as lysophosphatidylcholine, but not as free acid, enriches brain DHA and improves memory in adult mice. *Sci Rep.* 2017;7(1):11263.
27. Braverman NE, Moser AB. Functions of plasmalogen lipids in health and disease. *Biochim Biophys Acta.* 2012;1822(9):1442–1452.
28. Vakifahmetoglu-Norberg H, Ouchida AT, Norberg E. The role of mitochondria in metabolism and cell death. *Biochem Biophys Res Commun.* 2017;482(3):426–431.

29. Kuschner CE, Choi J, Yin T, et al. Comparing phospholipid profiles of mitochondria and whole tissue: Higher PUFA content in mitochondria is driven by increased phosphatidylcholine unsaturation. *J Chromatogr B Analyt Technol Biomed Life Sci*. 2018;1093-1094:147–157.
30. Pamplona R, Portero-Otin M, Ruiz C, Gredilla R, Herrero A, Barja G. Double bond content of phospholipids and lipid peroxidation negatively correlate with maximum longevity in the heart of mammals. *Mech Ageing Dev*. 2000;112(3):169–183.
31. Winczura A, Zdzalik D, Tudek B. Damage of DNA and proteins by major lipid peroxidation products in genome stability. *Free Radic Res*. 2012;46(4):442–459.
32. Kim J, Lampe JW, Yin T, Shinozaki K, Becker LB. Phospholipid alterations in the brain and heart in a rat model of asphyxia-induced cardiac arrest and cardiopulmonary bypass resuscitation. *Mol Cell Biochem*. 2015;408(1–2):273–281.
33. Kim J, Villarroel JP, Zhang W, et al. The responses of tissues from the brain, heart, kidney, and liver to resuscitation following prolonged cardiac arrest by examining mitochondrial respiration in rats. *Oxid Med Cell Longev*. 2016;2016:7463407.
34. Drgova A, Likavcanova K, Dobrota D. Changes of phospholipid composition and superoxide dismutase activity during global brain ischemia and reperfusion in rats. *Gen Physiol Biophys*. 2004;23(3):337–346.
35. Hamazaki K, Kim HY. Differential modification of the phospholipid profile by transient ischemia in rat hippocampal CA1 and CA3 regions. *Prostaglandins Leukot Essent Fatty Acids*. 2013;88(4):299–306.
36. van der Vusse GJ, Glatz JF, Stam HC, Reneman RS. Fatty acid homeostasis in the normoxic and ischemic heart. *Physiol Rev*. 1992;72(4):881–940.
37. Arnsdorf MF, Sawicki GJ. The effects of lysophosphatidylcholine, a toxic metabolite of ischemia, on the components of cardiac excitability in sheep Purkinje fibers. *Circ Res*. 1981;49(1):16–30.
38. Man RY. Lysophosphatidylcholine-induced arrhythmias and its accumulation in the rat perfused heart. *Br J Pharmacol*. 1988;93(2):412–416.
39. Bonventre JV. The 85-kD cytosolic phospholipase A2 knockout mouse: a new tool for physiology and cell biology. *J Am Soc Nephrol*. 1999;10(2):404–412.
40. Sun D, Gilboe DD. Ischemia-induced changes in cerebral mitochondrial free fatty acids, phospholipids, and respiration in the rat. *J Neurochem*. 1994;62(5):1921–1928.
41. Kim J, Yin T, Yin M, et al. Examination of physiological function and biochemical disorders in a rat model of prolonged asphyxia-induced cardiac arrest followed by cardio pulmonary bypass resuscitation. *PLoS One*. 2014;9(11):e112012.

CHAPTER

3

The Impact of Omega-3 Fatty Acids on Quality of Life

Ondine van de Rest, M Yulianto Kurniawan, Lisette CPGM de Groot

Division of Human Nutrition and Health, Wageningen University and Research, Wageningen, The Netherlands

ABBREVIATIONS

AA	Arachidonic acid
AUDIT	Alcohol use disorders identification test
BMI	Body mass index
DHA	Docosahexaenoic acid
EPA	Eicosapentaenoic acid
FFQ	Food Frequency Questionnaire
HRQoL	Health-related quality of life
LA	Linoneic acid
MCS	Mental component summary
PCS	Physical component summary
QoL	Quality of life
SF-12	Short-form 12 health survey
SF-36	Short-form 36 health survey
SF-8	Short-form 8 health survey
WHO	World Health Organization
WHOQOL	World Health Organization Quality-of-Life questionnaire
WHOQOL-BREF	Short version of the World Health Organization Quality-of-Life questionnaire

INTRODUCTION

Current evidence indicates that a low dietary intake as well as low levels of omega-3 fatty acids, such as eicosapentaenoic acid (EPA, C20:5n-3) and docosahexaenoic acid (DHA, C22:6n-3), may be associated with mental health problems, which are among the leading causes of impaired quality of life (QoL) in old age. A possible role for omega-3 fatty acids in

https://doi.org/10.1016/B978-0-12-815238-6.00003-1

mental health is supported by multiple plausible biochemical pathways, which are nicely described and explained by.[1] These include the role of omega-3 fatty acids as structural component of the brain cell membranes, increased brain phospholipid synthesis, increased neurite growth, improved membrane fluidity, more effective neurotransmission, effects on eicosanoid synthesis and its anti-inflammatory and vasodilatory properties, and protection against neuronal loss and neurodegeneration.[1]

QoL has been defined by the World Health Organization (WHO) as an individual's perception of their position in life in the context of culture and value systems in which they live and in relation to their goals, expectations, standards, and concerns.[2] It is a broad ranging concept affected in a complex way by a person's physical health, psychological state, level of independence, social relationships, and his/her relationship to salient features of their environment.[3] Depression, anxiety, and the presence of diseases such as cancer, cardiac disease, and diabetes are highly associated with a reduced QoL.[4,5] Therefore, studies investigating these associations have also mostly been performed in diseased populations and much less in the general population. However, improving QoL is becoming an increasingly important outcome in elderly research. The fact that we are living longer is a good thing for many people, but particularly if these years could be lived well, which is hampered by inevitable age-related deterioration of physical and mental health and an increased risk of developing chronic diseases. These age-related problems are likely to affect the QoL of individuals in later years, as well as adding pressure to infrastructures such as health and social care services. Also for studies in health promotion, generic outcomes are even more relevant than disorder-specific outcomes. Such data are sorely needed since caregivers and professionals may wish to improve the patient's QoL, especially for depression, dementia, or cognitive impairment for which pharmacological treatment is often ineffective. In this chapter first the assessment of QoL will be shortly addressed and subsequently the chapter provides an overview of the current observational and trial evidence with respect to the impact of omega-3 fatty acids on QoL.

ASSESSMENT OF QOL

There are several valid and reliable questionnaires to assess QoL. The most commonly used questionnaire is the Short-Form 36 (SF-36) Health Survey and another generic QoL questionnaire is the WHO QoL (WHOQOL) questionnaire. These questionnaires are generic health-related QoL surveys, because they can be used across age (18 years and older), disease, and treatment group, as opposed to disease-specific health surveys, which focus on a particular condition or disease. Disease-specific QoL measures are not included in this chapter, because this would introduce too much heterogeneity to the results and the main focus is on the general, nonphysically diseased population.

The SF-36 Health Survey consists of 36 items, grouped into 8 scales; each 0–100 scale measuring a different aspect of health.[6] Two summary scores can be calculated: a psychometrically based physical component summary (PCS) score and a mental component summary (MCS) score. Physical health (PCS) is assessed with scales on physical functioning, role limitations due to physical health, bodily pain, and general health. The MCS score is calculated with scales on vitality (energy/fatigue), role limitations due to emotional well-being, social functioning, and mental health. The reference period is the past 4 weeks for the scales that have a recall period. Higher scale scores represent better self-reported health and the scales have a good validity for measuring physical and mental health constructs.[7] There are also two shortened versions: the Short-Form 12 (SF-12) Health Survey and the Short-Form 8 (SF-8) Health Survey that measure the same eight health domains.[8,9]

The WHOQOL is a 100-item QOL instrument developed by the WHO to facilitate cross-cultural comparisons in QOL research.[10] The WHOQOL instruments focus on individual's own views of their well-being. The short version of the World Health Organization quality-of-life questionnaire (WHOQOL-BREF) was developed to enable a brief, but accurate, assessment of QOL in routine clinical work, epidemiological studies, and clinical trials.[11] The questionnaire comprises 26 items, including two general items and 24 items covering four domains: (1) physical health, (2) psychological health, (3) social relationships, and (4) environment. The scores of the two general items range from 1 to 5 and the domain scores range from 7 to 35, 6 to 30, 3 to 15, and 8 to 40 for the domains 1 through 4 consecutively. The total score range of the WHOQOL-BREF is 26–130 with higher scores indicating a favorable condition. The reference period is the previous 2 weeks and questions have a five-point Likert scale. Studies have shown good content validity, discriminant validity, and test–retest reliability.[12,13]

CURRENT EVIDENCE ON OMEGA-3 FATTY ACIDS AND QOL

Evidence From Observational Studies

Three observational studies have investigated the association between omega-3 fatty acids and QoL (Table 1).[14–16]

Schiepers et al. observed a significant positive trend of fish consumption in relation to physical well-being in 470 Dutch individuals aged 24–81 years[16] (P for trend <.01; $R^2=0.187$). However, no association was observed between fish consumption and the mental health component.

Contrary to the first study, Silvers and Scott observed a significant positive association between fish intake and self-reported scores on the mental health part of the SF-36 and a negative association between fish consumption and physical health (P for trend = .045) in 4644 New Zealand adults.[15]

The third observational study was also cross-sectional and based on data from the same New Zealand population-based survey as described by Silvers and Scott.[14] This time fatty acid composition of serum phospholipids was used as an indicator of omega-3 status, which resulted in a smaller sample size ($n=2416$) than in the earlier analysis that used fish intake. A significant positive trend (P for trend = .009) was observed across quintiles of the proportion of EPA in relation to self-reported physical well-being of the SF-36. The ratio of EPA to arachidonic acid (AA) was also positively associated with self-reported physical well-being (P for trend = .012). Neither the proportion of EPA nor that of DHA was associated with the mental health component, however, there was a positive association between the EPA/AA ratio and self-reported mental well-being (P for trend = .044).

TABLE 1 Summary of Observational Studies Assessing the Association Between Omega-3 Fatty Acids and Quality of Life

Author, Year	Study Population (*n*)	Design	Nutritional Status	Measure(s) of QoL	Covariate(s)	Results
Schiepers et al., 2010	Population-based sample of the Maastricht Aging study (MAAS), aged 24–81 years ($n=470$)	Cross sectional	Fish consumption measured by means of a validated short self-report questionnaire	SF-36	Age, sex, level of education, marital status, alcohol consumption, smoking, and BMI	Significant positive association between fish consumption and physical well-being (P for trend <.01; $R^2=0.187$). No significant association between fish consumption and the mental health component ($R^2=0.053$)
Crowe et al., 2007	Population-based survey in New Zealanders, aged ≥15 years ($n=2416$)	Cross sectional	Fatty acid composition in serum phospholipids	SF-36	Age category, sex, BMI category, season of blood collection, urban or rural dwelling, socioeconomic status, marital status, employment, smoking status, diabetes, asthma, use of blood pressure medication, and AUDIT category	Positive association between the proportion of EPA and the ratio of EPA/AA and self-reported physical well-being, P for trend = .009 and 0.012 respectively. No associations between the proportion of EPA or DHA with mental health. However, positive association between the EPA/AA ratio and self-reported mental health (P for trend = .044).
Silvers and Scott, 2002	Population-based survey in New Zealanders, aged ≥15 years ($n=4644$)	Cross sectional	Fish consumption assessed with FFQ	SF-36	Age, annual household income, smoking status, AUDIT category, and eating patterns	Significant association between more fish intake and better self-reported mental health status (P for trend = .005) and between more fish intake and impaired physical functioning ($P=.045$).

Abbreviations: *QoL*, Quality of life; *SF-36*, Short Form-36 Health Survey; *BMI*, Body Mass Index; *AUDIT*, Alcohol Use Disorders Identification Test; *EPA*, eicosapentaenoic acid; *DHA*, docosahexaenoic acid; *AA*, arachidonic acid; *FFQ*, Food Frequency Questionnaire.

Evidence From Clinical Trials

Three intervention studies have been performed in nonphysically diseased populations and are summarized in Table 2.

The first trial has been performed in 50 elderly aged >50 years with mild cognitive impairment who were randomly allocated to daily supplements containing DHA-rich oil (1.55 g DHA + 0.40 g EPA), EPA-rich oil (1.67 g EPA + 0.16 g DHA), or a control oil [safflower oil providing 2.2 g linoleic acid (LA), a n-6 fatty acid].[17] After 6 months of supplementation there

TABLE 2 Summary of Intervention Studies Assessing the Effect of Omega-3 Fatty Acid Supplementation on Quality of Life

Author, Year	Study Population (*n*)	Intervention and Design	Measure(s) of QoL	Results
Sinn et al., 2012	Population-based prospective cohort of elderly, 70–89 years ($n=937$)	DHA-rich oil (1.55 g DHA + 0.40 g EPA); EPA-rich oil (1.67 g EPA + 0.16 g DHA), or 2.2 g LA for 6 months	SF-36	No treatment effects on QoL. Increased erythrocyte DHA was associated with improved self-reported physical health
Rondanelli et al., 2010	Depressed women living in a nursing home, aged 66–95 years ($n=46$)	2.5 g omega-3 fatty acids (1.67 g DHA + 0.83 g EPA) or placebo for 8 weeks	SF-36	Significant difference between the omega-3 and the placebo group on the physical function score (difference (95% CI) 21.1 (10.8, 31.3), $P<.001$) and on the mental function score (difference (95% CI) 25.2 (16.3, 34.1), $P<.001$)
Van de Rest et al., 2009	Individuals without depression or dementia, aged >65 years ($n=302$)	1800 mg EPA + DHA (EPA:DHA ratio = 3:2); 400 mg EPA + DHA, or placebo for 26 weeks	WHOQOL-BREF	The mean difference (95% CI) compared to placebo was −1.42 (−3.40–0.57) for the 1800 mg EPA + DHA and 0.02 (−1.95–1.99) for the 400 mg EPA + DHA group, which was not significantly different

Abbreviations: *DHA*, docosahexaenoic acid; *EPA*, eicosapentaenoic acid; *QoL*, Quality of life; *SF-36*, Short Form-36 Health Survey; *WHOQOL-BREF*, World Health Organization Quality-of-life questionnaire, short version.

were no significant treatment effects on QoL parameters. However, increased erythrocyte DHA levels were significantly associated with improved self-reported physical functioning of the SF-36 scale. With respect to depressive symptoms a beneficial effect was shown in both the EPA and the DHA group and these improved scores correlated with increased EPA + DHA erythrocyte levels.

Another trial was performed in 46 depressed women aged 66–96 years and living in a nursing home in Italy.[18] These women were supplemented with 2.5 g omega-3 fatty acids (1.67 g DHA + 0.83 g EPA) or placebo during 8 weeks. Significant differences between the omega-3 and the placebo group were found on both the physical function score [difference (95% CI) 21.1 (10.8, 31.3), $P<.001$] and the mental function score [difference (95% CI) 25.2 (16.3, 34.1), $P<.001$] of the SF-36. Results on the total QoL score were not reported.

The third trial was performed in healthy individuals ($n=302$, mean age 70 years), who were supplemented with either 1800 mg EPA + DHA (EPA:DHA ratio = 3:2; ≈1093 mg EPA + 847 mg DHA), 400 mg EPA + DHA (226 mg EPA + 176 mg DHA), or placebo capsules for a period of 6 months.[19] No treatment effects on total QoL or any of the separate domains as measured with the WHOQOL-BREF were found.

Additional Observations From Other Clinical Trials

Some other clinical trials have been performed; however, they either did not use a validated QoL questionnaire or were performed in physically diseased patients. Therefore, they are not included in our Results, Table 2, and the conclusion of this chapter; however, they are still worth mentioning.

One such trial was conducted in 34 hemodialysis patients with a major depressive disorder who were supplemented with capsules containing either 180mg EPA and 120mg DHA or placebo for a period of 4months. A significant difference between the omega-3 and the placebo group was found on physical functioning [difference (95% CI) 23.59 (9.62, 40.15), $P=.003$].[20]

Patients with Meibomian gland dysfunction received a capsule of Brudysec 1.5g: a supplement containing a combination of omega-3 fatty acids and antioxidants (TG-DHA 350mg + EPA 42.5mg + DPA 30mg; vitamin; essential trace elements; other components) or a placebo supplement (500mg sunflower oil).[21] After 3months treatment significantly higher mean PCS and MCS scores were observed in patients treated with the active omega-3 dietary supplement as compared with controls [mean (standard deviation) PCS score 53.3 ± 5.6, $P=.008$]; [mean (standard deviation) MCS score 54.6 ± 5.6, $P=.0005$].

Another trial was performed in 100 Alzheimer's Disease patients who received SR-3: a compound comprising a 1:4 ratio of omega-3 to omega-6 fatty acids for a period of 4weeks.[22] Twelve areas of behavior were rated and beneficially improved in this study. The authors' state that based on improvements in these areas a beneficial effect on QoL was observed, however, QoL has not been assessed with a validated QoL questionnaire.

Some other recent trials did not find a significant effect of omega-3 treatment on QoL. Andreeva et al. performed an intervention trial on the effect of omega-3 fatty acids on health-related quality of life (HRQoL), measured with the SF-36, in 2029 French patients aged 45–80years and having acute myocardial infarction, unstable angina, or ischemic stroke.[23] During 5years intervention, all participants received either 600mg EPA and DHA in a 2:1 ratio; or B vitamins and omega-3 fatty acids combined. No significant effects between medium/long-term supplementation of n-3 fatty acids on HR-QoL were observed, that is, mean difference (95% CI)–0.0 (−0.9, 0.8) on the physical function score and mean difference (95% CI) 0.3 (−0.6, 1.2) on the mental function score.

Another study which was performed in Dallas (Texas) investigated the effect of fish oil on QoL in 50 systemic lupus erythematosus patients aged 18–64years.[24] The patients ingested either fish oil (2.25g EPA and 2.25g DHA) or placebo (olive oil) for 6months. There was an insignificant difference between the fish oil and the placebo group on physical functioning [median (IQR) −2.5(−10.0 to 16.3) (fish oil); 5.0 (−10.0 to 21.3) (placebo), $P=.47$] and on emotional well-being [median (IQR) 16.0 (−1.0 to 33.0) (fish oil); 4.0 (−5.0 to 14.0) (placebo), $P=.07$]. Sabour et al. executed an intervention with omega-3 on HRQoL in 104 patients with spinal cord injury.[25] Each individual was administered two capsules daily with each capsule containing either omega-3 fatty acids (465mg DHA and 63mg EPA) or placebo for a period of 14months. A nonsignificant difference between the omega-3 and the placebo group was found on the physical function score [difference−0.91 (omega-3); −0.76 (placebo), $P=.90$] and on the mental function score [difference 3.59 (omega-3); 2.67 (placebo), $P=.40$].

DISCUSSION

With only three observational and three intervention trials the number of studies that investigated omega-3 fatty acids in relation to QoL in nonphysically diseased populations is limited. Some results from the observational studies are favorable, but not consistent for mental and physical health components. The results from the trials only showed an effect of omega-3 fatty acid supplementation in the trial that included depressed individuals.

Dementia, depression, and coronary heart disease, constitute the main causes of impaired QoL in the Netherlands.[26] It is likely that QoL in general, a non-diseased population is already close to optimal and that further improvement could not be achieved with a higher omega-3 fatty acid intake or supplementation. The total body of evidence from observational as well as clinical studies on omega-3 and depression, one of the strongest determinants of QoL, is much larger and is addressed in another chapter. However, the results observed in these studies support the current results for QoL, that is, the evidence provides some support of a beneficial effect of omega-3 fatty acids in depressed individuals, but not in individuals without a diagnosis of depression.[27] If true, it can be concluded that extra omega-3 fatty acids are of little added value to mental health in a generally healthy population, but may have a role as a complementary treatment for depression or dementia[22] and as such also improve QoL. The role of omega-3 fatty acids as adjuvant treatment is supported by some other trials in patients with physical diseases where active omega-3 dietary supplementation significantly improved the physical health of hemodialysis patients with depressive symptoms[20] and physical and mental health of patients with Meibomian gland dysfunction.[21]

Of the included studies in this chapter, five used the SF-36 and one used the WHOQOL-BREF. Both questionnaires have shown good internal consistency, excellent discriminant validity, and good sensitivity.[7,12,13] Although the SF-36 is also generally the most widely used QoL questionnaire, the WHOQOL includes a strong mental health component and emphasizes the perception of the individual. Because of the implications for an association between omega-3 fatty acids and mental well-being, the WHOQOL questionnaire is also a very suitable questionnaire to be used in future studies in the field. Despite the availability of several valid and reliable QoL instruments it is a pity that they have also seldom been applied in efficacy trials that address mental well-being.[28]

The aspects included in QoL questionnaires have been generated by systematic qualitative and quantitative research among different cultures and populations and the psychometric properties have been validated in several studies. However, because mood state, physical pain or disease and QoL are strongly correlated, QoL is a subjective concept to measure as it depends on persons' self-report and an individual's perception of status. This possible lack of objectivity has to be taken into account.

Dietary DHA can be incorporated in the brain not only during development, but also later in life. The half-life of DHA in the brain is about 21 days, so studies should at least last 3–4 weeks, before the additional DHA will be incorporated into the brain.[29] Furthermore, the underlying mechanism that is targeted should also be taken into account, since different plausible mechanisms have been proposed and could be classified as either short term (i.e., weeks to months) or long term (months to years). The three intervention studies had a study duration of either 8 or 26 weeks and a beneficial effect was only found in the shortest study that lasted 8 weeks. However, this probably has more to do with the fact that in that study more sensitive subjects, that is, depressed patients were investigated rather than that a study duration of 8 weeks would be more preferable than 26 weeks.

CONCLUSION AND RECOMMENDATIONS

Based on the current evidence we conclude that the limited and conflicting total body of evidence hinders drawing a firm conclusion with respect to the impact of a higher intake of omega-3 fatty acids via the diet or via supplementation on QoL. The results from the few observational studies are partly favorable but conflicting. The results from the clinical trials are not favorable, except in depressed participants.

Because causality cannot be inferred from cross-sectional studies, which are three of the six studies that have been performed, at least some prospective studies should be performed before a valid conclusion about the role of omega-3 fatty acids in QoL can be drawn. Moreover, more intervention studies investigating the effects of omega-3 fatty acids on QoL are warranted. These intervention studies should be of sufficient study duration and particularly in subjects with an impaired QoL and/or impaired omega-3 fatty acid status offering room for improvement.

References

1. Parletta N, Milte CM, Meyer BJ. Nutritional modulation of cognitive function and mental health. *J Nutr Biochem*. 2013;24:725–743.
2. The WHOQOL group. The World Health Organization quality of life assessment (WHOQOL): Position paper from the World Health Organization. *Soc Sci Med*. 1995;41:1403–1409.
3. WHO. Study protocol for the World Health Organization project to develop a quality of life assessment instrument (WHOQOL). *Qual Life Res*. 1993;2:153–159.
4. Ali S, Stone M, Skinner TC, Robertson N, Davies M, Khunti K. The association between depression and health-related quality of life in people with type 2 diabetes: a systematic literature review. *Diabetes Metab Res Rev*. 2010;26:75–89.
5. Baumeister H, Hutter N, Bengel J, Harter M. Quality of life in medically ill persons with comorbid mental disorders: a systematic review and meta-analysis. *Psychother Psychosom*. 2011;80:275–286.
6. Ware Jr JE, Sherbourne CD. The MOS 36-item short-form health survey (SF-36). I Conceptual framework and item selection. *Med Care*. 1992;30:473–483.
7. McHorney CA, Ware Jr JE, Raczek AE. The MOS 36-item short-form health survey (SF-36): II. Psychometric and clinical tests of validity in measuring physical and mental health constructs. *Med Care*. 1993;31:247–263.
8. Ware Jr J, Kosinski M, Keller SD. A 12-item short-form health survey: construction of scales and preliminary tests of reliability and validity. *Med Care*. 1996;34:220–233.
9. Ware JE, Kosinski M, Dewey J, Gandek B. *How to Score and Interpret Single-Item Health Status Measures: A Manual for Users of the of the SF-8 Health Survey: (With a Supplement on the SF-6 Health Survey)*. Boston, MA/Lincoln, RI: Quality Metric, Inc; 2001.
10. The WHOQOL group. The World Health Organization quality of life assessment (WHOQOL): development and general psychometric properties. *Soc Sci Med*. 1998;46:1569–1585.
11. The WHOQOL group. Development of the World Health Organization WHOQOL-BREF quality of life assessment. The WHOQOL Group. *Psychol Med*. 1998;28:551–558.
12 Nelson CB, Lotfy M.1999. *The World Health Organization's WHOQOL-Bref quality of life assessment of psychometric qualities*. Results of field trial. WHO, Report No.: MNH/MHP/99.7.
13. Trompenaars FJ, Masthoff ED, Van Heck GL, Hodiamont PP, De Vries J. Content validity, construct validity, and reliability of the WHOQOL-Bref in a population of Dutch adult psychiatric outpatients. *Qual Life Res*. 2005;14:151–160.
14. Crowe FL, Skeaff CM, Green TJ, Gray AR. Serum phospholipid n 3 long-chain polyunsaturated fatty acids and physical and mental health in a population-based survey of New Zealand adolescents and adults. *Am J Clin Nutr*. 2007;86:1278–1285.

15. Silvers KM, Scott KM. Fish consumption and self-reported physical and mental health status. *Public Health Nutr*. 2002;5:427–431.
16. Schiepers OJ, de Groot RH, Jolles J, van Boxtel MP. Fish consumption, not fatty acid status, is related to quality of life in a healthy population. *Prostaglandins Leukot Essent Fatty Acids*. 2010;83:31–35.
17. Sinn N, Milte CM, Street SJ, et al. Effects of n-3 fatty acids, EPA v. DHA, on depressive symptoms, quality of life, memory and executive function in older adults with mild cognitive impairment: a 6-month randomised controlled trial. *Br J Nutr*. 2012;107:1682–1693.
18. Rondanelli M, Giacosa A, Opizzi A, et al. Effect of omega-3 fatty acids supplementation on depressive symptoms and on health-related quality of life in the treatment of elderly women with depression: a double-blind, placebo-controlled, randomized clinical trial. *J Am Coll Nutr*. 2010;29:55–64.
19. van de Rest O, Geleijnse JM, Kok FJ, et al. Effect of fish oil supplementation on quality of life in a general population of older Dutch subjects: a randomized, double-blind, placebo-controlled trial. *J Am Geriatr Soc*. 2009;57:1481–1486.
20. Dashti-Khavidaki S, Gharekhani A, Khatami MR, et al. Effects of omega-3 fatty acids on depression and quality of life in maintenance hemodialysis patients. *Am J Ther*. 2014;21:275–287.
21. Olenik A, Mahillo-Fernandez I, Alejandre-Alba N, et al. Benefits of omega-3 fatty acid dietary supplementation on health-related quality of life in patients with meibomian gland dysfunction. *Clin Ophthalmol*. 2014;8:831–836.
22. Yehuda S, Rabinovtz S, Carasso RL, Mostofsky DI. Essential fatty acids preparation (SR-3) improves Alzheimer's patients quality of life. *Int J Neurosci*. 1996;87:141–149.
23. Andreeva VA, Latarche C, Hercberg S, Briancon S, Galan P, Kesse-Guyot E. B vitamin and/or n-3 fatty acid supplementation and health-related quality of life: ancillary findings from the SU.FOL.OM3 randomized trial. *PLoS One*. 2014;9:e84844.
24. Arriens C, Hynan LS, Lerman RH, Karp DR, Mohan C. Placebo-controlled randomized clinical trial of fish oil's impact on fatigue, quality of life, and disease activity in systemic lupus Erythematosus. *Nutr J*. 2015;14:82.
25. Sabour H, Latifi S, Soltani Z, et al. The effect of Omega-3 on health-related quality of life in individuals with spinal cord injury: A double-blinded randomized clinical trial. *Clin Med J*. 2015;1:6–12.
26. RIVM. *RIVM Mortality, Morbidity and Quality of Life*. Bilthoven: RIVM; 2006. 29 May 2007.
27. Appleton KM, Rogers PJ, Ness AR. Updated systematic review and meta-analysis of the effects of n-3 long-chain polyunsaturated fatty acids on depressed mood. *Am J Clin Nutr*. 2010;91(3):757–770.
28. Scholzel-Dorenbos CJ, van der Steen MJ, Engels LK, Olde Rikkert MG. Assessment of quality of life as outcome in dementia and MCI intervention trials: a systematic review. *Alzheimer Dis Assoc Disord*. 2007;21:172–178.
29. Connor WE, Neuringer M, Lin DS. Dietary effects on brain fatty acid composition: the reversibility of n-3 fatty acid deficiency and turnover of docosahexaenoic acid in the brain, erythrocytes, and plasma of rhesus monkeys. *J Lipid Res*. 1990;31:237–247.

CHAPTER

4

The Balance Between n-6 and n-3 and its Relation to Executive Function

Kelly W. Sheppard[*,†], *Carol L. Cheatham*[‡,§]

*Nationwide Children's Hospital, Center for Biobehavioral Health, Columbus, OH, United States †The Ohio State University, Department of Pediatrics, Columbus, OH, United States ‡University of North Carolina at Chapel Hill, Nutrition Research Institute, Kannapolis, NC, United States §University of North Carolina at Chapel Hill, Department of Psychology and Neuroscience, Chapel Hill, NC, United States

INTRODUCTION

Fatty acids (FAs) are integral to every cell in the body. They are especially important in the brain where they fulfill structural and functional roles. The omega-6 (n-6) and omega-3 (n-3) FA are both important to neural tissues. Linoleic (LA; 18:2n-6) and linolenic (LNA; 18:3n-3) acids are essential; longer chain FA can be obtained from the diet or endogenously through a series of enzymatic steps. The latter method is thought to be relatively inefficient,[1] but important factors such as genetics, epigenetics, and balance were not considered when that conclusion was drawn. Importantly, the balance between the n-6 and n-3 metabolic pathways is complex and not well understood. As detailed further, the n-6 and n-3 metabolic pathways compete for the elongases and desaturases necessary to produce the longer-chain FAs (see Fig. 1). These enzymes are under control of the *FADS* gene complex, and the delta-6 desaturase from the *FADS2* gene is considered the rate-limiting enzyme. To predict the effects of an n-6 and n-3 imbalance, one must understand the genetics, the inherent differential affinities, and the various environmental influences, including diet; much research is needed to complete the understanding of this multifaceted system. In this chapter, we discuss research in which scientists utilized the n-6 to n-3 ratio. We provide the rationale for a change in terminology to "balance" rather than "ratio." After providing an evolutionary perspective, we discuss in detail the relation between the n-6/n-3 balance and the development of higher cognitive abilities known collectively as executive functions (EFs).

Omega Fatty Acids in Brain and Neurological Health
https://doi.org/10.1016/B978-0-12-815238-6.00004-3

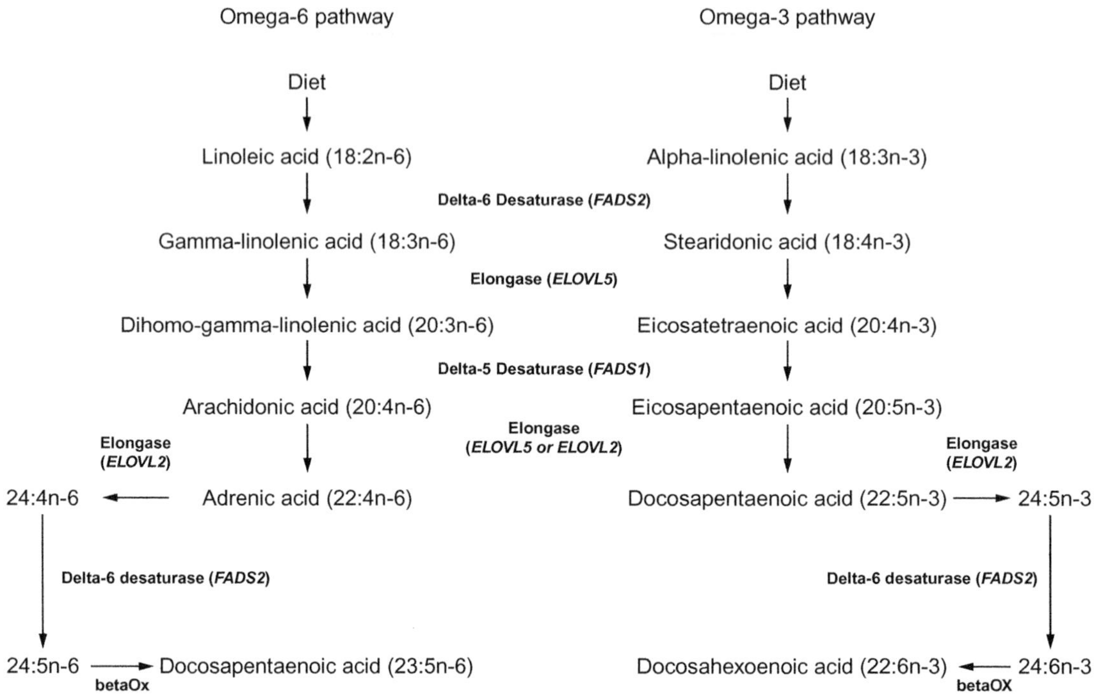

FIG. 1 The metabolic pathways of n-3 and n-6 fatty acids. *Adapted from Sheppard KW, Cheatham CL. Omega-6 to omega-3 fatty acid ratio and higher-order cognitive functions in 7- to 9-y-olds: a cross-sectional study. Am J Clin Nutr 2013;98(3):659–667.*

EVOLUTIONARY PERSPECTIVE

The human genome changes very slowly, and thus, humans today have the same or at least, a genome very similar to their Paleolithic ancestors. It has been suggested that we need to eat commensurate with our genes.[2] How do we then know what the appropriate genomic diet would be? What was the ancestral diet? Distal theories are always laden with supposition, and thus, are only an approximation, at best, of reality. Although great care has gone into determining the diet of our Paleolithic ancestors,[3–6] the fact is year-to-year climate and soil quality, in part, determine nutrient content of plant foods, and we cannot be certain what the weather and soil quality were thousands of years ago. Nonetheless, Eaton and his colleagues provide a well-researched and characterized estimate of micronutrient and energy intake in the ancestral diet.[3,6] Given the nature of the hunter-gather life, we assume that gathered plant foods were eaten immediately with little processing or even cooking, which provides the best nutritive value from plants. Hunting was most likely only randomly successful, and thus, meat consumption was not a major source of energy. Anthropologists set the intake at 65:35 plant- to animal-based calories.[7] Moreover, wild meat is generally quite lean. Thus, exogenous intake of the longer chain FA such as docosahexaenoic acid (DHA; 22:6n-3) was most likely negligible in non-coastal tribes.

It is equally difficult to determine the ancestral FA intake ratio. Nonetheless, it has oft times been posited that the Paleolithic diet contained an n-6/n-3 ratio of 1:1[2,8,9] and as such, 1:1 is put forth as a gold standard goal for eating to match ones' genes. The ratio purported to be found in the contemporary non-fish-eating Western diet is 20:1.[9–11] However, it is important to note that researchers who have collected data on consumption in non-fish-eating populations in developed countries find an n-6/n-3 ratio of closer to 10:1[12–18] Adding to the complexity, it is now widely accepted that genes are not deterministic, but rather are switched on and off by epigenetics. This programming occurs, in part, in utero. With respect to nutrition, the fetus' system is set based on maternal diet. The first evidence of this epigenetic programming of nutritional set points came from famine research: infants born to women who were pregnant during a famine have nutritional needs that have been adjusted to a world with few calories.[19–23] Thus, whereas the genome may have not changed much since the Paleolithic Era, gene expression is individualized, and nutritional needs are based on familial experience rather than distal theories.

METABOLIC PATHWAYS

Endogenous synthesis of the vital longer-chain FA arachidonic acid (AA; 20:4n-6) and DHA from their precursors LA and LNA, respectively, has two limiting factors: competition for enzymatic resources and genetically controlled production of those enzymes. Specifically, FA metabolism is limited by the delta-5 and delta-6 desaturase coded by the *FADS* genes.[24–26] The polyunsaturated FAs (PUFAs) rely on delta-5 and delta-6 desaturase to synthesize the long-chain FAs (LC-PUFAs) AA and DHA from their precursors LA and LNA, respectively. As mentioned, LNA and LA are essential nutrients: they must be obtained from the diet. The n-3, n-6, and n-9 pathways compete for a limited supply of delta-5 and delta-6 desaturase coded for by the *FADS* genes and of elongases coded for by the *ELOVL* genes. Beyond the competition for resources, the supply of the delta-5 and delta-6 desaturase (and possibly the elongase) is further limited by an individual's genotype. Single nucleotide polymorphisms (SNPs) in the *FADS* gene complex have been shown by others and us to be related to lower levels of FA in plasma [27,28] and in human milk.[29–31] It follows that the need for exogenous DHA is dependent, in part, on an individual's ability to synthesize endogenous long-chain FA from the essential parent FA: if an individual has the SNP, consumption of exogenous FA would be crucial. There is no consensus regarding the recommendations for the converse: if one does not have the SNP. Some researchers contend the endogenous conversion rate is so low, the DHA produced is inconsequential with the mean LNA:DHA rate ~0.047%,[32,33] but results of studies have been, again, inconsistent[34–36] and as mentioned, it is difficult to account empirically for all the potential influences on this complex system.

IMPORTANCE OF n-6/n-3 BALANCE

One of the influences on this complex metabolic system is the balance between the n-6 and n-3 pathways. This balance has become known as the "ratio." Use of this term by others and us has obfuscated, albeit unintentionally, other important data points. The ratio is merely a mathematical way to quantify the balance between the two pathways: it is a datapoint, not an ecologically relevant number. Indeed, given the complexity of the pathways and the many different intermediary FA, it is difficult to imagine that collapsing one's data into a number that reflects total n-6 FA/total n-3 FA would allow one to draw valid conclusions, and its use may in fact obscure the actual predictors. Harris and colleagues detail this issue and provide an empirical example of the inclusion of n-6 in the variable of interest diminishing the actual and more salient finding that n-3 levels are related to risk of cardiovascular disease.[37] That is, the ratio is a composite variable that should not be used on its own, but rather as part of an investigation of the individual FAs. In work detailed further, we have found that the relation of the ratio to cognitive data is further qualified by n-3 status.[17,18] Thus, the ratio is a good statistical starting point if the focus remains on balance as the true predictor and if individual FAs are also assessed for their independent contributions to the outcome of interest.

The balance between the two pathways (n-6 and n-3) has been empirically related to several disease states. As mentioned in the section Evolutionary Perspectives, it is thought that across time, the human diet has changed from an assumed nearly equal intake between the two pathways to a diet that is skewed to much higher n-6 intake (by as much as 10–20:1 n-6:n-3). Coincident with the change to higher n-6 intake is an increase in diseases of inflammation. All diseases have some degree of inflammation in their symptomology. Diseases specifically related to inflammation include allergies, some cancers, obesity, arthritis, and auto-immune disorders—all of which have been on the rise in Western countries. Metabolites of the n-6 pathway are pro-inflammatory; metabolites of the n-3 pathway are anti-inflammatory. Imbalance of dietary intake of these FA in favor of n-6 FAs, which would result in an increase in pro-inflammatory eicosanoids, has been implicated in atherosclerosis, obesity, diabetes, and autoimmune disorders[38,39] Increasing dietary n-3, and thereby, improving the balance, has been shown to improve the symptomology of lupus[40] and rheumatoid arthritis.[41] Associations with psychological functions have also been shown. For example, the relation between the balance and depressive mood states[42] and other psychological diseases such as the early stages of Alzheimer's[43] have been established. Interestingly and beyond dietary intake, the balance of FAs stored in adipose tissue is related to cognitive function: lower n-6 to n-3 content of tissue is related to better cognitive function.[44] These are only a few examples of the importance of an n-6 to n-3 balance in the prevention of disease states.

It is difficult to determine what the proper balance should be in any one individual. Individual differences exist as a function of fetal programming, dietary intake, genetics, epigenetics, and even, mitochondrial function. As mentioned previously, some clues are available from the research done on the ancestral diet. There are also potential indications of the optimal balance that can be gleaned from the system itself. The genes that code for the desaturases involved in the metabolism of FA are regulated by dietary LC-PUFAs such as AA and DHA. The *FADS* genes that code for the desaturases delta-5 and delta-6 are upregulated when the diet does not contain enough PUFA. Importantly, those same genes are downregulated when the balance between the pathways equalizes.[45] From this animal work, we can extrapolate to humans a homeostatic predisposition for a one-to-one balance of the n-6 and n-3 FAs. Therefore, those consuming the typical American diet or even, the diets in most developed non-coastal societies (1, 10–20) are at a much higher risk for disease based on diet alone.

Necessarily, the research regarding the interactions between the n-6 and n-3 pathways have been carried out for, the most part, animal studies. It is becoming clear that this system is very complex and that there are several genes, differential affinities, and individual differences involved. Work with humans is ongoing. Because the brain areas that contain the highest levels of n-3 FAs subserve the cognitive abilities memory and EFs (hippocampus and frontal areas, including prefrontal cortex), associations among the FAs and these cognitive functions are studied most often. Importantly, longitudinal supplementation studies that have not found effects of the supplementation early on have found effects when the EF abilities begin to develop at preschool age[46] attesting to the specificity of the effects of FAs on cognitive development.[47]

In what follows, we begin with the definition of EFs, their developmental trajectories, and the neural substrates that support them. We then detail the research that has been completed exploring the balance between the FA pathways and the human ability to perform higher order cognitive functions such as working memory, planning, and inhibitory control.

EXECUTIVE FUNCTIONS

EFs are higher-order cognitive processes that control behavior, emotion, and cognition[48–51] and are subserved by the frontal lobes.[52] EFs develop from early childhood into adulthood,[53,54] and are linked to theory of mind (understanding that others may have a differing viewpoint from one's own)[55–57] and math and reading abilities.[49,58] EFs develop over a long time-span, with research starting as early as 2 years of age and extending well into adulthood.[53,59,60] They are important functions for daily life in adulthood, and as such, it is important to understand what nutritional support is necessary for optimal development and function.

Definition of EF

The definition of EF has been approached from brain-based research and behavior-based research. The central role of the frontal cortex in EF has been emphasized in brain-based research for several decades,[52,61,62] but recent work has expanded to studying EF as subserved by networks, all of which include frontal areas, but also incorporate areas such as the amygdala, hippocampus, and cerebellum depending on the EF of interest.[61,63–66] Behavior-focused research has often defined EFs as control of goal-directed behaviors[51,67,68] or regulatory processes that control automatic responses.[50,56,69,70] In these investigations, EF are usually divided into discrete functions, such as inhibitory control, working memory, planning, set-shifting, updating, and attentional control.

Attempts have been made to bridge these two research approaches for a cohesive definition of EF that incorporates both brain and behavioral aspects and acknowledges the complexity of EF, which develop across a long-time span.[53,59,60] These attempts are important for understanding the role of nutrients (such as n-3 and n-6 FA) in EF. First, the role of complex neural networks in EF requires understanding the different developmental trajectories of the brain areas involved and how nutrients are taken up and used in those brain areas. Second, the distinct EFs studied have unique elements, but their similarities (e.g., in cognitive flexibility) are important for a unified definition and for understanding the role of specific nutrients. Third, the overarching concept of control or regulation that is seen in most approaches to EF is important for understanding how EF are measured, and therefore, what might be affected by changes in nutrient availability.

Neural Substrates Subserving EFs

The frontal cortex consists of Brodmann areas 4, 6, 8–11, and 44–47. It is structurally defined as the area in front of the central sulcus and above the lateral sulcus (Sylvian fissure). The frontal cortex is large and complex, and its structural and functional organization are still being studied today.[71] It is generally divided into three main areas when studied: a lateral section, a medial section, and an orbital section. These areas have direct connections to other brain areas, which provide clues as to their roles in various cognitive functions. There are two functional gradients that are important when studying EF: the medial-lateral gradient and the dorsal-ventral gradient. The connectivity of each area of the frontal cortex helps us understand why various nutritional deficiencies or supplements may affect some functions and not others, and helps us understand where we need to focus efforts if attempting to support development or prevent decline. Fig. 2 shows the general connections and flow of information in the neural substrates that support EF.

Medial-Lateral Gradient

Medial areas of the frontal cortex (including the medial prefrontal cortex) are typically found to be involved in action monitoring,[72] and connect to areas such as the anterior cingulate cortex (ACC), which is related to error monitoring and correction; the amygdala, which is connected to processing threats and fear-inducing stimuli; and the temporal lobe, including the hippocampus,[73] which is a clearinghouse of sorts for memory information. Action monitoring and error correction are important elements of daily life that are under the purview of EF.[74] Lateral frontal areas tend to be involved in broader cognitive control, with significant

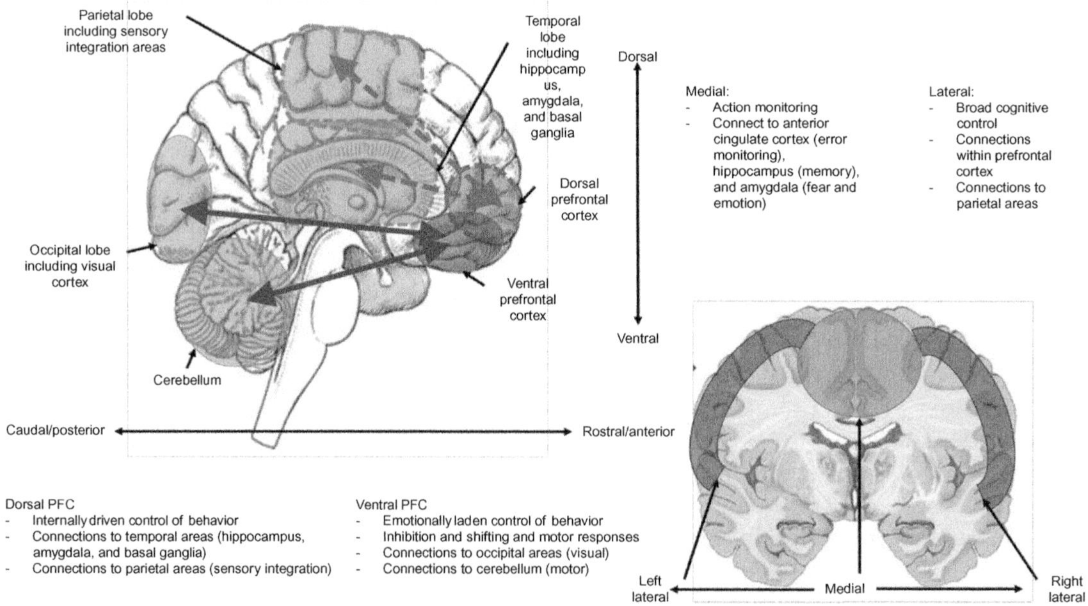

FIG. 2 Diagram of information flow in the brain for executive functions. Sagittal view of brain from Clipart Library 2016.

connections to other prefrontal areas and the parietal cortex.[71] These connections support findings that lateral frontal areas underlie working memory and perception.[75]

Dorsal-Ventral Gradient

The dorsal-ventral gradient in the frontal cortex is less well-understood and studied than the medial-lateral gradient, but ventral areas of the frontal cortex are thought to be more involved in emotionally laden information processing (sometimes called "hot" EF). These ventral areas connect to visual cortices, occipital areas, and fusiform areas.[71] More ventral areas have been connected to language abilities[76] and to inhibition and shifting performance,[77] which are important EF. Dorsal areas of the frontal cortex have been heavily linked to working memory and updating,[77] and specifically dorsolateral areas have been linked to internally driven control of behavior.[71,78]

Functional Connections

It is important to highlight two additional brain areas functionally connected to the frontal cortex that are important for EF. The hippocampus, a brain area in the temporal lobe linked to spatial processing and memory, is physically connected to frontal areas through the dentate gyrus and fornix. Functionally, the hippocampus has been found to be involved in working memory and inhibition tasks through its connections to areas involved in the medial-lateral and dorsal-ventral gradients.[79–81] It holds a seat of importance that has not been fully elucidated, but the hippocampus is involved in mature use of EF. In addition, the cerebellum, the brain area responsible for executing motor movements, is functionally connected to the frontal cortex and thus, is implicated in EF. In both laboratory EF tasks and daily life, EF often require motor movements, such as responding according to task rules (e.g., pressing a button) or executing a planned action (e.g., walking through a store picking up groceries). Evolutionarily newer areas of the cerebellum have also been linked to complex motor sequences that require coordination by an executive.[66] The cerebellum and prefrontal cortex develop along similar trajectories (physical and temporal) [61,82] and become increasingly connected as we age.[82] The cerebellum also connects to the frontal cortex across the medial-lateral and dorsal-ventral gradients, implying its high level of importance to all EF.[83]

Development

These functionally distinct gradients in the frontal cortex have led to some consistent conclusions regarding the role of the frontal cortex in EF, such as the role of the dorsolateral prefrontal cortex (DLPFC) in working memory and the ventromedial prefrontal cortex (VMPFC) and orbitofrontal prefrontal cortex (OPFC) in inhibitory control.[84] It is important to note that these frontal areas (and the functions they subserve) have a protracted development period. The prefrontal cortex is the last area of the brain to mature, and performance on EF tasks continues to improve well into adulthood.[84] The functional connectivity and segregation of function mentioned briefly earlier, therefore, develops slowly. Two main changes characterize the development of a mature frontal cortex and related EF: a change from diffuse to focal activity and increased segregation and integration of brain areas.

Diffuse to Focal Activity

Studies of functional brain development have often found that early, immature brain activity when completing a task is more diffuse than later, mature brain activity, especially with EF.[65,85–89] The development of EF involves increasingly distinct pathways of activation that depend on task demands[90] such that immature EF (i.e., failure or inadequate performance often seen in younger children) activate frontal regions generally, and mature EF (i.e., reliably successful performance often seen in adults) activate distinct neural networks.[87,89] There is evidence that the increase in focal activity is related to attenuation of activation in areas no longer required for task performance.[85] This increase in focal activity is seen across domains, including mentalizing, attentional control, motor control, and emotion regulation.[89]

Integration and Segregation of Function

Linked to the movement from diffuse to focal activity across development is increased in segregation and integration of function.[83,91] As EF develops, specific brain areas are used for specific functions, as in examples offered earlier. However, there is also increased integration among brain areas, such that the frontal cortex connects to more areas in the brain (e.g., the cerebellum), allowing the formation of more complex networks to support more complex functions. This segregation and integration mirrors the overall developmental trajectory in the brain, which includes pruning of gray matter (segregation, more efficient use of specific brain areas for specific tasks) and increases in white matter tracts (integration, myelinated axons used for long-distance communication between brain areas).[92–99] Increased segregation and integration are generally linked to improved behavioral performance on EF tasks.[83,89] Researchers also find that EF task performance clusters on a single factor in early childhood, but can be dissociated into three or more factors from later childhood onward.[50,100,101]

In models of EFs in adults, an executive component is seen as the seat of control of other functions [102,103] that produces a coherent response through representing rules and rule structures,[70,104] inhibiting distracting or incorrect information,[105,106] updating information while completing a task,[87] and finally, providing a correct response or series of responses to the situation. These distinct functions are fairly easily dissociated, and the most commonly studied EFs include working memory, inhibitory control, set shifting, and planning. The picture of EF in children is less clear. In some models, EFs are considered to consist of three distinct functions in children[67,100] generally comprising working memory, inhibitory control, and set shifting, whereas other models have still found EF to be a single factor in young children.[50] Developmentally, EFs are seen as becoming increasingly differentiated both in terms of behavioral function and cortical activation.

Fig. 3 illustrates the inter-dependencies and complexities of EF.

Research on EF development has focused on where children differ from adults—what errors they make in tasks that require an executive. As Smith et al.[107] explain, errors on EF tasks in childhood can be seen as problems with action, problems with inhibition, or problems with working memory. For instance, children and infants can correctly indicate where an object belongs in a sorting task through words or eye gaze.[108,109] and only fail at search tasks when

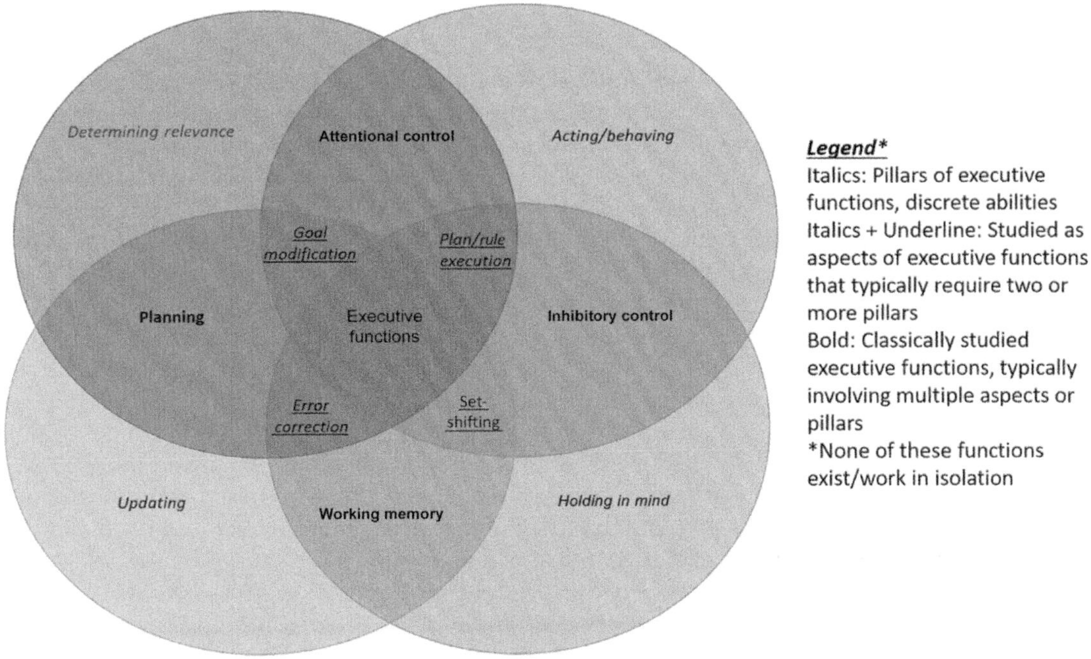

FIG. 3 Diagram of executive functions and their relations.

asked to execute that search.[107,110–112] In the brain, the prefrontal cortex and cerebellum are developing connections throughout the first 2 years of life, and those connections are thought to support knowledge of the correct sorting or reaching behavior to guide actual actions.[61] There is an inherent development of the connectivity between distant brain areas and an implied differentiation of function that allows specificity in the use of rules and actions to correctly respond to task demands.[113]

The executive is the organizer of information from brain areas required for control of behavior, emotion, and cognition, and the increases in coordination and efficiency required to develop mature EF are going to be constrained by the brain's structural and functional development. Neuronal structure and function are directly related to the availability of nutrients needed to produce neurons, neurotransmitters, and receptors. Broad changes across development include increases in gray matter until age 4 when it begins to decline and increases in white matter steadily into adulthood.[97,98,114,115] Later, changes in gray matter development differ between brain areas. Peaks in gray matter occur around adolescence and differ between the frontal, parietal, temporal, and occipital lobes.[116] The changes in white and gray matter are related to changes in the composition of the lipid bilayer and increased myelination that changes the overall balance of cholesterol, proteins, and fats in the brain.[99] When studying FA, it is important to understand how the brain is developing during a given period and what the optimal balance of FA may be to support structure and function.[117]

BALANCE OF FAS AND EFS

n-3 FA have been the main focus of investigations into FA, and yet n-6 FAs, as described earlier, are not only also prevalent in the brain, they also play integral and complementary roles to n-3 FA.[118–121] Consumption of FA should support the brain's need for both n-6 and n-3 FAs. Moreover, an imbalance of these two families of FA may prevent the beneficial effects of supplementation. EFs in particular are more susceptible to imbalances in n-6 and n-3 FA consumption because of the large networks involved in EF and their extended developmental trajectory. For instance, Innis and de la Presa Owens[122] found that supplementing rats with fish oil (high in n-3 FA) was negatively correlated with dopamine and serotonin levels in the rat brain. Dopamine and serotonin are two highly prevalent monoamine neurotransmitters in the frontal cortex, and alterations in these monoamines have been shown to affect EF.[123–125] Interestingly, serotonin levels were highest in the brains of rats in high n-6 group, followed by the brains of rats in the high n-3, and the lowest serotonin levels were found in the brains of rats in the medium n-6 group. The groups who consumed the two unbalanced ratios (371.5 and 0.12) had the highest levels of serotonin despite one being very high in n-6 and one being very high in n-3. These findings indicate that the balance of FA plays an important role in the ultimate outcome in the brain, and simply the amounts of n-6 or n-3 cannot be the sole focus of research. Additionally, the clear role for FA in neuronal growth, a process that would be particularly important during periods of considerable brain growth like the first 2 years of life and the adolescent years, implies that the balance of n-6 and n-3 FA will be important for cognitive functions that develop during those periods and functions that require large brain networks such as EFs.

Inhibitory Control

Inhibitory control is defined as the ability to "suppress inappropriate but prepotent responses of various kinds."[56] It represents the ability to take the large amounts of information we get on a daily basis and select the relevant pieces for use while inhibiting irrelevant or distracting information. Research has indicated that inhibitory control undergoes considerable development in early childhood, with particularly rapid improvement from 2 to 5 years old.[57,126,127] Inhibitory control tasks reliably activate the frontal cortex, and the development of inhibitory control is related to changes in the use of neural resources as the complexity of the inhibitory control requirements increases. Children will become proficient at simpler inhibitory control tasks before more complex tasks. Brain activation clearly distinguishes between children who have developed the ability to control inhibition and those who have not. For instance, adults had greater ACC activation across inhibitory control tasks (area of the brain involved in attentional control), but they only had different DLPFC activation during the more complex inhibitory control tasks compared to adolescents.[128] Supporting connections between the ACC and DLPFC are important for supporting EF development.

Similarly, mature inhibitory control not only means knowing the correct response but carrying out the correct action in accordance with the correct response. Inhibitory control requires efficient coordination of a frontal network and connections to cerebellar areas for motor responses.[61] Rule-switching, a complex form of inhibitory control, in adults involves coordination of the ventrolateral prefrontal cortex (VLPFC) and the pre-supplementary motor areas (pre-SMA). Children reach adult levels of activation of the pre-SMA by adolescence, but even adolescents show different activation of the VLPFC compared to adults.[104] Understanding the nutrient needs of each of the brains areas involved in EF is essential to supporting their optimal development.

The frontostriatal network frequently discussed with inhibitory control is generally considered to include the PFC, ACC, striatum, and cerebellar areas (see Fig. 2). All of these areas are affected by n-3 FA deficiency and changes in the balance of n-6 and n-3 FA.[129–131] However, each brain area tends to respond differently. In the face of n-3 deficiency, the frontal cortex tends to become depleted of n-3 FA, and levels of n-6 FA only increase somewhat, whereas the striatum and cerebellum tend to accumulate more n-6 FA and not lose as much DHA.[132,133] The best accretion of ARA and DHA in the frontal cortex and cerebellum occurs in ratios close to 4:1 and not diets focused on one or the other FA family.[130,131] These alterations in the FA composition of membranes affects dopamine receptor density in the frontal cortex[129] and serotonin receptor density in the frontal cortex and ACC,[134] which affects communication within the frontostriatal network as monoamine neurotransmitters are highly prevalent in these brain areas. Disruption to dopamine and serotonin receptors has been shown to disrupt EF task performance in animals.[123,124] The full picture of neurotransmitter function that supports EF is, of course, complex, but these monoamines have been consistently implicated[125] such that the role of FA in monoamine receptors and neurotransmission is likely an important element of their role in EF.

Working Memory

Working memory is defined as a system for temporarily holding information and using it for complex cognitive tasks.[102] Baddeley and Hitch's widely accepted tri-partite model of adult working memory[135] is composed of the central executive and two subsystems. The central executive and subsystems of the tri-partite model of working memory develop at different rates and in different ways. The central executive coordinates the activity of the two subsystems, the phonological loop and visuospatial sketchpad, which are directly responsible for the information being stored and processed. Similar to inhibitory control, mature working memory cannot be achieved until the frontal areas are connected to response elements (visual, motor, or verbal).

Investigations into the development of working memory highlight increases in the roles of distinct brain areas and increased connectivity between frontal, parietal, temporal, and occipital areas. Success on working memory tasks across development involves increasing specificity in the brain areas involved. Successful infants had greater electroencephalography (EEG) coherence across the whole brain in a simple working memory task, whereas children 4.5 years old who were successful had increased coherence only between the medial frontal and posterior temporal areas and the medial frontal and occipital areas.[65] EEG coherence is thought to be indicative of the connections between brain areas, and greater coherence likely indicates more connectivity and organization between brain areas.[136]

In general, improvements in speed and accuracy of working memory proceed linearly across childhood and adolescence.[53,137,138] However, investigations that use a variety of tasks often find that patterns of performance begin to diverge. Performance on phonological and visuospatial tasks was highly correlated among 4- and 5-year olds, but those correlations decreased over time until there were clearly distinct domains with phonological working memory tasks correlated with one another and distinct from visuospatial working memory tasks.[138] Across middle childhood and adolescence, there are increases in activity in frontal, parietal, and temporal areas that correspond to increases in accuracy and speed of processing.[139–141] There is also evidence of a lateralization across development that is related to improved performance[87] and reflects the type of stimulus being held in memory and manipulated (spatial vs object number).[142] The development of working memory involves a process of differentiation (behaviorally and in use of cortical resources) and increased involvement of the frontal cortex.

Evidence points to the important role of n-6 and n-3 FA. First, the hippocampus, a brain area recruited during memory and spatial processing tasks, is another area of the brain that accumulates considerable n-3 and n-6 FA.[143] The hippocampus is also sensitive to deficiency and imbalance of FA.[130,131] Working memory tasks, particularly spatial tasks, recruit both the frontal cortex and hippocampus. This recruitment of two brain areas sensitive to alterations in FA balance makes working memory a function of specific interest in understanding the role of FA. Alterations in monoamine vesicles in the presynaptic terminal,[144] neuron size and amount of dendritic branching,[145] and basal acetylcholine activity and muscarinic receptor density[146] have been found in the hippocampus in response to n-3 deficiency. All these alterations relate to the ability of the hippocampus to communicate with other brain areas. However, adding DHA at levels similar to other studies did not replete n-3 FA levels in the brain or improve working memory performance in previously deficient rats. Instead, increasing the level of DHA and lowering n-6 FA to lower the ratio repleted brain levels of n-3 FA and improved performance.[147]

The working memory literature also highlights the role of differentiation in the development of mature EF. Over time, brain function becomes more distinct in response to specific stimuli and specific processing requirements, and with aging, brain structure and function reduce in differentiation. This process of integration and differentiation through the growth of neurons, synapses, and dendritic branching combined with synaptic pruning supports the development of necessary connections and decreases unused or excess connections. Typically, researchers have demonstrated that n-3 FA increase neuronal growth and dendritic branching, and n-6 FA decrease neuronal growth and dendritic branching in both the hippocampus and frontal cortex.[145,148,149] A balance between n-6 and n-3 FA could therefore support a system that is changing neuronal connections and structure in support of newly developing functions. It would be expected that balanced FA pathways would support the process of integration and differentiation associated with mature working memory.

Planning

Planning is an EF characterized by assessing a goal, determining how to reach the goal, executing the steps, and then evaluating errors and goal attainment.[150,151] Planning recruits similar cortical resources to working memory and inhibitory control. The dorsolateral and rostrolateral PFC,[152] inferior frontal gyrus[153] and superior frontal areas[154] are found to be activated during planning tasks. Planning develops well into the adult years, and even studies with adults demonstrate considerable variability in performance.[59,155] For instance, when investigating EF in children 3–12 years old and a group of adults, researchers found that children reached adult performance on the simpler problems by age 6 but did not reach adult performance until age 12 on the more difficult problems.[156]

The contribution of n-6 and n-3 FA to planning is similar to the role of FA in the development of inhibitory control and working memory, which set the groundwork for the development of planning. The appropriate balance of FA during the development of inhibitory control and working memory will likely set the stage for optimal planning development. The balance between n-6 and n-3 FA predicts planning task performance, particularly number of moves and time required to solve complex problems.[17,18] Additionally, there is evidence that the optimal balance that supports planning changes across development. In one study, younger children (7–9 years old) performed best with balanced n-6 and n-3 FA levels measured in blood and through dietary intake, but older children (10–12 years old) were best supported by higher n-3 FA, and less balance between n-3 and n-6 FA.[17] The findings were mirrored by the balance (or imbalance) of n-3 and n-6 FA that supported brain activity during optimal performance of the planning tasks.

Across these ages, children improve in their inhibitory control,[105,128] working memory,[53,137] and planning[155,156] abilities. However, these changes reflect both a movement toward adult-level performance and brain activation and distinct differences.[88,105] Some of these differences are related to pubertal changes in hormones and concomitant changes in brain structure and function. n-3 FA have been found to increase neurite outgrowth and dendritic branching[145,157] that is important during periods of brain growth. High n-6 intake may inhibit the metabolism and use of n-3 FA that are important to brain growth. Younger children, whose brains are not undergoing the rapid growth associated with puberty, may require a balanced diet because both n-6 and n-3 FA are important for neuronal communication.[129,158,159]

Conversely, puberty brings with it changes in brain structure and function that could be best supported with a balance of n-6 and n-3 FA that provide more n-3 FA. n-3 FA support neurite growth in the frontal cortex and hippocampus,[122,157] and DHA has been found to affect monoamine neurotransmission during puberty in rats.[160] Monoamine neurotransmission has been shown to be affected by hormones[161,162] and n-6 and n-3 FA.[129,144,163] These developmental considerations make FA intake recommendations more difficult, but also highlight the importance of considering the balance of n-6 and n-3 FA and being precise in examining when to supplement.

CONCLUSION

The balance of n-6 and n-3 FAs is likely particularly important to the frontal cortex because of the unique demands of EF. The need to coordinate between many brain areas and to coordinate both incoming information and outgoing information to produce a coherent response would likely be susceptible to imbalances in FA because of the integral nature of both n-6 and n-3 FA to neuronal growth and communication. The frontal cortex is usually the focus of neuroimaging investigations of EF, but EFs are actually subserved by additional functional cortical networks that incorporate areas outside the frontal cortex. There are different networks related to different elements of EF, such as trial-by-trial updating and sustained maintenance of overall goal-directed behaviors in adults [78], and those networks undergo a process of segregation (decreases in short-range connections) and integration (increases in long-range connections) across development.[83] Therefore, whereas the needs of the frontal cortex are important, understanding the FA requirements of different structures within the EF networks and how those structures communicate is necessary. It is useful to approach EF in terms of the role of the frontal cortex[52,156] and from the lens of the inherent qualities of the functions that fall under the umbrella term "EFs."[70,109]

References

1. Brenna JT, Salem N, Sinclair AJ, Cunnane SC. α-Linolenic acid supplementation and conversion to n-3 long-chain polyunsaturated fatty acids in humans. *Prostaglandins Leukot Essent Fatty Acids*. 2009;80(2–3):85–91.
2. Simopoulos AP. The importance of the ratio of omega-6/omega-3 essential fatty acids. *Biomed Pharmacother*. 2002;56(8):365–379.
3. Eaton SB, Eaton III SB, Konner MJ. Paleolithic nutrition revisited: a twelve-year retrospective on its nature and implications. *Eur J Clin Nutr*. 1997;51:207–216.
4. Eaton SB, Konner M. Paleolithic nutrition: a consideration of its nature and current implications. *N Engl J Med*. 1985;312(5):283–289.
5. Eaton SB, Eaton III SB, Konner MJ. An evolutionary perspective enhances understanding human nutritional requirements. *J Nutrition*. 1996;126:1732–1740.
6. Eaton SB, Eaton IIISB, Sinclair AJ, Cordain L, Mann NJ. Dietary intake of long-chain polyunsaturated fatty acids during the paleolithic. In: Simopoulos A, ed. *World Review of Nutrition & Dietetics*. Basel, Switzerland: Karger; 1998:12–23. Vol. 83.
7. Lee RB. What hunters do for a living, or, how to make out on scarce resources. In: Lee RB, DeVore I, eds. *Man the Hunter*. Chicago: Aldine; 1968:30–48.
8. Simopoulos AP. Evolutionary aspects of diet, the omega-6/omega-3 ratio and genetic variation: Nutritional implications for chronic diseases. *Biomed Pharmacother*. 2006;60(9):502–507.
9. Simopoulos AP. Evolutionary aspects of diet: the omega-6/omega-3 ratio and the brain. *Mol Neurobiol*. 2011;44(2):203–215.
10. Innis SM. Omega-3 fatty acid biochemistry: perspectives from human nutrition. *Mil Med*. 2014;179(11):82–87.
11. Elmadfa I, Kornsteiner M. Fats and fatty acid requirements for adults. *Ann Nutr Metab*. 2009;55(1–3):56–75.
12. Fleith M, Clandinin MT. Dietary PUFA for preterm and term infants: review of clinical studies. *Crit Rev Food Sci Nutr*. 2005;45(3):205–229.
13. Friesen RW, Innis SM. Dietary arachidonic acid to EPA and DHA balance is increased among Canadian pregnant women with low fish intake. *J Nutr*. 2009;139(12):2344–2350.
14. Ervin RB, Wright JD, Wang CY, Kennedy-Stephenson J. Dietary intake of fats and fatty acids for the United States population: 1999–2000. *Adv Data*. 2004;(348):1–6.
15. Keim SA, Branum AM. Dietary intake of polyunsaturated fatty acids and fish among US children 12–60 months of age. *Matern Child Nutr*. 2015;11(4):987–998.
16. Sheppard KW, Cheatham CL. Omega-6/omega-3 fatty acid intake of children and older adults in the U.S.: Dietary intake in comparison to current dietary recommendations and the healthy eating index. *Lipids Health Dis*. 2018;17(1):43.
17. Sheppard KW, Cheatham CL. Executive functions and the omega-6-to-omega-3 fatty acid ratio: a cross-sectional study. *Am J Clin Nutr*. 2017;105(1):32–41.
18. Sheppard KW, Cheatham CL. Omega-6 to omega-3 fatty acid ratio and higher-order cognitive functions in 7- to 9-y-olds: a cross-sectional study. *Am J Clin Nutr*. 2013;98(3):659–667.
19. Ravelli AC, van der Meulen JH, Michels RP, et al. Glucose tolerance in adults after prenatal exposure to famine. *Lancet*. 1998;351(9097):173–177.
20. Ravelli GP, Stein ZA, Susser MW. Obesity in young men after famine exposure in utero and early infancy. *N Engl J Med*. 1976;295(7):349–353.
21. Roseboom TJ, Painter RC, van Abeelen AF, Veenendaal MV, de Rooij SR. Hungry in the womb: What are the consequences? Lessons from the Dutch famine. *Maturitas*. 2011;70(2):141–145.
22. Susser M, Stein Z. Timing in prenatal nutrition: a reprise of the Dutch famine study. *Nutr Rev*. 1994;52(3):84–94.
23. Tobi EW, Lumey LH, Talens RP, et al. DNA methylation differences after exposure to prenatal famine are common and timing- and sex-specific. *Hum Mol Genet*. 2009;18(21):4046–4053.

24. Rzehak P, Heinrich J, Klopp N, et al. Evidence for an association between genetic variants of the fatty acid desaturase 1 fatty acid desaturase 2 (FADS1 FADS2) gene cluster and the fatty acid composition of erythrocyte membranes. *Br J Nutr*. 2009;101(1):20–26.
25. Schaeffer L, Gohlke H, Muller M, et al. Common genetic variants of the FADS1 FADS2 gene cluster and their reconstructed haplotypes are associated with the fatty acid composition in phospholipids. *Hum Mol Genet*. 2006;15(11):1745–1756.
26. Nwankwo JO, Spector AA, Domann FE. A nucleotide insertion in the transcriptional regulatory region of FADS2 gives rise to human fatty acid delta-6-desaturase deficiency. *J Lipid Res*. 2003;44(12):2311–2319.
27. Steer CD, Hibbeln JR, Golding J, Davey Smith G. Polyunsaturated fatty acid levels in blood during pregnancy, at birth and at 7 years: Their associations with two common FADS2 polymorphisms. *Hum Mol Genet*. 2012;21(7):1504–1512.
28. Lupu DS, Cheatham CL, Corbin KD, Niculescu MD. Genetic and epigenetic transgenerational implications related to omega-3 fatty acids. Part I: Maternal FADS2 genotype and DNA methylation correlate with polyunsaturated fatty acid status in toddlers: an exploratory analysis. *Nutr Res*. 2015;35(11):939–947.
29. Xie L, Innis SM. Genetic variants of the FADS1 FADS2 gene cluster are associated with altered (n-6) and (n-3) essential fatty acids in plasma and erythrocyte phospholipids in women during pregnancy and in breast milk during lactation. *J Nutr*. 2008;138(11):2222–2228.
30. Molto-Puigmarti C, Plat J, Mensink RP, et al. FADS1 FADS2 gene variants modify the association between fish intake and the docosahexaenoic acid proportions in human milk. *Am J Clin Nutr*. 2010;91(5):1368–1376.
31. Cheatham CL, Sheppard KW. Synergistic effects of human milk nutrients in the support of infant recognition memory: an observational study. *Nutrients*. 2015;7(11):9079–9095.
32. Pawlosky RJ, Hibbeln JR, Novotny JA, Salem Jr N. Physiological compartmental analysis of alpha-linolenic acid metabolism in adult humans. *J Lipid Res*. 2001;42(8):1257–1265.
33. Lin YH, Llanos A, Mena P, Uauy R, Salem Jr N, Pawlosky RJ. Compartmental analyses of (2)H(5)-alpha-linolenic acid and (13)C-U-eicosapentaenoic acid toward synthesis of plasma labeled 22:6-23 in newborn term infants. *Am J Clin Nutrit*. 2010;92(2):284–293.
34. Li D, Sinclair A, Wilson A, et al. Effect of dietary alpha-linolenic acid on thrombotic risk factors in vegetarian men. *Am J Clin Nutrit*. 1999;69(5):872–882.
35. Fokkema MR, Brouwer DA, Hasperhoven MB, Martini IA, Muskiet FA. Short-term supplementation of low-dose gamma-linolenic acid (GLA), alpha-linolenic acid (ALA), or GLA plus ALA does not augment LCP omega 3 status of Dutch vegans to an appreciable extent. *Prostaglandins Leukot Essent Fatty Acids*. 2000;63(5):287–292.
36. Ezaki O, Takahashi M, Shigematsu T, et al. Long-term effects of dietary alpha-linolenic acid from perilla oil on serum fatty acids composition and on the risk factors of coronary heart disease in Japanese elderly subjects. *J Nutr Sci Vitaminol (Tokyo)*. 1999;45(6):759–772.
37. Harris WS, Assaad B, Poston WC. Tissue omega-6/omega-3 fatty acid ratio and risk for coronary artery disease. *Am J Cardiol*. 2006;98(4A):19i–26i.
38. Simopoulos AP. An increase in the omega-6/omega-3 fatty acid ratio increases the risk for obesity. *Nutrients*. 2016;8(3):128.
39. Simopoulos AP. The importance of the omega-6/omega-3 fatty acid ratio in cardiovascular disease and other chronic diseases. *Exp Biol Med (Maywood)*. 2008;233(6):674–688.
40. Borges MC, Santos Fde M, Telles RW, Correia MI, Lanna CC. Polyunsaturated omega-3 fatty acids and systemic lupus erythematosus: What do we know? *Rev Bras Reumatol*. 2014;54(6):459–466.
41. Calder PC. Fatty acids and inflammation: the cutting edge between food and pharma. *Eur J Pharmacol*. 2011;668(Suppl 1):S50–S58.
42. Ryback R. Bioelectrical modulators and the cell membrane in psychiatric medicine. *Psychopharmacol Bull*. 2001;35(4):5–44.
43. Freund-Levi Y, Eriksdotter-Jonhagen M, Cederholm T, et al. Omega-3 fatty acid treatment in 174 patients with mild to moderate Alzheimer disease: OmegAD study: a randomized double-blind trial. *Arch Neurol*. 2006;63(10):1402–1408.
44. Stilling F, Wolk A, Religa D, et al. Adipose tissue fatty acid composition and cognitive impairment. *Nutrition*. 2018;54:153–157.
45. Wijendran V, Downs I, Srigley CT, et al. Dietary arachidonic acid and docosahexaenoic acid regulate liver fatty acid desaturase (FADS) alternative transcript expression in suckling piglets. *Prostaglandins Leukot Essent Fatty Acids*. 2013;89(5):345–350.

46 Helland IB, Smith L, Saarem K, et al. Maternal supplementation with very-long-chain n-3 fatty acids during pregnancy and lactation augments children's IQ at 4 years of age. Pediatrics 2003;111(1):e39–44.
47. Cheatham CL. Omega-3 fatty acids and the development of cognitive abilities: a review of DHA supplementation studies. *CAB Rev Perspect Agric Vet Sci Nutr Nat Res*. 2008;3(1):1–15.
48. Carlson SM. Developmentally sensitive measures of executive function in preschool children. *Dev Neuropsychol*. 2005;28:595–616.
49. Bull R, Espy KA, Wiebe SA. Short-term memory, working memory, and executive functioning in preschoolers: longitudinal predictors of mathematical achievement at age 7 years. *Dev Neuropsychol*. 2008;33(3):205–228.
50. Wiebe SA, Sheffield T, Nelson JM, Clark CA, Chevalier N, Espy KA. The structure of executive function in 3-year-olds. *J Exp Child Psychol*. 2011;108(3):436–452.
51. Zelazo PD, Carter A, Reznick JS, Frye D. Early development of executive function: a problem-solving framework. *Rev Gen Psychol*. 1997;1:198–226.
52. Roberts RJ, Pennington BF. An interactive framework for examining prefrontal cognitive processes. *Dev Neuropsychol*. 1996;12(1):105–126.
53. Conklin HM, Luciana M, Hooper CJ, Yarger RS. Working memory performance in typically developing children and adolescents: behavioral evidence of protracted frontal lobe development. *Dev Neuropsychol*. 2007;37:103–128.
54. Pelphrey KA, Reznick JS. Working memory in infancy. *Adv Child Dev Behav*. 2003;31:173–227.
55. Carlson SM, Moses LJ. Individual differences in inhibitory control and children's theory of mind. *Child Dev*. 2003;72:1032–1053.
56. Carlson SM, Moses LJ, Breton C. How specific is the relation between executive function and theory of mind? Contributions of inhibitory control and working memory. *Infant Child Develop*. 2002;11(2):73–92.
57. Carlson SM, Moses LJ, Claxton LJ. Individual differences in executive functioning and theory of mind: An investigation of inhibitory control and planning ability. *J Exp Child Psychol*. 2004;87(4):299–319.
58. Bull R, Espy KA, Senn TE. A comparison of performance on the towers of London and Hanoi in young children. *J Child Psychol Psychiatry*. 2004;45:743–754.
59. Luciana M, Nelson CA. The functional emergence of prefrontally-guided working memory systems in four- to eight-year-old children. *Neuropsychologia*. 1998;36:273–293.
60. Wiebe SA, Morton JB, Buss AT, Spencer JP. The emergence of executive function. *Monogr Soc Res Child Dev*. 2014;79(2):1–132.
61. Diamond A. Close interrelation of motor development and cognitive development and of the cerebellum and prefrontal cortex. *Child Dev*. 2000;71:44–56.
62. Chandler DJ, Waterhouse BD, Gao WJ. New perspectives on catecholaminergic regulation of executive circuits: evidence for independent modulation of prefrontal functions by midbrain dopaminergic and noradrenergic neurons. *Front Neural Circuits*. 2014;8:53.
63. Logue SF, Gould TJ. The neural and genetic basis of executive function: attention, cognitive flexibility, and response inhibition. *Pharmacol Biochem Behav*. 2014;123:45–54.
64. Rohr CS, Dreyer FR, Aderka IM, et al. Individual differences in common factors of emotional traits and executive functions predict functional connectivity of the amygdala. *Neuroimage*. 2015;120:154–163.
65. Bell MA, Wolfe CD. Changes in brain functioning from infancy to early childhood: evidence from EEG power and coherence during working memory tasks. *Dev Neuropsychol*. 2007;31:21–38.
66. Riva D, Cazzaniga F, Esposito S, Bulgheroni S. Executive functions and cerebellar development in children. *Appl Neuropsychol Child*. 2013;2(2):97–103.
67. Garon N, Bryson SE, Smith IM. Executive function in preschoolers: a review using an integrative framework. *Psychol Bull*. 2008;134(1):31–60.
68. Carlson SE, Werkman SH, Rhodes PG, Tolley EA. Visual-acuity development in healthy preterm infants: effect of marine-oil supplementation. *Am J Clin Nutr*. 1993;58(1):35–42.
69. Miyake A, Friedman NP. The nature and organization of individual differences in executive functions: four general conclusions. *Curr Dir Psychol Sci*. 2012;21(1):8–14.
70. Zelazo PD, Frye D. Cognitive complexity and control: II. The development of executive function in childhood. *Curr Direct Psychol Sci*. 1998;7(4):121–128.
71. Orr JM, Smolker HR, Banich MT. Organization of the human frontal pole revealed by large-scale DTI-based connectivity: implications for control of behavior. *PLoS One*. 2015;10(5):e0124797.
72. Bonini F, Burke B, Liegeois-Chauvel C, Regis J, Chauvel P, Vidal F. Action monitoring and medial frontal cortex: leading role of supplementary motor area. *Science*. 2014;343:888–891.

73. Gilbert SJ, Gonen-Yaacovi G, Benoit RG, Volle E, Burgess PW. Distinct functional connectivity associated with lateral versus medial rostral prefrontal cortex: a meta-analysis. *Neuroimage*. 2010;53(4):1359–1367.
74. de la Vega A, Chang LJ, Banich MT, Wager TD, Yarkoni T. Large-scale meta-analysis of human medial frontal cortex reveals tripartite functional organization. *J Neurosci*. 2016;36(24):6553–6562.
75. Bludau S, Eickhoff SB, Mohlberg H, et al. Cytoarchitecture, probability maps and functions of the human frontal pole. *Neuroimage*. 2014;93(Pt 2):260–275.
76. Neubert FX, Mars RB, Thomas AG, Sallet J, Rushworth MF. Comparison of human ventral frontal cortex areas for cognitive control and language with areas in monkey frontal cortex. *Neuron*. 2014;81(3):700–713.
77. Collette F, Hogge M, Salmon E, Van der Linden M. Exploration of the neural substrates of executive functioning by functional neuroimaging. *Neuroscience*. 2006;139(1):209–221.
78. Dosenbach NU, Fair DA, Miezin FM, et al. Distinct brain networks for adaptive and stable task control in humans. *Proc Natl Acad Sci U S A*. 2007;104(26):11073–11078.
79. Toepper M, Markowitsch HJ, Gebhardt H, et al. Hippocampal involvement in working memory encoding of changing locations: an fMRI study. *Brain Res*. 2010;1354:91–99.
80. Wu JQ, Peters GJ, Rittner P, Cleland TA, Smith DM. The hippocampus, medial prefrontal cortex, and selective memory retrieval: evidence from a rodent model of the retrieval-induced forgetting effect. *Hippocampus*. 2014;24(9):1070–1080.
81. Mala H, Andersen LG, Christensen RF, et al. Prefrontal cortex and hippocampus in behavioural flexibility and posttraumatic functional recovery: reversal learning and set-shifting in rats. *Brain Res Bull*. 2015;116:34–44.
82. Ramnani N. The primate cortico-cerebellar system: anatomy and function. *Nat Rev Neurosci*. 2006;7(7):511–522.
83. Fair DA, Dosenbach NU, Church JA, et al. Development of distinct control networks through segregation and integration. *Proc Natl Acad Sci U S A*. 2007;104(33):13507–13512.
84. Casey BJ, Giedd J, Thomas KM. Structural and functional brain development and its relation to cognitive development. *Biol Psychol*. 2000;54:241–257.
85. Durston S, Davidson M, Tottenham N, et al. A shift from diffuse to focal cortical activity with development. *Dev Sci*. 2006;9(1):1–20.
86. Moriguchi Y, Hiraki K. Prefrontal cortex and executive function in young children: a review of NIRS studies. *Front Hum Neurosci*. 2013;7:867.
87. Tsujii T, Yamamoto E, Masuda S, Watanabe S. Longitudinal study of spatial working memory development in young children. *Neuroreport*. 2009;20(8):759–763.
88. Tsujimoto S. The prefrontal cortex: functional neural development during early childhood. *Neuroscientist*. 2008;14(4):345–358.
89. Kelly AM, Di Martino A, Uddin LQ, et al. Development of anterior cingulate functional connectivity from late childhood to early adulthood. *Cereb Cortex*. 2009;19(3):640–657.
90. Buss AT, Fox N, Boas DA, Spencer JP. Probing the early development of visual working memory capacity with functional near-infrared spectroscopy. *Neuroimage*. 2014;85(Pt 1):314–325.
91. Hwang K, Velanova K, Luna B. Strengthening of top-down frontal cognitive control networks underlying the development of inhibitory control: a functional magnetic resonance imaging effective connectivity study. *J Neurosci*. 2010;30(46):15535–15545.
92. Barnea-Goraly N, Menon V, Eckert M, et al. White matter development during childhood and adolescence: a cross-sectional diffusion tensor imaging study. *Cereb Cortex*. 2005;15(12):1848–1854.
93. Lenroot RK, Giedd JN. Brain development in children and adolescents: insights from anatomical magnetic resonance imaging. *Neurosci Biobehav Rev*. 2006;30(6):718–729.
94. Stevens MC, Skudlarski P, Pearlson GD, Calhoun VD. Age-related cognitive gains are mediated by the effects of white matter development on brain network integration. *Neuroimage*. 2009;48(4):738–746.
95. Giedd JN, Blumenthal J, Jeffries NO, et al. Brain development during childhood and adolescence: a longitudinal MRI study. *Nat Neurosci*. 1999;2:861–863.
96. Pfefferbaum A, Mathalon DH, Sullivan EV, Rawles JM, Zipursky RB, Lim KO. A quantitative magnetic resonance imaging study of changes in brain morphology from infancy to late adulthood. *Arch Neurol*. 1994;51:874–887.
97. Toga AW, Thompson PM, Sowell ER. Mapping brain maturation. *Trends Neurosci*. 2006;29(3):148–159.
98. Sowell ER, Trauner DA, Gamst A, Jernigan TL. Development of cortical and subcortical brain structures in childhood and adolescence: a structural MRI study. *Develop Med Child Neurol*. 2002;44:4–16.
99. Barkovich AJ, Kjos BO, Jackson DE, Norman D. Normal maturation of the neonatal and infant brain: MR imaging at 1.5 T. *Neuroradiology*. 1988;166:173–180.

100. Miyake A, Friedman NP, Emerson MJ, Witzki AH, Howerter A, Wager TD. The unity and diversity of executive functions and their contributions to complex "frontal lobe" tasks: a latent variable analysis. *Cogn Psychol*. 2000;41(1):49–100.
101. Hughes C, Ensor R, Wilson A, Graham A. Tracking executive function across the transition to school: a latent variable approach. *Dev Neuropsychol*. 2010;35(1):20–36.
102. Baddeley A. Working memory. *Science*. 1992;255:556–559.
103. Baddeley A. Exploring the central executive. *Q J Exp Psychol*. 1996;49A:5–28.
104. Crone EA, Donohue SE, Honomichl R, Wendelken C, Bunge SA. Brain regions mediating flexible rule use during development. *J Neurosci*. 2006;26(43):11239–11247.
105. Schroeter ML, Zysset S, Wahl M, von Cramon DY. Prefrontal activation due to Stroop interference increases during development—an event-related fNIRS study. *Neuroimage*. 2004;23(4):1317–1325.
106. Casey BJ, Richards JE. A refractory period for the heart rate response in infant visual attention. *Dev Psychobiol*. 1991;24(5):327–340.
107. Smith LB, Thelen E, Titzer R, McLin D. Knowing in the context of acting: The task dynamics of the A-not-B error. *Psychol Rev*. 1999;106:235–260.
108. Zelazo PD, Reznick JS, Pinon DE. Response control and the execution of verbal rules. *Dev Psychol*. 1995;31:508–517.
109. Zelazo PD. The development of conscious control in childhood. *Trends Cogn Sci*. 2004;8(1):12–17.
110. Cathro DM, Forsyth CC, Cameron J. Adrenocortical response to stress in newborn infants. *Arch Dis Childhood*. 1969;44:88–95.
111. Diamond A. Development of the ability to use recall to guide action, as indicated by infants' performance on AB. *Child Dev*. 1985;56:868–883.
112. Diamond A, Doar B. The performance of human infants on a measure of frontal cortex function, the delayed response task. *Dev Psychobiol*. 1989;22:271–294.
113. Morton JB, Munakata Y. Active versus latent representations: a neural network model of perseveration, dissociation, and declage. *Dev Psychobiol*. 2002;40:255–265.
114. Carlson SE, Goldfarb S. A sensitive enzymatic method for determination of free and esterified tissue cholesterol. *Clin Chim Acta*. 1977;79(3):575–582.
115 Carskadon MA, Dement WC. Normal sleep and its variations. In M.H. Kryger, T. Roth, & W.C. Dement (Eds.), Principles and Practice of Sleep Medicine, 5th ed., (pp 16-26). St Louis: Elsevier Saunders.
116. Carlson SE, Werkman SH, Montalto MB, Tolley EA. Visual acuity development of preterm (PT) infants fed docosahexaenoic acid (DHA) and arachidonic acid (ARA): effect of age at supplementation. *Pediatr Res*. 1999;45:279A.
117. Wainwright PE. Dietary essential fatty acids and brain function: a developmental perspective on mechanisms. *Proc Nutr Soc*. 2007;61(01):61–69.
118. Carlson SE. Docosahexaenoic acid and arachidonic acid in infant development. *Semin Neonatol*. 2001;6(5):437–449.
119. Carlson SE, Werkman SH, Peeples JM, Cooke RJ, Tolley EA. Arachidonic acid status correlates with first year growth in preterm infants. *Proc Natl Acad Sci U S A*. 1993;90:1073–1077.
120. Connor S, Tenorio G, Clandinin MT, Sauve Y. DHA supplementation enhances high-frequency, stimulation-induced synaptic transmission in mouse hippocampus. *Appl Physiol Nutr Metab*. 2012;37(5):880–887.
121. Richards DA, Bliss TVP, Richards CD. Differential modulation of NMDA-induced calcium transients by arachidonic acid and nitric oxide in cultured hippocampal neurons. *Eur J Neurosci*. 2003;17(11):2323–2328.
122. Innis SM, De La Presa Owens S. Dietary fatty acid composition in pregnancy alters neurite membrane fatty acids and dopamine in newborn rat brain. *J Nutr*. 2001;131:118–122.
123. Puig MV, Miller EK. Neural substrates of dopamine D2 receptor modulated executive functions in the monkey prefrontal cortex. *Cereb Cortex*. 2015;25(9):2980–2987.
124. Maiti P, Gregg LC, McDonald MP. MPTP-induced executive dysfunction is associated with altered prefrontal serotonergic function. *Behav Brain Res*. 2016;298(Pt B):192–201.
125. Robbins TW, Roberts AC. Differential regulation of fronto-executive function by the monoamines and acetylcholine. *Cereb Cortex*. 2007;17(Suppl 1):i151–i160.
126. Diamond A, Kirkham NZ, Amso D. Conditions under which young children can hold two rules in mind and inhibit a prepotent response. *Dev Psychol*. 2002;38:352–362.
127. Zelazo PD, Reznick JS. Age-related asynchrony of knowledge and action. *Child Dev*. 1991;62:719–735.
128. Rubia K, Smith AB, Woolley J, et al. Progressive increase of frontostriatal brain activation from childhood to adulthood during event-related tasks of cognitive control. *Hum Brain Mapp*. 2006;27(12):973–993.

129. Delion S, Chalon S, Herault J, Guilloteau D, Besnard J-C, Durand G. Chronic dietary alpha-linolenic acid deficiency alters dopaminergic and serotoninergic neurotransmission in rats. *J Nutr*. 1994;124:2466–2476.
130. Jumpsen J, Lien EL, Goh YK, Clandinin MT. During neuronal and glial cell development diet n-6 to n-3 fatty acid ratio alters the fatty acid composition of phosphatidylinositol and phosphatidylserine. *Biochim Biophys Acta*. 1997;1347:40–50.
131. Jumpsen J, Lien EL, Goh YK, Clandinin MT. Small changes of dietary (n-6) and (n-3)/fatty acid content ratio alter phosphatidylethanolamine and phosphatidylcholine fatty acid composition during development of neuronal and glial cells in rats. *J Nutr*. 1997;127:724–731.
132. Carlson SE, Salem Jr. N. Essentiality of omega 3 fatty acids in growth and development of infants. *World Rev Nutr Diet*. 1991;66:74–86.
133. Levant B, Ozias MK, Carlson SE. Specific brain regions of female rats are differentially depleted of docosahexaenoic acid by reproductive activity and an (n-3) fatty acid-deficient diet. *J Nutr*. 2007;137:130–134.
134. du Bois TM, Deng C, Bell W, Huang XF. Fatty acids differentially affect serotonin receptor and transporter binding in the rat brain. *Neuroscience*. 2006;139(4):1397–1403.
135. Gathercole SE. The development of memory. *J Child Psychol Psychiatry*. 1998;39:3–27.
136. Thatcher RW, Krause PJ, Hrybyk M. Cortico-cortical associations and EEG coherence: a two-compartmental model. *Electroencephalogr Clin Neurophysiol*. 1986;64:123–143.
137. Luciana M, Conklin HM, Hooper CJ, Yarger RS. The development of nonverbal working memory and executive control processes in adolescents. *Child Dev*. 2005;76:697–712.
138. Gathercole SE, Pickering SJ, Ambridge B, Wearing H. The structure of working memory from 4 to 15 years of age. *Dev Psychol*. 2004;40(2):177–190.
139. Blain-Briere B, Bouchard C, Bigras N, Cadoret G. Development of active control within working memory: active retrieval versus monitoring in children. *Int J Behav Dev*. 2013;38(3):239–246.
140. Klingberg T, Forssberg H, Westerberg H. Increased brain activity in frontal and parietal cortex underlies the development of visuospatial working memory capacity during childhood. *J Cogn Neurosci*. 2002;14:1–10.
141. Nelson CA, Monk CS, Lin J, Carver LJ, Thomas KM, Truwit CL. Functional neuroanatomy of spatial working memory in children. *Dev Psychol*. 2000;36:109–116.
142. Smith EE, Jonides J, Koeppe RA, Awh E, Schumacher EH, Minoshima S. Spatial versus object working memory: PET investigations. *J Cogn Neurosci*. 1995;7:337–356.
143. Martinez M. Tissue levels of polyunsaturated fatty acids during early human development. *J Pediatr*. 1992;120:S129–S138.
144. Zimmer L, Delpal S, Guilloteau D, Aioun J, Durand G, Chalon S. Chronic n-3 polyunsaturated fatty acid deficiency alters dopamine vesicle density in the rat frontal cortex. *Neurosci Lett*. 2000;284:25–28.
145. Ahmad A, Murthy M, Moriguchi T, Salem N, Greiner RS. A decrease in cell size accompanies a loss of docosahexaenoate in the rat hippocampus. *Nutr Neurosci*. 2002;5(2):103–113.
146. Aid S, Vancassel S, Poumes-Ballihaut C, Chalon S, Guesnet P, Lavialle M. Effect of a diet-induced n-3 PUFA depletion on cholinergic parameters in the rat hippocampus. *J Lipid Res*. 2003;44(8):1545–1551.
147. Ikemoto A, Ohishi M, Sato Y, et al. Reversibility of n-3 fatty acid deficiency-induced alterations of learning behavior in the rat: level of n-6 fatty acids as another critical factor. *J Lipid Res*. 2001;42:1655–1663.
148. Ahmad A, Momenan R, van Gelderen P, Moriguchi T, Greiner RS, Salem Jr. N. Gray and white matter brain volume in aged rats raised on n-3 fatty acid deficient diets. *Nutr Neurosci*. 2004;7(1):13–20.
149. Ikemoto A, Kobayashi T, Watanabe S, Okuyama H. Membrane fatty acid modifications of PC12 cells by arachidonate or docosahexaenoate affect neurite outgrowth but not norepinephrine release. *Neurochem Res*. 1997;22:671–678.
150. Lezak MD. The problem of assessing executive functions. *Int J Psychol*. 1982;17:281–297.
151. Welsh MC, Pennington BF. Assessing frontal lobe functioning in children: views from developmental psychology. *Dev Neuropsychol*. 1988;4(3):199–230.
152. Baker SC, Rogers RD, Owen AM, et al. Neural systems engaged by planning: a PET study of the tower of London task. *Neuropsychologia*. 1996;34:513–526.
153. Fincham JM, Carter S, van Veen V, Stenger VA, Anderson JR. Neural mechanisms of planning: a computational analysis using event-related fMRI. *Proc Natl Acad Sci U S A*. 2002;99(5):3346–3351.
154. Morris RG, Ahmed S, Syed GM, Toone BK. Neural correlates of planning ability: frontal lobe activation during the tower of London test. *Neuropsychologia*. 1993;31(12):1367–1378.

155. Luciana M, Collins PF, Olson EA, Schissel AM. Tower of London performance in healthy adolescents: the development of planning skills and associations with self-reported inattention and impulsivity. *Dev Neuropsychol.* 2009;34(4):461–475.
156. Welsh MC, Pennington BF, Groisser DB. A normative-developmental study of executive function: a window on prefrontal function in children. *Dev Neuropsychol*. 1991;7(2):131–149.
157. Calderon F, Kim HY. Docosahexaenoic acid promotes neurite growth in hippocampal neurons. *J Neurochem*. 2004;90(4):979–988.
158. Fang YJ, Zhou MH, Gao XF, Gu H, Mei YA. Arachidonic acid modulates Na+ currents by non-metabolic and metabolic pathways in rat cerebellar granule cells. *Biochem J*. 2011;438(1):203–215.
159. Kim SJ, Zhang Z, Saha A, et al. Omega-3 and omega-6 fatty acids suppress ER- and oxidative stress in cultured neurons and neuronal progenitor cells from mice lacking PPT1. *Neurosci Lett*. 2010;479(3):292–296.
160. Weiser MJ, Wynalda K, Salem Jr. N, Butt CM. Dietary DHA during development affects depression-like behaviors and biomarkers that emerge after puberty in adolescent rats. *J Lipid Res*. 2015;56(1):151–166.
161. Rubinow DR, Schmidt PJ, Roca CA. Estrogen-serotonin interactions: implications for affective regulation. *Biol Psychiatry*. 1998;44:839–850.
162. Epperson CN, Wisner KL, Yamamoto BY. Gonadal steroids in the treatment of mood disorders. *Psychosom Med*. 1999;61:676–697.
163. Zimmer L, Vancassel S, Cantagrel S, et al. The dopamine mesocorticolimbic pathway in affected by deficiency in n-3 polyunsaturated fatty acids. *Am J Clin Nutr*. 2002;75:662–667.

CHAPTER

5

Arachidonic Acid, Docosahexaenoic Acid, and Ethanol

Robert R. Miller, Jr, Marissa Owen†, Christine Scanlan*, Kathryn VanEgmond‡, Taylor K. Wessel§, Charles N. Zawatsky*,¶*

***Hillsdale College, Biology Dept., Hillsdale, MI, United States †Michigan State University College of Veterinary Science, East Lansing, MI, United States ‡Penn State Health, Milton S. Hershey Medical Center, Hershey, PA, United States §Pfizer Incorporated, Kalamazoo, MI, United States ¶National Institute on Alcohol Abuse and Alcoholism, Bethesda, MD, United States**

INTRODUCTION

The predominant omega-3 (*n*-3) fatty acid and omega-6 (*n*-6) fatty acid within brain membranes is docosahexaenoic acid (DHA; 22:6, *n*-3) and arachidonic acid (AA; 20:4, *n*-6). These fatty acids comprise approximately 20% of all adult mammalian brains fatty acids.[1,2] When hydrolyzed from phospholipids, AA is the precursor of prostaglandin 2, thromboxane 2, and leukotriene 4 and are pro-inflammatory, leading to increased production of interleukin-1 (IL_{-1}), interleukin-6 (IL_{-6}), NF-κB (nuclear factor-kappa light chain enhancer of activated B cells), and tumor necrosis factor-α (TNF-α). AA is ultimately synthesized from γ-linolenic acid (18:3, *n*-6).[3] DHA is the elongation and desaturation product of dietary α-linolenic acid (18:3, *n*-3) and eicopentaenoic acid (EPA; 20:5, *n*-3). DHA supplementation blocks AA metabolites from initiating inflammation and apoptosis.[3]

Phospholipase A_2, AA, and Ethanol in Adult Rat Brains

Phospholipases A_2 (PLA_2s; EC 3.1.1.4) are a superfamily of enzymes that catalyze the hydrolysis and release of fatty acids from the *sn-2* position (second carbon of glycerol) within phospholipids and glycolipids. The PLA_2 superfamily includes 30 different isozymes that fall into groups including secreted (s)PLA_2; cytosolic (c)PLA_2; calcium-independent (i)PLA_2;

https://doi.org/10.1016/B978-0-12-815238-6.00005-5

platelet-activating factor (PAF) acetylhydrolase; and oxidized lipoprotein-associated (Lp) PLA_2.[4,5] Chronic binge ethanol (EtOH) exposure within adult rats promotes increased levels of $cPLA_2$ IVA, phospho-$cPLA_2$, and $sPLA_2$ IIA coupled with decreased sPLA2 levels within the entorhinal-hippocampal cortex as compared to controls.[6,7] This same rat brain area correlates with the increased levels of aquaporin-4 (AQ4P), poly (ADP-ribose) polymerase-1 (PARP-1), and 4-hydroxynonenal (4-HNE)-conjugated proteins.[6–8] AQP4 is water channel abundantly found on astrocyte surfaces and cause EtOH-induced brain swelling (edema); increased PLA_2 activities; AA hydrolysis; and cyclooxygenase (EC1.14.99.1)-mediated conversion of AA to IL-1, IL-6; and NF-κB involvement coupled with increased TNF levels, and is ameliorated by DHA supplementation.[6–9] EtOH-induced neurodegeneration, neuroinflammation, and brain edema was first reported by Grace Y Sun and Albert Y Sun.[7,10,11] NF-κB involvement and increased TNFα levels, PARP-1, 4-hydroxynonenal (4-HNE)-conjugated proteins are markers of lipid peroxidation, apoptosis, and oxidative stress.[7,8,12] Binge EtOH treatments cause increased Ca^{+2}-independent cytosolic PLA_2 ($iPLA_2$) and Ca^{+2}-dependent cytosolic PLA_2 ($cPLA_2$) levels and catalyze the release of membrane bound AA, the formation of 4-HNE-conjugated proteins, and other oxidative stress markers within neonatal rat brains (2–2.5 weeks).[5,13]

Besides depleting AA levels, EtOH-treatments caused depleted brain DHA levels in adult cats and primates, and compensatory increased docosapentaenoic acid (DPA: 22:5, *n*-6) levels.[14–16] EtOH-induced decreased DHA levels and compensatory increases in DPA (22:5, *n*-6) levels were also reported in rat glioma cells and brains[17] and DHA augmentation alleviates apoptosis within DHA-depleted brains.[18] EtOH-induced decreased hepatic AA (20:4, *n*-6) and DHA (22:6, *n*-3) levels have also been observed in adult guinea pigs and miniature swine.[19,20] EtOH-induced DHA reductions in phosphatidylserine involve PS synthase[18] and DHA peroxidation.[21] Lipid peroxidation mechanisms and subsequent apoptosis effects and oxidative stress markers have been reviewed.[12] Calcium-independent PLA_2 ($iPLA_2$) prefers DHA hydrolysis within phosphatidylserine in astrocytes and $cPLA_2$ mediates hydrolysis of DHA within phosphatidylethanolamine and phosphatidylcholine in both astrocytes and neurons.[5,20,22–24] While AA hydrolysis is unaffected by 10 μM forskolin, a stimulator of adenylate cyclase, DHA hydrolysis is strongly amplified.[22] Hence, differences in cAMP levels, Ca^{20} levels, or PLA_2 isozymes present can selectively impact hydrolysis of either AA or DHA.[20] Astrocytes supply neurons with AA (20:4, *n*-6), DPA (22:5, *n*-6), and DHA (22:6, *n*-3) synthesized from α-linolenic acid (18:3, *n*-3) and eicosapentaenoic acid (EPA; 20:5, *n*-3).[21,25]

DHA and Adult Neuroprotection D1

Free DHA may either undergo lipid peroxidation and be converted into reactive aldehydes, which perpetuate oxidative stress and apoptosis,[12] or be converted into neuroprotection D1 (10,17-dehydroxydocosahexaenoic acid)[5,23,24,26,27] via 15-lipoxygenase-1. In retinal pigment epithelial cells, neuroprotectin D1 upregulates antiapoptotic proteins Bcl-2 and Bcl-x_L by increasing protein phosphatase 2A activity and dephosphorylating Bcl-x_L.[24] Pro-apoptotic proteins, Bax and Bad, are activated by H_2O_2/TNFα-induced oxidative stress and destabilize and insert into mitochondrial membranes.[24,28] Once mitochondrial membranes are destabilized, cytochrome *c* enters the cytoplasm, activates Apaf-1, which then activates procaspases 3 and 7.[12,24,28] Caspase-3 and caspase-7 are killer caspases. Hence, neuroprotection D1 protects retinal pigment epithelial cells from oxidative stress and apoptosis.

DHA is concentrated within adult brain synapses and DHA and neuroprotectin D1 levels decrease during Alzheimer's disease (AD).[29–31] Early AD onset is associated with inflammatory signaling, senile plaques, and accumulation of extracellular amyloid-β (Aβ40 and Aβ42) peptides derived from the transmembrane glycoprotein β-amyloid precursor protein via βγ-secretase cleavage.[26,27,31,32] Neuroprotection D1 represses Aβ40 and Aβ42 release and apoptosis by upregulating the antiapoptotic protein Bcl-2.[22] DHA decreases are promoted by free-radical-mediated lipid peroxidation,[33–36] decreased dietary *n*-3 intake and subsequent inhibited DHA synthesis, and/or impaired DHA transport from liver to brain.[33] While decreased DHA synthesis, and/or impaired DHA transport from liver to brain[36] is possible, the loss of brain DHA correlates with increased brain reactive aldehyde levels (4-hydroxynonenal and acrolein), both markers of apoptosis, and the appearance of isoprostanes within the cerebral spinal fluid of AD patients.[3]

Although DHA-derived neuroprotection D1 is linked to the amelioration of oxidative-stress and apoptosis within adult retinal pigment epithelia1 cells, AD, and most recently to ischemic stroke recovery,[37] it is unknown if neuroprotection D1 ameliorates ethanol-induced oxidative stress in adults. Nor is it known if neuroprotectin D1 exists within vertebrate embryos. However, it is well known that vertebrate embryos exhibit ethanol-induced oxidative stress.

AA, Prostaglandins, and Implantation

In rodents and humans, $cPLA_{2\alpha}$ activity catalyzes AA hydrolysis and AA conversion to PGI_2, $PGF_{2\alpha}$, PDE_2, and PGD_2 via cyclooxygenases 1 and 2; and interleukin 1 ($IL\text{-}_{1\alpha}$) and $IL\text{-}_6$, are secreted by blastocysts approaching uterine epithelium.[38–40] In fact, $cPLA_2$, cyclooxygenase 1, and cyclooxygenase 2 transcripts have been detected in human two-cell blastulas prior to implantation.[41,42] During mammalian implantation, blastocysts secrete $IL\text{-}_{1\alpha}$, TNF-α, platelet-derived growth factor (PDGF), and epithelial growth factor (EGF), which are potent inducers of leukemia inhibitory factor, an interleukin-6 related cytokine.[40,43] In response, uterine epithelium secretes TNF-α and LIF (inhibitory factor), while LIF receptors reside on trophoblast cells.[43] TNF-α activate inducible nitrous oxide synthase (iNOS) and both iNOS and endothelial NOS are found in placental tissue and catalyze the synthesis of nitrous oxide (NO).[42,43] Because of an unpaired electron, NO is a highly reactive free radical and can damage proteins, carbohydrates, lipids, and inflammatory mediators. NO can also promote oxidative stress and relax arterial and venous smooth muscles acting as a vasodilator within placental blood vessels.[44] Placentas of both humans and rats possess superoxide dismutase, xanthine oxidase/dehydrogenase.[45–47] Human placentas and fetuses also possess catalase and glutathione peroxidase.[48,49] The relative abundance and enzymatic activities of these antioxidant enzymes increase with gestational age and appear to signal the transition from relatively anaerobic metabolism during early development and implantation to an aerobic environment during later developmental stages.[45–49] Excessive oxidative stress has been linked to impaired female fertility.[42,44]

EMBRYONIC OXIDATIVE STRESS, APOPTOSIS, AND POLYUNSATURATED FATTY ACIDS

Reports of EtOH-induced changes within prenatal brain membrane fatty acid composition have been inconsistent.[12,50,51] Yet, prenatal EtOH-induced reductions in membrane DHA levels and reduced *n*-3/*n*-6 ratios have been reported in embryonic mice and rat brains especially within phosphatidylcholine fractions.[50–52] Similarly, EtOH administration to pregnant guinea pig dames caused lower fetal brain DHA levels.[53]

Because EtOH metabolism within placental mammals is the result of maternal metabolism, fetal metabolism, and transport mechanisms between mother and child, we have employed a simpler vertebrate system. Because chick embryos are self-contained (cleidoic egg), EtOH can be injected into the air sac, diffuse into the yolk sac, and is taken into embryos as yolk sac vascularization proceeds. Consequently, avian eggs solely represent embryonic transport and metabolism.

Approximately 25 μL of 95% EtOH was injected into each air sac of fertile eggs (blood type SC; Hy-Vac International, Adel, IA) at 0 days of development (E_0) and each hole sealed with molten Paraplast. At either 18 days of development (stage 44),[54] or 2-day neonatal development, chick brains were excised, fixed, stained, and brain neuron densities measured within the avian *lateral pallium* (homologous to frontal lobes within mammalian cerebral hemispheres). Brain membrane (phospholipid fractions) fatty acids were measured via gas chromatography (GC).[55,56] From a morphology standpoint, a single injection of exogenous 6 mmol EtOH/kg egg, on E_0, caused decreased neuron densities within several regions of the avian brain at 18 days of development (stage 44; late embryogenesis) and in 2-day neonatal brains.[55,56] From a membrane fatty acid standpoint, EtOH exposure caused reduced brain membrane DHA (22:6, *n*-3) and AA (20:4, *n*-6) levels coupled with increased myristic acid (14:0) and palmitic acid (16:0) levels at 18 days of development (stage 44; late embryogenesis) and 2-day neonatal brains.[55,56] EtOH-induced increased brain membrane myristic acid (14:0) and palmitic acid (16:0) levels may be compensatory for EtOH-induced reductions in AA and DHA levels and may promote EtOH-induced increases in fatty acid synthase (EC 2.3.1.85) activities. Exogenous treatments of 2.5 μmol α-tocopherol/kg egg or 2.5 μmol γ-tocopherol/kg egg partially ameliorated EtOH-induced reductions in brain AA and DHA levels; EtOH-induced increases in brain myristic acid (14:0) and palmitic acid (16:0) levels; and EtOH-induced reductions in brain neuron densities at 18 days of development (stage 44; late embryogenesis) and in 2-day neonatal brains.[55,56] EtOH-induced increases in brain and hepatic caspase 3 activities were observed and were partially ameliorated by α-tocopherol or γ-tocopherol.[55,56]

DHA and AA decreases may be caused by EtOH-induced free-radical-mediated lipid peroxidation,[43,44] decreased hepatic DHA and AA synthesis, and/or impaired DHA and AA transport from liver to brain.[35,36] While decreased hepatic DHA and AA synthesis, and/or impaired DHA and AA transport from liver to brain[36] are possible, use of these two well-known antioxidants (α-tocopherol or γ-tocopherol) ameliorated EtOH-induced decreased DHA and AA and partially ameliorated EtOH-induced increased brain and hepatic caspase-3 activities.[55,56] Hence, we suspect EtOH-induced decreased brain DHA and AA levels within avian embryos are partially caused by EtOH-induced lipid peroxidation. Possible sources of reactive oxygen species have been reviewed.[12,57]

PLA_2 Activities in Chick Embryos

Chronic EtOH effects on PLA_2 activities have been reported in embryonic chick brains, heart, and livers.[58] PLA_2, activities have been reported in chick brains at 14–16 days of development (stages 40–42)[54,59] and PLA_2 receptors were isolated from retinal progenitor cells from 6-day to 8-day-old embryos (stages 20–22).[54,60] Specifically, $cPLA_2$ activities exist within dorsal root ganglions of 10- and 11-day-old chick embryos (stages 36–37)[54,61] and iPA_2 (Ca^{+2}-independent phospholipase A_2) activity from brain glia isolated from cerebral hemisphere homologs in 11-to-14-day-old chick embryos has been reported (stages 37–40).[54,62] So far, all stages of chick PLA_2 activity precede 18 days of chick brain development (stage 44).[54–56] So, EtOH-induced release of AA and DHA from brain phospholipids and subsequent lipid peroxidation is possible. In addition, $cPLA_2$ presence in cardiomyocytes isolated from 13-day chick embryos (stage 39) were reported.[54,63,64]

EtOH-Induced Changes in Brain Membrane Fatty Acids During Late Chick Embryogenesis

EtOH-induced changes in brain membrane fatty acid levels during late chick embryogenesis have been studied. Control eggs were injected 3 times with approximately 25 μL of H_2O into air sacs on 0, 1, and 2 days of incubation ($E_{0–2}$; theoretical stages 1, 7, and 12).[54] Experimental eggs were injected 3 times with approximately 25 μL of either a 1:1 (*v*/v) mixture of 95% EtOH and H_2O (3.0 mmol EtOH/kg egg); or 95% EtOH (6 mmol/kg egg) into the air sac of each egg. The first 3 days of chick development represent an early embryogenesis (stage 12/46 total stages = 26%) and an attempt is made to model human binge drinking before a woman knows that she is pregnant.[54] After each injection, the punctures were sealed with molten Paraplast.

After injections, all embryos were incubated in a forced air incubator at 37.5°C and turned every 4 h with relative humidity ranging from 80% to 90%. Embryos were harvested at 7 days of development (stage 31; stage 31/stage 46 = 67.4% of development); 9 days of development (stage 35; stage 35/stage 46 = 76.1% of development); 11 days of development (stage 37/stage 46 = 80.4% of development); and 15 days of development (stage 41/stage 46 = 89.1% of development).[54] After excision, embryonic brains were stored at –80°C until subsequent analyses.

Embryonic brains were homogenized in 400 μL of ice-cold 50 mM 4-(2-hydroxyethyl)-1-piperazineethanesulfonic acid (HEPEs) buffer (pH 7.4), 1 mM dithiothreitol (DTT), 0.1 mM of the disodium salt of ethylenediaminetetraacetic acid (EDTA), and total lipids were extracted.[65,66] Samples were concentrated by solvent evaporation under N_2 and stored overnight at −40°C. Neutral lipids were separated from membrane lipids (phospholipids) on silicic acid columns and dried under N_2. Membrane lipid samples were saponified and liberated fatty acids methylated.[66–68] Fatty acid methyl esters (FAMEs) were then separated by GC as discussed.[67–70]

Chick Brain DHA Levels

The primary polyunsaturated fatty acids within brain phospholipids are DHA (22:6, *n*-3) and AA (20:4, *n*-6). Control DHA levels in stage 35 (E_9; 9 days of development), stage 37 (E_{11}; 11 days of development), and stage 41 (E_{15}; 15 days of development) were approximately 2.0-fold, 2.2-fold, and 2.3-fold greater than stage 31 controls (E_7),[54] respectively (Fig. 1). DHA levels within stages 35–41 brains were significantly greater than stage 31 controls (56; $P < 0.0001$). Because DHA is ultimately synthesized from dietary α-linolenic (18:3, *n*-3) within the liver

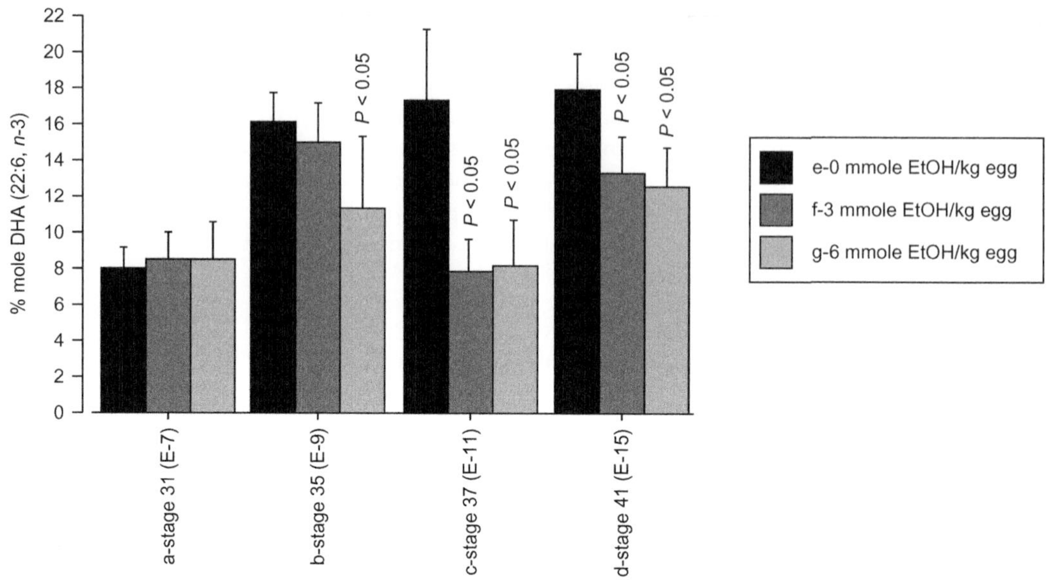

ANOVA table for %mole DHA

	DF	Sum of squares	Mean square	F-Value	P-Value	Lambda	Power
Dev stage	3	774.146	258.049	38.529	<.0001	115.586	1.000
mmole EtOH/kg egg	2	492.742	246.371	36.785	<.0001	73.570	1.000
Dev stage * mmole EtOH/kg egg	6	433.980	72.330	10.799	<.0001	64.797	1.000
Residual	118	790.311	6.698				

Note: $P < 0.05$ denotes statistical significance as compared to stage-specific controls.

FIG. 1 Effect of EtOH on brain membrane (phospholipid fraction) docosahexaenoic acid (DHA) levels during late chick embryogenesis.

and transported to the brain,[12,36,70] low DHA levels within stage 31 chick brains may be due to less differentiated hepatic tissues as compared to brains and/or reduced fatty acid transport proteins capabilities.[36]

While EtOH-induced reductions in brain DHA levels were observed during late embryogenesis stages, EtOH failed to promote reduced DHA levels in stage 31 (E_7) embryos. Later developmental stages have greater DHA levels as compared to stage 31 (E_7) embryos (Fig. 1). In stage 35 (E_9) embryos, EtOH caused dose-dependent decreased DHA levels. Although 3 mmol EtOH/kg egg dosages caused insignificant decrease in brain DHA levels as compared to controls, 6 mmol EtOH/kg egg dosages caused significant decrease in DHA levels as compared to stage 35 (E_9) controls ($P < 0.05$). In stage 37 embryos (E_{11}), both 3 mmol EtOH/kg egg[67,68] and 6 mmol EtOH/kg egg dosages[69] caused reduced brain DHA levels as compared to stage 37 controls (E_{11}; $P < 0.05$).[54] In stage 41 (E_{15}) embryos, both 3 mmol EtOH/kg egg and 6 mmol EtOH/kg egg dosages caused decreased DHA levels as compared to stage 41 controls ($P < 0.05$).[54] These EtOH-induced decreased DHA levels in stage 41 embryos were less severe as compared to EtOH-treated stage 37 embryos ($P < 0.05$).[54]

Chick Brain AA Levels

EtOH failed to promote significant brain AA reductions in stage 31 (E_7) embryos (Fig. 2).[54] Control brain AA levels in stage 35 (E_9), stage 37 (E_{11}),[54,67–69] and stage 41 (E_{15})[54] were approximately 1.5-fold, 1.4-fold, and 1.5-fold, respectively, greater as compared to stage 31 controls (E_7) ($P<0.0001$) (Fig. 2). Because AA is ultimately synthesized from dietary γ-linolenic (18:3, *n*-6) within the liver and transported to the brain,[12,36,70] low AA levels within stage 31 chick brains may be due to less differentiated hepatic tissues as compared to brains and/or reduced fatty acid transport proteins capabilities.[36]

Besides later stages of embryogenesis having greater brain AA levels as compared to stage 31 (E_7) embryos, exogenous EtOH also reduced AA levels (Fig. 2). In stage 35 (E_9) embryos, exogenous EtOH caused dose-dependent decreases in brain AA levels. Although 3 mmol EtOH/kg egg dosages caused insignificant brain AA decreases as compared to controls,[67,68] 6 mmol EtOH/kg egg dosages caused significant AA reductions[69] as compared to stage 35 (E_9) controls (56; $P<0.05$). In stage 37 embryos (E_{11}), both 3 mmol EtOH/kg egg[67,68] and

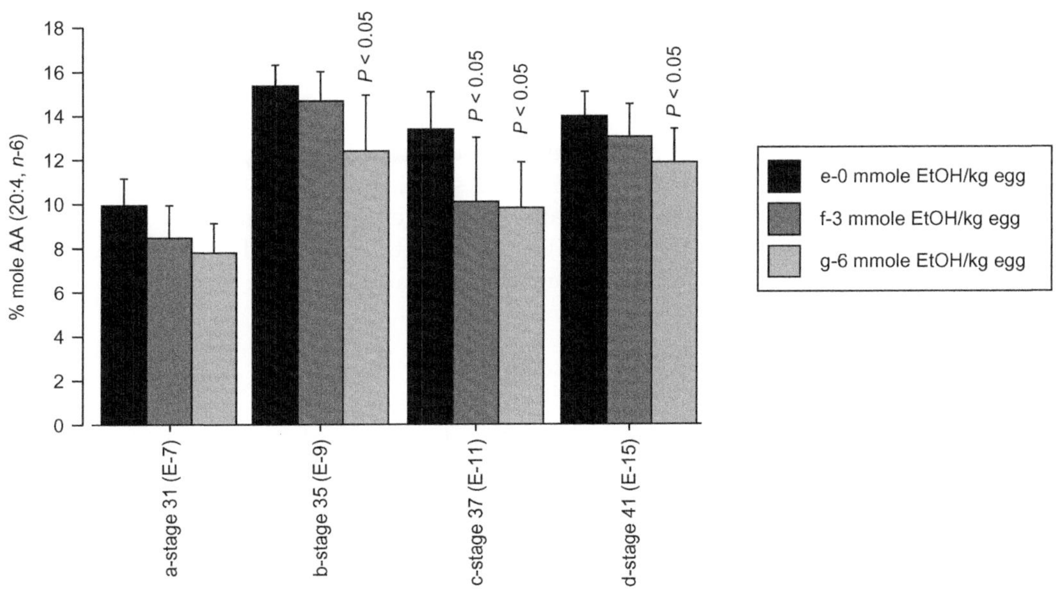

ANOVA table for arcsin AA

	DF	Sum of squares	Mean square	*F*-Value	*P*-Value	Lambda	Power
Dev stage	3	374.897	124.966	41.777	<.0001	125.332	1.000
mmole EtOH/kg egg	2	389.735	194.867	65.146	<.0001	130.293	1.000
Dev stage * mmole EtOH/kg egg	6	318.768	53.128	17.761	<.0001	106.568	1.000
Residual	118	352.965	2.991				

Note: $P < 0.05$ denotes statistical significance as compared to stage-specific controls.

FIG. 2 Effect of EtOH on brain membrane (phospholipid fraction) arachidonic acid (AA) levels during late chick embryogenesis.

6 mmol EtOH/kg egg[69] dosages caused reduced brain AA levels as compared to stage 37 controls (E_{11}; $P<0.05$). In stage 41 (E_{15}) embryos, 3 mmol EtOH/kg egg dosages caused insignificant brain AA reductions as compared to controls. However, 6 mmol EtOH/kg egg dosages caused significant AA reductions as compared to stage 41 (E_{15}) controls (56; $P<0.05$). These EtOH-induced decreased brain AA levels in stage 41 embryos were less severe as compared to EtOH-treated stage 37 embryos ($P<0.05$). While EtOH-induced decreased brain AA levels were observed during late embryogenesis stages, EtOH failed to cause significant brain AA reductions in stage 31 (E_7) embryos.

EtOH-Induced Increases in Short Saturated Membrane Fatty Acids

Besides causing decreased brain DHA and AA levels, exogenous EtOH caused increased levels of saturated brain membrane fatty acids in stage 35 (E_9), stage 37 (E_{11}),[67–69] and stage 41 embryos. EtOH-induced alterations in saturated brain membrane fatty acid levels were not observed in stage 31 (E_7) chick brains. Control membrane palmitic acid (16:0) levels in stage 35 chick brains were approximately 7.78% mol and exposure to 6 mmol EtOH/kg egg caused approximately a 1.6-fold increase ($P<0.05$). Control membrane palmitic acid levels in stage 37 chick brains were 9.87% mol[67–69] and 6 mmol EtOH/kg egg exposure[69] caused approximately a 1.4-fold increase ($P<0.05$). Control membrane stearic acid levels in stage 41 (E_{15}) brains were approximately 1.30% mol. Exposure to 3 mmol EtOH/kg egg caused approximately a 1.6-fold increase ($P<0.05$) and 6 mmol EtOH/kg egg exposure caused approximately a 1.8-fold increase as compared to controls ($P<0.05$). These EtOH-induced increased brain membrane palmitic acid (16:0) and stearic acid (18:0) levels may be compensatory for EtOH-induced reductions in AA and DHA levels and may indicate EtOH-induced increases in fatty acid synthase (EC 2.3.1.85) activities. EtOH-induced increases in saturated membrane fatty acids may impact membrane fluidity and/or serve as energy sources via gluconeogenesis.[12,70]

EtOH-Induced Increases in Brain LPO Levels

EtOH-induced decreased brain DHA and AA levels may be due to inhibited hepatic DHA synthesis from dietary α-linolenic (18:3, *n*-3) and inhibited AA synthesis from dietary γ-linolenic (18:3, *n*-6) coupled with impaired DHA and AA transport from liver to brain.[36,70] However, we suspect EtOH-induced reductions in brain DHA and AA levels are partially caused by lipid peroxidation of fatty acids.[12,55–57] DHA and AA may be primary oxidative targets because they are not only highly polyunsaturated, but are predominant fatty acids within brain phospholipids, and are rivaled only by the monounsaturated fatty acid, oleic acid (18:1, *n*-9).[55,56] Relative membrane oleic acid (18:1, *n*-9) levels within controls were 7.91± 1.42% mol in stage 31 (E_7); 12.17±1.83% mol in stage 35 chick (E_9); 19.36±7.74% mol in stage 37 (E_{11})[67–69]; and 24.70±3.25% mol in stage 41 (E_{15}) chick brains. Relative phospholipid AA (20:4, *n*-6) levels within controls were 10.54±1.99% mol in stage 31 (E_7); 15.00±1.38% mol in stage 35 chicks (E_9); 13.58±1.57% mol in stage 37 (E_{11})[67,68]; and 14.06±0.98% mol in stage 41 (E_{15}) chick brains (Fig. 2). Relative phospholipid DHA (22:6, *n*-3) levels within controls were 7.58±1.36% mol in stage 31 (E_7); 15.97±1.71% mol in stage 35 (E_9); 17.64±5.46% mol in stage 37 (E_{11})[67,68]; and 17.50±2.42% mol in stage 41 (E_{15}) chick brains (Fig. 1). While oleic acid, AA, and DHA are abundant during late chick brain embryogenesis, only AA and DHA are highly polyunsaturated making them primary fatty acid peroxidation targets.

If EtOH-induced decreased DHA and AA levels are partially due to EtOH-induced fatty acid peroxidation, then DHA and AA decreases should be associated with EtOH-induced increased brain lipid hydroperoxide (LPO) levels. LPOs are fatty acid peroxidation intermediates and the mechanism and structures have been reviewed.[12] During lipid peroxidation, the electron donated by either a Fe^{+2} ion or hydroxyl radical (HO–˙) is absorbed by a hydrogen atom attached to a saturated carbon adjacent to a carbon–carbon double bond within the fatty acid. During the second step, INITIATION, the absorption of an electron causes the formation of an alkyl radical and promotes a rearrangement of double bonds to form the alkyl radical. In the third step, a reaction with oxygen causes the formation of lipid peroxyl radicals and leads to PROPAGATION. In the fourth step, lipid peroxyl radicals remove an electron from a nearby alkyl radical and form LPOs which promote TERMINATION. In the fifth step, the Fe^{+2} (ferrous)-dependent cleavage of a LPO forms an alkoxyl radical. Finally, alkoxyl radical cleavage by β-(beta) scission is observed. This creates a shorter, less unsaturated fatty acid, and a number of reactive cytotoxic aldehydes.[12]

In order to test if exogenous EtOH caused increased brain LPO levels, control eggs were injected three times with approximately 25 μL of H_2O into each air sac on 0, 1, and 2 days of incubation (E_{0-2}; theoretical stages 1, 7, and 12).[54] Experimental eggs were injected three times with approximately 25 μL of either a 1:1 (*v*/v) mixture of 95% EtOH and H_2O (3.0 mmol EtOH/kg egg); or 95% EtOH (6 mmol/kg egg) into the air sac of each egg and sealed with Paraplast wax. Brains from stages 31(E_7), 35 (E_9), 37 (E_{11}), and 41(E_{15}) were isolated and LPO levels measured.[67–69]

LPOs were extracted by homogenizing brains in 500 μL of ice-cold 50 mM HEPES buffer (pH 7.4), 1 mM DTT, 0.1 mM EDTA.[67,68] Samples were deproteinated by adding 500 μL of *meta*-phosphoric acid (20 mg/mL in methanol) to each sample and protein-containing precipitates removed by centrifugation (5000 ×*g* for 5 min). Nonpolar LPOs were extracted from each supernatant by adding 2 mL of chloroform, followed by centrifugation (5 min at 2000 ×g). A 1 mL aliquot was removed from each organic layer and spectrophotometrically assayed for LPOs by using LPO assay kits (Cayman Chemical Co., Ann Arbor, MI, catalog no. 705003). LPOs are highly unstable and readily promote the conversion of Fe^{2+} to Fe^{3+} and were detected by ChromaGen using thiocyanate ions.[71]

EtOH failed to alter DHA, AA levels, and LPO levels (Fig. 3) within stage 31 (E_7) brains. However, EtOH-induced dose-dependent increases in brain LPO levels were observed in stages 35 (E_9) and 37 (E_{11}; 69–71) as compared to their respective controls ($P < 0.05$). In stage 41 (E_{15}) chick brains, the 6 mmol EtOH/kg egg dosage caused approximately a 4.6-fold increase in brain LPO levels as compared to stage 41 controls ($P < 0.05$) and was significantly greater than both 3 and 6 mmol EtOH/kg egg dosages in stages 35 (E_9) and stage 37 (E_{11}).[67–69] Because LPOs are a marker of fatty acid peroxidation, EtOH-induced increased lipid peroxidation rates increase during late embryogenesis.

EtOH-Induced Decreases in Brain Glutathione Levels

EtOH-induced increased brain LPO levels during later stages of chick embryogenesis appeared surprising. Peter Surai reported changing glutathione (reduced GSH), glutathione peroxidase (GSH-Px; EC 1.11.1.9), and other antioxidant enzymes levels/activities in chick brain, liver, and other tissues at 10 days of embryonic development (stage 36; E_{10}) through

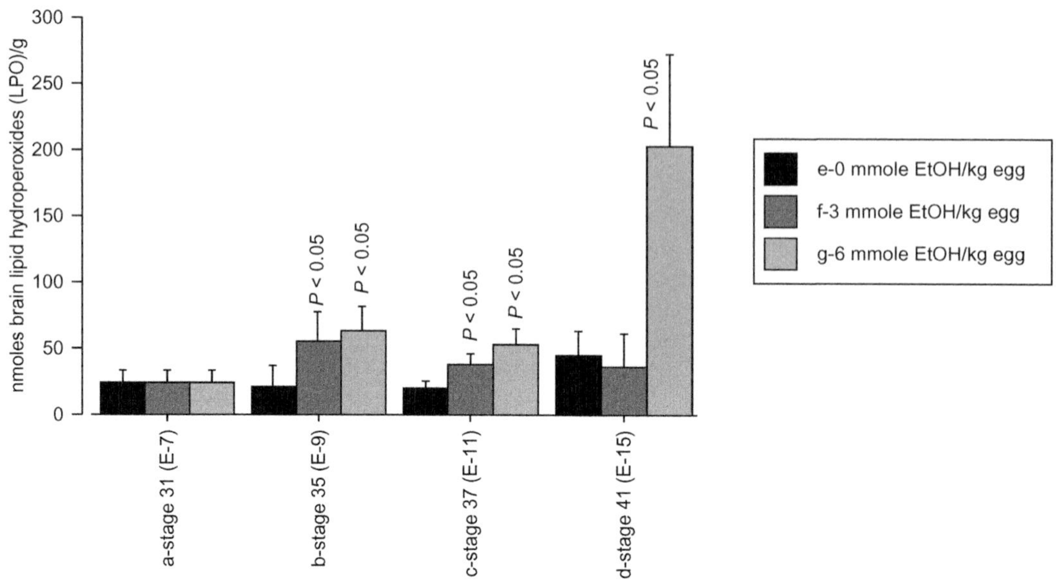

ANOVA table for nmoles LPO/g brain

	DF	Sum of squares	Mean square	*F*-Value	*P*-Value	Lambda	Power
Dev stage	3	82673.568	27557.856	42.641	<.0001	127.924	1.000
mmole EtOH/kg egg	2	81062.688	40531.344	62.716	<.0001	125.432	1.000
Dev stage * mmole EtOH/kg egg	6	128605.034	21434.172	33.166	<.0001	198.996	1.000
Residual	115	74320.982	646.269				

Note: $P < 0.05$ denotes statistical significance as compared to stage-specific controls.

FIG. 3 Effect of EtOH on brain lipid hydroperoxide (LPO) levels during late chick embryogenesis.

1-day of neonatal development.[72] Brain GSH-Px activities increased from stage 37 (E_{11}) to stage 41 (E_{15}) and then gently decreased from stage 43 (E_{17}) to 1-day of neonatal development.[71] GPx is a GSH-dependent enzyme that catalyzes the degradation of LPOs (ROOH; organic peroxides) into stable fatty acid alcohols (ROH) and forms oxidized glutathione dimer (GSSG) and is illustrated as follows[73–75]:

$$\text{ROOH} + 2\text{GSH} \xrightarrow{\text{GPx}} \text{ROH} + \text{GSSG} + \text{H}_2\text{O}$$

This reaction predicts that brain LPO levels should decrease from stage 37 (E_{11}) to stage 41 (E_{15}) as GPx activity increases. LPOs are cleaved into shorter, less unsaturated fatty acids, and a number of reactive aldehydes including 4-hydroxynonenal (HNE) and 4-oxo-2-nonenal (ONE) from *n*-6 peroxidation and *trans*-4-hydroxy-2-hexanal (HHE) from *n*-3 peroxidation.[12,75] HNE, ONE, and HHE are known to promote lipid peroxidation, apoptosis, and form GSH-adducts.[12,75] ETOH-induced GSH-adducts deplete GSH pools and inhibit GSH-dependent antioxidant enzymes, such as GPx.[12,75] Consequently, we measured brain GSH levels, in the presence and absence of EtOH in stage 37 (E_{11}) and stage 41 (E_{15}) brains.[68,73,74]

As Surai reported,[72] brain GSH levels were higher in stage 37 brains as compared to stage 41 brains (Fig. 4; $P<0.0001$). Early exposure to 3 mmol EtOH/kg egg[67,68] and 6 mmol EtOH/kg egg (E_{0-2}) caused decreased GSH levels as compared to their stage specific controls. This is consistent with the hypothesis that EtOH promoted GSH-adducts and decreased the activity of GSH-dependent enzymes, such as GPx, and further promoted neural lipid peroxidation (increased LPO levels) and apoptosis within late embryogenesis.[12,75] We have verified EtOH-induced decreases in brain GSH (Fig. 4), Se-dependent, and non-Se-independent brain GPx activities in stages 37 and 41 (data not presented).

EtOH-Induced Increases in Apoptosis Rates

Apoptosis rates in embryonic chick brains were measured in stage 31 (E_7), stage 35 (E_9), stage 37 (E_{11}), and stage 41 (E_{15}) chicks by measuring caspase-3 (casp-3; EC 3.4.22.56) activities.[55,56,67–69] While there are many methods used to determine apoptosis rates, we measured caspase-3 activities because caspase-3 and caspase-7 are "killer" caspases used to finish apoptosis whether cellular attack begins at cell membrane surfaces (extrinsic pathway) through the presence/absence of ligand/survival factor receptor interactions; or the attack of a cell

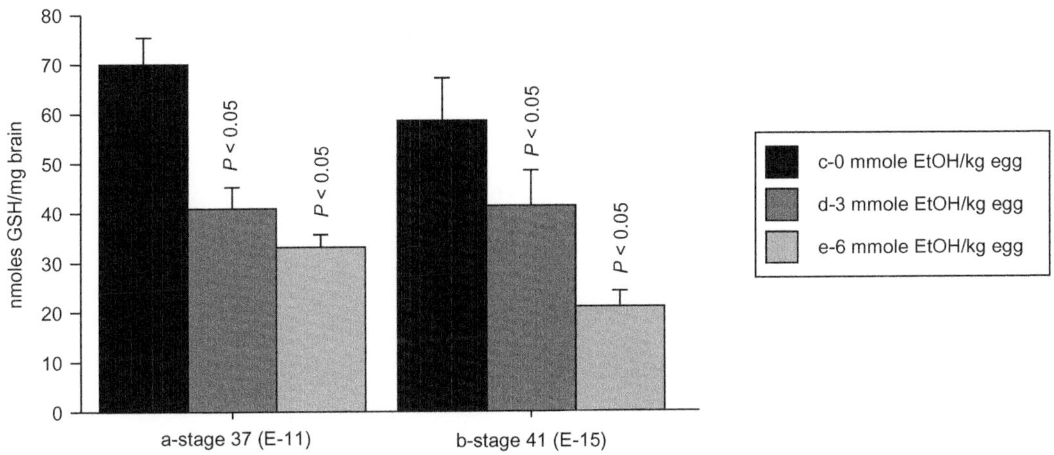

ANOVA table for nmoles GSH/mg brain

	DF	Sum of squares	Mean square	F-Value	P-Value	Lambda	Power
Dev stage	1	936.018	936.018	24.834	<.0001	24.834	1.000
mmole EtOH/kg egg	2	15820.386	7910.193	209.870	<.0001	419.740	1.000
Dev stage * mmole EtOH/kg egg	2	599.834	299.917	7.957	.0008	15.915	.959
Residual	66	2487.603	37.691				

Note: $P < 0.05$ denotes statistical significance as compared to stage-specific controls.

FIG. 4 Effect of EtOH on brain glutathione (GSH) levels in stage 37 and stage 41 chick brains.

membrane by reactive oxygen species; or attack of mitochondrial membranes by insertion of pro-apoptotic proteins or reactive oxygen species that spread to the cytoplasm (intrinsic pathway).[12,27,57] Once cytoplasmic procaspase 3 and procaspase 7 are converted into casp-3 and casp-7, they cleave any protein that has a DEVD (aspartate–glutamate-valine-aspartate) domain.[12,57] They are a final killing mechanism.

Fig. 5 illustrates that exogenous EtOH fails to alter casp-3 activities in stage 31 (E_7) chick brains. However, chick brains isolated from stage 37 (E_{11})[67–69] and stage 41 (E_{15}) have significantly greater casp-3 activities as compared to stage 31 and stage 35 controls [$F=(3, 110)$ 13.24; $P<0.0001$]. Hence, EtOH-induced brain apoptosis rates increased during later embryogenesis stages. Exogenous EtOH caused increased brain casp-3 activities in stage 35 (E_9), stage 37 (E_{11}), and stage 41 (E_{15}) embryos (Fig. 5) [$F=(3, 110)$ 37.41; $P<0.0001$]. In stage 35 (E_9) embryos, exogenous 6 mmol EtOH/kg egg dosages during the first 3 days of development ($E_{0\text{–}2}$), caused approximately a 1.5-fold increase in brain casp-3 activities as compared to controls ($P<0.05$). In stage 37 (E_{11}) embryos, exogenous 3 mmol EtOH/kg egg[67,68] and 6 mmol EtOH/kg egg dosages[69] during the first 3 days of development ($E_{0\text{–}2}$),

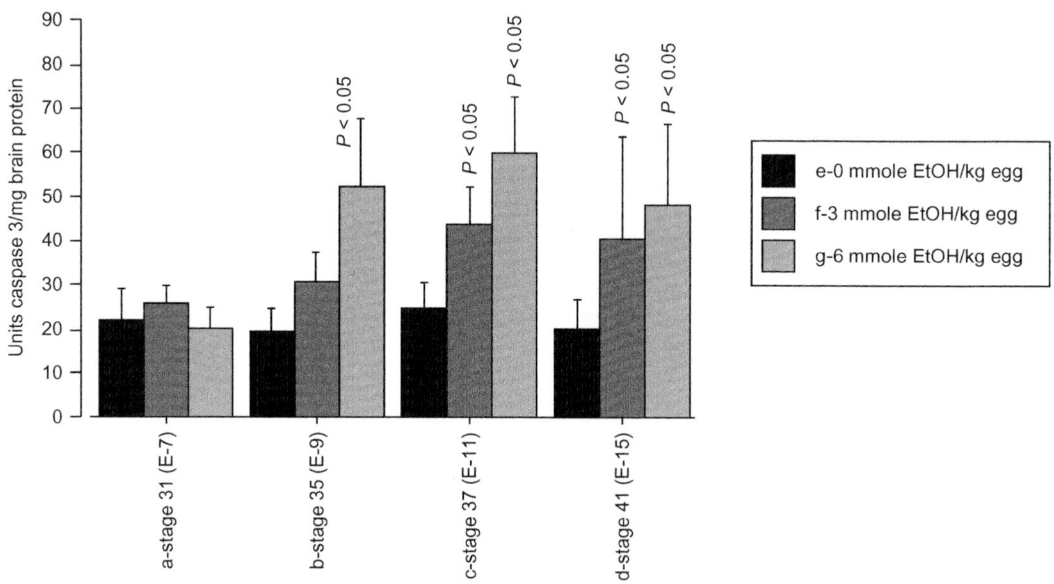

ANOVA table for units casp-3/mg brain protein

	DF	Sum of squares	Mean squares	*F*-Value	*P*-Value	Lambda	Power
Dev stage	3	5793.600	1931.200	13.239	<.0001	39.716	1.000
mmole EtOH/kg egg	2	10913.452	5456.726	37.406	<.0001	74.813	1.000
Dev stage * mmole EtOH/kg egg	6	4617.980	769.663	5.276	<.0001	31.657	0.996
Residual	110	16046.423	145.877				

Note: $P<0.05$ denotes statistical significance as compared to stage-specific controls.

FIG. 5 Effect of EtOH on brain caspase-3 activities during late embryogenesis.

caused approximately 1.8-fold to 2.5-fold increases in brain casp-3 activities as compared to controls, respectively ($P<0.05$). In stage 41 (E_{15}) embryos, exogenous 3 mmol EtOH/kg egg and 6 mmol EtOH/kg egg dosages during the first 3 days of development (E_{0-2}) caused approximately 2.0-fold to 2.4-fold increases in brain casp-3 activities as compared to controls, respectively ($P<0.05$) and were dose-dependent brain casp-3 activities increases in stages 37 and 41 embryos.

ALCOHOL DEHYDROGENASE AND XANTHINE OXIDASE EPISTASIS

Exogenous EtOH failed to cause decreased DHA and AA levels; increased LPO levels, and increased casp-3 activities within stage 31 (E_7) chick brains and stage 31 predates alcohol dehydrogenase expression.[76,77] EtOH is predominantly metabolized by alcohol dehydrogenase isozymes (ADH; EC 1.1.1.1) to acetaldehyde and acetaldehyde to acetate by aldehyde dehydrogenase isozymes (ALDH; EC 1.2.1.3), and both enzymes are NAD^+-dependent.[12,57] ADH isozymes are first seen in embryonic chick brains and livers at 9–10 days of development (theoretical stage 35–36) and were undetected at earlier stages.[76,77] Two polymorphic chick ADH forms exist.[77] Once injected during early chick embryogenesis, EtOH resides in the egg until ADH expression.[76,77] In mouse embryos, Class I ADH (Adh-1), the major EtOH metabolizing isozyme, was undetected until after implantation and cleavage.[78]

ADH and ALDH isozymes may be in epistasis with xanthine dehydrogenase (XDH; EC 1.17.1.4, 1.17.3.2) and xanthine oxidase (XO; EC 1.17.3.2).[12,57,68] We have measured primarily brain XO activities at stages 36 and 37. Allopurinol is a known XDH inhibitor and EtOH-induced increased XO activities are/were unaffected by exogenous 184 nmol allopurinol/kg egg. NAD^+-dependent XDH is normally the preferred enzyme in the conversion of hypoxanthine to xanthine. However, as NAD^+ levels are depleted during EtOH metabolism, XO activity increases promoting the formation of urate and H_2O_2 as follows:

$$\text{Hypoxanthine} + H_2O + NAD^+ \xrightarrow[\text{dehydrogenase}]{\text{Xanthine}} \text{Xanthine} + NADH + H^+$$

$$\text{Hypoxanthine} + H_2O + 2O_2 \xrightarrow{\text{Xanthine oxidase}} \text{Xanthine} + O_2^{-\cdot} + H_2O_2$$

In the presence of Fe^{+2} ions, H_2O_2 is converted to hydroxyl ions ($HO^{\cdot}$) and hydroxyl radicals initiate lipid peroxidation and LPOs are observed in stage 35 chick brains and later (Figs. 1–5).[12] Through the transfer of a single electron from NADPH, xanthine oxidase can also form less reactive superoxide radicals ($O_2^{-}.$), which are utilized during later lipid peroxidation, by transferring a single electron to molecular oxygen.[12] Through the transfer of a single electron from NADPH, xanthine oxidase can also form less reactive superoxide radicals ($O_2^{-}.$), which are utilized during later lipid peroxidation, by transferring a single electron to molecular oxygen.[12] Hence, ADH expression may initiate reactive oxygen species formation.

Acknowledgments

We wish to acknowledge and thank David Congdon and Benjamin J. Hooker. David aided in the collection of some LPO and Casp-3 data and Ben Hooker collected XO data in this paper.

ETHICS APPROVAL

Fertile chick eggs were purchased from Hy-Vac International (Adel, IA) and Charles Rivers Avian Services (North Franklin, CT) and animal care was in compliance with the Guide for the Care and Use of Laboratory Animals, 8th Edition (National Academies Press, Washington DC, USA, 2011). Because embryonic chicks cannot self-regulate temperatures and are not self-reliant, chick embryo usage is not currently under the scrutiny of US Institutional Animal Care and Use Committees.

References

1. Contreras MA, Chang MC, Myers CS, Salem Jr N, Rapoport SI. Nutritional deprivation of alpha-linolenic acid decreases but does not abolish turnover and availability of unacetylated docosahexaenoic acid and docosahexaenoyl-CoA in rat brain. *J Neurochem*. 2000;75(6):2392–2400.
2. Rapoport SI. Arachidonic acid the brain. *J Nutr*. 2008;138(12):2515–2520.
3. Simopoulos AP. An increase in the Omega-6/Omega-3 fatty acid ratio increases the risk of obesity. *Nutrients*. 2016;128. https://doi.org/10.3390/nu8030128.
4. Burke JE, Dennis EA. Phospholipase A_2 structure/function, mechanism, and signaling. *J Lipid Res*. 2009; [April Supplement: S237–S242].
5. Lee JC-M, Simonyl A, Sun AY, Sun GY. Phospholipases A2 and neural membrane dynamics: implications for Alzeimer's disease. *J Neurochem*. 2011;116(5):813–819.
6. Tajuddin N, Moon K-H, Marshall SA, et al. Neuroinflammation and neurodegeneration in adult rat brain from binge ethanol exposure: abrogation by docosahexaenoic acid. *Plos One*. 2014;9(7):e101223. https://doi.org/10.1371/journal.pone.0101223.
7 Collins MA, Tajuddin N, Moon K-H, Kim H-Y, Nixon K, Neafsey EJ. Alcohol, phospholipase A_2 associated neuroinflammation, and ω3 docosaehexaenoic acid protection. Mol Neurobiol 214; 50: 239–245. doi: https://doi.org/10.1007/s12305.014-8690-0.
8. Collins MA, Neafsey EJ. Neuroinlflammatory pathways in binge alcohol-induced neural degeneration: oxidative stress cascade involving aquaporin brain edema, and phospholipase A2 activation. *Neurotox Res*. 2012;21:70–78. https://doi.org/10.1007/s12640-011-9276-5.
9. Collins MA, Moon K-H, Tajuddin N, Neafsey EJ, Kim H-J. Docosahexaenoic acid (DHA) prevent binge ethanol-dependent aquaporin-4 elevations while inhibiting neurodegeneration: experiments in rat adult-age entorhino-hippocampal slice cultures. *Neurotox Res*. 2013;23(1):105–110.
10. Sun AY, Tumbleseson ME, Dexter JO, Sun GY. Miniature swine as an animal model. In: Sun GY, Rudden PK, Wood WG, Wei YH, Sun AY, eds. *Molecular Mechanisms of Alcohol: Neurobiology and Metabolism*. Boca Raton: Humana Press; 1989:371–397.
11. Sun AY, Ingelman-Sundberg M, Neve E, et al. Ethanol and oxidative stress. *Alcohol Clin Exp Res*. 2001;25(5). 237S–43S.
12. Miller Jr RR. Ethanol-induced lipid peroxidation and apoptosis in embryopathy. In: Watson RR, Preedy VR, Zibadi S, eds. *Alcohol, Nutrition, and Health Consequences*. New York: Springer Science; 2013:35–62.
13. Moon K-H, Tajuddin N, Brown III J, Neafsey EJ, Kim H-Y, Collins MA. Phospholipase A_2, ethanol-treated organotypic slice cultures of developing rat brain. *Alcohol Clin Exp Res*. 2014;38(1):161–169.
14. Pawlowsky RJ, Salam Jr N. Ethanol exposure causes a decrease in docosahexaenoic acid and an increase in docosapentaenoic acid in feline brains and retina. *Am J Clin Nutr*. 1995;61(6):1284–1289.

15. Pawlosky RJ, Bacher J, Salem Jr N. Ethanol consumption alters electroretinograms and depletes neural tissues of docosahexaenoic acid in rhesus monkeys: nutritional consequences of a low *n*-3 fatty acid diet. *Alcohol Clin Exp Res*. 2001;25(12):1758–1765.
16. Kim HY, Hamilton J. Accumulation of docosahexaenoic acid in phosphatidylserine is selectively inhibited by chronic ethanol exposure in C-6 glioma cells. *Lipids*. 2000;35(2):187–195.
17. Kim HY, Akbar M, Lau A, Edsall L. Inhibition neuronal apoptosis by docosahexaenoic acid (22:6, ω-3). Role of phosphatidylserine in antiapoptotic effect. *J Biol Chem*. 2000;275(45):35215–35223.
18. Kim HY. Biochemical and biological functions of docosahexaenoic acid in the nervous system: modulation by ethanol. *Chem Phys Lipids*. 2008;153(1):34–46.
19. Hidiroglou N, Madere R. Effect of chronic ethanol dosing on hepatic triglyceride and phospholipid profile and fatty acids in the Guinea pig. *Alcohol*. 1999;19(3):229–233.
20. Nakamura MT, Tang AB, Villnueva J, Halsted CH, Phinney SD. Reduced tissue arachidonic acid concentration with chronic ethanol feeding in miniature pigs. *Am J Clin Nutr*. 1992;56:467–474.
21. Shikano M, Masuzawa Y, Yazawa K, Takayama K, Kudo I, Inoue K. Complete discrimination of docosahexaenoic acid from arachidonate by 85 kDa cytosolic phospholipase A_2 during hydrolysis of diacyl- and alkenylacylglycerophosphoethanloamine. *Biochim Biophys Acta*. 1994;1212(2):211–216.
22. Moore SA. Cerebral endothelial and astrocytes cooperate in supplying docosahexaenoic acid to neurons. *Adv Exp Med Biol*. 1993;331:229–233.
23. Moore SA. Polyunsaturated fatty acid synthesis and release by brain-derived cells in vitro. *J Mol Neurosci*. 2001;16:195–200 [discussion 115-21].
24. Ghosh M, Tucker DE, Burchett SA, Tucker DE. Properties of group IV phospholipase A_2 family. *Prog Lipid Res*. 2006;45:487–510.
25. Strokin M, Sergeeva M, Reiser G. Docosahexaenoic acid and arachidonic acid release in rat brain astrocytes is mediated by separate isoforms of phospholipase A_2 and is differently regulated by cyclic AMP and Ca^{2+}. *Br J Pharmacol*. 2003;139:1014–1022.
26. Lukiw WJ, Cui JG, Marcheselli VL, et al. A role of docosahexaenoic acid-derived neuroprotection D1 in neural cell survival and Alzheimer disease. *J Clin Invest*. 2005;115(10):2774–2783.
27. Rajee A, Lukiw WJ, Bazan NG. Neuroprotectin D1 induces dephosphorylation of Bcl-x_L in a PP2A-dependent manner during oxidative stress and promotes retinal pigment epithelial cell survival. *J Biol Chem*. 2010;285(24):18301–18308.
28. Asatryan A, Bazan NG. Molecular mechanisms of signaling via docosanoid neurprotectin D1 for cellular homeostasis and neuroprotection. *J Biol Chem*. 2017;292(30):12390–12397.
29. Butovick IA. On the structure and synthesis of neuroprotection D1, a novel anti-inflammatory compound of the docosahexaenoic acid family. *J Lipid Res*. 2005;46:2311–2314.
30. Mukerjee PK, Marchesseli VL, Serhan CN. Bazan Neuroprotectin D in retinal pigment epithelia1: a docosahexaenoic acid-derived docosatriene protects human retinal pigment epithelial cells from oxidative stress. *Proc Natl Acad Sci U S A*. 2004;101(22):8491–8496.
31. Soderberg M, Edlund C, Kristensson K, Dallner G. Fatty acid composition of brain membrane phospholipids in aging and in Alzheimer' disease. *Lipids*. 1991;26(6):421025.
32. Prasad MR, Lovell MA, Yatin M, Dhillon H, Markesbery G. Regional membrane phospholipid alterations in Alzheimer's disease. *Neurochem Res*. 1998;23(1):81–88.
33. Bazan NG, Lukiw WJ. Cyclooxygenase-2 and presenlin-1 gene expression induced by interleukin beta and amyloid beta 42 peptide is potentiated by hypoxia in primary human neural cells. *J Biol Chem*. 2002;277(31):30359–30367.
34. Nourooz-Zadeh J, Liu EH, Yhlen B, Angaard EE, Halliwell B. F4-isoprostanes in specific marker of docosahexaeonic peroxidation in Alzeimer's disease. *J Neurochem*. 1999;72(2):734–740.
35. Montine TJ, Neeley MD, Quinn JF, et al. Lipid peroxidation in aging brain and Alzheimer's disease. *Free Rad Biol Med*. 2002;33(5):620–626.
36. Scott BL, Bazan NG. Membrane docosahexaenoate is supplied to the developing brain and retina by liver. *Proc Natl Acad Sci U S A*. 1989;86(8):2903–2907.
37. Belayev L, Mukherjee PK, Belaszczuk V, et al. Neuroprotectin D1 upregulates Iduna expression and provides protection in cellular uncompensated oxidative stress and in experimental ischemic stroke. *Cell Death Differ*. 2017;24:1091–1099. https://doi.org/10.1038/cdd.2017.55 April 2017.

38. Lim H, Gupta RA, Ma WM, et al. Cyclo-oxygenase-2-derived prostacyclin mediates embryo implantation in the mouse via. *PPARδ Genes Dev.* 1999;13(2):1561–1574.
39. Song H, Lim H, Paria BC, et al. Cytosolic phospholipase $A_{2\alpha}$ is crucial for "on-time" embryo implantation that directs subsequent development. *Devel.* 2002;129:2879–2889.
40. Achache H, Revel A. Endometrial receptivity markers, the journey to successful embryo implantation. *Hum Reprod Update.* 2006;12(6):731–746.
41. Wang H, Wen Y, Mooney S, Behr B, Polan ML. Phospholipase A_2 and cyclooxygenase gene expression in human preimplantation embryos. *J Clin Endocrinol Metab.* 2002;87(6):2629–2634.
42. Agarwal A, Gupta S, Sharma RK. Role of oxidative stress in female reproduction. *Reprod Biol Endocrinol.* 2005;3:28. https://doi.org/10.1186/1477-7827-3-28.
43. Duc-Goiran P, Mignot TM, Bourgeois C, Ferre' F. Embryo-maternal interactions at the implantation site: a delicate equilibrium. *Eur J Obstet Gynecol Reprod Biol.* 1999;83:85–100.
44. Krause BJ, Hanson MA, Casanello P. Role of nitric oxide in placental vascular development and function. *Placenta.* 2011;32(11):797–805.
45. Many A, Westerhausen-Larson A, Kanour-Shakir A, Roberts JM. Xanthine oxidase/dehydrogenase is present in human placenta. *Placenta.* 1996;17:361–365.
46. Telfer JF, Thomson AJ, Cameron IT, Greer IA, Norman JE. Expression of superoxide dismutase and xanthine oxidase in myometrium, fetal membranes and placenta during normal human pregnancy and parturition. *Hum Reprod.* 1997;12(10):2306–2312.
47. Kane AD, Camm EJ, Richter HG, et al. Maternal-to-fetal allopurinol transfer and xanthine oxidase suppression in the late gestation pregnant rat. *Physiol Rep.* 2013;1(6):e00156e00158. https://doi.org/10.1002/phy2.156.
48. Watson AL, Skepper JN, Jauniax E, Burton GJ. Changes in concentration, localization, and activity of catalase within human placenta during early gestation. *Placenta.* 1998;19(1):27–34.
49. Pristov JB, Spasojevic I, Mikovic Z, Mandic V, Cerovic N, Spasic M. Antioxidative defense in placenta protect placent and fetus in inherited thrombophilia from hydrogen peroxide. *Oxid Med Cell Longev.* 2009;2(1):14–18.
50. Salem Jr N, Ward GR. The effects of ethanol on polyunsaturated fatty acid composition. In: Alling C, Diamond I, Leslie SW, Sun GY, Gibbson Wood W, eds. *Alcohol, Cell Membranes, and Signal Transduction in Brain.* New York: Plenum Press; 1993:33–46.
51. Wainwright PE, Huang Y-S, Simmons V, Ward GR, Winfield D, McCutcheon D. Effects of prenatal ethanol and long-chain *n*-3 fatty acid supplementation on development in mice. 2. Fatty acid composition of brain membrane phospholipids. *Alcohol Clin Exp Res.* 1990;14(3):413–420.
52. Ward GR, Xing H-C, Wainwright PE. Effect of postnatal ethanol exposure on brain growth and lipid composition in n-3 fatty acid-deficient and adequate-rats. *Lipids.* 1999;34(11):1177–1186.
53. Burdge GC, Postle AD. Effect of maternal ethanol consumption during pregnancy on the phospholipid composition of fetal Guinea-pig brain, liver, and plasma. *Biochim Biophys Acta.* 1995;1256:346–352.
54. Hamburger V, Hamilton HL. Series of normal stages in the development of the chick embryo. *J Morphol.* 1951;88:49–52.
55. Miller Jr RR, Slathar JR, Luvisotto ML. α-Tocopherol and γ-tocopherol attenuate ethanol-induced changes in brain membrane fatty acid composition in embryonic chick brains. *Teratology.* 2000;62(1):26–35.
56. Miller Jr RR, Olson BM, Rorick N, Wittingen AL, Bullock M. Embryonic exposure to exogenous α- and γ-Tocopherol partially attenuates ethanol-induced changes in brain morphology and brain membrane fatty acid composition. *Nut Neurosci.* 2003;6(4):201–212.
57. Miller Jr RR. Alcohol-induced membrane lipid peroxidation. In: Preedy VR, Watson RR, eds. *Alcohol in Disease: Nutrient Interactions and Dietary Intake.* Boca Raton, FL: CRC Press; 2004:336–339.
58. Natsuki R. Effect of chronic ethanol on phospholipase A2- and C-activity in chick embryo brain, heart, and liver. *Nihon Arukoru Kenkyo to Yakubutsa Izon (Japanese Journal of Alcohol Studies & Drug Dependence).* 1995;30(5):348–357.
59. Socodato R, Brito R, Calaza KC, Paes-de-Carvalho R. Developmental regulation of neuronal survival by adenosine in the *in vitro* and *in vivo* avian retina depends on a shift of signaling pathways leading to CREB phosphorylation or dephosphorylation. *J Neurochem.* 2011;116(2):227–239.
60. Seifert J. A tentative mechanism of solubilization of neuropathy target esterase from chicken embryo brain by phospholipase A_2. *Sci World J.* 2008;8:346–349. https://doi.org/10.1100/tsw.2008.51.
61. Fioretti B, Catacuzzeno L, Tata AM, Franciolini F. Histamine activates a background, arachidonic acid-root sensitive K channel in embryonic chick dorsal root ganglion neurons. *Neurosci.* 2004;125:119–127.

62. Daniele JJ, Fidelio GD, Bianco ID. Calcium dependency of arachidonic acid incorporation into cellular phospholipids of different cell types. *Prostaglandins Other Lipid Mediat*. 1999;57:341–350.
63. Pavoine C, Magne S, Sauvadet A, Pecker F. Evidence for a β_2-adrenergic/arachidonic acid pathway in ventricular cardiomyocytes. *J Biol Chem*. 1999;274(2):628–637.
64. Magne S, Couchiee D, Pecker F, Pavoine C. β_2-Adrenergic receptor agonists increase intracellular free Ca^{2+} concentration cycling in ventricular cardiomyocytes through p38 and p42/44 MAPK-mediated cytosolic phospholipase A_2 activation. *J Biol Chem*. 2001;276(43):39539–39548.
65. Folch J, Lees M, Sloane Stanley GH. A simple method for the isolation and purification of total lipids from animal tissues. *J Biol Chem*. 1957;226(1):497–509.
66. Dittmer JC, Wells MA. Quantitative and qualitative analysis of lipids and lipid components. *Methods Enzymol*. 1969;14:482–530.
67. Berning EJ, Bernharson N, Coleman K, et al. Ethanol- and/or Taurine-induced oxidative stress in chick embryos. *J Amino Acids*. 2003. https://doi.org/10.1155/2013/240537. Article ID 240537, 11 pages.
68. Bauer AK, Fitzgerald M, Ladzinsky AT, et al. Dual behavior of *N*-acetylcysteine during ethanol-induced oxidative stress in embryonic chick brains. *Nutr Neurosci*. 2017;20(8):478–488.
69. Hancock ML, Miller Jr RR. Reveratrol can only partially attenuate ethanol-induced stress in embryonic brains. *Nutr Neurosoci*. 2006;9(3.4):121–129.
70. Reitz RC. Effects of dietary fatty acids and alcohol on fatty acid composition in cellular membranes. In: Watson RR, Watzl B, eds. *Nutrition and Alcohol*. Boca Raton, FL: CRC Press; 1992:191–204.
71. Mihaljevic B, Katuson-Razem B, Razem D. The reevaluation of the ferric thiocyanate assay for lipid hydroperoxides with special considerations of the mechanical aspects of the response. *Free Radic Biol Med*. 1966;21(1):53–63.
72. Surai PF. Tissue-specific changes in the activities of antioxidant enzymes during the development of the chicken embryo. *Br Poultry Sci*. 1999;40(3):397–405.
73. Mannervik B. Glutathione peroxidase. *Methods Enzymol*. 1985;113:490–495.
74. Ursini F, Maiorino M, Brigelius-Flohe R, et al. Diversity of glutathione peroxidases. *Methods Enzymol*. 1995;252:38–53.
75. Maffi SK, Rathinam ML, Cherian PP, Hamby-Mason R, Schenker S, Hendersen GI. Glutathione content as a potential mediator of the vulnerability of cultured fetal cortical neurons to ethanol-induced apoptosis. *J Neurosci Res*. 2008;86(5):1064–1076.
76. Wilson RW, Kalamus GW, Pennington SN. Alcohol dehydrogenase activity in the developing chick embryo. *Comp Biochem Physiol*. 1984;77B(1):191–196.
77. Miller Jr RR, Touney EJ, Vandivier WJ, Raymond FJ. A developmental profile of the effects of ethanol on the levels of chick brain phospholipids. *Comp Biochem Physiol*. 1998;120C:91–98.
78. Rout UK, Armant DR. Expression of genes for alcohol and aldehyde metabolizing enzymes in mouse oocytes and preimplanatation embryos. *Reprod Toxicol*. 2002;16(3):253–258.

Further Reading

79. Strokin M, Sergeeva M, Reiser G. Prostaglandin synthesis in rat brain astrocytes is under the control of *n*-3 docosahexaenoic acid, release by group VIB calcium-independent phospholipase A2. *J Neurochem*. 2007;102:1771–1782.
80. Bazan NG. Neuroprotectin D1 (NDPD1) a DHA-derived mediator that protects brain and retina against cell-induced oxidative stress. *Brain Pathol*. 2005;15(2):159–166.
81. Freeman BM, Manning AC. Glyconeogenesis and lipolysis in *Gallus domesticus* during the perinatal period. *Comp Gen Pharmacol*. 1971;2(6):198–204.

PART B

OMEGA FATTY ACIDS: BRAIN AND NEUROLOGICAL DEVELOPMENT

CHAPTER

6

Prenatal n-3 Long-Chain Polyunsaturated Fatty Acids and Children's Executive Functions

Jacqueline F. Gould[*,†], *Lisa G. Smithers*[‡,§]

[*]Healthy Mothers, Babies and Children, South Australian Health and Medical Research Institute, North Adelaide, SA, Australia [†]School of Psychology, University of Adelaide, Adelaide, SA, Australia [‡]School of Public Health, University of Adelaide, North Terrace, SA, Australia [§]Robinson Research Institute, School of Medicine, University of Adelaide, SA, Australia

INTRODUCTION

The prenatal period hosts a crucial window for brain development. The fetus relies on the placental supply of a range of nutrients. An inadequate supply of nutrients can have irreparable long-term consequences for brain development, such as the case with folate deficiency and neural tube defects. Recently in developed countries, interest has focused on nutrition to ensure optimal brain development, rather than to avoid serious abnormal development. Long-chain polyunsaturated fatty acids, particularly docosahexaenoic acid (DHA) is one such nutrient that is of interest to researchers. The DHA exposure (in the absence of a deficit in DHA supply) during fetal development is thought to primarily influence the development of the frontal lobes, the area of the brain responsible for higher-order cognitive skills known as the executive functions. The frontal lobes have the most protracted development and maturity. In this chapter, we will briefly cover brain development with a focus on the structures responsible for the executive functions, and then review the evidence for the effect of DHA exposure during pregnancy and the development of child executive functions. We will focus on the results of randomized controlled trials (RCTs) as the methodology of RCT's minimizes much of the bias associated with other methodologies.

https://doi.org/10.1016/B978-0-12-815238-6.00006-7

BRAIN DEVELOPMENT

Brain development starts after conception and continues throughout childhood into adolescence. The main cells in the brain are neurons (or neural cells) that consist of a cell body with a long projection known as an axon on one end, and a branch of projections known as dendrites at the other end of the cell body. Neurons have the unique feature of communicating with one another through these structures using electrical impulses and neurotransmitters (chemical messengers).[1]

Prenatal Brain Development

Lasting from conception until the end of week 8 of pregnancy, the *Embryonic Period* involves proliferation and differentiation of neural stem cells.[2] The neural stem cells form the neural tube at approximately three and a half weeks' postconception and subsequently differentiate into the precursors of the preliminary structures for the central and peripheral nervous system by 9 weeks postconception.[3]

Beginning at the postconception week 9 and lasting until term (week 40 postconception), *fetal brain development* is characterized by neural migration and dramatic growth. The neural cells differentiate and migrate to their appropriate locality[3,4] so that the various structures and regions in the brain become increasingly recognizable. As neurons migrate, their axonal projections branch out laying the foundation for later development of communication pathways throughout the brain.[3] As the brain increases in size, the outer layer of the brain (the cerebral cortex) develops distinctive pattern of sulci (surface folds) and gyri (ridges). By the end of the second trimester (13–28 weeks' postconception) most structures are present and recognizable, in an underdeveloped form.

During the last trimester (29–40 weeks postconception) the fetal brain undergoes a rapid growth spurt.[5,6] The volume of the brain increases at a rate of approximately 22 mL/week[7] so that by the end of the 11-week period, the brain has more than doubled in size from approximately 125–400 g.[5] The sulci and gyri on the surface of the brain as well as the various neurological structures become increasingly refined over the last trimester, particularly the frontal lobes which undergo an intense period of growth.[5,6]

At term, defined as 37–41 weeks' postconception, an infants' brain is 25% of the size of an adult brain. Over 100 billion neurons are present, the majority of which will be present in adulthood,[8] although only a small fraction of the connections between neurons are formed and functional. Neurological structures responsible for autonomic functions necessary for survival outside the womb (such as breathing and regulating heart rate) are fully functional. The regions responsible for processing various sensory input (i.e., visual, auditory, tactile) and motor control are capable of basic processing; however, regions responsible for more complex domains, such as the frontal lobes responsible for high-order cognitive skills, are underdeveloped.

Infant and Child Brain Development

Following birth at term gestation (approximately 40 weeks), another period of rapid brain development occurs during the first two postnatal years. By 24 months of age the brain reaches 80% of its adult weight,[9] and along with a vast increase in volume, there are two other key developmental processes underway; myelination and synaptogenesis. *Myelination* is the production and distribution of myelin, a fatty sheath, around neuronal axons. Axons that are myelinated transmit electrical impulses much faster than axons that are not yet myelinated; hence, myelination increases the rate of information processing. *Synaptogenesis* is the formation of a connection (a synapse) between the dendrite of one neuron and the axon of another neuron.[3] Each neuron can share synapses with over 1000 other neurons to form communication networks throughout the brain. The formation of synapses is somewhat dependent on stimulation from the surrounding environment and occurs continually during the first few years (up to 700 formed per second[1]) so that an excess of synapses are created. By 3 years of age there are more synapses formed within the brain than will be present in adulthood.[10] Over time, synapses that are used repeatedly are reinforced and strengthened, as are networks that are reinforced through use, while networks and synapses that are not used will be pruned.

Myelination and synaptogenesis enable new abilities and refinement of existing abilities throughout early childhood. Early infancy is characterized by developmental gains in neurosensory and motor areas; visual and auditory processing enhances rapidly in the first months after birth, followed by increasing motor control, allowing infants to recognize the faces and voices of their caregivers, orient toward a person or object, smile, and stabilize their core muscles to support their own bodies without assistance. A dependent new-born infant develops basic abilities to walk unsupported, speak in short phrases, and understand others' speech over the first 24 months after birth. The areas of the brain responsible for memory, cognition, executive functions, and emotion are similarly undergoing developmental processes during infancy, although at a slower rate than the sensory, motor, and language areas.

After 24 months of age, brain development continues but at a somewhat slower rate than during pregnancy and infancy. Both myelination and synaptogenesis continue throughout the brain at a rapid rate so that by around 4 years of age, synaptogenesis has peaked and synapses that are not reinforced are pruned to increase efficiency. Importantly, new synapses continue to be formed across childhood and into early adulthood.

Motor and language skills, along with social and emotional development progresses gradually across childhood so that abilities in these domains become more refined and coordinated, for example, catching or kicking a ball with increasing precision, expanding vocabulary, and writing or drawing neatly. By around 5 years, the brain reaches 90% of its adult size and children have typically mastered emotional regulation sufficiently to form friendships with other children. By late adolescence motor and language skills are refined, and social and emotional skills are well developed. The executive functioning cognitive abilities on the other hand, are governed by one of the last areas of the brain to mature-the frontal lobes, and abilities continue to mature into adulthood.

The Frontal Lobes

The frontal lobes, also referred to as the prefrontal cortex, are unique in that they are one of the most recently evolved structures in the brain, they are proportionally largest in humans compared to other mammals (see reviews[11,12]) and are one of the slowest developing regions of the brain (see reviews[9,12,13]). During the last trimester of pregnancy, the frontal lobes undergo an intense growth spurt, followed by a major period of growth shortly before 24 months of age and 6 years of age, and a less intense period of development in early adolescence.[9,14] The frontal lobes do not reach full maturity until early adulthood, however after the adolescent growth spurt, development proceeds gradually.[9,14] The emergence of executive functioning skills coincides with the major periods of frontal lobe growth.[15]

The Executive Functions

The executive functions are generally considered to be a set of higher-order cognitive skills that are governed by the frontal lobes. A summary of selected cognitive abilities considered to be executive functions are detailed in Table 1.[16,17] The executive functioning processes are responsible for selecting a personal goal (such as doing well in a test, or mastering a new skill) and self-monitoring and regulating behavior in order to achieve that goal.[18] The executive functions are responsible for regulating attention, controlling impulses, planning and organizing, problem solving, making complex judgments, and flexibility to adapt to novel challenges or situations. The executive functions are partially responsible for integrating and processing information from various neurological domains, and hence rely on other less complex neurological functions. Early executive functioning is evident as infants learn to control their attention to focus on stimuli of interest without reflexively attending or reacting to distracting stimuli. Some precursors to later executive functions emerge at around 12 months as infants develop the ability to perform very basic problem solving. It is around 3–4 years that executive function skills that more closely resemble adult skills emerge and can be more reliably assessed; however, executive functions continue to develop across childhood and into adolescence with ongoing emergence of new abilities.[9]

Appropriate development of the executive functions are critical for other aspects of child development as they are implicated in knowledge acquisition, planning, decision-making, moderating behavior, solving novel problems, and adapting to new situations. There is some evidence that variation within the hypothesized normal range of early childhood executive function abilities can be predictive of meaningful outcomes such as children's Theory of Mind,[17,19] the ability to reason about the mental states of one self and of others, which is implicated in social competence during preadolescence.[20] Impaired executive functions include inability to maintain or focus attention, impulsivity, lack of inhibition, lack of self-monitoring and failure to learn from mistakes. Such impairments are thought to contribute to low school attainment, reduced vocational opportunities, poor coping skills, and lower quality of life (see review[21]).

TABLE 1 An Overview of the Selected Executive Functioning Abilities

Executive Function	Description
Attention	Specifically, endogenous (consciously self-directed and controlled) attention The direction, control and maintenance of attention toward a target stimulus, with inhibition of sensory input from nontarget (or distracting) stimuli
Working Memory	Also known as "Updating" Short term retention, processing, and manipulation of information
Inhibition	Self-control, suppression of impulses, pre-potent responses, and inappropriate responses
Mental Flexibility	Also known as "Shifting" Ability to switch between thinking about two different concepts or tasks

FATTY ACIDS AND BRAIN DEVELOPMENT

The brain is composed of approximately 10% lipid, so that lipids make up 60% of the brains dry weight, including omega-3, omega 6, and omega-9 fatty acids, and sterols.[22–24] Arachidonic acid (AA, 20:4n-6) is the predominant omega-6, and docosahexaenoic acid (DHA, 22:6n-3) is the predominant omega-3 long-chain polyunsaturated fatty acid. Both are rich in the membranes of neurones and both accumulate rapidly during the last trimester of pregnancy.[25,26] DHA has been of particular interest due to the role in fetal neural tissue development due to links with some functional outcomes.

The DHA is present in all cell membranes as a structural component of the phospholipid bilayer where it plays an integral role in membrane fluidity[27]; however, it is highly concentrated in the brain.[28,29] DHA has active roles in neurogenesis, signal transduction, and neurotransmission,[30] and accumulates in neural tissues at the greatest velocity during the last trimester of pregnancy.[26] This coincides with the peak period of rapid and intense brain growth.[5,6]

Both DHA and AA can be synthesized by humans from their respective precursor omega-3 and omega-6 fatty acids. However, during pregnancy the fetus receives preformed DHA and AA via placental transfer from maternal circulation.[31,32] The DHA and AA from the diet, along with DHA and AA synthesized from precursor omega-3 and omega-6 fatty acids and released from adipose stores circulate in the maternal blood stream. The amount of fatty acids released from adipose stores or synthesized is not clear; however, a link between maternal DHA intake during pregnancy and maternal circulating DHA suggests that diet may be a major source of DHA for the developing fetus [33–37] Observational studies of fish (a primary source of dietary DHA) consumption during pregnancy have linked higher levels of intake with better cognitive,[38,39] language,[38–42] and motor scores[38,40,43] as well as social,[40,43] and behavioral outcomes.[42] However, observational studies may be subject to unknown confounding factors for fish consumption that also effect child development.[44]

DHA has been shown to accumulate in high concentrations in the cerebral cortex and synapses of baboons, mice, rats, and other mammals.[45–47] Animal models of severe omega-3 fatty acid deprivation demonstrate reduced levels of DHA in neural tissue,[48–52] with compensatory increases in the level of omega-6 fatty acids.[52,53] Interestingly, deprivation of all omega-3 fatty acids in rats results in changes to the frontal lobes.[54,55] Resulting functional deficits in offspring include decreased learning,[50,55–57] poorer motor skills,[58] lower performance on memory tasks,[55,57] behavioral disturbances,[56] poorer orientation,[58] and poorer spatial memory,[50] as well as the suggestion of deficits or suboptimal functioning of the frontal lobes.[50,55,57] Another model of omega-3 fatty acid deprivation in rats demonstrated lower dopamine, a neurotransmitter used at the synapse to communicate between neurons, in the frontal lobes.[59] This important animal work has linked the consequences of DHA deficits to functional outcomes and spurred an entire field of research among humans, which we will now examine more closely.

DHA and Brain Development of Infants Born Preterm

Infants born preterm are born before 37 weeks' completed gestation, but can survive birth as early as 22 weeks' gestation, and are denied the maternal supply of DHA during the fetal brain growth spurt. As discussed previously, from week 29 gestational age to 40 weeks, the fetal brain more than doubles in volume.[5] For infants born preterm, much of this rapid growth takes place ex utero, rather than in utero, when infants no longer have access to the placental transfer of nutrients such as preformed DHA. All requirements for DHA must be met through milk, which has variable DHA content. Consequently, infants born preterm and not fed dietary DHA have lower levels of neural tissue DHA than infants born full term.[26]

Children born preterm have a higher risk than their term born counterparts of poor developmental outcomes such as problems in cognition,[60–62] behavior,[60,63–67] and attention[68] as well as executive functioning.[69] This has led to much research activity to determine if increasing the supply of DHA to infants born preterm will result in improved cognitive outcomes.

Early preterm infant formula did not contain DHA, so that initial interventions in preterm infants were restricted to formula-fed infants and formula containing some DHA was compared with formula containing no DHA.[70–77] No consistent differences to brain development were evident on global measures, such as the Bayley series of assessments[78–80] However, improvements to preterm infant visual acuity led to the addition of DHA to preterm infant formula (DHA as 0.3% of total fatty acids to match breast milk levels in Western populations) as part of standard clinical practice.[75,76,81] Further work to clarify whether the brain development of infants born preterm can be optimized through DHA enriched milks is still needed.

Prenatal DHA RCT's and Brain Development

The human fetus is estimated to acquire ~70 mg of omega-3 long-chain polyunsaturated fatty acid per day during the last trimester, with DHA being the predominant fatty acid.[82] It has been hypothesized that these fetal DHA requirements exceed dietary intake of DHA for pregnant women consuming Western diets and therefore, this may be a factor limiting optimal brain development. Observational studies linking fish consumption during pregnancy to developmental gains support the hypothesis.[38–43]

There are now 12 published RCT's of DHA supplementation during pregnancy with child development outcome measures, some with multiple waves of follow-up.[37,83–102] Together these trials have demonstrated negligible effects on neurodevelopment.[103–106]

A hypothesized reason for the lack of demonstrated benefit of prenatal DHA supplementation is the sensitivity of the measures used to assess neurodevelopment. The majority of prenatal DHA RCT's have evaluated efficacy using global psychometric assessments that attempt to comprehensively capture a range of cognitive abilities. These assessments have the advantages of standardized administration and scoring; however, they are designed for use in clinical practice to detect developmental problems. Hence, these tests may not be suitable for detecting small differences in abilities within the normal range of development. Furthermore, by combining performance of multiple cognitive abilities, global psychometric tests may mask the sensitivity of specific neurological domains, such as the frontal lobes.[107] Given that single-nutrient interventions in generally healthy populations are likely to have a subtle effect on early brain development, it is recommended that researchers use assessments that target the appropriate neurological domain hypothesized to be influenced by the nutrient of interest.[107–109] Thus, to adequately evaluate a prenatal DHA intervention, assessments should target the frontal lobes through measures of executive functions.

One RCT that provided DHA to pregnant women in the form of fortified chicken eggs (treatment: 135 mg DHA/egg, control: 35 mg DHA/egg) conducted an experimental measure of early executive functioning.[110,111] Due to varied intake of DHA across and within the groups, the authors combined the intervention and control groups and dichotomized the sample according to median maternal (erythrocyte membrane) blood DHA status at delivery. Using repeat measures in $n=50$ children at multiple ages, they reported that children born to mothers with high erythrocyte membrane DHA at delivery had accelerated development of attention from 4 to 18 months of age compared with mothers who had low-DHA status.[110,111] Although this study could not determine whether the DHA intervention effected child executive functioning, the results suggest benefit.

The following section summarizes the evidence from RCT's of DHA supplementation during pregnancy that includes a measure of child executive functioning.

PRENATAL DHA RCT'S WITH EXECUTIVE FUNCTIONING ASSESSMENTS

A total of six RCT's of prenatal DHA supplementation have included an assessment of executive functioning as an outcome.[37,83–94] All but two trials[86,88] were restricted to women with a singleton pregnancy, and one of these trials only followed up the firstborn child from a twin pregnancy.[88] This RCT was further restricted to women with a history of allergic disease as the primary outcome of the trial was reduction of child allergic disease.[88,89] The interventions of all RCT's involved oral supplementation, with majority using capsules,[86,88–90,94] and one using milk-based sachets.[37] The results of these RCT's are summarized in Table 2 and will be discussed according to the specific domain of executive function assessed, age of the children when executive functioning was assessed and dose and duration of the DHA intervention.

Executive Functioning Domains

In this section, the executive functioning assessment outcomes will be discussed according to the specific ability they target, specifically covering attention, inhibition, and working memory. Parent-rated measures will be discussed separately from performance-based measures. While performance-based measures are generally considered more objective and reflective of an individual's capacity of executive functioning abilities, parent-ratings are also important because they reflect children's functioning in real-world settings.

The abstract by Karlsson et al. reports a benefit of DHA supplements on executive and working memory task(s) of the Wechsler Preschool and Primary Scale of Intelligence version III; however, the test battery contains no official measure of executive function or working memory and authors did not specify the task(s) they found a benefit on.[89]

Attention: The most commonly assessed domain of executive functioning was attention, with five trials reporting measures of attention among children whose mothers were involved in antenatal DHA supplementation trials.[83,85,87,91,93,94] As with Colombo et al.'s original exploration of maternal DHA status and attention,[110,111] the new RCT involved repeated behavioral measures of visual attention conducted at three ages in the first year of life to capture the developmental trajectory of early attention.[94] The measure[110] involves showing infants a stimuli (images of adult faces) and measuring infants looking at repeated stimuli whilst also recording heart rate variability and defining infants' state of attention. Authors reported that although there were no differences between the DHA and control group infants behavioral or heart-rate attentional outcomes, infants from mothers supplemented with DHA tended to habituate (learn the stimulus) slightly faster and maintained a consistent pattern of sustained attention (state of attention

TABLE 2 A Summary of Randomized Controlled Trials of Docosahexaenoic acid (DHA) Supplementation during Pregnancy and Assessments of Child Executive Functioning

Author and Reference	Setting and Participants	Intervention	Follow-up and Assessment Result	Comments
Colombo 2004 110 Kannass 2009 111	Country: USA. Setting: medical center N enrolled: 350 70 selected for follow-up Majority receiving government medical care	Period: 24–28 wk. preg-birth Form: DHA enriched eggs (135 mg DHA/egg) vs normal egg (35 mg DHA/egg) DHA/d: NA Comparison groups dichotomized into high and low exposure from median maternal and infant DHA status at delivery	Age: 4, 6, and 8 mo, $n=50$ Outcome: infant control habituation-no diff Age: 12 and 18 mo, $n=47$ Outcome: toddler free-play attention (single object, distractibility) -no diff Age: 12 and 18 mo, $n=47$ Outcome: toddler free-play attention (multiple object)-no diff	Data not analyzed as an RCT Maternal DHA-accelerated pattern of longitudinal development with high-DHA
Decsi, 2005 37 Catena, 2015 83	Country: Germany, Hungary, Spain Setting: Antenatal care clinic N enrolled: 315[1] Singleton pregnancy 136 selected for follow-up Excluded preterm birth, congenital abnormalities and deafness	Period: 20 wk. preg-birth (infant formula provided until 6 months of age if not breastfed) Form: milk-based sachets Omega-3 LCPUFA/d: 650 mg DHA/d: 500 mg	Age: 8.5 yr, $n=130$ Outcome: Attention Network Test (child version, measuring executive, alerting, orienting)-no diff -DHA slight improvement in speed + accuracy score	
Innis, 2008 84 Mulder, 2018 85	Country: Canada Setting: Hospital N enrolled: NR Singleton pregnancy Excluded if taking lipid supplement	Period: 16 wk. preg-birth (infant formula provided 60 d if not breastfed) Form: 2× capsules Omega-3 LCPUFA/d: 0 DHA/d: 400 mg	Age: 5–6 yr, n = 89 Outcome: Test of Variables Of Attention (TOVA, attention, impulsivity)-no diff	Authors testing whether sample was DHA deficient rather than whether supplements improve outcome Maternal DHA status not associated with TOVA

Ramakrishnan, 2010 [86] Ramakrishnan, 2016 [87]	Country: Mexico Setting: Hospital (routine prenatal) N enrolled: 1094 Medium-low SES Excluded if taking fish oil or DHA	Period: 18–22 wk. preg-birth Form: 2× capsules Omega-3 LCPUFA/d: 0 DHA/d: 400 mg	Age: 5 yr, $n=797$ Outcome: Conners' Kiddie Continuous Performance Test (K-CPT, attention) -no diff for incorrect responses, response time or impulsive responses -DHA fewer omissions Outcome: Behavioral Assessment System for Children, 2nd ed. (BASC-2, parent-report executive functioning subscale)-no diff	K-CPT also examined in terms of cut-offs for possible deficit-no diff between proportions in each group
Furuhjelm, 2009 [88] Karlsson, 2010 [89]	Country: Sweden Setting: Antenatal departments and local newspaper advertisements N enrolled: 145 All had allergic disease (mother, father, previous child) Excluded if taking omega-3 LCPUFA or soy/fish allergy	Period: 25 wk. preg-3.5 mo Form: 9× capsules Omega-3 LCPUFA/d: 2700 mg DHA/d: 1100 mg	Age: 46 mo (3.8 yr), n = NR Outcome: Wechsler Preschool and Primary Scale of Intelligence, 4th ed. Executive task-trt > ctrl Visuo-spatial task-trt > ctrl Memory task-no diff	Abstract only Not available on a clinical trial registry
Colombo, 2016 [94]	Country: USA Setting: Most recruited in antenatal clinics N enrolled: 350 Singleton pregnancy Excluded women with health conditions, post randomization excluded preterm birth	Period: <20 wk. preg-birth Form: 3× capsules/d DHA/d: 600 mg	Age: 4, 6, and 9 mo, $n=220$ Outcome: infant control habituation-No diff to heart rate and behavioral measures of attention -Possible DHA benefit for longitudinal quality of state attention, speed, and habituation	Did not exclude DHA supplements in control group

Continued

TABLE 2 A Summary of Randomized Controlled Trials of Docosahexaenoic acid (DHA) Supplementation during Pregnancy and Assessments of Child Executive Functioning—cont'd

Author and Reference	Setting and Participants	Intervention	Follow-up and Assessment Result	Comments
Makrides, 2010 ,[90] Gould, 2014 ,[91] Makrides, 2014,[92] Gould, 2017[93]	Country: Australia Setting: Hospital (routine antenatal) N enrolled: 726 Pre-specified subgroup for neurodevelopmental follow-up Singleton pregnancy Excluded if taking DHA	Period: 18–21 wk. preg-birth Form: 3× capsules Omega-3 LCPUFA/d: 900 mg DHA/d: ~800 mg	Age: 27 mo, $n=154$ Outcome: toddler free-play attention (single object, multiple object, distractibility) –No diff overall -Control better on one Multiple Object outcome Outcome: A-not-B task (working memory and inhibitory control)-no diff Age: 4 yr, $n=646$ Outcome: Stroop (inhibition)-no diff Outcome: digit recall (working memory)-no diff Outcome: picture recognition (working memory)-no diff Outcome: Behavior Rating Inventory of Executive Functioning-Preschool edition (BRIEF-P; parent-reported behavioral manifestations of executive function) -no diff overall - Emergent Metacognition and Plan/Organize subscales DHA worse Age: 7 yr, $n=543$ Outcome: Stroop (inhibition)-no diff Outcome: digit recall (working memory)-no diff Outcome: Rey Auditory Verbal Learning Test (working memory)-no diff Outcome: Test of Everyday Attention for Children (attention)-no diff Outcome: Rey Complex Figure (visuospatial memory)-no diff Outcome: BRIEF (parent-reported behavioral manifestations of executive function)-no diff overall - The Global Executive Composite, Behavioral Regulation Index and Metacognition Index DHA worse	Prespecified subgroup $n=185$ at 27 months only (excluded preterm, low birth weight, neurological pathologies) No association between cord blood DHA and outcomes at 27 months

Abbreviations: DHA, docosahexaenoic acid; LCPUFA, long-chain polyunsaturated fatty acid; mg, milligrams; mo, months; RCT, randomized controlled trial.

associated with processing visual information).[94] Authors also noted less "fussiness," or uncooperative behavior from the children leading to termination of the assessment or invalid data in children from the DHA group compared with control group children, although this was not a planned outcome of the trial.[94] Another RCT using Colombo et al.'s[110,111] original attention measure assessed older children.[91] As appropriate for their age, these children underwent free-play assessments involving a toy to play with freely under the following conditions (1) When there was no distraction or competition for attention (Single Object), (2) there was competition (multiple toys) for attention (Multiple Object), and (3) there was a distraction (a television plying pseudo-random intermittent bursts of children's programs, Distractibility).[91] For each of these conditions there was a series of outcomes, none of which suggested any group differences in performance with the exception that there was a negligible difference in the number of times control group children looked away from all toys in the Multiple Object condition.[91] This difference may be construed as an advantage for the DHA group children, who looked away from all of the stimuli two times fewer than control group children; however, children in both groups spent the same total amount of time not looking at the toys. Follow-up of children in this trial to 7 years of age also indicated no effect of DHA intervention on attention as assessed with four tasks from the Test of Everyday Attention for Children.[93] The tasks involved (1) Identifying target stimuli whilst ignoring nontarget stimuli, (2) Maintaining attention over a long duration and counting auditory stimuli, (3) Doing both task 1 and task 2 at the same time, and (4) counting stimuli, whilst following directions to switch between counting backwards and forwards.[112]

Ramakrishnan et al.[87] assessed attention at 5 years using the Conners' Kiddie Continuous Performance Test. The assessment involves presenting children with stimuli on a screen and asking children to respond quickly to specific target stimuli without responding to nontarget stimuli. Scores are standardized to a mean of 50 (SD = 10) and can be used to screen for attention disorders. Children in the DHA group were less likely to fail to respond to target stimuli and indicated some advantages for components of vigilance, although there were no differences in incorrect responses to nontarget stimuli, response time for correct responses or indices of impulsive responses.[87] Authors also report that Bonferroni correction to account for multiple comparisons diminished these results.[87]

Another RCT used the Test of Variables of Attention at 5–6 years of age, which similarly measures attention, impulsivity (via incorrect responses to nontarget stimuli and failing to respond to target stimuli) as well as response time.[85] In this RCT, no differences in performance were detected in any attention outcome.[85] Results may differ to Ramakrishnan et al.'s[87] due to the smaller sample ($n = 89$ vs $n = 797$) potentiality limiting the power to detect an effect, in addition to differences in the assessments used.

Catena et al.[83] used a child version of the Attention Network Test at 8.5 years of age, whilst electroencephalography data was collected. In the computerized test, children are presented with visual stimuli in a number of conditions (such as with or without a cue) and had to indicate as quickly as possible the orientation of the stimulus. The DHA group had a slight advantage for their speed and accuracy, but no other outcomes.[83]

Of the six assessments of attention from five trials of prenatal DHA, either all outcomes or the majority of outcomes were null in five measures,[83,85,91,93,94] whilst one showed a benefit of intervention for most outcomes.[87] There were two assessments which suggested a benefit in one attentional outcome,[83,94] and one that found a possible detriment from supplementation[91]; however, all of these attention assessments contained multiple outcomes and statistically significant differences reported may be due to chance. It is not clear from these studies as to whether prenatal DHA can enhance development of attention.

Working Memory: The earliest assessment of working memory was based on Smith et al.'s "A-not-B" task[113] and tested both working memory and inhibitory control.[91] Toddlers watched as a toy was buried in a specific location (location A) in a large pit of lentils, and asked to retrieve it several times. Then toddlers watched as the toy was buried in another location (location B) and were distracted briefly before being asked to retrieve the toy. Correctly locating the toy requires the toddler to remember the new location and inhibit the impulse to reach for the original location. In RCTs involving prenatal DHA supplementation, no differences were found between the groups during the test condition, although control group children were slightly more accurate when locating the toy during the training trials than the DHA group children.[91] The functional significance of this finding is unclear and may be due to chance.[91,114]

Other assessments of working memory in these children included the digit recall test, a picture recognition task, the Rey Complex Figure and the Rey Auditory Verbal Learning Test. Recall of digits requires listing increasingly long strings of numbers that the child must correctly repeat, in the same order and/or backwards order. There were no group differences in recall of digits at 4 years (forwards only),[92] or at 7 years (forwards and backwards).[93] The Rey Auditory Verbal Learning Test similarly involves listing a series of words that children must try to recall on a number of trials, following by an unprompted recall after a delay and then recognition of the stimulus words from a larger list of words. No difference in the outcomes at 7 years of age on the Rey Auditory Verbal Learning Test have been noted following prenatal DHA intervention.[93] The picture recognition task involves presenting children with a pictured object (such as a ball, a car or a boat) and then asking them to identify the picture they have just seen from a larger selection of pictured objects. There are a number of trials with an increasing number of pictured objects to recall. Performance in this task at 4 years did not differ between intervention groups.[93] The Rey Complex Figure task requires children to copy a complex drawing, and then recall and draw the figure again after a delay, and again suggested no effects of prenatal DHA supplementation.[93]

Inhibition: The earliest assessment of inhibition revealed no effect of prenatal DHA supplementation, as discussed earlier.[91] These same children underwent subsequent assessments of inhibition at 4 and 7 years of age using the Stroop test.[92,93] The Day/Night Stroop requires children to say "day" when presented with a picture of the moon, and "night" when presented with a picture of the sun. At 4 years of age, there were no group differences.[92] The Fruit Stroop at 7 years requires children to recall the correct colors of fruits that are presented incorrectly colored (for example a picture of a blue tomato must be correctly identified as red) and similarly did not result in any group differences in performance.[93]

Mental Flexibility: Two subtests of the Test of Everyday Attention for Children attention assessment used by Gould et al.[93] can be said to capture mental flexibility. Creature counting involves switching between counting backwards and forwards, and Sky Search DT requires the child to simultaneously identify visual target stimuli from nontarget stimuli and count auditory stimuli.[112] No group differences in performance of either of these tasks were identified.[93] No other prenatal DHA RCT's included an assessment of mental flexibility.

Parent-rated executive functioning: Parent-rated measures involve an informant who knows the child well to rate difficulties the child has with undertaking a series of everyday tasks and activities (usually with a Likert scale). The Behavioral Assessment System for Children, Second Edition is a behavioral measure that includes an optional subscale for executive functioning. Ramakrishnan et al. used the Spanish version, and found no differences between children whose mothers received prenatal DHA or control supplements.[87]

The Behavior Rating Inventory of Executive Functioning (BRIEF) is a parent-rated measure that captures behavioral manifestations of executive dysfunction within everyday life. There is a preschool version (BRIEF-P) and a version for older children (BRIEF). Both target executive functioning only and provide an overall score (the Global Executive Composite) as well as age-appropriate subscales. The BRIEF-P was used in a 4-year follow-up, and group scores were compared for overall functioning as well as an Inhibitory Self-Control Index, Flexibility Index, Emergent Meta-Cognition Index, Inhibition Scale, Shift Scale, Emotional Control Scale, Working Memory Scale, and Plan/Organize Scale.[92] The majority of scores did not differ between the groups, although the Plan/Organize Scale and the Emergent Meta-Cognition Index scores were slightly higher in the DHA group than the control group, indicating that parents perceived more executive functioning problems in the DHA group children.[92] However, findings were similar when these same children were assessed with the BRIEF at 7 years.[93] The Global Executive Composite, Behavioral Regulation Index and Metacognition Index were all slightly worse in the DHA group.[93] Individual scale scores were not reported.

Summary of the domains assessed: The vast majority of executive functioning assessments have thus far focused on the construct of attention. This may be partially due to the availability of attention measures to evaluate intervention efficacy during infancy,[91,94] as well as during the school years.[83,85,87,93] It may also be because children with attention problems often have lower levels of omega-3 long-chain polyunsaturated fatty acids.[115,116] No assessment of executive functioning was used in more than one study, although two follow-ups from the same trial used age-appropriate versions of digit recall, Stroop, BRIEF, and an assessment of attention.[91–93]

There were no apparent domains where DHA intervention had a consistent effect. Working memory, Inhibition, and Mental Flexibility tasks did not suggest a benefit of supplementation. The results of attention and parent-rated assessments were mixed. The attention assessments were largely null[83,85,91,93,94] although three studies found some evidence of benefit on some of the attentional outcomes.[83,87,94] An abstract by Karlsson et al. reports a benefit of DHA supplements on executive and working memory task(s) of the Wechsler Preschool and Primary Scale of Intelligence version III; however, the test battery contains no official measure of executive function or working memory, and authors did not specify the task(s) they found a benefit on.[89] One study with a parent-rated behavioral assessment included an executive function subscale that suggested no group differences, although a RCT with a full parent-rated executive function assessment at two ages found scores were slightly worse in the DHA group children compared with controls.[92,93] This is one of the first reports of a possible negative effect of DHA supplementation.

There is debate as to whether parent-rated measures capture the same executive functioning constructs as performance-based measures, which are highly controlled in their administration procedures by a psychologist (or other examiner) in a clinic setting.[117] The fact that performance-based measures of effective function suggest positive or null effects of DHA, while parent-rated assessments have suggested negative effects, may actually be due to differences in the underlying constructs reflected in each measure.[117] However, another layer of complexity is that there is evidence of selective reporting.[89] Selective reporting is where it is possible that there are other data which have not yet been published or made publicly available. For example, Karlsson reported findings in abstract form in 2010, but there is still no peer-reviewed full paper describing the results of this follow-up and we were unable to locate a report of the trial on any trial registry.[88,89] The authors of the trial reporting a possible negative outcome advise that it may be a chance finding, and there are currently no known mechanisms of DHA to explain an adverse effect.[92,93] Future trials should include some of the above discussed measures, particularly the BRIEF, to allow direct comparison of findings.

Age at Executive Function Assessment

It is useful to consider children's age when investigating effects of prenatal DHA supplementation on executive functioning. Not just the neural structures, but the domains of executive functions themselves take time to develop. It is not possible to test for an effect on executive functioning until children reach an age at which the domain has developed to be tested reliably, otherwise we risk seeing null effects. As the executive functions mature over time, we might expect that any effect of prenatal DHA would be manifested consistently at older ages. Age at testing could reveal important insights such as prenatal DHA accelerating the developmental trajectory of the executive functions. On the contrary, it is also possible that with sufficient time, the executive functioning of control group children might eventually 'catch up' to their supplemented counterparts. Thus, we look at the evidence for the effects of prenatal DHA on executive functioning, according to age bands that reflect major periods of executive function development in childhood.

Infancy and toddlerhood (≤2 years): Only one study assessed executive functioning (attention) during infancy, although authors conducted multiple repeat measures with the same assessment at 4, 6, and 9 months of age in order to analyze trajectory of development.[94] No differences were found on many of the planned outcomes; however, authors concluded that overall findings indicated that prenatal DHA intervention conferred benefits to the development of attention in the first year of life.[94] Further studies would need to test and confirm this finding.

Preschool age (between 2 and 5 years): Two trials included assessments during the preschool period, and one of which had multiple assessments at 27 months and 4 years of age.[89,91,92] One RCT assessed executive functioning in children at 27 months of age, and again at 4 years.[91,92] No effect of the DHA intervention was detectable at 27 months, or 4 years on several performance-based measures; however, there were some instances in the parent-rated measure at 4 years where the DHA group scored slightly worse than the control group children.[91,92] Conversely, Karlsson et al. reported benefits of DHA supplementation for some tasks tapping executive functioning skills at 3.8 years of age, although actual data and details of the assessment are not available.[89]

School age (≥5 years): Of the four RCT's that included an assessments of executive functions in children over 5 years old,[83,85,87,93] one reported a possible adverse effect on global executive functioning and null effects on other outcomes and two reported positive effects.[83,87,93] For the majority of outcomes reported in this age group, particularly the performance-based measures, there were no group differences.[83,85,93]

Summary of the ages assessed: The results of these trials can be explored with respect to age at assessment and type of outcome assessed. However, no two trials conducted assessments at the same age and all used different assessments of executive functioning. This results in considerable heterogeneity in the evidence. Furthermore, it is not possible to ascribe the heterogeneity to the age at assessment, the domain of executive functioning or the task used to measure that domain as there are simply too few studies within each age band, or too few studies with the same outcome measurement to draw any robust conclusions on the effects of prenatal DHA supplementation might be manifested in childhood executive functions.

Dose of DHA

Most trials commenced supplementation around mid-pregnancy (20 weeks postconception, range 16 weeks to 28 weeks). Four trials ceased supplementation at birth,[37,84,86,90] one trial continued the intervention until 3.5 months post-birth,[88,89] and one trial provided formula to non-breastfed infants for the first 6 months, including a formula free of DHA and AA for the control group.[37] There does not seem to be any clear patterns in the results of the RCT's according to duration of the intervention.

The question of what is the appropriate dose of prenatal DHA has been argued by health agencies. The Global Organization for Eicosapentaenoic acid and DHA (GOED) recommend that pregnant women consume around 200 mg of DHA per day.[118] None of the trials in this chapter used a dose that was less than this and therefore the absence of effects on executive functioning cannot be due to not meeting minimum recommended intakes according to GOED. In fact, most RCTs have used much higher daily doses of DHA, ranging from 400 mg/day[85,87] (500 mg,[83] 600 mg,[94] 800 mg[90]) up to 1100 mg.[88] The evidence from observational studies suggests that higher doses lead to better outcomes and the highest doses used in these RCTs are in the upper limits of what could be possible to consume through diet alone. Under the concept that 'more is better,' such high doses of DHA supplements should maximize the potential to detect effects of DHA on executive functions. However, the 'more is better' mantra may not be the case with DHA. The 800 mg/day supplements of DHA in one trial resulted in marginally poorer executive functioning scores at 4 and 7 years of age.[92,93] No recommendations can be made on the basis of one trial, but this evidence underscores the importance of long-term executive functioning follow-up from other large RCTs.

The supplement taken by the control group also has potential to influence the findings of a RCT. No RCTs used a control supplement containing DHA, most were corn or soy oil based.[84,86,89,90,94] Interestingly, one trial intervening with a dose of 600 mg DHA/day,[94] modified eligibility criteria to allow women taking prenatal supplements containing DHA due to the high prevalence of supplement use preventing interfering with recruitment into the trial.[119] One trial involved four groups (a placebo group, a folate group, a fish oil group, and a folate + fish oil group) and results discussed refer to the comparison of the two fish oil supplemented groups vs the two non-fish oil groups and the fish oil group compared to the placebo group.[37]

More work may be needed to link circulating DHA levels to optimal executive functioning performance. Obviously maternal DHA status reflects maternal diet and mobilization of DHA from stores. In contrast, cord blood DHA also includes placental transfer processes to reflect the amount of DHA reaching the fetus. Detailed information on links between cord blood concentrations most closely correlated with executive function performance would help elucidate the 'ideal' dose of DHA to test in future RCTs. The original work in this area explored prenatal DHA and executive functioning using maternal DHA status at birth.[110,111] Since then, only one RCT has included an investigation of cord blood DHA status and executive functioning.[91] Fortunately, one of the assessments used in this RCT was the Colombo et al. method of assessing early attention.[110,111] Children underwent the attention assessment, as well as an assessment of working memory and inhibitory control at 27 months of age.[91] Unlike Colombo et al., associations detected between performance in the tests and cord blood DHA status were inconsistent.[91] The source of the different findings between the study by Colombo et al.[110,111] and by Gould et al.[91] are unclear. These differences included maternal vs cord blood DHA, a single measure of attention vs repeat measures and analysis of trajectories, and the ages of the children assessed. The more recent trial by Colombo et al. did not include an assessment of DHA status during pregnancy and the developmental trajectory of attention, although cord blood DHA levels were measured so further analyses may be possible.[94]

Only one other RCT included data for DHA status, where maternal DHA was measured at 20 weeks' gestation, 30 weeks' gestation, and delivery, when cord blood and umbilical blood were also sampled.[83] Some of the timed-outcomes were associated with some of DHA status outcomes; however, there was no consistency between outcomes of time points when samples were taken, and the majority of outcomes tested had no association with DHA status.[83] With the evidence currently available, it is unclear whether there is a dose–response relationship between prenatal DHA and executive functioning. Where data are available from current RCT's and observational studies, it would be useful to determine whether DHA status during pregnancy or at birth is associated with child executive functioning abilities.

SUMMARY

All together there were six RCT's with 18 assessments of executive functioning, most of which contained multiple outcomes.[83,85,87,89,91–94] Twelve of the assessments came from one RCT that assessed children at 2, 4, and 7 years of age.[91–93] Assessments used were a combination of performance-based measures targeting attention, working memory and inhibitory control, and parent-rated measures capturing attention and general executive functioning in everyday life. Although the majority of the outcomes revealed a null effect of prenatal DHA intervention on executive functions, the findings were mixed with beneficial, null, and adverse effects detected. Heterogeneity between the RCT's and the executive functioning assessments prevents combination of data in a meta-analysis.

There are important factors to consider when interpreting the results of these studies. Executive functions have only recently become of interest to nutritional researchers and included as outcome measures on RCT's, as global measures of cognition have consistently resulted in null findings.[103–106] Many of the published prenatal DHA RCT's assessing brain development have not included any measure of executive functioning.[95–102] Of the prenatal DHA RCT's that have measured executive functioning, only one follow-up had an executive functioning assessment as the primary outcome[91] and this follow-up only included a predefined subsample of the children in the trial. Future work needs to focus on areas of deposition of DHA in the frontal lobes and assessments of executive functioning to adequately evaluate effects on brain development or functioning. Furthermore, some harmonization of methods and ages at follow-up would go some considerable way toward generating effects that could be meta-analyzed.

All but one of the RCT's took place in developed, high-income countries, and the other RCT was conducted in a low-to-middle income country in a population of healthy women.[120] Thus all women in these RCTs are likely to have been well-nourished and are possibly less likely to demonstrate a benefit of supplementation. Interestingly, the Colombo et al. RCT that linked maternal DHA status to the developmental trajectory of attention was also conducted in a developed country (USA) although the majority of the women were receiving government care and were likely to have been socially disadvantaged.[110,111] Further RCT's with a focus on women in developing countries or within disadvantaged population subgroups as well as populations with low DHA intakes are needed to determine whether there are specific individuals that would benefit from increased DHA. Currently, there is little evidence to support prenatal DHA supplementation in well-nourished pregnant women.

As a single nutrient intervention, the effect of prenatal DHA supplementation is likely to be modest and require a sizeable sample to be powered to detect an effect. Large samples also increase the probability than the unknown potential confounders are balanced across the randomization group. There were only two RCT's that were powered to detect a smaller effect ($n > 500$),[87,92,93] as most other RCTs had small samples ($n < 100$ follow-up per group)[83,85,89,91] and were powered only to detect large effect sizes. Furthermore, the two largest trials also had follow-ups with attrition <20%,[87,92] which is the minimum follow-up considered acceptable for limiting attrition bias.[121] Loss to follow-up and declined appointments were the main reasons for attrition in each study. However, several RCTs excluded children after randomization had taken place for reasons such as complications during the pregnancy,[87,120] and infant anomalies or disorders likely to affect the development.[83–85,94] Post-randomization exclusions such as these are problematic for conducting intention-to-treat analyses. Furthermore, the majority of the follow-up studies discussed here actively excluded infants who were born preterm.[83,85,89,91,94] These children missed the benefit of the intrauterine supply of DHA during the peak period of foetal DHA accrual[26,122] and hence, may have been the most likely to benefit from supplementation. Furthermore, there have been suggestions that prenatal DHA intervention may increase the length of gestation and therefore reduce the prevalence of preterm birth.[90] Only two RCT's did not exclude children born preterm from the follow-up assessments.[87,92,93] Interestingly one of these RCT's used a single measure of executive functioning and found a benefit of supplementation,[87] while the other used multiple measures at two ages and found mostly no group differences and some possible adverse outcomes of prenatal DHA intervention.[92,93] With the current available body of evidence, it is unclear as to whether development of infants born preterm can be improved through prenatal DHA supplementation. Due to the idiopathic nature of the majority of preterm births,[123] identifying women likely to experience preterm labour for targeted DHA intervention is not yet feasible.

Given the lack of supporting evidence for prenatal DHA supplementation on executive functioning, it is important to question whether the prenatal period is the most effective window for DHA supplementation to have a demonstrable effect on brain development. Prenatal DHA supplementation may not have demonstrable benefits because brain development is protected in utero. There is evidence that DHA synthesis from precursor fatty acids is upregulated during pregnancy,[124] and mobilization of maternal DHA adipose stores together with preferential transfer of DHA across the placenta[125,126] may protect against insufficient DHA supply to the fetus. This is particularly likely in the generally healthy, well-nourished populations included in the RCT's discussed in this chapter.

CONCLUDING REMARKS

There is evidence from both animal[107–109] and human samples[110,111] that DHA exposure may have an effect on frontal lobe functioning, which is one key area of the brain that participates in executive functioning. However, RCT's of prenatal DHA supplementation among well-nourished pregnant women suggest that the effects on children's executive functioning are mixed. The reasons for heterogeneity could be due to the different types of executive functioning assessments, doses tested, and ages at follow -up.

Future research exploring the effect of DHA on brain development should focus on (1) identifying populations with low DHA status for DHA efficacy trials; (2) postnatal DHA interventions with follow-up into childhood when executive function assessments can be more reliably assessed; (3) trials designed and powered for executive functioning assessments as a primary outcome; and (4) selecting executive functioning assessments from the above discussed studies to facilitate direct comparison of results, with specific inclusion of the parent-rated outcomes that have indicated possible adversity of supplementation.

References

1. Hoffnung M, et al. *Lifespan Development: A Chronological Approach (Second Australasian Edition)*. Milton, Queensland, Australia: John Wiley & Sons; 2013.
2. Bystron I, Blakemore C, Rakic P. Development of the human cerebral cortex: Boulder Committee revisited. *Nat Rev Neurosci*. 2008;9(2):110–122.
3. Stiles J, Jernigan TL. The basics of brain development. *Neuropsychol Rev*. 2010;20(4):327–348.
4. Kostovic I, Jovanov-Milosevic N. The development of cerebral connections during the first 20–45 weeks' gestation. *Semin Fetal Neonatal Med*. 2006;11(6):415–422.
5. Dobbing J, Sands J. Quantitative growth and development of human brain. *Arch Dis Child*. 1973;48:757–767.
6. Volpe, J.J., Neuronal proliferation, migration, organization, and myelination, in Neurology of the Newborn, W. B. Saunders Staff, Editor. 1987, W. B. Saunders Company: Philadelphia. p. 33–68.
7. Huppi PS, et al. Quantitative magnetic resonance imaging of brain development in premature and mature newborns. *Ann Neurol*. 1998;43(2):224–235.
8. Johnson MH. Functional brain development in humans. *Nat Rev Neurosci*. 2001;2(7):475–483.
9. Casey BJ, Giedd JN, Thomas KM. Structural and functional brain development and its relation to cognitive development. *Biol Psychol*. 2000;54(1–3):241–257.
10. Innocenti GM, Price DJ. Exuberance in the development of cortical networks. *Nat Rev Neurosci*. 2005;6(12):955–965.
11. Fuster JM. Frontal lobe and cognitive development. *J Neurocytol*. 2002;31(3–5):373–385.
12. Dempster F. The rise and fall of the inhibitory mechanism: toward a unified theory of cognitive development and aging. *Develop Rev*. 1992;12:45–75.
13. Benes F. The development of frontal cortex: the maturation of neurotransmitter systems and their interactions. In: Nelson C, Luciana M, eds. *Handbook of Developmental Cognitive Neuroscience*. Cambridge, MA: MIT Press; 2001; 79-92.
14. Gathercole SE, et al. The structure of working memory from 4 to 15 years of age. *Dev Psychol*. 2004;40(2):177–190.
15. Bell MA, Fox NA. The relations between frontal brain electrical activity and cognitive development during infancy. *Child Dev*. 1992;63(5):1142–1163.
16. Blair C. School readiness: integrating cognition and emotion in a neurobiological conceptualization of children's functioning at school entry. *Am Psychol*. 2002;57(2):111–127.
17. Carlson SM, Moses LJ. Individual differences in inhibitory control and children's theory of mind. *Child Dev*. 2001;72(4):1032–1053.
18. Anderson V, Jacobs R, Anderson P. *Executive Functions and the Frontal Lobes. A Lifespan Perspective*. New York: Taylor & Francis; 2008.

19. Carlson SM, Mandell DJ, Williams L. Executive function and theory of mind: stability and prediction from ages 2 to 3. *Dev Psychol*. 2004;40(6):1105–1122.
20. Bosacki S, Astington JW. Theory of mind in preadolescence: relations between social understanding and social competence. *Social Develop*. 1999;8(2):237–255.
21. Garon N, Bryson SE, Smith IM. Executive function in preschoolers: a review using an integrative framework. *Psychol Bull*. 2008;134(1):31–60.
22. Sastry PS. Lipids of nervous tissue: composition and metabolism. *Prog Lipid Res*. 1985;24(2):69–176.
23. McIlwain H, Bachelard HS. *Biochemistry and the Central Nervous System*. Edinburgh: Churchill Livingstone; 1985.
24. Uauy R, Dangour AD. Nutrition in brain development and aging: role of essential fatty acids. *Nutr Rev*. 2006;64(5 Pt 2). p. S24–33; discussion S72–91.
25. Clandinin MT, et al. Intrauterine fatty acid accretion rates in human brain: implications for fatty acid requirements. *Early Hum Dev*. 1980;4(2):121–129.
26. Martinez M. Tissue levels of polyunsaturated fatty acids during early human development. *J Pediatr*. 1992;120(4 Pt 2): S129–S138.
27. Feller SE, Gawrisch K, MacKerell Jr AD. Polyunsaturated fatty acids in lipid bilayers: intrinsic and environmental contributions to their unique physical properties. *J Am Chem Soc*. 2002;124(2):318–326.
28. Bourre JM, et al. The effects of dietary à-linolenic acid on the composition of nerve membranes, enzymatic activity, amplitude of electrophysiological parameters, resistance to poisons and performance of learning tasks in rats. *J Nutr*. 1989;119:1880–1892.
29. Lands WE, Morris A, Libelt B. Quantitative effects of dietary polyunsaturated fats on the composition of fatty acids in rat tissues. *Lipids*. 1990;25(9):505–516.
30. Innis SM. Dietary (n-3) fatty acids and brain development. *J Nutr*. 2007;137(4):855–859.
31. Dutta-Roy AK. Transport mechanisms for long-chain polyunsaturated fatty acids in the human placenta. *Am J Clin Nutr*. 2000;71(suppl):315S–322S.
32. Haggarty P. Placental regulation of fatty acid delivery and its effect on fetal growth—a review. *Placenta*. 2002;23. Suppl A: p. S28–38.
33. Innis SM. Essential fatty acid transfer and fetal development. *Placenta*. 2005;26. Suppl A: p. S70–5.
34. Elias SL, Innis SM. Infant plasma trans, n-6, and n-3 fatty acids and conjugated linoleic acids are related to maternal plasma fatty acids, length of gestation, and birth weight and length. *Am J Clin Nutr*. 2001;73(4):807–814.
35. Helland IB, et al. Supplementation of n-3 fatty acids during pregnancy and lactation reduces maternal plasma lipid levels and provides DHA to the infants. *J Mater Fetal Neonatal Med*. 2006;19(7):397–406.
36. van Houwelingen AC, et al. Essential fatty acid status in neonates after fish-oil supplementation during late pregnancy. *Br J Nutr*. 1995;74(5):723–731.
37. Decsi T, Campoy C, Koletzko B. Effect of N-3 polyunsaturated fatty acid supplementation in pregnancy: the Nuheal trial. *Adv Exp Med Biol*. 2005;569:109–113.
38. Mendez MA, et al. Maternal fish and other seafood intakes during pregnancy and child neurodevelopment at age 4 years. *Public Health Nutr*. 2009;12(10):1702–1710.
39. Hibbeln JR, et al. Maternal seafood consumption in pregnancy and neurodevelopment outcomes in childhood (ALSPAC study): an observational cohort study. *Lancet*. 2007;369(9561):578–585.
40. Daniels JL, et al. Fish intake during pregnancy and early cognitive development of offspring. *Epidemiology*. 2004;15(4):394–402.
41. Oken E, et al. Maternal fish intake during pregnancy, blood mercury levels, and child cognition at age 3 years in a US cohort. *Am J Epidemiol*. 2008;167(10):1171–1181.
42. Gale CR, et al. Oily fish intake during pregnancy—association with lower hyperactivity but not with higher full-scale IQ in offspring. *J Child Psychol Psychiatry*. 2008;49(10):1061–1068.
43. Oken E, et al. Associations of maternal fish intake during pregnancy and breastfeeding duration with attainment of developmental milestones in early childhood: a study from the Danish National Birth Cohort. *Am J Clin Nutr*. 2008;88(3):789–796.
44. Lawlor DA, et al. Those confounded vitamins: What can we learn from the differences between observational versus randomised trial evidence? *Lancet*. 2004;363(9422):1724–1727.
45. Bazan NG, Scott BL. Dietary omega-3 fatty acids and accumulation of docosahexaenoic acid in rod photoreceptor cells of the retina and at synapses. *Ups J Med Sci*. 1990;48:97–107.

46. Bowen RA, Clandinin MT. Dietary low linolenic acid compared with docosahexaenoic acid alter synaptic plasma membrane phospholipid fatty acid composition and sodium-potassium ATPase kinetics in developing rats. *J Neurochem*. 2002;83(4):764–774.
47. Sarkadi-Nagy E, et al. The influence of prematurity and long chain polyunsaturate supplementation in 4-week adjusted age baboon neonate brain and related tissues. *Pediatr Res*. 2003;54(2):244–252.
48. Lozada LE, et al. Perinatal brain docosahexaenoic acid concentration has a lasting impact on cognition in mice. *J Nutr*. 2017;147(9):1624–1630.
49. Abedin L, et al. The effects of dietary alpha-linolenic acid compared with docosahexaenoic acid on brain, retina, liver and heart in the Guinea pig. *Lipids*. 1999;34:475–482.
50. Moriguchi T, Greiner RS, Salem Jr N. Behavioral deficits associated with dietary induction of decreased brain docosahexaenoic acid concentration. *J Neurochem*. 2000;75(6):2563–2573.
51. Levant B, Radel JD, Carlson SE. Decreased brain docosahexaenoic acid during development alters dopamine-related behaviors in adult rats that are differentially affected by dietary remediation. *Behav Brain Res*. 2004;152(1):49–57.
52. Pawlosky RJ, et al. Retinal and brain accretion of long-chain polyunsaturated fatty acids in developing felines: the effects of corn oil-based maternal diets. *Am J Clin Nutr*. 1997;65:465–472.
53. Galli C, Trzeciak HI, Paoletti R. Effects of dietary fatty acids on the fatty acid composition of brain ethanolamine phosphoglyceride: reciprocal replacement of n-6 and n-3 polyunsaturated fatty acids. Biochimica et biophysica acta. *Lipids Lipid Metabol*. 1971;248(3):449–454.
54. Aid S, et al. Effect of a diet-induced n-3 PUFA depletion on cholinergic parameters in the rat hippocampus. *J Lipid Res*. 2003;44(8):1545–1551.
55. Catalan J, et al. Cognitive deficits in docosahexaenoic acid-deficient rats. *Behav Neurosci*. 2002;116(6):1022–1031.
56. Takeuchi T, Fukumoto Y, Harada E. Influence of a dietary n-3 fatty acid deficiency on the cerebral catecholamine contents, EEG and learning ability in rat. *Behav Brain Res*. 2002;131(1–2):193–203.
57. Ahmad A, et al. A decrease in cell size accompanies a loss of docosahexaenoate in the rat hippocampus. *Nutr Neurosci*. 2002;5(2):103–113.
58. Champoux M, et al. Fatty acid formula supplementation and neuromotor development in rhesus monkey neonates. *Pediatr Res*. 2002;51(3):273–281.
59. Zimmer L, et al. Modification of dopamine neurotransmission in the nucleus accumbens of rats deficient in n-3 polyunsaturated fatty acids. *J Lipid Res*. 2000;41(1):32–40.
60. Bhutta AT, et al. Cognitive and behavioral outcomes of school-aged children who were born preterm: a meta-analysis. *JAMA*. 2002;288(6):728–737.
61. Aarnoudse-Moens CS, et al. Executive function and IQ predict mathematical and attention problems in very preterm children. *PLoS One*. 2013;8(2):e55994.
62. Kerr-Wilson CO, et al. Meta-analysis of the association between preterm delivery and intelligence. *J Public Health (Oxf)*. 2012;34(2):209–216.
63. Aarnoudse-Moens CS, et al. Meta-analysis of neurobehavioral outcomes in very preterm and/or very low birth weight children. *Pediatrics*. 2009;124(2):717–728.
64. Johnson S. Cognitive and behavioural outcomes following very preterm birth. *Semin Fetal Neonatal Med*. 2007;12(5):363–373.
65. Lindstrom K, Lindblad F, Hjern A. Preterm birth and attention-deficit/hyperactivity disorder in schoolchildren. *Pediatrics*. 2011;127(5):858–865.
66. Arpi E, Ferrari F. Preterm birth and behaviour problems in infants and preschool-age children: a review of the recent literature. *Dev Med Child Neurol*. 2013;55(9):788–796.
67. Spittle AJ, et al. Early emergence of behavior and social-emotional problems in very preterm infants. *J Am Acad Child Adolesc Psychiatry*. 2009;48(9):909–918.
68. van de Weijer-Bergsma E, Wijnroks L, Jongmans MJ. Attention development in infants and preschool children born preterm: a review. *Infant Behav Dev*. 2008;31(3):333–351.
69. Anderson PJ, Doyle LW, V.I.C.S. Group. Executive functioning in school-aged children who were born very preterm or with extremely low birth weight in the 1990s. *Paediatrics*. 2004;114:50–57.
70. Clandinin MT, et al. Growth and development of preterm infants fed infant formulas containing docosahexaenoic acid and arachidonic acid. *J Pediatr*. 2005;146:461–468.
71. Fewtrell MS, et al. Double-blind, randomized trial of long-chain polyunsaturated fatty acid supplementation in formula fed to preterm infants. *Pediatrics*. 2002;110(1 Pt 1):73–82.

72. Fewtrell MS, et al. Randomized, double-blind trial of long-chain polyunsaturated fatty acid supplementation with fish oil and borage oil in preterm infants. *J Pediatr*. 2004;144(4):471–479.
73. O'Connor DL, et al. Growth and development in preterm infants fed long-chain polyunsaturated fatty acids: a prospective, randomized controlled trial. *Pediatrics*. 2001;108(2):359–371.
74. Vanderhoof J, et al. Evaluation of a long-chain polyunsaturated fatty acid supplemented formula on growth, tolerance, and plasma lipids in preterm infants up to 48 weeks postconceptional age. *J Pediatr Gastroenterol Nutr*. 1999;29(3):318–326.
75. Birch EE, et al. Dietary essential fatty acid supply and visual acuity development. *Invest Ophthalmol Vis Sci*. 1992;33(11):3242–3253.
76. Carlson SE, Werkman SH, Tolley EA. Effect of long-chain n-3 fatty acid supplementation on visual acuity and growth of preterm infants with and without bronchopulmonary dysplasia. *Am J Clin Nutr*. 1996;63(5):687–697.
77. Carlson SE, et al. First year growth of preterm infants fed standard compared to marine oil n-3 supplemented formula. *Lipids*. 1992;27(11):901–907.
78. Moon K, et al. Longchain polyunsaturated fatty acid supplementation in preterm infants. *Cochrane Database Syst Rev*. 2016;12. p. Cd000375.
79. Schulzke SM, Patole SK, Simmer K. Long-chain polyunsaturated fatty acid supplementation in preterm infants. *Cochrane Database Syst Rev*. 2011;2:. p. CD000375.
80. Smithers LG, et al. Effect of long-chain polyunsaturated fatty acid supplementation of preterm infants on disease risk and neurodevelopment: a systematic review of randomized controlled trials. *Am J Clin Nutr*. 2008;87(4):912–920.
81. Carlson SE, et al. Visual-acuity development in healthy preterm infants: effect of marine-oil supplementation. *Am J Clin Nutr*. 1993;58:35–42.
82. Clandinin MT, et al. Fatty acid utilization in perinatal de novo synthesis of tissues. *Early Hum Dev*. 1981;5:355–366.
83. Catena A, et al. Folate and long-chain polyunsaturated fatty acid supplementation during pregnancy has long-term effects on the attention system of 8.5-y-old offspring: a randomized controlled trial. *Am J Clin Nutr*. 2015;.
84. Innis SM, Friesen RW. Essential n-3 fatty acids in pregnant women and early visual acuity maturation in term infants. *Am J Clin Nutr*. 2008;87(3):548–557.
85. Mulder KA, Elango R, Innis SM. Fetal DHA inadequacy and the impact on child neurodevelopment: a follow-up of a randomised trial of maternal DHA supplementation in pregnancy. *Br J Nutr*. 2018;119(3):271–279.
86. Ramakrishnan U, et al. Effects of docosahexaenoic acid supplementation during pregnancy on gestational age and size at birth: randomized, double-blind, placebo-controlled trial in Mexico. *Food Nutr Bull*. 2010;31(2 Suppl): S108–S116.
87. Ramakrishnan U, et al. Prenatal supplementation with DHA improves attention at 5 y of age: a randomized controlled trial. *Am J Clin Nutr*. 2016;104(4):1075–1082.
88. Furuhjelm C, et al. Fish oil supplementation in pregnancy and lactation may decrease the risk of infant allergy. *Acta Paediatr*. 2009;98(9):1461–1467.
89. Karlsson T, et al. *LC-PUFA supplemented to mothers during pregnancy and breast-feeding improves cognitive performance in the children four years later-an rct study*. In: *ISSFAL*; 2010. Maastricht. p. 113.
90. Makrides M, et al. Effect of DHA supplementation during pregnancy on maternal depression and neurodevelopment of young children: a randomized controlled trial. *JAMA*. 2010;304(15):1675–1683.
91. Gould JF, et al. Randomized controlled trial of maternal omega-3 long-chain PUFA supplementation during pregnancy and early childhood development of attention, working memory, and inhibitory control. *Am J Clin Nutr*. 2014;99(4):851–859.
92. Makrides M, et al. Four-year follow-up of children born to women in a randomized trial of prenatal DHA supplementation. *JAMA*. 2014;311(17):1802–1804.
93. Gould JF, et al. Seven-year follow-up of children born to women in a randomized trial of prenatal DHA supplementation. *JAMA*. 2017;317(11):1173–1175.
94. Colombo J, et al. Prenatal DHA supplementation and infant attention. *Pediatr Res*. 2016;80(5):656–662.
95. Helland IB, et al. Similar effects on infants of n-3 and n-6 fatty acids supplementation to pregnant and lactating women. *Pediatrics*. 2001;108(5):1–10.
96. Helland IB, et al. Effect of supplementing pregnant and lactating mothers with n-3 very-long-chain fatty acids on children's IQ and body mass index at 7 years of age. *Pediatrics*. 2008;122(2):e472–e479.

97. Helland IB, et al. Maternal supplementation with very-long-chain n-3 fatty acids during pregnancy and lactation augments children's IQ at 4 years of age. *Pediatrics*. 2003;111(1). p. e39–44.
98. Tofail F, et al. Supplementation of fish-oil and soy-oil during pregnancy and psychomotor development of infants. *J Health Popul Nutr*. 2006;24(1):48–56.
99. Judge MP, Harel O, Lammi-Keefe CJ. Maternal consumption of a docosahexaenoic acid-containing functional food during pregnancy: benefit for infant performance on problem-solving but not on recognition memory tasks at age 9 mo. *Am J Clin Nutr*. 2007;85(6):1572–1577.
100. Dunstan JA, et al. Cognitive assessment of children at 2 1/2 years following fish oil supplementation in pregnancy: a randomized controlled trial. *Arch Dis Fetal Neonatal Ed*. 2008;93(1):F45–F50.
101. van Goor SA, et al. The influence of supplemental docosahexaenoic and arachidonic acids during pregnancy and lactation on neurodevelopment at eighteen months. *Prostaglandins Leukot Essent Fatty Acids*. 2011;84(5–6):139–146.
102. Brei C, et al. Impact of the n-6:n-3 long-chain PUFA ratio during pregnancy and lactation on offspring neurodevelopment: 5-year follow-up of a randomized controlled trial. *Eur J Clin Nutr*. 2017;.
103. Gould JF, Smithers LG, Makrides M. The effect of maternal omega-3 (n-3) LCPUFA supplementation during pregnancy on early childhood cognitive and visual development: a systematic review and meta-analysis of randomized controlled trials. *Am J Clin Nutr*. 2013;97(3):531–544.
104. Campoy C, et al. Omega 3 fatty acids on child growth, visual acuity and neurodevelopment. *Br J Nutr*. 2012;107(Suppl 2):S85–106.
105. Jasani B, et al. Long chain polyunsaturated fatty acid supplementation in infants born at term. *Cochrane Database Syst Rev*. 2017;3.p. Cd000376.
106. Delgado-Noguera MF, et al. Supplementation with long chain polyunsaturated fatty acids (LCPUFA) to breastfeeding mothers for improving child growth and development. *Cochrane Database Syst Rev*. 2015;7.p. CD007901.
107. Wainwright PE, Colombo J. Nutrition and the development of cognitive functions: interpretation of behavioral studies in animals and human infants. *Am J Clin Nutr*. 2006;84(5):961–970.
108. Cheatham, C.L., J. Colombo, and S.E. Carlson, n-3 Fatty acids and cognitive and visual acuity development: methodologic and conceptual considerations. Am J Clin Nutr, 2006. 83 (supplement): p. 1458S–1466S.
109. Georgieff MK. Nutrition and the developing brain: nutrient priorities and measurement. *Am J Clin Nutr*. 2007;85(suppl):614S–620S.
110. Colombo J, et al. Maternal DHA and the development of attention in infancy and toddlerhood. *Child Dev*. 2004;75(4):1254–1267.
111. Kannass KN, Colombo J, Carlson SE. Maternal DHA levels and toddler free-play attention. *Dev Neuropsychol*. 2009;34(2):159–174.
112. Gould JF, et al. Does n-3 LCPUFA supplementation during pregnancy increase the IQ of children at school age? Follow-up of a randomised controlled trial. *BMJ Open*. 2016;6(5):e011465.
113. Spencer JP, Smith LB, Thelen E. Tests of a dynamic systems account of the A-not-B error: the influence of prior experience on the spatial memory abilities of two-year-olds. *Child Dev*. 2001;72(5):1327–1346.
114. Althouse AD. Adjust for multiple comparisons? It's not that simple. *Ann Thorac Surg*. 2016;101(5):1644–1645.
115. Milte CM, et al. Increased erythrocyte Eicosapentaenoic acid and docosahexaenoic acid are associated with improved attention and behavior in children with ADHD in a randomized controlled three-way crossover trial. *J Atten Disord*. 2015;19(11):954–964.
116. Parletta N, Niyonsenga T, Duff J. Omega-3 and Omega-6 polyunsaturated fatty acid levels and correlations with symptoms in children with attention deficit hyperactivity disorder, autistic Spectrum disorder and typically developing controls. *PLoS One*. 2016;11(5):e0156432.
117. Toplak ME, West RF, Stanovich KE. Practitioner review: Do performance-based measures and ratings of executive function assess the same construct? *J Child Psychol Psychiatry*. 2013;54(2):131–143.
118. GOED. *Global Recommendations for EPA and DHA Intake*. [Rev 19 November 2014]; 2014.
119. Carlson SE, Colombo J. KUDOS Trial. Clinical Trials Registry [cited 2012 Jan 19]; Available from: http://clinicaltrials.gov/ct2/show/NCT00266825?term=DHA+and+pregnancy&rank=5.
120. Ramakrishnan U, et al. *Effect of prenatal supplementation with docosahexanoic acid on child size and development at 18 mo: randomized placebo-controlled trial in Mexico*. In: *ISSFAL*; 2010. Maastricht, p. 112.
121. Higgins J, Green S, eds. *Cochrane Handbook for Systematic Reviews of Interventions Version 5.0.2*. The Cochrane Collaboration; 2009. Available from, www.cochrane-handbook.org.

122. Clandinin MT, et al. Extrauterine fatty acid accretion in infant brain: implications for fatty acid requirments. *Early Hum Dev*. 1980;4(2):131–138.
123. Bennett P. Preterm Labour. In: *Dewhurst's Textbook of Obstetrics & Gynaecology*. Blackwell Publishing; 2008:177–191.
124. Burdge GC, Calder PC. Conversion of alpha-linolenic acid to longer-chain polyunsaturated fatty acids in human adults. *Reprod Nutr Dev*. 2005;45(5):581–597.
125. Hanebutt FL, et al. Long-chain polyunsaturated fatty acid (LC-PUFA) transfer across the placenta. *Clin Nutr*. 2008;27(5):685–693.
126. Haggarty P, et al. Long-chain polyunsaturated fatty acid transport across the perfused human plactenta. *Placenta*. 1997;18(8):635–642.

CHAPTER

7

Omega-3 and Omega-6 Fatty Acid Metabolism: Modeling Growth and Disease Using *Caenorhabditis elegans*

Andrea J. Connor, Jennifer L. Watts

School of Molecular Biosciences, Washington State University, Pullman, WA, United States

INTRODUCTION

The roundworm *Caenorhabditis elegans* is a powerful model organism for elucidating fundamental biological processes. The quick generation time and easy lab maintenance of *C. elegans* provides a high-throughput way to address research questions that may be too complex to approach in higher order organisms. Many genes in *C. elegans* are conserved from worms to humans, thus nematodes provide a way to address questions in the context of a whole organism that otherwise must be done *in vitro* or in isolated cell culture systems.[1] *C. elegans* consume *Escherichia coli* as food in the laboratory setting, which enables gene knockdown by RNAi to be carried out simply by feeding *E. coli* engineered to produce double-stranded RNA corresponding to *C. elegans* genes.[2] *C. elegans* are also transparent, allowing the use of fluorescent markers or dyes, and transgenic technology has allowed for the engineering of disease model strains of *C. elegans*. These aspects of *C. elegans* biology, together with conserved lipid biosynthesis pathways, make *C. elegans* an attractive model for functional studies of fatty acids and their derivatives.

Fatty acids are carboxylic acids that have long aliphatic chains of 14–24 carbons and are important energy sources and components of membranes. They also act as precursors to signaling molecules such as prostaglandins and eicosanoids. Mammals must obtain the essential fatty acids from their diets because they lack the enzymes necessary to convert monounsaturated fatty acids to polyunsaturated fatty acids (PUFAs). *C. elegans*, however, does not have this limitation and is able to synthesize a range of essential monounsaturated fatty acids (MUFAs), branched-chain fatty acids (BCFAs) and PUFAs starting from acetyl-CoA or isobutyryl-CoA and ending with several species of C20 PUFAs.[3,4]

Omega Fatty Acids in Brain and Neurological Health
https://doi.org/10.1016/B978-0-12-815238-6.00007-9

Fatty acids are characterized by the degree of saturation of the aliphatic chain. This degree of saturation is important not only for membrane dynamics, such as adaptation to temperature changes,[5,6] but also for enzymatic activity and the formation of lipid raft domains, which often influence cellular signal transduction.[7] Long-chain PUFAs are categorized as omega-6 or omega-3, the difference being the position of the final double bond is relative the methyl-end of the aliphatic chain. Mammals, in particular, lack the ability to convert omega-6 to omega-3 and vice versa, and the ratio of these two is determined by dietary intake. Functional differences between omega-3 and omega-6 PUFAs have implications for human health, but the mechanisms behind these functional differences are not fully understood.

C. *ELEGANS* SYNTHESIZES OMEGA-6 AND OMEGA-3 FATTY ACIDS

C. elegans can synthesize omega-6 and omega-3 PUFAs de novo, without the need for essential dietary fatty acids.[3] This ability depends on the presence of plant-like Δ12 desaturase encoded by the *fat-2* gene and the omega-3 desaturase, encoded by the *fat-1* gene[3,8–10] (Fig. 1). While these activities are not present in vertebrates, a growing body of evidence demonstrates the ability to synthesize omega-6 and omega-3 PUFAs by certain species of terrestrial invertebrates including insects[11] and the Collembola, commonly known as springtails.[12,13] Thus, synthesis of omega-6 and omega-3 fatty acids by nematodes and other invertebrates may constitute a larger source of dietary PUFAs in the soil food web than previously recognized.[13]

C. elegans synthesizes a complement of C18 and C20 PUFAs using seven fatty acid desaturases, including three partially redundant Δ9 desaturates (FAT-5, -6, and -7), a single copy of the plant-like Δ12 and omega-3 desaturases (FAT-2 and FAT-1), and a single gene each for Δ6 and Δ5 desaturases (FAT-3 and FAT-4)[9,10,14] (Fig. 1). The fatty acid desaturases require the 3-ketoacyl-CoA reductase (LET-767)[15] and several cytochrome *b*5 reductases encoded by *hpo-19* and T05H4.4.[16] In addition to the fatty acid desaturases, fatty acid elongases ELO-1 and ELO-2 are required for C18 and C20 PUFA production[10,17]; however, *C. elegans* lacks the elongase activity to produce C22 PUFAs. Thus, capacity for PUFA synthesis makes *C. elegans* an excellent system to genetically dissect the role of various C18 and C20 PUFAs with resolution not possible in mammals.

Even though *C. elegans* has all the enzymes for de novo fatty acid synthesis, they rely heavily on dietary fats from their *E. coli* diet for fatty acids and precursors, such as palmitic acid (16:0), cis-vaccenic acid (18:1n-7) and cyclopropane fatty acids. Using a mixture of C12 and C13 isotopes, mass spectrometry analysis revealed that 80%–90% of *C. elegans* fatty acids are derived from dietary *E. coli* fatty acids, either directly incorporated 16:0, or modified, such that the 16:0 from *E. coli* is elongated and desaturated. The remaining 10%–20% of fatty acids are synthesized de novo from malonyl CoA.[18] Remarkably, *C. elegans* produces nearly its own weight in progeny each day during peak reproduction, thus fatty acids and other dietary nutrients exhibit high turnover. Pulse chase experiments using a mixture of *E. coli* grown in C12 or C13 isotopes revealed that approximately 4.5% of membrane fatty acid acyl components of phospholipids are turned over every hour, and 2.7% of neutral (storage) lipids are turned over every hour, leading to a nearly complete turnover of nematode membrane fatty acids daily.[19]

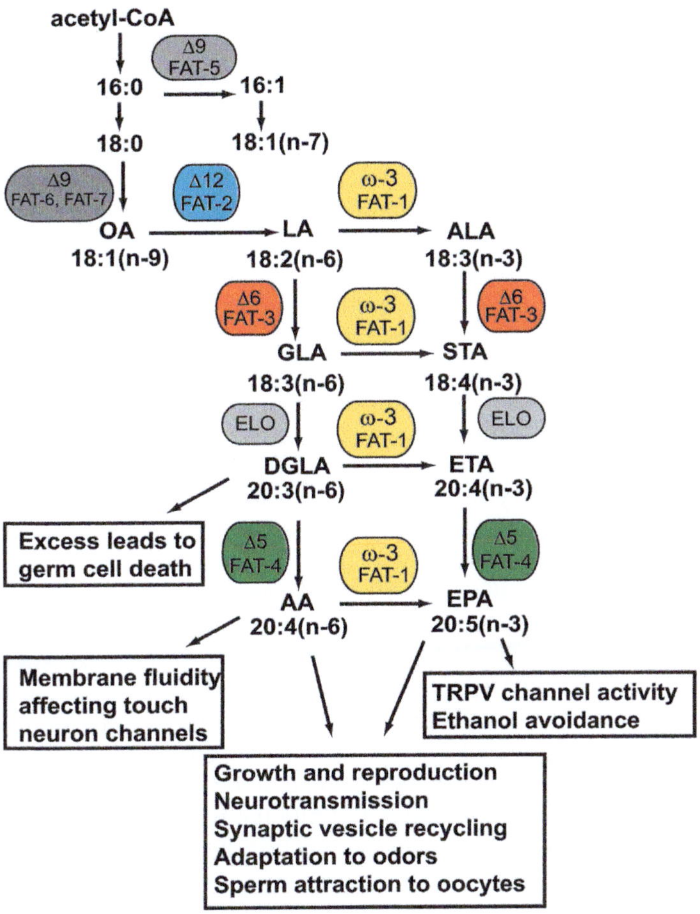

FIG. 1 Pathway of polyunsaturated fatty acid synthesis in *C. elegans* and functions of omega-3 and omega-6 PUFAs. Enzyme names and activities are enclosed in ovals. Abbreviations: ELO, elongase; OA, oleic acid; LA, linoleic acid; ALA, alpha linoleic acid; GLA, gamma linoleic acid; STA, stearidonic acid; DGLA, dihomo-gamma-linoleic acid; ETA, eicosatetraenoic acid; AA, arachidonic acid; EPA, eicosapentaenoic acid.

ROLES FOR OMEGA-3 AND OMEGA-6 PUFAS IN GROWTH AND DEVELOPMENT

Studies of *C. elegans* mutants defective in fatty acid desaturation and PUFA elongation reveal roles for PUFAs in growth and development of nematodes. PUFA synthesis is severely disrupted in the *fat-2* Δ12 desaturase mutants and in the *fat-6;fat-7* Δ9 desaturase double mutants, which are unable to synthesize conventional 18-carbon and 20-carbon PUFAs. These mutants grow very slowly, have reduced fat stores and display severe reproductive defects, producing only 10% of the number of live progeny compared to wild type.[10,20] Similarly, depletion of *elo-1* and *elo-2*, which cooperate to elongate palmitic acid (16:0) to stearic acid (18:0) and 18-carbon PUFAs to 20-carbon PUFAs, result in increased saturated fatty acid, decreased PUFA, slow growth, and disrupted reproduction.[17] The *fat-3* Δ6 desaturase mutants produce 18-carbon PUFAs but very little 20-carbon PUFAs. They also display defects in growth rate and reproduction, although not as severe as the *fat-2* or *fat-6;fat-7* mutants.[21]

Interestingly, the *fat-1* mutants, which lack omega-3 PUFAs and the *fat-4* mutants, which lack arachidonic acid (AA, 20:4n-6), show no obvious developmental or reproductive defects when grown in laboratory conditions.[10] Because the dietary *E. coli* readily incorporate PUFAs into their membranes, dietary supplementation of PUFAs is possible simply by including PUFAs in the growth media.[22] These types of supplementation experiments indicate that individual omega-6 fatty acids, such as AA, or omega-3 fatty acids, such as eicosapentaenoic acid (EPA, 20:5n-3), are equally capable of rescuing the developmental and neurological defects in PUFA-deficient *fat-3* mutants.[21] This demonstrates that under laboratory conditions, omega-3 and omega-6 PUFAs perform redundant functions in *C. elegans* for growth and reproduction.

Recently, PUFAs have been shown to be important in developing embryos, and are required for proper formation of the embryonic permeability barrier when they are not provided by the mother.[23] Both PUFA deficiency and mutations in the biotin ligating protein *bpl-1*, which is necessary for the biotinylation of acetyl-CoA carboxylase and for de novo fatty acid biosynthesis, are necessary for proper polarity and permeability of embryos.[23] Decreased de novo fatty acid biosynthesis leads to depletion of embryonic PUFAs, which may be modified in the embryonic membrane and incorporated into the hydrophobic permeability barrier layer of the eggshell. Proper eggshell formation is necessary for the establishment of polarity of the embryo.[24]

While either omega-6 or omega-3 PUFAs generally rescue *C. elegans* PUFA deficiencies to a similar extent, a role for omega-6 PUFAs for oocyte development has been suggested.[25] Researchers used coherent anti-Stokes Raman scattering (CARS) microscopy to examine yolk lipoprotein delivery to oocytes. The PUFA-deficient *fat-2* strain was not efficient at delivering yolk lipoproteins to developing oocytes, and supplementation with 18:3n-6 omega-6 PUFAs rescued this defect more efficiently than supplementation with 18:3n-3.[25] While the researchers concluded that omega-6 PUFAs were essential for oocyte development, it must be noted that the dietary 18:3n-6 will enter the PUFA biosynthesis pathway, and due to the presence of the FAT-1 omega-3 desaturase, this dietary precursor is likely converted into both omega-6 and omega-3 C20 PUFAs.

In spite of the requirement for omega-6 PUFAs during growth and development, excess dietary omega-6 PUFA leads to detrimental effects in the germ line. Specifically, dihomo-gamma-linolenic acid (DGLA, 20:3n-6) and AA supplementation resulted in death of germ cells during larval and adult development.[26] This cell death is independent of the physiological apoptosis that occurs in the germ line, because *ced-3* and *ced-4* mutants lacking the apoptotic machinery also become sterile when feeding on DGLA.[27] Furthermore, both male and hermaphrodite germ cells are affected by DGLA, and physiological apoptosis only occurs in hermaphrodites. The cell death in the germ line is likely mediated by oxidized metabolites of DGLA and AA generated by cytochrome P450 enzymes,[28] and mutants with enhanced stress resistance are resistant to the effects of dietary DGLA.[27]

ROLES FOR EICOSANOIDS AND ENDOCANNABINOIDS IN GROWTH AND DEVELOPMENT OF C. *ELEGANS*

The relative composition of omega-6 and omega-3 PUFAs have been suggested to influence inflammatory responses in mammals due to the synthesis of short range signaling molecules collectively called eicosanoids. These molecules are enzymatically produced, oxygenated signaling molecules derived from C20 omega-6 and omega-3 PUFAs, which are cleaved from the phospholipid membrane by phospholipases and act as precursors for eicosanoids such as prostaglandins, leukotrienes, and hydroxy-, epoxy-, and hydroperoxy-derived molecules.[29] Cyclooxygenases, lipoxygenases, and cytochrome P450 enzymes are responsible for eicosanoid formation by acting on the liberated PUFAs.[30,31] In general, eicosanoids produced from omega-6 precursors are considered to be pro-inflammatory, whereas eicosanoids produced from omega-3 precursors are considered to be antiinflammatory.[32] These eicosanoids are important signaling molecules for immune responses and reproduction.[33] PUFAs and eicosanoids derived from them can also act as transcription factors and ligands for endocannabinoid responses, such as appetite and pain.[34,35]

In *C. elegans,* the genes encoding the main enzymes for prostaglandin biosynthesis, cyclooxygenase 1 and 2, are not present in the genome. However, evidence for prostaglandin formation and function comes from experiments from the Miller lab that found that wild-type sperm did not chemotax toward oocytes from *fat-2* mutants, which lack C20 PUFAs.[36] However, dietary supplementation of either omega-6 or omega-3 PUFAs rescued the phenotype. The signal produced by oocytes is likely to be a prostaglandin produced from C20 PUFA precursors, since injection of F-series prostaglandins into the *fat-2* germline rescued the sperm migration defect.[37] In *C. elegans,* F-series prostaglandins derived from PUFAs have been identified by mass spectrometry.[38] Interestingly, these prostaglandins appear to be formed using cyclooxygenase-independent mechanisms, which are likely conserved because similar F-series prostaglandins were also identified in cyclooxygenase double knockout mutant mice.[39]

A lipidomic study of *C. elegans* dauer larvae, the ecologically important long-lived larval stage induced by crowding and low-nutrient conditions, revealed that dauer larvae sequester relatively high levels of PUFAs in their phospholipids.[40] After 3 weeks of dauer endurance, nonesterified PUFAs appear, as do eicosanoids derived from PUFAs, especially from EPA. The authors suggest that sequestration of PUFAs in phospholipids during the first weeks of dauer diapause contributes to longevity of the nonfeeding dauer larvae, in part by suppressing eicosanoid formation that may signal reproductive growth.

A recent study using *C. elegans* suggests that oxylipins derived from a specific omega-3 PUFA may play a role in adult longevity. The researchers found that exposure to dietary α-linoleic acid (ALA, 18:3n3) led to increased lifespan in *C. elegans* in a dose-dependent manner.[41] This may be mediated by oxylipin derivatives of ALA, which lead to activation of SKN-1/Nrf2, a transcription factor responsible for upregulation of genes required for oxidative stress responses.[42]

Endocannabinoids derived from C20 PUFAs and other fatty acids are produced in *C. elegans*. The endocannabinoid eicosapentaenoyl ethanolamide (EPEA), derived enzymatically from PE molecules containing EPA, promotes larval growth and may act as a sensor of nutrient availability and energy balance.[43] This metabolite is detected in fed, but not starved larvae, and depletion of EPEA by overexpression of the fatty acid amide hydrolase FAAH-1 leads to delayed larval growth. In adults, caloric restriction leads to lifespan extension, and application of exogenous EPEA reduces the lifespan extension, providing further evidence that this endocannabinoid acts as a signal to couple nutrient availability to regulate growth and lifespan.[43]

ROLES FOR OMEGA-6 AND OMEGA-3 PUFAS IN NEURONAL FUNCTION

PUFAs are vital for neuronal health. It is well known that omega-6 and omega-3 PUFAs aid in neurotransmission.[44] Neuron membranes contain a large amount of long-chain PUFAs and altering that composition leads to neuronal diseases. *C. elegans* possesses a very simple nervous system consisting of 302 neurons and 56 glial cells. In spite of this simplicity, these nematodes exhibit a range of movement and behaviors. A combination of laser ablation of single cells and genetic analysis has resulted in the elucidation of neuronal circuits and signal transduction pathways activating in specific neurons.[45]

C. elegans fat-3 mutants that lack C20 PUFAs exhibit slow movement compared to wild-type worms.[10,21] The movement defects can be rescued by expression of *fat-3* in neurons or supplementation of long-chain PUFAs into the diet of *C. elegans*.[21,46] Interestingly, synaptic vesicles are depleted at the neuromuscular junction in *fat-3* mutants, leading to reduced neurotransmitter levels.[46] This depletion is likely due to defects in synaptic vesicle endocytosis and recycling in *fat-3* mutants, consistent with the observation that the synaptic vesicle component synaptobrevin is mislocalized and the phosphoinositide phosphatase synaptojanin is depleted at synapses.[47]

Free fatty acids directly or indirectly modulate the activity of several types of ion channels.[48] Studies in *C. elegans* demonstrated that C20 PUFAs are specifically required for optimal activity of transient resistance potential (TRP) cation channels in sensory neurons.[49] In *C. elegans*, volatile odorants are sensed by two distinct neurons, AWA and AWC, which possess unique signal transduction pathways and sense distinct odors.[50] The involvement of C20 PUFAs in TRP channels was uncovered by demonstrating that *C. elegans* mutants lacking AA or EPA, such as *fat-3* and *fat-1;fat-4* double mutants, displayed defective chemotaxis behavior in response to odorants sensed by AWA neurons. The AWA neurons require TRPV channels to function, while AWC neurons do not require TRPV channels. Interestingly, PUFA-deficient worms are able to chemotax to odorants sensed by AWC neurons nearly as efficiently as wild-type worms.[49] Another ciliated sensory neuron, ASH, uses TRPV channels to stimulate an escape response to stimuli such as heavy metals, high osmolarity, and nose touch.[51] Behavioral responses to these stimuli are muted in mutants lacking C20 PUFAs. Both the chemotaxis and the avoidance behaviors were rescued in mutants supplemented with AA and EPA. Membrane formations and integrity of sensory neurons were not affected, suggesting that C20 PUFAs act as regulators of the TRP channel rather than membrane components in these neurons. Finally, calcium imaging of ASH neuronal response in *fat-3* mutants showed diminished calcium responses, while exogenous EPA elicited calcium responses and avoidance responses in *fat-3* mutants, bypassing the PUFA biosynthesis defects and providing evidence for 20-carbon omega-3 PUFAs as regulators of in vivo TRPV channel activity.[49]

Further studies in *C. elegans* revealed that an epoxide derivative of EPA, 17, 18-epoxyeicosatrienoic acid (17,18-EEG) is sufficient for TRPV channel function in ASH neurons in mutants lacking PUFAs.[52] In this study, both EPA and EEG were sufficient to rescue ASH-induced behavioral responses in PUFA-deficient mutants, while AA only rescued partially. Carrying their findings to human epithelial cells, the researchers found that dietary omega-3 fatty acids led to more TRPV channels in membranes, as well as increased membrane fluidity and enhanced TRPV function.[52]

While PUFAs are not as vital to the immediate sensory response to odorants detected by AWC neurons, they were shown, along with TRPV channels, to modulate the adaptation process in AWC neurons.[53] Adapting to changes in odors is crucial to survival and EGL-4, a protein kinase, is necessary to adapt to changes. EGL-4 accumulates in the nucleus in response to AWC activation, and PUFAs act downstream of this signal, together with TRPV channels, to dampen the behavioral response to long-term odor exposure.[53]

As opposed to a key role for omega-3 PUFAs in TRPV channels, a key role for the omega-6 PUFA AA was found for optimal activity of mechanoelectrical transduction channels.[54] *C. elegans* mutants lacking AA, including *fat-3*, *fat-1;fat-4*, and *fat-4* mutants, exhibit decreased responses to prolonged touch. Dietary supplementation of AA, but not EPA, was able to rescue touch response. Supplementation of a nonmetabolizable form of AA, eicosatetraenoic acid (ETA), was sufficient to restore touch response in *fat-1;fat-4* double mutants, suggesting that AA composition in phospholipids, rather than an eicosanoid signal produced from liberated AA, is mediating the effect. Using atomic force microscopy, the researchers found that PUFA deficiency yielded more rigid membranes, suggesting that PUFAs are necessary for optimal mechanical properties of membranes that are crucial for touch sensation.[54]

Neuronal plasticity, particularly ethanol avoidance, is affected when *C. elegans* are deficient in long-chain PUFAs, specifically EPA.[55] Using *fat-1*, *fat-3*, and *fat-4* desaturase mutants, it was determined that ethanol avoidance responses were being affected by the fatty acid composition. Because *fat-1* was unable to develop ethanol avoidance responses, it was hypothesized that EPA may be the PUFA necessary for normal behavior to ethanol. Supplementation of EPA in all three desaturase mutants was sufficient to rescue this activity, whereas AA was not. This study not only found that long-chain PUFAs were necessary for ethanol avoidance behaviors, but also that long-chain PUFAs, particularly EPA, are crucial during development for normal movement. *C. elegans* strains deficient in EPA during development were slow moving in adulthood and post developmental-supplementation of EPA was not adequate for movement rescue. These studies imply that genetic variation leading to altered PUFA composition, as well as dietary intake of fats, could influence behavioral responses to ethanol in humans.[55]

Many aspects of omega-3 and omega-6 PUFA biology have not yet been examined in *C. elegans*. For example, omega-3 PUFAs have been suggested to protect against neurodegenerative disease.[56] Several neurodegenerative diseases have been modeled in *C. elegans* by expressing human disease proteins associated with neurodegenerative diseases in the nematodes.[57] For example, to model Parkinson's disease in *C. elegans*, a mutant form of the protein α-synuclein fused to YFP was expressed in the body wall muscle.[58,59] These worms have decreased movement with age, increased YFP signaling with age, and reduced lifespan. Alzheimer's disease models in which Aβ peptide[60] and Tau proteins[61] were expressed in neurons and other tissues show similar decreases in movement and short lifespans. Other models include a Huntington's disease model in which human polyglutamine repeats are expressed in the body wall muscle of *C. elegans*,[62,63] as well as amyotrophic lateral sclerosis models, in which mutant forms of TDP43 and SOD1 are expressed in various tissues.[64–66] These models have been used to identify genetic suppressors and enhancers of the induced phenotypes, leading to a more clear understanding of the various disease mechanisms. It has not yet been determined whether dietary PUFA or PUFA deficiency would improve or worsen the outcomes of worms expressing any of these human disease mutations.

CONCLUSIONS

Fatty acids are important as energy sources and signaling molecules. The PUFAs, specifically the omega-3 PUFAs and omega-6 PUFAs, play important roles in many aspects of human physiology. Studies in *C. elegans* demonstrate requirements for PUFAs and PUFA-derived signaling molecules for growth, development, and optimal lifespan. Furthermore, even though *C. elegans* possesses a very simple nervous system, genetic analysis has revealed specific roles for PUFAs as mediators of TRP channels, mechanoelectrical transduction, and neuronal plasticity.

References

1. Corsi AK, Wightman B, Chalfie M. A transparent window into biology: a primer on *Caenorhabditis elegans*. *Genetics*. 2015;201(1):387–407.
2. Kamath RS, Fraser AG, Dong Y, et al. Systematic functional analysis of the *Caenorhabditis elegans* genome using RNAi. *Nature*. 2003;421(6920):231–237.
3. Wallis JG, Watts JL, Browse J. Polyunsaturated fatty acid synthesis: what will they think of next? *Trends Biochem Sci*. 2002;27(9):467.
4. Watts JL, Ristow M. Lipid and carbohydrate metabolism in *Caenorhabditis elegans*. *Genetics*. 2017;207(2):413–446.
5. Gracey AY, Fraser EJ, Li W, et al. Coping with cold: an integrative, multitissue analysis of the transcriptome of a poikilothermic vertebrate. *Proc Natl Acad Sci USA*. 2004;101(48):16970–16975.
6. Murray P, Hayward SA, Govan GG, Gracey AY, Cossins AR. An explicit test of the phospholipid saturation hypothesis of acquired cold tolerance in *Caenorhabditis elegans*. *Proc Natl Acad Sci USA*. 2007;104(13):5489–5494.
7. Turk HF, Chapkin RS. Membrane lipid raft organization is uniquely modified by n-3 polyunsaturated fatty acids. *Prostag Leukot Essn Fatty Acids*. 2013;88(1):43–47.
8. Peyou-Ndi MM, Watts JL, Browse J. Identification and characterization of an animal delta(12) fatty acid desaturase gene by heterologous expression in Saccharomyces cerevisiae. *Arch Biochem Biophys*. 2000;376(2):399–408.
9. Spychalla JP, Kinney AJ, Browse J. Identification of an animal omega-3 fatty acid desaturase by heterologous expression in Arabidopsis. *Proc Natl Acad Sci USA*. 1997;94(4):1142–1147.
10. Watts JL, Browse J. Genetic dissection of polyunsaturated fatty acid synthesis in *Caenorhabditis elegans*. *Proc Natl Acad Sci USA*. 2002;99(9):5854–5859.
11. Zhou XR, Horne I, Damcevski K, Haritos V, Green A, Singh S. Isolation and functional characterization of two independently-evolved fatty acid Delta 12-desaturase genes from insects. *Insect Mol Biol*. 2008;17(6):667–676.
12. Malcicka M, Ruther J, Ellers J. De novo synthesis of linoleic acid in multiple Collembola species. *J Chem Ecol*. 2017;43(9):911–919.
13. Menzel R, Geweiler D, Sass A, Simsek D, Ruess L. Nematodes as important source for omega-3 long-chain fatty acids in the soil food web and the impact in nutrition for higher trophic levels. *Front Ecol Evol*. 2018;6:96.
14. Brock TJ, Browse J, Watts JL. Genetic regulation of unsaturated fatty acid composition in *C. elegans*. *PLoS Genet*. 2006;2(7):e108.
15. Entchev EV, Schwudke D, Zagoriy V, et al. LET-767 is required for the production of branched chain and long chain fatty acids in *Caenorhabditis elegans*. *J Biol Chem*. 2008;283(25):17550–17560.
16. Zhang YR, Wang HZ, Zhang JJ, et al. The cytochrome b5 reductase HPO-19 is required for biosynthesis of polyunsaturated fatty acids in *Caenorhabditis elegans*. *BBA-Mol Cell Biol L*. 2016;1861(4):310–319.
17. Kniazeva M, Sieber M, McCauley S, Zhang K, Watts JL, Han M. Suppression of the ELO-2 FA elongation activity results in alterations of the fatty acid composition and multiple physiological defects, including abnormal ultradian rhythms, in *Caenorhabditis elegans*. *Genetics*. 2003;163(1):159–169.
18. Perez CL, Van Gilst MR. A 13C isotope labeling strategy reveals the influence of insulin signaling on lipogenesis in *C. elegans*. *Cell Metab*. 2008;8(3):266–274.
19. Dancy BC, Chen SW, Drechsler R, Gafken PR, Olsen CP. 13C- and 15N-labeling strategies combined with mass spectrometry comprehensively quantify phospholipid dynamics in *C. elegans*. *PLoS ONE*. 2015;10(11):e0141850.

20. Brock TJ, Browse J, Watts JL. Fatty acid desaturation and the regulation of adiposity in *Caenorhabditis elegans*. *Genetics*. 2007;176(2):865–875.
21. Watts JL, Phillips E, Griffing KR, Browse J. Deficiencies in C20 polyunsaturated fatty acids cause behavioral and developmental defects in *Caenorhabditis elegans* fat-3 mutants. *Genetics*. 2003;163(2):581–589.
22 Deline ML, Vrablik TL, Watts JL. Dietary supplementation of polyunsaturated fatty acids in *Caenorhabditis elegans*. J Visual Exp. 2013(81), e50879.
23. Watts JS, Morton DG, Kemphues KJ, Watts JL. The biotin-ligating protein BPL-1 is critical for lipid biosynthesis and polarization of the *Caenorhabditis elegans* embryo. *J Biol Chem*. 2018;293(2):610–622.
24. Olson SK, Greenan G, Desai A, Muller-Reichert T, Oegema K. Hierarchical assembly of the eggshell and permeability barrier in *C. elegans*. *J Cell Biol*. 2012;198(4):731–748.
25. Chen WW, Yi YH, Chien CH, et al. Specific polyunsaturated fatty acids modulate lipid delivery and oocyte development in *C. elegans* revealed by molecular-selective label-free imaging. *Sci Rep*. 2016;6:32021.
26. Watts JL, Browse J. Dietary manipulation implicates lipid signaling in the regulation of germ cell maintenance in *C. elegans*. *Dev Biol*. 2006;292(2):381–392.
27. Webster CM, Deline ML, Watts JL. Stress response pathways protect germ cells from omega-6 polyunsaturated fatty acid-mediated toxicity in *Caenorhabditis elegans*. *Dev Biol*. 2013;373(1):14–25.
28. Deline M, Keller J, Rothe M, Schunck WH, Menzel R, Watts JL. Epoxides derived from dietary Dihomo-gamma-Linolenic acid induce germ cell death in *C. elegans*. *Sci Rep*. 2015;5:15417.
29. Buczynski MW, Dumlao DS, Dennis EA. Thematic review series: proteomics. An integrated omics analysis of eicosanoid biology. *J Lipid Res*. 2009;50(6):1015–1038.
30. Funk CD. Prostaglandins and leukotrienes: advances in eicosanoid biology. *Science*. 2001;294(5548):1871–1875.
31. Spector AA, Kim HY. Cytochrome P450 epoxygenase pathway of polyunsaturated fatty acid metabolism. *Biochim Biophys Acta*. 2015;1851(4):356–365.
32. Innes JK, Calder PC. Omega-6 fatty acids and inflammation. *Prostaglandins Leukot Essent Fatty Acids*. 2018;132:41–48.
33. Nebert DW, Wikvall K, Miller WL. Human cytochromes P450 in health and disease. *Philos Trans R Soc Lond B Biol Sci*. 2013;368(1612):20120431.
34. Xu JY, Chen C. Endocannabinoids in synaptic plasticity and neuroprotection. *Neuroscientist*. 2015;21(2):152–168.
35. Maccarrone M, Bab R, Biro T, et al. Endocannabinoid signaling at the periphery: 50 years after THC. *Trends Pharmacol Sci*. 2015;36(5):277–296.
36. Kubagawa HM, Watts JL, Corrigan C, et al. Oocyte signals derived from polyunsaturated fatty acids control sperm recruitment in vivo. *Nat Cell Biol*. 2006;8(10):1143–1148.
37. Edmonds JW, Prasain JK, Dorand D, et al. Insulin/FOXO signaling regulates ovarian prostaglandins critical for reproduction. *Dev Cell*. 2010;19(6):858–871.
38. Hoang HD, Prasain JK, Dorand D, Miller MA. A heterogeneous mixture of F-series prostaglandins promotes sperm guidance in the *Caenorhabditis elegans* reproductive tract. *PLoS Genet*. 2013;9(1):e1003271.
39. McKnight K, Hoang HD, Prasain JK, et al. Neurosensory perception of environmental cues modulates sperm motility critical for fertilization. *Science*. 2014;344(6185):754–757.
40. Lam SM, Wang Z, Li J, Huang X, Shui G. Sequestration of polyunsaturated fatty acids in membrane phospholipids of *Caenorhabditis elegans* dauer larva attenuates eicosanoid biosynthesis for prolonged survival. *Redox Biol*. 2017;12:967–977.
41. Qi W, Gutierrez GE, Gao X, et al. The omega-3 fatty acid alpha-linolenic acid extends *Caenorhabditis elegans* lifespan via NHR-49/PPARalpha and oxidation to oxylipins. *Aging Cell*. 2017;16(5):1125–1135.
42. Oliveira RP, Porter Abate J, Dilks K, et al. Condition-adapted stress and longevity gene regulation by *Caenorhabditis elegans* SKN-1/Nrf. *Aging Cell*. 2009;8(5):524–541.
43. Lucanic M, Held JM, Vantipalli MC, et al. N-acylethanolamine signalling mediates the effect of diet on lifespan in *Caenorhabditis elegans*. *Nature*. 2011;473(7346):226–229.
44. Bazinet RP, Laye S. Polyunsaturated fatty acids and their metabolites in brain function and disease. *Nat Rev Neurosci*. 2014;15(12):771–785.
45. Sengupta P, Samuel AD. *Caenorhabditis elegans*: a model system for systems neuroscience. *Curr Opin Neurobiol*. 2009;19(6):637–643.
46. Lesa GM, Palfreyman M, Hall DH, et al. Long chain polyunsaturated fatty acids are required for efficient neurotransmission in C. elegans. *J Cell Sci*. 2003;116(Pt 24):4965–4975.

47. Marza E, Long T, Saiardi A, et al. Polyunsaturated fatty acids influence synaptojanin localization to regulate synaptic vesicle recycling. *Mol Biol Cell*. 2008;19(3):833–842.
48. Antollini SS, Barrantes FJ. Fatty acid regulation of voltage- and ligand-gated ion channel function. *Front Physiol*. 2016;7:573.
49. Kahn-Kirby AH, Dantzker JL, Apicella AJ, et al. Specific polyunsaturated fatty acids drive TRPV-dependent sensory signaling in vivo. *Cell*. 2004;119(6):889–900.
50. Bargmann CI, Hartwieg E, Horvitz HR. Odorant-selective genes and neurons mediate olfaction in *C. elegans*. *Cell*. 1993;74(3):515–527.
51. Colbert HA, Smith TL, Bargmann CI. OSM-9, a novel protein with structural similarity to channels, is required for olfaction, mechanosensation, and olfactory adaptation in *Caenorhabditis elegans*. *J Neurosci*. 1997;17(21):8259–8269.
52. Caires R, Sierra-Valdez FJ, Millet JRM, et al. Omega-3 fatty acids modulate TRPV4 function through plasma membrane remodeling. *Cell Rep*. 2017;21(1):246–258.
53. O'Halloran DM, Altshuler-Keylin S, Lee JI, L'Etoile ND. Regulators of AWC-mediated olfactory plasticity in *Caenorhabditis elegans*. *PLoS Genet*. 2009;5(12):e1000761.
54. Vasquez V, Krieg M, Lockhead D, Goodman MB. Phospholipids that contain polyunsaturated fatty acids enhance neuronal cell mechanics and touch sensation. *Cell Rep*. 2014;6(1):70–80.
55. Raabe RC, Mathies LD, Davies AG, Bettinger JC. The omega-3 fatty acid Eicosapentaenoic acid is required for normal alcohol response behaviors in *C. elegans*. *PLoS One*. 2014;9(8).
56. Kerdiles O, Laye S, Calon F. Omega-3 polyunsaturated fatty acids and brain health: preclinical evidence for the prevention of neurodegenerative diseases. *Trends Food Sci Technol*. 2017;69:203–213.
57. Kraemer B, Schellenberg GD. Using *Caenorhabditis elegans* models of neurodegenerative disease to identify neuroprotective strategies. *Int Rev Neurobiol*. 2007;77:219–246. https://doi.org/10.1016/S0074-7742(06)77007-6.
58. van Ham TJ, Thijssen KL, Breitling R, Hofstra RM, Plasterk RH, Nollen EA. *C. elegans* model identifies genetic modifiers of alpha-synuclein inclusion formation during aging. *PLoS Genet*. 2008;4(3):e1000027.
59. Cooper JF, Van Raamsdonk JM. Modeling Parkinson's disease in *C. elegans*. *J Parkinsons Dis*. 2018;8(1):17–32. https://doi.org/10.3233/JPD-171258.
60. Link CD, Taft A, Kapulkin V, et al. Gene expression analysis in a transgenic *Caenorhabditis elegans* Alzheimer's disease model. *Neurobiol Aging*. 2003;24(3):397–413.
61. Kraemer BC, Zhang B, Leverenz JB, Thomas JH, Trojanowski JQ, Schellenberg GD. Neurodegeneration and defective neurotransmission in a *Caenorhabditis elegans* model of tauopathy. *Proc Natl Acad Sci USA*. 2003;100(17):9980–9985.
62. Satyal SH, Schmidt E, Kitagawa K, et al. Polyglutamine aggregates alter protein folding homeostasis in *Caenorhabditis elegans*. *Proc Natl Acad Sci USA*. 2000;97(11):5750–5755.
63. Machiela E, Dues DJ, Senchuk MM, Van Raamsdonk JM. Oxidative stress is increased in C. elegans models of Huntington's disease but does not contribute to polyglutamine toxicity phenotypes. *Neurobiol Dis*. 2016;96:1–11.
64. Oeda T, Shimohama S, Kitagawa N, et al. Oxidative stress causes abnormal accumulation of familial amyotrophic lateral sclerosis-related mutant SOD1 in transgenic *Caenorhabditis elegans*. *Hum Mol Genet*. 2001;10(19):2013–2023.
65. Liachko NF, Guthrie CR, Kraemer BC. Phosphorylation promotes neurotoxicity in a *Caenorhabditis elegans* model of TDP-43 proteinopathy. *J Neurosci*. 2010;30(48):16208–16219.
66. Veriepe J, Fossouo L, Parker JA. Neurodegeneration in *C. elegans* models of ALS requires TIR-1/Sarm1 immune pathway activation in neurons. *Nat Commun*. 2015;6:7319.

CHAPTER

8

Fatty Acids and the Hippocampus

H.M. Francis, R.J. Stevenson

Department of Psychology, Macquarie University, Sydney, NSW, Australia

INTRODUCTION

The typical Western diet is high in saturated fatty acids (SFAs) and has reduced intake of polyunsaturated fatty acids (PUFAs). It is becoming recognized that the poor dietary practices typical of the modern world may have important consequences for cognition, and perhaps especially learning and memory, and may be a risk factor for development of neurodegenerative diseases such as Alzheimer's. Our focus here is on long-term memory function and its impairment, which in turn is closely tied to the integrity of the hippocampus. In particular, we explore two claims in this chapter. First, higher levels of dietary SFA can adversely impact learning and memory function, by impairing synaptic plasticity, elevating oxidative stress (OS) and causing neuroinflammation, which in the longer term may then contribute to the onset of Alzheimer's disease. Second, diets rich in n-3 fatty acids [or omega-3 PUFA; i.e., eicosapentaenoic acid (EPA) and docosahexaenoic acid (DHA)] may be able to remediate some of the harms of SFA consumption, as well as having a neuroprotective effect.

In structuring this review we have tried to draw as generally as we can on the relevant animal and human literature. Because this is an emerging literature, especially in humans, we have chosen to include all of the studies that utilize measures likely to be sensitive to changes in hippocampal function. We have arranged the review into two basic sections, the first addressing the evidence for the claim that SFA can impair memory function and how these effects may be instantiated in the brain, the second evaluating the evidence that n-3 fatty acids may have a neuroprotective effect.

SFAS, MEMORY, AND THE HIPPOCAMPUS–EVIDENCE FROM ANIMAL MODELS

There is a substantial amount of animal research demonstrating an effect of high levels of dietary SFA on performance on tests of hippocampal-dependent learning and memory function. In one of the earliest studies in this field, Greenwood and Winocur[1] fed rats a diet

Omega Fatty Acids in Brain and Neurological Health
https://doi.org/10.1016/B978-0-12-815238-6.00008-0

consisting of 20% lard as a source of SFA. This was designed to represent the upper limits of typical Western consumption patterns. Rats maintained on this diet for 3 months were impaired on a series of learning and memory-related tasks: radial arm maze, Hebb-Williams maze, and a variable interval delayed alternation task.

Spatial memory tasks are a widely accepted model for assessing hippocampal dysfunction in rodents, with the most common being the Morris water maze (MWM). The MWM involves placing the rodent in a pool of opaque water that contains an escape platform hidden beneath the surface. Over several trials, the animal should learn and remember the location of the platform based on spatial cues. The task was originally developed by Morris[2] who showed that memory for the platform location was impaired in rats with hippocampal lesions. Performance on this task has been compared after rats have been maintained on diets that are high in SFA or low fat standard chow. Impairment on the MWM task following high SFA diet consumption is now considered a robust effect as the overwhelming majority of studies that have been published to date have shown this effect (see Ref. 3 for a review). Impairment on the MWM has been reported in rodents exposed to a high SFA diet from 1 month[4] to 8 months.[5] These studies typically use lard, corn oil, and vegetable oil as a source of saturated fat, ranging from 21% fat[6] to 60% fat.[7]

Animal studies have also investigated whether exposure to a high SFA diet impairs performance on other tests of learning and memory that are known to be hippocampal dependent. Findings have been equivocal regarding the Barnes Maze. Mice fed a 60% fat diet for 6 months were impaired on the Barnes Maze compared to controls fed a 10% fat diet.[8] However, another study showed mice fed a 45% fat diet showed no difference in Barnes Maze performance compared to controls fed standard chow.[9] Several studies have shown impairment on the radial arm maze following consumption of a 25% fat diet, and a high SFA, high cholesterol diet.[10,11] We note that most studies report on fat in general, however in one study that compared two high fat diets; rats fed a high PUFA diet were intact whereas those fed a high SFA diet were impaired on the radial arm maze.[12]

There are also a number of studies that demonstrate hippocampal-dependent memory impairment following maintenance on diets that are high in SFA but also high in refined sugars.[4,5] While these studies are more representative of the typical Western diet, it is difficult to attribute the effects to the separate dietary components. Nevertheless, these results do lend weight to the general finding that diets high in SFA have a detrimental impact on memory functioning.

While most studies have observed increases in body weight of the animals concurrently with impaired memory performance, there are some studies that have demonstrated high SFA-related diet impairment without an increase in body weight[13,14] or in the absence of metabolic syndrome.[15] Thus, it appears likely that the effect of high SFA exposure on memory function may be independent of effects on body weight or metabolic disorders often associated with SFA consumption.

Neurological Impact of Saturated Fat Consumption

There are several neurological changes that have been shown to occur as a result of high SFA consumption and a number of studies have linked these neurological changes to the dietary-induced memory impairments described in the previous section. Such changes include reductions in brain-derived neurotrophic factor (BDNF), OS, neuroinflammation, and impaired blood brain barrier integrity.

Brain-Derived Neurotrophic Factor

Learning and memory requires synaptic plasticity: activity dependent changes in the strength of synaptic connections. BDNF is a neurotrophin that is abundant in the hippocampus and cerebral cortex[16] and is a major player in the molecular mechanisms underlying synaptic plasticity.[17] In particular, BDNF has been shown to facilitate synaptic transmission[18] and is essential for long-term potentiation.[19] Animals with deletion in one copy of the BDNF gene show impaired performance on the MWM.[19] Additionally, the BDNF val66met polymorphism, which affects the release of BDNF, results in impaired fear extinction learning[20] and episodic memory function[21] and reduces hippocampal activity during memory encoding and retrieval processes.[22]

Several studies link diet-induced impaired memory function to reductions in BDNF in the hippocampus. Molteni et al.[4] showed that rats maintained on a high SFA and refined sugar diet for 2 years had decreased levels of BDNF mRNA in the hippocampus after 2 months and levels decreased, with the lowest levels of BDNF mRNA observed after 2 years. Importantly, the results of MWM testing performed at 2 months correlated with the hippocampal BDNF levels. Stranahan et al.[5] found that rats maintained on a high SFA and refined sugar diet for 8 months had reduced hippocampal BDNF, which was accompanied by impaired MWM learning, as well as reduced long-term potentiation and dendritic spine density. This research provides further support for the notion that diet-induced impairment to hippocampal function is due to BDNF-mediated effects on synaptic plasticity. In another study, decreased expression of BDNF and its receptor TrkB in diet-induced obese mice was normalized when mice were placed on a low fat or energy-restricted diet.[23]

Notably, in the above studies the effects of SFA compared to refined sugars cannot be separated. However, Wu et al.[24] showed that 2 months maintenance on a diet differing only in regards to higher SFA content impaired learning on MWM and this was accompanied by reductions of hippocampal BDNF, as well as its downstream effectors synapsin I and cyclic AMP-response element-binding protein (CREB). BDNF-dependent phosphorylation of these proteins can affect synaptic plasticity through promotion of synaptic vesicle exocytosis,[25] the establishment and maintenance of functional synapses[26] and gene expression.[27,28]

Oxidative Stress

Oxidative phosphorylation is the process through which energy, in the form of adenosine triphosphate (ATP), is generated.[29] While oxidative phosphorylation is essential for energy metabolism, it produces reactive oxygen species (ROS) such as superoxide and hydrogen peroxide,[29] which can cause damage to many cellular components including DNA, proteins, and lipids.[30] Typically, antioxidants such as superoxide dismutase, vitamin E, vitamin C, and glutathione, reduce the levels of ROS to prevent free-radical-mediated damage.[30] An imbalance between levels of ROS and the rate at which the damage caused by ROS can be repaired causes oxidative stress (OS).

Several studies have observed increased OS following high SFA diets. Rats fed a diet comprised of 18% lard as a source of SFA for a period of 5 months showed increased levels of ROS in the cerebral cortex.[30] NADPH oxidase is one of the predominant sources of ROS and the above study found high dietary SFA resulted in increased cerebral cortex levels of NADPH oxidase subunits. This suggests high levels of SFA might cause OS via the NADPH pathway. In support of this, endothelial cell dysfunction in obese Zucker rats resulted from OS due to overactivation of NADPH oxidase.[31]

Diet-induced OS has also been associated with cognitive impairment. For example, markers of OS were increased in rats consuming a HF diet and accompanied by impaired MWM performance, with this effect exacerbated in offspring of dams who had also consumed a HF diet.[32] Wu et al.[24] showed that consumption of a high SFA diet for 2 months caused an increase in two markers of OS, oxidized protein levels and nucleic acids, in the rat hippocampus. SFA-fed rats were impaired on the MWM task and a strong negative relationship was observed between oxidized protein levels and MWM performance, suggesting OS contributed to cognitive impairment. Interestingly, OS in this study was associated with decreased levels of BDNF, synapsin I, and CREB in the hippocampus. The authors therefore suggested that OS mediates the effects of diet on cognition by reducing levels of BDNF and its downstream effectors, subsequently impairing synaptic plasticity and cognitive function.

Neuroinflammation

The neurophysiological mechanisms underlying learning and memory, such as long-term potentiation, are susceptible to the effects of neuroinflammation.[33,34] Obesity has been associated with chronic inflammation in the brain, with studies finding elevated levels of inflammatory markers in obese compared to lean individuals.[35,36] There is a quantity of evidence from animal studies that consumption of dietary SFA plays a role in driving this difference. For example, mice consuming a high SFA diet showed microglial reactivity and elevated levels of cytokines (markers of inflammation), which were associated with reduced BDNF and impaired cognitive performance on the T-stone maze task.[37] High SFA-fed rats that were mothered by high SFA-fed dams showed increases in inflammatory signaling in the brain.[32] Rats maintained on a high SFA diet for 5 months had increased cortical levels of markers of inflammation, and this was associated with impaired performance on the MWM.[38] Importantly, the two aforementioned studies were also mentioned above as having also observed increased OS. Thus, these two factors are likely to work together as mechanisms underlying the effects of high SFA diet on learning and memory.

SFAS, MEMORY, AND THE HIPPOCAMPUS: HUMAN DATA

Several studies in a human population have also provided evidence that high SFA intake is implicated in hippocampal impairment. These studies show impairment on tasks that assess long-term memory function and are therefore assumed to reflect hippocampal functioning. While these studies also tended to assess functioning of other cognitive domains (e.g., attention and executive function), they are not the focus of this chapter therefore they are not described in detail.

Devore et al.[39] found that in women with a mean age of 74 years and with type 2 diabetes, higher intake of SF since midlife was associated with significantly worse performance on cognitive tests measuring immediate and delayed memory, working memory, and verbal fluency. In a population of men and women aged 45–70 years old, higher rates of SFA intake were associated with a trend toward greater impairment of memory.[40] Higher SFA intake was also associated with poorer global cognitive function and memory in older adults.[41]

In a longitudinal study, high SFA intake over a period of 6 years was associated with poorer performance on delayed memory tasks, as well as other aspects of cognitive functioning.[42] Importantly, this relationship was present after adjusting for age, sex, race, education, and antioxidant intake. A study of women aged >45 found that verbal memory performance assessed over the course of 4 years was associated with higher SFA intake, but not total fat or PUFA intake.[43] These results suggest that the effects on memory function are specific to SFA and not fat intake overall. However, a general weakness in the literature is a difficulty in separating the effects of specific nutrients given that individuals who consume a high SFA diet will also tend to have an overall greater fat intake, as well as greater sugar intake. Indeed, Francis and Stevenson[44] showed that hippocampal-dependent memory function was impaired in individuals who self-reported high intake of SFA and sugar compared to individuals who self-reported low intake of these nutrients.

There are few intervention studies examining the effects of increased dietary SFA intake. Elderly adults who consumed a high SFA (>25%) diet for 4 weeks showed poorer visuospatial memory performance than those on a low SFA (<7%) and high glycemic index diet,[45] though there was no difference between groups on measures of verbal long-term memory. Young adults who consumed a Western-style breakfast across 4 days showed significant reductions in hippocampal-dependent learning and memory compared to those consuming a healthier breakfast.[41] While the breakfast was both high in sugar and SFA, diet diaries completed across the intervention showed that those consuming the breakfast compensated throughout the rest of the day by consuming fewer carbohydrates, whereas SFA intake remained elevated compared to the control group. These findings support the animal literature where there is greater ability to control dietary factors such as specific nutrient intake.

OMEGA-3 FATTY ACIDS, MEMORY, AND THE HIPPOCAMPUS: EVIDENCE FROM ANIMAL STUDIES

Omega-3 fatty acids are prominent in neural tissue and have profound influences in the CNS.[46,47] For instance, DHA is the predominant structural n-3 PUFA in the brain and is an important component of the phospholipid bilayer that surrounds all neurons.[48] DHA is essential for normal CNS development and is vital to brain structure and function.[49] EPA is detected as a trace component in the brain and is used to synthesize DHA.[50] The brain is unable to efficiently synthesize these PUFAs by itself, therefore, they must be obtained through dietary sources.[51] Fish oil is a rich source of DHA and EPA and its use as a dietary supplement has therefore been the focus of much research in this field. Other good sources of n-3 fatty acids include krill oil, nuts (notably walnuts), and plant seeds and oils such as perilla and flaxseed.[52–55]

There is a wealth of evidence from animal studies showing that n-3 enriched diets can enhance learning and memory function. Rats fed a diet with 1.25% DHA showed improved MWM performance compared to controls after only 7 days on the diet.[56] Similarly, mice treated with 50 and 100 mg/kg/day for 7 weeks had significantly improved spatial memory on the passageway water maze test.[57] Improved spatial memory performance was also demonstrated in rats administered a combination of fish oil and soybean oil for 10 weeks and in rats-fed diets rich in alternative sources of dietary n-3 fatty acids such as walnuts[58] and perilla oil.[59] Thus, diets with high levels of n-3 fatty acids from a variety of sources that are particularly rich in DHA and EPA have consistently been shown to be beneficial to hippocampal-dependent memory function.

The role of n-3 fatty acids in learning and memory has been demonstrated through various studies showing changes to performance on spatial memory tasks following manipulation of dietary n-3 availability. Methods of inducing dietary DHA deficiency typically involve removing all n-3 fatty acids from the diet throughout gestation and lactation so that the offspring then has lower levels of DHA. Employing this approach, Moriguchi et al.[60] demonstrated this procedure reduced DHA by over 80% in second- and third-generation DHA-deprived rats and that deficiency was accompanied by impairment in learning and memory performance on the MWM task. These findings have been replicated in subsequent studies in the same laboratory.[61–63] Adding weight to the argument that it is the deprivation of DHA specifically that is responsible for the effects, substitution of DHA with omega-6 PUFAs resulted in impairment of MWM performance.[64] Further, the effect of n-3 fatty acid deficiency on MWM performance can be reversed by provision of an adequate diet containing DHA.[60,65]

Given the evidence for a role of n-3 fatty acids in hippocampal learning and memory, n-3 supplementation has therapeutic potential in remediating memory deficits induced by various conditions. The conditions that will be the focus of this chapter are associated with neuroinflammation, synaptic plasticity, and OS.

DHA and EPA have antiinflammatory properties that have been extensively documented in nonneural tissues and more recently in neural tissues.[66] In the hippocampus specifically, decrease in dietary DHA has been shown to cause increased production of thepro-inflammatory cytokine interleukin-6.[67] Therefore, DHA and EPA may be relevant candidates for reversing impairment in learning and memory due to conditions associated with neuroinflammation. For instance, age-related impairments in learning and memory are associated with neuroinflammatory processes in the hippocampus.[68] Labrousse et al.[69] demonstrated that 2 months supplementation with an n-3 PUFA diet resulted in increased levels of DHA and EPA in the brain, as well as reduction in hippocampal levels of the pro-inflammatory cytokines interleukin-6 and tumor necrosis factor α. These hippocampal changes were accompanied by remediation of age-related deficits in spatial memory on the Y maze task.

N-3 fatty acids, particularly DHA, are implicated in synaptic plasticity: changes in the strength of synaptic connections that are necessary for memory formation and storage. Of particular relevance to this chapter is the role of n-3 fatty acids on synaptic plasticity in the hippocampus specifically. DHA supplementation enhances processes related to synaptic plasticity in the hippocampus, such as increased glutamatergic synaptic activity, synaptogenesis, and expression of proteins known to be involved in synaptic plasticity.[56,59,70] Similarly, EPA supplementation improved long-term potentiation in the hippocampus.[71] There is also evidence for involvement of n-3 fatty acids in regulation of synaptic plasticity-related genes such as BDNF and CREB. For instance, n-3 deficiency reduced the levels of BDNF, BDNF signaling through its TrkB receptor and activation of its downstream effector CREB.[72] Conversely, DHA supplementation and krill oil supplementation have both been shown to increase expression of BDNF.[57,73] In light of these findings, it has been suggested that n-3 fatty acids have potential as strategies for preventing memory impairments related to disorders that reduce synaptic plasticity in the hippocampus. Indeed, fish oil supplementation ameliorated spatial learning deficits induced by diabetes by increasing neuronal excitability.[74] Fish oil supplementation also normalized levels of BDNF, synapsin I, and CREB following TBI in rats, and counteracted learning deficits on the MWM task.[24] Agrawal and Gomez-Pinilla[75] showed that high fructose consumption impacted on synaptic plasticity by decreasing phosphorylation of CREB, synapsin I, and synaptophysin levels. All of these parameters were attenuated by the presence of dietary n-3 fatty acid.

The antioxidant properties of n-3 fatty acids render them an effective means of neuroprotection against the neuronal damage caused by ROS. For instance, fish oil supplementation has been shown to protect against OS and mitochondrial dysfunction induced by rotenone and 3-nitropropionic acid.[76,77] In the hippocampus specifically, fish oil supplementation reduced OS and counteracted learning impairment in a rat model of traumatic brain injury (TBI).[24] OS is also characteristic of aging and is associated with a deficit in cognitive function.[78] Dietary n-3 fatty acid supplementation has the ability to modulate the age-related increase in oxidative changes in the hippocampus, as well as the associated decrease in spatial learning.[79]

In conclusion, there is evidence that n-3 fatty acids are involved in hippocampal-dependent learning and memory function and that they are effective in remediating deficits caused by TBI, aging, high fructose intake, and diabetes. In light of the fact that the underlying mechanisms for their ability to remediate cognitive symptoms (i.e., neuroinflammation, synaptic plasticity, and OS) are the same mechanisms underlying deleterious effects of high SFA diet on learning and memory (see *Section* Neurological Impact of Saturated Fat Consumption), it is possible that n-3 fatty acids could have a therapeutic use in remediating damage to the hippocampus caused by SFAs.

CAN N-3 FATTY ACIDS REMEDIATE DAMAGE TO THE HIPPOCAMPUS IN HUMANS?

Omega-3 Fatty Acids and Their Effect on Memory and Cognition

Dietary n-3 fatty acid consumption can be measured indirectly by assessing oily fish consumption using food frequency questionnaires or by measuring plasma levels of DHA, EPA, ALA, and total n-3 fatty acids. There are several epidemiological studies which have investigated whether dietary intake of n-3 fatty acids (assessed using self-report questionnaires, or plasma levels) is associated with enhanced performance on neuropsychological tests known to tap hippocampal function. In one study, Inuit children had plasma n-3 levels established at birth from cord blood, and when aged 11.[80] Performance on the California Verbal Learning Test (CVLT) correlated with DHA levels from cord blood and from the most recent plasma sample, controlling for a range of factors. Importantly, the CVLT recognition score has been shown in a number of studies to be sensitive to hippocampal integrity (e.g., Ref. 81,82). While this cross-sectional and longitudinal study suggests that higher plasma n-3s are beneficial for memory, not all studies support this conclusion.

Several studies have looked for an association between performance on hippocampal-dependent neuropsychological tasks and estimation of n-3 intake. In a cross-sectional study,[40] n-3 intake was assessed from dietary recall in a large Dutch sample of 45- to 70-year-old participants, with memory performance obtained using a delayed verbal recall measure. In this case, no relationship was observed between performance on this memory test and n-3 intakes, although general cognitive impairment was found to be associated with n-3 intake. A cross-sectional study in healthy middle-aged adults examined the relationship between plasma estimates of n-3 fatty acids and performance on a range of neuropsychological tests.[83] Memory measures known to be associated with hippocampal integrity—logical memory and verbal paired associates[84,85]—were found to correlate positively with plasma estimates of DHA, but not EPA or ALA. These relationships withstood control for confounding variables including intelligence, blood pressure, age, and gender. These controls are important, especially as higher n-3 intake may reflect *generally* more healthful dietary choices as well as being a proxy for intelligence (i.e., following health-related guidelines). Andruchow et al.[86] showed that a higher ratio of n-3 to n-6 fatty acid intake correlated with better hippocampus-dependent spatial memory and faster learning on a virtual navigation task.

Other studies have investigated whether self-reported n-3 intake is associated with structural differences in hippocampal volume. Fifty-five healthy middle-aged adults underwent a structural brain scan, so that gray matter volume in three structures (the amygdala, hippocampus, and anterior cingulate cortex) could be correlated with estimates of their n-3 fatty acid intake taken from two dietary recall interviews.[87] Even after controlling for confounding variables (e.g., sex, age, total gray matter volume), higher n-3 intake was associated with a larger gray matter volume in the hippocampus, amygdala, and anterior cingulate cortex.

There have been several trials to investigate whether n-3 supplementation can improve memory function. A meta-analysis of 15 intervention trials showed that n-3 supplementation had no effect on episodic memory among adults with no cognitive impairment at baseline. However, in adults with mild memory complaints, DHA supplementation, alone or in combination with EPA, significantly improved performance on tests of episodic memory.[88] Improvements were observed from at least 1 g/day of DHA and EPA.

In sum, epidemiological studies show that habitual intake of n-3 fatty acids through diet is associated with a healthier hippocampus. This is reflected in better performance on memory measures known to be sensitive to damage to this structure, and in one case on a measure of hippocampal size obtained by neuroimaging. There is also evidence that supplementation with n-3 fatty acids can remediate memory function for individuals suffering mild memory problems, but there does not appear to be any benefit for individuals with intact cognition.

Omega-3 Fatty Acids and Depression

Abnormalities of the hippocampus accompany depression. Several studies have found that a smaller hippocampus is associated with depression, although there is uncertainty as to whether this may precede the onset of depressive symptoms or may be a consequence of depression (e.g., 89). Whatever the case, there is good evidence that antidepressant medications act to increase neurogenesis in the hippocampus and that this happens over a time course that parallels remission of symptoms.[90] Together, this would suggest that any agent that acts to reduce the likelihood of depression or to act in a curative manner, is likely to be affecting the hippocampus.

Using a food frequency approach to assess self-reported intake of foods high in n-3 fatty acids has provided only weak evidence that a diet rich in n-3 fatty acids is protective against depression.[91] More support comes from studies using plasma estimates. Conklin et al.[92] found that plasma n-3 fatty acid levels in otherwise healthy hypercholesterolemic individuals were associated with scores on a standardized depression inventory and on related measures of affect regulation. In particular, higher plasma levels of EPA, DHA, and overall n-3s were associated with better mood regulation. Conklin et al.[93] also found that higher EPA levels were associated with better affect regulation in healthy individuals. The significance of these observations in healthy people is that plasma levels of n-3 fatty acids are depleted in the acute phase of a major depressive episode (e.g., 94), suggesting that low levels of n-3 fatty acids may be a risk factor for the onset of depression.

A meta-analysis of the use of n-3 fatty acids to treat major depression indicates that they exert a significant curative effect.[95] This appears to be mediated by EPA and it seems to work best when combined with an antidepressant drugs. A more recent meta-analysis, correcting for publication bias, came to the same conclusion.[96] Higher EPA dose and concurrent antidepressant use were significantly associated with a better outcome of n-3 supplementation.

In sum, lower n-3 fatty acid levels seem to be associated with a greater risk for depression, and elevating these levels seems to assist recovery. As the hippocampus is significantly involved in depression and the drugs which target depression seem to exert some of their therapeutic effect by upregulating neurogenesis in this structure, we tentatively suggest that n-3 fatty acids exert some of their beneficial effects by targeting this structure.

CONCLUSION

This chapter explored two claims. First, SFA have an adverse effect on hippocampal-dependent learning and memory. Second, n-3 fatty acids have a remedial or neuroprotective effect. Both claims are generally supported in the literature, especially by the animal data, where it is much easier to directly test causality. In addition, the animal literature has also provided a number of plausible candidate pathways by which SFA and n-3 fatty acids could exert their effects on the hippocampus. Currently, these include OS, reduced levels of BDNF, and dietary-induced neuroinflammatory responses. While the human data are much less developed, there is clear epidemiological data linking elevated SFA intake to poorer hippocampal-dependent memory performance, and some albeit indirect evidence suggesting that n-3 fatty acids may exert protective effects on cognition, including memory.

Two conclusions emerge from this work. First, there is an important need to establish the causal effects of SFA and n-3 fatty acids on human learning and memory. This needs to employ both short-term dietary intervention studies, where diet can be carefully manipulated over several days, as well as longer-term dietary interventions lasting months or years. If the animal studies are correct there should be reasonably large and readily detectable effects on cognition following changes in SFA and n-3 fatty acid consumption. Second, and again in humans, we need to establish whether any of the proposed causal pathways in animals are triggered by similar diets in people. This would help to establish the case in humans that SFA and n-3 fatty acids do exert causal effects on hippocampal-dependent memory. This is important, because as pointed out before, adverse changes to the hippocampus may themselves predispose people to over consume the very foods that initiated this neural impairment, leading to weight gain and further damage to the hippocampus (see Ref. 3 for a review).

References

1. Greenwood CE, Winocur G. Learning and memory impairment in rats fed a high saturated fat diet. *Behav Neural Biol*. 1990;53:74–87.
2. Morris RG. Spatial localisation does not depend on the presence of local cues. *Learn Motiv*. 1981;12:239–260.
3. Kanoski SE, Davidson TL. Western diet consumption and cognitive impairment: links to hippocampal dysfunction and obesity. *Physiol Behav*. 2011;103:59–68.
4. Molteni R, Barnard RJ, Ying Z, Roberts CK, Gomez-Pinilla F. A high-fat, refined sugar diet reduces hippocampal brain-derived neurotrophic factor, neuronal plasticity, and learning. *Neuroscience*. 2002;112:803–814.
5. Stranahan AM, Norman ED, Lee K, et al. Diet-induced insulin resistance impairs hippocampal synaptic plasticity and cognition in middle-aged rats. *Hippocampus*. 2008;18:1085–1088.
6. Xia SF, Xie ZX, Qiao Y, et al. Differential effects of quercetin on hippocampus-dependent learning and memory in mice fed with different diets related with oxidative stress. *Physiol Behav*. 2015;138:325–331.
7. Lu J, Wu DM, Zheng YL, et al. Ursolic acid improves high fat diet-induced cognitive impairments by blocking endoplasmic reticulum stress and IκB kinase β/nuclear factor-κB-mediated inflammatory pathways in mice. *Brain Behav Immun*. 2011;25:1658–1667.
8. Anderson NJ, King MR, Delbruck L, Jolivalt CG. Role of insulin signaling impairment, adiponectin and dyslipidemia in peripheral and central neuropathy in mice. *Dis Model Mech*. 2014;7:625–633.
9. Li L, Wang Z, Zuo Z. Chronic intermittent fasting improves cognitive functions and brain structures in mice. *PLoS One*. 2013;8:e66069.
10. Alzoubi KH, Abdul-Razzak KK, Khabour OF, Al-Tuweiq GM, Alzubi MA, Alkadhi KA. Vitamin E prevents high-fat high-carbohydrates diet-induced memory impairment: the role of oxidative stress. *Physiol Behav*. 2013;119;72–78.
11. Alzoubi KH, Abdul-Razzak KK, Khabour OF, Al-Tuweiq GM, Alzubi MA, Alkadhi KA. Adverse effect of combination of chronic psychosocial stress and high fat diet on hippocampus-dependent memory in rats. *Behav Brain Res*. 2009;204:117–123.
12. Greenwood CE, Winocur G. Cognitive impairment in rats fed high-fat diets: a specific effect of saturated fatty-acid intake. *Behav Neurosci*. 1996;110:451–459.
13. Francis HM, Mirzaei M, Pardey MC, Haynes PA, Cornish JL. Proteomic analysis of the dorsal and ventral hippocampus of rats maintained on a high fat and refined sugar diet. *Proteomics*. 2013;(13):3076–3191.
14. Kanoski SE, Davidson TL. Different patterns of memory impairments accompany short- and longer-term maintenance on a high-energy diet. *J Exp Psychol Anim B*. 2010;36:313–319.
15. Valladolid-Acebes I, Stucchi P, Cano V, et al. High-fat diets impair spatial learning in the radial-arm maze in mice. *Neurobiol Learn Mem*. 2011;95:80–85.
16. Leibrock J, Lottspeich F, Hohn A, et al. Molecular cloning and expression of brain-derived neurotrophic factor. *Nature*. 1989;341:149–152.
17. Lu B, Gottschalk W. Modulation of hippocampal synaptic transmission and plasticity by neurotrophins. *Prog Brain Res*. 2000;128:231–241.
18. Tyler WJ, Pozzo-Miller LD. BDNF enhances quantal neurotransmitter release and increases the number of docked vesicles at the active zones of hippocampal excitatory synapses. *J Neurosci*. 2001;21:4249–4258.
19. Linnarsson S, Bjorklund A, Ernfors P. Learning deficit in BDNF mutant mice. *Eur J Neurosci*. 1997;9:2581–2587.
20. Soliman F, Glatt CE, Bath KG, et al. A genetic variant BDNF polymorphism alters extinction learning in both mouse and human. *Science*. 2010;327:863–866.
21. Egan MF, Kojima M, Callicott JH, et al. The BDNF val66met polymorphism affects activity-dependent secretion of BDNF and human memory and hippocampal function. *Cell*. 2003;112:257–269.
22. Hariri AR, Goldberg TE, Mattay VS, et al. Brain-derived neurotrophic factor val66met polymorphism affects human memory-related hippocampal activity and predicts memory performance. *J Neurosci*. 2003;23:6690–6694.
23. Yu Y, Wang Q, Huang XF. Energy-restricted pair-feeding normalizes low levels of brain-derived neurotrophic factor/tyrosine kinase B mRNA expression in the hippocampus, but not ventromedial hypothalamic nucleus, in diet-induced obese mice. *Neuroscience*. 2009;160:295–306.
24. Wu A, Ying Z, Gomez-Pinilla F. The interplay between oxidative stress and brain-derived neurotrophic factor modulates the outcome of a saturated fat diet on synaptic plasticity and cognition. *Eur J Neurosci*. 2004;19:1699–1707.

25. Jovanovic JN, Czernik AJ, Fienberg AA, Greengard P, Sihra TS. Synapsins as mediators of BDNF-enhanced neurotransmitter release. *Nat Neurosci*. 2000;3:323–329.
26. Melloni Jr RH, Apostolides PJ, Hamos JE, DeGennaro LJ. Dynamics of synapsin I gene expression during the establishment and restoration of functional synapses in the rat hippocampus. *Neuroscience*. 1994;58:683–703.
27. Finkbeiner S. CREB couples neurotrophin signals to survival messages. *Neuron*. 2000;25:11–14.
28. Ying SW, Futter M, Rosenblum K, et al. Brain-derived neurotrophic factor induces long-term potentiation in intact adult hippocampus: requirement for ERK activation coupled to CREB and upregulation of Arc synthesis. *J Neurosci*. 2002;22:1532–1540.
29. Wallace DC. Diseases of the mitochondrial DNA. *Annu Rev Biochem*. 1992;61:1175–1212.
30. Bayir H. Reactive oxygen species. *Crit Care Med*. 2005;33:S498–S501.
31. Chinen I, Shimabukuro M, Yamakawa K, et al. Vascular lipotoxicity: endothelial dysfunction via fatty-acid-induced reactive oxygen species overproduction in obese Zucker diabetic fatty rats. *Endocrinology*. 2007;148:160–165.
32. White CL, Pistell PJ, Purpera MN, et al. Effects of high fat diet on Morris maze performance, oxidative stress, and inflammation in rats: contributions of maternal diet. *Neurobiol Dis*. 2009;35:3–13.
33. Bellinger FP, Madamba SG, Campbell IL, Siggins GR. Reduced long-term potentiation in the dentate gyrus of transgenic mice with cerebral overexpression of interleukin-6. *Neurosci Lett*. 1995;198:95–98.
34. Jankowsky JL, Patterson PH. Cytokine and growth factor involvement in long-term potentiation. *Mol Cell Neurosci*. 1999;14:273–286.
35. Cazettes F, Cohen JI, Yau PL, Talbot H, Convit A. Obesity-mediated inflammation may damage the brain circuit that regulates food intake. *Brain Res*. 2011;1373:101–109.
36. Vachharajani V, Granger DN. Adipose tissue: a motor for the inflammation associated with obesity. *IUBMB Life*. 2009;61:424–430.
37. Pistell PJ, Morrison CD, Gupta S, et al. Cognitive impairment following high fat diet consumption is associated with brain inflammation. *J Neuroimmunol*. 2010;219:25–32.
38. Zhang X, Dong F, Ren J, Driscoll MJ, Culver B. High dietary fat induces NADPH oxidase-associated oxidative stress and inflammation in rat cerebral cortex. *Exp Neurol*. 2005;191:318–325.
39. Devore EE, Stampfer MJ, Breteler MM, et al. Dietary fat intake and cognitive decline in women with type 2 diabetes. *Diabetes Care*. 2009;32:635–640.
40. Kalmijn S, van Boxtel MP, Ocke M, Verschuren WM, Kromhout D, Launer LJ. Dietary intake of fatty acids and fish in relation to cognitive performance at middle age. *Neurology*. 2004;62:275–280.
41. Attuquayefio T, Stevenson RJ, Oaten M, Francis HM. A four-day Western-style dietary intervention causes reductions in hippocampal-dependent learning and memory and interoceptive sensitivity. *PLoS One*. 2017;12:e0172645.
42. Morris MC, Evans DA, Bienias JL, Tangney CC, Wilson RS. Dietary fat intake and 6-year cognitive change in an older biracial community population. *Neurology*. 2004;62:1573–1579.
43. Okereke OI, Rosner BA, Kim DH, et al. Dietary fat types and 4-year cognitive change in community-dwelling older women. *Ann Neurol*. 2012;72:124–134.
44. Francis HM, Stevenson RJ. Higher reported saturated fat and refined sugar intake is associated with reduced hippocampal-dependent memory and sensitivity to interoceptive signals. *Behav Neurosci*. 2011;125:943–955.
45. Bayer-Carter JL, Green PS, Montine TJ, et al. Diet intervention and cerebrospinal fluid biomarkers in amnestic mild cognitive impairment. *Arch Neurol*. 2011;68:743–752.
46. Chalon S, Delion-Vancassel S, Belzung C, et al. Dietary fish oil affects monoaminergic neurotransmission and behavior in rats. *J Nutr*. 1998;128:2512–2519.
47. Kaplan RJ, Greenwood CE. Dietary saturated fatty acids and brain function. *Neurochem Res*. 1998;(23):615–626.
48. Salem Jr N, Litman B, Kim HY, Gawrisch K. Mechanisms of action of docosahexaenoic acid in the nervous system. *Lipids*. 2001;36:945–959.
49. Gomez-Pinilla F. Brain foods: the effects of nutrients on brain function. *Nat Rev Neurosci*. 2008;9:568–578.
50. Moore SA, Yoder E, Murphy S, Dutton GR, Spector AA. Astrocytes, not neurons, produce docosahexaenoic acid (22:6 omega-3) and arachidonic acid (20:4 omega-6). *J Neurochem*. 1991;56;518–524.
51. Sinclair AJ. Incorporation of radioactive polyunsaturated fatty acids into liver and brain of developing rat. *Lipids*. 1975;10:175–184.
52. Asif M. Health effects of omega-3,6,9 fatty acids: *Perilla frutescens* is a good example of plant oils. *Orient Pharm Exp Med*. 2011;(1):51–59.

53. Tou JC, Jaczynski J, Chen YC. Krill for human consumption: nutritional value and potential health benefits. *Nutr Rev*. 2007;65:63–77.
54. Willis LM, Shukitt-Hale B, Cheng V, Joseph JA. Dose-dependent effects of walnuts on motor and cognitive function in aged rats. *Br J Nutr*. 2009;101:1140–1144.
55. Winther B, Hoem N, Berge K, Reubsaet L. Elucidation of phosphatidylcholine composition in krill oil extracted from Euphausia superba. *Lipids*. 2011;46:25–36.
56. Chytrova G, Ying Z, Gomez-Pinilla F. Exercise contributes to the effects of DHA dietary supplementation by acting on membrane-related synaptic systems. *Brain Res*. 2010;1341:32–40.
57. Jiang LH, Shi Y, Wang LS, Yang ZR. The influence of orally administered docosahexaenoic acid on cognitive ability in aged mice. *J Nutr Biochem*. 2009;20:735–741.
58. Haider S, Batool Z, Tabassum S, et al. *Effects of walnuts (Juglans regia)* on learning and memory functions. *Plant Foods Hum Nutr*. 2011;66:335–340.
59. Lee J, Park S, Lee JY, Yeo YK, Kim JS, Lim J. Improved spatial learning and memory by perilla diet is correlated with immunoreactivities to neurofilament and alpha-synuclein in hilus of dentate gyrus. *Proteome Sci*. 2012;10:72.
60. Moriguchi T, Greiner RS, Salem Jr N. Behavioral deficits associated with dietary induction of decreased brain docosahexaenoic acid concentration. *J Neurochem*. 2000;75:2563–2573.
61. Catalan J, Moriguchi T, Slotnick B, Murthy M, Greiner RS, Salem Jr N. Cognitive deficits in docosahexaenoic acid-deficient rats. *Behav Neurosci*. 2002;116:1022–1031.
62. Lim GP, Calon F, Morihara T, et al. A diet enriched with the omega-3 fatty acid docosahexaenoic acid reduces amyloid burden in an aged Alzheimer mouse model. *J Neurosci*. 2005;25:3032–3040.
63. Lim SY, Hoshiba J, Moriguchi T, Salem Jr N. n-3 fatty acid deficiency induced by a modified artificial rearing method leads to poorer performance in spatial learning tasks. *Pediatr Res*. 2005;58:741–748.
64. Lim SY, Hoshiba J, Salem Jr N. An extraordinary degree of structural specificity is required in neural phospholipids for optimal brain function: n-6 docosapentaenoic acid substitution for docosahexaenoic acid leads to a loss in spatial task performance. *J Neurochem*. 2005;95:848–857.
65. Fedorova I, Salem Jr N. Omega-3 fatty acids and rodent behavior. *Prostaglandins Leukot Essent Fatty Acids*. 2006;75:271–289.
66. Calder PC. Immunomodulation by omega-3 fatty acids. *Prostaglandins Leukot Essent Fatty Acids*. 2007;77:327–335.
67. Mingam R, Moranis A, Bluthe RM, et al. Uncoupling of interleukin-6 from its signalling pathway by dietary n-3-polyunsaturated fatty acid deprivation alters sickness behaviour in mice. *Eur J Neurosci*. 2008;28:1877–1886.
68. Kadish I, Thibault O, Blalock EM, et al. Hippocampal and cognitive aging across the lifespan: a bioenergetic shift precedes and increased cholesterol trafficking parallels memory impairment. *J Neurosci*. 2009;29:1805–1816.
69. Labrousse VF, Nadjar A, Joffre C, et al. Short-term long chain omega3 diet protects from neuroinflammatory processes and memory impairment in aged mice. *PLoS One*. 2012;7:e36861.
70. Cao D, Kevala K, Kim J, et al. Docosahexaenoic acid promotes hippocampal neuronal development and synaptic function. *J Neurochem*. 2009;111:510–521.
71. Kawashima A, Harada T, Kami H, Yano T, Imada K, Mizuguchi K. Effects of eicosapentaenoic acid on synaptic plasticity, fatty acid profile and phosphoinositide 3-kinase signaling in rat hippocampus and differentiated PC12 cells. *J Nutr Biochem*. 2010;21:268–277.
72. Bhatia HS, Agrawal R, Sharma S, Huo YX, Ying Z, Gomez-Pinilla F. Omega-3 fatty acid deficiency during brain maturation reduces neuronal and behavioral plasticity in adulthood. *PLoS One*. 2011;6:e28451.
73. Wibrand K, Berge K, Messaoudi M, et al. Enhanced cognitive function and antidepressant-like effects after krill oil supplementation in rats. *Lipids Health Dis*. 2013;12:6.
74. Yang RH, Wang F, Hou XH, et al. Dietary omega-3 polyunsaturated fatty acids improves learning performance of diabetic rats by regulating the neuron excitability. *Neuroscience*. 2012;212:93–103.
75. Agrawal R, Gomez-Pinilla F. 'Metabolic syndrome' in the brain: deficiency in omega-3 fatty acid exacerbates dysfunctions in insulin receptor signalling and cognition. *J Physiol*. 2012;590:2485–2499.
76. Denny Joseph KM, Muralidhara. Enhanced neuroprotective effect of fish oil in combination with quercetin against 3-nitropropionic acid induced oxidative stress in rat brain. *Prog Neuropsychopharmacol Biol Psychiatry*. 2013;40:83–92.
77. Denny Joseph KM, Muralidhara M. Fish oil prophylaxis attenuates rotenone-induced oxidative impairments and mitochondrial dysfunctions in rat brain. *Food Chem Toxicol*. 2012;50:1529–1537.

78. O'Donnell E, Vereker E, Lynch MA. Age-related impairment in LTP is accompanied by enhanced activity of stress-activated protein kinases: analysis of underlying mechanisms. *Eur J Neurosci*. 2000;12:345–352.
79. Kelly L, Grehan B, Chiesa AD, et al. The polyunsaturated fatty acids, EPA and DPA exert a protective effect in the hippocampus of the aged rat. *Neurobiol Aging*. 2011;32:e2311–e2315.
80. Boucher O, Burden M, Muckle G, et al. Neuropsychologic and neurobehavioral evidence of beneficial effects of prenatal omega-3 fatty acid intake on memory function at school age. *Am J Clin Nutr*. 2011;93:1025–1037.
81. Beyer M, Bronnick K, Hwang K, et al. Verbal memory is associated with structural hippocampal changes in newly diagnosed Parkinson's disease. *J Neurol Neurosurg Psychiatry*. 2012;84. https://doi.org/10.1136/jnnp-2012-303054.
82. Libo D, Bogdanoff B, Cloud B, et al. Declarative and procedural learning, quantitative measures of the hippocampus and subcortical white alterations in Alzheimer's disease and ischaemic vascular dementia. *J Clin Exp Neuropsychol*. 1998;20:30–41.
83. Muldoon M, Ryan C, Sheu L, Yao J, Conklin S, Manuck S. Serum phospholid docosahexaenonic acid is associated with cognitive functioning during middle adulthood. *J Nutr*. 2010;140:848–853.
84. Saling M, Berkovic F, O'Shea F, Kalnins M, Darby G, Bladin F. Lateralization of verbal memory and unilateral hippocampal sclerosis: evidence of task-specific effects. *J Clin Exp Neuropsychol*. 1993;15:608–618.
85. Sass J, Sass K, Westerveld M, et al. Specificity in the correlation of verbal memory and hippocampal neuron loss: dissociation of memory, language, and verbal intellectual ability. *J Clin Exp Neuropsychol*. 1992;14:662–672.
86. Andruchow ND, Konishi K, Bohbot VD. A lower ratio of omega-6 to omega-3 fatty acids predicts better hippocampus-dependent spatial memory and cognitive status in older adults. *Neuropsychol*. 2017;31:724–734.
87. Conklin S, Harris J, Manuck S, Yao J, Hibbeln J, Muldoon M. Serum omega-3 fatty acids are associated with variation in mood, personality and behavior in hypercholesterolemic community volunteers. *Psychiatry Res*. 2007;152:1–10.
88. Yurko-Mauro K, McCarthy D, Rom D, et al. Beneficial effects of docosahexaenoic acid on cognition in age-related cognitive decline. *Alzheimers Dement*. 2010;6:456–464.
89. den Heijer T, Tiemeier H, Luijendijk H, et al. A study of the bidirectional association between hippocampal volume on MRI and depression in the elderly. *Biol Psychiatry*. 2011;70:191–197.
90. Santarelli L, Saxe M, Gross C, et al. Requirement of hippocampal neurogenesis for the behavioral effects of antidepressants. *Science*. 2003;301:805–809.
91. Sanhueza C, Ryan L, Foxcroft D. Diet and the risk of unipolar depression in adults: systematic review of cohort studies. *J Hum Nutr Diet*. 2012;26:56–70.
92. Conklin S, Manuck S, Yao J, Flory J, Hibbeln J, Muldoon M. High w-6 and low w-3 fatty acids are associated with depressive symptoms and neuroticism. *Psychosom Med*. 2007;69:932–934.
93. Conklin S, Gianaros P, Brown S, et al. Long-chain omerga-3 fatty acid intake is associated positively with corticolimbic gray matter volume in healthy adults. *Neurosci Lett*. 2007;421:209–212.
94. Tiemeier H, van Tuijl H, Hofman A, Kiliaan A, Breteler M. Plasma fatty acid composition and depression are associated in the elderly: the Rotterdam study. *Am J Clin Nutr*. 2003;78:40–46.
95. Martins J. EPA but not DHA appears to be responsible for the efficacy of Omega-3 long chain polyunsaturated fatty acid supplementation in depression: evidence from a meta-analysis of randomized controlled trials. *J Am Coll Nutr*. 2009;28:525–542.
96. Mocking RJT, Harmsen I, Assies J, Koeter MWJ, Ruhe HG, Schene AH. Meta-analysis and meta-regression of omega-3 polyunsaturated fatty acid supplementation for major depressive disorder. *Transl Psychiatry*. 2016;6(3):e756.

CHAPTER

9

Energy Metabolism of Neural Cells Under the Control of Phospholipases A_2 and Docosahexaenoic Acid

Peter Schönfeld, Georg Reiser†*

*Institute for Biochemistry and Zell Biology, Medical Faculty, Otto-von-Guericke-University Magdeburg, Magdeburg, Germany †Institute for Inflammation and Neurodegeneration (Neurobiochemistry), Medical Faculty, Otto-von-Guericke-University Magdeburg, Magdeburg, Germany

ABBREVIATIONS

ALA	α-linolenic acid
ANT	ADP/ATP-antiporter
CL	cardiolipin
DHA	docosahexaenoic acid
IMM	inner mitochondrial membrane
$iPLA_2$	Ca^{2+}-independent phospholipase
LCFA	long-chain fatty acids
PPAR	peroxisome-proliferator-activated receptor
PTP	permeability transition pore
PUFA	polyunsaturated fatty acids
RC	respiratory chain
RET	reversed electron transport
ROS	reactive oxygen species
$\Delta\psi$	mitochondrial membrane potential

Omega Fatty Acids in Brain and Neurological Health
https://doi.org/10.1016/B978-0-12-815238-6.00009-2

ENERGY METABOLISM OF NEURAL CELLS AT A GLANCE

Human brain tissue has a high energy demand, mostly needed for synaptic signaling. Consequently, according to a previously published detailed estimate,[1] neurons contribute up to 85% (maybe even 95%) to the total energy required, although the cell number of astrocytes in brain tissue is much higher. ATP is generated in neurons predominantly by the oxidative phosphorylation machinery of mitochondria, whereas astrocytes provide the necessary ATP demand mainly by anaerobic glycolysis via substrate-level phosphorylation. Nevertheless, astrocytes are crucial for the fulfillment of the neuronal energy requirement (see for recent reviews Refs. 2–4). Chemical energy is predominantly supplied to the brain as glucose, taken up from the circulation by a concerted cooperation of the transporters Glut1 (from circulation) and Glut1 (from astrocytes) or Glut1 (from circulation) and Glut3 (from neurons). A growing number of reports have uncovered prominent differences in the use of glucose as fuel in astrocytes and neurons (see references in reviews[2–4]). Contrary to astrocytes, neurons do not store glucose as glycogen. Furthermore, even though astrocytes have high glycolytic capacity and are equipped with mitochondria, the end-product pyruvate becomes only poorly oxidized by the citric acid cycle for delivering reducing equivalents (NADH, $FADH_2$) to mitochondria. This surprising finding is a result of maintaining the astrocytic pyruvate dehydrogenase in an inactive, phosphorylated state, achieved by the high enzymatic capacity of pyruvate dehydrogenase kinase 4 in astrocytes. Inactivation of the pyruvate dehydrogenase by phosphorylation enables astrocytes to generate lactate, and to donate it to neurons as fuel. Otherwise, neurons would be unable to respond to an enhanced energy requirement by upregulation of the glycolytic glucose degradation, because the enzyme 6-phosphofructo-2-/fructo-2,6-bisphosphatase, a positive key regulator of the glycolysis, is constantly degraded in proteasomes. In contrast, this enzyme is fully active in astrocytes, and, therefore allows the upregulation of glycolysis. In addition, methylglyoxal (reduced pyruvate) is formed as side-product during glycolytic glucose degradation.[3] This highly reactive dicarbonyl compound methylglyoxal has been associated with the development of neurodegeneration. Contrary to neurons, astrocytes contain high capacities of glyoxalase enzymes for detoxification of methylglyoxal. This further enables astrocytes to deliver lactate to neurons.

In conclusion, astrocytes generate through glucose degradation mostly lactate, whereas neurons use glucose preferentially for generation of NADPH by the pentose-phosphate pathway, and thereby protect their antioxidative status (e.g., by keeping glutathione in the reduced form). Furthermore, the energy requirement of neurons seems mostly based on the use of astrocytic lactate as energy-delivering substrate for the oxidative phosphorylation.

According to the astrocyte-neuron lactate shuttle hypothesis, lactate as the final product of the astrocytic glycolysis is transferred to neurons by means of the monocarboxylate transporters MCT1/4 (astrocyte) and MCT2 (neuron).[5,6] Neurons stimulate the astrocytic glycolysis through their synaptic activity via the combined release of glutamate from neurons and its subsequent uptake by astrocytes. Since glutamate uptake is coupled with co-uptake of Na^+, the astrocytic Na^+/K^+-ATPase has the task to bring back Na^+ from the cytosolic into the extracellular environment. Consequently, the ATP consumption by the Na^+/K^+-ATPase and, in addition, the ATP-dependent conversion of glutamate to glutamine, stimulates the glycolytic lactate generation by astrocytes. Two possibilities have been discussed for the fate of astrocytic lactate in neurons. Firstly, the cytosolic lactate dehydrogenase (LDH1) oxidizes lactate to pyruvate and NADH, and NADH is transferred into mitochondria by the malate–aspartate-shuttle (see for review Ref. 7). Alternatively, lactate is transported from the cytosol via a monocarboxylate transporter into mitochondria, where mitochondrial LDH oxidizes lactate to pyruvate plus NADH.[4]

Moreover, in sharp contrast to other organs with high-energy turnover, such as heart, kidney or liver, mitochondrial β-oxidation seems to play no role as hydrogen delivering process in brain. Nevertheless, the discussion on the capability of the brain tissue to degrade long-chain fatty acids (LCFAs) is still going on.[8–10] Nevertheless, older (as we reviewed in Ref. 11) and more recent studies exclude mitochondrial β-oxidation in brain mitochondria and astrocytes.[12,13] In addition, the high susceptibility of brain to reactive oxygen species (ROS)[14] and the fact that β-oxidation would increase cellular oxidative stress, are further arguments against the use of fatty acid degradation for the hydrogen supply in brain tissue.[15]

However, brain tissue uses LCFA as energy source indirectly, namely in the form of liver-derived ketone bodies (β-hydroxybutyrate and acetoacetate) during maturation in young age and upon prolonged fasting.[16] Oxidation of liver-derived ketone bodies can replace glucose to a high degree because it provides up to 60% of the required energy.[17] Being highly water soluble, the permeation of ketone bodies across the blood–brain barrier is mediated by the monocarboxylate transporters, MCT1–2.[16] With respect to fatty acid oxidation, the use of ketone bodies as fuel is advantageous for the cerebral energy metabolism. This statement is substantiated by a lower oxygen consumption for oxidation of NADH derived from ketone body oxidation and, in addition, a lesser ROS generation due to the non-usage of the electron transfer flavoprotein-ubiquinone oxidoreductase in this pathway.[18]

DHA AND CA^{2+}-INDEPENDENT PHOSPHOLIPASES A_2

Docosahexaenoic acid (DHA; 22:6, n-3) is synthesized in the mammalian liver from α-linolenic acid (ALA; 18:3, n-3) by a series of desaturations, elongations, and one step of β-oxidation.[19] These reactions are located in the endoplasmic reticulum (desaturations, elongations) and in peroxisomes (β-oxidation). Hepatic conversion of ALA to DHA is considered to be sufficient for healthy individuals to meet the demand of DHA in the brain.[19, 20] In addition, liver takes up dietary DHA and uses it for the production of phospholipids, which are released into the blood circulation. Brain has the ability to extract lipoprotein-bound DHA or its precursor ALA from the blood.[19,21–23] Similarly to liver, brain has the enzymatic equipment to convert ALA into DHA. Moreover, DHA esterified to membrane glycerophospholipids is a further source for donating free DHA into the cerebral metabolism. In the glycerophospholipids, DHA is fixed to a high degree in the sn-2 position, from where it is liberated by the enzymatic activities of distinct phospholipases A_2.

Phospholipases A_2 which are known for a long time to be enzymes were recognized already for two decades to have wide importance for many neurochemical processes.[24] Briefly, phospholipases A_2 form a superfamily, consisting of 16 groups (I–XVI).[25] Group VI summarizes the so-called Ca^{2+}-independent phospholipases A_2 ($iPLA_2$s), enzymes that exhibit some catalytic activity in the absence of Ca^{2+}, and, in addition, have a serine residue in their catalytic center. It is believed that $iPLA_2$ accounts for more than 70% of PLA_2 activity in brain[26] and, especially for the release of DHA.[27,28] Two of them, the membrane-associated $iPLA_2\beta$ (PLA2G6) and $iPLA_2\gamma$ (PNPLA8) that are attached to mitochondria, have attracted particular attention. Moreover, in mammalian mitochondria $iPLA_2\gamma$ is the predominant phospholipase activity, and this phospholipase becomes markedly activated by divalent cations (Ca^{2+}, Mg^{2+}), by depolarization of the inner mitochondrial membrane (IMM) and by membrane-bound, negatively charged cardiolipin.[29–31] Ca^{2+}-induced $iPLA_2\gamma$ activation was completely inhibited by the CoA-derivates of LCFA.[29] In addition, there is also a group of cytosolic PLA_2 ($cPLA_2$) in brain.[32] $cPLA_2$ is activated after translocating to the plasma membrane from the cytosol, where it remains then constitutively membrane bound.

In contrast to $iPLA_2$, $cPLA_2$ preferentially catalyzes the release of arachidonic acid[28] and, is thereby involved in the production of arachidonic acid-derived lipid mediators, such as prostaglandins, thromboxanes, and leukotrienes, for brain metabolism.[32]

MULTIPLE CELLULAR PROCESSES UNDER THE CONTROL OF DHA

As a lipid molecule consisting of 22 carbon atoms with 6 double bonds, DHA can exists in multiple configurations, and so it can change after its incorporation into cellular membranes, the membrane properties. With respect to other LCFA, the DHA molecule has a greater molecular volume (355 $Å^3$). This feature results in increased membrane fluidity, when DHA becomes esterified with membrane phospholipids.[33] As a consequence, various membrane-associated functions such as transmembranal drug permeability, carrier-mediated metabolite transport, activities of membrane-bound enzymes, ion channels and receptors, and neurotransmission are modulated by the increased membrane fluidity.[33–37]

Brain has a particularly high demand for DHA during pre- and peri-natal development. In this period, synaptogenesis, dendrite formation, and other neural membrane building processes take place, to form brain circuit structures.[20] Moreover, similar to other members of the PUFA family, DHA can be active in diverse intracellular functions, such as being an effector of intracellular signaling, of immune response, and in homeostasis of the metabolism in tissues.[38] In more detail, DHA and other PUFAs are metabolized to resolvins (dihydroxy or trihydroxy metabolites of DHA), which have anti-inflammatory activity.[39] Furthermore, DHA operates as endogenous ligand of PPAR-γ and PPAR-α. Thus activation of PPAR-γ by DHA is involved in the microglial activation, myelination, and heat shock protein response.[38] Finally, DHA displays also anti-inflammatory properties by activation of $cPLA_2$. The underlying mechanism is mediated by the GPR120 receptor, and results in the production of prostaglandin A_2.[40]

DHA operates either directly as signaling molecule or serves as precursor of docosanoids, which are sustaining neural cell integrity.[20,32,33,41] It has been also proposed that a DHA-related intracellular Ca^{2+}-mediated signaling contributes to a regulation of a brain-specific prostaglandin formation.[42] Importantly, despite their function in brain tissue, DHA and other PUFA have the potential to counteract insulin resistance, probably by modulating mitochondrial bioenergetics.[43]

AMBIGUOUS ROLE OF DHA IN OXIDATIVE STRESS EVENTS

Basically, high cerebral capacity of oxygen-dependent energy turnover combined with low intrinsic capacity of antioxidant defense and high content of polyunsaturated lipids requires an effective protection of the nonregenerating neurons against the harmful action of ROS. For that purpose, neurons receive strong antioxidative support, particularly from astrocytes, which supply neurons with precursors for the synthesis of glutathione, to allow the inactivation of ROS.[44]

Generally, depending on the metabolic condition, LCFAs can either increase or decrease the intracellular ROS generation. There are three mechanisms accounting for the possibility that DHA modulates mitochondrial ROS generation (see Fig. 1 for mitochondrial targets of DHA, which contribute to ROS generation): Firstly, as a potential substrate of the mitochondrial and peroxisomal β-oxidation in noncerebral tissues, DHA is able to donate electrons to the ROS generation.[45–48] Secondly, nonesterified DHA or their activated acyl-CoA derivative are able to impair the electron flux within the respiratory chain (RC), and thereby enhance ROS generation (for review, see Refs. 18, 49, 50). This has been shown with rat brain mitochondria[51] and rat pheochromocytoma cells (PC12), a model of neural cells.[52] Since PC12-associated ROS generation was paralleled by an increase of glycolytically generated lactate and, in addition, negatively correlated with the mitochondrial NADH-cytochrome *c* reductase activity, it is highly suggestive to attribute the ROS generation to an impairment of the electron transport in the RC by DHA.

Thirdly, slight depolarization of the IMM by "mild-uncoupling" by nonesterified DHA dramatically decreased a ROS generation, which was driven by reversed electron transport (RET).[51] This strong dependency of RET-dependent ROS generation on the mitochondrial membrane potential ($\Delta\psi$) has attracted attention to $iPLA_2$ as a tool to liberate DHA from membrane lipids.[53] Additional evidence showed that the uncoupling protein-2 reduces in concert with an H_2O_2-activated mitochondrial $iPLA_2$ oxidative stress in the cell. According to this view, fatty acids released from membrane phospholipids initiate uncoupling protein-2-mediated "mild-uncoupling." It is also worth to mention that a diminished activity of $iPLA_2\gamma$ correlates reciprocally with increased mitochondrial peroxidation and mitochondrial dysfunction in a rotenone-induced model of Parkinson's disease.[54] This finding suggests that $iPLA_2\gamma$ acts in the midbrain as a protective enzyme throughout oxidative stress.

Moreover, being a polyunsaturated fatty acid, DHA is highly susceptible to peroxidative damage. Therefore, it was surprising that, despite of its abundance in brain tissue, DHA decreased lipid peroxidation (see for review Ref. 33). This discrepancy is explained by the ability of DHA to increase the capacity of antioxidative enzymes, such as catalase, glutathione peroxidase, and glutathione reductase. DHA-associated neuroprotection has been demonstrated with studies, where DHA was esterified in membrane phospholipids.[55]

A further likely mechanism for the neuroprotection by DHA is the stimulation of plasmalogen synthesis, driven by the DHA-regulated expression of enzymes involved in their synthesis.[56,57] Remarkably, in plasmalogens, the double bond of the vinyl-ether bound side-chain at the sn-1 position is highly reactive to ROS, thereby transforming plasmalogens into natural scavengers for ROS. In addition, a derivative of DHA, called neuroprotectin D1 (10,17S-docosatriene), exhibits neuroprotection after episodes of ischemia–reperfusion. This activity of neuroprotectin D1 is ascribed to the upregulation of the antiapoptotic proteins Bcl-2 and BclxL and the decrease of the proapoptotic proteins Bax and Bad.

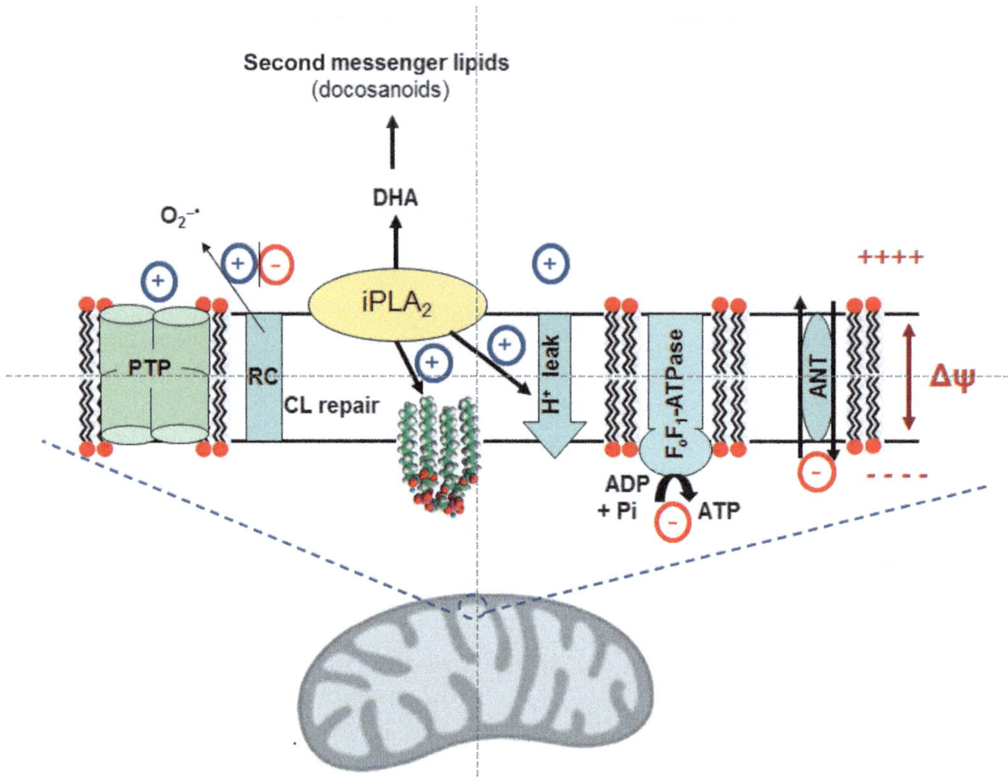

FIG. 1 Modulation of mitochondrial energy-related activities connected with the inner mitochondrial membrane (IMM) by docosahexaenoic acid (DHA) and $iPLA_2$. Binding of nonesterified DHA to the mitochondrial permeability transition pore (PTP) facilitates its opening and, consequently, DHA reduces cell viability. Moreover, elevated concentrations of DHA exert a dual action on ROS generation by the respiratory chain (RC). Interaction of DHA with components of the RC slows down the forward electron transport, thereby stimulating the generation of superoxide ($O_2^{-\bullet}$). Otherwise, the uncoupling activity of nonesterified DHA, being a characteristic feature of long-chain fatty acids, diminishes the mitochondrial membrane potential ($\Delta\psi$) across the IMM by enhancing the back-transport of H^+ from the cytosolic compartment (H^+ leak). Modulation of H^+ leak has been also attributed to $iPLA_2$. In conclusion, due to the decline of $\Delta\psi$, the $O_2^{-\bullet}$ generation driven by reverse electron transport becomes abolished. In addition, DHA partly inhibits phosphorylation of ADP to ATP by interaction with constituents of the F_0F_1-ATPase and, has the capacity to partly reduce the export of matrix-formed ATP to the cytosol via the ADP/ATP-antiporter (ANT). Besides these activities of DHA on the IMM and their protein components, DHA serves as precursor for the formation of second messenger lipids (docosanoids). Finally, $iPLA_2$ itself is able by its enzymatic activity to remove peroxidatively damaged fatty acid chains from cardiolipin (CL) and thereby, helps to regenerate CL. This feature of $iPLA_2$ is essential for sustaining energy-driven mitochondrial functions in cells.

Finally, cardiolipin is nearly exclusively present in the IMM, and it is located in the proximity of ROS-generating sites. For that, and due to multiple bis-allylic hydrogen atoms (-HC=CH-C**H**$_2$-HC=CH-) in cardiolipin, this phospholipid is highly susceptible to oxidative stress. There is evidence that $iPLA_2\gamma$ released peroxidized fatty acid chains from cardiolipin, and thereby, pathological products which are able to disturb energy conversion by mitochondria.[31]

ALTERATION OF MITOCHONDRIAL ENERGY-DEPENDENT FUNCTIONS

A growing number of reports revealed an important role of $iPLA_2\gamma$ in animal energy metabolism, particularly for maintaining bioenergetic functions of mitochondria. An early study reported that mitochondria isolated from the heart of $iPLA_2\gamma^{-/-}$ mice exhibits a 65% decrease in the ascorbate-mediated activity of Complex IV.[58] More importantly, skeletal muscle mitochondria[59] and liver mitochondria[60] from $iPLA_2\gamma$-ablated mice have a dramatically decreased ADP-stimulated respiration. This decrease was independent of the type of substrates (pyruvate plus malate or succinate or pyruvate plus glutamate) used for donation of reducing equivalents (NADH or $FADH_2$) to the RC. Moreover, the reduced capacity of oxidative phosphorylation results probably from a slight uncoupling of the ATP-synthase reaction and the electron transport within the RC. Since in mitochondria from $iPLA_2\gamma^{-/-}$ mice the cardiolipin content is also lower, it has been speculated that the uncoupling results from a greater H^+ leakage of the IMM. Taken together, $iPLA_2\gamma$ is an obligatory upstream enzyme essential for oxidative ATP production. In line with this conclusion are the findings that $iPLA_2\gamma^{-/-}$ mice suffer from growth retardation, cold intolerance, reduced exercise endurance, and are resistant to high fat diet-induced weight gain.[58,59] Notably, ablation of $iPLA_2\gamma$ is also causal for mitochondrial degeneration in brain, as indicated by an alteration of their lipid metabolism and membrane structure.[61] These findings are accompanied by cognitive dysfunctions.

Similar to $iPLA_2\gamma$, $iPLA_2\beta$ also contributes to the maintenance of mitochondrial respiration and ATP synthesis.[62] In addition, with brain mitochondria isolated from mice it was demonstrated that a defective activity of $iPLA_2\beta$ decreased the ability of mitochondria to take up and retain Ca^{2+}.[63] Since Ca^{2+} is known to be a universal second messenger (for review see Ref. 64), this finding suggests a function of mitochondria-bound $iPLA_2$ as regulator of cytosolic Ca^{2+} concentration. This view is supported by their ability to sense micropools of high Ca^{2+} concentrations created by the $InsP_3$-mediated opening of the Ca^{2+} store near of the endoplasmic reticulum.[65] In addition, by means of the reversible Ca^{2+} uptake and release, mitochondria control the activity of pyruvate dehydrogenase and two dehydrogenases of the citric acid cycle (tricarboxylic cycle).[66]

Moreover, there are arguments suggesting that $iPLA_2\gamma$ plays a role in the opening of the mitochondrial permeability transition pore (PTP) and, thereby acts as a trigger initiating apoptotic cell death (see Fig. 1). Firstly, it has been reported that activated $iPLA_2\gamma$ is involved in the release of apoptogenic factors, such as cytochrome *c*, Bax, or Bid.[67,68] Secondly, an involvement of the $iPLA_2\gamma$ in the intrinsic apoptotic cell death is substantiated by their capacity to release nonesterified LCFAs or their acyl-CoA derivatives, to trigger the Ca^{2+}-induced opening of PTP.[60] Thirdly, genetic ablation of mt-$iPLA_2\gamma$ attenuates Ca^{2+}-induced opening of PTP in liver mitochondria and thus, the release of the apoptogenic cytochrome *c*.[60] Since liver mitochondria from $iPLA_2\gamma^{-/-}$ mice were much more resistant to Ca^{2+}-induced swelling in the presence or absence of phosphate in comparison with wild-type littermates, $iPLA_2\gamma$ is considered to be a critical mechanistic participant in the Ca^{2+}-induced opening of PTP.

It is highly suggestive that some of the $iPLA_2\gamma$-associated changes in the mitochondrial functions are directly mediated by nonesterified DHA. Fig. 1 shows some of the main targets of DHA. Isolated mitochondria exposed to low micromolar concentrations of DHA exhibit slight depolarization of the IMM, a partly impaired ADP/ATP-antiporter and a slightly impaired electron transport within the RC.[51,52] In addition, DHA most likely inhibits the phosphorylation of ADP to ATP by an interaction with constituents of the F_0F_1-ATPase (see Fig. 1). In summary, these activities can explain a lowered efficacy for oxidative ATP regeneration in the presence of micromolar concentrations of DHA.

Moreover, as also indicated in Fig. 1, notably, the activation of iPLA$_2\gamma$ by cardiolipin (CL) has consequences for the mitochondrial bioenergetics during oxidative stress and intracellular signaling.[31] Firstly, CL-activated iPLA$_2\gamma$ mediates the hydrolysis of arachidonic acid from phosphatidylcholine, thereby integrating the production of lipid messengers from different lipid classes in mitochondria. Secondly, CL-activated iPLA$_2\gamma$ removes peroxidized fatty acids from CL, thereby offering the way for the reesterification of CL with intact fatty acids. Finally, it is also worth to recall that CL itself exerts an important role in the mitochondrial bioenergetics and mitochondria-associated signaling.[69–71] In spite of many other proteins of IMM, CL is required for the optimal activity of the complexes I, III, and IV of the RC as well as complex V (ATPase). One example for the bioenergetic consequences of CL-deficiency is clearly illustrated by the pathomechanisms found in Barth syndrome.[72] There, the underlying changes in composition of CL-aliphatic chains due to a defective CL remodeling result in mitochondria with decreased quality control.

CONCLUSIONS

Brain differs greatly from other tissues with respect to its distinct energy metabolism and susceptibility to oxidative stress. For maintaining the supply of energy, brain uses mostly the energy of glucose, whereas that of hydrogen-rich fatty acids, including DHA, is spurned. Nevertheless, besides a multiplicity of physiological functions in tissues, DHA protects and modulates the energy metabolism of neural cells. One strategy is "mild-uncoupling," thereby attenuating attacks of ROS on the system of a myriad of proteins, enzymes, unsaturated sidechains of membrane lipids and metabolite carriers involved in oxidative ATP generation and, the integrity of cerebral membranes. DHA-associated protection against oxidative stress is additionally supported by two further features of the molecule DHA. Firstly, DHA is a main constituent of cerebral membrane phospholipids and, secondly, it acts due to its highly unsaturated nature as an antioxidant. iPLA$_2$s account for more than 70% of the total phospholipase A_2 activity. Besides others, their physiological role comprises the release of DHA for the synthesis of second messenger lipids and the membrane remodeling. The latter contributes to the removal of peroxidatively damaged fatty acid chains, and, thus, allows their subsequent substitution by nondefective fatty acid chains in CL. Moreover, studies mimicking deficiency of the iPLA$_2$ have uncovered the role of DHA in sustaining effective oxidative phosphorylation (iPLA$_2\gamma$) and in the intracellular Ca^{2+} homeostasis (iPLA$_2\beta$). Lack of activity of iPLA$_2\beta$ in mitochondria decreased their ability to take up and retain Ca^{2+}. In addition to the iPLA$_2$-related functions, DHA directly modulates several activities of protein complexes of IMM and, thereby, affects the cellular energy metabolism. In conclusion, by targeting mitochondria, DHA severely affects the cellular energy metabolism.

References

1. Attwell D, Laughlin SB. An energy budget for signaling in the grey matter of the brain. *J Cereb Blood Flow Metab*. 2001;21:1133–1145.
2. Bolaños JP. Bioenergetics and redox adaptations of astrocytes to neuronal activity. *J Neurochem*. 2016;139(Suppl 2):115–125.
3. Magistretti PJ, Allaman I. A cellular perspective on brain energy metabolism and functional imaging. *Neuron*. 2015;86:883–901.
4. Schurr A. Cerebral glycolysis: a century of persistent misunderstanding and misconception. *Front Neurosci*. 2014;8:360.
5. Bélanger M, Allaman I, Magistretti PJ. Brain energy metabolism: focus on astrocyte-neuron metabolic cooperation. *Cell Metab*. 2011;14:724–738.
6. Mächler P, Wyss MT, Elsayed M, et al. In vivo evidence for a lactate gradient from astrocytes to neurons. *Cell Metab*. 2016;23:94–102.
7. Dienel GA. Brain lactate metabolism: the discoveries and the controversies. *J Cereb Blood Flow Metab*. 2012;32:1107–1138.
8. Chen CT, Liu Z, Ouellet M, Calon F, Bazinet RP. Rapid beta-oxidation of eicosapentaenoic acid in mouse brain: an in situ study. *Prostaglandins Leukot Essent Fatty Acids*. 2009;80:157–163.
9. Chen CT, Bazinet RP. β-Oxidation and rapid metabolism, but not uptake regulate brain eicosapentaenoic acid levels. *Prostaglandins Leukot Essent Fatty Acids*. 2015;92:33–40.
10. Panov A, Orynbayeva Z, Vavilin V, Lyakhovich V. Fatty acids in energy metabolism of the central nervous system. *Biomed Res Int*. 2014;2014:472459.
11. Schönfeld P, Reiser G. Why does brain metabolism not favor burning of fatty acids to provide energy? Reflections on disadvantages of the use of free fatty acids as fuel for brain. *J Cereb Blood Flow Metab*. 2013;33:1493–1499.
12. Schönfeld P, Reiser G. Inhibition of β-oxidation is not a valid therapeutic tool for reducing oxidative stress in conditions of neurodegeneration. *J Cereb Blood Flow Metab*. 2017;37:848–854.
13. Schönfeld P, Reiser G. Brain energy metabolism spurns fatty acids as fuel due to their inherent mitotoxicity and potential capacity to unleash neurodegeneration. *Neurochem Int*. 2017;109:68–77.
14. Cobley JN, Fiorello ML, Bailey DM. 13 reasons why the brain is susceptible to oxidative stress. *Redox Biol*. 2018;15:490–503.
15. Speijer D. Oxygen radicals shaping evolution: why fatty acid catabolism leads to peroxisomes while neurons do without it: $FADH_2$/NADH flux ratios determining mitochondrial radical formation were crucial for the eukaryotic invention of peroxisomes and catabolic tissue differentiation. *Bioessays*. 2011;33:88–94.
16. Morris AA. Cerebral ketone body metabolism. *J Inherit Metab Dis*. 2005;28:109–121.
17. Cahill Jr GF. Fuel metabolism in starvation. *Annu Rev Nutr*. 2006;26:1–22.
18. Schönfeld P, Wojtczak L. Fatty acids as modulators of the cellular production of reactive oxygen species. *Free Radic Biol Med*. 2008;45:231–241.
19. Domenichiello AF, Kitson AP, Bazinet RP. Is docosahexaenoic acid synthesis from α-linolenic acid sufficient to supply the adult brain? *Prog Lipid Res*. 2015;59:54–66.
20. Bazan NG, Musto AE, Knott EJ. Endogenous signaling by omega-3 docosahexaenoic acid-derived mediators sustains homeostatic synaptic and circuitry integrity. *Mol Neurobiol*. 2011;44:216–222.
21. Haast RA, Kiliaan AJ. Impact of fatty acids on brain circulation, structure and function. *Prostaglandins Leukot Essent Fatty Acids*. 2015;92:3–14.
22. Kitson AP, Metherel AH, Chen CT, et al. Effect of dietary docosahexaenoic acid (DHA) in phospholipids or triglycerides on brain DHA uptake and accretion. *J Nutr Biochem*. 2016;33:91–102.
23. Ouellet M, Emond V, Chen CT, et al. Diffusion of docosahexaenoic and eicosapentaenoic acids through the blood-brain barrier: an in situ cerebral perfusion study. *Neurochem Int*. 2009;55:476–482.
24. Kudo I, Murakami M. Phospholipase A2 enzymes. *Prostaglandins Other Lipid Mediat*. 2002;68-69:3–58.
25. Dennis EA, Cao J, Hsu YH, Magrioti V, Kokotos G. Phospholipase A2 enzymes: physical structure, biological function, disease implication, chemical inhibition, and therapeutic intervention. *Chem Rev*. 2011;111:6130–6185.
26. Yang HC, Mosior M, Johnson CA, Chen Y, Dennis EA. Group-specific assays that distinguish between the four major types of mammalian phospholipase A2. *Anal Biochem*. 1999;269:278–288.
27. Shikano M, Masuzawa Y, Yazawa K, Takayama K, Kudo I, Inoue K. Complete discrimination of docosahexaenoate from arachidonate by 85 kDa cytosolic phospholipase A2 during the hydrolysis of diacyl- and alkenylacylglycerophosphoethanolamine. *Biochim Biophys Acta*. 1994;1212:211–216.

28. Strokin M, Sergeeva M, Reiser G. Docosahexaenoic acid and arachidonic acid release in rat brain astrocytes is mediated by two separate isoforms of phospholipase A_2 and is differently regulated by cyclic AMP and Ca^{2+}. *Br J Pharmacol*. 2003;139:1014–1022.
29. Moon SH, Jenkins CM, Liu X, Guan S, Mancuso DJ, Gross RW. Activation of mitochondrial calcium-independent phospholipase $A_2\gamma$ (iPLA$_2\gamma$) by divalent cations mediating arachidonate release and production of downstream eicosanoids. *J Biol Chem*. 2012;287:14880–14895.
30. Rauckhorst AJ, Pfeiffer DR, Broekemeier KM. The iPLA$_2\gamma$ is identified as the membrane potential sensitive phospholipase in liver mitochondria. *FEBS Lett*. 2015;589:2367–2371.
31. Liu GY, Moon SH, Jenkins CM, et al. The phospholipase iPLA2γ is a major mediator releasing oxidized aliphatic chains from cardiolipin, integrating mitochondrial bioenergetics and signaling. *J Biol Chem*. 2017;292:10672–10684.
32. Rosa AO, Rapoport SI. Intracellular- and extracellular-derived Ca^{2+} influence phospholipase A_2-mediated fatty acid release from brain phospholipids. *Biochim Biophys Acta*. 2009;1791:697–705.
33. Hashimoto M, Hossain S, Al Mamun A, Matsuzaki K, Arai H. Docosahexaenoic acid: one molecule diverse functions. *Crit Rev Biotechnol*. 2017;37:579–597.
34. Fernstrom JD. Effects of dietary polyunsaturated fatty acids on neuronal function. *Lipids*. 1999;34:161–169.
35. Lee JY, Plakidas A, Lee WH, et al. Differential modulation of toll-like receptors by fatty acids: preferential inhibition by n-3 polyunsaturated fatty acids. *J Lipid Res*. 2003;44:479–486.
36. Chalon S. Omega-3 fatty acids and monoamine neurotransmission. *Prostaglandins Leukot Essent Fatty Acids*. 2006;75:259–269.
37. Itokazu N, Ikegaya Y, Nishikawa M, Matsuki N. Bidirectional actions of docosahexaenoic acid on hippocampal neurotransmissions in vivo. *Brain Res*. 2000;862:211–216.
38. Bazinet RP, Laye S. Polyunsaturated fatty acids and their metabolites in brain function and disease. *Nat Rev Neurosci*. 2014;15:771–785.
39. Farooqui AA, Horrocks LA, Farooqui T. Modulation of inflammation in brain: a matter of fat. *J Neurochem*. 2007;101:577–599.
40. Liu Y, Chen LY, Sokolowska M, et al. The fish oil ingredient, docosahexaenoic acid, activates cytosolic phospholipase A_2 via GPR120 receptor to produce prostaglandin E_2 and plays an anti-inflammatory role in macrophages. *Immunology*. 2014;143:81–95.
41. Kuda O. Bioactive metabolites of docosahexaenoic acid. *Biochimie*. 2017;136:12–20.
42. Strokin M, Sergeeva M, Reiser G. Role of Ca^{2+}-independent phospholipase A_2 and n-3 polyunsaturated fatty acid docosahexaenoic acid in prostanoid production in brain: perspectives for protection in neuroinflammation. *Int J Dev Neurosci*. 2004;22:551–557.
43. Lepretti M, Martucciello S, Burgos Aceves MA, Putti R, Lionetti L. Omega-3 fatty acids and insulin resistance: focus on the regulation of mitochondria and endoplasmic reticulum stress. *Nutrients*. 2018;10.
44. Baxter PS, Hardingham GE. Adaptive regulation of the brain's antioxidant defences by neurons and astrocytes. *Free Radic Biol Med*. 2016;100:147–152.
45. St-Pierre J, Buckingham JA, Roebuck SJ, Brand MD. Topology of superoxide production from different sites in the mitochondrial electron transport chain. *J Biol Chem*. 2002;277:44784–44790.
46. Hoffman DL, Brookes PS. Oxygen sensitivity of mitochondrial reactive oxygen species generation depends on metabolic conditions. *J Biol Chem*. 2009;284:16236–16245.
47. Perevoshchikova IV, Quinlan CL, Orr AL, Gerencser AA, Brand MD. Sites of superoxide and hydrogen peroxide production during fatty acid oxidation in rat skeletal muscle mitochondria. *Free Radic Biol Med*. 2013;61:298–309.
48. Cortassa S, Sollott SJ, Aon MA. Mitochondrial respiration and ROS emission during β-oxidation in the heart: an experimental-computational study. *PLoS Comput Biol*. 2017;13:e1005588.
49. Schönfeld P, Reiser G. Rotenone-like action of the branched-chain phytanic acid induces oxidative stress in mitochondria. *J Biol Chem*. 2006;281:7136–7142.
50. Schönfeld P, Wojtczak L. Fatty acids decrease mitochondrial generation of reactive oxygen species at the reverse electron transport but increase it at the forward transport. *Biochim Biophys Acta*. 2007;1767:1032–1040.
51. Nordmann C, Strokin M, Schönfeld P, Reiser G. Putative roles of Ca^{2+}-independent phospholipase A_2 in respiratory chain-associated ROS production in brain mitochondria: influence of docosahexaenoic acid and bromoenol lactone. *J Neurochem*. 2014;131:163–176.
52. Schönfeld P, Schlüter T, Fischer KD, Reiser G. Non-esterified polyunsaturated fatty acids distinctly modulate the mitochondrial and cellular ROS production in normoxia and hypoxia. *J Neurochem*. 2011;118:69–78.

53. Jaburek M, Jezek J, Zelenka J, Jezek P. Antioxidant activity by a synergy of redox-sensitive mitochondrial phospholipase A_2 and uncoupling protein-2 in lung and spleen. *Int J Biochem Cell Biol*. 2013;45:816–825.
54. Chao H, Liu Y, Fu X, et al. Lowered iPLA2γ activity causes increased mitochondrial lipid peroxidation and mitochondrial dysfunction in a rotenone-induced model of Parkinson' disease. *Exp Neurol*. 2018;300:74–86.
55. Strokin M, Chechneva O, Reymann KG, Reiser G. Neuroprotection of rat hippocampal slices exposed to oxygen-glucose deprivation by enrichment with docosahexaenoic acid and by inhibition of hydrolysis of docosahexaenoic acid-containing phospholipids by calcium independent phospholipase A_2. *Neuroscience*. 2006;140:547–553.
56. Jump DB. Dietary polyunsaturated fatty acids and regulation of gene transcription. *Curr Opin Lipidol*. 2002;13:155–164.
57. Horrocks LA, Farooqui AA. Docosahexaenoic acid in the diet: its importance in maintenance and restoration of neural membrane function. *Prostaglandins Leukot Essent Fatty Acids*. 2004;70:361–372.
58. Mancuso DJ, Sims HF, Han X, et al. Genetic ablation of calcium-independent phospholipase $A_2γ$ leads to alterations in mitochondrial lipid metabolism and function resulting in a deficient mitochondrial bioenergetic phenotype. *J Biol Chem*. 2007;282:34611–34622.
59. Mancuso DJ, Sims HF, Yang K, et al. Genetic ablation of calcium-independent phospholipase A2γ prevents obesity and insulin resistance during high fat feeding by mitochondrial uncoupling and increased adipocyte fatty acid oxidation. *J Biol Chem*. 2010;285:36495–36510.
60. Moon SH, Jenkins CM, Kiebish MA, Sims HF, Mancuso DJ, Gross RW. Genetic ablation of calcium-independent phospholipase $A_2γ$ ($iPLA_2γ$) attenuates calcium-induced opening of the mitochondrial permeability transition pore and resultant cytochrome *c* release. *J Biol Chem*. 2012;287:29837–29850.
61. Mancuso DJ, Kotzbauer P, Wozniak DF, et al. Genetic ablation of calcium-independent phospholipase $A_2γ$ leads to alterations in hippocampal cardiolipin content and molecular species distribution, mitochondrial degeneration, autophagy, and cognitive dysfunction. *J Biol Chem*. 2009;284:35632–35644.
62. Kinghorn KJ, Castillo-Quan JI, Bartolome F, et al. Loss of PLA2G6 leads to elevated mitochondrial lipid peroxidation and mitochondrial dysfunction. *Brain*. 2015;138:1801–1816.
63. Strokin M, Reiser G. Mitochondria from a mouse model of the human infantile neuroaxonal dystrophy (INAD) with genetic defects in VIA $iPLA_2$ have disturbed Ca^{2+} regulation with reduction in Ca^{2+} capacity. *Neurochem Int*. 2016;99:187–193.
64. Carafoli E, Krebs J. Why calcium? How calcium became the best communicator. *J Biol Chem*. 2016;291:20849–20857.
65. Rizzuto R, Brini M, Murgia M, Pozzan T. Microdomains with high Ca^{2+} close to IP_3-sensitive channels that are sensed by neighboring mitochondria. *Science*. 1993;262:744–747.
66. Denton RM, McCormack JG. On the role of the calcium transport cycle in heart and other mammalian mitochondria. *FEBS Lett*. 1980;119:1–8.
67. Brustovetsky T, Antonsson B, Jemmerson R, Dubinsky JM, Brustovetsky N. Activation of calcium-independent phospholipase a (iPLA) in brain mitochondria and release of apoptogenic factors by BAX and truncated BID. *J Neurochem*. 2005;94:980–994.
68. Gadd ME, Broekemeier KM, Crouser ED, Kumar J, Graff G, Pfeiffer DR. Mitochondrial iPLA2 activity modulates the release of cytochrome *c* from mitochondria and influences the permeability transition. *J Biol Chem*. 2006;281:6931–6939.
69. Patil VA, Greenberg ML. Cardiolipin-mediated cellular signaling. *Adv Exp Med Biol*. 2013;991:195–213.
70. Paradies G, Paradies V, De Benedictis V, Ruggiero FM, Petrosillo G. Functional role of cardiolipin in mitochondrial bioenergetics. *Biochim Biophys Acta*. 2014;1837:408–417.
71. Tyurina YY, Poloyac SM, Tyurin VA, et al. A mitochondrial pathway for biosynthesis of lipid mediators. *Nat Chem*. 2014;6:542–552.
72. Gaspard GJ, McMaster CR. The mitochondrial quality control protein Yme1 is necessary to prevent defective mitophagy in a yeast model of Barth syndrome. *J Biol Chem*. 2015;290:9284–9298.

CHAPTER

10

Omega-3 and Cognition in Children with Malnutrition

Veronica Portillo Reyes, John William Capps†, Yolanda Loya Méndez*, Gilberto Reyes Leal*, Graciela Avitia**

*Universidad Autónoma de Ciudad Juárez, Cd. Juárez, Chih, Mexico †The University of Texas at El Paso, El Paso, TX, United States

OMEGA-3

Malnutrition generally refers to deficiencies, excesses, or imbalances in energy and/or nutrient intake and encompasses three categories including undernutrition, overweight/obesity, and micronutrient malnutrition.[1,2] The consequences of malnutrition have significant and lasting effects which impact economic, social, and healthcare systems globally.[3] The magnitude of the negative impact of malnutrition may be greatest when considering one of the most affected populations, children. According to the World Health Organization, 224 million children under the age of 5 are under nourished, where 41 million are overweight or obese.[4] Likewise, 45% of deaths of this group are linked to undernutrition and obesity is in the rise in almost every country worldwide.[4] Malnutrition is not only related to a range of noncommunicable diseases but is considered to cause a range of issues which negatively impact the individual over their entire lifespan[5,6] and future generations via epigenetic mechanisms.[7] One of the primary concerns with child malnutrition is the negative impact it has on brain development and cognitive functioning altering physical and psychological characteristics and wellbeing of the child and their future self.

During development, the central nervous system (CNS) and brain are susceptible to a variety of factors which influence brain development and cognitive functioning.[8] These factors include both environmental and biological factors (i.e., genes and cellular mechanisms) and their interaction.[9] Nutrition is an especially significant environmental factor because of the influence it has on gene expression and epigenetics[10] and the specific cellular and developmental mechanisms which depend on essential nutrients for proper function and growth.[11] Likewise, with increasing duration of malnutrition, these problems are exacerbated increasing the long-term and more serious consequences and lifelong residual effects.[12]

https://doi.org/10.1016/B978-0-12-815238-6.00010-9

Specific nutrients are critical during different developmental epochs and for specific cellular mechanisms which underlie brain development and cognitive functioning. One nutrient of significant importance to both development and day-to-day functioning is omega-3 fatty acids. The following chapter reviews the specifics of how malnutrition in children can impact development and cognitive performance. Details are provided on the specific influence of both undernutrition and overweight/obesity on cognitive performance. Moreover, the bulk of the following specifically highlights the empirical evidence demonstrating the effect that Omega-3 fatty acid consumption or supplementation has on cognition and neuropsychological performance in children with malnutrition.

MALNUTRITION AND COGNITION IN CHILDREN

The World Health Organization (WHO) defines malnutrition as "the cellular imbalance between the supply of nutrients and energy and the body's demand to ensure growth, maintenance and specific functions"[13] resulting in accumulated energy, protein, or micronutrient deficits that may negatively affect growth, development, and other pertinent outcomes.[14] It has been demonstrated that children who experience retardation in growth during early childhood have increased probabilities of cognitive, motor, and language deficits along with poorer outcomes in academic performance and slowed cognitive development.[15]

In comparison to the body, the brain develops rapidly during the first years of life reaching 80% of adult weight by age 2,[16] but this rapid growth increases vulnerabilities to the negative effects of nutritional deficiencies.[17] Animal models have demonstrated that malnutrition during this period of rapid development can alter the number of cell,[18] the migration of cells,[19] neuronal myelination,[20] formation of synapses,[21] hippocampal formation,[22] and neurotransmission.[23] In addition, undernourished children may have less energy which may result in less exploratory behaviors and interest in learning which also impacts cognitive development.[24] This association between malnutrition and development of cognition is well established by cross-sectional studies, however, only a few longitudinal studies have been conducted and no international statistics exist which demonstrate the actual progress of cognitive development in children.[25]

Various findings which demonstrate the influence malnutrition has on child development and cognition have arisen from specific geographical incidents, such as famine or lacking a specific nutrient in a local area. However, it is important to consider that nutrients do not act in isolation but act within complex system and sequences which are integral to specific and general mechanisms which require the provision of other micronutrients.[15] This is especially true when considering the impact of nutrition on brain development and cognitive functioning.

IMPACT OF UNDERNUTRITION ON COGNITIVE FUNCTION

The impact of undernutrition, that is, lacking in energy and/or specific micronutrients, had been demonstrated to impact brain development and cognitive function starting during pregnancy through adolescence. The nutritional guidelines for pregnant mothers and infants have been greatly influenced by findings showing theses critical effect of nutrition. Malnutrition may begin before the child is even born, when the mother is undernourished, as is often the case in low-income countries.[26] However, in western countries, delays in fetal growth is often the result of a medical conditions for example with severe hypertension,[27] or if the mother consumes higher levels of alcohol.[28]

Also undernourishment during this period is associated with decreased birth weight, delays in cognitive development and with a decrease of 4–8 points in IQ scores compared to children with normal birth weight.[29] Frequently malnutrition during pregnancy is also followed by poor nutrition during the first months and years of life further resulting in severely stunted growth.[30] Stunting of growth in early childhood caused by a nutritional deficiency is associated with poorer cognitive development, academic performance in later childhood,[25] and specifically protein-energy malnutrition can cause delays in growth of the hippocampus and cortex.[31] Furthermore, reviews of the both animal and human studies have shown that mild but persistent malnutrition during the first 2 years of life negatively affects reasoning, visuospatial functions, language development, attention, academic achievement, IQ, and learning.[32] Research on brain development and cognition have demonstrated that some nutrients may more influential than others, including proteins, energy, certain fats, iron, zinc, copper, iodine, selenium, vitamin A, choline, and folic acid. Micronutrient deficiencies are a major concern among children in developing countries, which affect their growth and cognitive development.[33] The specific profiles or lack of resources with regional diets which have little variation may result in lack of important micronutrients essential for development.[34]

IMPACT OF OVERWEIGHT/OBESITY ON COGNITIVE FUNCTION

Obesity is a special form of malnutrition considered as over nutrition, as opposed to undernutrition, but it is considered malnutrition because it is likely that the diet has a low density of nutrients along with a high content of fats and carbohydrates.[35] Obesity represents a rapidly growing problem greatly affecting developed regions and is steadily increasing in developing countries[36] and is considered a world pandemic[37]. Childhood obesity and the associated burden of associated diseases and disorders are greatly impacting healthcare systems in the United States and abroad, and the increase in prevalence continues to rise.[38] In 2010, it was estimated that 43 million children worldwide, including 35 million in developing countries, were overweight or obese, and it is expected that this number will continue to grow to 60 million by 2020.[39]

Childhood obesity is not only associated with many major adverse health consequences but predisposes them as adults to hypertension, dyslipidemia, diabetes, cardiovascular disease (CVD), stroke, and many types of cancer.[39] Correspondingly, childhood obesity is associated with poorer cognitive function and future outcomes such as academic performance and cognitive functioning.[40] Dietary components such as high saturated fats and sugars may also have a detrimental impact on brain function although much of this evidence is derived from animal models.[41] Brain-derived neurotrophic factor (BDNF) appears to function at the crossroads of cognitive and metabolic regulation.[42] BDNF modulates insulin resistance and glucose metabolism and deletion or polymorphisms of BDNF are related to abnormal hippocampal function among rodent and human studies.[43] Exposure to a diet high in saturated fat and sucrose, independent of obesity, was related to decreased BDNF in the hippocampus and poorer learning and memory performance.[44]

A recent review of the literature concluded that overweight and obesity may result in lower academic performance as measured as literacy, numeracy, and school grades,[45] but only a few studies have investigated a possible connection between being obesity and/or overweight and cognitive performance in children. Associations have been demonstrated between overweight status and poor school performance among all children from 7th to 9th grade, but not from 3 to 6.[46] It has been found that obese children have a worse performance in mental updating, understanding of language and memory in comparison with normal weight children.[47] Likewise, others have observed that, among Chinese elementary school children, children with severe obesity had a significantly lower IQ than controls.[48] An association between the BMI percentile category and cognitive impairment in the visuospatial organization and general mental capacity, the importance of these findings can be judged by noting that the adverse effects of increasing body weight status on the functioning of cognition begin to play as early as in childhood.[49]

DEFICIENCIES OF SPECIFIC MICRONUTRIENTS AND BRAIN DEVELOPMENT

It is well established that nutritional deficiencies can affect the brain and alter development and subsequent behavior permanently as seen with the prime examples of iron, folic acid, zinc, and iodine deficiency. Studies of these micronutrients have established evidence which has been used to inform public policy and community interventions to better health outcomes and promote healthy development in children. Furthermore, this area of research demonstrates the increased need to better understand the influence of malnutrition, especially considering the influence of specific nutrients, on important outcomes which impact may affect the individual over their whole lifespan and even future generations.

The importance of nutrition to development is highlighted by the following specific micronutrients and their impact on cognitive functioning, likewise, it is important to consider the implications of malnutrition where the child or the pregnant mother is lacking basic sustenance. Any of the micronutrient deficiencies may occur alone or concurrently in individuals lacking sufficient caloric intake exacerbating the impact of malnutrition on development and cognition, especially in the epochs of development where specific micronutrients and energy are critical.

Iron deficiency anemia is the most prevalent nutritional deficiency worldwide, and it is estimated that up to half of all children worldwide are deficient in iron.[50] In children, the relationship between iron and cognitive development has been well researched, however, there are mixed results.[15] Meta-analysis has demonstrated that short-term iron treatment for anemia in children under 3 did not improve cognitive development, however in older children, there have been results indicated that supplementation did increase IQ.[51–53] Iron appears to modify the development processes in hippocampal neurons by altering dendritic growth[54] which may cause long-lasting effects on neural structure and behavior.[55] If a deficiency of iron occurs in the 6–12 months of life, the damage may be irreversible.[56]

Folic acid is another example of a nutrient vital to brain development. Deficiency of folic acid between 21 and 28 days after embryo is fertilized and implanted, when the neural tube is closed, predisposes the fetus to a congenital malformation, called a neural tube defect.[57] If there is not enough folic acid, in this critical period, there is an irreversible change in the structure and function of the brain, supplementation of folic acid can prevent this and associated fetal mortality.[58] Furthermore, vitamin B12 is often given alongside folic acid because of its critical and connected roles in metabolism mechanisms. Since folate and vitamin B12 are metabolically associated, deficiency of one can influence the functioning of the other.[59] Like folic acid, B12 deficiency has been showed to cause brain atrophy and demyelination of nerve cells that directly affect cognitive and language development in the long term.[60] An important function of B12 in the nervous system is the metabolism of the fatty acids necessary for the maintenance of the myelin sheath surrounding the axon and with long-term deficiency can cause irreversible brain damage.[59]

Another major problem worldwide is zinc deficiency, affecting 40% of the world's population.[61] Zinc is involved in the process of neurogenesis,[62] maturation and migration of neurons,[63] and overall in motor functioning[64] and cognitive development.[65] Likewise it is plays a critical role in in synaptic transmission[66] especially in hippocampal neurons, involved in learning and memory[67] and is known to modulate neurotransmitters, including glutamate and gamma-aminobutyric acid receptors (GABAs).[68] Evidence has demonstrated changes motor development and cognitive performance,[69–71] however some have demonstrated no benefits[72,73] and the exact mechanism which zinc is acting on behavior and cognitive development is undecided.[74]

Additionally, iodine deficiency is a significant global problem, which is especially important in developing children and during pregnancy[75] and can result in hypothyroidism, causing an insufficient production of thyroid hormones.[76] These hormones play a critical role in brain development and neurological processes, including differentiation of neuronal cells, maturation and migration, myelination, neurotransmission, and synaptic plasticity.[77,78] Past meta-analysis of the importance of iodine to cognitive development has shown significant effects on measures of cognition.[79] Similarly, a more recent meta-analysis using more advanced techniques and only studies from China, where the soil has a severe iodine deficiency, show similar results. It was found that there was a decrease of 12.3 points in the IQ of those children whose mothers lived in regions with iodine deficient soil as compared to those who lived in regions with sufficient soil iodine concentrations.[80] However, the association between mild to moderate maternal iodine deficiency and cognitive development is not as clear as it is when iodine deficiency is severe.[81]

As seen with iron, folic acid, zinc, and iodine micronutrients deficiencies of essential fatty acids are now considered to play a major role in development and cognitive function. Of interest are omega-3 polyunsaturated fatty acids (PUFAs) and its importance in biological mechanisms, brain structure, and behavior. The primary role that omega-3 plays in brain and behavior can further be demonstrated by the findings from studies in which supplement of omega-3 in children with malnutrition.

OMEGA-3, BRAIN DEVELOPMENT, AND COGNITION IN MALNUTRITION

The role of omega-3 fatty acids has been the focus for a broad range psychological issues and pathologies in children ranging from behavioral problems, attention deficit hyperactive disorder (ADHD), psychotic mania, autism spectrum disorder, and mood disorders.[82]

The basis of this area of research is that omega-3 polyunsaturated fatty acids (PUFAs) are important for physiological functioning especially in the CNS and have been shown to associate with decreased neuroinflammation.[83] Like the previously mentioned micronutrients, humans depend on dietary sources for essential fatty acids including omega-3 and omega-6 fatty acids. Linoleic acid (LA), the omega-6 precursor, and α-linoleic acid (ALA), the omega-3 precursor, are major essential fatty acids. ALA can be modified by the body into eicosatetraenoic acid (EPA) and docosahexaenoic acid (DHA), however, these conversions are slow and inefficient.[84] The main dietary sources of EPA and DHA are fatty fish such as salmon, herring, tuna and halibut, while ALA comes mainly from plant sources such as canola oil, walnuts, and flaxseed.[85]

Both omega-3 and omega- 6 are beneficial for health, however omega-6 is more prevalently consumed in western diets, while omega-3 is consumed less[86] and children who consume high levels of omega-6 are at risk of consuming low levels of omega-3.[87] The ratio of omega-6 to omega-3 consumption is of concern in westernized diets, as ratios are as high as 10:1 to 20:1 and are a large departure from the 1:1 ratio proposed by evolutionary perspectives.[88] This ratio is thought to be important to proper balance of lipids in brain tissues and brain functioning.[89] Increased blood serum level ratios of omega-6 to omega-3 have also been found to associate with increased risk for obesity, increased symptoms of ADHD, and increased symptoms of Autism Spectrum Disorder.[90] Lack of omega-3 from dietary sources produces deficiencies in the biological systems because of the inefficient conversion, which promotes inflammation, pathogenesis of disease, and hinders optimal brain functioning.[91]

The role of fatty acids in development and cognition is important as apparent when looking at brain tissue composition and the role that fatty acids play in brain cellular functioning and structure.[92] Sixty percent of brain tissue is composed of lipids[93] and is unique in its levels of specific fatty acids, with high levels of AA and DHA and lower levels of EPA.[94] The consumption of these essential fatty acids helps to conserve the neuronal membranes of this unique lipid composition which vital for the proper functioning of the CNS.[95] Specifically, 30%–40% of the lipids which comprise the cellular membranes in gray matter are DHA and higher concentrations are found synaptic membranes.[96] Omega-3 PUFAs have been associated with the integrity of the CNS and are important for development and functioning of the CNS.[97]

Essential fatty acids, especially DHA, contribute to the optimal functioning of neuronal membranes and neurotransmission, including the influence on inflammation, deficiencies of fatty acids are associated with problems with behavior, learning, and cognition.[98,99] Likewise, studies have shown that deficiency of omega-3 is associated with problems of the endocannabinoid system which helps regulate neuronal activities associated with mood and behavior.[100] Animal models demonstrate that maternal deficiency of omega-3 during gestation affects the development of the brain in the offspring and was associated with offspring spatial memory problems.[101] Omega-3 deficiency has also be associated with problems in learning and brain functioning and these behavioral deficits are thought to be the result of problems with the synthesis of phospholipids (which make up brain cell membranes), transportation and utilization of glucose, and decreased neuron size in the hippocampus, hypothalamus, and cerebral cortex.[98]

The regulation of membrane fluidity and gene expression is believed to be particularly important during infant and childhood development.[102] Likewise, problems in the fatty acid metabolism are also related with neurodevelopmental disorders such as attention deficit hyperactivity disorder, dyslexia, dyspraxia, and autistic spectrum disorder.[103] However, most research on the association between omega-3 intake and cognitive development has focused on preterm or low-birth-weight babies.[8]

The accumulation of DHA in the brain begins in the uterus, mainly in the second half of pregnancy, when gray matter growth accelerates.[104] In the last trimester of gestation and the first months of life DHA storage in the brain continues to increase, this is when neurogenesis is very active and neuroblast migration, differentiation and synaptogenesis occur.[105] Therefore preterm infants, who do not receive the intrauterine supply of DHA in the third trimester, they may have DHA deficiencies which will make them vulnerable to reductions in neuronal growth and decreased regional cortical gray matter volumes.[105]

As realized with other micronutrients, Omega-3 deficiency has been the focus of investigations to improve health and psychological functioning associated with the bulk of these studies have centered around the first couple years of life when development is rapidly occurring. However, it is important to consider that development is an ongoing process and is not uniform across individuals, especially when considering cognitive development.[106] Likewise, effects from early malnutrition are seen to cause lasting effects on cognitive abilities into adolescence and even young adulthood.[106,107]

When considering normal cognitive development, research has demonstrated beneficial effects from nutritional supplementation of omega-3. A recent systematic review and meta-analysis conducted by Shulkin and colleagues demonstrated that omega-3 supplementation improves child cognitive development.[108] Their meta-analysis included 15 randomized controlled trials including 2525 children with 20 different intervention types (supplementation with different combinations of DHA, EPA, AA at different time periods). They demonstrated that these studies, which had an average supplementation duration of about 7 months and measures taken after an average of 16 months, that both material and infant supplementation improved cognitive development as measured by the Bayley Scales of Infant Development (BSID) having a standard mean difference (SMD) of 0.21 and 0.24, respectively. The effect size was greatest (SMD = 0.40), on the psychomotor development subscale of the BSID when DHA and/or EPA was used. However, the combined use of DHA and AA had the greatest effect size (SMD = 0.17) on the mental development index used. The use of high-quality RCTs and differential intervention types aids in bettering the understanding of the effects of omega-3 supplementation in normal development, showing that there are clear benefits. Likewise, it is equally important to explore the differences in children with malnutrition, as this population represents one that is highly susceptible to cognitive deficits which are caused by lack of energy and nutrients to ensure optimal growth and development.

SUPPLEMENTATION AND IMPACT ON COGNITION WITH UNDERNUTRITION

In respect to undernutrition, it is important to explore avenues to aid children with the associated health and cognitive issues which may arise, especially in respect to cognitive functioning. This large, growing, and underserved population characterizes a unique group where early intervention aids development and functioning throughout the lifespan.

However, there is limited research on the direct effects of only omega-3 supplementation on cognitive function in undernourished children, especially when it comes to studies which use high-quality methodology such as randomized controlled trials. Furthermore, many of the results from studies connecting changes in cognition are controversial.

The use of other types of supplementation methods has been explored as well. Dalton et al. used a fish-flour spread to deliver omega-3 PUFAs to children to better understand the influence on cognitive outcomes.[109] One hundred and eight three children ages 7–9 who attended a primary school which served low SES community in Northern Cape Province of South Africa were randomized into two groups. Of the children who were participated in the study, anthropometric indices indicated that malnutrition was present, approximately 26% were classified as having stunting, 23.6% were classified as underweight, and 9.7% were classified as having wasting. No significant differences were seen between control or treatment groups in any demographic or anthropometric measurements. Both groups received two slices of bread with a 25 g of spread one which had marine fish flour and was designed to deliver 892 mg of DHA per week. This was provided to the children over a 6-month period for 104 days only taking place when school in session. Children were evaluated using the Hopkins Verbal Learning test (HVLT), a Reading test, and a Spelling test. This single-blind study found that for two of the subtests of the HVLT, that there were marginally significant treatment effects and for two others significant treatment effects, demonstrating that the treatment group was able to recall and recognize more words from the learning phase of the test. Likewise, it was seen that the treatment group had a significant increase in the Reading test, however, for the Spelling test did not change significantly from baseline. However, the control groups scores on the Spelling test actually declined significantly from baseline, which may have been due to high temperatures experienced during the follow-up testing. The authors indicate that this increase in cognitive performance is likely due to the community have no intake of fatty fish and very low consumption of lean fish and the supplementation increasing blood levels of DHA in the treatment group. However, these findings come from children with and without undernutrition, so it makes it difficult to indicate the true effect that omega-3 is having on children with undernutrition. About 20% of their low SES community sample could be classified as having undernutrition but similar studies have found mixed results of omega-3 using such samples.

Muthayya and colleagues conducted research with children ages 6–10 in two primary schools serving low SES communities in Bangalore, India. This double-blind randomized controlled trial used four different treatment groups to determine the effects of high and low micronutrient supplementation and high and low omega-3 PUFA supplementation. A total of 598 children participated and intervention lasted 12 months. After the intervention all four groups had significant improvements in weight and height, however, there were no significant differences between low and high omega-3 PUFA supplements. Likewise, measures of cognitive performance increased in all treatment groups as well. Low micronutrient treatment was the most beneficial on cognitive performance at 12 months, however, no significant effects or interactions were found when exploring the differences between low and high omega-3 PUFA treatments. The authors purport that the use of only 100 mg of omega-3 might be too low to see effects and that the use of a placebo control group would aid in making comparisons. Likewise, the children from this study were not all malnourished, inclusion criteria stated that they must not be severely malnourished, which makes it difficult to make clear conclusions about the effects of omega-3 on cognition in malnourished children. Only a small percentage of children who were included in the study could be classified as undernourished with 15% being classified as stunting and 20% being classified as underweight. The effect of omega-3 supplementation on cognition has likewise been studied in combination with iron supplementation, in similar samples of children, who have iron deficiency and come from low-income villages in eastern South Africa.

To better understand the combined effect of iron and omega-3 supplementation, Baumgartner and colleagues employed a randomized placebo-controlled double-blind design.[110] They used this design to investigate the effects of iron supplementation, DHA/EPA, and the combination of Iron and DHA/EPA supplementation in 321 children ages 6–11 who had iron deficiency. Four treatment groups were used Iron and placebo, Placebo and DHA/EPA, Iron and DHA/EPA, and Placebo and Placebo. The intervention lasted for 8.5 months and cognitive performance was assessed using four subtests of the Kaufman Assessment Battery for children (KABC-II) and the HVLT at baseline and conclusion of treatment. The groups were assessed using anthropometric indices, but all different categories were included in the study. Notably, their sample had a wider variation in the sample proportion classified as having malnutrition. The largest proportion of the sample, 31.5% had mild stunting (height for age z-scores less than −1 SD but greater than −2 SD), next 21.5% were overweight, 7.2% obese, 6.2% stunting, and 2.1% as underweight. Mixed results were found treatment effects regarding changes in cognitive performance, iron supplementation showed small significant effects on recall for HVLT, however, DHA/EPA supplementation had a large negative effect for working memory and for girls with iron deficiency a large negative effect on long-term memory subtests. Boys in the DHA/EPA group tended to benefit from supplementation having higher scores on the long-term memory subtest. The authors support this finding by the example of an animal study which found that using DHA/EPA or iron alone in rats which are deficient in both leads to poorer performance on working memory and memory tasks. Furthermore, they state that both the mechanisms which iron and PUFAs play in neurotransmission and in inflammation may influence their findings. However, again their study did not specifically determine the impact of omega-3 on undernutrition, their sample had a large mixture of children with differential nutritional status and anthropometric categorization. With this in consideration, it is important to acknowledge that their findings may be imprecise when trying to make conclusions about the effects of omega-3 on cognition in children with undernutrition.

Portillo Reyes et al. addressed this gap in the literature by determining the effects of omega-3 supplementation in children ages 8–12 classified as having mild to moderate levels of malnourishment as defined by being 85%–95% in height/age and 70%–90% in weight/height.[111] The study used a randomized double-blind placebo-controlled and case-controlled designed which supplemented children over a 3-month period using a daily dose of 180 mg of DHA and 270 mg of EPA. The children who received the omega-3 supplementation at conclusion showed significant changes in tests of processing speed, visuoperceptual integration, visuoconstructive integration, attention, and executive functioning. When comparing the treatment group with the placebo group, 63% (12 of 19) of the neuropsychological measures showed changes indicating moderate to large effect sizes, as measured by Cohen's d, whereas the placebo group only showed changes in 10.5% (2 of 19) of measures. Likewise, when unique change for each participant was calculated as an index of clinical significant, it was found that 70% or more of children in the omega-3 supplementation treatment group had large improvements ($d > 0.80$) in processing speed (Symbol Search), coordination visuomotor (Block Design), perceptual integration (Embedded Figures), attention (Letter Cancellation), and executive functioning (Letter/Number Sequencing & Matrix Reasoning). Even though blood serum levels of omega-3 or the omega-6/3 ratio were not obtained, nutritional questionnaires indicated that this sample of mild to moderate malnourished children was not frequently consuming foods rich in omega-3; 8% indicated consuming a portion of fish two or more times per week, 39% one portion per week, and 19% one portion per 2 weeks. Likewise, typical Mexican diets, especially in northern regions, are high in omega-6. Considering this, the supplementation of omega-3 in this sample of children and the effects which are seen in cognitive functioning are in line with studies demonstrating the importance to physiological brain functions.

The improvements in visuoperception may be due to the effect of omega-3 that has on biological mechanisms underlying vision and visual processing, such as seen with newborn children.[112] Furthermore, electrophysiology studies also demonstrate that supplementation of omega-3 improves direct measures of neuro functioning underlying visual processing.[113] Findings of increased processing speed with omega-3 may point to the effect which it has on brain structure, brain cell membranes and myelin, and neurotransmission. Witte et al. found that omega-3 supplementation improved the structural integrity of white matter in older adults which others have found as related to the cognitive speed of processing.[114] Likewise, the importance of omega-3 to the composition of neuronal membranes may be driving the enhancement seen in processing speed via its influence on membrane fluidity in membrane protein incorporation and function,[115] signal transduction speed, and neurotransmission.[102] These same physiological benefits of omega-3 supplementation also may be driving improves in attention and executive functioning which rely heavily on the integration of information from different brain areas and quick processing speed to produce improvements in cognitive measures. These improvements are in line with findings from studies showing that omega-3 improves symptoms of attention deficit disorder[116,117] and reductions in decline seen in older adults.[118]

The effect of omega-3 consumption in undernourished children is substantial to aid in the development process and lifelong functioning when it comes to both physiological and psychological well-being. Early intervention with omega-3 or nutritional education may be able to aid these children's functional role in society. This large and underserved population is important however, it is also important to understand how omega-3 consumption may influence the deficits which are seen with overnutrition. Childhood obesity is a growing problem and the long-term physiological, psychological, and societal consequences are just being revealed, however, the small number of studies available does seem to indicate that omega-3 is just as important to the overweight and obese children as it is with undernourished child populations in the same and other unique ways.

SUPPLEMENTATION AND IMPACT ON COGNITION WITH OVERWEIGHT/OBESE CHILDREN

The consumption of more omega-3 fatty acids has been associated with enhancements in cognition (depending on the population) while on the other hand consumption of a westernized diet (high in saturated fats and calories) has been associated with brain physiology[119,120] and is associated with poorer cognitive function.[121] Researchers have started to fill the gap on how omega-3 fatty acids influence cognitive functioning, especially concerning childhood obesity, where early prevention and treatment can maximize individual and societal benefits by deterring lifetime concurrency of disease and comorbid conditions. However, there are large gaps in the literature concerning the effects of omega-3 fatty acids or the ratio of omega-6 to omega-3 with respect to cognitive performance in children who are overweight or obese. Furthermore, the evidence is now growing which demonstrates that increased dietary intake of omega-3 fatty acids may aid in prevention and treatment of obesity via actions on the mechanisms underlying appetite.[122]

Regarding the relationship of omega-3 and cognitive functioning in childhood obesity, very limited evidence exists. Haapala and colleagues conducted a study with 386 normal weight and 58 overweight children ages 6–8 years old and found that higher levels of EPA and DHA in blood plasma of the obese children were associated with higher scores on the nonverbal reasoning (as measured by Raven's Colored Progressive Matrices).[122] However, they did not find any significant associations between normal-weight children's level of blood EPA and DHA and scores. They point out that earlier studies which investigated the levels of EPA and DHA have had mixed results in respect to cognition in normal weight children, only combined levels of blood EPA and DHA had an association with working memory in one study[123] whereas another study found no association.[124]

They highlight that their study included child weight status as a modifying factor, which have a stronger influence on cognitive performance via reductions in inflammation, which is known to be associated with excess body fat. However, their study only found associations with levels of EPA and DHA in plasma triacylglycerols, which reflect recent fatty acid consumption and not with fatty acids in plasma cholesteryl esters or phospholipids, which reflect consumption over the prior months. However, their findings of these direct associations do reflect findings from Harris and colleagues which found that children who had increased levels of palmitic acid and arachidonic acid had associated biomarkers of increased low-grade inflammation, whereas omega-6 and total polyunsaturated fatty acids were associated with decreased inflammation.[125] Furthermore, the findings from Haapala et al. also agree with findings from studies using animal models, adult populations, and aging populations.[126–128]

CONCLUSIONS

The investigation of cognitive development and the relationship between diet and cognition are both emerging fields. However, some clear connections have been made concerning normal development and nutrition which has arisen from observations of the consequences when diets are restricted, deficient, or have excesses. The role that essential fatty acids play in the brain and cognition, especially concerning omega-3, biological are critical to structure and function; however, mixed results have been seen with normal populations. Even though the research on child development is large, the exact timeline of cognitive development has no standards making it hard to compare and indicate how omega-3 is influencing developmental mechanisms and the observable cognitive performance.

However, when examining the studies which explore populations with malnutrition, the small amount of evidence seems to indicate a large benefit to development and cognition. However, there is a definite lack of research which uses high-quality designs and methodologies and specific populations of malnourished to determine the differential effects of omega-3 in this varied umbrella category. Future investigations should consider looking at specific profiles of malnourished children (i.e., mild undernourished, severely undernourished) to better determine the influence of omega-3 consumption on cognitive functioning. Many of the studies try to use omega-3 supplementation as a community intervention with low SES children to better cognitive outcomes, however, fail to make specific comparisons within the different profiles of these children.

Public policy and nutritional agencies are taking heed to the evidence with suggestions about omega-3 consumption as expert committees have started recommending it during pregnancy[129] and for adult populations for cardiovascular health.[130] Likewise, nutritional agencies are including omega-3 suggestions.[131] However, the greatest magnitude of impact may be in the growing number of children who consume a western diet or those who live in low SES communities. With the growing rates of childhood obesity, due partly to the consumption of calorie-dense but nutritionally deprived foods,[132] there is a great concern, especially when considering the effects of malnutrition on cognitive functioning. The deficits in cognitive function along with long-term epigenetic consequences may be impossible to factor in when calculating the overall impact malnutrition is having globally.

References

1. Elia M. Defining, recognizing, and reporting malnutrition. *Int J Low Extrem Wounds*. 2017;16(4):230–237.
2. Cederholm T, Barazzoni R, Austin P, et al. ESPEN guidelines on definitions and terminology of clinical nutrition. *Clin Nutr*. 2017;36(1):49–64.
3. Alderman H. Long term consequences of early childhood malnutrition. *Oxford Econ Pap*. 2006;58(3):450–474.
4. World Health Organization. Malnutrition. http://www.who.int/news-room/fact-sheets/detail/malnutrition. Published February 16, 2018. Accessed April 4, 2018.
5. Roseboom TJ, Meulen JHVD, Ravelli AC, Osmond C, Barker DJ, Bleker OP. Effects of prenatal exposure to the Dutch famine on adult disease in later life: an overview. *Mol Cell Endocrinol*. 2001;185(1–2):93–98.
6. Susser E, St Clair D. Prenatal famine and adult mental illness: interpreting concordant and discordant results from the Dutch and Chinese famines. *Soc Sci Med*. 2013;97:325–330.
7. Szutorisz H, Hurd YL. Feeding the developing brain: the persistent epigenetic effects of early life malnutrition. *Biol Psychiatry*. 2016;80(10):730–732.
8. Rosales FJ, Reznick JS, Zeisel SH. Understanding the role of nutrition in the brain and behavioral development of toddlers and preschool children: identifying and addressing methodological barriers. *Nutr Neurosci*. 2009;12(5):190–202.
9. Bryan J, Osendarp S, Hughes D, Calvaresi E, et al. Nutrients for cognitive development in school-aged children. *Nutr Rev*. 2004;62(8):295–306.
10. Attig L, Gabory A, Junien C. Early nutrition and epigenetic programming: chasing shadows. *Curr Opin Clin Nutr Metab Care*. 2010;13(3):284–293.
11. Hibbeln JR, Gow RV. Omega-3 fatty acid and nutrient deficits in adverse neurodevelopment and childhood behaviors. *Child Adolesc Psychiatr Clin N Am*. 2014;23(3):555–590.
12. Barrett DE, Radke-Yarrow M, Klein RE. Chronic malnutrition and child behavior: effects of early caloric supplementation on social and emotional functioning at school age. *Dev Psychol*. 1982;18(4):541–556.
13. De Onís M, Monteiro C, Akré J, Glugston G. The worldwide magnitude of protein-energy malnutrition: an overview from the WHO global database on child growth. *Bull World Health Organ*. 1993;71(6):703–712.
14. Mehta NM, Corkins MR, Beth L, et al. Defining pediatric malnutrition. *J Parenter Enteral Nutr*. 2013;37(4):460–481.

15. Nyaradi A, Li J, Hickling S, Foster J, Oddy WH. The role of nutrition in children's neurocognitive development, from pregnancy through childhood. *Front Hum Neurosci*. 2013;7:.
16. Bryan J, Osendarp S, Hughes D, Calvaresi E, Baghurst K, Klinken J-W. Nutrients for cognitive development in school-aged children. *Nutr Rev*. 2004;62(8):295–306.
17. Benton D. The influence of dietary status on the cognitive performance of children. *Mol Nutr Food Res*. 2010;54(4):457–470.
18. Babu S, Sinha RA, Mohan V, et al. Effect of hypothyroxinemia on thyroid hormone responsiveness and action during rat postnatal neocortical development. *Exp Neurol*. 2011;228(1):91–98.
19. Gilbert ME, Ramos RL, McCloskey DP, Goodman JH. Subcortical band heterotopia in rat offspring following maternal Hypothyroxinaemia: structural and functional characteristics. *J Neuroendocrinol*. 2014;26(8):528–541.
20. Yusuf HKM, Haque Z, Mozaffar Z. Effect of malnutrition and subsequent rehabilitation on the development of mouse brain myelin. *J Neurochem*. 1981;36(3):924–930.
21. Velasco I, Bath S, Rayman M. Iodine as essential nutrient during the first 1000 days of life. *Nutrients*. 2018;10(3):290.
22. Granados-Rojas L, Larriva-Sahd J, Cintra L, Gutiérrez-Ospina G, Rondán A, Díaz-Cintra S. Prenatal protein malnutrition decreases mossy fibers-CA3 thorny excrescences asymmetrical synapses in adult rats. *Brain Res*. 2002;933(2):164–171.
23. Debassio W, Kemper T, Tonkiss J, Galler J. Effect of prenatal protein deprivation on postnatal granule cell generation in the hippocampal dentate gyrus. *Brain Res Bull*. 1996;41(6):379–383.
24. Mathangi DC, Namasivayam A. Effect of chronic protein restriction on motor co-ordination and brain neurotransmitters in albino rats. *Food Chem Toxicol*. 2001;39(10):1039–1043.
25. Engle PL, Fernández PD. INCAP studies of malnutrition and cognitive behavior. *Food Nutr Bull*. 2010;31(1):83–94.
26. Grantham-McGregor S, Cheung YB, Cueto S, Glewwe P, Richter L, Strupp B. Developmental potential in the first 5 years for children in developing countries. *The Lancet*. 2007;369(9555):60–70.
27. Henriksen T, Clausen T. The fetal origins hypothesis: placental insufficiency and inheritance versus maternal malnutrition in well-nourished populations. *Acta Obstet Gynecol Scand*. 2002;81(2):112–114.
28. Carter RC, Jacobson JL, Molteno CD, Dodge NC, Meintjes EM, Jacobson SW. Fetal alcohol growth restriction and cognitive impairment. *Pediatrics*. 2016;138(2):e20160775.
29. Pallotto EK, Kilbride HW. Perinatal outcome and later implications of intrauterine growth restriction. *Clin Obstet Gynecol*. 2006;49(2):257–269.
30. Dewey KG, Begum K. Long-term consequences of stunting in early life. *Matern Child Nutr*. 2011;7:5–18.
31. Lee K-H, Calikoglu AS, Ye P, D'Ercole AJ. Insulin-like growth factor-I (IGF-I) ameliorates and IGF binding Protein-1 (IGFBP-1) exacerbates the effects of undernutrition on brain growth during early postnatal life: Studies in IGF-I and IGFBP-1 transgenic mice. *Pediatr Res*. 1999;45(3):331–336.
32 Laus MF, Duarte Manhas Ferreira vales L, Braga Costa TM, Sousa Almeida S. Early postnatal protein-calorie malnutrition and cognition: A review of human and animal studies. Int J Environ Res Public Health 2011;8(2):590–612.
33. Díaz JR, Cagigas A, de las Rodríguez R. Micronutrient deficiencies in developing and affluent countries. *Eur J Clin Nutr*. 2003;57(S1):S70–S72.
34. Teodosio N, Lago E, Romani S, Guedes R. A regional basic diet from Northeast Brazil as a dietary model of experimental malnutrition. *Arch Latinoam Nutr*. 1990;40(4):533–547.
35. Tanumihardjo SA, Anderson C, Kaufer-Horwitz M, et al. Poverty, obesity, and malnutrition: an international perspective recognizing the paradox. *J Am Diet Assoc*. 2007;107(11):1966–1972.
36. Abarca-Gómez L, Abdeen ZA, Hamid ZA, et al. Worldwide trends in body-mass index, underweight, overweight, and obesity from 1975 to 2016: a pooled analysis of 2416 population-based measurement studies in 128·9 million children, adolescents, and adults. *The Lancet*. 2017;390(10113):2627–2642.
37. Ludwig DS. Epidemic childhood obesity: not yet the end of the beginning. *Pediatrics*. 2018;141(3):e20174078.
38. Malecka-Tendera E, Mazur A. Childhood obesity: a pandemic of the twenty-first century. *Int J Obes (Lond)*. 2006;30(S2):S1–S3.
39. Freemark M. Childhood obesity in the modern age: global trends, determinants, complications, and costs. *Pediatr Obes*. 2017;3–24.
40. Eichen DM, Appleton-Knapp S, Boutelle KN. Childhood obesity and cognitive function. *Pediatr Obes*. 2017;539–551.
41. Das UN. Obesity: genes, brain, gut, and environment. *Nutrition*. 2010;26(5):459–473.

42. Gómez-Pinilla F. Brain foods: the effects of nutrients on brain function. *Nat Rev Neurosci*. 2008;9(7):568–578.
43. Egan MF, Kojima M, Callicott JH, et al. The BDNF val66met polymorphism affects activity-dependent secretion of BDNF and human memory and hippocampal function. *Cell*. 2003;112(2):257–269.
44. Molteni R, Barnard R, Ying Z, Roberts C, Gómez-Pinilla F. A high-fat, refined sugar diet reduces hippocampal brain-derived neurotrophic factor, neuronal plasticity, and learning. *Neuroscience*. 2002;112(4):803–814.
45. Burkhalter TM, Hillman CH. A narrative review of physical activity, nutrition, and obesity to cognition and scholastic performance across the human lifespan. *Adv Nutrition*. 2011;2(2):201S–206S.
46. Mo-suwan L, Lebel L, Puetpaiboon A, Junjana C. School performance and weight status of children and young adolescents in a transitional society in Thailand. *Int J Obes (Lond)*. 1999;23(3):272–277.
47. Portillo V, Puente A, Pérez M, Loya Y. Deterioro neuropsicológico en niños mexicanos con estados nutricionales alterados: Desnutrición leve a moderada vs. obesidad. *Rev Neurophys Neurophys Neuroci*. 2011;11(2):133–146.
48. Li X. A study of intelligence and personality in children with simple obesity. *Int J Obes Relat Metab Disord: J Int Assoc Study Obesity*. 1995;19(5):355–357.
49. Li Y, Dai Q, Jackson JC, Zhang J. Overweight is associated with decreased cognitive functioning among school-age children and adolescents. *Obesity*. 2008;16(8):1809–1815.
50. World Health Organization. *Iron deficiency anemia: Assessment, prevention and control: a guide for program managers*; 2001.
51. Gauntlett E, Hugman R, Kenyon P, Logan P. A meta-analysis of the impact of community-based prevention and early intervention action. *Policy Res Paper*. 2001;11.
52. Sachdev H, Gera T, Nestel P. Effect of iron supplementation on mental and motor development in children: systematic review of randomized controlled trials. *Public Health Nutr*. 2005;8(02).
53. Falkingham M, Abdelhamid A, Curtis P, Fairweather-Tait S, Dye L, Hooper L. The effects of oral iron supplementation on cognition in older children and adults: a systematic review and meta-analysis. *Nutr J*. 2010;9(1).
54. Jorgenson LA, Wobken JD, Georgieff MK. Perinatal iron deficiency alters apical dendritic growth in hippocampal CA1 pyramidal neurons. *Dev Neurosci*. 2003;25(6):412–420.
55. Lozoff B, Beard J, Connor J, Felt B, Georgieff M, Schallert T. Long-lasting neural and behavioral effects of Iron deficiency in infancy. *Nutr Rev*. 2008;64:S34–S43.
56. Beard JL. Why Iron deficiency is important in infant development. *J Nutr*. 2008;138(12):2534–2536.
57. Botto LD, Moore CA, Khoury MJ, Erickson JD. Neural-tube defects. *N Engl J Med*. 1999;341(20):1509–1519.
58. Blencowe H, Cousens S, Modell B, Lawn J. Folic acid to reduce neonatal mortality from neural tube disorders. *Int J Epidemiol*. 2010;39(1):110–121.
59. Benton D. Neurodevelopment and neurodegeneration: Are there critical stages for nutritional intervention? *Nutr Rev*. 2010;68:S6–S10.
60. Dror DK, Allen LH. Effect of vitamin B12 deficiency on neurodevelopment in infants: current knowledge and possible mechanisms. *Nutr Rev*. 2008;66(5):250–255.
61. Maret W, Sandstead HH. Zinc requirements and the risks and benefits of zinc supplementation. *J Trace Elem Med Biol*. 2006;20(1):3–18.
62. Levenson CW, Morris D. Zinc and neurogenesis: making new neurons from development to adulthood. *Adv Nutr*. 2011;2(2):96–100.
63. Frederickson CJ, Suh SW, Silva D, Frederickson CJ, Thompson RB. Importance of zinc in the central nervous system: the zinc-containing neuron. *J Nutr*. 2000;130(5):1471S–1483S.
64. Black MM. The evidence linking zinc deficiency with Children's cognitive and motor functioning. *J Nutr*. 2003;133(5):1473S–1476S.
65. Bhatnagar S, Taneja S. Zinc and cognitive development. *Br J Nutr*. 2001;85(S2):S139.
66. Huang EP. Metal ions and synaptic transmission: think zinc. *Proc Natl Acad Sci*. 1997;94(25):13386–13387.
67. Frederickson CJ, Danscher G. Chapter 6 zinc-containing neurons in hippocampus and related CNS structures. *Prog Brain Res*. 1990;83:71–84.
68. Hosie AM, Dunne EL, Harvey RJ, Smart TG. Zinc-mediated inhibition of GABAa receptors: discrete binding sites underlie subtype specificity. *Nat Neurosci*. 2003;6(4):362–369.
69. Penland JG, Sandstead HH, Alcock NW. A preliminary report: effects of zinc and micronutrient repletion on growth and neuropsychological function of urban Chinese children. *J Am Coll Nutr*. 1997;16(3):268–272.
70. Sandstead HH, Penland JG, Alcock NW. Effects of repletion with zinc and other micronutrients on neuropsychologic performance and growth of Chinese children. *Am J Clin Nutr*. 1998;68(2):470S–475S.

71. Penland J, Sanstead H, Egger N, et al. Zinc, iron and micronutrient supplementation effects on cognitive and psychomotor function of Mexican-American school children. *FASEB J*. 1999;13:A921.
72. Hamadani JD, Fuchs GJ, Osendarp SJ, Khatun F, Huda SN, Grantham-McGregor SM. Randomized controlled trial of the effect of zinc supplementation on the mental development of Bangladeshi infants. *Am J Clin Nutr*. 2001;74(3):381–386.
73. Castillo-Durán C, Perales CG, Hertrampf ED, Marín VB, Rivera FA, Icaza G. Effect of zinc supplementation on development and growth of Chilean infants. *J Pediatr*. 2001;138(2):229–235.
74. Penland JG. Behavioral data and methodology issues in studies of zinc nutrition in humans. *J Nutr*. 2000;130(2):361S–364S.
75. Zimmermann MB, Hess SY, Molinari L, et al. New reference values for thyroid volume by ultrasound in iodine-sufficient schoolchildren: a World Health Organization/nutrition for health and development iodine deficiency study group report. *Am J Clin Nutr*. 2004;79(2):231–237.
76. Delange F. The disorders induced by iodine deficiency. *Thyroid*. 1994;4(1):107–128.
77. Skeaff S. Iodine deficiency in pregnancy: the effect on neurodevelopment in the child. *Nutrients*. 2011;3(2):265–273.
78. Cao X-Y, Jiang X-M, Dou Z-H, et al. Timing of vulnerability of the brain to iodine deficiency in endemic cretinism. *New England J Med*. 1994;331(26):1739–1744.
79. Bleichrodt N, Born MP. *A meta-analysis of research on iodine and its relationship to cognitive development. The damaged brain of iodine deficiency*. New York: Cognizant Communication; 1994:195–200.
80. Qian M, Wang D, Watkins WE, Gebski V, Yan YQ, Li M. The effects of iodine on intelligence in children: a meta-analysis of studies conducted in China. *Asia Pac J Clin Nutr*. 2005;14(1):32.
81. Zimmermann MB. Iodine deficiency. *Endocr Rev*. 2009;30(4):376–408.
82. Kidd PM. Omega-3 DHA and EPA for cognition, behavior, and mood: clinical findings and structural-functional synergies with cell membrane phospholipids. *Altern Med Rev*. 2007;12(3):207.
83. Mori TA, Beilin LJ. Omega-3 fatty acids and inflammation. *Curr Atheroscler Rep*. 2004;6(6):461–467.
84. Davidson MH. Omega-3 fatty acids. *Curr Opin Lipidol*. 2013;24(6):467–474.
85. Giles GE, Mahoney CR, Kanarek RB. Chapter 25- omega-3 fatty acids and cognitive behavior. In: Watson RR, Meester FD, eds. *Omega-3 Fatty Acids in Brain and Neurological Health*. Boston: Academic Press; 2014:303–325.
86. Huffman SL, Harika RK, Eilander A, Osendarp SJM. Essential fats: How do they affect growth and development of infants and young children in developing countries? A literature review. *Matern Child Nutr*. 2011;7:44–65.
87. Briend A, Dewey KG, Reinhart GA. Fatty acid status in early life in low-income countries - overview of the situation, policy and research priorities. *Matern Child Nutr*. 2011;7:141–148.
88. Simopoulos A. An increase in the Omega-6/Omega-3 fatty acid ratio increases the risk for obesity. *Nutrients*. 2016;8(3):128.
89 Yehuda S. Omega-6/omega-3 ratio and brain-related functions. Omega-6/Omega-3 Essential Fatty Acid Ratio: The Scientific Evidence. 2003;92:37–56.
90. Parletta N, Niyonsenga T, Duff J. Omega-3 and omega-6 polyunsaturated fatty acid levels and correlations with symptoms in children with attention deficit hyperactivity disorder, autistic spectrum disorder and typically developing controls. Lakshmana MK, ed. *PLoS One*. 2016;11(5):e0156432.
91. Calder PC. Marine omega-3 fatty acids and inflammatory processes: Effects, mechanisms and clinical relevance. *Biochim Biophys Acta (BBA) – Mol Cell Biol Lipids*. 2015;1851(4):469–484.
92. Luchtman DW, Song C. Cognitive enhancement by omega-3 fatty acids from child-hood to old age: Findings from animal and clinical studies. *Neuropharmacology*. 2013;64:550–565.
93. O'Brien JS, Sampson EL. Lipid composition of the normal human brain: gray matter, white matter, and myelin. *J Lipid Res*. 1965;6(4):537–544.
94. Crawford MA, Williams G, Hassam AG, Whitehouse WL. Essential fatty acids and fetal brain growth. *The Lancet*. 1976;307(7957):452–453.
95. Brenna JT, Diau G-Y. The influence of dietary docosahexaenoic acid and arachidonic acid on central nervous system polyunsaturated fatty acid composition. *Prostaglandins Leukot Essent Fatty Acids*. 2007;77(5–6):247–250.
96. Pottala JV, Yaffe K, Robinson JG, Espeland MA, Wallace R, Harris WS. Higher RBC EPA + DHA corresponds with larger total brain and hippocampal volumes: WHIMS-MRI study. *Neurology*. 2014;82(5):435–442.
97. Dyall SC. Long-chain omega-3 fatty acids and the brain: a review of the independent and shared effects of EPA, DPA and DHA. *Front Aging Neurosci*. 2015;7.

98. McNamara RK, Carlson SE. Role of omega-3 fatty acids in brain development and function: potential implications for the pathogenesis and prevention of psychopathology. *Prostaglandins Leukot Essent Fatty Acids*. 2006;75(4–5):329–349.
99. Yehuda S, Rabinovitz S, Mostofsky DI. Essential fatty acids and the brain: From infancy to aging. *Neurobiol Aging*. 2005;26(1):98–102.
100. Lafourcade M, Larrieu T, Mato S, et al. Nutritional omega-3 deficiency abolishes endocannabinoid-mediated neuronal functions. *Nat Neurosci*. 2011;14(3):345–350.
101. Labrousse VF, Leyrolle Q, Amadieu C, et al. Dietary omega-3 deficiency exacerbates inflammation and reveals spatial memory deficits in mice exposed to lipopolysaccharide during gestation. *Brain Behav Immun*. 2018;.
102. Innis SM. Dietary (n-3) fatty acids and brain development. *J Nutr*. 2007;137(4):855–859.
103. Richardson AJ, Ross MA. Fatty acid metabolism in neurodevelopmental disorder: a new perspective on associations between attention-deficit/hyperactivity disorder, dyslexia, dyspraxia and the autistic spectrum. *Prostaglandins, Leukotrienes and Essential Fatty Acids (PLEFA)*. 2000;63(1–2):1–9.
104. Koletzko B, Lien E, Agostoni C, et al. The roles of long-chain polyunsaturated fatty acids in pregnancy, lactation and infancy: review of current knowledge and consensus recommendations. *J Perinat Med*. 2008;36(1).
105. McNamara RK. DHA deficiency and prefrontal cortex neuropathology in recurrent affective disorders. *J Nutr*. 2010;140(4):864–868.
106. Branca F. Nutritional solutions to major health problems of preschool children: How to optimise growth and development. *J Pediatr Gastroenterol Nutr*. 2006;43(Suppl 3):S4–S7.
107. Wang C, An Y, Yu H, et al. Association between exposure to the Chinese famine in different stages of early life and decline in cognitive functioning in adulthood. *Front Behav Neurosci*. 2016;10.
108. Shulkin ML, Pimpin L, Bellinger D, et al. Effects of omega-3 supplementation during pregnancy and youth on neurodevelopment and cognition in childhood: a systematic review and meta-analysis. *FASEB J*. 2016;30(1 Supplement):295.5.
109. Dalton A, Wolmarans P, Witthuhn RC, van Stuijvenberg ME, Swanevelder SA, Smuts CM. A randomised control trial in schoolchildren showed improvement in cognitive function after consuming a bread spread, containing fish flour from a marine source. *Prostaglandins Leukot Essent Fatty Acids*. 2009;80(2–3):143–149.
110. Baumgartner J, Smuts CM, Malan L, et al. Effects of iron and n-3 fatty acid supplementation, alone and in combination, on cognition in school children: A randomized, double-blind, placebo-controlled intervention in South Africa. *Am J Clin Nutr*. 2012;96(6):1327–1338.
111. Portillo-Reyes V, Pérez-García M, Loya-Méndez Y, Puente AE. Clinical significance of neuropsychological improvement after supplementation with omega-3 in 8–12 years old malnourished Mexican children: A randomized, double-blind, placebo and treatment clinical trial. *Res Dev Disabil*. 2014;35(4):861–870.
112. Kirby A, Woodward A, Jackson S. Benefits of omega-3 supplementation for schoolchildren: Review of the current evidence. *Br Educ Res J*. 2010;36(5):699–732.
113. Bauer I, Crewther DP, Pipingas A, Rowsell R, Cockerell R, Crewther SG. Omega-3 fatty acids modify human cortical visual processing—a double-blind, crossover study. *PLoS One*. 2011;6(12):e28214.
114. Turken U, Whitfield-Gabrieli S, Bammer R, Baldo JV, Dronkers NF, Gabrieli JDE. Cognitive processing speed and the structure of white matter pathways: convergent evidence from normal variation and lesion studies. *Neuroimage*. 2008;42(2):1032–1044.
115. Horrocks LA, Farooqui AA. Docosahexaenoic acid in the diet: its importance in maintenance and restoration of neural membrane function. *Prostaglandins Leukot Essent Fatty Acids*. 2004;70(4):361–372.
116. Weber WA, Tallman MJ, Blom TJ, Schurdak JD, Patino LR, McNamara RK. Effects of docosahexaenoic acid supplementation on cortical network integrity in medication-free children with attention-deficit/hyperactivity disorder: A preliminary multimodal neuroimaging trial. *J Nutr Therapeut*. 2018;6(4):92–106.
117. Artukoglu BB, Bloch MH. Editorial: Can omega-3 fatty acids improve executive functioning? Will this reduce ADHD and depression? *J Child Psychol Psychiatry*. 2018;59(6):615–617.
118. Sydenham E, Dangour AD, Lim W-S. Omega 3 fatty acid for the prevention of cognitive decline and dementia. *Sao Paulo Med J*. 2012;130(6):419.
119. Farooqui AA. Neurochemical effects of long term consumption of high fat diet. *High Calorie Diet and the Human Brain*. 2015;29–76.
120. Fu Z, Wu J, Nesil T, Li MD, Aylor KW, Liu Z. Long-term high-fat diet induces hippocampal microvascular insulin resistance and cognitive dysfunction. *Am J Physiol-Endocrinol Metab*. 2017;312(2):E89–E97.

121. Freeman LR, Haley-Zitlin V, Rosenberger DS, Granholm A-C. Damaging effects of a high-fat diet to the brain and cognition: a review of proposed mechanisms. *Nutr Neurosci*. 2013;17(6):241–251.
122. Haapala EA, Viitasalo A, Venäläinen T, et al. Plasma polyunsaturated fatty acids are directly associated with cognition in overweight children but not in normal weight children. *Acta Paediatr*. 2016;105(12):1502–1507.
123. Montgomery P, Burton JR, Sewell RP, Spreckelsen TF, Richardson AJ. Low blood long chain omega-3 fatty acids in UK children are associated with poor cognitive performance and behavior: a cross-sectional analysis from the DOLAB study. *PLoS One*. 2013;8(6):e66697.
124. Boucher O, Burden MJ, Muckle G, et al. Neurophysiologic and neurobehavioral evidence of beneficial effects of prenatal omega-3 fatty acid intake on memory function at school age. *Am J Clin Nutr*. 2011;93(5):1025–1037.
125. Harris C, Demmelmair H, von Berg A, et al. Associations between fatty acids and low-grade inflammation in children from the LISAplus birth cohort study. *Eur J Clin Nutr*. 2017;71(11):1303–1311.
126. Fourrier C, Sauvant J, Aubert A, Layé S, Joffre C, Castanon N. Effect of an omega-3/antioxidants supplemented diet on emotional and cognitive alterations and neuroinflammatory processes associated with obesity. *Glia*. 2015;63:E327.
127. Issa AM, Mojica WA, Morton SC, et al. The efficacy of omega–3 fatty acids on cognitive function in aging and dementia: a systematic review. *Dement Geriatr Cogn Disord*. 2006;21(2):88–96.
128. Lukaschek K, von Schacky C, Kruse J, Ladwig K-H. Cognitive impairment is associated with a low Omega-3 index in the elderly: results from the KORA-age study. *Dement Geriatr Cogn Disord*. 2016;42(3–4):236–245.
129. Greenberg JA, Bell SJ, Ausdal WV. Omega-3 fatty acid supplementation during pregnancy. *Rev Obstet Gynecol*. 2008;1(4):162–169.
130. Siscovick DS, Barringer TA, Fretts AM, et al. Omega-3 polyunsaturated fatty acid (fish oil) supplementation and the prevention of clinical cardiovascular disease. *Circulation*. 2017;135(15):e867–e884.
131. Institute of Medicine. *Dietary Reference Intakes for Energy, Carbohydrate, Fiber, Fat, Fatty Acids, Cholesterol, Protein, and Amino Acids*. Washington, DC: The National Academies Press; 2005.
132. Anderson PM, Butcher KF. Childhood obesity: trends and potential causes. *Future Child*. 2006;16(1):19–45.

Further Reading

133. Muthayya S, Eilander A, Transler C, et al. Effect of fortification with multiple micronutrients and n–3 fatty acids on growth and cognitive performance in Indian schoolchildren: The CHAMPION (Children's health and mental performance influenced by optimal nutrition) study. *Am J Clin Nutr*. 2009;89(6):1766–1775.
134. Witte AV, Kerti L, Hermannstädter HM, et al. Long-chain Omega-3 fatty acids improve brain function and structure in older adults. *Cereb Cortex*. 2013;24(11):3059–3068.
135. Campbell S, Bello N. Omega-3 fatty acids and obesity. *J Food Nutr Disord*. 2012;01(02).

CHAPTER

11

Neuroprotective Effects of Melatonin and Omega-3 on the Central Nervous System Exposed to Electromagnetic Fields in the Pre- and Postnatal Periods

Kıymet Kübra Yurt, Elfide Gizem Kıvrak*, Gamze Altun*, Abit Aktas†, Arife Ahsen Kaplan‡, Süleyman Kaplan**

***Department of Histology and Embryology, Faculty of Medicine, Ondokuz Mayis University, Samsun, Turkey †Department of Histology and Embryology, Faculty of Veterinary, Istanbul University Cerrahpasa, Istanbul, Turkey ‡Department of Histology and Embryology, Faculty of Medicine, İstanbul Medipol University, Istanbul, Turkey**

ABBREVIATIONS

AA	arachidonic acid
AC	adenylyl cyclase
DAG	diacylglycerol
CNS	central nervous system
BBB	blood-brain barrier
BDNF	brain derived neurotrophic factor
bHLH	Basic helix-loop-helix
CaMKII	Ca^{+2}/CaM kinase II
cGMP	cyclic guanosinemonophosphate
DHA	docosahexaenoic acid
EFA	essential fatty acids
EMF	electromagnetic field

Omega Fatty Acids in Brain and Neurological Health
https://doi.org/10.1016/B978-0-12-815238-6.00011-0

EPA	eicosapentaenoic acid
Et-DHA	ethyl docosahexaenoate
ω-3	omega-3
GFAP	glial fibrillary acidic
GSH	glutathione
HSPs	Heat-shock proteins
IP3	inositol 1,4,5-trisphosphate
MAP-2	microtubule-associated protein 2
MT	melatonin
NAT	N-acetyltransferase
NO	nitric oxide
Nrf-2	nuclear factor erythroid 2-related factor 2
NSC	neural stem cell
PIP2	phosphatidylinositol-4,5-bisphosphate
PKC	protein kinase C
PLC	phospholipase C, Alpha linoleic acid
PUFAs	polyunsaturated fatty acids
RF-EMF	radiofrequency-electromagnetic field
ROS	reactive oxygen species
TD-SCDMA	time-division synchronous code division multiple access

INTRODUCTION

The use of mobile phones in close proximity to the brain causes indisputable electromagnetic field (EMF) effects on it. The nervous system is sensitive to various physical signals during and after development. The nervous system acts like an irradiated organ, and electrical activity in the brain changes.[1] Exposure to EMF can affect brain development and cause changes in neuronal functions, such as neuron-surface interactions, neuron growth,[2] differentiation,[3–5] and the direction of neurite growth.[6] The duration and intensity of EMF play an important role in these effects.[7] While numerous studies have reported a therapeutic role of low-frequency EMF, others have discussed the health-threatening effects of EMF and have emphasized the need for precautionary measures against these. Melatonin acts as a powerful radical scavenger against oxidative stress.[8] It also has neuroprotective, and neuron-regulating properties, as well as exhibiting antioxidant activity. Melatonin has a positive effect on neuronal survival, neurogenesis, and neuronal differentiation in the central nervous system (CNS). The function of melatonin in the mammal brain is commonly mediated through the MT1 and MT2 receptors linked to the G protein.[9] The cytoskeletal structure remains undamaged while melatonin exhibits these effects, and many benefits such as the protection of mitochondrial homeostasis occur [10,11]. Although the underlying mechanisms of the effects of melatonin effects have not yet been fully explained, melatonin is known to exhibit a neuroprotective effect against EMF exposure in the brain and cerebellum.[12,13] Omega-3 (ω-3) plays an important role in the development of the CNS in mammals. Maternal ω-3 status in pregnancy is of critical importance.[14] Receiving adequate amounts of eicosapentaenoic acid (EPA) and docosahexaenoic acid (DHA), the biologically active forms of ω-3 fatty acids, can improve cognitive, and memory functions and visual performance in children.[15–17] ω-3 fatty acids also retard oxidative stress and inhibit neuroinflammation, and may also autooxidize and induce in vivo lipid peroxidation.[18] Oxidative stress necrosis may lead to neurological disorders associated with neuron losses in the brain.[19] Ω-3 reduces the neuronal damage caused by the oxidative stress induced by EMF.[12,20] Ω-3 has neuroprotective properties and is the sole method for treating neurodegenerative disorders.

Cellular and Structural Impacts of EMF-Induced Toxicity

Although the biological effects of EMF have been demonstrated in animals, the exact impacts of EMF on human health are still unclear. The basic mechanisms of the interaction between EMF and living matter are also unclear, for which reason studies have reported conflicting results. In particular, there have been a number of studies on the genomic and cellular effects of EMF, such as cell cycle regulation, cell proliferation, specific gene expression and signal transduction pathways, and DNA damage. Exposure to EMF can also result in alterations in some cellular processes, DNA breaks, and other chromosomal aberrations. Subjects are well known to be more susceptible to exposure during the developmental process. Even a low level of EMF exposure, especially during neurogenesis in fetal life, may have a profound effect on brain cells.[21] There have been a number of studies investigating cellular functions,[22] genotoxicity,[23] gene/protein expression changes,[24] neurogenesis,[25] protein misfolding, and DNA breaks[26] for the purpose of determining the structural effects of EMF in neurogenic cells. Epidemiological and animal studies have reported that EMF may cause neurodevelopmental disorders, damages hippocampal neuronal plasticity, changes blood-brain barrier (BBB) permeability, and increases the risk of neurodegenerative diseases.[27,28] The calcium-mediated signaling pathways play a major role in the regulation of neuronal development, such as neuronal survival and migration, and the differentiation of neurons in the nervous system.[29] Recent studies have confirmed the adverse effects of EMF exposure on Ca^{+2}, which plays a major role in regulating numerous neuronal processes.[30] EMF leads to an increase in cellular Ca^{+2} levels, as well as in free radicals. DNA damage, protein misfolding, and inhibition of cell growth may therefore occur.[26] Nitric oxide-peroxynitrite-oxidative stress may be effective against EMF exposure. EMF may lead to increased nitric oxide levels by affecting voltage-gated calcium channels (causing an increase in intracellular Ca^{+2}). This, a precursor of peroxynitrite, results in free-radical products. Pilla[31] reported a rapid increase in intracellular Ca^{+2} levels after EMF exposure, while the level of Ca^{+2}/ calmodulin-dependent nitric oxide increased in less than 5 min.[31] Researchers have reported that Ca^{+2} homeostasis disorders resulting from EMF exposure can affect normal hippocampal functions, thereby compromising neuronal connections and integration.[32]

DNA Breakage and Repair

DNA damage can have deleterious effects on cell functions and lead to cell death. The comet assay, a microgel electrophoresis technique, is very commonly used to see the effects of EMF on DNA damage, single- and double-strand breaks. Most studies conducted using the comet assay have reported an increase in DNA strand breaks in cells after EMF exposure.[33–37] This was confirmed by Lai and Singh, who demonstrated single- and double-strand DNA breaks in the neurons of the rat brain after 2-ho exposure to 2450-MHz RFR. This side effect can also be blocked by antioxidants, indicating that free radicals play an important role in this process.[38–40] In addition, EMF exposure at 0.5 and 7.5 μT leads to an increase in DNA double-strand breaks in mouse brain neurons.[34,35] Additionally, Sarkar et al. suggested that 2450-MHz microwave energy leads to breakages in DNA strands following EMF exposure to the testis and brain.[41] The most important of the possible mechanisms of these harmful effects is that the free radicals that form in the cells damage the DNA structure. EMF increases the free-radical activity of the cells through the Fenton reaction. This is a process in which hydrogen peroxide is converted to hydroxyl-free radicals and produces cytotoxic products. Free radicals also subsequently affect many macromolecules, such as DNA and membrane lipids.[40,42–44] Free radicals increase in a brain cell exposed to a magnetic field of 60 Hz through an iron-mediated process, which triggers an increase in DNA-DNA cross-links.[45] EMF exposure also affects the membranes of cell bodies such as residual body lysosomes and mitochondria, leading to the production of reactive oxygen species (ROS) and heavy metals in the neurons.[46,47] It therefore appears that ROS play a crucial role in EMF-induced tissue damage. Human SH-SY5Y has been found to cause secondary DNA damage by increasing ROS production at 900 and 872 MHz EMF in neuroblastoma cells [48]. Various molecular pathways such as caspase-3-dependent pathway,[49,50] cAMP/PKA pathway,[50] ATM-Chk2-p21 pathway,[51] and ERK pathway lead to the pathogenesis of EMF-induced neurological diseases.[52,53] One study of the effects of 900 MHz EMF on the rat brain reported neural edema and neuronal cell degeneration after exposure, resulting in impaired spatial memory and BBB permeability. The study suggested that EMF exposure increased the expression of mkp-1, resulting in ERK dephosphorylation.[53]

Researchers have also determined changes in protein-folding after EMF exposure. Heat-shock proteins (HSPs) are very sensitive and can be impaired by EMF exposure.[26,54] Studies have determined an effect of 900 MHz EMF on HSPs in the human endothelial cell line, such as upregulation of HSP27.[54] Exposure to 2.45 GHz EMF elicits a stress response associated with HSP in the rat hippocampus. Study results have revealed significant HSP27 and HSP70 upregulation. This effect has mainly been seen in the CA3 and dentate gyrus regions of the hippocampus[55] (Figs. 1 and 2). Another study reported that acute exposure to 2.45 GHz EMF caused an imbalance in HSP-90 levels in rats.[56] Similarly, expression changes in HSP22, HSP68, HSP70bc, HSP70, and HSP60d were detected following exposure to 50 Hz EMF in *Drosophila*.[57]

Electromagnetic Field and Gene Expression

Radiation can directly affect gene expression.[58] Studies have observed the effects of EMF on genes that regulate the expression of the protein at the molecular level, regulating the branching, external growth, and sprouting phenomena of the protein that plays a role in neural patterning.[59,60] Short-term exposure to 900 MHz EMF has been shown to cause an increase in Egr-1 mRNA level I in human neuroblastoma cells. This is evidence of the direct effect of electromagnetic radiation.[58,61] Exposure to low frequency EMF may affect the cell adhesion system that changes E-cadherin gene expression. The developing neural cell can thus affect the determination of the shape of neurons by altering the adhesion of the substrate.[7] There is also some evidence for the involvement of EMF in the regulation of synaptic plasticity-related genes, such as Egr-1. EMF influences cell growth and apoptosis by inducing Egr-1 gene activation.[61,62] The Egr family plays a crucial role in the regulation of gene expression that manages neuronal plasticity,[63–65] synaptic strength,[66,67] and learning[68] in the nervous system. It has also been reported that 1800 MHz EMF can disrupt the

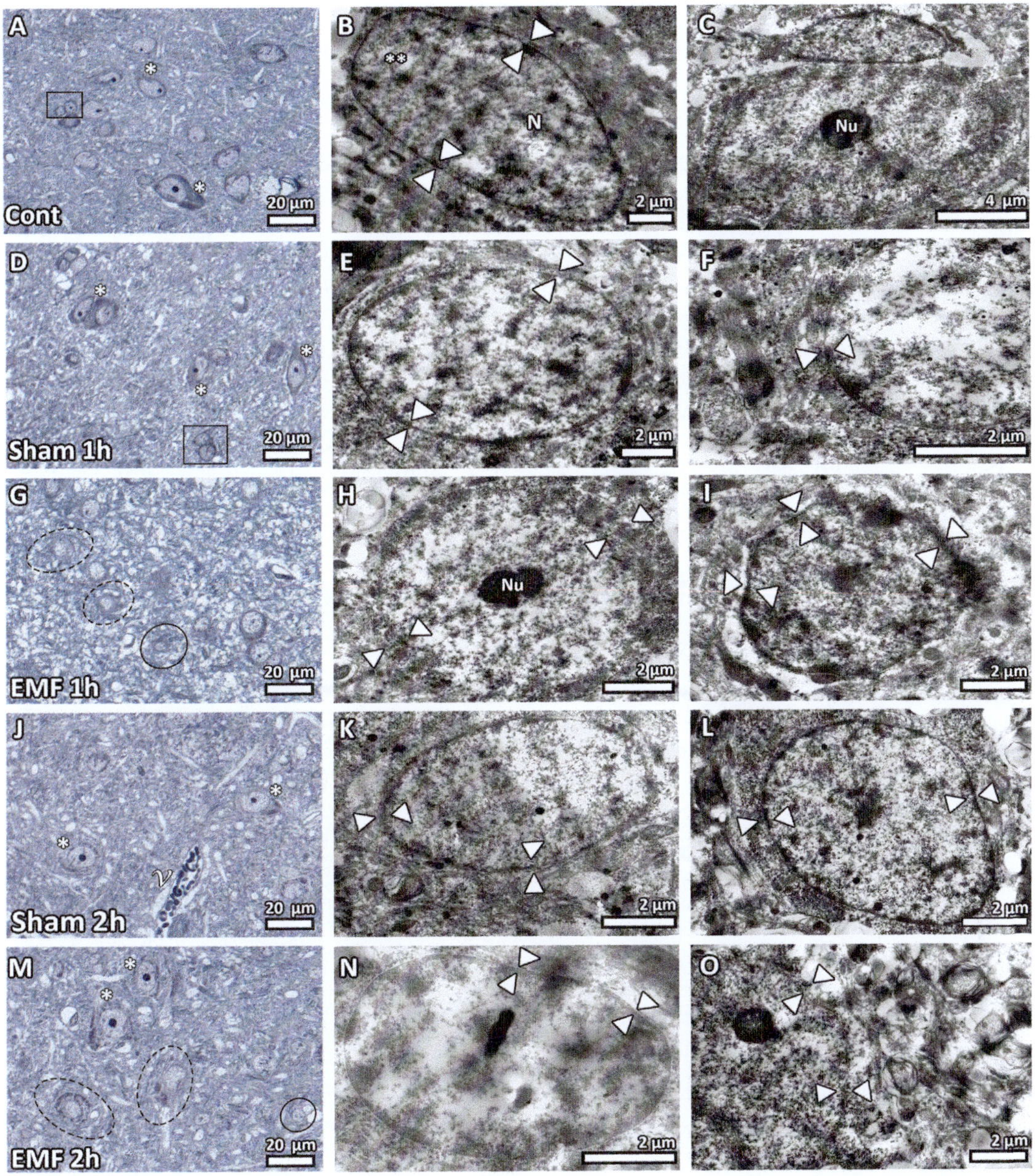

FIG. 1 Photomicrographs of semithin and ultrathin sections of the cerebral cortex of the control, the sham, and EMF-exposed rats. (A) In the control group neurons displayed having a prominent nucleus (N) with an apparent rounded nucleolus (Nu) and an abundant nuclear content (**). (D, J) Several normal neurons and granule cells have been seen in the sham groups. 900–MHz EMF exposures (during 1 and 2 hours) cause not only loss of cell boundaries but also affect the nucleus of cells. (G, M) Nucleus is dense and delimitation of it is difficult. Neuroglia (rectangle) with smaller and lighter nuclei is easily visible. In the semithin sections (dashed lines) have showed the pyramidal neurons, which have lost cell boundaries, and the similar impairment can be seen in the neuroglia that have been shown by a continuous line. Electron microscopic views have confirmed the EMF side effects on the neuronal structure. This can be seen when EMF exposed electron microscopic images (H, I, N, O) are compared to control and sham groups (B, C, E, F, K, L). It is clearly seen that the borders of the nucleus (between arrow heads) in the EMF exposed (1- and 2-hours-EMF exposed) groups are irregular and unclear.

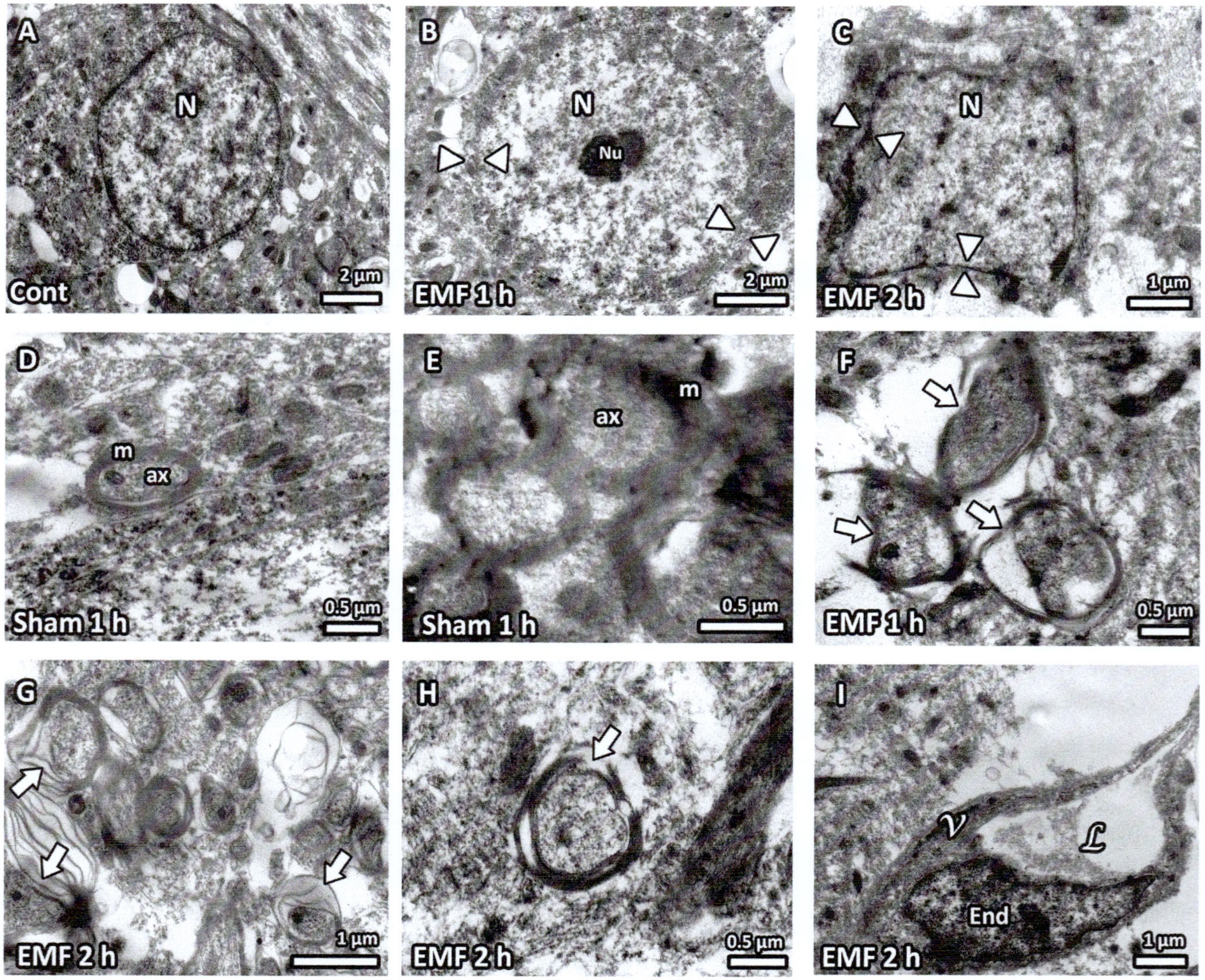

FIG. 2 Electron microscopic micrographs were taken from the ultrathin cross-sections of the rat brain cortex. A, B, and C illustrate the nucleus structure and boundaries. As seen in these pictures that the nuclear boundaries are unclear in the EMF exposed groups. Also, the structure of myelin sheath and its integrity has impaired in the EMF groups (it was shown by arrows in the F, G, H). Intact myelin sheet (m) and axons (ax) are observed in the sham group (D, E). Exposure to EMF affects nervous tissue as well as vascularization. Evidence of this effect is that the impaired structure of the blood vessel (V) in the 2-hours-EMF exposed group seen in the electron microscopic image (I), the wall of vessel is separated from around tissue, and lumen (L) of it was collapsed. N; nucleus, Nu; nucleolus, arrows; impaired myelin sheet, End; endothelium.

expression of the basic helical-loop-helix (bHLH) and thus influence neural differentiation and neurite growth.[69] One study investigated the effects of neurite overgrowth inhibition after EMF exposure at the gene level. EMF exposure was significantly associated with genes closely related to the control of neurite outgrowth. The expression of Hes1 also significantly increased, downregulating the expression of Ngn1 and NeuroD. These results suggest that the potential mechanism of neurite outgrowth leads to the impairment of bHLH expression.[69] EMF can regulate cell cycle progression and activation of transcriptional factors of genes ultimately responsible for inducing apoptosis.[61] Functional classes of genes affected by EMF in human HL-60 cells exposed to 2.45 GHz EMF were established in one study, and apoptosis-related genes were found to be composed of upregulated and downregulated cell-cycle genes.[70] In addition, gene products such as TNFa, p53, and Bax, apoptosis activators, mediate the response to inhibition of growth against the effect of ionizing radiation.[61] Tumor suppressor gene p53 can be activated with ionizing radiation. This gene regulates the cell cycle, DNA repair and apoptosis, autophagy, differentiation, and development. In utero exposure to irradiation can affect neuronal differentiation and mitotic spindle assembly by inducing p53-dependent genes.[71]

One study investigated the molecular mechanisms of the p53 gene on prenatal exposure to EMF in the brain. The results showed that prenatal EMF exposure causes apoptosis and an increase in p53 gene expression and induces cell death in the developing rat brain.[72]

EMF and BBB Permeability

EMF may disrupt the permeability of the biological barrier that protects mammalian brain from potentially harmful compounds in the body.[73] The BBB plays a crucial role in maintaining cerebral homeostasis. Many studies have shown that EMF enhances the permeability of the BBB.[74–76] Disturbance of the structure of the BBB induces brain albumin leaks and causes degeneration of neuronal cells.[77] High concentrations of albumin also lead to astrocyte proliferation.[78] The effects of albumin concentration on brain tissue is a study requiring future investigation.[79] Albumin extravasation has been shown to increase following 2-week exposure to EMF. However, this is correlated with the prevalence of dark neurons 4 weeks after EMF exposure.[46] EMF exposure causes an increase in temperature in the brain of more than 1°C. A reversible increase in BBB permeability may occur as a result. Changes in BBB permeability may also be due to the thermal effects of EMF.[80–82]

EMF and Melatonin Production

EMF can affect melatonin and its metabolite, 6-sulfatoxymelatonin. Decreased melatonin production has been reported in both adult humans and animals following exposure to EMF.[83] Age-related changes in serum melatonin and pineal N-acetyltransferase (NAT) activity have been observed following exposure to a 50-Hz magnetic field, and it has been suggested that elderly rats are less sensitive to EMF exposure than younger animals.[84] Human studies have also reported disruption of melatonin even when exposed to weak EMFs,[85] and that magnetic fields generated by incubators cause changes in the production of melatonin by newborns, which may lead to long-term health effects in humans.[86] EMF can reduce the production of melatonin by affecting several parameters in the indole cascade. For example, changes in pineal cAMP levels suppress the pineal gland, resulting in a decrease in blood melatonin concentrations.[87]

CURRENT KNOWLEDGE OF THE NEUROPROTECTIVE EFFECTS OF MELATONIN

Melatonin, a neurohormone secreted in darkness by the epiphyseal gland, has many physiological functions and plays a crucial role in the control of the circadian rhythm.[88,89] Melatonin is also produced in the gastrointestinal tract, bone marrow, and retinas[90] and is important to the prevention of mitochondrial homeostasis because of its anti-inflammatory and antioxidant properties.[91] This hormone provides mitochondrial homeostasis and increases mitochondrial glutathione (GSH) levels. Free-radical formation is thus reduced. Melatonin is also used as a therapeutic agent against free-radical damage in the brain.[11] Melatonin has also recently been as a mitochondrial protector.[92] The biological effects of melatonin are mediated by MT1 and MT2 receptors linked to G protein.[91,93] In addition, melatonin also exhibits a neuroprotective effect due to its antioxidant and anti-inflammatory properties. This neuroprotective effect can be used for the treatment of neurodegenerative diseases.[94] The acute and chronic administration of melatonin has been shown to protect dopaminergic neurons against neurotoxicity in the treatment of Parkinson's disease.[95]

The pathogenesis of neurodegenerative diseases is closely related to oxidative stress mechanisms.[96] Endogenous antioxidant levels in the CNS are very low.[97] Oxidative damage resulting from the formation of free radicals leads to degenerative effects in the CNS. In addition, high levels of unsaturated fatty acids in the neurons strengthen their defense mechanisms against oxidative stress. Similarly, melatonin secreted from the pineal gland exhibits a strong antioxidant effect against this oxidative damage. Melatonin is thus capable of preventing the progression of neurodegenerative diseases. This hormone scavenges superoxide, hydroxyl, peroxyl and peroxynitrite anion radicals, induces the antioxidative potentials of cells, and assists with the synthesis of antioxidative enzymes. These mechanisms confer a neuroprotective effect on melatonin.[96]

The cells of the CNS consume a considerable amount of oxygen and are therefore highly susceptible to oxidative stress. They also produce more reactive oxygen species than other tissues. In addition, many of the Fenton reactions involve unsaturated fatty acids and transition metals.[97] The expression of antioxidant enzymes is increased by non-receptor-mediated mechanisms of melatonin against neuronal injury caused by the Fenton reaction.[98–100] Exogenous melatonin has been shown to cause an increase in NO and cGMP production after transient ischemia.[101] It is also capable of protecting the physiological levels of γ-glutamylcysteine synthetase. This enzyme also plays a crucial role in limiting GSH.[102] Decreased GSH levels, especially in cerebral ischemia, can be significantly restored with melatonin therapy.[103,104]

Neuronal Protection

The fact is that melatonin is a cytoskeletal regulator and free-radical scavenger makes it suitable for use in the treatment of neurodegenerative diseases. Proliferative activities of melatonin have particularly been reported in the dentate gyrus.[105] Erdem Koç et al. showed that melatonin exhibits a neuroprotective effect against oxidative damage resulting from EMF exposure by preserving neuron numbers.[12] Melatonin may modulate neuronal functions[106] since several melatonin-binding sites have been identified in the hippocampal regions. This hormone is also effective in the elimination of mitochondrial damage and enhances mitochondrial complexs I and IV activity; thereby preventing electron leakage from the mitochondria.[107] Mitochondrial dysfunction is an important stage in the pathophysiology of neurodegenerative disorders. During the progression of these diseases, mitochondrial dysfunction increases and induces oxidative stress, leading to apoptosis. This disturbance then leads to changes in the cellular environment. Due to its strong antioxidant effects, exposure to melatonin can result in protection of the mitochondria. The protective effect of melatonin on neurons is also related to its role in the mitochondrial membrane. Damaged mitochondrial membrane permeability leads to cell apoptosis. At this point, cardiolipin is thought to be an important mediator in the membrane-protective effect of melatonin. Erdem Koç et al. observed that the side effects of prenatal exposure to 900MHz EMF in the hippocampus can be ameliorated by melatonin exposure.[12]

Glial Protection

Glial cells are responsible for the homeostatic regulation of the CNS, and protective effects against oxidative stress. Furthermore, astrocytes play a major role in metabolic support for neurons and maintain brain homeostasis. Neurotoxic agents, tumors, brain trauma, and ischemia induce astrocytic activation, and this in turn induces neuronal damage.[108,109]

Apoptosis in astrocytes can lead to progression of neurodegenerative diseases.[110] High-dose radiofrequency-electromagnetic field (RF-EMF) exposure has been shown to lead to hippocampal, striatal, and cortex-reactive astrogliosis.[111,112] Activation of microglia and astrocytes has been observed after exposure to RF-EMF at different frequencies. Studies have reported that 835 and 900MHz RF-EMF lead to activation of astrocytes, whereas 915 and 2450MHz RF-EMF lead to microglial activation and pro-inflammatory cytokine production.[111,113–115] Increased expression of glial fibrillary acidic protein (GFAP) is regarded as a response to oxidative damage.[116]

Synaptic Protection

Melatonin has been shown to enhance neuron survival and neurogenesis in the dentate gyrus of the hippocampus.[117–119] Dendritogenesis is a complex process involving cytoskeletal arrangements in response to kinase activation of membrane receptors and Ca^{+2} signals.[120,121] Synaptic connectivity between the prefrontal cortex and the hippocampus has been reported in some neuropsychiatric disorders, such as major depression and schizophrenia.[122–126] Ca^{+2}/CaM kinase II (CaMKII) activated by calmodulin phosphorylates microtubule-associated protein 2 (MAP2) during dendrite formation.[127–129] Melatonin is capable of activating protein kinase C through MT 1/2 receptors and receptor-mediated pathways. In particular, non-receptor pathways for protein kinase C activation by melatonin have been described in neuroblastoma culture studies.[130] Protein kinase C provides the melatonin downstream, and CaMKII upstream. Both receptor-mediated and non-receptor mediated signaling pathways have been found in melatonin-induced dendrite formation[131] (Fig. 2).

The Role of Melatonin Receptors in Neuronal Damage Resulting from Exposure to EMF

Oxidative stress in the brain is a critical neurotoxic effect of RF-induced radiation. Melatonin exhibits a protective effect against oxidative damage at the mitochondrial level by increasing levels of 8-hydroxyguanine in the neuron mitochondria.[132] It is found at high concentrations in hippocampal and cortical cells due to its ability to cross the BBB.[133] Functions of melatonin, such as circadian rhythm regulation, occur through the activation of MT1 (Mel1a) and MT2 (Mel1b) receptors bound to G protein, which are present in the cortex, cerebellar cortex, thalamus, hippocampus, and suprachiasmatic nucleus in the CNS.[134–137] The CA1, CA3, and dentate gyrus regions in the hippocampus in particular contain binding sites for MT1 and MT2 receptors.[106] Additionally, MT1 and MT2 are found in the granular and pyramidal neurons of the hippocampus.[138] MT1 receptors are also present in such regions of the brain as hypothalamic nuclei, the nucleus accumbens, substantia nigra, and caudate nucleus (Fig. 3).[134–137] Endogenous melatonin released from the pineal

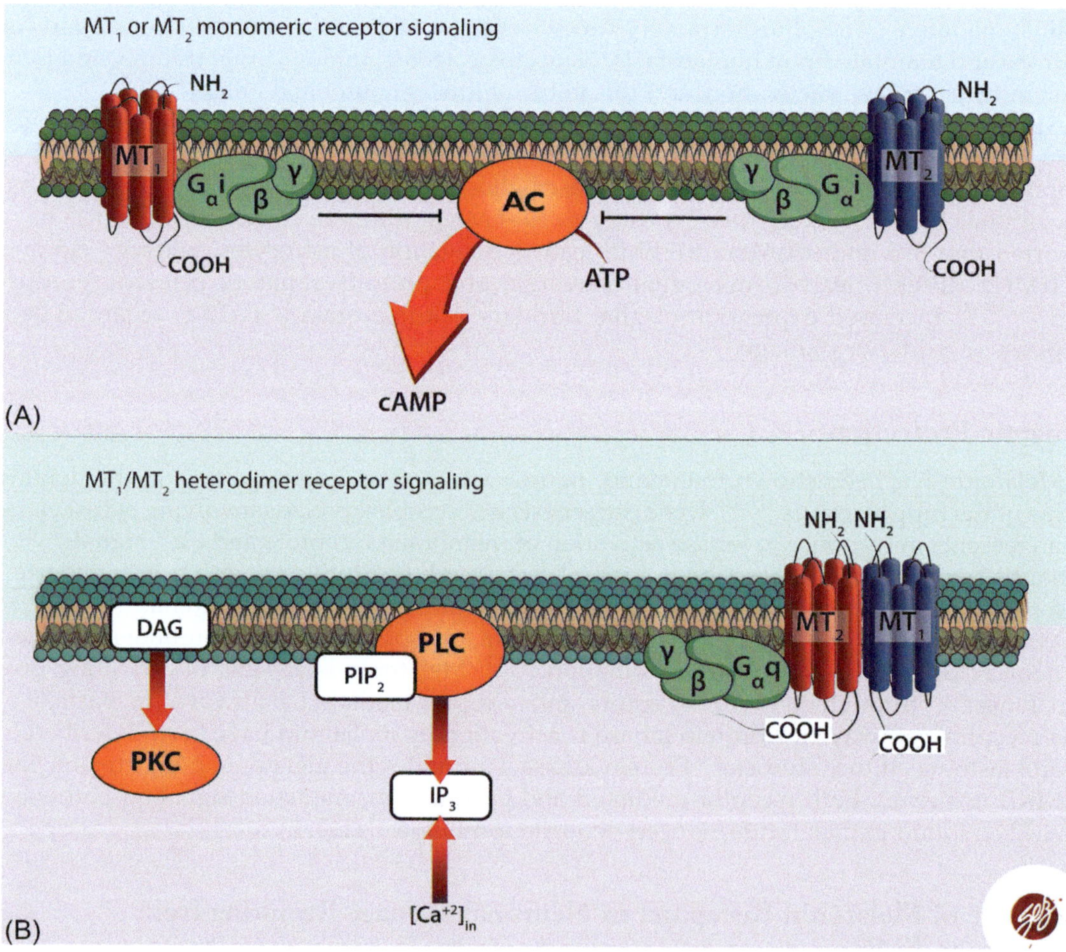

FIG. 3 MT1 and MT2 monomeric and heterodimeric receptors signal pathways. Abbreviations: AC, adenylyl cyclase; DAG, diacylglycerol; IP3, inositol 1,4,5-trisphosphate; PIP2, phosphatidylinositol-4,5-bisphosphate; PKC, protein kinase C; PLC, phospholipase C. This figure is modified from Ref. 135.

gland may stimulate MT1 and MT2 receptors.[135,139] In addition, the MT3 receptor is the quinone reductase enzyme against oxidative stress and inhibits the electron transfer reactions of quinones.[140]

Activation of melatonin receptors in the suprachiasmatic nucleus is important to the coordination of the circadian rhythm.[141] The role of these receptors in neurodegenerative diseases has been demonstrated histologically. Increased hippocampal MT1 receptors and decreased MT2 receptors have been determined in Alzheimer's disease.[142] Human MT1 and MT2 receptors have different pharmacological, molecular structures, and different chromosomal regions.[143–145] The protective function of the MT1 and MT2 receptors is related to control of the expression of antioxidant enzymes, such as superoxide dismutase, GSH peroxidase, and GSH reductase.[146] The effects of EMF and underlying mechanisms on the melatonin receptors are not yet fully understood.

The Effects of Melatonin on Hippocampal and Cerebellar Cells Survive Exposure to EMF

The effects of 900 MHz EMF on the hippocampus and cerebellum have been demonstrated in many experimental studies.[12,147–149] The effects of EMF may lead to various disorders in the electrophysiological properties of the Purkinje cells in the cerebellum. In particular, male and female fetuses become more susceptible to electrophysiological disorders after maternal exposure to electromagnetic radiation. Exposure to EMF during the prenatal period may affect the development of the cerebellum. This may also affect synaptic maturation in the cerebellum.[150] In addition to its radical scavenging properties, melatonin also modulates apoptosis in the neurons via voltage-gated Ca^{2+} channels and transient receptor potential induced by 2.45 GHz EMF.[151,152] It also regulates the delay in K^+ channels in order to protect cerebellar granular cells from apoptosis.[153,154] Studies have underlined the effect of melatonin on Na^+ channel activity and neuronal excitability in the central and peripheral nervous system after EMF exposure.[155] Melatonin may have toxic effects when used at doses of 100 and 50 mg/kg. Possible neuroprotective effect may even be observed at doses lower than 50 mg/kg.[156]

Melatonin provides an antioxidative response to ROS by activating the transition of the nuclear factor erythroid 2-related factor 2 (Nrf-2) from the cytosol to the nucleus. The Nrf-2-dependent antioxidant mechanism is thus induced, and neuroinflammation and gliosis are reduced. In summary, the role of the Nrf-2 pathway in the induction of the neuroprotective effects of melatonin is indisputable.[157]

Protective Effects of Melatonin on Neurogenesis in the Hippocampus Exposed to EMF

Disorders in memory and spatial learning may be related to decreased neurogenesis.[158,159] The hippocampus is the region of brain responsible for memory and spatial learning. The cells in CA1 and dentate gyrus of hippocampus are rich in glucocorticoid receptors.[160] Neurogenesis in the adult hippocampus assists life-time hippocampal functions.[161] The presence of neuroblast markers in the postmortem brain is evidence of neurogenesis in adulthood.[162] The role of neurogenesis in the functioning of healthy and diseased brains has been discussed in the literature. Understanding the relationship between neurogenesis levels and neuropathological parameters permit the development of new therapeutic options and strategies.[163] In this context, positive clinical results have been achieved with agomelatine, the prototypical substance of a new class of antidepressants.[164]

Melatonin regulates neurogenesis in the dentate gyrus and promotes the differentiation and proliferation of embryonic stem cells in early postnatal rats.[105,165] Similarly, it has been suggested that melatonin stimulates the differentiation of neuronal precursor cells.[166] Melatonin concentrations differ depending on different tissue and cell types.[167,168] In addition to the effects of physiological concentrations of melatonin, pharmacological concentrations also affect the regulation of neurogenesis. Both pharmacological and physiological concentrations of melatonin thus stimulate neurogenesis activation.[117] Administration of exogenous melatonin may activate adult neurogenesis through its antioxidant and antiapoptotic effects.[169] Some effects of melatonin, including membrane receptor activation, can even be seen at nanomolar concentrations. Regulation of neural stem cell (NSC) viability by the administration of 6–10 mM melatonin has been reported,[166] while 2–10 mM melatonin exhibits paracrine and autocrine effects.[167]

Neurite development is crucial for achieving differentiation and establishing intercellular communication. Following the establishment of the synaptic connection, the integration of neurons in the hippocampus increases.[170] The protective effect of melatonin on the cytoskeleton may also be related to neuronal survival, maturation, and differentiation in the adult brain.[10] These cytoskeletal effects may occur independently of membrane receptors. Microfilament rearrangement and expansion in microtubules are regulated through MT1 and MT2 membrane receptor antagonists.[171] This hormone inhibits microtubule degradation following increased lipid peroxidation.[172] One study showed that melatonin plays a role in precursor cell survival in the adult hippocampus.[117] Melatonin has also been shown to induce cell viability and to increase the numbers of neurospheres during neuronal differentiation. The MT receptor-mediated pathway induces phosphoinositide 3-kinase/Akt signaling, thus allowing melatonin to activate the neural differentiation of pluripotent stem cells.[173]

Glucocorticoids can lead to neuronal degeneration by suppressing cell proliferation and neurogenesis. In this context, in functional terms, melatonin protects neural cells from the potentially deleterious effects of glucocorticoids. This hormone may be considered an agent involved in neuronal differentiation[174] (Fig. 4).

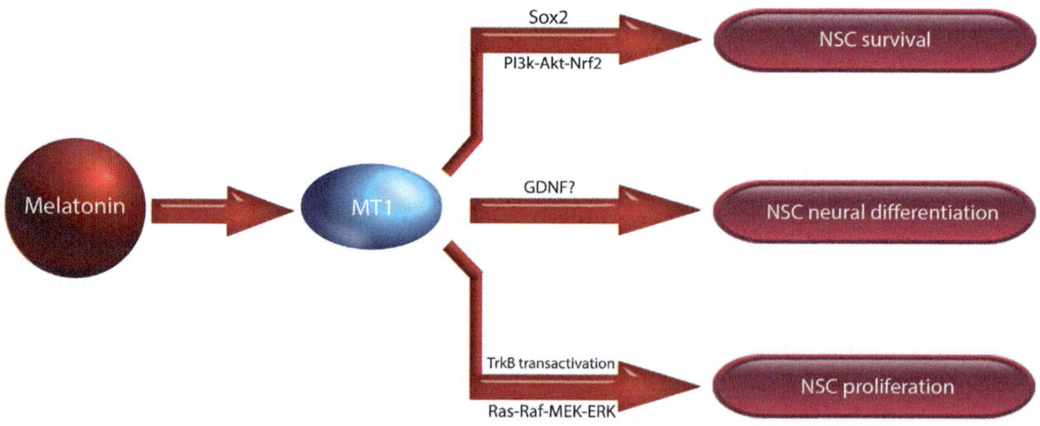

FIG. 4 Possible signaling pathways in the role of MT1 on the neuronal survival and differentiation. *NSC*, neural stem cell.

Substances, which impair glucocorticoid activity, may have a therapeutic effect in the treatment of neurodegenerative diseases. New molecular and dose-dependent studies are now required in order to clarify the underlying mechanisms involved in the association between melatonin and glucocorticoids.[175]

FUNCTION AND METABOLISM OF Ω-3 FATTY ACIDS IN THE BRAIN

Fatty acids play an essential lifetime role in many parts of the body. These include the regulation of neuronal functions and oxidative stress mechanisms necessary for proper neural and brain functions. Fatty acids, which play an important role in human metabolism, are involved in membrane formation and act as an energy substructure for brain development and function.[176,177] There are two types of polyunsaturated fatty acids (PUFAs) in the human body that cannot be produced in sufficiently large quantities since their conversion rates are very low. However, all forms of omega-6 can be produced from linoleic acid (LA), and all other forms of ω-3 from alpha linoleic acid (ALA), which is absorbed with diet. The two main forms of ω-3, EPA and DHA, are converted from ALA, but this process is insufficient for the body's requirements.[178] DHA represents 40% of total fatty acids in the brain, while EPA constitutes less than 1%.[179] EPA levels in the brain are typically 250–300 times lower than those of DHA, and the most significant ω-3 PUFA in the brain is therefore DHA. At the same time, DHA is an excellent raw material for well-functioning cell membranes.[180,181] The extension of ALA to DHA and the desaturation capacity are insufficient in the fetus and the newborn. Maternal ω-3 status in pregnancy is therefore of critical importance.[182] EPA and DHA are incorporated into the cell membrane structure and may affect membrane quality. Phospholipids containing these ω-3 regulate the action of membrane proteins such as receptors, ion channels, and membrane fluidity by altering the cholesterol/phospholipid ratio.[183] ω-3 also regulates both the inflammatory response and the release of inflammatory factors by affecting the synthesis of lipids. Due to these properties, the development of diseases such as multiple sclerosis, cerebral ischemia, Alzheimer's disease, and Parkinson's disease may be affected by ω-3.[184–186] At the same time, ω-3 has been shown to regulate inflammatory responses and immune functions in many studies.[187,188]

Ω-3 is generally assumed to regulate cell membrane organization, structure, and permeability[189]; it therefore has a significant effect on glucose uptake, neuronal function, neuronal development, and neurotransmission in the brain.[181,190–192] DHA is concentrated in the membranes of the synaptic button in the cerebral cortex, in the mitochondria, and in the photoreceptors of the retina.[193] DHA is unique in that it maintains and builds the flexible brain cell structure required for effective functioning. In particular, cell membranes require DHA for optimal functioning. DHA provides the flexibility needed for effective communication with cell membranes, a property that permits faster and better thinking. In case of insufficient DHA in the body, the brain will use saturated and trans fat, causing the cell membranes to harden and solidify.[193,194] EPA modulates synaptic plasticity similarly to DHA by activating the phosphatidylinositol-3 kinase/Akt pathway, but it increases membrane fluidity more strongly than EPA by changing the cholesterol content.[195,196] EPA exhibits important neuroprotective properties by regulating anti-oxidative activity and anti-inflammatory processes.

Studies have shown that exclusion of ω-3 from the diet leads to mental deficits such as learning and memory impairments [197–199]. It seems reasonable to assume that ω-3 plays a critical role in mental development and brain functioning.[189,200] For this reason, it has been investigated as an alternative treatment method in neurological diseases.

The Neuroprotective Role of ω-3 Against EMF Exposure

One study reported that even extremely low exposure to EMF emitted from mobile phones may cause health problems.[201] Several studies have reported findings such as stress, headache, tiredness, anxiety, decreased learning ability, impairment in cognitive functions, and poor concentration in case of exposure to microwave radiation emitted from mobile phones.[202,203] EMF influences metabolic processes in the human body and exert various biological effects on cells through a range of mechanisms. It disrupts the chemical structures of tissue, since high-degree electromagnetic energy absorption can change the tissue's electric current.

The ω-3 fatty acids EPA and DHA are essential for normal brain functioning and development throughout all stages of life. Recent studies have focused on whether these fatty acids can cure or prevent psychiatric and neurological disorders.[204–206] For example, the use of EPA has been reported to produce beneficial effects in behavioral disorders, [207] while DHA has been used in neurological disorders such as Alzheimer's disease.[208] These fatty acids are abundant in brain cell membranes, and preserve cell membrane functions while facilitating communication between cells.[209]

Antiapoptotic Benefits of ω-3

The use of mobile phone has increased considerably worldwide over the previous decade, and this has led to a high degree of concern about possible risks for human health. There has therefore been a substantial increase in studies focusing on the effects of exposure to RF on living cells.[210,211] Numerous studies have shown that EMF increases apoptosis, inhibits cell proliferation, induces DNA breaks, and alters gene expression in different cell types.[212,213] Various proteins control apoptosis, including caspase-3, Bax and Bcl-2.[214–216] Bcl-2 and Bax, which regulate mitochondrial functions, also play a critical role in the prevention of apoptosis [217]. Liu et al. investigated the effect of 1950-MHz time-division synchronous code division multiple accesses (TD-SCDMA) on the mitochondria of astrocytes. They observed that 48-h exposure to 1950 MHz TD-SCDMA damaged astrocytic mitochondria and apoptosis is induced by the caspase-3 dependent pathway.[49] At the same time, EMF can induce necrosis or apoptosis by altering the activity of calcium channels.[50,218,219]

Ω-3 protects neurons from apoptosis by preserving the structural integrity of cell membranes and controlling gene expression in the brain [220,221]. Lakshmi et al. investigated whether fish oil improves oxidative stress induced by acrylamide in the neurons in the cerebral cortex. They reported that fish oil protects the neurons from apoptosis by inhibiting acrylamide-induced oxidative stress.[222] Altun et al. investigated the effects of 900 MHz electromagnetic radiations on the CNS and demonstrated that EMF caused cellular damage in the hippocampus and cerebellum, leading to a decrease in neuron numbers. They also reported that ω3 produces significant neuroprotective effects against oxidative stress through its antiapoptotic effects.[20] Florent et al. investigated whether DHA inhibits neuronal apoptosis stimulated by soluble amyloid-beta oligomers. The findings of that study showed that DHA inhibits cytoskeleton perturbations and caspase activation that induces apoptosis in neurons.[223] Koc et al. evaluated the chronic effects of 900 MHz EMF on the brain in the prenatal period and showed that ω-3 protected cells against EMF damage.[12] Luchtman et al. suggested that EPA exhibits antiapoptotic effects by attenuating increases in pro-apoptotic Bax and caspase-3 mRNAs.[224] Another study investigated whether ω-3 improves mitochondrial dysfunction in brain aging. The results showed that ω-3 increased anti-apoptotic activity by raising Bcl-2 protein levels.[225] Numerous studies have also shown that DHA deficiency in the brain is responsible for behavioral, cognitive, and sensory dysfunctions.[220,226,227] Kelly et al. observed that ω-3 supplementation in diet reduced caspase-3 activity and suggested that it can eventually treat age-related spatial learning and long-term potential age-related decline.[228] Ahmad et al. observed that DHA deficiency leads to a decrease in hippocampal neuron sizes, and that neurons grow more slowly in young rats with DHA deficiency than in rats receiving ω-3 supplementation.[220]

Antioxidant Benefits of ω-3

Another feature of the ω-3 fatty acids is their effect on oxidative stress. Ω-3 PUFAs exhibit important neuroprotective properties in the regulation of anti-oxidative activity processes in various pathological conditions.[229] Increased oxidative stress due to EMF causes mitochondrial dysfunction, DNA damage, and changes in energy metabolism. Increased ROS under the effect of EMF have been shown to cause apoptosis in neurons and to increase inflammation. Similarly, nitrogen species also cause neurodegeneration in the CNS, damaging macromolecules such as DNA, proteins, and lipids.[230–232] Increasing antioxidant production or eliminating ROS prevents major depressive disorder and neurodegenerative diseases.

Several studies have suggested that EMF may cause the formation of ROS in exposed cells.[233–235] This occurs by means of an increase in free-radical activity in neurons, especially via the Fenton reaction, a catalytic course that transforms hydrogen peroxides, a product of oxidative respiration in the mitochondria, into the highly toxic hydroxyl free radical.[236] Since the levels of antioxidant enzymes are low in nerve tissue, this type of enzyme capacity has to be increased for protection against ROS damage. Recent studies have shown that a dietary supplement rich in PUFAs can eliminate oxidative stress by suppressing free-radical formation.[230,231] At the same time, ω-3 is thought to have anti-oxidative properties that repair long-term potentiation and improve memory in aging by reducing age-related activation of microglia and increased hippocampal IL-1β.[228] Several studies have shown that EMF induces lipid peroxidation under the effects of oxidative stress.[236–238] EMF-induced lipid peroxidation results in significant damage in cells, such as impairing cell membrane fluidity, and causing damage to protein receptors in membrane structures, defects in membrane permeability, and changes in the activity of cell enzymes. Awad and Hassan investigated the rat brain exposed to 900-MHz EMF emitted from mobile phones for 1 h/day for 1 week. They observed an increase in lipid peroxidation after exposure to mobile phones.[239] Hoyto et al. showed considerable induction of lipid peroxidation after exposure to EMF in the mouse SH-SY5Y cell and L929 fibroblast cells.[240] Ω-3 may make nerve tissues more susceptible to lipid peroxidation and strengthen antioxidant defense systems.[241] Avramovic et al. examined cerebral antioxidant and oxidant levels in 24-month-old Wistar rats exposed to ω-3. At the end of the experiment, they determined that ω-3 reinforcement reduced lipid peroxidation and increased SOD activity.[242] Another study investigated the antioxidative effects of essential fatty acids (EFAs), the ω-3 derivative, against apoptosis in the hippocampus exposed to cerebral ischemia. Their results indicated that EFA may be useful against cerebral vascular disease by reducing oxidative stress.[243] Cao et al. investigated the effects of ethyl docosahexaenoate (Et-DHA) on diabetic ischemia-induced arachidonic acid (AA) metabolites, ingested through diet, and evaluated Na, K-ATPase activity, cerebral edema, and brain damage. The experimental findings showed that Et-DHA, which reduces AA release and consequently Na, K-ATPase activity, may be useful against brain damage.[244] The dosage of ω-3 should be well adjusted in order to avoid adverse consequences. These alterations in the antioxidant balance in tissues are the result of the effects of DHA and EPA on lipids and enzymes. At this point, lipid peroxidation increases in the target tissues where ω-3 is overused, so the amount of ω-3 ingested is important.[245]

A Therapeutic Option in Neurological Disorders: Ω-3

Benefits of ω-3 in the Treatment of Alzheimer's and Parkinson's Diseases

The effects of prenatal EMF exposure on the development and apoptosis of neurons are striking. These pronounced effects are not only seen in the developing brain, but also in neurogenesis in the adult brain. A decrease in neuron numbers or a change in membrane lipid content under the effect of EMF may give rise to neurodegenerative diseases such as Parkinson's, amyotrophic lateral sclerosis, and Alzheimer's. The aging brain is more prone to neurodegenerative diseases, the most common of which is Alzheimer's, a disease that affects cognitive function, mood, and behavior. Parkinson's is the second most common neurological disorder in the world.[246] The common mechanisms of these diseases are mitochondrial dysfunction, neuro-inflammation, and oxidative stress.[247,248]

Both experimental and clinical studies have shown that ω-3 may cure important neurodegenerative and psychiatric diseases. PUFAs, important membrane components, play an important role in cerebral functioning by altering membrane functions, and by modulating apoptotic pathways and the neuroimmune system.[231,249]

EMF creates an oscillating force on each free ion on both sides of the cell membrane and causes them to cross that membrane. This movement of ions causes deterioration in the ion channels on the membrane, biochemical changes within the membrane, and thus impairment of cellular functions.[250] It has been suggested that EMF exposure disrupts the calcium ion balance in cells, and this change triggers increased production of amyloid beta. Amyloid beta gives rise to senile plaques in the brain that are responsible for Alzheimer's disease.[251] Numerous studies have observed that ω-3 PUFA intake reduces the risk of Alzheimer's disease.[205,252–254] Lim et al. observed that DHA reduces beta amyloid levels in transgenic Alzheimer's disease models [255]. Other studies have shown that DHA reduces β-amyloid peptide levels, which cause Alzheimer's disease.[256] Dietary administration of DHA and EPA to Alzheimer's patients reduces the genes that cause neurodegeneration.[257]

Loss of synaptic function is known to be capable of causing Alzheimer's disease. High levels of DHA are present in the synaptic regions of the cortex, and DHA is characterized by its fluidizing properties, which help to create new synapses.[258] Calon et al. observed that high levels of dietary DHA supplementation preserve synaptic components in Alzheimer's disease.[259] Parkinson's disease is a neurological disorder that occurs with a decrease in the amount of dopaminergic neurons in the substantia nigra. A-synuclein is a protein, which causes neuronal losses, an important predictor of this disease. This pathogenic factor is associated with EMF exposure and is thought to increase oxidative stress α-synuclein mutation in cells.[260] ω-3 is thought to be capable of providing significant benefits in Parkinson's disease. Numerous studies on this subject have focused on the effects of DHA.[261–263] However, when absorbed through diet, therapeutic effects have been observed in schizophrenia[264] and behavioral disorders. At the same time EPA can be used to treat Parkinson's disease symptoms, such as depression.[265]

Effects of ω-3 Intakes on Memory and Learning: Neurite Formation and Synaptogenesis

Rates of neurobehavioral disorders, low academic achievement, and poor school performance in children are currently all increasing. Some of this increase may derive from environmental stress conditions in prenatal life. The effect of EMF exposure should also be taken into account because of the increased use of mobile phones and similar devices. EMF exposure has adverse effects on cerebral functions, such as synaptic plasticity regulation, neurotransmitter release, neuronal survival, and learning and memory, by changing the structural (channel, protein and receptor) and functional organization of neuronal cells and reducing cell numbers in the brain. Exposure to such environmental stress factors during the fetal period may cause various postnatal neurological deficits. Prenatal exposure to EMF may also interfere with dentate neurogenesis and block the transformation of neural cells into mature cells.[201,266,267] In one study, exposure to EMF in the fetal period was found to cause impaired glutamatergic synaptic transmission in the V-pyramidal neurons in the prefrontal cortex, and this effect was found to lead to neurobehavioral disorders.[21]

Ω-3, a major component of the brain is essential for neuronal development and cerebral functions. Nevertheless, the mechanism of positive action in the brain is still unclear. Long-term potentiation is assumed to represent the initial stage of memory formation.[268,269] DHA deficiency is associated with memory impairment and emotional disturbances in rodents. The beneficial effect of ω-3 on cerebral functions has been associated with various mechanisms. [270]. Ω-3 has been shown to exhibit a pro-neurogenic effect and that neural stem cells induce differentiation into neurons.[271,272] The effects of ω-3 units (DHA and EPA) on neurogenesis may differ; EPA significantly increases levels of Hes1, a repressor-type transcription factor that hinders neuronal differentiation and increases neural stem cells, while DHA significantly reduces Hes1 levels. The effects of ω-3 on the regulation of neural stem cell flux may be related to the fine balance between its effects in basic coil-loop-coil transcription factors.[273] At the same time, decreased plasma ω-3 PUFA levels have been reported in children and adolescents with attention deficit/hyperactivity impairment. McNamara et al. reported a decreased signs of inattention in boys with and without attention deficit/hyperactivity in high-dose ω-3 long chain-PUFA rein fraction.[274] Ω-3 (especially DHA) regulates signal transduction and gene expression and protects neurons against death. Wu et al. investigated the protective effect of ω-3 on learning disability after traumatic brain injury in 4-week-old rats. Ω-3 has been shown to facilitate synaptic transmission and learning ability by regulating brain-derived neurotrophic factor (BDNF), synapsin I, and sensitive element binding protein levels.[275]

Synaptic plasticity is the main mechanism underlying learning and memory, which are important functions of the hippocampus. However, the cerebellum also plays a critical role in learning, balance, and impulse control and synapse formation. BDNF is a neurotrophin that facilitates synaptic conduction and is found in substantial quantities in the hippocampus and cerebral cortex.[276] BDNF val66met polymorphism, which affects the secretion of BDNF, results in learning impairment and episodic memory function, as well as reducing memory activity and hippocampal activity during recall processes.[277] Ω-3 has therapeutic potential in terms of preventing memory problems associated with disorders that reduce synaptic plasticity in the hippocampus. DHA reinforcement has been shown to increase BDNF expression.[278] Gamoh et al. examined whether chronic use of DHA improves learning memory related to reference memory in young rats. They reported that chronic use of DHA helps to improve learning ability related to reference memory.[226] The effects on visual, learning, and memory in healthy adults of DHA administration at 900 mg/day for 24 weeks were assessed by the Cambridge neuropsychological test. Analysis revealed improved learning and memory scores.[253] Patten et al. examined the effects of ω-3 on long-term deficits in hippocampal synaptic plasticity caused by ethanol exposure in the prenatal period. They determined that ω-3 reduces synaptic plasticity deficits in the dentate gyrus, and that ω-3 may therefore represent an alternative treatment for neurological deficits.[279]

ω-3 PUFAs also exhibit significant effects on the arrangement of glutamatergic synapses that are responsible for memory formation. Glutamate is the most important stimulating neurotransmitter in the cerebrum, and glutamatergic synapses are abundantly present in the hippocampus. Several studies have reported that EPA has direct neurite-supporting effects in primary sensory neuronal cultures in the young, adult, and elderly.[258,280] EPA and DHA together increase neurite growth during development. DHA also exhibits positive effects on tissues from elderly rats with significantly low degrees of plasticity. Another study showed that DHA increases maximum neurite length and the total number of neuritis in embryonic hippocampal and cortical neuronal cultures.[281] Ω-3 supplementation can prevent neurological deficits by inhibiting EMF-induced oxidative stress, inhibiting cellular apoptosis, and increasing membrane fluidity.

Relations Between ω-3 Fatty Acids and Depression

EMF can cause behavioral disturbances by affecting the BBB, intracellular calcium homeostasis, neuronal apoptosis, and extracellular signaling pathways.[21] Numerous studies of EMF have reported decreased numbers of neurons in the hippocampus and cerebellum.[12,20,282] Increased depressive disorders in humans are thought to be associated with severe EMF exposure. Abnormalities in the hippocampus may cause depression. The adverse effects of EMF on the hippocampus strengthen this thesis.[283] One study classified 130 electricians into five groups depending on severity of exposure to EMF. The results indicated a high probability of depression, obsessive-compulsive behavior, and anxiety following exposure to high EMF.[284] Many problems affecting mental health, such as schizophrenia and depression, may be due to changes in the phospholipid structure of neuronal membranes.[285] Due to the regulatory effect of ω-3 on metabolic structure, several epidemiological studies have suggested that dietary intake of ω-3 is associated with a decrease in depressive disorders.[286–288] A combination of EPA and selective serotonin reuptake inhibitors has been reported to exhibit a therapeutic effect against diseases such as depression.[289] Sivrioglu et al. observed that ω-3 improves various side effects caused by schizophrenia.[290] In another study, ω-3 was found to cause behavioral responses in a schizophrenia model and a reduction in various biochemical substances that may lead to brain oxidative damage in excess quantities. At this experiment, ω-3 in the hippocampus was observed to reduce oxidative stress.[291] Martin et al. showed that the use of ω-3 exhibited an important therapeutic effect in treating major depression. This, combined with the use of EPA with antidepressants, has important implications for treating depressive disorders.[292] Research indicates that ω-3 regulates neurogenesis and has beneficial effects on behavioral disorders such as depression by exhibiting a therapeutic effect against EMF.

CONCLUSION

The subject of whether or not EMF is harmful to humans is still a controversial one. However, in the light of the widespread use of mobile phones, their effects on the CNS need to be tested. In many experimental researches have showed that exposure to EMF in the prenatal period can lead to harmful effects on the nervous system in the fetus, and these potential neurotoxic effects should also be evaluated. Several factors are implicated in increased free-radical activity in brain. Iron particles (known as magnetite) are found in tissues, particularly in the brain. Magnetite may increase cellular free-radical activity, and the nervous system is particularly susceptible to this effect. While free radicals exert a common effect on all type of cells, high concentrations of iron ions in the brain may exacerbate this effect of EMF on neuronal DNA. There have been substantial reports that EMF impairs embryonic development and cornu amnion formation, which may lead to impaired behavioral and cognitive functions, as well as short-term memory and learning deficiency in postnatal life. Due to its antioxidant effects, melatonin exhibits a neuroprotective effect against EMF exposure. Although experimental studies on the effects of melatonin on EMF-induced neuronal loss are limited, the neuroprotective activity of melatonin is a proven fact. However, inducing effects of melatonin involve not only neuron survival, but also on neurogenesis. The mechanisms underlying these effects are associated with MT1 and MT2. The therapeutic role of melatonin in neurodegenerative diseases is currently being investigated in several clinical trials. The effects of various concentrations of melatonin may vary depending on the type of tissue. Ω-3 fatty acids are essential for the development of the infant brain in mammals. Maternal ω-3 status in pregnancy is therefore of crucial importance. Ω-3 appears to exhibit promising effects in terms of protecting the brain against oxidative damage and improving histopathological changes caused by EMF exposure. Further studies are now needed do to evaluate the effects of both prenatal and postnatal EMF exposure on the nervous system and to determine melatonin and ω-3 activity against the harmful effects of EMF.

Acknowledgments

We would like to thanks to Mr. *Carl Nino Rossini* for his excellent language revision and suggestion on the chapter and we are always grateful that we receive very good works and contribution to our academic studies. We are also thankful to Mr. *Sefa Ersan Kaya* for assistance with the editing of pictures.

References

1. Hong FT. Magnetic field effects on biomolecules, cells, and living organisms. *Biosystems*. 1995;36:187–229.
2. Shah JP, Midkiff P, Brandt PC, Sisken BF. Growth and differentiation of PC6 cells: the effects of pulsed electromagnetic fields (PEMF). *Bioelectromagnetics*. 2001;22:267–271.

3. Macias MY, Battocletti JH, Sutton CH, Pintar FA, Maiman DJ. Directed and enhanced neurite growth with pulsed magnetic field stimulation. *Bioelectromagnetics*. 2000;21:272–286.
4. McFarlane EH, Dawe GS, Marks M, Campbell IC. Changes in neurite outgrowth but not in cell division induced by low EMF exposure: influence of field strength and culture conditions on responses in rat PC12 pheochromocytoma cells. *Bioelectrochemistry*. 2000;52:23–28.
5. Zhang Y, Ding J, Duan W, Fan W. Influence of pulsed electromagnetic field with different pulse duty cycles on neurite outgrowth in PC12 rat pheochromocytoma cells. *Bioelectromagnetics*. 2005;26:406–411.
6. Rajnicek AM, Robinson KR, McCaig CD. The direction of neurite growth in a weak DC electric field depends on the substratum: contributions of adhesivity and net surface charge. *Dev Biol*. 1998;203:412–423.
7. Del Vecchio G, Giuliani A, Fernandez M, et al. Continuous exposure to 900 MHz GSM-modulated EMF alters morphological maturation of neural cells. *Neurosci Lett*. 2009;455:173–177.
8. Bandyopadhyay D, Biswas K, Bandyopadhyay U, Reiter RJ, Banerjee RK. Melatonin protects against stress-induced gastric lesions by scavenging the hydroxyl radical. *J Pineal Res*. 2000;29:143–151.
9. Bahna SG, Niles LP. Epigenetic regulation of melatonin receptors in neuropsychiatric disorders. *Br J Pharmacol*. 2018;175:3209–3219.
10. Benitez-King G. Melatonin as a cytoskeletal modulator: implications for cell physiology and disease. *J Pineal Res*. 2006;40:1–9.
11. Srinivasan V, Pandi-Perumal SR, Maestroni GJ, Esquifino AI, Hardeland R, Cardinali DP. Role of melatonin in neurodegenerative diseases. *Neurotox Res*. 2005;7:293–318.
12. Erdem Koc G, Kaplan S, Altun G, et al. Neuroprotective effects of melatonin and omega-3 on hippocampal cells prenatally exposed to 900 MHz electromagnetic fields. *Int J Radiat Biol*. 2016;92:590–595.
13. Liu DD, Ren Z, Yang G, Zhao QR, Mei YA. Melatonin protects rat cerebellar granule cells against electromagnetic field-induced increases in Na(+) currents through intracellular Ca(2+) release. *J Cell Mol Med*. 2014;18:1060–1070.
14. Dubnov-Raz G, Finkelstein Y, Koren G. omega-3 fatty acid supplementation during pregnancy—For mother, baby, or neither? *Can Fam Physician*. 2007;53:817–818.
15. Jordan RG. Prenatal omega-3 fatty acids: review and recommendations. *J Midwifery Wom Heal*. 2010;55:520–528.
16. Mulder KA, King DJ, Innis SM. Omega-3 fatty acid deficiency in infants before birth identified using a randomized trial of maternal DHA supplementation in pregnancy. *Plos One*. 2014;9(e83764):1–10.
17. Boucher O, Burden MJ, Muckle G, et al. Neurophysiologic and neurobehavioral evidence of beneficial effects of prenatal omega-3 fatty acid intake on memory function at school age. *Am J Clin Nutr*. 2011;93:1025–1037.
18. Freund-Levi Y, Vedin I, Hjorth E, et al. Effects of supplementation with omega-3 fatty acids on oxidative stress and inflammation in patients with Alzheimer's disease: the OmegAD study. *J Alzheimers Dis*. 2014;42:823–831.
19. Chen X, Guo C, Kong J. Oxidative stress in neurodegenerative diseases. *Neural Regen Res*. 2012;7:376–385.
20. Altun GKS, Deniz ÖG, Kocacan SE, Canan S, Davis D, Marangoz C. Protective effects of melatonin and omega 3 on the hippocampus and the cerebellum of adult Wistar albino rats exposed to electromagnetic fields. *JMAU*. 2017;5:230–241.
21. Aldad TS, Gan GL, Gao XB, Taylor HS. Fetal radiofrequency radiation exposure from 800–1900 MHz-rated cellular telephones affects neurodevelopment and behavior in mice. *Sci Rep-Uk*. 2012;2:312.
22. Pirozzoli MC, Marino C, Lovisolo GA, Laconi C, Mosiello L, Negroni A. Effects of 50 Hz electromagnetic field exposure on apoptosis and differentiation in a neuroblastoma cell line. *Bioelectromagnetics*. 2003;24:510–516.
23. Su L, Yimaer A, Wei X, Xu Z, Chen G. The effects of 50 Hz magnetic field exposure on DNA damage and cellular functions in various neurogenic cells. *J Radiat Res*. 2017;58:474–486.
24. Zhao TY, Zou SP, Knapp PE. Exposure to cell phone radiation up-regulates apoptosis genes in primary cultures of neurons and astrocytes. *Neurosci Lett*. 2007;412:34–38.
25. Piacentini R, Ripoli C, Mezzogori D, Azzena GB, Grassi C. Extremely low-frequency electromagnetic fields promote in vitro neurogenesis via upregulation of Ca(v)1-channel activity. *J Cell Physiol*. 2008;215:129–139.
26. Gye MC, Park CJ. Effect of electromagnetic field exposure on the reproductive system. *Clin Exp Reprod Med*. 2012;39:1–9.
27. Noonan CW, Reif JS, Yost M, Touchstone J. Occupational exposure to magnetic fields in case-referent studies of neurodegenerative diseases. *Scand J Work Environ Health*. 2002;28:42–48.
28. Stam R. Electromagnetic fields and the blood-brain barrier. *Brain Res Rev*. 2010;65:80–97.
29. Rosenberg SS, Spitzer NC. Calcium signaling in neuronal development. *Cold Spring Harb Perspect Biol*. 2011;3:a004259.

30. Pall ML. Electromagnetic field activation of voltage-gated calcium channels: role in therapeutic effects. *Electromagn Biol Med*. 2014;33:251.
31. Pilla AA. Electromagnetic fields instantaneously modulate nitric oxide signaling in challenged biological systems. *Biochem Biophys Res Commun*. 2012;426:330–333.
32. Maskey D, Kim M, Aryal B, et al. Effect of 835 MHz radiofrequency radiation exposure on calcium binding proteins in the hippocampus of the mouse brain. *Brain Res*. 2010;1313:232–241.
33. Ahuja YR, Vijayashree B, Saran R, Jayashri EL, Manoranjani JK, Bhargava SC. In vitro effects of low-level, low-frequency electromagnetic fields on DNA damage in human leucocytes by comet assay. *Indian J Biochem Biophys*. 1999;36:318–322.
34. Svedenstal BM, Johanson KJ, Mattsson MO, Paulsson LE. DNA damage, cell kinetics and ODC activities studied in CBA mice exposed to electromagnetic fields generated by transmission lines. *In Vivo*. 1999;13:507–513.
35. Svedenstal BM, Johanson KJ, Mild KH. DNA damage induced in brain cells of CBA mice exposed to magnetic fields. *In Vivo*. 1999;13:551–552.
36. Zmyslony M, Palus J, Jajte J, Dziubaltowska E, Rajkowska E. DNA damage in rat lymphocytes treated in vitro with iron cations and exposed to 7 mT magnetic fields (static or 50 Hz). *Mutat Res*. 2000;453:89–96.
37. Phillips JL, Campbell-Beachler, M., Ivaschuk, O., Ishida-Jones, T., Haggnen, W. Exposure of molt-4 lymphoblastoid cells to a 1 G sinusoidal magnetic field at 60-Hz: effects on cellular events related to apoptosis. Annual Review of Research on Biological Effects of Electric and Magnetic Fields from the Generation, Delivery, and Use of Electricity. Frederick, MD:W/L Associates, Ltd. In:1997.
38. Lai H, Singh NP. Acute low-intensity microwave exposure increases DNA single-strand breaks in rat brain cells. *Bioelectromagnetics*. 1995;16:207–210.
39. Lai H, Singh NP. Single- and double-strand DNA breaks in rat brain cells after acute exposure to radiofrequency electromagnetic radiation. *Int J Radiat Biol*. 1996;69:513–521.
40. Lai H, Singh NP. Melatonin and a spin-trap compound block radiofrequency electromagnetic radiation-induced DNA strand breaks in rat brain cells. *Bioelectromagnetics*. 1997;18:446–454.
41. Sarkar S, Ali S, Behari J. Effect of low power microwave on the mouse genome: a direct DNA analysis. *Mutat Res*. 1994;320:141–147.
42. Thomas JR, Schrot J, Banvard RA. Comparative effects of pulsed and continuous-wave 2.8-GHz microwaves on temporally defined behavior. *Bioelectromagnetics*. 1982;3:227–235.
43. Lai H, Singh NP. Melatonin and N-tert-butyl-alpha-phenylnitrone block 60-Hz magnetic field-induced DNA single and double strand breaks in rat brain cells. *J Pineal Res*. 1997;22:152–162.
44. Lai H, Singh NP. Magnetic-field-induced DNA strand breaks in brain cells of the rat. *Environ Health Perspect*. 2004;112:687–694.
45. Altman SA, Zastawny TH, Randers-Eichhorn L, et al. Formation of DNA-protein cross-links in cultured mammalian cells upon treatment with iron ions. *Free Radic Biol Med*. 1995;19:897–902.
46. Eberhardt JL, Persson BR, Brun AE, Salford LG, Malmgren LO. Blood-brain barrier permeability and nerve cell damage in rat brain 14 and 28 days after exposure to microwaves from GSM mobile phones. *Electromagn Biol Med*. 2008;27:215–229.
47. Narayanan SN, Kumar RS, Potu BK, Nayak S, Mailankot M. Spatial memory performance of Wistar rats exposed to mobile phone. *Clinics (Sao Paulo)*. 2009;64:231–234.
48. Luukkonen J, Hakulinen P, Maki-Paakkanen J, Juutilainen J, Naarala J. Enhancement of chemically induced reactive oxygen species production and DNA damage in human SH-SY5Y neuroblastoma cells by 872 MHz radiofrequency radiation. *Mutat Res*. 2009;662:54–58.
49. Liu YX, Tai JL, Li GQ, et al. Exposure to 1950-MHz TD-SCDMA electromagnetic fields affects the apoptosis of astrocytes via caspase-3-dependent pathway. *Plos One*. 2012;7:e42332.
50. He YL, Liu DD, Fang YJ, Zhan XQ, Yao JJ, Mei YA. Exposure to extremely low-frequency electromagnetic fields modulates Na+ currents in rat cerebellar granule cells through increase of AA/PGE2 and EP receptor-mediated cAMP/PKA pathway. *Plos One*. 2013;8:e54376.
51. Huang CY, Chang CW, Chen CR, et al. Extremely low-frequency electromagnetic fields cause G1 phase arrest through the activation of the ATM-Chk2-p21 pathway. *Plos One*. 2014;9:e104732.
52. Caraglia M, Marra M, Mancinelli F, et al. Electromagnetic fields at mobile phone frequency induce apoptosis and inactivation of the multi-chaperone complex in human epidermoid cancer cells. *J Cell Physiol*. 2005;204:539–548.

53. Tang J, Zhang Y, Yang L, et al. Exposure to 900 MHz electromagnetic fields activates the mkp-1/ERK pathway and causes blood-brain barrier damage and cognitive impairment in rats. *Brain Res.* 1601;2015:92–101.
54. Leszczynski D, Joenvaara S, Reivinen J, Kuokka R. Non-thermal activation of the hsp27/p38MAPK stress pathway by mobile phone radiation in human endothelial cells: molecular mechanism for cancer- and blood-brain barrier-related effects. *Differentiation.* 2002;70:120–129.
55. Yang X-S, He G-L, Hao Y-T, et al. Exposure to 2.45GHz electromagnetic fields elicits an HSP-related stress response in rat hippocampus. *Brain Res Bull.* 2012;88:371–378.
56. Jorge-Mora TAFM, Leiro-Vidal JM, Jorge-Barreiro FJ, Ares-Pena FJ, López-Martin E. Exposure to 2.45 GHz Microwave Radiation Provokes Cerebral Changes in Induction of Hsp-90 α/β Heat Shock Protein in Rat. *Prog Electromagn Res.* 2010;100:351–379.
57. Li SS, Zhang ZY, Yang CJ, Lian HY, Cai P. Gene expression and reproductive abilities of male Drosophila melanogaster subjected to ELF-EMF exposure. *Mutat Res.* 2013;758:95–103.
58. Remondini D, Nylund R, Reivinen J, et al. Gene expression changes in human cells after exposure to mobile phone microwaves. *Proteomics.* 2006;6:4745–4754.
59. Pollard TD, Blanchoin L, Mullins RD. Molecular mechanisms controlling actin filament dynamics in nonmuscle cells. *Annu Rev Biophys Biomol Struct.* 2000;29:545–576.
60. Dent EW, Gertler FB. Cytoskeletal dynamics and transport in growth cone motility and axon guidance. *Neuron.* 2003;40:209–227.
61. Buttiglione M, Roca L, Montemurno E, Vitiello F, Capozzi V, Cibelli G. Radiofrequency radiation (900 MHz) induces Egr-1 gene expression and affects cell-cycle control in human neuroblastoma cells. *J Cell Physiol.* 2007;213:759–767.
62. Sukhatme VP, Cao XM, Chang LC, et al. A zinc finger-encoding gene coregulated with c-fos during growth and differentiation, and after cellular depolarization. *Cell.* 1988;53:37–43.
63. Milbrandt J. A nerve growth factor-induced gene encodes a possible transcriptional regulatory factor. *Science.* 1987;238:797–799.
64. Sheng M, McFadden G, Greenberg ME. Membrane depolarization and calcium induce c-fos transcription via phosphorylation of transcription factor CREB. *Neuron.* 1990;4:571–582.
65. O'Donovan KJ, Tourtellotte WG, Millbrandt J, Baraban JM. The EGR family of transcription-regulatory factors: progress at the interface of molecular and systems neuroscience. *Trends Neurosci.* 1999;22:167–173.
66. Worley PF, Bhat RV, Baraban JM, Erickson CA, McNaughton BL, Barnes CA. Thresholds for synaptic activation of transcription factors in hippocampus: correlation with long-term enhancement. *J Neurosci.* 1993;13:4776–4786.
67. Williams AJ, Khachigian LM, Shows T, Collins T. Isolation and characterization of a novel zinc-finger protein with transcription repressor activity. *J Biol Chem.* 1995;270:22143–22152.
68. Revest JM, Di Blasi F, Kitchener P, et al. The MAPK pathway and Egr-1 mediate stress-related behavioral effects of glucocorticoids. *Nat Neurosci.* 2005;8:664–672.
69. Chen C, Ma Q, Liu C, et al. Exposure to 1800 MHz radiofrequency radiation impairs neurite outgrowth of embryonic neural stem cells. *Sci Rep-UK.* 2014;4:5103.
70. Lee BC, Johng HM, Lim JK, et al. Effects of extremely low frequency magnetic field on the antioxidant defense system in mouse brain: a chemiluminescence study. *J Photochem Photobiol B.* 2004;73:43–48.
71. Quintens R, Verreet T, Janssen A, et al. Identification of novel radiation-induced p53-dependent transcripts extensively regulated during mouse brain development. *Biol Open.* 2015;4:331–344.
72. Bolaris EBA, Benekou HPS. In utero radiation-induced apoptosis and p53 gene expression in the developing rat brain. *Int J Radiat Biol.* 2009;77:71–81.
73. Salford LGNH, Persson BR. Effects of EMF from wireless communication upon the blood–brain barrier. In: *BioInitiative 2012: A Rationale for a Biologically-Based Public Exposure Standard for Electromagnetic Fields (ELF and RF).* 2012:1–52.
74. Oscar KJ, Hawkins TD. Microwave alteration of the blood-brain barrier system of rats. *Brain Res.* 1977;126:281–293.
75. Nittby H, Brun A, Eberhardt J, Malmgren L, Persson BR, Salford LG. Increased blood-brain barrier permeability in mammalian brain 7 days after exposure to the radiation from a GSM-900 mobile phone. *Pathophysiology.* 2009;16:103–112.

76. Salford LG, Brun A, Sturesson K, Eberhardt JL, Persson BR. Permeability of the blood-brain barrier induced by 915 MHz electromagnetic radiation, continuous wave and modulated at 8, 16, 50, and 200 Hz. *Microsc Res Tech*. 1994;27:535–542.
77. Iwasaki Y, Ito S, Suzuki M, Nagahori T, Yamamoto T, Konno H. Forebrain ischemia induced by temporary bilateral common carotid occlusion in normotensive rats. *J Neurol Sci*. 1989;90:155–165.
78. Hassel B, Iversen EG, Fonnum F. Neurotoxicity of albumin in vivo. *Neurosci Lett*. 1994;167:29–32.
79. Szymas J, Hossmann KA. Determination of endogenous serum proteins in normal and oedematous brain tissue of cat by rocket and crossed immunoelectrophoresis. *Acta Neurochir (Wien)*. 1990;105:169–177.
80. Moriyama E, Salcman M, Broadwell RD. Blood-brain barrier alteration after microwave-induced hyperthermia is purely a thermal effect: I. Temperature and power measurements. *Surg Neurol*. 1991;35:177–182.
81. Kiyatkin EA, Sharma HS. Permeability of the blood-brain barrier depends on brain temperature. *Neuroscience*. 2009;161:926–939.
82. Hossmann KA, Hermann DM. Effects of electromagnetic radiation of mobile phones on the central nervous system. *Bioelectromagnetics*. 2003;24:49–62.
83. Bellieni CV, Nardi V, Buonocore G, Di Fabio S, Pinto I, Verrotti A. Electromagnetic fields in neonatal incubators: the reasons for an alert. *J Matern Fetal Neonatal Med*. 2019;32:695–699.
84. Selmaoui B, Touitou Y. Age-related differences in serum melatonin and pineal NAT activity and in the response of rat pineal to a 50-HZ magnetic field. *Life Sci*. 1999;64:2291–2297.
85. Halgamuge MN. Pineal melatonin level disruption in humans due to electromagnetic fields and ICNIRP limits. *Radiat Prot Dosimetry*. 2012;154:405–416.
86. Bellieni CV, Tei M, Iacoponi F, et al. Is newborn melatonin production influenced by magnetic fields produced by incubators? *Early Hum Dev*. 2012;88:707–710.
87. Reiter RJ. Static and extremely low frequency electromagnetic field exposure: reported effects on the circadian production of melatonin. *J Cell Biochem*. 1993;51:394–403.
88. Macchi MM, Bruce JN. Human pineal physiology and functional significance of melatonin. *Front Neuroendocrinol*. 2004;25:177–195.
89. Pandi-Perumal SR, BaHammam AS, Brown GM, et al. Melatonin antioxidative defense: therapeutical implications for aging and neurodegenerative processes. *Neurotox Res*. 2013;23:267–300.
90. Reiter RJ, Tan DX, Rosales-Corral S, Manchester LC. The universal nature, unequal distribution and antioxidant functions of melatonin and its derivatives. *Mini Rev Med Chem*. 2013;13:373–384.
91. Hardeland R, Cardinali DP, Srinivasan V, Spence DW, Brown GM, Pandi-Perumal SR. Melatonin—a pleiotropic, orchestrating regulator molecule. *Prog Neurobiol*. 2011;93:350–384.
92. Wongprayoon P, Govitrapong P. Melatonin as a mitochondrial protector in neurodegenerative diseases. *Cell Mol Life Sci*. 2017;74:3999–4014.
93. Reiter RJ. Melatonin: the chemical expression of darkness. *Mol Cell Endocrinol*. 1991;79:C153–C158.
94. Olakowska E, Marcol W, Kotulska K, Lewin-Kowalik J. The role of melatonin in the neurodegenerative diseases. *Bratisl Lek Listy*. 2005;106:171–174.
95. AcunaCastroviejo D, CotoMontes A, Monti MG, Ortiz GG, Reiter RJ. Melatonin is protective against MPTP-induced striatal and hippocampal lesions. *Life Sci*. 1996;60:PI23–PI29.
96. Srinivasan V. Melatonin oxidative stress and neurodegenerative diseases. *Indian J Exp Biol*. 2002;40:668–679.
97. Castroviejo DA, Lopez LC, Escames G, Lopez A, Garcia JA, Reiter RJ. Melatonin-mitochondria Interplay in Health and Disease. *Curr Top Med Chem*. 2011;11:221–240.
98. Barlowwalden LR, Reiter RJ, Abe M, et al. Melatonin stimulates brain glutathione-peroxidase activity. *Neurochem Int*. 1995;26:497–502.
99. Li HY, Wang Y, Feng DX, et al. Alterations in the time course of expression of the Nox family in the brain in a rat experimental cerebral ischemia and reperfusion model: effects of melatonin. *J Pineal Res*. 2014;57:110–119.
100. Reiter RJ, Paredes SD, Manchester LC, Tan DX. Reducing oxidative/nitrosative stress: a newly-discovered genre for melatonin. *Crit Rev Biochem Mol*. 2009;44:175–200.
101. Guerrero JM, Reiter RJ, Ortiz GG, Pablos MI, Sewerynek E, Chuang JI. Melatonin prevents increases in neural nitric oxide and cyclic GMP production after transient brain ischemia and reperfusion in the Mongolian gerbil (Meriones unguiculatus). *J Pineal Res*. 1997;23:24–31.
102. Rodriguez C, Mayo JC, Sainz RM, et al. Regulation of antioxidant enzymes: a significant role for melatonin. *J Pineal Res*. 2004;36:1–9.

103. Ozacmak VH, Barut F, Ozacmak HS. Melatonin provides neuroprotection by reducing oxidative stress and HSP70 expression during chronic cerebral hypoperfusion in ovariectomized rats. *J Pineal Res*. 2009;47:156–163.
104. Sinha K, Degaonkar MN, Jagannathan NR, Gupta YK. Effect of melatonin on ischemia reperfusion injury induced by middle cerebral artery occlusion in rats. *Eur J Pharmacol*. 2001;428:185–192.
105. Kim MJ, Kim HK, Kim BS, Yim SV. Melatonin increases cell proliferation in the dentate gyrus of maternally separated rats. *J Pineal Res*. 2004;37:193–197.
106. Musshoff U, Riewenherm D, Berger E, Fauteck JD, Speckmann EJ. Melatonin receptors in rat hippocampus: molecular and functional investigations. *Hippocampus*. 2002;12:165–173.
107. Martin M, Macias M, Escames G, et al. Melatonin-induced increased activity of the respiratory chain complexes I and IV can prevent mitochondrial damage induced by ruthenium red in vivo. *J Pineal Res*. 2000;28:242–248.
108. Blanc EM, Bruce-Keller AJ, Mattson MP. Astrocytic gap junctional communication decreases neuronal vulnerability to oxidative stress-induced disruption of Ca2+ homeostasis and cell death. *J Neurochem*. 1998;70:958–970.
109. Kang W, Hebert JM. Signaling pathways in reactive astrocytes, a genetic perspective. *Mol Neurobiol*. 2011;43:147–154.
110. Takuma K, Baba A, Matsuda T. Astrocyte apoptosis: implications for neuroprotection. *Prog Neurobiol*. 2004;72:111–127.
111. Brillaud E, Piotrowski A, de Seze R. Effect of an acute 900MHz GSM exposure on glia in the rat brain: a time-dependent study. *Toxicology*. 2007;238:23–33.
112. Mausset-Bonnefont AL, Hirbec H, Bonnefont X, Privat A, Vignon J, de Seze R. Acute exposure to GSM 900-MHz electromagnetic fields induces glial reactivity and biochemical modifications in the rat brain. *Neurobiol Dis*. 2004;17:445–454.
113. Maskey D, Kim HJ, Kim HG, Kim MJ. Calcium-binding proteins and GFAP immunoreactivity alterations in murine hippocampus after 1 month of exposure to 835 MHz radiofrequency at SAR values of 1.6 and 4.0 W/kg. *Neurosci Lett*. 2012;506:292–296.
114. Yang X, He G, Hao Y, et al. The role of the JAK2-STAT3 pathway in pro-inflammatory responses of EMF-stimulated N9 microglial cells. *J Neuroinflammation*. 2010;7:54.
115. Campisi A, Gulino M, Acquaviva R, et al. Reactive oxygen species levels and DNA fragmentation on astrocytes in primary culture after acute exposure to low intensity microwave electromagnetic field. *Neurosci Lett*. 2010;473:52–55.
116. Lu Y, He M, Zhang Y, et al. Differential pro-inflammatory responses of astrocytes and microglia involve STAT3 activation in response to 1800 MHz radiofrequency fields. *Plos One*. 2014;9:e108318.
117. Ramirez-Rodriguez G, Klempin F, Babu H, Benitez-King G, Kempermann G. Melatonin modulates cell survival of new neurons in the hippocampus of adult mice. *Neuropsychopharmacol*. 2009;34:2180–2191.
118. Rennie K, De Butte M, Pappas BA. Melatonin promotes neurogenesis in dentate gyrus in the pinealectomized rat. *J Pineal Res*. 2009;47:313–317.
119. Tocharus C, Puriboriboon Y, Junmanee T, Tocharus J, Ekthuwapranee K, Govitrapong P. Melatonin enhances adult rat hippocampal progenitor cell proliferation via ERK signaling pathway through melatonin receptor. *Neuroscience*. 2014;275:314–321.
120. Hamad MI, Jack A, Klatt O, et al. Type I TARPs promote dendritic growth of early postnatal neocortical pyramidal cells in organotypic cultures. *Development*. 2014;141:1737–1748.
121. Ramakers GJ, Avci B, van Hulten P, et al. The role of calcium signaling in early axonal and dendritic morphogenesis of rat cerebral cortex neurons under non-stimulated growth conditions. *Brain Res Dev Brain Res*. 2001;126:163–172.
122. Campbell S, Marriott M, Nahmias C, MacQueen GM. Lower hippocampal volume in patients suffering from depression: a meta-analysis. *Am J Psychiatry*. 2004;161:598–607.
123. Ebdrup BH, Glenthoj B, Rasmussen H, et al. Hippocampal and caudate volume reductions in antipsychotic-naive first-episode schizophrenia. *J Psychiatry Neurosci*. 2010;35:95–104.
124. Simic G, Kostovic I, Winblad B, Bogdanovic N. Volume and number of neurons of the human hippocampal formation in normal aging and Alzheimer's disease. *J Comp Neurol*. 1997;379:482–494.
125. Scheff SW, Price DA, Schmitt FA, Mufson EJ. Hippocampal synaptic loss in early Alzheimer's disease and mild cognitive impairment. *Neurobiol Aging*. 2006;27:1372–1384.
126. Sun D, Stuart GW, Jenkinson M, et al. Brain surface contraction mapped in first-episode schizophrenia: a longitudinal magnetic resonance imaging study. *Mol Psychiatry*. 2009;14:976–986.

127. Babu YS, Bugg CE, Cook WJ. Structure of calmodulin refined at 2.2 A resolution. *J Mol Biol*. 1988;204:191–204.
128. Diez-Guerra FJ, Avila J. MAP2 phosphorylation parallels dendrite arborization in hippocampal neurones in culture. *Neuroreport*. 1993;4:419–422.
129. Vaillant AR, Zanassi P, Walsh GS, Aumont A, Alonso A, Miller FD. Signaling mechanisms underlying reversible, activity-dependent dendrite formation. *Neuron*. 2002;34:985–998.
130. Rivera-Bermudez MA, Gerdin MJ, Earnest DJ, Dubocovich ML. Regulation of basal rhythmicity in protein kinase C activity by melatonin in immortalized rat suprachiasmatic nucleus cells. *Neurosci Lett*. 2003;346:37–40.
131. Dominguez-Alonso A, Valdes-Tovar M, Solis-Chagoyan H, Benitez-King G. Melatonin stimulates dendrite formation and complexity in the hilar zone of the rat hippocampus: participation of the Ca++/Calmodulin complex. *Int J Mol Sci*. 2015;16:1907–1927.
132. Xu S, Zhou Z, Zhang L, et al. Exposure to 1800 MHz radiofrequency radiation induces oxidative damage to mitochondrial DNA in primary cultured neurons. *Brain Res*. 2010;1311:189–196.
133. Cabrera J, Reiter RJ, Tan DX, et al. El-Sokkary G. Melatonin reduces oxidative neurotoxicity due to quinolinic acid: in vitro and in vivo findings. *Neuropharmacology*. 2000;39:507–514.
134. Dubocovich ML, Markowska M. Functional MT1 and MT2 melatonin receptors in mammals. *Endocrine*. 2005;27:101–110.
135. Liu JB, Clough SJ, Hutchinson AJ, Adamah-Biassi EB, Popovska-Gorevski M, Dubocovich ML. MT1 and MT2 melatonin receptors: a therapeutic perspective. *Annu Rev Pharmacol*. 2016;56:361–383.
136. Wu YH, Zhou JN, Balesar R, et al. Distribution of MT1 melatonin receptor immunoreactivity in the human hypothalamus and pituitary gland: colocalization of MT1 with vasopressin, oxytocin, and corticotropin-releasing hormone. *J Comp Neurol*. 2006;499:897–910.
137. Brunner P, Sozer-Topcular N, Jockers R, et al. Pineal and cortical melatonin receptors MT1 and MT2 are decreased in Alzheimer's disease. *Eur J Histochem*. 2006;50:311–316.
138. Leeboonngam T, Pramong R, Sae-Ung K, Govitrapong P, Phansuwan-Pujito P. Neuroprotective effects of melatonin on amphetamine-induced dopaminergic fiber degeneration in the hippocampus of postnatal rats. *J Pineal Res*. 2018;64.
139. Foulkes NS, Borjigin J, Snyder SH, Sassone-Corsi P. Rhythmic transcription: the molecular basis of circadian melatonin synthesis. *Trends Neurosci*. 1997;20:487–492.
140. Huang F, Yang Z, Li CQ. The melatonergic system in anxiety disorders and the role of melatonin in conditional fear. *Vitam Horm*. 2017;103:281–294.
141. Dubocovich ML. Melatonin receptors: role on sleep and circadian rhythm regulation. *Sleep Med*. 2007;8(Suppl 3):34–42.
142. Savaskan E, Ayoub MA, Ravid R, et al. Reduced hippocampal MT2 melatonin receptor expression in Alzheimer's disease. *J Pineal Res*. 2005;38:10–16.
143. Audinot V, Mailliet F, Lahaye-Brasseur C, et al. New selective ligands of human cloned melatonin MT1 and MT2 receptors. *Naunyn Schmiedebergs Arch Pharmacol*. 2003;367:553–561.
144. Reppert SM, Weaver DR, Godson C. Melatonin receptors step into the light: cloning and classification of subtypes. *Trends Pharmacol Sci*. 1996;17:100–102.
145. Slaugenhaupt SA, Roca AL, Liebert CB, Altherr MR, Gusella JF, Reppert SM. Mapping of the gene for the Mel1a-melatonin receptor to human chromosome 4 (MTNR1A) and mouse chromosome 8 (Mtnr1a). *Genomics*. 1995;27:355–357.
146. O'Neal-Moffitt G, Delic V, Bradshaw PC, Olcese J. Prophylactic melatonin significantly reduces Alzheimer's neuropathology and associated cognitive deficits independent of antioxidant pathways in A beta PPswe/PS1 mice. *Mol Neurodegener*. 2015;10.
147. Bas O, Odaci E, Kaplan S, Acer N, Ucok K, Colakoglu S. 900 MHz electromagnetic field exposure affects qualitative and quantitative features of hippocampal pyramidal cells in the adult female rat. *Brain Res*. 2009;1265:178–185.
148. Sonmez OF, Odaci E, Bas O, Kaplan S. Purkinje cell number decreases in the adult female rat cerebellum following exposure to 900 MHz electromagnetic field. *Brain Res*. 2010;1356:95–101.
149. Ikinci A, Odaci E, Yildirim M, et al. The effects of prenatal exposure to a 900 megahertz electromagnetic field on hippocampus morphology and learning behavior in rat pups. *Neuroquantology*. 2013;11:582–590.
150. Haghani M, Shabani M, Moazzami K. Maternal mobile phone exposure adversely affects the electrophysiological properties of Purkinje neurons in rat offspring. *Neuroscience*. 2013;250:588–598.

151. Naziroglu M, Celik O, Ozgul C, et al. Melatonin modulates wireless (2.45 GHz)-induced oxidative injury through TRPM2 and voltage gated Ca(2+) channels in brain and dorsal root ganglion in rat. *Physiol Behav.* 2012;105:683–692.
152. Celik O, Naziroglu M. Melatonin modulates apoptosis and TRPM2 channels in transfected cells activated by oxidative stress. *Physiol Behav*. 2012;107:458–465.
153. Jiao S, Wu MM, Hu CL, Zhang ZH, Mei YA. Melatonin receptor agonist 2-iodomelatonin prevents apoptosis of cerebellar granule neurons via K(+) current inhibition. *J Pineal Res*. 2004;36:109–116.
154. Liu LY, Hoffman GE, Fei XW, Li Z, Zhang ZH, Mei YA. Delayed rectifier outward K+ current mediates the migration of rat cerebellar granule cells stimulated by melatonin. *J Neurochem*. 2007;102:333–344.
155. Banasiak KJ, Burenkova O, Haddad GG. Activation of voltage-sensitive sodium channels during oxygen deprivation leads to apoptotic neuronal death. *Neuroscience*. 2004;126:31–44.
156. Rogerio F, de Souza Queiroz L, Teixeira SA, Oliveira AL, de Nucci G, Langone F. Neuroprotective action of melatonin on neonatal rat motoneurons after sciatic nerve transection. *Brain Res*. 2002;926:33–41.
157. Ali T, Rehman SU, Shah FA, Kim MO. Acute dose of melatonin via Nrf2 dependently prevents acute ethanol-induced neurotoxicity in the developing rodent brain. *J Neuroinflammation*. 2018;15:119.
158. Gould E, Beylin A, Tanapat P, Reeves A, Shors TJ. Learning enhances adult neurogenesis in the hippocampal formation. *Nat Neurosci*. 1999;2:260–265.
159. Shors TJ, Miesegaes G, Beylin A, Zhao M, Rydel T, Gould E. Neurogenesis in the adult is involved in the formation of trace memories. *Nature*. 2001;410:372–376.
160. McEwen BS, Macewen BS, Weiss JM. The uptake and action of corticosterone: regional and subcellular studies on rat brain. *Prog Brain Res*. 1970;32:200–212.
161. Kempermann G. The neurogenic reserve hypothesis: what is adult hippocampal neurogenesis good for? *Trends Neurosci*. 2008;31:163–169.
162. Knoth R, Singec I, Ditter M, et al. Murine features of neurogenesis in the human hippocampus across the lifespan from 0 to 100 years. *Plos One*. 2010;5:e8809.
163. Bergmann O, Spalding KL, Frisen J. Adult neurogenesis in humans. *Cold Spring Harb Perspect Biol*. 2015;7:a018994.
164. Eser D, Baghai TC, Moller HJ. Evidence of agomelatine's antidepressant efficacy: the key points. *Int Clin Psychopharmacol*. 2007;22(Suppl 2):S15–S19.
165. Moriya T, Horie N, Mitome M, Shinohara K. Melatonin influences the proliferative and differentiative activity of neural stem cells. *J Pineal Res*. 2007;42:411–418.
166. Kong X, Li X, Cai Z, et al. Melatonin regulates the viability and differentiation of rat midbrain neural stem cells. *Cell Mol Neurobiol*. 2008;28:569–579.
167. Bellon A, Ortiz-Lopez L, Ramirez-Rodriguez G, Anton-Tay F, Benitez-King G. Melatonin induces neuritogenesis at early stages in N1E-115 cells through actin rearrangements via activation of protein kinase C and Rho-associated kinase. *J Pineal Res*. 2007;42:214–221.
168. Reiter RJ, Tan DX, Leon J, Kilic U, Kilic E. When melatonin gets on your nerves: its beneficial actions in experimental models of stroke. *Exp Biol Med (Maywood)*. 2005;230:104–117.
169. Reiter RJ. Oxidative damage in the central nervous system: protection by melatonin. *Prog Neurobiol*. 1998;56:359–384.
170. Ge S, Yang CH, Hsu KS, Ming GL, Song H. A critical period for enhanced synaptic plasticity in newly generated neurons of the adult brain. *Neuron*. 2007;54:559–566.
171. Dubocovich ML, Masana MI, Iacob S, Sauri DM. Melatonin receptor antagonists that differentiate between the human Mel(1a), and Mel(1b) recombinant subtypes are used to assess the pharmacological profile of the rabbit retina ML(1) presynaptic heteroreceptor. *N-S Arch Pharmacol*. 1997;355:365–375.
172. Benitez-King G, Tunez I, Bellon A, Ortiz GG, Anton-Tay F. Melatonin prevents cytoskeletal alterations and oxidative stress induced by okadaic acid in N1E-115 cells. *Exp Neurol*. 2003;182:151–159.
173. Shu T, Wu T, Pang M, et al. Effects and mechanisms of melatonin on neural differentiation of induced pluripotent stem cells. *Biochem Biophys Res Commun*. 2016;474:566–571.
174. Yu X, Li Z, Zheng H, Ho J, Chan MT, Wu WK. Protective roles of melatonin in central nervous system diseases by regulation of neural stem cells. *Cell Prolif*. 2017;50.
175. Quiros I, Mayo JC, Garcia-Suarez O, et al. Melatonin prevents glucocorticoid inhibition of cell proliferation and toxicity in hippocampal cells by reducing glucocorticoid receptor nuclear translocation. *J Steroid Biochem Mol Biol*. 2008;110:116–124.

176. Cole GM, Ma QL, Frautschy SA. Omega-3 fatty acids and dementia. *Prostaglandins Leukot Essent Fatty Acids*. 2009;81:213–221.
177. Hooijmans CR, Kiliaan AJ. Fatty acids, lipid metabolism and Alzheimer pathology. *Eur J Pharmacol*. 2008;585:176–196.
178. Pu HJ, Guo YL, Zhang WT, et al. Omega-3 polyunsaturated fatty acid supplementation improves neurologic recovery and attenuates white matter injury after experimental traumatic brain injury. *J Cerebr Blood F Met*. 2013;33:1474–1484.
179. McNamara RK, Carlson SE. Role of omega-3 fatty acids in brain development and function: potential implications for the pathogenesis and prevention of psychopathology. *Prostaglandins Leukot Essent Fatty Acids*. 2006;75:329–349.
180. Chen CT, Liu Z, Ouellet M, Calon F, Bazinet RP. Rapid beta-oxidation of eicosapentaenoic acid in mouse brain: an in situ study. *Prostaglandins Leukot Essent Fatty Acids*. 2009;80:157–163.
181. Luchtman DW, Song C. Cognitive enhancement by omega-3 fatty acids from child-hood to old age: findings from animal and clinical studies. *Neuropharmacology*. 2013;64:550–565.
182. Crawford MA, Hassam AG, Williams G. Essential fatty acids and fetal brain growth. *Lancet*. 1976;1:452–453.
183. Wang Y, Botelho AV, Martinez GV, Brown MF. Electrostatic properties of membrane lipids coupled to metarhodopsin II formation in visual transduction. *J Am Chem Soc*. 2002;124:7690–7701.
184. Michael-Titus AT, Priestley JV. Omega-3 fatty acids and traumatic neurological injury: from neuroprotection to neuroplasticity? *Trends Neurosci*. 2014;37:30–38.
185. Hasadsri L, Wang BH, Lee JV, et al. Omega-3 fatty acids as a putative treatment for traumatic brain injury. *J Neurotrauma*. 2013;30:897–906.
186. Delattre AM, Carabelli B, Mori MA, et al. Maternal omega-3 supplement improves dopaminergic system in pre- and postnatal inflammation-induced neurotoxicity in Parkinson's disease model. *Mol Neurobiol*. 2017;54:2090–2106.
187. Ren H, Yang Z, Luo C, et al. Enriched endogenous omega-3 fatty acids in mice ameliorate parenchymal cell death after traumatic brain injury. *Mol Neurobiol*. 2017;54:3317–3326.
188. Scrimgeour AG, Condlin ML. Nutritional treatment for traumatic brain injury. *J Neurotrauma*. 2014;31:989–999.
189. Gorjao R, Azevedo-Martins AK, Rodrigues HG, et al. Comparative effects of DHA and EPA on cell function. *Pharmacol Ther*. 2009;122:56–64.
190. Freemantle E, Vandal M, Tremblay-Mercier J, et al. Omega-3 fatty acids, energy substrates, and brain function during aging. *Prostaglandins Leukot Essent Fatty Acids*. 2006;75:213–220.
191. Stonehouse W. Does consumption of LC omega-3 PUFA enhance cognitive performance in healthy school-aged children and throughout adulthood? Evidence from clinical trials. *Nutrients*. 2014;6:2730–2758.
192. Kim HY, Huang BX, Spector AA. Phosphatidylserine in the brain: Metabolism and function. *Prog Lipid Res*. 2014;56:1–18.
193. Bang HO, Dyerberg J, Hjoorne N. The composition of food consumed by Greenland Eskimos. *Acta Med Scand*. 1976;200:69–73.
194. Perlmutter D, Colman C. *The better brain book: the best tools for improving memory, sharpness, and preventing aging of the brain*. New York: Riverhead Books; 2004.
195. Reynolds C, Draper E, Loscher CE, Roche HM. Docosahexaenoic acid but not eicosapentaenoic acid mediates the anti-inflammatory effect in dendritic cells via peroxisome proliferator-activated receptor-gamma-dependent pathway(s). *P Nutr Soc*. 2006;65:61a.
196. Mercier R, Dominguez-Cuevas P, Errington J. Crucial role for membrane fluidity in proliferation of primitive cells. *Cell Rep*. 2012;1:417–423.
197. Reisbick S, Neuringer M, Gohl E, Wald R, Anderson GJ. Visual attention in infant monkeys: effects of dietary fatty acids and age. *Dev Psychol*. 1997;33:387–395.
198. Greiner RS, Moriguchi T, Hutton A, Slotnick BM, Salem N. Rats with low levels of brain docosahexaenoic acid show impaired performance in olfactory-based and spatial learning tasks. *Lipids*. 1999;34:S239–S243.
199. Catalan J, Moriguchi T, Slotnick B, Murthy M, Greiner RS, Salem N. Cognitive deficits in docosahexaenoic acid-deficient rats. *Behav Neurosci*. 2002;116:1022–1031.
200. Ratnayake WMN, Galli C. Fat and fatty acid terminology, methods of analysis and fat digestion and metabolism: a background review paper. *Ann Nutr Metab*. 2009;55:8–43.
201. Manikonda PK, Rajendra P, Devendranath D, et al. Influence of extremely low frequency magnetic fields on Ca2+ signaling and NMDA receptor functions in rat hippocampus. *Neurosci Lett*. 2007;413:145–149.

202. Soderqvist F, Carlberg M, Hardell L. Use of wireless telephones and serum S100B levels: A descriptive cross-sectional study among healthy Swedish adults aged 18–65 years. *Sci Total Environ*. 2009;407:798–805.
203. Megha K, Deshmukh PS, Banerjee BD, Tripathi AK, Abegaonkar MP. Microwave radiation induced oxidative stress, cognitive impairment and inflammation in brain of Fischer rats. *Indian J Exp Biol*. 2012;50:889–896.
204. Jazayeri S, Keshavarz SA, Tehrani-Doost M, et al. Effects of eicosapentaenoic acid and fluoxetine on plasma cortisol, serum interleukin-1beta and interleukin-6 concentrations in patients with major depressive disorder. *Psychiatry Res*. 2010;178:112–115.
205. Quinn JF, Raman R, Thomas RG, et al. Docosahexaenoic acid supplementation and cognitive decline in Alzheimer disease: a randomized trial. *JAMA*. 2010;304:1903–1911.
206. Lebbadi M, Julien C, Phivilay A, et al. Endogenous conversion of omega-6 into omega-3 fatty acids improves neuropathology in an animal model of Alzheimer's disease. *J Alzheimers Dis*. 2011;27:853–869.
207. Nimmerjahn A, Kirchhoff F, Helmchen F. Resting microglial cells are highly dynamic surveillants of brain parenchyma in vivo. *Science*. 2005;308:1314–1318.
208. Vines A, Delattre AM, Lima MMS, et al. The role of 5-HT1A receptors in fish oil-mediated increased BDNF expression in the rat hippocampus and cortex: a possible antidepressant mechanism. *Neuropharmacology*. 2012;62:184–191.
209. Dyall SC. Long-chain omega-3 fatty acids and the brain: a review of the independent and shared effects of EPA, DPA and DHA. *Front Aging Neurosci*. 2015;7:52.
210. Tkalec M, Malaric K, Pevalek-Kozlina B. Exposure to radiofrequency radiation induces oxidative stress in duckweed Lemna minor L. *Sci Total Environ*. 2007;388:78–89.
211. Hirose H, Sasaki A, Ishii N, et al. 1950 MHz IMT-2000 field does not activate microglial cells in vitro. *Bioelectromagnetics*. 2010;31:104–112.
212. Cleary SF, Du Z, Cao G, Liu LM, McCrady C. Effect of isothermal radiofrequency radiation on cytolytic T lymphocytes. *FASEB J*. 1996;10:913–919.
213. Ivancsits S, Diem E, Pilger A, Rudiger HW, Jahn O. Induction of DNA strand breaks by intermittent exposure to extremely-low-frequency electromagnetic fields in human diploid fibroblasts. *Mutat Res*. 2002;519:1–13.
214. Chao DT, Korsmeyer SJ. BCL-2 family: regulators of cell death. *Annu Rev Immunol*. 1998;16:395–419.
215. Franco R, Sanchez-Olea R, Reyes-Reyes EM, Panayiotidis MI. Environmental toxicity, oxidative stress and apoptosis: menage a trois. *Mutat Res*. 2009;674:3–22.
216. Vejux A, Lizard G. Cytotoxic effects of oxysterols associated with human diseases: Induction of cell death (apoptosis and/or oncosis), oxidative and inflammatory activities, and phospholipidosis. *Mol Aspects Med*. 2009;30:153–170.
217. Chen YB, Aon MA, Hsu YT, et al. Bcl-xL regulates mitochondrial energetics by stabilizing the inner membrane potential. *J Cell Biol*. 2011;195:263–276.
218. Pall ML. Electromagnetic fields act via activation of voltage-gated calcium channels to produce beneficial or adverse effects. *J Cell Mol Med*. 2013;17:958–965.
219. Ramundo-Orlando A, Mattia F, Palombo A, D'Inzeo G. Effect of low frequency, low amplitude magnetic fields on the permeability of cationic liposomes entrapping carbonic anhydrase: II. No evidence for surface enzyme involvement. *Bioelectromagnetics*. 2000;21:499–507.
220. Ahmad A, Moriguchi T, Salem N. Decrease in neuron size in docosahexaenoic acid-deficient brain. *Pediatr Neurol*. 2002;26:210–218.
221. Black KL, Hoff JT, Radin NS, Deshmukh GD. Eicosapentaenoic acid: effect on brain prostaglandins, cerebral blood flow and edema in ischemic gerbils. *Stroke*. 1984;15:65–69.
222. Lakshmi D, Gopinath K, Jayanthy G, Anjum S, Prakash D, Sudhandiran G. Ameliorating effect of fish oil on acrylamide induced oxidative stress and neuronal apoptosis in cerebral cortex. *Neurochem Res*. 2012;37:1859–1867.
223. Florent S, Malaplate-Armand C, Youssef I, et al. Docosahexaenoic acid prevents neuronal apoptosis induced by soluble amyloid-beta oligomers. *J Neurochem*. 2006;96:385–395.
224. Luchtman DW, Meng Q, Wang X, Shao D, Song C. Omega-3 fatty acid eicosapentaenoic acid attenuates MPP+-induced neurodegeneration in fully differentiated human SH-SY5Y and primary mesencephalic cells. *J Neurochem*. 2013;124:855–868.
225. Afshordel S, Hagl S, Werner D, et al. Omega-3 polyunsaturated fatty acids improve mitochondrial dysfunction in brain aging—impact of Bcl-2 and NPD-1 like metabolites. *Prostaglandins Leukot Essent Fatty Acids*. 2015;92:23–31.

226. Gamoh S, Hashimoto M, Sugioka K, et al. Chronic administration of docosahexaenoic acid improves reference memory-related learning ability in young rats. *Neuroscience*. 1999;93:237–241.
227. Greiner RS, Moriguchi T, Slotnick BM, Hutton A, Salem N. Olfactory discrimination deficits in n-3 fatty acid-deficient rats. *Physiol Behav*. 2001;72:379–385.
228. Kelly L, Grehan B, Chiesa AD, et al. The polyunsaturated fatty acids, EPA and DPA exert a protective effect in the hippocampus of the aged rat. *Neurobiol Aging*. 2011;32:2318 e2311–2315.
229. Castillo RL, Arias C, Farias JG. Omega 3 chronic supplementation attenuates myocardial ischaemia-reperfusion injury through reinforcement of antioxidant defense system in rats. *Cell Biochem Funct*. 2014;32:274–281.
230. Ruszkiewicz J, Albrecht J. Changes in the mitochondrial antioxidant systems in neurodegenerative diseases and acute brain disorders. *Neurochem Int*. 2015;88:66–72.
231. Corsi L, Dongmo BM, Avallone R. Supplementation of omega 3 fatty acids improves oxidative stress in activated BV2 microglial cell line. *Int J Food Sci Nutr*. 2015;66:293–299.
232. Wang JY, Wen LL, Huang YN, Chen YT, Ku MC. Dual effects of antioxidants in neurodegeneration: direct neuroprotection against oxidative stress and indirect protection via suppression of glia-mediated inflammation. *Curr Pharm Des*. 2006;12:3521–3533.
233. Yu BP. Approaches to anti-aging intervention: the promises and the uncertainties. *Mech Ageing Dev*. 1999;111:73–87.
234. Yao K, Wu W, Wang K, et al. Electromagnetic noise inhibits radiofrequency radiation-induced DNA damage and reactive oxygen species increase in human lens epithelial cells. *Mol Vis*. 2008;14:964–969.
235. Oktem F, Ozguner F, Mollaoglu H, Koyu A, Uz E. Oxidative damage in the kidney induced by 900-MHz-emitted mobile phone: protection by melatonin. *Arch Med Res*. 2005;36:350–355.
236. Aydin B, Akar A. Effects of a 900-MHz electromagnetic field on oxidative stress parameters in rat lymphoid organs, polymorphonuclear leukocytes and plasma. *Arch Med Res*. 2011;42:261–267.
237. Kim MJ, Rhee SJ. Green tea catechins protect rats from microwave-induced oxidative damage to heart tissue. *J Med Food*. 2004;7:299–304.
238. Moustafa YM, Moustafa RM, Belacy A, Abou-El-Ela SH, Ali FM. Effects of acute exposure to the radiofrequency fields of cellular phones on plasma lipid peroxide and antioxidase activities in human erythrocytes. *J Pharm Biomed Anal*. 2001;26:605–608.
239. Awad SM, Hassan NS. Health Risks of electromagnetic radiation from mobile phone on brain of rats. *J Appl Sci Res*. 2008;4:1994–2000.
240. Hoyto A, Luukkonen J, Juutilainen J, Naarala J. Proliferation, oxidative stress and cell death in cells exposed to 872 MHz radiofrequency radiation and oxidants. *Radiat Res*. 2008;170:235–243.
241. Cosar M, Songur A, Sahin O, et al. The neuroprotective effect of fish n-3 fatty acids in the hippocampus of diabetic rats. *Nutr Neurosci*. 2008;11:161–166.
242. Avramovic N, Dragutinovic V, Krstic D, et al. The effects of omega 3 fatty acid supplementation on brain tissue oxidative status in aged wistar rats. *Hippokratia*. 2012;16:241–245.
243. Bas O, Songur A, Sahin O, et al. The protective effect of fish n-3 fatty acids on cerebral ischemia in rat hippocampus. *Neurochem Int*. 2007;50:548–554.
244. Cao DH, Xu JF, Xue RH, Zheng WF, Liu ZL. Protective effect of chronic ethyl docosahexaenoate administration on brain injury in ischemic gerbils. *Pharmacol Biochem Behav*. 2004;79:651–659.
245. Song JH, Miyazawa T. Enhanced level of n-3 fatty acid in membrane phospholipids induces lipid peroxidation in rats fed dietary docosahexaenoic acid oil. *Atherosclerosis*. 2001;155:9–18.
246. Perluigi M, Swomley AM, Butterfield DA. Redox proteomics and the dynamic molecular landscape of the aging brain. *Ageing Res Rev*. 2014;13:75–89.
247. Franco-Iborra S, Vila M, Perier C. The Parkinson disease mitochondrial hypothesis: Where are we at? *Neuroscientist*. 2016;22:266–277.
248. Xie A, Gao J, Xu L, Meng D. Shared mechanisms of neurodegeneration in Alzheimer's disease and Parkinson's disease. *Biomed Res Int*. 2014;2014:648740.
249. Song C, Shieh CH, Wu YS, Kalueff A, Gaikwad S, Su KP. The role of omega-3 polyunsaturated fatty acids eicosapentaenoic and docosahexaenoic acids in the treatment of major depression and Alzheimer's disease: Acting separately or synergistically? *Prog Lipid Res*. 2016;62:41–54.
250. Alberts B, Johnson A, Lewis J, Raff M, Roberts K, Walter P. *Ion Channels and the Electrical Properties of Membranes*. Vol 4th ed; 2002.

251. Sobel E, Davanipour Z. Electromagnetic field exposure may cause increased production of amyloid beta and eventually lead to Alzheimer's disease. *Neurology*. 1996;47:1594–1600.
252. Grimm MO, Zimmer VC, Lehmann J, Grimm HS, Hartmann T. The impact of cholesterol, DHA, and sphingolipids on Alzheimer's disease. *Biomed Res Int*. 2013;2013:814390.
253. Yurko-Mauro K, McCarthy D, Rom D, et al. Beneficial effects of docosahexaenoic acid on cognition in age-related cognitive decline. *Alzheimers Dement*. 2010;6:456–464.
254. Irving GF, Freund-Levi Y, Eriksdotter-Jonhagen M, et al. Omega-3 fatty acid supplementation effects on weight and appetite in patients with Alzheimer's disease: the omega-3 Alzheimer's disease study. *J Am Geriatr Soc*. 2009;57:11–17.
255. Lim GP, Calon F, Morihara T, et al. A diet enriched with the omega-3 fatty acid docosahexaenoic acid reduces amyloid burden in an aged Alzheimer mouse model. *J Neurosci*. 2005;25:3032–3040.
256. Amtul Z, Uhrig M, Rozmahel RF, Beyreuther K. Structural insight into the differential effects of omega-3 and omega-6 fatty acids on the production of Abeta peptides and amyloid plaques. *J Biol Chem*. 2011;286:6100–6107.
257. Vedin I, Cederholm T, Freund-Levi Y, et al. Effects of DHA-rich n-3 fatty acid supplementation on gene expression in blood mononuclear leukocytes: the OmegAD study. *Plos One*. 2012;7:e35425.
258. Wurtman RJ, Cansev M, Ulus IH. Synapse formation is enhanced by oral administration of Uridine and Dha, the circulating precursors of brain phosphatides. *J Nutr Health Aging*. 2009;13:189–197.
259. Calon F, Lim GP, Yang F, et al. Docosahexaenoic acid protects from dendritic pathology in an Alzheimer's disease mouse model. *Neuron*. 2004;43:633–645.
260. Terro F, Magnaudeix A, Crochetet M, et al. GSM-900 MHz at low dose temperature-dependently downregulates alpha-synuclein in cultured cerebral cells independently of chaperone-mediated-autophagy. *Toxicology*. 2012;292:136–144.
261. Delattre AM, Kiss A, Szawka RE, et al. Evaluation of chronic omega-3 fatty acids supplementation on behavioral and neurochemical alterations in 6-hydroxydopamine-lesion model of Parkinson's disease. *Neurosci Res*. 2010;66:256–264.
262. Dyall SC, Michael-Titus AT. Neurological benefits of omega-3 fatty acids. *Neuromolecular Med*. 2008;10:219–235.
263. Sharon R, Bar-Joseph I, Frosch MP, Walsh DM, Hamilton JA, Selkoe DJ. The formation of highly soluble oligomers of alpha-synuclein is regulated by fatty acids and enhanced in Parkinson's disease. *Neuron*. 2003;37:583–595.
264. Horrobin DF, Jenkins K, Bennett CN, Christie WW. Eicosapentaenoic acid and arachidonic acid: collaboration and not antagonism is the key to biological understanding. *Prostaglandins Leukot Essent Fatty Acids*. 2002;66:83–90.
265. Peet M, Stokes C. Omega-3 fatty acids in the treatment of psychiatric disorders. *Drugs*. 2005;65:1051–1059.
266. Odaci E, Bas O, Kaplan S. Effects of prenatal exposure to a 900 MHz electromagnetic field on the dentate gyrus of rats: a stereological and histopathological study. *Brain Res*. 2008;1238:224–229.
267. Panagopoulos DJ, Chavdoula ED, Nezis IP, Margaritis LH. Cell death induced by GSM 900-MHz and DCS 1800-MHz mobile telephony radiation. *Mutat Res*. 2007;626:69–78.
268. Bliss TV, Collingridge GL. A synaptic model of memory: long-term potentiation in the hippocampus. *Nature*. 1993;361:31–39.
269. Fagnou DD, Tuchek JM. The biochemistry of learning and memory. *Mol Cell Biochem*. 1995;149–150:279–286.
270. Denis I, Potier B, Vancassel S, Heberden C, Lavialle M. Omega-3 fatty acids and brain resistance to ageing and stress: body of evidence and possible mechanisms. *Ageing Res Rev*. 2013;12:579–594.
271. Katakura M, Hashimoto M, Okui T, Shahdat HM, Matsuzaki K, Shido O. Omega-3 polyunsaturated Fatty acids enhance neuronal differentiation in cultured rat neural stem cells. *Stem Cells Int*. 2013;2013:490476.
272. Kawakita E, Hashimoto M, Shido O. Docosahexaenoic acid promotes neurogenesis in vitro and in vivo. *Neuroscience*. 2006;139:991–997.
273. Mormina ME, Thakur S, Molleman A, Whelan CJ, Baydoun AR. Cannabinoid signalling in TNF-alpha induced IL-8 release. *Eur J Pharmacol*. 2006;540:183–190.
274. McNamara RK, Able J, Jandacek R, et al. Docosahexaenoic acid supplementation increases prefrontal cortex activation during sustained attention in healthy boys: a placebo-controlled, dose-ranging, functional magnetic resonance imaging study. *Am J Clin Nutr*. 2010;91:1060–1067.
275. Wu A, Ying Z, Gomez-Pinilla F. Dietary omega-3 fatty acids normalize BDNF levels, reduce oxidative damage, and counteract learning disability after traumatic brain injury in rats. *J Neurotraum*. 2004;21:1457–1467.

276. Leibrock J, Lottspeich F, Hohn A, et al. Molecular cloning and expression of brain-derived neurotrophic factor. *Nature*. 1989;341:149–152.
277. Hariri AR, Goldberg TE, Mattay VS, et al. Brain-derived neurotrophic factor val66met polymorphism affects human memory-related hippocampal activity and predicts memory performance. *J Neurosci*. 2003;23:6690–6694.
278. Jiang LH, Shi Y, Wang LS, Yang ZR. The influence of orally administered docosahexaenoic acid on cognitive ability in aged mice. *J Nutr Biochem*. 2009;20:735–741.
279. Patten AR, Sickmann HM, Dyer RA, Innis SM, Christie BR. Omega-3 fatty acids can reverse the long-term deficits in hippocampal synaptic plasticity caused by prenatal ethanol exposure. *Neurosci Lett*. 2013;551:7–11.
280. He C, Qu X, Cui L, Wang J, Kang JX. Improved spatial learning performance of fat-1 mice is associated with enhanced neurogenesis and neuritogenesis by docosahexaenoic acid. *Proc Natl Acad Sci U S A*. 2009;106:11370–11375.
281. Calderon F, Kim HY. Docosahexaenoic acid promotes neurite growth in hippocampal neurons. *J Neurochem*. 2004;90:979–988.
282. Kıvrak EGAB, Alkan I, Yurt KK, Kocaman A, Önger ME. Effects of 900-MHz radiation on the hippocampus and cerebellum of adult rats and attenuation of such effects by folic acid and Boswellia sacra. *JMAU*. 2017;5:216–224.
283. Holewijn S, Heijer MD, Swinkels D, Stalenhoef A, Graaf JD. Gender Specific Impact of Inflammation and Obesity on Subclinical Atherosclerosis in the General Population. *Atherosclerosis Supp*. 2011;12:172.
284. Yousefi HANP. Psychological effects of occupational exposure to electromagnetic fields. *J Res Health Sci*. 2006;6:18–21.
285. Peet M. Omega-3 polyunsaturated fatty acids in the treatment of schizophrenia. *Isr J Psychiatry Relat Sci*. 2008;45:19–25.
286. Hoffmire CA, Block RC, Thevenet-Morrison K, van Wijngaarden E. Associations between omega-3 polyunsaturated fatty acids from fish consumption and severity of depressive symptoms: an analysis of the 2005–2008 National Health and Nutrition Examination Survey. *Prostaglandins Leukot Essent Fatty Acids*. 2012;86:155–160.
287. Beydoun MA, Fanelli Kuczmarski MT, Beydoun HA, Hibbeln JR, Evans MK, Zonderman AB. omega-3 fatty acid intakes are inversely related to elevated depressive symptoms among United States women. *J Nutr*. 2013;143:1743–1752.
288. Sanhueza C, Ryan L, Foxcroft DR. Diet and the risk of unipolar depression in adults: systematic review of cohort studies. *J Hum Nutr Diet*. 2013;26:56–70.
289. Nemets H, Nemets B, Apter A, Bracha Z, Belmaker RH. Omega-3 treatment of childhood depression: A controlled double-blind pilot study. *Int J Neuropsychoph*. 2006;9:S182.
290. Sivrioglu EY, Kirli S, Sipahioglu D, Gursoy B, Sarandol E. The impact of omega-3 fatty acids, vitamins E and C supplementation on treatment outcome and side effects in schizophrenia patients treated with haloperidol: an open-label pilot study. *Prog Neuropsychopharmacol Biol Psychiatry*. 2007;31:1493–1499.
291. Zugno AI, Chipindo HL, Volpato AM, et al. Omega-3 prevents behavior response and brain oxidative damage in the ketamine model of schizophrenia. *Neuroscience*. 2014;259:223–231.
292. Martins JG. EPA but not DHA appears to be responsible for the efficacy of omega-3 long chain polyunsaturated fatty acid supplementation in depression: evidence from a meta-analysis of randomized controlled trials. *J Am Coll Nutr*. 2009;28:525–542.

CHAPTER

12

Maternal Omega-3 Status: Effects of Cross-Fostering on Offspring Brain Lipids and Hippocampal Neurogenesis

Elise Maximin, Christine Heberden

Micalis Institute INRA, Jouy-en-Josas, France

INTRODUCTION

Omega-3 and -6 fatty acids (ω3 or n-3; ω6 or n-6) are necessary components of the brain, and diet is the main supplier of these essential lipids. Because the main source of n-3 fatty acids comes from seafood in western societies, a deficiency in n-3 is now highly widespread. It is considered that a balanced diet should supply a daily ratio of n-6/n-3 equal to 5, whereas nutritional studies show that the actual ratio in most developed countries is around 15. Many studies have focused on this problem, and preclinical studies have shown that n-3 in low amounts and n-6 in high proportions are detrimental to cerebral physiology and structures.[1]

This is especially important in infants. Indeed the synthesis capacities of the fetus and infants are very limited. In the early phases of life, the fetus and infants are entirely dependent on the maternal diet and the maternal stores during either pregnancy or the lactating period.[2]

Maternal supplementation in omega 3 translates into numerous benefits for the children. Indeed, the children born to supplemented mothers, or fed a diet supplemented in n-3 are more attentive and less hyperactive.[3] The reports showing a positive effect on IQ have been questioned though.[4–6]

Interestingly, the benefits of a n-3-rich diet on the children are not only immediate, but can be recorded for during several years in the childhood.[3] Preclinical models have shown consistent data: omega-3 supplementation proved protective toward emotional behavior, cognition, and neurogenesis in rodents.[7–10]

The supplementation is also necessary and beneficial to the mothers; indeed, under inadequate omega-3 supply, the maternal docosahexaenoic acid (DHA) stores are depleted under

Omega Fatty Acids in Brain and Neurological Health
https://doi.org/10.1016/B978-0-12-815238-6.00012-2

a single reproductive cycle in the rat.[11] In humans as well as in rodents, n-3 maternal supplementation improves the mood of the expectant mothers and prevents the development of postpartum depression.[12, 13]

The omega-3 molecular action mechanisms are not fully unveiled, but several studies have pointed to the modification of gene transcriptions[14] or DNA methylation.[15] The latter study suggests that omega 3 induce epigenetic effects, which have been also described in other studies.[16] For instance, we have shown that in vitro neural stem cells (NSCs) derived from pups born to deficient or supplemented mothers behave differently in vitro displaying a higher rate of proliferation, higher maturation, and modifications of gene expressions.[17]

In this report, we have questioned whether the dietary environment of pups born to omega-3 deficient or supplemented mothers could impact the proportions of PUFA in brain lipids and the rate of hippocampal neurogenesis in the adult offspring.

METHODS

Male Wistar rats bred in our own facilities were used for this experiment. Female rats were fed 2 weeks before mating with one of the two following diets: the "deficient" diet provided 0.01% of the total energy as linoleic acid (5 mg/100 g diet), and the "supplemented" diet provided high levels of DHA (1.0% of the total energy, 500 mg/100 g diet). The diets contained 6.6 g lipids/100 g and were prepared using mixtures of rapeseed, sunflower, and tuna fish oils, kindly provided by Lesieur (Asnières-sur-Seine, France).[18]

Before the pups were 1-week old, the litters were allocated to surrogate mothers, so that all the dams would adopt pups from another female. The sizes of the litters were similar for all the surrogate mothers ($n = 6$). Therefore, the pups were separated into four groups: (i) born to a supplemented mother, adopted by a supplemented mother (noted ω3-ω3), (ii) born to a supplemented mother, adopted by a deficient mother (ω3-ω6), (iii) born to a deficient mother, adopted by a deficient dam (ω6-ω6), and (iv) born to a deficient mother, adopted by a supplemented one (ω6-ω3).

After weaning, the offspring was fed the same diet as the surrogate dams up until 4 months of age, and were maintained in a comforting environment, at 2 per cage to avoid the stress of isolation. Three days before the sacrifice, they were intraperitoneally injected daily with 50 mg/kg BrdU dissolved in 0.007 N NaOH in PBS (10 mg/mL). They were euthanized by decapitation and the brains were rapidly removed. The cerebellum was excised and saved for lipid analysis and the rest was rapidly frozen in chilled isopentane and kept at −80 degree until analysis.

The fatty acid compositions of the two main classes of membrane PLs, the ethanolamine-phosphoglycerolipids (EPGs) and phosphatidylcholine (PC), were analyzed by gas chromatography, as previously described.[18]

Brains were sectioned at 16 μm with a cryostat and the sections were revealed for BrdU presence in the cells by immunohistochemistry as described before.[19] The cells positive for BrdU in the dorsal hippocampus were counted using Image J.

Statistical analyses for lipid analyses and neurogenesis were performed in GraphPad prism by one- or two-way ANOVA (followed by Tukey's multiple comparisons). The results are expressed as mean ± SD. Results were significantly different at $P < .05$.

RESULTS

The pups' weights at birth were similar (6 ± 0.5 g) regardless of the mothers' diet. At 4 months, the rats' weights were not significantly different even though the animals born to deficient dams and fed with high proportions of ω3 were a little heavier: ω3-ω3: 564 ± 22 g; ω3-ω6: 546 ± 41 g; ω6-ω6: 548 ± 38 g; ω6-ω3: 611 ± 62.9 g.

The cerebellum were excised from the brains, and analyzed for the fatty acid compositions in EPG and PC.

Figs. 1 and 2 show abbreviated data from the fatty acid analyses for EPG and PC, respectively: total n-6 and n-3 have been shown, along with 20:4 n-6 (arachidonic acid, AA) and 22:6 n-3 (DHA) since AA and DHA are the main fatty acid from each family.

In EPG, statistical analysis showed that ω3-ω3 and ω6-ω3 are not statistically different; the levels of total n-3 and DHA are restored in pups born to deficient mothers by higher proportions offered by the lactation and diet after weaning. This result shows that a diet

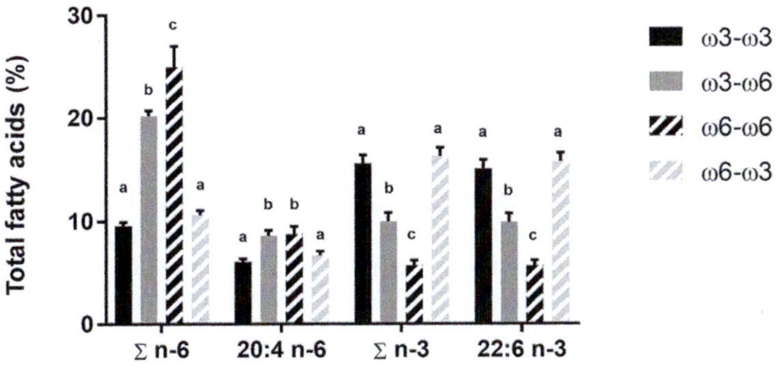

FIG. 1 Modifications of the percentages of polyunsaturated fatty acids in the EPG fraction. For each group, six cerebellums were collected and analyzed by gas chromatography. Different superscripts denote a significant difference at $P < .05$.

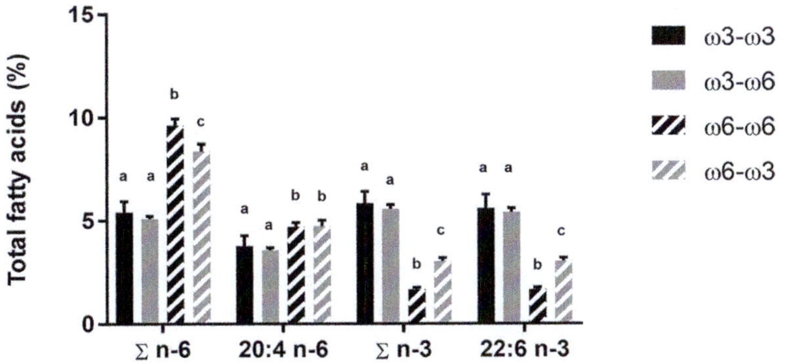

FIG. 2 Modifications of the percentages of polyunsaturated fatty acids in the PC fraction. For each group, six cerebellums were collected and analyzed by gas chromatography. Different superscripts denote a significant difference at $P < .05$.

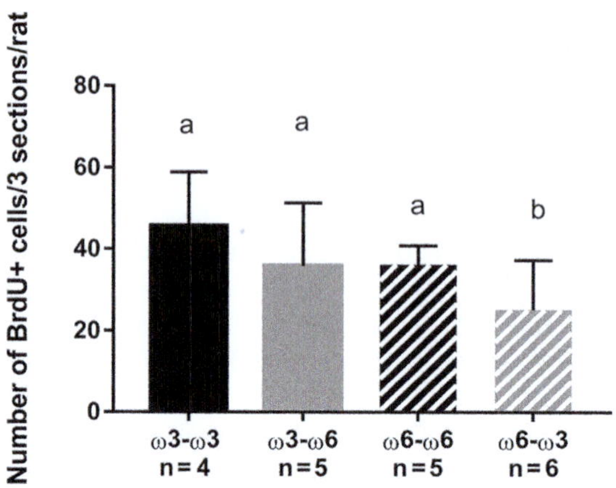

FIG. 3 Number of cells positive for BrdU in the dentate gyrus of the dorsal hippocampus. Immunohistochemistry was performed and for each animal, cells from three consecutive sections were counted. Different superscripts denote a significant difference at $P<.05$.

supplemented as early as the lactating period was able to reverse the deficiency noted in ω6-ω6 animals. Indeed this group was deeply depleted in ω3.

In PC, total ω3 and DHA percentages obtained in ω3-ω3 offspring were similar to the ones obtained in the ω3-ω6 animals, but interestingly, total ω3 and DHA in ω6-ω3 animals remained lower than in animals born to supplemented mothers and fed the supplemented diet (ω3-ω3). This shows that ω3 supply at lactation and further could enhance the proportions of 22:6 n-3, but not to a level equivalent to the one registered in EPG and could not totally counteract the deficit created in utero.

To analyze the rate of hippocampal neurogenesis, the rats were injected with BrdU, and cells positive for BrdU were counted.

Fig. 3 shows the data obtained in the dorsal hippocampus. Cells positive for BrdU tend to be less numerous in ω6-ω6 than in ω3-ω3, although the difference is not statistically significant. The ω6-ω3 animals had fewer positive cells than ω3-ω3 rats. This result shows that the level of dorsal hippocampal neurogenesis was not entirely restored by n-3 supplementation at lactation and by the diet.

DISCUSSION

Consistent with the data in the literature, we have shown that pups born to deficient mothers display a deficit in total ω3 and DHA in the PL, and this deficit is present in the early fetal phases.[19] This deficit can be totally (EPG) or partially (PC) erased by subsequent dietary supplementation. EPG and PC are the most abundant PL species in the cell membrane, with PC in higher quantities than EPG. Therefore, the composition of the membranes in deficient animals can be modified by a lifelong ω3 supplementation but the restoration of the levels registered in animals born to supplemented mothers is not entirely complete.

The animal body growth was not affected either.

Yet, in the hippocampus, the rate of neurogenesis was lower in animals ω3-depleted in utero, even though they were fed adequate levels of ω3 after birth.

This result seems to suggest that a critical event in the development in which ω3 are decisive, has been lacking in deficient animals.

On the other hand, ω3-ω6 rats that have been fed a deficient diet after birth but in utero have received an adequate level of ω3 are not impacted by the subsequent ω3 deficiency.

This is consistent with the idea that there seems to be a "window of opportunity" in the brain, as noted by Kodas et al.[20] Indeed the study showed that dopaminergic neurotransmission was restored in deficient pups when ω3 were provided during lactation but not after weaning. Our result suggests that as far as neurogenesis is concerned, this window could take place even earlier.

The fact that the restoration of the levels of neurogenesis was not complete, also suggests that epigenetic effects can eventually modulate the fate of NSCs. NSCs are known to be sensitive to epigenetic effects, and we have shown that NSCs derived from deficient or supplemented pups behave differently when cultured in vitro, with higher rates of proliferation and deeper maturation once allowed to differentiate.[17] Furthermore, numerous epigenetic effects have been attributed to ω3, whether they modify gene transcription through DNA methylation or miRNA synthesis.[16]

Therefore, our study shows that while the fatty acid compositions of membrane PL is restored by ω3 supplementation, hippocampal neurogenesis illustrates the fact that not all the consequences of ω3 deficiency in utero can be reversed by adequate levels provided after birth. Womb, birth, weaning: in the brain, these periods translate into different phases of development and "windows of opportunity." The ω3 family has proven essential in these different events.

This result argue for a very early ω3 maternal supplementation, prior to and during pregnancy.

Acknowledgments

This study was supported by a grant from the Groupe Lipides Nutrition (Neuilly-sur-Seine, France). We are indebted to Mr. P. Dahirel and Mrs. C. Maudet for excellent animal care.

References

1. Guesnet P, Alessandri JM. Docosahexaenoic acid (DHA) and the developing central nervous system (CNS)-implications for dietary recommendations. *Biochimie*. 2011;93:7–12.
2. Haggarty P. Fatty acid supply to the human fetus. *Annu Rev Nutr*. 2010;30:237–255.
3. Ramakrishnan U, Gonzalez-Casanova I, Schnaas L, et al. Prenatal supplementation with DHA improves attention at 5 y of age: a randomized controlled trial. *Am J Clin Nutr*. 2016;104:1075–1082.
4. Helland IB, Smith L, Saarem K, Saugstad OD, Drevon CA. Maternal supplementation with very-long-chain n-3 fatty acids during pregnancy and lactation augments children's IQ at 4 years of age. *Pediatrics*. 2003;111:e39–e44.
5. Gale CR, Robinson SM, Godfrey KM, et al. Oily fish intake during pregnancy–association with lower hyperactivity but not with higher full-scale IQ in offspring. *J Child Psychol Psychiatry*. 2008;49:1061–1068.
6. Gould JF, Treyvaud K, Yelland LN, et al. Does n-3 LCPUFA supplementation during pregnancy increase the IQ of children at school age? Follow-up of a randomised controlled trial. *BMJ Open*. 2016;6:e011465.
7. Lozada LE, Desai A, Kevala K, Lee JW, Kim HY. Perinatal brain docosahexaenoic acid concentration has a lasting impact on cognition in mice. *J Nutr*. 2017;147:1624–1630.

8. Fan C, Fu H, Dong H, Lu Y, Lu Y, Qi K. Maternal n-3 polyunsaturated fatty acid deprivation during pregnancy and lactation affects neurogenesis and apoptosis in adult offspring: associated with DNA methylation of brain-derived neurotrophic factor transcripts. *Nutr Res*. 2016;36(9):1013–1021.
9. Sakayori N, Tokuda H, Yoshizaki K, et al. Maternal nutritional imbalance between linoleic acid and alpha-Linolenic acid increases Offspring's anxious behavior with a sex-dependent manner in mice. *Tohoku J Exp Med*. 2016 Sep;240(1):31–37.
10. Gao J, Wu H, Cao Y, et al. Maternal DHA supplementation protects rat offspring against impairment of learning and memory following prenatal exposure to valproic acid. *J Nutr Biochem*. 2016;35:87–95.
11. Levant B, Radel JD, Carlson SE. Reduced brain DHA content after a single reproductive cycle in female rats fed a diet deficient in N-3 polyunsaturated fatty acids. *Biol Psychiatry*. 2006;60:987–990.
12. Harauma A1, Sagisaka T2, Horii T2, et al. The influence of n-3 fatty acids on maternal behavior and brain monoamines in the perinatal period. *Prostaglandins Leukot Essent Fatty Acids*. 2016;107:1–7.
13. Chen HF, Su HM. Fish oil supplementation of maternal rats on an n-3 fatty acid-deficient diet prevents depletion of maternal brain regional docosahexaenoic acid levels and has a postpartum anxiolytic effect. *J Nutr Biochem*. 2012;23:299–305.
14. Andrade-da-Costa B, Isaac A, Silva E, et al. Low omega 6/omega 3 fatty acid ratio in maternal diet favors leukemia inhibitory factor (LIF) gene transcription in the offspring's neural cells. *Glia*. 2017;65:E250.
15. Sable P, Randhir K, Kale A, et al. Maternal micronutrients and brain global methylation patterns in the offspring. *Nutr Neurosci*. 2013;18:30–36.
16. Heberden C, Maximin E. Epigenetic effects of N-3 polyunsaturated fatty acids. In: Preedy VR, Patel VB, eds. *Handbook of Nutrition, Diet, and Epigenetics*. Springer International Publishing AG; 2017:1–15.
17. Goustard-Langelier B, Koch M, Lavialle M, Heberden C. Rat neural stem cell proliferation and differentiation are durably altered by the in utero polyunsaturated fatty acid supply. *J Nutr Biochem*. 2013;24:380–387.
18. Langelier B, Linard A, Bordat C, et al. Long chain-polyunsaturated fatty acids modulate membrane phospholipid composition and protein localization in lipid rafts of neural stem cell cultures. *J Cell Biochem*. 2010;110:1356–1364.
19. Maximin E, Langelier B, Aioun J, et al. Fatty acid binding protein 7 and n-3 poly unsaturated fatty acid supply in early rat brain development. *Dev Neurobiol*. 2016;39:287–297.
20. Kodas E, Vancassel S, Lejeune B, et al. Reversibility of n-3 fatty acid deficiency-induced changes in dopaminergic neurotransmission in rats: critical role of developmental stage. *J Lipid Res*. 2002;43:1209–1219.

CHAPTER

13

Linking Omega-3 Fatty Acids and Depression

*Justyna Godos**, *Sabrina Castellano*†, *Fabio Galvano**, *Giuseppe Grosso**

*Department of Biomedical and Biotechnological Sciences, University of Catania, Catania, Italy

†Department of Educational Sciences, University of Catania, Catania, Italy

INTRODUCTION

Polyunsaturated fatty acids (PUFAs) are fatty acids with two or more double bonds between carbon atoms. PUFAs can be classified primarily as omega-3 or omega-6, depending on the position of the first double bond from the methyl end in the carbon chain. Omega-3 (also called as ω-3 fatty acids or n-3 fatty acids) refers to a group of PUFAs, in which the first double bond is found between the third and the fourth carbon atom from the methyl end of the molecule. The omega-6 (also called as ω-6 fatty acids or n-6 fatty acids) is a family of PUFAs, in which the first double bond is between the sixth and the seventh carbon atom, counting from the methyl end.[1]

Omega-3 PUFAs include α-linolenic acid (ALA), stearidonic acid (SDA), eicosapentaenoic acid (EPA), docosapentaenoic acid (DPA), and docosahexaenoic acid (DHA). The initial rate-limiting step of omega-3 PUFAs synthesis involves desaturation of ALA (18:3n-3) by Δ6-desaturase to SDA (18:4n-3), followed by an elongation to eicosatetraenoic acid (20:4n-3). Next, Δ5-desaturase catalyzes a second desaturation to form EPA (20:5n-3). EPA is then further elongated by elongase, first to DPA (22:5n-3) and then to tetracosapentaenoic acid (24:5n-3). Tetracosapentaenoic acid then undergoes a second Δ6 desaturation to form tetracosahexaenoic acid (24:6n-3). The initial steps occur in the endoplasmic reticulum. Nonetheless, the final stage of DHA synthesis occurs in the peroxisome following the translocation. In the peroxisome, during single cycle of β-oxidation, tetracosahexaenoic acid is shortened to DHA (22:6n-3) and then translocated back to the endoplasmic reticulum for subsequent esterification into aminophospholipids.[1] On the other hand, omega-6 PUFAs derive from linoleic acid (LA, 18:2n-6), which can be converted into gamma linolenic acid (GLA, 18:3n-6), dihomo-gamma-linolenic acid (DGLA, 20:3n-6), and arachidonic acid (AA, 20:4n-6). Both LA

Omega Fatty Acids in Brain and Neurological Health
https://doi.org/10.1016/B978-0-12-815238-6.00013-4

and ALA are considered essential fatty acids as mammalian cells are incapable of synthesizing these fatty acids from simpler precursors.[2]

The major dietary sources of PUFAs are aquatic organisms, vegetable oils (i.e., flaxseed, walnut, canola, soybean) and seeds (i.e., chia, hemp, flaxseed).[3] High amounts of ALA, the precursor of other omega-3 PUFAs, are mainly found in plant-based foods, including flaxseed, chia seeds, soybeans, and walnuts. Instead, the primary sources of EPA and DHA comprise liver of lean white fish such as cod and halibut, the body of oily fish such as salmon and mackerel, the blubber of marine mammals such as seals and whales, and marine algae.[4] Finally, similar to ALA, the precursor of omega-6 PUFAs may be found mostly in crop seeds and vegetables oils, including canola, soybean, and sunflower oil.[5] In fact, a joint expert consultation of the United Nations Food and Agriculture Organization (FAO) and the World Health Organization (WHO) recommend an intake of 1–2 servings of fish per week, considering that each serving is providing 200–500 mg DHA and EPA.[6] Moreover, the data support the inclusion of vegetable oils and food sources high in ALA, especially in a vegetarian diet.

It has been established that dietary ALA and LA play a crucial role in maintaining tissue omega-3 and omega-6 long-chain PUFAs (LC-PUFAs) levels, which are important components of cell membrane phospholipids. LC-PUFAs contribute not only to the membrane structure but also to membrane fluidity, functionality of membrane receptors, and cell signal transduction. Additionally, LA and ALA are critically responsible for producing various classes of pro-inflammatory and anti-inflammatory eicosanoids. Eicosanoids are signaling lipids that include prostaglandins (PGs), thromboxanes (TXs), leukotrienes (LTs), and hydroxyeicosatetraenoic acids (HETEs), which have been previously linked to various pathological processes such as inflammation. Omega-3-derived eicosanoids mostly promote anti-inflammatory response, while omega-6-derived eicosanoids are pro-inflammatory. Nonetheless, the synthesis of AA-derived eicosanoids depends on the concentration of DGLA, as DGLA competes with AA for cyclooxygenase (COX) and lipoxygenase (LOX), enzymes implicated in LC-PUFAs metabolism. The excess of DGLA inhibits the synthesis of AA-derived eicosanoids due to its higher affinity for the COX and LOX enzymes.[7] Taking into account above-mentioned mechanisms, an adequate and balanced intake of omega-3 and omega-6 should be considered in order to maintain homeostasis.

Over the last century, extreme nutritional changes, particularly in industrialized countries, caused a substantial decrease in dietary intake of omega-3 PUFAs.[8] Extensive refining and processing of foods, as well as cultural dietary choices have led to an increase in omega-6 PUFAs intake (i.e., associated with Western diet increased consumption of vegetable oils). Nowadays, fatty acids represent 30%–40% of total energy consumed by European populations, whereas, in ancestral nutrition, fatty acid consumption was limited to approximately 20%–30% of total energy intake. The ancestors, who ate a diet richer in omega-3 fatty acids, had an estimated ratio of omega-6:omega-3 of 1.5:1. However, according to the nutritional changes in the Western diet, this ratio has now increased, and the estimated range is from 10:1 to 20:1.[9,10]

Dietary omega-3 PUFAs have been considered of particular interest due to the association with human health.[11] In particular, many epidemiological and experimental studies emphasized their possible role in the prevention and treatment of depressive disorders. Due to evidence from animal and human studies reporting that omega-3 deficiency may lead to impaired neuronal function and altered inflammatory status, the biological plausibility of the effects of the omega-3 PUFAs raised several hypotheses, although merely speculative.[12]

The aim of this chapter is to review the current evidence regarding the association between omega-3 PUFAs and depression, taking into account both epidemiological and clinical studies. Finally, proposed molecular mechanisms underlying the beneficial effect of omega-3 PUFAs toward depression will be pointed out.

BURDEN OF DEPRESSION

Depression is a mental disorder characterized by sadness, lack of interest in activities, and anhedonia. Other symptoms include loss of confidence and self-esteem, inappropriate feeling of guilt or worthlessness, diminished concentration, disturbance of sleep and appetite, suicidal thoughts, and sometimes attempted or actual suicide.[13] Depending on the number and severity of symptoms, a depressive episode can be categorized as mild, moderate, or severe. There are different types of depression, including: (i) recurrent depressive disorder that involves repeated depressive episodes, during which, the person may experience depressed mood, loss of interest and enjoyment, reduced energy leading to diminished activity, anxiety symptoms, disturbed sleep and appetite, feeling of guilt or low self-worth, or poor concentration; (ii) bipolar affective disorders that typically consist of both manic and depressive episodes separated by periods of normal mood, manic episodes may involve elevated or irritable mood, over-activity, inflated self-esteem, and a decreased need for sleep.[13]

The Diagnostic and Statistical Manual of Mental Disorders (DSM–V) defines major depressive disorder (MDD) diagnosis based on the presence of primarily symptoms, either depressed mood or loss of interest or pleasure, during the 2-week period. In addition, the diagnostic criteria require the presence of at least five items listed in DSM–V, among which are (i) depressed mood, (ii) markedly diminished interest or pleasure in activities, (iii) significant weight loss when not dieting or weight gain, (iv) insomnia or hypersomnia, (v) psychomotor agitation or retardation, (vi) fatigue or loss of energy, (vii) feelings of worthlessness or excessive or inappropriate guilt, (viii) diminished ability to think or concentrate, or indecisiveness, (ix) recurrent thoughts of death, recurrent suicidal ideation, or a suicide attempt.[14]

The course of MDD is quite variable, some individuals may experience remission (rarely), while others may experience many years with few or no symptoms between discrete episodes. The risk of recurrence becomes progressively lower over time as the duration of remission increases, however, with higher rates in individuals whose preceding episode was severe, in younger individuals, and in individuals who have already experienced multiple episodes. The persistence of even mild depressive symptoms during remission is a powerful predictor of recurrence.[14]

Depression is a significant contributor to the global burden of disease and affects people in all communities across the world. According to the recent report of the Global Burden of Disease Study, MDD is the third leading cause of global disability.[15] By 2020, depression is projected to be the second leading cause of disease burden worldwide after heart disease. Currently, depression is estimated to affect 350 million people. The World Mental Health Survey conducted in 17 countries found that on average about 1 in 20 people reported having an episode of depression in the previous year. Nonetheless, despite the high prevalence of depression, the mental health budgets of the majority of countries represent <1% of their total health expenditures. >40% of countries have no mental health policy and over 30% have no mental health programs.[16] Moreover, both direct economic costs of depression in terms of cost of treatment and indirect costs through lost days of work and reduced productivity, represent a major issue for public health.[17]

EPIDEMIOLOGICAL EVIDENCE

Mental, physical, and social health represents fundamental elements of the general well-being of a person. These factors are closely interwoven and deeply interdependent. For instance, the increased prevalence of depression over the last decades in Western countries has been accompanied by increased prevalence of cardiovascular diseases (CVDs) and fundamental changes in dietary habits.[17,18] Based on the epidemiological evidence, several studies suggested that depression may share common pathophysiologic mechanisms with CVDs and their risk factors,[19] including increased production of pro-inflammatory cytokines,[20] endothelial dysfunction,[21] elevated plasma homocysteine levels,[22] blood flow abnormalities (i.e., in depression, hypoperfusion in the limbic system and prefrontal cortex),[23] and impaired glucose metabolism (i.e., low glucose utilization in a number of brain regions correlating negatively with severity of depression).[24] Consequently, it has been hypothesized that besides environmental, behavioral, and genetic factors,[25] nutritional factors could significantly contribute to the increase in prevalence of both depression and CVDs. Indeed, significant shift over the last century in the dietary intake of long-chain PUFAs toward an increased intake of saturated fats and change in the ratio of omega-6 to omega-3 fatty acids intake was annotated.[26]

The first observation of an inverse association between per capita fish consumption and national annual prevalence of major depression was reported about 15 years ago.[27] Since then, several epidemiological studies demonstrated a significant inverse association between dietary intake of oily fish and prevalence[28–33] and incidence[34–36] of depression and bipolar disorder,[37] setting a threshold of vulnerability of about 650 mg/day. Interestingly, regarding the Mediterranean countries, several studies reported a decreased prevalence[38] and incidence[39–41] of depression and/or depressive symptoms in subjects more adherent to the Mediterranean dietary pattern, which is characterized by a high consumption of fish. The favorable effects of the Mediterranean diet on mental health may depend on the synergic beneficial properties of a variety of foods with a high content of omega-3 PUFAs, such as oily fish, but also olive oil and varieties of nuts.[42] Such positive effects on mental health of long-chain fatty acids contained in the Mediterranean diet may also explain the evidence supporting protective effects of Mediterranean diet toward CVDs.[43] However, these conclusions are still not definitive, as other components of the Mediterranean dietary pattern may exert considerable positive effects on the brain function, and thereby undermining the evidence regarding omega-3 fatty acids.[44]

Recently, the association between dietary omega-3 PUFAs and EPA/DHA intake and depression was explored using meta-analytical approach. The study reported significant reduced risk of depression, when considering the highest versus the lowest category of dietary omega-3 PUFAs intake, even though several of the individual studies reported nonsignificant results and the amount of omega-3 PUFAs estimated across included studies varied to a great extent.

The dose–response analysis demonstrated a J-shaped decreased risk of depression up to 1.8 g/day of omega-3 PUFAs intake. Accordingly, dose–response calculated for studies on EPA/DHA showed significant decreased risk of depression up to 0.6 g/day of EPA/DHA intake, despite nonsignificant decreased risk was evident also for further increment in dietary intake.[45]

Nonetheless, the clarity and reliability of the results obtained from epidemiological studies on omega-3 PUFAs consumption and prevalence and incidence of depression may depend on the limitations characteristic for this type of studies. Indeed, in cross-sectional studies it is difficult to determine whether the outcome followed exposure in time or exposure resulted from the outcome. On the other hand, prospective studies may be affected by the misclassification of exposure, if dietary intake of omega-3 PUFAs is assumed to be constant over the entire follow-up period. Finally, in both types of studies, the use of food frequency questionnaires (FFQs) for the estimation of habitual dietary intake of omega-3 PUFAs may lead to recall bias.

CLINICAL EVIDENCE

Although epidemiological evidence supports an inverse association between omega-3 PUFAs and depression, the validity of the findings from clinical research is limited due to the methodological issues. Up to date, several meta-analyses on clinical studies have been conducted regarding the effect of omega-3 PUFAs toward depression, mostly demonstrating benefits of the administration in ameliorating symptoms of depression.[46–48] Nonetheless, inconsistent results have been reported in other systematic revisions of literature and meta-analysis.[49,50]

Systematic reviews and meta-analyses can provide evidence relevant to many aspects of human health; however, the conclusions are less clear and reliable when the included studies have differing results reflected in high heterogeneity. There are several factors that can account for the high heterogeneity across the studies, such as publication bias, unstandardized diagnosis (reliability of outcome measures), variability in treatment dose, and timing or duration of the trial. Other factors include features of the population such as severity of illness, comorbidities, age, and gender.[51]

In fact, among the limitations of pooled analysis, were study eligibility criteria, differences in omega-3 PUFAs dose, timing and duration of the trial, and overall quality of the study. However, it seems that unstandardized depression assessment (i.e., clinical diagnosis of MDD vs. self-reported depressive symptoms) and comorbidities (CVD, Alzheimer, and schizophrenia), were notably affecting the efficacy of the treatment with omega-3 PUFAs in these pooled analyses, as different mechanisms may underlie the pathophysiology of depression.[52]

Certain meta-analyses focused on the type of fatty acid used, resulting in a positive effect on depressive symptoms of EPA rather than DHA content of the regime.[53,54] Particularly, supplements containing EPA $\geq$ 60% of total EPA+DHA, in a dose range of 200–2200 mg/day of EPA in excess of DHA, were effective against primary depression.[55] It has been also reported that the more severe was the depression, the more likely omega-3 PUFAs supplementation would reduce depressive symptoms.

In summary, omega-3 PUFAs may exert antidepressant effects in patients with MDD but not "mood-improving" effects in individuals with only self-reported depressive symptoms, and however, the efficacy strongly depends on the origin of depression.

MOLECULAR MECHANISMS UNDERLYING POSSIBLE EFFECTS OF OMEGA-3 PUFAS

Susceptibility to depression is influenced by a variety of genetic, epigenetic, endocrine, environmental, and dietary risk factors. It has been established, that dysregulation of the hypothalamic–pituitary–adrenal (HPA) axis, and abnormal release of neurotrophic factors, sex steroids, metabolic, and/or inflammatory cytokines can lead to alterations in neurotransmitters, intracellular signaling, gene transcription, and epigenetic changes that in turn can contribute to short-term and long-lasting imbalances of neuronal function and behavior.[56] Therefore, the suggested preventive and therapeutic effects of omega-3 PUFAs toward depression could be underlined by the alteration of one or more of the above-mentioned mechanisms.

Neuro-Inflammation

A growing body of scientific evidence indicates that depression is associated with excessive production of pro-inflammatory cytokines such as interleukin-1beta (IL-1β), interleukin-12 (IL-12), and interferon-gamma (IFN-γ),[57] and reduced levels of anti-inflammatory cytokines, such as interleukin-4 (IL-4), interleukin-10 (IL-10), and tumor growth factor-beta1 (TGF-β1).[58,59] In fact, recent meta-analysis of experimental studies reported a significantly higher concentration of the pro-inflammatory cytokines, including tumor necrosis factor-alpha (TNF-α) and interleukin-6 (IL-6) in depressed subjects compared to healthy controls.[57] Furthermore, several studies have demonstrated a positive correlation between the severity of the symptoms of depression and the increase in the inflammatory status[58]. Dysregulation of inflammatory balance, partially caused by the activation of microglia in central nervous system (CNS),[60] may lower neurotransmitter precursor availability and therefore alter the metabolism of neurotransmitters, reduce synaptic plasticity and hippocampal neurogenesis, as well as elicit both sickness behavior and psychological pain.[58]

Chronic inflammation has been associated with the promotion of neurodegeneration,[61] a process characterized by progressive loss of structure or function of neurons. The key anti-inflammatory effect of omega-3 PUFAs has been long recognized to depend on their metabolites, namely eicosanoids. As described previously, eicosanoids are biologically active lipid mediators produced from PUFAs which play an important role in regulation of immune system function.[62] Several studies demonstrated the association between the severity of depression and the degree of incorporation of omega-3 PUFAs in erythrocyte membranes, which are decreased in more severe conditions.[63–66] Moreover, results from a case–control study conducted on 16 depressed and 22 non-depressed women recruited during the third trimester of pregnancy demonstrated that high DHA, high total omega-3 and a low omega-6:omega-3 ratio were associated with significantly lower odds of depression.[67] Similar findings were reported in some studies conducted on depressed patients with post-myocardial infarction[68] and acute coronary syndromes[69,70] in which, compared to control group, lower levels of long-chain omega-3 PUFAs as measured by a ratio of AA/EPA were found.

It is well established that depression is associated with immune system dysregulation. Nonetheless, it has not yet been elucidated whether this immune dysregulation plays a role in the pathophysiology of MDD or whether it increases the susceptibility of the depressed patient to immune-related diseases.[71] The peripheral immune activation observed in depressed patients, mediated by the release of pro-inflammatory cytokines, is responsible for the variety of behavioral, neuroendocrine and neurochemical alterations that are associated with the severity of the symptomology.[71] Omega-3 PUFAs have been reported to modulate the release of key pro-inflammatory and anti-inflammatory cytokines and growth factors.[72] At the cellular level, they have been demonstrated to regulate the expression of nuclear factor kappa-light-chain-enhancer of activated B cells (NF-κB), a key transcription factor involved in cytokine production and cell survival.[72] Recent studies have pointed out the possible role of omega-3 PUFAs in inducing an antidepressant-like effect by modulating oxidative stress response and inflammatory cytokine production in microglia and neuronal cells, through decreased expression of TNF-α, IL-6, nitric oxide synthase (NOS), and upregulated expression of (IFN-γ) and heme oxygenase-1 (HO-1).[73] However, results of experimental studies on immune system regulation after administration of omega-3 PUFAs are not univocal. For example, long-term administration of omega-3 PUFAs in a rat model increased plasma serotonin concentration and the expression of hippocampus c-AMP response element binding protein (CREB) and reduced release of IL-6, however, a clear dose-dependent effects and significant differences in expressions of IL-1β, TNF-α, brain-derived neurotrophic factors, and phosphorylated CREB were not observed.[74]

As mentioned above, MDD is characterized by increased levels of pro-inflammatory cytokines and reduced levels of anti-inflammatory cytokines such as IL-10 and TGF-β1.[58,75] It has been demonstrated that plasma TGF-β1 levels are reduced in MDD patients and show a significant negative correlation with the Hamilton Depression Rating Scale,[59,76,77] and that antidepressant treatment significantly increases TGF-β1 levels.[59] In example, SSRI drugs such as sertraline might exert immunomodulatory effects in vivo through a decrease in the release of the pro-inflammatory cytokine IL-12, and an increase in the anti-inflammatory cytokines such as IL-4 and TGF-β1.[77] Similarly, venlafaxine was shown to inhibit microglial activation, reduce pro-inflammatory cytokine secretion, and increase the release of TGF-β1 in an in vitro astroglia-microglia coculture model.[78] Finally, recent experimental studies suggested that omega-3 PUFAs can increase the synthesis of TGF-β1[79,80] and, in particular, in pregnant women,[81] although no studies have been yet conducted in MDD patients. Taking into consideration recent evidence, it is important to assess whether TGF-β1 signaling pathway is a common target for both omega-3 PUFAs and antidepressant drugs.

Neurogenesis

Neurogenesis is the process by which new neurons are generated from neural stem cells. Adult neurogenesis occurs in two specific regions of the brain, the subventricular zone (SVZ) of the olfactory bulb and the subgranular zone (SGZ) of the hippocampal dentate gyrus.[82,83] The rate of neurogenesis and survival of new neurons in adults can be enhanced by many factors, such as, growth factors and neurotransmitters,[84] exercise,[85] or enriched environment.[86] Interestingly, much evidence indicate that neurogenesis underlies specific dynamics and flexible features of learning and memory, and that there is a cross-talk between neurogenesis and synaptic plasticity.[87] Noteworthy, increased neurogenesis has been observed in rodents following ischemia,[88,89] stroke,[90,91] and after seizures,[92,93] suggesting implication in brain self-repair and possibility that enhancing neurogenesis and the subsequent survival of new neurons may have significant therapeutic potential.[83]

Neurogenesis is decreased in the hippocampus of animal models of MDD, and many therapeutic strategies for the disorder, including antidepressant medication and electroconvulsive therapy, increase hippocampal neurogenesis. In recent years, an interesting hypothesis regarding the role of hippocampal neurogenesis in depression has been proposed. Specifically, it has been suggested that insufficient hippocampal neurogenesis causally underlies depression,[94] and may be related to glucocorticoids and serotonin levels, which are also associated with a decreased and increased rate of neurogenesis, respectively.[95] Nevertheless, the associations are still uncertain.

Recent studies demonstrated the ability of DHA to induce the differentiation of neural stem cells into neurons, however, the evidence regarding other omega-3 PUFAs is limited. Several mechanisms may be involved in the induction of adult neurogenesis by DHA, including promotion of the proliferative activity of neural stem cells and increase in the number of newborn neurons. Consequently, it has been demonstrated that DHA effectively promotes the differentiation of neural stem cells into neurons through promoting cell cycle exit and suppressing cell death. Additionally, dietary administration of DHA was shown to significantly increase the number of newborn neurons in the granule cell layer of the dentate gyrus in adult rodents.[96]

In neural stem cells, neurogenesis is modulated by a large superfamily of transcriptional factors forming basic helix–loop–helix (bHLH).[97] Neurogenesis is promoted by activator-like bHLH transcription factors, such as neurogenin, NeuroD, and Ascl1, while Hes1 and Hes5 exert suppressor-like properties, and thereby prevent terminal differentiation and preserve a pool of stem cells. Accordingly, the effect of DHA administration on the expression of bHLH transcription factors in neural stem cells has been investigated. DHA was shown to significantly decrease the expression of Hes1 and increase expression of neurogenin1 and NeuroD. Interestingly, MAP2 expression, a neuron-specific protein involved in microtubule assembly, an essential step in neuritogenesis, was also significantly increased. Taking into consideration that MAP2 is activated by NeuroD and repressed by Hes1, it can be hypothesized that DHA stimulates neuronal differentiation by altering the expression of bHLH transcription factors.[98]

Brain-derived neurotrophic factor (BDNF), a member of the neurotrophin family of growth factors, involved in spatial learning and memory, acts on certain neurons of the CNS and the peripheral nervous system, supporting the survival of existing neurons, as well as the growth and differentiation of new neurons and synapses.[99] Interestingly, as demonstrated in in vivo study, PUFAs supplementation was associated with an increase in the volume of hippocampus, increased synaptogenesis and BDNF expression, and subsequent raise in newborn cells in dentate gyrus and possible differentiation into matured neurons.[100]

The effects of omega-3 PUFAs, in particular DHA, toward adult neurogenesis are promising. Nevertheless, further studies are needed in order to elucidate the retrieved associations.

Synaptic Plasticity

Synaptic plasticity represents one of the most fundamental and important functions of the brain, which is the ability of the neural activity generated by an experience to modify neural circuit function and thereby modify subsequent thoughts, feelings, and behavior[101]

Synaptic plasticity refers to the activity-dependent modification of the strength or efficacy of synaptic transmission at preexisting synapses, and has been proposed to play an essential role in the capacity of the brain to integrate transient experiences into persistent memory traces. Furthermore, synaptic plasticity is thought to play an important role in the early development of neural circuitry, and accumulating evidence suggests that impairments in synaptic plasticity may contribute to pathophysiology of multiple neuropsychiatric disorders, including depression.[101] Up to date, the most consistent findings regarding the implication of disrupted synaptic plasticity in depression, include the reduction of volume of the prefrontal cortex (PFC) and hippocampus,[102,103] reduced synapse number in PFC,[104] atrophy and loss of neurons, and glia in the PFC and hippocampus,[105–107] which have been associated with the length of illness, time of treatment, and the severity of depression.[101]

Several studies explored the potential effects of omega-3 PUFAs, mainly DHA and EPA, on neurite outgrowth and synaptogenesis in diverse in vitro and in vivo models. In vitro, DHA has been shown to promote hippocampal neurite outgrowth through increased individual neurite length and number of branches.[108] Although, DHA action was specific, supplementation with other PUFAs, such as AA and DPA, under identical culture conditions did not affect neurite growth.[108] Similarly, it has been demonstrated that moderate concentrations of DHA administrated in in vitro settings improve the cortical neurons survival and outgrowth.[109] Interestingly, as demonstrated in in vivo study, DHA deficiency during brain maturation reduces plasticity and compromises brain function in adulthood.[110] Therefore, adequate levels of dietary DHA seem crucial for building long-term neuronal resilience for optimal brain performance and protection against neurological disorders. In addition to neurite outgrowth, DHA promotes synaptogenesis and increases the levels of pre- and postsynaptic proteins involved in synaptic transmission and long-term potentiation (LTP), and thereby improves synaptic function.[111] Finally, a potential neuroprotective effect of EPA in hippocampus was observed, in particular, through the modulation of synaptic plasticity and activation of the PI3-kinase pathway possibly by its direct effects on neurons and glial cells and by its capacity to increase brain DHA.[112]

CONCLUSIONS

In conclusion, the epidemiological and clinical studies demonstrated the potential beneficial role of omega-3 PUFAs toward prevention and treatment of depression. Several molecular mechanisms were hypothesized to be implicated in the neuroprotective effects of omega-3 PUFAs, including regulation of neuro-inflammation, modulation of adult hippocampal neurogenesis, neurite outgrowth, and synaptogenesis. Overall, increasing scientific evidence supports the neuroprotective properties of omega-3 PUFAs, nevertheless there are still some controversies about their efficacy. Therefore, further research is needed in order to elucidate the role of omega-3 PUFAs toward both prevention and treatment of depression. Finally, a better understanding of the effects of omega-3 PUFAs would allow us to confront the difficulty of how to improve their inadequate dietary intake in Westernized dietary pattern, while maintaining low intake of pro-inflammatory omega-6 PUFAs, and develop evidence-based systems innovations and policies to effectively reduce the burden of depression.

Acknowledgments

Justyna Godos is a PhD student of the International PhD Program in Neuroscience at the Università di Catania.

References

1. Simopoulos AP. The importance of the ratio of omega-6/omega-3 essential fatty acids. *Biomedicine & pharmacotherapy = Biomedecine & pharmacotherapie*. 2002;56(8):365–379.
2. Dyall SC, Michael-Titus AT. Neurological benefits of omega-3 fatty acids. *Neuromolecular Med*. 2008;10(4):219–235.
3. Grosso G, Estruch R. Nut consumption and age-related disease. *Maturitas*. 2016;84:11–16.
4. Shahidi F, Ambigaipalan P. Omega-3 polyunsaturated fatty acids and their health benefits. *Annu Rev Food Sci Technol*. 2018;9:345–381.
5. Moghadasian MH. Advances in dietary enrichment with n-3 fatty acids. *Crit Rev Food Sci Nutr*. 2008;48(5):402–410.
6. Kris-Etherton PM, Harris WS, Appel LJ, American Heart Association. Nutrition C. Fish consumption, fish oil, omega-3 fatty acids, and cardiovascular disease. *Circulation*. 2002;106(21):2747–2757.
7. Barcelo-Coblijn G, Murphy EJ. Alpha-linolenic acid and its conversion to longer chain n-3 fatty acids: benefits for human health and a role in maintaining tissue n-3 fatty acid levels. *Prog Lipid Res*. 2009;48(6):355–374.
8. Simopoulos AP. Executive summary of the international conference on genetic variation and nutrition. *World Rev Nutr Diet*. 1990;63:1–13.
9. Simopoulos AP, Ordovas JM. *Nutrigenetics and Nutrigenomics*. Vol. 93; 2004. Washington DC.
10. Marventano S, Kolacz P, Castellano S, et al. A review of recent evidence in human studies of n-3 and n-6 PUFA intake on cardiovascular disease, cancer, and depressive disorders: Does the ratio really matter? *Int J Food Sci Nutr*. 2015;66(6):611–622.
11. Simopoulos AP. Evolutionary aspects of diet, the omega-6/omega-3 ratio and genetic variation: Nutritional implications for chronic diseases. *Biomedicine & pharmacotherapy = Biomedecine & pharmacotherapie*. 2006;60(9):502–507.
12. Sinclair AJ, Begg D, Mathai M, Weisinger RS. Omega 3 fatty acids and the brain: review of studies in depression. *Asia Pac J Clin Nutr*. 2007;16(Suppl 1):391–397.
13. Levant B. N-3 (omega-3) polyunsaturated fatty acids in the pathophysiology and treatment of depression: pre-clinical evidence. *CNS Neurol Disord Drug Targets*. 2013;12(4):450–459.
14. Battle DE. Diagnostic and statistical manual of mental disorders (DSM). *Codas*. 2013;25(2):191–192.
15. Disease GBD. Injury I, prevalence C. Global, regional, and national incidence, prevalence, and years lived with disability for 310 diseases and injuries, 1990–2015: a systematic analysis for the global burden of Disease study 2015. *Lancet*. 2016;388(10053):1545–1602.
16. Organization WH. *A public health approach to mental health*. Geneva 2001.
17. Sobocki P, Jonsson B, Angst J, Rehnberg C. Cost of depression in Europe. *J Ment Health Policy Econ*. 2006;9(2):87–98.
18. Tiihonen J, Lonnqvist J, Wahlbeck K, et al. No mental health without physical health. *Lancet*. 2011;377(9766):611.
19. Prince M, Patel V, Saxena S, et al. No health without mental health. *Lancet*. 2007;370(9590):859–877.
20. Sanchez-Villegas A, Martinez-Gonzalez MA. Diet, a new target to prevent depression? *BMC Med*. 2013;11:3.
21. Machado-Vieira R, Mallinger AG. Abnormal function of monoamine oxidase-a in comorbid major depressive disorder and cardiovascular disease: pathophysiological and therapeutic implications (review). *Mol Med Rep*. 2012;6(5):915–922.
22. Do DP, Dowd JB, Ranjit N, House JS, Kaplan GA. Hopelessness, depression, and early markers of endothelial dysfunction in U.S. adults. *Psychosom Med*. 2010;72(7):613–619.
23. Severus WE, Littman AB, Stoll AL. Omega-3 fatty acids, homocysteine, and the increased risk of cardiovascular mortality in major depressive disorder. *Harv Rev Psychiatry*. 2001;9(6):280–293.
24. Ito H, Kawashima R, Awata S, et al. Hypoperfusion in the limbic system and prefrontal cortex in depression: SPECT with anatomic standardization technique. *J Nucl Med*. 1996;37(3):410–414. official publication, Society of Nuclear Medicine.
25. Kimbrell TA, Ketter TA, George MS, et al. Regional cerebral glucose utilization in patients with a range of severities of unipolar depression. *Biol Psychiatry*. 2002;51(3):237–252.
26. Stapelberg NJ, Neumann DL, Shum DH, McConnell H, Hamilton-Craig I. A topographical map of the causal network of mechanisms underlying the relationship between major depressive disorder and coronary heart disease. *Aust N Z J Psychiatry*. 2011;45(5):351–369.

27. Council NHaMR. Nutrient Reference Values for Australia and New Zealand including recommended dietary intakes. In: *NHMRC;* 2006.
28. Tanskanen A, Hibbeln JR, Tuomilehto J, et al. Fish consumption and depressive symptoms in the general population in Finland. *Psychiatr Serv.* 2001;52(4):529–531.
29. Bountziouka V, Polychronopoulos E, Zeimbekis A, et al. Long-term fish intake is associated with less severe depressive symptoms among elderly men and women: The MEDIS (MEDiterranean ISlands elderly) epidemiological study. *J Aging Health.* 2009;21(6):864–880.
30. Timonen M, Horrobin D, Jokelainen J, Laitinen J, Herva A, Rasanen P. Fish consumption and depression: the northern Finland 1966 birth cohort study. *J Affect Disord.* 2004;82(3):447–452.
31. Suominen-Taipale AL, Partonen T, Turunen AW, Mannisto S, Jula A, Verkasalo PK. Fish consumption and omega-3 polyunsaturated fatty acids in relation to depressive episodes: a cross-sectional analysis. *PloS one.* 2010;5(5):e10530.
32. Appleton KM, Woodside JV, Yarnell JW, et al. Depressed mood and dietary fish intake: Direct relationship or indirect relationship as a result of diet and lifestyle? *J Affect Disord.* 2007;104(1–3):217–223.
33. Murakami K, Miyake Y, Sasaki S, Tanaka K, Arakawa M. Fish and n-3 polyunsaturated fatty acid intake and depressive symptoms: Ryukyus child health study. *Pediatrics.* 2010;126(3):e623–e630.
34. Li Y, Dai Q, Ekperi LI, Dehal A, Zhang J. Fish consumption and severely depressed mood, findings from the first national nutrition follow-up study. *Psychiatry Res.* 2011;190(1):103–109.
35. Sanchez-Villegas A, Henriquez P, Figueiras A, Ortuno F, Lahortiga F, Martinez-Gonzalez MA. Long chain omega-3 fatty acids intake, fish consumption and mental disorders in the SUN cohort study. *Eur J Nutr.* 2007;46(6):337–346.
36. Astorg P, Couthouis A, Bertrais S, et al. Association of fish and long-chain n-3 polyunsaturated fatty acid intakes with the occurrence of depressive episodes in middle-aged French men and women. *Prostaglandins Leukot Essent Fatty Acids.* 2008;78(3):171–182.
37. Noaghiul S, Hibbeln JR. Cross-national comparisons of seafood consumption and rates of bipolar disorders. *Am J Psychiatry.* 2003;160(12):2222–2227.
38. Skarupski KA, Tangney CC, Li H, Evans DA, Morris MC. Mediterranean diet and depressive symptoms among older adults over time. *J Nutr Health Aging.* 2013;17(5):441–445.
39. Sanchez-Villegas A, Delgado-Rodriguez M, Alonso A, et al. Association of the Mediterranean dietary pattern with the incidence of depression: the Seguimiento Universidad de Navarra/University of Navarra follow-up (SUN) cohort. *Arch Gen Psychiatry.* 2009;66(10):1090–1098.
40. Rienks J, Dobson AJ, Mishra GD. Mediterranean dietary pattern and prevalence and incidence of depressive symptoms in mid-aged women: results from a large community-based prospective study. *Eur J Clin Nutr.* 2013;67(1):75–82.
41 Hodge A, Almeida OP, English DR, Giles GG, Flicker L. Patterns of dietary intake and psychological distress in older Australians: benefits not just from a Mediterranean diet. Int Psychogeriatr/IPA. 2013;25(3):456–466.
42. Luciano M, Mottus R, Starr JM, et al. Depressive symptoms and diet: their effects on prospective inflammation levels in the elderly. *Brain Behav Immun.* 2012;26(5):717–720.
43. Antonogeorgos G, Panagiotakos DB, Pitsavos C, et al. Understanding the role of depression and anxiety on cardiovascular disease risk, using structural equation modeling; the mediating effect of the Mediterranean diet and physical activity: the ATTICA study. *Ann Epidemiol.* 2012;22(9):630–637.
44. Sanchez-Villegas A, Henriquez P, Bes-Rastrollo M, Doreste J. Mediterranean diet and depression. *Public Health Nutr.* 2006;9(8A):1104–1109.
45. Grosso G, Micek A, Marventano S, et al. Dietary n-3 PUFA, fish consumption and depression: A systematic review and meta-analysis of observational studies. *J Affect Disord.* 2016;205:269–281.
46. Freeman MP, Hibbeln JR, Wisner KL, et al. Omega-3 fatty acids: evidence basis for treatment and future research in psychiatry. *J Clin Psychiatry.* 2006;67(12):1954–1967.
47. Lin PY, Su KP. A meta-analytic review of double-blind, placebo-controlled trials of antidepressant efficacy of omega-3 fatty acids. *J Clin Psychiatry.* 2007;68(7):1056–1061.
48. Grosso G, Pajak A, Marventano S, et al. Role of omega-3 fatty acids in the treatment of depressive disorders: a comprehensive meta-analysis of randomized clinical trials. *PLoS One.* 2014;9(5):e96905.
49. Appleton KM, Hayward RC, Gunnell D, et al. Effects of n-3 long-chain polyunsaturated fatty acids on depressed mood: systematic review of published trials. *Am J Clin Nutr.* 2006;84(6):1308–1316.
50. Rogers PJ, Appleton KM, Kessler D, et al. No effect of n-3 long-chain polyunsaturated fatty acid (EPA and DHA) supplementation on depressed mood and cognitive function: a randomised controlled trial. *Br J Nutr.* 2008;99(2):421–431.

51. Fletcher J. What is heterogeneity and is it important? *BMJ*. 2007;334(7584):94–96.
52. Lin PY, Mischoulon D, Freeman MP, et al. Are omega-3 fatty acids antidepressants or just mood-improving agents? The effect depends upon diagnosis, supplement preparation, and severity of depression. *Mol Psychiatry*. 2012;17(12):1161–1163 [author reply 1163-1167].
53. Ross BM, Seguin J, Sieswerda LE. Omega-3 fatty acids as treatments for mental illness: Which disorder and which fatty acid? *Lipids Health Dis*. 2007;6:21.
54. Martins JG. EPA but not DHA appears to be responsible for the efficacy of omega-3 long chain polyunsaturated fatty acid supplementation in depression: evidence from a meta-analysis of randomized controlled trials. *J Am Coll Nutr*. 2009;28(5):525–542.
55. Sublette ME, Ellis SP, Geant AL, Mann JJ. Meta-analysis of the effects of eicosapentaenoic acid (EPA) in clinical trials in depression. *J Clin Psychiatry*. 2011;72(12):1577–1584.
56. Russo SJ, Murrough JW, Han MH, Charney DS, Nestler EJ. Neurobiology of resilience. *Nat Neurosci*. 2012;15(11):1475–1484.
57. Wichers M, Maes M. The psychoneuroimmuno-pathophysiology of cytokine-induced depression in humans. *Int J Neuropsychopharmacol/Off Scient J Collegium Int Neuropsychopharmacol*. 2002;5(4):375–388.
58. Wager-Smith K, Markou A. Depression: A repair response to stress-induced neuronal microdamage that can grade into a chronic neuroinflammatory condition? *Neurosci Biobehav Rev*. 2011;35(3):742–764.
59. Maes M, Yirmyia R, Noraberg J, et al. The inflammatory & neurodegenerative (I&ND) hypothesis of depression: leads for future research and new drug developments in depression. *Metab Brain Dis*. 2009;24(1):27–53.
60. Krishnan V, Nestler EJ. Linking molecules to mood: new insight into the biology of depression. *Am J Psychiatry*. 2010;167(11):1305–1320.
61. Myint AM, Leonard BE, Steinbusch HW, Kim YK. Th1, Th2, and Th3 cytokine alterations in major depression. *J Affect Disord*. 2005;88(2):167–173.
62. Perry VH, Cunningham C, Holmes C. Systemic infections and inflammation affect chronic neurodegeneration. *Nat Rev Immunol*. 2007;7(2):161–167.
63. Calder PC. N-3 polyunsaturated fatty acids, inflammation, and inflammatory diseases. *Am J Clin Nutr*. 2006;83(6 Suppl):1505S–1519S.
64. Peet M, Murphy B, Shay J, Horrobin D. Depletion of omega-3 fatty acid levels in red blood cell membranes of depressive patients. *Biol Psychiatry*. 1998;43(5):315–319.
65. Edwards R, Peet M, Shay J, Horrobin D. Omega-3 polyunsaturated fatty acid levels in the diet and in red blood cell membranes of depressed patients. *J Affect Disord*. 1998;48(2–3):149–155.
66. Maes M, Christophe A, Delanghe J, Altamura C, Neels H, Meltzer HY. Lowered omega3 polyunsaturated fatty acids in serum phospholipids and cholesteryl esters of depressed patients. *Psychiatry Res*. 1999;85(3):275–291.
67. Rizzo AM, Corsetto PA, Montorfano G, et al. Comparison between the AA/EPA ratio in depressed and non depressed elderly females: omega-3 fatty acid supplementation correlates with improved symptoms but does not change immunological parameters. *Nutr J*. 2012;11:82.
68. Rees AM, Austin MP, Owen C, Parker G. Omega-3 deficiency associated with perinatal depression: case control study. *Psychiatry Res*. 2009;166(2–3):254–259.
69. Schins A, Crijns HJ, Brummer RJ, et al. Altered omega-3 polyunsaturated fatty acid status in depressed post-myocardial infarction patients. *Acta Psychiatr Scand*. 2007;115(1):35–40.
70. Frasure-Smith N, Lesperance F, Julien P. Major depression is associated with lower omega-3 fatty acid levels in patients with recent acute coronary syndromes. *Biol Psychiatry*. 2004;55(9):891–896.
71. Mamalakis G, Kiriakakis M, Tsibinos G, Kafatos A. Depression and adipose polyunsaturated fatty acids in an adolescent group. *Prostaglandins Leukot Essent Fatty Acids*. 2004;71(5):289–294.
72. Dowlati Y, Herrmann N, Swardfager W, et al. A meta-analysis of cytokines in major depression. *Biol Psychiatry*. 2010;67(5):446–457.
73. Calder PC. N-3 fatty acids, inflammation and immunity: new mechanisms to explain old actions. *Proc Nutr Soc*. 2013;72(3):326–336.
74. Lu DY, Tsao YY, Leung YM, Su KP. Docosahexaenoic acid suppresses neuroinflammatory responses and induces heme oxygenase-1 expression in BV-2 microglia: implications of antidepressant effects for omega-3 fatty acids. *Neuropsychopharmacol: Off Publ Am College Neuropsychopharmacol*. 2010;35(11):2238–2248.
75. Navarro C, Gonzalez-Alvarez I, Gonzalez-Alvarez M, et al. Influence of polyunsaturated fatty acids on cortisol transport through MDCK and MDCK-MDR1 cells as blood-brain barrier in vitro model. *Eur J Pharm Sci*. 2011;42(3):290–299.

76. Prescott SM, Stenson WF. Fish oil fix. *Nat Med*. 2005;11(6):596–598.
77. Kim Y, Lee C. The gene encoding transforming growth factor beta 1 confers risk of ischemic stroke and vascular dementia. *Stroke; J Cereb Circ*. 2006;37(11):2843–2845.
78. Sutcigil L, Oktenli C, Musabak U, et al. Pro- and anti-inflammatory cytokine balance in major depression: effect of sertraline therapy. *Clin Dev Immunol*. 2007;2007:76396.
79. Vollmar P, Haghikia A, Dermietzel R, Faustmann PM. Venlafaxine exhibits an anti-inflammatory effect in an inflammatory co-culture model. *Int J Neuropsychopharmacol/Off Sci J Collegium Int Neuropsychopharmacol*. 2008;11(1):111–117.
80. Hida M, Fujita H, Ishikura K, Omori S, Hoshiya M, Awazu M. Eicosapentaenoic acid inhibits PDGF-induced mitogenesis and cyclin D1 expression via TGF-beta in mesangial cells. *J Cell Physiol*. 2003;196(2):293–300.
81. Thienprasert A, Samuhaseneetoo S, Popplestone K, West AL, Miles EA, Calder PC. Fish oil n-3 polyunsaturated fatty acids selectively affect plasma cytokines and decrease illness in Thai schoolchildren: a randomized, double-blind, placebo-controlled intervention trial. *J Pediatr*. 2009;154(3):391–395.
82. Ehninger D, Kempermann G. Neurogenesis in the adult hippocampus. *Cell Tissue Res*. 2008;331(1):243–250.
83. Dyall SC. Long-chain omega-3 fatty acids and the brain: a review of the independent and shared effects of EPA, DPA and DHA. *Front Aging Neurosci*. 2015;7:52.
84. Cameron HA, Hazel TG, McKay RD. Regulation of neurogenesis by growth factors and neurotransmitters. *J Neurobiol*. 1998;36(2):287–306.
85. van Praag H, Shubert T, Zhao C, Gage FH. Exercise enhances learning and hippocampal neurogenesis in aged mice. *J Neurosci*. 2005;25(38):8680–8685.
86. Lazarov O, Robinson J, Tang YP, et al. Environmental enrichment reduces Abeta levels and amyloid deposition in transgenic mice. *Cell*. 2005;120(5):701–713.
87. Stuchlik A. Dynamic learning and memory, synaptic plasticity and neurogenesis: an update. *Front Behav Neurosci*. 2014;8:106.
88. Takagi Y, Nozaki K, Takahashi J, Yodoi J, Ishikawa M, Hashimoto N. Proliferation of neuronal precursor cells in the dentate gyrus is accelerated after transient forebrain ischemia in mice. *Brain Res*. 1999;831(1–2):283–287.
89. Wang C, Zhang M, Sun C, et al. Sustained increase in adult neurogenesis in the rat hippocampal dentate gyrus after transient brain ischemia. *Neurosci Lett*. 2011;488(1):70–75.
90. Darsalia V, Heldmann U, Lindvall O, Kokaia Z. Stroke-induced neurogenesis in aged brain. *Stroke*. 2005;36(8):1790–1795.
91. Woitke F, Ceanga M, Rudolph M, et al. Adult hippocampal neurogenesis poststroke: More new granule cells but aberrant morphology and impaired spatial memory. *PLoS One*. 2017;12(9):e0183463.
92. Jessberger S, Parent JM. Epilepsy and adult neurogenesis. *Cold Spring Harb Perspect Biol*. 2015;7(12).
93. Parent JM, Yu TW, Leibowitz RT, Geschwind DH, Sloviter RS, Lowenstein DH. Dentate granule cell neurogenesis is increased by seizures and contributes to aberrant network reorganization in the adult rat hippocampus. *J Neurosci*. 1997;17(10):3727–3738.
94. Jacobs BL, van Praag H, Gage FH. Adult brain neurogenesis and psychiatry: a novel theory of depression. *Mol Psychiatry*. 2000;5(3):262–269.
95. Brezun JM, Daszuta A. Depletion in serotonin decreases neurogenesis in the dentate gyrus and the subventricular zone of adult rats. *Neuroscience*. 1999;89(4):999–1002.
96. Kawakita E, Hashimoto M, Shido O. Docosahexaenoic acid promotes neurogenesis in vitro and in vivo. *Neuroscience*. 2006;139(3):991–997.
97. Jones S. An overview of the basic helix-loop-helix proteins. *Genome Biol*. 2004;5(6):226.
98. Katakura M, Hashimoto M, Shahdat HM, et al. Docosahexaenoic acid promotes neuronal differentiation by regulating basic helix-loop-helix transcription factors and cell cycle in neural stem cells. *Neuroscience*. 2009;160(3):651–660.
99. Numakawa T, Odaka H, Adachi N. Actions of brain-derived neurotrophic factor and glucocorticoid stress in Neurogenesis. *Int J Mol Sci*. 2017;18(11).
100. Venna VR, Deplanque D, Allet C, Belarbi K, Hamdane M, Bordet R. PUFA induce antidepressant-like effects in parallel to structural and molecular changes in the hippocampus. *Psychoneuroendocrinology*. 2009;34(2):199–211.
101. Citri A, Malenka RC. Synaptic plasticity: multiple forms, functions, and mechanisms. *Neuropsychopharmacology*. 2008;33(1):18–41.
102. MacQueen GM, Yucel K, Taylor VH, Macdonald K, Joffe R. Posterior hippocampal volumes are associated with remission rates in patients with major depressive disorder. *Biol Psychiatry*. 2008;64(10):880–883.

103. Savitz J, Drevets WC. Bipolar and major depressive disorder: neuroimaging the developmental-degenerative divide. *Neurosci Biobehav Rev*. 2009;33(5):699–771.
104. Kang HJ, Voleti B, Hajszan T, et al. Decreased expression of synapse-related genes and loss of synapses in major depressive disorder. *Nat Med*. 2012;18(9):1413–1417.
105. Duman RS, Aghajanian GK. Synaptic dysfunction in depression: potential therapeutic targets. *Science*. 2012;338(6103):68–72.
106. McEwen BS, Eiland L, Hunter RG, Miller MM. Stress and anxiety: structural plasticity and epigenetic regulation as a consequence of stress. *Neuropharmacology*. 2012;62(1):3–12.
107. Morrison JH, Baxter MG. The ageing cortical synapse: hallmarks and implications for cognitive decline. *Nat Rev Neurosci*. 2012;13(4):240–250.
108. Calderon F, Kim HY. Docosahexaenoic acid promotes neurite growth in hippocampal neurons. *J Neurochem*. 2004;90(4):979–988.
109. Cao D, Xue R, Xu J, Liu Z. Effects of docosahexaenoic acid on the survival and neurite outgrowth of rat cortical neurons in primary cultures. *J Nutr Biochem*. 2005;16(9):538–546.
110. Bhatia HS, Agrawal R, Sharma S, Huo YX, Ying Z, Gomez-Pinilla F. Omega-3 fatty acid deficiency during brain maturation reduces neuronal and behavioral plasticity in adulthood. *PLoS One*. 2011;6(12):e28451.
111. Cao D, Kevala K, Kim J, et al. Docosahexaenoic acid promotes hippocampal neuronal development and synaptic function. *J Neurochem*. 2009;111(2):510–521.
112. Kawashima A, Harada T, Kami H, Yano T, Imada K, Mizuguchi K. Effects of eicosapentaenoic acid on synaptic plasticity, fatty acid profile and phosphoinositide 3-kinase signaling in rat hippocampus and differentiated PC12 cells. *J Nutr Biochem*. 2010;21(4):268–277.

CHAPTER

14

Omega-3 Fatty Acids and the Cardiovascular System

Andrew Elagizi, Carl J. Lavie

Department of Cardiovascular Diseases, John Ochsner Heart and Vascular Institute, Ochsner Clinical School-the University of Queensland School of Medicine, New Orleans, LA, United States

ABBREVIATIONS

AHA	American Heart Association
ALA	Alpha linolenic acid
CHD	Coronary heart disease
CV	Cardiovascular
CVD	Cardiovascular disease
DHA	Docosahexaenoic acid
EPA	Eicosapentaenoic acid
HF	Heart failure
LDL-C	Low-density lipoprotein cholesterol
MI	Myocardial infarction
n-3 PUFA	Omega-3 polyunsaturated fatty acid
n-6 PUFA	Omega-6 polyunsaturated fatty acid
RCT	Randomized controlled trial
RLP	Remnant lipoprotein
SCD	Sudden cardiac death
SDA	Stearidonic acid
TG	Triglyceride

INTRODUCTION

Because of the abundance of epidemiological and randomized controlled trial (RCT) evidence suggesting beneficial effects of Omega-3 polyunsaturated fatty acids (n-3 PUFAs) on cardiovascular (CV) health, the American Heart Association (AHA) has endorsed a nutritional supplement for the secondary prevention of CV events in patients with documented coronary heart disease (CHD) for the first time.[1] n-3 PUFA have been studied extensively for decades since Sinclair's observation of the rarity of CHD in Greenland Eskimos, who consumed a diet

Omega Fatty Acids in Brain and Neurological Health
https://doi.org/10.1016/B978-0-12-815238-6.00014-6

high in whale, seal, and fish.[2] More than 30 years ago, Bang and Dyberg reported that despite a diet low in fruits, vegetables, and complex carbohydrates but high in saturated fat and cholesterol, serum cholesterol, and triglyceride (TG) levels were lower in Greenland Inuit than in age-matched residents of Denmark, and also had lower risk of myocardial infarction (MI).[3]

More prospective observational studies and large RCTs have investigated the potential effects of fish or n-3 PUFA consumption on CV disease (CVD) outcomes compared to any other food or nutrient.[4] The benefits of n-3 PUFA have also been investigated with regard to malignancy, dementia, depression, military combat, and cognitive development, among others. Trials such as GISSI-P, which demonstrated that 1 g/day n-3 PUFA decreased the risk of death, nonfatal MI, and stroke in patients surviving recent MI generated more interest regarding the relationship between n-3 PUFA and CV benefit.[5] However, later trials have not found as optimistic results as GISSI-P, and have had conflicting results. There remains a controversy regarding the role of n-3 PUFA in the treatment and/or prevention of CVD and CVD events. The focus of this chapter is to provide a background of n-3 PUFA types, uses, and ongoing developments with a focus on CV implications.

OMEGA-3 AND OMEGA-6

In 1929, biochemists Evans and Burr discovered the essential fatty acids, showing that mammals do not possess enzymes to synthesize double bonds at the n-3 and n-6 positions of the carbon chain of a fatty acid.[6] Therefore, humans must obtain these essential fatty acids linoleic acid (C18:2n-6) and alpha linolenic acid (ALA, C18:3n-3), as well as some of their elongation products from dietary sources; the lack of these nutrients leads to a syndrome of deficiency of essential fatty acids.[6,7] In humans, this deficiency is usually characterized by desquamative rashes and hyperkeratotic dermatoses.[7] n-3 PUFA supplementation may be most beneficial for those who consume inadequate amounts of n-3 PUFA and are therefore deficient, similarly to how vitamin C bestows dramatic life-saving benefits in people with scurvy, but is no better than placebo in people who are replete with vitamin C.[8]

ALA can be extended into eicosapentaenoic acid (EPA, C20:5n-3) and docosahexaenoic acid (DHA, C22:6n-3) through elongation and saturation,[6] although this is a limited and complex conversion in humans. ALA is found in abundance in flaxseed, soybean, linseed, and to a lesser extent in canola and olive oil, walnuts, and other tree nuts, and in trace amounts in green leafy vegetables.[3,7] In Western diets the main source of EPA and DHA is fish, particularly oily fish.[7] Estimates of minimum requirements for n-6 and n-3 PUFA in adults are 1% and 0.2% of daily energy intake, respectively.[7]

It has been suggested that the modern human diet contains significantly more n-6 PUFA uptake than ancestral humans ingested.[9,10] Western diets are typically deficient in n-3 PUFA and have excessive n-6 PUFA,[7] and this high ratio of n-6:n-3 PUFA in the diet has been said to promote the pathogenesis of many diseases, including CVD, cancer, and inflammatory/autoimmune diseases, whereas increased levels of n-3 PUFA (diets with a lower n-6:n-3 PUFA ratio) may express suppressive effects.[9] Because arachidonic acid is an n-6 PUFA, and is the substrate for the synthesis of a variety of proinflammatory molecules, some individuals and groups have advocated for lower n-6 PUFA intake to potentially lower inflammation and the risk for CHD.[11] However, the AHA has concluded that the consumption of at least 5%–10%

of energy from n-6 PUFA reduces the risk of CHD relative to lower intakes.[11] The data also suggest that higher intakes appear to be safe and may be even more beneficial (as part of a low saturated-fat, low cholesterol diet), and that to reduce n-6 PUFA intake would be more likely to increase than to decrease risk for CHD.[11]

DIETARY N-3 PUFA INTAKE

The AHA recommends dietary fish consumption of preferably oily fish at least twice per week.[12] Fish oil is obtained in the human diet by eating oily fish, such as herring, mackerel, trout, salmon, albacore tuna, and sardines, or by consuming fish oil supplements.[3,7] n-3 PUFA are not synthesized in fish de novo but stem from unicellular organisms at the base of their natural food chain.[3, 13] Although farm fed fish are typically grain fed, in the last month or months before harvest, the fish diet is changed to a higher content of n-3 PUFA to meet the values seen in typical wild fish of a given species.[14]

Recognizing the benefits of n-3 PUFA consumption, an expert panel for the US military from the Nutritional Armor for the Warfighter Conference unanimously agreed that a daily recommended n-3 PUFA intake should be established to receive CV, immunological, and surgical benefits.[15] It was also agreed that preloading with n-3 PUFA before combat may be beneficial, in addition to improved outcomes regarding depression. The panel felt that n-3 PUFA supplements, however, would be ineffective.

It is certainly possible that 1 day the n-3 PUFA status of those who do not consume adequate marine sources of n-3 PUFA will improve. "Stealth health" (Fig. 1) is a concept that

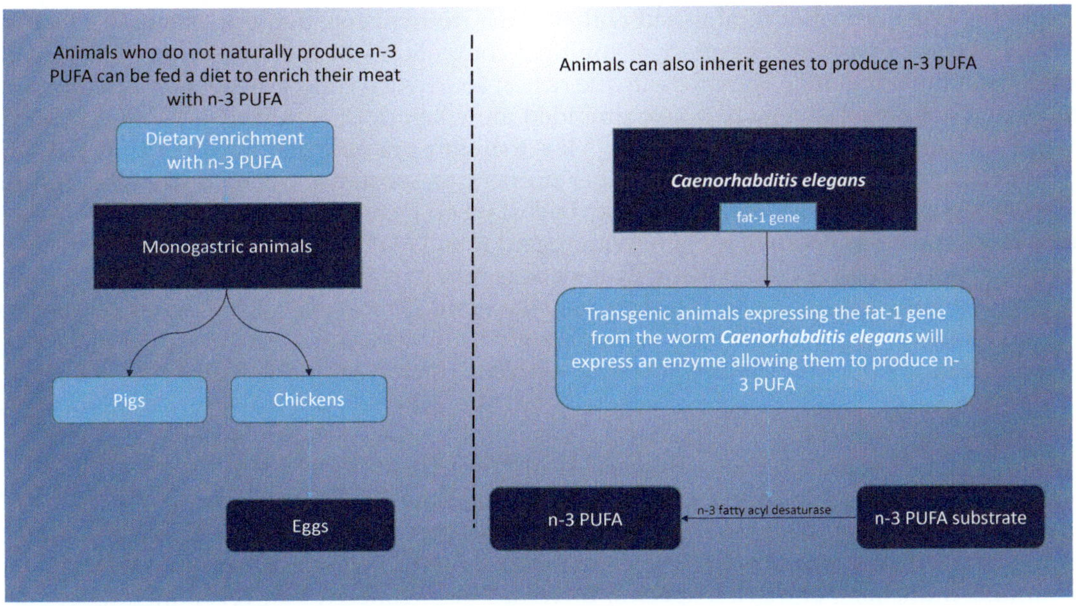

FIG. 1 Stealth Health. Methods which may be used to enrich mammalian cells with n-3 PUFA and enhance n-3 PUFA status in people without dietary change. n-3 PUFA, Omega-3 Polyunsaturated fatty acid.

involves intrinsically increasing the n-3 PUFA content of various foods, including the meat from chickens and other monogastric animals such as pork, using a similar dietary strategy as eggs, which seems readily achievable.[15] Also, transgenic animals that express the fat-1 gene from the worm *Caenorhabditis elegans* have been raised, and this gene encodes a n-3 fatty acyl desaturase enzyme that converts 18- and 20-carbon substrates into n-3 PUFA; the supplementation with meat from these transgenic animals might permit a marked enrichment of mammalian cells with n-3 PUFA.[7] These stealth health foods could be an ideal mechanism for delivering n-3 PUFA without changing dietary habits and could improve nutritional status among those with the worst dietary habits.[15]

While these methods, in addition to n-3 PUFA supplements may improve an individual's n-3 PUFA status, there is evidence that would suggest CV benefit from fish consumption (opposed to fish oil supplementation). Kromhout et al.[16] studied 852 middle-aged men in the Netherlands and followed them for 20 years, finding that people consuming at least 30 g of fish/day had 50% lower mortality from CHD. The DART trial found that participants who received advice on increasing fatty fish intake decreased all-cause mortality by 29% compared to those receiving no advice.[17] Some would argue that the benefits of whole food consumption cannot be met or replaced sufficiently by taking isolated nutrient supplements; the "whole food versus nutrient" debate.

WHOLE FOOD VERSUS NUTRIENT CONSUMPTION

There are potential benefits of whole-food consumption rather than isolated nutrient supplementation. Many nutrients may exert their health benefits through consumption of the whole food, which may be lost or diminished with isolated nutrient consumption, suggesting a possible synergistic or dependent effect of multiple components of the whole food for a given health benefit. For example, tomatoes are the richest source of lycopene in the Western diet, and studies have shown that tomato intake provided more favorable results on CV risk endpoints than did lycopene supplementation alone.[18] It is therefore possible that the tomato, in addition to containing lycopene, contains some other nutrient or substance that is responsible for the lycopene's potential CV benefit, which may be lost when lycopene is isolated as a supplement and not consumed as part of a tomato. With regard to n-3 PUFA, similar arguments have been made that fish contain several potentially beneficial nutrients not contained in fish oil (vitamin D, selenium, naturally occurring antioxidants),[1] therefore, clinical and policy recommendations that focus on dietary n-3 PUFA from fish rather than supplements are justified, but those who cannot consume sufficient fish or want more supplementation should use fish oil.[4]

SUSTAINABILITY

Considering the rapid decline of global fish stocks over past decades, with some commercially popular fish species collapsing to only 10% of their historic maximum and over 100 confirmed extinctions of marine species, there has been concern that recommendations for the population as a whole to increase fish consumption may put further strain on our overburdened fish stocks.[1] The supply of seafood from global capture fisheries is at or near the

TABLE 1 Sources of n-3 PUFA

SOURCE
FISH AND MARINE LIFE
– Mackerel
– Trout
– Salmon
– Herring
– Albacore tuna
– Sardines
– Oysters
PLANT SOURCES
– Flaxseed
– Soybean
– Walnuts
– Tree nuts
– Olive oil
– Canola oil
SINGLE CELL OILS
– Yeast
– Microalgae
FISH OIL SUPPLEMENTS
"STEALTH HEALTH" FOODS
– Dietary enhancement
– Transgenic animals

biological maximum and is unlikely to supply adequate amounts of additional seafood to the world's growing population, considering that 90% of the fish stocks globally are rated as fully or overexploited.[14] Options other than fish consumption need to further be explored in order to meet global demands using sustainable methods.

Rather than fish consumption, some alternatives for n-3 PUFA sources include single-cell oils from yeast or microalgae, which come as a sustainable alternative capable of supplementing the global demand, with less environmental impact.[19] However, there are drawbacks to these alternate methods; plants rich in the n-3 PUFA stearidonic acid (SDA) are expensive, metabolic engineering is unfavorably accepted by consumers in many countries, cultivation of microalgae is very expensive, and Antarctic Krill harvesting must be restricted.[20] These sources, in addition to the stealth health methods mentioned previously, may provide more sustainable methods to meet global demands of n-3 PUFA with less environmental impact. See Table 1 for a list of sources of n-3 PUFA.

N-3 PUFA INDEX

The n-3 PUFA index measures the content of n-3 PUFA in red blood cell membranes and is an objective measure of an individual's n-3 PUFA status.[8] Epidemiologic studies have shown that higher blood levels of EPA and DHA were consistently associated with lower risk of endpoints such as mortality, SCD, or major adverse CV events.[21] It has been suggested that an n-3 PUFA index <4% has been associated with increased CHD risk, particularly SCD, whereas an index >8% is considered low risk and 4%–8% is intermediate risk (Fig. 2).[8] Many previous n-3 PUFA intervention trials have yielded neutral results, and it has been proposed that incorporating the n-3 PUFA index into trial designs and treating to a target (e.g., 8%–11%) will make more efficient trials possible.[21] The n-3 PUFA index has been proposed as a novel CV risk factor, can be used to reclassify individuals from intermediate CV risk into high or low risk categories,[21] and can help standardize results of future studies as it can be used to assess baseline n-3 PUFA status prior to treatment, measure adherence to treatment, and serve as a therapeutic target.[21–23]

N-3 PUFA INTERVENTIONS

GISSI-P was a landmark trial which prompted many further research endeavors following their demonstration of CV benefits with an n-3 PUFA intervention of 1 g/day following acute MI. There are multiple factors that need to be taken into consideration when discussing the results of trials such as GISSI-P. First, only 5% of patients in GISSI-P received statin therapy at baseline,[24] which is a critical component of modern treatment for atherosclerotic CVD. Reduced CV events in patients receiving modern medical therapy for CVD (including statins) may explain why some modern trials do not show beneficial effects of n-3 PUFA therapy on CV events. Also, GISSI-P was performed in Italy, where fish consumption and baseline serum n-3 PUFA levels are higher than those of Western populations. The lack of some modern trials ability to demonstrate findings similar to those of GISSI-P may at least in part be due to the differences in patient populations with regard to baseline medical therapy, n-3 PUFA status, diet and geographic location. 1 g/day may simply be too low a dose to demonstrate CV benefits in Western populations (with lower baseline n-3 PUFA status), particularly in those receiving modern medical therapy.

The modern use of maximal medical therapy in addition to falling smoking rates could lower the potential population level benefit from n-3 PUFA supplementation because of lower underlying risk.[25] This notion is supported by the results of the OMEGA trial which concluded that the guideline directed medical treatment of acute MI results in a low rate of SCD and other clinical events within 1 year follow-up, which could not be shown to be further reduced by n-3 PUFA.[26] It is therefore important to assess the effects of n-3 PUFA supplementation in adequately powered studies, using higher intervention doses (>1 g/day), and to account for confounders of baseline fish consumption and n-3 PUFA status by using an objective measurement, such as the n-3 PUFA index, to standardize future trials. It has been argued that many discrepancies in the results of various trials can be explained on the basis of faulty study design (short follow-up, low baseline CV risk, lack of power, high baseline n-3 PUFA intake), inadequate dose of n-3 PUFA used, and differences in study populations.[17]

PHYSIOLOGIC EFFECTS

n-3 PUFA demonstrate many physiologic benefits, including anti-arrhythmic effects, benefits related to cardiac structure and function (fibrosis, myocardial oxygen demand), and

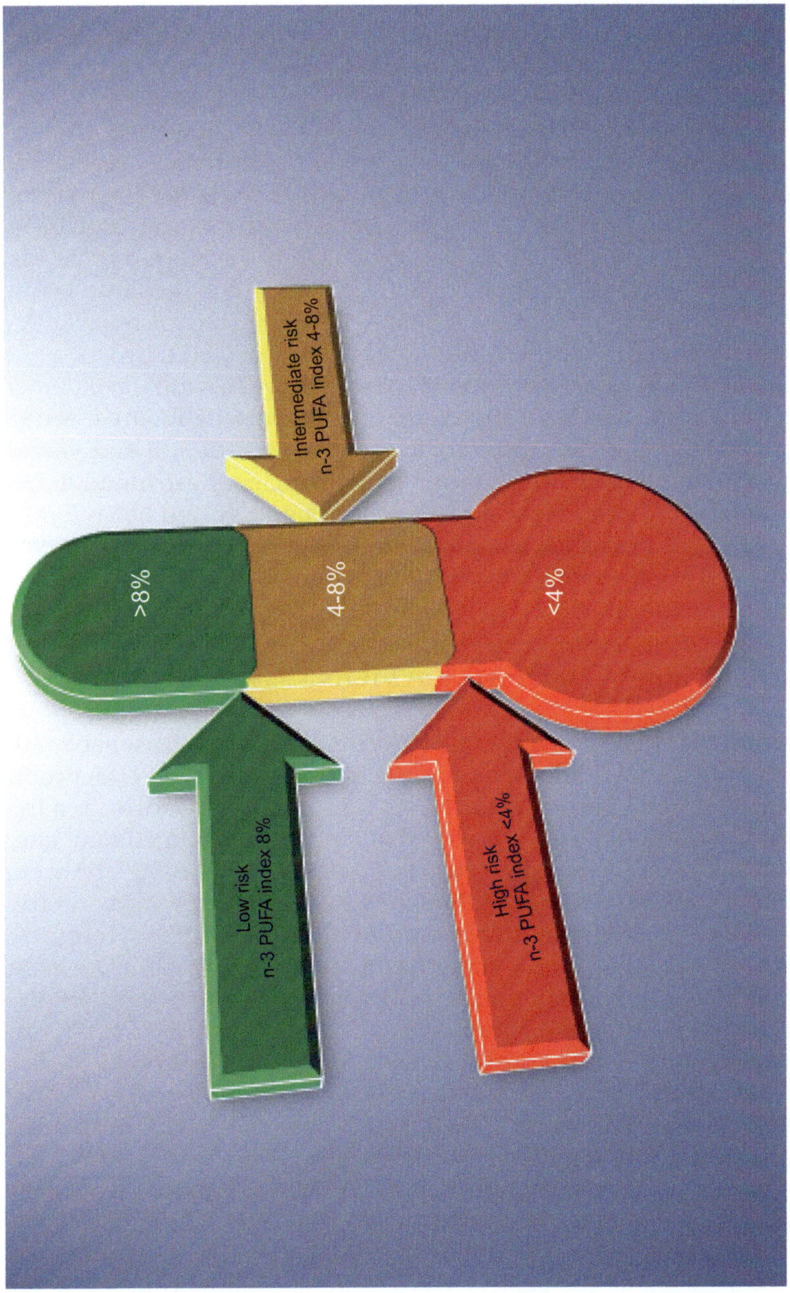

FIG. 2 n-3 PUFA index. Proposed risk stratification using the n-3 PUFA index as a measure of cardiovascular risk. n-3 PUFA, Omega-3 Polyunsaturated fatty acid.

improvements in arterial and endothelial function, autonomic function, elevated TG levels, thrombosis, blood pressure, inflammation, lipoprotein metabolism, atherosclerotic plaque, and reduced platelet aggregation.[3, 4, 7, 27–29] n-3 PUFA are known to lower TG levels, although doses >1.5 g/day are typically needed for a clinically meaningful change.[27] A recent meta-analysis showed that TG lowering therapies showed an overall benefit on CV event risk of 12%, and was even more significant (18%) in subsets with elevated TG (≥150 mg/dL), especially if accompanied by low high-density lipoprotein.[27] TG lowering doses of DHA and EPA of 3–4 mg/day typically reduce TG levels by 30%–40%, and have been shown to reduce severely elevated TG (≥500 mg/dL) by 45%.[3] Typically, there is no improvement in low-density lipoprotein cholesterol (LDL-C) levels with fish oil therapy, and in fact there is often an increase.[3] However, this n-3 PUFA enriched LDL-C develops a molecular structure described as "larger and fluffier" (pattern A LDL-C), which is potentially less atherogenic than the smaller, denser particles (pattern B LDL-C).[3] The effects of n-3 PUFA on platelet activity and antiplatelet therapy are discussed in the section of CHD.

n-3 PUFA also have a well-established role in cognitive and neural development, particularly in early life. Clinical studies have shown that n-3 and n-6 PUFA are essential nutrients for fetal and infant development, and that deficiencies lead to neural and vascular complications of preterm infants.[30] Newborns with n-3 PUFA deficiency are found to have reduced light sensitivity of retinal rod photoreceptors as well as cognitive and behavioral disorders.[30] While it seems that DHA is important for visual and neurologic development, studies examining the effects of DHA supplemented formulas for infants have had mixed results.[30]

FORMULATIONS

Although it seems that fish oil supplementation is a safe and reasonable alternative to dietary fish consumption, some studies would suggest that not all n-3 PUFA products are created equal. EPA and DHA may be present in an ethyl ester form, TG form, as a free fatty acid or as phospholipids, and it has been suggested that the form consumed may affect bioavailability.[31] Superior bioavailability of TG versus ethyl ester forms of n-3 PUFA were identified by several researchers over 30 years ago, although there have been some contrary findings.[31] Further study is required to determine the efficacy of different n-3 PUFA formulations.

Currently approved n-3 PUFA supplement formulations by the US FDA include Lovaza (ethyl ester formulation) and Vascepa (icosapent ethyl, containing >96% EPA ethyl ester), and are approved for the treatment of very high TG levels (≥500 mg/dL).[3, 32] In a phase 3 trial, icosapent ethyl 4 g/day significantly decreased the placebo-corrected median TG levels by 33.1%, had favorable effect on other lipid parameters without increasing LDL-C and was tolerated similarly to placebo.[32] Icosapent ethyl was used in the MARINE (TG 500–2000 mg/dL) and ANCHOR (TG 200–500 mg/dL) phase III studies, and found reductions of Remnant Lipoprotein (RLP) by 29.8% (P = .004) and 25.8% (P = .0001), respectively.[33] RLP exerts atherogenic effects and is considered an important CVD risk factor.[6] The ongoing reduction of cardiovascular events with icosapent ethyl-intervention trial (REDUCE-IT) has randomized approximately 8000 patients at 470 centers worldwide and aims to reveal if treatment for moderate TG elevation decreases CV events, particularly among patients already receiving statin therapy.[34]

CURRENTLY ACCEPTED REGARDING N-3 PUFA

In 2002, the AHA published a scientific statement recommending that patients with documented CHD consume ~1 g/day EPA + DHA, preferably from oily fish.[35] In an updated 2017 statement, the AHA concluded that n-3 PUFA "treatment is reasonable" for (1) secondary prevention of CHD and SCD among patients with prevalent CHD; and (2) secondary prevention of adverse outcomes in patients with heart failure (HF).[35] Based on a single, large RCT which demonstrated total mortality reduction of 9% and risk of CV-related hospitalizations or death reduction of 8%, treatment with n-3 PUFA is reasonable among patients with HF with reduced ejection fraction.

In spite of many anti-arrhythmic properties of n-3 PUFA, the cumulative evidence would not suggest that n-3 PUFA have any role in the primary or secondary prevention of atrial fibrillation.[35] It has also been concluded that the available evidence does not support the use of n-3 PUFA in the general population who are not at high CVD risk, although there was a lack of consensus.[35] As in 2002, in consultation with a physician, n-3 PUFA supplementation is reasonable for secondary prevention of CHD in patients with a recent CHD event such as MI.[35]

The risk of recurrent events among contemporary patients with CHD events is much lower than the corresponding risk of post-MI patients from 15 to 20 years ago.[35] It is therefore possible that n-3 PUFA has little benefit in the presence of maximal medical treatment for ischemic heart disease, which may account for later studies to not reproduce as significant beneficial results on CV events compared to GISSI-P. However, even if the benefit is quantitatively modest, a potential reduction in CHD death of 10% would justify the use of a relatively safe therapy such as n-3 PUFA, according to the AHA (Table 2).[35]

CORONARY HEART DISEASE

As mentioned previously, multiple studies have shown improved CHD morbidity and mortality associated with increased fish consumption. A large, high-powered meta-analysis performed by Alexander et al.[36] analyzed the effects of EPA + DHA on CHD and CHD risk. Data were analyzed from 93,000 subjects in RCTs and 732,000 subjects in prospective cohort studies. Among RCTs, a nonstatistically significant reduction in CHD risk with EPA + DHA was found, while subgroup analysis showed statistically significant CHD risk reduction with

TABLE 2 Recommendations From the AHA for n-3 PUFA Use

AHA Recommendation	Patient Population
Treatment is reasonable (minority of authors)	Prevention of CHD in patients with high CVD risk
Treatment is reasonable	Secondary prevention of CHD and SCD in CHD patients
Treatment is reasonable	Secondary prevention of outcomes in patients with HF with reduced ejection fraction

CHD, Coronary heart disease; CVD, Cardiovascular disease; HF, Heart failure; SCD, Sudden cardiac death.

EPA + DHA among higher-risk populations (those with elevated TG or LDL-C levels).[36] Their analysis of prospective cohort studies demonstrated a statistically significant risk reduction of any CHD event for higher intakes of EPA + DHA.

Antiplatelet therapy using aspirin, clopidogrel, or both, is a cornerstone in the primary prevention and treatment of atherosclerotic CVD, including CHD. However, there are inter-individual differences in the biological response to these agents, and there is at least moderate incidence of resistance to both aspirin and clopidogrel.[37] This hyporesponsiveness is associated with poor compliance, gene polymorphisms, intrinsic high platelet reactivity, variability of drug metabolism, and drug interactions, and patients with resistance to therapy have been shown to be at higher risk of ischemic events.[37] Triple therapy with the addition of cilostazol has also been reported to improve the biological effects of clopidogrel.[37]

n-3 PUFA are also known to reduce platelet aggregation, and studies have shown that their addition to antiplatelet therapy following percutaneous coronary intervention significantly potentiates platelet response to clopidogrel.[37] Aspirin resistance is also associated with adverse CV events, and it has been demonstrated the n-3 PUFA use can overcome aspirin resistance and effectively reduce platelet activity in a substantial portion of patients.[38] Prolonged bleeding time has been noted in Greenland Eskimos and when giving PUFA >20 g/day to normal volunteers usually remain on the high end of normal, however, Harris[39] concluded that there was no increased risk of clinically significant bleeding noted with n-3 PUFA doses up to 7 g/day DHA + EPA, even when combined with antiplatelet therapy or warfarin.

n-3 PUFA supplementation has been shown to significantly lower platelet activity as measured by $P2Y_{12}$ reactivity index by 22% after 1 month compared to placebo ($P = .02$).[37] It has also previously been concluded that aspirin resistance can be overcome by either increasing the dosage, or with the addition of n-3 PUFA therapy. Therefore, there appear to be three major methods to overcome antiplatelet agent resistance: (1) addition of n-3 PUFA supplementation, (2) increasing aspirin dosage, or (3) triple therapy with cilostazol. The addition of n-3 PUFA supplementation may be the safest option. Despite the reduced platelet aggregation and thrombosis seen with n-3 PUFA consumption, bleeding risk does not seem to be significantly increased.

POST MI

Multiple trials have shown benefits with the use of n-3 PUFA therapy following acute MI, such as decreased risk of all-cause mortality and recurrent acute MI.[5] It is important to note that one trial that included patients with a history of MI up to 10 years prior to randomization showed no effect of n-3 PUFA on the rate of adverse CV events compared to placebo. Another trial found that high-dose n-3 PUFA therapy following MI was associated with a reduction of adverse left ventricular remodeling, non-infarct myocardial fibrosis, and serum biomarkers of systemic inflammation beyond current guideline-based standard of care.[23] The improvements seen in the latter trial were only modest for patients who began treatment 2–4 weeks post MI. Taken together, these results suggest that earlier initiation of n-3 PUFA therapy in the first days following MI may provide more CV protection. However, not all trials have shown benefits in patients receiving modern guideline directed therapy,[26] and therefore the evidence is not conclusive.

SUDDEN CARDIAC DEATH

Overall, the findings of many RCTs and observational studies indicate that consumption of fish or fish oil significantly reduces CHD mortality, including fatal MI and SCD, in populations with and without established CVD.[4] Aggregate RCT and observational evidence has also been fairly consistent in showing benefits of higher n-3 PUFA intake, particularly in higher risk subsets, for cardiac death.[27] Higher n-3 PUFA levels may alter the risk for SCD through several pathways, including anti-arrhythmic effects, effects related to cardiac structure and function, endothelial and autonomic function, thrombosis, blood pressure, inflammation, and lipoprotein metabolism.[4] The cumulative findings from RCTs suggest that n-3 PUFA supplementation may reduce CHD death through a reduction in ischemia-induced SCD, and given that the benefit of n-3 PUFA use likely outweighs any risk of treatment, its use is reasonable for secondary prevention of CHD death.[35]

ATRIAL FIBRILLATION

Multiple trials have assessed the relationship between n-3PUFA and AF, considering their antiarrhythmic properties and association with lower incidence of AF.[22] Various trials, both with and without trans-telephonic monitoring have found no reduction of AF recurrence.[22,40] There are also trials which suggest that n-3 PUFA therapy does not appreciably reduce post-operative AF, regardless of the type of cardiac surgery,[41] although there are further ongoing trials. Overall, n-3 PUFA do not appear to play any role in the primary or secondary prevention of AF.[35]

HEART FAILURE

HF is one of the CV conditions with the most promising evidence for n-3 PUFA therapy. In addition to reducing the risk of incident HF,[42] multiple studies have suggested that increased n-3PUFA intake can decrease inflammation and cachexia in advanced HF.[3] Patients with HF have also been shown to have improved cardiorespiratory fitness, diastolic function, and body composition when consuming diets rich in unsaturated fatty acids.[43]

Following MI, n-3PUFA have recently been shown to effect left ventricular remodeling and significantly improved measures of cardiac structure and function.[23] The GISSI-HF trial suggested that 1 g/day n-3 PUFA was associated with lower death from any cause (27% vs. 29%) and hospitalizations for CV reasons (57% vs. 59%) in the n-3 PUFA groups versus placebo, respectively.[17] The AHA now recommends that patients with HF and reduced ejection fraction be treated with n-3 PUFA therapy to reduce mortality and hospitalizations.[35]

Importantly, the benefits of n-3 PUFA therapy in HF have also been demonstrated in patients receiving modern optimal medical therapy. This is an important finding because many older studies describing reduced recurrence of CHD events with n-3 PUFA therapy have had inconsistent results when performed in patient populations receiving modern medical therapy for CHD. Therefore, the utility of modern medical therapy for CHD appears to have a "blunting effect" on the benefits of n-3 PUFA. This blunting effect may not be as significant or as prevalent in the HF population, which may suggest a larger benefit of n-3 PUFA use in HF patients. n-3 PUFA have also demonstrated benefits in heart transplant patients, regarding reduced heart rate, blood pressure, and treatment of cyclosporine-induced hypertension.[44]

Considering heart transplant patients receiving cyclosporine therapy, systolic blood pressure was decreased by 2 ± 4 mmHg in the n-3 PUFA treatment group and increased by 17 ± 4 mmHg in the placebo group ($P < .01$), whereas diastolic blood pressure increased by 10 mmHg ± 3 and 21 mmHg ± 2 in the treatment and placebo groups, respectively ($P = .01$).[44] Another trial using an intervention of n-3 PUFA at 3 g/day in orthotopic cardiac transplant patients receiving cyclosporine therapy also showed blood pressure and systemic vascular resistance reductions in the n-3 PUFA treatment group.[45]

ISCHEMIC STROKE AND PERIPHERAL ARTERIAL DISEASE

Although meta-analyses suggest that fish consumption reduces the risk of ischemic stroke, stroke incidence has not been significantly affected in fish oil trials, and there are currently not enough appropriately powered RCTs to evaluate the effects of n-3 PUFA on stroke.[4] Overall, there is no evidence that n-3 PUFA supplementation reduces the risk of recurrent stroke or other CVD in patients with prior stroke.[35] There is very little evidence regarding n-3 PUFA use and CV outcomes in peripheral arterial disease, and one meta-analysis found no evidence to suggest a protective effect of n-3 PUFA therapy regarding MI, CV death, angina, stroke, amputation, revascularization, pain-free walking distance, or quality of life.[46]

MALIGNANCY

Articles in popular media have suggested that n-3 PUFA may increase the risk of prostate cancer, and many of these commentaries advise against the use of supplemental fish oil.[47] It is important to note that in men who already have prostate cancer, a regular high intake of fish has been linked to a marked increase in survival.[47] Japanese men consume approximately eight times more fish than American men, yet Japan in the past decade has been characterized by a rate of prostate cancer mortality many-fold lower than Western nations.[47] Also, according to the American Cancer Society, most men diagnosed with prostate cancer do not actually die from it,[48] and 91% of prostate cancers are discovered at a local or regional stage, with 5-year survival approaching 100%.[49] Because SCD is the initial sign of CHD in a substantial fraction of cases, discouraging the use of fish or fish oil is highly inadvisable without substantial evidence of risk.[47] With regard to cancer in general, many large randomized trials of fish oil supplementation have never found an increase in cancer incidence or mortality in those receiving n-3 PUFA supplements.[47,50]

RECENT EVIDENCE

More recent, powerful studies have found neutral results regarding n-3 PUFA and CV health. A recent, large meta-analysis of 10 trials involving >77,000 high-risk participants (many of whom had prior CHD, stroke, or diabetes) found that n-3 PUFA had no significant association with fatal or nonfatal CHD or any major adverse CV events.[51] While this trial does not find evidence to support recommendations for 1 g/day n-3 PUFA supplementation, the authors acknowledge that further trials are needed to assess if higher doses of n-3 PUFA (3–4 g/day) may have significant effects on major CV events.

A recent Cochrane review has been published regarding the use of n-3 PUFA for the primary and secondary prevention of CVD, and is the most extensive systematic assessment of effects of n-3 PUFA on CV health to date. The authors conclude that most of the evidence would suggest that increasing n-3 PUFA intake has little or no effect on mortality or CV outcomes.[52] The authors concluded that n-3 PUFA do not have important positive or negative effects on mortality or CVD events and moderate quality evidence suggests little or no effect on other measures of CV health in primary or secondary prevention.[52] A benefit was found with the use of ALA for reducing CHD mortality and arrhythmia, however the overall effects were small with 143 needed to treat to prevent one person developing arrhythmia, and 1000 to prevent one CVD event or CHD death. Also, no effects were found based on dose of n-3 PUFA used on any outcomes, except for lowering serum TG levels.

CONCLUSION

Currently, there are no class I indications from the AHA for n-3 PUFA therapy, however, the AHA recently increased the number of recommendations for which n-3 PUFA therapy is reasonable. The best evidence for the use of n-3 PUFA to benefit CV health is for CHD mortality, SCD, following acute MI and for patients with HF and reduced ejection fraction. Some trials, using higher doses of n-3 PUFA in patients receiving modern guideline directed medical therapy for CVD, have shown favorable results, suggesting that the list of CV conditions benefited with n-3 PUFA may yet continue to grow. However, recent large publications including a Cochrane review, which is the most extensive assessment of n-3 PUFA effects relating to CVD to date, suggest that n-3 PUFA have little beneficial effect on CV health.

The use of an objective measurement of a patient's n-3 PUFA status, such as the n-3 PUFA index, may be used to standardize the results of future trials and study the efficacy of treating to a certain target. Although there is still controversy regarding the benefits of n-3 PUFA with regard to CV health, this therapy is largely free of adverse effects and is considered safe. Even a modest reduction of CVD risk should encourage the use of a largely innocuous therapy, such as fish oil.

While dietary consumption of n-3 PUFA in the form of oily fish is currently recommended by the AHA, fish oil supplementation is certainly a safe and reasonable alternative for those who cannot, or choose not, to consume the recommended amounts of fish in their diet. The current strain on global fish stocks has lead researchers to investigate alternative, sustainable methods of meeting the global demand for n-3 PUFA, including single cell oils and "stealth health" methods which require further development before becoming more accessible and widely implemented.

DISCLOSURES/CONFLICTS OF INTEREST

Dr. Lavie is a speaker and consultant for Amarin on Vascepa and for DSM Nutritional Products and for the Global Organization for EPA and DHA Omega-3 s. Dr. Elagizi reports no conflicts of interest.

References

1. Lee JH, O'Keefe JH, Lavie CJ, Harris WS. Omega-3 fatty acids: cardiovascular benefits, sources and sustainability. *Nat Rev Cardiol*. 2009;6(12):753–758. https://doi.org/10.1038/nrcardio.2009.188.
2. Sinclair HM. The diet of Canadian Indians and Eskimos. *Proc Nutr Soc*. 1953;12(1):69–82. https://doi.org/10.1079/pns19530016.
3. Lavie CJ, Milani RV, Mehra MR, Ventura HO. Omega-3 polyunsaturated fatty acids and cardiovascular diseases. *J Am Coll Cardiol*. 2009;54(7):585–594. https://doi.org/10.1016/j.jacc.2009.02.084.
4. Mozaffarian D, Wu JH. Omega-3 fatty acids and cardiovascular disease effects on risk factors, molecular pathways, and clinical events. *J Am Coll Cardiol*. 2011;58(20):2047–2067. https://doi.org/10.1016/j.jacc.2011.06.063.
5. Greene SJ, Temporelli PL, Campia U, et al. Effects of polyunsaturated fatty acid treatment on Postdischarge outcomes after acute myocardial infarction. *Am J Cardiol*. 2016;117(3):340–346. https://doi.org/10.1016/j.amjcard.2015.10.050.
6. Kromhout D, Yasuda S, Geleijnse JM, Shimokawa H. Fish oil and omega-3 fatty acids in cardiovascular disease: do they really work? *Eur Heart J*. 2012;33(4):436–443. https://doi.org/10.1093/eurheartj/ehr362.
7. De Caterina R. N–3 fatty acids in cardiovascular disease. *N Engl J Med*. 2011;364(25):2439–2450. https://doi.org/10.1056/nejmra1008153.
8. O'Keefe JH, Jacob D, Lavie CJ. Omega-3 fatty acid therapy: the tide turns for a fish story. *Mayo Clin Proc*. 2017;92(1):1–3. https://doi.org/10.1016/j.mayocp.2016.11.008.
9. Simopoulos AP. The importance of the Omega-6/Omega-3 fatty acid ratio in cardiovascular disease and other chronic diseases. *Exp Biol Med*. 2008;233(6):674–688. https://doi.org/10.3181/0711-mr-311.
10. Marik PE, Varon J. Omega-3 dietary supplements and the risk of cardiovascular events: a systematic review. *Clin Cardiol*. 2009;32(7):365–372. https://doi.org/10.1002/clc.20604.
11. Harris WS, Mozaffarian D, Rimm E, et al. Omega-6 fatty acids and risk for cardiovascular disease: a science advisory from the American Heart Association nutrition Subcommittee of the Council on nutrition, physical activity, and metabolism; council on cardiovascular nursing; and council on epidemiology and prevention. *Circulation*. 2009;119(6):902–907. https://doi.org/10.1161/circulationaha.108.191627.
12. Lichtenstein AH, Appel LJ, Brands M, et al. Diet and lifestyle recommendations revision 2006: a scientific statement from the American Heart Association nutrition committee. *Circulation*. 2006;114(1):82–96. https://doi.org/10.1161/circulationaha.106.176158.
13. Dinicolantonio JJ, McCarty MF, Chatterjee S, Lavie CJ, O'Keefe JH. A higher dietary ratio of long-chain Omega-3 to Total Omega-6 fatty acids for prevention of COX-2-dependent adenocarcinomas. *Nutr Cancer*. 2014;66(8):1279–1284. https://doi.org/10.1080/01635581.2014.956262.
14. Raatz SK, Silverstein JT, Jahns L, Picklo MJ. Issues of fish consumption for cardiovascular disease risk reduction. *Nutrients*. 2013;5(4):1081–1097. https://doi.org/10.3390/nu5041081.
15. Coulter ID. The response of an expert panel to nutritional armor for the Warfighter: can Omega-3 fatty acids enhance stress resilience, wellness, and military performance? *Mil Med*. 2014;179(11S):192–198. https://doi.org/10.7205/MILMED-D-14-00189.
16. Kromhout D, Bosschieter EB, de Lezenne Coulander C. The inverse relation between fish consumption and 20-year mortality from coronary heart disease. *N Engl J Med*. 1985;312(19):1205–1209.
17. DiNicolantonio JJ, Niazi AK, McCarty MF, O'Keefe JH, Meier P, Lavie CJ. Omega-3s and cardiovascular health. *Ochsner J*. 2014;14(3):399–412.
18. Burton-Freeman B, Sesso HD. Whole food versus supplement: comparing the clinical evidence of tomato intake and lycopene supplementation on cardiovascular risk factors. *Adv Nutr*. 2014;5(5):457–485.
19. Finco AMI, Mamani LDG, Carvalho JC, de Melo Pereira GV, Thomaz-Soccol V, Soccol CR. Technological trends and market perspectives for production of microbial oils rich in omega-3. *Crit Rev Biotechnol*. 2017;37(5):656–671. https://doi.org/10.1080/07388551.2016.1213221.
20. Delarue J, Guriec N. Opportunities to enhance alternative sources of long-chain n-3 fatty acids within the diet. *Proc Nutr Soc*. 2014;73(3):376–384. https://doi.org/10.1017/S0029665114000123.
21. Von Schacky C. Omega-3 index and cardiovascular health. *Nutrients*. 2014;6(2):799–814. https://doi.org/10.3390/nu6020799.
22. Nigam A, Talajic M, Roy D, et al. Fish oil for the reduction of atrial fibrillation recurrence, inflammation and oxidative stress. *J Am Coll Cardiol*. 2014;64(14):1441–1448. https://doi.org/10.1016/j.jacc.2014.07.956.

23. Heydari B, Abdullah S, Pottala JV, et al. Effect of Omega-3 acid ethyl esters on left ventricular Remodeling after acute myocardial infarction: the OMEGA-REMODEL randomized clinical trial. *Circulation*. 2016;134(5):378–391. https://doi.org/10.1161/CIRCULATIONAHA.115.019949.
24. Kromhout D, Giltay EJ, Geleijnse JM. N–3 fatty acids and cardiovascular events after myocardial infarction. *N Engl J Med*. 2010;363(21):2015–2026. https://doi.org/10.1056/nejmoa1003603.
25. Balk EM, Lichtenstein AH. Omega-3 fatty acids and cardiovascular disease: summary of the 2016 Agency of Healthcare Research and Quality Evidence Review. *Nutrients*. 2017;9:865. https://doi.org/10.3390/nu9080865.
26. Rauch B, Schiele R, Schneider S, et al. OMEGA, a randomized, placebo-controlled trial to test the effect of highly purified omega-3 fatty acids on top of modern guideline-adjusted therapy after myocardial infarction. *Circulation*. 2010;122(21):2152–2159.
27. Maki KC, Palacios OM, Bell M, Toth PP. Use of supplemental long-chain omega-3 fatty acids and risk for cardiac death: an updated meta-analysis and review of research gaps. *J Clin Lipidol*. 2017;11(5):1152–1160. https://doi.org/10.1016/j.jacl.2017.07.010.
28. Goodfellow J, Bellamy MF, Ramsey MW, Jones CJ, Lewis MJ. Dietary supplementation with marine omega-3 fatty acids improve systemic large artery endothelial function in subjects with hypercholesterolemia. *J Am Coll Cardiol*. 2000;35(2):265–270. https://doi.org/10.1016/s0735-1097(99)00548-3.
29. ORIGIN Trial Investigators, Bosch J, Gerstein HC, et al. N–3 fatty acids and cardiovascular outcomes in patients with dysglycemia. *N Engl J Med*. 2012;367(4):309–318. https://doi.org/10.1056/nejmoa1203859.
30. Fares H, Lavie CJ, Dinicolantonio JJ, O'Keefe JH, Milani RV. Omega-3 fatty acids: a growing ocean of choices. *Curr Atheroscler Rep*. 2014;16(2):389.
31. Laidlaw M, Cockerline CA, Rowe WJ. A randomized clinical trial to determine the efficacy of manufacturers' recommended doses of omega-3 fatty acids from different sources in facilitating cardiovascular disease risk reduction. *Lipids Health Dis*. 2014;13:99. https://doi.org/10.1186/1476-511x-13-99.
32. Kim ES, McCormack PL. Icosapent ethyl: a review of its use in severe hypertriglyceridemia. *Am J Cardiovasc Drugs*. 2014;14(6):471–478. https://doi.org/10.1007/s40256-014-0099-7.
33. Ballantyne CM, Bays HE, Philip S, et al. Icosapent ethyl (eicosapentaenoic acid ethyl ester): effects on remnant-like particle cholesterol from the MARINE and ANCHOR studies. *Atherosclerosis*. 2016;253:81–87. https://doi.org/10.1016/j.atherosclerosis.2016.08.005.
34. Bhatt DL, Steg PG, Brinton EA, et al. Rationale and design of REDUCE-IT: reduction of cardiovascular events with icosapent ethyl-intervention trial. *Clin Cardiol*. 2017;40(3):138–148. https://doi.org/10.1002/clc.22692.
35. Siscovick DS, Barringer TA, Fretts AM, et al. Omega-3 polyunsaturated fatty acid (fish oil) supplementation and the prevention of clinical cardiovascular disease: a science advisory from the American Heart Association. *Circulation*. 2017;135(15):e867–e884. https://doi.org/10.1161/cir.0000000000000482.
36. Alexander DD, Miller PE, Elswyk ME, Kuratko CN, Bylsma LC. A meta-analysis of randomized controlled trials and prospective cohort studies of Eicosapentaenoic and docosahexaenoic long-chain Omega-3 fatty acids and coronary heart disease risk. *Mayo Clin Proc*. 2017;92(1):15–29. https://doi.org/10.1016/j.mayocp.2016.10.018.
37. Gajos G, Rostoff P, Undas A, Piwowarska W. Effects of polyunsaturated Omega-3 fatty acids on responsiveness to dual antiplatelet therapy in patients undergoing percutaneous coronary intervention: the OMEGA-PCI (OMEGA-3 fatty acids after PCI to modify responsiveness to dual antiplatelet therapy) study. *J Am Coll Cardiol*. 2010;55(16):1671–1678. https://doi.org/10.1016/j.jacc.2009.11.080.
38. Lev EI, Solodky A, Harel N, et al. Treatment of aspirin-resistant patients with Omega-3 fatty acids versus aspirin dose escalation. *J Am Coll Cardiol*. 2010;55(2):114–121. https://doi.org/10.1016/j.jvs.2010.06.089.
39. Harris WS. Expert opinion: Omega-3 fatty acids and bleeding—cause for concern? *Am J Cardiol*. 2007;99(6A):44C–46C.
40. Macchia A, Grancelli H, Varini S, et al. Omega-3 fatty acids for the prevention of recurrent symptomatic atrial fibrillation: results of the FORWARD (randomized trial to assess efficacy of PUFA for the maintenance of sinus rhythm in persistent atrial fibrillation) trial. *J Am Coll Cardiol*. 2013;61(4):463–468. https://doi.org/10.1016/j.jacc.2012.11.021.
41. Mozaffarian D, Wu JH, Marcia C, et al. Fish oil and post-operative atrial fibrillation a meta-analysis of randomized controlled trials. *J Am Coll Cardiol*. 2013;61(21):2194–2196. https://doi.org/10.1016/j.jacc.2013.02.045.
42. Djousse L, Akinkuolie AO, Wu JH, Ding EL, Gaziano JM. Fish consumption, omega-3 fatty acids and risk of heart failure: a meta-analysis. *Clin Nutr*. 2012;6:846–853. https://doi.org/10.1016/j.clnu.2012.05.010.

43. Carbone S, Canada JM, Buckley LF, et al. Dietary fat, sugar consumption, and cardiorespiratory fitness in patients with heart failure with preserved ejection fraction. *JACC: Basic to Translational Science*. 2017;2(5):513–525. https://doi.org/10.1016/j.jacbts.2017.06.009.
44. Andreassen AK, Hartmann A, Offstad J, Geiran O, Kvernebo K, Simonsen S. Hypertension prophylaxis with Omega-3 fatty acids in heart transplant recipients. *J Am Coll Cardiol*. 1997;29(6):1324–1331. https://doi.org/10.1016/s0735-1097(97)82757-x.
45. Ventura HO, Milani RV, Lavie CJ, et al. Cyclosporine-induced hypertension: efficacy of omega-3 fatty acids in patients after cardiac transplantation. *Circulation*. 1993;88(5 Pt 2):11281–11285.
46. Enns JE, Yeganeh A, Zarychanski R, et al. The impact of omega-3 polyunsaturated fatty acid supplementation on the incidence of cardiovascular events and complications in peripheral arterial disease: a systematic review and meta-analysis. *BMC Cardiovasc Disord*. 2014;14:70. https://doi.org/10.1186/1471-2261-14-70.
47. McCarty MF, DiNicolantonio JJ, Lavie CJ, O'Keefe JH. Omega-3 and prostate cancer: examining the pertinent evidence. *Mayo Clin Proc*. 2014;89(4):444–450. https://doi.org/10.1016/j.mayocp.2013.10.029.
48. American Cancer Society. Key Statistics for Prostate Cancer. https://www.cancer.org/cancer/prostate-cancer/about/key-statistics.html. Accessed December 20, 2017.
49. American Cancer Society. Cancer Facts & Figures 2018.
50. Zhang YF, Gao HF, Hou AJ, Zhou YH. Effect of omega-3 fatty acid supplementation on cancer incidence, non-vascular death, and total mortality: a meta-analysis of randomized controlled trials. *BMC Public Health*. 2014;14:204. https://doi.org/10.1186/1471-2458-14-204.
51. Aung T, Halsey J, Kromhout D, et al. Associations of Omega-3 fatty acid supplement use with cardiovascular disease risks. *JAMA Cardiol*. 2018. https://doi.org/10.1001/jamacardio.2017.5205.
52. Abdelhamid AS, Brown TJ, Brainard JS, et al. Omega-3 fatty acids for the primary and secondary prevention of cardiovascular disease. *Cochrane Database Syst Rev*. 2018;7:CD003177. https://doi.org/10.1002/14651858.CD003177.pub3.

PART C

DIET AND BRAIN DISEASE TREATMENT AND PREVENTION

CHAPTER

15

Dietary Strategies and Supplements for the Prevention of Cognitive Decline and Alzheimer's Disease

Ligia J. Dominguez, Mario Barbagallo

Geriatric Unit, Department of Internal Medicine and Geriatrics, University of Palermo, Palermo, Italy

INTRODUCTION

Aging of the world population is an unquestionable current reality. The recently published World Report on Aging and Health states that today most people around the world can expect to live over 60 years.[1] The worrying side of this successful story is that persons with cognitive decline will increase exponentially in the coming future in parallel with aging of the world population. This entails remarkable finance and human burden being dementia one of the greatest causes of disability.[2] Worldwide, 47 million persons are now affected by dementia; 60%–75% have Alzheimer's disease (AD), followed by vascular and Lewy bodies dementia.[3] A recent systematic review of 51 trials with low-moderate risk of bias concluded that the available evidence does not support the use of pharmacological treatments (i.e., dementia medications, antihypertensive and diabetes drugs, anti-inflammatory medications, and estrogen/progestin agents) for the prevention of cognitive decline in persons with normal cognition or with mild cognitive impairment (MCI).[4] Brain damage develops years before the manifestation of cognitive decline in old age, which offers feasibility to preventive or delaying measures. Two systematic reviews, one on cognitive training (11 trials),[5] and another on exercise (16 trials),[6] concluded that evidence for prevention of cognitive decline or dementia with these strategies is insufficient. Conversely, a multidomain intervention including physical activity, diet, and cognitive training improved several cognitive outcomes.[7]

Dietary factors are plausible modifiable contributors to age-associated cognitive decline.[8] Various minerals, micronutrients, and vitamins with antioxidant/anti-inflammatory have been studied. However, even if data suggests biological plausibility evidence on neuroprotective actions of these compounds have been judged as insufficient.[9] Furthermore, various

Omega Fatty Acids in Brain and Neurological Health
https://doi.org/10.1016/B978-0-12-815238-6.00015-8

dietary patterns have been studied in relation to their association with cognitive effects suggesting that any benefit of nutritional factors may derive from the synergistic interactions of diverse foods/nutrients.[10]

In this chapter, we review the evidence for possible effects of dietary components, supplements, and dietary patterns on cognitive decline and dementia, as well as possible mechanisms involved in these actions.

DIETARY COMPONENTS AND SUPPLEMENTS WITH POTENTIAL EFFECTS ON COGNITIVE DECLINE AND AD

Caffeine

There is evidence of antioxidant[11] and amyloid-beta suppressive properties of caffeine in animal models of AD.[12] It is well-known that coffee and caffeine strengthen the short-term memory and cognition in human beings, but there is limited evidence regarding the long-term effects. High serum levels of caffeine were associated with lack of progression to dementia in a case–control study of 124 older adults with MCI.[13] Another study showed a 65% reduction of incident AD and dementia in persons consuming 3–5 cups of coffee/day.[14] In the Honolulu-Asia Aging Study, deceased men in the highest quartile of caffeine intake were less likely to have neuropathological dementia lesions vs. men in the lowest quartile. However, coffee and caffeine consumption in midlife were not associated with the risk of cognitive impairment, AD, vascular dementia, or moderate/high levels of neuropathological lesions.[15]

A meta-analysis of few and largely heterogeneous studies evaluating the relationship of caffeine consumption and cognitive decline showed nonsignificant effects.[16] A Portuguese study reported an association of caffeine consumption with slower cognitive decline,[17] whereas French[18] and Finnish[19] studies found no association. A longitudinal study reported less cognitive decline among coffee consumers without dose response.[20] There is no evidence from RCTs.

Cocoa and Cocoa-Derived Products

Cocoa has been associated with cardiovascular (CV) benefits attributed to its rich content of flavonoids. Small acute and short-term chronic trials have suggested neuroprotective properties. A small study ($n=18$) reported significant increases in regional brain perfusion following consumption of a high flavanol drink (494 mg), particularly in the anterior cingulate cortex and the central opercular cortex of the parietal lobe measured with functional MRI.[21] Another small study of 34 healthy older adults showed an increased blood flow velocity in the middle cerebral artery after 1 (8%) and 2 weeks (10%) of flavanol-rich cocoa consumption using transcranial Doppler ultrasound.[22] A small RCT enrolling 37 healthy persons aged 50–69 years showed enhanced dentate gyrus function, as measured by functional MRI and cognitive testing, in those who consumed a high cocoa flavanol-containing diet for 3 months.[23] A prospective study including 531 participants aged ≥65 years followed for median 48 months showed that chocolate consumption was associated with a 41% lower risk of cognitive decline after adjustment for confounders.[24]

The mechanisms proposed to explain the potential benefit of cocoa and derived products include direct interactions with cellular signaling that promote neurogenesis, neuronal function and brain connectivity, and improved brain blood-flow and angiogenesis. This preliminary evidence underlines the need for further well-design and large RCTs on these compounds.

Curcuminoids

Curcumin is a potent antioxidant polyphenolic compound from the rhizome of *Curcuma longa* contained in curry spice turmeric, widely used in traditional Indian medicine.[25] A study by Ganguli et al.[26] reported a lower prevalence of AD in the Indian population compared to the US population. Ng et al. found that older healthy people who consume more frequently curry had better cognitive performance.[27] Numerous studies in experimental models confirmed the potent antioxidant and anti-inflammatory properties of curcumin and its AD protection actions. Baum et al.[28] enrolled 34 patients with AD receiving either 1 or 4g/day of curcumin or placebo for 6months and reported no significant effects in Mini Mental State Examination (MMSE). Ringman et al.,[29] randomized 36 patients with dementia to 2 or 4g/day of Curcumin-C3-Complex (95% of curcuminoids) or placebo for 24weeks, extended to 48weeks as an open-label trial, reporting no significant difference between treatment groups and placebo in any of the scores or cerebrospinal fluid markers. Hishikawa et al.[30] showed that three dementia patients treated with 100mg/day of curcumin and donepezil had a lower NPI score after 12weeks of therapy. Supplements of curcumin have limited bioavailability, which has been suggested to improve with lipidated formulations.[31]

Garlic (*Allium sativum*)

In experimental models, garlic extracts have shown antioxidant properties and protection against amyloid-beta-induced neurotoxic effects.[32] Allicin, an organosulfur compound contained in garlic, inhibits cholinesterase enzymes and upregulated brain acetylcholine levels in vitro.[33] There are no RCTs for these compounds. In the Doetinchem cohort study, including 2613 participants aged 43–70years, higher consumption of allium (onion, garlic, and leek) was associated with worse scores on cognitive flexibility and speed of cognitive processes in cross-sectional analyses, while in longitudinal analyses, allium consumption was not associated with cognitive decline.[34]

Ginkgo biloba

An RCT included 3069 older adults who were cognitively intact or had MCI at baseline, and randomized them to *Ginkgo biloba* treatment or placebo. After a median follow-up of 6.1 years the results indicated no difference in the rate of cognitive decline for the two treatments[35].

Magnesium

There is evidence showing that magnesium (Mg) deficiency results in increased free radicals production in various tissues, increased oxidative tissue damage, increased superoxide anion production by inflammatory cells, decreased antioxidant enzyme expression and activity, decreased cellular and tissue antioxidant levels, and increased oxygen peroxide production[36,37]; Mg is essential for synaptic conduction, affects *N*-methyl-D-aspartate receptor response to excitatory amino acids,[38] inhibits calcium channels, calcium influx, and glutamate release, and has effects on stability and viscosity of cell membranes.[36] Mg insufficiency produces vasospasm in cerebral arteries that regresses with elevated Mg concentrations.[39] With this basis, the role of Mg in cognitive decline and dementia has been examined in recent years. Earlier studies showed low serum[40] and autoptic brain tissue Mg concentrations in AD patients.[41] We observed reduced ionized free Mg concentrations in plasma obtained from AD patients,[42] which were related to cognitive dysfunction severity. Cilliler et al.[43] reported inverse associations of serum Mg levels with 2 rating AD severity scales, further confirming a potential protective role of Mg on cognitive function. There are no RCTs with Mg supplements in the prevention or therapy of cognitive disorders. Therefore, Mg's protective effects against AD remains to be further examined in well-design trials.

Omega-3 Fatty Acids

Polyunsaturated fatty acids (PUFAs) are essential constituents of neuronal cell membranes, preserving membrane fluidity for neurotransmitter communication and synaptic vesicle fusion. They may be precursors for lipid messengers in signaling processes that promote neuroprotection or neuronal damage.[44] In the aged brain, especially in AD, PUFAs tend to decrease in the hippocampus, cortex, and cerebellum.[45] Omega-3 PUFAs are widely studied regarding cognitive decline with dissimilar results. A systematic review reported data supporting a role for long-chain omega-3 PUFAs in the reduction of cognitive decline in persons without dementia.[46] Other trials have reported negative results. The "Supplementation with Folate, vitamin B6 and B12 and/or OMega-3 fatty acids" trial including 1748 participants with a history of CVD reported no significant effects of vitamin B and omega-3 fatty acid supplementation on cognitive function.[47] An RCT involving 302 cognitively intact persons aged ≥65 years comparing 1.8 g/day eicosapentaenoic (EPA)-docosahexaenoic (DHA), 0.4 g/day EPA-DHA, or placebo for 26 weeks found no significant effects.[48] A meta-analysis of results from 3536 persons aged >60 years cognitively intact at baseline supplemented with omega-3 PUFAs reported no significant effects on cognitive function.[49] A Chinese study showed that persons aged ≥65 years who consumed ≥100 g/week of fish had a reduction of ~65% in the mean annual rate of global cognitive decline with no associations among adults aged 55–64 years.[50] Recently, the "Multidomain Alzheimer Preventive Trial" reported that multidomain intervention and omega-3 PUFAs supplementation, either alone or in combination, had no significant effects on cognitive decline over 3 years in 1525 participants.[51]

Phytoestrogens

Soy isoflavones usually comprise combinations of them, such as genistein and daidzein, which impede to discriminate the effects of specific components. Studies of soy consumption and phytoestrogen supplements on cognitive function have shown inconclusive results.[52] Incongruities are attributed to the use of diverse isoflavone supplements at variable doses in small short-term trials. Negative trials are mostly from European cohorts, with low dietary soy consumption. Conversely, studies in older Asian populations with higher consumption of soy-derived foods have shown lower rates of cognitive decline. After oral intake of conjugated isoflavones, daidzein is converted to S-equol by a gut biotransformation. About 70% of older Japanese persons are "equol producers,"[53] which may help explain the different results in European and Asian studies.

In the Women's Isoflavone Soy Health trial, 313 postmenopausal women aged 45–92 years were randomized to receive 25 g/day of isoflavone-rich soy protein or milk protein-matched placebo. After 2.5 years, there were no differences between the groups in cognition changes from baseline. However, women within 5–10 years of menopause in the isoflavone group showed a marginal trend toward cognitive improvement, and a significant improvement in verbal episodic memory.[54] Further studies are necessary to validate these findings. No RCTs of phytoestrogens for the prevention or treatment of AD are available.

Resveratrol

This phenolic compound has antioxidant and anti-inflammatory actions.[55] Most dietary resveratrol in humans derives from grapes and red wine.[56] In animal models of AD resveratrol treatment has shown reduced hippocampal neurodegeneration and increased memory performance, but human clinical trials are limited. A placebo-control study ($n=22$ healthy adults) showed a dose-dependent (250–500 mg/day) increase in cerebral blood flow in the prefrontal cortex during cognitive tasks.[57] Another small trial ($n=23$ overweight persons, aged 50–75 years, receiving 200 mg/day of resveratrol for 26 weeks vs a group receiving placebo) reported improved memory performance and higher functional connectivity of the hippocampus in neuroimaging with resveratrol treatment.[58] Several clinical trials are currently investigating resveratrol's potential effects on cognitive function, MCI and AD.[59] At present, there is no sufficient evidence for prescribing this supplement to improve cognitive function.

Tea and (−)-Epigallocatechin-3-gallate

In Asian cultures, tea is considered a cognitive stimulator. This effect has been linked to antioxidants contained in tea, such as epigallocatechin-3-gallate (EGCG), L-theanine, and caffeine.[60] Tea may also be neuroprotective by regulation of stress hormones and inhibition of acetylcholinesterase.[61] However, there is hitherto inconsistent evidence with no definitive conclusion on its neuroprotective actions.[62] L-theanine and EGCG have shown anti-amyloidogenic and antioxidant properties in vitro, but the evidence for their use as nutraceuticals are limited.

Vitamins

Short-term beta-carotene (provitamin A) treatment showed no significant effect on cognitive function, while longer-term (18-year) administration showed beneficial effects in 4052 participants from the Physicians' Health Study and Physicians' Health Study II.[63]

Studies on cognitive effects of B-vitamins have produced variable results. An RCT of 299 men >75 years showed no significant effect on cognitive functions after 2-year supplementation with folic acid, vitamin B6, and B12.[64] A meta-analysis of 9 RCTs ($n=2835$) indicated no significant effect of folic acid with or without other B-vitamins on cognitive function.[65] Conversely, an RCT of 900 persons aged 60–74 years showed that folic acid+vitamin B12 were significantly superior vs. placebo for improving cognitive tests.[66]

Numerous studies have evaluated the combined effects of vitamins C and E on cognitive function in healthy populations with inconsistent results. The women's health study (6377 women >65 years) showed no significant effects on cognitive function of vitamin E supplementation.[67] Analyses of 616 over 65-years-old participants of the Duke Established Populations for Epidemiologic Studies showed that vitamins C and/or E supplements had no influence on incident dementia or AD.[68] Conversely, the Canadian study of health and aging (894 participants aged ≥65 years) reported a significant decreased rate of cognitive decline with a combination of vitamins C and E.[69] A recent study reported the relations of alpha- and gamma-tocopherol brain concentrations with AD neuropathology among 115 deceased participants of the prospective Rush Memory and Aging Project. Gamma-tocopherol concentrations were associated with lower amyloid load and lower neurofibrillary tangle severity, while alpha-tocopherol concentrations were not associated with neuropathology.[70]

The Third National Health and Nutrition Examination Survey showed an association of vitamin D deficiency with an increased risk of cognitive decline in the United States.[71] However, the Women's Health Initiative showed no significant effect of vitamin D and calcium supplementation on cognitive function in 2034 women aged over 65 years, followed for a mean of 7.8 years.[72]

Results from the age-related eye disease study including 2166 participants followed for 6.9 years showed that a combination of antioxidant vitamins (E, C, beta-carotene), zinc, and copper had no significant effects on any of six tests of cognitive function.[73] Similarly, the Women's Antioxidant CV study including 2824 women with or at risk for CVD who received a combination of vitamins E, C, and beta-carotene or placebo showed no effects on cognition with the multivitamin supplementation.[74]

Dietary Patterns

There is no single dietary, nutritional, or lifestyle intervention proven in RCTs to prevent incident AD. However, epidemiological data suggest that adopting a healthy, balanced diet and lifestyle known to improve CV risk may help delay incident AD.[75,76].

Mediterranean Diet

The eating and lifestyle pattern traditionally followed by populations from the Mediterranean basin has been widely reported as a model of healthy eating. Table 1 shows the main characteristics of Mediterranean diet (MeDiet). A greater adherence to MeDiet has been associated with a reduced incidence of AD, Parkinson's disease, overall mortality, CV mortality and events, and cancer mortality and/or incidence.[77] A systematic review of longitudinal cohorts exploring MeDiet effects on cognition found that participants in the highest tertile of adherence to MeDiet had 33% less risk of MCI or AD vs those in the lowest tertile.[78] The RCT PREDIMED (Prevención con Dieta Mediterranea), including adults aged 55–80 years at high CV risk, showed that participants on a MeDiet supplemented with either EVOO or nuts, had a 30% reduced incidence of CV events vs participants following a low-fat diet over a 5-year

TABLE 1 Characteristics of the Mediterranean Dietary Pattern and Lifestyle

- A variety of fresh fruits and vegetables; legumes, nuts, seeds every day
- Bread and other grain products (pasta, rice), mostly whole
- Herbs and spices to season food
- Cold pressed extra-virgin olive oil as the main source of fat
- Fresh fruit everyday as dessert; sweets, cakes and dairy desserts only on occasion
- Frequent consumption of fish and seafood (2–3 times per week)
- Dairy products on a daily basis, mainly yogurt (less often small portions of local cheese)
- Eggs (high quality proteins) 2–4 per week
- Red and processed meat consumed unfrequently, in moderate portions, if possible as a part of stews and other recipes (1–2 every month)
- Drinking plenty of water
- Wine in moderation with meals (≤1drink/day for women and 1–2 drinks/day for men), (respecting beliefs of each community and former habits)
- Foods that have undergone minimal processing that are fresh and locally produced
- Direct connection with nature
- Tasty cooking
- Moderation in portion size
- Moderate physical active every day is just as important as eating well
- Meals and cooking in the company of others
- Adequate rest (enough night sleeping time and eventually for small periods of time during the day [siesta]).

follow-up.[79] Two subanalyses of this RCT have reported improved cognitive function with the MeDiet supplemented with either EVOO oil or nuts vs low-fat diet.[80,81]

The effects of MeDiet on AD have been examined in longitudinal studies from the United States,[82–86] and France.[87,88] Two reviews[77,89] suggested that even with the evidence that adherence to MeDiet is associated with a reduced risk of AD, additional confirmation in populations with diverse ethnicities and dietary habits is needed. The personality and total health (PATH) through life longitudinal study[90] of healthy participants from Australia found no protection of MeDiet for cognitive decline, while the Australian imaging, biomarkers, and lifestyle (AIBL) study[91] showed that participants with AD and MCI had lower adherence to MeDiet vs. healthy controls. A 7-year longitudinal study[92] evaluated adherence to HEI-2005 or to MeDiet in relation to cognitive change in 3790 adults aged >65years from the ongoing Chicago health and aging project longitudinal study. The results showed that white participants had higher energy-adjusted MeDiet scores but lower HEI-2005 scores compared to black participants. Higher MeDiet scores, adjusted for confounders, were associated with slower rates of cognitive decline, but no association was found for HEI-2005 scores. MeDiet embraces other cultural and lifestyle components, such as physical activity, adequate rest, social engagement, and culinary activities.[93] These lifestyle factors may also influence cognitive functions. Future studies should consider all MeDiet lifestyle factors into their design.

Dietary Approaches to Stop Hypertension

This dietary pattern (Table 2) has shown to effectively lower CV outcomes.[94] Higher adherence to dietary approaches to stop hypertension (DASH) and MeDiet and greater consumption of whole grains, nuts, and legumes were significantly associated with higher average MMSE scores in a prospective study including 3831 participants aged >65years, followed for

TABLE 2 Characteristics of the Dietary Approaches to Stop Hypertension (DASH)

- Eating vegetables, fruits, and whole grains
- Including fat-free or low-fat dairy products, fish, poultry, beans, nuts, and vegetable oils
- Limiting foods that are high in saturated fat, such as fatty meats, full-fat dairy products, and tropical oils such as coconut, palm kernel, and palm oils
- Limiting sugar-sweetened beverages and sweets

>11 years.[95] A combined dietary pattern of MeDiet and DASH (called MIND) in 960 participants of the Memory and Aging Project followed for a mean of 4.7 years was associated with slower decline in global and five single cognitive domains, after adjusting for confounders.[96] There was a significant reduction in incident AD for the second and third tertile of adherence scores to MIND diet vs. the lowest tertile in 923 participants, aged 58–98 years, followed for average 4.5 years.[97] Future studies in diverse populations are warranted to confirm these positive results with DASH and MIND. *insert Table 2 here.

Other Dietary Patterns With Unconfirmed or Conflicting Effects on Cognitive Function or Prevention of Cognitive Decline

Several dietary patterns with confirmed positive effects for diverse health outcomes (e.g., total mortality, incident cancer, CV events, body weight control) have no definitive evidence on the prevention of cognitive decline or dementia.

- *Healthy Eating Index-2005 (HEI-2005)*—Two cross-sectional studies[98,99] showed associations of HEI-2005 and improved cognition. However, two longitudinal studies[92,100] followed up to 7.6 years showed no association. A longer study[101] (11-year follow-up) showed less cognitive decline with increased HEI scores, indicating that longer follow-up is needed to disclose effects of diet on cognition.
- *Ketogenic diet*—This dietary pattern has shown neuroprotective effects in some forms of resistant epilepsy. There is no solid evidence of AD prevention with ketogenic diet. A small RCT of 20 patients with MCI or AD showed memory scores improvement after 90 min of medium-chain triglycerides intake in non-*APOE* ε4 patients but not in *APOE* ε4 positive patients[102]. A clinical trial of 152 participants receiving medium-chain triglycerides for 90 days showed similar results.[103] Most findings with ketogenic diet come from animal studies and validation in human trials is not yet available.
- *Low-copper diet*—A longitudinal study ($n=3718$) showed that high intakes of copper, mostly supplements, together with a high-fat diet resulted in accelerated cognitive decline after 5.5 years.[104] Among 32 patients with mild to moderate AD, those with low plasma Cu had higher ADAS-cog scores.[105] Squitti et al. have proposed a "copper phenotype" as a basis for a diet-gene interplay in the genesis of AD, and promote a low-Cu diet to reduce the risk of developing the disease.[106] Yet, there is no evidence that depletion or supplementation of Cu may modify AD incidence or pathology.
- *Okinawa diet*—There is still no solid evidence for the effects of Okinawa diet on the prevention of cognitive decline and AD despite this dietary pattern is followed by populations with some of the longest and healthiest life expectancies in the world. One study comparing 157 migrants from Okinawa to Brazil vs 2217 residents of Okinawa

showed a higher incident dementia in the migrants living in Brazil.[107]. The 9-year difference between the studies and dissimilar sample size may have affected the results.

- *Paleolithic diet*—A study of 20 overweight postmenopausal women with a 6-month follow-up compared the effects of Paleolithic diet vs Nordic Nutrition Recommendations on weight loss, functional MRI parameters, and episodic memory tasks. There was a significant improvement in episodic memory performance after both dietary interventions, in parallel with a decreased waist circumference, increased hippocampal activity, and reduced plasma free fatty acid without differences between the diets.[108] There are no studies on the effects of Paleolithic diet on incident cognitive decline or AD.
- *Vegetarian diets*—Two sub-studies of the adventist health study explored the incidence of dementia in those consuming vegetarian diets vs meat-eating diets.[109] One study showed that meat-eating persons were > twofold likely to become demented vs their vegetarian counterparts, when matched for age, sex, and zip code. A second analysis showed no significant difference in incident dementia between vegetarian and meat-eating persons in unmatched participants. There was no clear evidence of any standardized cognitive assessment during the study.

POSSIBLE MECHANISMS MEDIATING THE EFFECTS OF DIETARY/METABOLIC FACTORS ON COGNITIVE DECLINE AND AD

The precise mechanisms by which dietary/nutritional factors may modify cognitive function and potentially prevent cognitive decline and incident AD are not completely clear. However, a body of accumulated research suggests plausible mechanisms of these actions, as follows:

- *Overweight/obesity in adulthood and late-onset cognitive decline*: traditional CV and metabolic risk, frequently linked to overweight and obesity, are risk factors for the development of several chronic diseases, including cognitive decline and AD (Fig. 1). A systematic review of 28 longitudinal studies conducted from 2003 to 2013 with follow-up of 5–40 years reported a 144% increased risk of late-onset dementia in overweight and obese participants in midlife vs. lean controls.[110] Most analyses were adjusted for vascular confounders (i.e., hypertension, dyslipidemia, diabetes); hence, other mechanisms such as a higher level of plasma amyloid proteins,[111] and impaired blood brain barrier[112] may have contributed.

Macrophages in adipose tissue can secrete pro- and anti-inflammatory cytokines (Fig. 1), which maintain low-grade systemic chronic inflammation and neuroinflammation. Obesity may trigger a switch from M2 macrophages to the proinflammatory M1 phenotype.[113]

Dietary fats, particularly saturated fatty acids, induce microglia inflammatory responses, with local cytokine production and consequent apoptosis of neurons involved in body weight control, energy balance, blood pressure, and glucose homeostasis.[114] Instead, weight reduction and frequent consumption of antioxidant-rich foods have been associated with reduced markers of inflammation.[115] Diet and exercise interventions that improved insulin resistance have been associated with decreased inflammatory cytokines and improved cognition.[116]

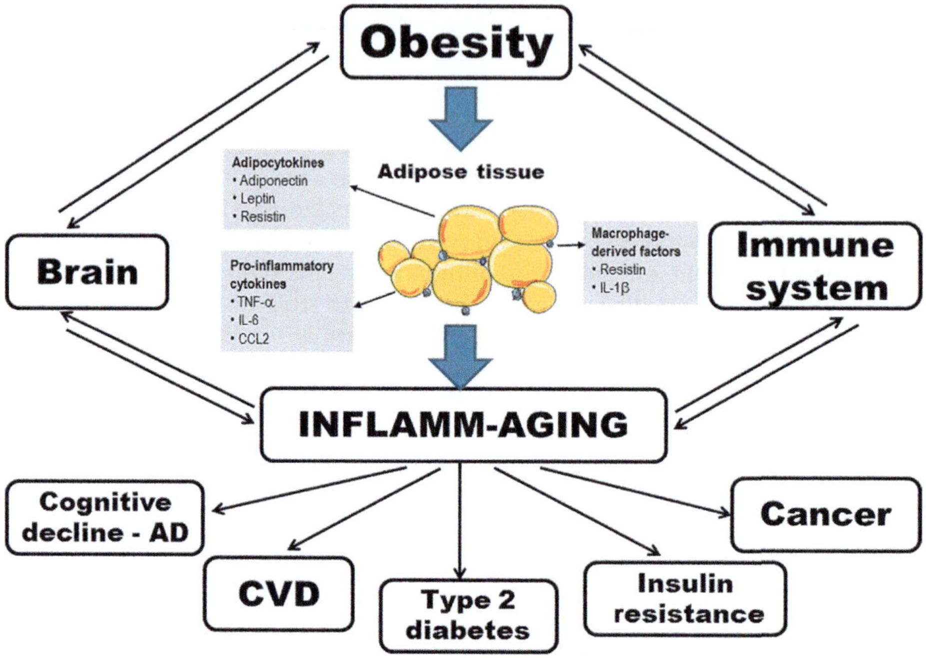

FIG. 1 Role of obesity in the genesis of inflamm-aging and related chronic diseases.

- *Oxidative stress and chronic metabolic inflammation*: a pro-oxidant, pro-inflammatory state called "inflammating," is characteristic of aging and age-related degenerative diseases. Brain tissue is particularly susceptible to oxidative damage and has low levels of antioxidant systems, rendering brain prone to damage from neurotoxic peptides such as amyloid-beta and neuroinflammation. Aggregated/misfolded proteins, such as amyloid, bind to microglia toll-like receptors and CD4, triggering innate immune responses with production of inflammatory mediators (e.g., cytokines, nitric oxide). Amyloid-beta accumulation induces neuroinflammation, which consecutively produces more amyloid-beta, propagating the injury[117] (Fig. 2). Gut microbiome and neuroinflammation are closely related with neurodegenerative diseases, although the specific mechanism is not yet completely clear (Fig. 3).
- *Autophagy and prothrombotic state*: the efficiency of degradation and recycling of cellular proteins, including amyloid and tau proteins, by autophagy decreases with age.

Several molecular mechanisms, such as mTOR pathway, neuroinflammation, and endocannabinoid system, among others, may be implicated in the pathogenesis of AD, which may be modulated by nutritional and metabolic factors.[118] Vascular factors and a prothrombotic state (e.g., amyloid beta fibrillization and platelet activation) may also contribute[119].

- *Epigenetic transformations*: these modifications occur over time in response to environmental effects. AD and other dementias have been linked to epigenetic transformations, which are latent until a pathological threshold is reached corresponding to the clinically apparent onset of dementia.[120] (Fig. 4) This implies that dementia is

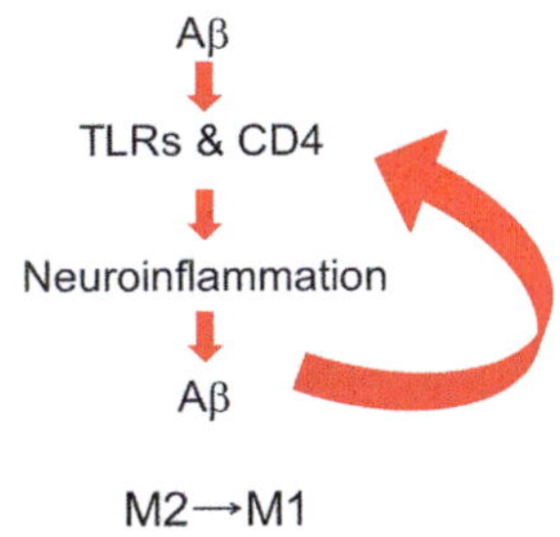

FIG. 2 Aggregated and misfolded proteins, such as amyloid beta (Aβ), bind to microglia toll-like receptors (TLRs) and cluster of differentiation (CD)4, initiating innate immune responses with production of inflammatory mediators. Amyloid-beta may lead to a preponderance of Macrophage 1 (pro-inflammatory cells) vs M2 (anti-inflammatory cells) and loss of the ability to switch phenotypes and lessen destruction.

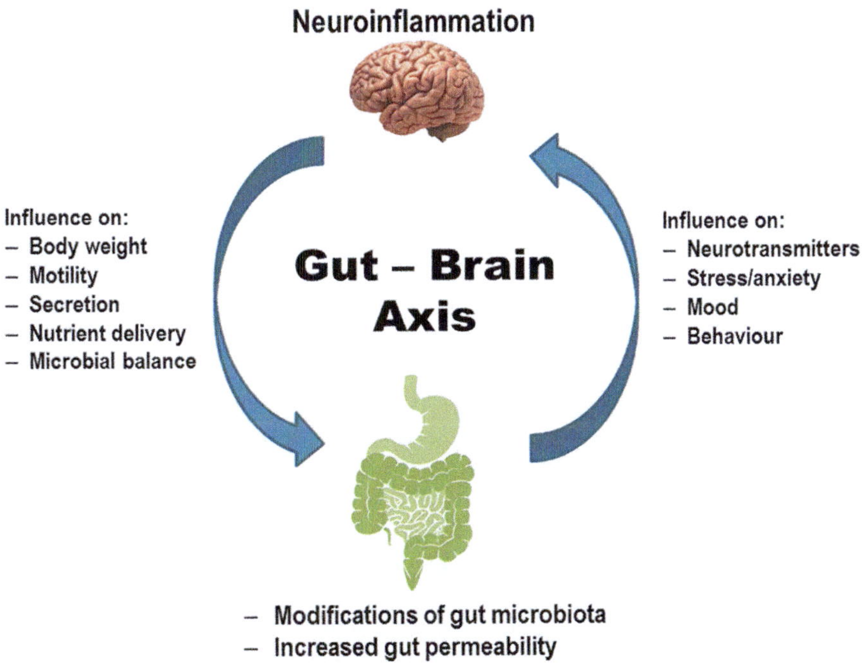

FIG. 3 Gut-brain axis is a proposed bidirectional neurohumoral connection, which maintains homeostasis and regulates function in both inter-related systems. Modifications of gut microbiota and increased gut permeability ("leaky gut") have been implicated in the genesis neuroinflammation.

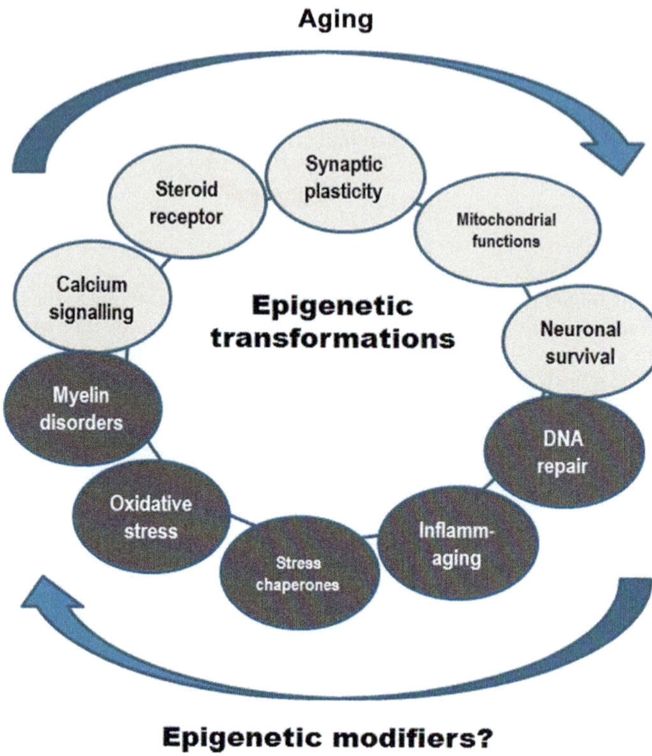

FIG. 4 Possible epigenetic transformations and hypothetical epigenetic modifiers.

not a rapidly occurring state, but rather a gradual change in crucial cellular pathways, transforming a healthy state, in a dysfunctional state as a result of neurodegeneration. This fascinating hypothesis still needs verification.

CONCLUSIONS

Currently available evidence from large-scale observational studies and clinical trials overall do not demonstrate an independent role for most nutritional and dietary factors in the prevention of cognitive decline or AD. Further research is still warranted to clarify the potential benefits of the single dietary/nutrient components.

Two healthy dietary patterns, MeDiet and DASH, and their combination have moderately convincing evidence that adherence to these eating patterns is associated with a reduced risk of AD. However, confirmation of this evidence in different populations with diverse dietary behaviors is still needed. At present, there is no definitive evidence to support the effects of HEI-2005, ketogenic, low-copper, Okinawa, Paleolithic, or vegetarian dietary patterns on the prevention of cognitive decline or AD.

CONFLICT OF INTEREST

None of the authors has a conflict of interest to report.

References

1. Beard JR, Officer A, de Carvalho IA, et al. The world report on ageing and health: a policy framework for healthy ageing. *Lancet*. 2016;387:2145–2154.
2. Black CM, Fillit H, Xie L, et al. Economic burden, mortality, and institutionalization in patients newly diagnosed with Alzheimer's disease. *J Alzheimers Dis*. 2018;61:185–193.
3. 2014 Alzheimer's disease facts and figures. *Alzheimers Dement*. 2014;10. e47–92.
4. Fink HA, Jutkowitz E, McCarten JR, et al. Pharmacologic interventions to prevent cognitive decline, mild cognitive impairment, and clinical Alzheimer-type dementia: a systematic review. *Ann Intern Med*. 2018;168:39–51.
5. Butler M, McCreedy E, Nelson VA, et al. Does cognitive training prevent cognitive decline? A systematic review. *Ann Intern Med*. 2018;168:63–68.
6. Brasure M, Desai P, Davila H, et al. Physical activity interventions in preventing cognitive decline and Alzheimer-type dementia: a systematic review. *Ann Intern Med*. 2018;168:30–38.
7. Ngandu T, Lehtisalo J, Solomon A, et al. A 2 year multidomain intervention of diet, exercise, cognitive training, and vascular risk monitoring versus control to prevent cognitive decline in at-risk elderly people (finger): a randomised controlled trial. *Lancet*. 2015;385:2255–2263.
8. Norton S, Matthews FE, Barnes DE, Yaffe K, Brayne C. Potential for primary prevention of Alzheimer's disease: an analysis of population-based data. *Lancet Neurol*. 2014;13:788–794.
9. Butler M, Nelson VA, Davila H, et al. Over-the-counter supplement interventions to prevent cognitive decline, mild cognitive impairment, and clinical alzheimer-type dementia: a systematic review. *Ann Intern Med*. 2018;168:52–62.
10. Jacobs Jr DR, Orlich MJ. Diet pattern and longevity: Do simple rules suffice? A commentary. *Am J Clin Nutr*. 2014;100(Suppl 1):313S–319S.
11. Azam S, Hadi N, Khan NU, Hadi SM. Antioxidant and prooxidant properties of caffeine, theobromine and xanthine. *Med Sci Monit*. 2003;9:BR325–330.
12. Arendash GW, Mori T, Cao C, et al. Caffeine reverses cognitive impairment and decreases brain amyloid-beta levels in aged Alzheimer's disease mice. *J Alzheimers Dis*. 2009;17:661–680.
13. Cao C, Loewenstein DA, Lin X, et al. High blood caffeine levels in mci linked to lack of progression to dementia. *J Alzheimers Dis*. 2012;30:559–572.
14. Eskelinen MH, Ngandu T, Tuomilehto J, Soininen H, Kivipelto M. Midlife coffee and tea drinking and the risk of late-life dementia: a population-based caide study. *J Alzheimers Dis*. 2009;16:85–91.
15. Gelber RP, Petrovitch H, Masaki KH, Ross GW, White LR. Coffee intake in midlife and risk of dementia and its neuropathologic correlates. *J Alzheimers Dis*. 2011;23:607–615.
16. Santos C, Costa J, Santos J, Vaz-Carneiro A, Lunet N. Caffeine intake and dementia: systematic review and meta-analysis. *J Alzheimers Dis*. 2010;20(Suppl 1):S187–S204.
17. Santos C, Lunet N, Azevedo A, de Mendonca A, Ritchie K, Barros H. Caffeine intake is associated with a lower risk of cognitive decline: a cohort study from Portugal. *J Alzheimers Dis*. 2010;20(Suppl 1):S175–S185.
18. Ritchie K, Carriere I, de Mendonca A, et al. The neuroprotective effects of caffeine: a prospective population study (the three city study). *Neurology*. 2007;69:536–545.
19. Laitala VS, Kaprio J, Koskenvuo M, Raiha I, Rinne JO, Silventoinen K. Coffee drinking in middle age is not associated with cognitive performance in old age. *Am J Clin Nutr*. 2009;90:640–646.
20. Arab L, Biggs ML, O'Meara ES, Longstreth WT, Crane PK, Fitzpatrick AL. Gender differences in tea, coffee, and cognitive decline in the elderly: the cardiovascular health study. *J Alzheimers Dis*. 2011;27:553–566.
21. Lamport DJ, Pal D, Moutsiana C, et al. The effect of flavanol-rich cocoa on cerebral perfusion in healthy older adults during conscious resting state: a placebo controlled, crossover, acute trial. *Psychopharmacology (Berl)*. 2015;232:3227–3234.
22. Sorond FA, Lipsitz LA, Hollenberg NK, Fisher ND. Cerebral blood flow response to flavanol-rich cocoa in healthy elderly humans. *Neuropsychiatr Dis Treat*. 2008;4:433–440.

23. Brickman AM, Khan UA, Provenzano FA, et al. Enhancing dentate gyrus function with dietary flavanols improves cognition in older adults. *Nat Neurosci*. 2014;17:1798–1803.
24. Moreira A, Diogenes MJ, de Mendonca A, Lunet N, Barros H. Chocolate consumption is associated with a lower risk of cognitive decline. *J Alzheimers Dis*. 2016;53:85–93.
25. Sikora E, Scapagnini G, Barbagallo M. Curcumin, inflammation, ageing and age-related diseases. *Immun Ageing*. 2010;7:1.
26. Ganguli M, Chandra V, Kamboh MI, et al. Apolipoprotein e polymorphism and Alzheimer disease: the indo-us cross-national dementia study. *Arch Neurol*. 2000;57:824–830.
27. Ng TP, Chiam PC, Lee T, Chua HC, Lim L, Kua EH. Curry consumption and cognitive function in the elderly. *Am J Epidemiol*. 2006;164:898–906.
28. Baum L, Lam CW, Cheung SK, et al. Six-month randomized, placebo-controlled, double-blind, pilot clinical trial of curcumin in patients with alzheimer disease. *J Clin Psychopharmacol*. 2008;28:110–113.
29. Ringman JM, Frautschy SA, Teng E, et al. Oral curcumin for alzheimer's disease: tolerability and efficacy in a 24-week randomized, double blind, placebo-controlled study. *Alzheimers Res Ther*. 2012;4:43.
30. Hishikawa N, Takahashi Y, Amakusa Y, et al. Effects of turmeric on alzheimer's disease with behavioral and psychological symptoms of dementia. *Ayu*. 2012;33:499–504.
31. Begum AN, Jones MR, Lim GP, et al. Curcumin structure-function, bioavailability, and efficacy in models of neuroinflammation and alzheimer's disease. *J Pharmacol Exp Ther*. 2008;326:196–208.
32. Ray B, Chauhan NB, Lahiri DK. The "aged garlic extract:" (age) and one of its active ingredients s-allyl-l-cysteine (sac) as potential preventive and therapeutic agents for alzheimer's disease (ad). *Curr Med Chem*. 2011;18:3306–3313.
33. Kumar S. Dual inhibition of acetylcholinesterase and butyrylcholinesterase enzymes by allicin. *Indian J Pharmacol*. 2015;47:444–446.
34. Nooyens AC, Bueno-de-Mesquita HB, van Boxtel MP, van Gelder BM, Verhagen H, Verschuren WM. Fruit and vegetable intake and cognitive decline in middle-aged men and women: the Doetinchem cohort study. *Br J Nutr*. 2011;106:752–761.
35. Snitz BE, O'Meara ES, Carlson MC, et al. Ginkgo biloba for preventing cognitive decline in older adults: a randomized trial. *JAMA*. 2009;302:2663–2670.
36. Barbagallo M, Belvedere M, Dominguez LJ. Magnesium homeostasis and aging. *Magnes Res*. 2009;22:235–246.
37. Barbagallo M, Dominguez LJ. Magnesium and aging. *Curr Pharm Des*. 2010;16:832–839.
38. Choi DW. Ionic dependence of glutamate neurotoxicity. *J Neurosci*. 1987;7:369–379.
39. Altura BT, Altura BM. Withdrawal of magnesium causes vasospasm while elevated magnesium produces relaxation of tone in cerebral arteries. *Neurosci Lett*. 1980;20:323–327.
40. Lemke MR. Plasma magnesium decrease and altered calcium/magnesium ratio in severe dementia of the alzheimer type. *Biol Psychiatry*. 1995;37:341–343.
41. Durlach J. Magnesium depletion and pathogenesis of Alzheimer's disease. *Magnes Res*. 1990;3:217–218.
42. Barbagallo M, Belvedere M, Di Bella G, Dominguez LJ. Altered ionized magnesium levels in mild-to-moderate Alzheimer's disease. *Magnes Res*. 2011;24:S115–S121.
43. Cilliler AE, Ozturk S, Ozbakir S. Serum magnesium level and clinical deterioration in alzheimer's disease. *Gerontology*. 2007;53:419–422.
44. Bazan NG. Lipid signaling in neural plasticity, brain repair, and neuroprotection. *Mol Neurobiol*. 2005;32:89–103.
45. Liu JJ, Green P, John Mann J, Rapoport SI, Sublette ME. Pathways of polyunsaturated fatty acid utilization: implications for brain function in neuropsychiatric health and disease. *Brain Res*. 2015;1597:220–246.
46. Fotuhi M, Mohassel P, Yaffe K. Fish consumption, long-chain omega-3 fatty acids and risk of cognitive decline or Alzheimer disease: a complex association. *Nat Clin Pract Neurol*. 2009;5:140–152.
47. Andreeva VA, Kesse-Guyot E, Barberger-Gateau P, Fezeu L, Hercberg S, Galan P. Cognitive function after supplementation with b vitamins and long-chain omega-3 fatty acids: ancillary findings from the su.Fol.Om3 randomized trial. *Am J Clin Nutr*. 2011;94:278–286.
48. van de Rest O, Geleijnse JM, Kok FJ, et al. Effect of fish oil on cognitive performance in older subjects: a randomized, controlled trial. *Neurology*. 2008;71:430–438.
49. Sydenham E, Dangour AD, Lim WS. Omega 3 fatty acid for the prevention of cognitive decline and dementia. *Cochrane Database Syst Rev*. 2012;6:CD005379.
50. Qin B, Plassman BL, Edwards LJ, Popkin BM, Adair LS, Mendez MA. Fish intake is associated with slower cognitive decline in chinese older adults. *J Nutr*. 2014;144:1579–1585.

51. Andrieu S, Guyonnet S, Coley N, et al. Effect of long-term omega 3 polyunsaturated fatty acid supplementation with or without multidomain intervention on cognitive function in elderly adults with memory complaints (mapt): a randomised, placebo-controlled trial. *Lancet Neurol.* 2017;16:377–389.
52. Soni M, Rahardjo TB, Soekardi R, et al. Phytoestrogens and cognitive function: a review. *Maturitas.* 2014;77:209–220.
53. Jackson RL, Greiwe JS, Schwen RJ. Emerging evidence of the health benefits of s-equol, an estrogen receptor beta agonist. *Nutr Rev.* 2011;69:432–448.
54. Henderson VW, St John JA, Hodis HN, et al. Long-term soy isoflavone supplementation and cognition in women: a randomized, controlled trial. *Neurology.* 2012;78:1841–1848.
55. Park EJ, Pezzuto JM. The pharmacology of resveratrol in animals and humans. *Biochim Biophys Acta.* 1852;2015:1071–1113.
56. Baur JA, Sinclair DA. Therapeutic potential of resveratrol: the in vivo evidence. *Nat Rev Drug Discov.* 2006;5:493–506.
57. Kennedy DO, Wightman EL, Reay JL, et al. Effects of resveratrol on cerebral blood flow variables and cognitive performance in humans: a double-blind, placebo-controlled, crossover investigation. *Am J Clin Nutr.* 2010;91:1590–1597.
58. Witte AV, Kerti L, Margulies DS, Floel A. Effects of resveratrol on memory performance, hippocampal functional connectivity, and glucose metabolism in healthy older adults. *J Neurosci.* 2014;34:7862–7870.
59. Tome-Carneiro J, Larrosa M, Gonzalez-Sarrias A, Tomas-Barberan FA, Garcia-Conesa MT, Espin JC. Resveratrol and clinical trials: the crossroad from in vitro studies to human evidence. *Curr Pharm Des.* 2013;19:6064–6093.
60. Einother SJ, Martens VE. Acute effects of tea consumption on attention and mood. *Am J Clin Nutr.* 2013;98:1700S–1708S.
61. Chen D, Milacic V, Chen MS, et al. Tea polyphenols, their biological effects and potential molecular targets. *Histol Histopathol.* 2008;23:487–496.
62. Song J, Xu H, Liu F, Feng L. Tea and cognitive health in late life: current evidence and future directions. *J Nutr Health Aging.* 2012;16:31–34.
63. Grodstein F, Kang JH, Glynn RJ, Cook NR, Gaziano JM. A randomized trial of beta carotene supplementation and cognitive function in men: the physicians' health study ii. *Arch Intern Med.* 2007;167:2184–2190.
64. Ford AH, Flicker L, Alfonso H, et al. Vitamins b(12), b(6), and folic acid for cognition in older men. *Neurology.* 2010;75:1540–1547.
65. Wald DS, Kasturiratne A, Simmonds M. Effect of folic acid, with or without other b vitamins, on cognitive decline: meta-analysis of randomized trials. *Am J Med.* 2010;123:522–527 [e522].
66. Walker JG, Batterham PJ, Mackinnon AJ, et al. Oral folic acid and vitamin b-12 supplementation to prevent cognitive decline in community-dwelling older adults with depressive symptoms--the beyond ageing project: a randomized controlled trial. *Am J Clin Nutr.* 2012;95:194–203.
67. Kang JH, Cook N, Manson J, Buring JE, Grodstein F. A randomized trial of vitamin e supplementation and cognitive function in women. *Arch Intern Med.* 2006;166:2462–2468.
68. Fillenbaum GG, Kuchibhatla MN, Hanlon JT, et al. Dementia and alzheimer's disease in community-dwelling elders taking vitamin c and/or vitamin e. *Ann Pharmacother.* 2005;39:2009–2014.
69. Maxwell CJ, Hicks MS, Hogan DB, Basran J, Ebly EM. Supplemental use of antioxidant vitamins and subsequent risk of cognitive decline and dementia. *Dement Geriatr Cogn Disord.* 2005;20:45–51.
70. Morris MC, Schneider JA, Li H, et al. Brain tocopherols related to alzheimer's disease neuropathology in humans. *Alzheimers Dement.* 2015;11:32–39.
71. Llewellyn DJ, Lang IA, Langa KM, Melzer D. Vitamin d and cognitive impairment in the elderly U.S. population. *J Gerontol A Biol Sci Med Sci.* 2011;66:59–65.
72. Rossom RC, Espeland MA, Manson JE, et al. Calcium and vitamin d supplementation and cognitive impairment in the women's health initiative. *J Am Geriatr Soc.* 2012;60:2197–2205.
73. Yaffe K, Clemons TE, McBee WL, Lindblad AS. Impact of antioxidants, zinc, and copper on cognition in the elderly: a randomized, controlled trial. *Neurology.* 2004;63:1705–1707.
74. Kang JH, Cook NR, Manson JE, Buring JE, Albert CM, Grodstein F. Vitamin e, vitamin c, beta carotene, and cognitive function among women with or at risk of cardiovascular disease: the women's antioxidant and cardiovascular study. *Circulation.* 2009;119:2772–2780.
75. Mucke L. Neuroscience: Alzheimer's disease. *Nature.* 2009;461:895–897.
76. The LN. Finding a cure for Alzheimer's disease starts with prevention. *Lancet Neurol.* 2016;15:649.

77. Sofi F, Abbate R, Gensini GF, Casini A. Accruing evidence on benefits of adherence to the mediterranean diet on health: an updated systematic review and meta-analysis. *Am J Clin Nutr*. 2010;92:1189–1196.
78. Singh B, Parsaik AK, Mielke MM, et al. Association of mediterranean diet with mild cognitive impairment and Alzheimer's disease: a systematic review and meta-analysis. *J Alzheimers Dis*. 2014;39:271–282.
79. Estruch R, Ros E, Salas-Salvado J, et al. Primary prevention of cardiovascular disease with a mediterranean diet. *N Engl J Med*. 2013;368:1279–1290.
80. Valls-Pedret C, Sala-Vila A, Serra-Mir M, et al. Mediterranean diet and age-related cognitive decline: A randomized clinical trial. *JAMA Intern Med*. 2015;175:1094–1103.
81. Martinez-Lapiscina EH, Clavero P, Toledo E, et al. Mediterranean diet improves cognition: the predimed-Navarra randomised trial. *J Neurol Neurosurg Psychiatry*. 2013;84:1318–1325.
82. Scarmeas N, Stern Y, Tang MX, Mayeux R, Luchsinger JA. Mediterranean diet and risk for Alzheimer's disease. *Ann Neurol*. 2006;59:912–921.
83. Scarmeas N, Luchsinger JA, Mayeux R, Stern Y. Mediterranean diet and alzheimer disease mortality. *Neurology*. 2007;69:1084–1093.
84. Scarmeas N, Stern Y, Mayeux R, Manly JJ, Schupf N, Luchsinger JA. Mediterranean diet and mild cognitive impairment. *Arch Neurol*. 2009;66:216–225.
85. Scarmeas N, Luchsinger JA, Schupf N, et al. Physical activity, diet, and risk of alzheimer disease. *JAMA*. 2009;302:627–637.
86. Gu Y, Luchsinger JA, Stern Y, Scarmeas N. Mediterranean diet, inflammatory and metabolic biomarkers, and risk of alzheimer's disease. *J Alzheimers Dis*. 2010;22:483–492.
87. Feart C, Samieri C, Rondeau V, et al. Adherence to a mediterranean diet, cognitive decline, and risk of dementia. *JAMA*. 2009;302:638–648.
88. Feart C, Torres MJ, Samieri C, et al. Adherence to a mediterranean diet and plasma fatty acids: data from the Bordeaux sample of the three-city study. *Br J Nutr*. 2011;106:149–158.
89. Solfrizzi V, Panza F, Frisardi V, et al. Diet and Alzheimer's disease risk factors or prevention: the current evidence. *Expert Rev Neurother*. 2011;11:677–708.
90. Cherbuin N, Anstey KJ. The mediterranean diet is not related to cognitive change in a large prospective investigation: the path through life study. *Am J Geriatr Psychiatry*. 2012;20:635–639.
91. Gardener S, Gu Y, Rainey-Smith SR, et al. Adherence to a mediterranean diet and Alzheimer's disease risk in an Australian population. *Transl Psychiatry*. 2012;2:e164.
92. Tangney CC, Kwasny MJ, Li H, Wilson RS, Evans DA, Morris MC. Adherence to a mediterranean-type dietary pattern and cognitive decline in a community population. *Am J Clin Nutr*. 2011;93:601–607.
93. Yannakoulia M, Kontogianni M, Scarmeas N. Cognitive health and mediterranean diet: Just diet or lifestyle pattern? *Ageing Res Rev*. 2015;20:74–78.
94. Tyson CC, Nwankwo C, Lin PH, Svetkey LP. The dietary approaches to stop hypertension (dash) eating pattern in special populations. *Curr Hypertens Rep*. 2012;14:388–396.
95. Wengreen H, Munger RG, Cutler A, et al. Prospective study of dietary approaches to stop hypertension- and mediterranean-style dietary patterns and age-related cognitive change: the cache county study on memory, health and aging. *Am J Clin Nutr*. 2013;98:1263–1271.
96. Morris MC, Tangney CC, Wang Y, et al. Mind diet slows cognitive decline with aging. *Alzheimers Dement*. 2015;11:1015–1022.
97. Morris MC, Tangney CC, Wang Y, Sacks FM, Bennett DA, Aggarwal NT. Mind diet associated with reduced incidence of Alzheimer's disease. *Alzheimers Dement*. 2015;11:1007–1014.
98. Huijbregts PP, Feskens EJ, Rasanen L, et al. Dietary patterns and cognitive function in elderly men in Finland, Italy and the Netherlands. *Eur J Clin Nutr*. 1998;52:826–831.
99. Correa Leite ML, Nicolosi A, Cristina S, Hauser WA, Nappi G. Nutrition and cognitive deficit in the elderly: a population study. *Eur J Clin Nutr*. 2001;55:1053–1058.
100. Shatenstein B, Ferland G, Belleville S, et al. Diet quality and cognition among older adults from the nuage study. *Exp Gerontol*. 2012;47:353–360.
101. Wengreen HJ, Neilson C, Munger R, Corcoran C. Diet quality is associated with better cognitive test performance among aging men and women. *J Nutr*. 2009;139:1944–1949.
102. Reger MA, Henderson ST, Hale C, et al. Effects of beta-hydroxybutyrate on cognition in memory-impaired adults. *Neurobiol Aging*. 2004;25:311–314.

103. Henderson ST, Vogel JL, Barr LJ, Garvin F, Jones JJ, Costantini LC. Study of the ketogenic agent ac-1202 in mild to moderate alzheimer's disease: a randomized, double-blind, placebo-controlled, multicenter trial. *Nutr Metab (Lond)*. 2009;6:31.
104. Morris MC, Evans DA, Tangney CC, et al. Dietary copper and high saturated and trans fat intakes associated with cognitive decline. *Arch Neurol*. 2006;63:1085–1088.
105. Pajonk FG, Kessler H, Supprian T, et al. Cognitive decline correlates with low plasma concentrations of copper in patients with mild to moderate alzheimer's disease. *J Alzheimers Dis*. 2005;8:23–27.
106. Squitti R, Siotto M, Polimanti R. Low-copper diet as a preventive strategy for Alzheimer's disease. *Neurobiol Aging*. 2014;35(Suppl 2):S40–S50.
107. Yamada T, Kadekaru H, Matsumoto S, et al. Prevalence of dementia in the older Japanese-Brazilian population. *Psychiatry Clin Neurosci*. 2002;56:71–75.
108. Boraxbekk CJ, Stomby A, Ryberg M, et al. Diet-induced weight loss alters functional brain responses during an episodic memory task. *Obes Facts*. 2015;8:261–272.
109. Giem P, Beeson WL, Fraser GE. The incidence of dementia and intake of animal products: preliminary findings from the adventist health study. *Neuroepidemiology*. 1993;12:28–36.
110. Emmerzaal TL, Kiliaan AJ, Gustafson DR. 2003–2013: a decade of body mass index, alzheimer's disease, and dementia. *J Alzheimers Dis*. 2015;43:739–755.
111. Jahangiri A, Wilson PG, Hou T, Brown A, King VL, Tannock LR. Serum amyloid a is found on apob-containing lipoproteins in obese humans with diabetes. *Obesity (Silver Spring)*. 2013;21:993–996.
112. Gustafson DR, Karlsson C, Skoog I, Rosengren L, Lissner L, Blennow K. Mid-life adiposity factors relate to blood-brain barrier integrity in late life. *J Intern Med*. 2007;262:643–650.
113. Harford KA, Reynolds CM, McGillicuddy FC, Roche HM. Fats, inflammation and insulin resistance: insights to the role of macrophage and t-cell accumulation in adipose tissue. *Proc Nutr Soc*. 2011;70:408–417.
114. Velloso LA, Folli F, Saad MJ. Tlr4 at the crossroads of nutrients, gut microbiota, and metabolic inflammation. *Endocr Rev*. 2015;36:245–271.
115. Calder PC, Ahluwalia N, Brouns F, et al. Dietary factors and low-grade inflammation in relation to overweight and obesity. *Br J Nutr*. 2011;106(Suppl 3):S5–78.
116. Kelly KR, Haus JM, Solomon TP, et al. A low-glycemic index diet and exercise intervention reduces tnf(alpha) in isolated mononuclear cells of older, obese adults. *J Nutr*. 2011;141:1089–1094.
117. Edison P, Brooks DJ. Role of neuroinflammation in the trajectory of Alzheimer's disease and in vivo quantification using pet. *J Alzheimers Dis*. 2018;64:S339–S351.
118. Uddin MS, Stachowiak A, Mamun AA, et al. Autophagy and Alzheimer's disease: From molecular mechanisms to therapeutic implications. *Front Aging Neurosci*. 2018;10:04.
119. Bacchetti T, Vignini A, Giulietti A, et al. Higher levels of oxidized low density lipoproteins in alzheimer's disease patients: Roles for platelet activating factor acetyl hydrolase and paraoxonase-1. *J Alzheimers Dis*. 2015;.
120. Maloney B, Lahiri DK. Epigenetics of dementia: Understanding the disease as a transformation rather than a state. *Lancet Neurol*. 2016;15:760–774.

CHAPTER

16

n-3 Long-Chain PUFA-Containing Phospholipids and Neuroprotection

Corinne Joffre[*,†], *Charlotte Rey*[*,†,‡], *Agnès Nadjar*[*,†], *Sophie Layé*[*,†]

*INRA, Nutrition et Neurobiologie Intégrée, Bordeaux, France †Université de Bordeaux 2, Bordeaux, France ‡ITERG, Institut des corps gras, Pessac, France

INTRODUCTION

n-3 Long-chain polyunsaturated fatty acids (n-3 LC-PUFAs), in particular docosahexaenoic acid (DHA) are essential for brain cell phospholipid (PL) membranes in which DHA constitutes up to 15% of total fatty acids.[1] n-3 LC-PUFAs have a key role in membrane fluidity and function and also act as signaling molecules. Furthermore, they are involved in neuronal structure and modulate the neurotransmission and the synaptic plasticity.[2,3] Moreover, DHA constitutes a neuroprotective agent against neurological disorders including Alzheimer's disease, Parkinson's disease that have an inflammatory component.[4] Indeed, n-3 LC-PUFAs have immunomodulatory properties as they are converted in bioactive lipid mediators called oxylipins. Oxylipins play a central role in the regulation of acute or chronic inflammation[5] and in the resolution of inflammation.[6] Among them, specialized pro-resolving mediators (SPMs) are synthetized after prostaglandins, leukotrienes and thromboxanes have first initiated, propagated inflammation, and stimulated their synthesis.[7,8] These SPMs are crucially involved in the resolution of inflammation in the brain.[9,10] They are good candidates to resolve neuroinflammation.[11] In this chapter, we present an overview of the anti-inflammatory role of n-3 LC-PUFAs and of the formation and action of SPMs in the brain, and the possible regulation of the resolution of inflammation by dietary means and by inflammation.

N-3 PUFA PLS IN THE BRAIN

n-3 PUFA Brain Content

The brain is one of the richest organs in lipids (50%–70%).[12] In the brain, fatty acids are mainly esterified in PLs, especially in phosphatidylcholine (PC, 42%–44%), phosphatidylethanolamine

Omega Fatty Acids in Brain and Neurological Health
https://doi.org/10.1016/B978-0-12-815238-6.00016-X

(PE, 36%–40%), and also in phosphatidylserine (PS, 11%–13%), phosphatidylinositol (PI, 2%–3%).[12] DHA is mostly esterified in PE and PS whereas the main n-6 LC-PUFA in the brain, arachidonic acid (AA), is esterified in PC and PI.[4,13,14] Human prefrontal cortex contains 30% PUFAs that are essentially DHA (14%) and AA (9%).[1] In rodents, the fatty acid composition is quite similar: 25% PUFAs with 12% DHA and 10% AA.[15–19] PUFAs are not homogenously distributed in the brain. In adult mice, hippocampus and prefrontal cortex contained the highest content of DHA and AA and hypothalamus has the highest n-6/n-3 ratio.[15,16,19] Moreover, PUFAs are heterogeneously distributed among brain cells. Brain glial cells comprise 70% astrocytes, 10%–15% oligodendrocytes, and 10%–15% microglial cells.[20] The main fatty acids of these cells are AA and DHA present in different proportions.[21,22]

Environmental factors such as nutrition can affect the fatty acid composition in all brain structures and cells. Indeed, the PUFA content is strongly impacted by the diet.[23] A dietary supplementation in n-3 PUFAs such as alpha-linolenic acid (ALA), the precursor of DHA, or ALA+DHA induces an increase in brain DHA in rodents.[24,25] A fish oil supplementation (rich in DHA and eicosapentaenoic acid, EPA, the precursor of DHA) increases the DHA content and decreases the content of AA and n-6 docosapentaenoic acid (n-6 DPA), a metabolite of AA, in rodent brain.[25–28] A DHA supplementation is more efficient than an ALA supplementation in increasing the DHA content in the brain.[29] In rodents, a DHA supplementation from 16 weeks to 16 months or from 20 to 22 months of age compensates a DHA decrease due to aging.[14,17,30] Contrary to the supplementation, a dietary n-3 PUFA deficiency induces a decrease in n-3 PUFAs in the brain. Indeed, an n-3 PUFA deficiency from weaning to 2–4 months decreases DHA content in all brain structures, the hippocampus being the most affected whereas hypothalamus the less.[19,31,32] This decrease in DHA is associated to an increase in n-6 DPA and AA.[19,31,33] However, it is possible to reverse the DHA decrease. Indeed, a fish oil-containing diet supply at weaning in animals deficient in n-3 PUFAs from the first day of gestation induces an increase in DHA and a decrease in n-6 DPA in the hippocampus after 2 months.[9] In a model of n-3 PUFA deficiency that induces a 75% depletion in DHA in the brain in the first generation, it is possible to restore DHA levels in all brain regions when animals are supplemented with ALA from P0, P7, and P14, weaning or 5 weeks of age for 12 weeks in the second generation.[15,34] Brain DHA metabolism depends on the diet to slow brain DHA loss. The half-life of DHA is increased when animal diet is deficient in n-3 PUFAs, reducing the loss of DHA in the brain.[35] Moreover, the activity of Δ6 desaturase and elongase that are responsible for the synthesis of LC-PUFAs from their precursors is increased in the liver of n-3 PUFA-deficient animals.[36–38] Brain cells are also impacted by dietary PUFA supply. An n-3 PUFA-deficient diet administered from gestation to 2-weeks-old rats induces in neurons, astrocytes, and oligodendrocytes a decrease in DHA levels that is compensated by an increase in n-6 DPA.[21] After the exchange of an n-6-rich diet for an n-3-rich diet, specific brain cells have a slow recovery of fatty acid composition as measured by either the increase of DHA or the decrease of n-6 DPA in neurons, astrocytes, and oligodendrocytes.[39] Moreover, an n-3 LC-PUFA supplementation administered during gestation and lactation increases PE-DHA and PS-DHA levels in glial cells.[40] A DHA+EPA supply induces a decrease in AA and n-6 DPA whereas a DHA supply leads to a reduction in n-6 DPA.[40,41] In line with these results, we recently showed for the first time in microglial cells, the immune cells of the brain, that modulating n-3 PUFA intake affects lipid profile in the developing brain.[22] Maternal n-3 LC-PUFA dietary supplementation during gestation and lactation increases DHA level and decreases n-6 DPA level in offspring's microglia of pups at P21.[22] Interestingly, microglial cells seem to be protected from an n-3 PUFA deficiency for DHA level: n-3 PUFA-deficient diet increases n-6 DPA as compared to n-3 PUFA balanced diet but does not impact DHA level in this group.[22]

n-3 PUFAs as Regulators of Neuroinflammation in Humans, Animals, and In Vitro

n-3 PUFAs have powerful anti-inflammatory properties.[42] Then they play an important role in the regulation of the synthesis and release of pro-inflammatory mediators.[31,43–46] Pro-inflammatory factors include interleukin-1 beta (IL-1β), interleukin-6 (IL-6), and tumor necrosis factor alpha (TNF-α) and play a role in neuronal plasticity.[31,46] If sustained, uncontrolled inflammation can lead to neuronal damage that is involved in many neuronal pathologies.[47–49] Hence, limiting the inflammation and enhancing the resolution of inflammation is of great interest.

Evidence in Human

In human, the anti-inflammatory properties of n-3 PUFAs have been first identified in epidemiological studies in Eskimos that eat a lot of n-3 LC-PUFAs through fishes.[50–52] Clinical studies have highlighted the beneficial effect of n-3 LC-PUFAs in chronic inflammatory and autoimmune diseases. Indeed, a fish oil supplementation induces a decrease in pro-inflammatory cytokine expression such as IL-1β in blood monocytes and improves the symptoms of patients suffering from rheumatoid arthritis[53–55] or multiple sclerosis.[56,57] Moreover, DHA supply for several months improves the working and long-term memories in patients with moderate cognitive alterations.[58,59] DHA supply also significantly decreases the circulating inflammatory markers and the oxidative stress.[58,60]

Evidence in Animals

In animals, numerous studies have demonstrated the anti-inflammatory properties of n-3 PUFAs in the brain. In a LPS-induced inflammation model, dietary n-3 LC-PUFA supplementation at adulthood prevents LPS-induced hippocampal increase of pro-inflammatory cytokines IL-1β and TNF-α.[10] A dietary supply in EPA decreases the production of LPS-induced pro-inflammatory cytokines IL-1β, IL-6, and TNF-α in hippocampus and increases the production of anti-inflammatory cytokines IL-10 and IL-4.[61–63] This modification is associated to a decrease in the phosphorylation of c-Jun and c-Jun N-terminal kinase (JNK) proteins and in nuclear factor-kappa B (NFκB) that regulate inflammation.[62,63] EPA dietary supply attenuates the activation of microglial cells and astrocytes triggered by an intracerebral IL-1β administration, increases the production of IL-10 in the hippocampus, and improves emotional behavior alteration and memory deficit.[64–66] Following a dietary DHA supplementation, DHA esterified in PLs is not sufficient to attenuate neuroinflammation, this pool is an important source of unesterified DHA that is necessary to attenuate neuroinflammation.[9] On the contrary, an n-3 PUFA deficiency from the first day of gestation to weaning induces an increase in the expression of pro-inflammatory markers in the hippocampus and an alteration in the motility and the phenotype of microglial cells and alters synaptic plasticity.[2,3,67] Moreover, when the deficiency continues at adulthood, it induces an alteration of GABAergic, dopaminergic and cholinergic neurotransmission[68–71] and anxiety-related, depressive-like behaviors that can be associated to autistic behavior in adults.[32,33,71,72] An n-3 PUFA dietary supply restores social and memory performances altered in autism models.[73–75] Following intraperitoneal (i.p.) injection of LPS, n-3 PUFA-deficient mice display altered hippocampal synaptic plasticity that likely contributes to spatial memory impairment and higher glucocorticoid levels.[31] Using the same model, we showed that Il-6 expression is strongly induced in n-3 PUFA-deficient mice whereas sickness behavior is downregulated via the impairment of IL-6 signaling in the brain.[33]

Another way to modulate neuroinflammation is to administer n-3 PUFAs directly in the brain or peripherally. Indeed, a 24-h-intracerebroventricularly (i.c.v.) DHA brain infusion attenuates hippocampal neuroinflammation initiated by i.c.v. LPS.[9] Moreover, intrathecal injection of DHA induces a decrease in microglial activation, MAPK phosphorylation and in the production of pro-inflammatory cytokines in spinal cord.[76] In addition, in a model of traumatic brain injury, n-3 PUFA i.p. injection attenuates microglial-induced inflammation by inhibiting NFκB pathway.[77]

Dietary n-3 LC-PUFA supplementation requires the use of fish oil. However, fish oil may provide confounding factors such as vitamins for example. Thus, the use of Fat-1 transgenic mice that convert n-6 to n-3 PUFAs through a desaturase from *C. elegans* is a model that allow the elimination of confounding factors provided by the diet.[78] These mice have higher DHA level in hippocampus and cortex.[9,79,80] They express less cyclooxygenase-2 (COX-2, that are involved in the production of pro-inflammatory oxylipins) in the cortex than wild-type mice.[79] They are protected against cognitive deficits induced by i.p. LPS injection through a decrease in neuroinflammation.[9,80] This is associated to a decrease in microglial activation.[9]

Evidence in In Vitro Studies

In vitro studies have shown that n-3 PUFAs have an inhibitory effect on the production of pro-inflammatory cytokines in microglial cells. Indeed, in microglial cell line or primary culture microglial cells, DHA prevents LPS-induced NKκB activation and then cytokine production by inhibiting LPS receptor presentation and decreases the oxidative stress, NO production and iNOS expression.[81–88] Moreover, DHA prevents LPS-induced MAPK phosphorylation (MAPK pathway playing essential roles in the expression of inflammatory molecules) and induces peroxisome proliferator-activated receptor (PPAR)-γ nuclear translocation that exerts anti-inflammatory effects.[83] Of interest, EPA also affects the production of TNF-α, IL-6 and NO by inhibiting NFκB phosphorylation via Sirt-1.[85,87,89] It is likely that DHA attenuates the inflammatory response in LPS-activated microglia by remodeling lipid bodies that are dynamic organelles in which DHA is incorporated and by altering their interplay with mitochondria and other associated organelles.[90] In addition, DHA and EPA are able to enhance myelin or Aβ peptide phagocytosis that is associated to a shift in microglial polarization toward the beneficial M2 phenotype and to a decrease in pro-inflammatory cytokine production.[85,91]

MECHANISMS OF ACTION FOR THE BENEFICIAL EFFECT OF N-3 PUFAS

Synthesis of SPMs

n-3 LC-PUFAs act in part through their conversion in oxidized lipid mediators called oxylipins. Two types of oxylipins are synthetized: those involved in the regulation of inflammation

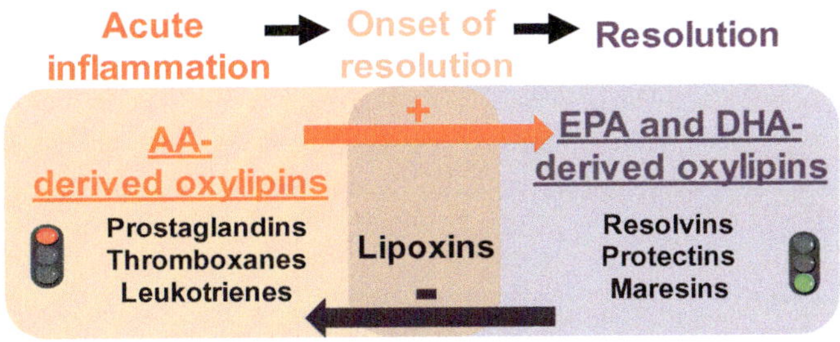

FIG. 1 Oxylipin contribution during the inflammatory process from the initiation to the resolution. AA: arachidonic acid; EPA: eicosapentaenoic acid; DHA: docosahexaenoic acid. *Adapted from Serhan CN. Pro-resolving lipid mediators are leads for resolution physiology. Nature 2014;510:92–101.*

(prostaglandins, thromboxanes, leukotrienes, hydroxyl fatty acids, epoxy fatty acids) and those implicated in the resolution of inflammation or SPMs (lipoxins, resolvins, protectins, and maresins)[6] (Fig. 1). The identification of these bioactive mediators makes evident that the resolution of inflammation is an active process that is finely tuned, orchestrating the decrease in pro-inflammatory cytokines and the increase in anti-inflammatory cytokines, and apoptosis and phagocytosis of cellular debris that permit a return to homeostasis.[92,93] Charles N. Serhan first identified these SPMs at the periphery and characterized their anti-inflammatory and pro-resolutive properties.[94] SPMs should fulfill some fundamental criteria among which the cessation of neutrophil tissue infiltration, the regulation of inflammation, the induction of apoptosis in neutrophils, and their subsequent efferocytosis by macrophages, the reprogramming of macrophages from classically activated to alternatively activated cells, and the induction of the return of the tissue to homeostasis.[93] SPMs orchestrate the inflammatory response in inflammatory diseases such as bacterial infections, colitis, peritonitis, arthritis or respiratory, renal or brain pathologies.[92,95] A failure in the resolution of inflammation is detrimental for tissue and contributes to chronic inflammatory response.

EPA is the precursor of resolvins E1 (RvE1), E2, and E3[94,96] It is converted by aspirin-triggered acetylated COX-2 or cytochrome P450 (CYP450) in 18-HEPE, that is transformed in RvE1 by 5-lipoxygenase (5-LOX)[97] (Fig. 2). RvE1 and 18-HEPE have been detected in hippocampus although EPA is very low in the brain.[9,98,99] RvE1 binds to two G protein-coupled receptors, chemokine-like receptor 1 (ChemR23) expressed on monocytes, macrophages and microglia, and leukotriene B4 receptor (BLT1) expressed on monocytes and neutrophils.[100–103] ChemR23 was identified as the RvE1 receptor in prefrontal cortex, hippocampus and cerebellum,[104] in microglial cells[105,106] and neurons.[107] In vitro, RvE1 inhibits microglial activation and pro-inflammatory cytokine release through NFκB pathway alteration.[106,108] In vivo it modulates the inflammatory profile and microglial activation.[108,109] RvE1 regulates phosphorylation of Akt and ribosomal protein rS6.[97]

DHA is the precursor of neuroprotectin D1 (NPD1), maresin 1 (Mar1) and RvDs that have pro-resolutive and anti-inflammatory properties.[96,110] DHA is converted in monohydroxy DHA (4-HDHA, 14-HDHA, and 17-HDHA) that serves as precursors for these resolving molecules.[96] For example, DHA is transformed by 15-LOX in 17S-hydroperoxy-DHA (17S-HpDHA) that is converted by 5-LOX in RvD1. RvD1 acts through lipoxin A4 receptor/formyl peptide receptor 2 (ALX/Fpr2) (in rodents) and G-protein-coupled receptor 32

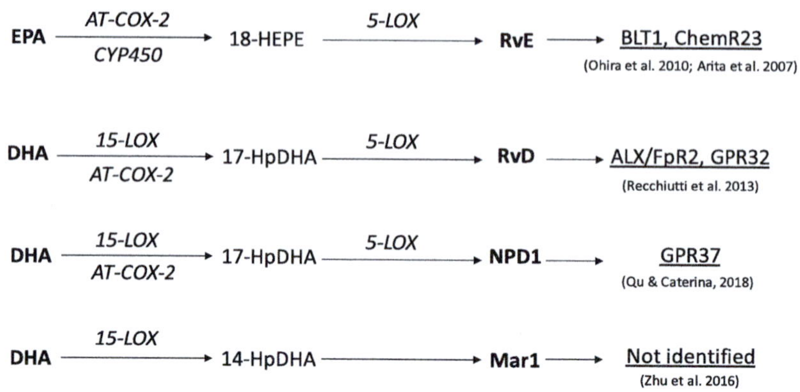

FIG. 2 Synthesis pathway of SPMs and their receptors. EPA: eicosapentaenoic acid; DHA: docosahexaenoic acid; 18-HEPE: 18-hydroxy-EPA; 17-HpDHA: 17-hydroperoxy-DHA; 14-HpDHA: 14-hydroperoxy-DHA; RvE: resolvin E; RvD: resolvin D; NPD1: neuroprotectin D1; Mar1: maresin 1; BLT1: leukotriene B4 receptor; ChemR23: chemokine-like receptor 1; ALX/Fpr2: N-formyl peptide receptor 2; GPR32: G protein-coupled receptor 32; GPR37: G protein-coupled receptor 37. In bold: SPMs and their precursors; Underlined: SPM receptors.

(GPR32) (in human)[111,112] at picomolar range[113] but induces biological effects at nanomolar range.[113,114] It regulates the expression of specific micro-RNA (miRNA) to control the intensity and the length of inflammation via the regulation of target genes such as inflammatory cytokine genes.[106,112,115,116] RvD1 was identified in vitro in DHA supplemented and zymosan A-treated glial cells[117] and in vivo in mouse after brain ischemia[118] Moreover, 5- and 15-LOX implicated in RvD1 synthesis are widely distributed in the brain suggesting the potential production of RvD1.[119] 15-LOX deletion induced a drastic decrease in RvD1 production in the brain.[119] Recent studies describe the anti-inflammatory properties of RvD1 in microglia and astrocyte cell cultures.[106,120,121] Indeed, in BV2 microglia cell culture, RvD1 promotes the IL-4-induced M2 phenotype[121] and inhibits LPS-induced pro-inflammatory cytokines.[106] In astrocytes cell culture, RvD1 attenuates LPS-induced TNF-α.[120] In vivo, in a model of remote damage, Bisicchia et al. (2018) recently show that RvD1 promotes functional recovery and reduces neuroinflammation via miRNA.[112]

Dysfunction of SPM production can be due to insufficient EPA and DHA supply leading to inadequate production of SPMs or to the polymorphism of the enzymes involved in their synthesis or to a defect in the binding of SPMs to their receptors.[93]

Free (unesterified) n-3 LC-PUFAs are released from membrane PLs through the action of phospholipases A2 (PLA2) in response to stimulation. DHA is hydrolyzed by iPLA2 from PLs and PlsEtn-PLA2 from plasmalogens.[122] After this step, n-3 LC-PUFAs undergo an enzymatic conversion to generate SPMs (Fig. 3).

Synthesis Pathways of n-3 PUFA-Derived Lipid Mediators

n-3 PUFA-derived lipid mediators are synthetized from EPA and DHA via COX, LOX, and CYP450. They include prostaglandins, leukotrienes, thromboxanes, and SPMs (resolvins, maresins, neuroprotectins).[6]

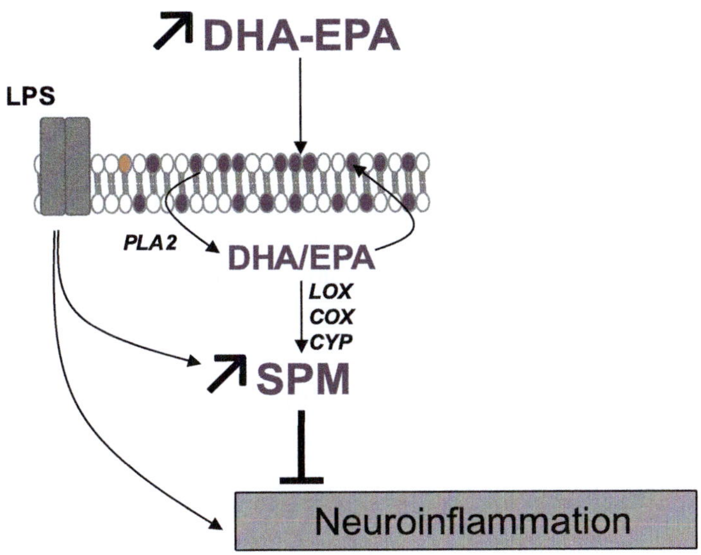

FIG. 3 Regulation of the resolution of inflammation by PUFA consumption and LPS.

Role of COX-2 in the Brain

There are two COX-2 isoforms: an inducible one and a constitutive one.[123] Indeed, COX-2 is an inducible enzyme expressed at sites of inflammation. In the absence of overt inflammation, COX-2 is also constitutively expressed in the brain.[79] In the brain, the inducible COX-2 is activated via NFκB pathway.[124] whereas the constitutive COX-2 is regulated by none of the two main transcriptional pathway NFκB and NFAT (nuclear factor of activated T cells) involved in kidneys.[123] COX-2 catalyzes the first step of the synthesis of prostaglandins and thromboxanes derived from n-6 PUFAs that contribute to the initiation of inflammation[125–127] COX-2 also produces hydroxyl fatty acids such as 13-hydroxy-DHA from DHA.[6] In response to LPS, COX-2 is rapidly expressed in the hippocampus.[10,128] In the central nervous system, it regulates blood arterial pressure,[129] pain hypersensitivity,[130,131] and neuronal plasticity[132,133] and is also involved in oxidative stress.[133] COX-2 inhibition has anti-inflammatory effect by inhibiting prostaglandin production and protects against neuronal degeneration.[134,135]

Role of 15-LOX and 5-LOX in the Brain

In the brain, 15-LOX and 5-LOX are the most abundant LOX. They are expressed in the second steps of inflammation in the hippocampus.[128,136] 15-LOX function in the brain is complex. It is involved in neurodegeneration and neurotoxicity due to the increased oxidative stress it generates in models of Alzheimer disease[137,138] and brain ischemia.[139] However, it is also implicated in neuroprotection.[140] The inhibition of 15-LOX induces hippocampus dependent cognitive alterations.[119] 5-LOX has pro- or anti-inflammatory properties. It enhances pro-inflammatory leukotriene synthesis when it is located in the nucleus whereas it induces the production of anti-inflammatory LxA4 when located in the cytoplasm.[141] In the brain, LPS induces an increase of 5-LOX expression in the late stage of inflammation.[10] The inhibition of 5-LOX is neuroprotective as it induces an inhibition of NFκB and a decrease in leukotriene production.[142–144]

Role of CYP450 in the Brain

CYP450 plays a complex role in inflammation by producing bioactive mediators such as epoxides that are anti-inflammatory in monocytes and macrophages.[145–148] 18-HEPE derived from EPA via ω-hydroxylase has an anti-cancer role by reducing pro-inflammatory cytokines and pro-proliferative factors possibly via conversion to *E*-series Rvs, enhancing the resolution of inflammation.[6]

REGULATION OF THE RESOLUTION OF INFLAMMATION IN HUMANS, ANIMALS, AND IN VITRO

Regulation of the Resolution of Inflammation by PUFA Consumption

We previously showed that PUFA consumption leads to modifications in brain PUFA levels in the brain. The oxylipin levels and their biosynthetic enzyme expression also depend on dietary PUFAs.

An n-3 PUFA-deficient diet induces an increase in COX-2 expression in prefrontal cortex, suggesting an increase in AA-derived pro-inflammatory mediator levels.[149] Conversely, a n-6 PUFA-deficient diet induces a decrease in COX-2 expression and an increase in 15-LOX expression.[150] Interestingly, Taha et al. (2017) showed that animals fed an n-6 PUFA-deficient diet are protected from an increase in COX-2 observed in LPS-treated n-3 PUFA-deficient animals.[151] However, we showed that n-3 LC-PUFA supplementation was not sufficient to modulate COX-2 expression in the brain.[10]

SPM levels in peripheral organs and brain are also modulated by different dietary supply in PUFAs (Fig. 3). In human, a dietary supplementation in DHA (1.7 g/j) and EPA (0.6 g/j) for 6 months induced an increase in RvD1 released by blood mononuclear cells of Alzheimer's disease patients as compared to controls.[138] Results from intervention studies show that EPA and DHA dietary supplementation induces an increase in EPA- and DHA-derived oxylipins although high inter-individual variances.[152] In animals, an n-3 PUFA dietary supplementation (140 mg EPA + 110 mg DHA/kg animal) in aged rats enhances the production of lipid mediators derived from DHA and from EPA in cerebral cortex.[153] DHA and EPA supply also decreases the n-6 PUFA-derived metabolites (that are mainly pro-inflammatory) and increases the n-3 PUFA-derived metabolites in the plasma and at the periphery[153–155] A diet enriched with fish oil (containing 9% EPA + 9% DHA) in mice induces an increase in plasmatic RvD1.[156] In Fat-1 mice, brain EPA increase is accompanied by an increase in EPA-derived HEPEs.[99] Conversely, an n-6 PUFA dietary supply from weaning to 15 weeks of age increases the n-6 PUFA metabolites and decreases the EPA-derived metabolites in the cortex and the plasma of rats.[157] Recently, we showed that dietary n-3 LC-PUFA supplementation induces n-3 LC-PUFA enrichment in the hippocampus and subsequently an increase in n-3 PUFA-derived SPMs and a decrease in n-6 PUFA-derived oxylipins.[10] Moreover, in response to LPS, n-3 LC-PUFA-deficient mice present a pro-inflammatory oxylipin profile whereas n-3 LC-PUFA supplemented mice display an anti-inflammatory oxylipin profile in the hippocampus.[10] These results differ from those of Trepanier et al. (2018) who found that in a model of Fat-1 mice, a higher increase in brain DHA induced by dietary supply has no effect on the resolution of inflammation after an i.c.v. LPS injection.[158]

Regulation of the Resolution by Inflammation

Numerous studies have highlighted that inflammation modulates oxylipin synthesis at the periphery and in the brain (Fig. 3).

Inflammation induces more drastic changes in the expression of biosynthetic enzymes than n-3 LC-PUFA dietary supply. In rat, traumatic brain injury induces an increase in COX-2, 5-LOX, 15-LOX, and CYP450 expression in hippocampus and cortex suggesting an alteration of all oxylipin biosynthesis pathways.[136] LPS increases COX-2 expression and activity in the brain.[10,151,159]

In human, Wang et al. showed that RvD1 level in cerebrospinal fluid (CSF) from Alzheimer's disease patients is positively correlated with cognitive function.[138]

In animals, brain ischemia induces an increase in the production of five of mono-, di-, and tri-hydroxy-DHA derivates.[160,161] However, LPS i.c.v. injection does not impact RvD1 level.[151,159]

In vitro, most studies have been performed in peripheral immune cells. The lipid mediator profile in rodent neutrophils and macrophages changes with macrophage phenotype M1 or M2, or with the inflammatory stimulus in neutrophils.[162] Indeed, M2 macrophages produce more resolvins, protectins, maresins, and less AA-derived pro-inflammatory mediators than M1 macrophages.[163]

CONCLUSION

Studies presented in this chapter show that dietary n-3 PUFA supply should affect neuroprotection likely by changing SPM profile. The drastic reduction in dietary n-3 PUFAs in Western societies leads to an imbalanced dietary n-6/n-3 PUFAs estimated at 12–20 instead of 5 recommended.[164] As we showed, this imbalance leads to enhanced neuroinflammatory responses that can be prevented by a balanced n-6/n-3 PUFA diet and solved by the use of SPMs as therapeutic compounds. SPMs are promising therapeutic compounds: they are of natural origin and are active at low concentrations (nM) as compared to their precursors (μM).[165,166] Orally RvD1 administration was shown to reduce acute inflammation and accelerate or initiate resolution,[167] suggesting a possibility to develop SPMs as a novel strategy for the treatment of central inflammatory diseases.

References

1. McNamara RK, Carlson SE. Role of omega-3 fatty acids in brain development and function: potential implications for the pathogenesis and prevention of psychopathology. *Prostaglandins Leukot Essent Fatty Acids*. 2006;75:329–349. https://doi.org/10.1016/j.plefa.2006.07.010.
2. Madore C, Nadjar A, Delpech J-C, et al. Nutritional n-3 PUFAs deficiency during perinatal periods alters brain innate immune system and neuronal plasticity-associated genes. *Brain Behav Immun*. 2014;41:22–31. https://doi.org/10.1016/j.bbi.2014.03.021.

3. Lafourcade M, Larrieu T, Mato S, et al. Nutritional omega-3 deficiency abolishes endocannabinoid-mediated neuronal functions. *Nat Neurosci*. 2011;14:345–350. https://doi.org/10.1038/nn.2736.
4. Bazinet RP, Layé S. Polyunsaturated fatty acids and their metabolites in brain function and disease. *Nat Rev Neurosci*. 2014;15:771–785. https://doi.org/10.1038/nrn3820.
5. Serhan CN, Savill J. Resolution of inflammation: the beginning programs the end. *Nat Immunol*. 2005;6:1191–1197. https://doi.org/10.1038/ni1276.
6. Gabbs M, Leng S, Devassy JG, Monirujjaman M, Aukema HM. Advances in our understanding of Oxylipins derived from dietary PUFAs. *Adv Nutr*. 2015;6:513–540. https://doi.org/10.3945/an.114.007732.
7. Levy BD, Clish CB, Schmidt B, Gronert K, Serhan CN. Lipid mediator class switching during acute inflammation: signals in resolution. *Nat Immunol*. 2001;2:612–619. https://doi.org/10.1038/89759.
8. Serhan CN. Pro-resolving lipid mediators are leads for resolution physiology. *Nature*. 2014;510:92–101.
9. Orr SK, Palumbo S, Bosetti F, et al. Unesterified docosahexaenoic acid is protective in neuroinflammation. *J Neurochem*. 2013;. August, http://www.ncbi.nlm.nih.gov/entrez/query.fcgi?cmd=Retrieve&db=PubMed&dopt=Citation&list_uids=23919613.
10. Rey C, Delpech JC, Madore C, et al. Dietary n-3 long chain PUFA supplementation promotes a pro-resolving oxylipin profile in the brain. *Brain Behav Immun*. August 2018. https://doi.org/10.1016/j.bbi.2018.07.025.
11. Laye S, Nadjar A, Joffre C, Bazinet RP. Anti-inflammatory effects of omega-3 fatty acids in the brain: physiological mechanisms and relevance to pharmacology. *Pharma Rev*. 2017;.
12. Sastry PS. Lipids of nervous tissue: Composition and metabolism. *Prog Lipid Res*. 1985;24:69–176.
13. Garcia MC, Ward G, Ma YC, Salem Jr N, Kim HY. Effect of docosahexaenoic acid on the synthesis of phosphatidylserine in rat brain in microsomes and C6 glioma cells. *J Neurochem*. 1998;70:24–30.
14. Bascoul-Colombo C, Guschina IA, Maskrey BH, Good M, O'Donnell VB, Harwood JL. Dietary DHA supplementation causes selective changes in phospholipids from different brain regions in both wild type mice and the Tg2576 mouse model of Alzheimer's disease. *Biochim Biophys Acta*. 1861;2016:524–537. https://doi.org/10.1016/j.bbalip.2016.03.005.
15. Xiao Y, Huang Y, Chen ZY. Distribution, depletion and recovery of docosahexaenoic acid are region-specific in rat brain. *Br J Nutr*. 2005;94:544–550.
16. Carrie I, Clement M, de Javel D, Frances H, Bourre JM. Specific phospholipid fatty acid composition of brain regions in mice. Effects of n-3 polyunsaturated fatty acid deficiency and phospholipid supplementation. *J Lipid Res*. 2000;41:465–472.
17. Little SJ, Lynch MA, Manku M, Nicolaou A. Docosahexaenoic acid-induced changes in phospholipids in cortex of young and aged rats: a lipidomic analysis. *Prostaglandins Leukot Essent Fatty Acids*. 2007;77:155–162. https://doi.org/10.1016/j.plefa.2007.08.009.
18. Chung WL, Chen JJ, Su HM. Fish oil supplementation of control and (n-3) fatty acid-deficient male rats enhances reference and working memory performance and increases brain regional docosahexaenoic acid levels. *J Nutr*. 2008;138:1165–1171.
19. Joffre C, Grégoire S, De Smedt V, et al. Modulation of brain PUFA content in different experimental models of mice. *Prostaglandins Leukot Essent Fatty Acids*. 2016;114:1–10. https://doi.org/10.1016/j.plefa.2016.09.003.
20. Renaud J, Therien HM, Plouffe M, Martinoli MG. [Neuroinflammation: Dr Jekyll or Mr Hyde?]. Med Sci (Paris) 2015;31:979–988. https://doi.org/10.1051/medsci/20153111012.
21. Bourre JM, Pascal G, Durand G, Masson M, Dumont O, Piciotti M. Alterations in the fatty acid composition of rat brain cells (neurons, astrocytes, and oligodendrocytes) and of subcellular fractions (myelin and synaptosomes) induced by a diet devoid of n-3 fatty acids. *J Neurochem*. 1984;43:342–348.
22. Rey C, Nadjar A, Joffre F, et al. Maternal n-3 polyunsaturated fatty acid dietary supply modulates microglia lipid content in the offspring. *Prostaglandins Leukot Essent Fatty Acids*. 2018;133:1–7. https://doi.org/10.1016/j.plefa.2018.04.003.
23. Alashmali SM, Hopperton KE, Bazinet RP. Lowering dietary n-6 polyunsaturated fatty acids: interaction with brain arachidonic and docosahexaenoic acids. *Curr Opin Lipidol*. 2016;27:54–66. https://doi.org/10.1097/MOL.0000000000000255.
24. Murthy M, Hamilton J, Greiner RS, Moriguchi T, Salem Jr N, Kim HY. Differential effects of n-3 fatty acid deficiency on phospholipid molecular species composition in the rat hippocampus. *J Lipid Res*. 2002;43:611–617.
25. Hiratsuka S, Koizumi K, Ooba T, Yokogoshi H. Effects of dietary docosahexaenoic acid connecting phospholipids on the learning ability and fatty acid composition of the brain. *J Nutr Sci Vitaminol (Tokyo)*. 2009;55:374–380.

26. de Theije CG, van den Elsen LW, Willemsen LE, et al. Dietary long chain n-3 polyunsaturated fatty acids prevent impaired social behaviour and normalize brain dopamine levels in food allergic mice. *Neuropharmacology*. 2015;90:15–22. https://doi.org/10.1016/j.neuropharm.2014.11.001.
27. Skorve J, Hilvo M, Vihervaara T, et al. Fish oil and krill oil differentially modify the liver and brain lipidome when fed to mice. *Lipids Health Dis*. 2015;14:88. https://doi.org/10.1186/s12944-015-0086-2.
28. Kitson AP, Metherel AH, Chen CT, et al. Effect of dietary docosahexaenoic acid (DHA) in phospholipids or triglycerides on brain DHA uptake and accretion. *J Nutr Biochem*. 2016;33:91–102. https://doi.org/10.1016/j.jnutbio.2016.02.009.
29. Lacombe RJS, Giuliano V, Colombo SM, Arts MT, Bazinet RP. Compound-specific isotope analysis resolves the dietary origin of docosahexaenoic acid in the mouse brain. *J Lipid Res*. 2017;58:2071–2081. https://doi.org/10.1194/jlr.D077990.
30. Labrousse VF, Nadjar A, Joffre C, et al. Short-term long chain omega3 diet protects from neuroinflammatory processes and memory impairment in aged mice. *PLoS ONE*. 2012;7:e36861. https://doi.org/10.1371/journal.pone.0036861.
31. Delpech J-C, Thomazeau A, Madore C, et al. Dietary n-3 PUFAs deficiency increases vulnerability to inflammation-induced spatial memory impairment. *Neuropsychopharmacology*. 2015;40:2774–2787. https://doi.org/10.1038/npp.2015.127.
32. Manduca A, Bara A, Larrieu T, et al. Amplification of mGlu5-endocannabinoid signaling rescues behavioral and synaptic deficits in a mouse model of adolescent and adult dietary polyunsaturated fatty acid imbalance. *J Neurosci*. 2017;37:6851–6868. https://doi.org/10.1523/JNEUROSCI.3516-16.2017.
33. Mingam R, Moranis A, Bluthé R-M, et al. Uncoupling of interleukin-6 from its signalling pathway by dietary n-3-polyunsaturated fatty acid deprivation alters sickness behaviour in mice. *Eur J Neurosci*. 2008;28:1877–1886. https://doi.org/10.1111/j.1460-9568.2008.06470.x.
34. Kodas E, Vancassel S, Lejeune B, Guilloteau D, Chalon S. Reversibility of n-3 fatty acid deficiency-induced changes in dopaminergic neurotransmission in rats: critical role of developmental stage. *J Lipid Res*. 2002;43:1209–1219.
35. Rapoport SI, Rao JS, Igarashi M. Brain metabolism of nutritionally essential polyunsaturated fatty acids depends on both the diet and the liver. *Prostaglandins Leukot Essent Fatty Acids*. 2007;77:251–261. https://doi.org/10.1016/j.plefa.2007.10.023.
36. Cho HP, Nakamura MT, Clarke SD. Cloning, expression, and nutritional regulation of the mammalian Delta-6 desaturase. *J Biol Chem*. 1999;274:471–477.
37. Wang Y, Botolin D, Christian B, Busik J, Xu J, Jump DB. Tissue-specific, nutritional, and developmental regulation of rat fatty acid elongases. *J Lipid Res*. 2005;46:706–715. https://doi.org/10.1194/jlr.M400335-JLR200.
38. Igarashi M, Ma K, Chang L, Bell JM, Rapoport SI. Dietary n-3 PUFA deprivation for 15 weeks upregulates elongase and desaturase expression in rat liver but not brain. *J Lipid Res*. 2007;48:2463–2470. https://doi.org/10.1194/jlr.M700315-JLR200.
39. Bourre JM, Durand G, Pascal G, Youyou A. Brain cell and tissue recovery in rats made deficient in n-3 fatty acids by alteration of dietary fat. *J Nutr*. 1989;119:15–22.
40. Bowen RA, Clandinin MT. Maternal dietary 22: 6n-3 is more effective than 18: 3n-3 in increasing the 22: 6n-3 content in phospholipids of glial cells from neonatal rat brain. *Br J Nutr*. 2005;93:601–611.
41. Destaillats F, Joffre C, Acar N, et al. Differential effect of maternal diet supplementation with alpha-Linolenic adcid or n-3 long-chain polyunsaturated fatty acids on glial cell phosphatidylethanolamine and phosphatidylserine fatty acid profile in neonate rat brains. *Nutr Metab (Lond)*. 2010;7:2. https://doi.org/10.1186/1743-7075-7-2.
42. Calder PC. Polyunsaturated fatty acids and inflammation. *Biochem Soc Trans*. 2005;33:423–427. https://doi.org/10.1042/BST0330423.
43. Cunningham C, Sanderson DJ. Malaise in the water maze: untangling the effects of LPS and IL-1beta on learning and memory. *Brain Behav Immun*. 2008;22:1117–1127. https://doi.org/10.1016/j.bbi.2008.05.007.
44. Hanisch UK, Kettenmann H. Microglia: active sensor and versatile effector cells in the normal and pathologic brain. *Nat Neurosci*. 2007;10:1387–1394. https://doi.org/10.1038/nn1997.
45. Pascual O, Ben Achour S, Rostaing P, Triller A, Bessis A. Microglia activation triggers astrocyte-mediated modulation of excitatory neurotransmission. *Proc Natl Acad Sci U S A*. 2012;109:E197–E205. https://doi.org/10.1073/pnas.1111098109.
46. Yirmiya R, Goshen I. Immune modulation of learning, memory, neural plasticity and neurogenesis. *Brain Behav Immun*. 2011;25:181–213. https://doi.org/10.1016/j.bbi.2010.10.015.

47. Blais V, Rivest S. Role of the innate immune response in the brain. *Med Sci (Paris)*. 2003;19:981–987. https://doi.org/10.1051/medsci/20031910981.
48. Layé S. Polyunsaturated fatty acids, neuroinflammation and well being. *Prostaglandins Leukot Essent Fatty Acids*. 2010;82:295–303. https://doi.org/10.1016/j.plefa.2010.02.006.
49. Solito E, Sastre M. Microglia function in Alzheimer's disease. *Front Pharmacol*. 2012;3:14. https://doi.org/10.3389/fphar.2012.00014.
50. Dyerberg J, Bang HO. Haemostatic function and platelet polyunsaturated fatty acids in Eskimos. *Lancet*. 1979;2:433–435.
51. Kromann N, Green A. Epidemiological studies in the Upernavik district, Greenland. Incidence of some chronic diseases 1950–1974. *Acta Med Scand*. 1980;208:401–406.
52. Simopoulos AP. The importance of the omega-6/omega-3 fatty acid ratio in cardiovascular disease and other chronic diseases. *Exp Biol Med (Maywood)*. 2008;233:674–688. https://doi.org/10.3181/0711-MR-311.
53. Kremer JM, Lawrence DA, Jubiz W, et al. Dietary fish oil and olive oil supplementation in patients with rheumatoid arthritis. Clinical and immunologic effects. *Arthritis Rheum*. 1990;33:810–820.
54. James SY, Williams MA, Kelsey SM, Newland AC, Colston KW. The role of vitamin D derivatives and retinoids in the differentiation of human leukaemia cells. *Biochem Pharmacol*. 1997;54:625–634.
55. Kremer JM. n-3 Fatty acid supplements in rheumatoid arthritis. *Am J Clin Nutr*. 2000;71:. 349S–51S.
56. Stewart TM, Bowling AC. Polyunsaturated fatty acid supplementation in MS. *Int MS J*. 2005;12:88–93.
57. Weinstock-Guttman B, Baier M, Park Y, et al. Low fat dietary intervention with omega-3 fatty acid supplementation in multiple sclerosis patients. *Prostaglandins Leukot Essent Fatty Acids*. 2005;73:397–404. https://doi.org/10.1016/j.plefa.2005.05.024.
58. Freund-Levi Y, Eriksdotter-Jonhagen M, Cederholm T, et al. Omega-3 fatty acid treatment in 174 patients with mild to moderate Alzheimer disease: OmegAD study: A randomized double-blind trial. *Arch Neurol*. 2006;63:1402–1408. https://doi.org/10.1001/archneur.63.10.1402.
59. Lee LK, Shahar S, Chin AV, Yusoff NA. Docosahexaenoic acid-concentrated fish oil supplementation in subjects with mild cognitive impairment (MCI): a 12-month randomised, double-blind, placebo-controlled trial. *Psychopharmacology (Berl)*. 2013;225:605–612. https://doi.org/10.1007/s00213-012-2848-0.
60. Kiecolt-Glaser JK, Epel ES, Belury MA, et al. Omega-3 fatty acids, oxidative stress, and leukocyte telomere length: a randomized controlled trial. *Brain Behav Immun*. 2013;28:16–24. https://doi.org/10.1016/j.bbi.2012.09.004.
61. Kavanagh T, Lonergan PE, Lynch MA. Eicosapentaenoic acid and gamma-linolenic acid increase hippocampal concentrations of IL-4 and IL-10 and abrogate lipopolysaccharide-induced inhibition of long-term potentiation. *Prostaglandins Leukot Essent Fatty Acids*. 2004;70:391–397. https://doi.org/10.1016/j.plefa.2003.12.014.
62. Lonergan PE, Martin DS, Horrobin DF, Lynch MA. Neuroprotective actions of eicosapentaenoic acid on lipopolysaccharide-induced dysfunction in rat hippocampus. *J Neurochem*. 2004;91:20–29. https://doi.org/10.1111/j.1471-4159.2004.02689.x.
63. Shi Z, Ren H, Huang Z, et al. Fish oil prevents lipopolysaccharide-induced depressive-like behavior by inhibiting Neuroinflammation. *Mol Neurobiol*. November 2016. https://doi.org/10.1007/s12035-016-0212-9.
64. Song C, Horrobin D. Omega-3 fatty acid ethyl-eicosapentaenoate, but not soybean oil, attenuates memory impairment induced by central IL-1beta administration. *J Lipid Res*. 2004;45:1112–1121. https://doi.org/10.1194/jlr.M300526-JLR200.
65. Song C, Manku MS, Horrobin DF. Long-chain polyunsaturated fatty acids modulate interleukin-1beta-induced changes in behavior, monoaminergic neurotransmitters, and brain inflammation in rats. *J Nutr*. 2008;138:954–963.
66. Dong Y, Xu M, Kalueff AV, Song C. Dietary eicosapentaenoic acid normalizes hippocampal omega-3 and 6 polyunsaturated fatty acid profile, attenuates glial activation and regulates BDNF function in a rodent model of neuroinflammation induced by central interleukin-1beta administration. *Eur J Nutr*. May 2017. https://doi.org/10.1007/s00394-017-1462-7.
67. Thomazeau A, Bosch-Bouju C, Manzoni O, Layé S. Nutritional n-3 PUFA deficiency abolishes endocannabinoid gating of hippocampal long-term potentiation. *Cereb Cortex*. 2017;27:2571–2579. https://doi.org/10.1093/cercor/bhw052.
68. Zimmer L, Delpal S, Guilloteau D, Aioun J, Durand G, Chalon S. Chronic n-3 polyunsaturated fatty acid deficiency alters dopamine vesicle density in the rat frontal cortex. *Neurosci Lett*. 2000;284:25–28.

69. Aid S, Vancassel S, Poumes-Ballihaut C, Chalon S, Guesnet P, Lavialle M. Effect of a diet-induced n-3 PUFA depletion on cholinergic parameters in the rat hippocampus. *J Lipid Res*. 2003;44:1545–1551. https://doi.org/10.1194/jlr.M300079-JLR200.
70. Chalon S. Omega-3 fatty acids and monoamine neurotransmission. *Prostaglandins Leukot Essent Fatty Acids*. 2006;75:259–269. https://doi.org/10.1016/j.plefa.2006.07.005.
71. Larrieu T, Hilal ML, Hilal LM, et al. Nutritional omega-3 modulates neuronal morphology in the prefrontal cortex along with depression-related behaviour through corticosterone secretion. *Transl Psychiatry*. 2014;4:e437. https://doi.org/10.1038/tp.2014.77.
72. Larrieu T, Madore C, Joffre C, Layé S. Nutritional n-3 polyunsaturated fatty acids deficiency alters cannabinoid receptor signaling pathway in the brain and associated anxiety-like behavior in mice. *J Physiol Biochem*. 2012;68:671–681. https://doi.org/10.1007/s13105-012-0179-6.
73. Pietropaolo S, Goubran MG, Joffre C, et al. Dietary supplementation of omega-3 fatty acids rescues fragile X phenotypes in Fmr1-Ko mice. *Psychoneuroendocrinology*. 2014;49:119–129. https://doi.org/10.1016/j.psyneuen.2014.07.002.
74. Madore C, Leyrolle Q, Lacabanne C, et al. Neuroinflammation in autism: Plausible role of maternal inflammation, dietary omega 3, and microbiota. *Neural Plast*. 2016;2016:3597209. https://doi.org/10.1155/2016/3597209.
75. Weiser MJ, Mucha B, Denheyer H, et al. Dietary docosahexaenoic acid alleviates autistic-like behaviors resulting from maternal immune activation in mice. *Prostaglandins Leukot Essent Fatty Acids*. 2016;106:27–37. https://doi.org/10.1016/j.plefa.2015.10.005.
76. Lu Y, Zhao LX, Cao DL, Gao YJ. Spinal injection of docosahexaenoic acid attenuates carrageenan-induced inflammatory pain through inhibition of microglia-mediated neuroinflammation in the spinal cord. *Neuroscience*. 2013;241:22–31. https://doi.org/10.1016/j.neuroscience.2013.03.003.
77. Chen X, Wu S, Chen C, et al. Omega-3 polyunsaturated fatty acid supplementation attenuates microglial-induced inflammation by inhibiting the HMGB1/TLR4/NF-kappaB pathway following experimental traumatic brain injury. *J Neuroinflammation*. 2017;14:143. https://doi.org/10.1186/s12974-017-0917-3.
78. Kang JX, Wang J, Wu L, Kang ZB. Transgenic mice: fat-1 mice convert n-6 to n-3 fatty acids. *Nature*. 2004;427:504. https://doi.org/10.1038/427504a.
79. Boudrault C, Bazinet RP, Kang JX, Ma DW. Cyclooxygenase-2 and n-6 PUFA are lower and DHA is higher in the cortex of fat-1 mice. *Neurochem Int*. 2010;56:585–589. https://doi.org/10.1016/j.neuint.2009.12.022.
80. Delpech J-C, Madore C, Joffre C, et al. Transgenic increase in n-3/n-6 fatty acid ratio protects against cognitive deficits induced by an immune challenge through decrease of neuroinflammation. *Neuropsychopharmacology*. 2015;40:525–536. https://doi.org/10.1038/npp.2014.196.
81. De Smedt-Peyrusse V, Sargueil F, Moranis A, Harizi H, Mongrand S, Layé S. Docosahexaenoic acid prevents lipopolysaccharide-induced cytokine production in microglial cells by inhibiting lipopolysaccharide receptor presentation but not its membrane subdomain localization. *J Neurochem*. 2008;105:296–307. https://doi.org/10.1111/j.1471-4159.2007.05129.x.
82. Lu DY, Tsao YY, Leung YM, Su KP. Docosahexaenoic acid suppresses neuroinflammatory responses and induces heme oxygenase-1 expression in BV-2 microglia: Implications of antidepressant effects for omega-3 fatty acids. *Neuropsychopharmacology*. 2010;35:2238–2248. https://doi.org/10.1038/npp.2010.98.
83. Antonietta Ajmone-Cat M, Lavinia Salvatori M, De Simone R, et al. Docosahexaenoic acid modulates inflammatory and antineurogenic functions of activated microglial cells. *J Neurosci Res*. 2012;90:575–587. https://doi.org/10.1002/jnr.22783.
84. Pettit LK, Varsanyi C, Tadros J, Vassiliou E. Modulating the inflammatory properties of activated microglia with docosahexaenoic acid and aspirin. *Lipids Health Dis*. 2013;12:16. https://doi.org/10.1186/1476-511X-12-16.
85. Chen S, Zhang H, Pu H, et al. n-3 PUFA supplementation benefits microglial responses to myelin pathology. *Sci Rep*. 2014;4:7458. https://doi.org/10.1038/srep07458.
86. Corsi L, Dongmo BM, Avallone R. Supplementation of omega 3 fatty acids improves oxidative stress in activated BV2 microglial cell line. *Int J Food Sci Nutr*. 2015;66:293–299. https://doi.org/10.3109/09637486.2014.986073.
87. Inoue T, Tanaka M, Masuda S, et al. Omega-3 polyunsaturated fatty acids suppress the inflammatory responses of lipopolysaccharide-stimulated mouse microglia by activating SIRT1 pathways. *Biochim Biophys Acta*. 1862;2017:552–560. https://doi.org/10.1016/j.bbalip.2017.02.010.

88. Fourrier C, Remus-Borel J, Greenhalgh AD, et al. Docosahexaenoic acid-containing choline phospholipid modulates LPS-induced neuroinflammation in vivo and in microglia in vitro. *J Neuroinflammation*. 2017;14:170. https://doi.org/10.1186/s12974-017-0939-x.
89. Moon DO, Kim KC, Jin CY, et al. Inhibitory effects of eicosapentaenoic acid on lipopolysaccharide-induced activation in BV2 microglia. *Int Immunopharmacol*. 2007;7:222–229. https://doi.org/10.1016/j.intimp.2006.10.001.
90. Tremblay ME, Zhang I, Bisht K, et al. Remodeling of lipid bodies by docosahexaenoic acid in activated microglial cells. *J Neuroinflammation*. 2016;13:116. https://doi.org/10.1186/s12974-016-0580-0.
91. Hjorth E, Zhu M, Toro VC, et al. Omega-3 fatty acids enhance phagocytosis of Alzheimer's disease-related amyloid-beta42 by human microglia and decrease inflammatory markers. *J Alzheimers Dis*. 2013;35:697–713. https://doi.org/10.3233/JAD-130131.
92. Serhan CN, Chiang N, Dalli J. The resolution code of acute inflammation: Novel pro-resolving lipid mediators in resolution. *Semin Immunol*. 2015;27:200–215. https://doi.org/10.1016/j.smim.2015.03.004.
93. Sugimoto MA, Sousa LP, Pinho V, Perretti M, Teixeira MM. Resolution of inflammation: What controls its onset? *Front Immunol*. 2016;7:160. https://doi.org/10.3389/fimmu.2016.00160.
94. Serhan CN, Clish CB, Brannon J, Colgan SP, Chiang N, Gronert K. Novel functional sets of lipid-derived mediators with antiinflammatory actions generated from omega-3 fatty acids via cyclooxygenase 2-nonsteroidal antiinflammatory drugs and transcellular processing. *J Exp Med*. 2000;192:1197–1204.
95. Serhan CN. Treating inflammation and infection in the 21st century: New hints from decoding resolution mediators and mechanisms. *FASEB J*. 2017;31:1273–1288. https://doi.org/10.1096/fj.201601222R.
96. Halade GV, Black LM, Verma MK. Paradigm shift-metabolic transformation of docosahexaenoic and eicosapentaenoic acids to bioactives exemplify the promise of fatty acid drug discovery. *Biotechnol Adv*. 2018;36:935–953. https://doi.org/10.1016/j.biotechadv.2018.02.014.
97. Ohira T, Arita M, Omori K, Recchiuti A, Van Dyke TE, Serhan CN. Resolvin E1 receptor activation signals phosphorylation and phagocytosis. *J Biol Chem*. 2010;285:3451–3461.
98. Chen CT, Liu Z, Bazinet RP. Rapid de-esterification and loss of eicosapentaenoic acid from rat brain phospholipids: an intracerebroventricular study. *J Neurochem*. 2011;116:363–373. https://doi.org/10.1111/j.1471-4159.2010.07116.x.
99. Siegert E, Paul F, Rothe M, Weylandt KH. The effect of omega-3 fatty acids on central nervous system remyelination in fat-1 mice. *BMC Neurosci*. 2017;18:19. https://doi.org/10.1186/s12868-016-0312-5.
100. Samson M, Edinger AL, Stordeur P, et al. ChemR23, a putative chemoattractant receptor, is expressed in monocyte-derived dendritic cells and macrophages and is a coreceptor for SIV and some primary HIV-1 strains. *Eur J Immunol*. 1998;28:1689–1700. https://doi.org/10.1002/(SICI)1521-4141(199805)28:05<1689::AID-IMMU1689>3.0.CO;2-I.
101. Arita M, Ohira T, Sun YP, Elangovan S, Chiang N, Serhan CN. Resolvin E1 selectively interacts with leukotriene B4 receptor BLT1 and ChemR23 to regulate inflammation. *J Immunol*. 2007;178:3912–3917.
102. Arita M, Bianchini F, Aliberti J, et al. Stereochemical assignment, antiinflammatory properties, and receptor for the omega-3 lipid mediator resolvin E1. *J Exp Med*. 2005;201:713–722.
103. Ji A, Diao H, Wang X, et al. n-3 Polyunsaturated fatty acids inhibit lipopolysaccharide-induced microglial activation and dopaminergic injury in rats. *Neurotoxicology*. 2012;33:780–788. https://doi.org/10.1016/j.neuro.2012.02.018.
104. Guo X, Fu Y, Xu Y, et al. Chronic mild restraint stress rats decreased CMKLR1 expression in distinct brain region. *Neurosci Lett*. 2012;524:25–29. https://doi.org/10.1016/j.neulet.2012.06.075.
105. Graham KL, Zabel BA, Loghavi S, et al. Chemokine-like receptor-1 expression by central nervous system-infiltrating leukocytes and involvement in a model of autoimmune demyelinating disease. *J Immunol*. 2009;183:6717–6723. https://doi.org/10.4049/jimmunol.0803435.
106. Rey C, Nadjar A, Buaud B, et al. Resolvin D1 and E1 promote resolution of inflammation in microglial cells in vitro. *Brain Behav Immun*. 2016;55:249–259. https://doi.org/10.1016/j.bbi.2015.12.013.
107. Xu ZZ, Zhang L, Liu T, et al. Resolvins RvE1 and RvD1 attenuate inflammatory pain via central and peripheral actions. *Nat Med*. 2010;16:592–597. 1p following 597, https://doi.org/10.1038/nm.2123.
108. Xu ZZ, Berta T, Ji RR. Resolvin E1 inhibits neuropathic pain and spinal cord microglial activation following peripheral nerve injury. *J Neuroimmune Pharmacol*. 2013;8:37–41. https://doi.org/10.1007/s11481-012-9394-8.
109. Harrison JL, Rowe RK, Ellis TW, et al. Resolvins AT-D1 and E1 differentially impact functional outcome, post-traumatic sleep, and microglial activation following diffuse brain injury in the mouse. *Brain Behav Immun*. 2015;47:131–140. https://doi.org/10.1016/j.bbi.2015.01.001.

110. Spite M, Serhan CN. Novel lipid mediators promote resolution of acute inflammation: impact of aspirin and statins. *Circ Res*. 2010;107:1170–1184. https://doi.org/10.1161/CIRCRESAHA.110.223883.
111. Recchiuti A. Resolvin D1 and its GPCRs in resolution circuits of inflammation. *Prostaglandins Other Lipid Mediat*. 2013;107:64–76. https://doi.org/10.1016/j.prostaglandins.2013.02.004.
112. Bisicchia E, Sasso V, Catanzaro G, et al. Resolvin D1 halts remote Neuroinflammation and improves functional recovery after focal brain damage via ALX/FPR2 receptor-regulated MicroRNAs. *Mol Neurobiol*. 2018;55:6894–6905. https://doi.org/10.1007/s12035-018-0889-z.
113. Krishnamoorthy S, Recchiuti A, Chiang N, et al. Resolvin D1 binds human phagocytes with evidence for proresolving receptors. *Proc Natl Acad Sci U S A*. 2010;107:1660–1665. https://doi.org/10.1073/pnas.0907342107.
114. Sun Y-P, Oh SF, Uddin J, et al. Resolvin D1 and its aspirin-triggered 17R epimer. Stereochemical assignments, anti-inflammatory properties, and enzymatic inactivation. *J Biol Chem*. 2007;282:9323–9334. https://doi.org/10.1074/jbc.M609212200.
115. Fredman G, Serhan CN. Specialized proresolving mediator targets for RvE1 and RvD1 in peripheral blood and mechanisms of resolution. *Biochem J*. 2011;437:185–197.
116. Recchiuti A, Krishnamoorthy S, Fredman G, Chiang N, Serhan CN. MicroRNAs in resolution of acute inflammation: identification of novel resolvin D1-miRNA circuits. *FASEB J*. 2011;25:544–560.
117. Hong S, Gronert K, Devchand PR, Moussignac RL, Serhan CN. Novel docosatrienes and 17S-resolvins generated from docosahexaenoic acid in murine brain, human blood, and glial cells. Autacoids in anti-inflammation. *J Biol Chem*. 2003;278:14677–14687. https://doi.org/10.1074/jbc.M300218200.
118. Marcheselli VL, Hong S, Lukiw WJ, et al. Novel docosanoids inhibit brain ischemia-reperfusion-mediated leukocyte infiltration and pro-inflammatory gene expression. *J Biol Chem*. 2003;278:43807–43817. https://doi.org/10.1074/jbc.M305841200.
119. Shalini SM, Ho CF, Ng YK, et al. Distribution of Alox15 in the rat brain and its role in prefrontal cortical Resolvin D1 formation and spatial working memory. *Mol Neurobiol*. February 2017. https://doi.org/10.1007/s12035-017-0413-x.
120. Abdelmoaty S, Wigerblad G, Bas DB, et al. Spinal actions of lipoxin A4 and 17(R)-resolvin D1 attenuate inflammation-induced mechanical hypersensitivity and spinal TNF release. *PLoS One*. 2013;8:e75543. https://doi.org/10.1371/journal.pone.0075543.
121. Li L, Wu Y, Wang Y, et al. Resolvin D1 promotes the interleukin-4-induced alternative activation in BV-2 microglial cells. *J Neuroinflammation*. 2014;11:72. https://doi.org/10.1186/1742-2094-11-72.
122. Farooqui AA, Horrocks LA. Phospholipase A2-generated lipid mediators in the brain: the good, the bad, and the ugly. *Neuroscientist*. 2006;12:245–260. https://doi.org/10.1177/1073858405285923.
123. Kirkby NS, Chan MV, Zaiss AK, et al. Systematic study of constitutive cyclooxygenase-2 expression: role of NF-kappaB and NFAT transcriptional pathways. *Proc Natl Acad Sci U S A*. 2016;113:434–439. https://doi.org/10.1073/pnas.1517642113.
124. Nadjar A, Tridon V, May MJ, et al. NFkappaB activates in vivo the synthesis of inducible Cox-2 in the brain. *J Cereb Blood Flow Metab*. 2005;25:1047–1059. https://doi.org/10.1038/sj.jcbfm.9600106.
125. Davidson J, Abul HT, Milton AS, Rotondo D. Cytokines and cytokine inducers stimulate prostaglandin E2 entry into the brain. *Pflugers Arch*. 2001;442:526–533.
126. Salinas G, Rangasetty UC, Uretsky BF, Birnbaum Y. The cycloxygenase 2 (COX-2) story: it's time to explain, not inflame. *J Cardiovasc Pharmacol Ther*. 2007;12:98–111. https://doi.org/10.1177/1074248407301172.
127. Engstrom L, Ruud J, Eskilsson A, et al. Lipopolysaccharide-induced fever depends on prostaglandin E2 production specifically in brain endothelial cells. *Endocrinology*. 2012;153:4849–4861. https://doi.org/10.1210/en.2012-1375.
128. Czapski GA, Gajkowska B, Strosznajder JB. Systemic administration of lipopolysaccharide induces molecular and morphological alterations in the hippocampus. *Brain Res*. 2010;1356:85–94. https://doi.org/10.1016/j.brainres.2010.07.096.
129. Davidge ST. Prostaglandin H synthase and vascular function. *Circ Res*. 2001;89:650–660.
130. Masferrer JL, Zweifel BS, Manning PT, et al. Selective inhibition of inducible cyclooxygenase 2 in vivo is antiinflammatory and nonulcerogenic. *Proc Natl Acad Sci U S A*. 1994;91:3228–3232.
131. Seibert K, Masferrer JL. Role of inducible cyclooxygenase (COX-2) in inflammation. *Receptor*. 1994;4:17–23.
132. Murray HJ, O'Connor JJ. A role for COX-2 and p38 mitogen activated protein kinase in long-term depression in the rat dentate gyrus in vitro. *Neuropharmacology*. 2003;44:374–380.

133. Hewett SJ, Bell SC, Hewett JA. Contributions of cyclooxygenase-2 to neuroplasticity and neuropathology of the central nervous system. *Pharmacol Ther*. 2006;112:335–357. https://doi.org/10.1016/j.pharmthera.2005.04.011.
134. Cortet B, Duquesnoy B. Action of non-steroidal anti-inflammatory agents on the immune system. *Rev Rhum Mal Osteoartic*. 1991;58:379–386.
135. Takemiya T, Maehara M, Matsumura K, Yasuda S, Sugiura H, Yamagata K. Prostaglandin E2 produced by late induced COX-2 stimulates hippocampal neuron loss after seizure in the CA3 region. *Neurosci Res*. 2006;56:103–110. https://doi.org/10.1016/j.neures.2006.06.003.
136. Birnie M, Morrison R, Camara R, Strauss KI. Temporal changes of cytochrome P450 (Cyp) and eicosanoid-related gene expression in the rat brain after traumatic brain injury. *BMC Genomics*. 2013;14:303. https://doi.org/10.1186/1471-2164-14-303.
137. Pratico D, Zhukareva V, Yao Y, et al. 12/15-lipoxygenase is increased in Alzheimer's disease: possible involvement in brain oxidative stress. *Am J Pathol*. 2004;164:1655–1662. https://doi.org/10.1016/S0002-9440(10)63724-8.
138. Wang X, Zhu M, Hjorth E, et al. Resolution of inflammation is altered in Alzheimer's disease. *Alzheimers Dement*. 2015;11:40–50. e1–2, https://doi.org/10.1016/j.jalz.2013.12.024.
139. Yigitkanli K, Zheng Y, Pekcec A, Lo EH, van Leyen K. Increased 12/15-lipoxygenase leads to widespread brain injury following global cerebral ischemia. *Transl Stroke Res*. 2017;8:194–202. https://doi.org/10.1007/s12975-016-0509-z.
140. Sun L, Xu YW, Han J, Liang H, Wang N, Cheng Y. 12/15-Lipoxygenase metabolites of arachidonic acid activate PPARgamma: a possible neuroprotective effect in ischemic brain. *J Lipid Res*. 2015;56:502–514. https://doi.org/10.1194/jlr.M053058.
141. Fredman G, Ozcan L, Spolitu S, et al. Resolvin D1 limits 5-lipoxygenase nuclear localization and leukotriene B4 synthesis by inhibiting a calcium-activated kinase pathway. *Proc Natl Acad Sci U S A*. 2014;111:14530–14535. https://doi.org/10.1073/pnas.1410851111.
142. Jatana M, Giri S, Ansari MA, et al. Inhibition of NF-kappaB activation by 5-lipoxygenase inhibitors protects brain against injury in a rat model of focal cerebral ischemia. *J Neuroinflammation*. 2006;3:12. https://doi.org/10.1186/1742-2094-3-12.
143. Martel-Pelletier J, Lajeunesse D, Reboul P, Pelletier JP. Therapeutic role of dual inhibitors of 5-LOX and COX, selective and non-selective non-steroidal anti-inflammatory drugs. *Ann Rheum Dis*. 2003;62:501–509.
144. Lin HC, Lin TH, Wu MY, et al. 5-Lipoxygenase inhibitors attenuate TNF-alpha-induced inflammation in human synovial fibroblasts. *PLoS One*. 2014;9:e107890. https://doi.org/10.1371/journal.pone.0107890.
145. Bystrom J, Wray JA, Sugden MC, et al. Endogenous epoxygenases are modulators of monocyte/macrophage activity. *PLoS One*. 2011;6:e26591. https://doi.org/10.1371/journal.pone.0026591.
146. Fleming I. Cytochrome P450-dependent eicosanoid production and crosstalk. *Curr Opin Lipidol*. 2011;22:403–409. https://doi.org/10.1097/MOL.0b013e32834a9790.
147. Nebert DW, Wikvall K, Miller WL. Human cytochromes P450 in health and disease. *Philos Trans R Soc Lond B Biol Sci*. 2013;368:20120431. https://doi.org/10.1098/rstb.2012.0431.
148. Gilroy DW, Edin ML, De Maeyer RP, et al. CYP450-derived oxylipins mediate inflammatory resolution. *Proc Natl Acad Sci U S A*. 2016;113:E3240–E3249. https://doi.org/10.1073/pnas.1521453113.
149. Rao JS, Ertley RN, DeMar Jr JC, Rapoport SI, Bazinet RP, Lee HJ. Dietary n-3 PUFA deprivation alters expression of enzymes of the arachidonic and docosahexaenoic acid cascades in rat frontal cortex. *Mol Psychiatry*. 2007;12:151–157. https://doi.org/10.1038/sj.mp.4001887.
150. Kim HW, Rao JS, Rapoport SI, Igarashi M. Dietary n-6 PUFA deprivation downregulates arachidonate but upregulates docosahexaenoate metabolizing enzymes in rat brain. *Biochim Biophys Acta*. 2011;1811:111–117. https://doi.org/10.1016/j.bbalip.2010.10.005.
151. Taha AY, Blanchard HC, Cheon Y, et al. Dietary linoleic acid lowering reduces lipopolysaccharide-induced increase in brain arachidonic acid metabolism. *Mol Neurobiol*. 2017;54:4303–4315. https://doi.org/10.1007/s12035-016-9968-1.
152. Ostermann AI, Reutzel M, Hartung N, et al. A diet rich in omega-3 fatty acids enhances expression of soluble epoxide hydrolase in murine brain. *Prostaglandins Other Lipid Mediat*. June 2017. https://doi.org/10.1016/j.prostaglandins.2017.06.001.
153. Hashimoto M, Katakura M, Tanabe Y, et al. n-3 fatty acids effectively improve the reference memory-related learning ability associated with increased brain docosahexaenoic acid-derived docosanoids in aged rats. *Biochim Biophys Acta*. 2015;1851:203–209. https://doi.org/10.1016/j.bbalip.2014.10.009.

154. Balvers MG, Verhoeckx KC, Bijlsma S, et al. Fish oil and inflammatory status alter the n-3 to n-6 balance of the endocannabinoid and oxylipin metabolomes in mouse plasma and tissues. *Metabolomics*. 2012;8:1130–1147. https://doi.org/10.1007/s11306-012-0421-9.
155. Lazic M, Inzaugarat ME, Povero D, et al. Reduced dietary omega-6 to omega-3 fatty acid ratio and 12/15-lipoxygenase deficiency are protective against chronic high fat diet-induced steatohepatitis. *PLoS One*. 2014;9:e107658. https://doi.org/10.1371/journal.pone.0107658.
156. Shevalye H, Yorek MS, Coppey LJ, et al. Effect of enriching the diet with menhaden oil or daily treatment with resolvin D1 on neuropathy in a mouse model of type 2 diabetes. *J Neurophysiol*. 2015;114:199–208. https://doi.org/10.1152/jn.00224.2015.
157. Taha AY, Hennebelle M, Yang J, et al. Regulation of rat plasma and cerebral cortex oxylipin concentrations with increasing levels of dietary linoleic acid. *Prostaglandins Leukot Essent Fatty Acids*. May 2016; https://doi.org/10.1016/j.plefa.2016.05.004.
158. Trépanier M-O, Hopperton KE, Giuliano V, Masoodi M, Bazinet RP. Increased brain docosahexaenoic acid has no effect on the resolution of neuroinflammation following intracerebroventricular lipopolysaccharide injection. *Neurochem Int*. 2018;118:115–126. https://doi.org/10.1016/j.neuint.2018.05.010.
159. Rosenberger TA, Villacreses NE, Hovda JT, et al. Rat brain arachidonic acid metabolism is increased by a 6-day intracerebral ventricular infusion of bacterial lipopolysaccharide. *J Neurochem*. 2004;88:1168–1178.
160. Farias SE, Basselin M, Chang L, Heidenreich KA, Rapoport SI, Murphy RC. Formation of eicosanoids, E2/D2 isoprostanes, and docosanoids following decapitation-induced ischemia, measured in high-energy-microwaved rat brain. *J Lipid Res*. 2008;49:1990–2000. https://doi.org/10.1194/jlr.M800200-JLR200.
161. Hennebelle M, Zhang Z, Metherel AH, et al. Linoleic acid participates in the response to ischemic brain injury through oxidized metabolites that regulate neurotransmission. *Sci Rep*. 2017;7:4342. https://doi.org/10.1038/s41598-017-02914-7.
162. Dieter P, Scheibe R, Kamionka S, Kolada A. LPS-induced synthesis and release of PGE2 in liver macrophages: regulation by CPLA2, COX-1, COX-2, and PGE2 synthase. *Adv Exp Med Biol*. 2002;507:457–462.
163. Dalli J, Serhan CN. Specific lipid mediator signatures of human phagocytes: Microparticles stimulate macrophage efferocytosis and pro-resolving mediators. *Blood*. 2012;120:e60–e72. https://doi.org/10.1182/blood-2012-04-423525.
164. Simopoulos AP. n-3 fatty acids and human health: defining strategies for public policy. *Lipids*. 2001;36. Suppl:S83–89.
165. Ariel A, Serhan CN. Resolvins and protectins in the termination program of acute inflammation. *Trends Immunol*. 2007;28:176–183.
166. Bannenberg G, Serhan CN. Specialized pro-resolving lipid mediators in the inflammatory response: An update. *Biochim Biophys Acta*. 1801;2010:1260–1273. https://doi.org/10.1016/j.bbalip.2010.08.002.
167. Recchiuti A, Codagnone M, Pierdomenico AM, et al. Immunoresolving actions of oral resolvin D1 include selective regulation of the transcription machinery in resolution-phase mouse macrophages. *FASEB J*. 2014;28:3090–3102. https://doi.org/10.1096/fj.13-248393.

CHAPTER

17

Omega-3 Polyunsaturated Fatty Acid Supplementation as a Potential Protective Factor or Treatment for Neurodegenerative Diseases

Erika Rodriguez-Guzman

The University of Arizona, Tucson, AZ, United States

INTRODUCTION

Omega-3 fatty acids are essential polyunsaturated fatty acids. Humans are unable to synthesize omega-3 fatty acids and must obtain them from their diet or supplementation.[1] Omega-3 fatty acids are metabolic precursors to alpha linolenic acid (ALA) and two long-chain omega-3 polyunsaturated fatty acids, docosahexaenoic acid (DHA), and eicosapentaenoic acid (EPA).[2] ALA is derived from dietary sources including nuts and seeds, and DHA is derived primarily from fish consumption.[2] Omega-6 fatty acids are the other major class of essential fatty acids. These are obtained from the diet as linolenic acid and then are elongated to gamma-linoleic acid and arachidonic acid in the body.[1]

The eicosanoids derived from omega-3 fatty acid prostaglandin metabolism are anti- inflammatory, while those derived from the omega-6 dependent arachidonic acid pathway are pro-inflammatory.[1] The omega-3 to omega-6 consumption ratio can, therefore, impact systemic inflammatory processes and immune function. Increased omega-3 consumption is believed to provide potential health benefits. DHA, in particular, is implicated in neural and visual development.[1]

Neurodegeneration is the slow, progressive dysfunction and loss of axons and neurons in the central nervous system (CNS).[3] It is the primary pathological component in neurodegenerative diseases.[3] Some of the most common neurodegenerative diseases include Alzheimer's disease (AD), frontotemporal dementia, Parkinson's disease (PD), and multiple sclerosis (MS).[4] Huntington's disease (HD), is the most common inherited neurodegenerative disease.[5]

Omega Fatty Acids in Brain and Neurological Health
https://doi.org/10.1016/B978-0-12-815238-6.00017-1

Diseases such as frontotemporal dementia, AD, HD, PD, and MS share neurodegeneration as the primary pathological component.[4] Neurodegeneration leads to slow, progressive dysfunction and loss of axons and neurons in the CNS.[3]

Many neurodegenerative diseases share other multiple clinical, molecular, and pathologic characteristics, as well as familial histories that indicate the role of genes in progression and causation.[4] Though some neurodegenerative diseases occur solely due to genetics, the majority are considered to occur due to a multifactorial pattern of inheritance that includes both genetic and environmental factors.[6] Neurodegenerative diseases tend to cluster in families due to family members sharing genetic information and similar environmental exposures, making them more likely to experience similar gene–gene and gene–environment interactions that lead to disease.[6]

Neurodegenerative diseases are a significant threat to human health and are becoming increasingly prevalent.[7] These diseases have diverse effects, including cognitive impairment and progressively losing the ability to move, speak, and breathe. As diseases tend to be age-dependent, the need for effective treatment becomes increasingly dire as the population ages.[7]

OMEGA-3 FATTY ACIDS IN THE TREATMENT OF DEMENTIA

Dementia is a debilitating, progressive disease that causes memory loss, language problems, psychological changes, and difficulties in performing activities of daily living.[2] Dementia primarily affects older adults; however, it can affect younger adults as well.[2] It is estimated that 80 million people will be affected by dementia by 2040 and 70% of these individuals will live in low-income countries.[2]

The currently available pharmacological therapies offer some symptomatic improvements but are unable to stop dementia progression.[2] In addition, due to negative personal effects and the cost to society associated with dementia care, there is a need to identify modifiable risk factors that may prevent the onset of dementia. Omega-3 fatty acid supplementation is among the proposed dietary interventions.[2]

Omega-3 fatty acid consumption may be a protective factor against the onset of dementia.[2] The brain is rich in fatty acids. DHA is a key component of brain phospholipids. Therefore, omega-3 fatty acid consumption may contribute to the maintenance of neuronal functioning and membrane integrity.[2] In addition, omega-3 oxidative products are key mediators for immunity and allergy, inflammation, oxidative stress, vascular responses and thrombosis, and bronchial constriction. This indicates they may be particularly effective against vascular dementia.[2] Furthermore, DHA may modify gene expression for genes that regulate neurogenesis and neuronal function, as well as other biological functions that influence cognitive health.[2]

However, no randomized control trials (RCT) have investigated the effects of Omega-3 fatty acids on incident dementia.[2] The short-term studies with secondary outcomes investigating the effects of omega-3 fatty acid supplementation on cognitive decline have shown no benefits.[2] There is a need for longer-term studies that investigate the effects of omega-3 fatty acid on cognitive decline, particularly in older populations.[2]

OMEGA-3 FATTY ACIDS IN THE TREATMENT OF ALZHEIMER'S DISEASE

AD, an age-dependent neurodegenerative disease, is the most prevalent form of dementia.[8] The characteristics of AD include short-term memory loss and impairment in cognitive thinking, ability to learn, sense of direction, oral communication, and calculation. Individuals with late-stage AD may develop depression, language deficits, psychosis, and aggressive behavior.[8] Ultimately, AD results in irreversible loss of cognitive function and memory.[9] AD is expected to become increasingly prevalent in the future, highlighting the need for the development of effective interventions.[9]

Omega-3 fatty acids, particularly DHA, are abundant in brain tissue membranes.[10] Diets enriched with DHA have been found to protect against cognitive decline in mouse models.[10] This has led to increased interest in the potential role of dietary interventions for the prevention of AD.[10]

A recent meta-analysis of prospective cohort studies evaluating dietary omega-3 fatty acid intake and dementia and AD risk determined that increased dietary intake of omega-3 fatty acids was not associated with decreased AD risk.[10] DHA supplementation has not been found to slow cognitive and functional decline in patients with mild to moderate AD[10]; however, there may be a potential protective effect of omega-3 fatty acid intake against AD incidence.[10]

Wu et al. (2014) found slight statistical significance for a 36% decreased risk of AD development in groups with highest fish consumption when compared to groups with lowest fish consumption.[10] The highest fish intake category was 500 g or more.[10] A weekly intake of 100 g of fish was found to be linked to an 11% decrease in risk of AD development.[10] The high omega-3 fatty acid content in fish is thought to be responsible for the protective effect of fish consumption associated with a 36% decreased risk for AD development.[10] However, the high nutrient, vitamin, amino acid, and trace elements available in fish or the replacement of dietary saturated fat sources with fish may be the alternative explanation for this protective factor.[10]

Given the potential for a protective factor against AD and the paucity of statistically significant evidence, there is a need for future studies to investigate this potential dose–response relationship. In addition, there is a need for future studies that investigate adequate omega-3 fatty acid dosages in comparison to fish consumption.

OMEGA-3 FATTY ACIDS IN THE TREATMENT OF FRONTOTEMPORAL DEMENTIA

Frontotemporal dementia is the leading early-onset dementia and is the third most common form of dementia.[11] It is a neurodegenerative disorder designated by progressive linguistic, behavioral, and executive functioning deficits.[11] There are three clinical variations of frontotemporal dementia: behavioral-variant frontotemporal dementia, non-fluent variant primary progressive aphasia, and semantic-variant primary progressive aphasia.[11] As the condition progresses, patients develop motor deficits and global cognitive impairment. The end-stage patients experience difficulty moving, swallowing, and eating. Death typically occurs 8 years following the symptom onset.[11]

In spite of the prevalence of frontotemporal dementia, there is a scarcity of pharmacological treatment trials dedicated to the treatment of frontotemporal dementia.[11] Despite promising evidence that omega-3 polyunsaturated fatty acids may improve brain function and dementia progression, no RCTs have been conducted that investigate omega-3 polyunsaturated fatty acid supplementation or dietary enrichment in frontotemporal dementia patients.[12] Future studies that investigate the potential effects of omega-3 intake on the progress of frontotemporal dementia are needed.

OMEGA-3 FATTY ACIDS IN THE TREATMENT OF MULTIPLE SCLEROSIS

MS is a chronic, progressive neurodegenerative disease involving CNS demyelination.[13,14] MS is the most prevalent chronic, disabling CNS disease in both young and middle-aged adults.[15] In patients with MS, the immune system is involved in myelin and nerve cell destruction.[16] The pathologic trademark of MS is the inflammatory demyelination plaques in the CNS.[16] In addition, it appears that inflammation triggers pathogenesis in MS and eventually leads to oxidative damage and injury to the mitochondria.[17] In early stages, MS responds to anti-inflammatory therapies.[18]

Omega-3 supplementation and fish consumption as a treatment for MS is controversial but appears to be beneficial.[19] Jelinek et al. (2013) investigated the association between the quality of life, disability, and fish consumption or omega-3 fatty acid supplementation.[19] This study found evidence supporting the role of omega-3 intake and fish consumption in the pathogenesis and expression of MS.[19] Investigators determined that omega-3 supplementation and fish consumption led to improvements in all components of the MS Quality of Life-54, a validated health-related quality-of-life measure.[19,20]

A review conducted by Farinotti et al. (2012) noted that omega-3 polyunsaturated fatty acid supplementation does not influence MS disease progression.[21] However, disease relapse appeared to be less frequent in MS patients who supplemented their diets with omega-3 polyunsaturated fatty acids, but this was a weak effect that no longer appeared following a sensitivity analysis.[21] No adverse effects were reported in any of the six studies reviewed.[21]

MS patients who previously consumed very low-fat diets and begin supplementing with omega-3 fatty acids appear to improve their physical and mental assessments, and the severity of the disease appears to decrease.[21] However, there is a lack of statistical significance supporting this finding.[21] The current body of literature on the effects of omega-3 fatty acids on MS is insufficient.[21] Further research is needed to determine the potential effects of dietary supplementation in the treatment of MS.[21]

OMEGA-3 FATTY ACIDS IN THE TREATMENT OF PARKINSON'S DISEASE

PD is a highly prevalent neurodegenerative disease that typically develops late in life.[22] Though the loss of dopaminergic neurons in the substantia nigra pars compacta is known to occur in PD, the cause of PD is unknown in most cases. Environmental and genetic factors, particularly those which mediate inflammation, autosomal autophagy system abrogation, and mitochondrial dysfunction are believed to contribute to PD development. In addition, dietary factors may influence the development of PD.[22]

Epidemiological studies indicate that Mediterranean diet adherence is associated with decreased incidence of PD.[23] In addition, high consumption of omega-3 polyunsaturated fatty acids was found to be associated with decreased PD risk in a population-based prospective cohort study.[24]

Though omega-3 polyunsaturated fatty acids appear to be protective factors against the development of a number of neurodegenerative diseases, thus far no intervention studies have investigated whether they are neuroprotective against PD.[22] Interestingly, da Silva et al. (2008) determined that fish oil supplementation improves depressive symptoms in PD patients.[25] This is a significant finding, as depression occurs in approximately 50% of the PD patients, yet antidepressant and PD medication interactions can cause adverse effects.[24] In da Silva et al. (2008), fish oil supplementation was not associated with any significant adverse effect.[25] This is a promising finding and indicates a need for future studies examining fish oil supplementation in larger samples of PD patients.[25]

OMEGA-3 FATTY ACIDS IN THE TREATMENT OF HUNTINGTON'S DISEASE

HD is a progressive neurodegenerative disorder that is autosomal-dominant.[26] It is a rare disorder, with a prevalence of 5–10 per 100,000 in Caucasians.[27]

The HD phenotype includes incoordination, chorea, dystonia, behavioral issues, and cognitive decline.[26] In addition, individuals with HD typically suffer from unintended weight loss, autonomic nervous system dysfunction, and sleep- and circadian rhythm disturbance.[28] Symptom onset typically occurs in middle age after having children, but they can begin at any time point. Individuals with HD who are considered to be in the "healthy period" do not have measurable abnormalities.[26] Following the healthy period is the "prediagnostic phase," in which there are measurable abnormalities, but the individual may be unaware of them.[26] In this phase, individuals with HD may become irritable and disinhibited and may experience increasing anxiety and forgetfulness. Individuals then progress to the "diagnostic stage" in which they begin to demonstrate incoordination, chorea, and motor impersistence.[26] In addition, individuals in this stage undergo progressive cognitive changes that affect their executive functioning, language, perceptual and spatial skills, and memory.[29] As HD progresses, these symptoms become increasingly severe.[29] Predictive testing is available for at-risk individuals.[30]

Disease pathology for HD involves a loss of neurons.[30] This neuron loss is concentrated in the striatum and cortex. The primary neurons targeted are medium spiny neurons that typically project into the lateral globus pallidus and contain γ-aminobutyric acid and enkephalin.[30] As the disease progresses, degeneration disseminates to the basal ganglia, cortex, and substantia nigra.[30] Mitochondrial dysfunction resulting in apoptosis is believed to play a significant role in HD pathophysiology.[27]

Currently, there are no treatments for HD that can delay or change the course of HD, but symptomatic therapies are beneficial for some individuals.[30] Dopamine, glutamate, and γ-aminobutyric acid are the current pharmacotherapy focuses as they are believed to be the most affected.[30]

Preliminary studies indicate that ethyl-eicosapentaenoic acid (Ethyl-EPA) may have beneficial effects for individuals with HD.[31] Ethyl-EPA is an ethyl ester EPA prodrug involved in fatty acid metabolism and the neuroendocrine system.[32] Ethyl-EPA reduces protein kinase C (PKC) activity, inhibits the apoptotic pathway, and inhibits phospholipase A2 (PLA2) activity.[32] Ethyl-EPA appears to modify the NF-kappa B pathway and the c-Jun amino-terminal kinases (JNK) pathways, two important pathways in HD.[32] These pathways are believed to cause mitochondrial dysfunction by activating antiproliferative factor p53.[32]

The literature on ethyl-EPA as an HD treatment is mixed. Overall, ethyl-EPA appears not to improve any HD clinical features.[31,33,34] However, one study found that ethyl-EPA is associated with significant brain atrophy reduction, predominantly in the thalamus and caudate.[35] As so few studies have investigated the influence of ethyl-EPA on HD, there is a need for future studies.

DISCUSSION

The neuroprotective properties exhibited by omega-3 polyunsaturated fatty acids may indicate their potential as treatments or protective factors against neurodegenerative diseases. Thus far, no RCTs have investigated the effect of omega-3 fatty acids on incident dementia or frontotemporal dementia. Omega-3 intake has not been found to decrease AD risk. High levels of fish consumption may be a protective factor against AD incidence. Omega-3 supplementation and fish consumption appear to improve quality of life in individuals with MS, decrease PD occurrence, and improve depressive symptoms in individuals with PD. The influence of omega-3 supplementation on HD is unclear, but ethyl-EPA may have beneficial effects. Due to the prevalence of neurodegenerative diseases and the paucity of information regarding their efficacy, as well as conflicting evidence from various studies, there is a need for further research to determine the effects of omega-3 fatty acid supplementation on neurodegenerative diseases.

References

1. Zhang A, MacIsaac R, Roberts L, et al. Omega-3 polyunsaturated fatty acid supplementation for improving peripheral nerve health: protocol for a systematic review. *BMJ Open*. 2018;8(3):e020804. https://doi.org/10.1136/bmjopen-2017-020804.
2. Sydenham E, Dangour A, Lim W. Omega 3 fatty acid for the prevention of cognitive decline and dementia. *Cochrane Database Syst Rev*. 2012. https://doi.org/10.1002/14651858.cd005379.pub3.
3. Amor S, Puentes F, Baker D, van der Valk P. Inflammation in neurodegenerative diseases. *Immunology*. 2010;129(2):154–169. https://doi.org/10.1111/j.1365-2567.2009.03225.x.
4. Lill C, Bertram L. Towards unveiling the Genetics of neurodegenerative diseases. *Semin Neurol*. 2011;31(05):531–541. https://doi.org/10.1055/s-0031-1299791.
5. Finkbeiner S. Huntington's disease. *Cold Spring Harb Perspect Biol*. 2011;3(6):a007476. https://doi.org/10.1101/cshperspect.a007476.
6. Fenoglio C. Genetics and Epigenetics in the neurodegenerative disorders of the central nervous system. In: Galimberti G, Scarpini E, eds. *Neurodegenerative Diseases: Clinical Aspects, Molecular Genetics and Biomarkers*. 1st edSpringer International Publishing; 2018:1–20.
7. Gitler A, Dhillon P, Shorter J. Neurodegenerative disease: models, mechanisms, and a new hope. *Dis Model Mech*. 2017;10(5):499–502. https://doi.org/10.1242/dmm.030205.

8. Ghezzi L. Diagnosis of Alzheimer's disease typical and atypical forms. In: Galimberti G, Scarpini G, eds. *Neurodegenerative Diseases: Clinical Aspects, Molecular Genetics and Biomarkers*. 1st ed.Springer International Publishing; 2018:21–22.
9. Mazza M, Pomponi M, Janiri L, Bria P, Mazza S. Omega-3 fatty acids and antioxidants in neurological and psychiatric diseases: an overview. *Progr Neuro-Psychopharmacol Biol Psychiat*. 2007;31(1):12–26. https://doi.org/10.1016/j.pnpbp.2006.07.010.
10. Wu S, Ding Y, Wu F, Li R, Hou J, Mao P. Omega-3 fatty acids intake and risks of dementia and Alzheimer's disease: a meta-analysis. *Neurosci Biobehav Rev*. 2015;48:1–9. https://doi.org/10.1016/j.neubiorev.2014.11.008.
11. Bang J, Spina S, Miller B. Frontotemporal dementia. *The Lancet*. 2015;386(10004):1672–1682. https://doi.org/10.1016/s0140-6736(15)00461-4.
12. Burckhardt M, Herke M, Wustmann T, Watzke S, Langer G, Fink A. Omega-3 fatty acids for the treatment of dementia. *Cochrane Database Syst Rev*. 2016. https://doi.org/10.1002/14651858.cd009002.pub3.
13. Yamasaki T, Muranaka H, Kaseda Y, Mimori Y, Tobimatsu S. Understanding the pathophysiology of Alzheimer's disease and mild cognitive impairment: a mini review on fMRI and ERP studies. *Neurol Res Int*. 2012;2012:1–10. https://doi.org/10.1155/2012/719056.
14. Shinto L, Marracci G, Mohr D, et al. Omega-3 fatty acids for depression in multiple sclerosis: a randomized pilot study. *PLoS One*. 2016;11(1):e0147195. https://doi.org/10.1371/journal.pone.0147195.
15. Gold R, Wolinsky J. Pathophysiology of multiple sclerosis and the place of teriflunomide. *Acta Neurol Scand*. 2010;124(2):75–84. https://doi.org/10.1111/j.1600-0404.2010.01444.x.
16. Wu G, Alvarez E. The Immunopathophysiology of multiple sclerosis. *Neurol Clin*. 2011;29(2):257–278. https://doi.org/10.1016/j.ncl.2010.12.009.
17. Mahad D, Trapp B, Lassmann H. Pathological mechanisms in progressive multiple sclerosis. *The Lancet Neurol*. 2015;14(2):183–193. https://doi.org/10.1016/s1474-4422(14)70256-x.
18. Stadelmann C, Wegner C, Brück W. Inflammation, demyelination, and degeneration—recent insights from MS pathology. *Biochim Biophys Acta (BBA) - Mol Basis Dis*. 2011;1812(2):275–282. https://doi.org/10.1016/j.bbadis.2010.07.007.
19. Jelinek G, Hadgkiss E, Weiland T, Pereira N, Marck C, van der Meer D. Association of fish consumption and omega 3 supplementation with quality of life, disability and disease activity in an international cohort of people with multiple sclerosis. *Int J Neurosci*. 2013;123(11):792–801. https://doi.org/10.3109/00207454.2013.803104.
20. Moore F, Vickrey B, Fortin K, Lee L. Two multiple sclerosis quality-of-life measures: comparison in a National Sample. *Can J Neurol Sci/J Can des Sci Neurol*. 2015;42(01):55–63. https://doi.org/10.1017/cjn.2014.128.
21. Farinotti M, Vacchi L, Simi S, Di Pietrantonj C, Brait L, Filippini G. Dietary interventions for multiple sclerosis. *Cochrane Database Syst Rev*. 2012. https://doi.org/10.1002/14651858.cd004192.pub3.
22. Solaro C, Trabucco E, Messmer Uccelli M. Pain and multiple sclerosis: pathophysiology and treatment. *Curr Neurol Neurosci Rep*. 2012;13(1). https://doi.org/10.1007/s11910-012-0320-5.
23. Seidl S, Santiago J, Bilyk H, Potashkin J. The emerging role of nutrition in Parkinson's disease. *Front Aging Neurosci*. 2014;6. https://doi.org/10.3389/fnagi.2014.00036.
24. da Silva T, Munhoz R, Alvarez C, et al. Depression in Parkinson's disease: a double-blind, randomized, placebo-controlled pilot study of omega-3 fatty-acid supplementation. *J Affect Disord*. 2008;111(2–3):351–359. https://doi.org/10.1016/j.jad.2008.03.008.
25. Gao X, Chen H, Fung T, et al. Prospective study of dietary pattern and risk of Parkinson disease. *Am J Clin Nutr*. 2007;86(5):1486–1494. https://doi.org/10.1093/ajcn/86.5.1486.
26. de Lau L, Bornebroek M, Witteman J, Hofman A, Koudstaal P, Breteler M. Dietary fatty acids and the risk of Parkinson disease: the Rotterdam study. *Neurology*. 2005;64(12):2040–2045. https://doi.org/10.1212/01.wnl.0000166038.67153.9f.
27. Frank S. Treatment of Huntington's disease. *Neurotherapeutics*. 2013;11(1):153–160. https://doi.org/10.1007/s13311-013-0244-z.
28. Walker F. Huntington's disease. *The Lancet*. 2007;369(9557):218–228. https://doi.org/10.1016/s0140-6736(07)60111-1.
29. Roos R. Huntington's disease: a clinical review. *Orphanet J Rare Dis*. 2010;5(1):40. https://doi.org/10.1186/1750-1172-5-40.
30. Craufurd D, Snowden J. Neuropsychological and neuropsychiatric aspects of Huntington's disease. *Oxford Monogr Med Genet*. 2002;48:62–94.

31. Puri B, Leavitt B, Hayden M, et al. Ethyl-EPA in Huntington disease: a double-blind, randomized, placebo-controlled trial. *Neurology*. 2005;65(2):286–292. https://doi.org/10.1212/01.wnl.0000169025.09670.6d.
32. Murck H, Manku M. Ethyl-EPA in Huntington disease—potentially relevant mechanism of action. *Brain Res Bull*. 2007;72(2–3):159–164. https://doi.org/10.1016/j.brainresbull.2006.10.020.
33. Huntington Study Group TREND-HD Investigators. Randomized controlled trial of ethyl-eicosapentaenoic acid in Huntington disease. *Arch Neurol*. 2008;65(12). https://doi.org/10.1001/archneur.65.12.1582.
34. Ferreira J, Rosser A, Craufurd D, Squitieri F, Mallard N, Landwehrmeyer B. Ethyl-eicosapentaenoic acid treatment in Huntington's disease: a placebo-controlled clinical trial. *Mov Disord*. 2015;30(10):1426–1429. https://doi.org/10.1002/mds.26308.
35. Puri B, Bydder G, Manku M, Clarke A, Waldman A, Beckmann C. Reduction in cerebral atrophy associated with ethyl-eicosapentaenoic acid treatment in patients with Huntington's disease. *J Int Med Res*. 2008;36(5): 896–905. https://doi.org/10.1177/147323000803600505.

CHAPTER

18

Neurobiological Meaning of Omega-3 Fatty Acids and Their Potential Role in the Treatment of Schizophrenia

Marianna Mazza, Giuseppe Marano*, Gianandrea Traversi†,‡, Salvatore Mazza*, Luigi Janiri**

*Institute of Psychiatry and Psychology, Department of Geriatrics, Neuroscience and Orthopedics, Fondazione Policlinico Universitario A. Gemelli IRCCS, Università Cattolica del Sacro Cuore, Rome, Italy †Department of Science, University of Rome "Roma Tre", Rome, Italy ‡Department of Research, Advanced Diagnostics and Technological Innovation, Regina Elena National Cancer Institute, Rome, Italy

INTRODUCTION

Schizophrenia is considered an exhausting psychiatric disease affecting approximately 1% of the world population.[1–3]

In 2001, the World Health Organization stated that around 450 million people worldwide suffered from a mental health problem and that one in four people will experience mental illness.[4]

The disorder is often chronic and disabling in several clinical areas, including social and occupational functioning, which are representative elements and significant diagnostic criteria associated with this disorder.[5] Likewise, patients suffering from schizophrenia appear to die 10–20 years earlier in comparison with the general population.[6,7] Because of the frequent recurrence rate, high mortality, and poor prognosis, schizophrenia imposes a weighty burden on patients, caregivers, and society.[8]

Chronic mental illness frequently results in decreased quality of life and social and economic burdens as well as high rates of physical illness. Individuals suffering from severe mental disorders are more likely to be affected by physical health problems because of behavioral factors that prompt increased risk of poor lifestyle choices like smoking, diet, and low physical activity.[4,9,10] Although psychopharmacological treatment seems to play a role

Omega Fatty Acids in Brain and Neurological Health
https://doi.org/10.1016/B978-0-12-815238-6.00018-3

in representing a risk factor for cardiovascular disease, lifestyle factors have been recognized as independent contributors.[11–13] Lifestyle choices may involve into common underlying biological mechanisms for both physical and mental illness and their comorbidities.[14] Patients affected by schizophrenia have an almost threefold increase in risk of dying prematurely, of cardiovascular diseases, or experiencing cardiovascular adverse actions as a result of antipsychotic treatment.[15]

The symptoms of schizophrenia can be grouped into positive symptoms (defining distortion of normal functioning), negative symptoms (showing deficits in functioning), and cognitive impairment.[16] This major mental disorder is distinguished by hallucinations, delusions, paranoia, agitation, feelings of apathy, affective flattening, disharmony between mental activity and environment, and deficits in learning, memory, attention, and executive functions.[1, 17, 18] Schizophrenia is commonly managed with a combination of psychotherapy and secondary socialization used to improve patient's interactions in social settings, together with the administration of medications, comprising atypical and typical antipsychotics.[19, 20]

Like physical health, mental health is configured as a complex interaction of biological, psychological, and social factors.[4] Despite recent progress in understanding the molecular and cellular mechanisms contributing to the disease and the fact that these pathophysiological advances have opened new avenues of investigation, at present the exact cause of schizophrenia remains unidentified and elusive. Several mechanisms have been suggested in the etiology of schizophrenia, such as genetic factors and disturbances in neurotransmitters, especially dopamine, serotonin, abnormalities of cerebral structure/function, and environmental conditions.[21–26]

One of the leading hypotheses regarding multifactorial etiology of schizophrenia is the dysregulation in the levels and actions of dopamine in the brain.[27–29] The accessibility of dopamine in the brain is interconnected to its synthesis, release, reuptake, and metabolism.[30–32]

Because the etiology and pathogenesis of schizophrenia is poorly understood, conventional prevention and treatment of the illness generally has partial success; as additional side effects/symptoms manifest, further treatment options can be introduced. In recent years, growing attention has concentrated on the role of nutritional interventions in neurologic and mental diseases in order to maximize the effects of nutritional components on brain function, neural plasticity, and mental health.[8]

BIOLOGICAL CONNOTATIONS OF OMEGA-3 FATTY ACID CONSUMPTION

Omega-3 polyunsaturated fatty acids form one of the major classes of dietary fat, which cannot be synthesized in sufficient amounts required for maximum health, but it can be obtained from several dietary sources.[33] Polyunsaturated fatty acids are indispensable elements of cell membrane phospholipids and are involved in numerous biological processes, such as the functionality and sensitivity of neurotransmitter receptors, ion channels, or mitochondrial activity,[34–36] signal transduction, and the synthesis of active metabolites, which regulate inflammation, oxidative stress, and defense processes. Polyunsaturated fatty acids play a pivotal role in the central nervous system, where they are the focal elements of neuronal membrane phospholipids.[37] Correct quantities of omega-3 and omega-6 fatty acids, as well as numerous growth factors, are indispensable for brain expansion during the perinatal period and adolescence.[38, 39] The supplies of docosahexaenoic acid, arachidonic acid, and nervonic acid are necessary. Docosahexaenoic acid constitutes the main component of excitable membranes, representing up to 20% of the brain's dry weight. The docosahexaenoic acid and arachidonic acid proportion is vital in the preservation of a suitable level of biological membrane fluidity, which is in turn, critical for ion channel function, membrane receptor activity, and the release of neurohormones, the signaling processes of every cell. The buildup of these fatty acids begins throughout intrauterine brain growth and endures in the course of the first months, subsequent birth, and on to pubertal phases.[36] From an evolutionary standpoint, these stages are of specific importance for the development of schizophrenia and other neuropsychiatric diseases since it is assumed that anomalies in the cytoarchitecture of the phospholipid membrane could signify a factor in schizophrenia risk or susceptibility, as Horrobin suggested in the early 1990s, expressing the membrane hypothesis of schizophrenia.[40] Horrobin[41] postulated that alterations in the membrane phospholipid composition of the brain, as a direct consequence of variations in fatty acid levels, could be implicated in the etiology of schizophrenia. This argument is based on findings of decreased omega-3 and omega-6 polyunsaturated fatty acids and anomalies in phospholipid metabolism in schizophrenia.[42, 43]

Omega-3 fatty acids cannot be produced by the organism; thus it has been advocated that a reduced nutritional amount of these fatty acids, as happens in technologically advanced areas, could elucidate the larger severity of schizophrenia in industrialized countries.[36, 44] Deficiencies of polyunsaturated fatty acid levels have repeatedly been observed in schizophrenia, especially at early stages of disease,[45] and magnetic resonance imaging examinations have been useful to suggest possible neuroprotective effects of polyunsaturated fatty acids in patients suffering from schizophrenia.[46]

Schizophrenia is treated as a neurodevelopmental and neurodegenerative disorder exhibiting disruption in multiple neurotransmission pathways.[37] Meanwhile, polyunsaturated fatty acids are crucial for regular neurodevelopment, disorders of polyunsaturated fatty acids metabolism, and they may be implicated in the etiology of neurodevelopmental disorders such as schizophrenia.[47] Neuronal membranes are rich in decosahexaenoic acid, which is vital for normal brain development.[48] Several dysfunctional neurotransmission pathways have been detected in schizophrenia, which increases the possibility that membrane phospholipids that control the activity of both receptors are involved in signal transduction recognized as part of the etiology of schizophrenia.[37] Deficiencies of polyunsaturated fatty acids were defined in red blood cell membranes of patients with schizophrenia.[42, 49–51] Metabolism of disorders in polyunsaturated fatty acids metabolism has also been presented at early stages of schizophrenia preceding initiation of antipsychotic therapy[51] and in individuals at ultra-high risk of psychosis.[52] Postmortem brain studies have discovered declined levels of polyunsaturated fatty acids, specifically docosahexaenoic acid and arachidonic acid, in the frontal lobes of schizophrenia patients. Deficits of polyunsaturated fatty acids in neuronal membranes can be the product of the enlarged release of polyunsaturated fatty acids from membrane phospholipids, or its reduced incorporation within them. Both processes were identified as faulty in patients suffering from schizophrenia.[37]

Nutritional deficiencies are documented as a risk-factor for various psychiatric disorders.[53, 54] People with schizophrenia commonly follow low-quality diets[55] and a range of dietary deficits, even from disease onset.[56] In addition, decreased amounts of vitamins and polyunsaturated fatty acids are connected with different adverse outcomes in first-episode psychosis comprising greater symptom severity, decreased neural integrity, and neurocognitive impairments.[56–58]

Inflammation and oxidative stress reach high levels of expression in early illness[59, 60] and may guide the neurocognitive abnormalities that originate during this time.[61] Thus, first-episode schizophrenia may be the perfect timeframe for dispensing antioxidant/antiinflammatory nutrients as supplementary treatments to contrast adverse neurobiological processes involved in schizophrenia, thus decreasing the probability of lasting symptoms and cognitive dysfunction.[62, 63] Berger et al.[64] observed that patients receiving first antipsychotic treatment are more likely to obtain symptomatic reduction within 6 weeks when supplemented with eicosapentaenoic acid daily, rather than a placebo. Two neuroimaging substudies of these participants revealed modifications of neuroprotective metabolites in the treatment groups. Findings from these studies suggested that omega-3 supplementation enhanced hippocampal neuronal health[46] and expressively augmented brain indicators, which reduce oxidative stress and neural damage, such as glutathione.[65] Moreover, enhanced hippocampal neuronal health and increased glutathione are associated with reduced negative symptoms.[53]

Data from three double-blind randomized controlled trials of polyunsaturated fatty acids in first-episode psychosis[64, 66] ended without any major side effect. In addition, feasibility data showed elevated tolerability, as retention was consistently high and adherence was >80%. A recently published review[53] looking at omega-3 fatty acids as adjunctive treatments for first-episode psychosis showed encouraging findings concerning the efficacy, tolerability, and biological mechanisms of action for nutrient-based compounds.

The relationship between polyunsaturated fatty acids' consumption and brain health has yet to be fully confirmed[67]; Kim et al.[28] theorized that inappropriate dietary habits and insufficient dietary intake of specific nutrients are associated with schizophrenia based on a sample of Korean patients. The male patients in the schizophrenia group presented lower dietary consumptions of polyunsaturated fatty acids in comparison with the male subjects in the control group; specifically, low levels of polyunsaturated fatty acids are documented as a factor in the neuropathology of schizophrenia.[68, 69] In addition, a high proportion of schizophrenia patients (64.2%) are overweight or obese compared with control subjects (39.7%). The male schizophrenia patients displayed considerably lower nutritional consumptions of polyunsaturated fatty acids in comparison with the male control subjects. No significant alimentary patterns in women affected by schizophrenia have been observed, despite the women having heavier body weights and body mass index compared with the female controls. This was a surprising observation; a potential clarification for these outcomes is that the women with schizophrenia were older than the female controls and these variances in age and age-related physiological modifications might impact the explanation. Patients affected by schizophrenia were less likely to choose correct nutrients for themselves, so the authors proposed a key topic: conduct research to define dietary guidelines and develop nutritional programs, prevent metabolic syndrome, and the worsening of symptoms.

INFLUENCE OF OMEGA-3 FATTY ACIDS ON PARAMETERS OF OXIDATIVE STRESS

Oxidative stress represents a disequilibrium between reactive oxygen species and reactive nitrogen species production and the activity of oxidative defense systems. A relative excess of aforementioned agents levels causes oxidative damage to macromolecules like nucleic acids, proteins, and lipids, along with changes in gene expression, intracellular signaling pathway modulation, and deleterious changes in the activity of critical cell enzymes (i.e., caspases), which can induce cellular death. Lipid peroxidation act for a well-characterized effect of reactive oxygen species that results in damage to the cell membrane as well as to the membranes of cellular organelles, such as mitochondria. Notwithstanding that the brain comprehends about 2% of the human body weight because of its high energy demand; it employs more than 20% of the consumed oxygen.[70, 71] The brain is composed of polyunsaturated fatty acids, which is highly susceptible to oxidative damage and peroxide formation by reason of the increased number of double carbon bonds in the molecule. The antioxidative defense system is usually distinguished into enzymatic and nonenzymatic parts, which together include its total antioxidant capacity. The enzymatic section of the brain's antioxidative defense system is composed mostly by enzymes like glutathione peroxidase and reductase and its substrates: reduced glutathione and its oxidized form, glutathione disulphide. Other antioxidant enzymes localized in the brain, i.e., catalase and superoxide dismutase, work in concert with glutathione peroxidase. The nonenzymatic component of the antioxidative defense system consists of α-Tocopherol (vitamin E), bilirubin, albumin, uric acid, niacin, carotenoids, and flavonoids.[72] The antioxidative defense of the brain is evaluated as weak in relation to its energy needs and oxygen utilization.[71, 73]

Elevated oxygen consumption, high double bond content in polyunsaturated fatty acids molecules, and low content of antioxidative defense system, together with a high metal content (iron, zinc, copper, manganese), make the brain remarkably vulnerable to oxidative damage. Modifications in the antioxidative defense system and the enlarged production of free radicals have been frequently mentioned in schizophrenia, both in preclinical and in clinical studies.[70, 74–77] A metaanalysis, performed by Flatow and colleagues,[59] supported the facts that total plasma or serum antioxidant status, red blood cell catalase, and plasma nitrite levels may operate as state markers in schizophrenia. On the other hand, superoxide dismutase content in red blood cells can be weighed as a trait marker of schizophrenia, for the reason that it is found to be reduced in acutely relapsed inpatients, first-episode schizophrenia, and stable outpatients. Anomalies occurring in the oxidative balance of the brain, characteristic of patients affected by schizophrenia, move to elevated lipid peroxidation with complex biochemical repercussions.[78]

The effects of polyunsaturated fatty acids are fundamental for overall health.[79] The proportion of polyunsaturated fatty acids in serum and erythrocyte phospholipids, which depends on endogenous metabolism controlled by genetic polymorphisms and dietary intake, is an important determinant of both health and disease.[80] Kiecolt-Glaser and co-workers[81, 82] appraised the effects of 4-month supplementation with omega-3 fatty acids in healthy individuals on the indicators of inflammation and oxidative stress. Together with a mitigation in pro-inflammatory cytokine levels (TNFα and Il-6), an omega-3 fatty acids' regimen substantially lessened lipid peroxide levels. Balcerczyk and colleagues[83] exposed that a 12-week supplementation with omega-3 fatty acids, antioxidants, vitamins E, and D3 in 66 healthy women aged 35–55 is connected with intensifications in telomerase levels, along with promising variations in antioxidative defense capacity and oxidative stress indicators. The increased production of reactive oxygen species and decreased oxidative defense mechanisms recognized in many animal models, in vitro and in clinical studies of patients diagnosed with either early or chronic schizophrenia mean that oxidative stress plays a part in the pathophysiology of mental disorders, including schizophrenia.[74] The effect of polyunsaturated fatty acids and metabolism abnormalities has frequently been investigated in individuals at different stages of schizophrenia development, such as in ultra-high risk individuals, first-episode, and chronic patients.[49, 84, 85] A biochemical anomaly related to schizophrenia concerns the high action of phospholipase A2 (Phosphatidylcholin-2-acylhydrase), which releases fatty acids from membrane phospholipids.[86]

Behdani et al.[87] conducted a study in order to explore the influence of omega-3 polyunsaturated fatty acids on metabolic and inflammatory parameters by assessing changes in weight, body mass index, waist circumference, blood pressure, serum high sensitivity-C-reactive protein, and lipid profiles in a group of chronic schizophrenia patients. They found that omega-3 fatty acids as add-on treatment may be an interesting treatment option because of some beneficial effects improving anthropometric indices in patients suffering from schizophrenia in clozapine and sodium valproate pharmacotherapy.

Smesny and co-workers[88] reported that omega-3 fatty acids' intervention seems to sustain the antioxidative defense at membrane level; the effect of omega-3 fatty acids on total glutathione level could demonstrate antioxidative effects, resulting in decreased demand for glutathione. The authors hypothesize that the vitamin E and glutathione mechanisms may explain the observed clinical effectiveness of omega-3 fatty acids in high clinical risk individuals. A more recently published study[71] also supported the effectiveness of omega-3 fatty acids on improvements in oxidative stress indices in patients with first-episode schizophrenia. Findings from this study suggest that the efficacy of a 6-month intervention with omega-3 fatty acids may be linked to betterment in oxidative stress parameters.

The possible mechanisms in polyunsaturated fatty acids modulation of inflammatory response are investigated in a number of studies.[89–92] Based on their randomized, double-blind, placebo-controlled, parallel-group study Pawełczyk et al.[71] postulate that the potential underlying mechanism of the therapeutic action of omega-3 fatty acids may be neuroprotective, acting via changes in physical and biochemical properties of cellular membranes,[93, 94] modulation of the inflammatory responses,[75, 95, 96] and antioxidative intracellular defense system,[97] and reduction of dopamine toxicity; it may operate by reducing inflammation via modulation of cytokine production.[98] It is possible that the effect of polyunsaturated fatty acids' supplementation may depend on a direct interaction between eicosapentaenoic acid or decosahexaenoic acid and glutamatergic neurotransmission.[99, 100] A magnetic resonance spectroscopy study in first-episode subjects has verified that eicosapentaenoic acid augmentation regulates glutathione and the glutamine/glutamate cycle in early psychosis and some of the metabolic variations in the brain associated with improvements in negative symptoms.[65]

RELATIONSHIP OF NEUROANATOMICAL CHANGES AND POLYUNSATURATED FATTY ACIDS ON THE CLINICAL COURSE OF SCHIZOPHRENIA

Schizophrenia is a biological and behavioral disorder. The question of whether the brain abnormalities in schizophrenia are either neurodegenerative or neurodevelopmental in origin has been discussed for more than 100 years.[101] Some theories concerning the origins of schizophrenia suggest that structural brain changes are evidence of a neurodevelopmental basis for schizophrenia. The theories are based on intra-utero or early childhood events that are either genetic, environmental, or a combination of the two.[102] Another important reflection is whether specific brain regions are implicated in schizophrenia or whether there is a general involvement. Some authors have sustained that schizophrenia is a disorder of neural dysconnectivity[103] and that the brain modifications mirror comprehensive breakdown of neural connections.[104]

Studies addressing structural brain abnormalities on magnetic resonance imaging examination often reveal reduced brain volume and cortical thickness in patients affected by schizophrenia, specifically abnormalities of the frontotemporal and hippocampal regions of the brain and alike lateral ventricle enlargement, even in individuals at high clinical risk of schizophrenia[105] and at early stages of the disease.[106] Some studies have documented that these changes grow over time with diversified loss patterns at every stage of the schizophrenic disease.[107, 108] It has been discovered that brain volume loss is in some measure related to disease severity based on the number and frequency of relapses leading to excitotoxicity induced by glutamatergic over activity or by oxidative stress related to dopaminergic stimulation.[109] There is also evidence indicating a relation between brain tissue loss and intensive antipsychotic therapy.[110, 111]

Antipsychotic medications are widely prescribed in schizophrenia; the benefits of these agents in the initial treatment of psychotic symptoms are undisputed, as well as the effects of long-term use, including clinical outcomes and relapse prevention. Nevertheless, it is still a matter of discussion whether this treatment can impact brain volume changes or if these modifications may be due to the pathological processes related to principally to the disease.[112, 113] The mainstay of treatment for schizophrenia can have a significant influence on brain tissue mediating progressive brain volume decrement.[109–111, 113] Research also supports that the magnitude of brain volume loss is directly associated to worse clinical outcome.[114]

In 2018, Pawełczyk and colleagues[115] published a paper aimed to confirm whether the use of adjunctive polyunsaturated fatty acids therapy, with its neuroprotective effects, may preserve this cortical volume loss in patients with schizophrenia. The authors note that individuals who received polyunsaturated fatty acids showed reduced cortical thickness loss in the cortical areas that control functions known to be impaired in schizophrenia. These findings are also consistent with previous studies carried out in older individuals and patients with mild cognitive impairment, suggesting that polyunsaturated fatty acids may have neuroprotective properties, especially at early stages of neurodegenerative diseases, such as schizophrenia. In particular, the study conducted by Pawełczyk[115] observed decreased cortical thickness loss associated to polyunsaturated fatty acids, in comparison with placebo in schizophrenia patients treated with antipsychotics. The differences are observed in the left parieto-occipital cortex, specifically Brodmann areas 7 and 19, such as the cortical areas located near the temporo-parieto-occipital junction, which integrates information from both the external environment and from within the body.[116] The temporo-parieto-occipital junction is also implicated in other distinctions: the process damaged in schizophrenia and one of the key features of the disease. Brodmann area 7 is described as a somatosensory association cortex and is thought to play a role in visuomotor coordination, semantic categorization tasks, and temporal context recognition.[117] Brodmann area 19 is illustrated as associative visual cortex and involved in the detection of light intensity, feature attention and pattern detection, and is deputed to spatial working memory.[118, 119] Many of these activities, such as visuomotor coordination, attention, pattern detection, and working memory, are generally degraded in schizophrenia.[120]

It may be useful to remember that the literature has emerged that both areas of Brodmann are protected by polyunsaturated fatty acids and are involved in saccadic movement control.[121, 122] This function is contemplated as an endophenotype of schizophrenia, defective not only in patients diagnosed with schizophrenia, but also in their family members and people at increased clinical risk of schizophrenia development.[123] There is a substantial body of evidence from animal studies addressing the topic of polyunsaturated fatty acids' influence on cortical thickness; only one placebo-controlled study has been reported.[124] The authors[124] evaluate the effects of 8-week polyunsaturated fatty acids' supplementation on structural brain changes of aged mice in high-resolution magnetic resonance imaging examinations. A notable enlargement of gray matter volume has been recounted in the hippocampus, medial prefrontal cortex, and the retrosplenial cortex of the mouse brain. However, the reported increase in gray matter volume correlated significantly with improvement in cognitive task performance. In a previous study, the same group of researchers found that polyunsaturated fatty acids supplementation of aged mice is associated with augmented hippocampal cognitive functions which take place in the context of enhanced cellular plasticity and reduced neurodegeneration.[125]

Polyunsaturated fatty acids' supplementation elevated hippocampal neurogenesis and dendritic arborization of newborn neurons, increased neuronal volume and density, along with microglial cell number. In addition, it also reduced apoptosis, astrocytosis, and lipofuscin accumulation in the hippocampus. The increased levels of blood acetyl-L-carnitine and brain polyunsaturated fatty acids concentrations found in polyunsaturated fatty acids supplemented mice additionally described effective neuroprotection.[115] Findings from the study carried out by Pawełczyk et al.[115] seem to be supported even by recently published considerations that a Mediterranean-style diet is related to increased cortical thickness[126] and higher total brain volume[127] in a large Scottish cohort.

The goal of the study of Pawełczyk et al.[115] was to comprehensively evaluate polyunsaturated fatty acids neuroprotective properties at early stages of neurodegenerative diseases: these agents may be effective in attenuating cortical thickness loss when used as prophylaxis in healthy individuals or at early stages of neurodegenerative processes like mild cognitive impairment as a prodrome of Alzheimer's disease. The postulated mechanism for the neuroprotective effects of polyunsaturated fatty acids may be related to its known immunomodulatory properties,[98] including its ability to stimulate oxidative stress defense mechanisms[71, 128] and inhibit apoptosis.[129, 130]

Cognitive impairment is considered a core feature of schizophrenia, affecting up to 75% of patients.[131] Among different cognitive abilities, a wide range of cognitive functions can be overwhelmed, including attention, working memory, verbal learning and memory, and executive functions. For the identification of schizophrenia, some authors estimate that it is useful to consider the observation of the cognitive abilities of the subject,[131] such as a level of cognitive functioning suggesting a consistent severe impairment and/or a significant decline from premorbid levels considering the patient's educational, familial and socioeconomic background.[132–137] A review article depicts that generalized cognitive deficits are seen in patients suffering from schizophrenia similarly in different regions of the world despite linguistic and cultural differences.[138]

Cognitive function is powerfully linked with functional outcome in patients,[139–141] but the pathophysiology underlying cognitive impairment in schizophrenia is currently not well understood.[140] Correlation between omega-3 fatty acids and psychiatric symptoms have been stated in schizophrenia[43, 142–145]; conversely few studies have explored the role of omega-3 fatty acids in the pathophysiology of cognitive impairment. Previous studies established that omega-3 fatty acids' consumption improved cognitive performance in rats.[125, 146] Moreover, a metaanalysis explained that omega-3 fatty acids as add-on therapy is beneficial for memory in adults with mild memory complaints.[147] Condray and colleagues[148] found that low fatty acids' levels are connected to semantic memory and language dysfunctions in schizophrenia. Peters and co-workers[149] determined that omega-3 fatty acids concentration in the brain of schizophrenia patients are associated with anomalous fractional anisotropy in diffusion tensor imaging studies across extensive areas of working memory.

Satogami and co-workers[150] are the first to made evidence a direct association between omega-3 fatty acids and cognitive function as measured by neuropsychological assessment in patients suffering from schizophrenia. Their findings indicate that eicosapentaenoic acid and docosahexaenoic acid blood levels are significantly correlated with cognitive assessment, which is meaningfully associated with the social functioning. With this in mind, the authors recognized connectives that express a cause-effect relationship: diminished omega-3 fatty acid levels are linked to cognitive dysfunction, that consequently affects social functioning outcomes in patients affected by schizophrenia. In this study, it has been suggested a direct negative linear relationship between a daily dose of antipsychotic medications and omega-3 fatty acid levels together with a significant negative correlation between a daily dose of antipsychotic agents and the cognitive assessment. By adopting a multilevel metaanalysis framework, Knowles et al.[151] tried to shed lights on the fact that neurocognitive impairment in patients suffering from schizophrenia is substantially affected by a higher dosage of antipsychotics. Two studies pointed out a significant negative correlation between antipsychotic treatment and cognitive assessment.[152, 153] Montesinos-Rueda and colleagues[36] conducted an observational study investigating levels of omega 3 fatty acids in a sample of 31 patients affected by schizophrenia. Their exploratory study yielded some suggestive finding such as a relationship between low omega-3 levels and the use of long-acting antipsychotic agents (depot). The use of this formulation is also associated with worse cognitive performance in comparison with other types of antipsychotic drugs.[36]

Fang and co-workers[8] observed that polyunsaturated fatty acids attenuated cognitive impairment and hippocampal neurons loss in rats, by activating ERK and AKT to increase phosphorylation of CREB pathway. Thereby, polyunsaturated fatty acids may amply synaptic plasticity, decrease the loss of hippocampal neurons, and antagonize the cognitive dysfunction in schizophrenic rats induced by MK-801. These outcomes contribute to offer a theoretical basis for the adoption of polyunsaturated fatty acids in the treatment and prevention of schizophrenia.

INVESTIGATING THE ROLE OF OMEGA 3 FATTY ACIDS LEVELS IN SCHIZOPHRENIA

It is well known that polyunsaturated fatty acids are crucial for the structural and functional integrity of neural cells.[154] Literature data analysis informs that polyunsaturated fatty acids, such as omega-3, may reduce the symptoms of schizophrenia.[155]

The body of existing data, including rodent, primate and human studies, support the involvement of decosahexaenoic acid and eicosapentaenoic acid in neurodevelopment via modulation of oxidative defense, neuroinflammation, directly and indirectly stimulating both the neuroplastic and neuroprotective mechanisms of the brain.[92, 98, 156, 157] Among the possible underlying mechanisms, it is proposed that the biological action of polyunsaturated fatty acids might account for be neuroprotective,[158] via modulation of the antioxidative intracellular defense,[97] antiinflammatory via modulation of nuclear factor kappa B pathway,[98] neuroplastic via stimulation of synthesis of neurotrophins[159] or antiapoptotic.[160]

Fatty acids can additionally regulate the expression of genes involved in lipid homeostasis[161] and energy metabolism. Specifically, two transcription factors, sterol regulatory element binding protein-1c, SREBP-1c, and peroxisome proliferator activated receptor alpha, PPARalpha, represent a direct link between changes in the intracellular lipid milieu and the expression of the genome.[162]

A recent study provided[163] fascinating insight observing that polyunsaturated fatty acids deficiency may affect neurotransmission. At the molecular level, in a mouse behavioral study, a deprivation of dietary arachidonic acid and docosahexaenoic acid induce hyper-methylation of the promoters of the nuclear receptor genes Rxr and Ppar, eliciting schizophrenia-like phenotypes in mouse offspring at adulthood. Particularly, the expression of these nuclear receptor genes were similarly downregulated in hair-follicle cells from schizophrenia patients. Mouse[164, 165] and human[166, 167] genetic studies supported that susceptibility to schizophrenia might be most likely influenced by *RXR* or *PPAR*. Maekawa and co-workers[163] discovered that polyunsaturated fatty acids deficiency during the early neurodevelopmental period in mice could model the prodromal state of schizophrenia through changes in the epigenetic regulation of nuclear receptor genes.

Following a supplemental therapy with omega-3 fatty acids, first-episode schizophrenia patients experienced improvement in psychopathology and patient functioning. This observation raises hope that treatment with omega-3 fatty acids may increase efficacy of conventional therapies, alleviate schizophrenia symptoms, and offer other health benefits, such as improved cardiovascular status. Treatment with omega-3 fatty acids may reduce the antipsychotic cumulative dose needed to control psychotic symptoms and improve antipsychotic tolerability as well as causing less constipation and fewer extrapyramidal side effects[65]. Patients might see an improvement in cognitive performance[168] and in metabolic profile results.[169] Patients might also see a reduced chance of developing: diabetes mellitus,[170] cardiac arrhythmia,[171] sudden cardiac death,[172] and all-cause mortality.[173] In addition, polyunsaturated fatty acids may improve oxidative stress indices[71] and preserve cortical volume loss in patients affected by schizophrenia.[115]

According to Leucht,[174] a 50% decrease in rating scales could be adopted as a cut-off for response in schizophrenia trials for acutely ill and nonrefractory patients; a 25% cut-off could be employed for treatment-resistant patients in order to designate clinically significant advances. Qurashi and colleagues[175] provide evidence of the potential benefits of polyunsaturated fatty acids in the treatment of individuals at increased risk of developing psychosis. Cuéllar-Barboza and co-workers[176] described a case of 52-year-old man who was affected by schizophrenia since he was 20 years old and was treatment resistant. After add-on therapy with omega-3 fatty acids, the man reached remission. In addition, omega-3 supplementation proved useful as the patient benefited from significant growth of speech quantity and content and began taking care of himself and visiting church and restaurants with help.

A randomized controlled study observed that the addition of omega-3 fatty acids to pharmacotherapy with antipsychotics reduced the severity of symptoms. Notable improvements are also found in several secondary outcome measures, such as level of functioning, clinical global impression, and severity of depressive symptoms.

Data indicates that the persistence of depressive symptoms in patients suffering from schizophrenia could prognosticate a relapse in schizophrenia.[177, 178] The results obtained in the study conducted by Pawełczyka et al. regarding the improvement of depressive symptomatology could be used to design disease management strategies targeted at decreasing depressive signs in order to discourage recurrence and treatment of schizophrenic illness. Another double-blind, placebo-controlled study revealed the efficacy of 6-month add-on therapy with omega-3 fatty acids in decreasing the risk of relapse in first-episode schizophrenia patients in the course of the first year of therapy.[37]

Use of omega-3 fatty acids in chronic schizophrenia have shown variegated outcomes. Interestingly, some studies of chronic patients consistently showed that supplemental omega-3 could be helpful in decreasing symptoms of schizophrenia.[179–181] Yet two studies found no meaningful differences between the use of omega-3 fatty acid as a supplemental therapy versus a placebo.[182, 183] Subsequently, Emsley et al.[66] proposed a combination of omega-3 with alpha-linoleic acid to prevent relapse during antipsychotic discontinuation in first-episode schizophrenia. This trial was ineffective as results in relapse rates in both omega-3 and placebo conditions over 2 years were equivalent. A randomized placebo-controlled trial of an omega-3 fatty acid and vitamins E and C in schizophrenia patients was carried out by Bentsen et al.[184] The authors observed that schizophrenia patients do not benefit from additional EPA and vitamins E and C. Surprisingly, they found that these agents, administered separately at moderately high doses, produce psychotic symptoms in patients with low blood levels of polyunsaturated fatty acids. In general, the combination of these substances appears to be safe. A more recent multicenter trial investigated reactions to omega-3 fatty acids in 304 adults at ultra-high risk for psychotic disorders; the study results did not find omega-3 fatty acids beneficial for study recipients.[185, 186] In a metaanalysis, Fusar-Poli and Berger[187] did not find EPA augmentation effective in treating established schizophrenia. These various results could be related to the different stages of the disease. The intake of polyunsaturated fatty acids may be more effective in individuals at high risk of psychosis or early psychosis in comparison with patients affected by chronic schizophrenia. These supplements may offer a safe and effective prevention strategy in subjects with subthreshold psychotic states; in subjects susceptible to schizophrenia, the supplements might reduce the risk of developing the disorder or prevent transition to first-episode psychosis.[188] McEvoy and colleagues[189] have highlighted that changes in lipid metabolism occurred early in the course of schizophrenia according to Horrobin's original hypothesis[190] and literature data.[45, 84, 191] Intriguingly, early disease-related variations could reflect an accelerated breakdown in membrane phospholipids that may hold developmentally mediated synaptic pruning processes that may play a role in schizophrenia.[192] Moreover, metabolism disturbances occur early in the course of schizophrenia and decrease over time. Antipsychotic therapy could improve aberrant polyunsaturated fatty acids in first-episode schizophrenia patients, but not in chronic schizophrenia.[189] Antipsychotic medications combined with omega-3 fatty acids may offer a therapeutic benefit early in the course of psychosis when irreversible neurobiological changes are not yet established.

CONCLUSIONS AND RECOMMENDATIONS

Schizophrenia remains a serious psychotic mental illness. Despite pharmacological advancements, the treatment of schizophrenia represents a challenge. Understanding schizophrenia is a leading challenge for medical research today, as affirmed during the 2017 International Congress on Schizophrenia Research, held in San Diego, California.[193]

Polyunsaturated fatty acids play a role in maintaining normal physiological conditions and therefore in maintaining human health. There is accumulating evidence, in animal models and human adults, that omega-3 fatty acids intake may constitute a valuable add-on intervention in schizophrenic illness. This descriptive review explores the potential beneficial effect of omega-3 fatty acids as nutritional supplementation in individuals affected by schizophrenia, an illness in which associated impairment of the lipid metabolism was found.

Patients suffering from schizophrenia usually exhibit low levels of antioxidant and antiinflammatory agents such as omega-3 fatty acids. These compounds have beneficial effects as augmentation of pharmacotherapies reducing antipsychotic doses, accelerating stabilization of the disease, decreasing the duration of remission as well as comorbid somatic illnesses. The early interventions with omega-3 fatty acids may aid to interrupt the disease progression and improve outcomes reducing financial and social costs of schizophrenia.

However, recent evidence from several supplementation trials supports the idea that omega-3 fatty acids appear to be more effective in the ultra-high risk and first-episode psychosis stages of schizophrenia in comparison with the chronic stage of the disease. A possible use of omega-3 fatty acids may be as a prophylactic and continue treatment against the symptoms associated with schizophrenia.

Omega-3 fatty acids are generally recognized as safe and well tolerated. High doses of omega-3 fatty acids may protract bleeding time.[179, 182] Serious adverse effects like spontaneous bleeding are not common, but the decreased aggregation of thrombocytes should be noted particularly in persons prone to falls, with regard to planned surgeries, or in subjects with coagulation disturbances.[186]

The existing pharmacological treatments for schizophrenia are not perfect, and the lack of knowledge about the pathophysiology of the disorder complicates treatment strategies. Heimer suggests that "the path forward to better treatments, in the short term, is to use the treatment we have in a more efficacious way and make them available to more people".[193]

To revolutionize our understanding, prevention, and treatment of schizophrenia from its earliest stages, we can advance our knowledge of omega-3 fatty acids and exploit their potential positive benefits on the physical, psychiatric, and neurobiological outcomes of the disease.

References

1. El-Sayed El-Sisi A, Sokkar SS, El-Sayed El-Sayad M, Sayed Ramadan E, Osman EY. Celecoxib and omega-3 fatty acids alone and in combination with risperidone affect the behavior and brain biochemistry in amphetamine-induced model of schizophrenia. *Biomed Pharmacother*. 2016;82:425–431.
2. Gustavsson A, Svensson M, Jacobi F, et al. Cost of disorders of the brain in Europe 2010. *Eur Neuropsychopharmacol*. 2011;21(10):718–779.
3. Kay SR, Fiszbein A, Opler LA. The positive and negative syndrome scale (PANSS) for schizophrenia. *Schizophr Bull*. 1987;13(2):261–276.
4. World Health Organisation. *The World Health Report: Mental health—New Understandings, New hope*. Geneva: World Health Organisation; 2001.
5. Kim SW, Lee BJ, Kim JJ, et al. Design and Methodology of the Korean Early Psychosis Cohort Study. *Psychiatry Investig*. 2017;14(1):93–99.
6. Liu NH, Daumit GL, Dua T, et al. Excess mortality in persons with severe mental disorders: a multilevel intervention framework and priorities for clinical practice, policy and research agendas. *World Psychiatry*. 2017;16(1):30–40.
7. Sugai T, Suzuki Y, Yamazaki M, et al. High prevalence of obesity, hypertension, hyperlipidemia, and diabetes mellitus in Japanese outpatients with Schizophrenia: a nationwide survey. *PLoS ONE*. 2016;11(11):e0166429.
8. Fang MS, Li X, Qian H, et al. ω-3PUFAs prevent MK-801-induced cognitive impairment in schizophrenic rats via the CREB/BDNF/TrkB pathway. *J Huazhong Univ Sci Technol Med Sci*. 2017;37(4):491–495.
9. Parletta N, Aljeesh Y, Baune BT. Health behaviors, knowledge, life satisfaction, and Wellbeing in People with mental illness across four countries and comparisons with normative sample. *Front Psychiatry*. 2016;7:145.
10. Parletta N, Zarnowiecki D, Cho J, et al. People with schizophrenia and depression have a low omega-3 index. *Prostaglandins Leukot Essent Fatty Acids*. 2016;110:42–47.
11. Lai JS, Hiles S, Bisquera A, Hure AJ, McEvoy M, Attia J. A systematic review and meta-analysis of dietary patterns and depression in community-dwelling adults. *Am J Clin Nutr*. 2014;99(1):181–197.
12. Osborn DP, Nazareth I, King MB. Physical activity, dietary habits and coronary heart disease risk factor knowledge amongst people with severe mental illness: a cross sectional comparative study in primary care. *Soc Psychiatry Psychiatr Epidemiol*. 2007;42(10):787–793.
13. Peet M. Diet, diabetes and schizophrenia: review and hypothesis. *Br J Psychiatry Suppl*. 2004;47:S102–S105.
14. Peet M. International variations in the outcome of schizophrenia and the prevalence of depression in relation to national dietary practices: an ecological analysis. *Br J Psychiatry*. 2004;184:404–408.
15. Marano G, Traversi G, Romagnoli E, et al. Cardiologic side effects of psychotropic drugs. *J Geriatr Cardiol*. 2011;8(4):243–253.
16. van Os J, Kapur S. Schizophrenia. *Lancet*. 2009;374(9690):635–645.
17. Fervaha G, Foussias G, Agid O, Remington G. Motivational and neurocognitive deficits are central to the prediction of longitudinal functional outcome in schizophrenia. *Acta Psychiatr Scand*. 2014;130(4):290–299.
18. Tandon R, Gaebel W, Barch DM, et al. Definition and description of schizophrenia in the DSM-5. *Schizophr Res*. 2013;150(1):3–10.
19. Kapur S, Remington G. Dopamine D(2) receptors and their role in atypical antipsychotic action: still necessary and may even be sufficient. *Biol Psychiatry*. 2001;50(11):873–883.
20. Zugno AI, Canever L, Mastella G, et al. Effects of omega-3 supplementation on interleukin and neurotrophin levels in an animal model of schizophrenia. *An Acad Bras Cienc*. 2015;87(2 Suppl):1475–1486.
21. Birtwistle J, Baldwin D. Role of dopamine in schizophrenia and Parkinson's disease. *Br J Nurs*. 1998;7(14). 832-4, 836, 838-41.
22. Bleich A, Brown SL, Kahn R, van Praag HM. The role of serotonin in schizophrenia. *Schizophr Bull*. 1988;14(2):297–315.
23. Flashman LA, Green MF. Review of cognition and brain structure in schizophrenia: profiles, longitudinal course, and effects of treatment. *Psychiatr Clin North Am*. 2004;27(1):1–18. vii.
24. Levinson DF. Molecular genetics of schizophrenia: a review of the recent literature. *Curr Opin Psychiatry*. 2003;16:157–170.
25. McCarley RW, Wible CG, Frumin M, et al. MRI anatomy of schizophrenia. *Biol Psychiatry*. 1999;45(9):1099–1119.
26. Wright IC, Rabe-Hesketh S, Woodruff PW, David AS, Murray RM, Bullmore ET. Meta-analysis of regional brain volumes in schizophrenia. *Am J Psychiatry*. 2000;157(1):16–25.

27. Howes OD, McCutcheon R, Owen MJ, Murray RM. The role of genes, stress, and dopamine in the development of Schizophrenia. *Biol Psychiatry*. 2017;81(1):9–20.
28. Kim EJ, Lim SY, Lee HJ, et al. Low dietary intake of n-3 fatty acids, niacin, folate, and vitamin C in Korean patients with schizophrenia and the development of dietary guidelines for schizophrenia. *Nutr Res*. 2017;45:10–18.
29. Maia TV, Frank MJ. An integrative perspective on the role of dopamine in Schizophrenia. *Biol Psychiatry*. 2017;81(1):52–66.
30. Best JA, Nijhout HF, Reed MC. Homeostatic mechanisms in dopamine synthesis and release: a mathematical model. *Theor Biol Med Model*. 2009;6:21.
31. Healy-Stoffel M, Levant B. N-3 (Omega-3) fatty acids: effects on brain dopamine systems and potential role in the etiology and treatment of neuropsychiatric disorders. *CNS Neurol Disord Drug Targets*. 2018;17(3):216–232.
32. Vaughan RA, Foster JD. Mechanisms of dopamine transporter regulation in normal and disease states. *Trends Pharmacol Sci*. 2013;34(9):489–496.
33. McLaughlin V, Reeves KW, Bertone-Johnson E. Fatty acid consumption and the risk of osteoporotic fracture. *Nutr Rev*. 2013;71(9):600–610.
34. du Bois TM, Deng C, Huang XF. Membrane phospholipid composition, alterations in neurotransmitter systems and schizophrenia. *Prog Neuropsychopharmacol Biol Psychiatry*. 2005;29(6):878–888.
35. Horrobin DF, Bennet CN. In: Peet M, Glenn I, Horrobin DF, eds. *Phospholipids metabolism and the patophisiology of psychiatric and neurological disorders in Phospholipids spectrum disorders psychiatry and neurology*. Carnfordh: Marius Press; 2003:3–47.
36. Montesinos-Rueda L, Cañete-Crespillo J, Palma-Sevillano C, Giné-Serven E. Erythrocyte membrane polyunsaturated fatty acid (pufa) levels in a sample of patients with schizophrenia and relation with clinical and progression variables. *Actas Esp Psiquiatr*. 2015;43(5):170–176.
37. Pawełczyk T, Grancow M, Kotlicka-Antczak M, et al. Omega-3 fatty acids in first-episode schizophrenia—a randomized controlled study of efficacy and relapse prevention (OFFER): rationale, design, and methods. *BMC Psychiatry*. 2015;15:97.
38. Das UN. Biological significance of essential fatty acids. *J Assoc Physicians India*. 2006;54:309–319.
39. Das UN. Essential fatty acids—a review. *Curr Pharm Biotechnol*. 2006;7(6):467–482.
40. Horrobin DF, Glen AI, Vaddadi K. The membrane hypothesis of schizophrenia. *Schizophr Res*. 1994;13(3):195–207.
41. Horrobin DF. The membrane phospholipid hypothesis as a biochemical basis for the neurodevelopmental concept of schizophrenia. *Schizophr Res*. 1998;30(3):193–208.
42. Arvindakshan M, Ghate M, Ranjekar PK, Evans DR, Mahadik SP. Supplementation with a combination of omega-3 fatty acids and antioxidants (vitamins E and C) improves the outcome of schizophrenia. *Schizophr Res*. 2003;62(3):195–204.
43. Arvindakshan M, Sitasawad S, Debsikdar V, et al. Essential polyunsaturated fatty acid and lipid peroxide levels in never-medicated and medicated schizophrenia patients. *Biol Psychiatry*. 2003;53(1):56–64.
44. Christensen O, Christensen E. Fat consumption and schizophrenia. *Acta Psychiatr Scand*. 1988;78(5):587–591.
45. van der Kemp WJ, Klomp DW, Kahn RS, Luijten PR, Hulshoff Pol HE. A meta-analysis of the polyunsaturated fatty acid composition of erythrocyte membranes in schizophrenia. *Schizophr Res*. 2012;141(2-3):153–161.
46. Wood SJ, Cocchi L, Proffitt TM, et al. Neuroprotective effects of ethyl-eicosapentaenoic acid in first episode psychosis: a longitudinal T2 relaxometry pilot study. *Psychiatry Res*. 2010;182(2):180–182.
47. Assisi A, Banzi R, Buonocore C, et al. Fish oil and mental health: the role of n-3 long-chain polyunsaturated fatty acids in cognitive development and neurological disorders. *Int Clin Psychopharmacol*. 2006;21(6):319–336.
48. Singh M. Essential fatty acids, DHA and human brain. *Indian J Pediatr*. 2005;72(3):239–242.
49. Khan MM, Evans DR, Gunna V, Scheffer RE, Parikh VV, Mahadik SP. Reduced erythrocyte membrane essential fatty acids and increased lipid peroxides in schizophrenia at the never-medicated first-episode of psychosis and after years of treatment with antipsychotics. *Schizophr Res*. 2002;58(1):1–10.
50. Peet M, Shah S, Selvam K, Ramchand CN. Polyunsaturated fatty acid levels in red cell membranes of unmedicated schizophrenic patients. *World J Biol Psychiatry*. 2004;5(2):92–99.
51. Reddy RD, Keshavan MS, Yao JK. Reduced red blood cell membrane essential polyunsaturated fatty acids in first episode schizophrenia at neuroleptic-naive baseline. *Schizophr Bull*. 2004;30(4):901–911.
52. Amminger GP, Schäfer MR, Klier CM, et al. Decreased nervonic acid levels in erythrocyte membranes predict psychosis in help-seeking ultra-high-risk individuals. *Mol Psychiatry*. 2012;17(12):1150–1152.
53. Firth J, Rosenbaum S, Ward PB, et al. Adjunctive nutrients in first-episode psychosis: a systematic review of efficacy, tolerability and neurobiological mechanisms. *Early Interv Psychiatry*. 2018;12(5):774–783.

54. Sarris J, Logan AC, Akbaraly TN, et al. International Society for Nutritional Psychiatry Research. Nutritional medicine as mainstream in psychiatry. *Lancet Psychiatry*. 2015;2(3):271–274.
55. Dipasquale S, Pariante CM, Dazzan P, Aguglia E, McGuire P, Mondelli V. The dietary pattern of patients with schizophrenia: a systematic review. *J Psychiatr Res*. 2013;47(2):197–207.
56. Firth J, Carney R, Stubbs B, et al. Nutritional deficiencies and clinical correlates in first-episode psychosis: a systematic review and meta-analysis. *Schizophr Bull*. 2018;44(6):1275–1292.
57. Graham KA, Keefe RS, Lieberman JA, Calikoglu AS, Lansing KM, Perkins DO. Relationship of low vitamin D status with positive, negative and cognitive symptom domains in people with first-episode schizophrenia. *Early Interv Psychiatry*. 2015;9(5):397–405.
58. Shivakumar V, Kalmady SV, Amaresha AC, et al. Serum vitamin D and hippocampal gray matter volume in schizophrenia. *Psychiatry Res*. 2015;233(2):175–179.
59. Flatow J, Buckley P, Miller BJ. Meta-analysis of oxidative stress in schizophrenia. *Biol Psychiatry*. 2013;74(6):400–409.
60. Miller BJ, Buckley P, Seabolt W, Mellor A, Kirkpatrick B. Meta-analysis of cytokine alterations in schizophrenia: clinical status and antipsychotic effects. *Biol Psychiatry*. 2011;70(7):663–671.
61. Mondelli V, Cattaneo A, Murri MB, et al. Stress and inflammation reduce brain-derived neurotrophic factor expression in first-episode psychosis: a pathway to smaller hippocampal volume. *J Clin Psychiatry*. 2011;72(12):1677–1684.
62. Chaudhry IB, Hallak J, Husain N, et al. Minocycline benefits negative symptoms in early schizophrenia: a randomised double-blind placebo-controlled clinical trial in patients on standard treatment. *J Psychopharmacol*. 2012;26(9):1185–1193.
63. Meyer U, Schwarz MJ, Müller N. Inflammatory processes in schizophrenia: a promising neuroimmunological target for the treatment of negative/cognitive symptoms and beyond. *Pharmacol Ther*. 2011;132(1):96–110.
64. Berger GE, Proffitt TM, McConchie M, et al. Ethyl-eicosapentaenoic acid in first-episode psychosis: a randomized, placebo-controlled trial. *J Clin Psychiatry*. 2007;68(12):1867–1875.
65. Berger GE, Wood SJ, Wellard RM, et al. Ethyl-eicosapentaenoic acid in first-episode psychosis. A 1H-MRS study. *Neuropsychopharmacology*. 2008;33(10):2467–2473.
66. Emsley R, Chiliza B, Asmal L, et al. A randomized, controlled trial of omega-3 fatty acids plus an antioxidant for relapse prevention after antipsychotic discontinuation in first-episode schizophrenia. *Schizophr Res*. 2014;158(1-3):230–235.
67. Bos DJ, van Montfort SJ, Oranje B, Durston S, Smeets PA. Effects of omega-3 polyunsaturated fatty acids on human brain morphology and function: what is the evidence? *Eur Neuropsychopharmacol*. 2016;26(3):546–561.
68. Arroll MA, Wilder L, Neil J. Nutritional interventions for the adjunctive treatment of schizophrenia: a brief review. *Nutr J*. 2014;13:91.
69. Hamazaki K, Maekawa M, Toyota T, et al. Fatty acid composition and fatty acid binding protein expression in the postmortem frontal cortex of patients with schizophrenia: a case-control study. *Schizophr Res*. 2016;171(1-3):225–232.
70. Bitanihirwe BK, Woo TU. Oxidative stress in schizophrenia: an integrated approach. *Neurosci Biobehav Rev*. 2011;35(3):878–893.
71. Pawełczyk T, Grancow-Grabka M, Trafalska E, Szemraj J, Pawełczyk A. Oxidative stress reduction related to the efficacy of n-3 polyunsaturated fatty acids in first episode schizophrenia: secondary outcome analysis of the OFFER randomized trial. *Prostaglandins Leukot Essent Fatty Acids*. 2017;121:7–13.
72. Bansal M, Kaushal N. *Oxidative Stress Mechanisms and their Modulation*. New Delhi: Springer; 2014.
73. Snow PJ. *Free Radicals in the Brain: Aging, Neurological and Mental Disorders*. Berlin, Heidelberg: Springer; 2012.
74. Emiliani FE, Sedlak TW, Sawa A. Oxidative stress and schizophrenia: recent breakthroughs from an old story. *Curr Opin Psychiatry*. 2014;27(3):185–190.
75. Koga M, Serritella AV, Sawa A, Sedlak TW. Implications for reactive oxygen species in schizophrenia pathogenesis. *Schizophr Res*. 2016;176(1):52–71.
76. Wood SJ, Yücel M, Pantelis C, Berk M. Neurobiology of schizophrenia spectrum disorders: the role of oxidative stress. *Ann Acad Med Singapore*. 2009;38(5):396-6.
77. Yao JK, Keshavan MS. Antioxidants, redox signaling, and pathophysiology in schizophrenia: an integrative view. *Antioxid Redox Signal*. 2011;15(7):2011–2035.
78. Greco A, Minghetti L. Isoprostanes as biomarkers and mediators of oxidative injury in infant and adult central nervous system diseases. *Curr Neurovasc Res*. 2004;1(4):341–354.

79. Stupin A, Rasic L, Matic A, et al. Omega-3 polyunsaturated fatty acids-enriched hen eggs consumption enhances microvascular reactivity in young healthy individuals. *Appl Physiol Nutr Metab*. 2018;10:1–8.
80. Ristic-Medic D, Suzic S, Vucic V, Takic M, Tepsic J, Glibetic M. Serum and erythrocyte membrane phospholipids fatty acid composition in hyperlipidemia: effects of dietary intervention and combined diet and fibrate therapy. *Gen Physiol Biophys*. 2009. 28 Spec No:190-9.
81. Kiecolt-Glaser JK, Belury MA, Andridge R, Malarkey WB, Hwang BS, Glaser R. Omega-3 supplementation lowers inflammation in healthy middle-aged and older adults: a randomized controlled trial. *Brain Behav Immun*. 2012;26(6):988–995.
82. Kiecolt-Glaser JK, Epel ES, Belury MA, et al. Omega-3 fatty acids, oxidative stress, and leukocyte telomere length: a randomized controlled trial. *Brain Behav Immun*. 2013;28:16–24.
83. Balcerczyk A, Gajewska A, Macierzyńska-Piotrowska E, Pawelczyk T, Bartosz G, Szemraj J. Enhanced antioxidant capacity and anti-ageing biomarkers after diet micronutrient supplementation. *Molecules*. 2014;19(9):14794–14808.
84. Hoen WP, Lijmer JG, Duran M, Wanders RJ, van Beveren NJ, de Haan L. Red blood cell polyunsaturated fatty acids measured in red blood cells and schizophrenia: a meta-analysis. *Psychiatry Res*. 2013;207(1–2):1–12.
85. Rice SM, Schäfer MR, Klier C, Mossaheb N, Vijayakumar N, Amminger GP. Erythrocyte polyunsaturated fatty acid levels in young people at ultra-high risk for psychotic disorder and healthy adolescent controls. *Psychiatry Res*. 2015;228(1):174–176.
86. Knöchel C, Voss M, Grüter F, et al. Omega 3 fatty acids: novel neurotherapeutic targets for cognitive dysfunction in mood disorders and Schizophrenia? *Curr Neuropharmacol*. 2015;13(5):663–680.
87. Behdani F, Roudbaraki SN, Saberi-Karimian M, et al. Assessment of the efficacy of omega-3 fatty acids on metabolic and inflammatory parameters in patients with schizophrenia taking clozapine and sodium valproate. *Psychiatry Res*. 2018;261:243–247.
88. Smesny S, Milleit B, Schaefer MR, et al. Effects of omega-3 PUFA on the vitamin E and glutathione antioxidant defense system in individuals at ultra-high risk of psychosis. *Prostaglandins Leukot Essent Fatty Acids*. 2015;101:15–21.
89. Calder PC. Fatty acids and inflammation: the cutting edge between food and pharma. *Eur J Pharmacol*. 2011;668(Suppl 1):S50–S58.
90. Fritsche KL. The science of fatty acids and inflammation. *Adv Nutr*. 2015;6(3):293S–301S.
91. Khan SA, Vanden Heuvel JP. Role of nuclear receptors in the regulation of gene expression by dietary fatty acids (review). *J Nutr Biochem*. 2003;14(10):554–567.
92. Mazza M, Marano G, Traversi G, Di Nicola M, Catalano V, Janiri L. The complex interplay of depression, inflammation and omega-3: state of the art and progresses in research. *Clin Ter*. 2015;166(3):e242–e247.
93. Bradbury J. Docosahexaenoic acid (DHA): an ancient nutrient for the modern human brain. *Nutrients*. 2011;3(5):529–554.
94. Calder PC. Mechanisms of action of (n-3) fatty acids. *J Nutr*. 2012;142(3):592S–599S.
95. García-Bueno B, Bioque M, Mac-Dowell KS, et al. Pro-/anti-inflammatory dysregulation in patients with first episode of psychosis: toward an integrative inflammatory hypothesis of schizophrenia. *Schizophr Bull*. 2014;40(2):376–387.
96. Kirkpatrick B, Miller BJ. Inflammation and schizophrenia. *Schizophr Bull*. 2013;39(6):1174–1179.
97. Hammamieh R, Chakraborty N, Gautam A, et al. Transcriptomic analysis of the effects of a fish oil enriched diet on murine brains. *PLoS ONE*. 2014;9(3):e90425.
98. Calder PC. n-3 fatty acids, inflammation and immunity: new mechanisms to explain old actions. *Proc Nutr Soc*. 2013;72(3):326–336.
99. Keleshian VL, Kellom M, Kim HW, et al. Neuropathological responses to chronic NMDA in rats are worsened by dietary n-3 PUFA deprivation but are not ameliorated by fish oil supplementation. *PLoS ONE*. 2014;9(5):e95318.
100. Patten AR, Sickmann HM, Dyer RA, Innis SM, Christie BR. Omega-3 fatty acids can reverse the long-term deficits in hippocampal synaptic plasticity caused by prenatal ethanol exposure. *Neurosci Lett*. 2013;551:7–11.
101. Pino O, Guilera G, Gómez-Benito J, Najas-García A, Rufián S, Rojo E. Neurodevelopment or neurodegeneration: review of theories of schizophrenia. *Actas Esp Psiquiatr*. 2014;42(4):185–195.
102. Weinberger D. Schizophrenia as a neurodevelopmental disorder. In: Weinberger DR, Hirsch SR, eds. *Schizophrenia*. Oxford: Blackwell Science; 2003:326–348.

103. Andreasen NC. A unitary model of schizophrenia: Bleuler's "fragmented phrene" as schizencephaly. *Arch Gen Psychiatry*. 1999;56(9):781–787.
104. Buckley PF. Neuroimaging of schizophrenia: structural abnormalities and pathophysiological implications. *Neuropsychiatr Dis Treat*. 2005;1(3):193–204.
105. Benetti S, Pettersson-Yeo W, Hutton C, et al. Elucidating neuroanatomical alterations in the at risk mental state and first episode psychosis: a combined voxel-based morphometry and voxel-based cortical thickness study. *Schizophr Res*. 2013;150(2-3):505–511.
106. Hýža M, Kuhn M, Češková E, Ustohal L, Kašpárek T. Hippocampal volume in first-episode schizophrenia and longitudinal course of the illness. *World J Biol Psychiatry*. 2016;17(6):429–438.
107. Bois C, Levita L, Ripp I, et al. Longitudinal changes in hippocampal volume in the Edinburgh high risk study of Schizophrenia. *Schizophr Res*. 2016;173(3):146–151.
108. Torres US, Duran FL, Schaufelberger MS, et al. Patterns of regional gray matter loss at different stages of schizophrenia: a multisite, cross-sectional VBM study in first-episode and chronic illness. *Neuroimage Clin*. 2016;12:1–15.
109. Ho BC, Andreasen NC, Ziebell S, Pierson R, Magnotta V. Long-term antipsychotic treatment and brain volumes: a longitudinal study of first-episode schizophrenia. *Arch Gen Psychiatry*. 2011;68(2):128–137.
110. Ahmed M, Cannon DM, Scanlon C, et al. Progressive brain atrophy and cortical thinning in Schizophrenia after commencing clozapine treatment. *Neuropsychopharmacology*. 2015;40(10):2409–2417.
111. Fusar-Poli P, Smieskova R, Kempton MJ, Ho BC, Andreasen NC, Borgwardt S. Progressive brain changes in schizophrenia related to antipsychotic treatment? A meta-analysis of longitudinal MRI studies. *Neurosci Biobehav Rev*. 2013;37(8):1680–1691.
112. Andreasen NC, Liu D, Ziebell S, Vora A, Ho BC. Relapse duration, treatment intensity, and brain tissue loss in schizophrenia: a prospective longitudinal MRI study. *Am J Psychiatry*. 2013;170(6):609–615.
113. Goff DC, Falkai P, Fleischhacker WW, et al. The long-term effects of antipsychotic medication on clinical course in Schizophrenia. *Am J Psychiatry*. 2017;174(9):840–849.
114. van Haren NE, Cahn W, Hulshoff Pol HE, Kahn RS. Schizophrenia as a progressive brain disease. *Eur Psychiatry*. 2008;23(4):245–254.
115. Pawełczyk T, Piątkowska-Janko E, Bogorodzki P, et al. Omega-3 fatty acid supplementation may prevent loss of gray matter thickness in the left parieto-occipital cortex in first episode schizophrenia: a secondary outcome analysis of the OFFER randomized controlled study. *Schizophr Res*. 2018;195:168–175.
116. Abu-Akel A, Shamay-Tsoory S. Neuroanatomical and neurochemical bases of theory of mind. *Neuropsychologia*. 2011;49(11):2971–2984.
117. Burke L, Androutsos C, Jogia J, Byrne P, Frangou S. The Maudsley Early Onset Schizophrenia Study: the effect of age of onset and illness duration on fronto-parietal gray matter. *Eur Psychiatry*. 2008;23(4):233–236.
118. Jahshan C, Wynn JK, Mathalon DH, Green MF. Cognitive correlates of visual neural plasticity in schizophrenia. *Schizophr Res*. 2017;190:39–45.
119. Seymour K, Stein T, Sanders LL, Guggenmos M, Theophil I, Sterzer P. Altered contextual modulation of primary visual cortex responses in schizophrenia. *Neuropsychopharmacology*. 2013;38(13):2607–2612.
120. Zai G, Robbins TW, Sahakian BJ, Kennedy JL. A review of molecular genetic studies of neurocognitive deficits in schizophrenia. *Neurosci Biobehav Rev*. 2017;72:50–67.
121. Cieslik EC, Seidler I, Laird AR, Fox PT, Eickhoff SB. Different involvement of subregions within dorsal premotor and medial frontal cortex for pro- and antisaccades. *Neurosci Biobehav Rev*. 2016;68:256–269.
122. Parks NA, Mazzi C, Tapia E, et al. The influence of posterior parietal cortex on extrastriate visual activity: a concurrent TMS and fast optical imaging study. *Neuropsychologia*. 2015;78:153–158.
123. Ross RG, Olincy A, Harris JG, Radant A, Adler LE, Freedman R. Anticipatory saccades during smooth pursuit eye movements and familial transmission of schizophrenia. *Biol Psychiatry*. 1998;44(8):690–697.
124. Cutuli D, Pagani M, Caporali P, et al. Effects of omega-3 fatty acid supplementation on cognitive functions and neural substrates: a voxel-based morphometry study in aged mice. *Front Aging Neurosci*. 2016;8:38.
125. Cutuli D, De Bartolo P, Caporali P, et al. n-3 polyunsaturated fatty acids supplementation enhances hippocampal functionality in aged mice. *Front Aging Neurosci*. 2014;6:220.
126. Staubo SC, Aakre JA, Vemuri P, et al. Mediterranean diet, micronutrients and macronutrients, and MRI measures of cortical thickness. *Alzheimers Dement*. 2017;13(2):168–177.
127. Luciano M, Corley J, Cox SR, et al. Mediterranean-type diet and brain structural change from 73 to 76 years in a Scottish cohort. *Neurology*. 2017;88(5):449–455.

128. Liu Q, Wu D, Ni N, et al. Omega-3 polyunsaturated fatty acids protect neural progenitor cells against oxidative injury. *Mar Drugs*. 2014;12(5):2341–2356.
129. Sadli N, Ackland ML, De Mel D, Sinclair AJ, Suphioglu C. Effects of zinc and DHA on the epigenetic regulation of human neuronal cells. *Cell Physiol Biochem*. 2012;29(1-2):87–98.
130. Suphioglu C, De Mel D, Kumar L, et al. The omega-3 fatty acid, DHA, decreases neuronal cell death in association with altered zinc transport. *FEBS Lett*. 2010;584(3):612–618.
131. Talreja BT, Shah S, Kataria L. Cognitive function in schizophrenia and its association with socio-demographics factors. *Ind Psychiatry J*. 2013;22(1):47–53.
132. Aleman A, Hijman R, de Haan EH, Kahn RS. Memory impairment in schizophrenia: a meta-analysis. *Am J Psychiatry*. 1999;156(9):1358–1366.
133. Bokat CE, Goldberg TE. Letter and category fluency in schizophrenic patients: a meta-analysis. *Schizophr Res*. 2003;64(1):73–78.
134. Dickinson D, Ramsey ME, Gold JM. Overlooking the obvious: a meta-analytic comparison of digit symbol coding tasks and other cognitive measures in schizophrenia. *Arch Gen Psychiatry*. 2007;64(5):532–542.
135. Heinrichs RW, Zakzanis KK. Neurocognitive deficit in schizophrenia: a quantitative review of the evidence. *Neuropsychology*. 1998;12(3):426–445.
136. Keefe RS, Fenton WS. How should DSM-V criteria for schizophrenia include cognitive impairment? *Schizophr Bull*. 2007;33(4):912–920.
137. Keefe RS. Should cognitive impairment be included in the diagnostic criteria for schizophrenia? *World Psychiatry*. 2008;7(1):22–28.
138. Schaefer J, Giangrande E, Weinberger DR, Dickinson D. The global cognitive impairment in schizophrenia: consistent over decades and around the world. *Schizophr Res*. 2013;150(1):42–50.
139. Fett AK, Viechtbauer W, Dominguez MD, Penn DL, van Os J, Krabbendam L. The relationship between neurocognition and social cognition with functional outcomes in schizophrenia: a meta-analysis. *Neurosci Biobehav Rev*. 2011;35(3):573–588.
140. Green MF, Harvey PD. Cognition in schizophrenia: Past, present, and future. *Schizophr Res Cogn*. 2014;1(1):e1–e9.
141. Zaragoza Domingo S, Bobes J, García-Portilla M, Morralla C, EPICOG-SCH Study Group. Cognitive Performance associated to functional outcomes in stable outpatients with schizophrenia. *Schizophr Res Cogn*. 2015;2(3):146–158.
142. Bentsen H, Solberg DK, Refsum H, Bøhmer T. Clinical and biochemical validation of two endophenotypes of schizophrenia defined by levels of polyunsaturated fatty acids in red blood cells. *Prostaglandins Leukot Essent Fatty Acids*. 2012;87(1):35–41.
143. Sethom MM, Fares S, Bouaziz N, et al. Polyunsaturated fatty acids deficits are associated with psychotic state and negative symptoms in patients with schizophrenia. *Prostaglandins Leukot Essent Fatty Acids*. 2010;83(3):131–136.
144. Solberg DK, Bentsen H, Refsum H, Andreassen OA. Association between serum lipids and membrane fatty acids and clinical characteristics in patients with schizophrenia. *Acta Psychiatr Scand*. 2015;132(4):293–300.
145. Watari M, Hamazaki K, Hirata T, Hamazaki T, Okubo Y. Hostility of drug-free patients with schizophrenia and n-3 polyunsaturated fatty acid levels in red blood cells. *Psychiatry Res*. 2010;177(1–2):22–26.
146. Hajjar T, Meng GY, Rajion MA, et al. Omega 3 polyunsaturated fatty acid improves spatial learning and hippocampal peroxisome proliferator activated receptors (PPARα and PPARγ) gene expression in rats. *BMC Neurosci*. 2012;13:109.
147. Yurko-Mauro K, Alexander DD, Van Elswyk ME. Docosahexaenoic acid and adult memory: a systematic review and meta-analysis. *PLoS ONE*. 2015;10(3):e0120391.
148. Condray R, Yao JK, Steinhauer SR, van Kammen DP, Reddy RD, Morrow LA. Semantic memory in schizophrenia: association with cell membrane essential fatty acids. *Schizophr Res*. 2008;106(1):13–28.
149. Peters BD, Szeszko PR, Radua J, et al. White matter development in adolescence: diffusion tensor imaging and meta-analytic results. *Schizophr Bull*. 2012;38(6):1308–1317.
150. Satogami K, Takahashi S, Yamada S, Ukai S, Shinosaki K. Omega-3 fatty acids related to cognitive impairment in patients with schizophrenia. *Schizophr Res Cogn*. 2017;9:8–12.
151. Knowles EE, David AS, Reichenberg A. Processing speed deficits in schizophrenia: reexamining the evidence. *Am J Psychiatry*. 2010;167(7):828–835.

152. Elie D, Poirier M, Chianetta J, Durand M, Grégoire C, Grignon S. Cognitive effects of antipsychotic dosage and polypharmacy: a study with the BACS in patients with schizophrenia and schizoaffective disorder. *J Psychopharmacol*. 2010;24(7):1037–1044.
153. Hori H, Yoshimura R, Katsuki A, et al. Several prescription patterns of antipsychotic drugs influence cognitive functions in Japanese chronic schizophrenia patients. *Int J Psychiatry Clin Pract*. 2012;16(2):138–142.
154. Mazza M, Pomponi M, Janiri L, Bria P, Mazza S. Omega-3 fatty acids and antioxidants in neurological and psychiatric diseases: an overview. *Prog Neuropsychopharmacol Biol Psychiatry*. 2007;31(1):12–26.
155. Marano G, Traversi G, Nannarelli C, Mazza S, Mazza M. Omega-3 fatty acids and schizophrenia: evidences and recommendations. *Clin Ter*. 2013;164(6):e529–e537.
156. McNamara RK, Vannest JJ, Valentine CJ. Role of perinatal long-chain omega-3 fatty acids in cortical circuit maturation: Mechanisms and implications for psychopathology. *World J Psychiatry*. 2015;5(1):15–34.
157. Okereke OI, Reynolds 3rd CF, Mischoulon D, et al. The VITamin D and OmegA-3 TriaL-Depression Endpoint Prevention (VITAL-DEP): rationale and design of a large-scale ancillary study evaluating vitamin D and marine omega-3 fatty acid supplements for prevention of late-life depression. *Contemp Clin Trials*. 2018;68:133–145.
158. Pottala JV, Yaffe K, Robinson JG, Espeland MA, Wallace R, Harris WS. Higher RBC EPA + DHA corresponds with larger total brain and hippocampal volumes: WHIMS-MRI study. *Neurology*. 2014;82(5):435–442.
159. Gama CS, Canever L, Panizzutti B, et al. Effects of omega-3 dietary supplement in prevention of positive, negative and cognitive symptoms: a study in adolescent rats with ketamine-induced model of schizophrenia. *Schizophr Res*. 2012;141(2–3):162–167.
160. Brown I, Cascio MG, Rotondo D, Pertwee RG, Heys SD, Wahle KW. Cannabinoids and omega-3/6 endocannabinoids as cell death and anticancer modulators. *Prog Lipid Res*. 2013;52(1):80–109.
161. Eftekhari MH, Aliasghari F, Beigi MA, Hasanzadeh J. The effect of conjugated linoleic acids and omega-3 fatty acids supplementation on lipid profile in atherosclerosis. *Adv Biomed Res*. 2014;3:15.
162. Varga T, Czimmerer Z, Nagy L. PPARs are a unique set of fatty acid regulated transcription factors controlling both lipid metabolism and inflammation. *Biochim Biophys Acta*. 2011;1812(8):1007–1022.
163. Maekawa M, Watanabe A, Iwayama Y, et al. Polyunsaturated fatty acid deficiency during neurodevelopment in mice models the prodromal state of schizophrenia through epigenetic changes in nuclear receptor genes. *Transl Psychiatry*. 2017;7(9):e1229.
164. D'Agostino G, Cristiano C, Lyons DJ, et al. Peroxisome proliferator-activated receptor alpha plays a crucial role in behavioral repetition and cognitive flexibility in mice. *Mol Metab*. 2015;4(7):528–536.
165. Wietrzych-Schindler M, Szyszka-Niagolov M, Ohta K, et al. Retinoid x receptor gamma is implicated in docosahexaenoic acid modulation of despair behaviors and working memory in mice. *Biol Psychiatry*. 2011;69(8):788–794.
166. Costa M, Squassina A, Congiu D, et al. Investigation of endocannabinoid system genes suggests association between peroxisome proliferator activator receptor-α gene (PPARA) and schizophrenia. *Eur Neuropsychopharmacol*. 2013;23(7):749–759.
167. Feng J, Chen J, Yan J, et al. Structural variants in the retinoid receptor genes in patients with schizophrenia and other psychiatric diseases. *Am J Med Genet B Neuropsychiatr Genet*. 2005;133B(1):50–53.
168. Luchtman DW, Song C. Cognitive enhancement by omega-3 fatty acids from child-hood to old age: findings from animal and clinical studies. *Neuropharmacology*. 2013;64:550–565.
169. Savinova OV, Fillaus K, Harris WS, Shearer GC. Effects of niacin and omega-3 fatty acids on the apolipoproteins in overweight patients with elevated triglycerides and reduced HDL cholesterol. *Atherosclerosis*. 2015;240(2):520–525.
170. Takkunen MJ, Schwab US, de Mello VD, et al. Longitudinal associations of serum fatty acid composition with type 2 diabetes risk and markers of insulin secretion and sensitivity in the Finnish Diabetes Prevention Study. *Eur J Nutr*. 2016;55(3):967–979.
171. Rix TA, Joensen AM, Riahi S, et al. A U-shaped association between consumption of marine n-3 fatty acids and development of atrial fibrillation/atrial flutter-a Danish cohort study. *Europace*. 2014;16(11):1554–1561.
172. Nestel P, Clifton P, Colquhoun D, et al. Indications for omega-3 long chain polyunsaturated fatty acid in the prevention and treatment of cardiovascular disease. *Heart Lung Circ*. 2015;24(8):769–779.
173. Villegas R, Takata Y, Murff H, Blot WJ. Fish, omega-3 long-chain fatty acids, and all-cause mortality in a low-income US population: Results from the Southern Community Cohort Study. *Nutr Metab Cardiovasc Dis*. 2015;25(7):651–658.

174. Leucht S. Measurements of response, remission, and recovery in schizophrenia and examples for their clinical application. *J Clin Psychiatry*. 2014;75(Suppl 1):8–14.
175. Qurashi I, Chaudhry IB, Khoso AB, et al. A randomised, double-blind, placebo-controlled trial of minocycline and/or omega-3 fatty acids added to treatment as usual for at-risk mental states (NAYAB): study protocol. *Trials*. 2017;18(1):524.
176. Cuéllar-Barboza AB, Sánchez-Ruiz JA, Corral PM. Use of omega-3 polyunsaturated fatty acids as augmentation therapy in treatment-resistant Schizophrenia. *Prim Care Companion CNS Disord*. 2017;19(3): 16102040.
177. Jin H, Zisook S, Palmer BW, Patterson TL, Heaton RK, Jeste DV. Association of depressive symptoms with worse functioning in schizophrenia: a study in older outpatients. *J Clin Psychiatry*. 2001;62(10):797–803.
178. Tollefson GD, Andersen SW, Tran PV. The course of depressive symptoms in predicting relapse in schizophrenia: a double-blind, randomized comparison of olanzapine and risperidone. *Biol Psychiatry*. 1999;46(3):365–373.
179. Emsley R, Myburgh C, Oosthuizen P, van Rensburg SJ. Randomized, placebo-controlled study of ethyl-eicosapentaenoic acid as supplemental treatment in schizophrenia. *Am J Psychiatry*. 2002;159(9):1596–1598.
180. Jamilian H, Solhi H, Jamilian M. Randomized, placebo-controlled clinical trial of omega-3 as supplemental treatment in schizophrenia. *Glob J Health Sci*. 2014;6(7):103–108.
181. Peet M, Horrobin DF. A dose-ranging study of the effects of ethyl-eicosapentaenoate in patients with ongoing depression despite apparently adequate treatment with standard drugs. *Arch Gen Psychiatry*. 2002;59(10):913–919.
182. Emsley R, Niehaus DJ, Koen L, et al. The effects of eicosapentaenoic acid in tardive dyskinesia: a randomized, placebo-controlled trial. *Schizophr Res*. 2006;84(1):112–120.
183. Fenton WS, Dickerson F, Boronow J, Hibbeln JR, Knable M. A placebo-controlled trial of omega-3 fatty acid (ethyl eicosapentaenoic acid) supplementation for residual symptoms and cognitive impairment in schizophrenia. *Am J Psychiatry*. 2001;158(12):2071–2074.
184. Bentsen H, Osnes K, Refsum H, Solberg DK, Bøhmer T. A randomized placebo-controlled trial of an omega-3 fatty acid and vitamins E+C in schizophrenia. *Transl Psychiatry*. 2013;3:e335.
185. McGorry PD, Nelson B, Markulev C, et al. Effect of ω-3 polyunsaturated fatty acids in young people at ultra-high risk for psychotic disorders: the NEURAPRO randomized clinical trial. *JAMA Psychiatry*. 2017;74(1):19–27.
186. Mitra S, Natarajan R, Ziedonis D, Fan X. Antioxidant and anti-inflammatory nutrient status, supplementation, and mechanisms in patients with schizophrenia. *Prog Neuropsychopharmacol Biol Psychiatry*. 2017;78:1–11.
187. Fusar-Poli P, Berger G. Eicosapentaenoic acid interventions in schizophrenia: meta-analysis of randomized, placebo-controlled studies. *J Clin Psychopharmacol*. 2012;32(2):179–185.
188. Amminger GP, Schäfer MR, Papageorgiou K, et al. Long-chain omega-3 fatty acids for indicated prevention of psychotic disorders: a randomized, placebo-controlled trial. *Arch Gen Psychiatry*. 2010;67(2):146–154.
189. McEvoy J, Baillie RA, Zhu H, et al. Lipidomics reveals early metabolic changes in subjects with schizophrenia: effects of atypical antipsychotics. *PLoS ONE*. 2013;8(7):e68717.
190. Horrobin DF, Huang YS. Schizophrenia: the role of abnormal essential fatty acid and prostaglandin metabolism. *Med Hypotheses*. 1983;10(3):329–336.
191. Peet M, Laugharne J, Rangarajan N, Horrobin D, Reynolds G. Depleted red cell membrane essential fatty acids in drug-treated schizophrenic patients. *J Psychiatr Res*. 1995;29(3):227–232.
192. Keshavan MS, Anderson S, Pettegrew JW. Is schizophrenia due to excessive synaptic pruning in the prefrontal cortex? The Feinberg hypothesis revisited. *J Psychiatr Res*. 1994;28(3):239–265.
193. Heimer H. Highlights from the Biennial International Congress on Schizophrenia Research (ICOSR), March 24-March 28, 2017. *Clin Schizophr Relat Psychoses*. 2017;11(2):91–94.

CHAPTER

19

Neurodevelopmental and Behavioral Effects of Variations in Omega-3 Polyunsaturated Fatty Acids Levels in Vulnerable Populations

Danitsa Marcos Rodrigues[*,†], *Gisele Gus Manfro*[*,†], *Patrícia Pelufo Silveira*[*,‡,§,¶]

[*]Postgraduate Program in Neuroscience, Universidade Federal do Rio Grande do Sul (UFRGS), Porto Alegre, Brazil [†]Anxiety Disorders Program for Child and Adolescent Psychiatry (PROTAIA), Universidade Federal do Rio Grande do Sul (UFRGS), Porto Alegre, Brazil [‡]Department of Psychiatry, Faculty of Medicine, McGill University, Montreal, QC, Canada [§]Ludmer Centre for Neuroinformatics and Mental Health, Douglas Mental Health University Institute, McGill University, Montreal, QC, Canada [¶]Sackler Program for Epigenetics & Psychobiology at McGill University, Montreal, QC, Canada

ABBREVIATIONS

ADHD attention-deficit hyperactivity disorder
ALA α-linolenic acid
ARA arachidonic acid
BBB blood-brain barrier
DHA docosahexaenoic acid
EFA essential fatty acids
EPA eicosapentaenoic acid
FA fatty acids
FADS fatty acid desaturase gene
IUGR intrauterine growth restriction
LA linoleic acid
LC-PUFA long chain PUFA
PUFA polyunsaturated fatty acid
RCTs randomized controled trials

https://doi.org/10.1016/B978-0-12-815238-6.00019-5

INTRODUCTION

Lipids are a varied group of compounds which have many important biological functions such as working as structural components, energy storage, and participating in signaling pathways.[1] Lipids can be classified into five major subcategories including: fatty acids (FAs), triglycerides, phospholipids, sterol lipids, and sphingolipids. Five per cent of all human genes are devoted to lipid synthesis[2] and the brain makes use of all five classes of lipids, containing the second highest concentration of lipids in the human body.[3] Lipids are indispensable elements in the diet, both for the provision of energy (beta-oxidation of FAs), and for the supply of essential FAs. The brain is highly enriched in lipids and particularly long chain (>20 carbons) polyunsaturated fatty acids (LC-PUFAs), and this lipid composition substantially influences subjective perceptions, mood and emotional behavior, depending on a constant FAs supply.

The long chain omega-3 polyunsaturated fatty acids (LC n-3 PUFAs) play an important role in the brain development and functioning,[4–6] being essential in the modulation of neurochemical aspects and behavioral disorders related to stress response, mood regulation,[5,7] and aggressive and impulsive behaviors. It also modulates neurotransmission,[8] such as in the dopaminergic mesocorticolimbic pathway.[9,10] The PUFAs influence the metabolic status, decreasing cardiovascular risk and other chronic diseases.[5]

The most abundant PUFA in the Western diet is the linoleic acid (LA), the omega-6 precursor that in addition to its action in mammalian reproduction, is vital for brain function.[11] The LA is primarily sourced from practically every commercially manufactured food, mostly sourced from soybean oil, corn oil, and sunflower oil.[12] Although LA is a metabolic precursor of arachidonic acid (ARA), the main dietary sources of ARA are red meat and dairy products including eggs.[13] The α-linolenic acid (ALA), available in vegetable sources, is metabolized into eicosapentaeoic acid (EPA) and docosahexaenoic acid (DHA), which are considered the two major n-3 FAs. The EPA and DHA are associated with many important functions related to neural activity, such as cell membrane fluidity, neurotransmission, ion channels, enzyme regulation, gene expression, and myelination[14] and their richest sources are marine fish and seafood.[15] The omega-3 FAs mediate a variety of key neurotransmitter functions, including serotonergic responsivity, signal transduction, and phospholipid turnover.[16,17] DHA alone is essential for optimal neural function[18] composing approximately 30% of the phosphoglycerides[19] and preferentially accumulating in growth cones, astrocytes, synaptosomes, myelin, and microsomal and mitochondrial membranes of neural tissue.[20,21]

It is known that adequate dietary intake of LC-PUFA is important for proper neural development during prenatal and postnatal periods up to 2 years of age[22–24] Recent research suggests that dietary FAs, potentially influenced by genetic variations in enzymes involved in the FA metabolism, contribute to the central nervous system functioning in children with long-lasting outcomes. Children exposed to intrauterine growth restriction (IUGR) are at increased risk for chronic adult diseases demonstrating major effects on the incidence of chronic non-communicable disease.[25–27] In addition, research has shown that exposure to an adverse event in the fetal period is associated with childhood impulsivity and altered food preferences, with an increased consumption of palatable choices,[28–31] which in the long term seems to contribute for obesity and metabolic risks in adulthood.[32] An experimental study demonstrated that early stress associated with deficiency of n-3 PUFAs led, in adulthood, to obesity, insulin and leptin resistance, being preceded by hyperphagia in adolescence.[33] A recent research has shown that poor inhibitory control predicts childhood fussiness-type food intake and also suggests a protective effect of dietary n-3 PUFAs in vulnerable children.[34] Therefore, it may be assumed that neurocognitive functions could be nutritionally programed early in life, resulting in increased risk for maladaptive behaviors.

Understanding the importance of dietary FAs to human neurodevelopment is challenging because of the complexity of FAs metabolism, incomplete knowledge of the pathways of transfer and FA uptake in the brain, and the functional roles of DHA. Although lipids have multiple functions in the brain, including energy substrate and bioactive molecules, the most essential role for lipids in the brain is as components of cellular structural machinery. The lipid composition of the brain can be altered through long-term changes in the diet, which seems to produce consequences on cognitive and emotional behavior.[35–38] These results suggest that increased availability of FAs in the brain can have a beneficial effect on neurodevelopment reflected in mental health and behavioral outcomes. In this review, we discuss the LC-PUFAs structure, synthesis, and accumulation focusing on the DHA importance in the context of neurodevelopment and brain functions.

LC-PUFAS STRUCTURE AND SYNTHESIS

The FAs can be classified based on the number of double bonds present in side chains: saturated (SFAs, no double bonds), monounsaturated (MUFAs, a single double bond), and polyunsaturated (PUFAs, ≥2 double bonds); FAs can also be classified by their carbon chain length and the position of the first double bond on methyl terminal (omega; ω; or n − FAs).[39] The PUFAs are subdivided into the omega- 3 (n-3) series (the first double bond is 3 carbons from the end carbon atom of the molecule) that are synthetically derived from LA, and the omega-6 (n-6) series that are derived from ALA, both 18-carbon atom-containing FAs. These two essential fatty acids,[40] LA and ALA, cannot be synthesized in the body and must be obtained from diet and act as precursors of other FAs.[41]

In the liver, the LA and ALA fatty acids may be desaturated and elongated to become other LC-PUFAs. LA can be converted sequentially via a biosynthetic pathway into other omega-6 fatty acids, the gamma linolenic acid (GLA), and the ARA and dihomogammalinolenic acids (DGLA). Similarly, ALA is converted into LC n-3 PUFAs such as EPA and DHA.

The n-3 and n-6 PUFAs share the same enzymes and compete for their desaturation and elongation on the endoplasmic reticulum.[42] Synthesized fatty acids can be transported into the blood as lipoproteins. These two classes of essential fatty acids (EFA) are not interconvertible, are metabolically and functionally distinct, and often have important opposing physiological functions. The balance of EFA is important for good health and normal development: recent studies indicate that human beings evolved on a diet with equal amounts of n-6 and n-3 PUFAs[43,44]; a high ratio of around ~16: 1 is found in today's typical Western diets.[45] A high dietary intake of n-6 fatty acids as occurs today promotes many inflammatory and autoimmune diseases,[46] whereas increased levels of n-3 PUFAs (lower n-6/n-3 ratio) exert suppressive effects.[47,48] This balanced ratio of n-6 to n-3 is critical to human development during pregnancy and lactation in the prevention of chronic diseases and in their management.[49,50] Considering the proportion and the opposing effects of n − 3 and n−6 PUFAs, especially dietary LA and ALA play a significant role in regulating body homeostasis,[51] or by locally acting bioactive signaling lipids called eicosanoids derived from ARA, EPA, and DHA.[52]

In animals, primarily in the endoplasmic reticulum of liver cells, biosynthesis of LC-PUFAs starts with desaturation of LA and ALA, thus generating γ-linolenic acid (GLA) and stearidonic acid (SDA). By a specific elongase these desaturated FAs are elongated to yield dihomo-γ-linolenic acid (DGLA) and eicosatetraenoic acid (ETA) respectively. Finally, a desaturase, by adding a carbon double bond and carrying out one more desaturation produces ARA and EPA, respectively. By a desaturase, GLA, DHGLA, and ARA can be converted to SDA, ETA, and EPA, respectively. In mammals EPA undergoes two successive elongation cycles. First, docosapentaenoic acid (DPA, clupanodonic acid) is generated. Then tetracosanolpentaenoic acid is produced which then yields tetracosahexaenoic acid (THA) by desaturation. This PUFA is then subjected to β-oxidation by which its chain is shortened by two carbons to yield very-long-chain (VLC, FAs containing 22 or more carbon atom) PUFA DHA, the final product.[53] Since members of n-6 and n-3 PUFAs compete for corresponding desaturase and elongase enzymes, bioconversion of LA and ALA to their respective LC and VLC-PUFAs in humans depends on the ratio of ingested n-6 and n-3 FAs, with the highest rate of formation of EPA and DHA occurs when a 1:1 ratio of LA and ALA is present.[54] Compared to ALA and EPA, DHA supplementation is the most effective way to improve body DHA levels. It is important to note that genetic variation in genes encoding FAs desaturases also influences essential fatty acid metabolism and may increase requirements in some individuals. Recent studies have demonstrated that the metabolic capacity of the LC-PUFA biosynthetic pathway differs among individuals and populations, suggesting that genetic and epigenetic variations are highly associated with the levels of LC-PUFAs produced in human circulation, cells, and tissues.[55,56] Advances in nutrigenomics allowed to identify three factors involved in the regulation of the conversion of ALA and LA to n-3 LC-PUFAs and n-6 LC-PUFAs, such as (i) high heritability of LC-PUFA biosynthetic capacity (i.e., strong genetic regulation), (ii) marked epigenetic regulation (i.e., potential gene environment interplay in regulation), and (iii) differences in dietary PUFA intake (i.e., variability in the environment).[57] Understanding the relationship between these factors will enable the prediction of physiological and pathological outcomes of human nutrition and potentially guide nutritional recommendations and interventions to individuals and populations throughout the life span. In a recent study, our group created a multilocus genetic score from the fatty acid desaturase (FADS) gene cluster and investigated if these genes moderated the association between being born small for gestational age and having altered eating behaviors in childhood.[58] The fatty acid desaturase (FADS) 1 and 2 genes are important for PUFAs metabolism because they code for fatty acid desaturase enzymes D5D and D6D, which are responsible for the synthesis of highly unsaturated fatty acids such as eicosopentaenoic and docosahexaenoic acids (synthesized from α-linolenic acid), and arachidonic acid (synthesized from LA). Several SNPs in the FADS1–2 gene complex influence PUFAs short, medium, and long-term plasma availability.[59–61] We observed interactions between being born small for gestational age and the genetic score on obesogenic eating behaviors, suggesting that genetic markers associated with higher plasma PUFAs levels in childhood in these children relate to a more adaptive eating behavior with no effects on body mass index (BMI).[58] The findings of vulnerability and protection factors through eating behavior and PUFAs consumption have potential clinical applications in primary prevention of chronic metabolic diseases in this population.

The LC-PUFAs and their metabolites are present in practically all cells and tissues of the body and act as potent signaling molecules that impact a wide range of physiologic and pathophysiologic processes and can also be converted to a diverse family of metabolites including multiple forms of prostaglandins, thromboxanes, hydroxyeicosatetraenoic acids, epoxyeicosatrienoic acids, leukotrienes, lipoxins, resolvins, protectins, maresins, and endocannabinoids.[62–67]

The synthesized and/or dietary-derived FAs are taken up from circulating blood into the brain through the blood-brain barrier (BBB). The transport of fatty acids from blood into the brain is a complex mechanism due to the presence of tight junctions in the BBB and has been the subject of research.[68] Once FAs are within the brain, they can be subsequently converted into various metabolites, such as eicosanoids and endocannabinoids. Eicosanoids are hormone-like compounds that function locally and are produced from either ARA or EPA/DHA released from membrane phospholipids. Eicosanoids have important and complex functions in the brain and participate in the control of numerous physiological systems, involved in the regulation or halting of blood flow (prostacyclin and thromboxanes, respectively), in the anti-inflammatory processes (prostaglandins), and tissue homeostasis.[11,69] Eicosanoids have been involved in long-term potentiation, spatial learning, synapse plasticity, and sleep induction; they also reduce neuroinflammation and have neuroprotective properties.[16] The reports suggest that decresead n-3 PUFAs levels during critical stages of neurogenesis may alter parameters of cell signaling, including within neurotransmitter systems, resulting in impairments in behavior, learning, and cognition.[70–73] Low intake diets of n-3 PUFAs result in reductions of DHA in the brain with simultaneous increase in the turnover of ARA to eicosanoids,[74] which can be reverted by the supplementation of n-3 PUFAs. Endocannabinoids are another important class of PUFAs, which are derived from ARA, acting in the central nervous system to modulate synaptic plasticity, neurotransmitter release, and have protective actions.[75] It is well established that higher levels of ARA are associated with increased production of endocannabinoids,[76] but it remains unclear if the diet fat intake could alter brain endocannabinoid production.

LC-PUFAS BRAIN ACCUMULATION

About 50%–60% of the dry weight of an adult brain is comprised of lipid and at least 15%–30% of the lipid content is made up of LC-PUFAs.[77] These fatty acids are highly specialized with particular metabolic functions and unique biophysical properties. The lipid brain components consist mainly of phosphoglycerides and cholesterol rich in ARA and DHA.[78] The DHA is the most abundant n-3 fatty acid in the mammalian brain and its level in the brain membrane lipids is altered by the type and amount of fatty acids in the diet and with life stage, increasing with age.[79] The DHA accumulation in the brain starts during the brain growth spurt in the intrauterine stage and continues up to 2 years of age, then maintaining high levels throughout the life span.[80] The intrauterine LC-PUFAs supply derives mainly from the maternal circulation across the placenta[81] and postnatal accumulation on infant tissues is supported by maternal transfer of LC-PUFAs through breast milk.[82] The embryo requires approximately 40–60 mg of n-3 LC-PUFAs per kilogram of body weight per day during the third trimester of pregnancy.[83] The DHA and ARA, the most abundant LC-PUFAs in the brain and retina play a special and distinct role in neurodevelopment, requiring specific balance ratios for optimal mental health.[84] The last trimester of pregnancy appears to be a critical period of time for DHA and ARA accretion in the fetal brain.[85] Importantly, experimental evidence suggests that DHA and ARA accumulation in the neural tissue is regulated by distinct mechanisms, showing also that DHA levels vary with diet but ARA levels remain relatively constant during early childhood through adolescence.[86]

LC-PUFAS IN BRAIN DEVELOPMENT AND FUNCTION

Maternal diet is a major determinant of offspring health.[26,87,88] The dietary-induced modifications appear to directly affect the fetal and childhood neurodevelopment,[89,90] and this may have a long-term impact on the onset of neurodegenerative disease later in life.[91] Thus, the quantity and quality of the fatty acids ingested can significantly influence a wide range of brain functions, including maintence of axons and dendrites, cell shape, lipid raft formation, G protein-coupled signaling (leading to altered gene expression), polarity, neuronal plasticity, dopamine storage, vesicle formation and transport, glucose uptake, and hypothalamic regulations.[92]

The PUFAs have several important behavioral and neurochemical roles in the central nervous system, such as: (i) regulation of energy intake and glucose homeostasis, (ii) modulation of neuroinflammatory mechanisms, and (iii) support and regulation of neurotransmission.[41] DHA affects the blood–brain-barrier function, neuronal membrane fluidity, and also regulates neurotransmission systems,[93,94] with a significant effect on neuronal membrane dynamics and transporter, receptor, and neurotransmitter functions.[77] These mechanisms of action provide support for possible neuroprotective effects of n3-PUFAs on mental health outcomes.

Experimental evidence has investigated the nutritional effects of PUFAs, demonstrating that its consumption is associated with neurogenesis and neuroprotection in animals supplemmented with DHA.[95–97] Other animal studies examined declines in memory functioning related to deficits in n3-PUFA in the diet, which could be fully or partially restored by n3-PUFAs correction.[98] Recent work with n3-PUFAs supplementation in nonhuman primates has demonstrated reduction in anxiety levels and an enhanced cognitive performance.[99]

Prematurely born infants are at particular risk for LC-PUFA deficiency and its supplementation in breast milk and formula has been suggested to improve both cognitive and visual development in these children.[100] A recent review showed conflicting results in supplementation of DHA for preterm infants with inconsistent long-term improvements and supporting benefits only at the short-term.[101] Moreover, interventions linking early nutrition and cognitive development have shown ambiguous evidence of benefits associated with LC-PUFAs supplementation. A recent meta-analysis does not reveal that LC-PUFA supplementation has a significant impact on the level of intelligence in low birth weight infants.[102] Another meta-analysis concluded that there were no beneficial effects or harms of LC-PUFA supplementation on neurodevelopmental outcomes and no consistent benefits on visual acuity on full-term infants.[103] Recent reports have demonstrated that n-3 PUFAs exert a protective effect in IUGR children by limiting impulsive eating.[34,104] Thus, for the newborn infants, the recommendation that has an unequivocal benefit in relation to the functional effects of LC-PUFA supplementation in childhood is that which improves the rate of development of visual acuity.[105]

The brain DHA deposition continues to increase in childhood, and although the accretion rate declines, the incorporation of DHA is still high at least during the preschool years. Once high levels of DHA are achieved in the brain, these are maintained during later life and this presumably also depends on an optimal dietary supply, as dietary intake of DHA from fish in adults has been shown to be the dominant determinant of DHA levels in various lipid pools.[24] High-quality evidence demonstrates that increasing children's dietary intake of the n-3 LC- PUFAs may improve concentration, reduce disruptive behavior, and lead to better reading and spelling, with benefits in cognitive performances in children and adolescents with ADHD.[106,107] A randomized controled trial (the DOLAB I study) provided the first good evidence for the benefits of DHA, with significant improvement over placebo on behavior and learning among healthy but under-performing children, aged 7–9 years.[108] Surprisingly, the DOLAB II study did not replicate the original findings of significant, positive effects of omega-3 DHA on either learning or behavior.[109] A recent methylome-wide study suggests that altered early life methylation in the peroxisomal network, which is essential in DHA formation, could contribute to an impulsive phenotype, including Attention-deficit/hyperactivity disorder (ADHD) in children at age 7.[110] Despite all the contradictory results explained in part by the variety of interventions and part by the modulating effects of age and gender, it should be noted that children considered at risk, like IUGR individuals, might benefit the most from supplementation of LC n3-PUFA with regard to neurodevelopmental outcomes (e.g., memory, non-verbal cognitive development, processing speed, visual-perceptive capacity, attention, and executive function) and school achievement (e.g., reading and spelling). We have recently shown that a genetic variant of the PLIN4 gene (rs8887) - associated with enhanced sensitivity to the protective effects of dietary n-3 PUFAs on metabolic outcomes[111] – interacts with fetal growth and influences impulsivity in childhood.[112] In humans, the perilipin gene (*PLIN*) has been localized to chromosomal location 15q26,[113] a region previously linked to obesity, hypertriglyceridemia, and diabetes.[114,115] The PLIN4 protein, previously referred to S3–12, belongs to the PAT family.[116]

In adulthood, full brain development is already achieved. However, at this stage of life stressful events can alter mood states and cognition, increasing the susceptibility to psychopathologies.[117] Clinical trials on supplying n-3 PUFAs among healthy young adults report improved effects on decision-making[118] and attentional funtion,[93] contrasting to other with no consistent evidence for significant effects on memory and cognitive functioning[119] and for negative effect and impulsivity traits.[120]

The studies in the elderly population have contradicting results: a meta-analysis of fish consumption indicates that n-3 LCPUFAs was associated with protection against neurocognitive deficits[121] but other meta-analyses assessments have described no evidence for the efficacy of omega-3 supplements in the treatment of Alzheimer's disease[122] which was corroborated by another randomized study that found no improvement in the global cognitive function in institutionalized elderly people.[123]

Omega-3 FAs seem to play a role in the regulation of the stress-response, influencing the activity of the hypothalamic–pituitary–adrenal (HPA) axis. It was proposed that glucocorticoid receptor (GR) sensitivity contributes to n-3 PUFAs deficiency-related emotional problems.[124] Furthermore, data relates consequences of maternal low PUFA intake and the development of neuropsychiatric disorders. Maternal malnutrition has been linked to the development of different neuropsychiatric disorders in infancy [Autism Spectrum Disorder (ASD)],[125] in childhood [Attention deficit disorder (ADHD)],[126] in adolescence [Schizophrenia spectrum disorders (SSD)][127] or in adulthood (depression and anxiety).[128] Thus, it is possible to correlate altered stress response following maternal poor intake of *n*-3 PUFAs[129] with many psychiatric diseases, particularly depression[130] and psychosis.[131]

It has been shown in animal and human trials that n-3 PUFAs may improve memory performance. Patients with Major Depression Disorder (MDD) have decreased intake of n-3 PUFAs and limited metabolism when compared to healthy controls.[132] Affective disorder and Schizophrenia (SZ) are characterized by emotional and cognitive dysfunction encompassing several domains. Cognitive dysfunction is closely related to several impaired functioning and deteriorating health conditions. Epidemiological studies show that the dietary intake of *n*-3 PUFAs through fish consumption is inversely correlated with the prevalence of depression, thus suggesting that *n*-3 PUFAs have a protective effect.[133] Individuals with bipolar disorder (BD), MDD, and SZ are thought to have a disturbed lipid metabolism: the levels of DHA, ALA, and EPA in erythrocytes of these patients seem to be lower compared to healthy controls.[5] Importantly, there is evidence of a higher prevalence of bipolar spectrum disorders associated with lower fish consumption, added to the finding that first-degree relatives of these patients showed a trend toward lower levels of n-3 PUFAs, suggesting a connection between FAs with BD.[134] A recent prospective study showed that the *n*-6/3 PUFA ratio, used as a risk biomarker,[135] was associated with an increased ratio of mood disorders in young people exhibiting an high risk phenotype.[136] Abnormalities in phospholipid metabolism are suggested to be involved in the pathogenesis of SZ. Accordingly, there have been several investigations of the n3-PUFA content of membrane phospholipids in SZ. The literature shows that n3-FAs provide numerous health benefits and changes in their concentration are connected to a variety of psychiatric symptoms and disorders including stress, anxiety, cognitive impairment, mood disorders, and schizophrenia but results at this moment are non-conclusive.[137,138] Further studies are necessary to confirm n-3 FAs' supplementation as a potential rational treatment in psychiatric disorders.

CONCLUSIONS

The role of n-3 PUFAs in healthy brain development has been explored through intense scientific research and because of its importance for public health. Regarding the supplementation of n-3 PUFAs, there are some promises and nonconclusive evidence for clinical protective benefits for affective, cognitive disorders and for neural health, particularly in vulnerable individuals. Given the crucial role of DHA in neuropsychiatric diseases involved in altered prefrontal cortex (PFC) and subcortical circuitries, special attention should be paid to define the best dietary options. Investigations on mental health and disease highlighted the importance of lipidomics for the identification of risk biomarkers in subjects with high-risk phenotypes, revealing associations between membrane PUFA levels and mood disorders.

As shown, the available evidence is not always consistent, probably for the heterogeneity of the methods and samples in the original studies, which often had small and not homogeneous sample sizes, different selection criteria, diverse subtypes and dosage of LC-PUFAs as well as duration of supplementation. Another limiting methodological factor is the variable nature of the neuropsychological outcomes. It is noteworthy that the evidence is still emerging but that randomized controlled trials (RCTs) have been inconsistent with many design limitations. A major challenge is to conduct rigorous RCTs to provide the evidence for dietary recommendations for DHA.

Despite these limitations, results seem to highlight the beneficial role of n-3 PUFAs under certain conditions. Based on epidemiological evidence, the nutritional recommendations by health agencies are to reduce excessive intake of omega-6 PUFAs, and avoid the high omega-6/omega-3 ratio of the current western diets. Furthermore, intake of large amounts of fish combined to natural antioxidants is strongly recommended. This guideline can be translated into a weekly ingestion of 2–3 portions of oily fish or a daily 450–500 mg of EPA+DHA for general health improvements in the adult population.[139] The Food and Agriculture Organization (FAO) of the United Nations jointly with the World Health Organization (WHO) and the European Food Safety Authority (EFSA) recommend a daily EPA+DHA intake of at least 250 mg; increased amounts of up to 1 to 3–4 g/day of omega-3 LC-PUFAs should be ingested during pregnancy and lactation or to prevent most cardiovascular, neurodegenerative, and pro-inflammatory disorders[140].

Because of its differentiated characteristics of low conversion rates and participation in numerous neural processes (neurotransmission, neurogenesis, and protection against oxidative stress), DHA is one of the main nutrients capable of acting in neural programming, maximizing neurodevelopment, and minimizing neurodegeneration. Continued research into the need for adequate n-3 FAs for optimal brain development and function may better elucidate dietary conditions and diet-gene or diet-disease interactions that present a risk for brain damage.

CONFLICT OF INTEREST

The authors declare no conflict of interest.

FUNDING

Researchers were funded by Brazilian National Council for Technological and Scientific Development (CNPq), Coordination for the Improvement of Higher Education Personnel (CAPES), Ludmer Centre for Neuroinformatics and Mental Health, JPB Network on Toxic Stress.

References

1. Calder PC. Functional roles of fatty acids and their effects on human health. *JPEN Journal of Parenteral and Enteral Nutrition*. 2015;39(1 Suppl):18S–32S.
2. van Meer G, Voelker DR, Feigenson GW. Membrane lipids: where they are and how they behave. *Nat Rev Mol Cell Biol*. 2008;9(2):112–124.
3. Hamilton JA, Hillard CJ, Spector AA, Watkins PA. Brain uptake and utilization of fatty acids, lipids and lipoproteins: application to neurological disorders. *J Mol Neurosci*. 2007;33(1):2–11.
4. Schuchardt JP, Huss M, Stauss-Grabo M, Hahn A. Significance of long-chain polyunsaturated fatty acids (PUFAs) for the development and behaviour of children. *Eur J Pediatr*. 2010;169(2):149–164.
5. Balanza-Martinez V, Fries GR, Colpo GD, et al. Therapeutic use of omega-3 fatty acids in bipolar disorder. *Expert Rev Neurother*. 2011;11(7):1029–1047.

6. Rombaldi Bernardi J, de Souza Escobar R, Ferreira CF, Pelufo Silveira P. Fetal and neonatal levels of omega-3: effects on neurodevelopment, nutrition, and growth. *Scient World J*. 2012;2012:202473.
7. Ross BM, Seguin J, Sieswerda LE. Omega-3 fatty acids as treatments for mental illness: Which disorder and which fatty acid? *Lipids Health Dis*. 2007;6:21.
8. Long SJ, Benton D. A double-blind trial of the effect of docosahexaenoic acid and vitamin and mineral supplementation on aggression, impulsivity, and stress. *Hum Psychopharmacol*. 2013;28(3):238–247.
9. Zimmer L, Vancassel S, Cantagrel S, et al. The dopamine mesocorticolimbic pathway is affected by deficiency in n-3 polyunsaturated fatty acids. *Am J Clin Nutr*. 2002;75(4):662–667.
10. Zimmer L, Delion-Vancassel S, Durand G, et al. Modification of dopamine neurotransmission in the nucleus accumbens of rats deficient in n-3 polyunsaturated fatty acids. *J Lipid Res*. 2000;41(1):32–40.
11. Crawford MA, Bazinet RP, Sinclair AJ. Fat intake and CNS functioning: ageing and disease. *Ann Nutr Metab*. 2009;55(1–3):202–228.
12. Blasbalg TL, Hibbeln JR, Ramsden CE, Majchrzak SF, Rawlings RR. Changes in consumption of omega-3 and omega-6 fatty acids in the United States during the 20th century. *Am J Clin Nutr*. 2011;93(5):950–962.
13. Wood JD, Enser M, Fisher AV, et al. Fat deposition, fatty acid composition and meat quality: a review. *Meat Sci*. 2008;78(4):343–358.
14. Innis SM. Dietary omega 3 fatty acids and the developing brain. *Brain Res*. 2008;1237:35–43.
15. Brenna JT, Salem Jr. N, Sinclair AJ, Cunnane SC. International Society for the Study of fatty a, Lipids I. Alpha-Linolenic acid supplementation and conversion to n-3 long-chain polyunsaturated fatty acids in humans. *Prostaglandins Leukot Essent Fatty Acids*. 2009;80(2–3):85–91.
16. Yehuda S, Rabinovitz S, Mostofsky DI. Essential fatty acids are mediators of brain biochemistry and cognitive functions. *J Neurosci Res*. 1999;56(6):565–570.
17. Condray R, Yao JK, Steinhauer SR, van Kammen DP, Reddy RD, Morrow LA. Semantic memory in schizophrenia: association with cell membrane essential fatty acids. *Schizophr Res*. 2008;106(1):13–28.
18. Mitchell DC, Gawrisch K, Litman BJ, Salem Jr. N. Why is docosahexaenoic acid essential for nervous system function? *Biochem Soc Trans*. 1998;26(3):365–370.
19. Brenna JT, Diau GY. The influence of dietary docosahexaenoic acid and arachidonic acid on central nervous system polyunsaturated fatty acid composition. *Prostaglandins Leukot Essent Fatty Acids*. 2007;77(5–6):247–250.
20. Bourre JM, Bonneil M, Chaudiere J, et al. Structural and functional importance of dietary polyunsaturated fatty acids in the nervous system. *Adv Exp Med Biol*. 1992;318:211–229.
21. Jones CR, Arai T, Rapoport SI. Evidence for the involvement of docosahexaenoic acid in cholinergic stimulated signal transduction at the synapse. *Neurochem Res*. 1997;22(6):663–670.
22. Gibson RA, Muhlhausler B, Makrides M. Conversion of linoleic acid and alpha-linolenic acid to long-chain polyunsaturated fatty acids (LCPUFAs), with a focus on pregnancy, lactation and the first 2 years of life. *Matern Child Nutr*. 2011;7(Suppl 2):17–26.
23. Uauy R, Dangour AD. Nutrition in brain development and aging: role of essential fatty acids. *Nutr Rev*. 2006;64(5 Pt 2):S24–S33. [discussion S72-91].
24. Lauritzen L, Brambilla P, Mazzocchi A, Harslof LB, Ciappolino V, Agostoni C. DHA effects in brain development and function. *Nutrients*. 2016;8(1).
25. Barker DJ, Winter PD, Osmond C, Margetts B, Simmonds SJ. Weight in infancy and death from ischaemic heart disease. *Lancet*. 1989;2(8663):577–580.
26. Hales CN, Barker DJ. Type 2 (non-insulin-dependent) diabetes mellitus: the thrifty phenotype hypothesis. *Diabetologia*. 1992;35(7):595–601.
27. Lahti M, Eriksson JG, Heinonen K, et al. Late preterm birth, post-term birth, and abnormal fetal growth as risk factors for severe mental disorders from early to late adulthood. *Psychol Med*. 2015;45(5):985–999.
28. Lussana F, Painter RC, Ocke MC, Buller HR, Bossuyt PM, Roseboom TJ. Prenatal exposure to the Dutch famine is associated with a preference for fatty foods and a more atherogenic lipid profile. *Am J Clin Nutr*. 2008;88(6):1648–1652.
29. Barbieri MA, Portella AK, Silveira PP, et al. Severe intrauterine growth restriction is associated with higher spontaneous carbohydrate intake in young women. *Pediatr Res*. 2009;65(2):215–220.
30. Perala MM, Mannisto S, Kaartinen NE, et al. Body size at birth is associated with food and nutrient intake in adulthood. *PloS One*. 2012;7(9):e46139.
31. Kaseva N, Wehkalampi K, Hemio K, et al. Diet and nutrient intake in young adults born preterm at very low birth weight. *J Pediatr*. 2013;163(1):43–48.

32. Portella AK, Silveira PP. Neurobehavioral determinants of nutritional security in fetal growth-restricted individuals. *Ann N Y Acad Sci*. 2014;1331:15–33.
33. Bernardi JR, Ferreira CF, Senter G, et al. Early life stress interacts with the diet deficiency of omega-3 fatty acids during the life course increasing the metabolic vulnerability in adult rats. *PloS One*. 2013;8(4):e62031.
34. Reis RS, Bernardi JR, Steiner M, Meaney MJ, Levitan RD, Silveira PP. Poor infant inhibitory control predicts food fussiness in childhood—a possible protective role of n-3 PUFAs for vulnerable children. *Prostaglandins Leukot Essent Fatty Acids*. 2015;97:21–25.
35. Moriguchi T, Greiner RS, Salem Jr. N. Behavioral deficits associated with dietary induction of decreased brain docosahexaenoic acid concentration. *J Neurochem*. 2000;75(6):2563–2573.
36. Wainwright PE. Dietary essential fatty acids and brain function: a developmental perspective on mechanisms. *Proc Nutr Soc*. 2002;61(1):61–69.
37. Arsenault D, Julien C, Tremblay C, Calon F. DHA improves cognition and prevents dysfunction of entorhinal cortex neurons in 3xTg-AD mice. *PloS One*. 2011;6(2):e17397.
38. Coryell WH, Langbehn DR, Norris AW, Yao JR, Dindo LN, Calarge CA. Polyunsaturated fatty acid composition and childhood adversity: independent correlates of depressive symptom persistence. *Psychiatry Res*. 2017;256:305–311.
39. Saini RK, Keum Y-S. Omega-3 and omega-6 polyunsaturated fatty acids: Dietary sources, metabolism, and significance—a review. *Life Sci*. 2018;203:255–267.
40. Holman RT. Essential fatty acids. *Nutr Rev*. 1958;16(2):33–35.
41. Bazinet RP, Laye S. Polyunsaturated fatty acids and their metabolites in brain function and disease. *Nat Rev Neurosci*. 2014;15(12):771–785.
42. Innis SM. Perinatal biochemistry and physiology of long-chain polyunsaturated fatty acids. *J Pediatr*. 2003;143.(4 Suppl):S1-8.
43. Eaton SB, Konner M. Paleolithic nutrition. A consideration of its nature and current implications. *N Engl J Med*. 1985;312(5):283–289.
44. Eaton SB, Eaton 3rd SB, Sinclair AJ, Cordain L, Mann NJ. Dietary intake of long-chain polyunsaturated fatty acids during the paleolithic. *World Rev Nutr Diet*. 1998;83:12–23.
45. Simopoulos AP. Genetic variants in the metabolism of omega-6 and omega-3 fatty acids: Their role in the determination of nutritional requirements and chronic disease risk. *Exp Biol Med*. 2010;235(7):785–795.
46. Scaioli E, Liverani E, Belluzzi A. The imbalance between n-6/n-3 polyunsaturated fatty acids and inflammatory bowel disease: a comprehensive review and future therapeutic perspectives. *Int J Mol Sci*. 2017;18(12).
47. Wall R, Ross RP, Fitzgerald GF, Stanton C. Fatty acids from fish: the anti-inflammatory potential of long-chain omega-3 fatty acids. *Nutr Rev*. 2010;68(5):280–289.
48. Simopoulos AP. The importance of the omega-6/omega-3 fatty acid ratio in cardiovascular disease and other chronic diseases. *Exp Biol Med*. 2008;233(6):674–688.
49. Ailhaud G, Guesnet P. Fatty acid composition of fats is an early determinant of childhood obesity: a short review and an opinion. *Obes Rev*. 2004;5(1):21–26.
50. Simopoulos AP. Essential fatty acids in health and chronic disease. Am J Clin Nutr 1999;70(3 Suppl):560S–569S.
51. Kang JX, Liu A. The role of the tissue omega-6/omega-3 fatty acid ratio in regulating tumor angiogenesis. *Cancer Metastasis Rev*. 2013;32(1–2):201–210.
52. Dennis EA, Norris PC. Eicosanoid storm in infection and inflammation. *Nat Rev Immunol*. 2015;15(8):511–523.
53. Abedi E, Sahari MA. Long-chain polyunsaturated fatty acid sources and evaluation of their nutritional and functional properties. *Food Sci Nutr*. 2014;2(5):443–463.
54. Harnack K, Andersen G, Somoza V. Quantitation of alpha-linolenic acid elongation to eicosapentaenoic and docosahexaenoic acid as affected by the ratio of n6/n3 fatty acids. *Nutrition & Metabolism*. 2009;6:8.
55. Mathias RA, Pani V, Chilton FH. Genetic variants in the FADS gene: implications for dietary recommendations for fatty acid intake. *Curr Nutr Rep*. 2014;3(2):139–148.
56. Reynolds LM, Howard TD, Ruczinski I, et al. Tissue-specific impact of FADS cluster variants on FADS1 and FADS2 gene expression. *PloS One*. 2018;13(3):e0194610.
57. Chilton FH, Dutta R, Reynolds LM, Sergeant S, Mathias RA, Seeds MC. Precision nutrition and Omega-3 polyunsaturated fatty acids: a case for personalized supplementation approaches for the prevention and Management of Human Diseases. Nutrients Vol 92017.

58. Roman GTGE, Nguyen TT, Chakr VG, et al. *Multilocus genetic score from the fatty acid desaturase (FADS) gene cluster and eating behavior in children born after intrauterine growth restriction*. submitted, 2018.
59. Lattka E, Illig T, Koletzko B, Heinrich J. Genetic variants of the FADS1 FADS2 gene cluster as related to essential fatty acid metabolism. *Curr Opin Lipidol*. 2010;21(1):64–69.
60. Rzehak P, Heinrich J, Klopp N, et al. Evidence for an association between genetic variants of the fatty acid desaturase 1 fatty acid desaturase 2 (FADS1 FADS2) gene cluster and the fatty acid composition of erythrocyte membranes. *Br J Nutr*. 2009;101(1):20–26.
61. Malerba G, Schaeffer L, Xumerle L, et al. SNPs of the FADS gene cluster are associated with polyunsaturated fatty acids in a cohort of patients with cardiovascular disease. *Lipids*. 2008;43(4):289–299.
62. Serhan CN, Chiang N, Van Dyke TE. Resolving inflammation: dual anti-inflammatory and pro-resolution lipid mediators. *Nat Rev Immunol*. 2008;8(5):349–361.
63. Buczynski MW, Dumlao DS, Dennis EA. Thematic review series: Proteomics. An integrated omics analysis of eicosanoid biology. *J Lipid Res*. 2009;50(6):1015–1038.
64. Smith WL. The eicosanoids and their biochemical mechanisms of action. *Biochem J*. 1989;259(2):315–324.
65. Haeggstrom JZ, Funk CD. Lipoxygenase and leukotriene pathways: biochemistry, biology, and roles in disease. *Chem Rev*. 2011;111(10):5866–5898.
66. Smith WL, DeWitt DL, Garavito RM. Cyclooxygenases: structural, cellular, and molecular biology. *Annu Rev Biochem*. 2000;69:145–182.
67. Wilson RI, Nicoll RA. Endocannabinoid signaling in the brain. *Science*. 2002;296(5568):678–682.
68. Mitchell RW, Hatch GM. Fatty acid transport into the brain: of fatty acid fables and lipid tails. *Prostaglandins Leukot Essent Fatty Acids*. 2011;85(5):293–302.
69. Rett BS, Whelan J. Increasing dietary linoleic acid does not increase tissue arachidonic acid content in adults consuming Western-type diets: a systematic review. *Nutrition & Metabolism*. 2011;8:36.
70. Carlson SE. Docosahexaenoic acid and arachidonic acid in infant development. Semin Neonatol*: SN*. 2001;6(5):437–449.
71. Levant B, Radel JD, Carlson SE. Decreased brain docosahexaenoic acid during development alters dopamine-related behaviors in adult rats that are differentially affected by dietary remediation. *Behav Brain Res*. 2004;152(1):49–57.
72. McNamara RK, Vannest JJ, Valentine CJ. Role of perinatal long-chain omega-3 fatty acids in cortical circuit maturation: Mechanisms and implications for psychopathology. *World J Psychiatry*. 2015;5(1):15–34.
73. Montgomery P, Spreckelsen TF, Burton A, Burton JR, Richardson AJ. Docosahexaenoic acid for reading, working memory and behavior in UK children aged 7-9: a randomized controlled trial for replication (the DOLAB II study). *PloS One*. 2018;13(2):e0192909.
74. Tassoni D, Kaur G, Weisinger RS, Sinclair AJ. The role of eicosanoids in the brain. *Asia Pac J Clin Nutr*. 2008;17(Suppl 1):220–228.
75. Scheen AJ, Seutin V, Van Gaal LF. Endocannabinoid system in the brain...and elsewhere. *Rev Med Liege*. 2008;63(5–6):364–371.
76. Naughton SS, Mathai ML, Hryciw DH, McAinch AJ. Fatty acid modulation of the endocannabinoid system and the effect on food intake and metabolism. *Int J Endocrinol*. 2013;2013:361895.
77. Hallahan B, Garland MR. Essential fatty acids and mental health. Br J Psychiatry Vol 186. England2005:275–277.
78. Crawford MA. The role of dietary fatty acids in biology: their place in the evolution of the human brain. Nutr Rev 1992;50(4 (Pt 2)):3–11.
79. Innis SM. Dietary (n-3) fatty acids and brain development. *J Nutr*. 2007;137(4):855–859.
80. Carver JD, Benford VJ, Han B, Cantor AB. The relationship between age and the fatty acid composition of cerebral cortex and erythrocytes in human subjects. *Brain Res Bull*. 2001;56(2):79–85.
81. Herrera E. Implications of dietary fatty acids during pregnancy on placental, fetal and postnatal development--a review. *Placenta*. 2002;23(Suppl A):S9–19.
82. Jorgensen MH, Nielsen PK, Michaelsen KF, Lund P, Lauritzen L. The composition of polyunsaturated fatty acids in erythrocytes of lactating mothers and their infants. *Matern Child Nutr*. 2006;2(1):29–39.
83. Clandinin MT, Chappell JE, van Aerde JE. Requirements of newborn infants for long chain polyunsaturated fatty acids. *Acta Paediatr Scand Suppl*. 1989;351:63–71.
84. Gow RV, Hibbeln JR. Omega-3 fatty acid and nutrient deficits in adverse neurodevelopment and childhood behaviors. *Child Adolesc Psychiatr Clin N Am*. 2014;23(3):555–590.

85. Kuipers RS, Luxwolda MF, Offringa PJ, Boersma ER, Dijck-Brouwer DA, Muskiet FA. Fetal intrauterine whole body linoleic, arachidonic and docosahexaenoic acid contents and accretion rates. *Prostaglandins Leukot Essent Fatty Acids*. 2012;86(1–2):13–20.
86. Miller LR, Jorgensen MJ, Kaplan JR, et al. Alterations in levels and ratios of n-3 and n-6 polyunsaturated fatty acids in the temporal cortex and liver of vervet monkeys from birth to early adulthood. *Physiol Behav*. 2016;156:71–78.
87. Ravelli AC, van Der Meulen JH, Osmond C, Barker DJ, Bleker OP. Obesity at the age of 50 y in men and women exposed to famine prenatally. *Am J Clin Nutr*. 1999;70(5):811–816.
88. Barker DJ. The developmental origins of chronic adult disease. *Acta Paediatr Suppl*. 2004;93(446):26–33.
89. Widen EM, Kahn LG, Cirillo P, Cohn B, Kezios KL, Factor-Litvak P. Prepregnancy overweight and obesity are associated with impaired child neurodevelopment. *Matern Child Nutr*. 2018;14(1).
90. Murphy T, Dias GP, Thuret S. Effects of diet on brain plasticity in animal and human studies: mind the gap. *Neural Plast*. 2014;2014:563160.
91. Rong H, Xi Y, An Y, et al. The correlation between early stages of life exposed to Chinese famine and cognitive decline in adulthood: nutrition of adulthood plays an important role in the link? *Front Aging Neurosci*. 2017;9:444.
92. Kitajka K, Puskas LG, Zvara A, et al. The role of n-3 polyunsaturated fatty acids in brain: modulation of rat brain gene expression by dietary n-3 fatty acids. *Proc Natl Acad Sci U S A*. 2002;99(5):2619–2624.
93. Fontani G, Corradeschi F, Felici A, Alfatti F, Migliorini S, Lodi L. Cognitive and physiological effects of Omega-3 polyunsaturated fatty acid supplementation in healthy subjects. *Eur J Clin Invest*. 2005;35(11):691–699.
94. Kidd PM. Omega-3 DHA and EPA for cognition, behavior, and mood: clinical findings and structural-functional synergies with cell membrane phospholipids. *Altern Med Rev*. 2007;12(3):207–227.
95. Zhu W, Chi N, Zou P, Chen H, Tang G, Zhao W. Effect of docosahexaenoic acid on traumatic brain injury in rats. *Exp Ther Med*. 2017;14(5):4411–4416.
96. Schober ME, Requena DF, Abdullah OM, et al. Dietary Docosahexaenoic acid improves cognitive function, tissue sparing, and magnetic resonance imaging indices of edema and white matter injury in the immature rat after traumatic brain injury. *J Neurotrauma*. 2016;33(4):390–402.
97. Janssen CI, Zerbi V, Mutsaers MP, et al. Impact of dietary n-3 polyunsaturated fatty acids on cognition, motor skills and hippocampal neurogenesis in developing C57BL/6J mice. *J Nutr Biochem*. 2015;26(1):24–35.
98. Chung WL, Chen JJ, Su HM. Fish oil supplementation of control and (n-3) fatty acid-deficient male rats enhances reference and working memory performance and increases brain regional docosahexaenoic acid levels. *J Nutr*. 2008;138(6):1165–1171.
99. Vinot N, Jouin M, Lhomme-Duchadeuil A, et al. Omega-3 fatty acids from fish oil lower anxiety, improve cognitive functions and reduce spontaneous locomotor activity in a non-human primate. *PloS One*. 2011;6(6):e20491.
100. Fang PC, Kuo HK, Huang CB, Ko TY, Chen CC, Chung MY. The effect of supplementation of docosahexaenoic acid and arachidonic acid on visual acuity and neurodevelopment in larger preterm infants. *Chang Gung Med J*. 2005;28(10):708–715.
101. Smith SL, Rouse CA. Docosahexaenoic acid and the preterm infant. Matern Health Neonatol Perinatol Vol 32017.
102. Song Y, Liu Y, Pan Y, et al. The effect of long chain polyunsaturated fatty acid supplementation on intelligence in low birth weight infant during lactation: a meta-analysis. *PloS One*. 2018;13(4):e0195662.
103. Jasani B, Simmer K, Patole SK, Rao SC. Long chain polyunsaturated fatty acid supplementation in infants born at term. *Cochrane Datab Syst Rev*. 2017;3:CD000376.
104. Reis RS, Dalle Molle R, Machado TD, et al. Impulsivity-based thrifty eating phenotype and the protective role of n-3 PUFAs intake in adolescents. *Transl Psychiatry*. 2016;6:e755.
105. DHA and ARA and visual development. Scientific substantiation of a health claim related to docosahexaenoic acid (DHA) and arachidonic acid (ARA) and visual development pursuant to Article14 of Regulation (EC) No 1924/2006. *EFSA J*. 2009. Wiley Online Library.
106. Bloch MH, Qawasmi A. Omega-3 fatty acid supplementation for the treatment of children with attention-deficit/hyperactivity disorder symptomatology: Systematic review and meta-analysis. *J Am Acad Child Adolesc Psychiatry*. 2011;50(10):991–1000.
107. Chang JP, Su KP, Mondelli V, Pariante CM. Omega-3 polyunsaturated fatty acids in youths with attention deficit hyperactivity disorder: a systematic review and meta-analysis of clinical trials and biological studies. *Neuropsychopharmacology*. 2018;43(3):534–545.

108. Richardson AJ, Burton JR, Sewell RP, Spreckelsen TF, Montgomery P. Docosahexaenoic acid for reading, cognition and behavior in children aged 7-9 years: a randomized, controlled trial (the DOLAB study). *PloS one*. 2012;7(9):e43909.
109. Montgomery P, Spreckelsen TF, Burton A, Burton JR, Richardson AJ. Docosahexaenoic acid for reading, working memory and behavior in UK children aged 7–9: a randomized controlled trial for replication (the DOLAB II study). PLoS One Vol 132018.
110. Walton E, Pingault JB, Cecil CA, et al. Epigenetic profiling of ADHD symptoms trajectories: a prospective, methylome-wide study. *Mol Psychiatry*. 2017;22(2):250–256.
111. Richardson K, Louie-Gao Q, Arnett DK, et al. The PLIN4 variant rs8887 modulates obesity related phenotypes in humans through creation of a novel miR-522 seed site. *PloS one*. 2011;6(4):e17944.
112. Rodrigues DM, Manfro GG, Levitan RD, Steiner M, Meaney MJ, Silveira PP. Moderating effect of PLIN4 genetic variant on impulsivity traits in 5-year-old-children born small for gestational age. *Prostaglandins Leukot Essent Fatty Acids*. 2018;137:19–25.
113. Qi L, Corella D, Sorli JV, et al. Genetic variation at the perilipin (PLIN) locus is associated with obesity-related phenotypes in white women. *Clin Genet*. 2004;66(4):299–310.
114. Meirhaeghe A, Thomas S, Ancot F, et al. Study of the impact of perilipin polymorphisms in a French population. *J Negat Results Biomed*. 2006;5:10.
115. Qi L, Shen H, Larson I, et al. Gender-specific association of a perilipin gene haplotype with obesity risk in a white population. *Obes Res*. 2004;12(11):1758–1765.
116. Brasaemle DL. Thematic review series: Adipocyte biology. The perilipin family of structural lipid droplet proteins: stabilization of lipid droplets and control of lipolysis. *J Lipid Res*. 2007;48(12):2547–2559.
117. Cattaneo A, Macchi F, Plazzotta G, et al. Inflammation and neuronal plasticity: a link between childhood trauma and depression pathogenesis. *Front Cell Neurosci*. 2015;9:40.
118. Antypa N, Van der Does AJ, Smelt AH, Rogers RD. Omega-3 fatty acids (fish-oil) and depression-related cognition in healthy volunteers. *J Psychopharmacol*. 2009;23(7):831–840.
119. Jackson PA, Deary ME, Reay JL, Scholey AB, Kennedy DO. No effect of 12 weeks' supplementation with 1 g DHA-rich or EPA-rich fish oil on cognitive function or mood in healthy young adults aged 18–35 years. *Br J Nutr*. 2012;107(8):1232–1243.
120. Ginty AT, Muldoon MF, Kuan DCH, et al. Omega-3 supplementation and the neural correlates of negative affect and impulsivity: a double-blind, randomized, placebo-controlled trial in midlife adults. *Psychosom Med*. 2017;79(5):549–556.
121. Samieri C, Morris MC, Bennett DA, et al. Fish intake, genetic predisposition to Alzheimer disease, and decline in global cognition and memory in 5 cohorts of older persons. *Am J Epidemiol*. 2018;187(5):933–940.
122. Burckhardt M, Herke M, Wustmann T, Watzke S, Langer G, Fink A. Omega-3 fatty acids for the treatment of dementia. The Cochrane Database of Systematic Reviews. 2016;Vol. 4:CD009002.
123. Baleztena J, Ruiz-Canela M, Sayon-Orea C, et al. Association between cognitive function and supplementation with omega-3 PUFAs and other nutrients in ≥ 75 years old patients: A randomized multicenter study. *PloS One*. 2018;13(3):e0193568.
124. Larrieu T, Hilal ML, De Smedt-Peyrusse V, Sans N, Laye S. Nutritional Omega-3 deficiency alters glucocorticoid receptor-signaling pathway and neuronal morphology in regionally distinct brain structures associated with emotional deficits. *Neural Plast*. 2016;2016:8574830.
125. Morgese MG, Trabace L. Maternal malnutrition in the Etiopathogenesis of psychiatric diseases: role of polyunsaturated fatty acids. *Brain Sci*. 2016;6(3).
126. Monk C, Georgieff MK, Osterholm EA. Research review: Maternal prenatal distress and poor nutrition-mutually influencing risk factors affecting infant neurocognitive development. *J Child Psychol Psychiatry*. 2013;54(2):115–130.
127. Roseboom TJ, Painter RC, van Abeelen AF, Veenendaal MV, de Rooij SR. Hungry in the womb: What are the consequences? Lessons from the Dutch famine. *Maturitas*. 2011;70(2):141–145.
128. Sullivan EL, Riper KM, Lockard R, Valleau JC. Maternal high-fat diet programming of the neuroendocrine system and behavior. *Horm Behav*. 2015;76:153–161.
129. Larrieu T, Hilal ML, Fourrier C, et al. Nutritional omega-3 modulates neuronal morphology in the prefrontal cortex along with depression-related behaviour through corticosterone secretion. *Transl Psychiatry*. 2014;4:e437.
130. Stetler C, Miller GE. Depression and hypothalamic-pituitary-adrenal activation: a quantitative summary of four decades of research. *Psychosom Med*. 2011;73(2):114–126.

131. Colaianna M, Schiavone S, Zotti M, et al. Neuroendocrine profile in a rat model of psychosocial stress: relation to oxidative stress. *Antioxid Redox Signal*. 2013;18(12):1385–1399.
132. Park Y, Kim M, Baek D, Kim SH. Erythrocyte n-3 polyunsaturated fatty acid and seafood intake decrease the risk of depression: case-control study in Korea. *Ann Nutr Metab*. 2012;61(1):25–31.
133. Evans JM. Longitudinal associations between fish consumption and depression in young adults. *Am J Epidemiol*. 2014;180:557.
134. Noaghiul S, Hibbeln JR. Cross-national comparisons of seafood consumption and rates of bipolar disorders. *Am J Psychiatry*. 2003;160(12):2222–2227.
135. Davis J, Maes M, Andreazza A, McGrath JJ, Tye SJ, Berk M. Towards a classification of biomarkers of neuropsychiatric disease: from encompass to compass. *Mol Psychiatry*. 2015;20(2):152–153.
136. Berger ME, Smesny S, Kim SW, et al. Omega-6 to omega-3 polyunsaturated fatty acid ratio and subsequent mood disorders in young people with at-risk mental states: a 7-year longitudinal study. *Transl Psychiatry*. 2017;7:e1220.
137. Fusar-Poli P, Berger G. Eicosapentaenoic acid interventions in schizophrenia: meta-analysis of randomized, placebo-controlled studies. *J Clin Psychopharmacol*. 2012;32(2):179–185.
138. Agostoni C, Nobile M, Ciappolino V, et al. The role of Omega-3 fatty acids in developmental psychopathology: a systematic review on early psychosis, autism, and ADHD. Int J Mol Sci Vol 182017.
139. Zárate R, El Jaber-Vazdekis N, Tejera N, Pérez JA, Rodríguez C. Significance of long chain polyunsaturated fatty acids in human health. Clin Transl Med Vol 62017.
140. Lupton JR, Blumberg JB, L'Abbe M, et al. Nutrient reference value: non-communicable disease endpoints—a conference report. *Eur J Nutr*. 2016;55(Suppl 1):S1–10.

Further Reading

141. Moreira DKT, Ract JNR, Ribeiro APB, Macedo GA. Production and characterization of structured lipids with antiobesity potential and as a source of essential fatty acids. *Food Res Int*. 2017;99(Pt 1):713–719.

PART D

NUTRIENTS AND THE BRAIN

CHAPTER

20

Omega-3 Fatty Acids and Cognitive Behavior

Grace E. Giles,†, Caroline R. Mahoney*,†, Robin B. Kanarek‡*

*Cognitive Science Team, US Army Natick Soldier Research, Development, and Engineering Center, Natick, MA, United States †Center for Applied Brain and Cognitive Sciences, Medford, MA, United States ‡Department of Psychology, Tufts University, Medford, MA, United States

INTRODUCTION

Omega-3 (*n*-3) and omega-6 (*n*-6) fatty acids are the main dietary components of the family of polyunsaturated fatty acids (PUFAs). The *n*-3 fatty-acids include alpha linolenic acid (ALA), eicosapentaenoic acid (EPA), and docohexaenoic acid (DHA), while *n*-6 PUFAs include linoleic acid (LA) and arachidonic acid (ARA). Since our bodies cannot synthesize these compounds, they must be obtained from our diet. The primary dietary sources of EPA and DHA are fatty fish such as salmon, herring, tuna, and halibut, while ALA comes mainly from plant sources such as canola oil, walnuts, and flax seed. The *n*-6 PUFAs are most commonly consumed as LA, which is found predominantly in plant oils (e.g., corn oil, sunflower oil, and soybean oil), as well as in products made from these oils.[1]

Both *n*-3 and *n*-6 PUFAs are necessary for cells to maintain normal structure, function, and signal transduction. However, there is growing evidence that the ratio of *n*-3 to *n*-6 is more important than their absolute levels for these cellular processes.[2] For example, the ratio of *n*-3 to *n*-6 PUFA influences inflammation. Intake of *n*-6 PUFA increases pro-inflammatory cytokine production, whereas *n*-3 PUFAs reduce *n*-6 PUFA activity and thus decrease pro-inflammatory cytokine activity.[3]

The importance of *n*-3 PUFAs in brain function have been extensively studied in relation to psychological and neurological disorders. For instance, research indicates a beneficial influence of *n*-3 PUFAs on depressive symptoms in individuals with major depressive disorder although the relationship is far from understood.[4–6] The *n*-3 PUFAs may also benefit individuals with bipolar disorder,[7] anxiety disorders,[8] and attention deficit hyperactivity disorder.[9]

Omega Fatty Acids in Brain and Neurological Health
https://doi.org/10.1016/B978-0-12-815238-6.00020-1

Despite the reported benefits of *n*-3 PUFAs when brain functions go awry, as in psychological and neurological disorders, less research has examined the influence of *n*-3 PUFAs on cognitive function in healthy individuals. The recent research has begun to fill this gap by examining how *n*-3 PUFAs influence cognitive development in infants and children, cognitive performance in young adults, and age-related cognitive impairments in older adults. Just as the brain changes throughout the lifespan, so too may the influence of nutrition on cognition.[10] Thus, this chapter will review recent research examining the cognitive effects of *n*-3 intake across the lifespan.

INFANTS AND CHILDREN

n-3 PUFAs are critical for neurocognitive development from conception through early childhood. DHA, in particular, accumulates in the brain during this time, and influences gene expression, monoamine neurotransmitter activity, and neuron growth. Higher maternal dietary intake of DHA during pregnancy and lactation results in higher DHA levels in infants.[11]

Initial epidemiological studies explored the relationship between maternal intake of fish and other foods containing *n*-3 PUFAs and cognitive behavior in their offspring (see Table 1). In the majority of studies, intakes of fish by the mother during pregnancy and lactation were determined using food frequency questionnaires (FFQ) while cognitive performance in the offspring was measured with standardized tests such as the Fagan Test of Infant Intelligence, which measures novelty preference and is thought to predict IQ later in life,[25] the MacArthur Communicative Development Battery, which assesses infant and toddler linguistic abilities,[26] the Peabody Picture Vocabulary Test, which measures listening comprehension and vocabulary,[27] the Bayley Scales of Infant Development, which evaluates cognitive and motor development,[28] and intelligence tests including the Kaufman Brief Intelligence Test[29] and Wechsler Intelligence Scales for Children.[30] Most studies found that greater maternal fish intake and blood PUFA levels were associated with enhanced cognitive performance in infants and children, ranging from 6 months to 13 years.[12–15,18,21,22] Similarly, more frequent fish intake during adolescence was associated with enhanced intelligence and academic performance,[16,17] particularly up to the recommended intake levels.[19] However, it should be noted that not all studies have observed a positive relationship between early exposure to *n*-3 PUFAs and later cognitive functions.[20,23,24]

While the results of these epidemiological studies suggest that dietary exposure to *n*-3 PUFAs can have positive relationships with cognitive behavior, their correlational design make it impossible to assign causality. Moreover, a number of factors must be considered as possible confounds in such studies. For example, mothers who consume more fish may have a generally healthier diet, a greater level of education, and/or be older (or younger) than mothers who consume less fish. Another concern is that to quantify PUFA intake, these studies generally rely on FFQ which are subject to the underreporting of energy and food, including fats and fatty acids.[31]

To begin to address causality, randomized, double-blind, placebo-controlled trials (RCTs) on the effects of *n*-3 PUFAs on behavior are required. As a result, the following sections will concentrate on RCTs investigating the effects of *n*-3 PUFAs on cognitive behavior. These trials

TABLE 1 Epidemiological Studies Assessing the Association between Maternal and Child n-3 PUFAs and Child Cognition

Authors	*n* Mothers	*n*-3 PUFA Measurement	*n* Children (Female)	Age at Testing	Childhood Cognitive Measures	Results
Daniels et al.[12]	7421	FFQ[a] at 32 weeks' gestation, Infant intake fish and breastfeeding at 6 and 12 months	7421 (48%)	15 and 18 months	15 months: MCDI[b] 18 months: DDST[c]	Higher maternal fish intake during pregnancy was associated with enhanced vocabulary comprehension and social activity at 15 months, and higher DDST total and language scores at 18 months (no associations with social scores) Higher child fish intake at 6 months associated with enhanced vocabulary comprehension and social activity at 15 months, and higher DDST total, language, and social scores at 18 months Higher child fish intake at 12 months associated with enhanced vocabulary comprehension and social activity at 15 months, and higher DDST total and social scores at 18 months (no associations with language scores)
Hibbeln et al.[13]	11,875	FFQ at 32 weeks' gestation	8801 (nr[d])	6, 18, 30, 42, and 81 months	6, 18, 30, and 42 months: ALSPAC early development test subscales[e] 7 years: SDQ[f] 8 years: WISC-IIIUK[g]	Higher fish intake associated with enhanced fine motor skills (18 and 42 months, not 6 or 30 months) Higher fish intake associated with enhanced social development (30 and 42 months, not 6 or 18 months) Higher fish intake associated with greater communication (6 and 18 months) No associations with gross motor skills Greater fish intake associated with greater prosocial behavior No associations with hyperactivity, emotional symptoms, conduct problems, peer problems, or total difficulties Higher fish intake associated with higher verbal IQ No association performance or full-scale IQ

(*Continued*)

TABLE 1 Epidemiological Studies Assessing the Association between Maternal and Child n-3 PUFAs and Child Cognition—cont'd

Authors	*n* Mothers	*n*-3 PUFA Measurement	*n* Children (Female)	Age at Testing	Childhood Cognitive Measures	Results
Oken, Radesky, et al.[14]	341	FFQ and erythrocyte PUFA content in 2nd trimester	341 (51%)	3 years	PPVT[h] WRAVMA[i]	Maternal fish intake of >2 servings/week associated with higher WRAVMA drawing and total scores Higher maternal DHA[j] + EPA[k] associated with higher PPVT and WRAVMA total scores No associations maternal erythrocyte content and child cognition No interaction fish intake and breastfeeding
Oken, Osterdal, et al.[15]	101,042	FFQ at 25 weeks' gestation	25,446	6 and 18 months	Motor, social/cognitive, and total development	Higher maternal fish intake associated with higher motor, social/cognitive, and total development scores at 6 and 18 months for the highest quintile of fish intake
Åberg et al.[16]	n/a	Fish consumption at 15 years	18,158 (8898)	18 years	Intelligence tests	More frequent fish intake associated with enhanced combined intelligence, verbal, and visuospatial skills
Kim et al.[17]	n./a	Fish consumption at 15 years	18,158 (8898)	16 years	School grades	More frequent fish intake associated with higher school grades
Boucher et al.[18]	154	Umbilical cord PUFA content and plasma PUFA content at cognitive testing	154 (58.4%)	10–13 years	CRT[l] WISC-IV[m] CVLT[n] EEG[o]	Higher cord DHA associated with enhanced Digit span forward subtest and CVLT recognition memory Higher cord DHA associated with shorter FN400 latency and larger LPC amplitude Higher child plasma DHA associated with more negative FN400 amplitude No association cord or child plasma DHA and CRT performance
de Groot et al.[19]	n/a	FFQ at cognitive testing	700 (394)	12–18 years	Amsterdam Vocabulary Test Dutch version of the Youth Self-Report Academic performance	Quadratic relationship between fish intake and vocabulary and academic performance Higher DHA + EPA associated with enhanced vocabulary and academic performance (up to recommended amount) No association fish intake and attention problems
Keim et al.[20]	358	Breast milk PUFA content at 4 months postpartum	319	12 months	Mullen Scales of Early Learning	Breast milk DHA and AA content not associated with cognitive development

Steer et al.[21]	6249	Erythrocyte PUFA content during pregnancy	6249 (3133)	8 years	IQ[p]	Lowest quartile osbond acid, DHA, AA[q] content associated with lower IQ than higher quartiles Highest quartile adrenic, vaccenic, and osbond acid associated with lower IQ than lower quartiles No associations when FAs modeled as linear covariate
Rees et al.[22]	125	FFQ during 2nd and 3rd trimester	107 (55)	2 2 months	Object search task	Object search performance better with high than low third trimester DHA intake No differences in cognition with high vs low second trimester DHA intake, or second or third trimester EPA or AA intake
Strain et al.[23]	1265	Serum PUFAs at 28 weeks gestation	1265 (nr)	20 months	BSID-II MDI[r] BSID-II PDI[s] CDI[t] IBQ-R[u]	Higher serum DHA associated with lower MDI score Higher serum DHA associated with higher total gestures Higher n-6/n-3 ratio associated with higher MDI score and lower CDI scores (vocabulary produced, vocabulary understood, total gestures) No associations PUFAs and PDI or IBQ-R scores
Oken et al.[24]	1068	FFQ and erythrocyte PUFA content in 2nd trimester	1068 (534)	6–10 years	KBIT[v] WRAVMA[w] WRAML[x]	No associations with maternal fish intake or erythrocyte PUFA content and childhood cognition

[a] *Food Frequency Questionnaire (FFQ).*
[b] *Avon Longitudinal Study of Parents and Children (ALSPAC) adaption of the MacArthur Communicative Development Inventory (MCDI).*
[c] *ALSPAC adaptation of the Denver Developmental Screening Test (DDST).*
[d] *Not reported (nr).*
[e] *Gross motor, fine motor, communication, and social skills.*
[f] *Strengths and Difficulties Questionnaire (SDQ).*
[g] *Wechsler intelligence scale for children (WISC-IIIUK).*
[h] *Peabody Picture Vocabulary Test (PPVT).*
[i] *Wide Range Assessment of Visual Motor Abilities (WRAVMA).*
[j] *Docosahexaenoic acid (DHA).*
[k] *Eicosapentaenoic acid (EPA).*
[l] *Continuous recognition task (CRT).*
[m] *Wechsler Intelligence Scales for Children, 4th edition (WISC-IV).*
[n] *California Verbal Learning Test– Children's Version (CVLT).*
[o] *Electroencephalogram (EEG).*
[p] *Intelligence Quotient (IQ).*
[q] *Arachidonic acid (AA).*
[r] *Bayley Scales of Infant Development II (BSID-II) Mental Developmental Index (MDI).*
[s] *Psychomotor Developmental Index (PDI).*
[t] *MacArthur Bates Communicative Development Inventories (CDI): vocabulary produced, vocabulary understood, total gestures.*
[u] *Infant Behavior Questionnaire–Revised (IBQ-R).*
[v] *Kauffman Brief Intelligence Test (KBIT).*
[w] *Wide Range Assessment of Visual Motor Abilities (WRAVMA).*
[x] *Wide Range Assessment of Memory and Learning (WRAML).*

can generally be divided into three categories depending on the age at supplementation: (1) prenatally or infancy, via maternal supplementation, (2) infancy via formula supplementation, and (3) childhood.

MATERNAL SUPPLEMENTATION

As shown in Table 2, RCTs evaluating the influence of maternal *n*-3 PUFA supplementation on infant and child cognition have produced mixed results. For instance, cod liver oil supplementation during pregnancy through 3 months postpartum did not influence intelligence during the first year.[32] However, in the same sample, children whose mothers ingested cod liver oil during pregnancy and postpartum exhibited higher intelligence at age 4, but not at age 7.[33,48] In another cohort, pregnant women ingested DHA-rich fish oil from their second trimester until delivery, and children were followed through 7 years of age. At 18 months, maternal fish oil supplementation during pregnancy did not influence cognitive or language performance.[37] At just over 2 years, maternal fish oil supplementation during pregnancy did not influence attention, working memory, or inhibitory control.[39] At 4 years, maternal fish oil supplementation during pregnancy did not influence conceptual ability.[40] And most recently, at 7 years, maternal fish oil supplementation during pregnancy did not influence intelligence, language, academic performance, or executive function.[47] However, parent-rated behavior and executive function were poorer in the 4- and 7-year-olds whose mothers had consumed fish oil than the vegetable oil control. Similarly, DHA supplementation during pregnancy did not influence motor, cognitive, or behavioral development at 18 months,[43] or in the same cohort, global cognitive functioning at 5 years.[46] However, at both the time periods, DHA mitigated the association between home environment and cognitive development. DHA supplementation during pregnancy did not influence habituation during infancy, but attenuated a decline in sustained attention from 4 to 9 months, and reduced attrition due to crying at 6 and 9 months.[45] Other research has focused on preterm infants, that is, infants born before 37 weeks gestation. Breastfeeding mothers consumed DHA-rich fish oil or soy oil control, and formula-fed infants consumed DHA-rich or standard-DHA formula from delivery to their term date. The mental and psychomotor development did not differ at 18 months between the groups as a whole, but girls allocated to the high DHA diet scored higher on mental development than those to the low DHA diet. Further, fewer infants in the higher than the standard DHA group experienced delayed mental development.[36] However, no differences in intelligence were found in the same children at 7 years.[41]

The aforementioned work chose vegetable, corn, or soy oils as control oils. These oils, which are rich in *n*-6 PUFAs, may oppose the effects of *n*-3 PUFAs, and thus not act as true controls.[49] More recent work has chosen olive oil, which is composed of monounsaturated fatty acids. Relative to olive oil intake, fish oil intake in the postpartum period in primarily breastfed infants did not influence problem solving in the sample as a whole, but improved problem solving in girls. Further fish oil exerted few effects on language development, but impaired vocabulary comprehension, and sentence complexity, particularly in boys.[34] In the same sample, fish oil did not influence speed of processing or inhibitory control at 7 years. However, higher maternal *n*-3 PUFA intake during the initial postpartum period was associated with poorer speed of processing at age 7, and higher erythrocyte DHA levels at the age of 4 months

TABLE 2 Randomized Controlled Trials Assessing the Influence of Maternal n-3 PUFA Intake on Offspring Cognitive Development

Authors	*n* Mothers	Maternal age (years)	*n*-3 PUFA Manipulation (per day)	*n* Infants (Female)	Age at Testing	Childhood Cognitive Measures	Results
Helland et al.[32]	341	19–35	Cod liver oil (2632 mg/10 mL *n*-3, 235 mg/10 mL *n*-6) vs corn oil (350 mg/10 mL *n*-3, 4747 mg/10 mL *n*-6) 17–19 weeks gestation until three months postpartum	288 (175)	6, 9 months	FTII[a]	No differences
Helland et al.[33,b]	341	Cod liver oil: 28.5 ± 3.3, Corn oil: 28.0 ± 2.4	Cod liver oil (2494 mg/10 mL *n*-3) vs corn oil (4747 mg/10 mL *n*-6) 18 weeks gestation until 3 months postpartum	84 (43)	4 years	K-ABC[c]	Cod liver > Corn oil Mental Processing Composite Positive association with plasma [DHA] and K-ABC subscales Positive association with maternal DHA, EPA intake during pregnancy and K-ABC subscales Maternal DHA intake predicted Mental Processing Composite
Lauritzen et al.[34]	150	nr[d]	Fish oil (FO)[e] (1500 mg *n*-3 PUFA: 900 mg DHA[f]) vs OO[g] 9 ± 3 days postdelivery until 4 months postpartum vs high habitual fish intake	148 (65)	9 months 1, 2 years	Motor function (9 months) Means-end problem solving (9 months) MCDI[h] (1, 2 years)	No differences in motor function Higher problem-solving scores in FO than OO in girls only Lower vocabulary comprehension in FO than OO at 1 year (no differences at 2 years) and sentence complexity in boys

(*Continued*)

TABLE 2 Randomized Controlled Trials Assessing the Influence of Maternal n-3 PUFA Intake on Offspring Cognitive Development—cont'd

Authors	*n* Mothers	Maternal age (years)	*n*-3 PUFA Manipulation (per day)	*n* Infants (Female)	Age at Testing	Childhood Cognitive Measures	Results
Dunstan et al.[35]	98	FO: 30.9 ± 3.7, Control: 32.6 ± 3.6	FO (2200 mg DHA, 1100 mg EPA) vs OO 20 weeks gestation until delivery	72 (39)	2.5 years	GMDS[i] PPVT[j] CBCL[k]	Higher hand eye coordination score on GMDS in FO than OO Hand eye coordination positively associated with erythrocyte [EPA] and [DHA] and negatively correlated with [ARA] No differences in PPVT or CBCL
Makrides et al.[36]	657	High DHA Diet: 29.9 ± 5.8 Standard DHA Diet: 30.2 ± 5.4	3000 mg DHA-rich FO + (if necessary) high-DHA formula to infant vs 3000 mg soy oil capsules + (if necessary) standard preterm infant formula Preterm birth to term due date	657	18 months	MDI[l] of BSID-II PDI[m] of BSID-II	High > Standard DHA diet MDI scores (girls only) No differences in PDI scores High < Standard DHA diet number of infants with delayed mental development
Makrides et al.[37]	2320	FO: 28.9 ± 5.7, Control: 28.9 ± 5.6	FO (800 mg DHA, 100 mg EPA[n]) vs vegetable oil 19–20 weeks gestation until delivery	1196 (600)	18 months	BSID-III[o]	Mean scores did not differ by maternal treatment group FO < Control group for language score, language development, adaptive behavior (girls only)
Cheatham et al.[38]	107	nr	FO (1500 mg/day *n*-3 PUFA: 790 mg DHA, 620 mg EPA) vs OO vs high habitual fish intake	98 (44)	7 years	Woodcock Johnson Tests of Cognitive Abilities III Day/Night Stroop SDQ[p]	No differences in speed of processing, Stroop Score, SDQ Lower prosocial functioning in FO than OO (boys only) Total maternal *n*-3 PUFA intake negatively associated with speed of processing scores

Gould et al.[39,q]	158	DHA + EPA: 29.7 ± 5.3 Control: 29.4 ± 5.0	800 mg DHA + 100 mg EPA vs vegetable oil 18–21 weeks gestation until delivery	158 (82)	27 ± 2 months	Attention WMIC[r]	No differences in primary attention, working memory, or inhibitory control outcomes
Makrides et al.[40,ρ]	2320	FO: 28.9 ± 5.7, Control: 28.9 ± 5.6	FO (800 mg DHA) vs vegetable oil 19–20 weeks gestation until delivery	646 (nr)	4 years	GCA of DAS-II[s]	No differences in GCA Scores DHA < Control parent-rated executive function and behavior
Collins et al.[41,t]	657	High DHA Diet: 29.9 ± 5.8 Standard DHA Diet: 30.2 ± 5.4	3000 mg DHA-rich FO + (if necessary) high-DHA formula to infant vs 3000 mg soy oil capsules + (if necessary) standard preterm infant formula Preterm birth to term due date	604 (280)	7 years	Full Scale IQ and Vocabulary, Similarities, Block Design and Matrix Reasoning subtest of WASI[u]	No differences in IQ
Meldrum et al.[42]	98	FO: 30.9 ± 3.7, Control: 32.6 ± 3.6	FO (2200 mg DHA, 1100 mg EPA) vs OO 20 weeks gestation until gestation	50 (28)	12 years	WISC-IV[v] TVMI[w] CBCL[x] CCC-2[y]	No differences in cognition between children whose mothers had been supplemented with FO vs OO during pregnancy Erythrocyte DHA at 12 years associated with higher WISC-IV perceptual reasoning, working memory, processing speed, and full IQ (not any other cognitive measure) No association with cord blood PUFA content and cognition at 12 years
Ramakrishnan et al.[43]	730	DHA: 26.3 ± 4.6, Control: 26.5 ± 49	400 mg DHA vs soy oil from 18 to 22 weeks gestation until delivery	730 (336)	18 months	MDI of BSID-II PDI of BSID-II BRS[z] of BSID-II	No differences in cognition DHA < Control association between home environment at 12 months and psychomotor development at 18 months

(Continued)

TABLE 2 Randomized Controlled Trials Assessing the Influence of Maternal n-3 PUFA Intake on Offspring Cognitive Development—cont'd

Authors	*n* Mothers	Maternal age (years)	*n*-3 PUFA Manipulation (per day)	*n* Infants (Female)	Age at Testing	Childhood Cognitive Measures	Results
Hurtado et al.[44]	110	FO: 29.9 ± 4.7 Control: 30.5 ± 4.8	FO-enriched dairy drink (320 mg DHA, 54 mg EPA) vs control dairy drinks from 28 weeks gestation until 4 months lactation	90 (45)	12 months	MDI of BSID-II PDI of BSID-II	No differences in cognition
Colombo et al.[45]	350	DHA: 25.3 ± 4.9 Control: 24.8 ± 4.7	600 mg/day DHA vs soybean + corn oil from 8 to 20 weeks gestation until delivery	230	4, 6, and 9 months	Visual habituation	No effect DHA on look duration or habituation DHA < Control decrease in sustained attention from 4 to 9 months DHA < Control attrition due to fussiness on habituation task at 6 and 9 months
Ramakrishnan et al.[46,aa]	797	DHA: 26.3 ± 4.9, Control: 26.3 ± 4.7	400 mg DHA vs soy oil from 18 to 22 weeks gestation until delivery	797 (367)	5 years	MSCA[bb] BASC-2[cc] K-CPT[dd]	No effect in DHA on global cognitive functioning (MSCA) or behavior (BASC-2) Effect of home environment on MSCA scores lower in DHA than control group DHA < Control K-CPT omissions
Gould et al.[47,o]	2320	FO: 28.9 ± 5.7, Control: 28.9 ± 5.6	FO (800 mg DHA) vs vegetable oil 19–20 weeks gestation until delivery	543	7 years	BSID-III, WISC-II RCF,[ee] F-Stroop,[ff] CELF-4,[gg] BRIEF,[hh] TEACh,[ii] RAVLT,[jj] SDQ, WRAT-4[kk]	No differences in IQ, language, academic performance, executive function performance DHA > Control parent-rated behavior problems and executive dysfunction

[a] *Fagan Test of Infant Intelligence (FTII).*
[b] *Same cohort as 32.*
[c] *Kaufman Assessment Battery for Children (K-ABC).*
[d] *Not reported (NR).*
[e] *Fish oil (FO).*
[f] *Docosahexaenoic acid (DHA).*
[g] *Olive oil (OO).*
[h] *MacArthur Communicative Development Inventory (MCDI).*
[i] *Griffiths Mental Development Scales (GMDS).*
[j] *Peabody Picture Vocabulary Test (PPVT) IIIA.*
[k] *Child Behavior Checklist (CBCL).*
[l] *Mental Development Index (MDI) of the Bayley Scales of Infant Development, Second Edition (BSID-II).*
[m] *Psychomotor Development Index (PDI).*
[n] *Eicosapentaenoic acid (EPA).*
[o] *Bayley Scales of Infant and Toddler Development Third Edition (BSID-III).*
[p] *Strengths and Difficulties Questionnaire (SDQ).*
[q] *Same cohort as 35.*
[r] *Working memory and inhibitory control (WMIC).*
[s] *General Conceptual Ability (GCA) or the DAS, second edition (DAS II).*
[t] *Same cohort as 42.*
[u] *Wechsler Abbreviated Scale of Intelligence (WASI).*
[v] *Wechsler Intelligence Scale for Children-IV (WISC-IV).*
[w] *Beery-Buktenica Developmental Test of Visual-Motor Integration (TVMI).*
[x] *Child Behaviour Checklist, both parent and child reports.*
[y] *Children's Communication Checklist (CCC-2).*
[z] *Behavior Rating Scale (BRS).*
[aa] *Same cohort as 39.*
[bb] *Spanish language version of the McCarthy Scales of Children's Abilities (MSCA).*
[cc] *Behavioral Assessment System for Children, Second Edition (BASC-2).*
[dd] *Conners' Kiddie Continuous Performance Test (K-CPT).*
[ee] *Rey Complex Figure (RCF).*
[ff] *Fruit Stroop test (F-Stroop).*
[gg] *Number Repetition Subtest of the CELF, 4th Edition (CELF-4).*
[hh] *Behaviour Rating Inventory of Executive Function (BRIEF).*
[ii] *Everyday Attention for Children (TEACh).*
[jj] *Rey Auditory Verbal Learning Test (RAVLT).*
[kk] *Wide Range Achievement Test, Fourth edition (WRAT-4).*

were associated with poorer inhibitory control at age 7.[38] Fish oil intake during pregnancy did not influence vocabulary or behavioral problems when children turned 2.5 years old,[35] or intelligence, fine motor skills, or behavioral problems when the same sample of children turned 12 years old.[42] Finally, a fish oil-enriched dairy drink consumed by mothers from their second trimester pregnancy through fourth month lactation did not influence motor, cognitive, or behavioral development at 12 months.[44]

Thus, trials evaluating the influence of maternal supplementation with *n*-3 PUFA on their children's cognition vary widely in the type, dose, and duration of *n*-3 PUFA supplementation, control conditions, cognitive domains assessed, and age of assessment. The trials found largely null results, though several methodological limitations may account for limited differences, including small sample size compounded with high attrition rates, as well as a lack of clear reporting of randomization and intention-to-treat analysis.[50]

SUPPLEMENTATION DURING INFANCY

Rates of breast feeding have increased in recent years in the United States. However, approximately half of infants still receive formula by the age of 6 months.[51] Thus, research has evaluated the influence of *n*-3 PUFA supplementation in formula on infant and child cognitive development (Table 3). In full-term born infants, breastfed or randomized to DHA-, DHA and AA-, or non-supplemented formulas did not differ in mental or motor development at 12 or 14 months. Infants randomized to receive DHA-only-supplemented formula scored poorer on measures of vocabulary than those breastfed or fed the control formula.[52] The same children showed no cognitive differences at age two and half years.[55] Similarly, infants receiving DHA-, DHA and AA-, or non-supplemented formulas did not differ in mental or motor development at 18 months. However, analysis of the DHA-supplemented formulas together relative to the non-supplemented formula showed enhancement of mental development and language following DHA-relative to control.[57] Additional studies have found that fish oil-supplemented formula did not influence infants' mental or motor development, but enhanced language-associated measures, such as vocabulary expression and the use of gestures.[53,59] However, fish oil-supplemented formula and the use of PUFA-rich cooking oils and margarines for the first 5 years of life did not influence literacy or numeracy from 18 months to 8 years [61].

Other research has evaluated the influence of *n*-3 PUFA-supplemented formula on infants born preterm, that is, before 37 weeks of pregnancy was completed.[62] The DHA and AA-supplemented formula enhanced problem solving skills and recognition memory at 6 months.[56] Few differences were found between fish and fungal-oil supplemented formula and control formula on measures of mental and motor development, language, and intelligence, though fish oil enhanced motor development in low birth weight infants and vocabulary comprehension in non-Spanish speaking households.[54] Other research has specifically focused on infants born at very low birthweights and found that DHA and AA-supplemented formula enhanced sustained attention, but not the duration of focused attention or mental development at 20 months.[58] No differences in cognition were found in the same children at 8 years.[60]

TABLE 3 Randomized Controlled Trials Assessing the Influence of Infant n-3 PUFA Supplementation on Cognitive Development

Authors	Sample *n* (Female) Term	Gestational Age	*n*-3 PUFA Manipulation	Supplement Duration	Age at Testing (Months Unless Otherwise Noted)	Cognitive Measures	Results
Scott et al.[52]	274 (nr[a]) term	nr	0.2% DHA[b] formula vs 0.12% DHA + 0.43% AA[c] formula vs control formula vs BF[d]	14 months	12, 14	BSID[e] MCDI[f]	No differences in BSID DHA < BF vocabulary comprehension DHA < control vocabulary production
Auestad et al.[53]	404 (203) term	Egg-DTG[d]: 39.0 ± 1.3, FO[e]-Fungal: 39.3 ± 1.2, Control: 39.4 ± 1.2, Egg-DTG BF: 39.6 ± 1.3, Control BF: 39.2 ± 1.2	FO[g] + fungal oil formula (0.46 g/100 g AA, ≤0.04 g/100 g EPA,[h] 0.13 g/100 g DHA) vs Egg-DTG formula (0.45 g/100 g AA, 0.14 g/100 g DHA) vs control formula vs BF + Egg-DTG[i] vs BF control	12 months	1, 2, 4, 6, 9, 12, 14	BSID MCDI FTII[j] Infant Behavior Questionnaire	No differences in BSID, FTII FO-Fungal > Egg-DTG Formula vocabulary expression Egg-DTG Formula < Control Formula smiling and laughter (Infant Behavior Questionnaire)
O'Connor et al.[54]	470 (214) preterm	Egg-DTG: 29.7 ± 2.0, FO-Fungal: 29.8 ± 2.1, Control: 29.6 ± 1.9, Breastfed: 29.7 ± 2.1	FO + fungal oil formula (0.43 g/100 g AA, 0.08 g/100 g EPA, 0.27 g/100 g DHA) vs Egg-DTG formula (0.41 g/100 g AA, 0.24 g/100 g DHA) vs Control formula vs BF	12 months	2, 4, 6, 9, 14	BSID MCDI FTII	No differences in BSID, except FO-Fungal > Control motor score in infants ≤1250 g birth weight Egg-DTG > Fish-Fungal, Control groups novelty preference No differences in MCDI, except Egg-DTG, FO-Fungal < Control vocabulary comprehension after removing infants from Spanish-speaking families

(Continued)

TABLE 3 Randomized Controlled Trials Assessing the Influence of Infant n-3 PUFA Supplementation on Cognitive Development—cont'd

Authors	Sample *n* (Female) Term	Gestational Age	*n*-3 PUFA Manipulation	Supplement Duration	Age at Testing (Months Unless Otherwise Noted)	Cognitive Measures	Results
Auestad et al.[55,k]	157 (71) term	All: 39 ± 1	ARA + DHA (0.43 g/100 g ARA + 0.12 g/100 g DHA) vs DHA (0.23 g/100 g DHA) vs control formula vs BF	52 months	14, 39	SBIS[l] PPVT-R[m] BVM[n] MLU[o]	No differences
Henriksen et al.[56]	141 (64) preterm	*n*-3: 28.4 ± nr Control: 28.9 ± nr	48 mg/kg day DHA + 48 mg/kg day AA vs soy oil + medium-chain triglyceride oil	2.25 months (median)	6	Ages and Stages Questionnaire Recognition memory via ERP[p]	*n*-3 > Control problem solving *n*-3 > Control recognition memory
Drover et al.[57]	131 (52) term	All: 37–42	0.32% DHA (17 mg/100 kcal), 0.64% DHA (34 mg/100 kcal) or 0.96%DHA (54 mg/100 kcal) + 0.64% ARA (54 mg/100 kcal) vs control formula	12 months	18	BSID BRS[q]	No differences between groups analyzed separately Combined DHA groups > Control mental development, language Combined DHA groups > Controls BRS emotion regulation
Westerberg et al.[58]	92 (44) VLBW[r]	DHA + AA: 28.6 ± 2.9 Control: 28.7 ± 2.7	0.5 mL oil (32 mg DHA + 31 mg AA) or placebo per 100 mL milk	8 weeks	20	Free play sessions for Duration of Focused Attention and Summary Attention Rating BSID MDI[s]	*N*-3 > Control sustained attention No differences in duration of focused attention or MDI
Meldrum et al.[59]	420 (139) term	FO: 39.1 ± 1.1, Control: 29.4 ± 1.3	FO (230 mg DHA, 110 mg EPA) vs OO[t]	6 months	12, 18	BSID MCDI CBCL[u]	FO > Control MCDI later developing gestures, and total number gestures No differences in BSID FO > Control CBCL Anxious/depressed behaviors

Almaas et al.[60,v]	98 (49) VLBW	DHA + AA: 28.6 ± 2.9 Control: 28.6 ± 2.6	0.5 mL oil (32 mg DHA + 31 mg AA) or placebo per 100 mL milk	8 weeks	8 years	Full-Scale IQ, Verbal IQ, Performance IQ of WASI[w] Digit span forward Digit span backward CVLT-II[x] Grooved Pegboard test	No differences
Brew et al.[61]	239 (113) Term	All: 39.55 ± 1.30	500 mg FO (37% n-3 PUFA: 135 mg DHA, 32 mg EPA; 6% n-6 PUFA) + canola-based cooking oils/margarines (6% n-3 PUFA, 16% n-6 PUFA) vs 500 mg Sunola oil (0.3% n-3 PUFA, 6% n-6 PUFA)	5 years	18 months, 3, 5 and 8 years	NAPLAN[y]	No difference in total, reading or numeracy scores at any age Higher total plasma n-3 PUFA levels associated with higher NAPLAN scores at 8 years only

[a] *Not reported (nr).*
[b] *Docosahexaenoic acid (DHA).*
[c] *Arachidonic acid (AA).*
[d] *Breastfed (BF).*
[e] *Bayley Scales of Infant Development (BSID).*
[f] *MacArthur Communicative Development Inventories (MCDI).*
[g] *Fish oil (FO).*
[h] *Eicosapentaenoic acid (EPA).*
[i] *Egg-derived triglyceride (Egg-DTG).*
[j] *Fagan Test of Infant Intelligence (FTII).*
[k] *Same cohort as 136.*
[l] *Stanford-Binet Intelligence Scale (SBIS) Form L-M.*
[m] *Peabody Picture Vocabulary Test-Revised (PPVT-R).*
[n] *Beery Visual-Motor Index Test (BVM).*
[o] *Mean length of utterance (MLU).*
[p] *Event Related Potential (ERP).*
[q] *Behavior Rating Scale (BRS).*
[r] *Very low birthweight (VLBW).*
[s] *Mental Development Index (MDI).*
[t] *Olive oil (OO).*
[u] *Achenbach Child Behavior Checklist (CBCL).*
[v] *Same cohort as 61.*
[w] *Wechsler Abbreviated Scale of Intelligence (WASI).*
[x] *California Verbal Learning Test II (CVLT-II).*
[y] *National Assessment Program—Literacy and Numeracy (NAPLAN).*

In summary, while some studies have found beneficial influence of *n*-3 PUFAs on measures of language, attention, problem solving, and memory,[53,54,56–59] improvements do not appear to persist beyond infancy.[55,60] Like maternal *n*-3 PUFA supplementation, infant *n*-3 PUFA supplementation does not seem to influence global indices of cognitive development but may have more circumscribed effects on specific cognitive domains.

SUPPLEMENTATION DURING CHILDHOOD

The RCTs have also evaluated *n*-3 PUFA supplementation and cognition in 13-year-old children (Table 4). In children of age group 8–12 years, fish oil supplementation enhanced multiple measures of processing speed, visuomotor coordination, and executive function, but not attention, memory, language, or academic performance.[68] In 4- to 12-year-old children, DHA and DHA with ALA supplementation for 8 weeks to one year did not influence cognition across a range of cognitive domains.[63–65] The DHA supplementation enhanced reading ability in 6- to 10-year-old children below the 22nd centile in reading but not reading ability or working memory across a larger sample of children below the 33rd centile in reading.[66] Similarly, DHA and EPA supplementation did not influence reading or spelling in 3- to 13-year-old children, but enhanced cognitive ability measured using a nonverbal cognitive test.[67] Thus, the evidence is quite mixed as to whether *n*-3 PUFA supplementation benefits cognitive performance in school-age children.

YOUNG ADULTS

To date only a handful of studies have assessed the cognitive effects of *n*-3 PUFA supplementation in young adults and yielded inconsistent results (Table 5). Among the few, Fontani et al.[69] found speeded response time on tasks of response inhibition (Go/No-Go) and sustained attention (complex Go/No-Go), but not purer measures of response time (Simple and Choice Response Time). The electromyography (EMG) activation showed that reduced response time was not due to increased muscle contraction but instead due to enhanced central processing.[69] Other studies have found that fish oil improved verbal learning,[74] and that DHA supplementation improved episodic memory in women and working memory in men,[77] but found few other effects of *n*-3 PUFA supplementation across a range of tasks measuring inhibitory control, facial expression recognition, attention, and memory.[70,74,77] Similarly, 3 PUFA supplementation did not influence stress-induced changes to emotion-associated cognitive processing.[76]

Inconsistent results among studies may be due to the difference in DHA content of fish oil (e.g., 0.8 g/day in Fontani et al.[69] compared to 0.25 g/day in Antypa et al.[70] and 0.24 g/day in Karr et al.[74]). Subsequent research has compared DHA-rich fish oil to EPA-rich fish oil to better understand whether the DHA or EPA content moderates the differences in cognitive performance. The DHA-rich fish oil lowered reaction time on the Stroop Task relative to olive oil and EPA-rich fish oil lowered self-reported fatigue during high cognitive demand. Both DHA- and EPA-rich fish oil impaired episodic memory on the Names-to-Faces task, but this task was one of the five tasks measuring episodic memory, the other four of which did not generate differences indicating that the influence of fish oil on episodic memory is not entirely reliable.[71]

TABLE 4 Randomized Controlled Trials Assessing the Influence n-3 PUFA Supplementation in Childhood on Cognitive Development

Authors	Sample *n* (Female)	Age Range (mean ± SD) years	*n*-3 PUFA Manipulation (per day)	Duration (weeks)	Cognitive Measures	Results
Kennedy et al.[63]	88 (42)	10–12 (400 mg DHA[a]: 11.11 ± 0.79, 1000 mg DHA: 10.70 ± 0.79 Control: 10.87 ± 1.1)	400 mg DHA vs 1000 mg DHA vs vegetable oil	8	IB[b] CDR[c]	Control > 400, 1000 mg DHA relaxation ratings at baseline Post > Pre-DHA treatment relaxation ratings 400 mg < Placebo word recognition reaction time before and after breakfast 1000 mg DHA > Placebo word recognition reaction time before breakfast
Ryan and Nelson[64]	175 (79)	4.0–4.67 (*n*-3: 4.3 ± 0.2, Control: 4.3 ± 0.2)	400 mg DHA vs sunflower oil	16	Leiter-R[d] PPVT[e] Day-Night Stroop Test kCPT[f]	No differences
Muthayya et al.[65]	598 (204)	6–10 (All: 8.7 ± 1.2)	Micronutrient (100% RDA[g]: high vs 15% RDA: low) × *n*-3 PUFA (900 mg ALA[h] + 100 mg DHA: high vs 140 mg ALA: low)	52	KABC-II[i] WISC-R, WISC-4[j] RAVLT[k] NEPSY[l] Number Cancellation	High > Low micronutrient short-term memory (6 months only) Low > High micronutrient fluid reasoning (6 and 12 months) No differences in *n*-3 groups
Richardson et al.[66]	362 (170) below 33rd centile in reading	6–10 (*n*-3: 8.6 ± 0.8, Control: 8.7 ± 0.8)	Algal oil (600 mg DHA) vs corn/soybean oil	16	BAS-II[m] Recall of Digits Forward Backward CTRS-L, CPRS-L[n]	No changes working memory or reading scores across entire sample DHA > Control reading score in children below 22nd centile in reading DHA > Control anxiety, restless-impulsive, emotional labiality, global index from parent but not teacher ratings

(*Continued*)

TABLE 4 Randomized Controlled Trials Assessing the Influence n-3 PUFA Supplementation in Childhood on Cognitive Development—cont'd

Authors	Sample *n* (Female)	Age Range (mean ± SD) years	*n*-3 PUFA Manipulation (per day)	Duration (weeks)	Cognitive Measures	Results
Parletta et al.[67]	409 (191)	3–13 (n-3: 8.74 ± 2.20, Control: 8.60 ± 2.18)	750 mg DHA + EPA[o] + 60 mg/ GLA[p] vs palm oil	20	WRAT4[q] DAP[r] CBRS[s]	*n*-3 > Control nonverbal cognitive ability (largest effect in 7–Table 12-year-old indigenous children) No differences in reading and spelling
Portillo-Reyes et al.[68]	59 (nr[t])	8–12 (FO: 9.37 ± 1.173, Control: 9.09 ± 0.985)	FO (180 mg DHA, 270 mg EPA) vs soybean oil	13	Processing speed, visuomotor coordination attention, memory, language executive function, academic performance	FO > control processing speed (Symbol search, embedded figures, visual closure), visuomotor coordination (block design), and executive function (stroop-color and color-word, and matrix reasoning) No differences in attention, memory, language, academic performance

[a] *Docosahexaenoic acid (DHA).*
[b] *Internet Battery (IB).*
[c] *Cognitive Drug Research (CDR) Battery.*
[d] *Leiter-R Test of Sustained Attention (Leiter-R).*
[e] *Peabody Vocabulary Test (PPVT).*
[f] *Conner's Kiddie Continuous Performance Test (kCPT).*
[g] *Recommended Dietary Allowance (RDA).*
[h] *α-Linolenic acid (ALA).*
[i] *Kaufman Assessment Battery for Children, second edition (KABC-II).*
[j] *Wechsler Intelligence Scales for Children (WISC-R and WISC-4).*
[k] *Rey Auditory Verbal Learning Test (RAVLT).*
[l] *Neuropsychological Assessment Tool (NEPSY).*
[m] *British Ability Scales (BAS-II).*
[n] *Conner's Rating Scales (CTRS-L, CPRS-L).*
[o] *Eicosapentaenoic acid (EPA).*
[p] *Gamma linolenic acid (GLA).*
[q] *Wide Range Achievement Test: Fourth Edition (WRAT4).*
[r] *Draw-A-Person (DAP).*
[s] *Conners' Comprehensive Behavior Rating Scales (CBRS).*
[t] *Not reported (nr).*

TABLE 5 Randomized Controlled Trials Assessing the Influence of n-3 PUFA Supplementation in Young Adults on Cognitive Performance

Authors	Sample *n* (Female)	Age Range (years)	Mean Age ± SD	*n*-3 PUFA Manipulation (per day)	Duration (weeks)	Cognitive Measures	Results
Fontani et al.[69]	36 (32)	22–51	FO[a]: 33 ± 7, Control: 33 ± 3	4 g FO (0.8 g DHA,[b] 1.6 g EPA[c]) vs OO[d]	5	POMS,[e] SRT,[f] CRT,[g] GNG,[h] Complex GNG (sustained attention)	FO > Control vigor FO < Control anger, anxiety, fatigue, depression, confusion FO < Control GNG reaction time, number of errors No differences SRT, CRT
Antypa et al.[70]	54 (44)	18–27	FO: 22.2 ± 3.6, Control: 22.6 ± 4.1	3 g FO (0.25 g DHA, 1.74 g EPA)	4	Affective GNG, Attentional GNG; Facial Expression Recognition Task, Decision-Making (Gambling) Task, M.I.N.I.[i]; BDI-II[j]; POMS; BIS/BAS[k]; LEIDS-R[l]	FO < Control fatigue, LEIDS-R Control/Perfectionism, Risk Aversion and total score FO > Control Gambling Task risk-seeking decision-making in gains only trials No differences in GNG, Facial Expression Recognition
Jackson, Deary, et al.[71]	140 (94)	18–35	DHA-rich FO: 21.96 ± 0.54, EPA-rich FO: 22.74 ± 0.61, Control: 21.94 ± 0.50	1 g DHA-rich FO (0.45 g DHA, 0.09 g EPA) vs EPA-rich FO (0.2 g DHA, 0.3 g EPA) vs OO	12	COMPASS[m] CDB[n] Bond-Lader VAS[o] DASS[p]	DHA-rich FO < Control Stroop reaction time EPA-, DHA-rich FO < Control Names-to-Faces Task (episodic memory) EPA-rich FO < Control CBD self-reported fatigue No differences in mood, other cognitive measures
Jackson, Reay, et al.[72]	22 (13)	nr[q]	All: 21.96 ± nr	1 g DHA-rich FO (0.45 g DHA, 0.09 g EPA) vs EPA-rich FO (0.2 g DHA, 0.3 g EPA) vs OO	12	Stroop Task Peg-and-Ball Task 3-Back Task Wisconsin Card Sort Task NIRS[r]	DHA-rich FO > Control oxy-HB following DHA-rich FO during Stroop Task DHA-rich FO > Control total-HB during Stroop Task, Peg-and-Ball, 3-Back Tasks
Jackson, Reay, et al.[73]	65 (49)	18–29	1 g FO: 20.5 ± 0.43, 2 g FO: 19.95 ± 0.34. Control: 21.35 ± 0.62	1 g FO (0.45 g DHA, 0.09 g EPA) vs 2 g/day FO (0.9 g DHA, 0.18 g EPA) vs OO	12	COMPASS	1, 2 g FO > Control oxy-HB during all tasks 2 g FO > Control total-HB during all tasks 1 and 2 g FO < Control RT in CRT 2 g < Placebo RT in RVIP[s] task

(Continued)

TABLE 5 Randomized Controlled Trials Assessing the Influence of n-3 PUFA Supplementation in Young Adults on Cognitive Performance—cont'd

Authors	Sample *n* (Female)	Age Range (years)	Mean Age ± SD	*n*-3 PUFA Manipulation (per day)	Duration (weeks)	Cognitive Measures	Results
Karr et al.[74]	41 (29)		FO: 19.90 ± 18.3, Control: 20.43 ± 1.63	1 g FO (240 mg DHA, 360 mg EPA) vs coconut oil	4	RAVLT,[t] Stroop Test, TMT,[u] PANAS[v]	FO > Control final stages (6 and 7) RAVLT FO < Control TMT No differences in Stroop Test, PANAS
Dretsch et al.[75]	78 (6)[w]	20–54	All: 31 ± 7.4	2.5 g EPA (47%) + DHA (38%) vs corn oil	8.5	CNS-VS,[x] PSQI,[y] ESS,[z] ZDS and ZAS,[aa] NSI[bb]	No differences
Giles et al.[76]	72 (45)	18–29	FO: 20.80 ± 2.386, Control: 20.49 ± 1.704	2.8 g FO (1680 mg EPA, 1120 mg DHA) vs olive oil X acute stress/ no-stress	5	POMS, STICSA,[cc] EIT,[dd,ee]	OO > FO Stress-induced increase in rated anger and confusion No other cognition or mood difference

[a] *Fish oil (FO).*
[b] *Docosahexaenoic acid (DHA).*
[c] *Eicosapentaenoic acid (EPA).*
[d] *Olive oil (OO).*
[e] *Profile of Mood States (POMS).*
[f] *Simple Reaction Time (SRT).*
[g] *Choice Reaction Time (CRT).*
[h] *Go/No-Go (GNG).*
[i] *Mini International Neuropsychiatric Interview (M.I.N·I).*
[j] *Beck Depression Inventory-II (BDI-II).*
[k] *Behavioral Inhibition/Behavioral Activation Scales (BIS/BAS).*
[l] *Leiden Index of Depression Sensitivity—Revised (LEIDS-R).*
[m] *Computerized Mental Performance Assessment System (COMPASS).*
[n] *Cognitive Demand Battery (CDB).*
[o] *Visual Analogue Scales (VAS).*
[p] *Depression, Anxiety, and Stress Scales (DASS).*
[q] *Not reported (nr).*
[r] *Near-infrared spectroscopy (NIRS).*
[s] *Rapid Visual Information Processing (RVIP).*
[t] *Rey Auditory Verbal Learning Test (RAVLT).*
[u] *Trail Making Test (TMT), Parts A and B.*
[v] *Positive and Negative Affect Schedule (PANAS).*
[w] *U.S. deployed soldiers.*
[x] *Central Nervous SystemVital Signs (CNS-VS).*
[y] *Pittsburgh Sleep Quality Index (PSQI).*
[z] *Epworth Sleepiness Scale (ESS).*
[aa] *Zung Depression Scale (ZDS) and Zung Anxiety Scale (ZAS).*
[bb] *Neurobehavioral Symptom Inventory (NSI).*
[cc] *State–Trait Inventory for Cognitive and Somatic Anxiety (STICSA).*
[dd] *Emotional Interference Task (EIT).*
[ee] *Morphed Faces Task (MFT).*

Near-infrared spectroscopy (NIRS) is a brain imaging method that measures light absorbance to calculate oxy-hemoglobin (oxy-HB) and deoxy-hemoglobin (deoxy-HB), which provides an indirect measure of brain activity, particularly in the frontal cortex. Previous research had found increased prefrontal activation following DHA treatment.[78] The DHA- but not EPA-rich fish oil increased oxy-HB and total-HB in participants performing tasks measuring executive function, including selective attention, cognitive flexibility, and working memory.[72] These effects were replicated in a subsequent study that compared two doses of DHA-rich fish oil (1 and 2 g/day) to olive oil on a number of cognitive tasks, including those measuring episodic memory, psychomotor performance, executive function, and working memory. Both doses increased oxy-HB during all of the cognitive tasks relative to olive oil and whereas 2 g/day fish oil increased total-HB during all tasks, 1 g/day increased total-HB only during the Stroop Task and a rapid visual information processing task, which measured sustained attention.[73] Thus *n*-3 PUFA supplementation appears to influence prefrontal cortical oxygenation, but whether such changes can be reflected in behavioral improvements remains unclear.

OLDER ADULTS

Older adulthood poses an additional critical period in cognitive development, as aging is associated with a number of cognitive changes, including decline in episodic and working memory (for review, see Nyberg et al.[79]). Although cognitive decline often occurs with normal aging, it may also serve as an early indicator of Alzheimer's disease.[80] The age-related cognitive impairments range from mild cognitive impairments to intermediate degrees of dementia to more severe cases as in Alzheimer's Disease. One in nine individuals aged 65 years and older and one third of individuals aged 85 years and older have Alzheimer's disease.[81] The well-established risk factors for Alzheimer's disease include advancing age, family history of Alzheimer's disease, and carrying the apolipoprotein E ε4 (*APOE ε4*) gene as well as modifiable conditions, such as hypertension, diabetes, and smoking.[82] A number of epidemiological studies as well as RCTs have assessed the relationship between *n*-3 PUFA intake and age-related cognitive decline, dementia, and Alzheimer's disease. We will begin with epidemiological studies and then move to RCTS in (1) healthy older adults and (2) individuals with mild cognitive impairment and Alzheimer's disease.

EPIDEMIOLOGICAL STUDIES: THE ASSOCIATION BETWEEN N-3 PUFA INTAKE, N-3 PUFA LEVELS, AND COGNITIVE DECLINE

Epidemiological studies assessing the influence of *n*-3 PUFAs on cognitive decline in older adults generally measure dietary intake of fatty acids and/or plasma or erythrocyte levels of the fatty acids. Associations with cognitive performance are then determined at one or more time points (Table 6). The most common behavioral measure utilized in such designs is the Mini Mental State Exam (MMSE),[112] which measures general cognitive impairment and is often used to characterize mild cognitive impairment or cognitive decline on a discrete yes-or-no basis.[112] For instance, intake of fish and various *n*-3 PUFAs, primarily EPA and DHA,

TABLE 6 Epidemiological Studies Assessing the Association between n-3 PUFAs and Age-Related Cognitive Decline

Authors	Sample n (Female)	Age Range	Follow-up Duration (years)	PUFA Measure	Cognitive Measures	Results
Kalmijn et al.[83]	476 (0)	69–89	3	Fish and FA[a] intake	MMSE[b]	Baseline: higher intake total fat, PUFAs and LA[c] and lower intake total energy, fish, EPA,[d] and DHA[e] associated with cognitive impairment (no association with total *n*-3 PUFA) 3-year: No association with FA intake and cognitive decline in sample as whole, but higher intake LA and lower intake fish associated with cognitive decline in individuals free of cognitive decline at baseline
Heude et al.[84]	246 (143)	63–74	4	Erythrocyte FA (baseline only)	MMSE	Lower DHA, *n*-3:*n*-6, DHA:AA,[f] and higher total *n*-6 in individuals showing cognitive decline than no decline
Laurin et al.[85]	174 (117)	≥65	5	Plasma FA (baseline only), *APOE ε4* genotype	MMSE, DSM-III-R[g] dementia diagnosis	No association with dementia risk and plasma FA Lower *n*-6 and total PUFAs in individuals with dementia than without (*APOE ε4* carriers only) Higher DHA in individuals with dementia than without (*APOE ε4* noncarriers only) *Prospective analysis (only participants free of dementia at baseline):* Higher EPA in individuals who developed CIND[h] than who did not (free of dementia at baseline only) Higher DHA, *n*-3 PUFAs, and total PUFAs in individuals who developed dementia than who did not (free of dementia at baseline only)
Whalley et al.[86]	350 (171)	64	53	Oily fish and FA intake, Erythrocyte FA, *APOE ε4* genotype	MHT[i] (11 years old), MMSE, RPM,[j] RAVLT,[k] Uses of Common Objects Test, Digit Symbol Test, Block Design Test	No association with childhood IQ and supplement use Higher IQ and digit symbol score for fish oil, vitamin and other supplement users than nonusers Higher block design scores in fish oil users than nonusers Higher erythrocyte *n*-3 PUFA, EPA, DHA, and *n*-6:*n*-3 associated with higher IQ in childhood and 64 years Higher erythrocyte AA:DHA associated with higher block design and RPM scores Higher erythrocyte DHA and DHA:AA associated with higher digit symbol scores

Huang et al.[87]	2233 (1306)	≥65	0.1–8.4 (mean 5.4)	Lean fried fish, fatty fish, or total fish intake, *APOE ε4* genotype	MMSE, Digit Symbol Test, Benton Visual Retention Test, CES-D,[l] ADLs,[m] TICS,[n] IQCoDE[o]	No association between risk of dementia or AD[p] and fried fish intake Reduced risk dementia and AD in individuals consuming ≥4 servings/week tuna or non-fried fish Reduced risk dementia in ≥2 servings/week fatty fish in individuals (*APOE ε4* non-carrier only)
Morris et al.[88]	3178 (2306)	≥65	3, 6	Fish and FA intake	MMSE, East Boston Tests of Immediate and Delayed Recall, Symbol Digit Modalities Test	Less cognitive decline in individuals who consumed ≥1 serving fish/week No association with cognitive decline and *n*-3 PUFA intake
Beydoun et al.[89]	2251 (1141)	45–64	N/A	Plasma FA	Delayed Word Recall, Digit Symbol Substitution, Word Fluency	Higher plasma ARA[q] and lower LA associated with GCI[r] No associations with plasma *n*-3 and GCI Higher *n*-3 PUFAs associated with reduced decline in verbal fluency (individuals with higher hypertensive and dyslipidemic markers, and lower depressive symptoms only)
Dullemeijer et al.[90]	807 (226)	50–70	3	Plasma FA	Concept Shift Test, Stroop Test, Word Learning Test, Letter Digit Substitution Test, Verbal Fluency Test	Reduced 3-year cognitive decline in participants with higher *n*-3 PUFAs (placebo group in RCT[s]) No differences in memory, information processing speed, word fluency No associations *n*-3 PUFAs and cognitive performance (all participants in RCT)
van Gelder et al.[91]	210 (0)	70–89	5, 10	Fish and EPA, DHA intake	MMSE	Greater cognitive decline in men who consumed no fish than men who consumed fish Greater cognitive decline in lowest than highest tertile of EPA + DHA intake
Samieri et al.[92]	1214 (748)	≥65	2, 4	Plasma FA (baseline only), *APOE ε4* genotype	Dementia diagnosis, CES-D	Higher EPA and total *n*-3 PUFA associated with reduced risk dementia Association between AA:DHA ratio and dementia risk higher in subjects with depression than without
Whalley et al.[93]	120 (68)	63.8–65.3	64, 66, 68 years age	Erythrocyte FA (64 years only), *APOE ε4* genotype	MHT (11 years old) RPM, RAVLT, Uses of Common Objects Test, Digit Symbol Test, Block Design Test	Positive association between total *n*-3 PUFA and cognitive performance at 11 and 64 years (*APOE ε4* noncarriers only)

(Continued)

TABLE 6 Epidemiological Studies Assessing the Association between n-3 PUFAs and Age-Related Cognitive Decline—cont'd

Authors	Sample n (Female)	Age Range	Follow-up Duration (years)	PUFA Measure	Cognitive Measures	Results
Devore et al.[94]	5395 (3185)	≥55	10	Fish and FA intake, *APOE ε4* genotype	MMSE, GMS,[t] DSM-III-R dementia diagnosis	No association with fish or *n*-3 PUFA intake and risk of dementia
Kroger et al.[95]	663 (401)	≥65	5, 10	Plasma FA, *APOE ε4* genotype	MMSE, DSM-III-R dementia diagnosis,	No association with FA and dementia
van de Rest et al.[96]	1025 (0)	68 ± X	3, 6	Fish and FA intake	MMSE, Tests of memory, language, perceptual speed, and attention	No associations between cognitive performance and *n*-3 PUFA intake
Gonzalez et al.[97]	304 (177)	X (75.3 ± 6.7)	N/A	FA intake	MMSE	Higher cognitive score associated with higher intake *n*-3 PUFA, EPA, DHA, ALA[u] and lower *n*-6:*n*-3 ratio (no differences in total energy, lipids, SFA,[v] MUFA,[w] PUFA, or *n*-6 PUFA) In regression, higher intake EPA and DHA and lower intake *n*-6:*n*-3 PUFA predicted lower cognitive impairment
Roberts et al.[98]	1223 (592) free of dementia	70–89	N/A	FA intake	CDR[x] Scale, Neuropsychological test battery, DSM-IV dementia diagnosis, Modified Block 1995 Revision of the Health Habits History Questionnaire	MCI[y] associated with lower intake PUFA, *n*-6 PUFA, *n*-3 PUFA, fatty acids, LA, ALA, and (MUFA + PUFA):SFA ratio Lower risk MCI associated with higher intake MUFA (men only)
Gao et al.[99]	1475 (969) free of dementia	≥55	1.5	*n*-3 PUFA supplement intake	MMSE	Lower cognitive decline associated with *n*-3 supplement use

Milte et al.[100]	79 (29) With MCI: 50 (16), HC: 27 (13)	≥65	N/A	Erythrocyte FA	MMSE, Memory Functioning Questionnaire, SF-36 Health Survey, GDS,[z] RAVLT, Stroop Test, Boston Naming Test, Digits forward, digits backward, letter-number sequencing, Trail Making Task	MCI < No MCI EPA MCI > No MCI *n*-6 PUFAs Higher *n*-6 PUFAs associated with poorer mental health
Chiu et al.[101]	132 (96) With history MDD without cognitive impairment	≥60	N/A	Erythrocyte and plasma membrane FA, *APOE ε4* genotype	WAIS-III,[aa] CTT,[bb] Semantic verbal fluency	Higher erythrocyte ALA, total *n*-3 PUFA associated with greater immediate verbal memory
Ronnemaa et al.[102]	2009 (0)	50	35	Serum FA (baseline only), *APOE ε4* genotype	Dementia diagnosis	Higher serum LA in individuals who developed dementia than did not (no difference in *n*-3 PUFAs)
Ammann et al.[103]	2157 (2157)	65–80	5.9	RBC[cc] DHA and EPA	Finger Tapping Test, Card Rotations Test, Benton Visual Retention Test, California Verbal Learning Test, Primary Mental Abilities Vocabulary test, letter and categories fluency tests, Digit Span Forward and Backward Test	High > low tertile DHA + EPA levels fine motor speed, verbal knowledge, verbal fluency No differences after adjustment for risk factors for cognitive impairment/vascular disease
Leckie et al.[104]	344 (183)	30–54	N/A	Serum DHA and AA levels	N-Back, Trailing Making, Logical Memory	No association with serum DHA or AA and cognition Serum AA:DHA and PA[dd] interaction on n-back and trail making: Higher AA:DHA ratios and lower PA associated with poorer n-back and trail making performance Serum DHA and PA interaction on n-back: higher DHA reduced association between lower PA and cognitive performance

(*Continued*)

TABLE 6 Epidemiological Studies Assessing the Association between n-3 PUFAs and Age-Related Cognitive Decline—cont'd

Authors	Sample n (Female)	Age Range	Follow-up Duration (years)	PUFA Measure	Cognitive Measures	Results
Danthiir et al.[105]	390 (209)	65–91	N/A	Erythrocyte FAs, fish consumption frequency	Tasks of vocabulary, perceptual speed/, simple/choice RT,[ee] speed of memory scanning, reasoning speed, inhibition, psychomotor speed, reasoning, working memory, short-term memory, retrieval fluency	More frequent current total fish intake associated with slower inhibition, simple/choice RT, reasoning, memory scanning More frequent current oily fish intake associated with poorer inhibition, simple/choice RT More frequent historical total fish intake in childhood associated with poorer perceptual speed, simple/choice RT No associations with historical total fish intake in adulthood/middle-age and cognition No associations erythrocyte DHA, DPA, EPA and cognition Higher percent erythrocyte EPA associated with higher speed scores in men, lower speed scores in women
Baierle et al.[106]	45 (30)	≥60	N/A	Serum FA concentration	MMSE, CERAD[ff] battery	With > without cognitive impairment SFAs myristic acid and palmitic acid, MUFA *n*-7 palmitoleic acid With < without cognitive impairment SFA behenic acid, MUFA *n*-9 nervonic acid, DHA, total PUFA Higher SFAs myristic acid associated with poorer visual attention Higher MUFA *n*-7 palmitoleic acid associated with lower word memory and delay recall Higher DHA associated with higher word recognition and constructional praxis
Otsuka et al.[107]	430 (198)	60–79	10	Serum DHA, EPA	MMSE	Higher serum DHA associated with lower prevalence cognitive decline No association serum EPA and cognitive decline
Daiello et al.[108]	819 (342)	55–90	2	FO supplement use	MMSE, ADAS-cog,[gg] MRI cerebral cortex gray matter, ventricle, and hippocampus volumes	FO supplement use associated with lower ADAS-cog and higher MMSE scores (remained significant in *APOE ε4* noncarriers only and only in individuals with normal cognition at baseline) FO supplement use associated with increased hippocampal and cerebral cortex gray matter volume (remained significant in *APOE ε4* noncarriers only) and decreased ventricular volume (not when stratified by *APOE ε4*)
Nishihira et al.[109]	185 (142)	≥80	N/A	Serum *n*-3 PUFA	MMSE, executive function	Higher serum EPA and DHA + EPA associated with higher MMSE scores No associations executive function
Del Brutto et al.[110]	307 (170)	≥60	N/A	Oily fish servings per week	MoCA[hh]	Greater servings oily fish per week associated with enhanced cognitive performance

van de Rest et al.[111]	915 (686)	81.4 6 7.2	4.9	Seafood and PUFA intake	21 tests of global cognitive function, episodic memory, semantic memory, working memory, perceptual speed, and visuospatial ability	No associations with seafood or PUFA intake global cognitive decline One or more seafood meals per week associated with slower decline in semantic memory and perceptual speed than fewer meals No association with ALA or *n*-3 PUFA intake and cognition FO supplement intake associated with slower global cognitive decline and episodic memory than nonconsumers Among *APOE ε4* carriers only, weekly seafood intake associated with slower decline in global cognition, episodic memory, semantic memory, and perceptual speed Among *APOE ε4* carriers only, increased *n*-3 PUFA intake associated with slower declines in semantic memory and perceptual speed

[a] *Fatty acid (FA).*
[b] *Mini Mental State Exam (MMSE).*
[c] *Linoleic acid (LA).*
[d] *Eicosapentaenoic acid (EPA).*
[e] *Docosahexaenoic acid (DHA).*
[f] *Arachidonic acid (AA).*
[g] *Diagnostic & Statistical Manual of Mental Disorders–3rd Edition Revised (DSM-III-R).*
[h] *Cognitive impairment-no dementia (CIND).*
[i] *Moray House Test (MHT).*
[j] *Raven's Standard Progressive Matrices (RPM).*
[k] *Rey's Auditory Verbal Learning Test (RAVLT).*
[l] *Center for Epidemiological Studies of Depression Scale (CES-D).*
[m] *Activities of daily living scale (ADLs).*
[n] *Telephone Interview for Cognitive Status (TICS).*
[o] *Informant Questionnaire for Cognitive Decline in the Elderly (IQCoDE).*
[p] *Alzheimer's Disease (AD).*
[q] *Arachidonic acid (ARA).*
[r] *Global cognitive decline (GCI).*
[s] *Randomized controlled trial (RCT).*
[t] *Geriatric Mental State schedule (GMS).*
[u] *Alpha-linolenic acid (ALA).*
[v] *Saturated fatty acid (SFA).*
[w] *Monounsaturated fatty acid (MUFA).*
[x] *Clinical Dementia Rating (CDR) Scale.*
[y] *Mild cognitive impairment (MCI).*
[z] *Geriatric Depression Scale (GDS).*
[aa] *Wechsler Memory Scale (III) (WAIS-III).*
[bb] *Color Trail Test (CTT).*
[cc] *Red blood cell (RBC).*
[dd] *Physical activity (PA).*
[ee] *Response time (RT).*
[ff] *Consortium to Establish a Registry for Alzheimer's Disease (CERAD).*
[gg] *Alzheimer's Disease Assessment Scale (ASAS-cog).*
[hh] *Montreal Cognitive Assessment (MoCA).*

were associated with reduced cognitive impairment in cross-sectional studies[83,97,110] and reduced decline over 5–10 year in prospective studies.[87,91] Similarly, individuals who consumed *n*-3 PUFA supplements experienced slower cognitive decline over 1–2 years.[99,108] Individuals with mild cognitive impairment consumed less PUFAs, including *n*-3 and *n*-6 PUFAs, than those without mild cognitive impairment.[98]

However, other studies have found no associations between fish or *n*-3 PUFA intake and global cognitive function over 3 to 10 years.[94,96,111] Across 6 years, individuals who consumed at least one serving per week of fish showed less cognitive decline, but no association was found between total *n*-3 PUFA intake and cognitive decline.[88] Thus although data from FFQ give us some indication that increased *n*-3 PUFA intake is associated with reduced cognitive decline, the reliability of such measures is questionable, given the inconsistencies in many studies, for example, correlations between cognitive function and fish but not *n*-3 PUFA intake.[88]

The blood levels of *n*-3 PUFAs may provide more objective measures of *n*-3 PUFA intake. The serum PUFAs have been associated with reduced cognitive impairment.[100,106,109] Increased risk of cognitive decline over 4 years has been associated with lower erythrocyte DHA, EPA, *n*-3:*n*-6 ratio, DHA:AA ratio and higher total *n*-6 PUFA levels[84] as well as lower plasma EPA and total *n*-3 PUFAs.[92] Higher serum DHA, but not EPA, was associated with lower prevalence of cognitive decline.[107] Older adults who experienced cognitive decline over a 4-year period had lower erythrocyte DHA, EPA, *n*-3:*n*-6 ratio, DHA:AA ratio and higher total *n*-6 PUFA levels than those who did not. However, the higher erythrocyte DHA levels were associated with lower self-reported mental health, suggesting that enhanced cognitive performance does not necessarily equate to enhanced subjective well-being.[100]

Other research has not found any association between blood *n*-3 PUFAs and cognitive decline or incidence of dementia[95,102] but found opposite associations: in individuals initially free of dementia, plasma EPA levels were higher in those who developed cognitive impairment than those who did not, and plasma DHA, total *n*-3 PUFA and total PUFA levels were higher in those who developed dementia than those who did not.[85]

In addition to the studies utilizing the MMSE or other global measures of cognitive function, the same and other studies have assessed more specific aspects of cognition. For instance, higher plasma *n*-3 PUFA levels were associated with reduced global cognitive decline over a 3-year period, though no differences were found in specific cognitive domains such as executive function, memory, and visual information processing.[90] Other research has found that more frequent fish intake was associated with slower response time and poor inhibitory control, but found no associations between erythrocyte DHA and EPA and cognition.[105] Higher erythrocyte ALA and total *n*-3 PUFAs were associated with enhanced verbal memory, but in the same study no associations were found between fatty acid levels and other cognitive measures including intelligence, attention, psychomotor speed, and executive function.[101]

Using retrospective data, Walley and colleagues (2004, 2008) correlated intelligence scores of individuals measured when they were 11 years of age with their intelligence scores, cognitive behavior, and erythrocyte fatty acid levels at 64 years of age. Higher erythrocyte total *n*-3 PUFAs, EPA, DHA, and *n*-3: *n*-6 ratio at 64 years of age were associated with higher IQ at both 11 and 64 years of age. Moreover, (1) a higher erythrocyte AA:DHA ratio was associated with higher constructional ability and nonverbal reasoning at 64 years of age and (2) older adults who consumed fish oil supplements had higher IQ, psychomotor performance, and constructional ability than those who did not use supplements.[86] In a subsequent study, the positive association between erythrocyte total *n*-3 PUFA levels and cognitive performance at 11 and 64 years was found to be stronger for individuals who did not carry the *APOE ε4* gene than for *APOE ε4* carriers.[93]

The association between plasma fatty acid levels and cognitive decline may also depend on other health-related factors, including cardiovascular and psychological impairments. For instance, plasma *n*-3 PUFAs were associated with reduced decline in verbal fluency in individuals with higher hypertensive and dyslipidemic markers, and in individuals with lower depressive symptoms.[89] Similarly, individuals in the high relative to low tertiles for red blood cell DHA + EPA levels exhibited enhanced fine motor speed, verbal knowledge, and verbal fluency, but the associations were no longer significant after adjusting for risk factors for cognitive impairment and vascular disease.[103] Other research failed to find associations between serum DHA or AA and cognition, but found that higher DHA reduced the association between low physical activity and cognitive performance, indicating that DHA intake may confer some protective factor against physical inactivity.[104]

The results suggest that *n*-3 PUFA intake and levels may have greater beneficial effects for individuals not already at increased risk for Alzheimer's disease by carrying the *APOE ε4* gene. For instance, *APOE ε4* noncarriers who consumed at least two servings per week of fish were at reduced risk for dementia, but no such association was found for *APOE ε4* carriers.[87] Similarly, higher *n*-3 PUFA intake and supplement use was associated with enhanced cognitive function in *APOE ε4* noncarriers, but not carriers.[93,108] Higher plasma DHA was associated with dementia in *APOE ε4* noncarriers only.[85] However, similar results have also been found for *APOE ε4* carriers only: lower plasma *n*-6 and total PUFAs was associated with dementia,[85] and seafood and *n*-3 PUFA intake was associated with slower cognitive decline in *APOE ε4* carriers only.[111]

Thus evidence from epidemiological studies is quite mixed, with some studies reporting associations between higher *n*-3 PUFA intake and/or blood levels and lower risk of age-related cognitive decline and other reporting no associations.

RCTS: HEALTHY OLDER ADULTS

Conclusions drawn from epidemiological studies are limited to possible associations between *n*-3 PUFAs and cognitive decline, whereas RCTs are necessary to determine whether increased *n*-3 PUFA intake plays a causal role in reducing cognitive decline. A number of studies have assessed the influence of *n*-3 PUFA supplementation on cognitive function in healthy older adults (Table 7). DHA improved verbal fluency, memory, and rate of learning but not mental processing speed or accuracy.[113] DHA-rich fish oil improved visual and verbal learning and memory but not working memory or executive function.[115] Similarly, a DHA-rich multi-nutrient supplement improved verbal memory but not visual memory or executive function.[121] Other research found that EPA-rich *n*-3 PUFA supplementation improved executive function but not memory.[117]

Similar to results from epidemiological studies in which *n*-3 PUFA levels were differentially associated with cognitive decline between *APOE ε4* carriers and noncarriers,[86,87] low and high doses of EPA and DHA enhanced attention relative to placebo only in *APOE ε4*

TABLE 7 Randomized Controlled Trials Assessing the Influence of n-3 PUFA Supplementation on Cognitive Function in Healthy Older Adults

Authors	Sample *n* (Female)	Age Range mean ± SD)	*n*-3 PUFA Manipulation	Duration (weeks)	Cognitive Measures	Results
Johnson et al.[113]	49 (49)	60–80 DHA: 68.5 ± 1.3 Lutein: 66.7 ± 1.9 DHA + Lutein: 68.6 ± 1.3 Control: 68.0 ± 1.2	800 mg/day DHA[a] vs 12 mg/day Lutein vs DHA + Lutein vs placebo	16	Verbal Fluency, Digit Span Forward, and Backward, Shopping Test Task, Word List Memory Test, Memory in Reality Apartment Test, Stroop Test, NES2 Mood Scales: self-reported mood	DHA, Lutein, DHA + Lutein > Control verbal fluency DHA + Lutein < Control response time on Shopping List Memory Test, delayed recall in Memory in Reality Apartment Test No differences in mental processing speed, accuracy
van de Rest et al.[114]	196 (88)	> 65 High EPA-DHA: 69.9 ± 3.4 Low EPA-DHA: 69.5 ± 3.2 Control: 70.1 ± 3.7	900 mg/day fish oil low EPA-DHA (226 ± 3 mg EPA, 176 ± 4 mg DHA) vs high EPA-DHA (1093 ± 17 mg EPA, 847 ± 23 mg DHA) vs sunflower oil	26	Verbal Fluency, Word Learning Test, Digit Span Forward and Backward, Trail Making Test Version A and B, Stroop Test	Low EPA-DHA < Control memory at 13 but not 26 weeks Low and High EPA-DHA > Control attention at 26 weeks in *APOE ε4* carriers only Low EPA-DHA > Control attention at 26 weeks in men only
Yurko-Mauro et al.[115]	485 (282)	≥ 55 DHA: 70 ± 9.3 Control: 70 ± 8.7	Algal triglyceride oil (900 mg/day DHA) vs corn + soy oil	24	WAIS-III[b] logical memory, MMSE,[c] CANTAB,[d] Frequency of Forgetting-10 Scale, ADCS-ADL PI,[e] GDS[f]	DHA > Control CANTAB PAL,[g] VRM[h]
Stough et al.[116]	74 (43)	45–77 DHA: 55.08 ± 8.70 Control: 57.66 ± 8.67	1 g/day tuna oil (252 mg DHA, 60 mg EPA) vs soybean oil	90 days	STAI,[i] Cognitive Drug Research (CDR)[j]	No differences in CDR factors
Witte et al.[117]	65 (30)	50–75 *n*-3 PUFA: 65 ± 6.3 Control: 62.9 ± 6.8	2.2 g/day *n*-3 PUFA (1320 EPA, 880 DHA) vs sunflower oil	26	Verbal fluency, TMT[k] Stroop Test, AVLT,[l] Forward and backward digit spans, MRI[m]	*n*-3 PUFA > Control executive function No differences in memory, sensorimotor speed
Chew et al.[118]	3073 (1767)	*n*-3 PUFA: 72.7 ± 7.7 No *n*-3 PUFA: 72.7 ± 7.7	350 mg DHA + 650 mg EPA vs 10 mg lutein + 2 mg zeaxanthin vs 350 mg DHA + 650 mg EPA + 10 mg lutein + 2 mg zeaxanthin vs placebo	5 years	CES-D,[n] TICS-M,[o] Animal category, Letter fluency, Alternating fluency, WAIS-III, Digits backward, Delayed recall, TICS-M recall	No differences

Pase et al.[119]	160 (85)	50–70 59.27 ± 5.72	Multivitamin + 3 g (240 mg EPA + 240 mg DHA) FO[p] vs multivitamin + 6 g (480 mg EPA + 480 mg DHA) FO vs placebo multivitamin + 6 g FO vs placebo multivitamin + sunola oil	16	Swinburne University Computerized Cognitive Assessment Battery: Reaction time, Cognitive processing speed, short-term memory, visual memory	No differences
Jackson et al.[120]	84 (52)	EM[q]: 59.87 ± 4.70 FO: 60.31 ± 4.89 Control: 59.64 ± 5.28	2 g FO (896 mg DHA, 128 mg EPA) vs 2 g FO (946.4 mg DHA, 160 mg EPA) + Multinutrient vs sunflower oil (with 120 mg FO)	26	CBD,[r] NIRS[s]	No differences
Strike et al.[121]	27 (27)	60–84 Supplement: 67.7 ± 6.2 Control: 65.7 ± 4.5	Multinutrient supplement including 1 g DHA, 160 mg EPA vs isocalorific oil	26	MOT,[t] VRM,[u] PAL,[v] Executive function test	FO < Placebo MOT response latencies FO > Placebo VRM words remembered No differences in MOT accuracy, PAL words recognized, executive function

[a] *Docosahexaenoic acid (DHA).*
[b] *Wechsler Memory Scale (III) (WAIS-III).*
[c] *Mini Mental State Exam (MMSE).*
[d] *Cambridge Neuropsychological Test Automated Battery (CANTAB).*
[e] *Alzheimer's Disease Cooperative Study-Activities of Daily Living Prevention Instrument (ADCS-ADL PI).*
[f] *Geriatric Depression Scale (GDS).*
[g] *Paired Associative Learning (PAL).*
[h] *Verbal Recognition Memory (VRM).*
[i] *State Trait Anxiety Inventory (STAI).*
[j] *Cognitive Drug Research (CDR) assessment.*
[k] *Trail making test (TMT).*
[l] *Auditory verbal learning task (AVLT).*
[m] *Magnetic Resonance Imaging (MRI).*
[n] *Center for Epidemiologic Studies' Depression Scale (CES-D).*
[o] *Telephone Interview Cognitive Status-Modified (TICS-M).*
[p] *Fish oil (FO).*
[q] *Multinutrient (EM).*
[r] *Cognitive Demand Battery' (CDB).*
[s] *Functional near infrared spectroscopy (NIRS).*
[t] *Psychomotor response latency 'motor screening task' (MOT).*
[u] *Verbal Recognition Memory (VRM).*
[v] *Paired Associates Learning (PAL).*

carriers.[114] However, an equal number of trials have failed to find improvements in cognition after *n*-3 PUFA supplementation.[116,118–120] Such trials included a range of cognitive tasks, participant sample sizes (from 74 to over 3000), *n*-3 PUFA doses (from 250 to 900 mg/day DHA and 60 to 1320 mg/day EPA), and durations (from 90 days to 5 years), suggesting that *n*-3 PUFA supplementation does not reliably benefit cognition in healthy older adults.

RCTS: OLDER ADULTS WITH MILD COGNITIVE IMPAIRMENT OR ALZHEIMER'S DISEASE

Evidence suggests *n*-6 PUFAs exacerbate β-amyloid deposition, a hallmark outcome of Alzheimer's disease, and *n*-3 PUFAs (or low *n*-6 to *n*-3 ratios) may reduce the effects.[122] A number of RCTs have evaluated whether *n*-3 PUFA supplementation results in behavioral changes in mild cognitive impairment and Alzheimer's disease and results are largely inconsistent (Table 8). Among individuals with mild cognitive impairment or similar degrees of cognitive aging, DHA-rich fish oil improved verbal fluency relative to *n*-6-rich oil but not tests of verbal learning, memory, or inhibitory control.[126] Similar doses of fish oil improved working memory and some tests of verbal memory but not other tests of verbal memory, executive function, psychomotor speed, visuospatial skills, or MMSE scores.[127] Fish oil higher in EPA than DHA improved psychomotor speed and working memory but not verbal memory.[133] In individuals with subjective memory impairment, fish oil improved working memory at the highest working memory load but not under less challenging conditions.[131] However, the other researcher found no differences in cognitive decline over 36 months across a number of cognitive domains.[132]

Other research has looked at *n*-3 PUFA intake across a range of levels of cognitive impairment. The *n*-3 PUFA intake lessened the severity of cognitive symptoms in individuals with mild cognitive impairment but not Alzheimer's disease. Further, higher red blood cell membrane EPA was associated with improved cognitive Alzheimer's disease symptoms.[124] However, *n*-3 PUFA intake did not influence MMSE scores or a range of cognitive tasks across individuals with non-dementia cognitive impairment and Alzheimer's disease.[129] The level of cognitive impairment in individuals with Alzheimer's disease may influence the efficacy of *n*-3 PUFA supplementation, as *n*-3 PUFA intake improved memory in individuals with very mild and severe Alzheimer's disease only and ameliorated global cognitive impairment in individuals with very mild Alzheimer's disease.[123] However, other researchers failed to find cognitive effects of *n*-3 PUFA supplementation in individuals with Alzheimer's disease,[125,130] but greater increases in plasma DHA and EPA levels were associated with slower cognitive decline.[130] The *n*-3 PUFA with alpha lipoic acid, a compound with antioxidant and anti-inflammatory actions, improved MMSE scores more than placebo but no differences were found between *n*-3 PFA alone and placebo or for other measures of cognitive decline.[128]

Thus, there is some evidence to suggest that *n*-3 supplementation slows cognitive aging, perhaps most so in individuals with less initial cognitive impairment, or lower likelihood of developing dementia. Indeed, a recent meta-analysis in individuals free of dementia suggests that DHA intake improves certain aspects of memory, notably episodic memory in individuals with memory complaints with no such effects in semantic or working memory, or in individuals with no memory complaints,[134] However, the results are far from conclusive, as

TABLE 8 Randomized Controlled Trials Assessing the Influence on n-3 PUFA Supplementation on Cognitive Function in Older Adults with Mild Cognitive Impairment or Alzheimer's Disease

Authors	Sample *n* (Female)	Age Range (mean ± SD)	*n*-3 PUFA Manipulation	Duration (weeks)	Cognitive Measures	Results
Freund-Levi et al.[123]	204 (110) AD[a]	*n*-3 PUFA: 72.6 ± 9.0, Placebo: 72.9 ± 8.6	*n*-3 PUFA (150 mg EPA, 430 mg DHA) vs corn oil	24 (+ 24 weeks open label *n*-3 for all participants)	MMSE,[b] ADAS-cog,[c] CDR[d]	No differences in MMSE or ADAS-cog across all participants *n*-3 > Control MMSE delayed recall, attention (very mild AD) *n*-3 > Control ADAS-cog delayed recall (severe AD)
Chiu et al.[124]	43 (20) mild-moderate AD (21) or MCI[e] (21)	*n*-3 PUFA: 70.1–77.8 (74.0), Placebo: 71.8–81.1 (76.5)	*n*-3 PUFA (1080 mg EPA, 720 mg DHA) vs olive oil	24	ADAS-cog, CIBIC,[f] HDRS[g]	*n*-3 > Control CIBIC *n*-3 > Control ADAS-cog in MCI, not AD No differences in ADAS-cog, MMSE, HDRS across all participants
Quinn et al.[125]	402 (210) mild-moderate AD	*n*-3 PUFA: 76 ± 9.3, Placebo: 76 ± 7.9	2 g Algal DHA (0.9–1.1 g DHA) vs corn or soy oil	72	MMSE (baseline only), ADAS-cog, CDR, ADCS-ADL,[h] Quality of Life of Alzheimer's Disease Scale	No differences
Sinn et al.[126]	50 (16) MCI	≥ 65 EPA-rich FO:74.88 ± 5.06 DHA-rich FO: 74.22 ± 7 Control: 73 ± 3.96	EPA-rich fish oil (1.67 g EPA, 0.16 g DHA) vs DHA-rich fish oil (1.55 g DHA, 0.4 g EPA) vs safflower oil (2.2 g LA[i])	24	SF-36 Health Survey: health and quality of life, RAVLT,[j] WAIS-III,[k] Trail-Making Task, Stroop Test, Verbal Fluency	EPA, DHA > LA depressive symptoms DHA > LA verbal fluency No differences in quality of life

(*Continued*)

TABLE 8 Randomized Controlled Trials Assessing the Influence on n-3 PUFA Supplementation on Cognitive Function in Older Adults with Mild Cognitive Impairment or Alzheimer's Disease—cont'd

Authors	Sample *n* (Female)	Age Range (mean ± SD)	*n*-3 PUFA Manipulation	Duration (weeks)	Cognitive Measures	Results
Lee et al.[127]	36 (27) Low SES[l] MCI	≥ 60 FO: 66.4 ± 5.1 Control: 63.5 ± 3.0	FO (1.3 g DHA, 0.45 g EPA) vs corn oil	52	MMSE, GDS, memory (Visual reproduction I and II WMS[m] subtests, RAVLT, Digit span backward), executive function and attention (CDT,[n] Digit span forward), psychomotor speed (Digit symbol substitution), and visuospatial skills (Matrix reasoning, Block design)	FO > Placebo Digit span, visual reproduction I, RAVLT delayed recall No differences in MMSE, visual reproduction II, immediate recall, block design, digit symbol substitution, matrix reasoning, CDT, or GDS scores
Shinto et al.[128]	39 (15) AD	≥ 55 *n*-3 PUFA: 75.9 ± 8.1 *n*-3 PUFA + LA: 76.7 ± 10.6 Control: 75.2 ± 10.8	3 g/day FO (675 mg DHA, 975 mg EPA) vs 3 g/day FO + 600 mg ALA[o]) vs soybean oil (with 5% FO)	52	MMSE, ADAS-cog	*n*-3 PUFA + LA > Placebo MMSE No differences *in n*-3 PUFA and placebo for MMSE No differences in ADAS-cog
Phillips et al.[129]	76 (42) (57 CIND,[p] 19 CE)	*n*-3 PUFA: 71.1 ± 8.6 Control: 71.1 ± 9.5	*n*-3 PUFA (625 mg DHA, 600 mg EPA) vs OO	17	MMSE, HVLT-R,[q] BASDEC,[r] tests of executive function, language, verbal reasoning, visual memory, BADLS[s]	No differences
Eriksdotter et al.[130]	165 (86) AD	72.5 ± 8.9	*n*-3 PUFA (1720 mg DHA, 600 mg EPA) vs corn oil	26	MMSE, ADAS-cog	No differences Greater increases in plasma DHA levels associated with slower cognitive decline on ADAS-cog, particularly verbal recall (DHA and EPA) and orientation No associations in plasma *n*-3 FA and MMSE

Boespflug et al.[131]	21 (13) Subjective memory impairment	62–80 FO: 70.1 ± 6.12 Placebo: 66.4 ± 3.75	FO (1.6 g DHA, 0.8 g EPA) vs corn oil	24	N-back working memory, fMRI	FO > Placebo accuracy 2-back (not 0- or 1-back) No differences in RT
Andrieu et al.[132]	1525 (978) Memory complaint, daily living limitation, or slow gait speed	≥ 70 75.3 ± 4.4	Multidomain (cognitive, physical, nutrition training) + Placebo (paraffin oil) vs Multidomain + *n*-3 PUFA (800 mg DHA, 225 mg EPA) vs *n*-3 PUFA (800 mg DHA, 225 mg EPA) vs paraffin oil	36	Free and total recall of the Free and Cued Selective Reminding Test, MMSE orientation items, Digit Symbol Substitution Test, Category Naming Test	No differences
Bo et al.[133]	86 (35) MCI	≥ 60 *n*-3 PUFA: 71.75 ± 5.68 Control: 70.45 ± 6.82	*n*-3 PUFA (480 mg DHA, 720 mg EPA) vs OO	26	BCAT[t]	*n*-3 PUFA > Placebo perceptual speed, space imagery efficiency, working memory, total score No differences in mental arithmetic efficiency, recognition memory

[a] *Alzheimer's disease (AD).*
[b] *Mini Mental State Exam (MMSE).*
[c] *Cognitive portion of Alzheimer's Disease Assessment Scale (ADAS-cog).*
[d] *Clinical Dementia Rating (CDR) Scale.*
[e] *Mild cognitive impairment (MCI).*
[f] *Clinician's Interview-Based Impression of Change (CIBIC) Scale.*
[g] *Hamilton Depression Rating Scale (HDRS).*
[h] *ADCS Activities of Daily Living (ADCS-ADL) Scale.*
[i] *Linoleic acid (LA).*
[j] *Rey Auditory Verbal Learning Test (RAVLT).*
[k] *Wechsler Memory Scale (III) (WAIS-III).*
[l] *Socioeconomic status (SES).*
[m] *Wechsler Memory Scale-Revised (WMS-R).*
[n] *Clock drawing test (CDT).*
[o] *Alpha lipoic acid (ALA).*
[p] *Cognitive impairment no dementia (CIND).*
[q] *Hopkins Verb Learning Test—Revised (HVLT-R).*
[r] *Brief Assessment Schedule Depression Cards (BASDEC).*
[s] *Bristol's Activities of Daily Living Scale (BADLS).*
[t] *Basic Cognitive Aptitude Tests.*

an equal number of trials failed to find benefits of *n*-3 intake. A recent review of the effects of components of the Mediterranean on cognitive aging, including *n*-3 PUFA, summarized similar inconclusive evidence, and pointed to high saturated- and trans- fatty acid intake as a primary impediment to *n*-3 PUFA efficacy in cognitive aging.[135] Thus, further research is necessary to determine whether *n*-3 PUFA supplementation alone may slow cognitive decline or whether its combination reduced saturated-, trans-, and/or *n*-6 polyunsaturated-fatty acid intake yields greater cognitive benefits.

CONCLUSION

Epidemiological research and RCTS have evaluated *n*-3 PUFA intake and cognition across the lifespan with the majority of research focused on either end: at infancy and older adulthood. Indeed, a number of studies have found positive associations between maternal *n*-3 PUFA intake and infant cognitive development, particularly in problem solving, memory, and language development. However, others have found no differences suggesting that although *n*-3 PUFA supplementation may not relate to global cognitive development among infants; it may aid particular cognitive functions. However, these relationships are not reflected in the findings of RCTS, which largely found no effects of maternal and childhood *n*-3 PUFA supplementation on childhood cognition.

The opposite may be true in older adulthood, as prospective and cross-sectional studies suggest that higher *n*-3 PUFA intake and plasma levels are associated with reduced overall cognitive decline with less evidence for specific cognitive domains. The epidemiological results are not entirely supported by RCTs, which generally failed to show cognitive benefits of *n*-3 PUFA intake on cognition in healthy older adults. However, there is some evidence to suggest that *n*-3 PUFA intake slows cognitive decline in individuals with mild cognitive impairment and Alzheimer's disease. Alzheimer's disease risk factors included history of depression and carrying the *APOE ε4* gene may influence *n*-3 PUFAs' efficacy in preventing cognitive decline in older adults. Research in young adults remains fairly limited and although some data suggest positive effects of *n*-3 PUFA supplementation on mood and executive function, other studies have failed to replicate these effects.

Thus, though the evidence to date points to a beneficial influence of *n*-3 PUFAs on cognitive performance across the lifespan, the conclusions remain tenuous, and additional research is necessary to completely determine the dietary fat composition—*n*-3 PUFA but also saturated-, trans-, and/or *n*-6 polyunsaturated-fatty acid intake—is most beneficial cognition at each stage of life, as well as completely understand the mechanism by which *n*-3 PUFAs may influence the brain function.

References

1. Kris-Etherton PM, Taylor DS, Yu-Poth S, et al. Polyunsaturated fatty acids in the food chain in the United States. *Am J Clin Nutr*. 2000;71(1 Suppl). 179S–88S, http://www.ncbi.nlm.nih.gov/pubmed/10617969; (Accessed January 23, 2018).
2. Simopoulos AP. Importance of the omega-6/omega-3 balance in health and disease: evolutionary aspects of diet. *World Rev Nutr Diet*. 2011;102:10–21. https://doi.org/10.1159/000327785.

3. Calder PC. Omega-3 fatty acids and inflammatory processes. *Nutrients*. 2010;2(3):355–374. https://doi.org/10.3390/nu2030355.
4. Giles GE, Mahoney CR, Kanarek RB. Omega-3 fatty acids influence mood in healthy and depressed individuals. *Nutr Rev*. 2013;71(11):727–741. https://doi.org/10.1111/nure.12066.
5. Lin PY, Huang SY, Su KP. A meta-analytic review of polyunsaturated fatty acid compositions in patients with depression. *Biol Psychiatry*. 2010;68(2):140–147. https://doi.org/10.1016/j.biopsych.2010.03.018.
6. Parker G, Gibson NA, Brotchie H, Heruc G, Rees AM, Hadzi-Pavlovic D. Omega-3 fatty acids and mood disorders. *Am J Psychiatry*. 2006;163(6):969–978. https://doi.org/10.1176/appi.ajp.163.6.969.
7. Turnbull T, Cullen-Drill M, Smaldone A. Efficacy of omega-3 fatty acid supplementation on improvement of bipolar symptoms: a systematic review. *Arch Psychiatr Nurs*. 2008;22(5):305–311. https://doi.org/10.1016/j.apnu.2008.02.011.
8. Ross BM. Omega-3 polyunsaturated fatty acids and anxiety disorders. *Prostaglandins, Leukot Essent Fat Acids*. 2009;81(5–6):309–312. https://doi.org/10.1016/J.PLEFA.2009.10.004.
9. Frensham LJ, Bryan J, Parletta N. Influences of micronutrient and omega-3 fatty acid supplementation on cognition, learning, and behavior: methodological considerations and implications for children and adolescents in developed societies. *Nutr Rev*. 2012;70(10):594–610. https://doi.org/10.1111/j.1753-4887.2012.00516.x.
10. Benton D. Neurodevelopment and neurodegeneration: are there critical stages for nutritional intervention? *Nutr Rev*. 2010;68(Suppl 1):https://doi.org/10.1111/j.1753-4887.2010.00324.x. S6–10.
11. Innis SM. Dietary (n-3) fatty acids and brain development. *J Nutr*. 2007;137(4):855–859. http://www.ncbi.nlm.nih.gov/pubmed/17374644; (Accessed January 23, 2018).
12. Daniels JL, Longnecker MP, Rowland AS, Golding J. Fish intake during pregnancy and early cognitive development of offspring. *Epidemiology*. 2004;15(4):394–402. https://doi.org/10.1097/01.ede.0000129514.46451.ce.
13. Hibbeln JR, Davis JM, Steer C, et al. Maternal seafood consumption in pregnancy and neurodevelopmental outcomes in childhood (ALSPAC study): an observational cohort study. *Lancet*. 2007;369(9561):578–585. https://doi.org/10.1016/S0140-6736(07)60277-3.
14. Oken E, Radesky JS, Wright RO, et al. Maternal fish intake during pregnancy, blood mercury levels, and child cognition at age 3 years in a US cohort. *Am J Epidemiol*. 2008;167(10):1171–1181. https://doi.org/10.1093/aje/kwn034.
15. Oken E, Osterdal ML, Gillman MW, et al. Associations of maternal fish intake during pregnancy and breastfeeding duration with attainment of developmental milestones in early childhood: a study from the Danish National Birth Cohort. *Am J Clin Nutr*. 2008;88(3):789–796.
16. Åberg MA, Åberg N, Brisman J, Sundberg R, Winkvist A, Torén K. Fish intake of Swedish male adolescents is a predictor of cognitive performance. *Acta Paediatr*. 2009;98(3):555–560. https://doi.org/10.1111/j.1651-2227.2008.01103.x.
17. Kim JL, Winkvist A, Åberg MAI, et al. Fish consumption and school grades in Swedish adolescents: a study of the large general population. *Acta Paediatr Int J Paediatr*. 2010;99(1):72–77. https://doi.org/10.1111/j.1651-2227.2009.01545.x.
18. Boucher O, Burden MJ, Muckle G, et al. Neurophysiologic and neurobehavioral evidence of beneficial effects of prenatal omega-3 fatty acid intake on memory function at school age. *Am J Clin Nutr*. 2011;93(5):1025–1037. https://doi.org/10.3945/ajcn.110.000323.
19. de Groot RH, Ouwehand C, Jolles J. Eating the right amount of fish: inverted U-shape association between fish consumption and cognitive performance and academic achievement in Dutch adolescents. *Prostaglandins Leukot Essent Fatty Acids*. 2012;86(3):113–117. https://doi.org/10.1016/j.plefa.2012.01.002.
20. Keim SA, Daniels JL, Siega-Riz AM, Herring AH, Dole N, Scheidt PC. Breastfeeding and long-chain polyunsaturated fatty acid intake in the first 4 post-natal months and infant cognitive development: an observational study. *Matern Child Nutr*. 2012;8(4):471–482. https://doi.org/10.1111/j.1740-8709.2011.00326.x.
21. Steer CD, Lattka E, Koletzko B, Golding J, Hibbeln JR. Maternal fatty acids in pregnancy, FADS polymorphisms, and chil intelligence quotient at 8 y of age. *Am J Clin Nutr*. 2013;98(6):1575–1582. https://doi.org/10.3945/ajcn.112.051524.
22. Rees A, Sirois S, Wearden A. Maternal docosahexaenoic acid intake levels during pregnancy and infant performance on a novel object search task at 22 months. *Child Dev*. 2014;85(6):2131–2139. https://doi.org/10.1111/cdev.12280.
23. Strain JJ, Yeates AJ, van Wijngaarden E, et al. Prenatal exposure to methyl mercury from fish consumption and polyunsaturated fatty acids: Associations with child development at 20 months of age in an observational study in the Republic of Seychelles. *Am J Clin Nutr*. 2015;101(3):530–537. https://doi.org/10.3945/ajcn.114.100503.

24. Oken E, Rifas-Shiman SL, Amarasiriwardena C, et al. Maternal prenatal fish consumption and cognition in mid childhood: Mercury, fatty acids, and selenium. *Neurotoxicol Teratol*. 2016;57:71–78. https://doi.org/10.1016/j.ntt.2016.07.001.
25. Fagan JF, Detterman DK. The Fagan test of infant intelligence: a technical summary. *J Appl Dev Psychol*. 1992;13:173–193.
26. Fenson L, Dale PS, Reznick JS, et al. *The MacArthur Communicative Development Inventories: User's Guide and Technical Manual*. San Diego, CA: Singular Publishing Group; 1993.
27. Dunn L, Dunn L. *Examiners Manual for the Peabody Picture Vocabulary Test. III*. Circle Pines, MN: American Guidance Service; 1997.
28. Bayley N. *Bayley Scales of Infant Development*. San Antonio, TX: Psychological Corporation; 1969.
29. Kaufman AS, Kaufman NL. Kaufman Brief Intelligence Test. In: *Encyclopedia of Special Education*. 2nd ed2014:https://doi.org/10.1002/9781118660584.ese1325.
30. Wechsler D. The Wechsler intelligence scale for children. *Wechsler Intell scale Child Ed*. 2003;.
31. Schaefer EJ, Augustin JL, Schaefer MM, et al. Lack of efficacy of a food-frequency questionnaire in assessing dietary macronutrient intakes in subjects consuming diets of known composition. *Am J Clin Nutr*. 2000;71(3):746–751.
32. Helland IB, Saugstad OD, Smith L, et al. Similar effects on infants of n-3 and n-6 fatty acids supplementation to pregnant and lactating women. *Pediatrics*. 2001;108(5):E82.
33. Helland IB, Smith L, Saarem K, Saugstad OD, Drevon CA. Maternal supplementation with very-long-chain n-3 fatty acids during pregnancy and lactation augments children's IQ at 4 years of age. *Pediatrics*. 2003;111(1):e39–e44.
34. Lauritzen L, Jorgensen MH, Olsen SF, Straarup EM, Michaelsen KF. Maternal fish oil supplementation in lactation: effect on developmental outcome in breast-fed infants. *Reprod Nutr Dev*. 2005;45(5):535–547. https://doi.org/10.1051/rnd:2005044.
35. Dunstan JA, Simmer K, Dixon G, Prescott SL. Cognitive assessment of children at age 2(1/2) years after maternal fish oil supplementation in pregnancy: a randomised controlled trial. *Arch Dis Child Fetal Neonatal Ed*. 2008;93(1):F45–F50. https://doi.org/10.1136/adc.2006.099085.
36. Makrides M, Gibson RA, McPhee AJ, et al. Neurodevelopmental outcomes of preterm infants fed high-dose docosahexaenoic acid. *JAMA*. 2009;301(2):175. https://doi.org/10.1001/jama.2008.945.
37. Makrides M, Gibson RA, McPhee AJ, et al. Effect of DHA supplementation during pregnancy on maternal depression and neurodevelopment of young children: a randomized controlled trial. *JAMA*. 2010;304(15):1675–1683. https://doi.org/10.1001/jama.2010.1507.
38. Cheatham CL, Nerhammer AS, Asserhoj M, Michaelsen KF, Lauritzen L. Fish oil supplementation during lactation: effects on cognition and behavior at 7 years of age. *Lipids*. 2011;46(7):637–645. https://doi.org/10.1007/s11745-011-3557-x.
39. Gould JF, Makrides M, Colombo J, Smithers L. Randomized controlled trial of maternal omega-3 long-chain PUFA supplementation during pregnancy and early childhood development of attention, working memory, and inhibitory control. *Am J Clin Nutr*. 2014;99:851–859. https://doi.org/10.3945/ajcn.113.069203.1.
40. Makrides M, Gould JF, Gawlik NR, et al. Four-year follow-up of children born to women in a randomized trial of prenatal DHA supplementation. *JAMA*. 2014;https://doi.org/10.1001/jama.2016.21303.
41. Collins CT, Gibson RA, Anderson PJ, et al. Neurodevelopmental outcomes at 7 years' corrected age in preterm infants who were fed high-dose docosahexaenoic acid to term equivalent: a follow-up of a randomised controlled trial. *BMJ Open*. 2015;5(3):1–12. https://doi.org/10.1136/bmjopen-2014-007314.
42. Meldrum SJ, Dunstan JA, Foster JK, Simmer K, Prescott SL. Maternal fish oil supplementation in pregnancy: a 12 year follow-up of a randomised controlled trial. *Nutrients*. 2015;7(3):2061–2067. https://doi.org/10.3390/nu7032061.
43. Ramakrishnan U, Stinger A, Digirolamo AM, et al. Prenatal docosahexaenoic acid supplementation and offspring development at 18 months: Randomized controlled trial. *PLoS One*. 2015;10(8):1–14. https://doi.org/10.1371/journal.pone.0120065.
44. Hurtado JA, Iznaola C, Pena M, et al. Effects of maternal [OMEGA]-3 supplementation on fatty acids and on visual and cognitive development. *J Pediatr Gastroenterol Nutr*. 2015;61(4):472–480. https://doi.org/10.1097/MPG.0000000000000864.
45. Colombo J, Gustafson KM, Gajewski BJ, et al. Prenatal DHA supplementation and infant attention. *Pediatr Res*. 2016;80(5):656–662. https://doi.org/10.1038/pr.2016.134.

46. Ramakrishnan U, Gonzalez-Casanova I, Schnaas L, et al. Prenatal supplementation with DHA improves attention at 5 y of age: a randomized controlled trial. *Am J Clin Nutr*. 2016;104(4):1075–1082. https://doi.org/10.3945/ajcn.114.101071.
47. Gould JF, Treyvaud K, Yelland LN, et al. Seven-year follow-up of children born to women in a randomized trial of prenatal DHA supplementation. *JAMA*. 2017;317(11):1173. https://doi.org/10.1001/jama.2016.21303.
48. Helland IB, Smith L, Blomen B, Saarem K, Saugstad OD, Drevon CA. Effect of supplementing pregnant and lactating mothers with n-3 very-long-chain fatty acids on children's IQ and body mass index at 7 years of age. *Pediatrics*. 2008;122(2):https://doi.org/10.1542/peds.2007-2762. e472–9.
49. Kiecolt-Glaser JK. Stress, food, and inflammation: psychoneuroimmunology and nutrition at the cutting edge. *Psychosom Med*. 2010;72(4):365–369. https://doi.org/10.1097/PSY.0b013e3181dbf489.
50. Gould JF, Smithers LG, Makrides M. The effect of maternal omega-3 (n-3) LCPUFA supplementation during pregnancy on early childhood cognitive and visual development: a systematic review and meta-analysis of randomized controlled trials. *Am J Clin Nutr*. 2013;97(3):531–544. https://doi.org/10.3945/ajcn.112.045781.
51. Center for Chronic Disease Prevention N, Promotion H, of Nutrition D, Activity P. Breastfeeding Report Card, Progressing Toward National Breastfeeding Goals: United States 2016; 2016. https://www.cdc.gov/breastfeeding/pdf/2016breastfeedingreportcard.pdf. (Accessed February 6, 2018).
52. Scott DT, Janowsky JS, Carroll RE, Taylor JA, Auestad N, Montalto MB. Formula supplementation with long-chain polyunsaturated fatty acids: Are there developmental benefits? *Pediatrics*. 1998;102(5):E59. http://www.ncbi.nlm.nih.gov/pubmed/9794989; (Accessed February 6, 2018).
53. Auestad N, Halter R, Hall RT, et al. Growth and development in term infants fed long-chain polyunsaturated fatty acids: a double-masked, randomized, parallel, prospective, multivariate study. *Pediatrics*. 2001;108(2):372–381.
54. O'Connor DL, Hall R, Adamkin D, et al. Growth and development in preterm infants fed long-chain polyunsaturated fatty acids: a prospective, randomized controlled trial. *Pediatrics*. 2001;108(2):359–371.
55. Auestad N, Scott DT, Janowsky JS, et al. Visual, cognitive, and language assessments at 39 months: a follow-up study of children fed formulas containing long-chain polyunsaturated fatty acids to 1 year of age. *Pediatrics*. 2003;112(3 Pt 1):e177–e183.
56. Henriksen C, Haugholt K, Lindgren M, et al. Improved cognitive development among preterm infants attributable to early supplementation of human milk with docosahexaenoic acid and arachidonic acid. *Pediatrics*. 2008;121(6):1137–1145. https://doi.org/10.1542/peds.2007-1511.
57. Drover JR, Hoffman DR, Castaneda YS, et al. Cognitive function in 18-month-old term infants of the DIAMOND study: a randomized, controlled clinical trial with multiple dietary levels of docosahexaenoic acid. *Early Hum Dev*. 2011;87(3):223–230. https://doi.org/10.1016/j.earlhumdev.2010.12.047.
58. Westerberg AC, Schei R, Henriksen C, et al. Attention among very low birth weight infants following early supplementation with docosahexaenoic and arachidonic acid. *Acta Paediatr*. 2011;100(1):47–52. https://doi.org/10.1111/j.1651-2227.2010.01946.x.
59. Meldrum SJ, D'Vaz N, Simmer K, Dunstan JA, Hird K, Prescott SL. Effects of high-dose fish oil supplementation during early infancy on neurodevelopment and language: a randomised controlled trial. Br J Nutr. 2012;108(8):1443–1454. doi:https://doi.org/10.1017/S0007114511006878; https://doi.org/10.1017/S0007114511006878
60. Almaas AN, Tamnes CK, Nakstad B, et al. Long-chain polyunsaturated fatty acids and cognition in VLBW infants at 8 years: an RCT. *Pediatrics*. 2015;135(6):972–980. https://doi.org/10.1542/peds.2014-4094.
61. Brew BK, Toelle BG, Webb KL, Almqvist C, Marks GB. Omega-3 supplementation during the first 5 years of life and later academic performance: a randomised controlled trial. *Eur J Clin Nutr*. 2015;69(4):419–424. https://doi.org/10.1038/ejcn.2014.155.
62. World Health Organization. Preterm birth. WHO. http://www.who.int/mediacentre/factsheets/fs363/en/. Published 2017. (Accessed February 6, 2018).
63. Kennedy DO, Jackson PA, Elliott JM, et al. Cognitive and mood effects of 8 weeks' supplementation with 400 mg or 1000 mg of the omega-3 essential fatty acid docosahexaenoic acid (DHA) in healthy children aged 10–12 years. *Nutr Neurosci*. 2009;12(2):48–56. https://doi.org/10.1179/147683009X388887.
64. Ryan AS, Nelson EB. Assessing the effect of docosahexaenoic acid on cognitive functions in healthy, preschool children: a randomized, placebo-controlled, double-blind study. *Clin Pediatr (Phila)*. 2008;47(4):355–362. https://doi.org/10.1177/0009922807311730.

65. Muthayya S, Eilander A, Transler C, et al. Effect of fortification with multiple micronutrients and n-3 fatty acids on growth and cognitive performance in Indian schoolchildren: the CHAMPION (Children's Health and Mental Performance Influenced by Optimal Nutrition) study. *Am J Clin Nutr*. 2009;89(6):1766–1775. https://doi.org/10.3945/ajcn.2008.26993.
66. Richardson AJ, Burton JR, Sewell RP, Spreckelsen TF, Montgomery P. Docosahexaenoic acid for reading, cognition and behavior in children aged 7–9 years: a randomized, controlled trial (the DOLAB Study). *PLoS One*. 2012;7(9):https://doi.org/10.1371/journal.pone.0043909. e43909.
67. Parletta N, Cooper P, Gent DN, Petkov J, O'Dea K. Effects of fish oil supplementation on learning and behaviour of children from Australian indigenous remote community schools: a randomised controlled trial. *Prostaglandins Leukot Essent Fat Acids*. 2013;89(2–3):71–79. https://doi.org/10.1016/j.plefa.2013.05.001.
68. Portillo-Reyes V, Pérez-García M, Loya-Méndez Y, Puente AE. Clinical significance of neuropsychological improvement after supplementation with omega-3 in 8-12 years old malnourished Mexican children: a randomized, double-blind, placebo and treatment clinical trial. *Res Dev Disabil*. 2014;35(4):861–870. https://doi.org/10.1016/j.ridd.2014.01.013.
69. Fontani G, Corradeschi F, Felici A, Alfatti F, Migliorini S, Lodi L. Cognitive and physiological effects of Omega-3 polyunsaturated fatty acid supplementation in healthy subjects. *Eur J Clin Invest*. 2005;35(11):691–699. https://doi.org/10.1111/j.1365-2362.2005.01570.x.
70. Antypa N, Van der Does AJ, Smelt AH, Rogers RD. Omega-3 fatty acids (fish-oil) and depression-related cognition in healthy volunteers. *J Psychopharmacol*. 2009;23(7):831–840. https://doi.org/10.1177/0269881108092120.
71. Jackson PA, Deary ME, Reay JL, Scholey AB, Kennedy DO. No effect of 12 weeks' supplementation with 1 g DHA-rich or EPA-rich fish oil on cognitive function or mood in healthy young adults aged 18–35 years. *Br J Nutr*. 2012;107(8):1232–1243. https://doi.org/10.1017/S000711451100403X.
72. Jackson PA, Reay JL, Scholey AB, Kennedy DO. DHA-rich oil modulates the cerebral haemodynamic response to cognitive tasks in healthy young adults: a near IR spectroscopy pilot study. *Br J Nutr*. 2012;107(8):1093–1098. https://doi.org/10.1017/S0007114511004041.
73. Jackson PA, Reay JL, Scholey AB, Kennedy DO. Docosahexaenoic acid-rich fish oil modulates the cerebral hemodynamic response to cognitive tasks in healthy young adults. *Biol Psychol*. 2012;89(1):183–190. https://doi.org/10.1016/j.biopsycho.2011.10.006.
74. Karr JE, Grindstaff TR, Alexander JE. Omega-3 polyunsaturated fatty acids and cognition in a college-aged population. *Exp Clin Psychopharmacol*. 2012;20(3):236–242. https://doi.org/10.1037/a0026945.
75. Dretsch MN, Johnston D, Bradley RS, MacRae H, Deuster PA, Harris WS. Effects of Omega-3 fatty acid supplementation on neurocognitive functioning and mood in deployed U.S. soldiers: a pilot Study. *Mil Med*. 2014;179(4):396–403. https://doi.org/10.7205/MILMED-D-13-00395.
76. Giles GE, Mahoney CR, Urry HL, Brunyé TT, Taylor HA, Kanarek RB. Omega-3 fatty acids and stress-induced changes to mood and cognition in healthy individuals. *Pharmacol Biochem Behav*. 2015;132:10–19. https://doi.org/10.1016/j.pbb.2015.02.018.
77. Stonehouse W, Conlon CA, Podd J, et al. DHA supplementation improved both memory and reaction time in healthy young adults: a randomized controlled trial. *Am J Clin Nutr*. 2013;97(5):1134–1143. https://doi.org/10.3945/ajcn.112.053371.
78. McNamara RK, Able J, Jandacek R, et al. Docosahexaenoic acid supplementation increases prefrontal cortex activation during sustained attention in healthy boys: a placebo-controlled, dose-ranging, functional magnetic resonance imaging study. *Am J Clin Nutr*. 2010;91(4):1060–1067. https://doi.org/10.3945/ajcn.2009.28549.
79. Nyberg L, Lovden M, Riklund K, Lindenberger U, Backman L. Memory aging and brain maintenance. *Trends Cogn Sci*. 2012;16(5):292–305. https://doi.org/10.1016/j.tics.2012.04.005.
80. Britton GB, Rao KS. Cognitive aging and early diagnosis challenges in Alzheimer's disease. *J Alzheimers Dis*. 2011;24(Suppl 2):153–159. https://doi.org/10.3233/JAD-2011-110239.
81. Alzheimer's Association. 2015 Alzheimer's disease facts and figures. *Alzheimers Dement*. 2015;11:332–384. https://doi.org/10.1016/j.jalz.2015.02.003.
82. Reitz C, Tang MX, Schupf N, Manly JJ, Mayeux R, Luchsinger JA. A summary risk score for the prediction of Alzheimer disease in elderly persons. *Arch Neurol*. 2010;67(7):835–841. https://doi.org/10.1001/archneurol.2010.136.
83. Kalmijn S, Feskens EJ, Launer LJ, Kromhout D. Polyunsaturated fatty acids, antioxidants, and cognitive function in very old men. *Am J Epidemiol*. 1997;145(1):33–41.

84. Heude B, Ducimetiere P, Berr C, Study EVA. Cognitive decline and fatty acid composition of erythrocyte membranes—the EVA Study. *Am J Clin Nutr*. 2003;77(4):803–808.
85. Laurin D, Verreault R, Lindsay J, Dewailly E, Holub BJ. Omega-3 fatty acids and risk of cognitive impairment and dementia. *J Alzheimers Dis*. 2003;5(4):315–322.
86. Whalley LJ, Fox HC, Wahle KW, Starr JM, Deary IJ. Cognitive aging, childhood intelligence, and the use of food supplements: possible involvement of n-3 fatty acids. *Am J Clin Nutr*. 2004;80(6):1650–1657.
87. Huang TL, Zandi PP, Tucker KL, et al. Benefits of fatty fish on dementia risk are stronger for those without APOE epsilon4. *Neurology*. 2005;65(9):1409–1414. https://doi.org/10.1212/01.wnl.0000183148.34197.2e.
88. Morris MC, Evans DA, Tangney CC, Bienias JL, Wilson RS. Fish consumption and cognitive decline with age in a large community study. *Arch Neurol*. 2005;62(12):1849–1853. https://doi.org/10.1001/archneur.62.12.noc50161.
89. Beydoun MA, Kaufman JS, Satia JA, Rosamond W, Folsom AR. Plasma n-3 fatty acids and the risk of cognitive decline in older adults: The atherosclerosis risk in communities Study. *Am J Clin Nutr*. 2007;85(4):1103–1111.
90. Dullemeijer C, Durga J, Brouwer IA, et al. N 3 fatty acid proportions in plasma and cognitive performance in older adults. *Am J Clin Nutr*. 2007;86(5):1479–1485.
91. van Gelder BM, Tijhuis M, Kalmijn S, Kromhout D. Fish consumption, n-3 fatty acids, and subsequent 5-y cognitive decline in elderly men: the Zutphen elderly Study. *Am J Clin Nutr*. 2007;85(4):1142–1147.
92. Samieri C, Feart C, Letenneur L, et al. Low plasma eicosapentaenoic acid and depressive symptomatology are independent predictors of dementia risk. *Am J Clin Nutr*. 2008;88(3):714–721.
93. Whalley LJ, Deary IJ, Starr JM, et al. N-3 fatty acid erythrocyte membrane content, APOE varepsilon4, and cognitive variation: an observational follow-up study in late adulthood. *Am J Clin Nutr*. 2008;87(2):449–454.
94. Devore EE, Grodstein F, van Rooij FJ, et al. Dietary intake of fish and omega-3 fatty acids in relation to long-term dementia risk. *Am J Clin Nutr*. 2009;90(1):170–176. https://doi.org/10.3945/ajcn.2008.27037.
95. Kroger E, Verreault R, Carmichael PH, et al. Omega-3 fatty acids and risk of dementia: the Canadian Study of Health and Aging. *Am J Clin Nutr*. 2009;90(1):184–192. https://doi.org/10.3945/ajcn.2008.26987.
96. van de Rest O, Spiro 3rd A, Krall-Kaye E, Geleijnse JM, de Groot LC, Tucker KL. Intakes of (n-3) fatty acids and fatty fish are not associated with cognitive performance and 6-year cognitive change in men participating in the Veterans Affairs Normative Aging Study. *J Nutr*. 2009;139(12):2329–2336. https://doi.org/10.3945/jn.109.113647.
97. Gonzalez S, Huerta JM, Fernandez S, Patterson AM, Lasheras C. The relationship between dietary lipids and cognitive performance in an elderly population. *Int J Food Sci Nutr*. 2010;61(2):217–225. https://doi.org/10.3109/09637480903348098.
98. Roberts RO, Cerhan JR, Geda YE, et al. Polyunsaturated fatty acids and reduced odds of MCI: the Mayo clinic study of aging. *J Alzheimers Dis*. 2010;21(3):853–865. https://doi.org/10.3233/JAD-2010-091597.
99. Gao Q, Niti M, Feng L, Yap KB, Ng TP. Omega-3 polyunsaturated fatty acid supplements and cognitive decline: Singapore longitudinal aging studies. *J Nutr Health Aging*. 2011;15(1):32–35.
100. Milte CM, Sinn N, Street SJ, Buckley JD, Coates AM, Howe PR. Erythrocyte polyunsaturated fatty acid status, memory, cognition and mood in older adults with mild cognitive impairment and healthy controls. *Prostaglandins Leukot Essent Fatty Acids*. 2011;84(5–6):153–161. https://doi.org/10.1016/j.plefa.2011.02.002.
101. Chiu CC, Frangou S, Chang CJ, et al. Associations between n-3 PUFA concentrations and cognitive function after recovery from late-life depression. *Am J Clin Nutr*. 2012;95(2):420–427. https://doi.org/10.3945/ajcn.111.015784.
102. Ronnemaa E, Zethelius B, Vessby B, Lannfelt L, Byberg L, Kilander L. Serum fatty-acid composition and the risk of Alzheimer's disease: a longitudinal population-based study. *Eur J Clin Nutr*. 2012;66(8):885–890. https://doi.org/10.1038/ejcn.2012.63.
103. Ammann EM, Pottala JV, Harris WS, et al. Omega-3 fatty acids and domain-specific cognitive aging: secondary analyses of data from WHISCA. *Neurology*. 2013;81(17):1484–1491. https://doi.org/10.1212/WNL.0b013e3182a9584c.
104. Leckie RL, Manuck SB, Bhattacharjee N, Muldoon MF, Flory JM, Erickson KI. Omega-3 fatty acids moderate effects of physical activity on cognitive function. *Neuropsychologia*. 2014;59(1):103–111. https://doi.org/10.1016/j.neuropsychologia.2014.04.018.
105. Danthiir V, Hosking D, Burns NR, et al. Cognitive performance in older adults is inversely associated with fish consumption but not erythrocyte membrane n-3 fatty acids. *J Nutr*. 2014;144(3):311–320. https://doi.org/10.3945/jn.113.175695.

106. Baierle M, Vencato PH, Oldenburg L, et al. Fatty acid status and its relationship to cognitive decline and homocysteine levels in the elderly. *Nutrients*. 2014;6(9):3624–3640. https://doi.org/10.3390/nu6093624.
107. Otsuka R, Tange C, Nishita Y, et al. Serum docosahexaenoic and eicosapentaenoic acid and risk of cognitive decline over 10 years among elderly Japanese. *Eur J Clin Nutr*. 2014;68(4):503–509. https://doi.org/10.1038/ejcn.2013.264.
108. Daiello LA, Gongvatana A, Dunsiger S, Cohen RA, Ott BR. Association of fish oil supplement use with preservation of brain volume and cognitive function. *Alzheimers Dement*. 2015;11(2):226–235. https://doi.org/10.1016/j.jalz.2014.02.005.
109. Nishihira J, Tokashiki T, Higashiuesato Y, et al. Associations between serum Omega-3 fatty acid levels and cognitive functions among community-dwelling octogenarians in Okinawa, Japan: The KOCOA Study. *J Alzheimers Dis*. 2016;51(3):857–866. https://doi.org/10.3233/JAD-150910.
110. Del Brutto OH, Mera RM, Gillman J, Zambrano M, Ha J. Eun. Oily fish intake and cognitive performance in community-dwelling older adults: The Atahualpa project. *J Community Health*. 2016;41(1):82–86. https://doi.org/10.1007/s10900-015-0070-9.
111. van de Rest O, Wang Y, Barnes LL, Tangney C, Bennett DA, Morris MC. APOE ε4 and the associations of seafood and long-chain omega-3 fatty acids with cognitive decline. *Neurology*. 2016;86(22):2063–2070. https://doi.org/10.1212/WNL.0000000000002719.
112. Folstein MF, Folstein SE, Mchugh PR. "Mini-Mental State" a practical method for grading the cognitive state of patients for the clinician. J Psychiatr Res. 1975;12(3):189–198. http://home.uchicago.edu/~tmurray1/research/articles/printed and read/mini mental state_a practical method for grading the cognitive state of patients for the clinician.pdf. Accessed March 1, 2018.
113. Johnson EJ, McDonald K, Caldarella SM, Chung HY, Troen AM, Snodderly DM. Cognitive findings of an exploratory trial of docosahexaenoic acid and lutein supplementation in older women. *Nutr Neurosci*. 2008;11(2):75–83. https://doi.org/10.1179/147683008X301450.
114. van de Rest O, Geleijnse JM, Kok FJ, et al. Effect of fish oil on cognitive performance in older subjects: a randomized, controlled trial. *Neurology*. 2008;71(6):430–438. https://doi.org/10.1212/01.wnl.0000324268.45138.86.
115. Yurko-Mauro K, McCarthy D, Rom D, et al. Beneficial effects of docosahexaenoic acid on cognition in age-related cognitive decline. *Alzheimers Dement*. 2010;6(6):456–464. https://doi.org/10.1016/j.jalz.2010.01.013.
116. Stough C, Downey L, Silber B, et al. The effects of 90-day supplementation with the omega-3 essential fatty acid docosahexaenoic acid (DHA) on cognitive function and visual acuity in a healthy aging population. *Neurobiol Aging*. 2012;33(4):https://doi.org/10.1016/j.neurobiolaging.2011.03.019. 824.e1–824.e3.
117. Witte AV, Kerti L, Hermannstädter HM, et al. Long-chain omega-3 fatty acids improve brain function and structure in older adults. *Cereb Cortex*. 2014;24(11):3059–3068. https://doi.org/10.1093/cercor/bht163.
118. Chew EY, Clemons TE, Agrón E, Launer LJ, Grodstein F, Bernstein PS. Effect of omega-3 fatty acids, lutein/zeaxanthin, or other nutrient supplementation on cognitive function: The AREDS2 randomized clinical trial. *JAMA J Am Med Assoc*. 2015;314(8):791–801. https://doi.org/10.1001/jama.2015.9677.
119. Pase MP, Grima N, Cockerell R, et al. The effects of long-chain Omega-3 fish oils and multivitamins on cognitive and cardiovascular function: a randomized, controlled clinical trial. *J Am Coll Nutr*. 2015;34(1):21–31. https://doi.org/10.1080/07315724.2014.880660.
120. Jackson PA, Forster JS, Gordon Bell J, Dick JR, Younger I, Kennedy DO. Dha supplementation alone or in combination with other nutrients does not modulate cerebral hemodynamics or cognitive function in healthy older adults. *Nutrients*. 2016;8(2):https://doi.org/10.3390/nu8020086.
121. Strike SC, Carlisle A, Gibson EL, Dyall SC. A high Omega-3 fatty acid multinutrient supplement benefits cognition and mobility in older women: a randomized, double-blind, placebo-controlled pilot Study. *J Gerontol Ser A Biol Sci Med Sci*. 2016;71(2):236–242. https://doi.org/10.1093/gerona/glv109.
122. Corsinovi L, Biasi F, Poli G, Leonarduzzi G, Isaia G. Dietary lipids and their oxidized products in Alzheimer's disease. *Mol Nutr Food Res*. 2011;55(Suppl 2):https://doi.org/10.1002/mnfr.201100208. S161–72.
123. Freund-Levi Y, Eriksdotter-Jonhagen M, Cederholm T, et al. Omega-3 fatty acid treatment in 174 patients with mild to moderate Alzheimer disease: OmegAD study: a randomized double-blind trial. *Arch Neurol*. 2006;63(10):1402–1408. https://doi.org/10.1001/archneur.63.10.1402.
124. Chiu CC, Su KP, Cheng TC, et al. The effects of omega-3 fatty acids monotherapy in Alzheimer's disease and mild cognitive impairment: a preliminary randomized double-blind placebo-controlled study. *Prog Neuropsychopharmacol Biol Psychiatry*. 2008;32(6):1538–1544. https://doi.org/10.1016/j.pnpbp.2008.05.015.

125. Quinn JF, Raman R, Thomas RG, et al. Docosahexaenoic acid supplementation and cognitive decline in Alzheimer disease: a randomized trial. *JAMA*. 2010;304(17):1903–1911. https://doi.org/10.1001/jama.2010.1510.
126. Sinn N, Milte CM, Street SJ, et al. Effects of n-3 fatty acids, EPA v. DHA, on depressive symptoms, quality of life, memory and executive function in older adults with mild cognitive impairment: a 6-month randomised controlled trial. *Br J Nutr*. 2012;107(11).
127. Lee LK, Shahar S, Chin AV, Yusoff NA. Docosahexaenoic acid-concentrated fish oil supplementation in subjects with mild cognitive impairment (MCI): a 12-month randomised, double-blind, placebo-controlled trial. *Psychopharmacology (Berl)*. 2013;225(3):605–612. https://doi.org/10.1007/s00213-012-2848-0.
128. Shinto L, Quinn J, Montine T, et al. A randomized placebo-controlled pilot trial of omega-3 fatty acids and alpha lipoic acid in Alzheimer's disease. J Alzheimers Dis. 2014;38(1):111–120. doi:https://doi.org/10.3233/JAD-130722 [doi]
129. Phillips MA, Childs CE, Calder PC, Rogers PJ. No effect of omega-3 fatty acid supplementation on cognition and mood in individuals with cognitive impairment and probable Alzheimer's disease: a randomised controlled trial. *Int J Mol Sci*. 2015;16(10):24600–24613. https://doi.org/10.3390/ijms161024600.
130. Eriksdotter M, Vedin I, Falahati F, et al. Plasma fatty acid profiles in relation to cognition and gender in Alzheimer's disease patients during Oral Omega-3 fatty acid supplementation: The OmegAD Study. *J Alzheimers Dis*. 2015;48(3):805–812. https://doi.org/10.3233/JAD-150102.
131. Boespflug EL, McNamara RK, Eliassen JC, Schidler MD, Krikorian R. Fish oil supplementation increases event-related posterior cingulate activation in older adults with subjective memory impairment. *J Nutr Health Aging*. 2016;20(2):161–169. https://doi.org/10.1007/s12603-015-0609-6.
132. Andrieu S, Guyonnet S, Coley N, et al. Effect of long-term omega 3 polyunsaturated fatty acid supplementation with or without multidomain intervention on cognitive function in elderly adults with memory complaints (MAPT): a randomised, placebo-controlled trial. *Lancet Neurol*. 2017;16(5):377–389. https://doi.org/10.1016/S1474-4422(17)30040-6.
133. Bo Y, Zhang X, Wang Y, et al. The n-3 polyunsaturated fatty acids supplementation improved the cognitive function in the Chinese elderly with mild cognitive impairment: a double-blind randomized controlled trial. *Nutrients*. 2017;9(1):1–11. https://doi.org/10.3390/nu9010054.
134. Yurko-Mauro K, Alexander DD, Van Elswyk ME. Docosahexaenoic acid and adult memory: A systematic review and meta-analysis. *PLoS Biol*. 2016;10(3):e0120391. https://doi.org/10.1371/journal.pone.0120391.
135. Huhn S, Kharabian Masouleh S, Stumvoll M, Villringer A, Witte AV. Components of a Mediterranean diet and their impact on cognitive functions in aging. *Front Aging Neurosci*. 2015;7:132. https://doi.org/10.3389/fnagi.2015.00132.

CHAPTER

21

Link Between Sugar Intake and Fatty Acid Metabolism in Brain

Tomislav Mašek

Department of Animal Nutrition and Dietetics, Faculty of Veterinary Medicine, University of Zagreb, Zagreb, Croatia

INTRODUCTION

Over the course of human history, dietary carbohydrates consisted primarily of starch. However, the availability of sugar has greatly increased and average sugar intake has been increasing since the 18th century until today. The average sugar (sucrose) intake per individual was 1.8 kg in 1700, with an increase during the 18th and 19th centuries and after the introduction of high-fructose corn syrup in the 1970s, sugar intake drastically increased to 67.6 kg in 2000.[1]

Over recent decades, the relation between sugar intake and metabolic disorders has been intensively investigated in the scientific community and has simultaneously received a great deal of attention in public. Especially the consumption of sugar-sweetened beverages has emerged as a possible contributor to the development of insulin resistance, obesity, and diabetes type 2. Although the molecular mechanisms may not yet be completely understood, the parallel epidemic of diabetes mellitus type 2 and sugar consumption should be considered as relevant.[1,2] Nevertheless, additional studies are required to resolve completely the remaining questions, including risk assessment of the sugar levels commonly-consumed[3] as well as sugar consumption in relation to individual differences in nutrient metabolism.[4]

Simple sugars, glucose, fructose, and sucrose are commonly used for dietary induction of hyperglycemia in animal translational models for diabetes or metabolic syndrome. The protocols significantly differ in the amounts used and route of administration (water or feed), as well as in the combination with other nutrients (high fat diet) or pharmacological treatments (low dosage of streptozotocin). The diversity of protocols is necessary to mimic the complexity of pathogeneses of diabetes and metabolic syndrome in humans.

Omega Fatty Acids in Brain and Neurological Health
https://doi.org/10.1016/B978-0-12-815238-6.00021-3

GLUCOSE AND FRUCTOSE: SIMILARITIES AND DIFFERENCES

The sweeteners most consumed today include sucrose and high-fructose corn syrup (HFCS). These sugars differ in the ratio of fructose to glucose (sucrose, 50:50; HFCS, 55:42) and monosaccharides are "free" in HFCS-55 while they are "bound" within the sucrose.[5] High levels of fructose are nowadays considered responsible for the high incidence of metabolic syndrome in modern people consuming a "western diet," high in saturated fat and simple sugars,[2,6,7] although there is still ongoing debate.[3,8]

There are several key differences between fructose and glucose, which influence their metabolism (Fig. 1). The enzymes involved in glucose metabolism, glucokinase, and phosphofructokinase are regulated by downstream metabolites, which prevent excessive phosphorylation. In contrast, the key enzyme for fructose metabolism is fructokinase, which is poorly regulated and will phosphorylate all fructose rapidly with the depletion of ATP, which can cause a decrease in protein synthesis and pro-inflammatory effects.[9,10] Fructose is converted to fatty acids in the liver at a greater rate than glucose; it stimulates triglyceride synthesis and increases the fat deposition in the liver.[11,12] An additional characteristic of fructose is that it upregulates its transporter (Glut 5) and fructokinase by a positive feedback system.[13,14] Insulin and hepatic energy needs regulate the hepatic glucose metabolism, which allows much of the ingested glucose, arriving via the portal vein to bypass the liver and reach the systemic circulation.[3] In contrast, fructose does not induce insulin release and its metabolism is not regulated by hepatic energy status. Therefore, most ingested fructose is metabolized in the liver and very little reaches the systemic circulation.[15]

ANIMAL MODELS AND LIPOGENESIS

The fatty acids present in the living organism can originate from food or they can be de novo synthesized in lipogenic tissues (mainly the liver and adipose tissue depending on the species). In addition, fatty acids, by a series of steps of desaturation and elongation, can be bioconverted to longer-chain fatty acids with more double bonds, or shortened by β-oxidation steps and recycled between peroxisomes and the endoplasmatic reticulum.[16,17] Regulation of these steps involves different desaturation and elongation enzymes and transcriptional factors, as well as substrate availability, competition for enzymes, and partitioning into oxidation.[18–20]

The investigation of lipogenesis during metabolic disorders in animal models is challenging due to the highly complex regulation of lipogenesis, as well as the diversity in animal species and strains used. An additional challenge is the diversity of nutritional and

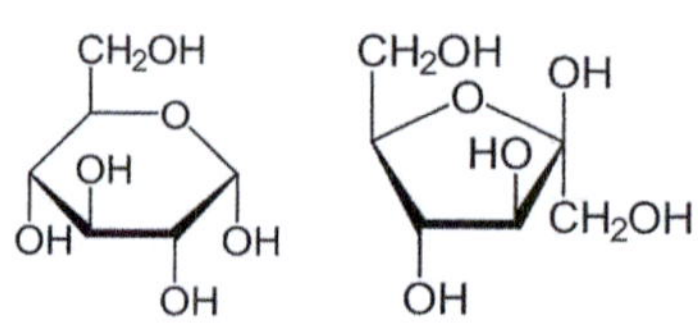

FIG. 1 Glucose (left) and fructose molecule (right).

pharmacological interventions used to induce diabetes type 1 (DM1) or 2 (DM2) or metabolic syndrome (MS). Today, different rodent models are available to scientists for investigation of lipogenesis in DM1, DM2, and MS. These models give us an insight into the changes in the fatty acid profile of different organs during the progression of these diseases. The most intensively investigated is liver lipogenesis, while there have been considerably fewer trials on the brain fatty acid profile. Considering the fact that the liver is the primary site of PUFA production for extrahepatic tissues such as the brain, differences in the liver metabolism of important fatty acids (e.g., DHA) in DM1 and DM2 or MS could have significant nutritional and clinical implications for brain health.

The brain is an organ with high lipid content consisting of complex lipids: glycerophospholipids, sphingolipids (sphingomyelin and cerebrosides), gangliosides, and cholesterol. Moreover, the brain possesses a unique polyunsaturated fatty acid (PUFA) profile with the highest proportion belonging to two important fatty acids: arachidonic acid (ARA, 20:4n6) and docosahexaenoic acid (DHA, 22:6n3). These fatty acids, together with their bioactive derivatives (DHA and ARA-derived mediators) regulate the important processes, such as neurotransmission, cell survival and neuroinflammation,[21] and endocannabinoid system.[22]

The enzymes necessary for fatty acid synthesis are expressed in the brain and saturated and monounsaturated fatty acids can be synthesized de novo within the brain. In contrast, only small amounts of linoleic and linolenic acid are converted to ARA and DHA in the brain.[23,24] In n3 deprived rats, conversion of 18:3n3 to DHA is low in rat brains and is similar in normal and n3 deprived rats,[25] which means that PUFA synthesis in the brain is not responsive to low dietary n3 supply, in contrast to the liver. These data suggest that the brain depends on a constant supply of ARA and DHA from the blood.[21]

HYPERGLYCEMIA INDUCED BY EXPERIMENTAL DIABETES TYPE 1 AND BRAIN FATTY ACIDS

The insulin-dependent diabetes mellitus (type 1 diabetes, DM1) is a chronic disease caused by deficiency in insulin production due to the autoimmune destruction of the pancreatic beta cells.[26] In DM1, changes in lipogenesis and fatty acid concentrations in different tissues occur due to the insulin deficiency. The brain is not an exception, although the brain tissue is the most resistant to changes in the fatty acid composition in cases of DM1.[27] One of the important features of DM1 is the decrease in the hepatic content of palmitic acid and de novo lipogenesis due to lack of insulin action. [28–30] Nevertheless, it is common finding that palmitic acid increases in the brain's total phospholipids,[31] or specifically phosphatidylcholine.[27]

Tissues differ in their ability to retain the ARA content in phospholipids in DM1. While the liver shows a rapid decline in ARA content, the brain retains normal ARA content for a longer period.[27] Several studies reported a complete lack of effect of streptozotocin induced DM1 on brain ARA values.[27,32] Nevertheless, other studies observed a decrease in ARA values.[33–35] Keeping in mind that depletion of brain ARA is evidently time dependent, the absence of ARA decrease in short-term studies could be expected. The rate of ARA synthesis within the brain is very low,[24] moreover, typical rodent feed has a very low ARA content which demonstrates the role of the liver in maintaining the brain ARA content in low Δ5-desaturase diabetic conditions. The decrease in Δ5-desaturation as an important reason for ARA decrease could specifically be seen in the increase in its precursors, linoleic and dihomo-gamma-linolenic acids, which is a regular finding in rat models of DM1.[36] In aged rats suffering from DM1, the brain content of ARA content further decreases,[31] probably in relation to the age-related decline in desaturase activity.[37] Another interesting aspect of ARA metabolism in DM1 is the partitioning of phospholipids into neutral lipids [29,36,38]. Whether similar partitioning is present in the brain and especially the possible biological effects are currently not known.

The brain content of DHA is not usually decreased during DM1, or it is even increased,[33] which may also be seen in the liver [39,40]. In addition, DHA content in the rat brain during DM1 could also depend on n6/n3 ratio.[36] In the long-term DM1 studies on rats, the content of DHA in the brain could decrease but only after its content starts to decrease in the liver (12 months).[31] This could mean that the DHA preserving mechanism in the brain is more susceptible to age than to other factors present in short-term diabetes, such as desaturation, elongation or differences in n6/n3 ratio.

HIGH SUGAR INTAKE AND BRAIN FATTY ACIDS

The consequences of diabetes mellitus type 2 (DM2) or metabolic syndrome (MS) on the fatty acid profile of different organs are much more complex compared to DM1. Initial studies with glucose or fructose in the diet have mostly shown the opposite influence on desaturases and consequently on the liver fatty acid composition, compared to DM1. The most consistent changes in the liver, as a source of the brain's fatty acids, are an increase in desaturase activity and in the content of oleic acid, and a decrease in the content of linoleic acid and DHA in the liver and plasma.[28,39–43] After the introduction of high fat diets as an experimental model, a difference was observed between high fat and high carbohydrate diets. Comparison of high fat and high carbohydrate diets reveals that a high carbohydrate diet leads to more intense lipid deposition, particularly of saturated fatty acids and monounsaturated fatty acids, as well as to higher de novo lipogenesis and Δ9-desaturase and Δ6-desaturase activities.[44] Consequently, the high carbohydrate diet group exhibits low levels of n-3 and n-6 PUFA, due to the more intense reduction of linoleic acid, eicosapentaenoic acid (EPA), and DHA levels during short-term trials (56 days).[44] The introduction of a high-fat low-streptozotocin model introduced an additional variable into the investigation of liver lipogenesis. Using that model, decreased desaturase expression along with increased liver DHA content could be found,_ENREF_37[45] which resembles the results found in DM1. These data point to the conclusion that the fatty acid profile of the brain could also vary, depending on the experimental model.

Consistent with the slow response of the brain to dietary manipulations, short-term trials with fructose in the drinking water or food result in weak and variable effects on the brain fatty acid profile. During a 6-week trial, rodents were supplemented with 15% fructose via their drinking water without any influence on the brain fatty acid profile.[46] Nevertheless, the authors found that n-3 deficiency increases vulnerability to the effects of fructose. In an investigation with very high fructose content (50 g per 100 mL water, duration 6 weeks) a significant increase in the content of ARA and DHA could be seen in free fatty acids.[47]

There has been limited investigation of the long-term effects of high fructose or sucrose dietary content on the brain's fatty acid profile. A long-term investigation with sucrose overload (20 weeks) showed increased brain weight, brain total lipid content and total fatty acid content[48] The insulin influence on lipogenic genes is mediated by the sterol regulatory element binding protein 1c (SREBP-1c).[49] The expression of SREBP-1c may significantly increase in the brain tissue after chronic sucrose treatment.[48] Although the brain SREBP-1c is involved in the lipogenesis and homeostasis of fatty acids within the brain, the SREBP-1c increase due to the higher nutritional conditions in the liver tissue compared to the brain.[50] Consequently, a much higher lipogenic capacity is found in the liver and adipose tissue of rodents than in the brain.[51] The brain fatty acid uptake and accumulation increase in metabolic syndrome.[52] Therefore, the increase in brain lipid content during chronic treatment is probably the result of the increased quantity of lipids that enter the brain. In rodent streptozotocin diabetes, the expression of SREBP1c is decreased in the liver but increased in kidney tissue indicating tissue specific regulation by hyperglycemia and/or insulin deficiency.[53,54] Increased expression of renal SREBP-1 induces ectopic lipid synthesis and the expression of profibrotic growth factors, which contributes to the kidney damage.[54] The regulation of SREBP1c expression in the brain and its possible influence on brain pathology should be further evaluated.

The brain fatty acid profile is characteristic and it contains high concentrations of DHA and ARA, which are accreted during brain development. This fatty acid profile is preserved in the adult brain due to the specific metabolism of fatty acids (Fig. 2). The fatty acids are transported as unesterified fatty acids or as lysophospholipids and lipoproteins and after entering the brain, DHA and ARA are esterified to phospholipids, while other PUFAs are β-oxidized [21]. After 20 weeks of sucrose intake, the content of n6 fatty acids decreases, whereas the n3 line fatty acids is not affected.[48] A decrease in n6 fatty acids is visible in different phospholipid classes, including phosphatidylethanolamine, phosphatidylcholine, and cardiolipin. The fatty acid composition of cardiolipin is organ-specific and the most abundant fatty acid in the heart and liver is linoleic acid, while oleic acid is predominant in the brain.[55,56] Cardiolipin is a constituent of the mitochondrial membrane with the highest abundance in the inner membrane,[57] which emphasizes its importance. It is interesting that a decrease in the linoleic acid content of cardiolipin may already be observed in the early stage of diabetes.[58]

During chronic sucrose intake, the content of n3 fatty acids could remain constant, which significantly decreases the n6/n3 ratio and the content of ARA relative to DHA. The liver's capacity for DHA synthesis is much higher than the brain's, and it can be upregulated by a deficient diet, which indicates that the brain's DHA originates from the circulation.[25] High sucrose intake trials with different duration have shown perturbations in lipogenesis with a decrease in liver and serum ARA and a constant or increased content of DHA.[41–43] The decrease in brain n6 fatty acids are the consequence of the decreased content of linoleic acid and other n6 fatty acids in the hepatic tissue and circulation. In these conditions ARA and DHA have different behavior, which could be explained by the different synthesis of these fatty acids. Meanwhile, ARA is relatively simply synthesized (delta-6-desaturation, one elongation and delta-5-desaturation), and the synthesis of DHA involves desaturation and elongation and also one cycle of peroxisomal β-oxidation.[16] The functional significance of the n6 fatty acid decrease in the brain caused by chronic sucrose intake is not known. A trial performed on growing Δ-6-desaturase knockout and wild-type mice demonstrated the essential nature of the combination of ARA and DHA, because only supplementation with both acids sustained

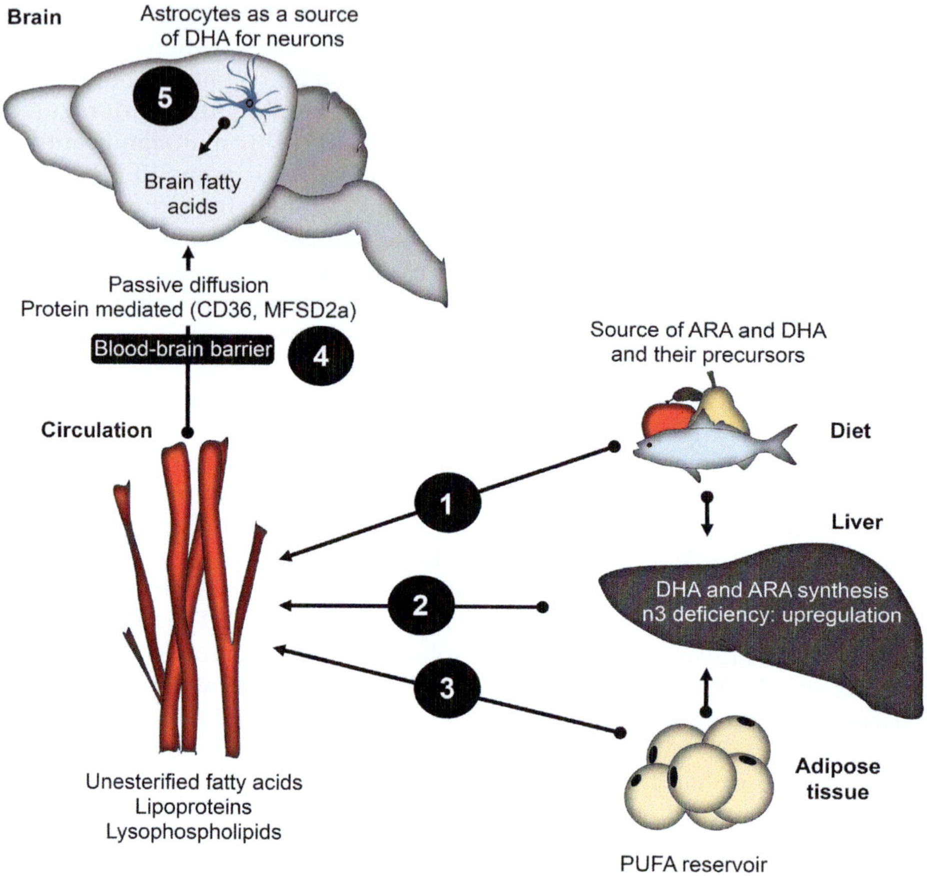

FIG. 2 The influence of hyperglycemia in DM1 and high sugar intake on the fatty acid profile of the brain. (1) Diet is a source of DHA and ARA that can be efficiently used by the brain, as well as DHA and ARA precursors (linolenic and linoleic fatty acids) that can be converted into ARA and DHA in the liver and fat tissue. The deficiency in n3 fatty acids increases the vulnerability to the effects of fructose used to induce metabolic syndrome.[46] (2) The liver tissue can synthesize DHA and ARA from its precursors and liver is important source of the brain's DHA in rats fed the adequate. In n3 deficient rats, liver synthesis of DHA is upregulated.[25] During chronic sucrose intake liver is depleted of linoleic and linolenic acid, which influences brain fatty acid profile.[41,42] In hyperglycemia caused by experimental DM1 liver fatty acid profile is quickly changed due to desaturation inhibition, while the brain is affected after longer period.[27,31] (3) Adipose tissue can synthesize DHA and ARA from its precursors, and it serves as a reservoir for both DHA and ARA and its precursors for the brain.[21,60] Similar to the liver, fat tissue is also depleted in linoleic and linolenic acid after chronic sucrose treatment which could influence brain fatty acid profile.[42,48] (4) Fatty acids can pass through blood–brain barrier either by passive diffusion or by protein mediated uptake.[21,61] The expression of different proteins involved in the transport of fatty acids across the blood brain barrier and possible modulation by hyperglycemia could have influence on the fatty acid profile of the brain. (5) Astrocytes can produce DHA from precursor and deliver it to neurons.[62]

normal growth and motor activity.[59] In that study, only the n6/n3 ratio in a narrow range (0.55) obtained optimal results, while other ratios failed to achieve results comparable to the wild-type mice. These data verify the significance of the n6/n3 ratio in addition to the n6 and n3 fatty acid content.

CONCLUSIONS

Compared to other organs, the rodent brain tissue is more resistant to changes in fatty acid profile during hyperglycemia caused by experimental diabetes or high sugar intake used for dietary induction of diabetes or metabolic syndrome. Therefore, the magnitude of the changes in fatty acid composition in the brain is smaller and it occurs much later than in other organs. When the changes occur, they are variable depending on numerous factors including the animal species and strain, as well as diet characteristics, such as the carbohydrate used and the duration of the dietary treatment. The importance of the complexity of various rodent models is that they give us an insight into the changes to the fatty acid profile and lipogenesis in the brain during the progression of different metabolic diseases and mimic the complex origin of metabolic diseases in humans.

References

1. Johnson RJ, Segal MS, Sautin Y, et al. Potential role of sugar (fructose) in the epidemic of hypertension, obesity and the metabolic syndrome, diabetes, kidney disease, and cardiovascular disease. *Am J Clin Nutr.* 2007;86(4):899–906.
2. Johnson RJ, Perez-Pozo SE, Sautin YY, et al. Hypothesis: Could excessive fructose intake and uric acid cause type 2 diabetes? *Endocr Rev.* Feb 2009;30(1):96–116.
3. Stanhope KL. Sugar consumption, metabolic disease and obesity: The state of the controversy. *Crit Rev Clin Lab Sci.* 2016;53(1):52–67.
4. Archer E. In defense of sugar: a critique of diet-centrism. *Prog Cardiovasc Dis.* 2018;61(1):10–19.
5. Hsu TM, Konanur VR, Taing L, et al. Effects of sucrose and high fructose corn syrup consumption on spatial memory function and hippocampal neuroinflammation in adolescent rats. *Hippocampus.* 2015;25(2):227–239.
6. Ferder L, Ferder MD, Inserra F. The role of high-fructose corn syrup in metabolic syndrome and hypertension. *Curr Hypertens Rep.* 2010;12(2):105–112.
7. Miller A, Adeli K. Dietary fructose and the metabolic syndrome. *Curr Opin Gastroenterol.* 2008;24(2):204–209.
8. Rippe JM, Angelopoulos TJ. Sucrose, high-fructose corn syrup, and fructose, their metabolism and potential health effects: what do we really know? *Adv Nutrition.* 2013;4(2):236–245.
9. Maenpaa PH, Raivio KO, Kekomaki MP. Liver adenine nucleotides: fructose-induced depletion and its effect on protein synthesis. *Science (New York, NY).* 1968;161(3847):1253–1254.
10. Cirillo P, Gersch MS, Mu W, et al. Ketohexokinase-dependent metabolism of fructose induces proinflammatory mediators in proximal tubular cells. *J Am Soc Nephrol: JASN.* Mar 2009;20(3):545–553.
11. Hirahatake KM, Meissen JK, Fiehn O, Adams SH. Comparative effects of fructose and glucose on Lipogenic gene expression and intermediary metabolism in HepG2 liver cells. *PLoS One.* 2011;6(11):e26583.
12. Stanhope KL, Havel PJ. Fructose consumption: Potential mechanisms for its effects to increase visceral adiposity and induce dyslipidemia and insulin resistance. *Curr Opin Lipidol.* Feb 2008;19(1):16–24.
13. Burant CF, Saxena M. Rapid reversible substrate regulation of fructose transporter expression in rat small intestine and kidney. *Am J Physiol.* Jul 1994;267(1 Pt 1):G71–G79.
14. Korieh A, Crouzoulon G. Dietary regulation of fructose metabolism in the intestine and in the liver of the rat. Duration of the effects of a high fructose diet after the return to the standard diet. *Arch Int Physiol Biochim Biophys.* Dec 1991;99(6):455–460.
15. Teff KL, Grudziak J, Townsend RR, et al. Endocrine and metabolic effects of consuming fructose- and glucose-sweetened beverages with meals in obese men and women: Influence of insulin resistance on plasma triglyceride responses. *J Clin Endocrinol Metab.* May 2009;94(5):1562–1569.
16. Sprecher H. Metabolism of highly unsaturated n-3 and n-6 fatty acids. *Biochim Biophys Acta.* 2000;1486(2–3):219–231.
17. Yin FQ, Chen Q, Sprecher H. A comparison of the metabolism of [3-C-14]-labeled 22- and 24-carbon (n-3) and (n-6) unsaturated fatty acids by rat testes and liver. *BBA-Mol Cell Biol L.* 1999;1438(1):63–72.
18. Jump DB, Tripathy S, Depner CM. Fatty acid-regulated transcription factors in the liver. *Ann Rev Nutrition.* 2013;33(1):249–269.

19. Tu WC, Cook-Johnson RJ, James MJ, Mühlhäusler BS, Gibson RA. Omega-3 long chain fatty acid synthesis is regulated more by substrate levels than gene expression. *Prostaglandins Leukot Essent Fatty Acids*. 2010;83(2):61–68.
20. Wang Y, Botolin D, Xu J, et al. Regulation of hepatic fatty acid elongase and desaturase expression in diabetes and obesity. *J Lipid Res*. 2006;47(9):2028–2041.
21. Bazinet RP, Laye S. Polyunsaturated fatty acids and their metabolites in brain function and disease. *Nat Rev Neurosci*. 2014;15(12):771–785.
22. Naughton SS, Mathai ML, Hryciw DH, McAinch AJ. Fatty acid modulation of the endocannabinoid system and the effect on food intake and metabolism. *Int J Endocrinol*. 2013;2013:11.
23. DeMar JC, Ma K, Chang L, Bell JM, Rapoport SI. α-Linolenic acid does not contribute appreciably to docosahexaenoic acid within brain phospholipids of adult rats fed a diet enriched in docosahexaenoic acid. *J Neurochem*. 2005;94(4):1063–1076.
24. DeMar Jr JC, Lee H-J, Ma K, et al. Brain elongation of linoleic acid is a negligible source of the arachidonate in brain phospholipids of adult rats. *Biochim Biophys Acta*. 2006;1761(9):1050–1059.
25. Igarashi M, DeMar JC, Ma K, Chang L, Bell JM, Rapoport SI. Docosahexaenoic acid synthesis from α-linolenic acid by rat brain is unaffected by dietary n-3 PUFA deprivation. *J Lipid Res*. 2007;48(5):1150–1158.
26. Katsarou A, Gudbjörnsdottir S, Rawshani A, et al. Type 1 diabetes mellitus. *Nat Rev Dis Primers*. 2017;3. 17016.
27. Hu Q, Ishii E, Nakagawa Y. Differential changes in relative levels of arachidonic acid in major phospholipids from rat tissues during the progression of diabetes. *J Biochem*. Mar 1994;115(3):405–408.
28. Comte C, Bellenger S, Bellenger J, Tessier C, Poisson JP, Narce M. Effects of streptozotocin and dietary fructose on delta-6 desaturation in spontaneously hypertensive rat liver. *Biochimie*. 2004;86(11):799–806.
29. Ghebremeskel K, Bitsanis D, Koukkou E, Lowy C, Poston L, Crawford MA. Liver triacylglycerols and free fatty acids in streptozotocin-induced diabetic rats have atypical n-6 and n-3 pattern. *Comp Biochem Physiol C: Toxicol Pharmacol*. 2002;132(3):349–354.
30. Chanussot B, Asdrubal P, Huang YS, Poisson JP. Adrenic acid delta4 desaturation and fatty acid composition in the liver of marine-oil fed streptozotocin diabetic rats. *Prostaglandins Leukot Essent Fatty Acids*. Dec 1997;57(6):539–544.
31. Mašek T, Filipović N, Hamzić LF, Puljak L, Starčević K. Long-term streptozotocin diabetes impairs arachidonic and docosahexaenoic acid metabolism and Δ5 desaturation indices in aged rats. Exp Gerontol. 2014;60(0):140–146.
32. Levant B, Ozias MK, Guilford BL, Wright DE. Streptozotocin-induced diabetes partially attenuates the effects of a high-fat diet on liver and brain fatty acid composition in mice. *Lipids*. 2013;48(9):939–948.
33. Celik S, Baydas G, Yilmaz O. Influence of vitamin E on the levels of fatty acids and MDA in some tissues of diabetic rats. *Cell Biochem Funct*. Mar 2002;20(1):67–71.
34. Pari L, Venkateswaran S. Protective role of Phaseolus vulgaris on changes in the fatty acid composition in experimental diabetes. *J Med Food*. Summer 2004;7(2):204–209.
35. Kumar G, Banu S, Murugesan AG. Influence of Helicteres isora administration for diabetes mellitus: The effect on changes in tissue fatty acid composition. *Food Chem Toxicol*. Aug 2009;47(8):1826–1830. an international journal published for the British Industrial Biological Research Association.
36. Starcevic K, Filipovic N, Galan A, Micek V, Gudan Kurilj A, Masek T. Hepatic lipogenesis and brain fatty acid profile in response to different dietary n6/n3 ratios and DHA/EPA supplementation in Streptozotocin treated rats. *Mol Nutr Food Res*. May 2018;62(9):e1701007.
37. Bourre JM, Piciotti M. Delta-6 desaturation of alpha-linolenic acid in brain and liver during development and aging in the mouse. *Neurosci Lett*. 1992;141(1):65–68.
38. Takahashi R, Horrobin DF, Watanabe Y, Kyte V, Billard V. Short-term diabetes increases triacylglycerol arachidonic acid content in the rat liver. *Biochim Biophys Acta*. 1987;921(1):151–153.
39. Brenner RR. Hormonal modulation of Δ6 and Δ5 desaturases: case of diabetes. *Prostaglandins Leukot Essent Fatty Acids*. 2003;68(2):151–162.
40. Montanaro MA, Bernasconi AM, González MS, Rimoldi OJ, Brenner RR. Effects of fenofibrate and insulin on the biosynthesis of unsaturated fatty acids in streptozotocin diabetic rats. *Prostaglandins Leukot Essent Fatty Acids*. 2005;73(5):369–378.
41. El Hafidi M, Cuéllar A, Ramírez J, Baños G. Effect of sucrose addition to drinking water, that induces hypertension in the rats, on liver microsomal Δ9 and Δ5-desaturase activities. *J Nutr Biochem*. 2001;12(7):396–403.
42. Mašek T, Filipović N, Vuica A, Starčević K. Effects of treatment with sucrose in drinking water on liver histology, lipogenesis and lipogenic gene expression in rats fed high-fiber diet. *Prostaglandins Leukot Essent Fatty Acids*. 2017;116:1–8.

43. Perez I, Hafidi M, Carvajal K, Baños G. Castration modifies aortic vasoreactivity and serum fatty acids in a sucrose-fed rat model of metabolic syndrome. *Heart Vessels*. 2009;24(2):147–155.
44. da Silva-Santi LG, Antunes MM, Caparroz-Assef SM, et al. Liver fatty acid composition and inflammation in mice fed with high-carbohydrate diet or high-fat diet. *Nutrients*. 2016;8(11).
45. Yao M, Hou L, Xie T, et al. The biosynthesis of DHA is increased in the liver of diabetic rats induced by high-fat diets and STZ, in correlation with increased activity of peroxisomal β-oxidation. *Eur J Lipid Sci Technol*. 2015;118(2):137–146.
46. Agrawal R, Gomez-Pinilla F. 'Metabolic syndrome' in the brain: deficiency in omega-3 fatty acid exacerbates dysfunctions in insulin receptor signalling and cognition. *J Physiol*. 2012;590(10):2485–2499.
47. Akgun B, Sari A, Ozturk S, Erol FS, Ozercan IH, Ulu R. Effects of Mucuna pruriens on free fatty acid levels and histopathological changes in the brains of rats fed a high fructose diet. *Med Princ practice: Int J Kuwait Univ, Health Science Centre*. 2017;26(6):561–566.
48. Mašek T, Starčević K. Chronic sucrose intake decreases concentrations of n6 fatty acids, but not docosahexaenoic acid in the rat brain phospholipids. *Neurosci Lett*. 2017;.
49. Owen JL, Zhang Y, Bae S-H, et al. Insulin stimulation of SREBP-1c processing in transgenic rat hepatocytes requires p70 S6-kinase. *Proc Natl Acad Sci U S A*. 2012;109(40):16184–16189.
50. Okamoto K, Kakuma T, Fukuchi S, Masaki T, Sakata T, Yoshimatsu H. Sterol regulatory element binding protein (SREBP)-1 expression in brain is affected by age but not by hormones or metabolic changes. Brain Res. 2006;1081(1):19–27.
51. Ferré P, Foufelle F. SREBP-1c transcription factor and lipid homeostasis: Clinical perspective. *Horm Res Paediatr*. 2007;68(2):72–82.
52. Karmi A, Iozzo P, Viljanen A, et al. Increased brain fatty acid uptake in metabolic syndrome. *Diabetes*. 2010;59(9):2171–2177.
53. Proctor G, Jiang T, Iwahashi M, Wang Z, Li J, Levi M. Regulation of renal fatty acid and cholesterol metabolism, inflammation, and fibrosis in Akita and OVE26 mice with type 1 diabetes. *Diabetes*. 2006;55(9):2502–2509.
54. Sun L, Halaihel N, Zhang W, Rogers T, Levi M. Role of sterol regulatory element-binding Protein 1 in regulation of renal lipid metabolism and glomerulosclerosis in diabetes mellitus. *J Biol Chem*. 2002;277(21):18919–18927.
55. Aoun M, Fouret G, Michel F, et al. Dietary fatty acids modulate liver mitochondrial cardiolipin content and its fatty acid composition in rats with non alcoholic fatty liver disease. *J Bioenerg Biomembr*. 2012;44(4):439–452.
56. McGree CD, Lieberman P, Greenwood CE. Dietary fatty acid composition induces comparable changes in cardiolipin fatty acid profile of heart and brain mitochondria. *Lipids*. 1996;31(6):611–616.
57. He Q, Han X. Cardiolipin remodeling in diabetic heart. Chem Phys Lipids. 2014;179(0):75–81.
58. Han X, Yang J, Yang K, Zhao Z, Abendschein DR, Gross RW. Alterations in myocardial cardiolipin content and composition occur at the very earliest stages of diabetes: a shotgun lipidomics study. *Biochemistry*. 2007;46(21):6417–6428.
59. Harauma A, Yasuda H, Hatanaka E, Nakamura MT, Salem N, Jr., Moriguchi T. The essentiality of arachidonic acid in addition to docosahexaenoic acid for brain growth and function. Prostaglandins Leukot Essent Fatty Acids.
60. Lefkowitz W, Lim SY, Lin Y, Salem Jr N. Where does the developing brain obtain its docosahexaenoic acid? Relative contributions of dietary alpha-linolenic acid, docosahexaenoic acid, and body stores in the developing rat. *Pediatr Res*. Jan 2005;57(1):157–165.
61. Murphy EJ. The blood-brain barrier and protein-mediated fatty acid uptake: Role of the blood-brain barrier as a metabolic barrier: An editorial comment for 'The blood-brain barrier fatty acid transport protein 1 (FATP1/SLC27A1) supplies docosahexaenoic acid to the brain, and insulin facilitates transport'. *J Neurochem*. 2017;141(3):324–329.
62. Williard DE, Harmon SD, Kaduce TL, et al. Docosahexaenoic acid synthesis from n-3 polyunsaturated fatty acids in differentiated rat brain astrocytes. *J Lipid Res*. Sep 2001;42(9):1368–1376.

CHAPTER

22

n-3 and n-6 Fatty Acids in Fish: A Focus on Non-Marine Species

Semra Kaçar

Mardin Artuklu University, Department of Nutrition and Dietetics, School of Health, Mardin, Turkey

INTRODUCTION

The fish that are the largest and the widest group of aquatic vertebrates constitute for approximately half of the vertebrates. The 39% of the fish species are freshwater fish. Most of the freshwater fish with a wide home range are tropical and subtropical, and rarely found in Polar Regions.[1]

Proteins and lipids are the main components that determine the quality of fish meat, which plays an integral role in a healthy human diet. The live weight of fish is 70%–80% water, 17%–20% protein, and 2%–10% fat. Fish are rich in the water-soluble vitamins thiamine, riboflavin, niacin, vitamin B_6, and vitamin B_{12}. They are also rich in vitamin A and D, which are fat-soluble vitamins.

Fish, unlike mammals that store lipids in the fat tissue, store lipids in the skeletal muscle and liver.[2] Some of these stored lipids are sent to body parts as needed according to the physiological condition of the animal and changes in environmental conditions.[3] Fish completely digest lipids, and they prefer them rather than carbohydrates as a source of energy.[4] The main sources of these fatty acids (FA) are the planktonic organisms that form the first ring of the food chain in the water.

THE STRUCTURE OF FATTY ACIDS

FAs with no double bonds are saturated fatty acids (SFAs); those with a double bond are monounsaturated fatty acids (MUFAs), and those with more than one double bond are polyunsaturated fatty acids (PUFAs). The number of carbon atoms, the number of double bonds, and the places where double bonds are located differ in FAs.

Omega Fatty Acids in Brain and Neurological Health
https://doi.org/10.1016/B978-0-12-815238-6.00022-5

The number one carbon atom of the FAs chain contains COOH, and the last carbon atom contains the CH_3 group. In PUFAs, when regarded as the methyl group, the FAs containing double bonds between the third and fourth carbon atoms are named as omega-3, and those containing double bonds between the sixth and seventh carbon atoms are named as omega-6 FAs. FAs that are rarely found freely in organisms are included in the structures of the triacylglycerols (TAGs), which are used as a power source, and the phospholipids (PLs) functioning in membranes.

Triacylglycerol

Triacylglycerols are formed by linking fatty acids with an ester linkage to three alcohol groups in glycerol. Triacylglycerols are the form in which fat energy is stored in adipose tissue. Triacylglycerols are sometimes referred to as triglycerides.

Phospholipid

In the structure of phospholipids found in the cell and organelles membranes of all living organisms, the saturated fatty acid binds to the number 1 carbon atom of glycerol; unsaturated fatty acid binds to the number 2 carbon atom, and the phosphate group binds to the number 3 carbon atom. Different alcohol groups bind to the phosphate group.

TOTAL LIPID CONTENT

Fish are usually classified as fat-free, medium-fat, and oily based on their fat content. Those with fat content less than 5% are considered fat free, those having between 5% and 10% are considered medium fat, and those having 10% and more are considered fatty fish.[5–7] Most freshwater fish are free of fat as they contain less than 5% lipids. Feeley et al.[8] stated that low-fat fish contain more water, and their meats are whiter. Fat-free fish, except the trout, usually store lipids in muscles and the liver, but mammals and trout store lipids in fat tissue. Muscle tissue is fat free.[9] The tissues where the oil is stored vary depending on the fish species. While the active fish store the lipids in muscles, the inactive fish living in the bottom of water store the lipids in the liver.[10]

The main factors affecting total lipid content in fish are whether or not the fish is a natural or farm fish, its climatic conditions, season, gonad maturation, sex, size of fish, age, and geographical region. Age variation and sex-linked maturity within the same species may also affect total lipid content.

Factors Affecting Total Lipid Content

Influence of Reproduction on Total Lipid Content

The total lipid amount in tissues of fish in their muscles, liver, and gonads increases in the prereproductive period and decreases in the postreproductive period. The need for protein, carbohydrate, and lipid for the development of gonads before the reproductive cycle is extensive. The lipids in the liver and muscle are transferred to the gonads for gonadal development, especially during the reproduction period.[11] The liver stores most of the lipids to be used during gonadal development and gametogenesis. However, the energy required for the reproduction is mostly derived from muscle lipids.[12] When fish find enough nutrients, they can control their reproduction and lipid storage period. The lipid storage cycle is directly related to the abundance of nutrients. If the nutrient amount in their surroundings is low, the lipid variation is low; if the nutrient is abundant, the lipid variation is high throughout the year.[13, 14]

Influence of the Seasons on Total Lipid Content

The important things in seasonal variation are the active feeding of fish and the time of egg-laying. Seasonal variations in lipid levels of fish mainly relate to the reproductive cycle. Fish accumulate lipid reserves, which are in excess before the gonad development. Lipid content in the interior is inversely related to the gonadosomatic index, according to most of the studies.[1] Total lipid content of fish generally increases in the early stages of summer but decreases in winter.[15]

The Influence of Geographic Region and Locality on Total Lipid Content

The total lipid content of the same type of fish living in different lakes was found to vary. For instance, it was found that three fish species living in six different lakes in Ethiopia have different amounts of total lipids. While the amount of lipid was high in fish living in some lakes, the amount of lipids in the same fish species living in other lakes was found to be low.[16]

TOTAL FATTY ACID

Fish oil and fatty acid composition are the biochemical compounds that change the most according to ecological factors and physiological conditions of fish. Saturated fatty acids and MUFAs generally function as energy sources, and PUFAs function structurally in organs. PUFA and MUFA ratios are high and SFA is low in fish living in cold regions; while in tropical regions, PUFA ratio is low. SFA and MUFA are higher in fish that live in both temperate[17] and tropical waters.[18]

The 16:0 from SFAs, 18:1n-9 from MUFAs, and EPA and DHA from PUFAs are found to be the highest in percentage distribution in fish. The 14:0 and 18:0 from the SFAs, 16:1n-7 from MUFAs, 18:2n-6 and 18:3n-3 from the PUFAs, and 20:3n-6 from the eicosanoids and AA were detected in a small percentage.[19–25]

Fatty acids such as 16:0, 16:1n-7, 18:1n-9, and 20:1n-9 do not have the essential features; they are FAs that can be synthesized from the precursors of carbohydrates and amino acids in tissues as they can be taken by means of nutrients. The proportion of these FAs, which are FA components of depot lipids, can vary from few to many. Ackman et al.[26] stated that the 16:0 is the key metabolite in fish, and its quantity is not influenced by nutrients. It is reported that the total amount of SFA in bony fish is 16:0, which is the most abundant fatty acid. Even its proportion can be up to 60%.[27, 28]

High amounts of 18:1n-9, 16:1n-7, and AA are characteristic for freshwater fish.[29] The 18:1n-9, one of these components, is the most abundant compound in fish (45%–58% of total MUFAs).

Linoleic and ALA essential FAs cannot be synthesized in fish tissue due to the absence of Δ^{12} and Δ^{15} desaturase enzymes. These FAs are taken up through nutrients and sent to other parts of the body. The other 20-carbon unsaturated long chain FAs, such as 20:3n-6, 20:4n-6, 20:5n-3, 22:5n-3, and 22:6n-3, are synthesized in tissues by the help of Δ^6, Δ^5, and Δ^4 desaturase enzymes from 18:2n-6 and 18:3n-3, as a result of the chain elongation and increasing the degree of unsaturation (desaturation). Fish species can provide these FAs from phytoplanktons, which also constitute the first ring of the food chain. Ackman[30] suggested that having C18 PUFAs at a high rate is a typical characteristic for freshwater fish.

The ratio of AA is high in fish living in tropical waters. The reason for the high amount of AA in freshwater fish stems from the lower solubility of oxygen in hotter waters.[31] However, the rate of this FA is usually low in some fish. Freshwater fish living in the dam contain AA and DHA in significant proportions.[32] Arachidonic acid is the precursor of eicosanoids, which have negative cardiovascular effects such as vasocontraction and platelet aggregation. Many researchers state that AA consumption should be reduced.[33, 34]

The most important factor affecting the quality of fish meat is the percentage of n-3 FAs present in lipids, EPA, and DHA. These are components that have functionality in vision, and enhance learning ability. Eicosapentaenoic acids are the precursors of eicosanoids, which are physiologically active substances, and have vasodilator effects on inhibiting platelet aggregation. Eicosapentaenoic acid and DHA have positive effects on cardiovascular diseases, arthritis, nephritis, skin diseases, and cancer.[35] Since fish need the n-3 PUFA much more nutritionally, the amount of DHA and EPA in their tissues is more than AA.[18]

Factors Affecting Total Fatty Acid Content

The Influence of Season on Total Fatty Acid Content

Seasonal water temperature changes and the nutrients in the environment affect the chemical composition of fish meat. Fatty acid composition of fish lipids is also affected by nutrient composition.

SFA and MUFA are generally abundant in hot regions and seasons, whereas PUFAs are more prevalent in cold regions and seasons.[36, 37]

The Influence of Reproduction on Total Fatty Acid Content

The amount of lipids in fish can vary according to temperature, season, geographical region, species, sex, different organs in the same species, and food. However, the most significant changes in the lipid content of other organs, such as gonads, notably both in the muscle and the liver, are seen during the reproductive period. As the gonads begin to develop, the deposit lipids in muscles, liver, and other organs become mobilized to gonads; while the amount of lipids in gonads increases, the content of muscle and liver lipids decreases.[38] The use of lipids for gonad development is less in males, and they are affected by reproductive activities less than females.

PUFAs are essential in the formation of the gamete; lack of these fatty acids causes infertility.[39] Changes in lipid metabolism, in proportion to sexual maturation, coincide to the same period, and depot fats are used for ovagenesis and spermatogenesis.[30, 40]

C20 PUFAs show a decrease during reproduction due to their use in the development of gonads and gametogenesis during the reproductive period. Fish need 20C PUFAs such as EPA, DHA, and AA for normal development and growth as well as for reproduction.[41] For example, PGE_2, which is derived from AA, stimulates ovarian and testicular steroidogenesis in the gonads of bony fish.[42] Eicosanoids function in hatching and early larval period during ovulation and embryonic development. Tocher and Sargent[43] stated that fatty acids accumulate in the ovule for embryo growth, and PUFAs are used in embryo growth, in particular.

While females use mainly SFA for energy needs of gonad maturation, males use MUFAs. But, the females mobilize n-3 and n-6 FAs, which are stored for gonad maturation during vitellogenesis.[44, 45] Sargent et al.[46] stated that MUFAs are catabolized to meet metabolic energy during gonad development.

Comparison of Total Fatty Acids of Sea and Freshwater Fish

Ackman[30] stated that the biochemistry of freshwater fish is the same as that of sea fish. However, some researchers have suggested that the fatty acid composition of freshwater and sea fish differs.[17]

Freshwater fish have high amounts of C18 PUFA and low amounts of n-3, EPA, and DHA compared to sea fish.[30] Freshwater fish are characterized by high amounts of n-6 PUFA, especially 18:2n-6 and AA.[47, 48] The FAs form C20 and C22; PUFAs are more abundant in sea fish.[30] The 20:5n-3, 22:2n-3 and 22:6n-3 of PUFAs with 20C are synthesized by the 18:3n-3. The reason why PUFAs with 20 and 22C are low in freshwater fish compared to those living at sea may be due to the lesser activity of the enzyme on the pathway synthesized from 18:3n-3. The water temperature and the nutrients should not be ignored. Freshwater fish contain more SFAs, especially the 16:0, than sea fish.[49] Freshwater fish can convert the short-chain 18:2n-6 and 18:3n-3, synthesized by algae or plants, to longer and more unsaturated EPA and DHA acids in comparison with those living in the sea.[1] Since freshwater fish contain less useful long-chain PUFAs, such as EPA and DHA, than those living at sea, they have lower nutritional value. This view is based on the findings that were obtained from studies during 1930-60.[30, 50, 51] The basic reason for this finding is seen in nutrients. Freshwater fish are usually fed with terrestrial or aquatic insects and benthic organisms. These nutritional organisms usually have the n-6 PUFAs at a high ratio.[1, 52] Lipids of freshwater fish have the SFA and 18C PUFA at higher ratios, and have the PUFA with 20 and 22C and n-3/n-6 at lower ratio.[1]

Freshwater fish are rich in 20:4n-6 in comparison with sea fish, and sea fish are rich in n-3PUFAs, such as EPA and DHA, which contain surplus plankton-forming nutrients. While the ratio of 20:4n-6 is 2%–3% in freshwater fish oils, it is generally lower than 1% in sea fish. The AA ratio in fatty acids of tropical fish is higher than cold water sea fish.[19]

One of the factors affecting the fatty acid distribution is the salinity of the water. The ratio of n-3/n-6 and n-3PUFA in sea fish is higher than freshwater fish. This difference in lipid composition is due to salinity.

Freshwater fish usually contain lesser amounts of n-3 FAs than sea fish.[53] However, some freshwater fish have been found to contain n-3 PUFA in a significant proportion, even more than some sea fish.[40, 54] Cold water fish are also important sources in terms of n-3 FAs.

The Influence of Nutrients on Total Fatty Acid Content

Nutrition is one of the most important factors affecting the FA composition of fish. Fatty acid composition of the body lipids of a fish are significantly affected by nutritional lipids. Freshwater algae and aquatic insect larvae, which are generally rich in FAs such as 18:2n-6, 18:3n-3 and EPA, are important nutrients for fish.[55, 56] The sea planktons have high amounts of 16:1n-7, 20:5n-3, and 22:6n-3. Diatoms are rich in the 20:5n-3. For this reason, the ratio of 20:5n-3 and 22:6n-3 in fish oils is based on nutrient.[57] Planktons, which are rich in EPA and DHA but poor in terms of n-6 PUFA, are the main nutrients of sea fish.[58, 59] Sea fish that are fed with plankton have a higher level of n-3. Freshwater algae, Crustaceans, and aquatic insect larvae are usually rich in 18:2n-6, 18:3n-3, and 20:5n-3. The 22:5n-3 and 22:6n-3 of the FAs have been found in Crustaceans instead of aquatic insects. Therefore, the 18:2n-6 and 20:4n-6 percentages are high in the fatty acid composition of freshwater fish.[1, 47] The diversity of FAs and tissue lipids in herbivorous fish is seen more than in carnivorous fish.[16]

The fatty acid profile of lipids in natural fish reflects the FA content of the aquatic nutrient chain. Algae are apparently present in the nutrients in the early-life stages of some freshwater fish. The lipids of freshwater algae contain more C18 PUFAs and C20 and C22 PUFAs in general. The algae include both 18:2n-6 and 18:3n-3. Eicosapentaenoic acid is found in large quantities in some algal species, especially in diatomaeas. However, 22:6n-3 is rarely found in them. Adults, larvae, or nymphs of aquatic insects are major nutrient sources for freshwater fish, especially for salmonids. Linoleic acid and 18:3n-3 are found in the PUFAs of these insects in excess amounts. The 20:4n-6 ratio increases up to 7% and 20:5n-3 ratio up to 25% in total FAs of these aquatic nutrients. Lipids of predator fish are influenced by the FA composition of the small fish they eat. On the other side, the lipids of phytoplankton and zooplankton forming the food chain in the sea are rich in the 18:3n-3, 20:5n-3, and 22:6n-3, but poor in terms of the n-6 PUFA. As a result, the main differences between freshwater and sea fish in terms of PUFA composition arise from differences in the FA composition of their nutrients.[1]

Herbivorous fish that have lots of PUFAs having 18-carbon excessively, but contain small quantities of C20 and C22 PUFAs and are fed with algae, contain lots of n-6 PUFAs and short-chain n-3 PUFAs. Carnivorous fish are rich in long-chain n-3 PUFAs but not in terms of 18:3n-3 because they consume fish that complete elongation and desaturation. While omnivorous are rich in 18:3n-3, they are poor in terms of the long-chain n-3 PUFAs, and this situation show that the components of fish lipids are directly related to the lipids in nutrients. Fatty acids of cold water and hot water fish are quantitatively different due to long-chain polyunsaturated n-3 FAs in cold water fish species; in hot water species, n-6 and n-9 FAs are found in greater amounts.[60]

When compared with culture fishing, SFA and PUFA ratios are high in natural fish whereas MUFA is low.[27] The reason for this could be due to the high monoenoic content in the nutrient of culture fish.[61]

The Influence of Temperature on Total Fatty Acid Content

Hot water fish contain more saturated FAs than cold water fish.[49, 62]

According to Kayam,[63] the cold and deep sea fish have lots of n-3 FAs due to the fact that the melting temperature of n-3 FAs is lower than that of n-6 FAs, and they participate more in the membrane structure of fish.

Freshwater fish living in temperate and hot regions have higher n-6 FAs, especially in those with a higher melting temperature. In addition, since culture freshwater fish are fed predominantly with fishmeal supplemented with vegetable oils, the amount of n-6 FAs in their tissues increases further.[18, 53]

PHOSPHOLIPID AND TRIACYLGLYCEROL FATTY ACID

Phospholipids and TAG have different roles in fish metabolism. Phospholipids are a key component of cell membrane and structure, and their C20 PUFAs serve as precursors of eicosanoids. The TAGs, on the other hand, are mainly stored in the adipose tissue and function as energy reserves.[46, 64]

Neutral lipids used for energy purposes store important physiological PUFAs temporarily along with significant amounts of saturated and monounsaturated fatty acids.[65, 66]

Total PLs in the tissues of natural fish contain more PUFAs and less MUFA than TAGs characteristically. The content of PUFAs in the PL fraction can increase up to 58%, but the rate of saturation is similar in both fractions. Additionally, PUFAs of PL have longer chains than neutral lipids. Since PLs are richer in PUFAs than neutral lipids, the changes that occur during temperature adaptation can be seen in this fraction. The unsaturated fatty acids in PLs are increasing as response to the decrease in ambient temperature. Generally, the most abundant PUFA in the PLs is 22:6n-3.[1]

PL fractions contain high levels of n-3HUFA in comparison with TAG fraction, and TAGs have MUFAs, especially 18:1n-9 at higher ratio.[67]

Since the excessively unsaturated fatty acids with twenty and twenty-two carbon are found in the PL fraction and the C18 PUFAs are found more in the TAG fraction, the ratio of C20 + C22 PUFAs to C18 PUFAs in the PL is 4.3-9.8 times higher than neutral lipids. The amounts of 20:5n-3 and 20:4n-6, which are general components of total PL in temperate species, vary depending on the species. The greater presence of MUFAs in TAG may be due to the use of these components as a preferential substrate for catabolism events. The PUFA and the DHA, EPA, and AA with 20 carbons are found in the PL fraction at a high ratio compared to TAG, which indicates that these FAs are important as components of membrane phospholipids.

As the fish grows, its lipid content increases, and more TAG is stored in the tissue. The amount of MUFA increases in the total lipid fraction as a result. Monounsaturated fatty acids are used as preferential substrates for catabolism events.[1]

Ackman[30] and Standsby[68] found that n-3/n-6 ratio ranged from 1.7 to 3.5 in the freshwater species in PL.

The ratio of n-3/n-6 in TAGs of freshwater fish is higher than sea fish, and it is in the range of 1–3.[35]

Factors Affecting the Triacylglycerol and Phospholipid Fatty Acid Content

TAG fatty acids, which constitute a significant proportion of the neutral lipids and are used as energy reserves, are influenced by factors, notably nutrient and reproductive status (spawning period). PLs found in its cell and organelle membranes structurally affect the preservation of the fluid-effluent properties of the membranes and the permeability. The ambient temperature generally affects the content of FAs in this fraction. PLs in cell and organelle membranes undergo more rapid changes than neutral lipids due to their biological activity.[69]

The Influence of Nutrients on the Content of Triacylglycerol and Phospholipid Fatty Acids

Neutral lipid and FA composition in PL of fish vary within the annual cycle.

The TAGs, which are the main components of neutral lipids, may change depending on physiological changes apart from nutrients.[67] Fatty acid profiles of nonpolar neutral lipids are more similar to the nutritional lipids in comparison with the FA content of structural polar PL.[1, 70, 71]

The effect of nutritional lipids on FA composition of body lipids differs according to the TAG and PL. It was determined that the FA composition of PLs is affected from TAGs to a greater extent. The unsaturation levels of freshwater fish are increased by subjecting the 18:2n-6 and 18:3n-3 taken as nutrients into the chain-length using enzymes. However, it was found that these FAs are stored in TAGs without alteration.[72, 73]

The Influence of Temperature on the Content of Triacylglycerol and Phospholipid Fatty Acids

The influence of temperature is more effective on PLs than on neutral lipids.[74] When temperature decreases, the quantity of unsaturated fatty acids increases in the neutral and PL.[73, 75]

As a consequence of exposure to low temperatures, the ratio of unsaturated fatty acids to saturated fatty acids in PLs increased. This is one of the mechanisms of thermal acclimatization (physiological adaptation to cold of cold-blooded animals), which maintain membrane fluidity, permeability, and flexibility (protection).[60, 74] Poikilotherm organisms ensure adaption of their membrane lipids physically along with changes in ambient temperature.

The SFA proportion reduces along with exposure to low temperatures, whereas the proportion of unsaturated fatty acids increases.

Increased unsaturated fatty acids are those included in either monoenes or polyenes class.[76] Increased ambient temperature reduces n-3 accumulation in the PL.[77] The salinity of the water and the change in its temperature affect the length and unsaturation of the fatty acids in the membrane PL of poikilotherms.[78] Gallagher et al.[79] suggested that adaptations of females and males of the same species to physiological events may change.

n-3/n-6 RATIO

PUFAs are classified as ω3 (n-3) and ω6 (n-6) within fatty acids of lipids according to the first double bond position, which is lined up after the last carbon atom ω (n). Fatty acids ω6, such as linoleic acid 20:3n-6, 20:4n-6, are known as compounds ω3, such as the 18:3n-3, 20:5n-3, 22:5n-3, and 22:6n-3.

When considering the many metabolic functions in the human body, the balance between the uptake of n-6 and n-3 PUFA is more important than the uptake of individual FAs. Experimental studies have shown that the uptake of n-6 at high ratios in terms of nutrition and n-3 at low are main risk factors for cancer, coronary heart disease, and cerebrovascular disease.[80, 81]

The WHO and FAO recommended that the n-6/n-3 PUFA ratio should be less than 4 in nutrients. Compared to other animal and vegetable lipids, the fish lipids having lower proportions of the n-6/n-3 PUFA are more important nutritionally. Additionally, eicosanoid balance in the body is also important to protect against heart disease and other chronic diseases. The ratio of 20:4n-6 and 20:5n-3 to each other, which are precursors of eicosanoid, gives the eicosanoid balance measurement of the body.[82]

Increasing the ratio of n-3/n-6 FAs in human nutrients reduces plasma lipids, prevents coronary heart disease, and reduces cancer risk.[83] The n-3/n-6 ratio is a good index to compare the nutritional value of fish oils.[84] Simopoulos[85] suggests that the ratio of n-3/n-6 should be 1:1. Some nutritionists indicate that n-6/n-3 ratio should be 4 at maximum. The amount of n-3 in the nutrient increases nutrient quality and prevents many chronic diseases.[86] Ackman[30] and Standsby[68] stated that this ratio is in the range of 1.7–3.5 in freshwater fish and Wang et al.[54] stated that it is in the range of 0.5–3.8. This ratio varies in the range of 2–3 in carp species.[35]

Factors Affecting n-3/n-6 Ratio

Aside from natural or culture fish, seasons, and spawning, the water environment in which the fish live, namely sea or freshwater environments, has the most effect on the n-3/n-6 ratio in fish.

n-3/n-6 Ratio in Sea and Freshwater Fish

Desaturase enzyme activity and chain extension (elongation) reactions increase n-6 PUFA and decrease n-3 PUFA in freshwater fish compared to sea fish; hence, the n-3/n-6 ratio is lower in freshwater fish characteristically.[87, 88]

When compared to sea fish, freshwater fish usually contain a great amount of C18 PUFAs, but at the same time, they also have significant amounts of EPA and DHA.[30, 89] In addition, the FA composition of freshwater fish is characterized by high rates of n-6 PUFAs, especially 18:2n-6 and 20:4n-6. For this reason, the ratio of total n-3/n-6 fatty acids is much lower in freshwater fish than sea fish and is in the range of 1–4. The lipids of sea fish are characterized by low levels of 18:2n-6 and 18:3n-3 and high level long-chain PUFAs. Eicosapentaenoic acid and DHA are the most abundant n-3 fatty acids.[90, 91] Due to the low content of n-6 fatty acids in sea fish, the total n-3/n-6 ratio is high and is around 5-10. Henderson and Tocher[1] stated that the n-3/n-6 ratio of total lipids in freshwater fish is mostly in the range of 0.5–3.8, and it is in the range of 4.7–14.4 in sea fish.

As freshwater fish contain long-chains of n-3PUFA in low amounts compared to sea fish,[18] the rate of n-3/n-6 in sea fish is higher than in freshwater fish, and this ratio may be higher than 5–10 or more than 10. Ackman[30] suggested that the proportion of n-3/n-6 in freshwater fish nutrients is lower than in sea fish since fatty acids such as 18:2n-6 and 18:3n-6 are more abundant than 18:3n-3.

The Influence of Nutrients on n-3/n-6 Ratios

The highest n-3/n-6 was found in Pike, which have a very low total lipid content with a rate of 8–9 in a study where 56 freshwater fish species were examined. However, very low ratios such as 1.1 and 1.8 were found in serpent fish with a higher lipid content. Fish having a high n-3/n-6 ratio are usually carnivores, which are fed with fish and insect larvae.[92]

Diatoms are rich in 20:5n-3. For this reason, the ratio of 20:5n-3 and 22:6n-3 in fish oils is dependent on the nutrient.[57]

*Sander lucioperca*s carnivorous fish are rich in long-chain n-3PUFAs, but poor in terms of 18:3n-3 because they consume other fish that complete chain elongation and desaturation. Carnivorous fish take up n-3 PUFA directly from nutrients.

In March, phytoplankton, which are a main nutrient source for fish, reach maximum levels when air temperature starts to increase. In autumn and winter, the amount of algae decreases due to the inadequate temperature of the water, sun, and air.

Carp, fed with phytoplankton and zooplankton, are rich in n-3 PUFA. However, it was also determined that these fish contain significant amounts of AA.[93] The phytoplankton is the main origin of the n-3 and n-6 series FAs, and therefore the FA composition of the fish is significantly affected by the nutrient's lipid content.

The n-3/n-6 Ratio in Culture and Natural Fish

The n-3/n-6 ratio of natural fish is higher than that of culture fish. The reason for this is that culture fish contain less n-3 FAs than natural ones. This ratio varies in the range of 1–4 for culture freshwater fish.[61]

The Influence of Temperature on n-3/n-6 Ratio

Fish that are found in warm water ecosystems have lower concentrations of polyunsaturated fatty acids than those found in cold water environments. Fish need high PUFA to provide a tolerance to low water temperatures, and this causes fish in cold environments to have a higher PUFA content than those in warm waters tropical areas. There are a number of experiments demonstrating the effect of environmental temperatures on FA composition of aquatic animals.[94] The general trend toward higher content of long chain PUFA at lower temperatures is quite clear. The n-6/n-3 ratio decreases with a decrease in temperature.[95] Fish raised in warmer waters, such as common carp, channel catfish, and tilapia, may do better with a mixture of n-6 and n-3 fatty acids.[95]

References

1. Henderson RJ, Tocher DR. The lipid composition and biochemistry of freshwater fish. *Prog Lipid Res*. 1987;26:281–347.
2. Neuhaus OW, Halver JC. *Fish in Research*. New York: Academic Press; 1969.
3. Johansson L, Kiessling A, Kiessling KH, Berglund L. Effects of altered ration levels on sensory characteristics, lipid content and fatty acid composition of rainbow trout (Oncorhynchus mykiss) their intrinsic variation and practical implications. *Food Qual Prefer*. 2000;11(3):247–254.
4. Cho CY, Cowey CB, Watanabe T. *Finfish Nutrition in Asia. Methodological Aproaches to Reseach and Devolopment*. 1. International Development Research Centre; 1985:26–33.
5. Ackman RG, Ratnayake WMN, Macpherson EJ. EPA and DHA contents of encapsulated fish oil products. *J Am Oil Chem Soc*. 1989;66(8):1162–1164.
6. Ackman RG. Seafood lipids and fatty acids. *Food Rev Int*. 1990;6(4):617–646.
7. Dean LM. Nutrition and preparation. In: Martin RE, Flick GJ, eds. *The Seafood Industry*. New York: Published Van Nostrand Rainhold; 1990:255–267.
8. Feeley RM, Criner DEC, Watt BK. Cholesterol content of foods. *J Am Diet Assoc*. 1972;61:134–148.
9. Huss HH. *Fresh Fish: Quality and Quality Changes*. Copanhagen, Denmark: Ministery of Fisheries Technical Universty Press, No: 29; 1988.

10. Castell JD, Sinnhuber RO, Wales JH, Lee DJ. Essential fatty acids in the diet of rainbow trout (*Salmo gairdneri*): growth, feed conversion and some gross deficiency symptoms. *J Nutr*. 1972;102:77–86.
11. Castell JD, Lee DJ, Sinnhuber RO. Essential fatty acids in the diet of rainbow trout (*Salmo gairdneri*): Lipid metabolism and fatty acid composition. *J Nutr*. 1972;102:93–100.
12. Atchison GJ. Fatty acid levels in developing brook trout (*Salvenus fontinalis*) eggs and fry. *J Fish Res Board Can*. 1975;32:2513–2515.
13. Kinsella JE, Shimp JL, Mai J. Fatty acid content and composition of fresh water finfish. *Am Oil Chem Soc*. 1977;54:424–429.
14. Mute P, Agren JJ, Lindovist OV, Hanninen O. Fatty acid composition of vendace (*Coregonus albula* L.) muscle and its plankton feed. *Comp Biochem Physiol*. 1989;92B:75–79.
15. Osaka K, Yamaguchi A, Kurokawa T, Kuwahara K, Saito H, Nozaki Y. Chemical components and body color of horse mackerel caught in different areas. *Fish Sci*. 2002;68:587–594.
16. Zenebe T, Ahlgren G, Boberg M. Fatty acid content of some freshwater fish of commercial importance from tropical lakes in the Ethiopian Rift Valley. *J Fish Biol*. 1998;53:987–1005.
17. Aggelousis G, Lazos ES. Fatty acid composition of the lipids from eight freshwater fish species from Greece. *J Food Compos Anal*. 1991;4:68–76.
18. Rahman SA, Huah TS, Hassan O, Daud NM. Fatty acid composition of some Malaysian freswater fish. *Food Chem*. 1995;54:45–49.
19. Ackman RG. Freshwater fish lipids-an overlooked source of benefical long- chain n-3 fatty acids. *Eur J Lipid Sci Technol*. 2002;104(5):253–254.
20. Kayhan H, Başhan M, Kaçar S. Seasonal variations in the fatty acid composition of phospholipids and triacylglycerols of brown trout. *Eur J Lipid Sci Technol*. 2015;117(5):738–744.
21. Kaçar S, Başhan M. Seasonal variations on the fatty acid composition of phospholipid and triacylglycerol in gonad and liver of *Mastacembelus simack*. *J Am Oil Chem Soc*. 2015;92(9):1313–1320.
22. Kaçar S, Başhan M. Comparison of lipid contents and fatty acid profiles of freshwater fish from the Ataturk Dam Lake. *Turk J Biochem*. 2016;41(3):150–156.
23. Kaçar S, Başhan M. Oymak. AS. Effect of seasonal variation on lipid and fatty acid profile in muscle tissue of male and female *Silurus triostegus*. *J Food Sci Technol*. 2016;53(7):2913–2922.
24. Kaçar S, Başhan M. 2017. Variations in the fatty acid compositions of the liver and gonad tissue of spiny eel (*Mastacembelus mastacembelus*) from Atatürk Dam Lake. *Turk J Biochem*. 2017;42(6):617–623.
25. Kaçar S, Başhan M, Oymak AS. Effect of seasons on the fatty acid profile of total lipid, phospholipid and triacylglycerol in *Mastacembelus mastacembelus* (Atatürk Dam Lake, Turkey). *Grasas Aceites*. 2018; https://doi.org/10.3989/gya.1108172.
26. Ackman RG, Eaton CA, Linne BA. Differentiation of freshwater characteristics of fatty acids in marine specimens of the Atlantic sturgeon (*Acipenser oxyrhynchus*). *Fish Bull*. 1975;73:838–845.
27. Alasalvar C, Taylor KDA, Zubcov E, Shahidi F, Alexis M. Differentiation of cultured and wild sea bass (*Dicentrarchus labrax*): total lipid content, fatty acid and trace mineral composition. *Food Chem*. 2002;79(2):145–150.
28. Wheeler SC, Morrissey MT. Quantification and distribution of lipid, moisture, and fatty acids of West Coast Albacore Tuna (*Thunnus alalunga*). *J Aquat Food Prod Technol*. 2003;12(2):3–16.
29. Andrade AD, Rubira AF, Matsushita M, Souza NE. Omega-3 fatty acids in freshwater fish from South of Brazil. *J Am Oil Chem Soc*. 1995;72(10):1207–1210.
30. Ackman RG. Characteristics of the fatty acid composition and biochemistry of some fresh-water fish oils and lipids in comparison with marine oils and lipids. *Comp Biochem Physiol*. 1967;22:907–922.
31. Smith MW, NGA M. In: Giles R, ed. *Animals and Environmental Fitness*. Oxford, UK: Pergamon Press; 1980.
32. Ackman RG. Fatty acid analysis of freshwater fish lipids. *J Am Oil Chem Soc*. 1996;73(4):537–538.
33. Leaf A, Weber PC. Cardiovascular effects of n-3 fatty acids. *N Engl J Med*. 1988;318:549–557.
34. Nordoy A, Marchioli R, Arnesen H, Videbaek J. N-3 polyunsaturated fatty acids and cardiovascular diseases. *Lipids*. 2001;36:127–129.
35. Steffens W, Wirth M. Freshwater fish-an important source of n-3 polyunsaturated fatty acids. *Arch Pol Fish*. 2005;13(1):5–16.
36. Skuladottir GV, Schiöth HB, Gudmundsottır E, Richards B, Gardarsson F, Jonsson L. Fatty acid composition of muscle, heart and liver lipids in Atlantic salmon, *Salmo salar*, at extremely low environmental temperature. *Aquaculture*. 1990;84:71–80.

37. Dey I, Buda C, Wiik H, Halver JE, Farkas T. Molecular and structural composition of phospholipid membranes in livers of marine and freshwater fish in relation to temperate. *Proc Natl Acad Sci U S A*. 1993;90:7498–7502.
38. Uysal K. *Eğirdir Gölü Sudak (Stizostedion lucioperca Lin., 1758) balıklarının total lipid, total yağ asidi ve yağ asidi bileşiminin mevsimsel incelenmesi*; 2000. Fen Bil. Enst Su Ürünleri Temel Bilimleri ABD, Doktora Tezi, s. 66. Eğirdir-Isparta.
39. Soivio A, Nıemıstö M, Backsröm M. Fatty acid composition of *Coregonus muksun* Pallas: changes during incubation, hatching, feeding and starvation. *Aquaculture*. 1989;79:163–168.
40. Agren J, Mute P, Hanninen O, Herranen J, Penttila I. Seasonal variation of lipid fatty acids of Boreal freshwater fish species. *Comp Biochem Physiol*. 1987;88B:905–909.
41. Rodriguez C, Acosta C, Badia P, Cejas JR, Santamaria FJ, Lorenzo A. Assessment of lipid and essential fatty acids requirements of black seabream (*Spondyliosoma cantharus*) by comparison of lipid composition in muscle and liver of wild and captive adult fish. *Comp Biochem Physiol*. 2004;139B:619–629.
42. Wade MG, Van Der Kraak G. Regulation of prostaglandins E and F production in the goldfish testes. *J Exp Zool*. 1993;266:108–115.
43. Tocher DR, Sargent JR. Analyses of lipids and fatty acids in ripe roes of some Northwest Eurpean marine fish. *Lipids*. 1984;19(7):492–499.
44. Medford BA, Mackay WC. Protein and lipid content of gonads, liver and muscle of Northern pike (*Esox lucius*) in relation to gonad growth. *J Fish Res Board Can*. 1978;35:213–219.
45. Cejas JR, Almansa E, Villamandos JE, Badia P, Bolanos A, Lorenzo A. Lipid and fatty acid composition of ovaries from wild fish and ovaries and eggs from captive fish of white sea bream (*Diplodus sargus*). *Aquaculture*. 2003;216(1-4):299–313.
46. Sargent JR, Bell JG, Bell M, Henderson RJ, Tocher DR. Requirement criteria for essential fatty acids. *J Appl Ichthyol*. 1995;11:183–198.
47. Steffens W. Effects of variation in essential fatty acids in fish feeds on nutritive value of freshwater fish for humans. *Aquaculture*. 1997;151:97–119.
48. Özoğul Y, Özoğul F, Alagöz S. Fatty acid profiles and fat contents of commercially important seawater and freshwater fish species of Turkey: a comparative study. *Food Chem*. 2007;103:217–223.
49. Gopakumar K, Nair MR. Fatty acid composition of eight Indian marine fish. *J Sci Food Agric*. 1972;23:493–496.
50. Lovern JA. Fat metabolism in fishes. I. General survey of the fatty acid composition of the fats of a number of fishes, both marine and fresh-water. *Biochem J*. 1932;26:1974–1998.
51. Gruger EH, Nelson RW, Stansby ME. Fatty acid composition of oils from 21 species of marine fish, freshwater fish and shellfish. *J Am Oil Chem Soc*. 1964;41:662–667.
52. Pike IH. *The role of fish oil in feeds for farmed fish*. 25IAFMM, International Association of Fish Meal Manufactureres; 1990.1–12.
53. Vlieg P, Body DR. Lipid contents and fatty acid composition of some New Zealand freshwater finsfish and Marine finfish, shellfish and roes. *NZ J Mar Freshwater Res*. 1988;22:151–162.
54. Wang YJ, Miller LA, Perren M, Addis PB. Omega-3 fatty acids in Lake Superior fish. *J Food Sci*. 1990;55:71–73.
55. Takahashi H, Yamada M. Lipid composition of seven species of Crustacean plankton. *Bull Jpn Soc Sci Fish*. 1976;42:769–776.
56. Hanson BJ, Cummins KW, Cagill AS, Lowry RR. Lipid content, fatty acid composition, and the effect of diet on fats of aquatic insects. *Comp Biochem Physiol*. 1985;80B:257–276.
57. Sargent JR. Fish oils and human diet. *Br J Nutr*. 1997;78:5–13.
58. Kelly PB, Riser R, Hood DW. The origin of the marine polyunsatureted fatty acids compositions of marine plankton. *J Am Oil Chem Soc*. 1959;36:104–106.
59. Kayama M, Tsuchiya Y, Mead JF. A model experiment of aquatic food chain whit special significance in fatty acid conversion. *Bull Jpn Soc Sci Fish*. 1963;29:452–458.
60. Lovell RT. Nutrition of aquaculture species. *J Anim Sci*. 1991;69(10):4193–4200.
61. Van Vliet T, Katan MB. Lower ratio of n-3 to n-6 fatty acids in cultured than in wild fish. *Am J Clin Nutr*. 1990;51:1–2.
62. Gibson RA, Kneebone R, Kneebone GM. Comparative levels of arachidonic acid and eicosapentaenoic acid in Malaysian fish. *Comp Biochem Physiol*. 1984;78C:325–328.
63. Kayam M. Feed oil. *Abst Bull*. 1977;6:6–10.

64. Kiessling A, Pickova J, Johansson L, Asgard T, Storebakken T, Kiessling KH. Changes in fatty acid composition in muscle and adipose tissue of farmed rainbow trout (Oncorhynchus mykiss) in relation to ration and age. *Food Chem*. 2001;73:271–284.
65. Tocher DR, Faser AJ, Sargent JR, Gamble JC. Fatty acid composition of phospholipids and neutral lipids during embryonic and early larval development in Atlantic herring (*Clupea harengus* L.). *Lipids*. 1985;20:69–74.
66. Napolitano GE, Ratmayake WMN, Ackman RG. Fatty acid component of larval *Ostrea edulis* L.: importance of triacylglycerols as a fatty acid reserve. *Comp Biochem Physiol*. 1988;90B:875–883.
67. Christie WW. The lipid composition of animal tissues. In: Christie WW, ed. *HPLC and Lipids: A Pratical Guide*. Oxford, UK: Pergamon Press; 1987:55.
68. Standsby MM. Fatty acid patterns in marine, freshwater and anadromous fish. *J Am Oil Chem Soc*. 1967;44:64–74.
69. Parker RS, Selivonchick DP, Sinnhuber RO. Turnover of label from 1-14C linolenic acid in phospholipids of coho salmon, *Oncorhynchus kisutch*. *Lipids*. 1980;15:80–85.
70. Arts MT, Ackman RG, Holub BJ. Essential fatty acids in aquatic ecosystems: a crucial link between diet and human health and evolution. *Can J Fish Aquat Sci*. 2001;58:122–137.
71. Jump DB. The biochemistry of n-3 polyunsaturated fatty acids. *J Biol Chem*. 2002;277:8755–8758.
72. Takeuchi T, Watanabe T. Requirement of carp for essential fatty acids. *Bull Jpn Soc Sci Fish*. 1977;43:541–551.
73. Farkas T, Csenger I, Majoros F, Olah J. Metabolism of fatty acids in fish III. Combined effect of environmental temperature and diet on formation and deposition of fatty acids in the carp, *Cyprinus carpio* Linnaeus 1758. *Aquaculture*. 1980;(20):29–40.
74. Jobling M, Bendiksen EA. Dietary lipids and temperature interact to influence tissue fatty acid compositions of Atlantic salmon, *Salmo salar* L. parr. *Aquac Res*. 2003;34:1423–1441.
75. Farkas T. Adaption of fatty acid composition to temperature. A study on planktonic crustaceans. *Comp Biochem Physiol*. 1979;64B:71–76.
76. Logue JA, DeVries AL, Fodor E, Cossins AR. Lipid compositional correlates of temperature-adaptive interspeciecific differences in membrane physical structure. *J Exp Biol*. 2000;203:2105–2115.
77. Delgado A, Estevez A, Hortelano P, Alejandre MJ. Analyses of fatty acids from different lipids in liver and muscle of sea bass (*Dicentrarchus labrax* L.). Influence of temperature and fasting. *Comp Biochem Physiol*. 1994;108A:673–680.
78. Cordier M, Brichon G, Weber JM, Zwingelstein G. Changes in the fatty acid composition of phospholipids in tissues of farmed sea bass (*Dicentrarchus labrax*) during annual cycle. Roles of environmental temperature and salinity. *Comp Biochem Physiol*. 2002;133B:281–288.
79. Gallagher ML, Harrell ML, Rulfson RA. Variation in lipid and fatty acid contents of Atlantic croakers, striped mullet, and summer flounder. *Trans Am Fish Soc*. 1991;120:614–619.
80. Kris-Etherton PM, Harris WS, Appel LJ. Fish consumption, fish oil, omega-3 fatty acids and cardiovascular disease. *Circulation*. 2002;106:2747–2757.
81. Calder PC. Long-chain n-3 fatty acids and cardiovascular disease: further evidence and insights. *Nutr Res*. 2004;24(10):761–772.
82. Sargent JR, Bell JG, McEvoy LA, Tocher DR, Estevez A. Recent developments in the essential fatty acid nutrition of fish. *Aquaculture*. 1999;177:191–199.
83. Kinsella JE, Lokesh B, Stone RA. Dietary n-3 polyunsaturated fatty acids and amelioration of cardiovascular disease: Possible mechanisms. *Am J Clin Nutr*. 1990;52(1):1–28.
84. Piggott GM, Tucker BW. *Seafood Effects of Technology on Nutrition*. New York: Marcel Dekker Inc; 1990.
85. Simopoulos AP. Summary of NATO advanced research workshop on dietary n3 and n6 fatty acids: Biological effects and nutritional essentiality. *J Nutr*. 1989;199:512–528.
86. Moreira AB, Visentainer JV, de Souza NE, Matsushita M. Fatty acids profile and cholesterol contents of three Brazilian brycon freshwater fishes. *J Food Compos Anal*. 2001;14:565–574.
87. Borlongan IG, Benitez LV. Lipid and fatty acid composition of milkfish (*Chanos chanos*) grown in freshwater and seawater. *Aquaculture*. 1992;104(1-2):79–89.
88. Sheikheldin M, DeSilva SS, Anderson TA, Gooley G. Comparison of fatty acid composition of muscle, liver, mature oocytes, and diets of wild and captive macquarie perch (*Macquaria australasica*) broodfish. *Aquaculture*. 1996;144:201–216.
89. Steffens W. *Principles of Fish Nutrition*. Chichester: Ellis Horwood Limited; 1989.
90. Hilditch TP, Williams PN. *The Chemical Constitution of Natural Fats*. 4th ed. New York: Wiley; 1964.

91. Yamada M, Hayashi K. Fatty acid composition of lipids from 22 species of fish and mollusk. *Bull Jpn Soc Sci Fish.* 1975;41:1143–1152.
92. Ahlgren G, Blomqvist P, Bobergi M, Gustafsson IB. Fatty acid content of the dorsal muscle—an indicator of fat quality in freshwater fish. *J Fish Biol.* 1994;45(1):131–157.
93. Steffens W, Wirth M, Mieth G, Lieder U. Fresh water fish as a source of n-3 poly unsaturated fatty acids and their application to human nutrition. In: Kaushik SJ, Luquet P, eds. *Fish Nutrition in Practice.* INRA; 1993:469–474.
94. Rasoarahona RE, Bernathan G, Bianchini JP, Gaydou ME. Influence of seasons on the lipid content and fatty acids profiles of three tilapia species, (Oreochromis Niloticus, O. Macrochir and Tilapia Rendalli) from Madagascar. *Food Chem.* 2005;91(4):683–694.
95. Meed JF, Kayama M. Lipid metabolism in fish. In: Stansby ME, ed. *Fish Oils.* Westport, CT: Avi Publ. Co; 1967:289–299.

CHAPTER

23

Aspartame-Induced Neurodegeneration Versus Omega 3 Alleviation

Eyad M.T. Ali[*,†]

[*]Faculty of Medicine, Taibah University, Madinah, Saudi Arabia [†]Faculty of Medicine, Mansoura University, Mansoura, Egypt

ASPARTAME

Discovery

In 1965, a scientist called James Schlatter working at the G.D. Searle research laboratories accidentally discovered aspartame while working to develop a gastrin inhibitor drug. A white, odorless powder dripped onto his hands. After licking his fingers, he observed an extreme sugary taste. Later on, this sweet powder was named aspartame, and in 1970, scientists proposed its consumption as an artificial sweetener.[1]

Introduction

Aspartame (ASP) is one of the most common artificial sweeteners used all over the world. In 1983, its usage was approved by the Food and Drug Administration (FDA) for use in carbonated soft drinks.[2] ASP is used in more than 6000 foodstuffs, such as chewing gum, sweets, yogurts, vitamins, medications, and primarily in diet soft drinks.[3] It is 180–200 times sweeter than sucrose; therefore, it could be administrated in minimal doses and considered to be almost non-caloric.[4] Fear of dental caries, obesity, and diabetes, in addition to low-cost, gorgeous commercials, markedly increased the consumption of these synthetic sweeteners.[5]

Aspartame is reported as an extremely hazardous compound and raises a great deal of arguments regarding its safety and the circumstances of its approval by the FDA until now. Neurological or behavioral reactions ranging from headaches, convulsions, blindness, and

Omega Fatty Acids in Brain and Neurological Health
https://doi.org/10.1016/B978-0-12-815238-6.00023-7

intellectual changes were described during its consumption.[6] The European Food Safety Authority and FDA considered that ASP's acceptable daily intake is 40–50 mg/kg/day respectively.[7]

Physical Properties

It is a white, odorless crystalline, extremely sweet powder. It is an L-aspartyl-L-phenylalanine methyl ester and its molecular formula is $C_{14}H_{18}N_2O_5$. ASP molar mass, density, and melting point are documented to be 294.31 $g\,mol^{-1}$, 1.347 g/cm^3, between 246°C and 247°C.[8] The maximum stability of ASP occurs at pH 4.3 in aqueous solution with 300 days half-life; therefore, PH, together with temperature variations, greatly affects the ASP solubility. Acidic solutions and higher temperatures increase the solubility of ASP.[9] Heating causes loss of ASP integrity by hydrolyzing its peptide bonds, so it cannot be used in cooking, baking, or for prolonged storage in liquids due to release of its amino acid content (phenylalanine and aspartic acid). ASP is also slightly soluble in alcohol.[10]

Consequently, aqueous and dry forms are ideal for ASP storage. In the dry state, stability is mainly temperature dependent and decreases at temperatures below 30°C or above 80°C. In aqueous form, stability is maximal at a pH of 4.3; beyond this pH, it breaks down into its metabolites and loses some of its sweetness.[11]

Biochemistry of Aspartame

Ingestion and Metabolism

Upon ingestion, aspartame is absorbed from the intestinal lumen and metabolized by its enzymes (esterase and peptidase) into three metabolites (amino acid isolates), phenylalanine (50%), aspartic acid (40%), and methanol (10%).[12] In the liver, phenylalanine hydroxylase enzyme processes phenylalanine into L-tyrosine. Then, tyrosine hydroxylase enzyme converts L-tyrosine into L-dopa (L-3,4 dihydroxyphenylalanine).[13] L-Dopa by its turn is further transformed into the catecholamines (dopamine, norepinephrine, and epinephrine) by the amino acid decarboxylase enzyme.[14]

Aspartic acid (di-carboxyl amino acid) is also metabolized in the liver by aspartate kinase enzyme into L-lysine and L-methionine. Multiple enzymes (species dependent) are involved in methanol metabolism into formaldehyde: alcohol dehydrogenase in primates and alcohol catalase in rodents. Formic acid, resulting from oxidation of formaldehyde by formaldehyde dehydrogenase, is rapidly metabolized into carbon dioxide and water in rodents since rodents produce more folic acid (tetrahydrofolate) than primates. Humans are uniquely sensitive because of their low hepatic folate concentrations; therefore, excess formic acid may lead to metabolic acidosis and tissue injury.[15]

Effect of Aspartame on the Brain

Chronic ASP ingestion is already accepted to cause neurological, mental, and behavioral disorders; these disturbances are assumed to its direct effects or its metabolites, affecting both learning and emotional functions.[6]

Phenylalanine represents about 50% of ASP metabolites. It is transferred to the general pool of amino acids present in the blood and may share in protein synthesis. As previously mentioned, phenylalanine is partially metabolized to tyrosine, L-dopa, and then to catecholamines.[14] The residual portion of phenylalanine can cross the blood brain barrier to the brain by competitive binding with neutral amino acid transporters (NAAT). These transporters are also incorporated in the transport of numerous amino acids such as tyrosine, threonine, methionine, histidine, glutamine, serine and tryptophan, which are imperative for synthesis of neurotransmitters. The competitive binding and the affinity of NAAT is proportionate to the amino acids' concentration in blood.[7]

Excess plasma concentration of phenylalanine will combine and saturate the NAAT, resulting in its increased transfer to the brain. In turn, it modifies the brain's neurochemical composition through interfering tyrosine conversion to the biogenic amines dopamine, adrenaline and tryptophan to serotonin.[16] Decreased dopamine levels in the substantia nigra neurons of the midbrain reduces the mobility, balance, and walking as they are connected to the corpus striatum and cerebral cortex. These changes are the first symptoms of Parkinson's disease.[6]

Moreover, lower serotonin levels in the hippocampus, raphe nucleus, and substantia nigra cause modifications in memory, sleep, sensory perception, mood, and temperature regulation by affecting cell proliferation, differentiation, migration, and synaptogenesis.[17] Reduced serotonin levels minimize GABA transporters activity in astrocytes, which increases GABA neurotransmitter levels leading to sustained depolarization inhibition of postsynaptic membranes in nerve cells.[6]

Aspartic acid represents about 40% of the ASP's metabolites, and its blood and brain concentrations are amplified in constant ASP intake.[18] In CNS, aspartate, a nonessential amino acid, acts as a neurotransmitter incorporated in impulse transmission in-between neurons. The amino acids glutamine, asparagine, and glutamate are derived from aspartate. Increased levels of these amino acids, particularly glutamate, cause hyperexcitability of neurons that induces neurodegeneration. Maintenance of the appropriate homeostasis of these amino acids is a crucial role played by astrocytes.[6]

Excess glutamate and aspartate released from neurons are collected by the astrocytic excitatory amino acid transporters: EAAT-1 (GLAST), EAAT-2 (GLT-1) And EAAT-3. Supersaturation of these transporters prevents the absorption of glutamate and aspartate by the astrocytes. Glutamate in the glial cells is converted to glutamine, which can be reused again by nerve cells in glutamate resynthesis. The uptake and release processes of these amino acids are dependent on calcium ions.[19] The excess excitatory glutamate activates the astrocytes metabotropic glutamate receptors mGluRs group I (mGluR1 and mGluR5) leading to intracellular ion concentration disturbance, particularly Ca^{2+}, Na^{+} and K^{+}. The upsurge in intracellular calcium results in release of extensive amounts of glutamate.[6] This released alpha glutamate stimulates NR2B subunit of presynaptic and postsynaptic N-methyl-D-aspartate receptors (NMDA) and presynaptic mGluRs receptors, leading to neuronal hyperexcitability. The excessive activation of NMDA receptors by glutamate results in neurodegeneration.[20] This process is called excitotoxicity and the excess aspartate and glutamate are called excitotoxins. Continuous neuronal hyperexcitability causes depletion of the intracellular ATP store, and this leads to modifications in the calcium influx. The alterations in calcium uptake result in mitochondrial damage and subsequent energetic disturbance. Finally, damage of the cell membrane, increasing capillary permeability, and release of the reactive oxygen species

(ROS) occur, causing oxidative stress and further continuous neurodegeneration of nerve cells and glia.[21] Additionally, reactive nitrogen production and accumulation increase due to hyperstimulation of glutamate receptors, which leads to free radicals formation such as nitric peroxide. Both oxygen and nitrogen reactive species act together to induce proteins and DNA damage, predominantly mitochondrial DNA in the neurons.[22] Neurons are mainly susceptible to oxidative stress because of high concentrations of polyunsaturated fatty acids, which render them much more vulnerable to lipid peroxidation compared with other tissues.[23]

The excess free oxygen and nitrogen reactive species become attached to the fatty acids components of the neuronal cell membrane, thereby hindering the neuronal cell functions. The excess free radicals' production may also increase permeability of the blood-brain barrier in a time- and concentration-dependent manner.[24]

On the other hand, the excess glutamate activates astrocytes, and amplification of this amino acid neurotoxicity occurs. The neurodegeneration caused by excitotoxicity of glutamate and aspartate toxic substances like nitric oxide and pro-inflammatory cytokines, IL-1, IL-6, TNF-α, PGE 2, are released by the activated microglia and astrocytes.[20] As mentioned above, neurochemical alterations in neurons and astrocytes are parallel to those occurring in neuro-degenerative diseases such as Alzheimer's, epilepsy, Huntington's, and multiple sclerosis.[6]

Methanol constitutes about 10% of ASP breakdown. After ASP ingestion, methanol is oxidized to formaldehyde and then to formic acid.[7] The conversion of methanol to formaldehyde in the retina may cause blurring of vision and even blindness. In addition, metabolic acidosis results from formic acid accumulation, and this could eventually cause CNS depression, coma, and death from respiratory system paralysis. Methanol poisoning can occur after drinking chilled beverages containing ASP following physical exertion or after excessive fluid loss due to high temperature.[18]

OMEGA-3

Introduction

Polyunsaturated fatty acids (PUFAs), a family of unsaturated fatty acids, are reported to be essential as they cannot be synthetized by the human body and must be provided from dietary sources. Omega-3 fatty acids exert an important role in normal brain development, growth, and cognitive/behavioral functions as they are, extremely concentrated in brain tissues.[25] Additionally, PUFAs have great impact on cell migration, apoptosis,[26] synaptogenesis, and synaptic transmission.[27]

Chemistry

Three major types of PUFAs are identified: alpha-linolenic acid (ALA), eicosapentaenoic acid (EPA), and docosahexaenoic acid (DHA). Upon ingestion, EPA and DHA are more easily used by the human body.

Omega-3s (n3) are unsaturated fatty acids that have carbon chains in which the first double bond exists as the third carbon-carbon bond from the terminal methyl end (n) of the carbon chain. ALA, EPA, and DHA are structurally formed of 18, 20, and 22 carbon atoms with 3, 5, or 6 double bonds respectively. These bonds have the two hydrogen atoms on their sides, and this is known as cis-configuration.[25] Most naturally occurring fatty acids have an even number of carbon atoms in cis-configuration where they are more easily transformable. In addition, this configuration exposes the last double bond geometrically and electrically, rendering the omega 3 acids more fragile.[25]

Daily values: Acceptable Intake and Acceptable Macronutrient Distribution Range

Macronutrients have AI (Acceptable Intake) and AMDR (Acceptable Macronutrient Distribution Range). The AI is 1.6g/day for men and 1.1g/day for women. The AMDR is estimated to be 0.6% to 1.2% of total energy. EPA and/or DHA consumption yields about 10% of the AMDR because of their greater physiological potency than ALA.[25]

Dietary sources: The AI and AMDR could be derived from the following sources; *Fish*: Considered as one of the most important sources of omega 3, however, they obtain it from the algae or plankton in their diet and do not produce it. Oils from salmon, herring, mackerel, anchovies, and sardines, and to a lesser extent, tuna, are very rich in PUFAs, especially EPA and DHA. Although these fatty acids are widely available in fish oils, the accumulation of heavy metals and fat-soluble pollutants in fish should be considered.[25] *Flax seeds*: Linseed oil derived from flax seeds is reported to be six times richer than fish oils in omega 3 content and is possibly the most widely existing botanical source (approximately 55% ALA). *Eggs*: Contain higher levels (mostly ALA). *Meat*: Grass-fed meat has higher omega 3 content than grain-fed meat. *Vegetables, fruits, and nuts*: Strawberries, acai palm fruit, broccoli, and walnuts contain higher and appreciable amounts. *Microalgae*: DHA can be derived from *Crypthecodinium cohnii* and *Schizochytrium* and may be manufactured commercially in bioreactors. This is the solitary accepted source of DHA for vegetarians. Brown algae oil (kelp) is an EPA source.[25]

Neuroprotective role of omega-3s

Omega-3 fatty acids have a crucial role in neuroprotection particularly in neurodegenerative diseases as Parkinson's and Alzheimer's by supporting neuronal growth via enhancing membrane capabilities and myelin sheaths fortification to repair damage.[25] The brain contains high concentrations of omega-3 PUFAs, especially DHA and arachidonic acid (AA). DHA is the most abundant PUFA, representing about 30%–40% of the phospholipids components in neural plasma membrane and can be synthesized from ALA by cerebrovascular endothelial cells and astrocytes, but not the neurons.[28]

DHA exerts a significant role in the development and repair of the brain under numerous circumstance.[29] After birth, it accumulates rapidly in the growth spurt of the brain to enhance neuronal development as well as improve and reestablish cognitive functions.[30] By time, DHA plays a beneficial influence on oxidative stress cellular injury[31] and also increases neuronal resistance to several kinds of damage, such as amyloid-induced oxidative stress,[32] and also via neurotransmission by providing a precursor pool for signaling molecules and lipid-derived mediators.

PUFAs are essential components of neuronal membrane phospholipids, therefore directly affecting the membrane biophysical properties through increasing its fluidity[33] and reducing the cholesterol level that toughens membranes.[34] The normal signal transduction process depends mainly on the membrane structure, which is kept normal by omega-3 fatty acids.[35] They can also modify the assembly and action of neurotransmitters, membrane bound enzymes, the receptors number and affinity,[36] and the ion channels.

One of the most important roles of Omega-3 PUFAs is to modify dopaminergic, serotoninergic, and cholinergic neurotransmission. The formation of dopamine storage vesicle was significantly decreased due to Chronic omega-3 PUFA deficiency.[37] Preservation of dopamine levels and protection against losses in dopaminergic neurons in the substantia nigra occurred by providing diet containing 7.8 g/kg total omega-3 PUFAs, and 5.3 g/kg diet DHA.[38]

Omega-3 deficiency also reduces serotonin release in the hippocampus of rats.[39] However, this effect was reversed by omega-3 PUFA supplementation in the frontal cortex, striatum, and hippocampus.[40] Moreover, Omega-3 PUFA deficiency modifies cholinergic transmission by significantly increasing the basal acetylcholine release; however, 10% reduction in muscarinic acetylcholine receptor binding was observed.[41]

Omega-3 PUFAs, through changing membrane structure, can control signal transduction pathways. EPA and DHA have inhibitory effects on protein kinase C, cAMP-dependent kinase, and Ca^{2+}/calmodulin-dependent protein kinase,[42] 5-HT-induced activation of mitogen-activated protein kinase[43] and G protein-coupled signaling.[44] Phosphatidylserine biosynthesis in nervous tissues is positively modulated by membrane-bound phospholipid DHA.[45] Its accumulation to some degree is responsible for antiapoptotic effect of DHA[46] and prevention of the age-related deterioration in cognitive ability and memory.[47]

A key role for omega-3 PUFAs is in the synthesis of eicosanoids and other lipid-derived messengers and mediators involved in the inflammatory process.[48] Eicosanoids are oxygenated compounds with antiinflammatory properties, formed during DHA metabolism, such as 10-17S-docosatriene (also called neuroprotectin D-1 or NPD1)[49] and 7,17S-dihydroperoxy-DHA.[50] NPD1 has been reported to exert a neuroprotective impact in ischemia-reperfusion injury[51] and in a model of Alzheimer's disease, where it inhibits the Aβ-triggered stimulation of the pro-inflammatory genes such as Bax, Bik, Bad, and Bid, and activates the antiapoptotic genes encoding Bcl-2, Bcl-xL and Bfl-1(A1).[52]

Increased inflammation and oxidative stress together with alterations in metabolism are the major biochemical characteristics of normal brain aging.[53] Age-related modifications also include decreased weight and volume of the brain, ventriculomegaly, a reduction in enzymatic activity, such as Na+/K+ATPase and diminished membrane fluidity.[54]

Long-term potentiation (LTP) reduction is one of the cellular changes in the aging brain that may cause memory and learning ability loss.[34] However, these changes could be reversed by DHA or EPA[55] through modifications in the glutamate receptor expression, particularly the N-methyl-D-aspartate (NMDA) glutamate receptors that have an imperative effect on synaptic transmission of memory and learning.[56] The chief mediators of fast excitatory neurotransmission induced by glutamate are a-Amino-3-hydroxy-5-methyl-4-isoxazole propionic acid (AMPA) receptors.[57] The GluR2 subunit determines the permeability of AMPA receptors to Ca^{2+}. During the aging process, the GluR2 subunit levels decrease, contributing to Ca^{2+} homeostasis modifications.[58] These modifications in Ca^{2+} flux may augment the neurodegeneration accompanying the neurodegenerative disorders.[59] Moreover, the GluR2 subunit is also essential for dendritic spines' growth and maintenance. These spines have an imperative role in information processing.[60] The age-related reductions in the GluR2 subunit and its consequences are reversed by a diet rich in EPA and DHA.[61]

Another subunit termed NR2B incorporated in learning and memory is also decreased during the aging process.[62] And its deficit is partially retreated by a diet enriched in EPA and DHA.[61] The reversibility of the age-related NR2B reduction in the hippocampus is associated with improved cognition.[63]

Progressive neurodegeneration caused by oxidative stress-induced apoptosis is the main character of neurodegenerative diseases. Aggregation of specific proteins, inflammatory response, mitochondrial dysfunction, and neurotoxicity are closely associated with these diseases.[64] The imbalance between the pro-oxidant's synthesis and degradation and impaired antioxidant mechanisms causes oxidative stress, which by its turn leads to neuronal damage and apoptosis.[65] The brain is mainly vulnerable to oxidative stress, because it contains high levels of easily peroxidizable omega-3 fatty acids such as DHA and AA. A large amount of glucose is consumed by neuronal mitochondria to supply the brain with energy, and this is accompanied with relatively high liberation of free radicals.[66]

DHA is described to have a neuroprotective role against oxidative stress in astrocytes, oligodendroglia cells, and retinal ganglion cells by scavenging the liberated free intracellular radical productions persuaded by hydrogen peroxide, superoxide anion, and hydroxyl radical.[67] As previously mentioned, DHA can give rise to a key bioactive effectors docosanoids such as docosatrienes and resolvins, that can stimulate the antiinflammatory and neuroprotective properties and prevent oxidative stress-induced apoptosis.[52] Moreover, DHA increases the concentrations of the antioxidant enzymes such as catalase and glutathione peroxidase and decreases glutathione levels that lead to elimination of reactive oxygen species.[68]

Another mechanism explaining the antioxidative effects of DHA is to initiate transcription factor nuclear factor erythroid 2-related factor 2 (Nrf2) translocation from the cytoplasm to the nucleus and hence triggering the expression of a series of antioxidant and phase II detoxification fragments, such as hemeoxygenase-1 (HO-1), NAD(P)H quinine oxidoreductase 1 (NQO-1), glutamate-cysteine ligase modifier subunit (GCLM), and glutamate-cysteine ligase catalytic subunit (GCLC) as a response to oxidative stress.[69]

References

1. Magnuson BA, Burdock GA, Doull J, et al. ASP: a safety evaluation based on current use levels, regulations, and toxicological and epidemiological studies. *Crit Rev Toxicol.* 2007; https://doi.org/10.1080/10408440701516184.
2. Butchko HH, Stargel WW, Comer CP, et al. ASP: review of safety. *Regul Toxicol Pharmacol.* 2002;35:. S1-93.
3. Rycerz K, Jaworska-Adamu JE. Effects of aspartame metabolites on astrocytes and neurons. *Folia Neuropathol.* 2013; https://doi.org/10.5114/fn.2013.34191.
4. Rowe RC, Sheskey PJ, Cook WG, Fenton ME, American Pharmacists Association. *Handbook of Pharmaceutical Excipients.* 7th ed. London; Philadelphia: Pharmaceutical Press; 2012. Washington, DC: American Pharmacists Association.
5. Tandel K. Sugar substitutes: Health controversy over perceived benefits. *J Pharmacol Pharmacother.* 2011; https://doi.org/10.4103/0976-500X.85936.
6. Humphries P, Pretorius E, Naudé H. Direct and indirect cellular effects of aspartame on the brain. *Eur J Clin Nutr.* 2008; https://doi.org/10.1038/sj.ejcn.1602866.
7. Ali EMT, Sonpol HMA. Neuroprotective and ameliorating impacts of omega-3 against aspartame-induced neuronal and astrocytic degeneration. *Anat Rec.* 2017; https://doi.org/10.1002/ar.23536.

8. Zafar T, Qayoom Naik ABVKS. Aspartame: effects and awareness. *MOJ Toxicol.* 2017;3(2):23–26.
9. Sun J-P, Han Q, Zhang X-Q, Ding M-Y. Investigations on the degradation of aspartame using high-performance liquid chromatography/tandem mass spectrometry. *Chin Chem Lett.* 2014; https://doi.org/10.1016/j.cclet.2014.04.012.
10. Chattopadhyay S, Raychaudhuri U, Chakraborty R. Artificial sweeteners—a review. *J Food Sci Technol.* 2014; https://doi.org/10.1007/s13197-011-0571-1.
11. Paixão JA, Rodrigues JB, Esmerino EA, Cruz AG, Bolini HMA. Influence of temperature and fat content on ideal sucrose concentration, sweetening power, and sweetness equivalence of different sweeteners in chocolate milk beverage. *J Dairy Sci.* 2014; https://doi.org/10.3168/jds.2014-7995.
12. Singh M, Kumar A, Tarannum N. Water-compatible 'aspartame'-imprinted polymer grafted on silica surface for selective recognition in aqueous solution. *Anal Bioanal Chem.* 2013; https://doi.org/10.1007/s00216-013-6812-6.
13. Daubner SC, Melendez J, Fitzpatrick PF. Reversing the substrate specificities of phenylalanine and tyrosine hydroxylase: aspartate 425 of tyrosine hydroxylase is essential for L-DOPA formation. *Biochemistry.* 2000; https://doi.org/10.1021/bi000493k.
14. Choudhary AK, Pretorius E. Revisiting the safety of aspartame. *Nutr Rev.* 2017; https://doi.org/10.1093/nutrit/nux035.
15. Cederbaum AI. Alcohol metabolism. *Clin Liver Dis.* 2012; https://doi.org/10.1016/j.cld.2012.08.002\rS1089-3261(12)00085-2. [pii].
16. Tsakiris S, Giannoulia-Karantana A, Simintzi I, Schulpis KH. The effect of aspartame metabolites on human erythrocyte membrane acetylcholinesterase activity. *Pharmacol Res.* 2006; https://doi.org/10.1016/j.phrs.2005.07.006.
17. Daubert EA, Condron BG. Serotonin: A regulator of neuronal morphology and circuitry. *Trends Neurosci.* 2010; https://doi.org/10.1016/j.tins.2010.05.005.
18. Ekong M. Aspartame: Sweet Or Bitter. *Internet J Health.* 2008;9(1):5.
19. Pinheiro PS, Mulle C. Presynaptic glutamate receptors: physiological functions and mechanisms of action. *Nat Rev Neurosci.* 2008; https://doi.org/10.1038/nrn2379.
20. Hamilton NB, Attwell D. Do astrocytes really exocytose neurotransmitters? *Nat Rev Neurosci.* 2010; https://doi.org/10.1038/nrn2803.
21. Bowen J, Evangelista AM. *Brain Cell Damage From Amino Acid Isolates.* World Natural Health Organisation; 2002. http://www.wnho.net/aspartame_brain_damage.htm.
22. Mourad IM, Noor NA. Aspartame (a widely used artificial sweetener) and oxidative stress in the rat cerebral cortex. *Int J Pharm Biomed Sci.* 2011;2(1):4–10.
23. Mariani E, Polidori MC, Cherubini A, Mecocci P. Oxidative stress in brain aging, neurodegenerative and vascular diseases: an overview. *J Chromatogr B Anal Technol Biomed Life Sci.* 2005; https://doi.org/10.1016/j.jchromb.2005.04.023.
24. Haorah J, Ramirez SH, Schall K, Smith D, Pandya R, Persidsky Y. Oxidative stress activates protein tyrosine kinase and matrix metalloproteinases leading to blood-brain barrier dysfunction. *J Neurochem.* 2007; https://doi.org/10.1111/j.1471-4159.2006.04393.x.
25. Mirajkar RN, Jamadar SA, Patil AV, Mirajkar SN. Omega 3 fatty acids—clinical implications. *Int J ChemTech Res.* 2011;.
26. Arita K, Yamamoto Y, Takehara Y, et al. Mechanisms of enhanced apoptosis in HL-60 cells by UV-irradiated n-3 and n-6 polyunsaturated fatty acids. *Free Radic Biol Med.* 2003; https://doi.org/10.1016/S0891-5849(03)00310-1.
27. Haag M. Essential fatty acids and the brain. *Can J Psychiatr.* 2003; https://doi.org/10.1177/070674370304800308.
28. Moore SA. Polyunsaturated fatty acid synthesis and release by brain-derived cells in vitro. *J Mol Neurosci.* 2001;.
29. Chang YL, Chen SJ, Kao CL, et al. Docosahexaenoic acid promotes dopaminergic differentiation in induced pluripotent stem cells and inhibits teratoma formation in rats with Parkinson-like pathology. *Cell Transplant.* 2012; https://doi.org/10.3727/096368911X580572.
30. Gamoh S, Hashimoto M, Hossain S, Masumura S. Chronic administration of docosahexaenoic acid improves the performance of radial arm maze task in aged rats. *Clin Exp Pharmacol Physiol.* 2001.
31. Kim EJ, Park YG, Baik EJ, et al. Dehydroascorbic acid prevents oxidative cell death through a glutathione pathway in primary astrocytes. *J Neurosci Res.* 2005; https://doi.org/10.1002/jnr.20384.
32. Lim GP, Calon F, Morihara T, et al. A diet enriched with the omega-3 fatty acid docosahexaenoic acid reduces amyloid burden in an aged Alzheimer mouse model. *J Neurosci.* 2005; https://doi.org/10.1523/JNEUROSCI.4225-04.2005.

33. Stillwell W, Wassall SR. Docosahexaenoic acid: membrane properties of a unique fatty acid. *Chem Phys Lipids*. 2003; https://doi.org/10.1016/S0009-3084(03)00101-4.
34. Yehuda S, Rabinovitz S, Carasso RL, Mostofsky DI. The role of polyunsaturated fatty acids in restoring the aging neuronal membrane. *Neurobiol Aging*. 2002; https://doi.org/10.1016/S0197-4580(02)00074-X.
35. Horrobin DF, Jenkins K, Bennett CN, Christie WW. Eicosapentaenoic acid and arachidonic acid: collaboration and not antagonism is the key to biological understanding. *Prostaglandins Leukot Essent Fat Acids*. 2002; https://doi.org/10.1054/plef.2001.0338.
36. Lafourcade M, Larrieu T, Mato S, et al. Nutritional omega-3 deficiency abolishes endocannabinoid-mediated neuronal functions. *Nat Neurosci*. 2011; https://doi.org/10.1038/nn.2736.
37. Zimmer L, Delpal S, Guilloteau D, Aioun J, Durand G, Chalon S. Chronic n-3 polyunsaturated fatty acid deficiency alters dopamine vesicle density in the rat frontal cortex. *Neurosci Lett*. 2000;284(1-2):25–28.
38. Bousquet M, Saint-Pierre M, Julien C, Salem N, Cicchetti F, Calon F. Beneficial effects of dietary omega-3 polyunsaturated fatty acid on toxin-induced neuronal degeneration in an animal model of Parkinson's disease. *FASEB J*. 2008; https://doi.org/10.1096/fj.07-9677com.
39. Kodas E, Galineau L, Bodard S, et al. Serotoninergic neurotransmission is affected by n-3 polyunsaturated fatty acids in the rat. *J Neurochem*. 2004; https://doi.org/10.1111/j.1471-4159.2004.02401.x.
40. Vancassel S, Leman S, Hanonick L, et al. n-3 Polyunsaturated fatty acid supplementation reverses stress-induced modifications on brain monoamine levels in mice. *J Lipid Res*. 2008; https://doi.org/10.1194/jlr.M700328-JLR200.
41. Aïd S, Vancassel S, Poumès-Ballihaut C, Chalon S, Guesnet P, Lavialle M. Effect of a diet-induced n-3 PUFA depletion on cholinergic parameters in the rat hippocampus. *J Lipid Res*. 2003; https://doi.org/10.1194/jlr.M300079-JLR200.
42. Seung Kim HF, Weeber EJ, Sweatt JD, Stoll AL, Marangell LB. Inhibitory effects of omega-3 fatty acids on protein kinase C activity in vitro. *Mol Psychiatry*. 2001; https://doi.org/10.1038/sj.mp.4000837.
43. Mirnikjoo B, Brown SE, Kim HF, Marangell LB, Sweatt JD, Weeber EJ. Protein kinase inhibition by omega-3 fatty acids. *J Biol Chem*. 2001; https://doi.org/10.1074/jbc.M008150200.
44. Niu SL, Mitchell DC, Lim SY, et al. Reduced G protein-coupled signaling efficiency in retinal rod outer segments in response to n-3 fatty acid deficiency. *J Biol Chem*. 2004; https://doi.org/10.1074/jbc.M404376200.
45. Hamilton J, Greiner R, Salem NJ, Kim HY. n-3 fatty acid deficiency decreases phosphatidylserine accumulation selectively in neuronal tissues. *Lipids*. 2000; https://doi.org/10.1007/S11745-000-0595-x.
46. Akbar M, Kim HY. Protective effects of docosahexaenoic acid in staurosporine-induced apoptosis: involvement of phosphatidylinositol-3 kinase pathway. *J Neurochem*. 2002; https://doi.org/10.1046/j.1471-4159.2002.01015.x.
47. Kim HY. Novel metabolism of docosahexaenoic acid in neural cells. *J Biol Chem*. 2007; https://doi.org/10.1074/jbc.R700015200.
48. Calder PC. n-3 polyunsaturated fatty acids, inflammation, and inflammatory diseases. *Am J Clin Nutr*. 2006. http://dx.doi.org/ 16841861.
49. Mukherjee PK, Marcheselli VL, Serhan CN, Bazan NG. Neuroprotectin D1: a docosahexaenoic acid-derived docosatriene protects human retinal pigment epithelial cells from oxidative stress. *Proc Natl Acad Sci U S A*. 2004; https://doi.org/10.1073/pnas.0402531101.
50. Butovich IA, Lukyanova SM, Bachmann C. Dihydroxydocosahexaenoic acids of the neuroprotectin D family: synthesis, structure, and inhibition of human 5-lipoxygenase. *J Lipid Res*. 2006; https://doi.org/10.1194/jlr.M600280-JLR200.
51. Marcheselli VL, Hong S, Lukiw WJ, et al. Novel docosanoids inhibit brain ischemia reperfusion mediated leukocyte infiltration and pro-inflammatory gene expression. *J Biol Chem*. 2003; https://doi.org/10.1074/jbc.M305841200.
52. Lukiw WJ, Cui JG, Marcheselli VL, et al. A role for docosahexaenoic acid-derived neuroprotectin D1 in neural cell survival and Alzheimer disease. *J Clin Investig*. 2005; https://doi.org/10.1172/JCI25420.
53. T a P, Mattson MP. Molecular mechanisms of brain aging and neurodegenerative disorders: lessons from dietary restriction. *Trends Neurosci*. 2001; https://doi.org/10.1016/S0166-2236(01)00005-4.
54. Anderton BH. Ageing of the brain. *Mech Ageing Dev*. 2002; https://doi.org/10.1016/S0047-6374(01)00426-2.
55. Martin DSD, Spencer P, Horrobin DF, Lynch MA. Long-term potentiation in aged rats is restored when the age-related decrease in polyunsaturated fatty acid concentration is reversed. *Prostaglandins Leukot Essent Fat Acids*. 2002; https://doi.org/10.1054/plef.2002.0408.
56. Lynch MA. Long-term potentiation and memory. *Physiol Rev*. 2004;84(1):87–136.

57. Segovia G, Porras A, Del Arco A, Mora F. Glutamatergic neurotransmission in aging: a critical perspective. *Mech Ageing Dev*. 2001; https://doi.org/10.1016/S0047-6374(00)00225-6.
58. Hof PR, Duan H, Page TL, et al. Age-related changes in GluR2 and NMDAR1 glutamate receptor subunit protein immunoreactivity in corticocortically projecting neurons in macaque and patas monkeys. *Brain Res*. 2002; https://doi.org/10.1016/S0006-8993(01)03345-5.
59. Foster TC, Kumar A. Calcium dysregulation in the aging brain. *Neuroscientist*. 2002; https://doi.org/10.1177/107385840200800404.
60. Passafaro M, Nakagawa T, Sala C, Sheng M. Induction of dendritic spines by an extracellular domain of AMPA receptor subunit GluR2. *Nature*. 2003; https://doi.org/10.1038/nature01781.
61. Dyall SC, Michael GJ, Whelpton R, Scott AG, Michael-Titus AT. Dietary enrichment with omega-3 polyunsaturated fatty acids reverses age-related decreases in the GluR2 and NR2B glutamate receptor subunits in rat forebrain. *Neurobiol Aging*. 2007; https://doi.org/10.1016/j.neurobiolaging.2006.01.002.
62. Xu SJ, Chen Z, Zhu LJ, Shen HQ, Luo JH. Visual recognition memory is related to basic expression level of NMDA receptor NR1/NR2B subtype in hippocampus and striatum of rats. *Acta Pharmacol Sin*. 2005;26(2):177–180.
63. Mesches MH, Gemma C, Veng LM, et al. Sulindac improves memory and increases NMDA receptor subunits in aged Fischer 344 rats. *Neurobiol Aging*. 2004; https://doi.org/10.1016/S0197-4580(03)00116-7.
64. Barnham KJ, Masters CL, Bush AI. Neurodegenerative diseases and oxidative stress. *Nat Rev Drug Discov*. 2004; https://doi.org/10.1038/nrd1330.
65. Steele M, Stuchbury G, Munch G. The molecular basis of the prevention of Alzheimer's disease through healthy nutrition. *Exp Gerontol*. 2007; https://doi.org/10.1016/j.exger.2006.06.002.
66. FLOYD R, HENSLEY K. Oxidative stress in brain aging Implications for therapeutics of neurodegenerative diseases. *Neurobiol Aging*. 2002; https://doi.org/10.1016/S0197-4580(02)00019-2.
67. Liu Q, Wu D, Ni N, et al. Omega-3 polyunsaturated fatty acids protect neural progenitor cells against oxidative injury. *Mar Drugs*. 2014;12(5):2341–2356. https://doi.org/10.3390/md12052341.
68. Hashimoto M, Tanabe Y, Fujii Y, Kikuta T, Shibata H, Shido O. Chronic administration of docosahexaenoic acid ameliorates the impairment of spatial cognition learning ability in amyloid b-infused rats. *J Nutr*. 2005.
69. Ma Q. Role of Nrf2 in oxidative stress and toxicity. *Annu Rev Pharmacol Toxicol*. 2013; https://doi.org/10.1146/annurev-pharmtox-011112-140320.

CHAPTER

24

DHA-Enriched Formulations as a Promising Strategy for the Treatment of Hypoxic/Ischemic Brain Injury

Humberto Milani, Daniela Velasquez Oliveira

Department of Pharmacology and Therapeutics, Heath Sciences Center, State University of Maringá, Maringá, Brazil

INTRODUCTION

In the context of cerebrovascular diseases, hypoxic/ischemic brain injury (HI-BI) is one of the most common and serious health problems, the prevalence of which has increased as the elderly population has grown. A reduction or interruption of cerebral blood flow (CBF) can affect restricted brain regions (i.e., focal cerebral ischemia) or the entire brain (i.e., global cerebral ischemia). Focal cerebral ischemia is most often referred to as stroke. It is usually caused by the embolic or thrombotic occlusion of one or more cerebral arteries. Clinically, stroke is reflected by the abrupt onset of neurologic deficits, the nature and severity of which primarily depend on the localization and extent of brain damage (e.g., sensory and/or motor deficits in a limb or paralysis of the entire half of the body along with emotional and cognitive impairments). Stroke patients often become partially or totally disabled with regard to engaging in productive work. No less dramatic, global cerebral ischemia can lead to widespread brain damage and lesions in the hippocampus, thalamus, and various cortical regions that are tightly correlated with the severity of functional sequelae.[1] Global cerebral ischemia most often occurs after reversible cardiac arrest, the duration of which will determine a constellation of sequelae that characterize so-called hypoxic-ischemic encephalopathy. Even after successful resuscitation, most patients still die in the hospital because of neurological or cardiac complications. Patients who survive long after the cardiac arrest may experience a broad spectrum of dysfunctions, the most prominent of which are cognitive impairments, including anterograde and retrograde amnesia, attention deficits, verbal communication disability, spatial/temporal disorientation, and decision-making impairments.[2] Hypoxic-ischemic brain injury also occurs under conditions of chronic cerebral hypoperfusion (CCH)

Omega Fatty Acids in Brain and Neurological Health
https://doi.org/10.1016/B978-0-12-815238-6.00024-9

that results from several cardiovascular and metabolic diseases that are highly prevalent in the aged population, including coronary artery disease, myocardial infarction, atherosclerosis of the carotid arteries, hypertension, and diabetes. The role of CCH in the development of age-related dementias is well recognized.[3] Overall, HI-BI imposes economic costs that may reach hundreds of millions of dollars annually worldwide. With regard to stroke specifically, more than USD$65 billion was spent in 2008 solely in the United States.[4]

ANIMAL MODELS AND PATHOPHYSIOLOGY

The basic mechanisms that underlie the pathophysiology of ischemia-induced brain damage have been revealed by in vitro studies (e.g., the oxygen/glucose deprivation model) that have provided valuable mechanistic insights into potential neuroprotective candidates. Major advances have also been made from in vivo models that are associated with cellular, electrophysiological, biochemical/immunohistochemical, and more recently metabolomic analyses in wild type and genetically modified animals.[5] Two in vivo models are notable because of their broad use: (i) middle cerebral artery occlusion (MCAO) model of stroke (i.e., focal/regional cerebral ischemia), which has been largely applied using rats and mice,[5] and (ii) four-vessel occlusion (4-VO) model of transient, global cerebral ischemia (TGCI). The 4-VO model is the rodent model that is most used to reproduce cerebral outcomes of reversible cardiac arrest. For full details of these two models, in addition to other animal models, excellent reviews have been published that discuss their main differences, advantages, and disadvantages from the perspective of human conditions of cerebral ischemia.[6–9]

Although HI-BI can have different etiologies (e.g., thromboembolism, cardiac arrest, severe hypotension, internal carotid artery stenosis/atherosclerosis, and even traumatic brain injury), the main pathophysiological mechanisms that culminate in irreversible brain damage may be similar.[10,11] These mechanisms have been investigated using both in vitro and in vivo models, with an emphasis on processes that lead to neuronal death, which can be differentiated by four key phases: (i) excitotoxicity, (ii) oxidative stress, (iii) neuroinflammation, and (iv) apoptosis. These four phases can develop across a few minutes (e.g., excitotoxicity) or hours to days post-ischemia (e.g., inflammation and apoptosis).[12,13] Ischemia-induced excitotoxicity refers to the neurotoxic effect of the excitatory amino acid glutamate. When brain perfusion drops below 15%–20% of normal CBF, the already poor energetic reserve of the brain collapses. At this level of brain perfusion, the immediate consequence is the breakdown of ion homeostasis in the neuronal membrane, particularly failure of the Na^+/K^+-adenosine triphosphatase pump.[14] Intracellular Na^+ accumulation then leads to cytotoxic edema, anaerobic metabolism, and acidosis. This energetic failure also suppresses glial and neuronal mechanisms that remove glutamate from the synaptic cleft, thus resulting in the excessive activation of glutamatergic receptors, mainly *N*-methyl-D-aspartate and α-amino-3-hydroxy-5-methyl-4-isoxazolepropionic acid receptors. The activation of a series of downstream signaling pathways then culminates in an overload of intracellular calcium that in turn activates lytic enzymes (e.g., proteases, lipases, and endonucleases). In the following hours to days of reperfusion, the pathological cascade of ischemia advances, partly as a downstream consequence of excitotoxicity. Oxidative and nitrosative stress, membrane degradation, mitochondrial and DNA damage, inflammation, and apoptosis then occur, culminating in irreversible damage to neuronal membranes and cell death. This sequence of events described above has well been characterized in an effort to identify potential targets for the development of neuroprotective strategies. However, myriad other intermediate molecular/biochemical changes may also participate in the etiology of ischemic damage. For a detailed view of the intricate nature of neuronal and nonneuronal molecular events that are triggered by ischemia, we refer the reader to excellent previous reviews.[13,15–17] Importantly, these neurodegenerative processes (i.e., excitotoxicity, oxidative/nitrosative stress, neuroinflammation, and apoptosis) and other intermediary changes operate at different time points and locations in the extra- and intracellular milieu. This spatial-temporal pattern of ischemic brain damage makes the development of pharmacological neuroprotection a tremendous challenge.

Concomitantly with neurodegenerative processes, neuroregenerative or protective mechanisms may also be recruited in the reperfusion phase of cerebral ischemia, including angiogenesis, neurogenesis, regeneration, synaptic plasticity, and axonal rewiring. Although neurons that survive ischemia can respond with ultrastructural changes that can promote functional recovery to some extent,[15,18–20] permanent deficits that are often observed in patients with brain ischemia indicate that the capacity of these endogenous regenerative mechanisms is usurped by the impact of ischemia. Therefore, pharmacological interventions that can stimulate these neuroregenerative processes may be potential alternatives to promote functional recovery after HI-BI.

PHARMACOTHERAPY OF HYPOXIC/ISCHEMIC BRAIN INJURY

Although several drugs have shown promising neuroprotective effects in animal models of HI-BI, the translation of preclinical data to clinically effective medications has seen no such success, thus leaving recombinant tissue plasminogen activator (rtPA; Alteplase), a thrombolytic agent, as the only approved therapy for acute stroke. However, its narrow time window of efficacy (TWE) (4.5h) and the possible occurrence of hemorrhage limit its application to only a minority of patients.[15,21,22]

Ischemia-induced neurodegeneration develops across several processes that occur at different time points post-ischemia. Therefore, any drug that targets a single pathological process may have only limited efficacy or no efficacy at all. This may at least partially explain the disappointing results of clinical trials that have tested the neuroprotective efficacy of monotherapies that target single processes in the intricate cascade of ischemic damage. Our current understanding is that drugs with pleiotropic actions may confer functional recovery after ischemic brain damage more efficiently than drugs that target only a single mechanism. Moreover, endogenous, neuroregenerative processes (e.g., neurotrophicity, neurogenesis, axonal rewiring, and synaptogenesis) may represent suitable targets of pleiotropic substances.[23] Docosahexaenoic acid (DHA), an omega-3 (ω-3) polyunsaturated fatty acid (PUFA), has received increasing attention as a promising candidate for neuroprotection after brain ischemia.

DHA: NEUROPROTECTION AND FUNCTIONAL RECOVERY AFTER HYPOXIC/ISCHEMIC BRAIN INJURY

The PUFAs refer to long-chain fatty acids with more than one double bond in their carbon side chains. They can be divided into three major classes, based on the position of their final carbon bond at the methyl end: ω-3, ω-6, and ω-9. In the ω-3 series, the number of double bonds ranges from three (C18:3n-3, α-linolenic acid [ALA]) to six (C22:6n-3, DHA), with the final carbon-carbon double bond in the n-3 (or ω-3) position (i.e., the third bond from the methyl end). As an essential fatty acid, ALA cannot be synthesized in mammals because they lack the enzyme Δ^{15}-dessaturase. Major dietary sources of ALA include green leafy vegetables, flaxseed oil, canola oil, linseed oil, and walnuts. In mammals, ALA is sequentially converted to a series of other ω-3 intermediaries in the liver endoplasmic reticulum and finally to DHA by β-oxidation in peroxisomes. However, certain deep-sea marine algae, as well as oil that is extracted from certain cold-water fish species (e.g., salmon, mackerel, halibut, sardines, tuna, and herring), constitute major dietary sources of DHA and its immediate precursor, eicosapentaenoic acid (EPA; 20:5n-3).[24,25] The dietary deficiency of EPA or DHA has been linked to several human diseases, including chronic heart disease (CHD), vascular diseases, atherosclerosis, and neuropsychiatric diseases, among others.[26–28] The preventive, therapeutic value of ω-3 fatty acids (e.g., ALA, EPA, and DHA) has been systematically investigated in the setting of CHD,[27,28] but controversial results have been reported.[29]

In the field of cerebral ischemia, several studies carried out in animals indicate that DHA may have therapeutic potential as a neuroprotective agent that can facilitate functional recovery. As defined by Myron Ginsberg, "neuroprotection encompasses any strategy, or combination of strategies, that antagonizes, interrupts, or slows the sequence of injurious biochemical and molecular events that, if left unchecked, would eventuate in irreversible ischemic injury."[30] Most recently, however, this author refined this concept to include strategies that primarily target the cerebral circulation, which constitutes the obligatory substrate on which ischemic brain survival depends.[31] In addition, the concept of neuroprotection transcends the molecular, biochemical, and histological levels of ischemic injury to include enduring functional recovery (or preservation). Indeed, the most important end-point for assessing the neuroprotective potential of drugs is functional, despite the obvious importance of preserving as many neurons as possible after ischemic insult.[32]

The efficacy of ω-3 PUFAs to treat the outcomes of HI-BI in humans has not yet been evaluated. However, increasing evidence from animal models points to likely efficacy. The efficacy of DHA in alleviating the outcomes of cerebral ischemia was reported for the first time by Umemura et al.,[33] where DHA supplementation for 8 weeks before thombotic occlusion of the middle cerebral artery (MCA) enhanced the thrombolytic efficacy of rtPA and reduced infarct size in young rats. Okada et al.[34] used the 4-VO model of TGCI and found that per oral DHA administration for 21 days prior to ischemia improved retrograde memory performance in the radial maze model and apparently reduced hippocampal neurodegeneration. Subsequent studies with gerbil models of TGCI also demonstrated that DHA treatment for 4 weeks prior to ischemia improved brain perfusion, survival rate, and locomotor activity while reducing cerebral edema and arachidonic acid-derived mediators.[35–37] These early studies suggested that DHA given preventively is able to mitigate the outcomes of cerebral ischemia.

However, cerebral ischemia is a condition that usually occurs abruptly and unpredictably. This implies that any therapeutic intervention will be instituted after the occurrence of ischemia, and the window of opportunity for a supposed neuroprotective treatment represents the most important variable for the therapeutic success. With this in mind, the efficacy of DHA as a possible strategy to treat acute ischemic stroke was evaluated more systematically in a series of studies by the research groups of Myron Ginsberg and Ludmila Belayev using the MCAO model of stroke. For example, treatment with DHA combined with intravenous albumin administration soon after the beginning of reperfusion provided robust neuroprotection, reflected by improvements in neurobehavioral scores and significant reductions of cortical infarction and brain swelling 72h post-ischemia.[38] These data were confirmed and later extended to a condition in which DHA was given alone 3h after ischemia, suggesting that DHA has a clinically relevant TWE. As already mentioned, this variable is crucial for the successful treatment of acute ischemic stroke.[39] The therapeutic potential of DHA became more evident when its beneficial effects on behavior and brain damage were found to be sustained for up to 7 days after stroke, with a TWE of 5h.[40] Belayev's group extended their previous findings by demonstrating that neuroprotectin-D1 (NPD1), a 10,17*S*-docosatriene metabolite of DHA, may be responsible for the neuroprotective properties of DHA through activation of the PI3K/Akt/glycogen synthase kinase pathway and the consequent reduction of neuroinflammation, stimulation of cell survival, and restoration of synaptic circuits in the penumbra area.[41] Altogether, these findings support the efficacy of DHA treatment alone or combined with albumin, which was further advanced by a study that found that DHA treatment was also effective in aged rats that were subjected to the MCAO model of acute stroke.[42]

These results that were obtained with DHA in the MCAO model of acute ischemic stroke are consistent with a series of data from our laboratory. In these studies, the effects of a standardized, high-grade pharmaceutical formulation of fish oil (FO; Omega-3 DHA 250, Biotik, São Paulo, Brazil) were systematically investigated in the 4-VO model of TGCI. We used a modified version of the eight-arm radial maze, referred to as the aversive radial maze, combined with a behavioral protocol that allows the assessment of remote, retrograde memory as the long-term functional end-point. Retrograde amnesia implies that the memory trace that was formed and consolidated during the training phase (i.e., prior to ischemia) was severely and persistently lost after TGCI. In our first study that used this paradigm, the FO was given orally for 3 days prior to TGCI and continued up to 28 days after the beginning of reperfusion. Under this treatment regimen, FO almost completely reversed the amnesic effect of TGCI. Importantly, this memory-protective effect of the FO was maintained for at least 2 weeks after the cessation of treatment, suggesting sustained functional recovery.[43] This finding was then replicated, even when the treatment duration was shortened from 31 to 10 days, provided it was given in the acute phase of ischemia. In fact, FO was no longer effective in preventing ischemia-induced memory loss when treatment was delayed for 15 days after ischemia. These data suggest that the efficacy of FO in facilitating memory recovery (or preservation) may be related to some FO-mediated cellular/molecular actions that occur in the early phase of ischemia/reperfusion and not to the ability of FO to stimulate memory performance on its own.[44] This hypothesis received support from a subsequent study that showed that the FO treatment for only 3 days prior to ischemia effectively prevented both ischemia-induced memory loss and oxidative stress. In addition, this memory protection was very similar to the memory-protective effects that were observed after 10 days of treatment (i.e., 3 days before ischemia +7 days after ischemia), suggesting that the antioxidant action that occurred in the hyperacute phase of ischemia/reperfusion at least partially contributed to the beneficial effect of FO on long-term retrograde memory.[45] We further found that FO continued to effectively prevent ischemia-induced memory loss even when treatment began 4h after TGCI in aged rats.[46] This 4-h TWE was later extended to up to 8h post-ischemia but not to 12h.[47]

In this sense, our results observed with FO after global cerebral ischemia agree closely with those observed with DHA observed in the MCAO stroke model.

The findings that were observed with pure DHA in the MCAO model of stroke[38–42] and the findings that were observed with FO that was tested in the 4-VO model[43–47] are reinforced by observations that Omegaven, a FO-based lipid emulsion, which is used clinically for parenteral nutrition, reduced both brain damage and neurological deficits when administered intravenously after MCAO in mice.[48] Further, neurohistological protection was observed in neonatal mice that received intraperitoneal injections of a DHA-enriched triglyceride emulsion up to 2h after hypoxia/ischemia.[49]

Altogether, the findings strongly suggest that DHA as a pure substance, DHA/EPA in high-grade FO formulations (e.g., Omacor, Omega-3 DHA 250), and DHA/EPA as a FO-based lipid emulsion (e.g., Omegaven) may have therapeutic utility for the treatment of HI-BI and deserve further consideration in clinical trials. In this regard, the latter two studies[48,49] are very important because they indicate that the availability of high-grade, pharmaceutic formulations that contain DHA might make the parenteral administration of DHA/EPA feasible in humans. These findings, therefore, might be the first translational data for the early clinical management of ischemic stroke with DHA-enriched preparations.[48]

MECHANISMS OF DHA-MEDIATED NEUROPROTECTION

The highest concentration of DHA is found in the brain and retina, where it is incorporated into membrane phospholipids, mainly phosphatidylethanolamine and phosphatidylserine, from which it is mobilized to play an important role in the development and maintenance of photoreceptor and neuronal cell integrity and function. In the cell membrane, DHA modulates neurotransmitter release, membrane-bound enzyme and ion channel activity, gene expression, and synaptic plasticity. This broad biological activity characterizes DHA as a pleiotropic substance. Under pathological conditions (e.g., hypoxia-ischemia), DHA is released from membrane phospholipids by phospholipase A2 (PLA2) and converted to 10,17*S*-docosatriene (i.e., the metabolite NPD1). The pharmacological or nutraceutical actions of NPD1 include the inhibition of neutrophil infiltration, the inhibition of pro-inflammatory gene expression (e.g., nuclear factor-κB, interleukin-1β, cyclooxygenase-2, and the activation of apoptotic signaling pathways (e.g., Bax, Bad, Bid, Bik). Conversely, NPD1 stimulates the expression of antiapoptotic Bcl-2/Bcl-xL family proteins and activation of the mechanistic target of rapamycin pathway (i.e., PI3K/Akt/mTOR) that is related to autophagy, angiogenesis, and nerve regeneration.[50,51]

The exact mechanisms by which DHA confers neuroprotection in animal models of cerebral ischemia are not completely understood, partly because of its wide range of cellular/molecular actions (i.e., pleiotropism). One proposal is that the efficacy of DHA in protecting against ischemic damage results from cooperative or synergistic actions at different levels, including (i) the reduction of arachidonic acid metabolism with a consequent decrease in pro-inflammatory eicosanoids, (ii) the normalization of cell membrane ionic homeostasis, (iii) the reduction of glutamate neurotransmitter excitotoxicity, (iv) the prevention of apoptosis, (v) the preservation of mitochondrial function, and (vi) the reduction of reactive oxygen species-mediated oxidative stress and inflammatory mediators.[52] Together with its anti-inflammatory and antioxidant properties, DHA in FO has the ability to activate endothelial nitric oxide synthase with consequent generation of nitric oxide, which improves vascular function and may prevent cardiovascular disorders.[28] These actions of DHA may extend to the brain and contribute to neuroprotection by improving blood perfusion in the penumbra area of infarct.[53] Other DHA-mediated actions may also contribute to neuroprotection, including the stimulation of neurotrophic factors (e.g., brain-derived neurotrophic factor, BDNF) and neuroplasticity (e.g., synaptogenesis and neurite outgrowth), the maintenance of neuronal and synaptic membrane fluidity and lipid raft function, and improvements in the coupling of blood flow to glucose utilization.[54] Details of the intricate molecular pathways and mechanisms that have been implicated in DHA-mediated neuroprotection are beyond the scope of this chapter; we refer the reader to the articles cited above.[28,50–54] Altogether, the anti-inflammatory, antioxidant, antiapoptotic, and neurotrophic actions of DHA and its metabolite NPD-1 may act synergistically to play a key role in neuroprotection after HI-BI.

Before concluding this topic, we refer again to the results of studies from our laboratory[43–47] to stress the possibility that DHA-mediated neurotrophic actions may account for the effectiveness of FO in facilitating functional recovery after TGCI. The memory-protective effect of the FO has repeatedly been observed in the absence of pyramidal, neuronal rescue in the hippocampus[43–47] and cerebral cortex.[44–46] The FO-mediated learning and memory recovery without pyramidal cells rescue has also been reported by others,[55] suggesting that this is a consistent finding. Therefore, one unresolved issue is the way in which FO prevents memory loss in the presence of damaged hippocampal/cortical circuits because the integrity of these circuits is required for spatial memory performance.[56–58] One possibility is that DHA in FO acts on neurons that survive ischemia. As mentioned above, ischemia-induced retrograde amnesia reflects the loss of memory that was formed and consolidated prior to ischemia. We posit that the neuronal networks that underlie such preexisting memory (i.e., retrograde memory) survive ischemia but become functionally impaired afterward. If so, then we can further speculate that FO can reestablish (or preserve) the functionality of these neuronal networks that survive ischemia, thus facilitating memory recovery or preservation. Neurons that survive an ischemic event can respond with ultrastructural changes, including dendritic restructuring, reactive synaptogenesis, and growth-promoting processes that, in turn, promote functional recovery even in the absence of neuronal rescue.[20] One possibility, therefore, is that FO acts on neurons that survive ischemia and stimulates ultrastructural changes that may contribute to memory recovery (or preservation), an effect that has been consistently observed.[43–47]

Preliminary data from our laboratory support this possibility. The same FO treatment regimen that was found to improve memory recovery and preserve the expression of microtubule-associated protein 2 (MAP2) after TGCI,[47] also increased the density of dendritic spines not only in the hippocampus but also in the prefrontal and postrhinal cerebral cortices (Fig. 1).

Notably, MAP2 is a structural protein that is highly compartmentalized in dendrites where it plays a key role in dendritic plasticity, and the dendritic spines are a major site of synaptic input, the density of which determines the surface area of stimulation that a given neuron receives, that is, their receptive field. Although the function of dendritic spines is far from being fully understood, they might play a crucial role in information transmission and processing and ultimately on learning and memory storage.[59] On the other hand, extrahippocampal regions, including the postrhinal (visual) cortex, prefrontal cortex, entorhinal cortex,

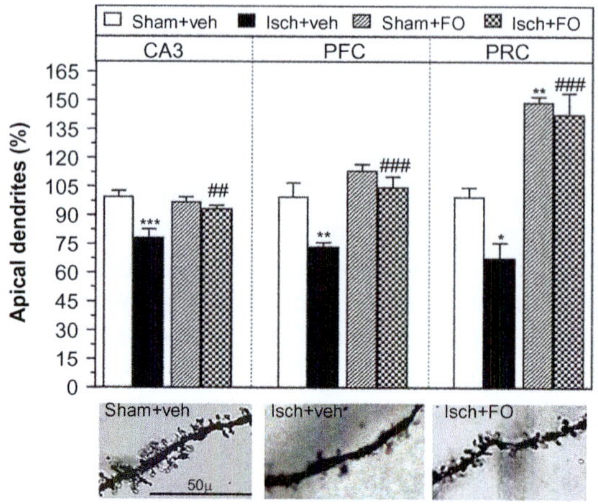

FIG. 1 (*Upper*) Effects of fish oil (FO) on the density of dendritic spines in a branch of apical dendrites in pyramidal neurons in the hippocampus (CA3 subfield), prefrontal cortex (PFC), and postrhinal cortex (PORC) in rats subjected to sham operation or global cerebral ischemia. (*Bottom*) Representative microphotographs of dendritic branches showing the density of spines in three experimental groups. Isch, ischemia; Veh, vehicle. The data are expressed as mean ± SEM. $^{*}p < 0.05$, $^{**}p < 0.01$, $^{***}p < 0.001$ vs. sham; $^{\#\#}p < 0.01$, $^{\#\#\#}p < 0.001$ vs. Isch/Veh.

and retrosplenial cortex, among others, are directly interconnected with the hippocampus and work together to play a key role in processing spatial memory formation, consolidation, and retrieval.[60,61] The postrhinal cortex, where greater efficacy of FO in increasing the density of dendritic spines was observed (Fig. 1), seems to be critically involved in contextual representations of the spatial layout of objects in the environment, a function that requires visual and spatial information processing.[62] Accordingly, this function of the visual cerebral cortex, together with other structures, may be important for resolution of the radial maze task that is used in our studies.

Together with the behavioral effects of FO[43–47] and its effects on MAP2,[47] the results that are presented in Fig. 1 suggest that the greater density of dendritic spines that is observed in hippocampal and extrahippocampal structures after FO treatment may represent a possible mechanism by which FO facilitates memory recovery (or preservation) after brain ischemia, even in the presence of neuronal death. Further support for the hypothesis that DHA-mediated dendritic neuroplasticity is correlated with functional recovery after brain injury comes from the studies that utilized models of peripheral nerve lesions. For example, a single dose of DHA enhanced the sprouting of uninjured corticospinal and serotonergic fibers, which was correlated with neurological improvement in rats that were subjected to cervical spinal cord injury.[63] Mice that were preventively treated with the high-grade FO formulation Omacor exhibited neurological improvements, a decrease in neuroinflammation, and an increase in the expression of growth-associated protein 43 after sciatic nerve damage.[64]

Finally, we further speculate that the FO-mediated dendritic plasticity and memory recovery after brain ischemia may combine with the antioxidant effect of FO that was also observed under similar experimental conditions.[45] The literature indicates that oxidative stress disrupts neuronal responses or neuroplastic changes that may otherwise cooperate to elicit functional recovery after brain damage. Altogether, we suggest that the antioxidant, anti-inflammatory, and neurotrophic actions of FO work together to confer functional recovery following HI-BI.

CONCLUSION

Several substances that have been evaluated in Phase III clinical trails in patients with HI-BI have failed to show clinical effectiveness. Accumulating evidence suggests that DHA may be a promising strategy for the treatment of HI-BI outcomes. Although DHA as a pure substance is not yet available for clinical administration in the setting of acute ischemic stroke, DHA-containing high-grade pharmaceutical formulations (e.g., Omacor, Omegaven, and Omega-3 DHA 250) may be used preventively in patients who have a high risk of stroke or curatively as a post-ischemic intervention to mitigate the structural and functional sequelae of ischemic brain damage. The findings that are discussed in this chapter create optimism for future clinical trials of DHA-based formulations in the setting of cerebral hypoxia/ischemia.

References

1. Horstmann A, Frisch S, Jentzsch RT, Müller K, Villringer A, Schroeter M. Resuscitating the heart but losing the brain: brain atrophy in the aftermath of cardiac arrest. *Neurology*. 2010;74(4):306–312.
2. Anderson CA, Arciniegas DB. Cognitive sequelae of hypoxic-ischemic brain injury: a review. *NeuroRehabilitation*. 2010;26(1):47–63.
3. Akinyemi RO, Mukaetova-Ladinska EB, Attems J, Ihara M, Kalaria RN. Vascular risk factors and neurodegeneration in ageing related dementias: Alzheimer's disease and vascular dementia. *Curr Alzheimer Res*. 2013;10:642–653.
4. Demaerschalk BM, Hwang HM, Leung G. US cost burden of ischemic stroke: a systematic literature review. *Am J Manag Care*. 2010;16(7):525–533.
5. Neuhaus AA, Couch Y, Hadley G, Buchan AM. Neuroprotection in stroke: the importance of collaboration and reproducibility. *Brain*. 2017;140(8):2079–2092.
6. Ginsberg MD, Busto R. Rodent models of cerebral ischemia. *Stroke*. 1989;20(12):1627–1642.
7. Traystman RJ. Animal models of focal and global cerebral ischemia. *ILAR J*. 2003;44(2):85–95.
8. Krafft PR, Bailey EL, Lekic T, et al. Etiology of stroke and choice of models. *Int J Stroke*. 2012;7(5):398–406.
9. Sommer CJ. Ischemic stroke: experimental models and reality. *Acta Neuropathol*. 2017;133(2):245–261.
10. Leker RR, Shohami E. Cerebral ischemia and trauma-different etiologies yet similar mechanisms: neuroprotective opportunities. *Brain Res Brain Res Rev*. 2002;39(1):55–73.
11 Hemphil JC. Critical care neurology, In: Kasper DL, Braunwald E, Fauci AS, Hauser SL, Longo DL, J. Jameson JL, eds. Harrison's Principle of Internal Medicine. New York: McGraw-Hill, 6th ed., 2005:1631–1637.
12. Nguemeni C, Gouix E, Bourourou M, Heurteaux C, Blondeau N. Alpha-linolenic acid: a promising nutraceutical for the prevention of stroke. *PharmaNutrition*. 2013;1:1–8.
13. Dirnagl U. Pathobiology of injury after stroke: the neurovascular unit and beyond. *Ann N Y Acad Sci*. 2012;1268:21–25.
14. Ginsberg MD. The new language of cerebral ischemia. *Am J NeuroRadiology*. 1997;18:1435–1445.
15. Dirnagl U, Endres M. Found in translation: preclinical stroke research predicts human pathophysiology, clinical phenotypes, and therapeutic outcomes. *Stroke*. 2014;45(5):1510–1518.
16. White BC, Sullivan JM, DeGracia DJ, et al. Brain ischemia and reperfusion: molecular mechanisms of neuronal injury. *J Neurol Sci*. 2000;179(1–2):1–33.
17. Mehta SL, Manhas N, Raghubir R. Molecular targets in cerebral ischemia for developing novel therapeutics. *Brain Res Rev*. 2007;54(1):34–66.
18. Benowitz LI, Carmichael ST. Promoting axonal rewiring to improve outcome after stroke. *Neurobiol Dis*. 2010;37(2):259–266.

19. Corbett D, Nguemeni C, Gomez-Smith M. How can you mend a broken brain? Neurorestorative approaches to stroke recovery. *Cerebrovasc Dis*. 2014;38(4):233–239.
20. García-Chávez D, González-Burgos I, Letechipía-Vallejo G, et al. Long-term evaluation of cytoarchitectonic characteristics of prefrontal cortex pyramidal neurons, following global cerebral ischemia and neuroprotective melatonin treatment, in rats. *Neurosci Lett*. 2008;48:148–152.
21. Chamorro Á, Dirnagl U, Urra X, Planas AM. Neuroprotection in acute stroke: targeting excitotoxicity, oxidative and nitrosative stress, and inflammation. *Lancet Neurol*. 2016;15(8):869–881.
22. Ginsberg MD. Current status of neuroprotection for cerebral ischemia: synoptic overview. *Stroke*. 2009;40 (3 Suppl):S111–S114.
23. Machicao F, Muresanu DF, Hundsberger H, Pflüger M, Guekht A. Pleiotropic neuroprotective and metabolic effects of Actovegin's mode of action. *J Neurol Sci*. 2012;322(1–2):222–227.
24. Kim HY. Novel metabolism of docosahexaenoic acid in neural cells. *J Biol Chem*. 2007;282(26):18661–18665.
25. Crupi R, Marino A, Cuzzocrea S. N-3 fatty acids: role in neurogenesis and neuroplasticity. *Curr Med Chem*. 2013;20(24):2953–2963.
26. Das UN. Essential fatty acids—a review. *Curr Pharm Biotechnol*. 2006;7(6):467–482.
27. Lavie CJ, Milani RV, Mehra MR, Ventura HO. Omega-3 polyunsaturated fatty acids and cardiovascular diseases. *J Am Coll Cardiol*. 2009;54:585–594.
28. Balakumar P, Taneja G. Fish oil and vascular endothelial protection: bench to bedside. *Free Radic Biol Med*. 2012;53(2):271–279.
29. Aung T, Halsey J, Kromhout D, et al. Associations of Omega-3 fatty acid supplement use with cardiovascular disease risks: meta-analysis of 10 trials involving 77 917 individuals. *JAMA Cardiol*. 2018;3(3):225–234.
30. Ginsberg MD. Neuroprotection for ischemic stroke: past, present and future. *Neuropharmacology*. 2008;55(3):363–389.
31. Ginsberg MD. Expanding the concept of neuroprotection for acute ischemic stroke: the pivotal roles of reperfusion and the collateral circulation. *Prog Neurobiol*. 2016;145(146):46–77.
32. Corbett D, Nurse S. The problem of assessing effective neuroprotection in experimental cerebral ischemia. *Prog Neurobiol*. 1998;54:531–548.
33. Umemura K, Toshima Y, Asai F, Nakashima M. Effect of dietary docosahexaenoic acid in the rat middle cerebral artery thrombosis model. *Thromb Res*. 1995;78(5):379–387.
34. Okada M, Amamoto T, Tomonaga M, Kawachi A, Yazawa K, Mine K, et al. The chronic administration of docosahexaenoic acid reduces the spatial congnitive deficit following transient forebrain ischemia in rats. *Neuroscience*. 1996;71:17–25.
35. Cao D, Xu JF, Xue RH, Zheng WF, Liu ZL. Protective effect of chronic ethyl docosahexaenoate administration on brain injury in ischemic gerbils. *Pharmacol Biochem Behav*. 2004;79:651–659.
36. Cao D, Li M, Xue R, Zheng W, Liu Z, Wang X. Chronic administration of ethyl docosahexaenoate decreases mortality and cerebral edema in ischemic gerbils. *Life Sci*. 2005;78:74–81.
37. Cao D, Zhou C, Sun L, Xue R, Xu J, Liu Z. Chronic administration of ethyl docosahexaenoate reduces gerbil brain eicosanoid productions following ischemia and reperfusion. *J Nutr Biochem*. 2006;17:234–241.
38. Belayev L, Marcheselli VL, Khoutorova L, et al. Docosahexaenoic acid complexed to albumin elicits high-grade ischemic neuroprotection. *Stroke*. 2005;36(1):118–123.
39. Belayev L, Khoutorova L, Atkins KD, Bazan NG. Robust docosahexaenoic acid-mediated neuroprotection in a rat model of transient, focal cerebral ischemia. *Stroke*. 2009;40(9):3121–3126.
40. Belayev L, Khoutorova L, Atkins KD, et al. Docosahexaenoic acid therapy of experimental ischemic stroke. *Transl Stroke Res*. 2011;2(1):33–41.
41. Eady TN, Belayev L, Khoutorova L, Atkins KD, Zhang C, Bazan NG. Docosahexaenoic acid signaling modulates cell survival in experimental ischemic stroke penumbra and initiates long-term repair in young and aged rats. *PLoS One*. 2012;7(10):e46151.
42. Eady TN, Khoutorova L, Obenaus A, Mohd-Yusof A, Bazan NG, Belayev L. Docosahexaenoic acid complexed to albumin provides neuroprotection after experimental stroke in aged rats. *Neurobiol Dis*. 2014;62:1–7.
43. Fernandes JS, Mori MA, Ekuni R, Oliveira RMW, Milani H. Long-term treatment with fish oil prevents memory impairments but not hippocampal damage in rats subjected to transient, global cerebral ischemia. *Nutr Res*. 2008;28:798–808.
44. Bacarin CC, Mori MA, Ferreira ED, Romanini CV, de Oliveira RM, Milani H. Fish oil provides robust and sustained memory recovery after cerebral ischemia: influence of treatment regimen. *Physiol Behav*. 2013;119:61–71.

45. Bacarin CC, de Sá-Nakanishi AB, Bracht A, et al. Fish oil prevents oxidative stress and exerts sustained Antiamnesic effect after global cerebral ischemia. *CNS Neurol Disord Drug Targets*. 2015;14(3):400–410.
46. Ferreira ED, Romanini CV, Oliveira JN, et al. Fish oil provides a sustained antiamnesic effect after acute, transient forebrain ischemia but not after chronic cerebral hypoperfusion in middle-aged rats. *Behav Brain Res*. 2014;265C:101–110.
47. Bacarin CC, Godinho J, de Oliveira RM, et al. Postischemic fish oil treatment restores long-term retrograde memory and dendritic density: an analysis of the time window of efficacy. *Behav Brain Res*. 2016;311:425–439.
48. Berressem D, Koch K, Franke N, Klein J, Eckert GP. Intravenous treatment with a long-chain Omega-3 lipid emulsion provides neuroprotection in a murine model of ischemic stroke—a pilot study. *PLoS One*. 2016;11(11):e0167329.
49. Williams JJ, Mayurasakorn K, Vannucci SJ, et al. N-3 fatty acid rich triglyceride emulsions are neuroprotective after cerebral hypoxic-ischemic injury in neonatal mice. *PLoS One*. 2013;8(2):e56233.
50. Bazan NG. Cell survival matters: docosahexaenoic acid signaling, neuroprotection and photoreceptors. *Trends Neurosci*. 2006;29(5):263–271.
51. Bazan NG, Molina MF, Gordon WC. Docosahexaenoic acid signalolipidomics in nutrition: significance in aging, neuroinflammation, macular degeneration, Alzheimer's, and other neurodegenerative diseases. *Annu Rev Nutr*. 2011;31:321–351.
52. Mayurasakorn K, Williams JJ, Ten VS, Deckelbaum RJ. Docosahexaenoic acid: brain accretion and roles in neuroprotection after brain hypoxia and ischemia. *Curr Opin Clin Nutr Metab Care*. 2011;14(2):158–167.
53. Zhang W, Wang H, Zhang H, et al. Dietary supplementation with omega-3 polyunsaturated fatty acids robustly promotes neurovascular restorative dynamics and improves neurological functions after stroke. *Exp Neurol*. 2015;272:170–180.
54. Frautschy SA, Cole GM. Why pleiotropic interventions are needed for Alzheimer's disease. *Mol Neurobiol*. 2010;41(2–3):392–409.
55. Plamondon H, Roberge MC. Dietary PUFA supplements reduce memory deficitsbut not CA1 ischemic injury in rats. *Physiol Behav*. 2008;95:492–500.
56. Keene CS, Bucci DJ. Damage to the retrosplenial cortex produces specific impairments in spatial working memory. *Neurobiol Learn Mem*. 2009;91:408–414.
57. Haijima V, Ichitani Y. Anterograde and retrograde amnesia of place discrimination in retrosplenial cortex and hippocampal lesioned rats. *Learn Mem*. 2008;15:477–482.
58. Ramos JM. Remote spatial memory and the hippocampus: effect of early andextensive training in the radial maze. *Learn Mem*. 2009;16:554–563.
59. Spruston N. Pyramidal neurons: dendritic structure and synaptic integration. *Nat Rev Neurosci*. 2008;9(3):206–221.
60. Ramos JM. Preserved learning about allocentric cues but impaired flexible memory expression in rats with hippocampal lesions. *Neurobiol Learn Mem*. 2010;93(4):506–514.
61. Wirt RA, Hyman JM. Integrating spatial working memory and remote memory: interactions between the medial prefrontal cortex and Hippocampus. *Brain Sci*. 2017;7(4):E43.
62. Heimer-McGinn VR, Poeta DL, Aghi K, Udawatta M, Burwell RD. Disconnection of the Perirhinal and Postrhinal cortices impairs recognition of objects in context but not contextual fear conditioning. *J Neurosci*. 2017;37(18):4819–4829.
63. Liu ZH, Yip PK, Priestley JV, Michael-Titus AT. A single dose of docosahexaenoic acid increases the functional recovery promoted by rehabilitation after cervical spinal cord injury in the rat. *J Neurotrauma*. 2017;34(9):1766–1777.

Further Reading

64. Silva RV, Oliveira JT, Santos BLR, et al. Long-chain omega-3 fatty acids supplementation accelerates nerve regeneration and prevents neuropathic pain behavior in mice. *Front Pharmacol*. 2017;8:723.

CHAPTER

25

Alpha-Linolenic Acid and Dementia in Epidemiological Studies

Kazumasa Yamagishi, Utako Murai*, Rie Usuki†, Hiroyasu Iso†*

*Department of Public Health Medicine, Faculty of Medicine, and Health Services Research and Development Center, University of Tsukuba, Tsukuba, Japan †Public Health, Department of Social Medicine, Osaka University Graduate School of Medicine, Suita, Japan

INTRODUCTION

Alpha-linolenic acid (ALA), unlike long-chain ω-3 polyunsaturated fatty acids of marine origin [i.e., eicosapentaenoic acid (EPA) or docosahexaenoic acid (DHA)], is a plant-derived ω-3 polyunsaturated fatty acid. Flaxseed oil and perilla oil are rich in ALA, but major sources of ALA are canola oil and soybean oil. The ALA is an essential fatty acid with neuroprotective and cerebrovascular-preventive functions. Several studies have reported that EPA and DHA as well as fish intake may be useful in prevention of dementia.[1–3] The ALA is also expected to have a similar preventive effect, but the evidence is limited. In this section, we summarize the existing epidemiological evidence regarding ALA and dementia.

OBSERVATIONAL EVIDENCE ON ALA AND DEMENTIA

Some cross-sectional studies showed the association between plasma ALA and dementia. The Invecchiare in Chianti (InCHIANTI) Study measured plasma fatty acids including ALA in 935 community-based older (aged ≥65 years) Italians. They found that patients with dementia, compared with participants with normal cognitive function, had significantly lower proportion of plasma ω-3 fatty acid levels, more specifically ALA. The proportion of plasma EPA or DHA did not differ significantly between dementia patients and participants with normal cognitive function. The odds ratio for 1-standard deviation lower plasma ALA were 1.65 (95% confidence interval: 1.17–2.33).[4] Another cross-sectional study of 57 Korean elderly

Omega Fatty Acids in Brain and Neurological Health
https://doi.org/10.1016/B978-0-12-815238-6.00025-0

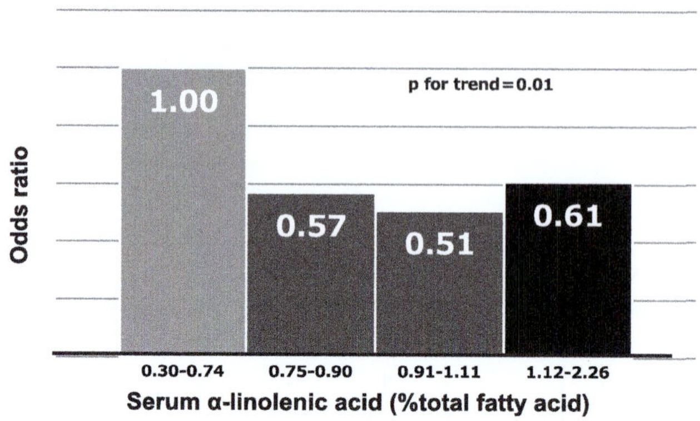

FIG. 1 Association between serum ALA and risk of disabling dementia in CIRCS in 315 dementia cases and 620 controls randomly chosen from 7586 community-based Japanese[6]. Odds ratios were age, sex, community and baseline year-matched, and adjusted for smoking, systolic blood pressure, diabetes mellitus, and medication of hypertension.

(aged ≥65 years) reported that human erythrocyte ALA, but not EPA and DHA, was correlated with cognitive decline.[5] These cross-sectional studies, however, did not assure the causality.

As a follow-up study, the authors performed a case–control study nested in the Circulatory Risk in Communities Study (CIRCS), involving 315 dementia cases and 630 controls randomly chosen from 7586 community-based Japanese aged 40–74 years.[6] There was a strong inverse association between serum ALA levels and disabling dementia (dementia diagnosed by attending physicians under the National Long-Term Care Insurance System, a compulsory insurance system for all Japanese individuals aged ≥40 years, with good validity[7]). As shown in Fig. 1, the association was virtually nonlinear and only the lowest quartiles had the higher risk of disabling dementia. On the other hand, two prospective studies did not find the association of ALA levels with risk of cognitive decline or dementia. One study from the Atherosclerosis Risk in Communities (ARIC) Study examined the plasma fatty acid composition and risk of cognitive decline in 2251 white men and women aged 50–65 years in Minnesota, United States. They found no association between plasma ALA levels and cognitive decline in their 9-year follow-up: the multivariable odds ratio for cognitive decline by 1-standard deviation increases in ALA was 0.97 (0.82–1.16) in plasma cholesteryl esters and 1.03 (0.84–1.22) in plasma phospholipids.[8] They did not find any association of EPA or DHA, either. Another study from the Three Cities Study, France, involving 1214 men and women aged ≥65 years, also did not find any association of plasma level of ALA with incident dementia in their 4-year follow-up: the multivariable hazard ratio of dementia for a 1-standard deviation increase in plasma ALA was 1.08 (0.93–1.26), although they found statistically significant inverse associations of plasma EPA and DHA.[9] The reasons for the discrepancy among the findings of these studies are unknown, but presumably, the nonlinear association mentioned above may in part explain the inconsistency. Further, the differences in follow-up period, number of cases, ethnicity, settings, age distribution, and endpoint determination across the studies may cause the discrepancy.

Several diet studies have also reported similar but weaker associations between ALA intake and risk of dementia. The Rotterdam Study, involving 5395 community-based Dutch men and women aged ≥55 years, did not find association between ALA intake and risk of dementia.[10] Another prospective study, the Chicago Health and Aging Project, involving 815 community-based American men and women aged ≥65 years, showed that ALA intake was inversely and linearly associated with risk of Alzheimer's disease in age-adjusted models. However, in a multivariable adjusted model including the *APOE* ε4 allele, the association was no longer statistically significant.[11] They also found an interaction of the *APOE* ε4 allele with ALA intake in relation to Alzheimer's disease; that is, a strong dietary ALA-Alzheimer's disease association was observed only among persons with the *APOE* ε4 allele: the multivariable adjusted odds ratio and *p*-value of Alzheimer's disease associated with a natural log 1 mg increase in dietary ALA was 0.08, $p=0.02$. Another study, the Rush Memory and Aging Project involving 915 community-based American men and women (mean age 81.4 years), showed that dietary intake of ALA had marginally linear protective relations with global cognitive score and perceptual speed among persons with *APOE* ε4 allele, but not total samples.[12] Similar protective effects (slower rates of decline in global cognition and in multiple cognitive domains) were also observed for seafood intake and long-chain ω-3 polyunsaturated fatty acids among persons with *APOE* ε4 allele.

INTERVENTION TRIALS ON ALA AND DEMENTIA

So far, two randomized controlled trials have tested this issue. Geleijnse et al. performed a randomized double-blind placebo-controlled trial among 2911 coronary patients.[13] The decline in scores of the mini-mental state examination did not differ significantly between the patients with 2 g/day supplementation of ALA and those taking the placebo after a 40-month intervention: the difference, however, was almost marginally significant (−0.74 vs −0.60 points, $p=0.12$). Another randomized double-blind placebo-controlled trial involving 100 Alzheimer's disease patients showed that, after the 4-week intervention, the percentage of patients whose short-term memory was improved were significantly higher among patients with the supplementation of a mixture of ALA and linolenic acids compared with patients who were prescribed placebo.[14]

UNDERLYING MECHANISMS OF ALA AND DEMENTIA

The ALA is an essential fatty acid and is converted to EPA and DHA by Δ-6 desaturase, Δ-5 desaturase, and elongase. However, because Δ-6 and Δ-5 desaturases are rate-limiting enzymes, typically it has been reported that <1% of food-derived ALA is converted to DHA.[15] On the other hand, studies showed conversion of ALA to DHA may be greater than expected and be sufficient to supply the adult brain.[15] Several animal studies showed that feeding ALA enhanced the proportion of DHA in brain phospholipids.[16–18] Therefore, beneficial impacts of EPA and DHA may also be applicable to ALA.

Another neuroprotective or cerebrovascular-preventive effects of ALA independent to EPA or DHA have been suggested by several animal studies[19]: impact on learning ability[16–18] and hyperactivity,[17] protection against ischemia-induced neuronal death,[20, 21] prevention of kainate-induced seizures and neuronal death,[20] neurofunctional recovery from ischemia[21] and injury,[22] and improvement of cerebrovascular flow[23] and vasodilation.[23] These mechanisms could explain the beneficial effect of ALA on both Alzheimer's and vascular dementias.

CONCLUSION

Existing evidence concerning ALA and dementia suggests its preventive effect of cognitive decline and dementia. However, more epidemiological evidence should be accumulated on this issue.

References

1. Kalmijn S, Launer LJ, Ott A, Witteman JC, Hofman A, Breteler MM. Dietary fat intake and the risk of incident dementia in the Rotterdam study. *Ann Neurol*. 1997;42:776–782.
2. Barberger-Gateau P, Letenneur L, Deschamps V, Peres K, Dartigues JF, Fish RS. Meat, and risk of dementia: cohort study. *Br Med J*. 2002;325:932–933.
3. Schaefer EJ, Bongard V, Beiser AS, et al. Plasma phosphatidylcholine docosahexaenoic acid content and risk of dementia and Alzheimer disease: The Framingham heart study. *Arch Neurol*. 2006;63:1545–1550.
4. Cherubini A, Andres-Lacueva C, Martin A, et al. Low plasma N-3 fatty acids and dementia in older persons: the InCHIANTI study. *J Gerontol A Biol Sci Med Sci*. 2007;62:1120–1126.
5. Kim M, Nam JH, Oh DH, Park Y. Erythrocyte α-linolenic acid is associated with the risk for mild dementia in Korean elderly. *Nutr Res*. 2010;30:756–761.
6. Yamagishi K, Ikeda A, Chei CL, et al. Serum α-linolenic and other ω-3 fatty acids, and risk of disabling dementia: Community-based nested case-control study. *Clin Nutr*. 2017;36:793–797.
7. Noda H, Yamagishi K, Ikeda A, Asada T, Iso H. Identification of dementia using standard clinical assessments by primary care physicians in Japan. *Geriatr Gerontol Int*. 2018;18:738–744.
8. Beydoun MA, Kaufman JS, Satia JA, Rosamond W, Folsom AR. Plasma n-3 fatty acids and the risk of cognitive decline in older adults: the atherosclerosis risk in communities study. *Am J Clin Nutr*. 2007;85:1103–1111.
9. Samieri C, Feart C, Letenneur L, et al. Low plasma eicosapentaenoic acid and depressive symptomatology are independent predictors of dementia risk. *Am J Clin Nutr*. 2008;88:714–721.
10. Devore EE, Grodstein F, van Rooij FJ, et al. Dietary intake of fish and omega-3 fatty acids in relation to long-term dementia risk. *Am J Clin Nutr*. 2009;90:170–176.
11. Morris MC, Evans DA, Bienias JL, et al. Consumption of fish and n-3 fatty acids and risk of incident Alzheimer disease. *Arch Neurol*. 2003;60:940–946.
12. van de Rest O, Wang Y, Barnes LL, Tangney C, Bennett DA, Morris MC. *APOE* ε4 and the associations of seafood and long-chain omega-3 fatty acids with cognitive decline. *Neurology*. 2016;86:2063–2070.
13. Geleijnse JM, Giltay EJ, Kromhout D. Effects of n-3 fatty acids on cognitive decline: a randomized, double-blind, placebo-controlled trial in stable myocardial infarction patients. *Alzheimers Dement*. 2012;8:278–287.
14. Yehuda S, Rabinovtz S, Carasso RL, Mostofsky DI. Essential fatty acids preparation (SR-3) improves Alzheimer's patients quality of life. *Int J Neurosci*. 1996;87:141–149.
15. Domenichiello AF, Kitson AP, Bazinet RP. Is docosahexaenoic acid synthesis from α-linolenic acid sufficient to supply the adult brain? *Prog Lipid Res*. 2015;59:54–66.
16. Yamamoto N, Saitoh M, Moriuchi A, Nomura M, Okuyama H. Effect of dietary α-linolenate/linoleate balance on brain lipid compositions and learning ability of rats. *J Lipid Res*. 1987;28:144–151.
17. Umezawa M, Kogishi K, Tojo H, et al. High-linoleate and high-α-linolenate diets affect learning ability and natural behavior in SAMR1 mice. *J Nutr*. 1999;129:431–437.
18. Lamptey MS, Walker BL. A possible essential role for dietary linolenic acid in the development of the young rat. *J Nutr*. 1976;106:86–93.
19. Devassy JG, Leng S, Gabbs M, Monirujjaman M, Aukema HM. Omega-3 polyunsaturated fatty acids and oxylipins in neuroinflammation and management of Alzheimer disease. *Adv Nutr*. 2016;7:905–916.

20. Lauritzen I, Blondeau N, Heurteaux C, Widmann C, Romey G, Lazdunski M. Polyunsaturated fatty acids are potent neuroprotectors. *EMBO J*. 2000;19:1784–1793.
21. Lang-Lazdunski L, Blondeau N, Jarretou G, Lazdunski M, Heurteaux C. Linolenic acid prevents neuronal cell death and paraplegia after transient spinal cord ischemia in rats. *J Vasc Surg*. 2003;38:564–575.
22. King VR, Huang WL, Dyall SC, Curran OE, Priestley JV, Michael-Titus AT. Omega-3 fatty acids improve recovery, whereas omega-6 fatty acids worsen outcome, after spinal cord injury in the adult rat. *J Neurosci*. 2006;26:4672–4680.
23. Blondeau N, Petrault O, Manta S, et al. Polyunsaturated fatty acids are cerebral vasodilators via the TREK-1 potassium channel. *Circ Res*. 2007;101:176–184.

CHAPTER

26

DHA A Key Supplement for Neurologic Health—An Insight

Vasanti Suvarna, Vikas Singh

Department of Pharmaceutical Chemistry and Quality Assurance, SVKM's Dr. Bhanuben Nanavati College of Pharmacy, Mithibai Campus, Mumbai, India

ABBREVIATIONS

AB	amyloid beta
AD	Alzheimer disease
ADAS-Cog	Alzheimer's disease Assessment Scale-Cognitive
ALA	α-linolenic acid
ChemR23	chemerin receptor 23
CREB	cAMP response element-binding protein
DHA	docosahexaenoic acid
DPA	docosapentaenoic acid
EPA	eicosapentaenoic acid
ETA	eicocosatetraenoic acid
GPCRs	G-protein coupled receptors
IFNg	interferon gamma
IL-1	interleukin-1
PD	Parkinson's disease
PLA2	phospholipase-A2
PPARs	peroxisome proliferator-activated receptors
RXRs	retinoid X receptors
SCI	spinal cord injury
SDA	stearidonic acid
TBI	traumatic brain injury
TBPA	thyroxine binding prealbumin
THA	tetracosahexaenoic acid

Omega Fatty Acids in Brain and Neurological Health
https://doi.org/10.1016/B978-0-12-815238-6.00026-2

TNF-α	tumour necrosis factor-α
TPA	tetracosapentaenoic acid
TREK-1	heat-activated background K+ channel
ω-3-PUFAs	Omega-3-poly unsaturated fatty acids

INTRODUCTION

The brain is one of the organs containing high amounts of lipids, which constitutes more than half of its dry weight.[1] It is comprised of long-chain omega-3 and omega-6 polyunsaturated fatty acids in large amounts, which suggests that poly-unsaturated fatty acids (PUFAs) play a major role in the optimum development, maturation, and aging of neuronal structure and their networks. Large number of evidences have highlighted the key role of omega-3-PUFAs supplementation in brain development, brain function, and mental performance. Optimum brain functioning demands optimal maintenance of its lipid composition.

Omega-3 PUFAs (ω-3-PUFAs) are comprised of stearidonic acid (SDA), α-Linolenic acid (ALA), eicosapentaenoic acid (EPA), docosapentaenoic acid (DPA) and docosahexaenoic acid (DHA). Among the ω-3 fatty acids, the well-studied fatty acid is DHA, which is the largest of the ω-3 fatty acids, having a 22-carbon chain and 6 double bonds.[2] Docosahexaenoic acid (DHA) is quantitatively the utmost significant omega-3 PUFA present in the brain, contributing to over 95% of ω-3 PUFAs and 10%–20% of total brain lipids.[1, 3] It is specifically present in grey matter in high concentrations. Primarily it is stored in membrane phospholipids like phosphatidylethanolamine (PA) and phosphatidylserine (PS). It is also found in phosphotidylcholine (PC). Membrane structures of synaptic terminus, mitochondria, and endoplasmic reticulum are enriched with DHA. Thus, DHA affects various cellular characteristics and physiological activities comprising membrane fluidity, transmembrane receptor function, gene expression, release of neurotransmitter, signal transduction, lipid raft function, myelination, neuroinflammation, and neuronal differentiation and growth.

DHA is synthesized denovo from ALA primarily within liver endoplasmic reticulum (ER). The rate of synthesis is typically below 5% in adult males compared the adult female population. DHA is necessary for growth and maturation of critical organs like the brain and retinas of foetus, infants, and children. The inefficient synthesis of DHA in the human body requires DHA dietary supplementation.[4] Many scientific studies report on the beneficial therapeutic outcomes of DHA in various brain disorders. The present review focuses on scientific findings supporting the role of DHA in neurological health.

SOURCES OF DHA

The major sources of DHA include oils from microalgae and oily fishes. Oily fishes include salmon, sardines, herring, menhaden, mackerel, and blubber of marine fish, such as whales and seals. DHA is also found in the livers of some lean fishes like cod, halibut, and skipjack tuna. Among seafood, salted mackerel contains large amounts of DHA. Marine algae are reported to be the major producers of DHA. Numerous algae species contain DHA. *Schizobytrium spp* and *Cryptobecodinium cobnii* are the two predominant algal sources of DHA with DHA content of 55% and 40% of total fatty acids respectively. DHA is majorly synthesized by phytoplankton and algae, further transferred via the food chain and deposited into the lipids of fish and marine mammals.[5, 6]

BIOSYNTHESIS OF DHA

DHA is synthesized denovo from α-Linolenic acid (ALA) through a series of desaturation and elongation steps occurring in endoplasmic reticulum and finally the β-Oxidation of tetracosahexaenoic acid (24:6n-3) that occurs in peroxisomes. The first step of the pathway is a rate limiting step, which involves desaturation of ALA to SDA by Δ^6 desaturase. This step is followed by an elongation step leading to formation of Eicocosatetraenoic acid (ETA) (20:4n-3); further desaturation by Δ^5 desaturase produces Eicosapentaenoic acid (EPA) (20:5n-3). Eicosapentaenoic acid is in turn elongated by elongase to Docosapentaenoic acid (DPA) (22:5n-3) followed by Tetracosapentaenoic acid (TPA) (24:5n-3). Tetracosapentaenoic acid further undergoes another desaturation step to form Tetracosahexaenoic acid (THA) (24:6n-3). All the above steps occur in endoplasmic reticulum. Later, translocation of THA to peroxisome leads to β-Oxidation catalyzed by enzyme acyl-CoA-oxidase, D-bifunctional enzyme and then peroxisomal thiolases in a sequential manner following translocation into endoplasmic reticulum[7, 8] (Fig. 26.1).

DHA IN VARIOUS NEUROLOGICAL DISORDERS

DHA and Cognition

Population aging is one of the major demographic trends observed in developed countries. Various health issues associated with aging, which include memory loss and behavioral alteration associated with brain functioning decline, have contributed to a significant impact on society and economy. Various degenerative brain diseases include Alzheimer's, dementia, and many chronic debilitating brain disorders. Therefore, nutrition is of paramount importance and can play a significant role in reducing the risk of cognitive function.[9] Cognitive decline during aging in elderly patients is associated with decreased DHA blood levels.[10]

In foetal and placental development, synthesis of DHA from its precursor is unable to meet the demands of rapidly developing neuronal tissue. This necessitates delivery of maternal DHA stores through placental transfer and mother's milk during pregnancy and postbirth respectively.[11] Thus, dietary intake of DHA by the mother greatly affects the DHA supply for brain and cognitive development of progeny. Therefore, daily consumption of 200mg of DHA is recommended for pregnant and nursing women.[12]

A study involving a larger population ($N = 11,875$) reported that a reduced intake of DHA-rich seafood during pregnancy was strongly correlated with risk of suboptimal neuronal development of the placenta. On the contrary, children born to women with higher consumption of DHA-rich seafood during pregnancy resulted in enhanced pro-social behavior, improved motor and social development scores, and greater verbal intelligence at the age of 8 years.[13] Several studies have reported the significance of DHA from mothers' milk in

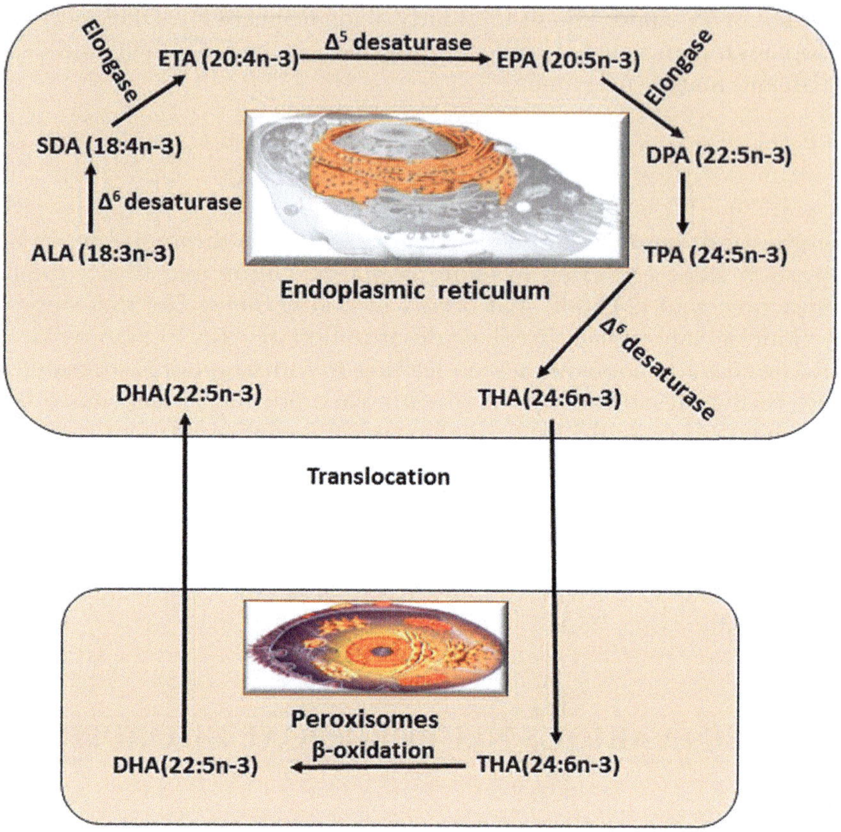

FIG. 1 Biosynthesis of docosahexaenoic acid.

postnatal development. Higher, DHA breast milk concentrations were associated with various brain-related beneficial health effects in infants. These beneficial effects include improved ability to adjust to changes in surroundings, mental development, better hand-eye coordination, improved attention scores, and memory performance in later life.[14] Reduced total ω-3-PUFA plasma levels in male children in the age range 6–12 years were strongly correlated with learning and behavioral problems. When administered with ω-3-PUFA-enriched diets for a duration of 6 months, children in the age range of 7–9 years showed significant enhancement of DHA levels, which was associated with enhanced verbal learning, reading, and spelling abilities.[15]

In normal adults of age 18 years and above, DHA supplementation was found to improve episodic memory significantly and semantic and working memory to a lesser extent.[16] A study evaluating cognitive enhancement of DHA in mainstream school children involved intake of milk fortified with PUFA for a duration of 5 months. Outcomes revealed that fortified milk significantly enhanced plasma DHA levels in addition to improving working memory and processing speed.[17]

A randomized clinical trial (RCT) of mentally healthy older adults aged greater than 55 years evaluated the effect of DHA on their cognitive performance. The supplementation of 900mg DHA from algal sources (*Shizochytrium* sp.) for a duration of 24 weeks resulted in significant reduction in associative learning errors.[18] Another RCT evaluating executive function and neuroemission in the healthy volunteers in the age group of 50 to 75 years showed that DHA causes significant improvement in executive functions like verbal fluency. In addition, alteration in white matter, microstructural integrity, and enhancement of gray matter volume in different brain areas like temporal, limbic, and parietal regions were observed.[19]

DHA and Alzheimer's Disease

Alzheimer's disease is one of the most common neurodegenerative diseases occurring in aging populations. A vast number of epidemiological studies have described association between increased risk of cognitive decline or dementia, especially Alzheimer's disease (AD) with less intake of ω-3-PUFA.[20] The present management of Alzheimer's disease largely involves symptomatic relief and improvement of cognitive dysfunction. Hypothesis for aetiology of disease based on lipid metabolism includes mitochondrial defects, altered phospholipid metabolism, and increased oxidative stress.[21, 22] AD is strongly associated with diminished ω-3-PUFA in brain and peripheral tissues. Alzheimer's patients have shown major decline in brain DHA levels compared to age-related counterparts.[23]

A study conducted by Hashimoto et al. in amyloid beta (AB) infused AD model rats, illustrated that dietary intake of DHA decreases amyloid burden for amyloid-β-infused brains, thus offering protection and causing improvement of learning impairment induced due to AB with consequent increase in DHA levels and decrease in lipid peroxide and reactive oxygen species levels in corticohippocampal tissues.[24] A metaanalysis by Wu et al. revealed that 100g per week increase of fish intake led to 11% lesser risk of AD.[25] Grimm et al. reported that brains of AD patients contain reduced levels of DHA. It is observed that DHA supplementation improves episodic memory, which includes semantic and working memory in normal adults. Thus DHA supplementation supports the impact of DHA on cognitive function.[26] Morris et al. showed that servings of fish containing about 200mg of DHA per week resulted in 60% lower risk of AD development. Therefore, proving DHA as a determinant of lower AD risk. A clinical trial conducted by Morris et al. involved daily supplementation of DHA over 6 months to AD patients; the trial revealed cognition preservation indicated by ADAS-cog scores. This effect was attributed to an increase in plasma level of DHA in these patients.[26]

A set of two clinical trials conducted by Nolan et al. demonstrated the beneficial effects of combining fish oil and xanthophyll carotenoid. The first trial involved treatment of 12 AD patients with xanthophyll carotenoid formulation for a duration of 18 months. The second trial involved treatment of 13 AD patients with a combination of xanthophyll carotenoid and fish oil containing 430mg DHA. Results indicated that a combination of xanthophyll carotenoids and fish oil exhibited slower progression of AD compared to xanthophyll treatment alone in the first trial.[27]

DHA and Parkinsonism

After Alzheimer's disease, Parkinson's disease (PD) is the second most common neurodegenerative disorder. It is associated with aging and has about 2% prevalence in older persons.

It typically affects basal ganglia in patients leading to slowed movement, tremor, muscle stiffness, and disturbed balance. Neuropathology of PD includes dopaminergic neuronal loss from substantia nigra and presence of intracellular inclusion called as Lewy bodies and α-Synuclein. There are a number of environmental risk factors involved in the development of PD including exposure to environmental toxins. Common clinical features of PD and AD include mitochondrial dysfunction, neuroinflammation, and oxidative stress. Though current treatment of PD is based on symptomatic correction of dopaminergic signaling with medications like carbidopa, levodopa and dopamine agonists; the efficacy of these treatments is limited due to serious treatment-associated side effects. Patients with early motor-stage and incidental PD are reported to exhibit significant reduction in DHA, but no difference was observed in postmortem analysis of cortex tissue from levodopa-treated patients. A decrease in DHA level was found to be associated with appearance of dyskinesia.[8] Preclinical models involved use of neurotoxins, such as methyl-4-phenyl-1,2,3,6-tetrahydropyridine (MPTP), leading to loss of dopaminergic cells mimicking PD pathogenesis. Levodopa administration in MPTP-treated monkeys was found to substantially increase arachidonic acid and decrease DHA concentration.[28] Further, short-term administration of DHA at a dose of 100 mg/kg was found to decrease the extent of levodopa-induced dyskinesia without altering its anti-Parkinsonism effect. Therefore, the study suggested that DHA leads to significant decline in severity of levodopa-induced dyskinesia. Inclusion of 5.3 g/kg DHA in the diet was found to maintain dopamine levels and hinder dopaminergic neuronal loss in substantia nigra.[29]

DHA and Stroke

According to 2015 statistics, stroke is the fifth leading cause of death in the United States. It is a leading cause of long-term disability in elderly populations worldwide. With prolonged life expectancy, there is a significant increase in strokes. Older populations are more likely to develop neurological disorders, large infarct volumes, higher morbidity and mortality rate following an ischaemic event.[30] Current treatment for acute ischaemic stroke includes thrombolytic therapy using recombinant tissue-plasminogen activator. But this benefits only a smaller population of stroke patients due to associated narrow therapeutic windows. Since stroke victims often develop physical or cognitive deficits, neurorestorative strategies have been greatly adopted recently to boost poststroke tissue repair. One of the potential treatments used for stroke are ω-3-PUFAs. Researchers have proven that ω-3-PUFA acts on multiple cell types in the brain, functions to inhibit pro-death or pro-inflammatory mechanisms, and exhibits cytoprotective responses. They are also proven to prevent blood brain barrier breakdown and brain dilemma development by maintaining endothelial tight junction proteins.[31]

A study exploring the effect of ω-3-PUFAs in neurological poststroke recovery suggested long-term functional improvement through stimulation of diverse endogenous-restorative mechanism following brain-injury. These mechanisms include neurogenesis, angiogenesis, and oligodendrogenosis.[32] An investigational study exploring prophylactic effects of DHA supplementation in combination with ω-3-PUFAs in ischaemic brain injury mice model revealed that the combination caused significant reduction in tissue atrophy and improved cognitive function. Thus the study demonstrated the ω-3-PUFA and DHA combination to be an effective therapeutic regimen for enhancement of neurovascular restoration and long-term cognitive recovery.[31] A study evaluated neurological protection exhibited by ω-3-PUFA in aging populations after chronic dietary supplementation in mice model. Results revealed that ω-3-PUFA supplementation for a period of 3 months resulted in significant increase in ratio of ω-3/ω-6 PUFA levels. Further mechanistic evaluation showed that ω-3-PUFA caused beneficial neurological effects through stimulation of postischaemic angiogenesis, neurogenesis, and white matter integrity enhancement.[30] Another study exploring neuroprotective potential of DHA in experimental ischaemic stroke illustrated that DHA caused significant improvement in short- and long-term behavioral functions and a decrease in cortical, subcortical, and total infarct volumes.[33]

DHA and Neurological Injury

Major neurological injuries include traumatic brain injury (TBI) and spinal cord injury (SCI). Both of these injuries are highly disabling and cause considerable healthcare expenditure. Injury management includes strategies like the protection of brain and spinal cord in the acute phase and stimulation of regeneration and neuroplasticity leading to restoration of function.[34] SCI further leads to long-lasting and debilitating motor, sensory, and bladder dysfunction. SCI indicates a complex cascade of damaging events, which include metabolic alteration, inflammation, axonal damage, cell-death, and demyelination.[35] These damaging effects result due to intrinsic balance between neurorestorative and neurodestructive signals. Extensive deregulation of membrane PUFA homeostasis is observed after SCI resulting in membrane derangements, arachidonic acid release, and DHA per-oxidation. These in turn lead to regulation in toxic fatty acids metabolism and decline in ω-3-PUFA levels. DHA plays a significant role in functional recovery of various SCI animal models through mechanisms like neuroprotection and antiinflammatory activity.

A study evaluating the efficacy of ω-3-PUFA enriched diet in reduction of dysfunction in SCI animal model was conducted for a duration of 8 weeks at a dose of 750 mg/kg/day. Outcomes have shown that pretreatment with ω-3-PUFA containing DHA resulted in better functional improvement including lower sensory deficits, faster improvement in locomotion, and autonomic bladder recovery. Since marked DHA deficiency caused significant alteration in cord-PUFA neurolipidome, enhancement of DHA level through ω-3-PUFA supplementation led to amelioration of the associated changes.[6] DHA injection after 30 min of SCI resulted in significant elevation of locomotor performance and neuroprotection through decreased lesion size, apoptosis, and increased neuronal and oligodendrocytes survival. The beneficial effect of DHA treatment in spinal cord compression injury was proposed to be due to significant antioxidant activity.

Plasma ω-3-PUFA level and pro-inflammatory marker levels are inversely co-related with each other. DHA administration through intravenous route at a dose of 250 mmol/kg for 30 min after SCI caused significant reduction in lesion size, oligodendrocyte, neuronal loss, and apoptotic death contributing to substantial neuroprotection. A study highlighting the therapeutic potential of DHA in TBI rat model has shown that ω-3-PUFA supplementation postinjury at a dose of 24 mg/kg resulted in significant improvement in neurological outcome, hind limb motor score, general balance, and limb co-ordination. Mechanism was proposed to involve decrease in caspase-3 level.[36] Wu et al. investigated diet supplementation with 8% fish oil as a pretreatment for a duration of 4 weeks and 1 week respectively in TBI model. Results indicated that ω-3-PUFA was found to reverse the decrease in trauma-induced brain-derived neurotropic factor, CREB protein, and synpaxin-1.ω-3-PUFA. Pretreatment was found to reduce oxidative stress induced by mild TBI and trauma-induced alterations in cellular energy metabolism.[37]

MECHANISM OF DHA

As a major component of phospholipid, incorporating DHA into the biological membrane changes the properties of specialized membrane micro domains like lipid rafts and caveolae. DHA is proven to influence a large number of membrane proteins such as receptors, ion channels, and enzymes. DHA is involved in activation of transcription factors like retinoid X receptors (RXRs), nuclear receptors, and G-protein coupled receptors (GPCRs). Further, it was found to activate Peroxisome proliferator-activated receptors (PPARs), which include three isotypes: PPARa, PPARg, and PPARd/b.[34] DHA exhibits functional improvement and tissue protective outcomes in SCI and TBI. These effects are mediated through macrophages activation and production and release of nitric oxide, IL-1, TNF-α, IFNg.[38]

The antiinflammatory effect of DHA is mediated through its binding to G-protein coupled receptor (GPCR) 120. This mechanism is involved in controlling the macrophage dependent chronic inflammation in tissue and antiapoptotic effects of fatty acids.[39] Binding of DHA to GPCR40 receptor present in brain and spinal cord is responsible for mediating the postinjury effects of DHA on neurogenesis. DHA serves as a precursor for inflammatory markers like resolvins, docosa-trienes, protectins, and maresins produced by macrophages.[40] Resolvins like resolvins D1 bind to specific receptor like formyl peptide receptor 2, GPCR 32 and lipoxinA4. ResolvinE1 exhibits its effect through binding to chemerin receptor23 (ChemR23) and leukotriene B4 receptor 1. An active metabolite named neuropectin D1 of DHA formed after postischaemic event in CNS is proposed to be the key mediator of the antioxidant and antiinflammatory activity of DHA following active CNS injury.[41]

Retinoid signalling pathway majorly regulates synaptic plasticity, learning, and memory. Thus contributing to pathophysiology of AD, DHA modulates dopaminergic, cholinergic, and serotonergic neurotransmitters. Further, DHA was found to inhibit cAMP-dependent kinase, Ca^{2+}/Calmodulin-dependent protein kinase, and protein kinase C.[42] DHA was found to upregulate expression of gene namely α-Synulein and trans-thyretin. Transthyretin (thyroxine binding prealbumin, TBPA) is an amyloid-β peptide scavenger[43] and α-Synuclein is a presynaptic protein with its mutation responsible for familial PD (Fig. 26.2).

CONCLUSION

DHA plays a crucial role in the development and maintenance of the nervous system, demonstrated by its predominant existence in neural tissues. Rich sources of DHA include marine fish and shellfish. DHA, one of the most commonly studied ω-3 fatty acids, is a key nutrient mandatory for growth and maturation of critical organs like the brain and retina of foetus, infants and children. Scientific evidence reported in medical journals and studies revealed the beneficial therapeutic outcome of DHA on various brain disorders. An overview of the clinical findings supports that dietary supplementation of DHA has significant impact on overall neuronal health.

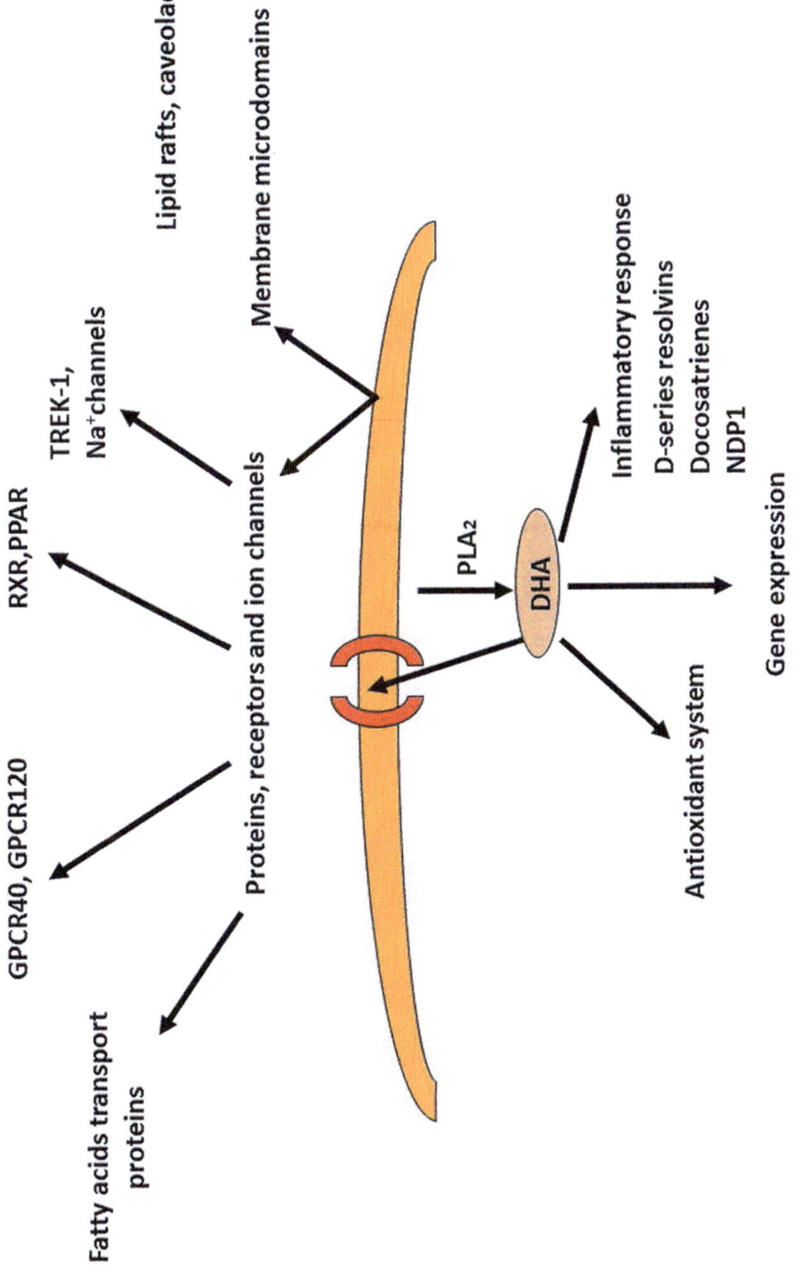

FIG. 2 Mechanism of action of DHA.

References

1. Weiser MJ, Butt CM, Mohajeri MH. Docosahexaenoic acid and cognition throughout the lifespan. *Nutrients.* 2016;8(2):99.
2. Lei E, Vacy K, Boon WC. Fatty acids and their therapeutic potential in neurological disorders. *Neurochem Int.* 2016;95:75–84.
3. Cunnane SC, Chouinard-Watkins R, Castellano CA, Barberger-Gateau P. Docosahexaenoic acid homeostasis, brain aging and Alzheimer's disease: can we reconcile the evidence? *Prostaglandins Leukot Essent Fatty Acids.* 2013;88(1):61–70.
4. Okubo R, Chen C, Sekiguchi M, Hamazaki K, Matsuoka YJ. Mechanisms underlying the effects of n-3 polyunsaturated fatty acids on fear memory processing and their hypothetical effects on fear of cancer recurrence in cancer survivors. *Prostaglandins Leukot Essent Fatty Acids.* 2018;.
5. Shahidi F, Ambigaipalan P. Omega-3 polyunsaturated fatty acids and their health benefits. *Annu Rev Food Sci Technol.* 2018;9:345–381.
6. Figueroa JD, Cordero K, Illán MS, De Leon M. Dietary omega-3 polyunsaturated fatty acids improve the neurolipidome and restore the DHA status while promoting functional recovery after experimental spinal cord injury. *J Neurotrauma.* 2013;30(10):853–868.
7. Dyall SC, Michael-Titus AT. Neurological benefits of omega-3 fatty acids. *Neuromolecular Med.* 2008;10(4):219–235.
8. Dyall SC. Long-chain omega-3 fatty acids and the brain: a review of the independent and shared effects of EPA, DPA and DHA. *Front Aging Neurosci.* 2015;7:52.
9. Andlin-Sobocki P, Jönsson B, Wittchen HU, Olesen J. Cost of disorders of the brain in Europe. *Eur J Neurol.* 2005;12:1–27.
10. Mohajeri MH, Troesch B, Weber P. Inadequate supply of vitamins and DHA in the elderly: implications for brain aging and Alzheimer-type dementia. *Nutrition.* 2015;31(2):261–275.
11. Chambaz J, Ravel D, Manier MC, Pepin D, Mulliez N, Bereziat G. Essential fatty acids interconversion in the human fetal liver. *Neonatology.* 1985;47(3):136–140.
12. EFSA Panel on Dietetic Products, Nutrition, and Allergies (NDA). Scientific opinion on dietary reference values for fats, including saturated fatty acids, polyunsaturated fatty acids, monounsaturated fatty acids, trans fatty acids, and cholesterol. *EFSA J.* 2010;8(3):1461. Cardoso C, Afonso C, Bandarra NM. Dietary DHA and health: cognitive function ageing. Nutr Res Rev 2016;29(2):281–94.
13. Hibbeln JR, Davis JM, Steer C, et al. Maternal seafood consumption in pregnancy and neurodevelopmental outcomes in childhood (ALSPAC study): an observational cohort study. *Lancet.* 2007;369(9561):578–585.
14. Belfort MB, Rifas-Shiman SL, Kleinman KP, et al. Infant feeding and childhood cognition at ages 3 and 7 years: effects of breastfeeding duration and exclusivity. *JAMA Pediatr.* 2013;167(9):836–844.
15. Dalton A, Wolmarans P, Witthuhn RC, van Stuijvenberg ME, Swanevelder SA, Smuts CM. A randomised control trial in schoolchildren showed improvement in cognitive function after consuming a bread spread, containing fish flour from a marine source. *Prostaglandins Leukot Essent Fatty Acids.* 2009;80(2–3):143–149.
16. Yurko-Mauro K, Alexander DD, Van Elswyk ME. Docosahexaenoic acid and adult memory: a systematic review and meta-analysis. *PLoS ONE.* 2015;10(3):e0120391.
17. Petrova D, Litrán MA, García-Mármol E, et al. Effects of fortified milk on cognitive abilities in school-aged children: results from a randomized-controlled trial. *Eur J Nutr.* 2018;7:1.
18. Wurtman RJ, Cansev M, Sakamoto T, Ulus I. Nutritional modifiers of aging brain function: use of uridine and other phosphatide precursors to increase formation of brain synapses. *Nutr Res.* 2010;68(suppl_2):S88–101.
19. Witte AV, Kerti L, Hermannstädter HM, et al. Long-chain omega-3 fatty acids improve brain function and structure in older adults. *Cereb Cortex.* 2013;24(11):3059–3068.
20. Cole GM, Ma QL, Frautschy SA. Omega-3 fatty acids and dementia. *Prostaglandins Leukot Essent Fatty Acids.* 2009;81(2–3):213–221.
21. Horrobin DF, Bennett CN. New gene targets related to schizophrenia and other psychiatric disorders: enzymes, binding proteins and transport proteins involved in phospholipid and fatty acid metabolism. *Prostaglandins Leukot Essent Fatty Acids.* 1999;60(3):141–167.
22. Markesbery WR. Oxidative stress hypothesis in Alzheimer's disease. *Free Radic Biol Med.* 1997;23(1):134–147.
23. Corrigan FM, Horrobin DF, Skinner ER, Besson JA, Cooper MB. Abnormal content of n− 6 and n− 3 long-chain unsaturated fatty acids in the phosphoglycerides and cholesterol esters of parahippocampal cortex from Alzheimer's disease patients and its relationship to acetyl CoA content. *Int J Biochem Cell Biol.* 1998;30(2):197–207.

24. Hashimoto M, Hossain S. Neuroprotective and ameliorative actions of polyunsaturated fatty acids against neuronal diseases: beneficial effect of docosahexaenoic acid on cognitive decline in Alzheimer's disease. m. *J Pharmacol Sci*. 2011;116(2):150–162.
25. Hashimoto M, Hossain S. Neuroprotective and ameliorative actions of polyunsaturated fatty acids against neuronal diseases: beneficial effect of docosahexaenoic acid on cognitive decline in Alzheimer's disease. *J Pharmacol Sci*. 2011;116(2):150–162.
26. Grimm MO, Zimmer VC, Lehmann J, Grimm HS, Hartmann T. The impact of cholesterol, DHA, and sphingolipids on Alzheimer's disease. *Biomed Res Int*. 2013;2013.
27. Nolan JM, Mulcahy R, Power R, Moran R, Howard AN. Nutritional Intervention to Prevent Alzheimer's Disease: Potential Benefits of Xanthophyll Carotenoids and Omega-3 Fatty Acids Combined. *J Alzheimers Dis*. 2018;64(2):367–378.
28. Samadi P, Grégoire L, Rouillard C, Bédard PJ, Di Paolo T, Lévesque D. Docosahexaenoic acid reduces levodopa-induced dyskinesias in 1-methyl-4-phenyl-1, 2, 3, 6-tetrahydropyridine monkeys. *Ann Neurol*. 2006;59(2):282–288.
29. Bousquet M, Saint-Pierre M, Julien C, Salem Jr N, Cicchetti F, Calon F. Beneficial effects of dietary omega-3 polyunsaturated fatty acid on toxin-induced neuronal degeneration in an animal model of Parkinson's disease. *FASEB J*. 2008;22(4):1213–1225.
30. Cai M, Zhang W, Weng Z, et al. Promoting neurovascular recovery in aged mice after ischemic stroke-prophylactic effect of omega-3 polyunsaturated fatty acids. *Aging Dis*. 2017;8(5):531.
31. Pu H, Jiang X, Hu X, et al. Delayed docosahexaenoic acid treatment combined with dietary supplementation of omega-3 fatty acids promotes long-term neurovascular restoration after ischemic stroke. *Transl Stroke Res*. 2016;7(6):521–534.
32. Wang J, Shi Y, Zhang L, et al. Omega-3 polyunsaturated fatty acids enhance cerebral angiogenesis and provide long-term protection after stroke. *Neurobiol Dis*. 2014;68:91–103.
33. Belayev L, Hong SH, Menghani H, et al. Docosanoids promote neurogenesis and angiogenesis, blood-brain barrier integrity, penumbra protection, and neurobehavioral recovery after experimental ischemic stroke. *Cell Mol Neurobiol*. 2018;1:1–7.
34. Michael-Titus AT, Priestley JV. Omega-3 fatty acids and traumatic neurological injury: from neuroprotection to neuroplasticity? *Trends Neurosci*. 2014;37(1):30–38.
35. Hulsebosch CE. Recent advances in pathophysiology and treatment of spinal cord injury. *Adv Physiol Educ*. 2002;26(4):238–255.
36. King VR, Huang WL, Dyall SC, Curran OE, Priestley JV, Michael-Titus AT. Omega-3 fatty acids improve recovery, whereas omega-6 fatty acids worsen outcome, after spinal cord injury in the adult rat. *J Neurosci*. 2006;26(17):4672–4680.
37. Wu A, Ying Z, Gomez-Pinilla F. Omega-3 fatty acids supplementation restores mechanisms that maintain brain homeostasis in traumatic brain injury. *J Neurotrauma*. 2007;24(10):1587–1595.
38. McTigue DM, Tripathi R, Wei P, Lash AT. The PPAR gamma agonist Pioglitazone improves anatomical and locomotor recovery after rodent spinal cord injury. *Exp Neurol*. 2007;205(2):396–406.
39. Talukdar S, Bae EJ, Imamura T, et al. GPR120 is an omega-3 fatty acid receptor mediating potent anti-inflammatory and insulin-sensitizing effects. *Cell*. 2010;142(5):687–698.
40. Ma D, Lu L, Boneva NB, et al. Expression of free fatty acid receptor GPR40 in the neurogenic niche of adult monkey hippocampus. *Hippocampus*. 2008;18(3):326–333.
41. Mukherjee PK, Marcheselli VL, Serhan CN, Bazan NG. Neuroprotectin D1: a docosahexaenoic acid-derived docosatriene protects human retinal pigment epithelial cells from oxidative stress. *Proc Natl Acad Sci*. 2004;101(22):8491–8496.
42. Mirnikjoo B, Brown SE, Kim HF, Marangell LB, Sweatt JD, Weeber EJ. Protein kinase inhibition by ω-3 fatty acids. *J Biol Chem*. 2001;276(14):10888–10896.
43. Link CD. Expression of human beta-amyloid peptide in transgenic Caenorhabditis elegans. *Proc Natl Acad*. 1995;92(20):9368–9372.

CHAPTER

27

Obesity, Cognitive Functioning, and Dementia: A Lifespan Prospective

Fayeza S. Ahmed, Olivia E. Bogucki*, Peter J. Dearborn*,†, Merrill F. Elias*,†*

***Department of Psychology, University of Maine, Orono, ME, United States †Graduate School of Biomedical Sciences and Engineering, University of Maine, Orono, ME, United States**

INTRODUCTION

The prevalence of obesity [body mass index (BMI $\geq$ 30 kg/m^2)] and overweight status (BMI of 25–29.9 kg/m^2) is rising dramatically. In 2016, the rates of individuals who were overweight or obese exceeded 1.9 billion and 650 million, respectively.[1] This is a major public health concern regarding general health, type 2 diabetes mellitus (T2DM), and heart disease[2,3] because cardiovascular diseases (CVDs) are associated with lower levels of cognitive performance and dementia[4–6] and lower cognitive performance earlier in life is a risk factor for dementia later in life.[7]

In this chapter, we review the literature on weight, including BMI, obesity, and central obesity, as it relates to dementia and to variations in cognitive ability in non-demented persons. We take a lifespan perspective, arguing that it is imperative to move backward in time from adult cognitive deficit and diagnosed dementia to examine its precursors. These precursors include weight and obesity, but also cognitive changes that occur prior to a diagnosis of mild cognitive impairment (MCI) or probable dementia. It is critical to consider the possible longitudinal covariation of weight-related variables and cognitive outcomes across the lifespan. Because we cannot review all the pertinent literature due to space limitations, we focus on prospective data and highlight selected studies that illustrate important issues of theory, design, methodology, and the need for a lifespan perspective. For other perspectives and emphases, see the following reviews.[8–15]

There has been some amount of emphasis on which specific cognitive abilities are related to obesity and overweight status and in this chapter, we comment on important methodological considerations with respect to this research goal. In addition, we discuss mechanisms underlying adiposity and cognitive decline. A comprehensive approach to the topic of

Omega Fatty Acids in Brain and Neurological Health
https://doi.org/10.1016/B978-0-12-815238-6.00027-4

mechanisms has been conducted by Gustafson.[12] Where necessary, we note findings with respect to underweight and related conditions, but the focus is on overweight status and obesity. Metabolic syndrome (MetS) is an area of research which is an important and integrative extension of studies of obesity in relation to metabolically-related CVD risk factors. This topic is examined in a review by Crichton and colleagues[16] and is thus briefly summarized in this chapter. Omega-3 fatty acids have shown promising results in the treatment of T2DM and may play a role in the mechanisms underlying the relation between overweight status, obesity, cognitive performance, and dementia.

Since the publication of the first edition of this book, there have been several new research publications, including systematic reviews and meta-analyses, on the topic of obesity, cognition, and dementia. In this chapter, we provide an update to the first version while following the same format of the original chapter for the reader's ease.

WEIGHT-RELATED VARIABLES AND DEMENTIA

Results of prospective investigations into the relationship between BMI and incident dementia are somewhat mixed. Some studies report no association between BMI and dementia.[17,18] Recent reviews and meta-analyses, however, have supported the notion that midlife obesity increases risk for dementia in late life.[19–26]

Moreover, lower BMI has also been related to increased risk for dementia, particularly Alzheimer's disease (AD).[8,19,21,24,27] Unfortunately, underweight status has not been properly examined across prospective studies, one reason being the low sample sizes of this subgroup.[17,21] The take-home message across these studies is that being underweight, overweight, or obese at midlife may increase the risk for dementia later in life. Of note, "may" is used to reflect some of the nonsignificant findings,[21] particularly with the relationship with overweight status and dementia risk.

Notably, there is evidence of the presence of an "obesity paradox" in which obesity in midlife, but not late life, is associated with dementia risk.[21,23,24,28] Using data from the Whitehall II Study, Singh-Manoux and colleagues[29] examined the association between BMI and dementia in participants at ages 50, 60, and 70. They found that being obese (BMI $\geq$ 30) was significantly related to dementia risk, but only at age 50 and not at ages 60 or 70. Adjusting for cardiovascular factors did not change this relationship. Further, examination of BMI trajectories using a backward timescale method revealed that BMI was high in midlife and then decreased 10 years prior to a dementia diagnosis. Examination of an average of 28.5-year follow-up in the Cardiovascular Risk Factors, Aging and Dementia (CAIDE) Study revealed dementia risk for those who had a reduction in their midlife BMI to their late-life BMI.[25] BMI change over time, as measured with data from the Health and Retirement Study and use of algorithms to predict memory decline, revealed that reductions in BMI from late 50s to mid-70s was related to predicted memory decline.[30] Using data from 19 studies (nine of which were cohort studies) for a total of over 500,000 participants, Albanese and colleagues[19] noted that when the follow-up between midlife and late-life assessment is >20 years, the relationship between midlife BMI and risk for dementia is weaker. The authors also noted high heterogeneity across studies, which further complicated analysis.

Kivimäki and colleagues[31] refer to a two-phase process, wherein obesity in midlife increases dementia risk in late-life, but weight does not necessarily remain stable the closer in time one is to dementia. Data were analyzed from 39 prospective cohort studies and results indicated when BMI is measured prior to the dementia process, there is an increased risk for dementia with increasing BMI; however, as participants near dementia, their BMI drops in the preclinical phase and therefore, the BMI-dementia risk relationship is affected. This means that it is not uncommon to find no significant relationship between increased BMI and dementia risk in late life.

The proposed mechanisms behind this seemingly paradoxical relationship are: (1) BMI is not a good indicator of obesity, especially in late life; (2) weight loss can be a precursor to dementia even before the presence of behavioral symptoms of dementia (e.g., memory loss); and (3) there is variation in the surveillance periods for exposure to weight gain or loss and obesity.[21,31] BMI is not an ideal measure of adiposity in the oldest old. Aging is characterized by a decline in lean body mass (i.e., sarcopenia) and an increase in adipose tissue without weight gain, a phenomenon that is not accurately captured by BMI.[28] Waist circumference (WCir) and waist-hip ratio (WHR) appear to be better adiposity measures in elderly people as these indices of central obesity reflect the abdominal distribution of body fat.[32–35] Whitmer and colleagues[35] make the following points: Central obesity is a more potent risk factor for CVD and mortality than total body obesity assessed by BMI. This phenomenon may be related in part to the role of intra-abdominal fat (i.e., visceral adiposity) on metabolic abnormalities associated with increased risk of T2DM and CVD. Visceral fat is more active metabolically than subcutaneous fat and is thought to have a stronger influence on insulin resistance and adipocytokine (i.e., cell to cell signaling protein secreted in adipose tissue) production. In dementia and stroke-free individuals, indicators of central obesity and indirect indicators of visceral fat (e.g., WHR and WCir) were more strongly associated with cognitive performance levels than BMI.[36] In a study by Whitmer and colleagues[35] in which participants were followed for 36 years from baseline, the highest quintile of sagittal abdominal diameter measured in midlife (40–45 years of age) was associated with a three-fold increased risk of dementia. In a cross-sectional study comparing subcutaneous versus visceral viscosity measured by abdominal computerized tomography (CT), Yoon and colleagues[37] reported that visceral adiposity (highest tertile) was associated with increased risk of low mini-mental state examination (MMSE) scores (−1 SD), with adjustment for age, gender, education, and T2DM, but that subcutaneous adipose tissue and MMSE scores were not significantly related. Findings were significant only for participants under age 70 ($n = 188$), but the over 70 groups had fewer subjects ($n = 62$) and thus, the negative finding may relate to relatively lower power. More studies with prospective and longitudinal designs that employ this type of quantification of visceral versus subcutaneous obesity are needed.

Luchsinger and colleagues[38] attempted to do this by examining not only BMI but also WCir as well as bioelectrical impedance analysis (BIA). Results indicated that across all three of these measures, increasing BMI, WCir, and fat mass was related to poorer performance in a global cognitive measure (100-point modified MMSE) and a visual scanning/processing speed task (Digit Symbol Substitution Test).

Regarding the second point of weight loss as a precursor to dementia, it has been found that among older age cohorts, associations between weight loss and dementia may reflect the fact that weight loss is a common consequence of dementia rather than a cause.[25,31] Importantly, however, weight loss often predates dementia onset by as many as 10 years,[39] and it is one of the principal manifestations of AD.[40] The third point about variation among length of follow-up time reflects the problem with heterogeneity among methodologies.[19]

Positive associations between being overweight or obese at midlife and incident dementia may be at least partially attributed to the emergence of cardiovascular risk factors and events that correlate with obesity at midlife. These include components of MetS such as central obesity, glucose intolerance, dyslipidemia, hypertension, development of T2DM, and clinical cardiovascular or cerebrovascular events.[28] However, at the end of the life cycle, the role of cardiovascular risk factors in cognitive performance may be diminished by attrition due to morbidity and mortality of subjects who are in the poorest health at midlife and who are likely the poorest performers on tests of cognitive ability. Interestingly, some recent studies have attempted to adjust for these factors still found a significant relationship between midlife BMI and dementia risk.[20,29]

It is also important to consider the changes in protective reproductive hormones and weight gain experienced during midlife by women undergoing the menopausal transition,[41] which has not been extensively explored in the literature. Data from the Cardiovascular Health Study (59% women) indicate that high BMI at midlife was related to increased risk of dementia in women, but it was associated with decreased risk for dementia after 65 years of age. The importance of this study is that weight variables were assessed both at midlife, age 50, and at 65 years or older.[28] Obesity at midlife (BMI > 30), as compared to normal weight (BMI 20–25), was associated with incident dementia (hazard ratio (HR) = 1.38; confidence interval (CI) = 1.03–1.95) when adjusted for demographic variables. These associations were attenuated, but not eliminated (HR = 1.36; CI = 0.94–1.95) with additional adjustments for CVD and risk factors, suggesting that they were among those mechanisms that mediated relations between obesity and cognition. However, in old age, risk estimates were reversed. Underweight study participants (BMI < 20) showed an increased risk of dementia (HR = 1.62; CI = 1.02–2.64) and presence of obesity was associated with a reduced risk for dementia (HR = 0.63; CI = 0.44–0.91) as compared to normal weight status.[28] The same results were obtained for the condition of overweight status (BMI 25–30; HR = 0.92; CI = 0.72–1.18) in comparison with normal weight status. These curvilinear associations between weight and cognition in old age have been widely reported.[8,42–44]

Prospective studies, in which risk factors are measured prior to onset of dementia, avoid confounding of precursors of dementia with the consequences of dementia such as weight loss and poor health. The length of the pre-dementia surveillance period used to assess weight, weight change, and obesity and its distance from the onset of dementia are both important design parameters. Clearly, more reliable assessments of risk factors are possible with more time to assess them and more time between risk exposure and dementia. The ideal study would start as early as possible in development with time points for measurement reasonably spaced, based on rate of development or decline, and would feature measurements of obesity, body weight, relevant comorbidity, confounders, and cognitive functioning at each of these points in time. It is difficult to achieve this ideal in a single study and these data will most likely evolve from an integration of data with surveillance periods beginning and ending at different points in the lifespan.

The preclinical phase of dementia may begin decades before dementia is diagnosed.[7,45,46] There is now a growing body of literature indicating that weight at midlife is a good predictor of cognition many years later and prospective studies are the design of choice. A detailed review of every study contributing to this growing literature is not possible and thus, we focus on studies in two categories based on these outcome measures in the next two sections: (1) dementia; and (2) cognitive performance, including studies that exclude persons with dementia and those that include them. We focus largely on studies with large sample sizes and prospective designs with reasonably long surveillance periods.

Prospective Studies with a Focus on Dementia

In a study examining whether the population attributable risk (PAR) on modifiable risk factors for dementia changed, researchers analyzed data from the Rotterdam Study and noted that while the PAR itself did not change, the risk factors with the highest PARs differed from those in the previous study.[47]

In the large sample ($N = 6538$) from the Kaiser-Permanente Study of Central Obesity,[35] standardized anthropometric measures were obtained when participants were between the ages of 40 and 45 years and related to medical records of dementia three decades later. Compared to those in the lowest quintile of sagittal abdominal diameter, participants in the highest quintile showed nearly a three-fold increase in risk of dementia (HR = 2.72; CI = 2.23–3.33). These associations were maintained when adjusted for demographic variables, T2DM, cardiovascular comorbidity, and BMI. The relation between increased sagittal abdominal diameter and risk for dementia was attenuated, but not eliminated, when adjusted for BMI (HR = 1.92; CI = 1.58–2.35). The number of years of follow-up and sample size are key strengths of the study, but the diagnosis of probable dementia was based on insurance records rather than comprehensive clinical and cognitive evaluation in a central laboratory or clinic setting.

In a study with the Baltimore Longitudinal Study of Aging (BLSA),[48] 2322 participants were followed for a median of 23.4 years with AD as the outcome variable. BMI and WCir increased with age at an annual rate of 0.07 kg/m^2 and 0.35 cm, respectively. For men, weight gain between ages 30 and 50 years for any five-year longitudinal interval was associated with an approximate three-fold increase in risk of AD. However, being an underweight male at ages 30, 40, or 45 years of age was associated with a more than five-fold increased risk for AD (HR = 5.76). Among older and midlife women, the joint presence of obesity and central obesity (WCir $>$80 percentile) at 30, 35, or 50 years increased the risk of developing AD by 6.6-fold. However, women who experienced weight loss between 30 and 35 years of age exhibited a two-fold increase in AD. A notable longitudinal increase in WCir was found to be protective against AD in women, but not men. The follow-up period is exemplary, but data on possible mediators of the sex differences among weight variables and AD outcomes were not available and thus, a substantive elaboration on the reasons for these differences was not possible. Later in this chapter, we discuss the possibility that sex differences may be a product of events that occur many years prior to AD. A discussion of possible causes of sex differences for obesity and other risk factors for dementia may be found in a review by Azad and colleagues.[49] A more recent study with the BLSA data showed that increasing BMI at midlife was significantly associated with increased neurofibrillary tangles at autopsy and a near-significant relationship with in-vivo amyloid deposition and cortical volume. Further, the authors found that the age of onset of AD was decreased by 6.7 months for every one-unit increase in BMI at midlife.[20]

The Italian Longitudinal Study on Aging (ILSA)[50] examined the progression from normal cognitive performance through two phases of clinically defined impairment, MCI, and dementia. The study is exemplary in terms of the prospective design, the population-based sample, the large number of participants ($N = 5632$, ages 65–84 years old), and adjustment for a large number of potential confounders including psychosocial variables such as depressed mood and undernutrition. The presence of MetS and its major components, central obesity, hypertension, and hypertriglyceridemia, were associated with a significantly higher incidence rate for progression to dementia in persons with MCI. However, obesity and the other MetS components did not predict progression from normal cognition to MCI. Thus, abdominal obesity did not predict incident MCI, but MCI modified the prospective relation of central obesity to dementia. In an explanation of negative results, the authors noted that the MMSE was a major source of dementia case definition and diagnosis of impairment, measures of executive function were not considered, and amnestic versus non-amnestic subtypes were not defined because of missing data on critical indices of these subtypes. Study participants were at the lower end of formal educational attainment (mean [M] = 7.5 years; SD = 4.9). This raises the possibility that cases of MCI and/or dementia may have been overestimated, albeit informant data and a neurologist's diagnosis were employed to make these classifications. Regardless of these issues, MCI is an important diagnosis based on measures of clinical impairment and other evidence. There is a pressing need for prospective studies that differentiate between subtypes of MCI as a critical step in the progression from normal cognition to dementia.

Yet another study found negative results when examining the relationship between midlife BMI and dementia risk. Using the Age, Gene/Environment Susceptibility-Reykjavik Study, Albanese and colleagues[17] analyzed data from 3864 participants. Results suggested that midlife BMI was not significantly associated with diagnosis of dementia, total brain volume, and measures of cerebrovascular damage. It was noted that a better marker for midlife obesity would be central adiposity.

Albanese and colleagues[19] analyzed data from 19 studies. Nine of these were cohort studies and the remaining were general observational studies. Using meta-regression, findings indicated that being underweight or obese at midlife was significantly related to dementia risk. Of note, they found that underweight status was only significantly associated when there was at least a 20-year follow-up, meaning that underweight status was not reflecting preclinical dementia. The authors also noted that only four studies attempted to adjust for cardiovascular factors and that there was notable heterogeneity among studies.

Sensitivity and specificity of cognitive measurement is best achieved by considering cognitive outcomes where the full range of test scores is employed. Accordingly, we now briefly review the literature on weight-related variables. In each of these studies, a range of cognitive test scores is the outcome measure or is used to create a binary outcome measure (i.e., good or poor performance).

Prospective Studies Focusing on Cognitive Functioning

Data from a French randomized control trial on dietary supplement and observational studies (SU.VI.MAX and SU.VI.MAX 2) were analyzed to examine the impact of BMI and central adiposity in older adults (mean ages across categories 65.1–66.1 years). After an average of 13.4 years, participants' performance on measures that included executive function components was lower with both increasing BMI and central adiposity.[51]

In the Swedish Adoption/Twin Study of Aging,[52] the association between BMI and cognitive decline was examined over 40 years of follow-up among participants initially aged

25–63 years ($N = 781$; M age = 41.6 years). Without excluding persons with probable dementia, BMI and global cognitive ability (a continuously distributed variable) were measured five times over a 20-year period beginning at age 25. Global cognitive ability was measured by taking the first principal component of 11 tests at baseline and standardizing it across times of measurement. Of note, specific tests in the composite were not reported. Latent growth curve analyses indicated that higher BMI was inversely related to the measure of global cognitive ability with longitudinal decline for both women and men. Results were similar when persons with probable dementia were excluded from the sample. Two study limitations were noteworthy: (1) BMI was calculated from self-report of height and weight at the earliest assessments (1963 and 1967) and no measures of WCir, WHR, and/or obesity were employed; and (2) control for education was less than ideal (i.e., low education ≤ 6 years and high education > 6 years).

This leaves a wide range of years of education that can have a major effect on cognitive outcomes. Aside from these issues, the long follow-up period adds important lifespan information to the literature. In a later study with mixed self-report and assessment of height and weight, these same investigators reported a negative effect on mean cognitive test performance across verbal and spatial abilities, memory, and perceptual speed for individuals who were overweight or obese in early midlife, late midlife, or across midlife.[52] With adjustment for age, education, physical activity, and vascular risk factors, overweight status (BMI 25–30) in midlife for the Origins of Variance in the Old-Old: Octogenarian Twin study subjects was related to lower performance on long-term memory, short-term memory, speed, and verbal and spatial ability 30 years later.[53]

Dahl and colleagues[54] examined both underweight and overweight or obese participants from the Swedish Adoption/Twin Studies during an average of 21.4 years follow-up period. Cognitive function was assessed with multiple validated measures. Using principal components analysis, researchers identified component scores for the cognitive functions. In general, being overweight or obese was associated with poorer cognitive performance. Notably, participants who remained overweight or obese throughout midlife had sharper declines in cognitive functioning. The authors noted that this was the first study to examine both change in cognition as well as the effects of stability of BMI over midlife, which, according to their findings, indicated an additive effect of years of overweight or obese status.

Employing the Whitehall II study data, Sabia and colleagues[55] examined the cumulative effect of BMI from early adulthood (25 years) to late midlife ($M = 61$ years) with respect to cognitive outcomes. In a sample restricted to white participants ($N = 9181$), BMI was self-reported at 25 years (early adulthood) but was measured at midlife ($M = 43.8$ years) and later life ($M = 60.8$ years). Persons with dementia were not excluded. Associations between obesity and two measures of cognitive ability (MMSE and executive functioning indexed by semantic and verbal fluency) became more pronounced over the follow-up period (M duration = 36 years). Although attenuated, these associations were observed after adjustment for health behaviors, blood pressure, and cholesterol at mid and late life.

An association between underweight and cognition, both at mid and late life, was also seen in the investigation by Sabia and colleagues.[55] The risk of cognitive deficit was enhanced with increasing levels of underweight status in a dose response manner, and this association was sustained with statistical adjustment for cardiovascular risk factors and events. The authors hypothesized that underweight status may be a product of poor health[56] and/or dysregulation in hormone secretion corresponding to that seen in anorexia nervosa, a disease associated with cognitive dysfunction.[57] Limitations of the study were that measures of central adiposity were not employed, weight was self-reported for the 25-year-old cohort, and an opportunity to compare results with the inclusion and exclusion of African Americans was missed. However, the sample size was impressive, as was the duration of longitudinal follow-up.

In a 2012 study using the Whitehall II cohort, Singh-Manoux and colleagues[58] analyzed data collected during a 10-year follow-up period. Notably, this study grouped participants into six categories: normal weight (as defined by BMI) and metabolically normal, normal weight and metabolically abnormal, overweight and metabolically normal, overweight and metabolically abnormal, obese and metabolically normal, and obese and metabolically abnormal. Average age ranged from a mean of 48.9 years (normal weight), 50.0 years (overweight), and 49.7 (obese). Both baseline and longitudinal analyses were conducted. Cognitive functions assessed included reasoning, word recall, semantic and letter fluency, and an overall global cognition score. At baseline, overweight and obese participants scored worse than normal weight participants. Further, normal weight but metabolically abnormal participants scored worse than normal weight and metabolically normal participants. Overweight participants performed worse, and the metabolically abnormal overweight participants had even lower scores than their metabolically normal counterparts. Interestingly, the obese participants did not reveal a significant difference in cognitive performance as a function of metabolic status. Longitudinal analyses revealed that decline was similar across all cognitive areas.

In the Basel Study, Brubacher and colleagues[42] related weight change and cognition for a sample of 445 men and 86 women described as optimally healthy and nonoptimally healthy community-dwelling persons. Baseline measurements of BMI were obtained in 1960, with follow-up examinations in 1965, 1971, 1990, and 2000, and multiple measures of cognitive performance were obtained at an average age of 69.4 years (SD = 7.8 years). Based on the full continuous distribution of test scores for 11 key variables extracted from the Consortium to Establish a Registry for Alzheimer's Disease-Neuropsychological Assessment Battery (CERAD-NAB) test scores, good and poor performance were defined as binary variables (i.e., two or more scores 1.5 SD below the mean was defined as poor performance). With zero annual weight change as the reference group, weight change in the positive direction was associated with progressively increasing odds for poor performance, but weight change in the negative direction from zero was also associated with increasing odds of poor performance, although risk did not reach the same level as change in the positive direction. The same curvilinear relation was seen for participants carrying the ApoE-ε4 allele but with modestly higher risk of poor performance at each point along the curve. The limitation in this study was that probable dementia was not defined by a dementia review but was estimated from telephone test scores and identified as "likely dementia."

Based on research with a large sample (2606 twin individuals) from the Finnish Twin Cohort Study, Laitala and colleagues[57] reported that midlife BMI (ages not specified) and other cardiovascular risk factors measured by self-report and medical registry records were related to cognitive functioning, as assessed via a validated phone interview technique. Height and weight estimates were based on self-report in 1975 and 1978 and related to cognition measured between 1999 and 2007. With adjustment for sex, education, birth year, and age, relations among other CVD risk factors and BMI at midlife were associated with a later-life cognition measure, based on the phone interview questions. Overweight status was associated with an increased risk of likely dementia. Most importantly, weight gain in excess of 1.7 kg/m^2 and weight loss of less than 2 kg/m^2 within an average of 5.6 years prior to cognitive assessment was related to lower telephone cognitive performance scores despite adjustment for BMI. Additive genetic correlation (i.e., the proportion of variance that two traits share due to genetic causes) appeared to mediate BMI and cognition, but this association was attenuated when adjusted for education. As the authors noted, telephone interviews cannot be definitive as a technique in assessing dementia, and, in our view, is less desirable than direct cognitive assessment. However, the study employed a very large sample, a necessity for genetic studies by making use of an emerging and important method of general cognitive assessment (e.g., telephone interview) where large samples are important. In 2013, 735 participants from the Finnish Twin Cohort Study were again examined, this time using a dementia risk score developed based on the CAIDE. Using this risk score, researchers examined risk for cognitive impairment as a result of cardiovascular factors at midlife. Notably, this was a 23-year follow-up period with a large sample size. After adjusting for APoE status, researchers found that, with the exception of hypercholesterolemia, cardiovascular factors during midlife were significantly related to lower cognitive function. Of the cardiovascular factors, obesity and low physical activity had the strongest effects; in fact, the authors noted that these two variables alone doubled the risk for cognitive impairment 23 years later.[59]

It should be noted that few to none of these studies examined weight in relation to a comprehensive set of cognitive functions, or multiple cognitive domains each indexed by one or more tests of ability, and several did not adjust for CVD risk factors other than weight. There are other prospective studies that we have not reviewed. To read further on these investigations and this topic, see the paper by Beydoun and colleagues[10] for a review and meta-analysis of 10 relevant and methodologically acceptable prospective studies employing adults between 40 and 80 years of age at baseline. This meta-analysis supports the general conclusion that obesity earlier in life is a risk factor for dementia and AD later in life. Two more recent reviews concluded that the longer one remains obese, the greater the cognitive dysfunction in late life.[60,61]

An area in need of further analysis is with non-White participants. Arvanitakis and colleagues[62] attempted to address this by examining late-life BMI and its impact on cognitive functioning in Black and White participants. Using data from the Minority Research Study and the Rush Memory and Aging Project, they found a sharper decrease in cognitive function in the areas of episodic memory and semantic fluency, as well as a trend toward decreased performance in working memory. There was no interaction from race. In addition to needing more research examining a more diverse set of participants within the United States, there is a greater need for more prospective cohort studies examining weight, adiposity, and cognition outside of the United States and Europe.

In the following section we focus on studies addressing two important questions: (1) do CVD risk factors and events that correlate highly with obesity account for relations between obesity and cognition; and (2) are there other variables that mediate and modify relations between weight-related variables and cognition?

WEIGHT AND COGNITIVE FUNCTION: ROLE OF CVD FACTORS

The earliest study reporting an association between the classification of overweight and cognitive functioning did not adjust for CVD correlates of obesity.[63] Extensive controls for CVD risk factors were employed for the first time in a prospective study employing data from the Framingham Heart Study.[64] Persons with probable dementia and prevalent acute stroke were excluded from the study. The surveillance period for examination of multiple CVD risk factors and events was 8 to 10 years prior to measurement of multiple cognitive abilities.[65] With statistical adjustment for basic demographics and CVD risk factor variables, obesity was associated with lower performance in measures of attention, visual–spatial ability, memory, executive function, and global performance for both men and women. Relations between obesity and cognitive performance were not as strong for women. Stronger and more consistent associations of obesity and cognitive performance for men[64] with control for CVD variables may be related to their higher risk for CVD-related mortality and morbidity.[66] See Elias and colleagues[64,65] for a more detailed elaboration of these hypotheses. Regardless of the reason for sex differences, this prospective study provides evidence for the conclusion that CVD risk factors play an important mediating role in relations between weight and cognitive functioning but does not entirely account for these associations.

Three years later, support for the generalization was provided by a cross-sectional study in which BMI and measures of central adiposity were related to multiple measures of cognitive ability with several statistical adjustments for health and medical variables.[67] Additionally, study participants with renal, hepatic, pulmonary, or neurologic disease, stroke, probable dementia, psychiatric disorder, heavy alcohol use, and low MMSE scores (<24) were excluded from the analyses. Despite these exclusions, both BMI and WCir were associated with lower levels of cognitive performance on measures of motor speed, manual dexterity, and executive functioning. Adjustment for demographic variables, fasting glucose, insulin, and specific lipid fractions attenuated, but did not eliminate, the associations between the measures of overweight and cognitive measures. Both studies by Elias and colleagues[64] and Waldstein and Katzel[67] indicate that the relation between overweight or T2DM statuses and prevalence of other CVD risk factors does not explain why persons who are overweight or diabetic perform at a lower level on tests of cognitive functioning. These data are generally consistent with reports that obese adults who are otherwise healthy perform more poorly on measures of memory and learning,[68,69] speed and attention,[70] and executive function.[71]

One may conclude from these studies that disease and risk factor correlates of overweight status and obesity partially moderate relations among weight and cognition, but they are insufficient to explain these associations. Other explanatory mechanisms are discussed in the following section on interactions.

Interactions of Obesity and CVD Risk Factors

Associations between obesity and cognition are modified by the presence of other CVD risk factors. In the Framingham Heart Study,[64] obesity and hypertension were independently associated with lower scores on multiple measures of cognitive functioning, but only among men were the adverse effects of obesity and hypertension additive with respect to measures of episodic memory and visual–spatial constructive abilities. The highest level of performance was observed in the absence of both obesity and hypertension. The second highest performance was observed for those with either hypertension or obesity, but not both. Lastly, the lowest performance was observed for those with both obesity and hypertension.

Interactions between T2DM and obesity were also found in the progeny of the Framingham Study participants. In a 12-year prospective study with 1184 men and women participating in the Framingham Offspring Study, the lowest levels of cognitive performance were seen for participants with both hypertension and central obesity as compared to those with either hypertension or obesity or neither of these risk factors.[36] The synergistic relation between T2DM and hypertension has been reported in at least three additional studies, two cross-sectional[67,72] and one longitudinal,[73] both featuring extensive controls for demographic variables, CVD, and CVD risk factor correlates of obesity.

Sakakura and colleagues,[74] employing the MMSE as an outcome variable, reported interactions between hypertension and weight in a sample of young-elder and elder adults with T2DM. There was no relation between obesity and performance, but the lean participants (BMI range = 14.5–20.3) with hypertension were the poorest performers on the MMSE. This finding suggests that hypertension enhances the magnitude of associations between low weight and cognition, and this is consistent with reports of higher mortality rate in lean hypertensive subjects.[75] An interaction study with important implications for individuals experiencing heart failure was performed by Alosco and colleagues.[76] Cerebral blood flow velocity (CBF-V), measured by transcranial Doppler sonography in 99 participants, was the index of cerebral perfusion and obesity and was defined by BMI. There were significant statistical main effects of CBF-V and BMI, but a combination of hypoperfusion and higher BMI showed a stronger relation to measures of attention and executive function than did either hypoperfusion or BMI alone. Weight reduction would appear to be advantageous, particularly in the early stages of heart failure.

Synergistic and additive relations among obesity and other CVD risk factors should come as no surprise. In general, risk factors are additive with respect to their impact on cognitive functioning. In a re-analysis of Framingham Heart Study data, obesity, T2DM, cigarette smoking, and hypertension were significantly related to lower cognitive performance; regardless of which risk factors were in the mix, risk for poor performance increased as the number of risk factors in the regression model increased.[4]

In the previous section, we discussed the fact that obesity-related CVD comorbidity provides an incomplete explanation of relations between overweight status or obesity and cognitive performance. In the following section, we turn to the issue of needing additional controls and variables that intervene between, and partially explain, relations between weight and cognitive functioning.

Controls and Potential Intervening Variables

Failure to adjust for physical activity or ability level, eating behaviors, and mood disorders is common in the weight and cognition literature we have reviewed, but the following studies indicate that these variables may intervene between weight and cognition. In a cross-sectional analysis using the Maine-Syracuse Longitudinal Study (MSLS) data ($N = 917$, ages 23–98 years, with stroke and dementia excluded), Dore and colleagues[77] found inverse associations between WHR and WCir for global cognition, scanning and tracking, and abstract reasoning domains with adjustment for age, other demographic variables, and multiple cardiovascular risk factors. But with further adjustment for self-reported physical activity expressed as metabolic energy expenditure, only the relation between central obesity and abstract reasoning was sustained, suggesting that activity and metabolic expenditure may be important mediators between central adiposity and cognition.

Dore and colleagues[77] statistically adjusted for demographic variables and multiple cardiovascular risk factors, including depressed mood as an estimated metabolic energy expenditure, but important nutrition and dietary habit variables were not considered. Annesi and Gorjala[78] found that exercise was associated with weight loss, but that this association was more related to improvements in mood state and improved eating and nutrition behaviors than by direct energy expenditure.

Mood state may be important in explaining sex differences with respect to relations between weight and cognitive functioning. Among younger adults, obesity tends to be more associated with depressive symptoms among women compared with men, but gender differences are less clear among older adults.[79,80] Among middle-aged and older adults, as few as three studies have systematically examined the bidirectional associations between obesity and depressive symptoms with adjustment for demographic and health variables. Two analyses, one from the MSLS[81] and another from the Alameda County Study,[82] found that for both men and women, BMI predicted changes in depressive symptoms over a period of 5 years but the reverse association was not found (i.e., depressive symptoms predicting change in BMI). In a third analysis with the Nurses Health Study cohort,[83] a bidirectional change association was observed over a 10-year period, but only among women. Although the associations between BMI and depressive risk as a function of gender are unclear, a recent study using the MSLS suggests that over periods up to 15 years, men experience a greater risk of CVD events as a function of depressive symptom level compared with women with adjustments for demographic variables, mean arterial pressure, and exclusion for the presence of CVD at baseline.[84] It is possible that the relative risks for CVD events as a function of obesity and depressive symptoms may result in higher risk for cognitive decline among men.

There are several potential variables that may explain associations between depressive symptoms and weight disturbance. Daily, monthly, or seasonal changes in mood are related to an excessive intake of carbohydrate-rich food and resistance to involvement in physical activity. Brain serotonin may also be involved in these mood and appetite disturbances.[85] Clearly, co-control of these mood state, dietary change, and nutrition variables is important when relating activity level to cognitive performance; unfortunately, these controls are not widely employed. Few, if any, studies have examined them as effect mediators using formal statistical methods such as path analysis or structural equation modeling (SEM).

INITIAL SUMMARY

Based on studies of weight and obesity where normal variation and a full range of cognitive test scores define the cognitive outcome variable, the following generalizations are possible: (1) cardiovascular risk factors, CVD, and CVD events play an important mediating role in relations between obesity and cognition, but are insufficient to fully explain these associations; (2) in large sample studies, central obesity shows a stronger relation to cognitive deficit than BMI; (3) effect sizes for weight and obesity are enhanced by the addition of other risk factors such as hypertension, T2DM, and ApoE-ε4 genotype; (4) adjustments for activity level and energy expenditure are associated with a reduced magnitude and number of relations between central obesity and cognitive performance; and (5) it is important that studies in which weight is adjusted for include further adjustment or control for diet and mood state. A review of overweight status and obesity is incomplete unless tied into the understanding of MetS.

METS

It is not possible to provide a comprehensive review of this rapidly growing area of the literature in this chapter, but a brief summary is important because central adiposity is the key component of MetS. We refer readers to a thorough review of this literature by Crichton and colleagues[16] and identify a few major studies. This review included nine studies that assessed MetS in relation to cognitive performance and 10 studies that used a combination of screening measures, neuropsychological testing, clinical evaluations, and brain imaging to assess MetS in relation to dementia. The authors conclude that despite the widespread variability in study design, statistical control of confounders, type of cognitive testing, and classification of tests, the presence of MetS increased the risk of poor cognitive test performance variously defined and of cognitive impairment in an absolute sense. For example, for cognitive function measured by neuropsychological testing, detriments in executive function, information processing speed, attention, and verbal memory were most frequently reported.[86–92] A more recent systematic review of 26 studies, almost all of which examined middle-aged to older adults, found that executive function was the only cognitive domain to consistently be negatively impacted by the presence of MetS.[93] Even with this reported consistency, performance on verbal fluency tasks that can be conceptualized as both a language and executive task was variable across studies.[93] MetS was associated with increased risk of AD,[94,95] and with an increased risk for both the development of vascular dementia (VaD) and the progression from MCI to dementia over a four-year period.[50,73,96]

The review by Crichton and colleagues[16] highlights some of the major issues facing MetS investigations. While studies have demonstrated relationships between the number of MetS components present and cognitive function,[86,89,97,98] it is often difficult to determine which among the metabolic risk factors included in the syndrome is driving the relation between MetS and cognitive performance. Two recent systematic reviews[93,99] also noted these methodological limitations; namely, that the co-occurrence of multiple illnesses that make up MetS is hard to parse apart[99] and the lack of consistent diagnostic criteria across studies further complicates this field of research.[93] It is important in future research to distinguish if the MetS relationship to cognition is driven by one or more major risk factors such as obesity or T2DM. Stated differently, the role of major metabolic risk factors such as obesity and T2DM must be separated from the cumulative negative force of MetS. Secondly, it will be important to determine if MetS is having a diffuse influence on global cognition or on specific cognitive abilities, which may help identify patterns of cognitive deficit prior to cognitive decline. As we discuss later, this goal is made difficult by the different definitions of various specific cognitive abilities and domains, especially where studies are examined in the context of meta-analysis. For example, a systematic review on MetS and cognition in an older adult population found that only 13 of 26 studies used more than a simple cognitive screen for cognitive assessment and upon further inspection of these studies, only three or four conducted evaluations that would be considered at least somewhat comprehensive.[100] We discuss the pitfall of using cognitive screeners later in this chapter. This is a very important area of future research because it is quite clear that risk factors do not act alone and that they cluster within the individual. We now turn to the consideration of biological mechanisms that may explain relations between weight, obesity, and cognition.

MECHANISMS: GENERAL

As discussed previously, cardiovascular risk factors and CVD at least partially mediate relations between obesity and cognition. Yet, relations between overweight status, obesity, and cognition remain despite control for these health variables. An emerging theme in the following section is that brain changes are occurring before weight gain,[101] before diagnosis of metabolic disease,[102] and before cognitive changes.[103]

There are other potentially important mediators between weight gain and cognitive performance. Inflammation has been proposed as one mechanism behind obesity and brain dysfunction. Whitmer and colleagues[35] found that there are toxic effects of visceral adiposity, a metabolically active endocrine tissue secreting several inflammatory cytokines and hormones, such as adiponectin and interleukin-6 (IL-6), that are associated with cognitive deficit.[104–108] Inflammation is a two-phase process; there is the short-lived inflammatory response, but for obesity, the pressing problem lies in a longer-term inflammatory response. In response to a high fat diet, a "prolonged inflammatory cascade" causes significant damage to the brain, as stress-mediated inflammatory responses (such as cytokines) impact the brain.[101] Further, this inflammatory cascade has been found to affect the hypothalamus and blood–brain barrier (BBB) and eventually, death of hypothalamic neurons. In addition to hypothalamic neurons, Jais and Brüning[101] explain that chronic high-calorie diets affect the integrity of endothelial cells, allowing openings in the BBB that can let in leukocytes. Of particular concern is that the hypothalamic inflammation begins before weight gain.[101] Whitmer and colleagues[35] also cite work indicating that leptin crosses the BBB, plays a role in neurodegeneration, and is involved in the depositing of beta 42, the main ingredient in AD-associated plaques.

Jais and Brüning[101] also explain that this inflammatory response also leads to insulin resistance. Insulin resistance occurs when cells in the body no longer respond to insulin and is considered another mechanism explaining the relation between obesity and cognitive decline. Ryu and colleagues[109] found that insulin resistant middle-aged and older adults who were otherwise healthy already displayed damage to white matter pathways.

There is evidence that total brain atrophy, and especially atrophy of the hippocampus, is related to obesity as there is increased white matter disease.[110,111] Research by Verstynen and colleagues[112] indicates that BMI is associated with a reduction of cerebral white matter throughout the brain. Walther and colleagues[113] reported a negative association between increased body fat and gray and white matter in older women, aged between 52 and 92 years, who underwent extensive high-resolution magnetic resonance imaging (MRI) scanning of the brain. Compared to normal weight women, obese women performed more poorly on tests of executive function, and reduction in gray matter in the left orbital frontal area was associated with executive functioning; while gray matter predicted performance on memory and visual-motor performance, obesity did not. This phenomenon may, or may not, reflect the fact that executive functioning is more sensitive to gray matter deficits than other cognitive measures or the relatively small sample size for a cognitive study as there are only 53 normal weight women, 22 overweight women, and 20 obese women in the study. Thus, we suggest that in studies combining MRI and cognitive measures, power should be determined on the basis of the cognitive measures.

Driscoll and colleagues[114] cite numerous cross-sectional studies in which obesity is related to whole brain, prefrontal regions, and the temporal lobe volumes, including the hippocampus, across a variety of age groups in non-demented study participants. But this phenomenon, as they point out, could be related to the fact that some of the individuals with normal cognitive functioning in midlife may progress to dementia and the volume reduction may be related to pathological processes inherent to the dementing process.

A review by Gustafson[12] addressed the issues of underlying mechanisms regarding adiposity-related cognitive decline in far more detail than is possible in this chapter. A brief summary of highlights is possible. Brain imaging studies have indicated that BMI is associated with multiple brain pathologies, including temporal lobe atrophy, white matter changes in the brain, and BBB disturbances that are more pronounced in late life. Importantly, Gustafson[12] concludes, based on the literature reviewed, that cross-sectional studies indicate that BMI is related to cerebrospinal fluid (CSF) markers and MRI markers in a way that is consistent with a relation between BMI and dementia. BMI is inversely correlated with CSF levels of total tau, is positively correlated with CSF αβ levels, and inversely related to amyloid deposits in the brain. Gustafson[12] points out that amyloid metabolism also occurs in peripheral adipose tissue and discusses the "potential of cross-talk between adipose tissue and the brain in relation to amyloid and perhaps in relation to dementia etiology"[12] and concludes that:

"Irrespective of the direction and temporality of associations, vascular and neurodegenerative pathologies in the brain occur against the background of peripheral metabolic correlates and consequences of overweight status and obesity and other vascular factors, and endocrine and neuroendocrine feedback loops. These alterations may lead to direct neuronal toxicity perhaps due to an altered blood brain barrier."[12(pS10)]

More recently, there has been an emphasis on studying microstructural brain changes related to obesity through diffusion tensor imaging (DTI). With DTI, damage to white matter pathways can be identified.[115] In a study of adults ages 51 to 81, researchers found increased BMI was related to evidence of white matter tract damage. Interestingly, these participants did not show cognitive deficits.[103] In a study of participants 21 to 37 years old, researchers examined both DTI as well as tissue factors (multiparametric mapping) and found that as BMI increased, there was evidence of changes in iron content, increased water, and decreased myelin; all of these are indicators of damage to white matter pathways. Of particular concern, none of the participants were diagnosed with a metabolic disease.[102] Both of these recent studies indicate that subtle brain changes are occurring before the onset of behavioral symptoms (i.e., cognitive dysfunction).

Therefore, it is even more important that longitudinal studies are conducted. The need for longitudinal studies is emphasized in a longitudinal study of midlife obesity in relation to trajectories of brain volume change in older adults conducted with 152 BLSA community-dwelling men and women, aged between 56 and 86.[114] The investigators predicted that the same brain regions that exhibit early pathological changes in association with AD would be related to obesity. There were no associations between central or global midlife obesity and brain volume changes when compared against a background of general age-related atrophy in their non-demented sample. They note that modest relations began to emerge when data up to the point of dementia were analyzed for those individuals who progressed to dementia. The BLSA study authors argue that cross-sectional studies, particularly those focusing on middle-aged individuals, may overstate the relation between obesity and brain volumes in non-demented individuals because they do not identify and/or do not exclude individuals who progress to dementia. Limitations of the BLSA study are that it is not population-based and most of the sample were White and highly educated. Nevertheless, the study makes a major contribution to the literature. It is clear that there is a need for further longitudinal work, especially work in the context of a background of pathological changes for those who will progress to dementia.

The impact of obesity on the brain should be considered in a lifespan perspective. Notably, Sbarbati and colleagues[116] found evidence of inflammation in adipose tissue of obese children. Obesity-related sleep disturbance, including obstructive sleep apnea, via tiredness and hypoxia, among other mechanisms, may play a role in obesity-related lowered cognitive performance in children and adults.[117–122] The study by Spruyt and Gozal[122] is especially noteworthy because they employed SEM, an extension of path analysis, to test whether sleep disordered breathing mediates between overweight and cognition in 351 community-living snoring and non-snoring children between 6 and 10 years of age. They found diminishing conceptualization and reasoning abilities with increasing BMI (direct association with weight), but they also found an increase (up to 0.55-fold) risk of poorer cognitive ability that was mediated by sleep disorder.

Frisardi and colleagues[123] and Gustafson[12] have provided a comprehensive and detailed integrative model and discussions of mechanism while Kumari and colleagues[124] have provided a mini-review on mechanisms relating overweight status and obesity to memory loss. It is clear that psychosocial variables such as activity level, dietary habits, and mood state disorders need to be included in theories relating obesity to cognitive ability and that sleep disturbance and related physiological changes play an important role as a mediating factor.

Spruyt and Gozal[122] note that the functional and structural differences in brains of children with sleep disorders, which are possible mediator between weight and lower cognitive performance, have yet to be examined. However, they do point to some preliminary evidence in other populations. More specifically, children with congenital central hypoventilation syndrome, a genetic disorder associated with sleep disordered breathing, show evidence of localized injury to the hippocampus; in adults with this condition, preliminary neuroimaging studies indicate that atrophy of the hippocampus and white matter lesions in the frontal lobes are seen in association with neural differences in motor, sensory, and autonomic regions of the brain. Thus, imaging studies in nonobese, overweight, and obese children could advance our understanding of the role of sleep disorders in cognitive deficits observed as a result of these conditions of overweight status.

EARLY INFLUENCES ON RELATIONS BETWEEN OBESITY AND COGNITION

The preclinical period for dementia is at least 10 years, but its time course has not yet been defined. Pathophysiological processes related to dementia may begin decades prior to the diagnosis of clinical dementia. Further, there may be different preclinical stages depending on how the neuropathological processes involved are characterized.[10,45] It is generally known that early experience and genotype alter brain structure and function.[125,126] Moreover, the early years of schooling are a critical period in acquisition of cognitive ability and motor development and these skills affect education and cognitive performance in adulthood.[127] Lower levels of occupation, education, and income are risk factors for cognitive deficits and lowered cognitive performance.[128] Lowered cognitive ability can lead to poorer health literacy including decisions about health practices, nutrition, and decreased activity level that increase the likelihood of sustaining or developing obesity later in life.[129–132] Li and colleagues[121] suggests a reverse causality model in which lower levels of cognition lead to being overweight and obese. It is very likely that multidirectional influences exist across the lifespan. More work with path and SEM analyses will be helpful in modeling these influences and in the development of relevant theory. It is well established that obesity in childhood can affect tests of cognition, learning, and academic performance at a very early age.[121] For example, Mond and colleagues[133] conducted a study of pre-school children in Germany to assess a wide array of cognitive domains important to development between 4.4 and 8 years of age. Boys exhibited more impairment than girls in motor development, development of speech, memory, abstraction, visual perception, arithmetic, and emotional behavior. After controlling for potential confounders, impairment in gross motor ability was greater in obese male children than normal weight male children, but impairment in attention skills was greater in obese female children than normal weight female children. These and other studies of children lead to the hypothesis that the basis of sex differences in relations between obesity, being overweight or underweight, and cognition in adults may relate to sex differences in psychosocial variables very early in life.

A study by Guxens and colleagues[134] with a US population summarizes the literature on obesity and cognition in childhood and is pertinent to the reverse cognition hypothesis (i.e., that lowered cognition predicts obesity). Early-life cognitive function scores were related to obesity and overweight classifications at two important developmental periods at 4 and 6 years of age. Higher executive functioning ability, verbal, quantitative skills, and memory skills scores at baseline (age 4) were associated with lower likelihood of being overweight at age 6 with adjustment for a significant number of socioeconomic variables including maternal BMI, sex, birth weight and height, gestational age, maternal smoking during pregnancy, maternal education, maternal age, pregnancy height and BMI, breastfeeding, maternal smoking at baseline, number of siblings at baseline, and child's consumption of sweetened beverages, sweets, and meat at baseline.

As is true in the adult literature on obesity and cognition, results with respect to obesity effect modification are conflicting and the precise mechanisms are unresolved. A systematic review attempting to examine whether there are differential cognitive effects of T2DM, hypertension, dyslipidemia, and obesity found the most consistent cognitive results for T2DM and hypertension; however, the significant variability among studies made it difficult to properly analyze results.[135] However, the literature on childhood obesity indicates that the answer to why we see sex by weight interactions in adults may lie in the literature on school children and prospective studies that span many years. Of particular interest is the emerging evidence that attention-deficit symptoms are associated with higher prevalence of obesity and hypertension in early adulthood;[136,137] thus raising the possibility of causal paths from attention disorder to obesity and to cognitive performance, particularly with regard to attention deficits. A recent review by Smith and colleagues[14] cites several prospective studies indicating that poor performance on measures of executive and motor performance predicts higher BMI in children from 6 to 11 years and discusses obesity as a cause and consequence of cognitive deficits.

Associations between weight and cognition may relate to a chain of events from infancy to dementia. Whalley and colleagues[46] emphasize the importance of a lifespan approach to the etiology of late-onset dementia, postulating that:

"Possible pathways from adverse developmental exposure to dementia include a chain of events triggered by fetal malnutrition, leading to low birth weight, low average childhood intelligence, low educational attainments, low occupational status, and an unhealthy lifestyle." [46(p93)]

Whalley and colleagues[46] hypothesize that malnutrition may trigger at least two pathways to lowered cognitive functioning, one adversely influencing neurodevelopment and the other affecting physiological regulation of blood pressure and glucose metabolism. Models with bidirectional relations between weight and cognitive performance make good sense but these hypotheses need to be tested further, especially using path analysis and SEM methods.

MORBID OBESITY AND CLINICALLY IMPORTANT COGNITIVE DEFICIT

Morbid obesity (BMI > 40) provides a good model for the study of the relationship between cognition and obesity in those circumstances where cognitive deficits are of major clinical importance.[138] Morbidly obese patients often seek bariatric surgery for this condition, thus allowing comparisons of cognitive performance before and after substantial weight reduction.[139,140] The rate of morbid obesity in the US has increased over time,[141] which is cause for concern given the disproportionate risk for obesity-related morbidity, mortality, physical disability, and psychological distress.[142,143] Morbidly obese individuals illustrate the wide range of biological mechanisms serving as mediators between obesity and cognition. Given that there is evidence that morbid obesity is associated with excessive activation of the hypothalamic–pituitary adrenal axis (HPAX),[144] examination of the role of HPAX activation in the morbidly obese provides an opportunity to examine the relationship between alteration of the counter-regulatory response to insulin-induced hypoglycemia and cognitive function after dramatic weight reduction following bariatric surgery.

In a study of HPAX response and cognition, Guldstrand and colleagues[145] evaluated visual–spatial problem solving and maze performance following weight loss from vertical banded gastroplasty. The mean weight loss was 39.8 kg (SD = 9.1), and after approximately 12 months, insulin sensitivity improved to an average increase of 376% (SD = 250%) of initial values. Cognitive functioning in persons with both normoglycemia and hyperglycemia was modified by weight reduction and changed from an accuracy-oriented approach to a speed-oriented approach with post-surgical improvements in processing rate, inspection rate, and correctly solved problems. Gunstad and colleagues[69] tested cognitive functioning across multiple domains after weight loss from bariatric surgery after 12 weeks. While there were no differences in attention, language, and executive functioning, bariatric surgery patients showed improved memory functioning compared to controls. While these results suggest that surgery-related weight loss is associated with cognitive benefits shortly after surgery, adherence to post-operative procedures is essential to sustained improvement in cognition.[146]

More recently, Galioto and colleagues[147] assessed different domains of cognitive functioning following weight loss and changes in glycemic control from bariatric surgery. After 12 months, significant reductions BMI, hyperglycemia, and insulin sensitivity were observed. Reductions in insulin sensitivity were associated with small but significant improvements in working memory, psychomotor speed, and cognitive flexibility while reductions in BMI and hyperglycemia were not associated with cognitive changes. While this provides some support for improved glycemic control contributing to improved cognitive functioning, there are inconsistencies in the literature. Additional studies exploring multiple physiological correlates and utilizing clinical trials with randomization to surgical treatment is needed.[148]

We now turn to a necessary question in relation to the causal associations between weight and cognitive performance: are there randomized-to-treatment clinical trials indicating an improvement in cognitive performance or lessening of dementia symptoms with weight reduction?

TREATMENT OF OVERWEIGHT STATUS AND OBESITY

The first question is whether therapies aimed at reducing weight are potentially useful in the restoration of cognitive performance or slowing of cognitive deficit. Short-term interventions of this nature often yield trivial to modest or no positive results.[6] Consequently, our view is that they are often ineffective or that improvement is modest.[5] It is possible that long-term interventions are necessary to lead to changes in cognitive performance; therefore, longitudinal findings are of interest and importance.

Lifestyle interventions have been explored as potential therapies for improving cognitive performance in cognitively healthy individuals. One noteworthy randomized clinical trial (RCT) is the Look AHEAD Ancillary Study, which investigated the long-term physical and cognitive impact of a lifetime intervention program designed to achieve and maintain weight loss through dietary and physical activity changes in overweight and obese individuals with T2DM. Data are currently available for over 1000 individuals who have participated in the lifestyle intervention program or control condition for up to 13 years. Multiple outcome measures have been collected including physical health measures and a comprehensive neuropsychological battery. The lifestyle intervention program was associated with improvements in multiple markers of physical health that have been identified as risk factors for cognitive decline, such as blood pressure, cholesterol, and hyperglycemia.[149,150] Despite these physiological changes, the lifestyle intervention program was not associated with changes in general or specific domains of cognitive functioning.[149–152] Of note, participants with obesity or a history of CVD who participated in the lifestyle intervention program exhibited higher rates of cognitive impairment and lower cognitive functioning despite engaging in similar levels of physical activity and attaining comparable weight loss.[151,152] It is possible that the cumulative effects of obesity and CVD present at baseline impacted the trajectory of cognitive decline. It may be that primary and secondary, rather than tertiary, prevention is necessary to mitigate the negative effects of obesity on cognition.

Lifestyle interventions have also been investigated as a way to slow the progression of cognitive decline in cognitively impaired individuals. Barnes and colleagues[153] point out that many of the intervention studies involving the goals of weight loss and improvement in cognitive performance in persons with MCI report that resistance training and aerobic exercise achieve modest improvements in cognitive functioning, particularly with respect to attention, processing speed, and executive function. Very often, the specifically targeted cognitive abilities improve, but the improvement does not appear to generalize to other cognitive abilities or general cognitive functioning. Despite these limitations, the Alzheimer's Association has concluded that there is sufficient evidence to conclude that risk for cognitive decline and dementia can be reduced through physical activity and a decrease in cardiovascular risk factors while the risk for cognitive decline may be reduced through cognitive training and dietary changes.[154]

There have been relatively few RCTs in which both weight reduction and cognitive training have been employed in the same trial for persons with MCI[153,155,156] or dementia.[157] In the most recent trial by Barnes and colleagues,[153] 126 inactive community-residing individuals were randomized to either mental activity (intensive computer work) or a mental activity control group (educational DVDs) plus exercise intervention (aerobic) or exercise control group (stretching and toning). Thus, there were four intervention groups and the outcome was global cognitive change based on a comprehensive neuropsychological test battery. Groups did not differ at baseline. The global composite cognitive score improved by 0.16 SD over time, but there were no differences among the intervention groups. As is very often seen in these types of studies, improvement due to the practice effect could not be separated from improvement due to specific interventions. Also, as noted in the editorial commentary on this paper by Lautenschlager and Cox,[158] the 12-week interval may have been too short for differences between the intervention strategies to be observed.

It is clear that long-term intervention studies relating weight loss to cognitive performance are necessary. While recent systematic reviews and a meta-analysis have concluded that combined weight reduction and cognitive training are associated with small but significant improvements in cognitive functioning,[159–161] the vast majority of studies have been conducted in healthy populations. Therefore, additional RCTs investigating the impact of combined weight reduction and cognitive training in individuals with cognitive impairments are warranted.

From a broader perspective, many interventions other than weight loss need to be empirically examined. The best approach to reducing cognitive deficits associated with excessive weight is to prevent, manage, and treat obesity as early as possible. Diet, exercise and physical activity, prescription-weight medication, and weight loss surgery are all important intervention methods for weight loss and this literature is far too extensive to review here. Behavioral and psychiatric treatment can be important but behavioral or psychological disorders may not be the primary issues in many cases of overweight status and obesity. Alternatively, they may not be an issue at all, as obesity involves a complex interaction of genetic, environmental, neuroendocrinological, psychosocial, and behavioral factors. While heredity is estimated to be involved in 30%–70% cases of obesity,[2] diet and exercise remain important factors in weight control as the prevalence and incidence of obesity increases.[141]

While weight reduction interventions are important, longitudinal data indicating that physical activity, while beneficial for many measures of physical health, is not associated with significant, if any, weight loss.[162] We now turn to a possible approach to the treatment of obesity, which is particularly relevant to the central theme of this chapter (i.e., the use of omega-3 polyunsaturated acids (PUFAs)).

TREATMENT OF OBESITY WITH OMEGA-3 FATTY ACIDS

Long-chain omega-3 PUFAs are essential nutrients that may potentially be of benefit to the management of several disease processes including obesity. Studies in animals have indicated promising early results in this area, in particular for eicosapentaenoic acid (EPA) and docosahexaenoic acid (DHA). Omega-3 fatty acids are necessary for normal cognitive development, playing an important role in brain function.[163] Sources of EPA and DHA include fatty fish such as salmon, albacore tuna, herring, mackerel, and anchovies, as well as fish oil supplements. Despite considerable evidence from a number of studies that long-chain omega-3 PUFAs can attenuate weight gain and reduce body fat in rodents,[164,165] relatively few studies have investigated such effects in humans.

The studies that have been conducted in overweight and obese people suggest that long-chain omega-3 PUFA supplementation may be associated with positive changes in body composition measurements. However, recent meta-analyses indicate that these changes appear to be negligible or nonsignificant when comparing an aggregate of studies. More specifically, Bender and colleagues[166] reported that participants ingesting long-chain omega-3 PUFAs showed greater reduction in weight (−0.59 kg), BMI (−0.24 kg/m^2), body fat percentage (−0.49%), and WCir (−0.81 cm). Du and colleagues[167] found participants who consumed fish oil supplements showed greater reductions in WHR (−0.52 cm), but not weight (−0.07 kg), BMI (−0.09 kg/m^2), or WCir (−0.16 cm). It has been suggested that long-chain omega-3 supplementation alone is not sufficient to result in marked weight loss; instead, the addition of alterations in dietary patterns and/or physical activity may be necessary for meaningful change to occur.[165]

Relatively few studies have examined the effects of long-chain omega-3 PUFA supplementation combined with energy restriction and/or physical activity on body composition. Results have been mixed. The majority of studies have failed to find any additional benefit from long-chain omega-3 PUFA supplementation on weight loss in addition to energy restriction and/or physical activity in overweight[168–171] and obese[172–181] samples. However, other studies have reported slightly better outcomes among individuals ingesting long-chain omega-3 PUFAs while restricting caloric intake and/or engaging in exercise.

Kunesova and colleagues[182] found that while weight loss did not significantly differ between groups, there was a modestly greater decrease in BMI among women supplementing a very low caloric diet (2200 kJ/day) with long-chain omega-3 PUFAs (2.8 g/day) and daily physical activity (60 min/day) for 3 weeks. Thorsdottir and colleagues[183] showed a modestly greater increase in weight loss with either long-chain omega-3 PUFA supplementation (6 × 1 g capsules yielding 1.5 g/day long-chain omega-3 PUFAs) or the inclusion of either lean or fatty fish, as part of an energy restricted diet over 8 weeks in males only. This differential with regard to gender is in contrast to the most recently conducted study, a double-blind RCT by Munro and Garg[184] investigating the effects of supplementation with long-chain omega-3 PUFAs alone, then consumed concomitantly with a very low energy diet on weight loss. In this study, the placebo group ($n = 19$) consumed 6 × 1 g capsules per day of monounsaturated oil, and the treatment group ($n = 20$) consumed 6 × 1 g capsules per day of long-chain omega-3 PUFAs (fish oil, 70 mg EPA and 270 mg DHA), while consuming their usual diet for 4 weeks. Each group continued with their supplements for another 4 weeks while following a very low energy diet. Despite no change at 4 weeks, there was a greater percentage decrease for females in the supplementation group compared to the placebo for both body weight (−7.21% vs 25.82%) and BMI (−7.43% vs 25.91%), respectively ($p < 0.05$ for both) at 8 weeks.

A recent meta-analysis focused primarily on studies that included long-chain omega-3 PUFA supplementation combined with a reduction in caloric intake and/or the addition of exercise regimes. Zhang and colleagues[185] reported that participants who consumed fish oil supplements showed greater reductions in WCir (−0.53 cm); however, differences in weight (0.00 kg) or BMI (−0.19 kg/m^2) were not detected between groups. This suggests that when examining an aggregate of currently available studies, the combination of omega-3 fatty acids, energy restriction, and/or physical activity does not appear to have a significant impact on body composition in overweight or obese individuals.

Much more work is necessary and, hopefully, will be inspired by the rapidly growing literature on overweight status, obesity, and cognitive functioning. The majority of studies have been of short duration, used small samples, and shown only modest effects.[164–167,185] Differences in study design, and the dosage, timing, and duration of supplement administration may also contribute to the conflicting results to date. Further, human studies are necessary as physiological and behavioral changes in animals may not be equivalent in humans.

In the longer term, large intervention studies in overweight and obese adults are needed to determine if long-chain omega-3 PUFA supplementation, either alone or in combination with changes in diet or physical activity, may be effective in reducing body fat, with the possibility of it being used as a public health strategy to combat the obesity epidemic.

OMEGA-3 FATTY ACID MECHANISMS FOR PROMOTING WEIGHT LOSS

A brief summary of the mechanisms by which the dietary intake of long-chain omega-3 PUFAs may facilitate body weight reduction may be helpful with respect to research hypotheses for further studies. Appetite reduction is an important possible mechanism. There is some evidence to support the hypothesis that long-chain omega-3 PUFA intake may increase postprandial satiety, thereby reducing subsequent food intake and impacting weight trajectory.[186,187] A smaller sub-study of the weight loss trial by Thorsdottir and colleagues[183] in overweight and obese volunteers showed that individuals who consumed higher long-chain omega-3 PUFA content meals (fatty fish and fish oil) had lower hunger sensations immediately after the test dinner and two hours after the meal.[188] However, recent randomized studies have failed to replicate this finding. While healthy-weight volunteers who consumed long-chain omega-3 and omega-6 PUFA supplements for seven days exhibited changes in physiological indicators of hunger (e.g., increased ghrelin, decreased peptide YY), ratings of satiety did not differ between the PUFA and control diet groups.[189]

In addition to influencing satiety, EPA and DHA may have a direct effect on specific brain structures and functions involved in appetite, food intake, and energy homeostasis.[186] More specifically, animal studies have shown that EPA and DHA may reduce brain endocannabinoids involved in overeating, and effect dopaminergic neurotransmission implicated in the regulation of mood, motivation, and reward (e.g., in response to food intake).[186]

Long-chain omega-3 PUFAs has been implicated in several biological processes that potentially interact. Long-chain omega-3 PUFAs has been associated with reduced adipocytes and central adipose tissue, resulting in more adaptive adipogenesis as well as reduced cytokines, leading to lower levels of inflammation.[165] These processes impact insulin sensitivity and the secretion of adipokines, which are implicated in obesity-related diseases.[165]

Another mechanism for operation of long-chain omega-3 PUFAs is possible alterations in the expression of genes involved in the regulation of fatty acid oxidation in adipose, liver, cardiac, intestinal, and skeletal muscle tissue.[165,190–192] These mechanisms may contribute to body fat loss by increasing fatty acid oxidation in these organs, reducing the capacity for deposition in adipose tissue, and impeding the storage of fat.[164,165] Finally, long-chain omega-3 PUFA supplementation may also improve vasodilator function and increase blood flow and energy utilization of fats.[165,174,187]

Further studies are needed to investigate all of these possible mechanisms, as they provide a guide to studies relating long-chain omega-3 PUFAs to weight loss and cognitive performance and relations between these phenomena.

METHODOLOGIES: GENERAL ISSUES

Measurement of Cognitive Function

A number of methodological issues deserve attention and it is important to define which cognitive abilities are adversely associated with the conditions of overweight status and obesity because the pattern of cognitive deficits provides an indication as to whether low cognitive performance will more likely progress to AD or VaD if conversion to dementia occurs. This goal is often frustrated by the fact that tests of different abilities may vary in difficulty level. Many of the clinical tests employed do not represent pure measures of the constructs they purport to measure because they assess related, and thus overlapping, abilities. See thorough and thoughtful discussions of these issues in Rabbitt[193] and Lyon and Krasnegor.[194]

In our interpretation of the literature, global cognitive ability and a wide range of specific abilities are adversely affected by the conditions of underweight and overweight status and by changes in weight and obesity. Studies specifically concerned with the pattern of cognitive deficiencies associated with obesity[54,58,59,68,70,71,195] indicate that memory, attention, psychomotor speed, and executive function are among those cognitive skills particularly vulnerable to obesity. These conclusions mainly result from consideration of amalgamating results from multiple studies because the memory studies we reviewed met our criteria for a comprehensive test battery. This is perhaps due to issues of participant burden or because investigators are making use of available data sets. In our view, a comprehensive battery should include a global measure and/or a composite of the individual measures in the battery, which would be, at minimum, attention (with separation of simple attention and working memory), verbal and nonverbal learning and memory (immediate and delayed), language, speed of performance, abstract reasoning, constructional ability, and some form of visual–spatial organization ability. Ideally, it would also have a measure of premorbid ability such as performance on a word-reading task, which is a good proxy for premorbid ability (for a review, see Graves, Carswell, & Snow[196]).

Obviously, test batteries with multiple neuropsychological tests measuring specific abilities are important. Ideally, we should be examining cognitive domains and those domains should be indexed by one or more specific abilities measured by specific tests, as there are both multiple cognitive domains involved in thinking/behavior and each domain should have at least two tests (for a detailed explanation, see Elias and colleagues[6]). For example, tests of attention and executive functioning are popular constructs but are often measured by one or a few test measures. More than one test purporting to measure these constructs should be used to index the executive functioning domain. For example, it is not uncommon to find studies that group simple attention and working memory into one variable when, in fact, they are measuring different abilities. In addition to including multiple measures per cognitive domain, researchers are encouraged to use factor analysis to identify different domains. This is particularly achievable in large cohort studies.[51,54] In two papers, Elias and colleagues recommend defining a domain by no less than two measures of that domain.[6,197]

Formal Definitions of MCI

Petersen and colleagues[198] described MCI as subjective complaints of memory loss, memory decline evidenced through formal assessment beyond what is typical for normal aging, and preservation of functional abilities. This has been considered the gold-standard among clinicians and researchers. This study, however, focused on amnestic MCI (i.e., where memory performance is notably lower other cognitive functions are relatively intact). While amnestic MCI has been considered a precursor to AD,[199] it does not necessarily apply to all dementia cases. Obesity, as aforementioned, is associated with cardiovascular, cerebrovascular, and metabolic disease and increases the risk for dementia (for a review, see Elias, Goodell, & Walstein[5]). Whether this is pure VaD or a mixed AD/VaD pathology is not always clear in the literature. Nevertheless, it is important that studies clearly describe and adhere to a formal diagnosis of MCI. A relatively recent working group from the NIA and Alzheimer's Association put forth the following set of criteria for MCI *in general*:

"Concern regarding change in cognition; impairment in one or more cognitive domains; preservation of independence in functional abilities; and not demented."[199p(3,4)]

Note that this definition does not state that memory must be impaired, thus broadening the concept of MCI. Additionally, as part of the aforementioned working group, Sperling and colleagues[45] emphasized the concept of a continuum for dementia, noting that MCI was one stop along this continuum for some people. They also described the significant contribution that health factors, such as obesity, have made on the pathophysiology of dementia.

But, what does a diagnosis of MCI mean in prospective studies? While amnestic MCI (i.e., MCI where memory decline is primary) appears to be the precursor for AD,[199] this is not true for all cases of MCI. In fact, Koepsell and Monsell[200] found that out of the 3020 people diagnosed with MCI at an AD Center, 16% of them were back to being cognitively normal after 1 year. Therefore, from a practical standpoint, diagnosing MCI alone does not necessarily imply a neurodegenerative process.

This leads us to the argument for multiphased diagnosis that involves case conference among interdisciplinary providers, which we cover in detail in the following section.

Diagnosis of Probable Dementia

We describe diagnosis of probable dementia, which is the closest to a diagnosis of dementia in vivo (from the AD literature, see McKhann et al.[201]). In recent years, efforts have been made to establish a more formal method for diagnosing probable dementia. In 2011, a series of papers were published from the National Institute on Aging (NIA) and Alzheimer's Association workgroups, in which they expanded on the diagnoses of AD but also of all-cause dementia. Our readers are encouraged to review this series of articles for the exact criteria.[202,203] Of particular significance in these criteria was the emphasis on neuropsychological assessment.

In addition to the aforementioned definition of probable dementia, we need to briefly review two major types of dementia: AD and VaD. The pathology of AD is considered to be one of beta amyloid plaques, neurofibrillary tangles, and neuronal loss that primarily targets temporal–parietal regions first.[201,202] VaD is defined as a dementia with vascular origins. This broad definition includes multiple possible etiologies, such as small vessel disease, leukoencephalopathy, and multiple infarcts, among other causes (for a review, see Jellinger[204]). Further complicating differential diagnosis is the prevalence of mixed AD/VaD, when both disease processes are co-occurring (for a review, see Jellinger[205]). Additionally, there is much evidence that vascular factors themselves are a pathway to AD.[206]

With respect to diagnosis of probable dementia and its employment as an outcome measure in research, differences in results may relate to differences in the way it is diagnosed. Studies often employ one or more of the following: comprehensive clinical dementia reviews, medical records, insurance records, and inferences from MMSE and Montreal Cognitive Assessment (MoCA) scores, both instruments that are good at screening for the presence of dementia but cannot inform on one's full cognitive profile. In our view, under no circumstances should a diagnosis of dementia be made solely on the basis of MMSE, MoCA, or other screening scores, although clearly they play a role as an important component of the diagnostic process.

However, researchers must not rely on one neuropsychological evaluation alone, as variables outside of cognition (e.g., effort) can also impact performance (see National Academy of Neuropsychology position paper on symptom validity[207]). Therefore, we recommend that researchers employ both a multiphase approach, in which participants who score below a cutoff on a cognitive screen such as the MMSE or MoCA go on for a comprehensive neuropsychological evaluation, and visit other medical and psychological providers so that information about relevant laboratory results, neuroimaging, and a thorough psychosocial evaluation are available. Lastly, these interdisciplinary providers should meet, review their findings, and arrive at a consensus for diagnosis. It is our hope that future longitudinal studies will adhere to these recommendations and therefore provide additional objective data for diagnosis of dementia. Readers are also encouraged to see recommendations for dementia studies by Elias and colleagues.[197]

Prospective Designs

Given that predictors and outcomes of dementia are confounded in studies conducted following a dementia diagnosis, the ideal designs are prospective, involving long surveillance periods prior to the onset of dementia. Additionally, prospective, and especially prospective-longitudinal designs, should include multiple weight and cognitive performance measurements over time. In the future, more lifespan longitudinal studies are needed, or at least the emergence of a connected body of literature that extends collectively over all periods of the life cycle. There are three important reasons why this is true: (1) bidirectional relations between obesity and cognition may begin in childhood;[5] (2) poorer cognitive performance in non-demented individuals is itself a risk factor for dementia; and (3) the direction of the relation between obesity and cognition is yet to be determined and further research is necessary to examine theoretical models suggesting that relations are bidirectional.[5]

Neuroimaging Studies

Self-reported height and weight is not an acceptable option for future studies of weight and cognitive performance due to inaccuracy in participant reporting, which has lead to underestimation of BMI across the literature.[208] However, it is important to note that anthropomorphic measurements of body weight, WCir, hip circumference (HCir), and WHR are also susceptible to inaccuracy, especially in overweight and obese persons.[209–211] Thus, more reliable CT, MRI, and dual-energy X-ray absorptiometry methods will hopefully be employed more frequently in future studies.[212] Functional and structural neuroimaging can play a very important role in studies relating potential mediating brain mechanisms (e.g., reductions in gray matter volume, white matter hyperintensities, and brain atrophy) to brain areas and specific cognitive abilities and major domains.[213,214]

Trials

RCTs with joint weight reducing methods and cognitive outcomes, such as bariatric surgery, dietary and physical activity regimes, cognitive training, and long-chain omega-3 PUFA supplementation, are in their infancy and hopefully will increase in number in future years. A comment on deficits in clinical trials with cognition and dementia outcomes has been published by Elias and colleagues.[197]

Clinical Implications

The previously reviewed research has suggested that treatments for obesity, including dietary and physical activity regimes and long-chain omega-3 PUFA supplementation, have modest effects in regard to anthropomorphic measurements such as body weight, WCir, HCir, and WHR. While the clinical significance of such interventions may be low at an individual level, the epidemiological significance has important implications at a population level. Small effects in a large population have a significant impact on the population health. Therefore, additional research on these and other treatments is warranted to reduce the impact of obesity.

FINAL SUMMARY AND CONCLUSIONS

From an empirical perspective, it is clear that overweight status and obesity are conditions related to lowered cognitive performance and to the dementias, and that deficits in cognition are enhanced by the presence of obesity with other risk factors, especially a constellation of correlated risk factors that we refer to as MetS. Measures of central obesity appear to be more strongly associated with dementia and cognitive deficit than BMI, and results are less conflicting when these measures are used as predictors. Visceral obesity, reflected in central obesity, is more strongly related to cognitive performance than subcutaneous obesity, but more studies with objective measurement of these weight parameters are needed in the future.

Less is not better than more with respect to risk of cognitive deficit or decline related to weight gain or loss. Especially with measures of central obesity as predictors, J- and U-shaped functions between weight and cognitive ability are seen. Where studies of weight loss have been undertaken, it is particularly important to have a long prospective surveillance period as weight loss may be precipitated by the dementing processes.

In the first controlled studies of weight and obesity, cardiovascular risk factors were viewed as mediators of relations between weight and cognition, but a rapidly evolving literature indicates that many other mediating mechanisms are possible. The next phase of research on weight and cognition will need to involve more studies of effect moderators and mediators. This phase of research could make much better use of path analyses and SEM statistical methods for evaluating and modeling the role of mediating variables in a systematic manner.

The mechanisms underlying sex modification of relations among overweight status, weight change, obesity, and cognition are not completely understood. It is our view that variables that will explain the role of sex in relations between obesity and cognition in later life are established at an early age. Fetal development, infancy, and childhood obesity is at the other end of the life-spectrum from dementia but understanding the path to dementia requires an understanding of the role of psychosocial variables, cognition, and obesity in childhood. This is especially important if we are to better understand possible reciprocal relations between cognition and weight.

As empirical research on dementia in general has moved backward in time from dementia diagnosis, we have achieved a better understanding of the predictors of dementia. The success achieved by these timescale studies reinforces arguments for lifespan approaches to dementia.[29,46,215,216] In the popular motion picture Back to the Future, a young man journeys into the past and influences, for the better, his parents' experiences and life in the future.[217] Our goal must be to influence the future of those who are at risk of dementia due to preventable and treatable risk factors, overweight status, and obesity. Among many other treatment strategies, supplementation with long-chain omega-3 PUFAs should be studied. Certainly, gaining a better understanding of the role of cognitive deficits in overweight status and obesity is as important as understanding the role of overweight status and obesity in cognitive deficits and the search for answers here leads us back to infancy, childhood, and possibly to neonatal development.

Acknowledgments

We would like to acknowledge Merrill F. Elias, Georgina E. Crichton, and Amanda L. Goodell, the authors of the first-edition chapter, for their significant contributions without which this current update would not be possible.

References

1 World Health Organization. Overweight and obesity. http://www.who.int/en/news-room/fact-sheets/detail/obesity-and-overweight. Published 2017. (Accessed 20 May 2018).
2. Eckel RH. Obesity and heart disease: A statement for healthcare professionals from the nutrition committee, American Heart Association. *Circulation*. 1997;96(9):3248–3250.
3. Sowers JR. Obesity and cardiovascular disease. *Clin Chem*. 1998;44(8):1821–1825.
4. Waldstein SR, Elias MF. Cardiovascular risk factors and cognitive functioning: an epidemiological perspective. In: *Neuropsychology of Cardiovascular Disease*. Psychology Press; 2001:99–120.
5. Elias MF, Goodell AL, Waldstein SR. Obesity, cognitive functioning and dementia: back to the future. *J Alzheimers Dis*. 2012;30(s2):S113–S125.
6. Elias MF, Goodell AL, Dore GA. Hypertension and cognitive functioning: a perspective in historical context. *Hypertension*. 2012;60(2):260–268.
7. Elias MF, Beiser A, Wolf PA, Au R, White RF, D'agostino RB. The preclinical phase of Alzheimer disease: a 22-year prospective study of the Framingham cohort. *Arch Neurol*. 2000;57(6):808–813.
8. Anstey KJ, Cherbuin N, Budge M, Young J. Body mass index in midlife and late-life as a risk factor for dementia: a meta-analysis of prospective studies. *Obes Rev*. 2011;12(5):e426–e437.
9. Barrett-Connor E. An introduction to obesity and dementia. *Curr Alzheimer Res*. 2007;4(2):97–101.
10. Beydoun MA, Beydoun HA, Wang Y. Obesity and central obesity as risk factors for incident dementia and its subtypes: a systematic review and meta-analysis. *Obes Rev*. 2008;9(3):204–218.
11. Gustafson D. A life course of adiposity and dementia. *Eur J Pharmacol*. 2008;585(1):163–175.
12. Gustafson DR. Adiposity and cognitive decline: underlying mechanisms. *J Alzheimers Dis*. 2012;30(s2):S97–S112.
13. Sellbom KS, Gunstad J. Cognitive function and decline in obesity. *J Alzheimers Dis*. 2012;30(s2):S89–S95.
14. Smith E, Hay P, Campbell L, Trollor JN. A review of the association between obesity and cognitive function across the lifespan: implications for novel approaches to prevention and treatment. *Obes Rev*. 2011;12(9):740–755.
15. Whitmer RA. The epidemiology of adiposity and dementia. *Curr Alzheimer Res*. 2007;4(2):117–122.
16. Crichton GE, Elias MF, Buckley JD, Murphy KJ, Bryan J, Frisardi V. Metabolic syndrome, cognitive performance, and dementia. *J Alzheimers Dis*. 2012;30(s2):S77–S87.
17. Albanese E, Davis B, Jonsson PV, et al. Overweight and obesity in midlife and brain structure and dementia 26 years later: the AGES-Reykjavik study. *Am J Epidemiol*. 2015;181(9):672–679.
18. Stewart R, Masaki K, Xue QL, et al. A 32-year prospective study of change in body weight and incident dementia: the Honolulu-Asia aging study. *Arch Neurol*. 2005;62(1):55–60.
19. Albanese E, Launer LJ, Egger M, et al. Body mass index in midlife and dementia: Systematic review and meta-regression analysis of 589,649 men and women followed in longitudinal studies. *Alzheimers Dement Diagn Assess Dis Monit*. 2017;8:165–178.
20. Chuang YF, An Y, Bilgel M, et al. Midlife adiposity predicts earlier onset of Alzheimer's dementia, neuropathology and presymptomatic cerebral amyloid accumulation. *Mol Psychiatry*. 2016;21(7):910–915.
21. Emmerzaal TL, Kiliaan AJ, Gustafson DR. 2003-2013: a decade of body mass index, Alzheimer's disease, and dementia. *J Alzheimers Dis*. 2015;43(3):739–755.
22. Kivipelto M, Ngandu T, Fratiglioni L, et al. Obesity and vascular risk factors at midlife and the risk of dementia and Alzheimer disease. *Arch Neurol*. 2005;62(10):1556–1560.
23. Loef M, Walach H. Midlife obesity and dementia: meta-analysis and adjusted forecast of dementia prevalence in the United States and China. *Obesity*. 2013;21(1):E51–E55.
24. Monda V, La Marra M, Perrella R, et al. Obesity and brain illness: from cognitive and psychological evidences to obesity paradox. *Diabetes Metab Syndr Obes Targets Ther*. 2017;10:473–479.
25. Tolppanen AM, Ngandu T, Kareholt I, et al. Midlife and late-life body mass index and late-life dementia: results from a prospective population-based cohort. *J Alzheimers Dis*. 2014;38(1):201–209.
26. Whitmer RA, Gunderson EP, Barrett-Connor E, Quesenberry CP, Yaffe K. Obesity in middle age and future risk of dementia: a 27 year longitudinal population based study. *BMJ*. 2005;330(7504):1360–1365.
27. Nourhashemi F, Deschamps V, Larrieu S, Letenneur L, Dartigues J-F, Barberger-Gateau P. Body mass index and incidence of dementia the PAQUID study. *Neurology*. 2003;60(1):117–119.

28. Fitzpatrick AL, Lopez OL, Kuller LH. Mid-and late life obesity: risk of dementia in the cardiovascular health cognition study. In: *Annals of Neurology*. Vol. 60. Hoboken, NJ: Wiley-Liss Div John Wiley & Sons Inc.; 2006. S4–S4.
29. Singh-Manoux A, Dugravot A, Shipley M, et al. Obesity trajectories and risk of dementia: 28 years of follow-up in the Whitehall II study. *Alzheimers Dement*. 2017;14(2):178–186.
30. Suemoto CK, Gilsanz P, Mayeda ER, Glymour MM. Body mass index and cognitive function: The potential for reverse causation. *Int J Obes (Lond)*. 2015;39(9):1383–1389.
31. Kivimäki M, Luukkonen R, Batty GD, et al. Body mass index and risk of dementia: analysis of individual-level data from 1.3 million individuals. *Alzheimers Dement*. 2018;14(5):601–609.
32. Allison DB, Faith MS, Heo M, Kotler DP. Hypothesis concerning the U-shaped relation between body mass index and mortality. *Am J Epidemiol*. 1997;146(4):339–349.
33. Janssen I, Katzmarzyk PT, Ross R. Waist circumference and not body mass index explains obesity-related health risk. *Am J Clin Nutr*. 2004;79(3):379–384.
34. Stevens J, Cai J, Thun MJ, Williamson DF, Wood JL. Consequences of the use of different measures of effect to determine the impact of age on the association between obesity and mortality. *Am J Epidemiol*. 1999;150(4):399–407.
35. Whitmer RA, Gustafson DR, Barrett-Connor E, Haan MN, Gunderson EP, Yaffe K. Central obesity and increased risk of dementia more than three decades later. *Neurology*. 2008;71(14):1057–1064.
36. Wolf PA, Beiser A, Elias MF, Au R, Vasan RS, Seshadri S. Relation of obesity to cognitive function: importance of central obesity and synergistic influence of concomitant hypertension. The Framingham heart study. *Curr Alzheimer Res*. 2007;4(2):111–116.
37. Yoon DH, Choi SH, Yu JH, Ha JH, Ryu SH, Park DH. The relationship between visceral adiposity and cognitive performance in older adults. *Age Ageing*. 2012;41(4):456–461.
38. Luchsinger JA, Biggs ML, Kizer JR, et al. Adiposity and cognitive decline in the cardiovascular health study. *Neuroepidemiology*. 2013;40(4):274–281.
39. Knopman DS, Edland SD, Cha RH, Petersen RC, Rocca WA. Incident dementia in women is preceded by weight loss by at least a decade. *Neurology*. 2007;69(8):739–746.
40. Gillette-Guyonnet S, Nourhashémi F, Andrieu S, et al. Weight loss in Alzheimer disease. *Am J Clin Nutr*. 2000;71(2):637S–642S.
41. Will MA, Randolph JF. The influence of reproductive hormones on brain function in the menopausal transition. *Minerva Ginecol*. 2009;61(6):469–481.
42. Brubacher D, Monsch AU, Stähelin HB. Weight change and cognitive performance. *Int J Obes (Lond)*. 2004;28(9):1163–1167.
43. Fewlass DC, Noboa K, Pi-Sunyer FX, Johnston JM, Yan SD, Tezapsidis N. Obesity-related leptin regulates Alzheimer's Aβ. *FASEB J*. 2004;18(15):1870–1878.
44. Sturman MT, de Leon CM, Bienias JL, Morris MC, Wilson RS, Evans DA. Body mass index and cognitive decline in a biracial community population. *Neurology*. 2008;70(5):360–367.
45. Sperling RA, Aisen PS, Beckett LA, et al. Toward defining the preclinical stages of Alzheimer's disease: Recommendations from the National Institute on Aging-Alzheimer's association workgroups on diagnostic guidelines for Alzheimer's disease. *Alzheimers Dement J Alzheimers Assoc*. 2011;7(3):280–292.
46. Whalley LJ, Dick FD, McNeill G. A life-course approach to the aetiology of late-onset dementias. *Lancet Neurol*. 2006;5(1):87–96.
47. de Bruijn R. *Emerging determinants of dementia*. 2015.
48. Beydoun MA, Lhotsky A, Wang Y, et al. Association of adiposity status and changes in early to mid-adulthood with incidence of Alzheimer's disease. *Am J Epidemiol*. 2008;168(10):1179–1189.
49. Azad NA, Al Bugami M, Loy-English I. Gender differences in dementia risk factors. *Gend Med*. 2007;4(2):120–129.
50. Solfrizzi V, Scafato E, Capurso C, et al. Metabolic syndrome, mild cognitive impairment, and progression to dementia. The Italian longitudinal study on aging. *Neurobiol Aging*. 2011;32(11):1932–1941.
51. Kesse-Guyot E, Andreeva VA, Touvier M, et al. Overall and abdominal adiposity in midlife and subsequent cognitive function. *J Nutr Health Aging*. 2015;19(2):183–189.
52. Dahl A, Hassing LB, Fransson E, et al. Being overweight in midlife is associated with lower cognitive ability and steeper cognitive decline in late life. *J Gerontol Ser Biomed Sci Med Sci*. 2009;65(1):57–62.

53. Hassing LB, Dahl AK, Pedersen NL, Johansson B. Overweight in midlife is related to lower cognitive function 30 years later: a prospective study with longitudinal assessments. *Dement Geriatr Cogn Disord*. 2010;29(6):543–552.
54. Dahl AK, Hassing LB, Fransson EI, Gatz M, Reynolds CA, Pedersen NL. Body mass index across midlife and cognitive change in late life. *Int J Obes (Lond)*. 2013;37(2):296–302.
55. Sabia S, Kivimaki M, Shipley MJ, Marmot MG, Singh-Manoux A. Body mass index over the adult life course and cognition in late midlife: The Whitehall II cohort study. *Am J Clin Nutr*. 2008;89(2):601–607.
56. Truett J, Cornfield J, Kannel W. A multivariate analysis of the risk of coronary heart disease in Framingham. *J Chronic Dis*. 1967;20(7):511–524.
57. Laitala V, Kaprio J, Koskenvuo M, Raiha I, O'Rinne J, Silventoinen K. Association and causal relationship of midlife obesity and related metabolic disorders with old age cognition. *Curr Alzheimer Res*. 2011;8(6):699–706.
58. Singh-Manoux A, Czernichow S, Elbaz A, et al. Obesity phenotypes in midlife and cognition in early old age the Whitehall II cohort study. *Neurology*. 2012;79(8):755–762.
59. Virta JJ, Heikkilä K, Perola M, et al. Midlife cardiovascular risk factors and late cognitive impairment. *Eur J Epidemiol*. 2013;28(5):405–416.
60. Dahl AK, Hassing LB. Obesity and cognitive aging. *Epidemiol Rev*. 2012;35(1):22–32.
61. Prickett C, Brennan L, Stolwyk R. Examining the relationship between obesity and cognitive function: a systematic literature review. *Obes Res Clin Pract*. 2015;9(2):93–113.
62. Arvanitakis Z, Capuano AW, Bennett DA, Barnes LL. Body mass index and decline in cognitive function in older black and white persons. *J Gerontol Ser A*. 2017;73(2):198–203.
63. Sørensen TI, Sonne-Holm S, Christensen U, Kreiner S. Reduced intellectual performance in extreme overweight. *Hum Biol*. 1982;765–775.
64. Elias MF, Elias PK, Sullivan LM, Wolf PA, D'agostino RB. Lower cognitive function in the presence of obesity and hypertension: the Framingham heart study. *Int J Obes (Lond)*. 2003;27(2):260–268.
65. Elias MF, Elias PK, Sullivan LM, Wolf PA, D'Agostino RB. Obesity, diabetes and cognitive deficit: the Framingham heart study. *Neurobiol Aging*. 2005;26(1):11–16.
66. Mosca L, Barrett-Connor E, Wenger NK. Sex/gender differences in cardiovascular disease prevention: What a difference a decade makes. *Circulation*. 2011;124(19):2145–2154.
67. Waldstein SR, Katzel LI. Interactive relations of central versus total obesity and blood pressure to cognitive function. *Int J Obes (Lond)*. 2006;30(1):201–207.
68. Gunstad J, Paul RH, Cohen RA, Tate DF, Gordon E. Obesity is associated with memory deficits in young and middle-aged adults. *Eat Weight Disord-Stud Anorex Bulim Obes*. 2006;11(1):e15–e19.
69. Gunstad J, Strain G, Devlin MJ, et al. Improved memory function 12 weeks after bariatric surgery. *Surg Obes Relat Dis*. 2011;7(4):465–472.
70. Cournot M, Marquie JC, Ansiau D, et al. Relation between body mass index and cognitive function in healthy middle-aged men and women. *Neurology*. 2006;67(7):1208–1214.
71. Gunstad J, Luyster F, Hughes J, Waechter D, Rosneck J, Josephson R. The effects of obesity on functional work capacity and quality of life in phase II cardiac rehabilitation. *Prev Cardiol*. 2007;10(2):64–67.
72. Kim E, Cho MH, Cha KR, et al. Interactive effect of central obesity and hypertension on cognitive function in older out-patients with type 2 diabetes. *Diabet Med*. 2008;25(12):1440–1446.
73. Solfrizzi V, Scafato E, Capurso C, et al. Metabolic syndrome and the risk of vascular dementia. The Italian Longitudinal Study on Aging. *J Neurol Neurosurg Psychiatry*. 2010;81:433–440.
74. Sakakura K, Hoshide S, Ishikawa J, et al. Association of body mass index with cognitive function in elderly hypertensive Japanese. *Am J Hypertens*. 2008;21(6):627–632.
75. Stamler R, Ford CE, Stamler J. Why do lean hypertensives have higher mortality rates than other hypertensives? Findings of the hypertension detection and follow-up program. *Hypertension*. 1991;17(4):553–564.
76. Alosco ML, Spitznagel MB, Raz N, et al. Obesity interacts with cerebral hypoperfusion to exacerbate cognitive impairment in older adults with heart failure. *Cerebrovasc Dis Extra*. 2012;2(1):88–98.
77. Dore GA, Elias MF, Robbins MA, Budge MM, Elias PK. Relation between central adiposity and cognitive function in the Maine–Syracuse study: attenuation by physical activity. *Ann Behav Med*. 2008;35(3):341–350.
78. Annesi JJ, Gorjala S. Relationship of exercise program participation with weight loss in adults with severe obesity: assessing psychologically based mediators. *South Med J*. 2010;103(11):1119–1123.
79. Atlantis E, Baker M. Obesity effects on depression: systematic review of epidemiological studies. *Int J Obes (Lond)*. 2008;32(6):881–891.

80. Xiang X, An R. Obesity and onset of depression among US middle-aged and older adults. *J Psychosom Res*. 2015;78(3):242–248.
81. Dearborn PJ, Robbins MA, Elias MF. Challenging the "jolly fat" hypothesis among older adults: high body mass index predicts increases in depressive symptoms over a 5-year period. *J Health Psychol*. 2018;23(1):48–58.
82. Roberts RE, Deleger S, Strawbridge WJ, Kaplan GA. Prospective association between obesity and depression: evidence from the Alameda County study. *Int J Obes (Lond)*. 2003;27(4):514–521.
83. Pan A, Sun Q, Czernichow S, et al. Bidirectional association between depression and obesity in middle-aged and older women. *Int J Obes (Lond)*. 2012;595:595–602.
84. Haigh EAP, Bogucki OE, Dearborn PJ, Robbins MA, Elias MF. Depressive symptoms prospectively predict cardiovascular disease among older adults: findings from the Maine-Syracuse longitudinal study. *J Health Psychol*. 2018;1–11.
85. Wurtman JJ. Depression and weight gain: the serotonin connection. *J Affect Disord*. 1993;29(2–3):183–192.
86. Cavalieri M, Ropele S, Petrovic K, et al. Metabolic syndrome, brain magnetic resonance imaging, and cognition. *Diabetes Care*. 2010;33(12):2489–2495.
87. Dik MG, Jonker C, Comijs HC, et al. Contribution of metabolic syndrome components to cognition in older individuals. *Diabetes Care*. 2007;30(10):2655–2660.
88. Knopman DS, Mosley TH, Catellier DJ, Coker LH, et al. Fourteen-year longitudinal study of vascular risk factors, APOE genotype, and cognition: the ARIC MRI study. *Alzheimers Dement J Alzheimers Assoc*. 2009;5(3):207–214.
89. Komulainen P, Lakka TA, Kivipelto M, et al. Metabolic syndrome and cognitive function: A population-based follow-up study in elderly women. *Dement Geriatr Cogn Disord*. 2007;23(1):29–34.
90. Segura B, Jurado MÁ, Freixenet N, Albuin C, Muniesa J, Junqué C. Mental slowness and executive dysfunctions in patients with metabolic syndrome. *Neurosci Lett*. 2009;462(1):49–53.
91. Van Den Berg E, Dekker JM, Nijpels G, et al. Cognitive functioning in elderly persons with type 2 diabetes and metabolic syndrome: the Hoorn study. *Dement Geriatr Cogn Disord*. 2008;26(3):261–269.
92. Yaffe K, Haan M, Blackwell T, Cherkasova E, Whitmer RA, West N. Metabolic syndrome and cognitive decline in elderly Latinos: findings from the Sacramento area Latino study of aging study. *J Am Geriatr Soc*. 2007;55(5):758–762.
93. Alcorn T, Hart E, Smith AE, et al. Cross-sectional associations between metabolic syndrome and performance across cognitive domains: a systematic review. *Appl Neuropsychol Adult*. 2017;1–14.
94. Razay G, Vreugdenhil A, Wilcock G. The metabolic syndrome and Alzheimer disease. *Arch Neurol*. 2007;64(1):93–96.
95. Vanhanen M, Koivisto K, Moilanen L, et al. Association of metabolic syndrome with Alzheimer disease a population-based study. *Neurology*. 2006;67(5):843–847.
96. Raffaitin C, Gin H, Empana JP, et al. Metabolic syndrome and risk for incident Alzheimer's disease or vascular dementia: the Three-City study. *Diabetes Care*. 2009;32(1):169–174.
97. Gatto NM, Henderson VW, St. John JA, McCleary C, Hodis HN, Mack WJ. Metabolic syndrome and cognitive function in healthy middle-aged and older adults without diabetes. *Aging Neuropsychol Cogn*. 2008;15(5):627–641.
98. Yaffe K, Weston AL, Blackwell T, Krueger KA. The metabolic syndrome and development of cognitive impairment among older women. *Arch Neurol*. 2009;66(3):324–328.
99. Alfaro FJ, Gavrieli A, Saade-Lemus P, Lioutas V-A, Upadhyay J, Novak V. White matter microstructure and cognitive decline in metabolic syndrome: a review of diffusion tensor imaging. *Metab-Clin Exp*. 2018;78:52–68.
100. Assuncao N, Sudo FK, Drummond C, de Felice FG, Mattos P. Metabolic syndrome and cognitive decline in the elderly: a systematic review. *PloS One*. 2018;13(3):e0194990.
101. Jais A, Brüning JC. Hypothalamic inflammation in obesity and metabolic disease. *J Clin Invest*. 2017;127(1):24–32.
102. Kullmann S, Callaghan MF, Heni M, et al. Specific white matter tissue microstructure changes associated with obesity. *Neuroimage*. 2016;125:36–44.
103. Bolzenius JD, Laidlaw DH, Cabeen RP, et al. Brain structure and cognitive correlates of body mass index in healthy older adults. *Behav Brain Res*. 2015;278:342–347.
104. Chaldakov GN, Stankulov IS, Hristova M, Ghenev PI. Adipobiology of disease: adipokines and adipokine-targeted pharmacology. *Curr Pharm Des*. 2003;9(12):1023–1031.
105. Kuller LH, Lopez OL, Newman A, et al. Risk factors for dementia in the cardiovascular health cognition study. *Neuroepidemiology*. 2003;22(1):13–22.

106. Landin K, Stigendal L, Eriksson E, et al. Abdominal obesity is associated with an impaired fibrinolytic activity and elevated plasminogen activator inhibitor-1. *Metabolism.* 1990;39(10):1044–1048.
107. Prins JB. Adipose tissue as an endocrine organ. *Best Pract Res Clin Endocrinol Metab.* 2002;16(4):639–651.
108. Yaffe K, Lindquist K, Penninx BW, et al. Inflammatory markers and cognition in well-functioning African-American and white elders. *Neurology.* 2003;61(1):76–80.
109. Ryu SY, Coutu JP, Rosas HD, Salat DH, et al. Effects of insulin resistance on white matter microstructure in middle-aged and older adults. *Neurology.* 2014;82(21):1862–1870.
110. Jagust W. What can imaging reveal about obesity and the brain? *Curr Alzheimer Res.* 2007;4(2):135–139.
111. Jagust W, Harvey D, Mungas D, Haan M. Central obesity and the aging brain. *Arch Neurol.* 2005;62(10):1545–1548.
112. Verstynen TD, Weinstein AM, Schneider WW, Jakicic JM, Rofey DL, Erickson KI. Increased body mass index is associated with a global and distributed decrease in white matter microstructural integrity. *Psychosom Med.* 2012;74(7):682–690.
113. Walther K, Birdsill AC, Glisky EL, Ryan L. Structural brain differences and cognitive functioning related to body mass index in older females. *Hum Brain Mapp.* 2010;31(7):1052–1064.
114. Driscoll I, Beydoun MA, An Y, et al. Midlife obesity and trajectories of brain volume changes in older adults. *Hum Brain Mapp.* 2012;33(9):2204–2210.
115. Basser PJ, Mattiello J, LeBihan D. MR diffusion tensor spectroscopy and imaging. *Biophys J.* 1994;66(1):259–267.
116. Sbarbati A, Osculati F, Silvagni D, et al. Obesity and inflammation: evidence for an elementary lesion. *Pediatrics.* 2006;117(1):220–223.
117. Amin RS, Kimball TR, Bean JA, et al. Left ventricular hypertrophy and abnormal ventricular geometry in children and adolescents with obstructive sleep apnea. *Am J Respir Crit Care Med.* 2002;165(10):1395–1399.
118. Bass JL, Corwin M, Gozal D, et al. The effect of chronic or intermittent hypoxia on cognition in childhood: a review of the evidence. *Pediatrics.* 2004;114(3):805–816.
119. Carvalho LBC, Prado LF, Silva L, et al. Cognitive dysfunction in children with sleep-disordered breathing. *J Child Neurol.* 2005;20(5):400–404.
120. Halbower AC, Mahone EM. Neuropsychological morbidity linked to childhood sleep-disordered breathing. *Sleep Med Rev.* 2006;10(2):97–107.
121. Li Y, Dai Q, Jackson JC, Zhang J. Overweight is associated with decreased cognitive functioning among school-age children and adolescents. *Obesity.* 2008;16(8):1809–1815.
122. Spruyt K, Gozal D. A mediation model linking body weight, cognition, and sleep-disordered breathing. *Am J Respir Crit Care Med.* 2012;185(2):199–205.
123. Frisardi V, Solfrizzi V, Seripa D, et al. Metabolic-cognitive syndrome: a cross-talk between metabolic syndrome and Alzheimer's disease. *Ageing Res Rev.* 2010;9(4):399–417.
124. Kumari M, Brunner E, Fuhrer R. Minireview: mechanisms by which the metabolic syndrome and diabetes impair memory. *J Gerontol A Biol Sci Med Sci.* 2000;55(5):B228–B232.
125. Dalton TC, Bergenn VW. *Early Experience, the Brain, and Consciousness: An Historical and Interdisciplinary Synthesis.* Psychology Press; 2007.
126. Eleftheriou BE, Elias MF, Castellano C, Oliverio A. Cortex weight: a genetic analysis in the mouse. *J Hered.* 1975;66(4):207–212.
127. Strauss RS, Pollack HA. Social marginalization of overweight children. *Arch Pediatr Adolesc Med.* 2003;157(8):746–752.
128. Scazufca M, Almeida OP, Menezes PR. The role of literacy, occupation and income in dementia prevention: the São Paulo Ageing & Health Study (SPAH). *Int Psychogeriatr.* 2010;22(8):1209–1215.
129. Datar A, Sturm R. Childhood overweight and parent-and teacher-reported behavior problems: evidence from a prospective study of kindergartners. *Arch Pediatr Adolesc Med.* 2004;158(8):804–810.
130. Datar A, Sturm R. Childhood overweight and elementary school outcomes. *Int J Obes (Lond).* 2006;30(9):1449–1460.
131. Datar A, Sturm R, Magnabosco JL. Childhood overweight and academic performance: national study of kindergartners and first-graders. *Obesity.* 2004;12(1):58–68.
132. Reilly JJ. Descriptive epidemiology and health consequences of childhood obesity. *Best Pract Res Clin Endocrinol Metab.* 2005;19(3):327–341.
133. Mond JM, Stich H, Hay PJ, Krämer A, Baune BT. Associations between obesity and developmental functioning in pre-school children: a population-based study. *Int J Obes (Lond).* 2007;31(7):1068–1073.

134. Guxens M, Mendez MA, Julvez J, et al. Cognitive function and overweight in preschool children. *Am J Epidemiol*. 2009;170(4):438–446.
135. Van den Berg E, Kloppenborg RP, Kessels RP, Kappelle LJ, Biessels GJ. Type 2 diabetes mellitus, hypertension, dyslipidemia and obesity: a systematic comparison of their impact on cognition. *Biochim Biophys Acta BBA-Mol Basis Dis*. 2009;1792(5):470–481.
136. Fuemmeler BF, Østbye T, Yang C, McClernon FJ, Kollins SH. Association between attention-deficit/hyperactivity disorder symptoms and obesity and hypertension in early adulthood: a population-based study. *Int J Obes (Lond)*. 2011;35(6):852–862.
137. Puder JJ, Munsch S. Psychological correlates of childhood obesity. *Int J Obes (Lond)*. 2010;34(S2):S37.
138. Chelune GJ, Ortega D, Linton JC, Boustany MM. Personality and cognitive findings among patients electing gastroplasty for morbid obesity. *Int J Eat Disord*. 1986;5(4):701–712.
139. Handley JD, Williams DM, Caplin S, Stephens JW, Barry J. Changes in cognitive function following bariatric surgery: a systematic review. *Obes Surg*. 2016;26(10):2530–2537.
140. Veronese N, Facchini S, Stubbs B, et al. Weight loss is associated with improvements in cognitive function among overweight and obese people: a systematic review and meta-analysis. *Neurosci Biobehav Rev*. 2017;72:87–94.
141. Sturm R, Hattori A. Morbid obesity rates continue to rise rapidly in the United States. *Int J Obes (Lond)*. 2013;37(6):889–891.
142. Dixon JB. The effect of obesity on health outcomes. *Mol Cell Endocrinol*. 2010;316(2):104–108.
143. Williams EP, Mesidor M, Winters K, Dubbert PM, Wyatt SB. Overweight and obesity: prevalence, consequences, and causes of a growing public health problem. *Curr Obes Rep*. 2015;4(3):363–370.
144. Rodriguez ACI, Epel ES, White ML, Standen EC, Seckl JR, Tomiyama AJ. Hypothalamic-pituitary-adrenal axis dysregulation and cortisol activity in obesity: a systematic review. *Psychoneuroendocrinology*. 2015;62:301–318.
145. Guldstrand M, Ahrén B, Wredling R, Backman L, Lins PE, Adamson U. Alteration of the counterregulatory responses to insulin-induced hypoglycemia and of cognitive function after massive weight reduction in severely obese subjects. *Metab-Clin Exp*. 2003;52(7):900–907.
146. Spitznagel MB, Galioto R, Limbach K, Gunstad PDJ, Heinberg PDL. Cognitive function is linked to adherence to bariatric postoperative guidelines. *Surg Obes Relat Dis*. 2013;9(4):580–585.
147. Galioto R, Alosco ML, Spitznagel MB, et al. Glucose regulation and cognitive function after bariatric surgery. *J Clin Exp Neuropsychol*. 2015;37(4):402–413.
148. Spitznagel MB, Hawkins M, Alosco M, et al. Neurocognitive effects of obesity and bariatric surgery. *Eur Eat Disord Rev*. 2015;23(6):488–495.
149. Espeland MA, Rapp SR, Bray GA, et al. Long-term impact of behavioral weight loss intervention on cognitive function. *J Gerontol Ser Biomed Sci Med Sci*. 2014;69(9):1101–1108.
150. Rapp SR, Luchsinger JA, Baker LD, et al. Effect of a long-term intensive lifestyle intervention on cognitive function: action for health in diabetes study. *J Am Geriatr Soc*. 2017;65(5):966–972.
151. Espeland MA, Carmichael O, Hayden K, et al. Long-term impact of weight loss intervention on changes in cognitive function: exploratory analyses from the action for health in diabetes randomized controlled clinical trial. *J Gerontol Ser A*. 2017;73(4):484–491.
152. Hayden KM, Baker LD, Bray G, et al. Long-term impact of intensive lifestyle intervention on cognitive function assessed with the National Institutes of Health toolbox: the look AHEAD study. *Alzheimers Dement Diagn Assess Dis Monit*. 2018;10:41–48.
153. Barnes DE, Santos-Modesitt W, Poelke G, et al. The mental activity and eXercise (MAX) trial: a randomized controlled trial to enhance cognitive function in older adults. *JAMA Intern Med*. 2013;173(9):797–804.
154. Baumgart M, Snyder HM, Carrillo MC, Fazio S, Kim H, Johns H. Summary of the evidence on modifiable risk factors for cognitive decline and dementia: a population-based perspective. *Alzheimers Dement J Alzheimers Assoc*. 2015;11(6):718–726.
155. Evans SC, Blossom JB, Canter KS, et al. Self-reported emotion reactivity among early-adolescent girls: evidence for convergent and discriminant validity in an urban community sample. *Behav Ther*. 2016;47(3):299–311.
156. Suzuki T, Shimada H, Makizako H, et al. Effects of multicomponent exercise on cognitive function in older adults with amnestic mild cognitive impairment: a randomized controlled trial. *BMC Neurol*. 2012;12(1):128–137.
157. Schwenk M, Zieschang T, Oster P, Hauer K. Dual-task performances can be improved in patients with dementia a randomized controlled trial. *Neurology*. 2010;74(24):1961–1968.

158. Lautenschlager NT, Cox KL. Can participation in mental and physical activity protect cognition in old age?: Comment on "The Mental Activity and eXercise (MAX) trial: a randomized controlled trial to enhance cognitive function in older adults.". *JAMA Intern Med*. 2013;173(9):805–806.
159. Lauenroth A, Ioannidis AE, Teichmann B. Influence of combined physical and cognitive training on cognition: a systematic review. *BMC Geriatr*. 2016;16(1):141–155.
160. Law LL, Barnett F, Yau MK, Gray MA. Effects of combined cognitive and exercise interventions on cognition in older adults with and without cognitive impairment: a systematic review. *Ageing Res Rev*. 2014;15:61–75.
161. Zhu X, Yin S, Lang M, He R, Li J. The more the better? A meta-analysis on effects of combined cognitive and physical intervention on cognition in healthy older adults. *Ageing Res Rev*. 2016;31:67–79.
162. Swift DL, Johannsen NM, Lavie CJ, Earnest CP, Church TS. The role of exercise and physical activity in weight loss and maintenance. *Prog Cardiovasc Dis*. 2014;56(4):441–447.
163. Luchtman DW, Song C. Cognitive enhancement by omega-3 fatty acids from child-hood to old age: findings from animal and clinical studies. *Neuropharmacology*. 2013;64:550–565.
164. Buckley JD, Howe PR. Long-chain omega-3 polyunsaturated fatty acids may be beneficial for reducing obesity—A review. *Nutrients*. 2010;2(12):1212–1230.
165. Albracht-Schulte K, Kalupahana NS, Ramalingam L, et al. Omega-3 fatty acids in obesity and metabolic syndrome: a mechanistic update. *J Nutr Biochem*. 2018;58:1–16.
166. Bender N, Portmann M, Heg Z, Hofmann K, Zwahlen M, Egger M. Fish or n3-PUFA intake and body composition: a systematic review and meta-analysis. *Obes Rev*. 2014;15(8):657–665.
167. Du S, Jin J, Fang W, Su Q. Does fish oil have an anti-obesity effect in overweight/obese adults? A meta-analysis of randomized controlled trials. *PLoS One*. 2015;10(11):e0142652.
168. Krebs JD, Browning LM, McLean NK, et al. Additive benefits of long-chain n-3 polyunsaturated fatty acids and weight-loss in the management of cardiovascular disease risk in overweight hyperinsulinaemic women. *Int J Obes (Lond)*. 2006;30(10):1535–1544.
169. Mori TA, Bao DQ, Burke V, Puddey IB, Watts GF, Beilin LJ. Dietary fish as a major component of a weight-loss diet: effect on serum lipids, glucose, and insulin metabolism in overweight hypertensive subjects. *Am J Clin Nutr*. 1999;70(5):817–825.
170. Paoli A, Moro T, Bosco G, et al. Effects of n-3 polyunsaturated fatty acids (omega-3) supplementation on some cardiovascular risk factors with a ketogenic Mediterranean diet. *Mar Drugs*. 2015;13(2):996–1009.
171. Ramel A, Parra D, Martinéz JA, Kiely M, Thorsdottir I. Effects of seafood consumption and weight loss on fasting leptin and ghrelin concentrations in overweight and obese European young adults. *Eur J Nutr*. 2009;48(2):107–114.
172. DeFina LF, Marcoux LG, Devers SM, Cleaver JP, Willis BL. Effects of omega-3 supplementation in combination with diet and exercise on weight loss and body composition. *Am J Clin Nutr*. 2010;93(2):455–462.
173. Harden CJ, Dible VA, Russell JM, et al. Long-chain polyunsaturated fatty acid supplementation had no effect on body weight but reduced energy intake in overweight and obese women. *Nutr Res*. 2014;34(1):17–24.
174. Hill AM, Buckley JD, Murphy KJ, Howe PR. Combining fish-oil supplements with regular aerobic exercise improves body composition and cardiovascular disease risk factors. *Am J Clin Nutr*. 2007;85(5):1267–1274.
175. Kratz M, Swarbrick MM, Callahan HS, Matthys CC, Havel PJ, Weigle DS. Effect of dietary n-3 polyunsaturated fatty acids on plasma total and high-molecular-weight adiponectin concentrations in overweight to moderately obese men and women. *Am J Clin Nutr*. 2008;87(2):347–353.
176. Lee HC, Cheng WY, Hsu YH, et al. Effects of calorie restriction with n-3 long-chain polyunsaturated fatty acids on metabolic syndrome severity in obese subjects: a randomize-controlled trial. *J Funct Foods*. 2015;19:929–940.
177. Méndez-Sánchez N, González V, Aguayo P, et al. Fish oil (n-3) polyunsaturated fatty acids beneficially affect biliary cholesterol nucleation time in obese women losing weight. *J Nutr*. 2001;131(9):2300–2303.
178. Munro IA, Garg ML. Dietary supplementation with n-3 PUFA does not promote weight loss when combined with a very-low-energy diet. *Br J Nutr*. 2012;108(8):1466–1474.
179. Munro IA, Garg ML. Dietary supplementation with long chain omega-3 polyunsaturated fatty acids and weight loss in obese adults. *Obes Res Clin Pract*. 2013;7(3):e173–e181.
180. Razny U, Kiec-Wilk B, Polus A, et al. Effect of caloric restriction with or without n-3 polyunsaturated fatty acids on insulin sensitivity in obese subjects: a randomized placebo controlled trial. *BBA Clin*. 2015;4:7–13.
181. Slivkoff-Clark KM, James AP, Mamo JC. The chronic effects of fish oil with exercise on postprandial lipaemia and chylomicron homeostasis in insulin resistant viscerally obese men. *Nutr Metab*. 2012;9(1):9–18.

182. Kunesova M, Braunerova R, Hlavatý P, Tvrzicka E. The influence of n-3 polyunsaturated fatty acids and very low calorie diet during a short-term weight reducing regimen on weight loss and serum fatty acid composition in severely obese women. *Physiol Res*. 2006;55(1):63–72.
183. Thorsdottir I, Tomasson H, Gunnarsdottir I, et al. Randomized trial of weight-loss-diets for young adults varying in fish and fish oil content. *Int J Obes (Lond)*. 2007;31(10):1560–1566.
184. Munro IA, Garg ML. Prior supplementation with long chain omega-3 polyunsaturated fatty acids promotes weight loss in obese adults: a double-blinded randomised controlled trial. *Food Funct*. 2013;4(4):650–658.
185. Zhang YY, Liu W, Zhao TY, Tian HM. Efficacy of omega-3 polyunsaturated fatty acids supplementation in managing overweight and obesity: a meta-analysis of randomized clinical trials. *J Nutr Health Aging*. 2017;21(2):187–192.
186. Golub N, Geba D, Mousa SA, Williams G, Block RC. Greasing the wheels of managing overweight and obesity with omega-3 fatty acids. *Med Hypotheses*. 2011;77(6):1114–1120.
187. Lorente-Cebrián S, Costa AG, Navas-Carretero S, Zabala M, Martínez JA, Moreno-Aliaga MJ. Role of omega-3 fatty acids in obesity, metabolic syndrome, and cardiovascular diseases: a review of the evidence. *J Physiol Biochem*. 2013;69(3):633–651.
188. Parra D, Ramel A, Bandarra N, Kiely M, Martínez JA, Thorsdottir I. A diet rich in long chain omega-3 fatty acids modulates satiety in overweight and obese volunteers during weight loss. *Appetite*. 2008;51(3):676–680.
189. Stevenson JL, Paton CM, Cooper JA. Hunger and satiety responses to high-fat meals after a high-polyunsaturated fat diet: a randomized trial. *Nutrition*. 2017;41:14–23.
190. Mori T, Kondo H, Hase T, Tokimitsu I, Murase T. Dietary fish oil upregulates intestinal lipid metabolism and reduces body weight gain in C57BL/6J mice. *J Nutr*. 2007;137(12):2629–2634.
191. Flachs P, Horakova O, Brauner P, et al. Polyunsaturated fatty acids of marine origin upregulate mitochondrial biogenesis and induce β-oxidation in white fat. *Diabetologia*. 2005;48(11):2365–2375.
192. Martínez-Fernández L, Laiglesia LM, Huerta AE, Martínez JA, Moreno-Aliaga MJ. Omega-3 fatty acids and adipose tissue function in obesity and metabolic syndrome. *Prostaglandins Other Lipid Mediat*. 2015;121:24–41.
193. Burgess PW, Rabbitt P. *Methodology of Frontal and Executive Function*. Psychology Press Hove; 1997.
194. Lyon G, Krasnegor NA. *Attention, Memory, and Executive Function*. Paul H Brookes Publishing; 1996.
195. Gunstad J, Lhotsky A, Wendell CR, Ferrucci L, Zonderman AB. Longitudinal examination of obesity and cognitive function: results from the Baltimore longitudinal study of aging. *Neuroepidemiology*. 2010;34(4):222–229.
196. Graves RE, Carswell LM, Snow WG. An evaluation of the sensitivity of premorbid IQ estimators for detecting cognitive decline. *Psychol Assess*. 1999;11(1):29–38.
197. Elias MF, Torres RV, Davey A. Clinical trials of blood pressure lowering and antihypertensive medication: is cognitive measurement state-of-the-art? *Am J Hypertens*. 2018;31(6):631–642.
198. Petersen RC, Smith GE, Waring SC, Ivnik RJ, Kokmen E, Tangelos EG. Aging, memory, and mild cognitive impairment. *Int Psychogeriatr*. 1997;9(S1):65–69.
199. Albert MS, DeKosky ST, Dickson D, et al. The diagnosis of mild cognitive impairment due to Alzheimer's disease: recommendations from the National Institute on Aging-Alzheimer's association workgroups on diagnostic guidelines for Alzheimer's disease. *Alzheimers Dement J Alzheimers Assoc*. 2011;7(3):270–279.
200. Koepsell TD, Monsell SE. Reversion from mild cognitive impairment to normal or near-normal cognition risk factors and prognosis. *Neurology*. 2012;79(15):1591–1598.
201. McKhann G, Drachman D, Folstein M, Katzman R, Price D, Stadlan EM. Clinical diagnosis of Alzheimer's disease report of the NINCDS-ADRDA work group* under the auspices of Department of Health and Human Services Task Force on Alzheimer's disease. *Neurology*. 1984;34(7):939–944.
202. McKhann GM, Knopman DS, Chertkow H, et al. The diagnosis of dementia due to Alzheimer's disease: recommendations from the National Institute on Aging-Alzheimer's association workgroups on diagnostic guidelines for Alzheimer's disease. *Alzheimers Dement J Alzheimers Assoc*. 2011;7(3):263–269.
203. Jack CR, Albert MS, Knopman DS, et al. Introduction to the recommendations from the National Institute on Aging-Alzheimer's association workgroups on diagnostic guidelines for Alzheimer's disease. *Alzheimers Dement J Alzheimers Assoc*. 2011;7(3):257–262.
204. Jellinger KA. The pathology of "vascular dementia": a critical update. *J Alzheimers Dis*. 2008;14(1):107–123.
205. Jellinger KA. Understanding the pathology of vascular cognitive impairment. *J Neurol Sci*. 2005;229:57–63.
206. Zlokovic BV. Neurodegeneration and the neurovascular unit. *Nat Med*. 2010;16(12):1370–1371.
207. Bush SS, Ruff RM, Tröster AI, et al. Symptom validity assessment: practice issues and medical necessity: NAN Policy & Planning Committee. *Arch Clin Neuropsychol*. 2005;20(4):419–426.

208. Shiely F, Hayes K, Perry IJ, Kelleher CC. Height and weight bias: the influence of time. *PloS One*. 2013;8(1):e54386.
209. Sebo P, Beer-Borst S, Haller DM, Bovier PA. Reliability of doctors' anthropometric measurements to detect obesity. *Prev Med*. 2008;47(4):389–393.
210. Sebo P, Herrmann FR, Haller DM. Accuracy of anthropometric measurements by general practitioners in overweight and obese patients. *BMC Obes*. 2017;4(1):23–30.
211. Seboe P, Haller DM, Pechere A, Bovier P, Herrmann F. Accuracy of doctors' anthropometric measurements in general practice. *Swiss Med Wkly*. 2015;145:w14115.
212. Hume PA, Kerr DA, Ackland TR. *Best Practice Protocols for Physique Assessment in Sport*. Springer; 2017.
213. Willette AA, Kapogiannis D. Does the brain shrink as the waist expands? *Ageing Res Rev*. 2015;20:86–97.
214. Bischof GN, Park DC. Obesity and aging: consequences for cognition, brain structure and brain function. *Psychosom Med*. 2015;77(6):697–709.
215. Launer LJ. The epidemiologic study of dementia: a life-long quest? *Neurobiol Aging*. 2005;26(3):335–340.
216. Breteler MM. Early life circumstances and late life Alzheimer's disease. *Epidemiology*. 2001;12(4):378–379.
217. Wikipedia. Back to the Future. http://en.wikipedia.org/wiki/Back_to_the_Future. Published 2011. (Accessed 1 September 2011).

CHAPTER

28

Fatty Acids and the Brain Health

Rafael Longhi

Department of Basic Health Sciences, Federal University of Health Sciences of Porto Alegre (UFCSPA), Porto Alegre, Brazil

INTRODUCTION

Trans Fatty Acids

In the 1960s, the food industry began using substantial amounts of partially hydrogenated vegetable oils in processed food following public health campaigns aimed at decreasing the use of animal fats. Thus began the public's increased consumption of trans fatty acids (TFAs).[1] In time, evidence showed that the increased consumption of partially hydrogenated vegetable oils (source of TFA) was associated with many comorbidities, and more recently, neurodegenerative diseases.[2–4] Following such research, most professional health organizations have recommended reduced consumption of foods containing TFA, including requiring the labelling of TFA on packaged foods. Despite the intense efforts to virtually eliminate this artificial fat, there is still a long way to go. From 2007 to 2011, a hundred brand-name food products in US supermarkets contained TFA but were cited as "TFA free," but due the quantity seen in each serving, TFA was available for consumption, based on the decisions made by food legislation.[5]

Despite important health concerns, companies continued to use TFA in manufactured products. The companies were influenced by the following commercial advantages: (a) enhanced consistency/texture, springiness, brittleness, and chewiness); (b) increased shelf life; (c) increased flavor stability; (d) decreased food's sensitivity to oxidation; (e) increased their stability against liquefaction; (f) increased storage stability at room temperature; (g) increased stability during frying; and (h) enhanced their emulsion stability.[6]

Hydrogenation Process

Obtaining TFA, by industrial processes, occurs through a process called hydrogenation. It was the partial hydrogenation of fats that introduced TFA into fats of vegetable origin.[7] During this process, hydrogen is bubbled through the fat at elevated temperature in the absence of oxygen and presence of a catalyst such as nickel; this partial hydrogenation rearranges the double bonds, converting some of them to the trans configuration and shifting the double

Omega Fatty Acids in Brain and Neurological Health
https://doi.org/10.1016/B978-0-12-815238-6.00028-6

bonds along the carbon chain,[8] the oil, liquid at room temperature, is converted to solid fat through the chemical process, producing TFA.[9] Among the oils available for consumption, soybean oil is one of the most used in the hydrogenation process. It is composed of saturated (SFA) and unsaturated (UFA) fatty acids, with the major SFAs being palmitic (C16:0; 10%–12%) and stearic (C18:0; 3%–5%); the major UFAS are oleic (C18:1; 24%), linoleic (C18:2; 54%), and linolenic (C18:3; 8.0%).[10] During soybean oil hydrogenation, the oleic acid (the largest percentage of the natural fatty acid in the human body, a cis configuration) goes through geometrical isomerization during hydrogenation to trans C18:1, 9t acid known as elaidic acid (EA). Thus the "natural" oleic acid is turned into elaidic acid during the hydrogenation process and becomes an "unnatural" fatty acid.[11]

In addition to the process cited above, thermal processes during refining or frying of vegetable oils are capable of producing TFA as follows:

1. *Refining*: Edible oils are commonly heated between 60°C and 100°C to remove certain impurities. The impurities include free fatty acids, phospholipids, carbohydrates, and proteins from degradation byproducts, which may alter the color, taste, and aroma; they are then subjected to deodorization, which aims to improve oil's sensorial characteristics; in this process, the temperature is raised to 180°C–270°C, which leads to the formation of TFA in the vegetable oil.[9]
2. *Frying*: During this process (150°C–190°C or more), edible oils/fats undergo various chemical reactions that include oxidation, hydrolysis, isomerization, polymerization, and cyclization. As a result, products like free fatty acids, mono- and diacylglycerols, oxidized monomers, dimers, and polymers are formed, which are then incorporated in the foods and are responsible for the appearance, aroma, and taste of the fried food. Studies with spectroscopy have shown a loss of cis double bonds along with an increase in trans unsaturation, thus confirming the formation of TFA.[9]

The chemical process mainly produces elaidic acid (EA),[12] which is a monounsaturated fat with 18 carbons in the chain. It is double bound in trans configuration in the position 9 (C18:1,9t), as you can see in Fig. 1.[9]

It is important to highlight that the presence of a trans double bond in a fatty acyl chain results in a smaller bond angle or kink than a cis double bond, resulting in a fatty acyl chain conformation more similar to a SFA than an UFA. Whereas the presence of cis double bonds impedes the fatty acyl chains from aligning (packing together), causing vegetable fats to be liquid at room temperature. The presence of trans double bonds makes it more likely that the fatty acyl chains will align, resulting in fats that are more viscous (solid) at room temperature.[13]

BIOCHEMICAL METABOLISM OF TRANS FATTY ACIDS

In relation to the metabolism, fats with trans configuration have different characteristics, which cause doubt in the scientific community. The degradation of unsaturated fatty

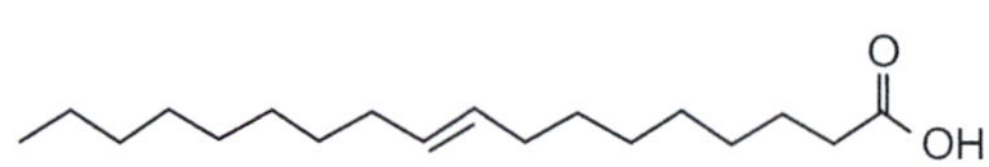

FIG. 1 Structure of elaidic acid.

acids having trans double bonds at even numbered positions yields the intermediate 2-trans, 4-trans-dienoyl-CoA. This can be further metabolized via two distinct routes. In the reductase-dependent pathway, 2-trans, 4-trans-dienoyl-CoA is a substrate for the 2,4-dienoyl-CoA reductase, generating 3-trans-enoyl-CoA, which is converted further by the D3,D2-enoyl-CoA isomerase to 2-trans-enoyl-CoA that re-enters the core β-oxidation cycle. An alternative pathway has also been described whereby the 2-trans,4-trans-dienoyl-CoA is directly metabolized by the enoyl-CoA hydratase II activity of the multifunctional enzyme (MFE) to 3- hydroxy, 4-trans-enoyl-CoA, which is further then degraded by the core β-oxidation cycle.[14]

General beta oxidation, which is seen in all lipids with trans configuration, disregards a possible incomplete beta oxidation, specifically from EA. EA is partially converted to 5-trans-tetradecenoic acid, generating an unusual leakage of an intermediate during beta oxidation. This 5-trans-tetradecenoyl-CoA is more rapidly formed than degraded by β-oxidation. The most obvious reason for the accumulation of 5-trans-tetradecenoyl-CoA would be its slower dehydrogenation by long-chain acyl-CoA dehydrogenase (LCAD), whereas the corresponding intermediates of more common dietary fatty acids, including 5-cis-tetradecenoyl-CoA derived from oleic acid, are degraded without accumulating to a significant extent. Therefore, these data support the conclusion that the presence of a 5-trans double bond in place of a 5-cis or no double bond in the substrate reduces the catalytic efficiency of LCAD. This generates a possible lower rate respiration or even adverse effects on other reactions of beta oxidation, as to move from the mitochondrial matrix into the cytosol where it can be activated by conversion to its CoA derivative and utilized for lipid synthesis.[15]

Currently, there is a global consensus to decrease consumption of TFAs produced by partial hydrogenation of vegetable oils. These recommendations are based on numerous studies demonstrating many adverse health effects of TFA, regardless of the amount consumed.[6]

ESSENTIAL FATTY ACIDS

Our diet has changed over the past 10,000 years, but our genes have not. Genetically speaking, humans today live in a nutritional environment that differs greatly from that of our ancient ancestors. Many studies on the evolutionary aspects of diet indicate that major changes have taken place in our diet, particularly in the type and amount of essential fatty acids (EFA).[16]

Omega 3 (n-3) and omega 6 (n-6) are both described as essential fatty acids (EFA). This nomenclature is related to their critical role in nutrition and health. They have different chemical structures and perform different roles. Linoleic acid (LA) is the precursor of the n-6 polyunsaturated fatty acids (PUFA) family, in which the first double bond in the molecule is six carbons from the methyl terminus. Alpha linolenic acid (ALA) is the precursor of the n-3 family of PUFA, in which the first double bond in the molecule is three carbons away from the methyl terminus.[17] The past 50 years brought recognition that dietary EFA accumulate as highly unsaturated fatty acids (HUFA) in tissues; they then form hormone-like eicosanoids that act on selective receptors, influencing physiological processes in nearly every tissue of the body.[18]

Actually, the typical Western diet is rich in n-6 and relatively lower in n-3 PUFA. The genetic constitution of human beings likely evolved on a diet with a ratio of n-6 to n-3 PUFAs of about 1:1. Whereas in Western diet, the typical ratio is 15:1.[19] This information is relevant because n-3 PUFAs appear to possess the most potent immunomodulatory activities in relation to n-6 PUFAs.[17, 20] For a better understanding, the term "Western Diet" refers to a diet including high-fat and cholesterol, high protein, high sugar, and excess salt intake, as well as the frequent consumption of processed and "fast foods".[21]

BIOCHEMICAL METABOLISM OF ESSENTIAL FATTY ACIDS

It is important to highlight that both n-3 and n-6 are EFA because mammalian cells lack the specific enzymes required for the de novo synthesis of these lipids; therefore, there are specific biochemical metabolism.

In relation to n-6, linoleic acid can be metabolized to other more unsaturated, long-chain members of the n-6 family by the insertion of additional double bonds during consecutive elongation and desaturation mechanisms. The initial rate limiting desaturation of LA to γ-linolenic acid (GLA) is catalyzed by the enzyme delta-6-desaturase (FADS2). Elongation then takes place to convert GLA (18:3n-6) to dihomo-γ-linolenic acid (DGLA—20:3n-6) by elongation of very long-chain fatty acids (ELOVL) 5. Finally, a cycle of elongation and desaturation by delta-5-desaturase (FADS1) generates arachidonic acid (AA—20:4 n-6),[22] see Fig. 2.

In reference to n-3, alpha-linolenic acid is metabolized to eicosapentaenoic acid (EPA). EPA is currently thought to occur predominately in liver cells by sequential desaturation and elongation: delta-6 desaturase, chain elongation, and delta-5 desaturase. EPA is elongated to DPAn-3 (22:5n-3 or docosapentaenoic acid). DPAn-3 is then metabolized to docosahexaenoic acid (DHA) in several relatively more complex steps. First, it is elongated; then it's desaturated via delta-6 desaturase to form the intermediate molecule 24:6n-3. It must then be taken into the peroxisomes to undergo limited β-oxidation to remove two carbon atoms in a process known as retro-conversion or the "Sprecher" pathway. These conversions have long been assumed to be rate-limited by hepatic delta-6 desaturase, although delta-6 desaturase has also been found in the brain and skeletal muscles, suggesting the possibility of alternative pathways for the maintenance of EPA and DHA. While brain delta-6 desaturase is demonstratively active in the neonate, it appears to rapidly decline with age,[23] justifying a diet consumption (see Fig. 2).

Findings suggest that industrially produced TFA can induce an increased risk of metabolic diseases and a possible risk for brain diseases in opposite of essential fatty acids that could be neuroprotector, when n-6/n-3 ratio is controlled. In the next paragraphs will study the relationship between the composition to be more important than quantity of fat consumed, focus on TFA, brain damage and a possible neuroprotector strategies with a correct consumption of n-3 and n-6 essential fatty acids.

TRANS FATTY ACIDS AND CENTRAL NERVOUS SYSTEM

In recent years, dietary trans fatty acids have raised health concerns largely from reports that high levels in the diet, compared with high levels of cis fatty acids, have resulted in unfavorable effects on lipid metabolism.[24, 25] Additionally, this has been associated with increased

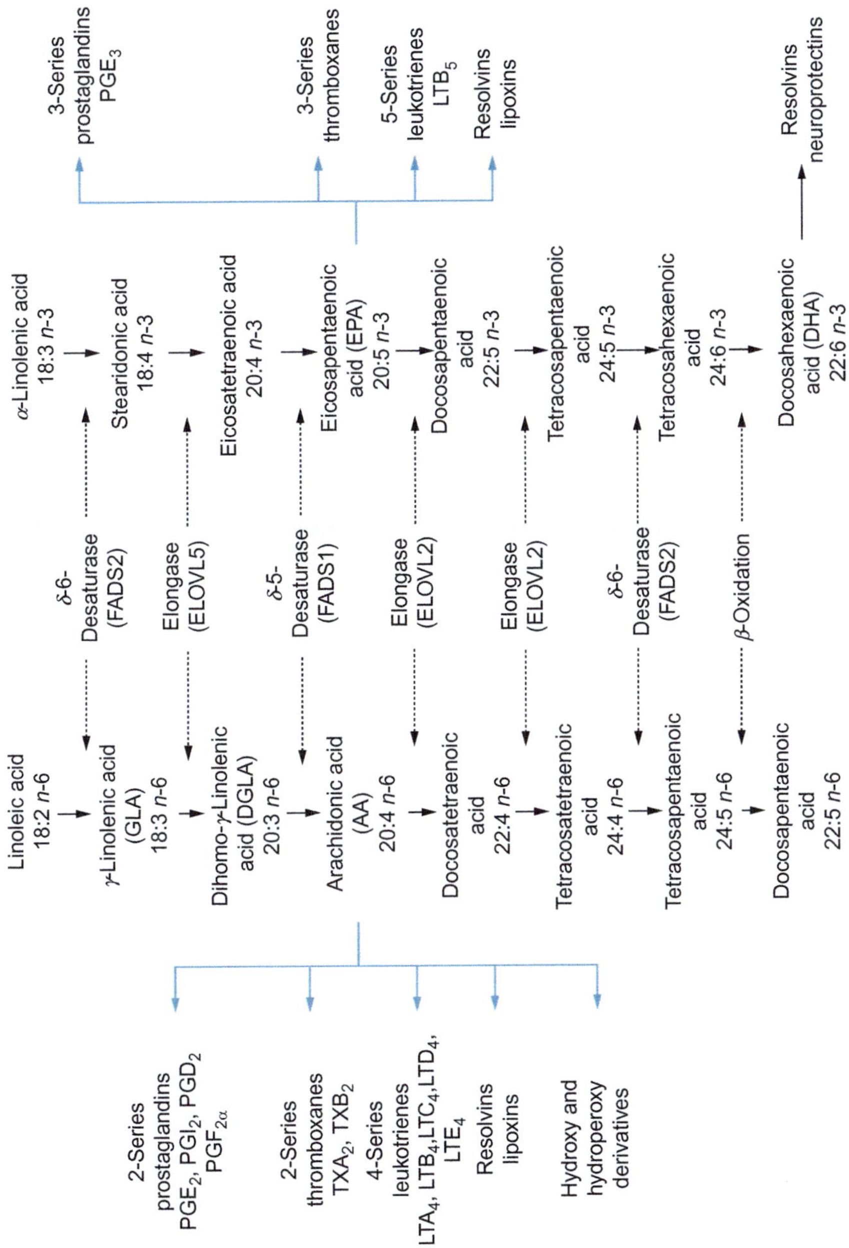

FIG. 2 Metabolism of n-6 and n-3 PUFA.

concentrations of inflammatory parameters,[26] insulin sensitivity, an increased risk and/or incidence of some types of cancer,[27] and neurodegenerative diseases.[28]

It´s important to emphasize that some research speculates that all fatty acids with one or more bonds in the trans configuration raise the ratio of LDL to HDL cholesterol irrespective of their origin or structure. However, it is important to note that animal fat with TFA, milk, and meat C18:1 trans-11 (vaccenic acid) are also major sources of saturated fatty acids in affluent diets and can be a confounding factor.[29] For example, there are varying reviews about the human health benefits of vaccenic acid (VA). Some authors argue that VA does not alter insulin sensitivity, and they have not found a convincing association between VA intake from animal fats and coronary heart disease. Some papers found benefits on immune function and inflammation.[30]

Specifically to the Central Nervous System (CNS), there is increasing evidence to suggest that diet, one of the most important modifiable environmental factors, may play a role in preventing or delaying cognitive decline and Alzheimer's disease (AD). AD, a major public health problem, has been estimated to affect 26,000,000 people worldwide; and the number of AD cases is expected to reach more than 100,000,000 by 2050.[31, 32] It is well known that diets rich in saturated fat increase oxidative stress on the brain, reduce neurogenesis, enhance neuroinflammation, and exert anxiety-like behavior.[33] Intake of TFA has been positively associated with risk of Alzheimer's disease[34] and dementia,[35] but it's important to note that the exact cause of AD remains elusive. Many possible risk factors and pathological alterations have been extensively studied in the elaboration of in vitro and in vivo models of this disease.[36] The relationship between TFA and brain damage will be discussed, as well as a focus on animal and human studies.

TRANS FATTY ACIDS AND ANIMAL STUDIES

Our research group has analyzed the cerebrospinal fluid (CSF) parameters in 60 adult male Wistar rats that were divided into four groups: normolipidic (control group) and hyperlipidic diet enriched with lard and the other two groups with the same composition but enriched with TFA. The treatment with TFA in both normo and hyperlipidic diets induced impairment of oxidized low-density lipoprotein (OxLDL) and LDL antibodies in CSF parameters (see Fig. 3). Specifically for the normolipidemic enriched with TFA, this concentration of lipids resulted in the same harmful effects as both of the hyperlipidic diets,[37] these results can suggest a possible inflammatory component to the TFA independent of concentration. Still in the same article, we found alterations in cytokines from CSF, Interleukin 1 (IL1), Interleukin 6 (IL6), Interleukin 10 (IL10), and Tumor necrosis factor alpha (TNF-α). Following the same trend, TFA diets consistently showed the worst results, supporting that inflammatory characteristics are due to the diet to which the animals were exposed (see Fig. 4).[37]

It was previously shown that dietary TFA may affect the CNS in Wistar rats by increased levels of cytokines in the CSF. In the next paper, we aimed at clarifying the impact of diets with different TFA concentrations on cerebral tissue, focusing on hippocampus and cortex and behavioral performance. Therefore, mitochondrial, inflammatory, and oxidative brain parameters, as well as performance in an open field task (OF), were evaluated after 90 days of treatment with the same diet composition.[28] TNF-α was altered in

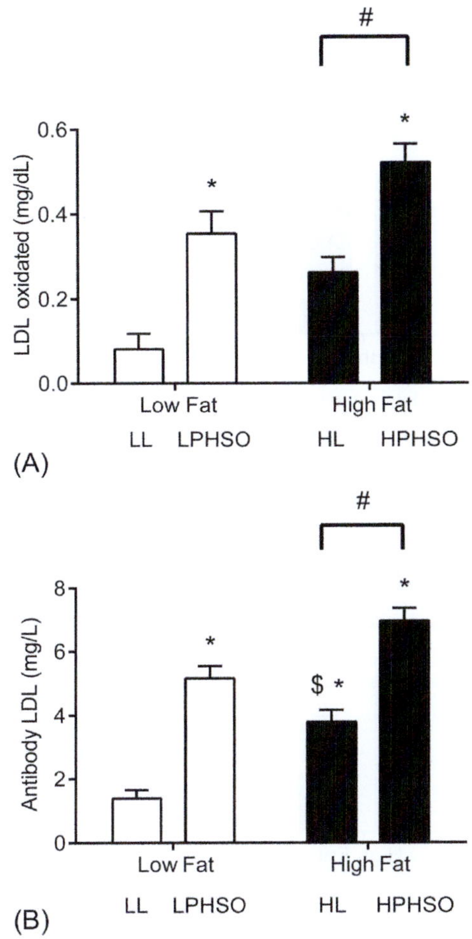

FIG. 3 Oxidized low-density lipoprotein (A) and low-density lipoprotein antibody (B) in CSF samples. $^{*}P<.05$ versus control (LL); $^{\#}P<.05$, difference between hyperlipidic diet; $^{\$}P<.05$, difference between LPHSO and HL only. All groups were analyzed using a two way ANOVA and Bonferroni post hoc test ($n=15$ animals/group). Data are expressed as the mean ± SEM and $P<.05$.

both the cortex and hippocampus, specifically in the hippocampus. In the normo diet with TFA, the values were higher than in the hyperlipidic diet from the control group (lard diet) (see Fig. 5).[28] In injured rat brains, upregulation of TNF-α was shown to precede leukocyte infiltration to the site of injury suggesting that TNF-α is produced early by resident brain cells in response to neuronal injury. Researchers suggest that TNF-α is only increased in rats subjected to a combined diffuse brain injury and hypoxia, but not in diffuse brain injury alone. In this case, only the diet altered this parameter.[38]

When cortex cells were evaluated (see Fig. 6), there were mitochondrial alterations and dysregulation of oxidative stress mediators in rats with TFA diets. Dietary TFAs provoked a reduction in mitochondrial content, a decrease in both cells with high mitochondrial mass, in cells with high membrane potential, and mean mitochondrial membrane potential. Likewise,

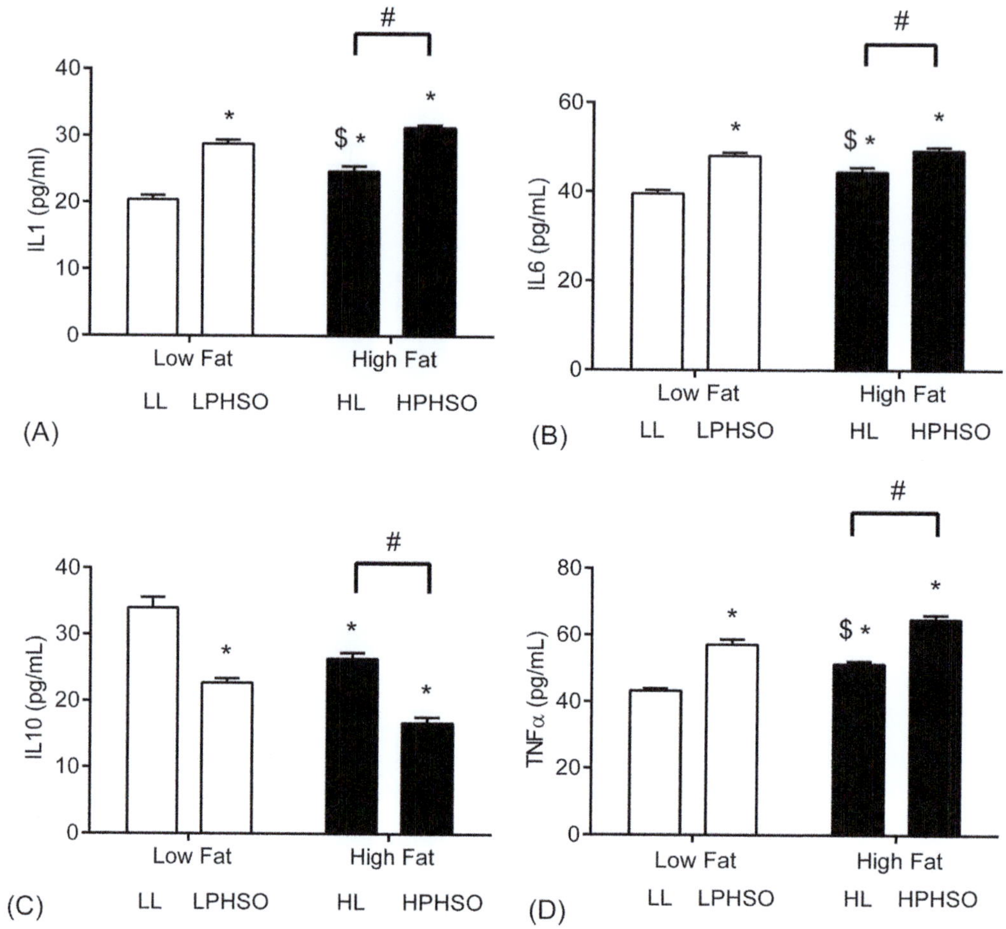

FIG. 4 IL1 (A), IL6 (B), IL10 (C) and TNF-α (D) in CSF samples. *$P<.05$ versus control (LL); #$P<.05$, difference between hyperlipidic diet; $$P<.05$, difference between LPHSO and HL only. All groups were analyzed using a two-way ANOVA and Bonferroni post hoc test ($n=15$ animals/group). Data are expressed as the mean ± SEM and $P<.05$.

the reactive oxygen species (ROS) were altered in TFA diet, independent of the concentration of the fat, reflecting a significant effect of diet on this parameter (see Fig. 7).[28] Changes in the parameters cited have been shown to be risk factors for a diminished bioenergetic activity; therefore it generated an important decrease in the capacity to generate ATP via oxidative phosphorylation in the cortex, which is a risk factor for neurodegeneration.[39]

Finally, the group with hyperlipidic diets enriched with TFA exhibited a behavioral aversion to the central zone, with less crosses in the central zone than the other groups in the open field task (see Fig. 8), showing a possible aversive response. This result can suggest a possible cognitive impairment as a result of the dietary treatment.[28] In another study, the influence of TFA on locomotor activity was evaluated for 60 weeks in 48 male Wistar rats (21 days old) using the quantification of orofacial dyskinesia (OD) and OF. In conclusion, this study showed that prolonged dietary intake of hydrogenated vegetable fat (source of TFA) allowed

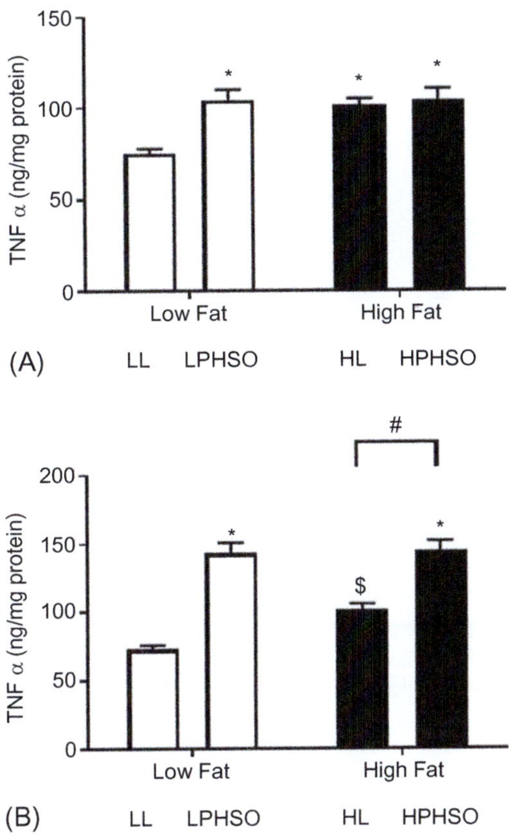

FIG. 5 Tumor necrosis factor-α (TNF-α) levels in the cortex (A) and in the hippocampus (B) were evaluated in the following groups: LL, LPHSO, HL, and HPHSO. *Represents a significant difference in comparison to the LL treatment. $Represents a significant difference between the LPHSO and HL treatments only. #Represents a significant difference between both hyperlipidic treatments after a two-way ANOVA followed by Tukey's post hoc tests. Data are expressed as the mean ± SEM and all analyses significant at $P < .05$ ($n = 15$ animals/group).

TFA incorporation of 0.33% in brain neuronal membranes, facilitating the development of movement disorders. This apparently slight incorporation can induce changes in memory function and anxiety-like symptoms, as well as in motor control, which was observed in this study by increased OD and decreased spontaneous locomotor activity in hydrogenated vegetable fat-fed rats.[40] The same authors have published a new research paper ratifying the previous conclusion; the hydrogenated vegetable fat was associated to higher incorporation of TFA in cortex, hippocampus, and striatum. Moreover, it was related to behavioral disturbances and weaker neuroprotection.[41] Two successive generations of female rats were supplemented with hydrogenated vegetable fat (19.8% of TFA), from pregnancy, lactation to adulthood at 90 days of age. In this study, hyperlocomotion (mania animal model), cortical fatty acid composition, and oxidative and molecular status were evaluated in a second generation of supplemented adult rats. The hydrogenated vegetable fat allowed TFA incorporation and increased omega 6 (n-6)/omega 3 (n-3) PUFA ratio in the brain-cortex and was

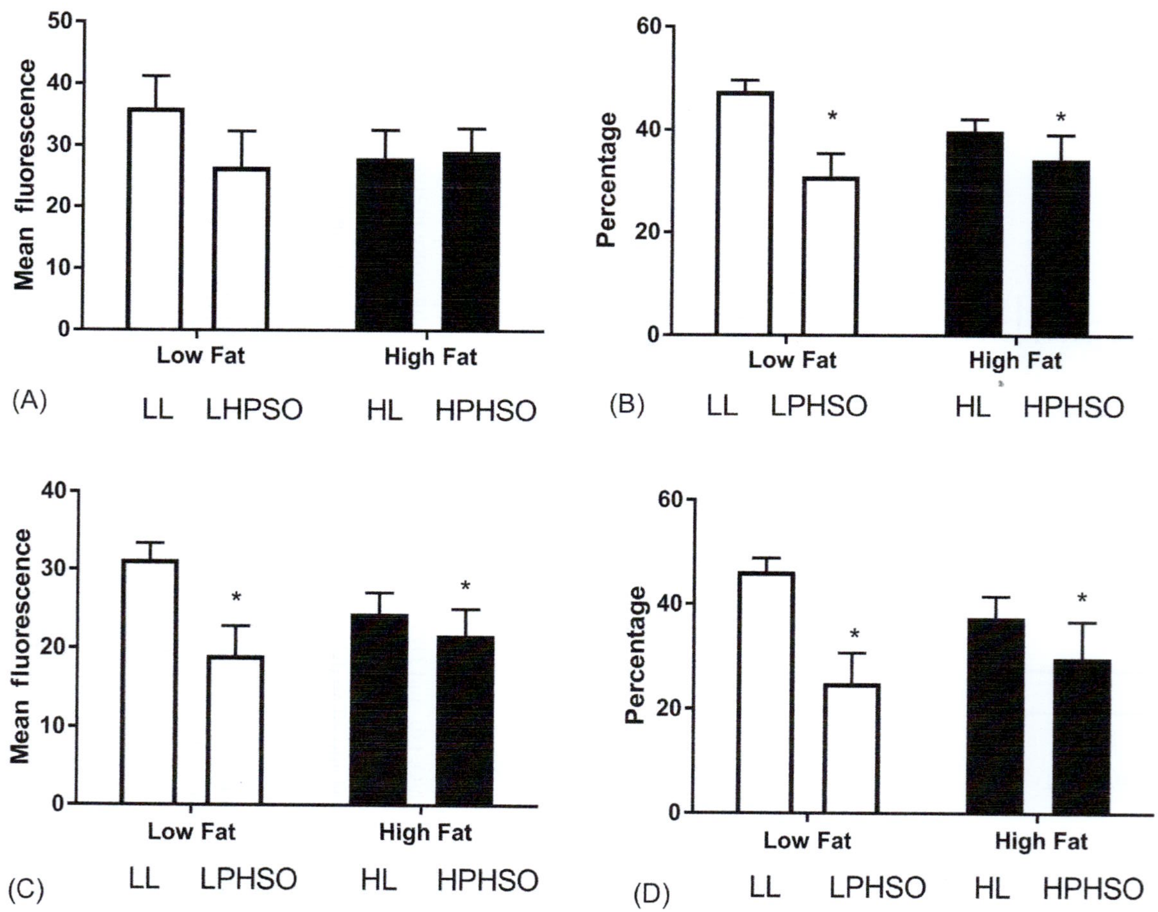

FIG. 6 Mean mitochondrial mass (A), the percentage of cells with high mitochondrial mass (B), the mean mitochondrial membrane potential (C), and the percentage of cells with high membrane mitochondrial potential (D) in cortex brain structure were evaluated in the following groups: LL, LPHSO, HL, and HPHSO. *Represents a significant difference in comparison to the LL treatment after a two way ANOVA followed by uncorrected Fisher's LSD post hoc tests. Data are expressed as the mean ± SEM and all analyses significant at $P < .05$ ($n = 15$ animals/group).

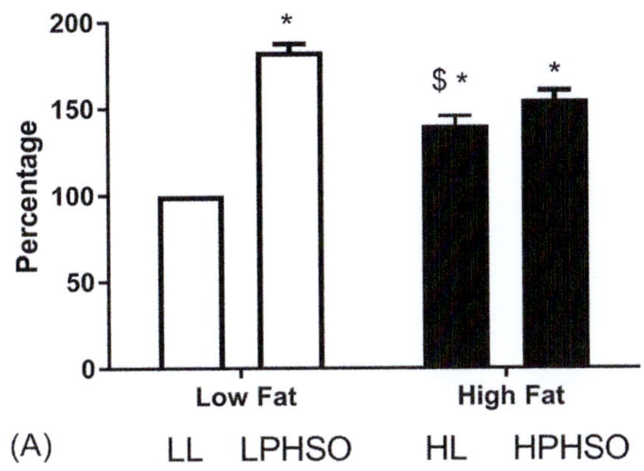

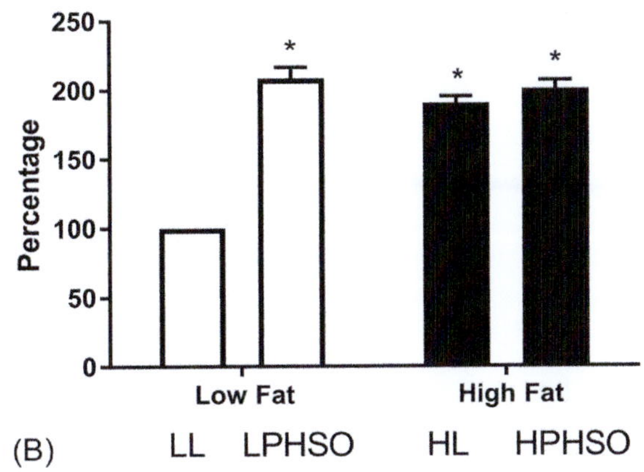

FIG. 7 Oxidative stress parameters in the cortex (A) in the hippocampus (B) in the following groups: LL, LPHSO, HL and HPHSO. *Represents a significant difference in comparison to the LL treatment. $Represents a significant difference between the LPHSO and HL treatments only, after a two-way ANOVA followed by Tukey's post hoc tests. Data are expressed as the mean ± SEM and all analyses significant at $P < .05$ ($n = 15$ animals/group).

associated with higher locomotion and exploration (hyperactivity behavior). Taken together, these events were able to change the composition of brain membranes, affecting the locomotor status, and facilitating the development of neuropsychiatric conditions, including bipolar disorder (BD).[42] The TFA incorporation into the brain is well studied. Wistar rat offspring whose mothers were fed with normolipidic diets containing soybean oil (control group) or hydrogenated vegetable oil (trans group) during gestation and lactation were used in this study. In 45 days, the rats received the same diets as their mothers. As result, TFA was incorporated in small amounts in the hippocampus, but it did not affect aversive memory or impair the cognitive behavior of animals in spatial memories.[43] In conclusion, the relationship between TFA and neurodegenerative disorders has been well documented.

In SH-SY5Y wild-type cells and cortical neuron from mice, TFA were added to culture to analyze whether the regulated intramembrane proteolysis of the amyloid precursor protein (APP) is affected by TFA. It's important to emphasize that AD disease has pathological

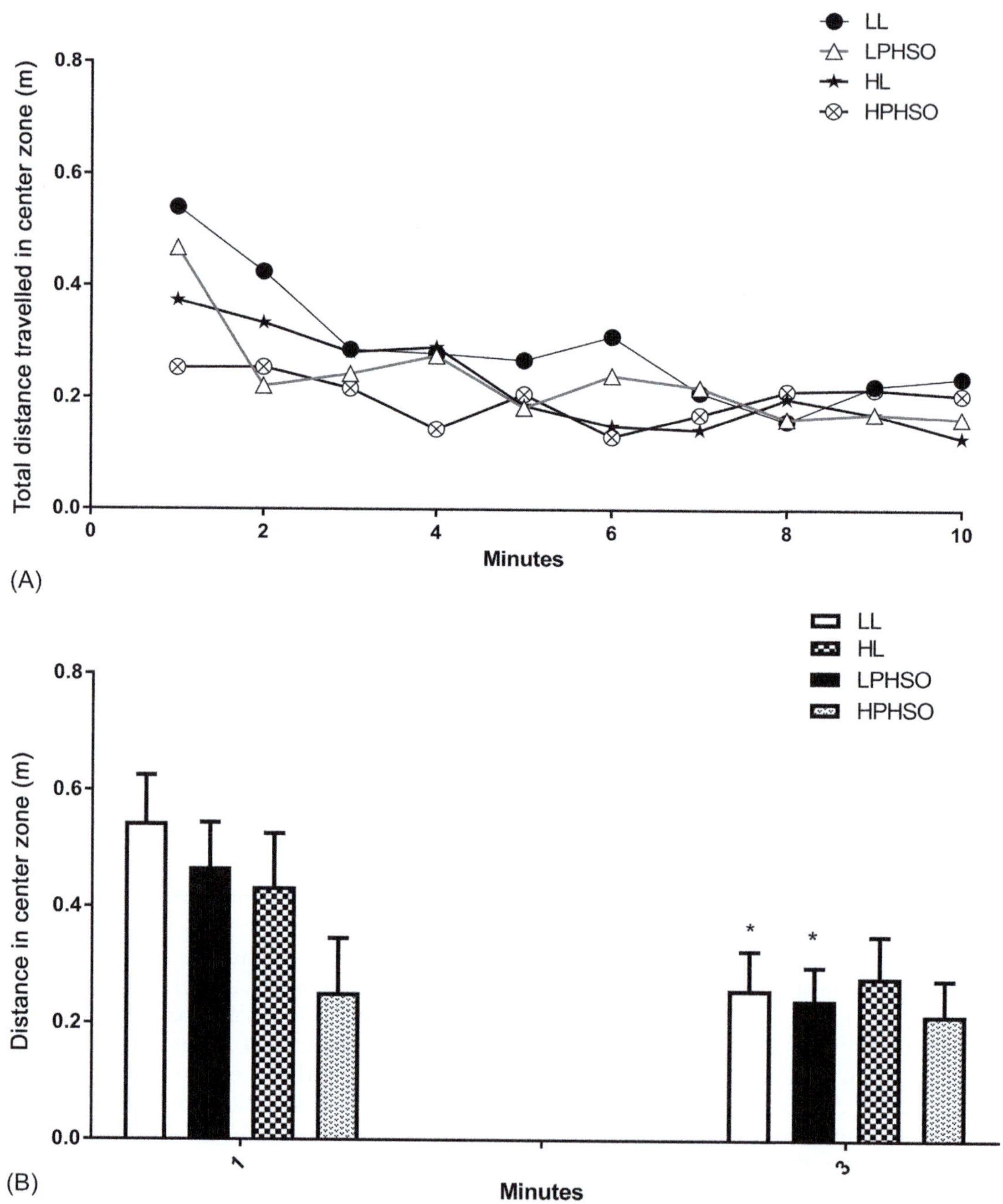

FIG. 8 Distance traveled in the center zone (A) and distance in the center zone traveled in the 1st and 3rd min in the OF task (B) in the following groups: LL, LPHSO, HL, and HPHSO. Data represent the activity in a 10-min session. A two-way ANOVA followed by Tukey's post hoc tests was used. *Represents a significant difference between the 1st and the 3rd min of the same treatment. Data are expressed as the mean ± SEM and all analyses significant at $P < .05$ ($n = 15$ animals/group).

hallmarks. The extracellular senile plaques composed of aggregated amyloid-beta (Aβ) peptides are released by sequential processing of the APP. For the generation of Aβ, APP is processed in the amyloidogenic pathway by β- and γ-Secretases. According to the authors, the experiments clearly showed that TFA increased amyloidogenic and decreased nonamyloidogenic processing of APP, resulting in an increased production of amyloid beta (Aβ) peptides, a neuropathological hallmark for AD.[44] Following the same trend, TFA were used to analyze their effects in nerve growth factor in differentiated PC12 cells with or without beta-amyloid peptide (Aβ). In the conclusions, the authors suggest that TFA was able to react with Aβ, which in turn enhanced oxidative stress and impaired neuronal cells; therefore, the results provided evidence to link TFA and AD.[45]

Likewise, in human trials, the relationship between TFA and brain damage are well documented.

TRANS FATTY ACIDS AND HUMAN TRIALS

Actually, many studies have investigated the effects of dietary fats on the development of AD. An important clinical evaluation was done on a stratified random sample of 815 community residents aged 65 years and older who were not affected by AD and were made a follow-up of 3.9 years. In the results, the authors found that 131 persons developed AD. The innovation in this case was that the intake of saturated fat and TFA were positively associated with increased risk, specifically TFA; they observed a strong increased risk.[34] Corroborating, Ginter and Simko,[46] commented in their review that TFA had an adverse effect on the brain and nervous system. They ratified the idea that this fat is incorporated into brain cell membranes and can alter the ability of neurons to communicate. Because of this alteration, it can diminish mental performance and therefore play a possible role in the development of AD and cognitive decline. Cognitive decline may be associated with the ingestion of TFA, but in this case, in conjunction with copper, according to Morris et al.[47]; a total of 6158 community residents 65 years and older participated in this project, and the data suggest that high dietary intake of copper in conjunction with TFA (and/or SFA) can accelerate the cognitive decline.

A systematic review (only with observational studies) evaluated the relationship between saturated and trans fat intake and risk of cognitive disorders. They used 12 articles in the review, and as a result, several (not all), prospective studies indicated a relationship between saturated and trans fat intake and risk of cognitive disorders. The possible mechanisms to link fat and dementia are to elevate total plasma and LDL cholesterol concentrations, which, in turn, may be associated with AD risk; cholesterol may play a key role in β-Amyloid production and deposition.[4]

Aggression and/or irritability were the aim of the study with 1018 male and female adults, minimum age 20 years (medium of 57.2). Nutrient data were collected using a food frequency questionnaire developed by the Nutrition Assessment Shared Resource of the Fred Hutchinson Cancer Research Center. In the results, TFA showed a significant association to behaviors. Indeed, TFA was more consistently predictive than that of other assessed and recognized aggression predictors, and the effect was robust to adjustment for potential confounders like age, education, smoking, and alcohol. In the conclusion, the authors comment that TFA consumption could serve as a marker for other untoward behaviors; n-3 may serve as a common protection against it.[48] Following the conclusions above, the same group found other results, which were published in a new paper. They found that TFA was significantly associated with worse word recall in younger adults (<45 years old), which is during years of high productivity. Therefore, these findings expand adverse associations of TFA to health generally and to central nervous system function in particular.[49]

Concluding, the excess in peripheral proinflammatory mediators, some of which can cross the blood-brain barrier (BBB), may trigger neuroinfammation, subsequently exacerbating neurodegeneration. Therefore, all conditions cited have oxidative stress as a risk factor.[28] Indeed, increased oxidative damage, mitochondrial dysfunction, accumulation of oxidized aggregated proteins, inflammation, and defects in protein clearance constitute complex intertwined pathologies that conspire to kill neurons, according to Halliwell.[50] Linking TFA and proinflammatory state, Bryk et al.[51] concluded that using human aortic endothelial cells, TFA has a direct proinflammatory effect, which promotes leukocyte adhesion to the endothelium through ROS-dependent NF-κB activation; thereby, the findings suggest a role of TFA in the development of a variety of vascular disorders.

ESSENTIAL FATTY ACIDS AND CENTRAL NERVOUS SYSTEM

Robust evidence shows us that public health may benefit from advice to eat less n-6 nutrients, more n-3 nutrients, and fewer calories per meal.[18] This concern is in relation to n-6, a polyunsaturated fatty acids (PUFA) (e.g., arachidonic acid [AA]). It is a precursor to potent lipid mediator signaling molecules, termed "eicosanoids," which have important roles in the regulation of inflammation; in general, eicosanoids derived from n-6 PUFA are proinflammatory.[22] Arachidonic acid and docosahexaenoic acid make up about 20% of fatty acids in the mammalian brain.[52]

In general, PUFA play an important role in the composition of all cell membranes where they maintain homeostasis for correct membrane protein function and influence membrane fluidity, thus regulating cell signaling processes, cellular functions, and gene expression.[22] According to Calder,[53] it has generally been assumed that all AA-derived eicosanoids are pro-inflammatory, however, this is an oversimplification since some actions of eicosanoids are antiinflammatory (e.g., prostaglandin (PGE^2) inhibits production of some inflammatory cytokines). It has been discovered quite recently that PGE^2 inhibits production of inflammatory leukotrienes and induces production of inflammation resolving lipoxin A.

In a robust review, the AA and its end products are involved in physiological processes in the central nervous system, such as synaptic signaling, neuronal firing, neurotransmitter release, nociception, neuronal gene expression, cerebral blood flow, sleep/wake cycle, and appetite.[54]

Dietary n-3 have a great relevance to brain health. It has been established that they are critical for proper infant growth and neurodevelopment. They are the most important n-3 with physiological significance for brain function, especially during prenatal brain development when it is incorporated into nerve growth cones during synaptogenesis.[55]

PUFAs can modulate many of the signal transduction mechanisms operating in neuronal membranes and, thus, in the synaptic cleft. This action is strongly influenced by ratio n-6 to n-3 that works on neurotransmission and prostaglandin formation, processes that are vital in the maintenance of normal brain function.[56] Both AA and DHA were originally established in infant neurodevelopment. These fatty acids are supplied by breast milk and rapidly accumulate in the brain during the first postnatal year; their impairment and/or deficiency have been linked with neurodevelopmental disorders. Studies have stressed the physiological importance of AA and DHA in the nervous systems,[57] for this reason. DHA and AA are nutrients that were first added to formulas in the United States in 2002.[58]

ESSENTIAL FATTY ACIDS AND ANIMAL STUDIES

The ratio between omega-3 and omega-6 polyunsaturated fatty acids (PUFAs), in particular, influences various aspects of serotoninergic and catecholaminergic neurotransmission, as shown by studies in animal models.[56] Moreira et al.[55] evaluated female rats that received isocaloric diets, either with n-3 (n-3 group) or n-3 deficient diet (D group), both containing 8% of total fat and differing only in the fatty acids composition. Female rats received their respective diets throughout the gestation as well as during suckling. After weaning (21 days old), the litters were maintained with the same diet of their dams until 60 days old. Ontogenetic experiments (hippocampal immunocontent of ionotropic NMDA and AMPA glutamatergic receptors subunits (NR2 A\B and GluR1, respectively) and the alpha isoform of the calcium-calmodulin protein kinase type II (aCaMKII) were done with litters at 02, 21 and 60 days old and the behavioral experiments were performed with male adult rats (60 days old). Their results support the evidence that dietary n-3 are crucially important to adequate brain development and point that n-3 dietary deficiency could affect the glutamatergic neurotransmission development, specifically by altering proteins involved in signal transduction, important to neuroplasticity.

Prokopiou et al.[59] evaluated the therapeutic effects of n-3 and n-6 in age-matched (9 months old) CCL2/ and C57BL/6 mice (3 months old and 9 months old, considered wild type, CY/EXP/101 breeding colony) of both sexes, with dry age-related macular degeneration (AMD), the animals were randomly treated with n-3 plus n-6 (100mg EPA+50mg DHA+30 mg ɣ-linolenic acid (GLA)) treated only n-3 (130mg EPA) and untreated. The animals in the treatment groups were administered fish oil daily by gavage for 3 months, and the oral dosing total volumes did not exceed 10mL/kg. The results showed an increase of EPA and a decrease of AA levels in both blood and retinas in the treatment groups. A decrease in NF-kB expression was noted in the treatment groups, and interleukin (IL) 18 protein levels demonstrated a significant reduction in the n-3-treated group only. In this way, the results suggest that treatment with n-3 PUFAs has an effect on the resolution of inflammation, but this effect is not the only determinant factor for disease regression. These results are relevant because the positive effects of n-3 PUFAs (e.g., EPA and DHA) is due to the formation of their metabolites, which possess antiinflammatory properties. This relation about n-3 metabolites and neuroprotection has been cited in an important review (with rats and mice) published by Orr et al.[60] This paper summarizes the effect of n-3 PUFA modulation in animal models of brain injury and disease where inflammation is also measured (sickness behavior, stroke, aging, depression, Parkinson's disease, diabetes, and cytokine- and irradiation-induced cognitive impairments). It is hypothesized that n-3 PUFA are antiinflammatory via their enzymatically-derived metabolites; however, the authors comment that these models provide persuasive but indirect evidence of antiinflammatory actions of n-3 PUFA in the brain, suggesting that a comprehensive lipidomics profiling during neuroinflammation has yet to be reported.

Kim et al.[61] fed male mice for 10 weeks that were divided into five groups (n = 7) according to the type of diet: control diet (C), control diet containing low-dose DHA-enriched fish oil (LD), control diet containing high-dose DHA-enriched fish oil (HD), control diet containing low-dose EPA-enriched fish oil (LE), and control diet containing high-dose EPA-enriched fish oil (HE). In summary, brown adipose tissue (BAT) activity is primarily regulated by the sympathetic nervous system (SNS) through the binding of noradrenaline (NA) to the β Adrenergic receptors, thereby inducing lipolysis and the activation of UCP1 to enhance thermogenesis. These mechanisms were found in the fish oil intake groups; in conclusion, fish oil intake induces UCP1 upregulation in brown and white adipose tissue via the sympathetic nervous system.

These discussions become relevant because the findings in the literature are still controversies as we can see in the paper from Islam et al.[62] This group has evaluated if n-3 could improve schizophrenia symptoms and delay the onset of first-episode psychosis in animal models of NMDA receptor hypofunction, NR1KD mice (mice express 10% of the normal levels of NMDA receptors due to a mutation in the gene that encodes the NR1 subunit (Grin1, GluN1)), to understand whether changes in glutamate neurotransmission could lead to changes in brain and serum fatty acids. In this research, we used the diet as follows: 1. Regular diet (10% fat with ratio of ω-6 to ω-3 = 10:1); 2. n-3 deficient diet (10% fat supplied from safflower oil with ratio of n-6 to n-3 = 760:1); 3. n-3 enriched diet (10% fat from: menhaden oil (2%) and safflower oil (8%), with ratio of n-6 to n-3 = 4:1). The findings showed that n-3 supplementation increased EPA, but did not increase DHA. They also did not improve any of the behavioral abnormalities of NR1KD mice, suggesting that n-3 supplementation is not effective, but n-3 depletion exacerbated their deficits in executive function. The depletion also caused extreme mortality among male mutant mice, with 75% mortality rate by 12 weeks of age. In this way, the question is: "Is there a difference between n-3 from food in comparison to supplementation?"

A robust systematic review and metaanalysis of the literature focused on effects of the relatively long-term n-3 supplementation on cognitive impairment, amyloid-β pathology, and neuronal loss in animal models of AD. Their results detailed a long-term n-3 supplementation that decreased the n-6/n-3 ratio and reduced the amount of amyloid-β in experimental animal models of AD. Omega-3 supplementation also improved cognitive function; this effect appeared more significant in rats compared to mice and in males compared to females. Moreover, n-3 supplementation diminished the amount of neuronal loss, especially in female animals. According to the authors, these benefits are more relevant during a minimum period of 10% of total life span.[63]

As we can see, the scientific community is looking for new approaches to offer focus on understanding the true impact of fat on human health. An intriguing paper presented an innovative technology for the production of both omega-6 and omega-3 essential fatty acids using transgenic mice (named "Omega mice"). Pai et al.[64] demonstrates the feasibility of producing mammals with the capability to convert SFA and MUFA to the essential n-6 and n-3 PUFA from transgenic mice, according to the authors. This transgenic technology (engineered to carry both optimized fat-1 and fat-2 genes from the roundworm *Caenorhabditis elegans* and are capable of producing essential n-6 and n-3 fatty acids from saturated fats or carbohydrates) serves as a new biotechnology for the production of essential fatty acids, especially n-3 EPA and DHA to meet increasing demand.

ESSENTIAL FATTY ACIDS AND HUMAN TRIALS

Important data provide the first prospective evidence that the n-6/n-3 PUFA ratio is associated with an increased risk for mood disorders in young people exhibiting an ultrahigh risk (UHR) phenotype. This study was a secondary analysis of a double-blind randomized controlled trial (RCT) of n-3 PUFAs for indicated prevention of psychosis transition in individuals at UHR for psychosis, with 12-week intervention with 1.2g per day of n-3 PUFAs (700mg EPA, 480mg docosahexaenoic acid (DHA)) or placebo was conducted with follow-up at 1 year and 7 years (median) after baseline. In conclusion, the data supported the hypothesis that higher levels of n-6 PUFAs and lower levels of n-3 PUFAs, possibly due to long-term changes in dietary intake or differential usage, pose a risk for depression.[65] Corroborating, Spencer et al.[66] found that low dietary intake of n-3 PUFA's can also contribute to depression through its effects on endocannabinoid and inflammatory pathways in specific brain regions leading to synaptic phagocytosis by microglia in the hippocampus, contributing to memory loss.

Taha et al.[67] evaluated 56 of the 67 randomized participants (with chronic headache) who completed the 12-week intervention phase, with 55 providing pre- and postintervention plasma samples, 28 in the low n-6 PUFA (L6 group) and 27 in the high n-3 plus low n-6 PUFA (H3-L6 group). The interventions were designed to be equally credible and to provide equivalent food and macronutrients, and the foods meeting nutrient targets were provided to participants for two meals and two snacks per day. After treatment, the H3-L6 group significantly reduced plasma AA concentration and produced significantly greater increases in n-3 PUFA concentrations in esterified and unesterified plasma lipids, suggesting that dietary n-6 fatty acid lowering increases bioavailability of n-3 polyunsaturated fatty acids in human plasma lipid pools. Following the same bioavailability theory, Jamilian et al.[68] evaluated 60 patients with documented schizophrenia randomly divided into two groups: n-3 (1000mg/day) ($n=30$) and placebo ($n=30$), in addition to their standard antipsychotic treatment in a double blind, placebo-controlled fixed-dose, add-on clinical trial conducted over 8 weeks. At the end of the trial, Positive and Negative Syndrome Scale (PANSS) score decreased significantly in both groups ($P<.05$) in comparison to baseline. Therefore using n-3 fatty acid as a supplement therapy could be helpful in the treatment of symptoms among schizophrenic patients as a potential supplement for reducing the total dose of these drugs, which could help the patients have a better life.

It's important to highlight that the studies were not conclusive regarding the real effects of n-3; for example, a metaanalysis of randomized, placebo-controlled trials of n-3 fatty acid treatment of major depressive disorder was searched. In 13 of them, the efficacy of n-3 fatty acids involving 731 participants was examined; metaanalysis demonstrated no significant benefit of n-3 fatty acid treatment compared to placebo, including a significant heterogeneity and publication bias, suggesting only a possible benefit in patients with more severe depression.[69]

Conversely, a randomized trial (VITACOG) with 266 participants 70 years and over with mild cognitive impairment (MCI) were randomized either to high dose B vitamins ($n=133$) or placebo for 2 years. The study found that B vitamin treatment to lower homocysteine slowed the rate of cognitive and clinical decline, but when n-3 fatty acid concentrations are low, B vitamin treatment has no effect on cognitive decline in MCI. Then, omega-3 levels are in the upper normal range. B vitamins interact to slow cognitive decline, and these effects were more significant for DHA alone than for EPA or combined n-3.[70] Following this same trend in healthy human volunteers, positron emission tomography (PET) was used to show that the brain consumes AA and DHA at respective rates of 17.8 and 4.6 mg/d, whereas in patients with AD, AA consumption is elevated.[52]

Lastly, the inflammation is the particular interest because aging is characterized by an increase in the concentration of a number of pro-inflammatory molecules in the circulation, a phenomenon that has been termed "inflammageing," an important risk factor for the most diseases and neurodiseases (Hypothalamic inflammation for example). In this way, the increased intake of n-3 offer a better incorporation of EPA and DHA into human inflammatory cells is partly at the expense of AA resulting in less substrate available for synthesis of the classic inflammatory eicosanoids like prostaglandin E_2, thus, antiinflammatory effects.[71]

FINAL CONSIDERATIONS

TFA has been shown to be toxic for our organs, independent of the dosage. This lipid exacerbates the inflammatory state and oxidative stress in this tissue in both animal and human trials. This suggests that this lipid may need to be eliminated from the diet. Following the same argument, the brain was shown to be vulnerable to the toxicity produced by this lipid, in both peripheral and local proinflammatory state. The brain generated important alterations in the metabolism, a possible link between TFA and brain damage was well documented throughout the chapter, including an increased risk of developing AD. Hence, it is possible that the first risk factor to AD is not the age but the fat dietary composition. However, the imbalance of n-3 and n-6 PUFA contribute to neurodevelopmental disorders by altering microglial activation resulting in abnormal formation of neuronal networks and activity, but, a correct proportion between n-3/n-6 ratio and consumption of fruits and vegetables high in polyphenolics can prevent and reverse age-related cognitive deficits by lowering oxidative stress and inflammation.[66] Obesity is clearly a risk factor to neurodegeneration. Alterations in the patterns of the fat consumption in the worldwide population are a public health issue and must be further discussed.

References

1. Downs SM, Thow AM, Leeder SR. The effectiveness of policies for reducing dietary trans fat: a systematic review of the evidence. *Bull World Health Organ*. 2013;91(4). 262-9H.
2. Machado RM, Stefano JT, Oliveira CP, et al. Intake of trans fatty acids causes nonalcoholic steatohepatitis and reduces adipose tissue fat content. *J Nutr*. 2010;140(6):1127–1132.
3. Mozaffarian D, Abdollahi M, Campos H, Houshiarrad A, Willett WC. Consumption of trans fats and estimated effects on coronary heart disease in Iran. *Eur J Clin Nutr*. 2007;61(8):1004–1010.
4. Barnard ND, Bunner AE, Agarwal U. Saturated and trans fats and dementia: a systematic review. *Neurobiol Aging*. 2014;35(Suppl 2):S65–S73.

5. Otite FO, Jacobson MF, Dahmubed A, Mozaffarian D. Trends in trans fatty acids reformulations of US supermarket and brand-name foods from 2007 through 2011. *Prev Chronic Dis*. 2013;10:E85.
6. Longhi IBaR. Trans fatty acids, does exist safety dosage? *Obes Weight Loss Therapy*. 2016;6(3):3.
7. Eckel RH, Borra S, Lichtenstein AH, Yin-Piazza SY. Trans Fat Conference Planning G. Understanding the complexity of trans fatty acid reduction in the American diet: American Heart Association Trans Fat Conference 2006: report of the Trans Fat Conference Planning Group. *Circulation*. 2007;115(16):2231–2246.
8. Priego-Capote R-JJ, de Castro MD. Identification and quantification of trans fatty acids in bakery products by gas chromatography–mass spectrometry after focused microwave Soxhlet extraction. *Food Chem*. 2007;100:859–867.
9. Bhardwaj S, Passi SJ, Misra A. Overview of trans fatty acids: biochemistry and health effects. *Diabetes Metab Syndr*. 2011;5(3):161–164.
10. Nacer Bellaloui AM, My AK. Effects of genetics and environment on fatty acid stability in soybean seed. *Food Nutr Sci*. 2013;4:165–175.
11. Kummerow FA. The negative effects of hydrogenated trans fats and what to do about them. *Atherosclerosis*. 2009;205(2):458–465.
12. Remig V, Franklin B, Margolis S, Kostas G, Nece T, Street JC. Trans fats in America: a review of their use, consumption, health implications, and regulation. *J Am Diet Assoc*. 2010;110(4):585–592.
13. Lichtenstein AH. Dietary trans fatty acids and cardiovascular disease risk: past and present. *Curr Atheroscler Rep*. 2014;16(8):433.
14. Robert J, Marchesini S, Delessert S, Poirier Y. Analysis of the beta-oxidation of trans-unsaturated fatty acid in recombinant Saccharomyces cerevisiae expressing a peroxisomal PHA synthase reveals the involvement of a reductase-dependent pathway. *Biochim Biophys Acta*. 2005;1734(2):169–177.
15. Yu W, Liang X, Ensenauer RE, Vockley J, Sweetman L, Schulz H. Leaky beta-oxidation of a trans-fatty acid: incomplete beta-oxidation of elaidic acid is due to the accumulation of 5-trans-tetradecenoyl-CoA and its hydrolysis and conversion to 5-trans-tetradecenoylcarnitine in the matrix of rat mitochondria. *J Biol Chem*. 2004;279(50):52160–52167.
16. Simopoulos AP. The importance of the ratio of omega-6/omega-3 essential fatty acids. *Biomed Pharmacother*. 2002;56(8):365–379.
17. McCowen KC, Bistrian BR. Essential fatty acids and their derivatives. *Curr Opin Gastroenterol*. 2005;21(2):207–215.
18. Lands B. Historical perspectives on the impact of n-3 and n-6 nutrients on health. *Prog Lipid Res*. 2014;55:17–29.
19. Arnold C, Konkel A, Fischer R, Schunck WH. Cytochrome P450-dependent metabolism of omega-6 and omega-3 long-chain polyunsaturated fatty acids. *Pharmacol Rep*. 2010;62(3):536–547.
20. Mantzioris E, James MJ, Gibson RA, Cleland LG. Differences exist in the relationships between dietary linoleic and alpha-linolenic acids and their respective long-chain metabolites. *Am J Clin Nutr*. 1995;61(2):320–324.
21. Manzel A, Muller DN, Hafler DA, Erdman SE, Linker RA, Kleinewietfeld M. Role of "Western diet" in inflammatory autoimmune diseases. *Curr Allergy Asthma Rep*. 2014;14(1):404.
22. Patterson E, Wall R, Fitzgerald GF, Ross RP, Stanton C. Health implications of high dietary omega-6 polyunsaturated Fatty acids. *J Nutr Metab*. 2012;2012:539426.
23. Bradbury J. Docosahexaenoic acid (DHA): an ancient nutrient for the modern human brain. *Nutrients*. 2011;3(5):529–554.
24. Hunter JE. Dietary trans fatty acids: review of recent human studies and food industry responses. *Lipids*. 2006;41(11):967–992.
25. Tokede OA, Petrone AB, Hanson NQ, et al. Plasma phospholipid trans fatty acids and risk of heart failure. *Am J Clin Nutr*. 2013;97(4):698–705.
26. Smit LA, Katan MB, Wanders AJ, Basu S, Brouwer IA. A high intake of trans fatty acids has little effect on markers of inflammation and oxidative stress in humans. *J Nutr*. 2011;141(9):1673–1678.
27. Wang Y, Proctor SD. Current issues surrounding the definition of trans-fatty acids: implications for health, industry and food labels. *Br J Nutr*. 2013;110(8):1369–1383.
28. Longhi R, Almeida RF, Pettenuzzo LF, et al. Effect of a trans fatty acid-enriched diet on mitochondrial, inflammatory, and oxidative stress parameters in the cortex and hippocampus of Wistar rats. *Eur J Nutr*. 2017;.
29. Brouwer IA, Wanders AJ, Katan MB. Effect of animal and industrial trans fatty acids on HDL and LDL cholesterol levels in humans--a quantitative review. *PLoS ONE*. 2010;5(3):e9434.
30. Field CJ, Blewett HH, Proctor S, Vine D. Human health benefits of vaccenic acid. *Appl Physiol Nutr Metab*. 2009;34(5):979–991.

31. Mosconi L, Murray J, Davies M, et al. Nutrient intake and brain biomarkers of Alzheimer's disease in at-risk cognitively normal individuals: a cross-sectional neuroimaging pilot study. *BMJ Open*. 2014;4(6):e004850.
32. Daulatzai MA. "Boomerang neuropathology" of late-onset Alzheimer's disease is shrouded in harmful "BDDS": breathing, diet, drinking, and sleep during aging. *Neurotox Res*. 2015;28(1):55–93.
33. Sharma S, Zhuang Y, Gomez-Pinilla F. High-fat diet transition reduces brain DHA levels associated with altered brain plasticity and behaviour. *Sci Rep*. 2012;2:431.
34. Morris MC, Evans DA, Bienias JL, et al. Dietary fats and the risk of incident Alzheimer disease. *Arch Neurol*. 2003;60(2):194–200.
35. Morris MC, Tangney CC. Dietary fat composition and dementia risk. *Neurobiol Aging*. 2014;35(Suppl 2):S59–S64.
36. Rodrigues L, Biasibetti R, Swarowsky A, et al. Hippocampal alterations in rats submitted to streptozotocin-induced dementia model are prevented by aminoguanidine. *J Alzheimers Dis*. 2009;17(1):193–202.
37. Longhi R, Almeida RF, Machado L, et al. Effect of a trans fatty acid-enriched diet on biochemical and inflammatory parameters in Wistar rats. *Eur J Nutr*. 2017;56(3):1003–1016.
38. Woodcock T, Morganti-Kossmann MC. The role of markers of inflammation in traumatic brain injury. *Front Neurol*. 2013;4:18.
39. Chacko BK, Kramer PA, Ravi S, et al. The Bioenergetic Health Index: a new concept in mitochondrial translational research. *Clin Sci*. 2014;127(6):367–373.
40. Teixeira AM, Dias VT, Pase CS, et al. Could dietary trans fatty acids induce movement disorders? Effects of exercise and its influence on Na(+)K(+)-ATPase and catalase activity in rat striatum. *Behav Brain Res*. 2012;226(2):504–510.
41. Trevizol F, Roversi K, Dias VT, et al. Influence of lifelong dietary fats on the brain fatty acids and amphetamine-induced behavioral responses in adult rat. *Prog Neuro-Psychopharmacol Biol Psychiatry*. 2013;45:215–222.
42. Trevizol F, Roversi K, Dias VT, et al. Cross-generational trans fat intake facilitates mania-like behavior: oxidative and molecular markers in brain cortex. *Neuroscience*. 2015;286:353–363.
43. de Souza AS, Rocha MS, Tavares do Carmo M. Effects of a normolipidic diet containing trans fatty acids during perinatal period on the growth, hippocampus fatty acid profile, and memory of young rats according to sex. *Nutrition*. 2012;28(4):458–464.
44. Grimm MO, Rothhaar TL, Grosgen S, et al. Trans fatty acids enhance amyloidogenic processing of the Alzheimer amyloid precursor protein (APP). *J Nutr Biochem*. 2012;23(10):1214–1223.
45. Tsai SJ, Liu WH, Yin MC. Trans fatty acids enhanced beta-amyloid induced oxidative stress in nerve growth factor differentiated PC12 cells. *Neurochem Res*. 2012;37(4):786–794.
46. Ginter E, Simko V. New data on harmful effects of trans-fatty acids. *Bratisl Lek Listy*. 2016;117(5):251–253.
47. Morris MC, Evans DA, Tangney CC, et al. Dietary copper and high saturated and trans fat intakes associated with cognitive decline. *Arch Neurol*. 2006;63(8):1085–1088.
48. Golomb BA, Evans MA, White HL, Dimsdale JE. Trans fat consumption and aggression. *PLoS ONE*. 2012;7(3):e32175.
49. Golomb BA, Bui AK. A fat to forget: trans fat consumption and memory. *PLoS ONE*. 2015;10(6):e0128129.
50. Halliwell B. Oxidative stress and neurodegeneration: where are we now? *J Neurochem*. 2006;97(6):1634–1658.
51. Bryk D, Zapolska-Downar D, Malecki M, Hajdukiewicz K, Sitkiewicz D. Trans fatty acids induce a proinflammatory response in endothelial cells through ROS-dependent nuclear factor-kappaB activation. *J Physiol Pharmacol*. 2011;62(2):229–238.
52. Rapoport SI. Arachidonic acid and the brain. *J Nutr*. 2008;138(12):2515–2520.
53. Calder PC. Polyunsaturated fatty acids and inflammatory processes: new twists in an old tale. *Biochimie*. 2009;91(6):791–795.
54. Bosetti F. Arachidonic acid metabolism in brain physiology and pathology: lessons from genetically altered mouse models. *J Neurochem*. 2007;102(3):577–586.
55. Moreira JD, Knorr L, Ganzella M, et al. Omega-3 fatty acids deprivation affects ontogeny of glutamatergic synapses in rats: relevance for behavior alterations. *Neurochem Int*. 2010;56(6-7):753–759.
56. Haag M. Essential fatty acids and the brain. *Can J Psychiatry*. 2003;48(3):195–203.
57. Kiso Y. Pharmacology in health foods: effects of arachidonic acid and docosahexaenoic acid on the age-related decline in brain and cardiovascular system function. *J Pharmacol Sci*. 2011;115(4):471–475.
58. Carlson SE, Colombo J. Docosahexaenoic acid and arachidonic acid nutrition in early development. *Adv Pediatr*. 2016;63(1):453–471.

59. Prokopiou E, Kolovos P, Kalogerou M, et al. Therapeutic potential of omega-3 fatty acids supplementation in a mouse model of dry macular degeneration. *BMJ Open Ophthalmol*. 2017;1(1):e000056.
60. Orr SK, Trepanier MO, Bazinet RP. n-3 Polyunsaturated fatty acids in animal models with neuroinflammation. *Prostaglandins Leukot Essent Fat Acids*. 2013;88(1):97–103.
61. Kim M, Goto T, Yu R, et al. Fish oil intake induces UCP1 upregulation in brown and white adipose tissue via the sympathetic nervous system. *Sci Rep*. 2015;5:18013.
62. Islam R, Trepanier MO, Milenkovic M, et al. Vulnerability to omega-3 deprivation in a mouse model of NMDA receptor hypofunction. *NPJ Schizophr*. 2017;3:12.
63. Hooijmans CR, Pasker-de Jong PC, de Vries RB, Ritskes-Hoitinga M. The effects of long-term omega-3 fatty acid supplementation on cognition and Alzheimer's pathology in animal models of Alzheimer's disease: a systematic review and meta-analysis. *J Alzheimers Dis*. 2012;28(1):191–209.
64. Pai VJ, Wang B, Li X, Wu L, Kang JX. Transgenic mice convert carbohydrates to essential fatty acids. *PLoS ONE*. 2014;9(5):e97637.
65. Berger ME, Smesny S, Kim SW, et al. Omega-6 to omega-3 polyunsaturated fatty acid ratio and subsequent mood disorders in young people with at-risk mental states: a 7-year longitudinal study. *Transl Psychiatry*. 2017;7(8):e1220.
66. Sarah J, Spencer AK, Layé S, Shukitt-Hale B, Barrientos RM. Food for thought: how nutrition impacts cognition and emotion. *Nat Sci Food*. 2017;1(7):8.
67. Taha AY, Cheon Y, Faurot KF, et al. Dietary omega-6 fatty acid lowering increases bioavailability of omega-3 polyunsaturated fatty acids in human plasma lipid pools. *Prostaglandins Leukot Essent Fat Acids*. 2014;90(5):151–157.
68. Jamilian H, Solhi H, Jamilian M. Randomized, placebo-controlled clinical trial of omega-3 as supplemental treatment in schizophrenia. *Global J Health Sci*. 2014;6(7):103–108.
69. Bloch MH, Hannestad J. Omega-3 fatty acids for the treatment of depression: systematic review and meta-analysis. *Mol Psychiatry*. 2012;17(12):1272–1282.
70. Oulhaj A, Jerneren F, Refsum H, Smith AD, de Jager CA. Omega-3 fatty acid status enhances the prevention of cognitive decline by B vitamins in mild cognitive impairment. *J Alzheimers Dis*. 2016;50(2):547–557.
71. Calder PC, Bosco N, Bourdet-Sicard R, et al. Health relevance of the modification of low grade inflammation in ageing (inflammageing) and the role of nutrition. *Ageing Res Rev*. 2017;40:95–119.

Further Reading

72. Menaa F, Menaa A, Treton J, Menaa B. Technological approaches to minimize industrial trans fatty acids in foods. *J Food Sci*. 2013;78(3):R377–R386.

CHAPTER

29

Epigenetic Effects of Omega-3 Fatty Acids on Neurons and Astrocytes During Brain Development and Senescence

Belmira Lara da Silveira Andrade-da-Costa, Alinny Rosendo Isaac*,†, Ricielle Lopes Augusto*, Raphael Fabricio de Souza*,‡, Hércules Rezende Freitas†, Ricardo Augusto de Melo Reis†*

***Laboratory of Neurophysiology, Department of Physiology and Pharmacology, Federal University of Pernambuco, Recife, Brazil †Laboratory of Neurochemistry, Institute of Biophysics Carlos Chagas Filho, Federal University of Rio de Janeiro, Rio de Janeiro, Brazil ‡Department of Physical Education, Federal University of Sergipe, São Cristovam, Brazil**

EPIGENETIC MECHANISMS RELATED TO NEURON AND ASTROCYTE DEVELOPMENT

Epigenetic can be defined by environmental influences on changes in gene expression (active versus inactive genes) that do not involve changes to the underlying DNA sequence. Therefore, epigenetic remodeling is required during early stages of brain development, especially for cell genesis, proliferation, and differentiation, inducing switching off or maintaining active genes necessary to convert pluripotent in specialized cells.[1–3] This process involves DNA methylation, chromatin remodeling, histone posttranslational modifications, and regulation mediated by noncoding RNAs, including microRNAs (miRNAs), and long noncoding RNAs (lncRNAs). These mechanisms act as modulatory players of programmed events, which are not directly dependent on the inherited DNA sequence.[4–6]

One of the most studied epigenetic mechanisms is DNA methylation, which can exert an important role in gene silencing during ontogeny. Such a process consists in the covalent

Omega Fatty Acids in Brain and Neurological Health
https://doi.org/10.1016/B978-0-12-815238-6.00029-8

addition of a methyl (CH3-) radical in a cytosine base, converting it to 5-methylcytosine. The methylation target cytosines are bound to guanine, which is located mainly in gene transcription promoter regions, constituting the so-called CpG islands.[7, 8] This reaction occurs through enzymes called DNA methyltransferases (DNMT), which catalyze the methylation of CpG islands, limiting the interaction with the transcription factors required for gene expression to start. DNMTs are divided into three subtypes: DNMT1, DNMT3A, and DNMT3B. DNMT1 is responsible for the maintenance of methylation patterns, preserving them during the process of cell division. In contrast, DNMT3A and DNMT3B are de novo methyltransferases, and their functional interaction represents a regulatory mechanism needed during neural cell differentiation.[1, 3]

The acquisition of multipotentiality in neural stem cells (NSCs) during development is strongly regulated by DNA methylation.[9] At the early and midgestational stage, astrocytic gene promoter, such as glial fibrillary acidic protein (GFAP), is hypermethylated in NSCs, impeding the differentiation of these cells.[10] At the embryonary stage of rodent central nervous system (CNS), DNMT1 is expressed in both proliferating neural progenitor cells (NPCs) and differentiated neurons.[11] On the other hand, DNMT3A is distributed in neural stem cells of the subventricular zone (SVZ) and in postnatal neurons of several brain regions, including olfactory bulb, cerebral cortex, hippocampus, striatum, and cerebellum.[12] DNMT3A activity regulates spine plasticity and emotional behavior in the nucleus accumbens.[13] In addition, both DNMT1 and DNMT3A also contribute to regulating synaptic function in adult forebrain neurons.[13]

A number of studies have also shown that high CpG methylation can occur in gene bodies (genomic loci spanning exons and introns), where it can increase gene transcription.[14] In the nervous system, a subset of genes that act at the synapse are highly methylated and expressed in immature neurons more than in the placenta.[15] Compared to nonneuronal cells, very low levels of methylation in CpG islands and very high methylation in gene bodies were reported for the human brain and neurons.[16]

Histone modifications also represent an important mechanism of epigenetic regulation during brain development and display high levels of diversity and complexity when compared to DNA methylation. Histones have a globular domain, where the DNA strand is involved. Flexible amino-terminal, also called tails, are rich in the amino acids lysine, arginine, serine, and threonine. Histone octamer composes the nucleosome unit, which is formed by dimers of four types of nuclear histones: H2A, H2B, H3, and H4.[2, 17] The binding between nucleosomes is given by the histone H1 that also is necessary for chromatin structuring and stabilization.[18] The long amino-terminal tails of H3 and H4 can be subjected to several posttranslational modifications, including methylation, acetylation, phosphorylation, ubiquitylation, sumoylation, glycosylation, biotinylation, carbonylation, and adenosine diphosphate (ADP)-ribosylation.[17, 19] Lysine (K) acetylation and methylation have been reported as essential steps during neural stem cell differentiation in neurons or astrocytes.[20]

Adequate levels of histone acetylation, which induces gene transcription, depend on a well-adjusted activity of the enzymes histone acetyltransferase (HAT) and histone deacetylases (HDAC). Histone methylation or demethylation is achieved by histone methyltransferases (HMT) and histone demethylases (HDM), respectively,[21] and the effect depends on which lysine or arginine residue is methylated: methylation at lysine residues H3K4, H3K9, H3K14 and H3K36 induce gene transcription. Whereas, methylated H3K9 and H3K27 can result in gene silence.[22–25]

Genes for different subclasses of these enzymes are more prevalent during brain development than in adulthood.[26] For example, histone deacetylation induced by one type of HDAC called HDAC11 plays an important role in the development of oligodendrocytes and some neurons, including hippocampal granule neurons.[27] In addition, other HDAC families were also associated with the differentiation and maturation processes of Schwann cells and oligodendrocytes, as well as the induction, regulation, and maintenance of myelination and axonal regeneration.[28, 29] Changes of H3K27Ac and H3K9Ac activities can repress genes related to neural stem cells and up-regulate genes associated with neuron differentiation.[30] H3K4Me3 and H3K9Ac activations are also involved in neuronal or glial functions during brain development,[30] and a number of genes affected by these modifications are related to cell-cell interactions and synaptic transmission.[30] When high global H3K27Me3 occurs simultaneously with low global H3K4 methylation, neuronal differentiation is repressed.[31, 32] H3K4 methylation is also important for orderly brain development and synaptic plasticity.[33]

The blockade of HDAC6 activity in cerebellar and hippocampal neuronal cultures is capable of promoting dendritic growth and arborization, suggesting the importance of this enzyme and acetylation levels on dendritogenesis.[34] In contrast, nonspecific inhibition of HDACs by valproic acid in hippocampal neural progenitor cells is able to induce cell differentiation, enhancing neurogenesis.[35]

It was shown that the increase in H3 acetylation around the STAT3 binding site after treatment of neural progenitor cells with an HDAC inhibitor (valproic acid), promotes a greater binding of this molecule to the GFAP promoter.[36] Majumder et al.[37] reported that the inhibition of HDAC by trichostatin increased the response of human neural progenitor cells to pro-astrocytic growth factors, such as BMP-2 and LIF.[37] Histone acetylation has also been shown to be crucial for the expression of GFAP isoforms as well as the organization of its intermediate filaments.[38]

The role of miRNAs as posttranscriptional regulators of gene expression has also been reported in several steps of CNS development, from the maintenance of the pluripotent state of embryonic stem cells to the differentiation of neural phenotypes.[39] Cell type-specific miRNA expression is necessary to maintain the control of gene expression within specific spatiotemporal windows. In this context, while miR-125b, for example, acts to keep NSC status, miR-9a has been detected exclusively in neural progenitor cells (NPCs).[40] miRNAs such as miR-124a, miR-128, miR-129, and miR-298, are expressed in neurons,[41] whereas miR-23 is restricted to the glial lineage.[42] miR-124, for example, is abundant miRNA within the CNS, and it regulates the transition from neural progenitor to mature neurons by inhibiting nonneuronal genes.[43] Others, miR-26 and miR-29 are more expressed in glia than in neurons.[42] On the other hand, miR-219 and miR-338 are involved in the oligodendrocyte differentiation pathway.[44] A number of studies have shown the global importance of miRNAs for CNS development through the deletion of key proteins implicated in their biogenesis and function. According to some of these studies, the absence of miRNAs can induce size reduction of the forebrain,[45] causing microcephaly in the cortex and hippocampus due to an impaired neuron dendritic branching and density.[46]

All these mechanisms described above are sensitive to environmental factors, which in the early stages of brain development can alter gene transcription and have a long-term repercussion on phenotypic markers.[47, 48] Nutritional changes that lead to an increase or deficiency of macro or micronutrients may modify the normal course of neural development as a product of changes in epigenetic mechanisms.[49–51] These changes in the epigenome may induce reprogramming and permanent modifications in a transgenerational manner, as a result of subtle alterations in cell differentiation or cumulative effects in the cell, compromising the normal function of the gene.[50, 52, 53]

IMPORTANCE OF OMEGA 6/OMEGA 3 RATIO FOR EPIGENETIC MODULATION DURING NEURAL DEVELOPMENT

Nutritional models using diets with high-fat content and low-protein levels, as well as intrauterine malnutrition, can lead to changes on several epigenetic mechanisms in different organs and systems, including liver, lung, heart, and CNS.[50, 54, 55] Animal or human studies have indicated that dietary omega 3 (n-3) alpha-linoleic and omega 6 (n-6) linoleic fatty acids, exert crucial roles during brain development. Special attention has been given to beneficial effects of the n-3 and n-6 long-chain polyunsaturated fatty acids (LC-PUFA) docosahexaenoic (DHA) and arachidonic acid (AA),[56, 57] respectively. However, accumulating evidence has reinforced the necessity of a balanced n-6/n-3 LC-PUFA to maximize the health benefits from gestation to senescence. The debate about an optimal n-6/n-3 ratio has been raised, considering that metabolic programming of long-term outcomes can be due to inadequate fatty acid nutrition in early life.[58–60] Thus, the dietary availability of these essential fatty acids represents a theme of increasing interest in public health.[61]

The elevated n-6/n-3 fatty acid ratio in Western diets (hyperlipidemic and hypercaloric) has been associated with epigenetic changes involved in the etiology of neurodegenerative diseases,[62] and impaired neocortical development.[63] DHA plays a key role in neural stem cells, astrocyte and neuron proliferation and differentiation,[64] intercellular communication[65] and also exert antiinflammatory effects,[66] in part due to a modulatory action on epigenetic players such as DNA methylation, histone posttranslational changes and miRNA expression, among others.[60, 62]

Gestational and early postnatal dietary deprivation of AA and DHA elicited schizophrenia-like phenotypes in mouse offspring at adulthood and the mechanisms involved in this effect were related with DNA hypermethylation of genes for nuclear receptors such as retinoid X receptor (RXR) and *peroxisome proliferator-activated receptors* (PPAR). As a consequence of this effect, a down-regulation of genes involved in oligodendrocyte integrity and the gamma-aminobutyric acid (GABA)-ergic system in the prefrontal cortex[67] was also detected. Maternal n-3 LC-PUFA deprivation during pregnancy and lactation also affects neurogenesis and apoptosis in adult offspring, and these effects have been associated with DNA methylation of brain-derived neurotrophic factor (BDNF) transcripts.[68] A similar epigenetic effect was also found when animals grown from gestation until 80 postnatal days under n-3 deficient diet were fed a hypercaloric and hyperlipidemic Western diet. However, the maternal supplementation of fish oil was able to reverse this effect.[69] N-3 LC-PUFA-enriched maternal diets were also able to induce higher DNA methylation in the offspring's gene body, which was linked to homeostasis in cellular and humoral immunity, and to the growth and development of children.[70] When n-3 supplementation was given to vitamin B12-deficient pregnant females, it was able to revert the global hypomethylation in the offspring cerebral cortex at birth and hypermethylation in adulthood, indicating that this effect could be involved in the neuroprotective actions in the brain.[71]

DHA supplementation in the maternal diet also induced long-term effects in the offspring neural stem cells kept in cultures for 25 days, reducing levels of HDAC as well as an increased expression of miRNA 210 and mRNA levels of trophic factors involved in neural cell proliferation.[64] When this dietary supplementation with fish oil occurs during gestation and lactation, a dose-dependent increase in GFAP protein expression was observed in the cerebral cortex of neonates at postnatal day (PND) 16.[72] Conversely, DHA deficiency in astrocyte primary culture reduced the expression of phosphorylated GFAP protein and increased neutral, basic, and nonphosphorylated isoforms. This finding was accompanied by an immature astrocyte phenotype.[73, 74] The importance of an adequate n-3 availability in the maternal diet was also reinforced by a recent study demonstrating that even under protein deficiency and low amount of lipids, the levels of DHA given by a n-6/n-3 ratio of 1.67 was enough to keep astrocyte and neuron differentiation in the neonate progeny.[75]

It is noteworthy that the evidence that GFAP expression and glial functions can be positively modulated by heterochromatin reprogramming related to H3K9Ac activation, increasing acetylation levels, either during development or in mature astrocytes.[38, 76]

Within a neuroprotection context in human neuroblastoma cell line, DHA supplementation in the culture medium was able to reverse deleterious effects of zinc on posttranslational modification of the histones 3 (H3) and 4 (H4) as well as reduced caspase-3 levels.[63, 64] These data reinforce the importance of adequate supplementation of PUFAs on epigenetic modulation in physiological or pathological conditions.

The schematic drawing in Fig. 1 illustrates some of the epigenetic mechanisms described here that are induced by n-3 fatty acids during neural development.

OMEGA-3 DIETARY AVAILABILITY INDUCES EPIGENETIC CHANGES IN THE SENESCENCE

During the aging process, a progressive dysfunction at the cellular level can occur due to the accumulation of molecular errors that increases disease susceptibility. It has been discussed that intrinsic and environmental factors may influence this process through epigenetic mechanisms.[77–79]

Several studies have indicated associations between low levels of DHA and/or unbalanced n-6/n-3 ratios and neurodegenerative and/or inflammatory conditions that occur during senescence.[80] Current evidence that n-3 fatty acids exert antiinflammatory therapeutic effects,[81] and alter gene expression within the cells has indicated that one of the mechanisms for this beneficial effect involves changes of epigenetic markers, such as DNA methylation, miRNAs, and histone posttranslational modifications.[66, 82]

A diet enriched with n-3 LC-PUFA was able to suppress the transcriptomic expression of APBA1, which is a key gene encoding the amyloid beta (A4) precursor protein-binding family.[83] Recently, this finding was supported by the evidence that a balanced diet containing a 1:1 n-6/n-3 ratio given to mice from three weeks of age till 25 weeks of age, induced overexpression of miR-381-3p, miR-30c-5p, and miR-103-3p that directly target APBA1.[60] This same dietary treatment also regulated the methylation status of a group of genes encoding proteins associated with spatial and working memory.[60] Such results indicated a long-term preventative effect provided by diets with an adequate proportion of n-3 and n-6 LC-PUFA on the

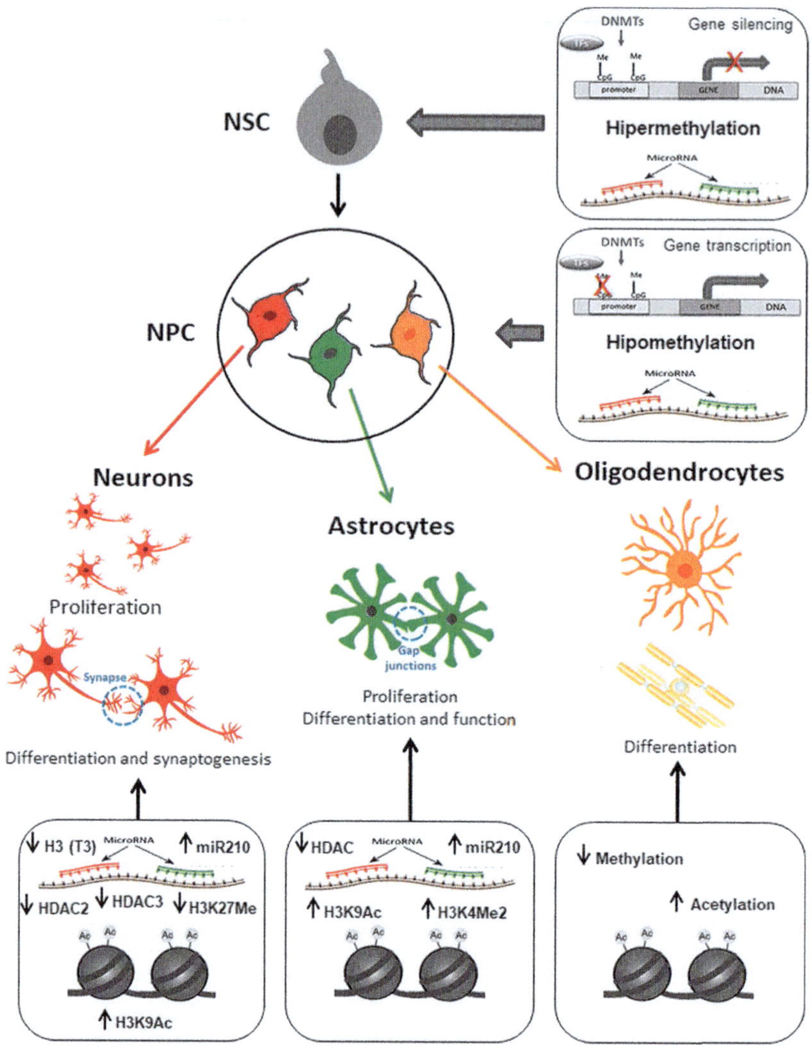

FIG. 1 Schematic drawing illustrating some of the epigenetic mechanisms that are induced by n-3 fatty acids during neural development. All data were obtained from the literature cited in this chapter. *NSC*, neural stem cells; *NPC*, neural progenitor cells; *HDAC2 AND HDAC3*, histone deacetylase 2 and 3; *miR210*, microRNA 210; *H3K9Ac*, histone 3 acetylated in lysine 9; *H3K27Me*, histone 3 methylated in the lysine 27; *H3K4Me2*, histone 3 dimethylated in lysine 4; *H3(T3)*, histone 3 phosphorylated.

development of Alzheimer's or another disease with impaired cognitive function. Another study in humans showed that dietary supplementation with n-3 LC-PUFA for six months was associated with global DNA hypomethylation in peripheral blood leukocytes in Alzheimer's disease patients.[84]

N-3 fatty acids are known to have positive effects on brain neurotrophins.[85] It has been reported that DHA supplementation in aged mice improved cognitive dysfunction through increased BDNF levels.[86] Conversely, a series of animal studies have demonstrated reduced levels of nerve growth factor (NGF) under conditions of decreased DHA levels.[87] Reduced levels of BDNF and NGF in the brain have been implicated in the pathophysiology of Alzheimer's,[88] Parkinson's,[89] and Huntington's diseases.[90] Association of memory with changes in DNA methylation on BDNF gene has been described,[91] and it was also demonstrated during development of the forebrain in mice.[92] Reduced levels of BDNF induced by DHA deficiency have been also associated with DNA methylation.[68]

During normal aging, the effect of LC-PUFA on prevention of cognitive impairment also involves its contribution in the prevention of brain hypoperfusion. Cerebral perfusion can be compromised as result of cerebrovascular disease, hypertension, or diabetes mellitus type 2, for example. Neurotrophins like BDNF and NGF are also implicated in the regulation of angiogenic markers in the brain[93, 94] stimulating the vascular endothelial growth factor (VEGF) transcription in neuronal tissue.[95] Maternal dietary supplementation with n-3 fatty acids was also able to increase VEGF gene expression in the offspring brain,[64] which can help promote and coordinate angiogenesis during development and adulthood. In addition, VEGF plays other multiple roles in the developing and adult nervous system by acting as a neuroprotective agent.[96]

Neuroinflammation induced by microglia activation has been associated with a number of neurodegenerative diseases that occur during senescence. Diabetes and obesity, for example, are recognized risk factors for dementia, and chronic neuroinflammation contributes to this increased risk. A growing body of evidence has indicated the role of n-3 LC-PUFA in shifting reactive microglia to the antiinflammatory phenotype, and these effects have been associated to an inhibitory effect on the NF-κB signaling pathway, involved in cytokine gene activation.[97, 98] A recent study demonstrated that such effects, plus the induction of autophagy promoted by n-3 supplementation, involve the SIRT1 protein deacetylase activity as well as increased expression of SIRT1 mRNA levels.[99] In a cross-sectional cohort of Greek preadolescents, associations between n-3 PUFAs and differential DNA methylation in the NFkB pathway were reported.[100]

The importance of an adequate dietary supplementation of n-3 fatty acids in the senescence is due to the fact that the aging brain seems to lose DHA, as reported for rodent models.[101] The proposition that an adequate n-3 PUFA supply may favor the resistance of the brain to aging, and stress is also supported by studies demonstrating the involvement of DHA in mechanisms related to synaptic function and plasticity necessary for memory formation as recently reviewed.[80, 102]

All the aspects described above were reinforced by a recent analysis using transcriptome and epigenome sequencing of the hypothalamus and hippocampus.[103] According to this latter study, a significant reprogramming of DNA methylation, transcript abundance, alternative splicing, and gene networks governing cell metabolism, cell communication, inflammation, and neuronal signaling were induced by 15% fructose consumption in the drinking water. These signals that converged with causal risks of metabolic, neurological, and psychiatric disorders, described in humans, was remarkably reversed by the simultaneous treatment with a diet rich in DHA.[103]

Altogether, the emerging experimental and epidemiological research supports the idea of a potentially helpful therapeutic use of n-3 fatty acids as a nutraceutical that can target epigenetic changes induced by neurodegenerative diseases during aging.

Fig. 2 illustrates deleterious neural effects induced by reduced levels of DHA during the senescence and how these effects can be reversed by n-3 supplementation through epigenetic mechanisms.

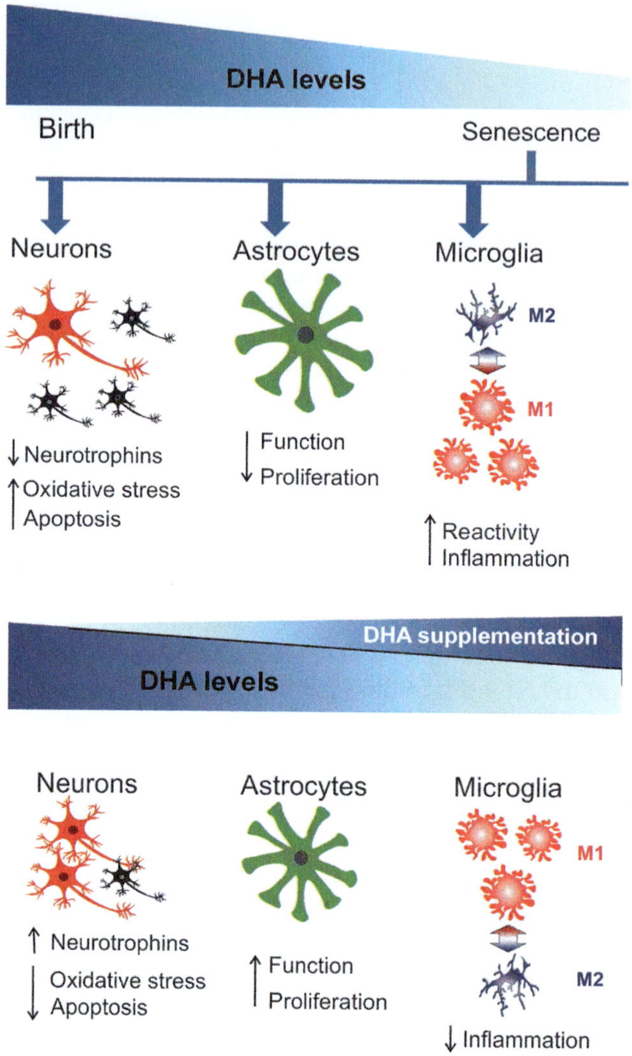

FIG. 2 Schematic drawing illustrating deleterious neural effects induced by reduced levels of DHA during the senescence and how these effects can be reversed by n-3 supplementation through epigenetic mechanisms. All data were obtained from the literature cited in this chapter. *M1*, microglia reactive phenotype; *M2*, microglia normal phenotype.

Acknowledgments

This study was financed in part by the Coordination for the Improvement of Higher Level or Education Personnel, Brazil (CAPES, Grant numbers: Finance Code 001 and PROCAD-2013), the Foundation for Science and Technology of the Estate of Pernambuco, Brazil (FACEPE; Grant number APQ 1037—2.07/15), the Foundation for Science and Technology of the Estate of Rio de Janeiro, Brazil (FAPERJ; Grant number E-26/210.878/2014), and the National Institute of Translational Neuroscience (INNT/CNPq Grant number 465346/2014-6).

References

1. LaSalle JM, Powell WT, Yasui DH. Epigenetic layers and players underlying neurodevelopment. *Trends Neurosci*. 2013;36(8):460–470.
2. Landgrave-Gomez J, Mercado-Gomez O, Guevara-Guzman R. Epigenetic mechanisms in neurological and neurodegenerative diseases. *Front Cell Neurosci*. 2015;9:58.
3. Yao B, Christian KM, He C, et al. Epigenetic mechanisms in neurogenesis. *Nat Rev Neurosci*. 2016; 17(9):537–549.
4. Lillycrop KA, Burdge GC. The effect of nutrition during early life on the epigenetic regulation of transcription and implications for human diseases. *J Nutrigenet Nutrigenom*. 2011;4(5):248–260.
5. Milagro FI, Mansego ML, De Miguel C, et al. Dietary factors, epigenetic modifications and obesity outcomes: progress and perspectives. *Mol Asp Med*. 2013;34(4):782–812.
6. Tammen SA, Friso S, Choi SW. Epigenetics: the link between nature and nurture. *Mol Asp Med*. 2013;34(4):753–764.
7. Goldberg M, Rummelt C, Laerm A, et al. Epigenetic silencing contributes to frequent loss of the fragile histidine triad tumor suppressor in basal cell carcinomas. *Br J Dermatol*. 2006;155(6):1154–1158.
8. Rodenhiser D, Mann M. Epigenetics and human disease: translating basic biology into clinical applications. *Can Med Assoc J*. 2006;174(3):341–348.
9. Juliandi B, Abematsu M, Nakashima K. Epigenetic regulation in neural stem cell differentiation. *Develop Growth Differ*. 2010;52(6):493–504.
10. Takizawa T, Nakashima K, Namihira M, et al. DNA methylation is a critical cell-intrinsic determinant of astrocyte differentiation in the fetal brain. *Dev Cell*. 2001;1(6):749–758.
11. Goto K, Numata M, Komura JI, et al. Expression of DNA methyltransferase gene in mature and immature neurons as well as proliferating cells in mice. *Differentiation*. 1994;56(1-2):39–44.
12. Feng J, Chang H, Li E, et al. Dynamic expression of de novo DNA methyltransferases Dnmt3a and Dnmt3b in the central nervous system. *J Neurosci Res*. 2005;79(6):734–746.
13. LaPlant Q, Vialou V, Covington 3rd HE, et al. Dnmt3a regulates emotional behavior and spine plasticity in the nucleus accumbens. *Nat Neurosci*. 2010;13(9):1137–1143.
14. Hellman A, Chess A. Gene body-specific methylation on the active X chromosome. *Science*. 2007;315(5815):1141–1143.
15. Schroeder DI, Lott P, Korf I, et al. Large-scale methylation domains mark a functional subset of neuronally expressed genes. *Genome Res*. 2011;21(10):1583–1591.
16. Schroeder JW, Conneely KN, Cubells JC, et al. Neonatal DNA methylation patterns associated with gestational age. *Epigenetics*. 2011;6(12):1498–1504.
17. Castillo J, Lopez-Rodas G, Franco L. Histone post-translational modifications and nucleosome organisation in transcriptional regulation: some open questions. *Adv Exp Med Biol*. 2017;966:65–92.
18. Chakravarthy S, Park YJ, Chodaparambil J, et al. Structure and dynamic properties of nucleosome core particles. *FEBS Lett*. 2005;579(4):895–898.
19. Strahl BD, Allis CD. The language of covalent histone modifications. *Nature*. 2000;403(6765):41–45.
20. Gavin DP, Grayson DR, Varghese SP, et al. Chromatin switches during neural cell differentiation and their dysregulation by prenatal alcohol exposure. *Genes*. 2017;8(5).
21. Bannister AJ, Kouzarides T. Regulation of chromatin by histone modifications. *Cell Res*. 2011;21(3):381–395.
22. Dulac C. Brain function, and chromatin plasticity. *Nature*. 2010;465(7299):728–735.
23. Hake SB, Xiao A, Allis CD. Linking the epigenetic 'language' of covalent histone modifications to cancer. *Br J Cancer*. 2004;90(4):761–769.
24. Pattaroni C, Jacob C. Histone methylation in the nervous system: functions and dysfunctions. *Mol Neurobiol*. 2013;47(2):740–756.
25. Staszewski O, Prinz M. Glial epigenetics in neuroinflammation and neurodegeneration. *Cell Tissue Res*. 2014;356(3):609–616.
26. Haberland M, Montgomery RL, Olson EN. The many roles of histone deacetylases in development and physiology: implications for disease and therapy. *Nat Rev Genet*. 2009;10(1):32–42.
27. Liu H, Hu Q, Kaufman A, et al. Developmental expression of histone deacetylase 11 in the murine brain. *J Neurosci Res*. 2008;86(3):537–543.

28. Marin-Husstege M, Muggironi M, Liu A, et al. Histone deacetylase activity is necessary for oligodendrocyte lineage progression. *J Neurosci*. 2002;22(23):10333–10345.
29. Jacob C, Lebrun-Julien F, Suter U. How histone deacetylases control myelination. *Mol Neurobiol*. 2011;44(3):303–312.
30. Ye Y, Li M, Gu L, et al. Chromatin remodeling during in vivo neural stem cells differentiating to neurons in early Drosophila embryos. *Cell Death Differ*. 2017;24(3):409–420.
31. Jepsen K, Solum D, Zhou T, et al. SMRT-mediated repression of an H3K27 demethylase in progression from neural stem cell to neuron. *Nature*. 2007;450(7168):415–419.
32. Park DH, Hong SJ, Salinas RD, et al. Activation of neuronal gene expression by the JMJD3 demethylase is required for postnatal and adult brain neurogenesis. *Cell Rep*. 2014;8(5):1290–1299.
33. Akbarian S, Huang HS. Epigenetic regulation in human brain-focus on histone lysine methylation. *Biol Psychiatry*. 2009;65(3):198–203.
34. Kim AH, Puram SV, Bilimoria PM, et al. A centrosomal Cdc20-APC pathway controls dendrite morphogenesis in postmitotic neurons. *Cell*. 2009;136(2):322–336.
35. Hsieh J, Nakashima K, Kuwabara T, et al. Histone deacetylase inhibition-mediated neuronal differentiation of multipotent adult neural progenitor cells. *Proc Natl Acad Sci U S A*. 2004;101(47):16659–16664.
36. Asano H, Aonuma M, Sanosaka T, et al. Astrocyte differentiation of neural precursor cells is enhanced by retinoic acid through a change in epigenetic modification. *Stem Cells*. 2009;27(11):2744–2752.
37. Majumder A, Dhara SK, Swetenburg R, et al. Inhibition of DNA methyltransferases and histone deacetylases induces astrocytic differentiation of neural progenitors. *Stem Cell Res*. 2013;11(1):574–586.
38. Kanski R, Sneeboer MA, van Bodegraven EJ, et al. Histone acetylation in astrocytes suppresses GFAP and stimulates a reorganization of the intermediate filament network. *J Cell Sci*. 2014;127(Pt 20):4368–4380.
39. Meza-Sosa KF, Valle-Garcia D, Pedraza-Alva G, et al. Role of microRNAs in central nervous system development and pathology. *J Neurosci Res*. 2012;90(1):1–12.
40. Le MT, Xie H, Zhou B, et al. MicroRNA-125b promotes neuronal differentiation in human cells by repressing multiple targets. *Mol Cell Biol*. 2009;29(19):5290–5305.
41. Lau PW, MacRae IJ. The molecular machines that mediate microRNA maturation. *J Cell Mol Med*. 2009;13(1):54–60.
42. Smirnova L, Grafe A, Seiler A, et al. Regulation of miRNA expression during neural cell specification. *Eur J Neurosci*. 2005;21(6):1469–1477.
43. Cheng LC, Pastrana E, Tavazoie M, et al. miR-124 regulates adult neurogenesis in the subventricular zone stem cell niche. *Nat Neurosci*. 2009;12(4):399–408.
44. Zhao X, He X, Han X, et al. MicroRNA-mediated control of oligodendrocyte differentiation. *Neuron*. 2010;65(5):612–626.
45. Makeyev EV, Zhang J, Carrasco MA, et al. The microRNA miR-124 promotes neuronal differentiation by triggering brain-specific alternative pre-mRNA splicing. *Mol Cell*. 2007;27(3):435–448.
46. Davis TH, Cuellar TL, Koch SM, et al. Conditional loss of Dicer disrupts cellular and tissue morphogenesis in the cortex and hippocampus. *J Neurosci*. 2008;28(17):4322–4330.
47. Keverne EB, Curley JP. Epigenetics, brain evolution and behaviour. *Front Neuroendocrinol*. 2008;29(3):398–412.
48. Lillycrop KA, Burdge GC. Epigenetic mechanisms linking early nutrition to long-term health. *Best Pract Res Clin Endocrinol Metab*. 2012;26(5):667–676.
49. Cortessis VK, Thomas DC, Levine AJ, et al. Environmental epigenetics: prospects for studying epigenetic mediation of exposure-response relationships. *Hum Genet*. 2012;131(10):1565–1589.
50. Jimenez-Chillaron JC, Diaz R, Martinez D, et al. The role of nutrition on epigenetic modifications and their implications on health. *Biochimie*. 2012;94(11):2242–2263.
51. Haggarty P. Epigenetic consequences of a changing human diet. *Proc Nutr Soc*. 2013;72(4):363–371.
52. van Otterdijk SD, Michels KB. Transgenerational epigenetic inheritance in mammals: how good is the evidence? *FASEB J*. 2016;30(7):2457–2465.
53. Ji Y, Wu Z, Dai Z, et al. Nutritional epigenetics with a focus on amino acids: implications for the development and treatment of metabolic syndrome. *J Nutr Biochem*. 2016;27:1–8.
54. Langie SA, Achterfeldt S, Gorniak JP, et al. Maternal folate depletion and high-fat feeding from weaning affects DNA methylation and DNA repair in brain of adult offspring. *FASEB J*. 2013;27(8):3323–3334.
55. Weng X, Zhou D, Liu F, et al. DNA methylation profiling in the thalamus and hippocampus of postnatal malnourished mice, including effects related to long-term potentiation. *BMC Neurosci*. 2014;15:31.

56. Freitas HR, Isaac AR, Malcher-Lopes R, et al. Polyunsaturated fatty acids and endocannabinoids in health and disease. *Nutr Neurosci*. 2017;1–20.
57. Novak EM, Innis SM. Impact of maternal dietary n-3 and n-6 fatty acids on milk medium-chain fatty acids and the implications for neonatal liver metabolism. *Am J Physiol Endocrinol Metab*. 2011;301(5):E807–E817.
58. Innis SM. Metabolic programming of long-term outcomes due to fatty acid nutrition in early life. *Matern Child Nutr*. 2011;7(Suppl 2):112–123.
59. Mennitti LV, Oliveira JL, Morais CA, et al. Type of fatty acids in maternal diets during pregnancy and/or lactation and metabolic consequences of the offspring. *J Nutr Biochem*. 2015;26(2):99–111.
60. Chakraborty N, Muhie S, Kumar R, et al. Contributions of polyunsaturated fatty acids (PUFA) on cerebral neurobiology: an integrated omics approach with epigenomic focus. *J Nutr Biochem*. 2017;42:84–94.
61. Novak EM, Keller BO, Innis SM. Metabolic development in the liver and the implications of the n-3 fatty acid supply. *Am J Physiol Gastrointest Liver Physiol*. 2012;302(2):G250–G259.
62. He F, Lupu DS, Niculescu MD. Perinatal alpha-linolenic acid availability alters the expression of genes related to memory and to epigenetic machinery and the Mecp2 DNA methylation in the whole brain of mouse offspring. *Int J Dev Neurosci*. 2014;36:38–44.
63. Sakayori N, Kikkawa T, Tokuda H, et al. Maternal dietary imbalance between omega-6 and omega-3 polyunsaturated fatty acids impairs neocortical development via epoxy metabolites. *Stem Cells*. 2016;34(2):470–482.
64. Goustard-Langelier B, Koch M, Lavialle M, et al. Rat neural stem cell proliferation and differentiation are durably altered by the in utero polyunsaturated fatty acid supply. *J Nutr Biochem*. 2013;24(1):380–387.
65. Champeil-Potokar G, Chaumontet C, Guesnet P, et al. Docosahexaenoic acid (22,6n-3) enrichment of membrane phospholipids increases gap junction coupling capacity in cultured astrocytes. *Eur J Neurosci*. 2006;24(11):3084–3090.
66. de la Rocha C, Perez-Mojica JE, Leon SZ, et al. Associations between whole peripheral blood fatty acids and DNA methylation in humans. *Sci Rep*. 2016;6:25867.
67. Maekawa M, Watanabe A, Iwayama Y, et al. Polyunsaturated fatty acid deficiency during neurodevelopment in mice models the prodromal state of schizophrenia through epigenetic changes in nuclear receptor genes. *Transl Psychiatry*. 2017;7(9):e1229.
68. Fan C, Fu H, Dong H, et al. Maternal n-3 polyunsaturated fatty acid deprivation during pregnancy and lactation affects neurogenesis and apoptosis in adult offspring: associated with DNA methylation of brain-derived neurotrophic factor transcripts. *Nutr Res (New York, NY)*. 2016;36(9):1013–1021.
69. Tyagi E, Zhuang Y, Agrawal R, et al. Interactive actions of Bdnf methylation and cell metabolism for building neural resilience under the influence of diet. *Neurobiol Dis*. 2015;73:307–318.
70. Lee HS, Barraza-Villarreal A, Biessy C, et al. Dietary supplementation with polyunsaturated fatty acid during pregnancy modulates DNA methylation at IGF2/H19 imprinted genes and growth of infants. *Physiol Genomics*. 2014;46(23):851–857.
71. Sable P, Randhir K, Kale A, et al. Maternal micronutrients and brain global methylation patterns in the offspring. *Nutr Neurosci*. 2015;18(1):30–36.
72. Tripathi S, Kushwaha R, Mishra J, et al. Docosahexaenoic acid up-regulates both PI3K/AKT-dependent FABP7-PPARgamma interaction and MKP3 that enhance GFAP in developing rat brain astrocytes. *J Neurochem*. 2017;140(1):96–113.
73. Joardar A, Sen AK, Das S. Docosahexaenoic acid facilitates cell maturation and beta-adrenergic transmission in astrocytes. *J Lipid Res*. 2006;47(3):571–581.
74. Joardar A, Das S. Effect of fatty acids isolated from edible oils like mustard, linseed or coconut on astrocytes maturation. *Cell Mol Neurobiol*. 2007;27(8):973–983.
75. Isaac AR, da Silva EAN, de Matos RJB, et al. Low omega-6/omega-3 ratio in a maternal protein-deficient diet promotes histone-3 changes in progeny neural cells and favors leukemia inhibitory factor gene transcription. *J Nutr Biochem*. 2018;55:229–242.
76. Ye Y, Gu L, Chen X, et al. Chromatin remodeling during the in vivo glial differentiation in early Drosophila embryos. *Sci Rep*. 2016;6:33422.
77. Huidobro C, Fernandez AF, Fraga MF. Aging epigenetics: causes and consequences. *Mol Asp Med*. 2013;34(4):765–781.
78. Gabbianelli R, Damiani E. Epigenetics and neurodegeneration: role of early-life nutrition. *J Nutr Biochem*. 2018;57:1–13.
79. Penner MR, Roth TL, Barnes CA, et al. An epigenetic hypothesis of aging-related cognitive dysfunction. *Front Aging Neurosci*. 2010;2:9.
80. Freitas HR, Ferreira GDC, Trevenzoli IH, et al. Fatty acids, antioxidants and physical activity in brain aging. *Nutrients*. 2017;9(11).

81. Nadjar A, Leyrolle Q, Joffre C, et al. Bioactive lipids as a new class of microglial modulators: when nutrition meets neuroimmunology. *Prog Neuro-Psychopharmacol Biol Psychiatry*. 2017;79(Pt A):19–26.
82. Ma Y, Follis JL, Smith CE, et al. Interaction of methylation-related genetic variants with circulating fatty acids on plasma lipids: a meta-analysis of 7 studies and methylation analysis of 3 studies in the Cohorts for Heart and Aging Research in Genomic Epidemiology consortium. *Am J Clin Nutr*. 2016;103(2):567–578.
83. Hammamieh R, Chakraborty N, Gautam A, et al. Transcriptomic analysis of the effects of a fish oil enriched diet on murine brains. *PLoS ONE*. 2014;9(3):e90425.
84. Karimi M, Vedin I, Freund Levi Y, et al. DHA-rich n-3 fatty acid supplementation decreases DNA methylation in blood leukocytes: the OmegAD study. *Am J Clin Nutr*. 2017;106(4):1157–1165.
85. Cai J, Hua F, Yuan L, et al. Potential therapeutic effects of neurotrophins for acute and chronic neurological diseases. *Biomed Res Int*. 2014;2014:601084.
86. Jiang LH, Shi Y, Wang LS, et al. The influence of orally administered docosahexaenoic acid on cognitive ability in aged mice. *J Nutr Biochem*. 2009;20(9):735–741.
87. Bhatia HS, Agrawal R, Sharma S, et al. Omega-3 fatty acid deficiency during brain maturation reduces neuronal and behavioral plasticity in adulthood. *PLoS ONE*. 2011;6(12):e28451.
88. Laske C, Stransky E, Leyhe T, et al. Decreased brain-derived neurotrophic factor (BDNF)- and beta-thromboglobulin (beta-TG)- blood levels in Alzheimer's disease. *Thromb Haemost*. 2006;96(1):102–103.
89. Scalzo P, Kummer A, Bretas TL, et al. Serum levels of brain-derived neurotrophic factor correlate with motor impairment in Parkinson's disease. *J Neurol*. 2010;257(4):540–545.
90. Baydyuk M, Xu B. BDNF signaling and survival of striatal neurons. *Front Cell Neurosci*. 2014;8:254.
91. Lester BM, Tronick E, Nestler E, et al. Behavioral epigenetics. *Ann N Y Acad Sci*. 2011;1226:14–33.
92. Dennis KE, Levitt P. Regional expression of brain-derived neurotrophic factor (BDNF) is correlated with dynamic patterns of promoter methylation in the developing mouse forebrain. *Brain Res Mol Brain Res*. 2005;140(1-2):1–9.
93. Kermani P, Hempstead B. Brain-derived neurotrophic factor: a newly described mediator of angiogenesis. *Trends Cardiovasc Med*. 2007;17(4):140–143.
94. Nico B, Mangieri D, Benagiano V, et al. Nerve growth factor as an angiogenic factor. *Microvasc Res*. 2008;75(2):135–141.
95. Calza L, Giardino L, Giuliani A, et al. Nerve growth factor control of neuronal expression of angiogenetic and vasoactive factors. *Proc Natl Acad Sci U S A*. 2001;98(7):4160–4165.
96. Mackenzie F, Ruhrberg C. Diverse roles for VEGF-A in the nervous system. *Development*. 2012;139(8):1371–1380.
97. Yang B, Li R, Michael Greenlief C, et al. Unveiling anti-oxidative and anti-inflammatory effects of docosahexaenoic acid and its lipid peroxidation product on lipopolysaccharide-stimulated BV-2 microglial cells. *J Neuroinflammation*. 2018;15(1):202.
98. Chen X, Chen C, Fan S, et al. Omega-3 polyunsaturated fatty acid attenuates the inflammatory response by modulating microglia polarization through SIRT1-mediated deacetylation of the HMGB1/NF-kappaB pathway following experimental traumatic brain injury. *J Neuroinflammation*. 2018;15(1):116.
99. Inoue T, Tanaka M, Masuda S, et al. Omega-3 polyunsaturated fatty acids suppress the inflammatory responses of lipopolysaccharide-stimulated mouse microglia by activating SIRT1 pathways. *Biochim Biophys Acta*. 2017;1862(5):552–560.
100. Voisin S, Almen MS, Moschonis G, et al. Dietary fat quality impacts genome-wide DNA methylation patterns in a cross-sectional study of Greek preadolescents. *Eur J Human Genet*. 2015;23(5):654–662.
101. Latour A, Grintal B, Champeil-Potokar G, et al. Omega-3 fatty acids deficiency aggravates glutamatergic synapse and astroglial aging in the rat hippocampal CA1. *Aging Cell*. 2013;12(1):76–84.
102. Denis I, Potier B, Vancassel S, et al. Omega-3 fatty acids and brain resistance to aging and stress: body of evidence and possible mechanisms. *Aging Res Rev*. 2013;12(2):579–594.
103. Meng Q, Ying Z, Noble E, et al. Systems nutrigenomics reveals brain gene networks linking metabolic and brain disorders. *EBioMedicine*. 2016;7:157–166.

Further Reading

104. Sadli N, Ackland ML, De Mel D, et al. Effects of zinc and DHA on the epigenetic regulation of human neuronal cells. *Cell Physiol Biochem*. 2012;29(1-2):87–98.
105. Suphioglu C, Sadli N, Coonan D, et al. Zinc and DHA have opposing effects on the expression levels of histones H3 and H4 in human neuronal cells. *Br J Nutr*. 2010;103(3):344–351.

Index

Note: Page numbers followed by *f* indicate figures and *t* indicate tables.

CPI Antony Rowe
Eastbourne, UK
November 20, 2019

PREFACE

It is best if the author could film the breathtaking moments, and record the turning points. The period of 2008-2009 is an uncommon period, during which a financial crisis not encountered for almost half a century broke out suddenly. What were the Chinese thinking about during this period? What have the authors done? And with what effects? This book aims to record all these things for us to remember forever.

However, what impressed us the most in 2009 is not disaster but cooperation. Statesmen all around the world are seeking cooperation, including the willingness to cooperate indicated as well as various achievements made by the leaders of China and the United States. The authors have witnessed unreserved appreciation from the European countries and the United States, if not the whole world, when the Chinese government took major measures in response to the crisis, which has rarely been seen for years.

This beautifully printed book in front of you is also a product of cooperation. The New York NOVA Press in the U.S. and the Encyclopedia of China Publishing House have, within the shortest time possible, reached a cooperation agreement with us to publish this book both in Chinese and English in China and the U.S. respectively. The act itself is full of the spirit of trust and cooperation, which has imbued my memory with pleasure and appreciation.

The writing process per se is even more of a cooperative process. Young professors happily cooperate with even younger graduate students; the scholars and students from Beijing Normal University, Southwestern University of Finance and Economics, and Harvard University and MIT efficiently cooperate with each other; doctors who have earned their degrees domestically and overseas cooperate complementarily; scholars from various specialized fields in and out of school also cooperate with exchange of ideas.

Why do people all agree to cooperate? Is it because of the need to unite to weather through the storm of financial crisis, or is it that the growing power of China attracts it more partners, or that greater complication of the issues the author are facing requires joint exploration? All in all, our consensus can be summarized as: only cooperation can bear more fruits, bring in more opportunities, and create a brighter future.

The authors still have a long way to go on the road of cooperation. *The Development of China's Economy under International Financial Crisis* is just the first volume of the series of *The Research Report on China's Economy and Resource Management.* The authors are looking forward to future cooperation with the publishing houses and the authors, as well as to meeting the demand of readers both at home and abroad, in order to bring on new volumes.

Let us bless the world an everlasting favor from cooperation.

Li Xiaoxi

Translator: Zhang Li(SUFE)
English Version Editors: Xiao Rong (SUFE), Chen Jinzhu (Harvard)

COLUMNS

GENERAL REMARKS: THE CHINA MODEL AND GREEN NEW DEAL

Li Xiaoxi *

"China's Economic and Resource Management Research Report 2010" launches the discussion on the theme of "China's economic development under the international financial crisis". As the title suggests, it is necessary for us not only to be aware of the international financial crisis, but also to expand analysis and exposition on China's current developing situation affected by the crisis. Undoubtedly the emphasis should be placed on "China's development", which includes many elements to be shared with readers, such as the background for Chinese government's introduction of "four trillion" economic stimulus package and its content and progress. After finishing some chapters, readers may inquire about what are the most obvious features of China's economy reflected in these issues. Or equally, about what are the most heated topics that are concerned by both China and the globe in China's current development context. Our answer is: the China Model and Green New Deal.

I. THE FINANCIAL CRISIS INTENSIFIES DISCUSSION ON "THE CHINA MODEL"

Within 30 years after its economic reform and opening up, China has witnessed its GDP per capita soaring from 381 Yuan in 1978 to 22640 Yuan in 2008, its GDP per capita growing nearly 60 times.[1] In 2007, former World Bank President Paul Wolfowitz said that about 67% of the world's achievements in poverty reduction should be attributed to China during the past

---

* Li Xiaoxi is Professor of economics as well as the Dean of Institute of Economics and Resource Management at Beijing Normal University, who earned his PhD degree at Graduate School of the Chinese Academy of Social Sciences. Email: lxx62205058@263.net

[1] 1978 data from *China Statistical Yearbook 2008*, detailed in http://www.stats.gov.cn/tjsj/ndsj/2008/indexch.htm; 2008 data from *National Economic and Social Development Statistical Communiqué 2008*, detailed in http://www.stats.gov.cn/tjgb/ndtjgb/qgndtjgb/t20090226_402540710.htm.

25 years.[2] As the world gets impressed by China's rapid economic development, appraises from scholars and other professionals all over the world have been given on "the China Model" that leads China to the world. In May 2004, Joshua Cooper Ramo, senior editor of "Time" magazine and senior consultant of Goldman Sachs, published a report entitled "the Beijing Consensus"[3] at UK Foreign Policy Center in London. The study points out that the Chinese have worked out a development model suited to their national conditions through hard work, innovation and bold initiative and independent practice. He addresses this model as "Beijing Consensus". The report says that the Chinese are moving in the direction of building a powerful country. Instead of a simple imitation of Western models, this effort is based on their own model, which has been studied by many other countries and regions as a role model. In its report entitled "New World Order, Starting from the London Summit", Du Ping, commentator of "Lianhe Zaobao" of Singapore, pointed out that "we have many different development models in the G20 groups. A particular one of them is the "China Model" that has already been going on for 30 years. As we ought to see, were there no China Model, the current global economy would appear much more spiritless, its prospects for recovery short of a hope and driving force."[4] A number of well-known scholars are also interested. Joseph Stiglitz, a professor in Columbia University, a famous Nobel Prize laureate in economics, published a monograph *China's New Economic Model.*[5] John Naisbitt, who enjoys a good reputation around the world for his once published book *Megatrends* predicating American society's moving direction, has published his new book "China Megatrends ".[6] Lin Yifu, World Bank Chief Economist and Senior Vice President, discusses the China Model from the perspective of the experience of China's development and transformation in his book entitled "China's Miracle: Development Strategy and Economic Reform" published in 1994.[7] There are many more papers and monographs in this area.[8] Now, as for European and American developed countries as well as Africa and Latin America countries; government officials, experts and scholars as well as corporate executives and business leaders; political scientists, economists as well as sociologists and historians, they frequently use the term "the China Model" when talking about China. As the Chinese people are celebrating the 30th Anniversary of reform and opening up as well as the 60th anniversary

---

[2] China Central Television: (How foreigners comment two meetings) interview with former World Bank president, "two observers: 67% of the world's achievements in poverty reduction owned by China", March 13th, 2007. Detailed in http://news.cctv.com/china/20070313/106619.shtml

[3] Joshua Cooper Ramo: The Beijing Consensus, detailed in http://fpc.org.uk/fsblob/244.pdf.

[4] Du Ping, April 30th,2009. *New World Order, Starting from the London Summit. Lian He Zao Bao.* Detailed in http://www.zaobao.com/special/forum/pages7/forum_zp090403.shtml

[5] Joseph E. Stiglitz. *China's New Economic Model.* Detailed in http://www.project-syndicate.org/ commentary/ stiglitz86

[6] Xxxx: Leading futurist John Naisbitt launches *China's Megatrends* in Beijing, November 9th, 2009. Detailed in http://chinaglobalspeakers.com/?p=2582

[7] Zhuang Juanju. *Review of Research on "the China Model".* www.People.Com.cn. Detailed in http:// theory. people.com.cn/GB/49154/49155/8992114.html

[8] Such as Wu Jinglian, Lin Yifu, Hu Angang, et al., research on the path of China's market economic reform from the economic perspective; Zheng Bijian, Yu Keping, Lin Shangli, et al., research on globalization and China's development model from the political perspective; Huang Ping, Li Peilin, Sun Liping, et al., study social transformation and China's experience from the sociological perspective; Luo Rongqu, Zhang Kaiyuan, Qian Chengdan, et al., conduct international comparative studies on the development path of China's modernization from the historical perspective. All of them have had a major impact.

of the founding of New China, the enthusiasm in discussion "the China Model" is being elevated, and the appraisal is being more affirmed.

Interestingly, at the same time, the international financial crisis intensified the discussion of the China Model. Scholars re-understand and judge "the China Model" from both positive and negative aspects. Considering the numerous different arguments, I summarize them as follows: For the positive evaluation, whether it is because China's more stringent financial supervision that prevented the international financial gamble's massive penetration, or it is the result of the style of Chinese government in concentrating power to hold the most important matters, China has been clearly immune from the adverse external effects, and the Chinese model has demonstrated high efficiency and strength in the face of international financial crisis. For the negative evaluation, the present financial crisis originates from the global economic imbalances, particularly the U.S. economy of high consumption and unregulated financial derivatives, coupled with China's high savings, high investment and high exports. Therefore, from the international economic perspective, the China Model is not sustainable. It is in this context that debate on "the China Model" is inspired among academicians and other professionals.

Facing the international financial crisis, the Chinese government takes powerful actions, which gains high affirmation from the international society. In particular, the announcement of Chinese government's "four trillion" economic stimulus plan has captured the world's attention and surprise. "Washington Post" published an article considering it as "A New Deal with Chinese characteristics".[9] "Economist" called this stimulus package "A New Deal, Chinese-style".[10] What role has this plan played? Will it trigger a new problem? Is it a typical case on "the China Model"? We will discuss these issues in the first chapter of this book. Similarly, the analysis of Chinese government's newly published "Top Ten Industries Stimulus Plan" in the fourth chapter is based on the same goal. Facing the international financial crisis which originates from the sub-prime loan crisis, how does China perform in the real estate industry which is tightly integrated with the financial development? We will have a thorough discussion on this topic in Chapter Five. How did the international financial crisis affect Hong Kong and Macau, the two most important special economic zones in China? Did the Chinese government implement corresponding support measures? These interesting issues will be discussed in the last chapter.

"The China Model" comes as a generalization among theorists. Then do the public approve the formulation of "the China Model"? A thematic investigation "How do you think the China Model?" was initiated by the People's Forum Magazine collaborated with People's Daily Online Magazine and the People's Forum Network, etc. A total of 4970 participated in the survey. In addition, the journalists from the People's Forum also interviewed 192 from the public on a random basis, summing it up to a total of 5162 participants. Results show that 75% of those surveyed believe that there exists "the China Model". 56% of those surveyed felt that "having become the fastest growing economy in the world" is the greatest

[9] Ariana Eunjung Cha, Maureen Fan: China Unveils $586 Billion Stimulus Plan. "The Washington Post" November 9,2008, detailed in http://www.washingtonpost.com/wp-dyn/content/article/2008/11/09/AR 2008 110900701. html

[10] Economist: Reflating the Dragon: Can the world's fastest-growing economy avoid a sharp downturn? "Economist" November 9, 2008, detailed in http://www.economist.com/ world/asia/PrinterFriendly.cfm? story _id=12606998

achievement of "the China Model". The top three answers to the main features of "the China Model" are: a strong government's dominance (2918 votes, 57% of the total), development strategy based on gradual reform (2424 votes, 48% of the total), domestic reform and opening-up at the same time (2276 votes, representing 45%). Except for its totally different emotion coloring, such perception coincides with the comment of Rowan Callick from the American Enterprise Institute, who summarizes "the China Model" as "economic freedom plus political repression."[11] It represents the mainstream views on "the China Model" from some western countries. The survey also shows that "the China Model" in the public's eyes mainly refers to "the market liberalization with Chinese characteristics" (3172 votes, 64%), with none of other replies surpassing 40% of those surveyed. The survey also shows why people don't think there is "the China Model". On one hand, they don't think China's development pattern can form a model. On the other hand, "the China Model" formulation has not been acknowledged worldwide.[12]

I understand that all the different summaries of "the China Model" have their own reasons, to which I have deep respect. However, the excessive diversity has easily resulted in the vagueness of the conception, which also makes "the China Model" a topic difficult to communicate. For example, is "the China Model" equivalent to "the Chinese pattern" including political reform? Is it similar to "the East Asian Model"? Is it the summary of China's 60 years of development experience? Should "the China Model" be an "economic growth model" or "social development model" or "the pattern of high investment, high exports and low consumption" ("two-high-and-one-low" Mode) summed up by some scholars? It is the summary of "two-high-and-one-low" that has intensified the debate on "the China Model" in the context of international financial crisis. Despite of the possible recommendations and conclusions it may render for future development, I believe the summary is hasty generalization and unacceptable. Clearly, the total denial of "the China Model" which turns China from being economically backward to a country that can affect the world economy is definitely not able to convince the Chinese people or even most of the world. A total acceptance is also considered to be troublesome, because there do exist serious problems including excessive consumption of resources and severe environmental degradation in China's rapid economic growth, which is emergent to be dealt with. Thus, while affirming "the China Model", I hope it becomes "greener" instead of being "gray" or even "black".

Combining the public's understanding, the scholars' various viewpoints, and my own thinking, I wonder if I can stress: the China Model mainly refers to China's economic development model since its reform and opening-up, the core of which is the socialist market economy. China's socialist market economy not only surmounts the capitalism, but also develops the good and discards the bad of the traditional socialism; it not only implements the government's macro-economic control, but also gives full play to the vitality of the market. As a result, the "invisible hand" and the "visible hand" functions properly and

---

[11] Rowan Callick: *The China Model*, The America, November/December 2007 Issue, by American Enterprise Institute. Detailed in http://www.american.com/archive/2007/november-december-magazine-contents/the-china-model/?searchterm=China Model

[12] "74.55% of the public approve 'the China Model". www.People.com.cn. January 13, 2009. Detailed in http://politics.people.com.cn/GB/143465/143480/8669715.html

simultaneously. Where do Chinese characteristics lie in? It lies exactly in combining "socialism" with "market economy". "The China Model" is a socio-economic development model which actively participates in economic globalization and development. She adheres to socialism, and emphasizes the national characteristics, promoting the harmony and mutual benefits among different social systems. The China Model is an unprecedented great creation by Chinese people. China has enriched and developed the development models of the world. At this point, I think the interpretations given by most foreign scholars and Chinese, although starting from a personal experience, has captured the essential characteristics of "the China Model."

II. "GREEN NEW DEAL" AGAINST THE FINANCIAL CRISIS

While people discussing "the China Model" heatedly, another term, "Green New Deal", has entered people's vision.

Let me briefly review: in 1971, the Canadian engineer Mctaggant initiated the "Green Peace Organization". In the summer of 1984, "Green Committees of Correspondence (GCOC)" was set up in Virginia, the United States, aiming to set up local "green groups".[13] "Green economy" was first proposed by the British economist David Pearce in his book "Blueprint for A Green Economy" published in 1989. Green economy advocates: the social and ecological conditions should be considered in the economic development. We should make them "bearable", i.e., bearable to the natural environment and mankind themselves. It will never cause social imbalance and ecological crisis for the blind pursuit of growth and production, and maintains a sustainable development without exhausting the natural resource.[14]

With the social development, "the green concept" has gradually become the mainstream thinking in the field of public policy,[15] and gradually becomes internationalized and globalized. The United Nations Environment Program (UNEP) proposed the "Global Green New Deal" in October 2008, suggesting the 21st century a historic opportunity for focusing on environmental protection and investment model, which will ensure economic prosperity and create job opportunities;[16] in April 2009, the agency published "Global Green New Deal Policy Summary" (A Global Green New Deal),[17] and described the six elements of "Global

[13] Green Committees of Correspondence—Green Politics
http://greenpolitics.wikia.com/wiki/Green_Committees_of_Correspondence

[14] UK Conservatives' *Blueprint for a Green Economy*
http://www.worldchanging.com/archives/007254.html

[15] Source: A Green New Deal for Asia: How Asia-Pacific countries are expanding the green sector in their economies. Detailed in
http://www.spireresearch.com/pdf/archive/ejournaljun09/A%20Green%20New%20Deal%20for%20the%20emerging%20world.pdf

[16] Source: UNEP: Global Green New Deal-Environmentally-focused Investment Historic Opportunity for 21st Century Prosperity and Job Generation. Detailed in
http://www.unep.org/Documents.Multilingual/Default.asp?DocumentID=548&ArticleID=5957&l=en

[17] Data source: UNEP: A Global Green New Deal, detailed in
http://www.unep.org/greeneconomy/docs/ggnd_Final%20Report.pdf

Green New Deal": (1) clean energy and clean technologies including recycling; (2) rural energy, including renewable and sustainable biomass; (3) sustainable agriculture , including organic agriculture; (4) ecosystem infrastructure; (5) reduced emissions from deforestation and forest degradation; (6) sustainable cities including planning, transportation and green building. The report calls for state leaders to implement the Green New Deal, which means to put within two years 1% of gross national product or equivalently about 750 billion U.S. dollars into the development of five key areas, including wind, solar, geothermal, biomass and other renewable energy sources within two years. The report notes that in the case of enhancing the energy efficiency of both old and new buildings, the use of current energy-saving technologies can reduce the present buildings' energy consumption by 80%. The related investments in the field will not only stimulate revitalization in construction industry and other related industries, but also create a large number of green employment opportunities. Within only Europe and the US, it is likely to create 2 to 3.5 million job opportunities. The potential of developing countries in this regard would be even greater. Till 2030, investment of 630 billion U.S. dollars in renewable energy will add at least 20 million job opportunities.[18] In June 2009, UN Secretary General Ban Ki-moon attended the World Environment Day activities on the theme of "Your planet needs you– Unite to address climate change" in Mexico. In his speech, Ban said that the world needs a "Green New Deal", focusing on investment in renewable energy, constructing eco-friendly infrastructure, improving energy efficiency, and investing in the green economy as part of the huge economic stimulus plan. As a result, we would be able to put today's crisis into tomorrow's sustainable growth, enable countries transiting to a low-carbon society to get more generous returns, and share new technologies with other countries in a superior position.[19] On the theme of "Development and Climate Change", the World Bank "World Development Report 2010"[20] suggests building "climate-smart world." Developing countries could promote development and poverty reduction by carbon-reduction model, which requires all countries work together to promote global sustainable economic development.

The wave of "Green New Deal" heats up in the context of global economic crisis. Countries are making increased efforts to promote green economic development. On one hand they take this opportunity to get rid of an economic recession. On the other hand they seek new development opportunities. Some experts have pointed out that if the climate, economy and financial system were looked as a whole, the global economy would face a "triple crunch":[21] a combination of a credit-fuelled financial crisis, an accelerating climate change and the soaring energy, particularly oil prices. It is increasingly clear that these three

---

[18] UNEP: "Global Green New Deal"-Environmentally-focused Investment Historic Opportunity for 21st Century Prosperity and Job Generation. Detailed in http://www.unep.org/Documents.Multilingual/Default.asp?DocumentID=548&ArticleID=5957&l=en

[19] UN's Report: Your Planet needs You—Unite to combat climate change. Detailed in http://www.unep.org/wed/2009

[20] Taken from the World Bank: "World Development Report 2010", November 2009. Detailed in http://econ.worldbank.org/WBSITE/EXTERNAL/EXTDEC/EXTRESEARCH/EXTWDRS/EXTWDR2010/0,,menuPK:5287748~pagePK:64167702~piPK:64167676~theSitePK:5287741,00.html

[21] The Green New Deal Group: UK needs 'Green New Deal' to tackle 'triple crunch' of credit, oil price and climate crises http://www.neweconomics.org/gen/greennewdealneededforuk210708.aspx

overlapping events threaten to develop into a perfect storm, which has not been seen since the Great Depression, with potentially devastating consequences.

"Green New Deal" launches rapidly in each country. First, looking at the United States, Obama's "Green New Deal" constitutes of promoting energy efficiency, developing new energy sources, and combating climate change and so on, among which the development of new energy is the core of "Green New Deal". In February 2009, President Obama signed the economic stimulus package with developing new energy sources as an essential component in Denver, the value of the package amounting to 787 billion U.S. dollars.[22] In April 2009, President Obama said in a speech that the United States must carry out a comprehensive reform, of which one important aspect was to establish a new economic growth driving force, i.e., Green Economy.[23] Since then, the U.S. government has introduced several policies: increasing investment in new energy sources; formulating strict vehicle exhaust emission standards; introduction of "The American Clean Energy and Security Act". Use the carbon emissions in 2005 as the base, "Clean Energy and Security Act"[24] proposed to reduce emissions by 17% in 2020, and 83% in 2050. In addition, a carbon trading market to promote alternative energy development will also be established. As a result, Obama have been addressed as "the first green US president", and the "New Deal" he signed is also regarded as "Green New Deal".[25][26]

Other countries have also launched "Green New Deal". In the UK, "Green New Deal" plays an important role in promoting employment, alternative energy development, sustainable developing transportation system and energy-saving efforts. It requires economic development to be transited to a low-carbon economy at the same time.[27] In New Zealand, using "a warm home and a cool planet" as the slogan, a stimulus plan named "Green New Deal" was introduced, providing a total of 3.3 billion U.S dollars in the next three years, maintaining balanced development and transformation between urban and rural areas.[28] Japan has the world's leading technology in solar power, low fuel consumption cars, electric cars and other aspects. However, the current problems Japan facing is the possibility of setting up industry with international competitiveness for next generation and making a great contribution to world's achievement in solving environmental and energy problems through the realization of key technology's practical application in low- carbon-emission society.[29]

---

22 www.news.cn Washington February 18, 2009: Obama signed 787 billion U.S. dollar economic stimulus package. Detailed in http://news.xinhuanet.com/photo/2009-02/18/content_10838401.htm

23 Gerald F. Seib, April 17, 2009. Obama's "House Upon a Rock". The Wall Street Journal. Detailed in http://online.wsj.com/article/SB123991149634726345.html

24 The American " Clean Energy and Security Act" Detailed in http://www.pewclimate.org/acesa

25 Anna Fifield, United States: Is Obama the first green US president? Detailed in http://www.ft.com/cms/s/0/81e7ed9a-a3e9-11de-9fed-00144feabdc0,dwp_uuid=c3def9ac-a3ec-11de-9fed-00144feabdc0.html?nclick_check=1

26 John M. Broder, February 17, 2009. Obama's Greenhouse Gas Gamble. New York Times. Detailed in http://www.nytimes.com/2009/02/28/science/earth/28capntrade.html?ref=business

27 Green New Deal Group: Put People First: Demand a Green New Deal. Detailed in http://www.greennewdealgroup.org/?p=125

28 A Green New Deal for New Zealand: A Warm Home and A Cool Planet. Detailed in http://www.greens.org.nz/greennewdeal

29 OZAKI Masahiko, April 20, 2009. Japanese version of the effect of green new deal — is helpful to the maintenance of technological superiority. Japan Economic News. Detailed in http://www.rietigo.jp/cn/papers/contribution/ozaki/01.html

China adheres to the road of sustainable development and firmly takes the green economic development model in the long course of economic development. Under the international financial crisis, the green imprint of China's policy was striking and clear: China's new exploration in energy development, emissions trading and green economies have become the new engine of economic development.

Firstly, let's look at the development of new energy sources. In order to cope with contradiction in energy supply and demand and environmental pressures in the course of economic and social development, China is stepping up its efforts in new energy and renewable energy development, and actively exploring an energy development path that is suitable for national conditions of China. China introduced the "The People's Republic of China Renewable Energy Law" and "renewable energy medium and long-term development plan" in 2005 and 2007 respectively. The Chinese government has also formulated a series of new energy and renewable energy policies which have greatly promoted the wind, solar, biomass and nuclear power [30] and other new energy sources and renewable energy development. In the sixth chapter of this book, the authors analyzed China's exploration in the development of new energy, and also pointed out barriers of technologies and institutions in the development of new energy, which is worth reading carefully. In the analysis we have also noted that China is constrained by shortage of independent innovation, out-of-date energy development and utilization system, poor financing channels and limited market space and many other restraints in current industrial development of new energy sources. How to break through a variety of technical and institutional barriers to boost the proportion of new energy consumption will be a core issue in China's energy development strategy.

Secondly, the issue of emission right trading, one of the focuses of this book, is the main discussion topic in Chapter Seven. Conducting emission right trading as an effective model for low-carbon development is necessary in China's realization of sustainable development in economy and society, active response to global climate change, and effective implementation of Chinese energy-saving and emission reduction development targets. The introduction of emission right trading has a certain foundation in China. In the 1990s, the state environmental protection departments have already begun research on SO_2 emission right trading of the U.S. electric power industry. Under the Clean Development Mechanism (CDM) required by "Kyoto Protocol", China has become the country who achieved the most reduction in certified CO_2 emission. Since 2008, China has set up a number of emission right exchanges in Beijing, Tianjin and Shanghai and some other cities, actively exploring the experimental sites of emission right trading. Of course, everything is just beginning, and there are still difficulties and obstacles such as weak legal basis, immature market conditions, and unstable global market framework. But looking to the future, China's emissions trading market is expected to form a more mature market framework on the basis of a clear legislation and policy, to set market model on the basis of an exact mid-term goals upon the global climate change, to expand the market space by improving emission-reduction technology and financial support efforts.

---

[30] In the strict sense, the nuclear power does not belong to new energy, but considering its importance in China's future energy structure adjustment and energy strategy choice, the authors take the nuclear power as a part of China's "new energy".

Both developing new energy and carrying out emission right trading are organic integral parts of the green economy. The development of green economy is the choice of this era, and the fundamental requirement of changing the growth pattern. It determines a country's future developing competitiveness. The development of the green economy brings a good opportunity to optimize the economic structure and enhance competitiveness of China. Both the Chinese government and different social communities pay much attention to and support the green industry, and actively participate in activities related to international organizations, signing a number of relevant agreements, such as "the Montreal Protocol about Depleting the Ozone Layer Substances" which entered into force in 1989, the Kyoto Protocol and the Bali Agreement and so on. China has also been actively involved in a number of global technical cooperation and extensively worked with other countries on issues such as clean-coal research. But as a large developing country, China faces a variety of contradictions in improving people's livelihood, enhancing well-being as well as developing green economy. Green economy is a complete system which needs support and action provided by all groups of the society. China is now committed to building and improving the system of protecting the green economic development, including the amendments to relevant laws and the establishment in green economic policy system involving public finance, finance and emission right trading. Among China's reported 1.18 trillion Yuan investment program, 210billion Yuan will be invested in environmental protection to cope with the international financial crisis. We will conduct detailed discussion on these elements in Chapter Eight "New Growth Engine –The Rising of Green Economy in China."

III. THE COUNTRYSIDE IS ORIGINALLY GREEN

When we use the green evaluation standards to measure, we find a new charm of the countryside. Our mountains, rivers and forests cross the motherland by passing through the vast field in rural areas. Our beloved prairie is closely linked with the survival of farmers and herdsmen. The numerous rivers and trees are closely intertwined with our small villages in rural areas. When big trees are replaced by skyscrapers and chimneys, when streams are turned into foul ditches by industrial wastewater, when blue sky and white clouds are darkened by the gray sky, we become nostalgic for rural areas, for the ecology in our childhood as well as the poor but fresh time. People have a new idea when comparing the lives between rural areas and urban cities. As is written by an internet user, the many shortcomings of urban life are full of risk, pollution, crowdedness and interference. We can say that city life is the least beneficial to health, and the least beneficial to life. Our human beings can maintain health and longevity to ensure high quality of life only when we live as close to nature and scatter settlements as possible as we can. If people focus on narrowing the differences between urban and rural areas as soon as possible but not on urbanization, if they make efforts to develop information and transportation means and achieve the rural-oriented development, then we can definitely promote our efficiency and degree of progress.[31] It is in this sense that I would like to call for attention that "the countryside is originally green"!

---

[31] Guang Di. "City, This Piece of Human Work". http://bbs.koubei.com/thread_178_3211_1.html

However, isn't the green countryside linked with poverty over the years? Isn't it because the hope of having a better life why people advocate urbanization and industrialization? For many years, people believe that urbanization is a worldwide trend and it is reasonable to narrow the rural area and to reduce the number of peasants. A country's powerfulness and the people's happiness depend on industrialization and urbanization. Nevertheless, we have been worrying about that too much emphasis on urbanization may result in the loss of rights and interests of farmers. For example, the government seeks to profit from the restructuring of land resources in the process of the implementation of urbanization.[32] Some experts criticize sharply, pointing out that one driving force in the process of the implementation of China's urbanization is that the local government would get extra non-taxation revenue by changing farm land into urban use, with another driving force being that the urbanization would help officials get promotion. Experts believe that the different systems and dynamic mechanism will result in very different urbanizations, and that we should adopt the urbanization of "free flow of the market portfolio", rather than that of "administrative planning and rent-driven powers".[33] Other experts support urbanization in which people will have the same national treatment and can obtain approximately the equal public service, but they don't think China has conditions for such comprehensive urbanization.[34] Another expert clearly expressed his own idea by saying that "I advocate a steady urbanization strategy with an urban-rural interaction, by which farmers can enter into city and return to their native villages freely, while oppose the advancement of an urbanization strategy by encouraging farmers to buy apartments or building a city slum. Otherwise, once world-wide economic crisis happen, under which exports will decline substantially and the coastal processing businesses will massively close down, there would be no escape route for large quantities of peasant laborers, as a result of which China would lose the ability to cope with major crises".[35] In short, the experts don't ignore the social prosperity produced by urbanization and industrialization, nor they object to urbanization or industrialization, they just object the tendency of sacrificing the interests of farmers and rejecting the practice of polluting the environment in industrialization.

To sum up, we have never forgotten "green" in the past, but we need to have new "green" means which is both profitable and environmentally friendly. Clearly, the "green" we are promoting is associated with the modern "green", which has gone beyond the traditional "green" in countryside and agriculture. However, the rural resources and the ecological and environmental problems have become increasingly prominent in the process of long-term economic development and construction. What should we do when we are losing the old "green" without the blossom of the new "green"? The countryside can neither be separated from the green nor isolated from the modern green civilization. Therefore, it's extremely important to propose to "build a new countryside". In this sense, we can deepen

32 Li Xiaoxi. China: The New Concept of Development. China Economic Press, 2009 edition, 16. It is a speech made in China Economic Forum on Urban and Rural Development, June 9, 2007

33 Zhou Qiren. "Urbanization, a by-product". November 8, 2001. From http://zhouqiren.org/archives/268.html

34 Wen Tiejun. " the Agricultural Policy, from Discussing Principle to Deal with Concrete Issues", http://www.caogen.com/blog/Infor_detail.aspx?ID=95&articleId=13467

35 He Xuefeng, May 16, 2008. Reflections On China's Urbanization Strategy. From "China with Three Rural Issues". http://www.chinavalue.net/Article/Archive/2008/5/16/115662.html

understanding about the significance of the ideas proposed by the Chinese government in 2006. The Chinese government sketched out a new blueprint of "the production development, the wealthy life, the local custom civilization, the clean village, and the democratic management" for the countryside, not only to raise farmers' income and improve their quality of life, but also to improve the overall appearance of the countryside and the ecological environment, and reproduce the beautiful rural scenery and landscapes. The problem of China's building a new countryside will be specially introduced and analyzed in the second chapter of this book.

It is necessary to point out that the urban and rural overall plan experimented in the city of Chengdu and Chongqing manifests a very valuable way of thinking. Urban and rural overall plan is different from the simple urbanization or the simple new rural reconstruction. It is an important way to simultaneously complete the urbanization and construction of the new countryside. The urban and rural overall plan should put equal importance on urban residents' and farmers' problems, with priority on the weak farmers first. Such action entitles with us a long-term basis for urbanization. Not only to attain the extra income from farmers' land, the urban and rural overall plan needs also to have the economic potentiality. Therefore, we must bring in the advantage of the market forces in the operation, which will facilitate the large promotion of government's ability of management and coordination.

But, till today, are the destitute local people able to get rid of poverty by only depending on the green economy? In the summer of 2009, I visited Qinghai, a big province in China's western area, the area of which is equal to that of four Guangdong Provinces, but with the population of only 5,000,000. When experts appeal the Qinghai people to develop green economy, we must answer a question "can the green plan support 5,000,000 Qinghai residents?" I reply self-confidently by saying that the answer is definitely yes. The reasons I have are easily understood as following. One is about the particularity of Qinghai's "green". Here locates Yangtze River and Yellow River's sources, which is called the Chinese water tank, and here has the snowy mountain and the plateau, which is called China's ecology shield. Therefore, Qinghai's ecology not only affects itself but also affects the whole nation. The second is Qinghai's "green" on the basis of vast land. The vast land offers us hope and space to afforest despite the heavy pressure from desertification and the frail ecology, with which many provinces in the eastern parts is hard to compare. The third is that Qinghai has green animal husbandry, green industry and ecological tourism. It has the Sanjiang-source ecosphere, Lake Qinghai ecosphere, Tsaidam ecosphere, South Qilianshan Mountain foothill ecosphere as well as Huangshui ecological preservation ring. Besides, there is the "new energy" which is becoming an important element in the development of Qinghai's ecologic economy. All of these provide material base for Qinghai's green economy. Lastly, Qinghai is also changing its non-green industries into "green" ones, which is a comprehensive development under the "green" idea. There are two state-level circulated economic bases and one batch of environment-friendly industrial enterprises. Certainly, the governments at all levels need to support the green industry development and the superior industry's greenization, to improve the implementation of ecology compensation and each kind of green projects (including returning farmland to pasture, wetland protection, desertification control and ecological immigration, etc.), to protect and strengthen popular base for the marvelous economic development (such as poverty reduction, social undertakings, livelihood projects, etc.). On this basis, I believe that Qinghai's economy not only needs to be "green" but also can be more "green", and that green economy will ensure the long lasting happiness for the people in Qinghai now and future![36] Here, of course, the words need to be repeated: Each local area has its own reasons for developing the green economy and can work out ways to gain happiness by the means of developing green economy despite of different conditions.

In response to the adverse effects caused by the financial crisis on China's rural development, the State Council executive meeting proposed ten policies and measures to expand domestic demand in November 2008. Among them, taking the new rural development as a breakthrough, one of the policies is to expand rural consumption and rural infrastructure construction. Since then, the Chinese government launched a series of policies in rural consumption, such as Home Appliances Going to the Countryside, Automobiles Going to the Countryside, and so on. These stimulus policies soon came into effect because of the great potential the China's rural consumer market has. During the first half of 2009, the growth rate of rural consumption kept higher than that of urban consumption for 6 months. This shows that it is quite necessary and correct for the Chinese government to implement the policy to stimulate rural consumption to cope with the international financial crisis.

The international financial crisis has also affected China's agricultural finance, which is mainly manifested in the declination in the share of agricultural loans. The proportion of farmers' loans in China to the total remained at about 4.2% until the third quarter of 2008, and it began to decline rapidly in the fourth quarter, and reached the lowest point of 3.84% in the first quarter of 2009. Since the third quarter of 2008, China's agricultural wholesale price index fell sharply, which directly affects the sustainability of agricultural capital chain, and then leads to the rise of the non-performing loan ratio of financial institutions and the increase of business risk. These effects will definitely decrease the loan delivered by financial institutions to agriculture. The Chinese government has intensified its policy support for the "the rural areas, agriculture and farmers", which has made significant progress in rural finance. Agricultural Development Bank (ADB) has increased its support on the construction of rural infrastructure. The Agricultural Bank of China (ABC) has developed various types of special credit products related to the "Three Agricultural Issues". Compared with the beginning of this year, agriculture-related loans has increased by 88.1 billion Yuan (RMB) in the first half of 2009, accounting for 40% in ABC's loan increment. Rural credit cooperatives have strengthened their services for the "Three Agricultural Issues", with asset quality improved significantly. By the end of June in 2009, the non-performing loans and its proportion in Rural Credit Cooperatives to the whole country has declined by 23.8 billion and 0.7% respectively. Within the first half of 2009, China's agricultural insurance achieved 7.003 billion Yuan (RMB) in premium income, a year-on-year increase of 59.81%. In the third chapter of this book, focusing on the difficulties of rural development, the authors deeply analyzed the plight and reform of rural finance, provided detailed and specific facts and data, and clearly described the potential achievements made by China's rural finance in the financial crisis.

---

[36] See Chapter Five of "Thinking of China's Market-Oriented Process". The People's Publishing House, 2009 Edition. It is a speech entitled "strategic thinking of the green economic development" made by the Beijing expert adviser group participating in the forum of "Great Western Development Review and Outlook" held in Qinghai on August 10, 2009.

IV. Green Vitality in the Minority-Concentrated Regions

The academic circle and community in European and American developed countries have many years of deep thinking about "green" and rich practice of green protection and development. These are well worthwhile for our study.

European and American ecology thoughts emphasize on integration, systematization and diversity.[37] This is very inspiring. We apparently do not only mean green when we advocate "green" ecology. We love flowers with profusion of colors .We love year-round snow surrounded by white mist. We love all the colors created by the nature. All of them can also be integrated into our green ecology thought. One day in 2007, I participated in a conference on the Chinese ecology modernization undertaken by the Chinese Academy of Sciences. At the meeting I proposed one question about the relation between ecology modernization and ecology primitivation. I said that we saw the blue sky, snowy mountains, green lakes and lawn, all of which were holy and serene when we visited Tibet and some sparsely inhabited area. Finding the great ecological conditions there, we felt obliged to keep this pure land. Then, is this kind of ecology primitivation better than the ecology modernization? Is it one of the goals that the ecology modernization needs to pursue? [38]

Let me extend this topic. The ecology condition and present situation of economic development in the national minority areas including Tibet and Xinjiang are extremely important components of Chinese green economy and development pattern. Without the analysis on the ethnic areas, our judgment will be only confined to half area of China. China is a multi-ethnic country which contains 56 ethnics. Up to the end of 2008, our ethnic areas possessed a population of 191.56 million, which was 14.42% of China's total population, with its distribution occupying the larger part in Chinese land area. For instance, the southwest and the northwest of China are the two most centralized regions in terms of the distribution of minorities. The west part of China containing 12 provinces, autonomous regions and municipalities accommodate nearly 70% of national minority population, and about 60% of national minorities are living in 9 provinces of the border provinces and the autonomous regions. Most Chinese minority community is of vast territory with a sparse population and abundant resources. The prairie area, forest and hydro-electric resources reserves in the multi-ethnic areas surpass or are close to half of the total possession in China. China has more than 22,000 kilometer land boundary line, of which 19,000 kilometers are in minority-concentrated regions. The national natural reserves in the minority regions cover more than 85% of the protection areas of the whole country. It is our nation's important ecology shield.[39] Therefore, when talking about the new green deal and "the China Model", we must emphasize specially on the analysis on China's multi-ethnic areas. Chapter Eleven in this book specially briefs and analyzes the present economic development situation of China's

---

[37] Wang Jianming, 2008. "Red" and "Green" reveal the new ecological, political and philosophical thinking of a new era of globalization (Chinese) [J]. Natural Diagnostic Method Research 12.

[38] My speech on China's Modernization Report sponsored by the Chinese Academy of Sciences, January 27, 2007.

[39] Taken from the White Paper, "China's ethnic policy and the common prosperity and development of all ethnic groups" published by the State Council Information Office, September 27, 2009 (Reporter Wei Wu, Wei Minli, Fu Shuanqi).

multi-ethnic areas It also gives thorough analysis on the characteristics of economic development in the multi-ethnic areas under the current financial crisis.

Let us continue to include the European and American scholars' ecology thoughts in the discussion on the economy and the development pattern in China's minority-concentrated regions. The European and American ecology thoughts value the relation between humanity and the nature, even up to the political level. To the Green Parties in the European and American countries, the ecological problems appear not only as general technical and scientific issues, nor just as economic and legal issues, but rather a great problem concerning healthy co-existence and sustainable development of human beings and ecology. Therefore, we must look at and deal with it as a major issue from the political perspective.[40] This really makes sense. The current ecological environment and the climate problem have already become big issues listed on the most important agenda by the global leaders. It is a critical issue about the harmony between man and man, man and nature, and even more related to whether the livelihood on the Earth could sustain human beings to continue to survive. In fact, the discussion about "Green New Deal" proposed by this article is exactly a response to this question.

Let's discuss the relation between human beings and the nature first. The God-like nature, as we see, has created a world filled with the green color no matter it is called the God or the Buddha that the Oriental believe in. If there were no human beings, the vast land on the earth would be like a green ocean. The places which have already become desert today were once the heaven of water and plant long ago. But nature is also showing his power, controlling the expansion of green through earthquake, thunder and lightning and so on. After their birth, human beings have obtained the basic survival condition under the protection of nature. In ancient China, the philosophers proposed the union of Heaven and Man, Tao (the way of nature) following Nature, emphasizing the harmonious co-existence between man and nature as well as human's reverence for the laws of nature. Since the industrialization, with formidable power, the humanity has employed the nature to create a modern world. However, the nature started to retaliate against the human beings at the same time. It is not until now that the humanity, truly united, has expressed good intentions to and cooperation with the nature through the "Green New Deal". China's minority regions fall behind comparatively in the economy, but they do the best in co-ordination between mans and nature as well as in natural protection. Regardless of where their philosophies come from, all deserves worth our respect.

Now we need to discuss the relation between man and man in the green ecology. In all human relationships, the most destructive one to nature is not others but war. The green is destroyed into pieces by gunfire, disappearing forever. Under the wrong concepts and policies, the natural color is also being destroyed by greedy claims and stupid plans. For the green, we need peace; for the green, we need ecological new deal, thus we need our country to play a better role. We realize that the modern country has assumed the leading responsibility in charge of ecological protection, green protection and climate protection. A series of international joint pledge to protect the Earth and the ecology and to prevent the air

[40] Wang Jianming, 2008. "Red" and "Green" reveal the new ecological, political and philosophical thinking of a new era of globalization (Chinese) [J]. Natural Dialectics Research 12. Reference on Cheng Rong Ning, 2003. "Analysis of green politics – A kind of new development concept". Realistic 5.

pollution are signed on an inter-country basis. That has given rise to a problem faced by all countries, which is how to keep harmony between their citizens and their ethnics, with neither hatred nor war. China's answer is that we need national equality and ethnic unity. To respect minorities, the Chinese government implements regional ethnic autonomy. That is the most concentrated expression of such goodwill. Of course, this kind of regional autonomy for ethnics should be implemented under the national unified leadership and takes the national approval as the premise. Why does it need the national approval? This may invite the history to reply, invite our older generation and national brothers' ancestors to reply; also may invite statesmen and the scholars from Europe and America to reply. They will tell you how a modern country maintains unification and realizes prosperity. As far as I am concerned, one is not qualified to enjoy the national autonomy right if he does not acknowledge his nation. In addition, might as well mention, the practice of national melting in the U.S. is also very worthy of thinking of China.

In recent years, in order to accelerate the development of ethnic minorities and minority areas, China has adopted the western development strategy, with "actions to realize prosperity for the border areas" and more focus on supporting the development of 22 nationalities with a small population. Through the training of ethnic minority cadres and developing the science, education, culture and public health undertakings, and full respect for and protection of minorities' freedom of religious belief, increasing their investment and quickening the pace of opening up in the ethnic minority areas, economic development in the ethnic minority areas shows a new vitality. Until 2008, the accumulative fixed assets in the western region has totaled 7.7899 trillion Yuan (RMB), of which 1.8453 trillion Yuan (RMB) was made in 2008, an increase of 5 times more than that in 2000, an average annual increase rate of 23.7%. The "West-to-East Natural Gas Transportation", "West-to-East Electricity Diversion Project" and a number of key projects have been completed. We have finished the construction of a number of airports, highways, water controls and other infrastructure projects. Our country also provides certain financial compensations to the ethnic areas which have contributed to the output of natural resources, national ecological balance and environmental protections. In 1994, the state adjusted the central government and autonomous regions' sharing of mineral resources compensation ratio to 4:6, other provinces and cities to 5:5. In 2004, our country began to establish the compensation mechanisms in ecological construction and environmental protection. In the exploration of the rich oil and natural gas resources, focusing on driving local development, the "West-to-East Gas Transportation Project" alone has been able to make an annual increase of 1billion Yuan (RMB) for Xinjiang fiscal revenue. In the meantime, we need to develop the green economy in the minority areas. For many years, our country has paid close attention to the ecological environmental protection and construction in the ethic areas. In particular, since the implementation of the western development strategy, a series of policy and measures have been introduced, including prohibiting forest harvesting in the upper reaches of major rivers, implementing the policy of returning farmland to forest and grassland, closing hillsides to facilitate afforesting and using grain to relief, coupled with others. The state resolves the issue of compensation for ecological construction properly. Food subsidies are provided for farmers and herdsmen whose farmlands have been returned to forest and grassland. Meanwhile, fiscal subsidy is offered to local government suffering from revenue reduction due to the prohibition of forest harvesting. In 1999, the Chinese government stressed that the grassland and farmland capital construction should be arranged to an equally important position, which will benefit more than a dozen of ethnic minorities including Mongolian, Tibetan, Kazakh, as China's main pastoral areas are places where these minorities live in, while the animal husbandry is their livestock business for generations and the foundation industry for survival and development.[41]

In 2008, the international financial crisis continued to spread and deepen, which had a serious influence on the Chinese economy. In such a macroscopic background, what are the characteristics of the economic development in the multi-ethnic areas? In the eleventh chapter of this book, the authors have induced the following characteristics. First, there has been a high GDP growth in the ethnic minority areas, which is higher than that of the national average level. Since 2008, GDP has grown by 9% compared with that of last year, but the economic growth rate in multi-ethic areas was averagely 12.15%, for some region such as Inner Mongolia, as high as 17.2%. Second, the national investment in fixed assets maintains a rapid growth in the ethic regions. The western region of China including the ethnic minority areas is one of the most important regions where the 4 trillion economic stimulus is to benefit. In 2008, the investment in the fixed assets in Tibet, Xinjiang and Inner Mongolia amounted to 30.333 billion Yuan (RMB), 231.4 billion Yuan (RMB) and 559.645 billion Yuan (RMB) respectively, which grew by 12.5%, 25.0% and 27.1% respectively compared to that of last year. Third, it maintains a steady import and export growth in the ethic areas. In 2008, the ethnic regions' total amount of imports and exports reached 60.75 billion U.S. dollars, an increase of nearly 37% compared with 2007, which is higher than the national growth rate. Fourth, the minority areas maintain a steady growth in consumer demand. For more details, please read our specific sections.

V. The Pursuit of "Green" Means Happiness for Generations of Mankind

There are some wonderful words I'd like to share with everyone. "The so-called green economy, according to experts, is defined as a new economic form oriented by the market, based on the traditional industrial economy, guided by the harmony of economy and environment, a state of development produced and shown by the industrial economy to meet human needs of environmental protection and health. The fundamental distinction between the green economy and the traditional industrial economy lies in that the former is no longer a loss-style characterized by the destruction of ecological balance, heavy users of energy and resources or damages to human health, but is a balance-like economy characterized by the maintenance of human survival environment, the rational protection of resources and energy, and benefits to human health. The green economy is not only a realistic choice for every country to defeat the financial crisis, but also a long-term strategy bound to protect the global environment. That's why the green economy is regarded as the fourth industrial revolution in the history, and its significance and impact is even greater than that of the last three industrial revolutions. In the coming 50 years to 100 years, the global economy will be confronted with

[41] Compiled according to "China's Ethnic Policy and the Common Prosperity and Development of All Ethnic Groups," 2009 White Paper, and other information et.

the transformation from industry civilization towards ecology civilization."[42] Those words give us clear understanding of the connotation and historical localization of developing green economy. But we still want to figure out a question: what is the pursuit of green all for?

The answer will take many forms. One of the more common answers goes like this: the pursuit of the green is to realize sustainable development of human beings, to achieve green development which will save resources, to protect human survival space and to benefit all mankind. Here, I would like to recommend an answer with a high level of generality to readers. That is to build a "harmonious society" and "harmonious world" proposed by the Chinese government. As we discuss the Green New Deal and the China Model, we often mention the concept of "Heaven and Man" in ancient Chinese civilization, that is, to respect nature and to live in harmony with nature. But the significance of "harmony" is not only confined to this. As early as in Fu Xi Era, human beings had understood that the harmony of the opposites, "yin" and "yang", is the root cause of all things flourishing. It is a great idea that the ancient Chinese civilization has contributed to contemporary world. Therefore, the "harmony" offers the idea of dealing with conflicts, that is, coordinating relations, resolving conflicts, balancing and integrating society, and striving to create a social atmosphere in which people can do their bests and live in harmony. It can be said that "harmony" is a potential equilibrium point for contradictory movement of everything, the new development of traditional virtue, a guide to improve the system, and a requirement of social civilization. So the faith and pursuit of harmony not only conform to the trend of world progress, but also reflect an assiduously social ideal of human.

What is the pursuit of harmony for? The final goal is for people's happiness. China takes "people-orientation" as the goal which contains five aspects as the basic contents, i.e., harmonious development of city and countryside, different regions, economic and social development, human and the nature, home and abroad. The nation takes the goal of "moving towards the harmonious society" as the highest ideal in which the pursuit of happiness for people is always the theme. The implementation of "Green New Deal" is to create more employment opportunities for people, to lead a new turn of economic growth, and to realize humanity's survival, happiness and development. The Green New Deal is distinguished from the environmental protection economic policy. The former pays close attention to spending money and materials in protecting and restoring the ecological environment, nevertheless, the latter emphasizes on propelling economic development by investment in the environmental projects.

Further more, we have realized that the pursuit of green was not only to deal with the financial crisis or to make happy life for people today, but also to pursue happiness for generations. Were our water and air polluted, were the earth dredged and natural resources over-consumed, how could our generations survive and develop? For the future, for our descendents, we must turn the development model into a "green" one and have a green lifestyle. Our Chinese have made a great resolution and much effort, and will continue to do so. In September 2009, during the United Nations General Assembly period, Chinese President Hu Jintao made a commitment to the international community at the UN summit on climate change: In the future, China would further integrate climate change into economic

[42] Source: Xie Li, September 12, 2009. Developing the Green Economy, China Has Been On the Road. Financial Times. Detailed in http://finance.sina.com.cn/g/20090912/04166740347.shtml

and social development plan, continuing to take strong measures, and strive to make the proportion of non-fossil energy in the primary energy consumption reach about 15% in 2020. By 2020, the forest area will have been 40 million hectares more than that of 2005, an increase of 1.3 billion cubic meters' forest reserves compared to 2005. These solemn and important commitments won high praise from the international community. Foreign dignitaries share the view that China's choice will have a decisive impact on other countries, the same as that from global rules, laws and changes in the market on China.[43] On the other hand, we also notice that experts worldwide are pondering over the future of the green economy and making many predictions. Nicholas Stern, the former chief economist in the World Bank, published "Stern Review Report" in November 2006, which showed that by 2050, the carbon content in the world energy industry will have been reduced by 60—75% to stabilize greenhouse gas emissions of carbon dioxide at the level equivalent to 550ppm or below, and then the annual output value of low-carbon energy products will have reached more than 500 billion U.S. dollars.[44] According to the prediction from McKinsey' report, from the present to 2030, China will raise a "green revolution", which includes green electricity, green transportation, green industry, green building and green eco-systems in five major fields.[45] Many predictions show that the green economy will create millions of job opportunities. But undoubtedly, one thing is that it is difficult to achieve a rapid change in certain fields of the green economy, especially the energy and transport systems which require massive input.

There is a topic need to be extended. When we talk about the happiness of people passionately, we shouldn't forget to integrate economic growth with social development. Here, people's health is one of the most important elements. The medical and health system is related to the maintenance of life and happiness of generations. At present, most countries regard the public health and basic medical care as an important livelihood issue. In the economic stimulus plan signed by President Obama, health care reform is one of the main events besides the energy and education. The public health reform and investment are also taken as one of the keys in the four trillion economic stimulus package introduced by the Chinese government. To illustrate the people-centered objective in Chinese Green New Deal, in Chapter Nine the authors deliberately brief the new health system reform started in 2009, which aims to not only solve the short-term problems, but also take the long-term arrangements for people's welfare. The Chinese government intends to invest 850 billion Yuan (RMB) to support the implementation of health reform program in three years, which expresses not only their positive attitude in determination to deal with the financial crisis, but also a concept of pursuing a harmonious society.

In such a big country as China, it is necessary to go forward step by step in the implementation of any major initiative. In the process of 30 years' reform and opening up, the pilot exploration has played a very important role. The pilot special economic zones in cities

---

[43] Tao Zhipeng, September 25, 2009. The International Community Spoke Highly of China's "Emission- Reduction" Commitments. Source: www.Xinhua.com.
Detailed in http://news.xinhuanet.com/world/2009-09/25/content_12109923.htm

[44] UK Ministry of Public Finance: Stern Review ReportDetailed in http://www.hm-treasury.gov.uk/stern_review _report.htm

[45] Martin Joerss, Jonathan Woetzel, March 2009. Green Opportunities in China. Mckinsey Quarterly.
Detailed in: http://china.mckinseyquarterly.com/Chinas_green_opportunity_2364

such as Shenzhen have made tremendous achievements grasping the attention of the whole world. In the new century, the pilot exploration becomes more important in taking a people-oriented development road with harmony and co-ordination. In Chapter Ten, from the new platform in system reform, the authors specially analyze a batch of new comprehensive pilot reform areas, including comprehensive pilot reform areas in Pudong New Area of Shanghai in 2005 and Binhai New Area of Tianjin in 2007, the national urban and rural comprehensive reform pilot areas approved to be established in Chongqing Municipality and Chengdu city in 2007, and the national comprehensive resource-saving, environment-friendly and society-building pilot reform areas in Wuhan City Circle and the Changsha-Zhuzhou-Xiangtan (CZX) City Group. Many provinces also launched comprehensive reform of the provincial-level test. Through comprehensive reform experiment, integrating the solution to practical problems in local areas and the exploration of national common problems, the breakthroughs in key areas and overall innovation, and integrating the reform of economic system and other reforms, emphasizing on conducting the reform safely, stably and orderly, we will achieve new progress in China's moving forward to a society with urban and rural co-coordinated, resource-saving, environment-friendly and sound system, and bring greater happiness to the people.

Green is the color of life, a symbol of hope, harmony and vitality. To protect "green" is to protect human beings themselves and the pursuit of the "green" is for the well-being of our generations. Today, when China's development model becomes "greener", we find that it conforms to the ideas proposed by the ancient sages and sacred laws of nature, in line with both the nature and people, and good for the succeeding generations. Therefore, let's strive together for the realization of the Green New Deal and the China Model, the two great goals!

Translator: Yu Chunmei (SUFE)
English Version Editor: Xiao Rong (SUFE), Liu Xuzhi (Harvard)

REFERENCES

[1] UNEP: Global trends in sustainable energy investment 2009 [EB/OL]. http://sefi.unep.org/fileadmin/media/sefi/docs/publications/Executive_Summary_2009_CN.pdf

[2] Geoff Mulgan , Omar Salem: Green economy: background, current situation and prospects [EB/OL]. http://www.shantoudialogues.cn/uploadfile/200812/22/124499025.pdf

[3] UNEP: Globalization and environment—Global crisis: international chaos [EB/OL]. http://www.unon.org/confss/doc/unep/unep_gc/gc_25/gc_25_16/k0843093.pdf

[4] State Council Information Office: "China's ethnic policy and the common prosperity and development of all ethnic groups," White Paper, November 27th,2009 [EB/OL], http://news.xinhuanet.com/politics/2009-09/27/content_12117333.htm

[5] Obama's " Green new deal". *International finance,* August 17th, 2009.

[6] Marty. Vanhanen, April 19th,2009. The "Green New Deal" led "green economy", theme luncheon speech in BoAo Asian Forum [EB/OL]. http://news.sina.com.cn/o/2009-04-20/064715494274s.shtml
[7] Rowan Callick. The China Model. The America. (November/December 2007 Issue, by American Enterprise Institute) [EB/OL]. http://www.american.com/archive/2007/november-december-magazine-contents/the-china-model/?searchterm=China Model
[8] UK Conservatives: Blueprint for a Green Economy[EB/OL], http://www.worldchanging.com/archives/007254.html
[9] A Green New Deal for Asia: How Asia-Pacific countries are expanding the Green sector in their economies [EB/OL], http://www.spireresearch.com/pdf/archive/ejournaljun09/A%20Green%20New%20Deal%20for%20the%20emerging%20world.pdf
[10] UNEP : Global Green New Deal-Environmentally-focused Investment Historic Opportunity for 21st Century Prosperity and Job Generation [EB/OL], http://www.unep.org/Documents.Multilingual/Default.asp?DocumentID=548&ArticleID=5957&l=en
[11] The World Bank, November 2009. *World Development Report 2010* [EB/OL], http://econ.worldbank.org/WBSITE/EXTERNAL/EXTDEC/EXTRESEARCH/EXTWDRS/EXTWDR2010/0,,menuPK:5287748~pagePK:64167702~piPK:64167676~theSitePK:5287741,00.html
[12] The American Clean Energy and Security Act of 2009 (ACES), http://www.pewclimate.org/acesa
[13] Nicholas Stern: *Stern Review Report* [EB/OL], http://www.hm-treasurygovuk/ stern_review_report.htm
[14] Martin Joerss, Jonathan Woetzel, March 2009. Green Opportunities in China [EB/OL]. Mckinsey Quarterly http://china.mckinseyquarterly.com/Chinas_green_opportunity_2364
[15] Hu Angang, 2008. "Green rise": China's new contribution to the world - China should make five contributions to human development (Chinese) [J]. The People's Forum 9.
[16] Chang Xiuze, 2007. China's development model: to explore the diversity in the development of human civilization (Chinese) [J]. *China Economic Tribune*, 22.
[17] Wang Jinnan, Li Xiaoliang, Ge Chazhong, 2009. The current development situation and prospects of China's Green Economy (Chinese) [J]. *Environment Protection*, 5.
[18] Xu Chongwen, 2009. Foreign comments on the China Model (Chinese) [J]. *Flag Presentations,* 8.
[19] Wang Jianming, 2008. "Red" and "Green": reveal the new thinking in ecology, politics philosophy in a new era of globalization (Chinese) [J]. Natural Dialectics Research 12
[20] Qi Si, 2009. The philosophical reflection on issues about Chinese Model development (Chinese) [J]. *Philosophical Trends,* 1.
[21] Fang Zhengzhu, 2009. The scientific concept of development to promote change in the pattern of economic development—Analysis on the transformation of economic growth mode for 30 years (Chinese) [J]. *Discussion on the Modern Economy*, 1.
[22] Zhuang Junju, Zhang Xili, 2009.: A Summary of Recent Research on the "China Model" (Chinese) [J]. *Red Flag Digest*, 2.

INTRODUCTION

The period of 2008-2009 is an uncommon period, during which a financial crisis not encountered for almost half a century broke out suddenly. What were we Chinese thinking about during this period? What have we done and with what effects? This book aims to record all these things for our references in future.

This book is covered with but not limited to the following fields: How the financial crisis intensifies discussion on "the China Model"? The comprehensive interpretation on the scale, structure and effects of the 4 trillion economic stimulus plan. The countryside is originally green and how the new countryside needs to be better constructed in China. How is the development of rural finance in China under circumstances of worldwide financial crisis? How the 10 Major Industries Stimulus Plan is favorable to the present and even more to the future? The Interpretation and analysis on the China Real Estate market in comparison with the Subprime Crisis in the U.S. Real Estate Market. What are the difficulties and prospects in China's new energy development? Market Mechanism and Low-carbon Economic Growth in China. The pursuit of 'green' means happiness for generations of mankind. How are the comprehensive and coordinated experimental reforms ongoing in China energetically and systematically? The one who fails to acknowledge his country won't deserve the autonomous right. Surviving the financial crisis, Hong Kong and Macao are bound to see a better tomorrow.

In: China's New Deal
Editors: Li Xiaoxi and Hu Biliang

ISBN: 978-1-61470-815-5

Chapter 1

Four-Trillion (RMB) Economic Stimulus Plan: China's Response to the International Financial Crisis

Zhang Shengling*[*], *Jin Sanlin, Liu Yimeng and Zhangshuai

Since the second half of 2008, the international financial crisis triggered by the U.S. sub-prime mortgage crisis has spread rapidly, and the world's major economies have stepped into recession. Due to the international financial crisis' effects on China's economic development, coupled with a number of accumulated structural problems in a long term, China's economy has experienced a substantial decline. The Chinese government made a resolute decision in time to adjust China's macroeconomic policies by implementing a proactive fiscal policy and loose monetary policy appropriately and quickly unveiling the "four trillion economic stimulus plan" which focuses on investment. These measures are being implemented and have made interim progress.

Section 1: The Background of "Four-Trillion-Yuan (RMB)" Economic Stimulus Plan

Since 2003, China's economy has remained more than 10% of the high-speed growth for years' running. However, the rapid growth in national economic strength brought such problems including increasing inflationary pressures, structural imbalances of economic growth and other issues. Since 2006, the Chinese government has begun to implement relatively tight economic policies in order to solve these problems. The main methods are controlling the two "brakes" and raising a "threshold", that is, to strictly examine and approve the use of land, to control money supply and credit growth, and to enhance the industry entry threshold. In 2007, the Chinese government continued to make great efforts on controlling by abolishing export tax rebate on "high energy consumption, high pollution, resource-based"

* Zhang Shengling is associate professor of the Institute of Economics and Resrouce Management, Beijing Normal Universiy and she earned her PhD degree from Beijing Normal University. Email: iermzh@hotmail.com

products, raising RMB deposit reserve ratio for 10 times and benchmark deposit and lending interest rates for six times, lowering the interest tax rate from 20% to 5%, putting more efforts on open-market operations and controlling the excessive growth of monetary credit. Despite taking these measures, China's economic growth, increasing price pressures and the structural imbalances of economic growth are still accelerating. Therefore, Central Economic Work Conference held by the end of 2007 proposed that the primary task of macroeconomic regulation and control should be to prevent relatively fast economic growth from becoming overheated, to prevent structural price increases from becoming obvious inflation for the purpose of promoting China's economic development with quality and speed. The conference also determined to implement steady fiscal policy and tight monetary policy in 2008. It was the first time the Chinese government switched monetary policy from "steady" to "tight" during the past 10 years.

However, in the second half of 2008, the international financial crisis spread to China and made the economic downward pressures increased. Foreign trade, industrial production and employment were greatly affected. The Government promptly adjusted macroeconomic policies by implementing the proactive fiscal policy and moderately loose monetary policy, and unveiling a large-scaled economic stimulus plan.

I. The Effects of the International Financial Crisis on China's Economy

In 2007, the outbreak of U.S. sub-prime mortgage crisis eventually evolved into a global financial crisis, which affected the real economy. It was the first time that the whole developed industrial countries plunged into recession since World War II. As China's economy depends on investment and exports, the international financial crisis has affected China's economic development through a variety of channels including finance, trade, price and psychological expectations in varying degrees.

1. A Substantial Decline in Exports

Financial crisis led to the global credit crunch, and made investors' confidence cracked down. The prices of global stock market, housing market, commodity and other assets declined sharply, which resulted in dramatic shrink of developed countries' demand. As a result of demand shrink, China's exports only increased 4.3% year-on-year, imports fell 8.8% year-on-year in the fourth quarter of 2008, which had been the lowest quarter growth rate since 2002. The growth rates of exports and imports between China and its major trading partners such as the United States, European Union, Japan, ASEAN were in decline. Bilateral trade between China and EU increased by 19.5% in the year, 8.2 percentage points lower than the growth in the first half of the year; Sino-US trade increased by 10.5%, 2.1 percentage points lower that growth in the first half of the year; Sino-Japanese trade increased by 13%, 4.7 percentage points lower than the growth in the first half of the year; China-ASEAN trade increased 14%, 11.8 percentage points lower than the growth in the first half of the year. As for China's exports, the exports to the United States were 252.3 billion US dollars, increasing by 8.4% year-on-year, 6 percentage points lower than the growth in the first half of the year; the exports to EU were 292.88 billion U.S. dollars, increasing by 19.5% year-on-year, a 9.6 percentage points lower than the growth in the first half of the year; the exports to ASEAN were 114.14 billion U.S. dollars, increased by 20.7% year-on-year, 11.7 percentage points lower than the growth in the first half of the year. The growth rate of exports to countries and regions mentioned above reached its lowest point since 2002. The only exception was the exports to Japan, which were 116.13 billion U.S. dollars, increased by 13.8% in 2008, 2.4 percentage points higher than that in the first half of the year.

In the first half of 2009, China's imports and exports to its major trading partners continued exhibiting a trend of sharp decline. China-EU trade was 159.97 billion U.S. dollars, decreased by 20.9%, while China exports to EU were 103.45 billion U.S. dollars, decreased by 24.5%, and the China imports from EU were 56.5 billion U.S. dollars, decreased by 13.2%. China-US trade were 132.09 billion U.S. dollars, decreased by 16.6%, of which, China exports to US were 97 billion U.S. dollars, decreased by 16.9%, and China imports from US was 35.01 billion U.S. dollars, decreased by 15.6%. China-Japan trade was 99.73 billion U.S. dollars, decreased by 23.1%, of which the exports to Japan were 44.03 billion U.S. dollars, decreased by 20.3%, and the imports from Japan were 55.7 billion U.S. dollars, decreased by 25.1%.China-ASEAN trade was 88.06 billion U.S. dollars, decreasing by 23.8%.

Table 1-1. The Trade Information of China and Its Trading Partners in the First Half Year of 2008-2009(Unit: billion dollars, %)

	01-06-2009		01-06-2008		01-12-2008	
	Imports & Exports	Year-on-year growth rate	Imports & Exports	Year-on-year growth rate	Imports & Exports	Year-on-year growth rate
Total trade volume	946.12	-23.5	1233.8	25.7	2561.6	17.8
China –EU (27countries)	159.97	-20.9	202.14	27.7	425.58	19.5
China-US	132.09	-16.6	158.31	12.6	333.74	10.5
China-Japan	99.73	-23.1	129.61	17.7	266.79	13.0
China- ASEAN	88.06	-23.8	115.79	25.8	231.12	14.0%

Sources: Ministry of Commerce of the People's Republic of China, http://www.mofcom.gov.cn/tongjiziliao/tongjiziliao.html

2. The Rapid Decline of Industrial Added Value

Affected by the international financial crisis, China's industrial growth has started its rapid decline since June 2008. The growth rate of industrial added value has fallen from 16% in June to 5.4% in November 2008, which was the lowest growth rate since March 2002. Since the industrial added value almost accounted for 43% of GDP, the industrial employment in the urban areas accounted for 35% of total employment, and the rapid decline of industrial growth directly led to the decline of GDP growth and the increase of the unemployment.

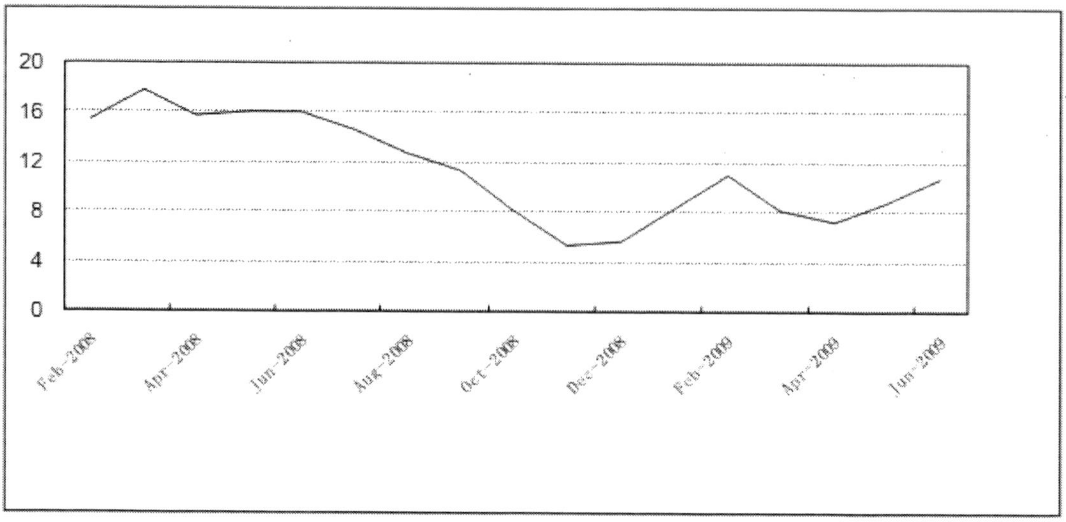

Figure 1-1 The Month-on-Month Growth Rate of China's Industrial Added Value in 2008 and the First Half of 2009.

3. The Labor Market Suffered a Big Blow

The Chinese economy depends on exports to large extent, and the exports decline results in many factories' temporary shutdown, or even bankruptcy, so unemployment is becoming increasingly prominent, particularly in a large number of migrant workers unemployed. At the end of 2008, China's registered urban unemployment rate was 4.2%, increasing by 0.2% compared with the end of 2007, a reversal of employment improvement in the past few years. By the end of 2008, up to 20 million rural migrant workers had returned home after losing their jobs due to the continuously deteriorating operating environment in the coastal areas, the decline of the growth rates and profits of importers and exporters as well as the failure of some small and medium sized export-oriented enterprises. Therefore, a great number of migrant workers lost job opportunities. In addition, the pressure of creating new jobs has been gradually increased. The employment rate of university graduates in 2008 was 65%, dropping by nearly 8% compared with that of 2005. The serious unemployment problem is directly related to the global financial crisis' short-term impact on China's labor market.[1]

4. The Rapid Decline in Asset Prices

The international financial crisis led to a rapid fall of asset prices in global stock market, housing market, bulk stock and other assets. The decline rate of stock market of the top10 countries was from 97% to 68.81%, and China's stock market decline leaped into the front rank of the world. In the fourth quarter of 2008, the decline of Shanghai Composite Index was 20.62%, and fell to 1664.93 points on October 28 which was the lowest point for nearly two years. Compared with 5522.78 points on January 14, the highest point of 2007, it fell to

[1] Chen Jiagui, 2009. *Analysis on the Prospect of China's Economy* (2009). Beijing: Social Sciences Academic Press, 4.

69.85% which was the biggest historical decline record within the same period time. The trading volume in Shanghai and Shenzhen stock market sharply shrank. The average daily trading volume declined from 210 billion Yuan (RMB) at the beginning of the year to 50 billion Yuan (RMB) in September and then went up to 80 billion Yuan (RMB) in the fourth quarter.

In 2008, China's real estate investments significantly declined from 32.27% in the first quarter to 9.97% in the fourth quarter. Prices had also continued the downward trend in the first three quarters, and dropped further. Particularly, in December, housing sales prices in 70 large and middle-sized cities fell 0.4% year-on-year, being 0.6% lower than the year-on-year growth rate of November, with a comparative decrease of 0.5% comparing with November, which was the first year-on-year decrease in housing prices since the release of the index in July 2005.[2]

5. Rapid Decline of Economic Growth Rate

China's exports decline and housing price decline resulted in the decline of investment growth rate. The rising unemployment rate had influenced residents' income and their expectation for income and then influenced the consumption. The effects of the international financial crisis on China have spread from export to the field of investment and consumption, resulting in a decline of aggregate demand. The economic growth rate has slid since the third quarter, and the growth rate in the fourth quarter was only 6.8%. China's GDP reached 30.067 trillion Yuan (RMB) in 2008, increased by 9.0% compared with the last year, and the growth rate fell by 4 percentage points compared with that in 2007. Considering the economic growth rate, the annual growth rate of 9% was a little bit lower than the average growth rate between 1979 and 2007, but it was not the lowest one. However, the economic growth rate dropped within a year by four percentage points, which was the biggest economic fluctuation over the past decade.

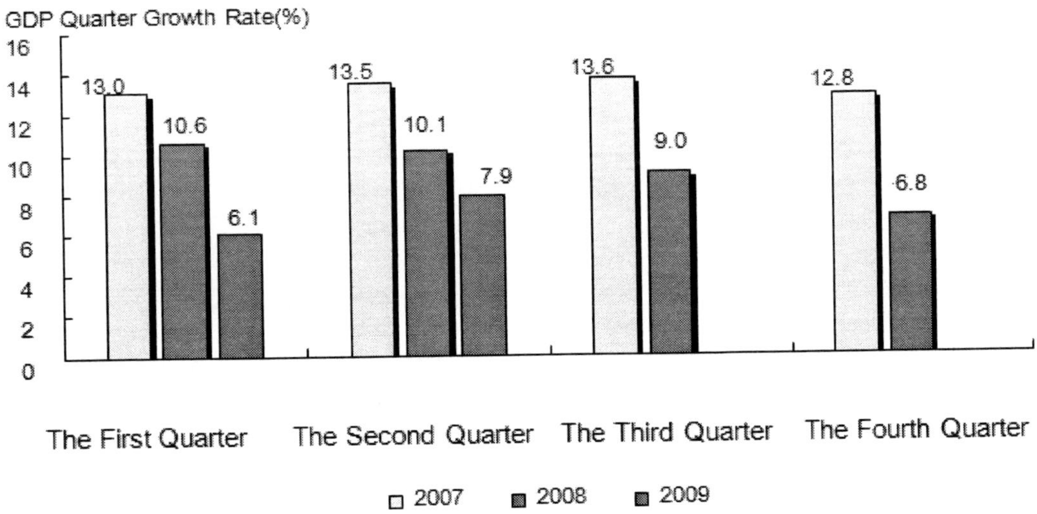

Figure 1-2. China's Quarterly GDP Growth Rate from 2007 to the First Half of 2009.

---

[2] Li Xiaoxi, 2009. China's Economic Trajectory in the New Century——Sub-quarter Economic Situational Analysis Report2007-2008. Beijing: People's Publishing House, 606,623.

As mentioned above, China's development has been faced with a more severe situation. In order to ensure sound operation of China's economy and avoid big ups and downs, we should implement effective economic stimulus plan through investment, stimulating domestic demand and foreign trade co-ordination for the purpose of achieving sound and rapid economic growth.

III. The Announcement of "Four Trillion" Economic Stimulus Plan

In light of major changes in the economic situation at home and abroad, the Chinese government responded quickly and unveiled a "four-trillion-Yuan (RMB)" economic stimulus plan timely.

1. The First Proposition of "Four-Trillion-Yuan (RMB)" Economic Stimulus Plan by the State Council Executive Meeting

In September 2008, the U.S. financial crisis fully deteriorated, and quickly spread to worldwide real economies. Major Western economies plunged into deep recession one after another. China's exports and industrial output growth sharply declined. In view of this increasingly grim situation, the State Council executive meeting decided to implement a proactive fiscal policy and moderately loose monetary policy in November 2008. The meeting also negotiated 10 measures to further expand domestic demand and promote economic growth:

(1) To accelerate the construction of housing projects by building more affordable and low-rent housing , speeding the reconstruction of slums, implementing projects in nomads settlement, and expanding the pilot projects in reconstructing dangerous buildings in rural areas;

(2) To speed up construction of rural infrastructure by enhancing the construction of rural marsh gas facilities, safe drinking water project, countryside roads, by completing rural power grids, accelerating the construction of the north-to-south-water-division project and reinforcing risky reservoirs and strengthening the water-saving transformation in large-scale irrigation districts; by reinforcing poverty alleviation efforts;

(3) To accelerate the construction of infrastructure including railways, highways and airports by constructing a number of passenger dedicated lines, projects of coal transportation corridors and western trunk railways; by improving high-way networks, planning the construction of trunk-line and feeder-line airports and strengthening the transformation of urban power grids ;

(4) To speed up health care, cultural and educational development by strengthening the health services and medical system at grass roots, by accelerating the reform of junior high schools in the central and western regions, and by promoting the construction of special educational schools and comprehensive cultural centers in towns;

(5) To improve ecological environment construction by enhancing the construction of sewage and rubbish disposal facilities and preventing water pollution in key areas, by strengthening the construction of both the natural and protective plantation forest

, by supporting the construction of key energy-saving and emission reduction projects;

(6) To enhance self-innovation and industrial restructuring by supporting the development of the high-tech and service industries;

(7) To accelerate the reconstruction in the areas hit by Wenchuan Earthquake;

(8) To raise average incomes in the rural and urban areas by raising the minimum purchasing price of grain in the next year and the standard of agricultural comprehensive direct subsidy, seed subsidies, farming machinery subsidies, by improving the social security for the low-income groups, increasing spending on allowances in the urban and rural areas, continuing to enhance the standards of basic pension for enterprise retirees and the special assistance subsidy for living;

(9) To extend reforms in value-added tax rules to all industries through encouraging technologically upgrading of enterprises and cutting the corporate taxes by 120 billion Yuan (RMB);

(10)To enhance financial support to maintain economic growth by removing the restriction on commercial bank credits and reasonably expanding the credit, by increasing the credit support to key project construction, the construction concerning agriculture, countryside and farmers, the credit support to middle-sized, small-sized enterprises, technological transformation, mergers and reorganization, by targeting to foster and consolidate consumer credit growth.

It is estimated that the government should invest four trillion Yuan (RMB) to implement the construction of those projects mentioned above at the end of 2010.

Column 1-1: Chinese Leaders Express Their Standpoints in Response to the Financial Crisis

China's state leaders have clearly expressed China's confidence, resolution and action in response to the financial crisis when participating in major international events on many occasions. On November 15th 2008, the G20 summit on financial markets and the world economy was held in Washington D.C. Chinese President Hu Jintao attended the meeting and delivered a speech entitled Making Joint Efforts to Tide over Difficulties. He said, "China has taken such measures like reducing bank deposit reserve ratio, cutting deposit and lending rates, cutting corporate taxes and other measures in order to promote the economic development. Recently, China announced another more forceful measure to expand domestic demand. China also decided to increase100 billion Yuan (RMB) investment to accelerate the construction of projects for people's livelihood, infrastructure and ecological environment, and post-disaster reconstruction, expecting to lead the community's total investment to reach 400 billion Yuan (RMB). From the fourth quarter in 2008 to the end of 2010, China has made an investment of 4 trillion Yuan (RMB) on the construction of these projects. The implementation of these measures will further promote China's economic development as well as the world's economic development.

Source: Summarized on basis of speech by Chairman Hu Jintao on G20 summit on financial markets and the world economy in Washington. People's Daily, November 16, 2008, http://politics.people.com.cn/GB/8346583.html

2. The Central Economic Conference Proposed to Maintain Growth, Expand Domestic Demand, and Adjust Structure

In December 2008, the Central Economic Conference clearly stated that the primary task of economic work for the next year should be to maintain stable and rapid economic development. Expanding domestic demand is the fundamental way to maintain growth. Speeding up the transformation of development patterns and adjusting structures are the main direction for maintaining growth. Opening up, deepening major sectors and reforming key links are driving force in maintaining growth. Improving livelihood for the people is the basis for maintaining growth.

The meeting proposed five key economic tasks in 2009:

(1) To strengthen and improve macro-economic control, implement proactive fiscal policy and moderately loose monetary policy, increase public spending at a relative large rate, implement structural tax reduction and promote the rational growth of monetary credit supply.
(2) To consolidate and develop the agricultural economy in a good shape, guarantee the efficient supply of agricultural products, put emphasis on the employment of peasant workers, continue to raise the income of farmers, and greatly increase the input in agriculture, villages and peasants.
(3) To speed up the transformation of development patterns, promote the strategic adjustment of economic structure, put emphasis on speeding up the transformation of development patterns, maintain the direction of promoting the strategic adjustment of economic structure, adjust the national income distribution pattern, optimize industrial structure, improve urban-rural structure, and adjust regional structure.
(4) To deepen reform and opening up, improve the mechanism and system benefited to scientific development, improve the price formation mechanism of resources and production factors, promote the reform of the public financial management system and speed up the reform of the financial system.
(5) To solve the issues concerning the interests of the masses, maintain social stability by implementing more active employment policies and promoting all-round employment growth.

3. The Government Work Report Proposed the Package Plan for Promoting Steady and Rapid Development of Economy

At the Second Session of the Eleventh National People's Congress in March 2009, Premier Wen Jiabao presented the government work report. He said that we should take the response to the international financial crisis and the promotion of steady and rapid economic development as the major task of government work in 2009. We should make overall plans and take all the factors into consideration, focus on the key point, and comprehensively implement the package plan for promoting the steady and rapid economic development. The tasks mainly consist of increasing government spending on a large scale, implementing and adjusting the industrial revitalization plan and increasing the level of the social security .

In 2009, the Chinese government will give high priority to the following work focusing on the implementation of the package plan:

(1) Improving and strengthening macroeconomic control and maintaining steady and rapid economic development. The Chinese government needs to adhere to a flexible and prudent control policy, implement proactive fiscal policy and moderately loose monetary policy. The government has set the central government deficit at 750 billion Yuan (RMB) in 2009 and the state council will allow local governments to issue 200 billion Yuan (RMB) government bonds. Moreover, the government will increase the broad money supply by about 17% and grant over five trillion Yuan (RMB) newly increased loans to ensure that the supply of money and credit meets the requirements of economic development.
(2) Actively boosting domestic demand, especially consumer demand, and strengthening the role of domestic demand in driving economic growth.
(3) Consolidating and strengthening the position of agriculture as the foundation of the economy, and promoting steady development of agriculture and continuous increase in rural incomes.
(4) Accelerating transformation of the pattern of development and energetically promoting strategic economic restructuring.
(5) Continuing to deepen the reform and opening up, and further improving systems and mechanisms conducive to developing in a scientific manner.
(6) Energetically developing social programs and vigorously improving people's livelihood.
(7) Intensifying government self-improvement and enabling itself to better manage overall economic and social development.

As for the launch of China's economic stimulus plan, some international media have made their comments. The American newspaper "Washington Post" published an article said that Beijing's new policy drive of upgrading infrastructure, rural land reforms, and expansion of social welfare is akin to a 'New Deal' with Chinese characteristics.[3] The British journal "Economist" called China's fiscal stimulus plan as "A New Deal, Chinese-style".[4] On November 11th, the French newspaper "Le Figaro" published an editorial believing that China were willing to make efforts to seek for the solutions of the international financial crisis.[5]

SECTION 2: THE INTRODUCTION TO CHINA'S ECONOMIC STIMULUS PLAN

Since the fourth quarter of 2008, Chinese Government adjusted the macro-economic policies timely and launched an economic stimulus plan in order to tackle the international

[3] Washington Post: China Unveils $586 Billion Stimulus Plan, 11-10-2008.. http://www.washingtonpost.com/wp-dyn/content/article/2008/11/09/AR2008110900701.html?nav=emailpage.
[4] http://theory.people.com.cn/GB/49154/49155/9679301.html, 07-19-2009.
[5] http://news.xinhuanet.com/world/2008-11/12/content_10348969.htm, 11-12-2008.

financial crisis. The package plan has been enriched and completed in practice. This plan consists of five areas.

I. The Implementation of Proactive Fiscal Policy Focusing on Four-trillion-Yuan (RMB) Investment

The main contents of the proactive fiscal policy are: the implementation of a two-year investment plan of a total amount of "four-trillion" Yuan (RMB) which includes a newly increased Chinese central government fund of 1.18 trillion Yuan (RMB), accounting for 30% of the total investment, the implementation of the structural tax reduction to expand of domestic demand, and the increase of export tax rebate to stabilize exports.

1. The Central Government Funds Guide the Local Funds

In the economic stimulus plan of four trillion Yuan (RMB), the central government provided 1.18 trillion Yuan (RMB), and the rest came from local government budget, local government bonds, policy loans, bank loans and non-government investment.[6] The investments will cover the projects concerning people's livelihood, the need of countryside, agriculture and farmers, the courses about healthcare, culture, education, the construction of ecological and environmental protection, self-innovation and economic structural adjustment.

The detailed outlay for the stimulus plan is following: 400 billion Yuan (RMB) will be spent on the construction of low-cost renting housing and the transformation of slums; 370 billion Yuan (RMB) on projects of people's livelihood and infrastructure including water supply, power grids, gas supply and housing in the rural areas; 1.5 trillion Yuan (RMB) on the construction of roads, railways, airports, water projects, and on the transformation of urban power grids; 150 billion Yuan (RMB) on developing the courses of health care, education, culture; 1 trillion Yuan (RMB) on post-disaster reconstruction. Of which, 37.5% will be spent on the construction of important infrastructures such as railways, roads, airports and water projects, which is the largest investment one. 25% will be spent on the post-earthquake reconstruction. 10% will be spent on the construction of indemnificatory housing.

2. Issuing Local Government Bonds by the Ministry of Finance

In March 2009, the two sessions, NPC and CPPCC were held. China's central government would issue bonds of 200 billion Yuan (RMB) on behalf of local governments, and the allowance for each province was also determined in the two sessions. The local government bonds will be issued by the Ministry of Finance through the present circulation of treasury bonds by bidding. The maturity period for the bonds would be 3 years. The interest rate will be the same as the interest rate of treasury bonds or a little bit higher than that of treasury bonds. In terms of solvency and economic strength, the eastern coastal areas occupy a distinct advantage, but the fund raised by issuing local government bonds will support the central and western regions to develop their infrastructure, environmental protection and people's livelihood.

[6] http://www.infzm.com/content/28912

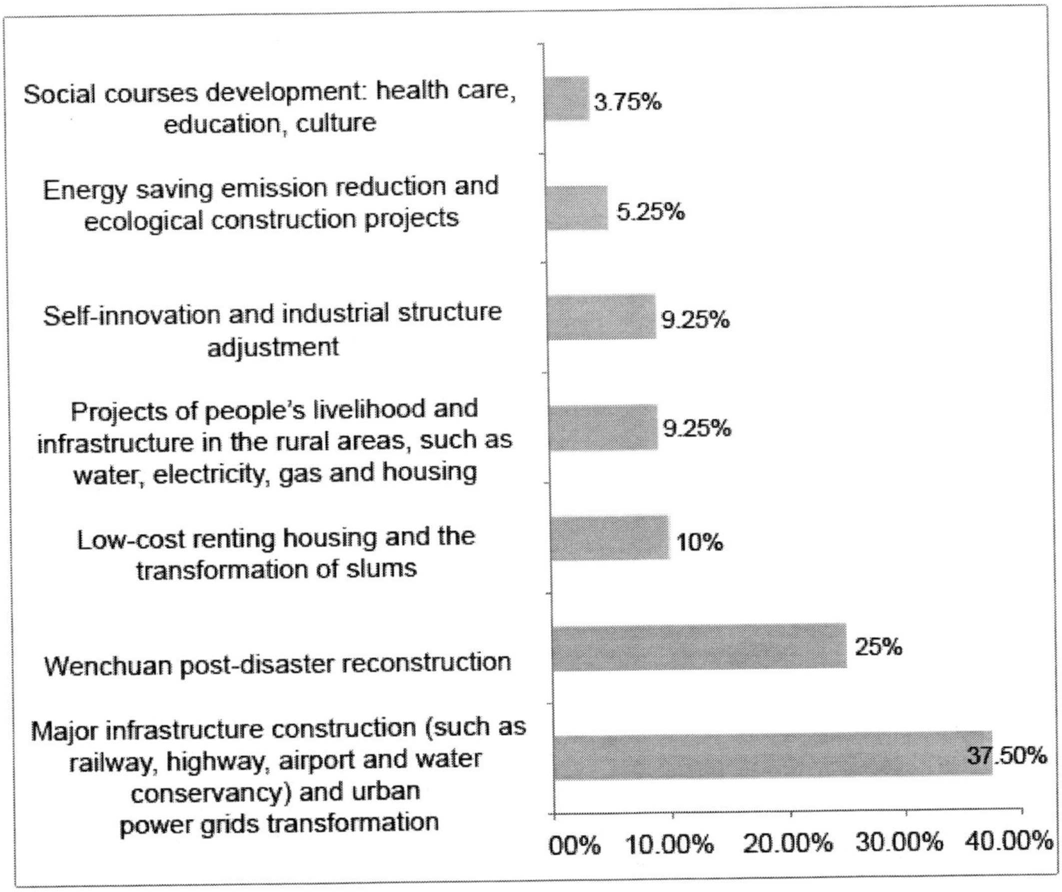

Figure 1-3. Major Industries Involved in China's Four-Trillion-Yuan Economic Stimulus Plan.

3. Implementation of Structural Tax Reduction Policy

The structural tax reduction policies are following:

(1) Fully implementing the consumption-based value-added tax, easing corporate tax burdens, promoting enterprises to increase the investment in self-innovation and technical transformation.
(2) Implementing the reform of refined oil tax, improving the tax-burden fairness, promoting energy saving and emission reduction.
(3) Increasing the salary standard deductions for individual income, lowering stamp duty on securities transactions with unilateral collection.
(4) Free from the savings deposits and securities transactions settlement funds on interest income.
(5) Reducing housing transaction tax.
(6) Substantially increasing the export tax rebate rate of some products, cancelling and reducing export tariffs on some products.

4. "Autos and Home Appliances Going to the Countryside" Program through Financial Subsidies

On May 19th 2009, the State Council executive meeting decided to implement autos and home appliances trade-in policy through financial subsidies. The Ministry of Finance will raise subsidies for auto replacements from 1 billion Yuan (RMB) to 5 billion Yuan (RMB), and allocate 2 billion Yuan (RMB) to encourage home appliance upgrades. On the basis of subsidy for auto replacement policy, the range and amount of subsidy will be expanded. Under the program, consumers who trade in their used mid- and small-sized trucks and some types of mid-sized passenger cars for new ones will receive a subsidy. Subsidies will also be given to consumers who sell automobiles that no longer meet the government's new emission standards but are still within life expectancy. The subsidy will be no more than the purchase tax of the automobile.

III. Implementation of Moderately Loose Monetary Policy

On the subject of moderately loose monetary policy, efforts will be made to promote the rational liquidity of monetary credit supply through interest rates and other monetary policy instruments. With the focus on the adjustment of credit structure, efforts will also be made to improve the quality and benefit of credit, to enhance the credit's support to the balance and sustainability of economic and social development.

1. Continuous Decline in Interest Rates

In order to ensure the liquidity of the banking system and support the economic growth, the People's Bank of China (PBC) cut the benchmark deposit and lending rates on five occasions within four months since the middle of September to the end of the year 2008. The benchmark one-year lending rate was lowered by 198 basis points, falling from 7.47% at the beginning of the year to 5.31% at the end of December. One-year deposit rate was reduced from 4.41% to 2.25%, of which, the decrease on November 26th was as high as 108 base points which is unprecedented in history. As the core meaning of moderately loose monetary policy, the successive reduction of interest rates is the traditional approach to promote investment of private assets, maintain the loose monetary environment and ensure the liquidity of monetary market.

2. Reduction of Reserve Requirement

As a quantitative instrument of monetary policy, the reserve requirement has direct controlling effects on liquidity. To meet the requirements of tight monetary policy, strengthen the liquidity management in the banking system and maintain the rational growth of credit, the PBC had raised the reserve requirement up to 17.5%, increased by 3 percentage points cumulatively on five occasions in the first half of 2008. In the wake of September, to cope with the gradually severe international economic and financial situation, and maintain the rapid growth of domestic economy, the PBC lowered RMB reserve requirement for financial institutions on September 25th, October 15th, December 5th and December 25th respectively. Specifically, the reserve requirement for large financial institutions decreased to 15.5%, cut by 4 percentage points cumulatively. Meanwhile, the reserve requirement for small and medium financial institutions decreased to 14.5%, cut by 3 percentage points cumulatively. Particularly, the PBC lowered the reserve requirement for large, small and medium financial institutions by 1 and 2 percentage points respectively on December 5th. The adjustment is unprecedented in history.

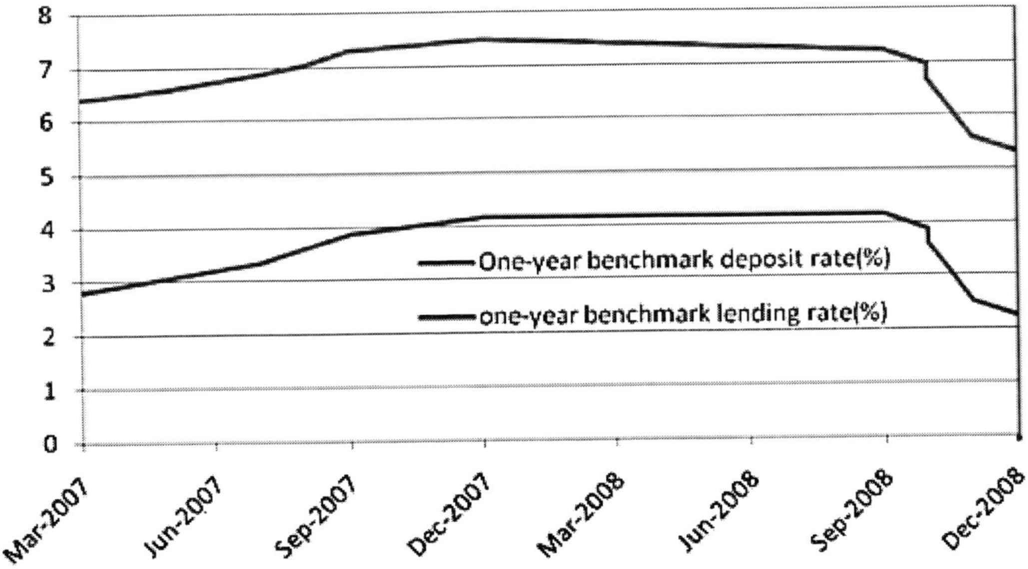

Figure 1-4. Readjustment of Interest Rate 2007—2008.

3. Increase in Money Supply

After deciding to implement the moderately loose monetary policy, at the end of March 2009, the outstanding balance of M2 money supply amounted to 53.06 trillion Yuan (RMB), increased 25.51% year-on-year, which was 7.69 percentage points higher than that in the end of 2008, and was 5.11 percentage points higher than that in February. The outstanding balance of M1 money supply reached 17.65 billion Yuan (RMB), increased 17.04% year-on-year, which was 7.98 percentage points higher than that in the end of 2008, and was 6.41 percentage points higher than that in February. The cash in circulation (M0) grew 10.88% year-on-year to 3.37 trillion Yuan (RMB).

In addition, the PBC granted 149.225 billion Yuan (RMB) by open market operations, breaking the practice of net funds withdrawal from circulation in one quarter. In the first half of 2009, the PBC granted 275.5 billion Yuan (RMB) through open market operations, which provides strong financial support to maintain economic growth to some extent.

4. RMB Settlement of Cross-border Trade Pilot Launched

On April 8th 2009, the State Council executive meeting decided on the pilot scheme of RMB settlement of cross-border transactions in five cities including Shanghai, Guangzhou, Shenzhen, Zhuhai and Dongguan. The meeting called for the relevant departments of the State Council to issue the administrative rules as soon as possible in order to standardize the RMB settlement of cross-border transactions and promote the pilot program. The pilot enterprises should have the right of pricing products and be the excellent enterprises with the ability to make the trading partners accept RMB. On June 29th, Zhou Xiaochuan, Governor of the PBC, and Joseph Yam, Chief Executive of the Hongkong Monetary Authority (HKMA), signed a supplementary Memorandum of Co-operation, to prepare for the implementation of the pilot scheme for the use of RMB in settling cross-border transactions between the enterprises in Hongkong and the enterprises in Shanghai, Guangzhou, Shenzhen, Zhuhai and Dongguan.

III. Large-scale Implementation of Industry Revitalization Plan

The industry revitalization plan is mainly concerned about promoting all-round adjustment and upgrading of industrial structure; implementing mergers and acquisitions, eliminating backward production capacity, preventing duplicate construction and promoting development of SMEs.

1. China Launched the Revitalization Plan for Major Ten Industries

In 2009, to enhance business investment information, and promote social investment growth, the State Council launched the adjustment and revitalization plan for major ten industries including steel industry, auto industry, shipbuilding industry, petrochemical industry, textile industry, light industry, non-ferrous metals industry, equipment manufacturing industry, electronics and information technology industry and logistics industry. The revitalization plan of major ten industries covers the following five areas:

(1) Expanding the domestic demand and export policies, opening up domestic and international markets for the industrial development and stability of market demand.
(2) Rationally adjusting corporate taxes, easing the tax burden of enterprises and industries, and lowering economic operating costs.
(3) Supporting the enterprises of production to establish their own brand, strive to achieve the industry leading role in the near future.
(4) Expanding the investment on advanced, technological equipments while eliminating backward production capacity and promoting the upgrading of industrial structure.
(5) Promoting mergers and acquisitions of enterprises and improving the large enterprises' international competitiveness in industries.

Besides the five areas, the Chinese government will study to launch mid-term or long-term plans for scientific technology development, and combine the implementation of major scientific projects with economic development in order to strengthen the function of science and technology to economy.

2. Efforts to Solve SMEs' Difficulties in Financing.

In the first half of 2009, the Central Government and the Ministry of Finance allocated 1 billion Yuan (RMB) of guarantee fund to 330 guarantee institutions, which will grant 250 billion Yuan (RMB) loans for 40,000 SMEs. Meanwhile, the Ministry of Industry and

Information, the People's Bank of China, China Banking Regulatory Commission and other departments have also taken a number of measures to promote SMEs' financing.

3. Adjustment of the Proportion of Capital Funds for Fixed Asset Investment Projects

The State Council decided to adjust capital ratio of fixed assets investment projects since May 25th 2009. The capital proportion of urban rail transit, coal mining, airports, ports were reduced while the capital proportion of project about calcium carbide and titanium alloy which is high-pollution, high-energy-consuming and resource-dependent. After adjustment, the minimum capital ratio of steel and electrolytic aluminum projects increased to 40%, of which, the capital ratio of electrolytic aluminum project increased five percentage points than that of 2005. The capital ratio of the cement project continued to maintain 35%. Meanwhile, the minimum capital ratio of coal mining, calcium carbide and inland navigation projects was lowered to 30%, of which, the capital ratio of coal mining was decreased by 5 percentage points than previous standard. The minimum capital ratio of railway and road projects decreased to 25%. The minimum capital ratio of ordinary commodity housing projects was reduced from 35% to 20%. Therefore, housing industry has become the greatest beneficiary of the capital ratio adjustment.

IV. Vigorous Promotion of Scientific and Technological Progress and Innovation

The State Council executive meeting held in May 2009, required that it should be further promoted to make science and technology play an important supporting role in expanding domestic demand, promoting economic growth, adjusting structures, upgrading and benefiting people's livelihood.

1. Setting Tasks

Efforts will be made to fully carry out the work concerning economic development and enterprises supported by science and technology, start to implement projects in scale application of self-innovation products, speed up the implementation of technological innovation projects and create newly economic growth point.

2. Increasing Investment

Besides the 32.8 billion Yuan (RMB) in 2009 and 30 billion Yuan (RMB) in 2010 allocated by the central government budget, enterprises will be encouraged to make investment.

3. Priority Setting

Efforts will be made to promote the implementation of 11 scientific and technological projects including advanced digital-control machines, basic manufacturing equipments, large aircrafts, next-generation broadband wireless mobile communication network, core electronic devices, advanced general chips and basic software products, large-scale integrated circuit manufacturing equipment and complete sets of technology, the development of large-scale oil and gas fields and coal seam gas, large-scale advanced pressurized water reactor and HTR nuclear power plant, water pollution control and management, new varieties of GMO cultivation, creation and production of new drugs, prevention and control of AIDS and viral hepatitis of major infectious diseases.

V. Substantially Improving Social Security Benefits

To improve social security benefits includes enhancing the basic pension for enterprise retirees, unemployment insurance, insurance for work-related injuries, increasing the margin of the minimum living allowance by the government to urban and rural residents and the "five guarantees" of food, clothing, medical care and burial expenses in the rural areas, launching a new agricultural insurance pilot project, and actively promoting the reform of medical and health system.

(1) Promoting the construction of systems. The Chinese government made efforts to improve the basic retirement insurance system, continue to carry out pilot projects in order to establish personal accounts so that funds can be accumulated, comprehensively promote provincial coordination and formulate the implementation schemes of retirement insurance for migrant workers. The new rural social security pilot project will cover 10% counties and cities around the whole country. China will launch "the transfer of pension insurance and its continuation methods", improve the insurance for work-related injuries and maternity insurance system, and the urban and rural social relief system.

(2) Expanding the social security coverage. The priority is to make non-state owned businesses, migrant workers, the peasants whose land has been requisitioned, people employed in a flexible manner and freelancers covered by the social insurances system. The Chinese government should help those low-income families cover their insurance, increase the national social security funds through various channels and earnestly strengthen the supervision of social security funds for the purpose of ensuring the safety of funds.

(3) Increasing the social security funds. From 2009 to 2010, the government will continue to raise the basic retirement pension for enterprises' retirees by an annual average of about 10% per capita, and increase unemployment insurance and industrial injury insurance pension. The guarantee of the minimum standard of living for urban and rural residents, the "five guarantees" in the rural areas will be further raised as well as the living allowances for key entitled groups. The government will vigorously develop social welfare and charity programs.

(4) Guarantee of capital investment. The Chinese government will allocate 850 billion Yuan (RMB) which is not included in 4 trillion Yuan (RMB), to carry out the pharmaceutical and health care reform.[7] The central government plans to spend 293 billion Yuan (RMB) on the social safety net this year, up 17.6% over the estimated figure for 2008.

[7] Premier Wen Jiabao's Press Conference Record, http://www.infzm.com/content/25438, 2008.

Column 1-2: The U.S Economic Stimulus Plan [1]

Establish-ment Time	Contents of Economic Stimulus Plan	Amount (Billion Dollars)	Implementation Period
Feb2008	Economic Stimulus of 2008: (1) nearly 120 billion dollars tax rebate to individuals with 0.13 billion families benefited; (2) nearly 48 trillion dollars of enterprises tax reduction (accelerated depreciation and new investment); (3) raising loan limits for single-parent families.	1680	2008—2009(152 billion in 2008,16 billion in 2009)
Mar2008	Emergency Economic Stabilization Act: (1) Troubled Asset Relief Program; (2) Raising the federal deposit insurance limit from 100,000 dollars to 250,000dollars.	7000	From issuance to December 31th 2009(550 billion dollars spent in June 2009)
Jul2008	Housing and Economic Recovery Act of 2008: (1) Purchasing debts and securities from enterprises invested by the government; (2) Hope for Homeowners Program; (3) Rebating the first-time home buyer's tax; (4) The new Mortgage Guarantee Scheme（10-01-2008—09-30-2011）。	3000	
Feb2009	U.S. Recovery & Reinvestment Act of 2009: (1) 37% for individual and corporate tax reduction; (2) 30% for health care, education and scientific research; (3) 20% for constructing infrastructure and investing new energy; (4) 13% for real estate foreclosure and vulnerable group relief (including the low-income earners, the unemployed, retirees, etc.).	7870	Over the next decade

Source: from information on official websites including http://www.usbudgetwatch.org, http://www.financialstability.gov, http://en.wikipedia.org/wiki, www.fdic.gov, http://www.newyorkfed.org, http://www.cbo.gov, http://www.cbo.gov, http://www.federalreserve.gov, etc.

SECTION 3: THE PROGRESS AND PRELIMINARY EFFECTS OF ECONOMIC STIMULUS PLAN

The progress, preliminary effect of the economic stimulus plan and the trend of China's macroeconomy has attracted great attention of governments, enterprises, media and research institutions at home and abroad.

I. The Progress of Economic Stimulus Plan

There are two years for China to implement its economic stimulus plan. Till the first half of 2009, 32% central government investment has been allocated to the right place, accounting for one third of 1.18 trillion Yuan (RMB). Funds were mainly channeled to major construction projects and livelihood projects, while the approval and process of construction projects has continued to accelerate, the combination of tax reduction measures and credit support has also been strengthened.

1. Rapid Allocation of Funds

The 100 billion Yuan (RMB) of new investment launched in the fourth quarter has played a significant role in the economic recovery, which was allocated to the right place in the first quarter in 2009. In the second half of 2009, the central government has allocated 280 billion Yuan (RMB) and the total funds allocated is 380 billion Yuan (RMB) in all, of which, 104.3 billion Yuan (RMB) for people's livelihood projects and infrastructure construction in the rural areas, 87.1 billion Yuan (RMB) for the construction of major infrastructure, 57.3 billion Yuan (RMB) for health care and education, 37.5 billion Yuan (RMB) for guaranteed housing construction, 39.6 billion Yuan (RMB) for energy-saving emission reduction and ecological environment construction, 54.2 billion Yuan (RMB) for self-innovation and structural adjustment. The funds channeled to the right place have gradually put into effects.[8]

Table 1-2. The Effect and the Availability of Investments from the Fourth Quarter of 2008 to the First Half of 2009（Unit: billion, RMB)

Investment Areas	Investment Amount	The Investment Achievements
People's livelihood projects and infrastructure construction in the rural areas	104.3	Solved the issue of drinking water safety for19.6 million residents in the rural areas; constructed1.78million rural biogas projects, transformed 7500 kilometers rural power grids
Guaranteed housing construction	37.5	250,000 low-rent housing buildings 105 million low-rent housing buildings started
Major infrastructure construction	87.1	Four major railway lines including Harbin-Dalian, Wuhan-Guangzhou, Nanning-Guangzhou, Guiyang-Guangzhou; South-to-North Water Diversion Project; constructed 1360 large and medium-sized water conservancy projects.
Health care, education	57.3	Constructed more than 7800 projects of primary health care services; completed 3 million square meters of rural junior high school
Energy-saving emission reduction ,ecological environment construction	39.6	Afforestation over 3700 mu, and also additional sewage treatment capacity of 4 million tons / day
Self-innovation, structural adjustment	54.2	176 high-tech industrialization projects and 146 industrial projects with technological advances; some of these projects have stepped into the major project construction and equipment installation and adjustment phase.

Source: website of the State Council Information Office of China, August7, 2009, http://gb.cri.cn/news/live/20090810.htm

[8] State Council Information Office, August 7th (15:00) 2009, http://gb.cri.cn/news/live/20090810.htm

2. Speeding up the Examination and Approval of Major Projects

Some projects included in the "four-trillion-Yuan (RMB)" investment were originally included in the "Eleventh Five-Year Plan", such as roads, railways and other infrastructure constructions. Made after full argument, these projects will be accelerated and promoted between 2009 and 2010. Moreover, the National Development and Reform Commission speeds up the examination and approval of new major projects. The year of 2008 witnessed the approval of 18 major projects of railways construction with a total length of 7389 kilometers and feasibility study reports, of which, 6 projects were approved in October and November, with a total length of 1992 kilometers which accounts for 30% of the total projects and length approved in the year. In addition, 45 major construction projects of roads were approved in 2008, with a total length of 4536.3 kilometers, of which, 18 projects were approved between October and December, with a total length of 1664.3 kilometers accounting for 40% of the total projects and length.

Table 1-3. Major Railway Projects Approved by China's National Development and Reform Commission between October and November in 2008

Number	Time for Approval	Project Name	Starting and End Points	Length (Kilometers)
1	October 2008	Xilinhot-Ulanhot Railway	Xilinhot-Ulanhot	651
2	October 2008	Litang- Guangzhou of Nanning-Guangzhou Raiway	Nanning-Guangzhou	471
3	October 2008	Chongqing-Lichuan and Feasibility Reports	Chongqing-Lichuan	261
4	October 2008	The feasibility report of building the Second Double Track between Jining and Baotou in Jining-Baotou railway	Jining-Baotou	263
5	November 2008	The project proposal of enlarging railway transportation capacity of Jinhua-Wenzhou railway	Jinhua-Wenzhou	188
6	November 2008	The project proposal of Shanghai-Hangzhou railway passenger line	Shanghai-Hangzhou	158

18 major railway construction projects were approved between January and July 2009, of which, three large projects are beyond 1000 kilometers: the Project Proposal of the Second Double Track Construction for Lanzhou-Xinjiang Railway, the Project Proposal of Newly-built Changsha-Kunming Railway Passenger Line, the Project Proposal of Newly-built Railway Corridor in Central and Southern Shanxi. All of these three large projects are located

in the central and western regions of China which represents the central government's strong support to the construction of these regions.[9]

Figure1-5. The Major Railway Projects Approved by the NDRC in the First Half of 2009.

From the examination and approval of major projects, it can be seen that the coordination of regional development is gradually enhanced. With the launch of a series of plans, policies and measures concerning the promotion of regional economic development, the Midwest and Northeast regions of China can seize the opportunities to strengthen the infrastructure construction, undertake industrial transfer and develop the industries with local features; the eastern region of China needs to overcome the impact of financial crisis and make new progress in promoting transformation, upgrading, self-innovation and competitiveness enhancement.

3. Substantial Enhancement of Structural Tax Reduction

In addition to increasing investment, the implementation of structural tax reduction as well as easing the tax burden on business and residents is one part of China's economic stimulus plan.

(1) Collecting the consumption-based value-added tax in an all-round way, easing corporate tax burdens, promoting enterprises to increase the investment in self-innovation and technical transformation.

[9] National Development and Reform Commission, http://zfxxgk.ndrc.gov.cn/PublicItemFrame.aspx

(2) Reducing the tax burdens on low-income individuals. With the tax adjustment strengthened in 2008, the standard of cost deduction in individual income increased to 2000Yuan (RMB) per month. Individual Income Tax on interest from savings deposits and interest form securities transactions settlement funds will be exempt. Collecting from only one side of the stock traders the stamp tax followed the tax reduction. The government will take measures to reduce residents' housing-purchasing burdens by declining property deed tax for the people buying their first home only if it is smaller than 90 square meters, exempting the stamp tax and land value-added tax.

(3) Increasing the export tax rebate rate of certain commodities. The adjustment policy of export tax rebate was implemented on August 1, 2008. The export tax rebate rate of some textiles and garments increased from 11% to 13%. The export tax rebate rate of some bamboo products increased to 11%. On November 1st, the export tax rebate rate of 3486 commodities of textiles and garments was readjusted. These commodities accounted for 25.8% of all commodities listed in China Customs Tariff. This adjustment of export tax rebate rate policy was the largest one which involved the most tariffs since 2004. Since December 1st, the government increased the export tax rebate rate and adjusted 3770 commodities' export tariffs, accounting for 27.9% of the total export commodities. In the wake of implementing the tax reduction measure, the growth rate of financial revenue declined to 19% in the second half of 2008. While the growth rate of financial revenue was 33% in the first half of 2008.

In the first half of 2009, the government implemented the tax reform of refined oil and promoted the energy-saving gas emission, cancelled 100 administrative fees in addition to continuously carry out measures of tax reduction and exemption. As for house transferring, tax payers can have access to business tax reduction or exemption for a year. Passenger vehicles with 1.6liters of displacement or below can be collected 5% vehicle purchase tax. From April 1st, the export tax rebate rates were increased for goods under tariff code no. 3802. The commodities are as follows: textiles, garments, light industry, electronic information, steel, nonferrous metals and petrochemical products. Meanwhile, the export tariff of some steel, chemicals, grains will be cancelled and the export tariff on certain fertilizer will be reduced. It is estimated that the tax reduction amount in 2009 will reach more than 500 billion Yuan (RMB).

According to the announcement of the Ministry of Finance of the People's Republic of China, the local government bonds worth 200 billion Yuan (RMB) had been issued completely till September 4th 2009, covered with all the provinces and autonomous regions except Hong Kong, Macau, and Tibet.

4. Substantially Relaxing Credit

Since the second half of 2008, the PBC began to implement loose monetary policy by relaxing commercial bank's credit, reducing interest rates on savings, loans and reserve requirement for commercial banks. At the beginning of December, with the deterioration of economic situation, the PBC declared the cancellation of credit line and provided window guidance to banks in terms of the credit extensions and channels. Since the announcement of "four-trillion-Yuan (RMB)" economic stimulus plan, the credit extension of financial institutions increased rapidly by the end of 2008. In the first half of 2009, the RMB loans reached 7.37 trillion Yuan (RMB), one point five times as much as the new loans of 4.91 trillion Yuan (RMB) in 2008, three times as much as the new loans of 2.45 trillion Yuan (RMB) in the first half of 2008. Thus, the accommodative monetary policy has provided significant additional liquidity into the economic field which plays an active role in promoting the economic recovery and stability.

II. The Preliminary Effects of Economic Stimulus Plan

With the implementation of economic stimulus plan, China's economy has developed in a good trend with positive factors increased. The situation is stabilized with a marked economic recovery.

1. Stable Economic Recovery in China

Based on the requirements of "four-trillion-Yuan (RMB)" package plan, China's economy has stepped into recovery in terms of the progress in expanding the investment and consumption, stabilizing exports, adjusting and optimizing economic structure, improving people's livelihood and farmers' income.

Firstly, China's GDP grows steadily. In the first half of 2009, the gross domestic product grew year on year by 7.1%, of which, 6.1% of growth in the first quarter, 7.9% of growth in the second quarter. It can be noticed that the GDP growth rate in the second quarter is 1.8 percentage points higher than that in the first quarter. During January and February, the above-scaled industrial production grew 3.8%, increased by 3.8% in March, declined in April, raised 8.9% in May, and increased by 10.7% in June.

Secondly, the domestic demand grows rapidly. Expanding the domestic demand is the driving force and focus task of the economic recovery. In the first half of 2009, the total retail sales of social consumer goods increased by 15%, an actual increase of 16.6% after deducting price factors, up year on year 3.7%. It should be noticed that the commercial housing sales and the automobile sales increased by 31.7% and 17.7% respectively.

Thirdly, the economic structural adjustment is actively promoted. To promote structural adjustment is the requirement and need of maintaining stable and rapid economic development. At first, the infrastructure and basic industries have been further strengthened. In the first half of 2009, the investment in infrastructure (power grid excluded) increased by 57.4%, of which, 126.5% increased in railroad transportation; 57.4% increased in road transportation; 54.5% increased in water conservancy, environment and public facility management industry; 71.3% increased in health care, social security and social welfare; 57.1% increased in culture, sports and entertainment. Secondly, the equipment industry has stepped into rapid recovery. It increased by 6.7% year-on-year in the first half of the year, 3 percentage points higher compared with that in the end of the first quarter. Finally, the coordination of regional development has been strengthened with the significant growth of investment and industrial production in the Midwest regions. In terms of fixed asset investment in the urban areas, the growth rate was 38.1% in the central region, 42.1% in the western region, 26.7% in the eastern region. As for industrial production, the above-scaled industrial added value increased by 6.8% in the central region, 13.2% in western region, and 5.9% in the eastern region.

Fourthly, people's livelihood continues to be improved. The overall employment maintained basic stability. Particularly, the population of rural migrant workers increased gradually. In the first half year, there were 5.69 million new jobs created, accounting for 63% of the yearly target. At the end of the second quarter, the rural migrant workers increased by 3.78 million compared with that of the first quarter, an increase of 2.6%. Among them, there were 0.56 million people increased in the eastern region, raised by 1.6%; 0.8 million people increased in central region by 1.8%; 2.42 million people increased in western region by 6.5%. Secondly, the residents' income has grown steadily. In the first half year, the disposable income of urban residents reached 8856 Yuan (RMB) per capita, increasing by year on year 9.8%, a real growth of 8.1%. Per capita cash income of rural residents was 2733 Yuan (RMB), climbing up by year on year 8.1%, a real growth of 8.1% after deducting price factors. Third, the social security expenditure further increased. With the large financial revenue pressures in the first half year, the fiscal expenditures for social security and employment increased by year on year 29.2%. The minimum living security allocated by the central government and the Ministry of Finance for urban residents and rural residents respectively increased by 49.9% and 140%.[10]

2. Market Confidence Rebounded

The government's economic stimulus plan is to convey a clear signal that the economy has to grow at least 8%. This signal is more important than any specific investment figures for it can help families and enterprises restore confidence and encourage enterprises to maintain the investment.

The news that the five indexes rebounded in the second quarter of 2009 was released on the web site of China National Bureau of Statistics, China Macroeconomic Information Network and China Federation of Logistics & Purchasing. Among the five indexes, China's Entrepreneur Confidence Index (ECI) gained 110.2 in the second quarter, promoted 9.1 points over previous quarter. The National Business Climate Index (BCI) declined since the third quarter in 2008. This index was 115.9 in the second quarter, rising by 10.3 points from the first quarter. China's Consumer Confidence Index (CCI) has continuously declined in 2009. Compared with the CCI in April, the CCI in May rebounded which implied that the consumer's confidence had stabilized and slightly recovered. That fact that China's Manufacturing Purchasing Managers Index (PMI) was 53.2%, above the critical point of 50%, demonstrating that the manufacturing economy as a whole had shown a steady upward trend due to the implementation of the central government's policies and measures concerning expanding domestic demand as well as promoting stable, rapid growth of national economy. The rebound of China's Macroeconomic Early-warning Index indicated that the current Chinese economic situation would develop well.

3. Reflections and Concerns on Sustainable Economic Growth

With the implementation of economic stimulus plan, China's package plan is already paying off. China's economy had shown the signs of economic growth in the second quarter of 2009. Besides the negative growth in foreign trade, other indexes have substantially rebounded. However, the investment-led economic recovery, inadequate consumption pulling

[10] http://www.mof.gov.cn/mof/zhengwuxinxi/caijingshidian/jjrb/200907/t20090717_182756.html

and too much credit loans have aroused people's reflections and concerns about inflation and economic growth sustainability.

First of all, the possibility of inflation will increase. The continued loose monetary policy has led to ample market liquidity which will enter the capital market with high return on investment so that the asset prices will be pushed up. Meanwhile, the revival of real economy is a relatively slow process. The process of increasing credit to increase investment to speed up economic growth requires a certain period of transmission time. Therefore, the price level would not change from deflation into inflation in a short time. As a result, the asset price and commodity price will be deviated from each other. Since the rise of asset price is independent from the growth of real economy, bubble risks will come into life. If the risks can not be controlled, the collapse of bubble economy will lead to a greater impact on economy.

Secondly, the stimulus effect of infrastructure construction on employment is limited. The new investments are highly channeled in capital-intensive heavy industry and infrastructure construction which is one important reason for non-further improvement in employment. However, the sectors mentioned above can create limited job opportunities. As a matter of fact, the maintenance of economic growth, in the final analysis, is to maintain the employment, people's livelihood and social security. Only solved the employment problems, coupled with a reasonable income distribution patterns and effective social security system, can we fundamentally solve the people's problems and achieve social harmony and stability.

Thirdly, the investment-led economy may make the income distribution deteriorated. As for the three demands of GDP growth in the first half of 2009, the economic growth contribution rate of the gross capital formation, mainly including fixed assets investment and inventory, was 87.6%, pulling GDP increase by 6.2 percentage points. The contribution rate of final consumption to economic growth was 53.4%, boosting GDP growth of 3.8 percentage points. The contribution rate of net exports to economic growth was -41%, decreasing GDP growth rate by 2.9 percentage points. The funds were mainly from the central government, local government and bank loans. Although the construction of major projects has stimulated the economic growth significantly, the benefits for most labors are still inadequate, which may result in further deterioration of income distribution.[11]

Finally, the investment in social security system is relatively smaller. Using Roosevelt's American New Deal for reference, the aim of the New Deal is striving to establish a series of economic security system in order to make people free from fear. It can be noticed that the core of the New Deal is to establish social security system and achieve balanced economic growth with stability in a long term rather than increase government spending to construct infrastructures. Although the projects of people's livelihood is included in the "four trillion" economic stimulus plan, the investment proportion in these projects is still much less than that in supporting infrastructure construction, boosting economic growth. If we continue the current pattern of investment-led economic growth, with the rapid economic growth, the structural contradictions will be more and more severe. The income proportion of household and workers will be much lower in the economic system, which will lead to low consumption willingness. Therefore, only by improving the income distribution pattern and establishing a sound social security system will China's economy maintain sustainable long-term growth.[12]

[11] Can Chinese Style Economic Recovery last? http://www.infzm.com/content/32207

[12] Huang Yiping. The most difficult time is the best chance for reform. http://www.infzm.com/content/32224

III. The Future Trend of National Economic Development

With the further implementation of the world economic stimulus plan, there have been positive changes in the world economy in the first half of 2009: the asset price increases and the external economic environment of China has also been improved. China's economic growth rate rebounded in the first quarter of 2009. The economic growth was close to the full-year target of 8% in the second quarter. The future fiscal and monetary policy will play a greater role in promoting stable and rapid economic development. As long as we can accurately predict the world economic development and recognize potential problems, diminish the unstable factors' impact on Chinese economy, it will move toward a sound development.

1. The Fixed Asset Investment's Stimulus Effects on Economic Growth Will Further Increase

For a long time, investment has played an important role in stimulating China's economic growth. Especially, the introduction and implementation of economic stimulus plan accelerated the investment growth further. The investment growth stamina is abundant, manifested as increased investment in infrastructure, the increasing new projects and the rapid growth of total investment in new projects. In particular, the construction of people's livelihood projects has been enhanced in the economic stimulus plan, which will solve the low-income group's problems, be beneficial to expand consumption and optimize economic growth structure.

2. Consumption Will Steadily Increase with Sustainability

Since the breakout of the financial crisis, to improve residents' income and expand the final consumption demand, China has unveiled relevant policies to adjust national income distribution pattern, improve the proportion of residents' income in national income, adjust the enterprise retirement pensions, improve residents' property income and vigorously promote residents' consumption in the urban and rural areas. The financial and fiscal measures adopted by Chinese government, such as issuing coupons, guiding home appliances and automobiles to the countryside, new-for-old program, increasing grain subsidies to farmers, and other policies and measures are conducive to the expansion of domestic consumption demand, thereby stimulating economic growth. These measures will lead to sustainable consumption growth and achieve economic growth co-stimulated by consumption and investment under the premise of optimizing investment structure.

3. Positive Changes in Imports and Exports

The international financial crisis not only lead to a drastic shrinkage of national wealth in the developed countries, decrease the current residents' consumption expenditure, affect the future economic growth prospects, but also aggravate the protectionism in many countries which causes negative impact on China' imports and exports. Although the situation of China's imports and exports is very grim, with negative contribution to the economic growth, we should pay attention to the positive changes. Firstly, the private-funded enterprises become much stronger than the state-owned enterprise and foreign-funded enterprises in terms of responding to crisis. In the first half of 2009, the rate of decline in imports and exports by private-funded enterprises was half of state-owned enterprises and foreign invested enterprises, and the imports and exports proportion of private-funded enterprises showed an upward trend in total national imports and exports. Secondly, the development momentum of domestic enterprises indicates the promising direction of improving the export enterprises' main structure of the market. The financial crisis has affected foreign-invested enterprises dealing with processing trade greater than domestic-funded enterprises. We should seize the opportunity to support domestic enterprises to cultivate export brands with their own intellectual property rights. With the implementation of national policy for supporting foreign trade growth, China's foreign trade will show a resumption of growth.

Based on the comprehensive research at home and abroad, China's economy is expected to achieve growth target of 8% in 2009 and is likely to maintain rapid growth in 2010. However, the ultimate performance of "four-trillion-Yuan (RMB)" investment remains unknown due to the uncertainty of domestic and international economic environment.

Translator: Luo Liyun (SUFE)
English Version Editors: Xiao Rong (SUFE), Jin Ji (Harvard)

REFERENCES

[1] Chen Jiagui, 2009. *Analysis on the Prospect of China's Economy* [M]. Beijing: Social Sciences Academic Press.

[2] Donald.J.Korn, 2009. Interactions between Monetary and Fiscal Policy in the Current Situation(Chinese)[J]. *China Finance*, 13, 33-35.

[3] Dwight.H.Perkins, 2009. Spectator perspective: China's existing and viable economic stimulus package (Chinese) [J]. *China Development Observation* 4, 46-47.

[4] Gao Peiyong, 2008. *The People's Well-being & Public Finance* [M]. Beijing: China Financial & Economic Publishing House.

[5] Gan Xingqiong, 2008. Tax Rebate and Tax Reduction - Comparison between Sino-US Economic Stimulus Plans (Chinese)[J]. Finance & Trade Economics 12, 69-72.

[6] Han Guangyu, 2009. *China's Development in the Context of International Financial Crisis* [M]. Beijing: China Renmin University.

[7] Jia Kang, 2009. Essentials of Fiscal Policy in 2009[J]. *China TodayForum*(z1), 25-26.

[8] Li Xiaoxi, 2009. *China's Economic Trajectory in the New Century——Sub-quarter Economic Situational Analysis Report2007-2008*[M]. Beijing: People's Publishing House.

[9] Wang Xiaolu, Fan Gang, Liu Peng, 2009. Transformation of Growth Pattern and Growth Sustainability in China(Chinese)[J]. *Economic Research Journal*, 1, 4-16.

[10] ANDREWS E L. Tracking the Bailout: Summary of U.S. bailout efforts. The New York Times:http://www.nytimes.com/2008/11/26/business/economy/26fed.html?_r=1.

[11] DOLLAR D. China's stimulus plan also aims to improve quality of life. November 12, 2008.http://eapblog.worldbank.org.

[12] LEE D. China's $586-billion stimulus plan could boost world economy. Los Angeles Times: http://articles.latimes.com/2008/nov/10/business/fi-china10.

[13] ROUBINI N. The Rising Risk of a Hard Landing in China: The Two Engines of Global Growth – U.S. and China – are Stalling. http://chineseculture.about.com/.

In: China's New Deal
Editors: Li Xiaoxi and Hu Biliang

ISBN: 978-1-61470-815-5

Comments on Chapter 1

KEYNES'S PHILOSOPHY IN CHINA: ASSESSING THE FOUR-TRILLION-YUAN ECONOMIC STIMULUS PACKAGE

Shu Yan *

Triggered by the U.S. sub-prime mortgage crisis in April 2007, the current financial crisis has engulfed the world. Although China did not suffer the western-style financial crisis, it was deeply affected by the collapse in external demand due to this crisis. China is a labor abundant country and has comparative advantage in the production of labor-intensive manufacturing goods. Over a long period of time, the export sector of these goods has been an important driver of China's high economic growth. But recently, when the financial crisis led to a dramatic fall on the international demand for goods, China's export sector has got into trouble. Thousands of export-oriented enterprises have been forced to shut down. The unemployment rate climbed fast in China inevitably. Compared with 2007, the GDP growth rate in 2008 decreased by four percentage points. China's economy experienced a real crisis to some extent. Fortunately, the Chinese government responded very quickly. In November 2008, the government unveiled a four-trillion-yuan economic stimulus package to be spent over the next two years (i.e. 2009 and 2010), aiming at stimulating domestic demand to compensate for the reduction in exports, maintaining high economic growth and boosting employment. It is a concrete manifestation of Keynesian counter-cyclical economic policy. The current situation in China shows that the stimulus policy is gradually playing a role.

The first chapter of China Economy & Resources Management Report 2010 describes in details the background, contents as well as developments and preliminary effects of this stimulus plan. Through the progressive implementation of the stimulus plan, China is showing signs of improvement. Official figures state that GDP growth hit 7.9 percent in the second quarter of 2009. This growth rate is very close to the target of "bao ba" or "preserve

* Shu Yan is associate professor in Research Institute of Economics and Management at Southwestern University of Finance and Economics, P. R. China. She earned her PhD degree in Economics from Florida International University, USA.

eight", which means maintaining a pace of economic growth at 8 percent. Those who are skeptical of the quality of China's macroeconomic statistics can look over some associated specific data: for example, the production of electricity raised 5.2 percent in June 2009, completely changing its declining situation in the previous eight months[1]. Nevertheless, it is too early to hail success of the stimulus plan. We should pay more attention to the problems that occurred or may occur during the implementation of the stimulus package. Among these are, for instance, possible inflation and asset bubbles, limited impact on boosting employment and consumption, etc. This comment on the first chapter of the Report will provide a brief discussion of some main concerns about the stimulus package, which attract a lot of domestic and foreign interests.

NEW INVESTMENT VERSUS OLD PROJECTS

China's four-trillion-yuan investment, the main component of the stimulus package, consists of several projects in the existing "Eleventh Five-Year Plan", especially those for the rest two years in the Plan. Thus, actual amount of new investment in the stimulus package is more likely to be less than the headline figure of "four-trillion-yuan". Although Keynesians tend to believe that only extra government investment can execute positive impact on the level of and growth rate of aggregate demand, China's declaration of "Four-trillion-yuan Economic Stimulus Package" is justified by the following reasons. First, when considering a stimulus plan at national level, we should not only ensure to achieve its short-term goals but also care about its effects on the long-term economic growth of the country. We believe only if it is in accord with the "Eleventh Five-Year Plan" can the four-trillion-yuan investment produce a long-standing constructive effect. Next, we must learn the lesson from the implementation of similar fiscal stimulus plan in China during the Asian Financial Crisis ten years ago. One of its key problems is that some investment projects were launched hastily at that time in order to have a fast influence on GDP growth[2]. As a result, there existed to some degree redundant construction and waste of resources. Since investment projects in the "Eleventh Five-Year Plan" survived full feasibility studies before the scope of the crisis became clear, it is appropriate to contain them in the current economic stimulus package. Last, the key to the success of a national stimulus package lies in its ability to boost market confidence. Through the declaration of "Four-trillion-yuan Stimulus Package", the Chinese government is sending the public such a signal that central government is capable of combating crisis and guaranteeing the sustainable and stable economic growth. From this perspective, the size of a stimulus package may not as important as that in theory. After all, it is difficult to know for sure how large the fiscal multipliers are and what scale of new investment is enough to rebound the economy from downturn or recession. If necessary, just as other countries, China can also add the second-round fiscal stimulus. But on top of that, we have to wait for further macroeconomic data to judge effects of the first stimulus package and then take corresponding measures.

---

1 YU Y. China's Stimulus shows the problem of success (August 25, 2009) Financial Times.

2 ZHENG Y., CHEN M. *How Effective will China's Four Trillion Yuan Stimulus Plan Be?* (March 2009) China Policy Institute, University of Nottingham.

SOURCES OF FUNDING

Another eye-catching question related to the four-trillion-yuan economic stimulus package is how China will come up with the cash. In May 2009, National Development and Reform Commission, China's macroeconomic management agency, first publicly disclosed detailed funding sources of the package. With regard to the four-trillion-yuan investment, central government will provide 1.18 trillion, which is about 30 percent of the total. The rest 2.82 trillion yuan will come from the provincial fiscal budget, local government bonds, policy loans (including subsidies on interest payments), corporate bonds and medium-term notes, as well as bank loans and private investment. However, the proportion of each fund was not published. Looking at the situation of the fourth quarter of 2008 and the first six months of 2009, we could know that central fiscal funds were issued very quickly and timely, which is the main force driving the total investment up. Over the past decade, China has had very low budget deficits, and even it ran a fiscal surplus in 2007. So the government has plentiful room to operate even if it needs to issue bonds to meet the requirement of the 1.18 trillion yuan central investment. In contrast, it seems likely that there is some uncertainty about the 2.82 trillion yuan "social matching funds". At the moment large contribution to these funds comes from bank loans, local government bonds issued by the Ministry of Finance, corporate bonds and policy loans. Among these bank loans are the most important part. It is noteworthy that private capital occupies a very small proportion in the four-trillion-yuan economic stimulus funds. The main reason is that private capital does not want to enter non-profitable projects while it has little chance of accessing to profitable projects. Hence, it is very necessary to set policies to stimulate enthusiasm in private investment, such as policies of tax breaks, fiscal subsidies, etc. Attracting foreign investment should also be an important consideration in raising funds in theory. But due to the liquidity shortage caused by the current global financial crisis and weak international real economies, foreign investment, as a source of funding, will be rather limited.

CREDIT FLOWING

To support the four-trillion-yuan economic stimulus package, China's central bank loosed control on credit scale. But things went far beyond that. In the first half of 2009, bank lending increased up to 7.37 trillion yuan, even greater than the official target of 5 trillion yuan for the whole year. Similarly, the broad money, M2, is growing at an exceptional rate.[3] Thus, the accommodative monetary policy has provided significant additional liquidity to our financial market and economy. There are more and more people worrying about inflation risks that may come out of this policy. Although overcapacity in many industries is delaying inflation in some sense, the inflation itself could happen at any time due to internal or external shocks. In particular, when liquidity was quickly built up and the use of funds was lack of effective supervision, there occurred that some companies who obtained credit funds did not engage in real investments to enhance production and employment, but speculate in the stock

[3] YU Y. China's Stimulus shows the problem of success (August 25, 2009) Financial Times.

and property markets which produce a higher return at the time being. Possible result of doing so is continuously rising asset prices ending with the formation of asset bubbles. Therefore, to avoid repeating mistakes which triggered the present western-style financial crisis, an urgent task for China is to strengthen the supervision of credit funds, thereby ensuring credit flowing into productive investment. Meanwhile, banks should pay more attention to their own risk management.

BOOSTING EMPLOYMENT

At present, one of important concerns to the Chinese government is that up to 20 million rural migrant workers lost their jobs or could not find jobs in 2008. Moreover, according to the report from China's Academy of Social Sciences, 7.8 million university graduates will enter the labor market in 2009. Among these, 40 percent (around 3 million) may not be able to find employment.[4] Right now, many people count on this four-trillion-yuan economic stimulus package to solve the employment problem. But indeed the package has very limited effects on creating jobs for the following two reasons. First, just as what the Report pointed out, the package focused on capital-intensive industry, for example infrastructure development, and will create relatively few jobs directly. Second, the vast majority of bank lending is flowing into large state-owned enterprises (SOEs). This exacerbates overcapacity in several sectors that are dominated by SOEs, such as steel, chemical, and telecommunication industries. In turn, their overcapacity squeezes hard private and especially small and medium-size enterprises (SMEs). As a result, SMEs faced difficulties to survive. Here comes the problem since these SMEs are the most important employment generators in the Chinese labor market. It is estimated that 75 percent of job positions in China comes from SMEs. Therefore, to boost employment, it is necessary for the government to provide SMEs financial guarantee and support their normal operations. On the other hand, China can also learn from the practice of Germany. The government could encourage enterprises to hire new workers or not dismiss original workers by means of fiscal subsidies.

STIMULATING CONSUMPTION

Over a long time, Chinese people's high rate of savings has made impossible for domestic household consumption to be the main drive force of economic growth. To stimulate the consumption, a popular approach, in Keynes's philosophy, is tax cuts targeted at low-income consumers. The reason of targeting at this group is that they tend to have a higher marginal propensity to consume than the rich. Given the features of China's tax system, this approach will not take effect as expected in China. Unlike the United States, China does not make the income tax as the main component of its tax system. The low-income consumers in China basically do not pay income taxes or pay very little tax. Therefore, China has to turn to other options than tax cuts. In effect, improving the social welfare system is the crux of stimulating household consumption because the extraordinarily high saving rates in China are

4 ERNST D. China's Stimulus Package: A Catalyst for Recovery? (June 2009) East-West Center.

not accidental but results of the low level of social security services. Only if their top concerns about, for example, education, healthcare, and pension insurance, are resolved, the ordinary Chinese people could feel safe to withdraw their savings and consume. This problem has already come to the Chinese government's attention for a long time. Despite the fact that only 3.75 percent of the four-trillion-yuan stimulus money is put on "social safety net" programs, the government is trying hard to change the status quo from other channels, such as investing additional 0.85 trillion yuan to carry out a three-year reform of medical and health system.

Opportunities for Foreign-Funded Enterprises

Although state-owned enterprises and domestic private companies will benefit most from the four-trillion-yuan stimulus package due to the characteristics of China's economic policy and the restrictions on foreign investment, foreign-funded enterprises can also find opportunities within the stimulus package.[5] It is not easy for them to obtain contacts of investment projects directly. But they can make full use of their advantage and perform as subcontractors to the large contracts that Chinese companies already got. Foreign-funded enterprises have the distinct advantage in converting advanced intellectual properties and providing value-added services, for example, their clean energy technologies, and project management services that will be useful to the rural infrastructure development or earthquake reconstruction, etc. These advantages create opportunities for them when faced with China's stimulus package.

Overall, from either political or economic perspective, China's four-trillion-yuan economic stimulus package is an important strategy to combat the global financial crisis. With the implementation of specific policies and measures, the stimulus package will gradually demonstrate its positive effects. At the same time, we should recognize problems and risks that may emerge from the implementation of the package and prevent and control those problems as soon as possible. Furthermore, we should fully seize opportunities within the stimulus plan and contribute to the realization of the goal of "maintaining growth, boosting employment and stimulating domestic demand".

References

ERNST D. China's Stimulus Package: A Catalyst for Recovery? (June 3, 2009) East-West Center.

JACKSON J K. The Financial Crisis: Impact on and Response by The European Union (June 24,2009). Congressional Research Service.

KWAN C. A Balanced Approach – China's Economic Stimulus Plan (November 2008). U.S. Chinese Services Group.

[5] REVELLE G., CHIANG J. *China Stimulus Package* (June 17, 2009) Stoel Rives LLP.

PRASAD E., SORKIN I. Assessing the G-20 Economic Stimulus Plans: A Deeper Look (March 2009). Brookings Series.

REVELLE G., CHIANG J. China Stimulus Package (June 17, 2009) Stoel Rives LLP.

YU Y. China's Stimulus shows the problem of success (August 25, 2009) Financial Times.

ZHENG Y., CHEN M. How Effective will China's Four Trillion Yuan Stimulus Plan Be? (March 2009) China Policy Institute, University of Nottingham.

In: China's New Deal
Editors: Li Xiaoxi and Hu Biliang

ISBN: 978-1-61470-815-5

Chapter 2

BUILDING A NEW COUNTRYSIDE: THE BREAKTHROUGH POINT OF STIMULATING DOMESTIC DEMAND

Li Jing[*], *Zhang Jiangxue, Feng Tao and Zhang Ning*

In order to effectively cope with the negative effect of the international financial crisis on China, executive meetings of the State Council initiated 10 measures to expand domestic demand on November 15th, 2008. One of the main focuses of these measures is to building a new countryside byincreasing countryside consumption as well as improving its infrastructural facilities. From then on, in order to stimulate consumption in the rural area, Chinese government has put forward a series of measures such as household electric appliances going to the countryside, automobile going to the countryside, etc., which have already make some achievements. However, to expand consumption in the rural area is a long-term and arduous task which asks for not only amelioration of the problem during the execution of measures but also the coordination of macro policies and the reform of the distribution of national income.

SECTION 1: NEW COUNTRYSIDE CONSTRUCTION AND NATIONAL ECONOMY

The primary problem existing in China's economic development for a long time is insufficient domestic demand. The potential consuming capacity in the rural area in China is huge, but the low income of farmers is the major factor that restricts consumption in the rural area. Therefore, the strategy of increasing farmers' income so as to build a new countryside is of great significance.

* Dr. Li Jing is a research fellow of the Chinese Academy of Social Sciences as well as a professor of the Graduate School of the Academy. Email: ljrdi@263.net

I. Rural Economy and National Economy

1. The Role of Rural Economy in National Economy

China's economic reform is initiated in the rural area. The reform and development of rural economy used to be an important driving force of China's economic growth. However, due to the restriction of urban-rural dual segmentation system, the economic development between the rural area, the city and the whole nation are asynchronous. Since 1990s, the proportion of rural economy in the national economy has been gradually decreasing as a result of a slowdown in agricultural growth and a decrease in farmers' income growth. Amongst all components of the decrease in rural economic growth proportion, the decrease of the proportion of primary industry in the national economy is the greatest, while that of the secondary industry has stagnated in recent years, and that of the tertiary industry has also shown some decline (see Table 2-1).

Table 2-1. The Urban-rural Structure Decomposition (Production Method) of Gross Domestic Product in the Three Industries (The National GDP is 100)

Year	Primary Industry	Secondary Industry			Tertiary Industry			Aggregate of Rural Area
	I	II	Urban Area II	Rural Area II	III	Urban Area III	Rural Area III	
1991	24.5	41.8	28.2	13.6	33.7	23.3	10.4	48.5
1993	19.7	46.6	24.9	21.7	33.7	23.1	10.6	52.0
1995	19.9	47.2	27.2	20.0	32.9	22.6	10.3	50.2
1997	18.3	47.5	27.1	20.4	34.2	23.4	10.8	49.5
1999	16.5	45.8	25.2	20.6	37.7	25.8	11.9	49.0
2001	14.4	45.1	24.3	20.8	40.5	28.0	12.5	47.7
2003	12.8	46.0	25.0	21.0	41.2	28.6	12.6	46.4
2005	12.2	47.1	25.9	21.8	40.1	27.9	12.2	46.2
2007	11.1	48.5	26.5	22.0	40.4	28.9	11.5	44.6
2008	11.3	48.6	26.6	22.1	40.1	28.8	11.3	44.7

Source: Quoted from "Analysis and Forecast on China's Rural Economy (2008-2009)" (P.40, Social Sciences Academic Press, April 2009) collaborated on by the Institute for Rural Development of the Chinese Academy of Social Sciences and the Rural Social and Economic Survey Division of the National Bureau of Statistics.

Table 2-1 shows that the decrease of the primary industry in rural economy is the most rapid — the percentage of the primary industry in GDP declines from 24.5% in 1991 to 11.3% in 2008; the secondary industry has reached a plateau and has stayed at around 40% of the national economy from 2001 to 2008, although it was still growing in the 90s of the last century; the proportion accounted for by the tertiary industry also shows a decrease from 12.6% in 2003, which is the year of the highest percentage, to 11.3% in 2008.

2. Fixed Asset Investment in the Rural Area

The driving force of China's economic growth is investment, which means low investment growth normally results in low economic growth. Therefore, the decrease in the proportion of rural economy in the national economy is directly related to the decreasing proportion of rural fixed asset investment in the whole country (see Table 2-2).

Table 2-2. The Volume of Fixed Asset Investment in the Rural Area and Its Proportion to the Whole Country (Unit: 100 million Yuan (RMB), %)

Year	Volume of fixed asset investment in the whole country	Volume of fixed asset investment in the rural area	The proportion of fixed asset investment in the rural area
1996	22913.5	5346.3	23.3
1998	28406.2	5914.8	20.8
2000	32917.7	6695.9	20.3
2002	43499.9	8011.1	18.4
2004	70477.4	11449.3	16.2
2006	109998.2	16629.5	15.1
2008	172291	24124	14.0

Source: website of the National Bureau of Statistics of China, http://www.stats.gov.cn.

Table 2-2 shows that from 1996 to 2008, the amount of fixed asset investment in the whole country has increased from 2291.35 billion to 17229.1 billion with a multiplier of 7.52 times; while the amount of fixed asset investment in the rural area has just increased from 534.66 billion to 2412.4 billion, or a 4.51 times increase. During the past 12 years, the proportion of rural fixed asset investment in the whole society has kept decreasing from 22.3% in 1996 to 14.0% in 2008.

3. Consumption in the Rural Area

For a long time, the consumption growth in the rural area has been slower than that in urban area. In the national retail market of consumer goods, the share of the rural area keeps decreasing continuously (see Table 3).

Table 2-3 shows that from 1978 to 2008, the total volume of retail sales of consumer goods increases from 155.86 billion to 10848.8 billion, with a growth of 69.6 times; the total volume of retail sales of consumer goods at and below county level increases from 105.34 billion in 1978 to 3475.3 billion in 2008, with a growth of 33 times, which is less than a half of the former. It is safe to draw the conclusion that the past 30 years since the reform and open-up policy have witnessed the decline of the proportion of rural residents' consumption in the country, because the proportion of the total volume of retail sales of consumer goods at and below county level in the country decreases from 67.6% in 1978 to 32% in 2008.

Table 2-3. Total Volume of Retail Sales of Consumer Goods at and below County Level and Its Proportion to the Country (Unit: 100 million Yuan (RMB), %)

Year	Aggregate retail sales of consumer goods nationwide	Total volume of retail sales of consumer goods at and below county level	Proportion
1978	1558.6	1053.4	67.6
1980	2140	1406.4	65.7
1985	4305	2430.5	56.5
1990	8300.1	4411.5	53.1
1995	23613.8	10634.4	45.0
1999	35647.9	13446.1	37.7
2005	67176.6	22082.3	32.9
2007	89210	28799.3	32.3
2008	108488	34753	32.0

Source: website of the National Bureau of Statistics of China, http://www.stats.gov.cn.

4. Disparity of Living Standards between Rural and Urban Residents

The reason for the consumption decrease in the rural area is that the growth rate of rural income is lower than that of the city. From 1978 to 2008, rural income increases from 133.6 Yuan (RMB) to 4761 Yuan (RMB), with an increase of 36 times, while the income of urban and township residents increases from 343.4 Yuan (RMB) to 15781 Yuan (RMB), with an increase of 46 times (see Table 2-4).

Numbers in Table 2-4 show that the income gap between rural residents and urban and township residents is widening, although the Chinese government keeps emphasizing the increase of rural income and has put forth a series of policies and measures with that aim. Due to the slow growth of per capita income of rural residents and the rapid growth of that of urban and township residents, the absolute income gap between urban and rural residents expands rapidly. In 2008, the income gap between urban and rural residents is 11020 Yuan (RMB). In 1978, per capita income of urban and township residents was 2.57 times of that of rural residents, while in 2008, the multiplier increased to 3.31. As the level of income determines the level of consumption, due to the low income of rural residents, the consumption level of rural residents is much lower than that of urban residents. The Engel coefficient (the proportion of residents' food expenses to living expenses) can be employed to comprehensively evaluate the gap in consumption and living standards between rural and urban residents. From Table 2-4, we can find that, in 2008, due to a large increase in food price, the Engel coefficient of both rural and urban households increased slightly. The Engel coefficient of rural households is 43.7, up by 0.6 percentage points over the previous year, while the Engel coefficient of urban and township households is 37.9, up by 1.6 percentage points. According to the Engel coefficient, the level of living standard and consumption of rural residents in 2007 was lower than that of urban and township residents in 1999, that is to say, the consumption level of rural residents is 8 years backward compared to urban residents.

Table 2-4. Per Capita Income of Rural and Urban Residents and the Engel Coefficient (Unit: Yuan (RMB), %)

Year	Per capita disposable income of urban and township residents	Per capita net income of rural residents	Engel coefficient of urban and township households	Engel coefficient of rural households
1978	343.4	133.6	57.5	67.7
1980	477.6	191.3	56.9	61.8
1985	739.1	397.6	53.3	57.8
1990	1510.2	686.3	54.2	58.8
1994	3496.2	1221.0	50.0	58.9
1998	5425.1	2162.0	44.7	53.4
2000	6280.0	2253.4	39.4	49.1
2003	8472.2	2622.2	37.1	45.6
2005	10493.0	3254.9	36.7	45.5
2007	13785.8	4140.4	36.3	43.1
2008	15781	4761	37.9	43.7

Source: website of the National Bureau of Statistics of China, http://www.stats.gov.cn.

II. New Countryside Construction Strategy

New countryside construction strategy, put forth by the Chinese government in 2006, is based on the following concerns:

1. The Widening Gap between Urban and Rural Areas

In 2005, Chinese GDP amounted to 18232.1 billion Yuan (RMB), equal to 2225.7 billion Dollar, and per capital GDP reached 1730 Dollar. Since then, China has entered into a phase of high frequency of conflicts, among which the one between urban and rural areas is one of the major conflicts that are prone to cause social differentiation and collision. Against this background, the Chinese government puts forth a strategy of the construction of a new socialist countryside, aiming at developing the rural economy and improving the income level of rural residents, thereby relieving the contradiction between urban and rural areas.

Column 2-1. Objectives and Tasks of Rural Reform and Development Before 2020

According to the new requirements of the objectives of building a well-off society in an all-round way and the claim of the construction of a new socialist countryside with developed production, well-off life, civilized atmosphere, neat appearance and democratic management, which are advanced by the 17th CPC National Congress, the objectives and tasks of the reform and development of the rural area before 2020 are as follows: the economic system in the rural area is sound; the social and economic development in the rural and urban areas has been integrated as a unified system; obvious progress has been made in modern agriculture construction; the comprehensive production capacity of agriculture rises notably; the national food security and the supply of main agricultural products are effectively guaranteed; net income of farmers is doubled compared to that of 2008, level of consumption is greatly improved, absolute poverty has been basically eliminated; organizations at grassroots level in the rural area have been further enhanced, the system of villagers' self-governance has been further perfected, and farmers' democratic rights have been effectively ensured; the equalization of public infrastructure between urban and rural areas has been notably enhanced, culture in the rural area flourishes, farmers' cultural rights and interests have been more effectively realized, all farmers have rights to good education, the system of basic living allowance and medical and public health becomes more comprehensive, and the social management system has been further improved; a resource-economical and environment-friendly agricultural production system has been basically formed, human settlement and ecological environment have been notably enhanced, and the capacity of sustainable development has been continuously enhanced.

Source: Excerpt from "CPC Central Committee's Resolutions on Some Important Issues Concerning Advancing the Reform and Development in the Rural Area"

2. The Necessity of Expanding Domestic Demand in Our Country

Before the financial crisis, the problem of apparently insufficient domestic demand already had emerged in China. However, because of the rapid growth of export, this problem has not foregrounded until the financial crisis, which forced China to face the predicament of insufficient domestic demand. In the domestic market of productive consumption, there is a severe lack of infrastructural construction in the rural area of China, while in the domestic market of personal consumption, the rural area whose population is 2/3 of that of the whole country only consumes 1/3 of the goods of the whole country; the daily per capita consumption in the rural area is only about 5 Yuan (RMB), which is almost 1/6 of that in the urban area. Therefore, by building a new countryside, the income of rural residents can be raised, which will in turn increase the consumption capacity of the rural area, and the problem of restrictions on the economic development of China imposed by insufficient domestic demand can be solved.

3. China's Economic Development Has Reached a Stage of "Industry Supports Agriculture, while the Urban Area Promotes the Rural Area"

In 2005, China's per capita GDP was 4.73 times that in 1989, and the national revenue exceeded 3 trillion. At the same time, the proportion of agriculture in GDP declined from 25.6% to 12.2%; the proportion of rural to total population declined from 73.79% to 58.24%. According to the international experience, China's economy has already reached a stage of "industry supports agriculture, while the urban area promotes the rural area". To implement the construction of new countryside is both the actual need and an urgent task of China's present social and economic development.

SECTION 2: GOVERNMENT'S POLICY OF EXPANDING RURAL DEMAND

Affected by the international financial crisis, China's economic development calls for the increase in domestic consumption demand, which makes it necessary to stimulate domestic consumption, particularly the rural consumer market. The population in China's rural area is more than 700 million, which constitutes the biggest consumption community in the world. Therefore, to develop the rural consumer market, whether for the purpose of coping with the influence brought by the international financial crisis or achieving sustainable development of the Chinese economy, is of great significance.

In order to deal with the international financial crisis, the Chinese government has successively put forth package plans and measures to expand domestic rdemand and stimulate the economy. Measures and plans related to the expansion of rural demand are as follows:

1. Construction of Rural Infrastructure

The 10 measures to expand domestic demand, promoted by the Central Government in November 2008, propose that the central government budget increases by 34 billion Yuan (RMB) to speed up the livelihood improvement program and infrastructure construction in the rural area. Departments like the Ministry of Agriculture and the Development and Reform Committee have completed the investment plan of newly added agriculture construction projects with a fund of 5.15 billion Yuan (RMB), which will be used in improving the agricultural production facility and farmers' living condition[1].

"The 1st document" put forth by the central government emphasizes five key fields: countryside electrical network construction, country road construction, countryside potable water safety construction, countryside methane construction, and countryside dangerous house transformation. The specific measures include: adjusting the countryside potable water

[1] The Ministry of Agriculture and other departments formulated the investment plan of newly added agriculture construction projects with a fund of 5.15 billion Yuan (RMB). Source : website of the Chinese Central Government, http://www.gov.cn

safety construction plan, increasing the investment and speeding up the construction, incorporating countryside schools, state-owned farms and tree farms into the construction plan; expanding the coverage of electric power network supply, speeding up the project of unified electric network and price between urban and rural areas, increasing the investment in countryside hydroelectric construction, expanding the substitution of small hydropower stations' electricity for fuel construction; speeding up the countryside highway construction so that by the end of 2010, basically all towns nationwide and qualified administrative villages in the middle east region will have asphalt (concrete) road, and qualified areas in the west will have roads, increasing the central budget allocation to countryside highway construction in the middle and west areas, establishing the policy-based subsidy system of countryside passenger transportation; increasing investment in the countryside methane construction project, expanding experimental demonstration sites for straw stalk solidification and gasification; developing countryside informatization; speeding up the transformation of national forest areas and shantytowns in reclamation areas, implementing the nomad settling project, and expanding the experiment site of countryside dangerous house transformation[2].

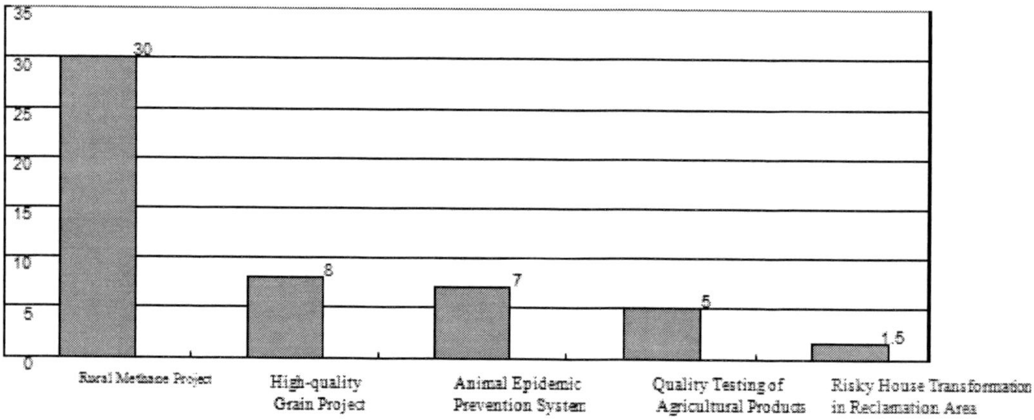

Source: website of the Chinese Central Government, www.gov.cn)

Figure 2-1. The Investment Plan of 5.15 billion Yuan (RMB) in Agriculture Construction Projects (Unit: 100 million Yuan).

2. Direct Agricultural Subsidy

In 2009, the central government budget will further increase subsidies for planting crops, for growing superior crop varieties, for purchasing farm machinery, as well as agricultural material synthesis subsidy, with a planned total subsidy of 123 billion Yuan (RMB), which is 20 billion Yuan (RMB) more than that of the previous year. The measure of direct agricultural subsidy will possibly be a long-term policy, and will be unceasingly improved.

---

[2] "The 1st document" of the central government (full text). Source : Xinhuanet, http://www.xinhuanet.com, 02-01-2009

The specific measures are: increasing the improved variety seed subsidy, raising the subsidy standard, incorporating paddy rice, wheat, corn and cotton into the subsidy, expanding the scope of superior seed variety subsidy for rape and soybean; increasing agricultural tools and machinery purchase subsidy massively, with a coverage across all the agriculture and husbandry counties/farms/pastures and a projected fund of 13 billion Yuan (RMB) from the central fiscal budget, or a 9 billion Yuan (RMB) increase from last year; increasing the agricultural material synthesis subsidy, consummating the dynamic adjusting mechanism of subsidy, and increasing the subsidy promptly in response to the rise of agricultural material price and the actual sown area of crops[3].

3. The Plan to Revive Textiles and other Light Industries

The small and medium-sized enterprises of light industry provide plenty of job opportunities for rural workers and increase the wage income of rural residents; 70% of light industry occupations and 50% of light industry output value are associated with the intensive processing of agricultural products and by-products, expansion of the agricultural product demand, enhancement of commercial and management income of rural residents; further expanding the state reserve of food and sugar, encouraging local governments to adopt measures like interest subsidy of liquid capital loan and so on, encouraging the enterprises to purchase and store agricultural product so as to solve the problem of agricultural product purchasing and storage; promoting the domestic consumption of textile clothing, actively developing the rural market, increasing sales in remote rural areas, making consumption for farmers more convenient[4].

4. The Policy of "Household Electric Appliances Going to the Countryside"

The Ministry of Public Finance and the Department of Commerce started the "household electric appliances going to the countryside" in December 2007 at the experiment sites in Henan, Shandong, and Sichuan, which provides a subsidy of 13% of the selling prices when farmers purchase color television, refrigerator, mobile phone, and washing machine. The local government assumes 20% and the central government 80% of the subsidy. From December 2008 on, the policy of "household electric appliances going to the countryside" has been launched in 14 provinces and cities, and will last for 4 years. According to farmers' wish and the development of the industry, more products will be eligible for subsidy under "household electric appliances going to the countryside", with guaranteed quality and satisfactory post-sale services[5]. In 2009, subsidy capital for the "household electric appliances going to the countryside" allocated by the central government budget is 20 billion Yuan

[3] Report on the work of the government to the Second Session of the Eleventh National People's Congress. Source : Xinhuanet, http://www.xinhuanet.com, 03-14-2009

[4] "Opinions on Reviving and Readjusting Ten Industries". Source： Xinhuanet, http://news.xinhuanet.com.

[5] Source: the information management system of "household electronic appliances going to the countryside", http://jdxx.zhs.mofcom.gov.cn.

(RMB), up by 9 times compared with 2008. According to surveys of the Ministry of Public Finance, during the four years of the implementation of the subsidy policy, the sales of household electric appliances is estimated to reach 700 million units, which will contribute 2100 billion Yuan (RMB) to domestic consumption[6].

5. The Policy of "Automobile Going to the Countryside"

According to the adjusting and reviving plan of automobile industry announced in January 2009, in 2009 the country will allocate 5 billion Yuan (RMB) to one-off subsidy for the scrap of three-wheeled motor vehicles and low-speed freight vehicles to exchange for light trucks as well as the purchase of miniature passenger buses whose displacement is below 1.3 liters. The subsidy standard is: the subsidy for the purchase of light truck with exchange or miniature passenger bus is 10% of the selling price, while that of the motorcycle is 13% of the selling price. The "automobile and motorcycle going to the countryside" subsidy is jointly funded by the central and provincial governments, with 80% of the subsidy born by the central fund and the rest 20% by provincial fund. Through the price subsidy, the consumption of farm vehicle and mini vehicle will be stimulated in a short time. The fiscal subsidy policy of scrap and renewal of old motor vehicles is adjusted. The subsidy fund of the scrap and renewal of old motor vehicles in 2009 will increase from 600 million Yuan (RMB) in 2008 to 1 billion Yuan (RMB). Through increase in subsidy support, rise of subsidy standard, and speeding up of elimination of old automobiles, consumption can be promoted[7].

6. The Policy of Logistics Construction in the Rural Area

In 2009, central finance will increase special funds for the development of logistics service system in the rural area and the promotion of service industry development. The method of award instead of subsidy and interest discount is adopted so as to arouse the initiative in local and social investment and support the development of a circulation system in the rural area[8]. In 2009, the Department of Commerce plans to build and transform 150,000 countryside stores and 1,000 countryside commodity allocation centers nationwide, covering 70% of villages and towns and 50% of administrative villages, and forming a modern countryside circulation network with urban stores taking the leading role, town stores as the mainstay, and village store as the basis, so as to improve the network's allocation ability and increase the commodity allocation rate of countryside stores up by more than 50%. Accelerate

---

6 The central government budget allocated a subsidy fund of 20 billion Yuan (RMB) to "household electronic appliances going to the countryside". Source: website of the central government, http://www.gov.cn, 03-07-2009.

7 "Plans to Revive and Readjust the Automobile Industry". Source: website of the central government, http://www.gov.cn

8 The Department of Commerce and the Ministry of Public Finance's Announcements of the Fund Management concerning Supporting the Circulation and Expanding Consumption. Source: Ministry of Public Finance, P. R. C., http://www.mof.gov.cnl, 02-05-2009

the implementation of the "countryside business information service project", advance the integration of the "thousands of villages and towns" network with the network of supply and sales, postal service, telecommunication and so on, in order to provide services like market information, matching of supply and demand, and so on. Expand the variety of centrally allocated commodity, reduce the operational cost, and guide manufacturing enterprises to develop farmer-oriented products, so as to better satisfy rural residents' demand as consumers[9].

7. The Policy to Guarantee the Supply of Agricultural Means of Production

Build suitable large-scale nitrogenous fertilizer process units in places with rich resources, optimize the distribution of phosphate fertilizer, and increase the domestic and foreign prospection and exploration of potassium ore resource. In 2011, the chemical fertilizer output will amount to 62.5 million ton; the proportion of chemical fertilizer produced in places with raw materials will increase to 60%. Improve the central and local (two levels) commercial reserve system of chemical fertilizer and enhance the allocation and transportation of chemical fertilizer in the off season, establish a scientific and reasonable adjusting system that "reserves in the off season and supplies in the high season", guarantee the supply, and stabilize the market price; adjust the product structure of pesticide products, develop those pesticides which are highly effective but minimally poisonous with low residue, promote the centralized production of technical material; unceasingly improve the county and township supply network of diesel oil for agricultural usage, so as to meet the seasonal consumption demand; implement the policy of subsidizing the procurement of agricultural tools and machinery to households as soon as possible[10].

8. The Policy of Providing Basic Pension, Medical and Healthcare Insurance to Rural People, while Improving Culture and Education Services in Rural Areas

New pension policy in the rural area: launch experiment sites of new countryside pension program in 10% of the counties (cities, districts) in 2009. The new countryside pension program adopts a basic model that unifies overall social planning and individual accounts and employs a mixed financing method incorporating personal payment, collectively pooled subsidy and government subsidy. Rural residents who are over 16 years old, not school students, and have not participated in the basic pension plan for urban workers are eligible for

9 The Department of Commerce and the Ministry of Public Finance's Announcements Concerning Speeding up the Implementation of "Market of Thousands of Villages and Towns Project". Source: public inquiry system of government information of the Department of Commerce, http://file.mofcom.gov.cn.

10 "Plans to Revive and Readjust the Petrochemical Industry". National Development and Reform Commission, http://zfxxgk.ndrc.gov.cn

the new countryside pension program . Those insured rural residents who are more than 60 years old and meet the relevant conditions can receive the basic pension[11].

The new healthcare reform package: the government put forth a new healthcare reform package in April 2009, which provides the public with a basic medical and health system as public good. 850 billion Yuan (RMB) will be invested in the implementation of five key reforms of the package, including: Speed up the construction of a basic medical insurance system, establish a basic national drug system, improve basic medical and health services at the grassroots level, promote the equalization of basic public health services, and advance the reform of public hospitals at experiment site. By 2011, the basic medical insurance system will benefit both urban and rural residents[12].

Compulsory education in the rural area: in 2009, the Ministry of Public Finance allocates 66.25 billion Yuan (RMB) to further add to the rural compulsory education funds, which is up by 16.1% compared to last year. Free secondary vocational education with an allocated subsidy fund of 4.5 billion Yuan (RMB), which is up by 115.3% compared to last year, will be gradually implemented. A construction fund of 13.7 billion Yuan (RMB) is allocated to enhance the construction of county centers for vocational education and demonstrative secondary vocational schools, speed up the transformation of junior high-school buildings in the midwest rural areas, and improve the ancillary facilities of countryside boarding schools. A fund of 24 billion Yuan (RMB) is allocated to the financial aid to students whose families are in financial trouble as well as to the student loan[13].

9. Credit Policy in the Rural Area

In 2009, the People's Bank of China will continue to encourage and guide various financial institutions to make use of small guaranteed loans, etc., so as to help capable immigrant workers start their own businesses in or out of their hometown, as well as vigorously develop consumption credit and enliven the consumer market in the rural area. "*Overall Working Plan of New Rural Finance Institution 2009 - 2011*" issued by the China Banking Regulatory Commission in August 2009 is a plan to establish 1294 financial institutions in 35 provinces (regions, cities, excluding Tibet) and cities specifically designated in the state plan from 2009 to 2011 all over the nation. Among these 1294 financial institutions, 1027 are villages and township banks, 106 loan companies, and 161 mutual-aid countryside fund institutions. Those financial institutions will be mainly established in those areas: counties with a larger agricultural industry than the national average, midwest regions, counties where the coverage of financial institutions is low, poverty stricken counties, and counties with active small- and medium-sized enterprises.

[11] Website of the Chinese central government, http://www.gov.cn, 06-24-2009.

[12] "*Recent Key Plans of Medical and Public Health Systems (2009-2011)* ". Website of the Chinese central government, http://www.gov.cn, 04-07-2009.

[13] The Ministry of Public Finance, "Draft of Central and Local Budget in 2009", 03-05-2009

Column 2-2. Experiment of New Rural Pension Reform

The national conference concerning the experiment site of new countryside social old-age pension insurance was held in August 2009 in Beijing, which decided that experiment sites of new countryside social old-age pension insurance would be established in 10% of all counties (cities, districts) in 2009, and later gradually spread to other areas so that the entire country will benefit from it by 2020. The basic principles of the new countryside social old-age pension insurance system are: First, start from the actual conditions of the countryside, start from a low level, and insist that the fund raising and treatment standard must comply with the economic development and various other capacities; Second, the responsibility is shared by the individual, the collective as well as the government, and rights and obligations should be matched; Third, combine government guidance and farmers' will, direct all the farmers to participate; Fourth, launch experiment sites first then gradually expand.

The project of experiment sites of new countryside social old-age pension insurance mainly includes two aspects; one is to provide old-age pension, which is a combination of the basic pension and individual account retirement pension, with the minimum standard basic pension entirely born by the central public finance. The other is to employ a fund raising means that unifies individual payment, collective subsidy and government subsidy, with local subsidy for farmers' individual payment. The central finance provides a subsidy for the total amount of minimum standard basic pension in the midwest area, and 50% of the total amount of minimum standard basic pension in the east area at the same time, and guarantees that farmers covered by the old-age insurance program in the same area will receive the same amount of basic old-age pension in the future, so as to manifest the fundamentality, fairness and widespread benefits of the insurance system. Rural residents who are over 16 years old, not school students, and have not participated in the basic old-age pension plan for urban workers may participate in the new countryside old-age pension insurance program. Insured rural residents who are more than 60 years old and meet the relevant conditions can receive the basic old-age pension.

On September 4, 2009, the State Council put forth the guideline on the experiment sites of new countryside social old-age pension insurance, which set the basic old-age pension for each person at 55 Yuan (RMB) a month. The local government may raise the standard of basic old-age pension according to the actual situation, and more basic old-age pension can be distributed to those rural residents who have paid for a long time. The extra pension payments should be born by the local government. Senior citizens over 60 years old with rural *hukou* and have not participated in the basic old-age pension plan for urban workers may receive the pension every month.

Rural residents who participate in the new countryside social old-age pension insurance should pay insurance premium according to the stipulation. Currently the insurance premium is graduated in five levels: 100 Yuan (RMB), 200 Yuan (RMB), 300 Yuan (RMB), 400 Yuan (RMB), and 500 Yuan (RMB) per year. The local government can add other premium levels based on the actual situation. The national government will adjust the premium standards according to the changes in per-capita net income of rural residents. The insured person can independently choose the premium standard. The more premium they pay, the more insurance pension they will receive. At the time of implementation of the new countryside social old-age pension insurance, rural residents who are 60 years old and above and did not participate in the basic old-age pension plan for urban workers can get the basic old-age pension monthly for free, but their qualified children should join the program and pay the insurance premium. For those who are expecting pension payments within 15 years, payment for the premium should be made annually, and makeup payments for the previous years are allowed, but with a total not exceeding 15 years. For those who can get the pension after 15 years, payment for the premium should be made annually for no less than 15 years.

Source:

1. *"The Conference on the Experiment Site of New Countryside Social Old-age Pension Insurance Held by the State Council on 18th", http://www.gov.cn, 08-18-2009.*
2. *"The State Council Puts forward Guidelines on Experiment Site of New Countryside Social Old-age Pension Insurance, and Encourages Farmer to Get Insured", http://www.xinhuanet.com, 09-04-2009.*

SECTION 3: THE PROGRESS AND PROBLEMS OF IMPLEMENTING THE NEW POLICIES OF EXPANDINGRURAL DEMAND

Until the first half of 2009, most of the Chinese government's policies to expand rural demand have been issued and actual effect can be traced. According to the statistical data from the National Bureau of Statistics, cash income per capita of Chinese rural residents is 2733 Yuan (RMB) in the first half of 2009, with an increase of 8.1% over that of the same period of last year; total retail sales of consumer goods at and below the county level increase by 16.4%, and the growth rate of rural consumption exceeds that of cities successively for six months in the first half of the year[14]. The contribution of countryside consumption to total consumption shows an upward tendency, which indicates that certain progress has been made in the implementation of the government's policy to expand domestic demand in rural areas. But in terms of the actual situation, some problems still exist in the implementation of these policies.

I. The Operation of the Policy to Expand Rural Demand

By the end of 2008, the number of counties (cities, districts) where the new rural cooperative medical care system has been established amount to 2792 throughout the country, with a cooperation rate of 91.5%, and the number of service items of family planning at the grassroots level amount to 1,140; the transformed area of junior high-school building sums approximately to 1.5 million square meters, and the constructed area of secondary vocational school buildings amounts approximately to 900,000 square meters; the planned allocation to direct grain production subsidy, synthetic agricultural material subsidy, improved variety of seeds subsidy and agricultural tools and machinery purchase subsidy is 111.6 billion Yuan (RMB); 31 provinces (cities, regions) have established the rural subsistence security system,

14 Source: National Bureau of Statistics of China, 07-24-2009.

which covers 43,060,000 rural residents[15]. By the end of April 2009, the central government investment project has solved the drinking water security problem for approximately 14.6 million rural population; 1.6 million rural households have started construction of methane for rural household usage; about 20,000 kilometers of countryside driving road and more than 40,000 kilometers of countryside electrical network have been completed; about 6500 items of medical and health service have been established. Up to June 2009, 7.81 billion Yuan (RMB) has been distributed as farm machinery purchase subsidy nationwide, the number of subsidized farm machines and tools surpasses 1.79 million units, the number of rural households who have benefited from the project exceeds 1.67 million, the rate of the implementation progress reaches 78.1%; the pre-allocated subsistence security allowance subsidy fund in the urban and rural areas is 27.447 billion Yuan (RMB), basic old-age pension insurance subsidy fund is 70 billion Yuan (RMB), special pension and subsidy fund for special-care recipients is 9.85 billion Yuan (RMB), and a lump sum living allowance of 9.067 billion Yuan (RMB) is granted to recipients of the subsistence security allowances.

According to statistics from the China Banking Regulatory Commission, from the end of 2006 to the end of June 2009, 118 new rural financial institutions have gone into operation in the country, among which 100 are village and small township banks and 7 are loan companies. Those financial institutions have collected a capital of 4.733 billion Yuan (RMB) and deposit of 13.1 billion Yuan (RMB), and the loans issued to rural residents total 5.5 billion Yuan (RMB). The development of new rural financial institutions has effectively shortened the financial gap between urban and rural areas, and improved financial service in the rural area; The financial support toward the reform of rural credit cooperatives has been basically put in places, up to the end of June 2009, special bills of 159.60 billion Yuan (RMB) have been cashed for rural credit cooperatives in 2,296 counties (cities) with a cashing progress rate of over 95%, and the issued special loans amount to 1.5 billion Yuan (RMB)[16].

The policy of "household electrical appliances, farm machinery, and automobiles (motorcycles) going to the countryside" and the halving of tax on passenger vehicle purchases with displacement of 1.6 liters and below have promoted consumption remarkably, among which the automobile industry is the most prominent one, for which nationwide production in the first 7 months accumulated to 7.11 million, up by 20.2% over the same period last year, and the production and sales volume exceeded 1 million for five successive month[17]. Since December 2007, notable successes have been achieved at the experiment sites of "household electric appliances going to the countryside" in Shandong, Henan, and Sichuan provinces, in 2008, the sales volume of household electric appliances increased by 40% over the same period last year, and the growth rate increased by 30 percentage points.

From Table 2-5, we can see that "household electric appliances going to the countryside" was officially launched nationwide since February 2009, and the sales volume of household electric appliances in the countryside rose from 338,000 units in January to 2.921 million units in June, with a growth rate of 764.2%; the sales value also rose from 479 million Yuan

---

15 The investment project of the central government runs smoothly. 214,000 low-priced apartments for rent have been constructed. Source: China Internet Information Center, http://www.china.com.cn, 05-27-2009

16 "*Report on the Implementation of Monetary Policy (the Second Quarter of 2009)*".

17 "Automobile Going to the Countryside" Policy Stimulates the Sales and Production. Source: Website of Market Research in China, http://www.cu-market.com.cn, 8-13-2009

(RMB) in January to 5.481 billion Yuan (RMB) in June, with a growth of more than 10 times (see Table 2-5). Due to the guaranteed post-sale service and a subsidy of 13% of the selling price, the project was warmly received by farmers.

Table 2-5. Statistics of the Sales of "Household Electric Appliances Going to the Countryside" form January to June in 2009

Month	Sales Volume (Unit)	Sales Value (Yuan (RMB))	Proportion of Sales Value (%)
January 2009	338135	479，418，972.40	2.95
February 2009	875390	1,305，471，067.66	8.04
March 2009	1484517	2,238，428，402.10	13.79
April 2009	1763304	2,778，494，194.41	17.12
May 2009	2227725	3,946，132，921.07	24.32
June 2009	2921096	5,481，072，926.46	33.77
Total	9610167	16，229，018，484.10	100

Source: Information Management System of "household electric appliances going to the countryside", http://jdxx.zhs.mofcom.gov.cn

II. Problems at the Present Stage of Implementation of the Policy to Expand Rural Demand

However, problems also exist in the operational process of the policy to expand rural demand. For example, in the process of "household electric appliances going to the countryside", some sellers infringed the relevant rules and cheated the consumers by faking it. Therefore, the Department of Commerce and other departments jointly promulgated the measures of "cracking down on counterfeit goods" on July 13, 2009, in order to regulate the household electric appliance market and protect the interest of consumers. Moreover, in terms of the construction of countryside methane, road, water safety, electricity, and so on, too much importance has been attached to quantity rather than quality. As a result, some completed facilities can hardly be used because of poor quality. With regard to funds invested to benefit farmers, problems of waste, leakage, corruption and so on also exist. Putting together situations in different areas, reasons for these mentioned problems can be summarized as follows.

1. Governments at the Grassroots Level do not Implement the Policy Effectively

Governments at the grassroots level are the main executive agents of the policy formulated by the central government and governments at all levels, and should bear the most direct responsibilities. The limitations of governments at the grassroots level will impede policy implementation. The quality of some cadres at the grassroots level in villages and towns is relatively low. Thus, they, on the one hand, do not have an adequate understanding of the regulations and policies, which leads to insufficient publicity, while on the other hand it is hard for them to entertain the idea of administration according to law, resulting in the frequent appearance of the phenomenon of power substituting law and people overruling law in policy execution. Moreover, due to information asymmetry between the local government and the central government, discrepancy may exist in their departure point, available resources, utility and objectives in policy formulation, which may lead to discrepancy, perfunctoriness, or even resistance in policy implementation by governments at the grassroots level, thereby negatively affecting the effectiveness of the policy.

2. Farmers are not Proactive Enough in Policy Implementation

Whether the execution of rural policy is successful or not also depends on its popularity with farmers and effective communication between policy subjects. If farmers are not proactive enough about the implementation of the policy, the evaluation on the result of rural policy implementation will be more difficult. On the one hand, the basic rural unit of livelihood and production is family in China, which makes it impossible for the central government to comprehensively understand each farmer's wish, and this kind of information acquisition is of great cost. All of these have led farmers to be passive in the central government's policy formulation process. On the other hand, in the face of poor or non-conforming enforcement by governments at the grassroots level of policies in favor of farmers formulated by the central government, farmers, in many cases, do not have a say, nor do they have access to an effective channel for feedback, which will also impede the implementation and execution of the policy. In addition, because of farmers' low education level and information asymmetry, they do not have an adequate understanding of many specific policies, thus they can not fully enjoy the preferential policy of the government.

3. Insufficient Material Conditions for Implementing the Policy

The continuous increase of farmers' income and the improvement of rural infrastructure is the foundation of expanding rural demand and the consumption of rural residents, but these problems can not be solved in a short time, which constrains the stability and continuity of the policy to expand domestic demand. Although governments of all levels have been continuously building up investments that will benefit farmers, the total fund is still severely insufficient, the rural social security system is still deficient, and the rural infrastructure is still wanting investment. These problems manifest themselves in the following aspects: the investment to rural compulsory education is insufficient; the rural medical service and health care system is still facing the problems of insufficient investment, low efficiency, lack of security, weak public health and so on, and pressure on the medical front is ever-mounting; Supply is still insufficient in the construction of small farmland irrigation and water conservancy, the construction of farm road, the dissemination of agricultural science and technology, the improvement of socialized rural service system, and the spread of market supply and demand information relating to agricultural products; the agricultural insurance system and the rural social security system still have not been completely established, which leads to tremendous market and natural risks against agriculture.

Column 2-3. Scrap Rate of Methane-generating Pits Worth Hundreds of Millions Yuan (RMB) Exceeds 90% in the Rural Area in Heilongjiang Province

"China Economic Weekly" Reporter Cui Xiaolin, Ma Yuzhong/Heilongjiang, reported in Beijing: the rural household methane project supported by the national debt has been advanced since 2003 in the nation. It is listed as "core technology to solve the pollution problem in the rural area" by the Ministry of Environment, "reproducible clean energy in the new countryside" by the Ministry of Agriculture and "comprehensive technology of the construction of a new socialist countryside" by the leading group of the central government in building a new countryside. The central government will subsidize 4/5 of the construction cost of a methane-generating pit in the rural area, which is definitely a beneficial project for farmers who are direct beneficiaries. But according to a survey published by "China Economic Weekly", the scrap rate of completed household methane-generating pits exceeds 90% in many villages in Heilongjiang Province, and the project leaves an extremely bad impression on farmers there.

The "Working Report of Energy Construction of the Province in 2008" provided by the general energy office of the provincial agricultural committee reads, "in 2008, 110,000 household methane-generating pits were built in the province, and the household methane project has covered all the cities (municipalities) in the province, and has been launched in 1,549 villages in 83 counties (cities, districts); in 2008, we received various national rural energy construction funds of 177 million Yuan (RMB), special funds at the provincial level of 108 million Yuan (RMB), aid-to-the-poor funds from the cities and counties of more than 40 million Yuan (RMB), and funds raised by farmers and enterprises themselves of 290 million Yuan(RMB), all of which have set new records." But in the same report, something is really disappointing: on the one hand, we see 110,000 newly constructed methane-generating pits and a total of special funds exceeding 600 million Yuan (RMB) while, on the other hand, a large number of methane-generating pits are scrapped.

On August 12th, the reporter of "China Economic Weekly" saw at the Quality Controlling Bureau of Heilongjiang Province a document titled "Investigation Report on the Construction and Usage of Brick and Concrete Household Methane-Generating Pits in the Rural Area of the Province", in which reasons for the scrap of methane-generating pits are listed as follows:

1. The construction is not in compliance with the technical standard, which leads to low quality of the project. Problem of inferior materials and shoddy work is prevalent in the project and somebody even use red bricks to build methane-generating pits. The reason for this lies in self-interest. In order to gain huge profits, the construction unit and some of the government officials collude to line their pocket with the invested funds, so that less than a half of the funds invested by the country and the province have actually been spent on the project.
2. Driven by local interest. Although some local official were perfectly aware that large-scale construction of red brick methane-generating pits is not suitable for their locality, they still construct several methane-generating pits in order to deal with inspection of their superiors, thus creating the illusion that the brick and concrete methane-generating pit is suitable for the entire province.

3. Instead of actually carrying out the work under the instructions of the provincial energy office, the agricultural committees and energy offices, even some leaders in a few counties and cities, illegally obtained, diverted, appropriated and even secretly divided the funds of the project by making up the roster, which leads to fund insufficiency and the untruthful construction reports.
4. Some energy office staff in a few cities and counties embezzled the project funds by using even viler method.

Source: http://news.QQ.com, August 31, 2009 00:47.

4. Deficiencies of the Surveillance and Management System

The central government's policy to expand rural demand involves the interests of many groups, which include the policy formulating organization, the implementing organization and the target group. Because of the imperfection of the surveillance and management system at various levels, decisions by different groups out of their own interests have led to discrepancies between the policy implemented by the government at the grassroots level and the policy formulated by the central government. At present, there exists the problem of the absence of a management and surveillance mechanism in the implementation of the central government's policy. The absence of a management mechanism is manifested in a low concentration rate of investment for assisting agriculture and low efficiency of the fund for assisting agriculture. The absence of a surveillance mechanism is manifested in the asymmetry between fund allocation and its management and surveillance, the ignorance of management and surveillance, the imperfect evaluation system of fund utilization and so on. Moreover, beneficiary groups like the farmers are comparatively disadvantaged against the executing agency, and can be effectively supervised only with great difficulty. Therefore, a sound surveillance and management mechanism must be established to smooth the implementation of the central government's policy in the rural area.

SECTION 4: GOVERNMENT'S NEW MACRO POLICIES' INFLUENCE ON NEW COUNTRYSIDE CONSTRUCTION

To cope with the international crisis, the Chinese government has launched packages of economic stimulus plans since November 2008. With the rural economy as an essential part of the national economy, the government's macro policies will absolutely have some influence on new countryside construction and expansion of rural consumption. Nevertheless, it is difficult to judge the exact influence due to limited time. Here, some simple deductions will be made according to the content of those policies and previous experience.

I. Proactive Fiscal Policies' Influence on New Countryside Construction

Central finance increases investment by 100 billion Yuan (RMB) in the fourth quarter of 2008, which is entirely distributed in the first quarter of 2009; meanwhile, in the first half year of 2009, 280 billion Yuan (RMB) is dispensed, making the total investment 380 billion

Yuan (RMB), among which funds for rural civil engineering and infrastructure together add up to 104.3 billion, or 27.4% of the total. Thanks to the investment, 19.60 million rural population's problem of safe drinking water has been solved; 1.78 million rural methane projects have been completed; 75 thousand miles of rural network lines have been upgraded. The above-mentioned data show that proactive fiscal policies are making a positive impact on new countryside construction. In addition, out of the 4,000 billion which is the core of the proactive fiscal policies, 370 billion, or 9.25%, is targeted at civil engineering and infrastructure such as rural water, electricity, road, gas, and housing; massive investments in other aspects will bring some indirect and beneficial impact, for instance, extensive infrastructure construction and post-disaster reconstruction will enlarge provide migrant workers more job opportunities; energy conservation and emission reduction will improve rural living conditions; investment and expenditure in culture, hygiene, education, and social security will bring positive impact on rural residents and children of rural migrant workers in cities.

However, judging from the implementation of the proactive fiscal policies of 4,000 billion Yuan (RMB), we can say that they are urban-oriented. Under recent proactive policies, the urban-oriented investment of 4,000 billion Yuan (RMB) may crowd out rural fixed asset investment. Although it seems that the amount of rural fixed asset investment in absolute terms is increasing continuously in 2009 and the following years, its proportion of it to overall social fixed asset investment remains decreasing.

II. Loose Monetary Policy's Influence on New Countryside Construction

Under the stimulus of loose monetary policy, credit supply of China's financial institutions is proliferating. During the first half of 2009, aggregate loans increase by 7,370 billion Yuan (RMB), or 1.5 times of the amount of newly-increased loans of year 2008. Nevertheless, according to the statistics, only a small fraction of the fund streams to rural economic areas, and medium and small enterprises still have limited access to credit. A large part of the fund flows into large state-owned enterprises and state-owned monopoly enterprises, even into the real estate market and the stock market, which leads to assets price inflation. In the first half of 2009, the Shanghai Stock Exchange Composite Index rises from 1,641 to 3,478, registering the largest growth rate in the world; at the same time, housing price in China has picked up from the downturn in the second half of last year. In July, housing price in 70 large and medium cities in China increases by 1.0% over that of the same period of last year, with an increase of 0.8 percentage points over that of June; increase in housing price over June this year is 0.9%, which is 0.1 percentage points higher than that of June[18]. Increase in asset price is conducive to wealth expansion for urban residents with capital income, yet it at the same time reduces the relative wealth of rural residents with little capital income. In addition, as credit and loan balloons, the evidence of inflation is more and more obvious. Once inflation occurs, it will represent a substantial plunder of farmers'

---

[18] National Statistics Bureau, 10th August in 2009. "The operation situation of the real estate market in China from January to July in 2009", China Information News.

wealth. Therefore, loose monetary policy has not brought positive impact on new countryside construction yet.

III. National Industrial Policies' Influence on New Countryside Construction

The State Council issued a series of plans for reviving and adjusting top ten key industries, including steel, automobile, ship, petrochemical engineering, textile, light industry, nonferrous metal, equipment manufacturing, electronic communication and logistics. However, agriculture is excluded from the list. Thus, the national industry reinvigorating plan does not have a direct positive impact on new countryside construction. However, from the perspective of indirect influence, it does. This is manifested mainly in the fact that industry reinvigorating policy promotes employment growth that further increases rural residents' wage income, which is the main source of rural income increase in recent years. Based on the statistics from the National Statistics Bureau, rural population going out for work has gone up in the first half of this year. Job opportunities in cities and towns increase by 5.69 million in the first half of the year, which achieves 63% of the target of the whole year. By the end of the second quarter, the population of rural labor out for work has risen by 3.78 million compared with that of the first quarter, or an increase of 2.6%.

Though agriculture is not included in the industry reinvigorating plan, it does not mean that agriculture does not need reinvigoration, because unfavorable factors against agriculture's sustainable and stable development like aging and lagging agricultural facility, low level of agricultural technology, soil degeneration, etc. have not seen fundamental improvement. The government can not ignore agricultural input and construction despite good crops for five successive years. The drought from south to north in China this year is a living example.

IV. The Regional Reinvigorating Plan's Influence on New Countryside Construction

The State Council's regional reinvigorating plan followed the 4,000-billion-Yuan (RMB) investment and the top ten industries reinvigorating plan. In the first half of 2009 alone, the State has already approved many regional reinvigorating plans for Haixi Area, Hengqin Island, the economic belt along the coast of Jiangsu, the Guanzhong—Tianshui economic zone development plan and the economic belt along the coast of Liaoning. Through the regional reinvigorating policy, the Chinese government hopes to cultivate the designated areas into new growth belts of our country's next round of economic and social development. The approved or to-be-approved reinvigorating plans are all directed to economically backward areas or underdeveloped provinces in the coastal areas, which provides a good opportunity for the economic development of the targeted areas. However, in terms of the content of existing regional reinvigorating policies, the main focus of it is still the cultivation of local leading industries and pillar industries, which has no direct positive impact on agriculture and new countryside construction, but is beneficial for enlarging local farmers' non-agriculture employment, enhancing local economic strength, and building up a material basis for new countryside construction.

To sum up, the Chinese government's policies for boosting rural consumption are both necessary and correct for coping with the international financial crisis. Though there are some problems in the process of implementation, the results are obviously positive on the whole. To sustain those positive results, it is not enough to solely rely on the existing policies, but other matching macro policies are much needed. More fundamentally, the national income distribution pattern has to be adjusted and farmers' income increased, which constitutes the foundation of expansion of rural domestic demand. Fortunately, the Chinese government has realized this, and the National Income Distribution Reform Suggestions drafted by the Development and Reform Commission has been handed to the State Council for a new round of advice-seeking. As is known to us, the core of it is to deepen the income distribution reform, to increase the proportion of household income to national income, and to improve the proportion of laborers' remuneration in primary distribution.

Translator: Yang Ke (SUFE)
English Version Editors: Xiao Rong (SUFE), Chen Jinzhu (Harvard)

REFERENCES

[1] Rural Development Institute of Chinese Academy of Social Sciences, Rural Social Economic Survey Department of National Statistics Institute, April 2009. Chinese Rural Economic Situation Analysis and Prediction 2008-2009. Society and Science Documents Press.

[2] Gao Huiqing, Xiong Yizhi, Hu Shaowei, 2008. World Financial Crisis and Its Impact on China (Chinese)[J]. *Studies of International Finance*, 11, 21-25.

[3] Chen Xiwen. *Why Do We Put forward New countryside construction Recently* [B]. www.jgny.net/news/200604/40633.htm

[4] Li Zheng, 2009. Reflections on Startup of Current Rural Consumption Market(Chinese) [J]. *Price Theory &Practice*, 2, 69-70.

[5] Research Team on Rural Consumption Issue, 2007. The Situation of Countryside Consumption and Relevant Suggestions (Chinese)[J]. *Finance & Trade Economics*, 2, 69-73

[6] Cai Yuezhou, 2009. Economic Stimulus Package and Start-up of Rural Consumption in China: Empirical Study Based on the Decomposition of Rural Household Income(Chinese) [J]. *Journal of Finance and Economics*, 7, 6-11.

[7] Cheng Yi, 2006. Discussion on Obstacles and Solutions of Implementation of the Policy Peasant, the rural areas and Agriculture(Chinese)[J]. *Truth Seeking*, 3, 83-86.

[8] Luo Ying, Yu Yanfeng, Daitian,2009. Problems and Solutions for Rural Policy Establishment and Implementation(Chinese)[J]. *Rural Economy*, 5, 20-22.

[9] Yin Ping, Tian Weimin, 2007. Discussion on Improvement of China's Fiscal Policies to Support Agriculture(Chinese) [J]. *Rural Economy*, 7, 94-95

[10] Wang Feng, 2009. Developing Situation of Chinese Rural Consumption Market, Constraints, and its Outdoor Development Countermeasures(Chinese)[J]. Rural Economy3, 66-68.

[11] Rural Development Institute of Chinese Academy of Social Sciences, Rural Social Economic Survey Department of National Statistics Institute, 2002. Chinese Rural Economic Situation Analysis and Prediction 1997-2002. Beijing, Society and Science Documents Press.

In: China's New Deal
Editors: Li Xiaoxi and Hu Biliang

ISBN: 978-1-61470-815-5

Comments on Chapter 2

FACTS BEHIND FIGURES: PRELIMINARY THOUGHTS ON BUILDING A NEW COUNTRYSIDE AND STIMULATING RURAL DEMAND IN CHINA

Hu Rong *

This chapter briefs the readers on the agricultural economic development in rural China. It gives particular attention to China's concept of building a new socialist countryside, and China's policies of expanding rural demand. Agriculture has been an important sector in national economy. The issues of agriculture, rural areas and farmers (referred to as the "three rural issues") are closely related with China's economic development and social stability under the current economic situation.

The position given to this chapter in the entire book shows that building a new countryside is of vital importance to promoting steady and sustainable development of China's national economy. This drive is also of special significance to stimulating rural demand and expanding domestic demand. Using numbers and figures, the authors of this chapter introduce the role of building a new countryside in national economy. They also detail the policies of expanding rural demand. They further brief the readers on the status of implementation of these policies, and the problems associated with carrying out these policies. They analyze the impact of government's macro-policies on building a new countryside. All data show that steady and sustainable development of China's economy and the realization of domestic demand expansion are closely related with agricultural revitalization, rural development and rising affluence in farmers.

The authors point out that despite the overall good trend of China's national economy and improved rural economic development in recent decade, agriculture obviously lags behind the development of other industries. The gap in development between agriculture and other industries is getting larger. The continuous widening in income disparity between urban and rural residents is the major obstacle to solving the "three rural issues". The income difference

* Hu Rong, Ph.D. in Agricultural Economics, Texas A&M University, Associate Professor, Research Institute of Economics and Management, Southwestern University of Finance and Economics

also restricts the sustainable growth of national economy, and directly affects social stability and harmony.

At the end of 2006, the 5th Plenary Session of the 16th Central Committee of the Communist Party of China put forward the development strategy of building a new countryside. Government officials at all levels as well as scholars showed great interest in this strategy. There appeared a plethora of research and exploration of this topic in terms of theory and practice. As the world economy falls into recession, there are increasing risks as a result of the international financial crisis. Therefore, it is timely and appropriate to launch the strategy of building a new countryside which aims at promoting agricultural production and rural economic development, and increasing farmers' income. It is the strategic option for China to accelerate economic development by expanding domestic demand and promoting economic growth under the circumstances of international financial crisis. And furthermore, the strategy of building a new countryside naturally becomes the breakthrough point for stimulating domestic demand, particularly rural demand.

The reform and opening up policy incorporates China with the world economy. Any changes in China's agricultural development and farmers' income growth will have impact on international agricultural trade. Any changes in China's agricultural production and rural consumption will bring about fluctuation to the world agricultural market (Lohmar et al., 2009). Not only the Chinese government and scholars give due emphasis to China's agricultural economic reform, but international organizations and scholars also show great interest in it. They have done extensive research on China's rural policies and the impact of these policies on China's agricultural economy. Examples include but not limited to the research on the price reform in grain market and market integration (Rozelle et al., 1997), rural labor market (DeBrauw et al., 2002) and rural income growth (Gale and Park, 2002).

The researchers who closely study China's agriculture at Economic Research Service of the United States Department of Agriculture pointed out that despite the tremendous changes in China's agricultural economy after 30 years reform and development, China's rural economic development still face many problems. In addition to the scarcity of rural farm land and water resources as well as difficulties in agricultural technology extension, other problems difficult to solve are developing modern marketing mechanism, infrastructure construction, and increasing farmers' income (Lohmar et al., 2009). Farmers' income growth falls below that of the urban residents. There is a huge income difference between urban and rural residents. International scholars hold that there is a need to stimulate farmers' income growth, which in turn will impact China's agricultural structure and the competitiveness of its agricultural products in international market (Gale and Park, 2002).

Because of the low level and slow growth of farmers' income, rural residents consume over 50% less than urban residents, though rural population in China constitutes the majority in the total population of the country. At present, the level and structure of consumption of rural residents are much less developed than those of urban residents. With social progress and economic development, consumption needs of rural residents have dramatically changed. They are no longer satisfied with the consumption style at the subsistence level. They are eager to have the consumption style of more luxurious content. They set higher requirements for the variety, quality and service of the commodities. Rural markets have greater potential. The environment for consumption has been improved compared with the one in the early years.

The Chinese government issues specific policies for expanding rural demand. These policies include those related with infrastructure construction, public service system, agricultural subsidy, and rural credit system. The improvements in social protection system and infrastructure will solve the most concerned issues for rural residents such as affordability of education, healthcare, pension, and adequacy of clean water, electricity, road and gas supply. These policies also improve the living environment and the conditions for agricultural production. They further enhance the farmers' willingness to consume and confidence of consuming. The project of "bringing-what-farmers-need-to- the-countryside" makes household electric appliances and vehicles affordable to rural residents. It enables rural residents to enjoy the cultural life, leisure activities and living conditions available only to the urban residents in the past.

Given the geographical locations, geological conditions and natural resources of the large rural areas, China should develop its agriculture into a sector full of efficiency, modernization and environment-friendliness, which will bring about growth in farmers' income, and result in stable and sustainable agricultural economy. The areas with appropriate conditions could start industrialized agricultural production, establish specialized economic cooperatives, and set up standardized production bases. There is a need to assist and set up "dragon head" enterprises of agricultural production and processing which have local characteristics and resource advantages. It is also necessary to speed up the process of urbanization, and remove the restrictions of resident permit system which will allow rational mobility of rural labor so that they could work and live in the cities. This will not only raise their income, improve their economic wellbeing of their families, but also will provide sufficient labor for the industry and service sector in the cities. These measures will transfer the surplus farm labor smoothly to non-agricultural jobs in the cities (Huang et al., 2008). Furthermore, it is essential to provide information, technology and training to the farmers who stay in production and processing in the rural areas. It is also essential to help them improve productivity and find markets for their agricultural products. The application, extension and transfer of agricultural technologies will greatly improve productivity and lower costs, which will be the key to enhancing competitiveness in the market.

We should learn the experience from other countries in terms of industrialization of agriculture, supply chain management, food safety system, rural community development and small town construction. For example, many developed countries implement supply chain management for agricultural production and processing. Processors establish long term contracts with farmers by which processors have steady supply of the inputs and guaranteed quality. On the other hand, farmers reduce the risk in income by securing their profits in the contract arrangement. Supply chain management makes it possible to use financial resources effectively and to share technology. It becomes easy for farmers to have access to credit and loans, and technical assistance (Gale and Park, 2002).

China has already started a production model of "company plus farmers". This model has resulted in some effects. However, it is different from the contract production in other countries in that government involvement can be seen more or less in the Chinese model. Production and processing are not totally privatized in the Chinese model (Frederick and Gale, 2003). International organizations and companies have rich experience in assisting developing countries and exploring overseas markets. China's agriculture should take active steps to seize opportunities to participate in international cooperation and joint venture projects. These international projects could accelerate local economic development.

Due to the small amount of land available to each farmer and the large number of small farms, intensive agriculture on large scale is not readily doable. The less intensive farming and processing cannot meet the demand of large super markets, hotels and luxury restaurants for products with consistent quality (Frederick and Gale, 2003). Small-scale production and marketing in rural China do not ensure food safety and quality (Lohmar et al., 2009). Many food products are still produced in small workshops not well-equipped. There is no quality guarantee of the inputs and final products, which in turn lead to health concerns in rural consumers with low income. If production, processing and distribution are vertically integrated, the food supply chain "from farm to table" will realize economies of scale. Integration will also ensure product quality. Moreover, distribution system and channels will be more efficient and smooth.

As mentioned by the authors of this chapter, there still exist many problems in building a new countryside at present. For example, there is no adequate publicity of this campaign. Local governments do not implement the policy effectively. Leaders in the townships and villages have limitations in understanding the policies, exercising proper leadership and identifying the problems. They may only see trees instead of the forest. They may have simple working style or even lack of enthusiasm at work.

Farmers are not proactive enough. Due to their limited knowledge and education, they do not have the same understanding of the policies as the government officials and scholars. Farmers at different age and in different regions have different needs. Leaders at the grass-roots level are farmers themselves. But they have different opinions and understanding of the policies from the ordinary farmers. Therefore, they differ in the effectiveness and outcomes when implementing the policies of expanding domestic demand. Other problems identified by the authors include insufficient material conditions for implementing the policy, and the imperfection of surveillance and management system.

The Chinese government has attached great importance to developing national economy. It shows great concern about the issues of agriculture, rural areas and farmers. However, there are many problems and challenges in the rural areas. Although market-based mechanisms can solve some of these problems, it is not easy and takes no short time to establish these mechanisms (Lohmar et al., 2009). Research and practice are needed. Since the strategy of building a new countryside was carried out not long ago, it is inevitable to experience biased understanding of the policies and ineffectiveness in implementing them. Governments and scholars should listen to the voices of the farmers, and communicate farmers' needs and wishes to the higher level. They should promptly solve the problems with which farmers are concerned most. Governments and scholars also need to make adjustment to the specific implementation plans according to local conditions.

There is a shortage of research from farmers' perspective on building a new countryside. Leaders at the grass-roots level may only report good stories to the upper level government. Therefore, it is necessary for the leaders and researchers to make field trips to rural areas to get first hand information. This helps to learn the real thoughts and needs of the farmers.

If space permits, this chapter could include detailed discussion of the difficulties and suggestions of expanding rural demand. It could also analyze the difficulties and breakthrough points of expanding rural demand from farmers' perspective. It could also give more discussion about the townships and villages which do a good job in expanding domestic demand. If supplied with more case studies and international views, this chapter could provide more information and enlightenment to the readers. It is a long-term and difficult task to turn the huge rural market potential into consumption reality. We anticipate more problems to emerge during the process. This is a great mission to fulfill. And there is a long way to go. Common concern and concerted efforts are demanded from the governments and all walks of life.

REFERENCES

DeBrauw, A., J. Huang, S. Rozelle, L. Zhang, and Y. Zhang. "The Evolution of China's Rural Labor Markets during the Reforms." working paper 02-003, Department of Agricultural & Resource Economics, University of California at Davis, 2002.

Frederick, H., and F. Gale. "China's Growing Affluence: How Food Markets are Responding." *Amber Waves, U.S. Department of Agriculture, Economic Research Service*, June 2003.

Gale, F., and A. Park. "Can Rural Income Growth Accelerate?" *China's Food and Agriculture: Issues for the 21st Century*, F. Gale, F. Tuan, B. Lohmar, H. Hsu, and B. Gilmour (eds.), AIB-775, *U.S. Department of Agriculture, Economic Research Service,* 2002.

Huang, J., K. Otsuka, and S. Rozelle. "Agriculture in China's Development." *China's Great Economic Transformation,* L. Brandt and T. Rawshi (eds.), Chapter 13, New York: Cambridge University Press, 2008.

Lohmar, B., F. Gale, F. Tuan, and J. Hansen. "China's Ongoing Agricultural Modernization Challenges Remain After 30 Years of Reform." EIB-51, U.S. *Department of Agriculture, Economic Research Service*, 2009.

Rozelle, S., A. Park, J. Huang, and H. Jin. "Liberalization and Rural Market Integration in China." *American Journal of Agricultural Economics* 79(2): 635-642, May 1997.

In: China's New Deal
Editors: Li Xiaoxi and Hu Biliang

ISBN: 978-1-61470-815-5

Chapter 3

SUPPORTING AGRICULTURAL AND RURAL DEVELOPMENT: THE DILEMMA AND REFORM IN RURAL FINANCE

Zeng Xuewen*, Qin Jianguo, Peng Linlin and Zhang Yifu

"The Three Rural Issues" concerning agriculture, countryside and farmers are universally considered as a hard nut, with tremendous challenges and difficulties in financial institution service for them. From the historical perspective, China's rural finance has gone through many arduous transformations, such as the phases of development, adjustment, stagnation and reform. The recent years have witnessed a leap-forward growth in the rural finance of China. Despite the impact of the international financial crisis, the rural finance of China has been rapidly developing with the support of government policy since 2008. And we can also see many institutional barriers and business risks, which call upon continuous efforts on the financial services concerning the "Three Rural Issues".

SECTION 1: RURAL FINANCE IN THE INTERNATIONAL FINANCIAL CRISIS

The impact of the financial crisis on the rural economy and finance has been increasingly intense, which primarily lies in the reduction of agricultural loans. Owing to the sharp decline in the demand from the foreign market, waves of farmer workers return to the countryside, a number of agriculture-related companies with export orientation and farmer workers as consumers, therefore, are short of labors, leading to the decreasing revenues, turnover slow-down and increasing loan risks. The undersupply of orders from agriculture-related enterprises, especially from the leading ones, would engender an even lower demand for agricultural raw materials eventually fewer orders and lower prices for farmers and less credit

* Dr. Zeng Xuewen is associate professor of the Institute of Economics and Resource Management, Beijing Normal University. Email: zxw@bnu.edu.cn

loans of financial institutions for agriculture-related enterprises and farmers. In this sense, some of the credit supply for farmers would be squeezed out to some extent in the event of relatively limited credit resources (See Figure 3-1).

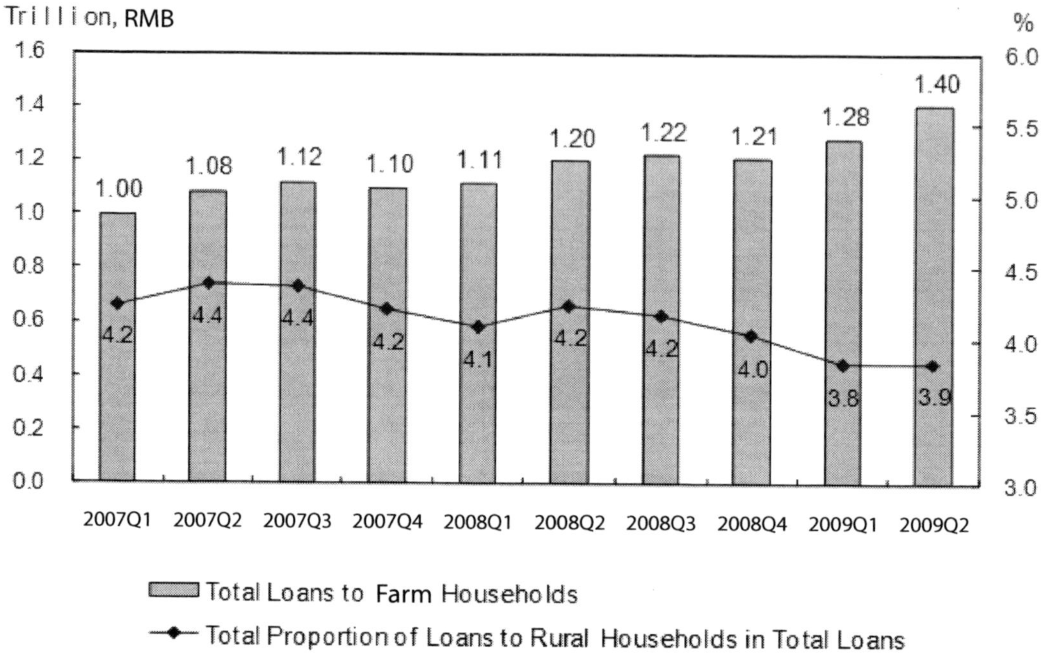

Source: RMB Credit Funds from Financial Institutions (by sector) 2007-2009, the People's Bank of China.

Figure 3-1. Comparison between the Total Amount of Loans and Loans to rural Households.

From Figure 3-1, the proportion of China's loans to rural households in the total amount of loans remains at 4.2% before the third quarter of 2008, but sharply declines in the fourth quarter of 2008, reaching the bottom at 3.84% in the first quarter of 2009. There has been a clear downward trend in loan issuing to rural households. It indicates financial institutions' reluctance to lend to farmers in the wake of the crisis. In reality, in every cycle of credit crunch of China, we can see considerable curtail in loans to agriculture, farmers and the rural areas due to the long production cycle, low profits and high risk of the agricultural industry. Take the capital chain of agricultural products for example. The international demand of agricultural products has apparently knocked down after the crisis, resulting in the continuous price cutting of agricultural products (See Figure 3-2).

In Figure 3-2 there is a sharp decline in the wholesale price index of agricultural products in China after the third quarter of 2008. The prices of rice and corn have been reduced by 5.1% and 11.5% in Feb2009, respectively. The continuous price reduction of agricultural products is just a strike against the sustainability of capital chain of agricultural products, indicating an increase of accounts receivables, the slow-down of turnover of capital and a decrease of repayment and profitability of agriculture-related enterprises and farmers. Consequently, the non-performing loan (NPL) ratio and business risks of financial institutions tend to rise up, forcing banks to tighten credit. The increase of NPL in rural financial institutions would inevitably lead to a reduction in the supply of agricultural loans.

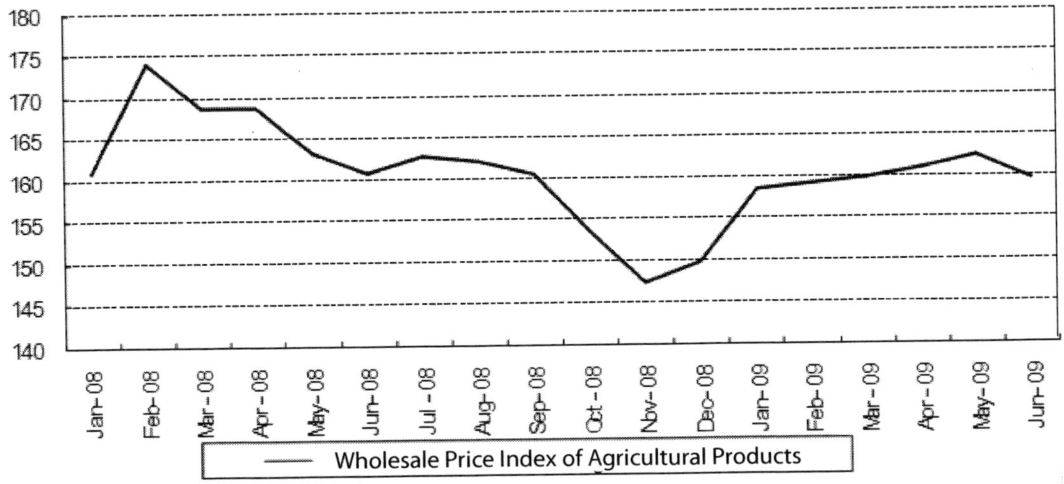

Figure 3-2. Trend Chart of Wholesale Price Index of Agricultural Products between Jan2008 and Jun2009.

Furthermore, the informal financial institutions have taken up a relatively large part in the rural financial market of China, which has played an active role in absorbing agricultural production and operation capital, stimulating rural financial market, raising financial efficiency and promoting the development of self-employed and private businesses in particular. Nonetheless, the operational risks can be aggravated in view of complicated debtor-creditor relationships and irregular ways of making loans, so that more and more disputes on private lending have come into existence. With the deepening impact of financial crisis on China, we are confronted with serious problems and increasing uncertainties, all imposing adverse effects on China's rural finance.

SECTION 2: NEW ADVANCES IN CHINA'S RURAL FINANCIAL SERVICES

On the whole, China has entered into an important phase of development, i.e., promoting agriculture through industry and promoting the development of countryside through city, transforming traditional agriculture, and moving toward the agricultural modernization with Chinese characteristics. We exert our strength on deconstructing the dual structure in order to integrate economic and social development in the urban and rural areas. Decision on Several Major Issues of Promoting Rural Reform and Development of the Central Committee of the Communist Party of China (CCCPC) in Oct2008 has unambiguously brought up the initiation of establishing modern rural financial system and supporting the development of rural finance with great efforts. By virtue of the urge and support of government policy, China's rural finance has made significant progress to albeit the adverse effects of the international financial crisis.

I. Agricultural Development Bank of China (ADBC) Keeping Expanding Service Domains

According to national laws, regulations and policies, ADBC, as the only agricultural policy bank approved by the State Council in China, raises fund, undertakes the policy-oriented financial service concerning agriculture, acts for the Ministry of Public Finance in raising funds to support agriculture, and offers service to the development of rural economy and agriculture using the state credits. Since 2008, ADBC has been expanding service domains and scopes of business in favor of stronger support for agriculture, farmers and the rural areas.

Firstly, ADBC issues the purchase loans of grains, cotton and oil and backs up the industrialization of agriculture. ADBC has taken up an active part in financially supporting the leading enterprises in agricultural industrialization and the agro-processing industry , which were seriously impinged in the financial crisis and encountered more setbacks and potential risks, increasing credit guarantee of the leading enterprises in agricultural industrialization and agro-product processing with no bad records but good fundamentals, credit history, market and orders and temporary operation difficulties, and ensuring the avalability of collateralized loan to business owners or principal shareholders of small businesses. In the first half of 2009, ADBC has cumulatively issued purchase loans of grain, cotton and oil of 198.31 billion Yuan(RMB), a year-on-year increase of 39.81 billion, loans to the leading enterprises in agricultural industrialization and agro-processing of 47.663 billion Yuan(RMB)[1], and supported the purchase of grain, cotton and grease by 193.8 billion *jin* (a unit of weight equal to half kilogram), 1.187 million *dan* (a unit of weight equal to 50 kilograms), and 5.2 billion *jin*, respectively, corresponding to an annual increase of 6.8 billion *jin*, 3.512 million dan, and 3.25 billion *jin*[2]. It has played an important role in stabilizing markets for grain, cotton and oil and standing up for the operation of agricultural industrialization.

Secondly, ADBC has strengthened the support for infrastructural construction, earthquake relief work and post-disaster reconstruction. Since 2008, it has placed priority on the major rural infrastructural constructions fulfilling the policy orientation to increase domestic demand, receiving the concerns of national and provincial governments, and already having obtained the matched financial funds. In the first half of 2009, the cumulative issuances of middle-and-long-term loans for agricultural infrastructural construction and integrated agricultural development have amounted to 89.28 billion Yuan(RMB), increasing year-on-year by 40.6 billion Yuan(RMB), and backing up 816 new projects[3]. Under the circumstance of limited credit scale, it has increased the credit funds and satisfied the need of credit for the purchase of grain, oil and non-staple foods in the disaster areas in the way of setting up green channels of credit, given more funding support to the enterprises and enhancing support for the reconstructions of road networks, water and power systems in the disaster-stricken areas. It has made such a great difference for the disaster areas in working

1 The Outstanding Loan of ADBC Increasing by 140 billion Yuan(RMB) at the First Half of the Year, Farmers' Daily, Aug 14, 2009.

2 ADBC Further Enhancing Agricultural Maintenance Support in Terms of Credit, People's Daily, Aug 16, 2009.

3 Agricultural Information Network of China (http://www.agri.gov.cn/gndt/t20090817_1332667.htm).

through difficulties that ADBC cumulatively input 61.98 billion Yuan(RMB) reconstruction loans into the disaster area in 2008[4].

Thirdly, it has deepened the internal reform and accelerated the internal mechanism construction of risk management and control. ADBC has made effective adjustments in helping businesses cope with the international financial crisis and strengthen the risk management by deepening the internal reform since 2008. On one hand, according to the principle of "dealing with each case on its merits and guaranteeing some expenditure while reducing others", it has intensified the support to the temporarily-in-trouble businesses of quasi-policy and commercial loans in terms of its degrees of loss, reasons of loss and potential risks; on the other hand, in light of different risk profiles, it has adopted distinct measures, such as limitation, contraction and evacuation, to take relevant precautions or remedial measures to manage and control risk, which has ensured the highest efficiency of credit in supporting the "New Countryside Construction".

By the means of deepening reforms, the lines of business of ADBC has been rapidly enlarged, switching from the unitary support for the purchases of grain, cotton and oil to the main business of purchases and loans of grain, cotton and oil accompanied by the intermediary business and the subsidiary business of the middle-and-long-term loans to the leading enterprises of agricultural industrialization and "New Countryside Construction". ADBC has been fulfilling the function of "supporting, strengthening and benefiting agriculture", and playing a more and more important role in rural finance and becoming an indispensable part in sustaining the development of rural economy and solving the "Three Rural Issues".

II. ABC Set Up the Division of the "Three Rural Issues" to Offer Service to Agriculture, Farmers and Rural Area

The Third Plenary Session of the 17th CPCCC in 2008 established the ideas of the development of rural finance i.e., focusing on business, cooperation and policy orientation. The emphasis of rural financial reform of China has started shifting from the single transformation of Rural Credit Cooperatives (RCCs) to the shared responsibility in both RCCs and Agriculture Bank of China (ABC). As one of the four state-owned commercial banks, ABC is an important component of the financial system in rural China. After the reform of establishing the corporate and share holding systems of ABC in Jan2009, in terms of "realizing commercialization in the face of the 'Three Rural Issues'", ABC has specially founded the financial division of "Three Rural Issues" and gradually served as the financial artery.

In the first place, ABC has innovated the management mode of financial division and explored new ways for large commercial banks to serve agriculture. After putting forward the thought of reform, namely, "line management, unit operation, responsibility delegation, independent accounting, positive incentive and effective constraint", ABC has successively commenced pilot programs of the financial division of the "Three Rural Issues" in seven sub-

---

[4] *Financial Bond Raising Prospectus of ADBC in 2009* (http://www. adbc.cn/1about/ detail. asp? channeled = 100250&page=2&id=5085)

branches including Gansu, Sichuan, Funjian, Guangxi, Zhejiang, Shandong and Chongqing, and founded management board of the sub-branch divisions in the senior management of head office. It has announced a succession of rules concerning credit and risk management of rural finance, such as basic instructions of credit business, framework policy of business risks, credit rating regulation of clients, credit authorization, credit line and collateral regulation, all in terms of the "Three Rural Issues".

In the second place, it has strengthened the credit support for agriculture, farmers and the rural areas. The aggregate input of agriculture-related loans in 2008 is 766.7 billion Yuan(RMB), and the outstanding loan is as high as 933 billion Yuan(RMB), including loans to 800,000 rural households worth 99.8 billion Yuan(RMB) and loans to more than 30,000 rural enterprises and organizations with outstanding volume of 630.4 billion Yuan(RMB)[5]. In the first half of 2009, ABC has altogether put in 378.9 billion Yuan(RMB) agriculture-related loans, taking up 26% of total loans of the bank; the outstanding of agriculture-related loans has reached 1.36 trillion Yuan(RMB) up to the end of April, with an increment of 88.1 billion Yuan(RMB) compared with the beginning of the year and a proportion of 40% in the increment of the bank loans as a whole, which has initially reverse the tide of low profitability in agriculture-related loans[6].

In the third place, ABC has urged the research on the financial products regarding agriculture, farmers and the rural areas. Since the reform of the share-holding system, ABC has introduced a large quantity of innovative products and integrated service solutions for customers like the leading enterprises of agricultural industrialization, construction projects of rural infrastructure and featured resources, developed simple and standardized micro-credit loans for large growers, cultivators and individually-owned businesses, exploited real estate mortgage, chattel financing, equipment leasing and factoring for small-and-medium-sized enterprises (SMEs) in the rural areas, and designed "Huinong" Card, a credit cards benefiting farmers with functions of wage distribution, remittance, micro-credit, deposit and consumption. Besides, ABC has gone further into the county investment banking service, supplying county government with finance consultancy and agriculture-related businesses with listing advice, downgrading the threshold for entry and raising property income of farmers. It has enormously enhanced the service function of rural financial products through a series of product innovations.

In the fourth place, ABC has complemented the risk management system of business concerning the "Three Rural Issues". First of all, it has taken up vertical management and independent operation to handle risks to adjust to the authority delegation and long management chain; in the next place, it has been engaged in the work of creating credit file database of agricultural clients and prioritized the clients of good credit on the loan list and cut red tapes for them; thirdly, it has carried out a positive incentive mechanism to primary banks, which made clear the risk tolerance of businesses, stipulated reasonable due diligence and exemption provisions for risk loss within the normal range, mitigated and exempted responsibility for those fulfilling due diligence, to motivate the primary banks in rural financial service.

---

[5] ABC Continuously Going Upward in Business Performance and Getting Ready for Listing, *Shanghai Securities News*, April 28, 2009

[6] ABC Emphasizing on County Networks (http://finance.qq.com/a/20080729/001884.htm)

ABC's attempt to offer service to agriculture, farmers and the rural areas has not only provided it with new profit growth point in the service for "New Countryside Construction", but also opened up new profit space and shown us a new way of moving away from competitive convergence to rural-urban linkage development, cultivating differentiated competitive advantage and exploring another broad market in counties and the countryside. More importantly, it has come up with a reference as to how the large commercial banks could offer service to agricultural sector.

Column 3-1. ABC Launching 'Huinong' Card

Based on the service platform of Golden Spike Debit Card, "Huinong" Card is a comprehensive bank card targeting at all rural households after ABC's reform of the share holding system and the announcement of its "Three Rural Issues" policy. After the one-year pilot project in seven provinces, ABC has set up the division of the "Three Rural Issues" in 2008, and officially launched the card.

In addition to all the functions of the debit card, "Huinong" Card is endowed with many other functions, such as social security, micro-credit and financial subsidies. Through the card, ABC has co-operated with Rural Pension Insurance and New Type of Rural Cooperative Medical Care System aimed to aid farmers in paying premiums and drawing pensions. The card has been used as an ID card for farmers to participate cooperative medical care system, helping them in paying medical expenses, correction fees and reimbursing medical expenses. The implement of the card has cut down the premium collection costs and payment costs of insurance benefits, facilitating farmers as well as improving the government management efficiency on the information of insurance participants. Furthermore, via the card, ABC has taken the agent release of financial subsidies, directly handing out the subsidy payments to the accounts according to the list of financial subsidy of the government, which has diminished of the cost of distributing financial subsidy to a large extent.

Different from other rural micro-credits, the card has made continuous usage of credit possible for rural households. Once acquiring credit, the farmers are authorized to conduct business within the credit line, such as lending, usage and repaying of small loans, in the bank branches or automatic teller machine (ATM) with the card, which has realized the idea of "crediting at a time, continuous usage, loaning and repaying whenever possible" and remarkably increased the efficiency of loan usage. The card has set up the sub-account for the lending of small loans, and the general preferential loans would be respectively distributed into card account and sub-account according to the agreed-on proportion. The fund in sub-account is limited to be used in the designated merchants only within the province or to repay the general preferential loans. The stipulation has guaranteed that the loans be primarily used in

Source: the website of ABC.

III. Rural Credit Cooperatives (RRCs) Deepening the Reform of Property Right System to Offer Service to Agriculture, Farmers and the Rural Areas

The second round of the reform of the property right system in RCCs has been officially initiated in Feb 2008 at the Regulatory Work Meeting of Small-and-medium-sized Financial Institutions in the Rural Area of China. The meeting has stated that the reform is supposed to focus on the transformation of property rights and direct towards shareholding system. With the substantial support of the country, conquering many disadvantages, such as the instable financial and economic environment home and abroad and the ponderous historical burdens, the reform has been steadily pushing forward and breaking new ground in many aspects, like the reforms in the shareholding system, provincial rural credit union(PRCU), and cross-regional operation of rural cooperative financial institutions.

Firstly, the reform of the shareholding system has made smooth progress. Through the reform of property rights in RCCs, there have been setting up 27 rural commercial banks and 174 rural cooperative banks in China until Jun2009, which justifies the high speed of development of rural finance. (See Figure 3-3)[7]

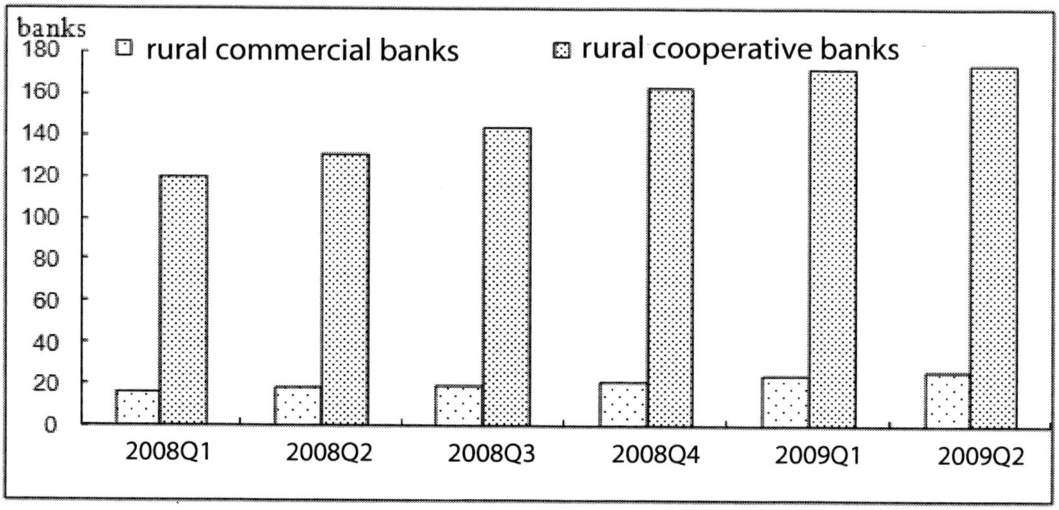

Source: *China Monetary Policy Implementation Report* (Quarter1, 2008 - Quarter2, 2009), the People's Bank of China.

Figure 3-3. The Reform of Shareholding System in RRCs since 2008.

In Figure 3-3, the reform in RCCs has not been stagnated by international financial crisis, but maintained a relatively high speed. Despite the slowdown since 2009, it is going upward as a whole and the process of reform has been going on successfully.

Secondly, the reform of PRCU has been carried out on a favorable condition. As the first rural commercial bank restructured from PRCU, Chongqing Rural Commercial Bank has started business in June 2008. It has generated revenue as much as 8.856 billion Yuan(RMB) in the same year, increasing by 2.392 billion Yuan(RMB) year-on-year, generated pre-provision profit 2.83 billion Yuan(RMB), growing by 1.17 billion Yuan(RMB) and 70.5% over the previous year.[8] The founding of Chongqing Rural Commercial Bank has realized transformations on management modes, switching from decentralized to unified operation, from indirect to direct intervention among juridical persons, which is in favor of the agglomeration effects of resource integration and management advantages and further

7 Source: China Monetary Policy Report, 2008Q1-2009Q2.

8 Source: PRCU Going from Reform to Transformation of Functions, 21st Century Business Herald (http://www.21cbh.com/) .

enhancing the core competitiveness. Thanks to the reform and development, the property relations of Chongqing Rural Credit Cooperative have come to be clearer, the financial condition and asset quality significantly improved, internal management effectively strengthened, corporate governance basically sound and service function evidently reinforced, which initially formed the sustainable long-term mechanism. In December of the same year, according to the principle of "diversified equity", the Rural Commercial Bank of Ningxia Hui Nationality Autonomous Region has completed reform into the shareholding system and established the Yellow River Rural Commercial Bank. The model of Rural Commercial Bank of Ningxia Hui Nationality Autonomous Region is to take capital as link to control the subordinate RCCs, which has retained the independence of sub-branches and modified the ambiguously defined property rights in county RRCs to some extent, advancing the reform of PRCU.

Thirdly, the cross-regional operation has been enforced in rural financial cooperatives. In Dec2008, marked as the preface of cross-regional operation of rural financial cooperatives, Changshu and Zhangjiagang Rural Commercial Banks of Jiangsu province have respectively opened Haimen and Tongzhou sub-branches in Nantong. The cross-regional operation and equity consolidation would broaden the channels of capital reflow into the rural areas, heating up the competition in rural financial market and promote the sound development of rural finance.

Fourthly, the asset quality of RRCs has been relatively bettered. In terms of the Loan Classification System, NPL outstanding and the ratio of China's RCCs are 320.3 billion Yuan(RMB) and 7.2% respectively, decreasing by 23.8 billion Yuan(RMB) and 0.7% compared with the end of 2008.

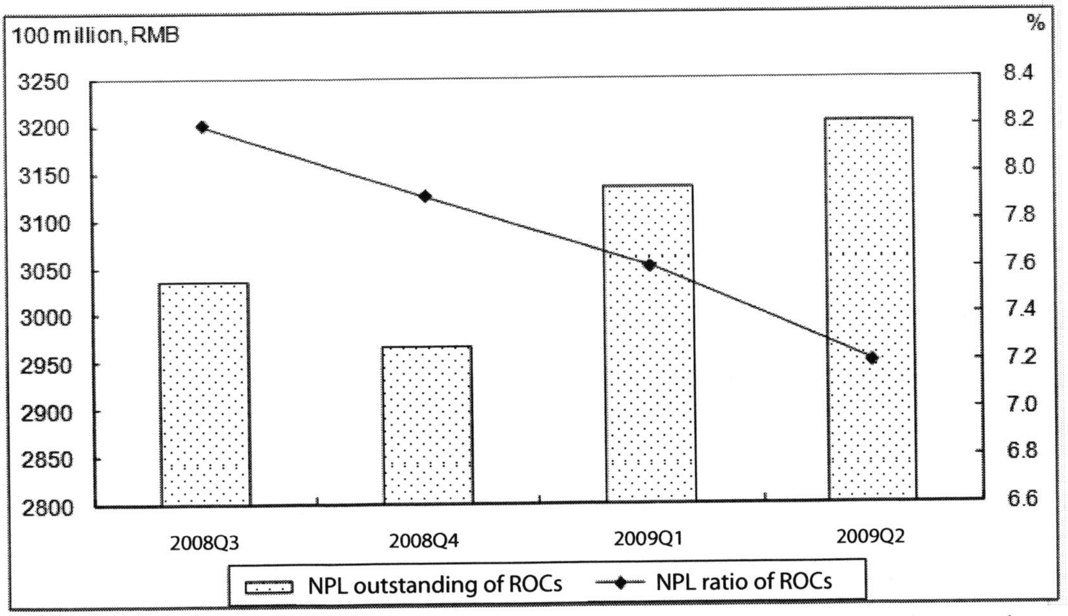

Source: Annual Report of China Banking Regulatory Commission 2008, China Banking Regulatory Commission; China Monetary Policy Implementation Report (Quarter4, 2008), the People's Bank of China.

Figure 3-4. NPL of RRCs since 2008Q2.

From Figure 3-4, the NPL ratio of RCCs tends to plummet, while the NPL outstanding has displayed the tendency of rebounding since the second half of 2008. The continuous dropping of NPL ratio is largely ascribed to central bank's cashing a great quantity of special notes to RRCs to strip bad assets and improve the asset quality; meanwhile, the rebounding NPL outstanding indicates the shock from the international financial crisis. In a word, the overall asset quality of RCCs has been improved. In 2008, the net profit after tax (NPAT) of rural financial institutions[9] has amounted to 39.59 billion Yuan(RMB), in which that of RCCs is 21.91 billion Yuan(RMB), rural commercial banks 7.32 billion Yuan(RMB) and rural cooperative banks 10.36 billion Yuan(RMB). The management has been further modified as well.[10]

The round of reform is another significant attempt for Chinese finance to resolve the "Three Rural Issues", in which two distinct differences have asserted themselves. Firstly, the emphasis of reform in property rights has gone from three modes, i.e. the systems of share-holding, joint-stock cooperative and cooperative to the ultimate demutualization; secondly, the management mode of PRCU founded in the first round of reform has become the target of the second round. By and large, the anti-risk capability and service function of RCCs [11] having been enhanced and the business vitality of RCCs has been further stimulated, which made a great difference in realizing the sound and healthy development of RCCs, wholly improving rural financial service and advancing the solution of the "Three Rural Issues" and the "New Countryside Construction".

IV. Postal Savings Bank Achieving in Capital Reflow to the Rural Areas

Established in 2007, the Postal Savings Bank of China (PSBC) is a financial institution of asset and liability established on the basis of postal service network of China Post and in succession to the postal financial services of the former State Post Bureau and China Post. The postal savings banking has gained ground and improved the service function for agriculture, farmers and the rural areas since 2008.

Firstly, PSBC offered wholesale funds to rural financial institutions. PSBC have conducted financial transactions in capital market and rural financial institutions to indirectly restitute funds to the countryside and support "New Countryside Construction" through the funding channels of rural financial institutions. At the end of Dec2008, the outstanding of funds for agriculture maintenance was 66.942 billion Yuan(RMB), taking up 4.23% of the autonomous asset balance of PSBC, in which the negotiated deposit on supporting agriculture was 19.23 billion Yuan(RMB) and the balance of subscribing bonds of ADBC amounted to 40.925 billion Yuan(RMB).[12] As for syndicated loan, by virtue of good asset sales and management capability of other financial institutions, PSBC has channeled a large volume of

9 Rural financial institutions include RCCs, rural commercial banks and rural cooperative banks restructured from RCCs.

10 Source: *Annual Report of CBRC 2008* and *China Monetary Policy Report 2008Q4*.

11 RCCs have been restructured into rural commercial banks or rural cooperative banks, yet the name of "RCCs" is still in use for the sake of correspondence in the context.

12 Kou Jian-ping. Li Ming, 2009. Capitalizing on Advantages and Giving Excellent Service to Farmers: An Interview with Tao Li-ming, the President of PSBC. C-thinking 4.

postal savings funds into the fields of key projects of the "Three Rural Issues", rural infrastructural constructions and comprehensive rural development by the means of participating in syndicated loan, indirectly leading capital flow back to the rural areas.

Secondly, PSBC has used micro-credit as a point of departure to upgrade rural credit service. In Jun2007, PSBC has commenced the pilot program of micro-credit. All the provincial and municipal branches have launched micro-credit business until Jan2009, with 1911 county sub-branches participating as well. As to business volume, PSBC has cumulatively issued 37.7 billion Yuan(RMB) small loan by Jan2009, with loan amount in the rural areas accounting for 70%.[13] In order to satisfy the need of individually-owned businesses, small business owners and operators, PSBC has initiated a great deal of retail asset businesses, such as loan secured by certificates of deposit (CDs), personal business loan, self-help housing loan and second-hand housing mortgage to enrich rural credit products.

Thirdly, PSBC has improved the serviceability of primary banks to supply all-round services to rural citizens. PSBC have been making efforts and achievements on strengthening service of primary network since 2008. Regarding deposit-taking, the savings balance of county networks of China's postal savings has come to 491.153 billion Yuan(RMB) by the end of 2008, making up 23.61% of the total deposits; the balance in rural networks was 840.499 billion Yuan(RMB), 40.41% of the total; concerning remittance, PSBC has set up the nation-wide networking including 350,000 outlets, greatly facilitating the deposit and withdrawal of farmer workers, individually-owned businesses, visitors and tourists; respecting various agency businesses, relying on the rural networks, PSBC has acted as the agent of collecting power charge, telephone fee and TV subscription, with a total of 8.5 billion Yuan(RMB) in 2008.[14] Moreover, PSBC has also aided in issuing food subsidies, funds of returning farmland to forest and family-planning grants, and even provided rural insurance service for farmers in need but without access to insurance service.

The founding of PSBC has put a full stop for its history of "deposit-only", to benefit the rural capital reflow and alleviate the imbalance between supply and demand, to consummate the service of rural and urban finance and encourage effective competition in rural financial market, to give CBRC a hand in supervising postal savings, forestalling and diversifying away postal financial risks, to further strengthen the financial support of postal funds for agriculture, increase satisfaction and capacity of rural financial service and accelerate the deepening of rural finance.

V. New-type Rural Financial Institutions Attaining Rapid Development

Since 2008, CBRC, the Ministry of Public Finance and the People's Bank of China have launched a series of measures and policies to confirm and ensure the legitimacy of New-type Rural Financial Institutions, give permission to the access of international and private capital into the rural financial market, and broaden the channels of capital reflow to the countryside in the way of substituting reward for subsidy, directed subsidies, and reducing deposit-reserve

13 PSBC Nurturing Rural Area to Serve Farmers. Nongbo website, March 16, 2009.

14 Kou Jian-ping. Li Ming. Capitalizing on Advantages and Giving Excellent Service to Farmers: An Interview

ratio in order to maintain sustainability of new rural financial institutions and to cope with the international financial crisis.

1. Village Bank Developing Rapidly

For CBRC has loosened the restriction on market access to rural banking, village banks, as a new rural financial institution, have achieved fast development. Depending on the existing financial institutions and flexible mechanisms, village bank has caught most attention in all types of new rural financial institutions. It has taken positive effects in changing the under-supply and deficient competition in the rural financial market since the first pilot program.

Column 3-2. The First Village Bank in Shanghai Founded

New achievement has been made in the reform of Shanghai rural financial institutions. On Feb19, 2009, the first village bank of Shanghai, Yangtze River Village Bank, started official operation in Chongming, with registered capital of 100 million Yuan(RMB). As the principal founder, Shanghai Rural Commercial Bank has put up capital of 51million Yuan(RMB), other five founders including Chongming asset management co. of Shanghai. The bank has agenda in "offering service to county, countryside and SMEs", stimulating the formation of financial service with diversified investment, various products, with wide coverage, flexible control and effective service in Chongming, raising capability and efficiency of financial service and settling the difficulty in obtaining loans for farmers and SMEs. It would play an active role in increasing farmers' income, adjusting the industrial structure of agriculture and backing up the "New Countryside Construction" in Chongming.

As the major initiator, Shanghai Rural Commercial Bank has endowed the village bank with management model, technology platform, business process, product development and recruitment through its own mature operation system and risk control and assisted the village bank in corporate governance, operation, development and risk prevention.

Source : The First Village Bank of Shanghai Launched, Shanghai Financial News, Feb20, 2009.

The capital source of village bank is mostly diversified, and the Chinese banks are initiated by one bank and joined by many corporations in most cases, while the foreign-owned banks remain independently-owned. The second half of 2008 has witnessed a number of village banks in the rural areas set up by ABC, Bank of Communications, Bank of Beijing, China Minsheng Bank and Bank of Nanjing. With regards to the foreign-funded banks, HSBC has opened the first village bank in Suizhou, Hubei in Dec2007 followed by others in Dazu Chongqing, Yongan Fujian, Miyun Beijing, and Enping Guangdong; Standard Chartered Bank set up its first village bank in Inner Mongolia in 2009 as well. The work arrangement of New Type Rural Financial Institutions from 2009 to 2011 issued by CBRC in 2009 has publicized the plan of establishing 1027 village banks all over the nation from 2009 to 2011. The action has appealed more to private capital and boosted development in rural finance. Despite difficulty in fund-raising and risk control for village bank, it has gradually turned into a profitable and effective model in rural finance to invigorate the rural financial market.

2. *Finance Company Expanding Steadily*

There are two types of finance company in China at present. Under the supervision of the Department of Industry and Commerce and excluded from financial institution, the first one is small-credit company ltd. or plc. set up by natural person, enterprise as legal person, and other social organization without public deposit. It is not classified into rural financial institutions until the issuance of CBRC's document allowing small-credit company's transformation into village bank. As non-banking financial institution and New-type Rural Financial Institution, the other one is finance company making loans to county farmers, agriculture and the rural areas which is set up by domestic commercial bank or rural cooperative bank with the permission of CBRC according to relevant laws and regulations.

Under the active impetus of local government, small-credit and finance companies have gained steady development after CBRC loosening the threshold of access to rural financial market. The nation-wide micro-credit companies have come into being and started operation up to June 2009, and seven finance companies have been founded as well. Through different innovations, the two types of institutions have exertcd active impact on spurring county and rural financial market and opened up a new path to handle the capital pressure concerning the "Three Rural Issues". Firstly, the two types of institutions have added and widened channels of capital flow back to the countryside, made up for the disadvantages of rural financial service and found a new method to fulfill the demand of farmers and small businesses for capital, which is of great importance to the development of rural economy, the increasing of farmers' income and pushing forward "New Countryside Construction". Secondly, by creating new method of loan and cutting off lending procedures, the institutions have put an end to the difficulty in making loans for some farm households. Thirdly, with the service chain of different levels and types, the institutions have cooperated with local commercial banks, village banks and RCCs to increase rural credit. Fourthly, the institutions have constrained the private lending rate and compressed the market of private lending by virtue of the demonstration effect of interest rate fixing.

As a loan organization to serve farmers, agriculture and the rural areas, finance company has progressively become an effective way of fund-raising for farmers and small businesses in the rural areas, dealing with the difficulty in financing, complementing rural financial system of various levels and actively promoting county economy and agricultural development.

3. *Rural Financial Credit Union (RFCU) Maintaining a Good Momentum*

CBRC has relaxed the restriction on the entry of financial institutions in the rural areas in 2007, and provided services of deposit, withdrawal and settlement to farmers and small businesses who voluntarily participated in the Financial Credit Union.

The present RFCU in China can be divided into three types as to degree of formality. The first type is approved by CBRC and registered at the Department of Industry and Commerce, with 11unions by the end of June 2009; the second one registered but not approved, with more than 30 by the end of June 2009; and the third one is the Mutual Aid Association of Development Fund in Poverty-stricken Village initiated by the State Council Leading Group

Office of Poverty Alleviation and Development, neither registered nor approved , established in 140 poverty-stricken villages in 28 counties, 14 provinces of China now.[15]

As for the rural areas as a whole, the establishment of RFCU has contributed to the development of diversified economies, yet due to the self-closed nature, the deposit and loan are confined among members, which would attract local farmers and small businesses to join in only for acquiring loans. In this sense, compared with RCC, RFCU is at a disadvantaged position on deposit-taking, which combined with the floor of interest rate leads to capital shortage. Therefore, the growth of RFCU cannot work out without the support of governments at all levels, besides pushing upward the floor of interest rate, enhancing savings absorption and working together with other banking institutions. The authority is supposed to relieve its tax burden since RFCU is a share-holding corporation founded by acquaintances in poverty-stricken area to eliminate poverty.

VI. Agricultural Insurance Continuously Going Upward

Since 2008, based on the precious experience of pilot program, the Ministry of Pubic Finance and China Insurance Regulatory Commission (CIRS) have taken several steps in enlarging the coverage of agricultural insurance and the range of premium subsidy and implementing the issuance of premium to build up long-term mechanism of agricultural security. The agricultural insurance has gone along well.

Firstly, the business scope of agricultural insurance has escalated rapidly and gradually come into function. In the first half of 2009, the former premium income of insurance of China's agricultural insurance has reached 7.003 billion Yuan(RMB), increasing by 59.81%, in which that of policy-oriented agricultural insurance is 6.714 billion Yuan(RMB), accounting for 95.87% of the former premium income of agricultural insurance; there have been 61.52 million insured rural households, growing by 49.2% year-on-year; the credit protection has been as high as 143.7 billion Yuan(RMB), up by 89.9% year-on-year; the indemnity to farm households arrived at 3.407 billion Yuan(RMB), increasing by 63.09% year-on-year, in which that of farming was 1.388 billion Yuan(RMB), up by 23.51% year-on-year, that of breeding industry was 2.019 billion Yuan(RMB), up by 10.923 billion Yuan(RMB) year-on-year. In the first half of 2009, the indemnity of agricultural insurance to farm households totaled 3.407 billion Yuan(RMB), growing by 63.09% year-on-year, mitigating the possibility for farmers to go back to poverty due to disaster to some extent and ensuring the agricultural production in the disaster area.[16]

Secondly, the coverage and service of agricultural insurance have been expanded. First of all, the operation network of agricultural insurance has preliminarily come into being. With 20 agricultural insurance companies in China up to the end of June 2009, the business network of agricultural insurance made up of nation-wide insurance companies and professional insurance companies has taken shape gradually. In the second place, the

---

15 Hao Yu-bin, 2009. On the Development of the Rural Financial Credit Union of China (Chinese) [J]. *Theoretical Exploration*, 2.

16 The website of CIRS: The Agricultural Insurance Maintaining Good Momentum of Development in the First Half of the Year (http://www.circ.gov.cn/tabid/106/InfoID/107057/frtid/3871/Default.aspx) .

responsibility scope of insurance has been magnified. In terms of farming, the companies have incorporated drought, insect pest and plant diseases into the insurance liability; as for breeding, all the animal diseases (plagues), natural disasters, accidents and government culling insurance liability have been covered. In the third place, the research on new insurances has been carried out. In 2009, the types of insurance receiving national financial support to premium subsidy have newly incorporated the insurances of pig farming and forest, with all 9 policy-oriented agricultural insurances; there have been more than 160 types of insurance all over China until June 2009, covering every aspect of the "Three Rural Issues".[17]

Column 3-3. Hainan Province Commencing A New Pilot of 'Linkage of Bank and Insurance'

As a part of the rural financial system, rural insurance is of great importance in financial support to agriculture, farmers and the rural areas. In the long run, there has been no reasonable risk-sharing mechanism in the countryside, which largely limits the growth of rural credit, collateral and futures, even the sound development of rural economy. On the strength of public finance of government, it is a new path to develop both rural insurance and credit by exploring the combination of them. As the pilot province of mutual insurance subsidy, Hainan has established itself as a model of the "Linkage of Bank and Insurance".

Since May 2009, Hainan has launched a new wave of rural insurance pilots, created the model of paying premium subsidies with financial funds and made some achievements, in which the financial department of the government would hand out 10% - 80% subsidy to insured farmers and agricultural or fishing enterprises. Hainan has planed to institute the Mutual Aid Association of Agricultural Insurance since 2009, selected 1 or 2 counties as pilots to probe into the models of "coinsurance management" and "mutual aid among farmers and farm households", based on agriculture trade associations, leading enterprises of industrialization and cooperative organizations with universal arrangement of start-up and operation funds by provincial financial department to reinforce the self-accumulation and anti-risk capability of rural insurance. In the meantime, Hainan has put emphasis on the research of agricultural insurance to stimulate the linkage of bank and insurance. By the means of "loan on the condition of insuring and pledged policy", Hainan has connected insurance with credit business and formed a positive interaction of "farm households taking insurance to be insured, and banking making loans to promote farmers' willingness for insurance", expanding the insurance business as well as promoting the development of credit.

The method of combining rural insurance and credit, joining with the public finance of government, and uniting different parts of rural finance as a whole is capable of pushing forward the industrialization of agriculture, stimulating agricultural production and consumption and guaranteeing the credit of rural financial market to resolve the "Three Rural Issues".

Source: 4 Measures of Hainan to Improve Rural Insurance, Hainan Journal, Feb. 10, 2009.
The People's Government of Hainan Province ： The Pilot Program of Agricultural Insurance of Hainan Province in 2009, the People's Government Office of Hainan Province(2009) No.92.

[17] The website of CIRS: The Agricultural Insurance Maintaining Good Momentum of Development in the First Half of the Year （http://www.circ.gov.cn/tabid/106/InfoID/107057/frtid/3871/Default.aspx） .

China's agricultural insurance has overcome double negative factors, the international financial crisis and Wenchuan Earthquake, with tremendous support of the government. The scope of agricultural insurance has kept growing, service domains widening, pilot programs starting in succession, which has assumed an active role in improving disaster and loss prevention and post-disaster recovery capacity, completing social protection system, ensuring rural stability and serving the "New Countryside Construction" and the "Construction of Harmonious Society".

SECTION 3: MAIN ISSUES IN RURAL FINANCIAL REFORM AND DEVELOPMENT AT PRESENT

Recently, China's reform in rural finance has made remarkable advancement, but we have to admit that there are still many new issues and contradictions as the tests and uncertainties for the development of rural finance, in particular under the profound impact of the international financial crisis.

I. Loan Risks concerning the "Three Rural Issues" Having Risen Up

To handle the international financial crisis, the credit scale kept inflating at a high speed. In the first half of 2009, there added 7.72 trillion Yuan(RMB) local and foreign currency loans, increasing by 32.8% year-on-year and tremendously outpacing the growth of local and foreign currency deposit. [18] The small-and-medium-sized financial institutions, represented by RCCs, have taken up an increasingly bigger proportion in incremental loans, especially since March 2009. The rapid expansion of credit scale indicated more liquidity consumed, the well-off level of which could be possibly changed due to the radical asset allocation. It is no less than a gamble with small odds in the case of unclear economic situation. At the same time, the loan risks in the rural areas have slightly ascended. Firstly, NPL has locally rebounded, demonstrated in 7 provinces and cities including Sichuan, Hunan, Zhejiang, Hebei, Gansu, Shanxi and Shanghai; for the new rural financial institutions with shareholding system, we have also witnessed the rebounding of NPL ratio in rural commercial banks in Wujiang, Changshu and Kunshan. Secondly, the total structure of NPL in rural financial institutions has deteriorated to some extent (See Figure 3-5).

From Figure 3-5, in the 593.9 billion Yuan(RMB) NPL, the proportion of sub-category, doubtful, and loss loans are respectively 28.7%, 51.5% and 19.8%. Compared with 2007, the ratio of doubtful and loss loans have increased by 2.3% and 0.5% respectively,[19] while NPL has tended to decline, and the loan risks increased slightly.

18 CBRC: 7 Characteristics of the Incremental Loans of Banking Institutions in the First Half of the Year (http://www.gov.cn/gzdt/2009-08/09/content_1387243.htm) The Portal Site of Central People's Government of People's Republic of China.

19 Source: Hope and Fear: The Performances of Small-and-medium-sized Rural Financial Institutions in 2008, *21st Century Business Herald* (http://www.21cbh.com/HTML/2009-2-25/HTML_14VYD 5XJI6 QU. html)

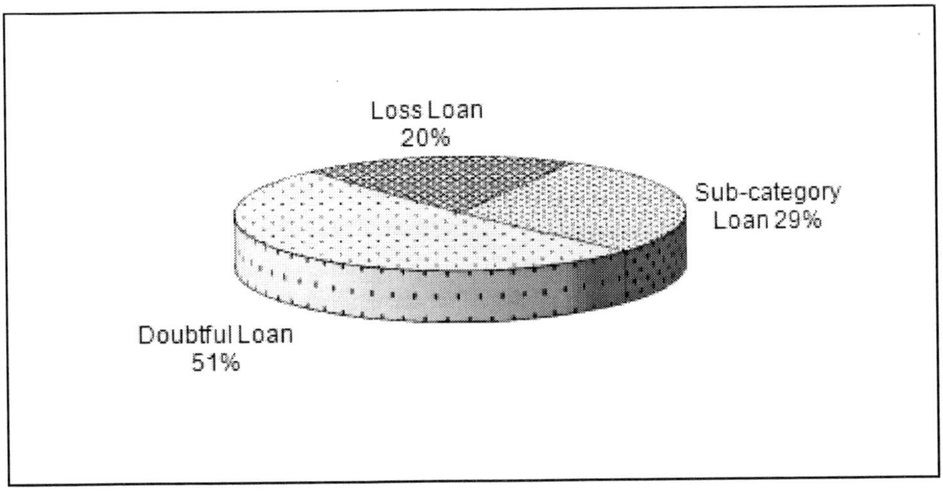

Figure 3-5. The Structure of NPL of Rural Financial Institutions in 2008.

II. Rural Credit Lacking in Effective Collateral

So far, the rural credit guarantee primarily consists of group lending of farm households and collateralized loan. The group lending of farm households has improved the rural credit environment, decreased the risk and partially handled the difficulty for farmers in making loans. However, we cannot ignore the limitations: firstly, there are cumbersome procedures and high thresholds; secondly, as a result of the joint liability of group lending, most farmers would not take part in the lending group unless they are in need of loans, which renders the members into both borrower and guarantor and diminishes the efficacy of the guarantee, even brings about wrangles and debt repudiations among members; thirdly, owing to the marginalization of the weak group, the rich households with great financial strength and little default risks are usually picked up rather than the poor ones, and the pecking order has obviously asserted itself here. In addition, the large amount and long term of the collateral loan can partly meet the demand of some farm households and rural businesses for large loans. Owing to the rural area lacking in effective collateral, the spread of collateral is at a very low speed, and the existing innovations of secured loans, such as forest-property mortgage and mortgage of usufruct of land contract, are at the stage of pilot and far from being popularized.

III. The System of Agricultural Policy Finance Being Incomplete

The development of China's agricultural policy-oriented finance lags behind that of agriculture, and the financial functions have not been brought into full play. The limitations have asserted themselves again in the financial crisis. Firstly, the business scale of policy-oriented finance is too limited to fulfill the demand; secondly, the boundary between policy-oriented and commercial finance is blurred since both ABC and RCCs have undertook some policy-type businesses; thirdly, there is not legislation regarding policy-oriented financial institutions, nor efficient regulation on the business scope, operation rules or penalties for violation, which confines the healthy growth of policy-oriented finance; fourthly, the mechanisms of risk prevention and control are not sound. Under the stress of financial crisis, financial risks have increased and the efficiency of policy-oriented finance would be seriously weakened by a quantity of issues like incomplete internal control mechanism, deficiency in risk forecast system, weak post-loan management and simplified solution to NPL.

IV. China Suffering the Absence of Agricultural Commercial Insurance

There has been a tendency for natural disasters to increase in recent years. According to the statistics, there have been 41.29 million hectares of affected crops on average from 2004 to 2008, in which the disaster- affected areas are 21.65 million hectares, taking up 53% of the affected areas and 14% of the sown areas.[20] Natural disasters have adversely affected the development of agriculture and crippled the competitiveness of agricultural products in the international market. Nowadays, the global economy has apparently decelerated due to the international financial crisis, which has called for further improvement on the quality of agricultural products and more caution in the aversion of operation risk. Nevertheless, China's agricultural insurance, commercial insurance in particular, is seriously under-developed and far from coping with the international financial crisis and developing agricultural industrialization even though the great leap-forward development in the late two years has been realized with the support of the central finance, in which the former premium income of agricultural policy-type insurance is 6.714 billion Yuan(RMB), taking up 95.87% of the former premium income of agricultural insurance, yet agricultural commercial insurance only accounting for 4.13%.[21]

V. China Being Deficient in the Channels for Private Capital to Flow Back

Impinged by the financial crisis, a large volume of big cultivators and rural SMEs are in the dilemma of broken funding chains because the loose monetary policy has not alleviated the deficiency of capital in essence. The formal financial institutions are reluctant to lend to the disadvantaged groups with low anti-risk capability who thus are forced to turn to the non-formal finance. Despite the opening-up of China's rural financial market in 2006, the degree is relatively small and the private capital can only go into rural funding cycle through finance company and new rural financial institution. As for the non-formal financial institutions, like private lending, pooled loan and mutual fund, they are still under strict regulation. A large quantity of capital is not able to flow back to the countryside, while the farmers and businesses have nowhere to make loans. Here comes a bizarre result: on one hand, the idle

---

[20] China Investment Consulting, ltd.: Investment analysis and Forecast of China's Agricultural Insurance Market from 2009 to 2012

[21] The website of CIRS: The Agricultural Insurance Maintaining Good Momentum of Development in the First Half of the Year (http://www.circ.gov.cn/tabid/106/InfoID/107057/frtid/3871/Default.aspx)

private funds find no path to be invested; on the other hand, many businesses have huge demand for but no access to capital. It has resulted in the low efficiency of capital allocation and posed a hurdle for the sound development of agriculture, farmers and the rural areas.

Section 4: Recommendations on Further Promoting the Financial Service concerning the "Three Rural Issues"

Though China has accomplished stabilized economic recovery, the international financial crisis has still had impact on our economy. We must make scientific prediction and research and take active precautions to enhance and deepen the policy support for the financial service concerning the "Three Rural Issues".

I. More Support for Agriculture, Farmers and the Rural Areas

The international financial crisis has brought about huge impacts on the economic and financial system of China. Its indirect and implicit influence on the rural economy and finance has gradually unveiled itself compared with that on the city. The international demand for main agricultural products is shrinking, price dropping off, and the export of some products have been hindered. The deceleration of domestic economy as a result of the international financial crisis and the weakening domestic demand have generated more pressure on the falling prices of agricultural products. The productivity of agricultural production is declining, while the difficulty of stabilizing agricultural development and farmers' income increment is increasing. Hence, the government ought to make a full plan on the comprehensive impact of the crisis on urban and rural areas. Firstly, it must enhance the support for agriculture to resolve the financial problems concerning people's welfare and satisfy the basic demand of farmers; secondly, it should upgrade the financial support for agricultural industrialization to protect farmers' willingness to produce, stabilize agricultural production, increase farmers' income and maintain the peace of rural area.

II. Accelerating the Establishment of Compensation Mechanisms of Financial Service regarding the "Three Rural Issues"

The rural financial market is characterized by segmentation and high risk, and the financial institutions in pursuit of profit maximization are reluctant to issue agriculture-related loans owing to the lacking of correspondent compensation mechanisms. The phenomenon is more apparent under the circumstance of the financial crisis. Therefore, we are supposed to speed up in the establishment of compensation mechanism of commercial financial service in terms of the "Three Rural Issues". Firstly, the financial service for the "Three Rural Issues" should be entitled with the generalized preference system. For example, we can hand out the proportional financial subsidies to relevant financial institutions according to the aggregate agriculture-related loans; for the commercial financial institutions making loans more than a certain ratio in total amount, they are given preference on taxation; and we could empower the rural financial institutions in the self-regulation of interest rate and permit them to absorb deposit and make loans in a wider range of interest rates. Secondly, the policy support framework of agricultural insurance is supposed to be completed. On the basis of completing the system of agricultural insurance and for the sake of sharing risks of commercial finance, we should probe into founding the system of agricultural policy-oriented insurance, including financial subsidies for catastrophe, the agricultural re-insurance system and the institution of third party vouching.

III. Completing the Rural Credit Model and Vouching Business Model

For the demand diversity has been the feature of rural finance, the supply of it is supposed to be well oriented. In the first place, we should develop new credit models suitable for the characteristics of rural economy as well as moving on popularizing the small-credit and group lending, such as the models of "company + farm household", "company + intermediary + farm household", and "company + professional market + farm household". In this sense, the models can take advantage of radiation and pulling effects of agricultural industrialization, promote the high-quality, high-efficiency agriculture with unique characteristics. In the second place, we could make innovations in vouching methods and add more effective collaterals. The research and promotion of new pledges, such as forest-property mortgage, CD, pledge of accounts receivable, chattel mortgage and agricultural product mortgage, would increase the satisfaction of rural financial service.

IV. Taking Full Advantage of Private Capital to Push forward the Development of New Rural Financial Institutions

We ought to step up the pilot program and promotion of new rural financial institutions, such as village bank, small-credit company and RFCU. It is of necessity to nurture and develop new rural financial institutions: firstly, to alleviate the undersupply of financial service in rural area; secondly, to raise the coverage of rural financial service; thirdly, to enhance the competition of rural financial market; and lastly, to foster the willingness of different parties to take part in the nurturing and development of such institutions. Governments at all levels and various types of capital have appealed to the regulators for their strong desires of establishing new rural financial institution, developing rural financial business, improving rural financial service and mitigating the difficulty of funding SMEs. Consequently, it is urgent and necessary to foster new rural financial institutions and make use of finance to support the construction of socialist new countryside and the development of SMEs. In the founding of new rural financial institutions, the government and regulators are supposed to fully mobilize private capital and gradually diversify the property rights of rural financial institutions.

V. Building up the Motivation System of Capital Reflow to Rural Area with Different Approaches

The capital outflow has long been the bottleneck of the development of agriculture, farmers and the rural areas. It becomes more severe since the financial crisis. The construction of effective mechanism for capital reflow to the rural areas is a key step of assisting and protecting agriculture, farmers and the rural area. We should focus on the following three aspects. Firstly, we are supposed to construct the mechanisms of financial input for capital reflow to the rural areas, increasing not only the aggregate of financial support to agriculture but also its ratio in the total expenditures, forming the steady mechanisms of national support for agriculture and giving more input to the rural areas. Secondly, we can build up the leverage incentives for capital reflow to the rural areas. The combination of taxation, interest rate and deposit-reserve ratio would encourage commercial financial institutions to operate in the rural financial market. We could also formulate more preference on deposit-reserve ratio for the rural financial institutions with higher percentage of agriculture-related loans and enforce more flexible policy on interest rate to raise the re-lending line. Thirdly, we are supposed to optimize the allocation of rural financial resources by the means of innovation and reform. Through the reform and restructuring of ADBC, ABC and PSBC, we can endow rural financial service with wider coverage, stronger anti-risk capacity and higher service quality, to further complete the financial environment in counties and lead more financial resources into the rural areas.

Translator: Li Hui (SUFE)

English Version Editors: Xiao Rong (SUFE), Du Wenxin (Harvard)

REFERENCES

[1] International Monetary Fund, 2009. *Global Financial Stability Report* [R]. Beijing: China Financial Publishing House.

[2] Rural Financial Services Research Group of the People's Bank of China, 2008. *China's Rural Financial Services Report 2008* [R]. Beijing: China Financial Publishing House.

[3] Rural Finance Academy of China,2008. *China's Rural Financial Reform and Development of Three Decades*[R]. Beijing: China Financial Publishing House.

[4] Han Jun, et al., 2009. *Survey on China's Rural Finance* [M]. Shanghai: Shanghai Yuandong Publishing House.

[5] Li Shu-sheng, He Guang-wen, 2008. *Study of Financial Innovations in Rural China* [M]. Beijing: China Financial Publishing House.

[6] Xiong De-ping,2009. *Study on Coordinated Development of Rural Finance and Rural Economy* [M]. Beijing: Social Sciences Academic Press.

[7] Zhang Xiao-shang, He An-nai, 2007. *Financial Transformation and Innovation in the Rural Area* [M]. Beijing: Social Sciences Academic Press.

[8] Jiang Ding-zhi, 2008. Reform and Development in Rural Finance for Three Decades[J]. *China Finance,* 23.

[9] Zang Jing-fan, 2009. The Reform and Development of Rural Finance [J]. *Rural Credit Cooperative of China,* 7.

[10] Gao Sheng-ping, Liu Ping, 2009. On the Credit Collateral in Rural Financial System: Predicament and Outlet [J]. *Journal of Financial Research* 2.

[11] Gong Jing, Liu Hong-yan, 2009. An Investigation on China's Rural Financial Innovation after Sub-prime Crisis[J]. *Rural Economy* 7.

[12] Guo Wu-yan, 2009. On the impact of Financial Crisis on China's Rural Financial Repression: A Reference to Deepening Rural Finance [J]. *Rural Economy* 5.

In: China's New Deal
Editors: Li Xiaoxi and Hu Biliang
ISBN: 978-1-61470-815-5

Comments on Chapter 3

IMPROVING RURAL FINANCE UNDER THE INDUSTRIAL STRUCTURE CHANGE

Zhang Tong *

The author of the chapter , "Confronting agriculture, rural area and farmers—the dilemma and reform of rural financial services (D & R)," points out that under the current circumstance of worldwide financial crisis, China Agriculture Bank creates new rural financial products and increases the credit loans to farmers due to the policy for aiding agriculture, rural area and farmers; Rural Commercial Bank, village banks, and small quantity loan companies develops gradually and the number of those banks for proving financial services increases progressively.

The author also points out that because of incompletion of rural credit system and because of the lack of the collateral and the lack of rural commercial insurance, the lending risk increases. Therefore, the author suggests that China should accelerate building up the mechanism of recompense for those rural financial institutions,

complete rural credit loan and vouching system, fully utilize nongovernmental capital, and construct incentive policy for attracting rural capital to reflow, then increasing the aid of rural financial system to Chinese agriculture, rural area and farmers.

The author of "D &R" describes the current situation of Chinese rural financial system and gives us some rational suggestions on future development of this system.

The description and suggestions are important and valuable for studying Chinese rural finance.

As the reviewer of "D & R", I deem that the Chinese rural financial system should transform from structural system to functional system in order to resolve the imbalance of supply and demand of rural capital and then provide better services for the development of rural area and for the construction of agriculture.

Chinese rural financial system and policy should meet the need of Chinese long term development, which is related to the change of production structure by the scalization and

* Zhang Tong, Ph. D in Agriculture Economics at Oklahoma State University, U. S. A, Associate Professor at Research Institute of Economics and Management, Southwestern University of Finance and Economics.

industrialization of agriculture, the construction of new rural area under the balance of urban and rural development, and the food security for such a large population.

Chinese rural financial system should provide multi-level services to satisfy the different needs of rural areas. Chinese government needs to clarify the service range of rural policy-type banks and complete the services of those rural policy-type banks,

in order to support the services of rural commercial banks for rural areas, to complete rural credit system, resolve the problems about protecting farmers' welfare induced by bankruptcy of those farmers.

Chinese government also needs to complete the relative laws to ensure the legitimacy of nongovernment financial institutions, such as agriculture funding, to develop Chinese agriculture wealthily and stably.

The adjustment of production structure induced by the development of rural area alters the rural demand of capital. At present, capital supplied by formal rural financial institutions is much less than the capital demand of smaller farmers and medium and small rural Enterprises.

During the 30 years since Reform and Opening up, the weight of Primary industry's GDP decreases from 28.2% in 1978 to 11.3% in 2008. The weight of Secondary Industry's GDP increases from 47.9% to 48.6%.

The weight of Tertiary Industry's GDP increases from 23.9% to 40.1%. Within the Primary Industry, weight of agriculture's GDP decrease from 80% to 50.4%, the weight of livestock's GDP increases from 15% to 30%, and the weight of Fishery's GDP increases from 1.6% to 9.1%.

Meanwhile, accompanying the economy development and the improvement of the urbanization, rural population decreases from 82.08% in 1978 to 55.06% in 2007. Moreover, the number of cities in China increases from 132 to 655.

During the process of urbanization, millions of rural people enter cities to find jobs, or immigrate into cities. During the economy construction of developing countries, people have to confront some issues such as adjustment of agricultural production structure, labor forces shirting from rural to urban areas, and urbanization.

Currently, the extent of industry concentration of Chinese Agro-processing is very low. Especially in recent 10 years, foreign multinational firms aim at China, the country with the largest population and agriculture production. Those foreign multinational firms merge the leading enterprises, joint venture with Chinese companies, or construct their offices in China, for the purpose of entering Chinese agricultural production market and occupying a large amount of market share.

Therefore, how to make sure the food security and protect local agricultural production, how to assist aquaculture and livestock and farming of Economical Crops, how to help small and medium rural enterprises to grow up and large rural enterprises to realize modernization, and how to serve rural people to urban industrial workers are the long term issues and focal points of Chinese government.

The main demanders of rural finance are farmers, rural enterprises, and local governments. Their corresponding demands are living expenses, capital for production requirements, and capital for construction of public infrastructures respectively.

From a structural point of view, current Chinese rural financial supply system is composed with policy-oriented financial institutions, commercial financial companies, and cooperative financial organizations. Different financial institutions satisfy different functions for different comities' demands.

However, most farmers and small and medium rural enterprises lack collateral and then are excluded by services of State-owned financial institutions. So, those farmers and rural enterprises search Non-formal financial services, and their lending behaviors make private lending rates higher and higher. To solve the problem China Agriculture Bank exploits "Huinong" card, one kind of credit cards for farmers, which partly recuperate the lack of credit loans for farmers. But rural credit loans are shot of large amount of funds. Thus, China Banking Regulatory Commission lowers the access standards for opening new rural financial institutions, encourages various types of capital to establish multi-objective and multi-level rural financial institutions, such as village banks, small loan companies, and Jointly Warranty organizations.

It is estimated that in future 15 years funding gap for new rural construction is around 1.4 trillion to 4.3 trillion RMBs.

To solve the problem of lack of capital for poor farmers, in some African countries such as Kenya and Eritrea, and in some Asian countries such as Bangladesh and India, Bangladesh rural Jointly Warranty model or India SHG-Bank Linkage model is very popular and provides capital supports to farmers, which cannot be obtained from other commercial financial systems.

Since 1994, when Jointly Warranty loan model is introduced to Chinese rural areas, this model does not work very well, and default rates are very high. Moreover, this model is based on the joint liability of members to solve the lack of credit loans, and it is commonly related to small amount loans and only suitable for very poor rural areas, where the formal financial institutions do not set up their branches.

Form the developing experiences of most countries in the world, we can find that most countries' governments apply policy-oriented financial methods to correct the failure of rural financial markets or to reach the targets of industrial distribution. Those countries' governments offer direct subsidies for agriculture producing, fund policy-oriented rural financial institutions, make favorable policies such as tax relief to encourage commercial financial institutions and investment companies to provide loans for agriculture, set up vouch agencies, and build agriculture production Insurance System to avoid the losses induced by disasters and the volatility of production prices.

For example, at the beginning of 1960's European Community (EC) starts The Common Agricultural Policy, which prescribes that EC provide funds to subsidy farmers for selling their agriculture productions, and then promote European modern large-scale agriculture business.

As large agriculture production export country, the U.S.A. has more valuable experiences for China to learn. American government provides support to agriculture development through investment and credit loan. In terms of investment, federal and local governments directly invest in the agricultural projects for public welfare through establishing budgets and borrowing debts to improve the construction of agriculture infrastructure. Meanwhile, federal and local governments relive some amount of taxes or provide some financial aids for profitable private projects and the investment in agriculture infrastructure. The long term and large amount of investment by American government gradually causes the developed and sound agriculture infrastructure. American government establishes Commodity Credit Company, Small Business Administration, Rural Electrification Board, and Farmers

Household Agency, four different kinds of policy-oriented rural financial institutions. Lending rates for agriculture projects commonly offered by policy-oriented rural financial institutions are one third to one half lower than the lending rates for industrial loans. Commercial banks whose more than 25% of total loans are agriculture loans can obtain tax relief and interest rate subsidies for their agriculture loans. American well-built agriculture insurance system plays an important role in dispersing bank risks. Federal programs also help initial farmers to buy commercial farms to realize large-scale production. America and Europe generally provide policy-related subsidies for agriculture, which cut down the prices of agriculture products that are exported to China, further decrease our country's prices of agriculture products, lower the profits of Chinese farmers, and eventually reduce farmers' production willingness. Therefore, the subsidies for agriculture services by commercial financial institutions are not sufficient to ensure agriculture production stability. We also need to provide policy-related subsidies directly to farmers and reinforce the construction of agriculture infrastructure. We should learn the lesson on our soybean industry, and ensure the market share for local agriculture products in order to avoid the monopoly of foreign-funded food enterprises in China and to avoid jeopardizing our food security.

In conclusion, under the circumstance of industrialization and internationalization of agriculture, Chinese rural financial system not only needs to complete its structure, but also needs to strengthen its functions for the purpose of fulfilling the requirements of better servicing multi-level new rural construction and Modern Agricultural Development.

REFERENCE

[1] Satyasai , K.J.S. "Rural Credit Delivery in India: Structural Constraints and Some Corrective Measures." *Agricultural Economics Research Review*, Vol. 21 (Conference Number) 2008 pp 387-394.

[2] George, O. "Is Micro-Finance Achieving Its Goal Among Smallholder Farmers in Africa? Empirical Evidence from Kenya Using Propensity Score Matching."XXV11 *International Conference of Agricultural Economists*, 16-22 August 2009, Beijing, China.

[3] Bahta ,Y.T., and Groenewald ,J.A. "Rural Credit for Resource-Poor Entrepreneurs: Lessons from the Eritrean Experience."Poster paper prepared for presentation at the International Association of Agricultural Economist Conference, Gold Coast, Australia, August 12-18, 2006.

[4] United States Department of Agriculture. "Is More Credit the Best Way to Assist Beginning Low-Equity Farmers? " *Agriculture Information Bulletin*, No. 724-04 August 1996。

[5] Matthew A. Diersen, Editor."Financing Agriculture and Rural America: Issues of Policy, Structure and Technical Change." Proceedings of the NC-221 Committee Annual Meeting Denver, Colorado. October 7-8, 2002.

In: China's New Deal
Editors: Li Xiaoxi and Hu Biliang

ISBN: 978-1-61470-815-5

Chapter 4

TEN MAJOR INDUSTRIES STIMULUS PLAN: NEW STRATEGIES FOR STRUCTURAL ADJUSTMENT

Han Jing*[*], *Xu Yan, Sun Peng, and Feng Rui

Since the end of 2008, an international financial crisis and economic recession have struck the whole world. In order to prevent China's economy from fast sliding, the State Council unveiled a package program, including ten major industries stimulus plan (the stimulus plan henceforth) to ensure economic growth. The stimulus plan involves logistics as well as nine key industries such as the textile industry, the steel industry, the automobile industry, the shipbuilding industry, the equipment manufacturing industry, the electronic information industry, the light industry, the petrol chemical industry, as well as the nonferrous metal industry, and more than 60 implementation rules have been established, which enlarges domestic consumption, stabilizes the production and operation of enterprises, and quickens the development of industrial technology. All in all, the implementation of the stimulus plan has given rise to some visible achievements.

SECTION 1: THE CONTEXT OF THE STIMULUS PLAN

At the beginning of 2009, the international financial crisis started to affect China's economy more profoundly and resulted in the shrink of market demand, difficulties in export industries due to difficulties in market sales, an increase in the number of the enterprises that suspended or partially suspended production, and an increase in unprofitable firms and the amount of losses, which as a whole led to a serious situation. In accordance with the deployment of the State Council, the State Development and Reform Commission and the Ministry of Industry and Information Technology together with other related departments of the State Council compiled the adjustment and stimulus plan of the ten key industries including the textile industry, the steel industry, the automobile industry, the shipbuilding

* Dr. Han Jing is associate professor at the Institute of Economics and Resource Management at Beijing Normal University, who earned her PhD degree at the Shool of economics ,Nankai University. Email: nankai_sky@sohu.com

industry, the equipment manufacturing industry, the electronic information, the light industry, the petrol chemical industry, the nonferrous metal industry and logistics, and they viewed the plan as the countermeasures against the international financial crisis and important strategies to maintain economic growth, enlarge domestic demand and adjust structures, which would last three years from 2009 to 2011. As for china's economy, the issue of the stimulus plan is both a self-rescuing action under unprecedentedly severe external economic environment ever since the reform and opening-up and a transforming strategy to explore development spaces in the new era.

I. The Stimulus Plan is Aiming at Relieving the Shocks from the International Financial Crisis

In November 2008, China's export growth rate fell from 19.2% in October to -2.2% and import growth rate from 15.7% to -17.9%. The fallback of export directly slowed down the growth rate of industries. The value added of the scaled industrial enterprises in China increased at the rate of 5.4% over the same period of the previous year, falling back by 11.9% over the same period of the previous year, and 2.8% over that of October. The production of pig iron, crude steel and rolled steel decreased by 16.2%、12.4% and 11% respectively. The production of automobiles amounted to 714 thousand, decreasing by 15.9%. The slide in the growth rate of industry resulted in the slide in generated electrical energy, which decreased by 9.6% and was the largest monthly decrease[1]. In the first quarter of 2009, the generated electrical energy of Chinese scaled electric power plants decreased by 2.0% over the same period of the previous year and social power consumption by 4%, among which industrial power decreased by 8.4%. The value added of Chinese scaled industries increased by 5.1% over that of the same period of previous year, much lower than that of 16.4% of the first quarter in 2008. GDP increased only 6.1% over that of the same period of previous year, much lower than the growth rate of same period of last year, that of the year of 2008 and that of the fourth quarter of 2008. Total export-import volume decreased by 24.9%, among which the export and the import volumes decreased by 19.7% and 30.9% respectively, the increasing range dropped by 41.1% and 59.5% respectively over that of the same period of last year.

As long as the downturn of economic growth is not changed, it would give rise to severe economic and social problems, considering that China is a large developing country. The slide of China's economic growth is out of expectation and the industries need revitalization. As indicated by experts, revitalizing industries aims to achieve the goal of "Maintaining Eight Percentage". If the economic growth rate is lower than 8%, unemployment problems would be aggravated and cause some social problems as well.[2] It is in such severe economic conditions that the nation released the major ten industries stimulus plan.

---

[1] Zhu Guodong, 01-16-2009. Battle of Maintaining the Growth Rate of 8% Fired, Stimulus Plan of the Ten Key industries Discharged [N]. *China Securities Journal.*

[2] Cao, Jianhai, 03-11-2009. A Roadmap for China's Industry Revitalization: Short-term Growth and Long-term Structure Adjustment, Economic Information Daily [N]. *Economic Information Daily*. The viewpoints in this book are quoted from Cao's essay and originally delivered by Yang Yiyong, the superintendent of Institution of Economic and Social Development of National Development & Reform Commission.

Viewing the ten selected industries, it can be inferred that the nation has been deliberate in making choices. Firstly, the ten key industries play important roles in the economy. Firstly, take economic aggregate as an example: in 2007, industrial value added of the nine key industries[3] takes up nearly 80% of total industrial value added and one third of GDP. In fact, the nine key industries play an irreplaceable role in national industry, finance, social employment and the livelihood of the people. Secondly, the nine key industries make great contribution to the state revenue. In 2007, total revenue in China amounted to 4.56 trillion Yuan(RMB), in which that of the scaled enterprises of the nine key industries amounted to 1.7 trillion Yuan(RMB), taking up 37.4% or so. Thirdly, stimulus plan of the nine key industries would promote social employment. Even with migrant workers out of consideration, employees of the nine key industries in towns and cities amount to 36.156 million, taking up 30% of total employment in urban China. Fourthly, the stimulus of the nine key industries is in favor of the solution to the issues concerning agriculture, countryside and farmers. Large amount of output value in such sub-industries of the light industry as food, paper making, furniture, household appliances, leather and household chemicals industry are generated from deep processing of agricultural products, which involve a 300-million-peasant market and take in nearly 60 million migrant workers in cities. The light industry alone can absorb 20 million migrant workers. In addition, medium and small enterprises per se are part of the nine key industries. And it is obvious that the stimulus of the nine key industries will revitalize related medium and small enterprises. Fifthly, the logistics industry is the compound service industry integrating the transportation industry, the storage industry, the freight forwarding industry and the information industry. It is an important component of national economy involving a large range of fields, creating a large amount of employment and playing an important role in expanding consumption. The value added of logistics industry takes up 16.5% of that of the whole service industry and 6.6% of GDP[4]. It can be said safely that if the stimulus plan of the ten key industries could be implemented smoothly as expected, China's economy would realize the goal of "Maintaining Eight Percentage" and quickly recover from the impact of the international financial crisis. The roadmap of the stimulus plan is displayed in table 4-1.

II. The Stimulus Plan Will Boost China's Economy Constantly

Through 30 years of hardships and fast development, China has become one of the great powers in terms of industrial scale and economic strength. In recent years, China's economy has kept growing at roughly a double-digit speed; such high growth rate conceals some limitations of the internal structure of China's economy, such as the high dependence of economic growth on external environment, low end of the majority of products evaluated in the international industrial division system, the dependence of core technology and sophisticated technique of certain industrial on other countries, and the insufficient

---

[3] As the nine industries are in the industrial sector while logistics is productive service industry, providing service for the previous nine industries, thus they are analyzed separately.

[4] National Development and Reform Commission. Background Information of the Stimulus Plan of Key Industries. [EB/OL]. (2009-02-27). http://www.china.com.cn/news/2009-02/27/content_17343584.htm

Table 4-1. Roadmap of the Stimulus Plan

Industry	Issuing Schedule and Main Contents	Major Impact
Steel Industry	Issued on January 14, the stimulus plan advocates the implementation of flexible export tax rebate policy, stimulation of export of high-end products, elimination of backwards capacity and support for merger and acquisition within the industry.	Although the stimulus plan has little impact on the situation of the decline of overseas market demand, its measures of elimination of backwards capacity and joint reorganization would postpone the industrial periodic sliding. In the long term, large preponderant steel enterprises will take the absolute lead in steel industry.
Automobile Industry	Issued on January 14, the stimulus plan stipulates that vehicle purchase tax of passenger car with displacement less than 1.6 liter be reduced to 5%, the amount of large automobile enterprises be reduced from 14% to 10% and that 10 billion Yuan(RMB) be invested in technological reform and development of new energy vehicles of the industry.	The stimulus plan focuses on structural adjustments with the attempt to promote the optimization and upgrading of automobile industry structures. Specifically, on the one hand, it accelerates the joint reorganization of enterprises, and on the other hand, it improves independent innovation and competitiveness of enterprises with new energy automobile as the breakthrough point.
Equipment Manufacturing Industry	Issued on February 4, the stimulus plan supports the joint reorganization of key enterprises. It is aiming at developing large groups which are capable of general project contracting, system integration of international trade and financing. It also encourages the preferential purchasing of domestically made production facility in case of large projects.	Joint reorganization will enable the related enterprises to expand its power through the integration of industrial chains. In addition, the demand for increasing the purchase ratio of domestically made equipment will expand the market share of large key enterprises.
Textile Industry	Issued on February 4, the stimulus plan increases export rebate rate of textile garment from 14% to 15% and enhances the support for loans and assurance.	Increase in export rebate rate will contribute to the ease of enterprise stresses. However in the context of shrinking global demand, it is difficult for the stimulus plan to expand export directly. Besides, the implementation of credit support is doubtful and investment amount of special funds is unclear. Thus, in the short term, the stimulus plan is hard to improve the ability of enterprises to make profits.

Industry	Issuing Schedule and Main Contents	Major Impact
Shipbuilding Industry	Issued on February 12, the stimulus plan accelerates the elimination of backwards capacity and encourages buyer credit under export both released by financial institution and by large watercraft enterprises, extending the present fiscal and financial support for domestically consumed ocean-going vessel to the year of 2012.	The orders in Chinese shipbuilding industry are mainly about bulk carrier, which takes up 70%. The stimulus plan is in favor of the expansion of domestic demand for watercraft and adjustment of capacities and structures. Since global orders in hand for oil tanker have already contained the newly increased demand resulting from the elimination of single-hull oil tanker, the stimulus plan will not make a great impact on an international scale.
Electronic Information Industry	Issued on February 12, the stimulus plan attaches great importance to the projects including upgrading the integrated circuit, 3G mobile communication industry, popularization of digital television, advancement of computer and next-generation internet.	The stimulus plan is beneficial for communication device industry (3G and broad band), electronic industry, colored TV industry and software industry. In addition, it will promote the growth of demand for domestic outsourcing services and domestic basic software enterprises.
Petrol Chemical Industry	Issued on February 19, the stimulus plan brings in 20 key construction projects in process and 20 new construction projects, the general goal of which is to achieve industrial value-added of 1.75 trillion Yuan(RMB) until 2011.	The stimulus plan is issued in an attempt to adjust and control the total amount and industrial layout and to accelerate technological upgrading. According to the stimulus plan, sub-industries such as oil refining, agrochemical and pesticide and new chemical materials will get strong supports.
Light Industry	Issued on February 19, the stimulus plan aims to expand the consumption of rural and urban markets and to accelerate technological reforms of such industries as paper making, household appliance and plastic.	Light industry totally involves around 45 large and small trades. “Home-Appliances-Going-to-the-Countryside” policy will surely expand the demand for major appliances.

Table 4-1. (Continued)

Industry	Issuing Schedule and Main Contents	Major Impact
Nonferrous Metal Industry	Issued on February 25, the stimulus plan aims to adjust taxes appropriately, promote merger and acquisition, implement national reserve system and develop 3 to 5 comprehensive nonferrous metal enterprise group with great competitive powers.	National reserve system plays an important role in the torsion of the predicament of some domestic key nonferrous metal enterprises with high inventory. Thus, the tension of market underselling would be eased and the price of metal would be sustained to some extent.
Logistics Industry	Issued on February 25, the stimulus plan focuses on the support for agricultural and village logistics and logistics for bulk production goods and consumer goods	The stimulus plan will relieve the shocks on logistics industry brought about by the international financial crisis.

Source: Collated from "Adjustment and Stimulus Plan of Industries [EB/OL]. (04-02-2009). http://www.gov.cn/zwgk/2009-04/02/content_1276054.htm"

development of the economy itself. The international financial crisis unveiled the deep structural problems in the development of China's economy. Even without the international financial crisis, it's obvious that the extensive developmental mode is not sustainable for China's economy. The international financial crisis enables China to reflect upon the developing road in the previous years after a 30-year high speed of growth of China's economy. With the stimulus plan, it becomes possible that the industries would gain long run competitive advantages through merger and acquisition and post-merger integration of enterprises within the ten key industries and the formation of certain leading roles of the ten key industries applying such methods.[1]

1. The Stimulus Plan Will Ensure the Promotion of Domestic Demand and a Steady and Sustained Increase of Economy

In the international financial crisis, European and American economies have experienced great recession, which resulted in the severe environment of China's foreign trade. In the case of weakened external demand, it is without doubt that we should stimulate the vast domestic market of the 1.3 billion of Chinese population to cope with the international financial crisis and better the economic situation. Thus, undoubtedly, the stimulus plan issued by the State Council will be the direct drive for the promotion of domestic demand; for example, Amendment I in the stimulus plan of Light Industry is to expand the consumption of city and countryside area; the opening chapter of that of Textile Industry and Steel Industry is to plan domestic and foreign markets in entirity, to enlarge domestic demand and to pull domestic consumption. And as to Automobile Industry which catches much attention, the stimulus plan points out the need to implement consumption policy actively and stabilize and enlarge the demand in order to accelerate the adjustment and stimulus of the industry.

2. The Stimulus Plan Will Promote Merge and Reorganization and Improve International Competitiveness of Chinese Enterprises

More than half of the ten key industries refer to the issue of merger and acquisition in their plans. For example, the stimulus plan of Steel Industry makes clear the intention to act as the leading role in all industries in the interest of promoting joint reorganization among enterprises, creating and developing large and ultra-large steel group with international competitiveness, optimizing industrial layout, and elevating concentration ratio.

In fact, before China entered WTO, there were experts who spared no effort to suggest the merger and acquisition in certain domestic industries against the shock from gargantuan multinational corporations. The reform of China's economy lasted only 30 years so far and the overall construction of market economy system only experienced a ten-year-or-so period. Chinese enterprises growing up under such circumstances are lacking in international operation experience, core technology and competitiveness compared to international enterprises alike. In addition, China has a vast territory, resulting in the scattering of many small companies of the same industry in all parts of country. Thus, considering the long-term development of the whole nation, the stimulus plan encourages merger and acquisition for the purpose of elevating international competitiveness of Chinese enterprises.

---

[1] Yuan Yuan, 03-05-2009. The Revitalization of the Ten Key Industries Brightens Chine's Economy [N]. 21st Century Business Herald

Column 4-1. Distributing Map of the Provincial Primacy in Annual Value-added of 9 Key Industries in 2008

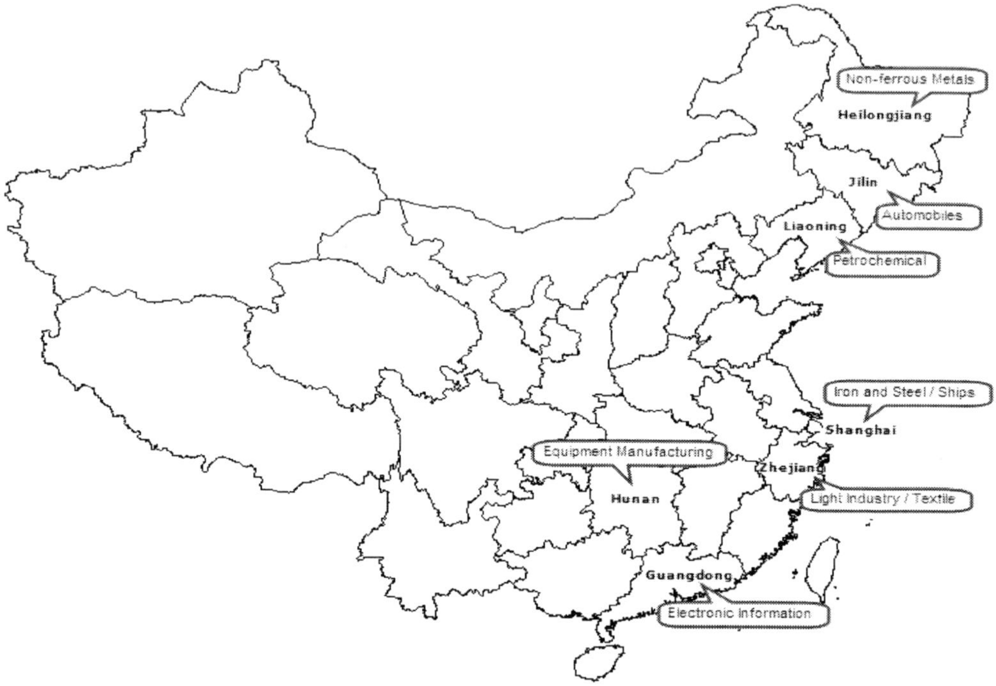

Column 4.2. Why Is Real Estate Industry Excluded in the Ten Major Industries Stimulus Plan?

Before the issue of the stimulus plan of Logistics, Cheng Siwei, the former vice chairperson of the Standing Committee, disclosed that real estate industry would substitute for energy industry and be counted one of the ten core industries, but it didn't come true. Liu Tienan, the Associate Director of State Development and Reform Commission, explained that the final selection of logistics is made on the basis of the characteristics of logistics for that it is an important branch of service industry and is closely related to the other nine key industries. Logistics is the important link of the nine key industries and between the industries and domestic and foreign markets. At the same time, the adjustment and stimulus of logistics as a service industry on the one hand meets the needs of logistics itself and on the other hand has a positive impact on the elevation of the competitiveness of the other nine key industries. Thus, the ten key industries are organically linked integrity because of logistics which brings the possibility of forming combinations in the adjustment and stimulus of industries. Comparatively, although real estate industry is also an important pillar industry in China, its contribution as a link is not remarkable.

Source: From "Explanation by the NBRC: Why is Real Estate Industry Excluded in the Stimulus Plan", http://news.xinhuanet.com/fortune/2009-02/27/content_10909094.htm

Section 2: Main Contents of the Stimulus Plan[2]

I. Raw Material Industry

1. The Stimulus Plan of Steel Industry

The output of crude steel in 2008 was 500 million tons, taking up 38% of the global production. The apparent domestic consumption amounted to 453 million tons and direct export, 60 million tons, accounting for 15% of the world steel trade volume. However, the problems resulting from the extensive development of steel industry in the past years now have become obvious, including the glut of capacity caused by blind investment, weakness in creativity, irrationality in industrial layout, low industrial concentration ratio, weakness in control of resources and disorder of circulation. Since the second half of 2008, with the expansion and extension of international financial crisis, China's steel industry is confronted with severe shocks and unprecedented challenges as for its rapidly sliding demand, dramatically falling price, difficulty in business operations and deficit of the whole industry.

The Stimulus Plan of Steel Industry is aiming at preventing steel industry from continuous sliding momentum and keeping the overall stability in 2009. And in 2011, the steel industry would hopefully change its extensive development modes and improve its technological level and creativity, which contributes to the elevation of overall competitiveness and enhancement of its role as a pillar industry and leads to the track of sound progress. The stimulus plan is designed to make the following achievements: (1) maintain the stability of domestic market and improve the export environment; (2) strictly control the total quantity of steel and accelerate the elimination of the unqualified; (3) expedite the reorganization of the enterprises and increase the concentration ratio of the industry; (4) intensify reforms of technology and promote the technological improvements; (5) optimize the layout of the steel industry and maintain a harmonious development; (6) adjust steel assortment and improve product quality; (7) keep the stability of imported iron ore resources and put the market in good order; (8) exploit domestic and foreign resources and ensure the safety of the industry. Policy measures include the adjustment of tariff rate of certain products, the implementation of fair trade policy, an increase in the investment in technology, perfection of the mechanism which enables the outdated production to withdraw and reorganizational policy of enterprises, amendment of steel industry policy, uplift of steel standard for construction engineering, realization of harmonious development with other related industries, implementation of financing policy with guarantee and curtailment and "to-be-global" strategy, establishment of disclosure system of industrial information and the exertion of industrial association.

2. The Stimulus Plan of Petrol Chemical Industry

Petrol chemical industry has developed at a high growth rate with its industrial scale enlarged and general strength improved. Industrial value added grew at an average annual increase of 20%, pulling up the growth of GDP at 1% or so. However, there are problems concerning the development of petrol chemical industry: low degree of intensive development

[2] This part in compiled according to the related sources from Stimulus Plan of Industrial Industries. [EB/OL]. (04-02-2009).http://www.gov.cn/zwgk/2009-04/02/content_1276054.htm

and decentralization of industrial layout; lack of innovation, which results in the dependence of the production technique and high-end products and large complete technique and equipment on import; irrationality of product mix with more mid- and low-end products; restrictions of resources and environment, which causes the contradiction between industrial development and environment protection; increasing demand of agricultural means of production, deficiency in capacity of low-cost products, and imperfection of market regulation and control system; and blind exploitation of coal chemical regardless of resources and environment and resource conversion efficiency. Since the second half of 2008, petrol chemical industry suffered from severe shocks from the international financial crisis, with the domestic market shrinking, production declining, inventory of enterprises increasing, price decreasing, and economic benefits of the industry declining. All in all, the production and operation of the industry is confronted with serious difficulties.

The goal of the Stimulus Plan of Petrol Chemical Industry is to maintain the stable and fast development of the industry. In 2009, a stable operation is the major concern and hopefully after three years' adjustment and revitalization, in 2011, the industrial structure will become reasonable, development modes transferred and general strength of the industry advanced. The major missions the stimulus plan include: (1) maintain the stable operation of the industry; (2) improve the indemnifying ability toward agricultural means of production; (3) demonstrate the coal chemical industry steadily; (4) implement significant projects; (5) plan the layout of significant projects as a whole; (6) promote the improvement of technology; (7) accelerate the elimination of the backwards capacity; (8) enhance environment protection; (9) support the joint reorganization of enterprises; (10) improve the indemnifying ability toward resources; (11) improve the management level of enterprises. Policy measures include the optimization of chemical fertilizer reserve mechanism, fulfillment of oil reserve, enhancement of credit policy support, optimization of product oil pricing mechanism, enlargement of investment in technological upgrading, support to exploitation of overseas resources, implementation of fair tax policy, promotion of merger and acquisition of enterprises, optimization of industrial development policy and reinforcement of anti-dumping anti-smuggling activities according to law.

3. The Stimulus Plan of Nonferrous Metal Industry

At the beginning of 21st century, China's nonferrous metal industry has developed fast with apparent achievements in the advancement of technology, improvement of the quality of varieties, elimination of backwards capacity and exploitation of overseas resources. Besides, the production and consumption scale is constantly enlarged, which makes China the largest nation in production and consumption of nonferrous metal. Since the second half of 2008, with more influence of the international financial crisis on the real economy, China's nonferrous metal industry has been overwhelmed by shocks with its product price plummeting, production declining, domestic consumption slumping, cash-flow of enterprises tightening and whole industry experiencing losses. In addition, the deep problems of China's nonferrous metal industry become evident, including the glut of productivity of some products, the irrationality of industrial layout, low degree of industrial intensification and indemnifying ability toward resources, lack of innovation, and low level of recycling ability.

The Stimulus Plan of Nonferrous Metal Industry is aiming at the goals of keeping the overall stability of industry and leading it to the track of sound progress in 2011, optimizing the industrial structure, changing its development modes, improving technological innovations and laying the foundation for the sustainable development of nonferrous metal industry. The stimulus plan is designed to make the following achievements: (1) maintain the stability of domestic market and improve the export environment; (2) strictly control the total quantity and accelerate the elimination of the backwards capacity; (3) intensify reforms of technology and promote the technological improvement; (4) expedite the reorganization of the enterprises and adjust the layout of the industry; (5) exploit domestic and foreign resources and improve the indemnifying ability toward resources; (6) develop a recycling economy and improve the recycling ability; (7) reinforce management and safety supervision of the enterprises and cultivate talented people. Policy measures include the optimization of export tax policy, establishment of national reserve system, enlargement of investment in technological upgrading, creation of pilot project for direct power-purchase, optimization of reorganizational policy of enterprises, encouragement of "to-be-global" strategies, amendment of industrial policy, proper resource allocation, implementation of financing policy with guarantee and curtailment, implementation of accountability system in energy conservation and emission reduction and elimination of the outdated capacities, establishment of communication and disclosure system of industrial information and the exertion of industrial association (chamber of commerce).

II. Equipment Manufacturing Industry

1. The Stimulus Plan of Automobile Industry

Automobile industry has developed at a high growth rate with the formation of the production and assembling system for multi-variety and all series of finished automobile and intermediate products, rise of industrial concentration ratio and technological level of products, which makes China the largest nation in producing automobiles. However, the problems of irrational industrial structure, low technological level, weak innovation and deficient consumption policy have become overwhelming in addition to the restrictions from resources, environment protection and urban transport. Since the second half of 2008, as the influence of the international financial crisis deepened and the international automobile markets dramatically shrank, China's automobile industry is overwhelmed by shocks which result in the negative growth of production and marketing of the whole industry, declination of the economic benefits of core enterprises, weakness of self-owned brand in competition, and the severe situation facing China's automobile industry.

The goal of the stimulus plan of Automobile Industry is to maintain the stable and fast increase in production and marketing of the industry, improve the environment of automobile consumption, optimize the structure of market demand, make progress in merger and acquisition, enlarge the market proportion of indigenous brands, form the scale of production and marketing of electric vehicles, improve R&D level of finished automobile, and become independent in the technology of designing and producing key components. The stimulus plan is designed to make the following achievements: (1) cultivate automobile market; (2) promote the reorganization of automobile industry; (3) support the independent innovation of enterprises; (4) intensify reforms of technology; (5) implement the strategies concerning new energy vehicles; (6) implement the strategies concerning indigenous brands; (7) implement the strategies concerning export of automobiles; (8) develop modern automobile service industry. Policy measures include: (1) cut down purchase tax of passenger vehicle; (2)

implement the policy of "Automobiles-Going-to-the-Countryside"; (3) accelerate the rejection and replacement of old automobiles; (4) cancel the irrational restriction on automobile purchase; (5) regulate automobile consumption credit; (6) regulate and promote second-hand automobile market; (7) accelerate the construction of urban transportation system; (8) perfect the reorganization policy of automobile industry; (9) enlarge the investment in technological improvement and reconstruction; (10) encourage the consumption of energy-saving automobiles and new energy automobiles; (11) perfect and implement *Automobile Industry Development policy*.

2. The Stimulus Plan of Shipbuilding Industry

Since 2003, China's shipbuilding industry has maintained fast development, with its industrial scale enlarged, production of watercrafts increased. In addition, the completions of shipbuilding, order quantity both newly received and held in hand have been in the front ranks of the world for many continuous years. However, some problems of shipbuilding industry become apparent, including the lack of independent innovation, extensive development modes, low-level overlapping investment, extreme surplus of productivity, sluggish development of corollary equipment used in watercrafts and equipment used in ocean engineering. Since the second half of 2008, the impact of the international financial crisis and dramatic sliding of the international shipping market have brought about great pressure on watercraft market with purchase orders for new watercraft declining rapidly, financing problems emerging and risks of performance of the contracts of delivering ships increasing, which as a whole leaves China's shipbuilding industry in the face of severe situation.

The Stimulus Plan of Shipbuilding Industry is aiming at the goals of maintaining the stable and fast development of the industry, enlarging the market share, enhancing abilities to provide the auxiliary items, making progress in adjustments of structure, uplifting the R&D level and improving the development quality. The major missions the stimulus plan include: (1) stabilize the production of watercraft enterprises; (2) enlarge the demand of watercraft market; (3) improve equipment of ocean engineering; (4) support the merger and acquisition of enterprises, (5) improve independent innovation; (6) enhance the reforms of technology; (7) develop ship repairing service; (8) develop international market; (9) enhance the management of watercraft enterprises. Policy measures include enhancement of the financing support for production and operation, increase in the release of buyer's credit in the export of watercraft, encouragement in the purchase of abandoned watercraft, expansion of domestic and international market demand, elimination of old watercrafts and single hulls, strict control on newly developed capacity, perfection in merger and acquisition strategies, and more investment in R&D and technological reforms.

3. The Stimulus Plan of Equipment Manufacturing Industry

China has become a powerful country in equipment manufacturing industry but the industry is still confronted with problems such as weakness in independent innovation, underdevelopment of basic manufacturing, low-level duplication of similar projects and difficulty in promoting independent innovation product. At the same time, influenced by the international financial crisis, domestic and markets shrinks dramatically, together with the slowing down of fast developing momentum of the industry, difficulty in production and operation of enterprises, declination of economic benefits and challenges toward sustainable development.

The goal of the Stimulus Plan of Equipment Manufacturing Industry is to maintain the stable development of the industry, enlarge the market share, make breakthrough in manufacturing major equipments, advance the basic corollary standard, optimize organizational structure and change the developing modes. The stimulus plan is designed to make the following achievements: (1) revitalize equipment manufacturing industry relying on the major projects of the ten fields, including electric power generation with high efficiency and cleanliness, power transmission and transformation of extra-high voltage, mining of coal and metal minerals, pipage of natural gas and storage and transportation of liquefied gas, high-speed railway, urban rail transportation, agriculture and rural area, infrastructure, ecological environment and the livelihood of the people and major science and technology project; (2) realize the automation of equipment with the support of the nine major industries, including steel industry, automobile industry, petrol chemical industry, shipbuilding industry, light industry, textile industry, nonferrous metal industry, electronic information industry and national defense and war industry; (3) advance the manufacturing standard of the four major matching products, including heavy casting and forging, basic components, processing assistive devices and specialty resources and lay a solid foundation for the development of the industry; (4) accelerate the implement of key work (including firstly, accelerate the adjustment of industrial organization and structure; secondly, improve independent innovation; thirdly, uplift the level of specialized production; fourthly, optimize product standard system; fifthly, exploit overseas resources and market; sixthly, develop modern manufacturing service industry and finally, enhance the management of enterprises and the development of talented people) and reform of industrial development modes. Policy measures include: (1) bring the VAT Reform Policy into effect; (2) enhance devices purchasing management of invested projects; (3) encourage the use of the first domestically made equipment; (4) intensify reforms of technology and promote the technological improvement; (5) support the export of equipment; (6) adjust preferential tax policies; (7) accelerate the merger and acquisition of enterprises; (8) implement the policy of subsidizing energy saving product and the purchase of agricultural equipment; (9) establish of the disclosure system of industrial information; (10) support the development of product inspection and certification authority.

III. Consumer Goods Industry

1. The Stimulus Plan of Light Industry

In 2008, the value added of light industry was 2.6235 trillion Yuan(RMB), taking up 8.7% of GDP with more than 100 varieties of products ranking the first place in the whole world, including household appliance, leather, plastic, food, furniture and hardware products. However, light industry is still confronted with some serious problems, including, weakness of independent innovation, irrationality of industrial structure, arduousness in energy saving and consumption reduction and quality problems of products. Since second half of 2008, the shocks from the international financial crisis have given rise to the unbalance of domestic and foreign markets, overstocks of products, difficulty in financing, production and operation. All in all, light industry is facing a severe situation.

The Stimulus Plan of Light Industry is aiming at the goals and tasks of maintaining the stable increase of production, making progress in independent innovation, optimizing industrial structure, reducing the emission of contaminations, eliminating the outdated capacities and enhancing quality and safety. Policy measures include: (1) increase the subsidized varieties for "Home-Appliances-Going-to-the-Countryside" policy; (2) uplift the export rebate rate of some light industry products; (3) adjust the catalogue of processing trade; (4) deal with the problem concerning purchase and reservation of agricultural products; (5) intensify reforms of technology and promote technological improvement; (6) enhance the financial support; (7) support medium and small enterprises; (8) give industrial policy instructions; (9) encourage merger and acquisition and eliminate the backwards capacity; (10) exert the power of industrial association (chamber of commerce).

2. The Stimulus Plan of Textile Industry

China has become a powerful country in the production of textile and garments. However, some problems concerning textile industry has risen, such as, weakness of independent innovation, irrationality of industrial layout, arduousness in energy saving and emission reducing and blindness in capacity expansion. Since the second half of 2008, the international financial crisis has brought about serious problems including the disequilibrium of demand and supply, difficulty in enterprise operation and increase in the loss of enterprises and declination of employment offered, which as a whole leaves China in the unprecedented predicament ever.

The goal of the Stimulus Plan of Textile Industry is to maintain the stable and fast growth of the industry, optimize industrial structure, improve independent innovation, technological and equipment level and quality of product varieties, adjust industrial layout, make breakthrough in the development of indigenous brands and eliminate the backwards capacity, which as a whole will hopefully lead China onto the track of a superpower country in textile industry. The stimulus plan is designed to make the following achievements: (1) stabilize domestic and foreign markets; (2) improve independent innovation; (3) accelerate the reforms of technology; (4) eliminate the backwards capacity; (5) optimize geographical distribution; (6) perfect public service system; (7) accelerate the development of indigenous brands; (8) improve the competitiveness of enterprises. Policy measures include: (1) keep increasing the export rebate rate of textile products; (2) enhance the purchase of cotton and steam filature; (3) intensify reforms of technology and promote the technological improvement; (4) enlarge domestic consumption; (5) encourage merger and acquisition of enterprises; (6) enhance the financial support towards textile industry; (7) reduce the burden of textile enterprises; (8) support medium and small enterprises; (9) give industrial policy instructions; (10) Exert the power of industrial association (chamber of commerce).

IV. Electronic Manufacturing Industry

Since the implementation of the reform and opening up policy, China's electronic information industry has maintained continuously fast development and China has become the largest base for manufacturing electronic information products. However, with the impact of the international financial crisis, since the second half of 2008, electronic information industry has been facing a serious situation with the export of electronic information products declining, growth rate of sales revenue dramatically falling, operation problems emerging in the key enterprises, and the exploitation of foreign capital declining. Some deep problems of the industry have risen and countermeasures have to be carried out in order to accelerate the adjustment of industrial structure, optimization of industry, enhancement of technological innovation and most importantly, to promote the stable and continuous development of electronic information industry.

The Stimulus Plan of Electronic Information Industry is aiming at promoting the growth, maintaining the stability, adjusting the structure, and changing the development modes of the industry. Over the next three years, electronic information industry should make the following achievements with the support of other nine key industries: maintenance of stable growth of the key industries, breakthrough in the strategically core industries and new growth resulting from new application, including: (1) the stable growth in computer, electronic components and audio and visual products; (2) breakthrough in the technology of core industries such as integrated circuit, display device, and software; (3) cultivation of new growth areas in communication apparatus, information service and application of information technology. Policy measures include: (1) expand domestic demand; (2) increase national investment; (3) enhance policy support; (4) improve investment and financing environment; (5) support the merge and organization of strong enterprise; (6) exploit foreign market; (7) develop independent innovation.

V. Logistics Industry

In 2008, the total value of social logistics goods in China amounts to 899 trillion Yuan(RMB), increasing by 4.2 times as against Year 2000, with an annual growth rate of 23%. The industry achieves value added of 20 trillion Yuan(RMB), increasing by 1.9 times as compared with that of 2000, with an annual growth rate of 14%. But the overall level of logistics industry is still underdeveloped and there arises some serious problems: (1) operating efficiency of social logistics is far from being satisfactory; (2) both socialized logistics demand and specialized logistics supply are insufficient; (3) logistics industry is still lacking in infrastructure construction with a more rationally distributed, smoothly linked, and comprehensive convenient transportation systems to be desired; (4) local blockade and industrial monopoly cause obstacles against resources integration and integrated operation, which results in the irregular logistics market; (5) logistics technology, personnel training and logistics standard can't meet current needs and logistics service is still weak in systematization and intensification. Since the second half of 2008, influenced by the international financial crisis, logistics industry are confronted with the following problems: shrinking of market demand, declination of profits as well as transportation price and warehousing charges, difficulty of the operation of some medium and small-size logistics enterprises, which endangered the traditional logistics enterprises which only provide simple services such as transportation and storage.

The Stimulus Plan of Logistics Industry is designed to achieve the following goals: (1) relieve the severe situation of logistics operation problem and maintain the stable development of the industry in 2009; (2) until 2011, develop a passel of large-scaled and comprehensive logistics enterprise groups which have international competitiveness and modern logistics service system with rational layout, advanced technology, energy and environment conservation ability, convenience, efficiency and safety. Hopefully the service ability and socialization and specialization of logistics industry would be improved, the proportion of third party logistics increased and logistics scale enlarged. The growth rate of logistics industry increases by 10% each year and the ratio of the total cost of social logistics against GDP would be lower than the present level. The major missions of the stimulus plan include: (1) enlarge the logistics market demand; (2) promote the socialization and specialization of logistics; (3) accelerate the merger and acquisition of logistics enterprises; (4) promote the development of logistics in key fields; (5) accelerate the development of international logistics and bonded logistics; (6) optimize the regional layout of logistics development; (7) enhance the link and coordination of logistics infrastructure construction; (8) improve logistics information standard; (9) perfect logistics standardization system; (10) enhance the development and application of new logistics technology. The major projects are as follows: (1) multimodal transport and transferring facility project; (2) logistics park project; (3) urban delivery project; (4) Bulk commodity and rural project; (5) joint development project of manufacturing industry and logistics industry; (6) project of logistics standard and promotion of technology; (7) logistics information platform project; (8) project for tackling key problems in logistics technology; and (9) emergency logistics projects. Policy measures include the enhancement of organization and coordination, reform of logistics management system, optimization of logistics laws and policies, formulation and implementation of subject planning, multiple-channel increase in investment in logistics, optimization of logistics statistical indicator system, promotion of opening up and international cooperation of domestic logistics, acceleration of personnel training, and the exertion of the power of industrial association.

Column 4-3. The "Most" of Ten Major Industrial Stimulus Plans [3]

Light Industry: Most Beneficial for the Stock Market

The Stimulus Plan of Light Industry benefits more than 40 sub-industries including household appliances, leather and food, etc. Thus, the stimulus plan is referred to as the most beneficial plan for the stock market. The emphasis of the stimulus plan is placed on supporting white goods while increasing the subsidized varieties of "Home-Appliances-Going-to-the-Countryside" policy and the amount of each variety that every family is allowed to purchase from one to two.

Automobile Industry: Most Yearning for Self-owned Brands

According to the stimulus plan of Automobile Industry, domestic market share of future self-owned brand of passenger car will hopefully rise up to 40% or more, in which 30% will be taken up by self-owned brand cars. In addition, the stimulus plan encourages the export of self-owned vehicles, the amount of which will take up 10% of total sales volume.

Shipbuilding Industry: Most Concerned about Purchase Order

Shipbuilding industry is marked by the long cycle of producing one single watercraft and its high cost, which makes the punctual performance of the purchase order very important. In the

[3] The "Most" in Ten Major Industries Stimulus Plans Issued by the State Council: Light Industry is the Most Beneficial for the Stock Market [N]. *International Finance News*. 02-26-2006

context of the financial crisis, established orders are frenquently withdrawn with less new orders signed. In accordance with the stimulus plan, the focus at present is to ensure the orders are performed on time and relieve operation risk through implementation of active credit guarantee measures.

Nonferrous Metal Industry: Rapidest in the Elevation of Rebate Rate

Export will become one of the most important strategies in the stimulus plan of Nonferrous Metal Industry. The government makes some relaxation of restrictions on the import and export of some metal products in tax adjustments. The stimulus plan further increases average export rebate rate of some products such as high-precision copper pipes from 5% to 13%, and that of other home-made highly processed products which are able to substitute for those imported from aboard could be raised to 17%.

Equipment Manufacturing Industry: Beneficial for the Largest Range of Upper and Lower Stream of Industrial Chain

Equipment manufacturing industry is an industry with a long upper and lower stream of industrial chain. The stimulus plan for the first time places basic components in a very important position. It can be predicted that government would issue more practical measures concerning financial support, tax preferences for purchasing home equipment, R&D financial support, and merger and acquisition to promote the development of the whole industrial chain.

Light Industry: Most Caring for the Livelihood of People

No other industries like light industry would have so long an industrial chain covering wide range of materials and products from cotton to plus material to clothing. And no other industries would provide the employment of more than 20 million positions except for light industry. It is obvious that employment security is the major task to ease the influence of the international financial crisis.

Logistics Industry: Luckiest in the Stimulus Plan

Logistics industry is the last to be listed in the stimulus plan, the process of which is dramatic with ups and downs. Even once, there was news saying that real estate industry would be chosen in the stimulus plan. As China's logistics industry is in its infancy, it is very lucky for it to be listed in the stimulus plan at the last moment. The reason might be that the government wouldn't allow too many shocks on the newly established fragile logistics system.

Electronic Information Industry: Most Effective to Get Instant Results

The Stimulus Plan of Electronic Information Industry stipulates the exemption of sales tax of software companies and increase in the export rebate rate of 25 varieties of key electronic products including colored tubes and glass envelope up to 17%. The plan also refers to the expansion of domestic demand, promotion of 3G mobile communication networks, and the investment scale of more than 600 billion Yuan(RMB). These measures would decrease the cost of enterprises and offer instant results as to promoting the achievements of companies.

Petrol Chemical Industry: Most Concerned about Structural Adjustments

The problems accumulated in the rising period of petrol chemical industry have begun to emerge, including the extreme surplus of capacity and out-of-order competition. On one hand, the capacity of primary chemical products is superfluous and dramatically dependent on export. On the other hand, the development of high-end fine chemical products is very weak and dependent on import. Thus, the stimulus plan focuses on structural adjustment of the industry, planning the developing trend of the whole industry from a macro perspective.

Steel Industry: Most Concerned about Reorganization

It is put forward in the stimulus plan of Steel Industry that steel industry should take the leading role of large groups, promote the joint reorganization of enterprises and develop large and ultra-large steel groups with international competitiveness. Specifically, until 2011, there should be several large groups with capacity of more than 50 million tons and international competitiveness such as Baosteel Group, Anben Iron and Steel Group and Wuhan Steel Group.

SECTION 3: IMPLEMENTATION EFFECTS OF THE STIMULUS PLAN

The stimulus plan has been implemented for almost half a year. The effects are evident. The policy effect of promoting economic stability and rapid development is gradually visualized; downturn trend in industry growth rate is restrained; recovery of enterprises' stable development tends is to become clear, and industry integrated operations tend to be sound.

I. Accelerative Recovery in Raw Material Industry

Raw Material Industry accelerates recovery as the fixed asset investment increases vigorously. From January to July of 2009, value added in raw material industry increased by 7.8% over that of the same period of last year, with an increase of 3.1% from January to February, and value added in March, April, May, June, July increases by 8.8%, 7%, 8.9%, 11%, 13% respectively.[4]

Recovery in production of metallurgy and nonferrous metal industry speeds up and the operation takes a favorable turn. From January to July of 2009, value added in metallurgy and nonferrous metal industry increased by 4.2% and 7.9% respectively over that of the same period of last year, with an increase of 14% and 11.5% in July. The output of crude steel of this period is 317 million tons, with an increase of 2.9% over that of the same period of last year, among which in July it reached 50.68 million tons with about 90% capacity restored. In July, the output of ten kinds of nonferrous metal increased by 2% over that of the same period of last year, which changed the situation of eight successive months' downturn in the growth rate of output from last November to this June. Also in July, the daily average output of electrolytic copper and electrolytic aluminum respectively increased by 17.8% and 29.5%. The domestic market prices of major products pick up in vibration trend; the performance of metallurgy and nonferrous metal industry has been improved. In the first half year of 2009, the total profit of 89 key large and medium enterprises amounted to 2.62 billion Yuan(RMB) with a profit of 6.64 billion Yuan(RMB) in June. They have gained profit in five successive

---

[4] The Ministry of Industry and Information Technology. Industry Economic Operation in the First Half Year of 2009—Part One: Material Industry [EB/OL].(07-23-2009). http://www.miit.gov.cn/ n11293472/ n11293832/ n11293907/ n11368223/12472674.html

months. Scaled nonferrous metal enterprises of 22 provinces earned gross profit of 13.9 billion Yuan(RMB), with a profit of 5 billion in June.[5]

Petroleum supply is steady on the whole, and chemical industry upturns strongly. From January to July of 2009, China's output of crude oil was 93.49 million tons, which decreased by 1% below that of the same period of last year, and the volume of imported crude oil in this period held the line with that of the same period of last year. The volume of crude oil processing was 175 million tons, with an increase of 1.5% above that of the same period of last year, 6% above that of the first quarter. In June, the output of crude oil was 15.71 million tons, with a decrease of 1.8% below that of the same period of last year; the volume of crude oil processing was 31.92 million tons, with an increase of 6%. In July, the growth rate of value added in chemical industry was 9.4%, with an increase of 14.7%; the increasing range increased by 0.1% above that of the same period of last year, with an increase of 3.5% above that of June, which is the highest increasing range since the second half year of last year. The output of soda ash and caustic soda in July rose by 6.5% and 13.2% respectively over the same period of last year, with an increase of 5.2 and 7.3 percentage points above that of June, which has maintained growth for two successive months. The output of ethylene has turned to increase by 2% from a decline of 3.3% in June. Owing to the policies to support and benefit agriculture, the rural areas and farmers, the growth of chemical fertilizer and pesticide production has been steady, with an increase of 10.4% and 9.2% in output in previous seven months.

II. Favorable Growth Situation in Equipment Manufacturing Industry

Equipment manufacturing industry appears to be in favorable growth situation, due to investment and policy supports. From January to July of 2009, value added in equipment manufacturing industry increased by 10.1% over that of the same period of last year, with an increase of 15.9% in July, maintaining double-digit growth rate in three successive months.[6]

The production and sales volume in automobile industry hit new records continuously and the increasing trend in agricultural machinery industry is good. Under the policies of half purchase tax for those passenger cars with displacement below 1.6L, "Automobiles-Going-to-the-Countryside", "Automobile trade-in", etc. from January to July of 2009, the output of automobile was 7.1 million, and the sales volume of automobile was 7.18 million, increasing by 20.2% and 23.4% respectively over that of the same period of last year (statistics by Automobile Association). The volume of automobile production and sales has exceeded 1 million for five successive months, among which the amount of those automobiles with displacement above 1.6 L increased by 39.8% over that of the same period of last year. The performance has been improved to some extent. In the first half year, 19 key auto enterprises gained gross profit of 42.2 billion, with a decrease of 1.4% and the decreasing range shrank

[5] The Ministry of Industry and Information Technology. Chinese Steel Industry Operation Features in the First Half Year of 2009[EB/OL].(07-29-2009).
http://www.miit.gov.cn/n11293472/n11293832/n11293907/n11368223/12481621.html

6 The Ministry of Industry and Information Technology. Industry Economic Operation—Part Two (July 2009): Equipment Manufacturing Industry [EB/OL].(08-27-2009).
http://www.miit.gov.cn/n11293472/n11293832/n11293907/n11368223/12537204.html

8.6 percentage points compared to that of the period from January to May. The policy of "Automobile-Going-to-the Countryside" triggers off farmers' initiative for purchasing agriculture machines which further promotes enterprises' production. Until the end of July, the central government has subsidized 9.04 billion Yuan(RMB) and 2.06 million agriculture machines, among which there are 194.7 thousand large and medium tractors, 104.3 thousand other machines like rice transplanting machines, corn harvesting machines, cereal harvesting machines (statistics by the Ministry of Agriculture). In the former seven months, the output of large and medium tractors is 228.1 thousand with an increase of 29.6% over the same period of last year, while the growth rates of production of crop harvesting machines and field work machines are 23.6% and 37.3% respectively.[7]

The situation of general equipments production has turned for the better; the increasing trend in special equipments production maintains steady. From January to July of 2009, value added in general equipment manufacturing industry increased by 7.7% over that of the same period of last year, among which the growth rate in July rose again to 11.3%; value added in special equipment manufacture industry increased by 12.1%, and it has kept a growth rate of more than 12% in the second quarter and July. For extensive construction of domestic engineering projects, domestic demand of construction machinery soars obviously. During the former seven months, the output of lifting equipments, conveying machinery, compaction machines, and concrete/brick machinery respectively increased by 9.9%, 12%, 9.5%, and 30.6% over that of the same period of last year; the production situation of forklifts, diggers, loaders fell into downturn due to the export slump; the output of special equipments for cement, refining and chemicals, grain, feed, printing increased by 15%—31.9% over that of the same period of last year.

Machine tool industry has accelerated its development towards being high-tech and large scale. From January to July of 2009, the output of metal-cutting machine tools and metal-forming machines decreased by 18.6% and 13.5% respectively, among which in July the output of metal cutting machine tools decreased by 10.6%, and the output of metal-forming machines increased by 11.3% compared to the same period of last year. The demand of general-purpose machine tools, low-grade machine tools shrank sharply; nevertheless the demand of large-sized, heavy and top-grade numerical control machines has been steady on the whole. Based on the statistics by China Machine Tool Industry Association about 179 key enterprises concerned, in the first half year, the numerical control rate of metal working machine tools reached 53.7%, with an increase of 7.7% over that of the same period of last year; the unit price of metal cutting machines rose by 36.7%; and the unit price of numerical control machines ascended 46.7% over that of the same period of last year. Recent investigation statistics of more than 130 enterprises by the association have shown that more than one third enterprises are preparing large-sized and heavy tools production.

Shipbuilding industry keeps a rapid growth rate, and the situation of newly received orders is improved. According to the statistics by China International Shipbuilding Industry Association, from January to July of 2009, the completed volume of Chinese ship manufacturing was 18.78 million DWT, with an increase of 78% over that of the same period of last year. In the former seven months, the number of new undertaking ship orders was 7.87

7 The Ministry of Industry and Information Technology. Analysis on the Operation of Chinese Automobile Industry in the First Half Year of 2009 [EB/OL].(08-05-2009). http://www.miit.gov.cn/n11293472/n11293832/n11293907/n11368223/12497380.html

million DWT, with a decrease of 78% over that of the same period of last year; the order situation has got better since June, among which in July the number of new undertaken orders was 1.99 million DWT. Until the end of July, the number of ship orders handed by watercraft enterprises is 192.35 million DWT, with a decrease of 6% below the beginning of the year. According to incomplete figures, China has canceled 75 ship orders and 3.88 million DWT.

III. Steady Operation in Consumer Goods Industry

Dependent on domestic demand, consumer goods industry is in the state of steady operations. From January to July of 2009, in main consumer goods trades, value added in light and scaled textile enterprises increased by 9.6% and 8.5% respectively over that of the same period of last year.[8]

Food production keeps a rapid growth rate, and the increasing trend of fitness equipments and electric bicycles industry is sound. From January to July of 2009, the growth rates of value added in farm and sideline food processing, food manufacture, beverage production are respectively 16.5%, 11.9%, and 12.9%. Among main products, the growth rates of refined edible oil production, meat, and freezing aquatic food production were respectively 25%, 36.5%, 34.8%; the growth rates of candy, quick-frozen rice vermicelli food, and can production ranged from 10.9% to 13.8%, and the growth rates of alcoholic drink and soft drink production were respectively 8.2% and19.2%. In the former seven months, the growth rates of indoor training fitness equipments and electric bicycles production were respectively 23.2% and13.5%, and in July the growth rates of them were 65% and 27.6% respectively.

Recovery in paper production accelerates, and export situation is in a favorable upturn. From January to July of 2009, value added in paper production and paper products trades increased by 6.5% over that of the same period of last year, among which in June and July the growth rate upturned to 10.9% and 9.4% respectively. On the other hand, according to the statistics by customs, from January to July of 2009, the export volume of paper and paper products decreased by 13.3% below that of the same period of last year, nevertheless the export volumes of June and July increased by 11.5% and 16.4% respectively, which changed the situation of continuous decline in export volume for former five months.

Home appliances production has been promoted more obviously by gradual implementation of the policy of "Home-appliances-Going-to-the-Countryside". From January to July of 2009, the accumulated sales volume of the policy was 13.88 million, with turnover of 25.1 billion, among which there were 7.68 million refrigerators, 1.16 million air conditioners, respectively taking the proportion of 62.9% and 24.6% of the turnover. In July, 4.27 million home appliances going to the countryside were sold, and the turnover was 8.8 billion, among which there were 2.56 million refrigerators and 928 thousand air conditioners. Under the policy of home appliance going to the countryside, the situation of home appliances production turns for the better, with an increase of 32.7% in the output of home refrigerators and an increase of 9.8% in the output of room air conditioner over that of the

8 The Ministry of Industry and Information Technology. Chinese Industry Economic Operation in the First half Year of 2009—Part Three: Consumer Goods Industry [EB/OL].(2009-08-27). http://www.miit.gov.cn/n11293472/n11293832/n11293907/n11368223/12537201.html

same period of last year. In the background of export slump, the overall situation of home appliance production is in downturn.

Textiles, clothing production is in smooth operation and the growth rate of chemical fiber production accelerates. From January to July of 2009, value added in textile industry, clothing, shoes and hats manufacturing industry increased by 7. 2% and 10.8% respectively over that of the same period of last year, among which in July the growth rates were 8.6% and 11.3% respectively; value added in chemical fiber industry increased by 5.7% over that of the same period of last year, among which in July the growth rate was 18.4%. From the perspective of main products output, from January to July, the output of yarn, cloth, clothes increased by 10%, 1.1% and 4.8% respectively over that of the same period of last year, with the increasing range falling back to 0.8, 5.6 and 1.4 percentage points respectively, among which in July the growth rates were 13.4%, 4.1% and 8.9% respectively that are higher than those of last July; the growth rate of chemical fiber production was 11.8%, with an increase of 7 percentage points over that of the same period of last year, among which the growth rate in July was 22.6%. Boost of export rebate rate has smoothened the decline trend of textile and clothes export. Export value of the former seven months decreased by 11.3%, with a declining range of 11.3% which was lower than the average of Chinese foreign trade export.

IV. Restrain of the Downward Trend in Electronic Manufacturing Industry

Owing to the policy of "Home-Appliances-Going-to-the-Countryside", and the boost in export rebate rate that encourages consumption, stabilizes export and 3G construction, the downward trend in production has been restrained. From January to July of 2009, value added in electronic manufacturing industry above the designated size increased by 0.8% over that of the same period of last year, which reversed the downward trend of the first half year. The growth rates in May, June, and July turned up to be 4.3%, 6.5% and5.3% respectively. The decreasing range of delivery value of exports shrank from 15.5% in the first quarter to 7.1%, 5.8% and 6.2%.[9]

Domestic market drives the upturn of electronic manufacturing industry. From January to July of 2009, output value of domestic market in electronic manufacturing industry increased by 12.0%, among which the growth rate of the second quarter was 19.6% with an increasing range of 2 percentage points more than that of May and June. The policies of "Home-Appliances-Going-to-the-Countryside" and "Substituting-the-Old-for-New" stimulated town and country market expenditure. From January to July, the accumulated sales volumes of color TV sets, computers, and mobile phones were 2.33 million, 249 thousand, 546 thousand, and the turnovers of them were 2.945 billion Yuan (RMB), 831 million Yuan (RMB), and 296 million Yuan (RMB) respectively, among which in July 488 thousand color TV sets, 139 thousand computers, 120 thousand mobile phones subject to the policy of "Home-Appliances-Going-to-the-Countryside" were sold.

9 The Ministry of Industry and Information Technology. Chinese Industry Economic Operation—Part Four (July 2009): Electronic Manufacturing Industry [EB/OL].(08-27-2009).
http://www.miit.gov.cn/n11293472/n11293832/n11293907/n11368223/12537198.html

The production in communication and network equipment industry has accelerated due to 3G construction and new business expansion. From January to July of 2009, the sales volumes of domestic market in communication equipment and computer network equipment industry increased by 10%, and 19.4% respectively. The output of main communication products increased rapidly, with an increase of 51.7% in optical cable production over that of the same period of last year in the former seven months; the output of basic station communication channels of mobile communication and program control exchanges increased by 159.1%, 3.3% respectively. The output of microcomputers was 88.49 million with an increase of 14.6% over that of the same period of last year, among which the quantity of computers was 74.26 million with an increase of 31.3% over that of the same period of last year.

Color TV industry has been gradually revived in the background of product mix readjustment. From January to July of 2009, the output of Chinese color TV was 50.18 million with an increase of 2% over that of the same period of last year, among which the output of LCD-TV increased by 77.1% that increases its proportion to the total output of color TV from 25.7% to 44.7%, and among which the output of CRT color TV decreased by 32.6% that descended its proportion to the total output of color TV from 46.2% to 30.5%. The export has turned for the better. According to the statistics by customs, from January to July, the accumulated export volume of color TV was 24.83 million with a decrease of 6.6% below that of the same period of last year, among which the accumulated export volume of color TV in June and July increased by 4.1% and 16.8% respectively over that of the same period of last year, which changed the continuous decline situation of monthly export volume in former five months.

V. Upturn in Logistics Industry against the Downward Trend

Updated statistics by National Statistics Institute shows that from January to June of 2009, Chinese freight volume totaled up to 12.935 billion tons, with an increase of 2.5% over that of the same period of last year; freight turnover amounted to 5602.256 billion T/KM, with an increase of 4.3% over that of the same period of last year; the volume of passenger transport added up to 14.767 billion with an increase of 3.7% over that of the same period of last year; passenger turnover volume amounted to 1210 billion P/K, with an increase of 4.5% over that of the same period of last year; the volume of scaled freight handled harbors in the coastal areas was 2.257 billion tons, with an increase of 1.9% over that of the same period of last year. The Statistics above amounts to the conclusion that logistics industry has upturned against the international financial crisis.[10]

The situation of the previous months shows that industrial economic operation has overcome the shock of the international financial crisis through the hardest period. Vigor of smooth upturn has congregated increasingly, and the operation condition has turned for the better on the whole.

[10] National Statistics Institute of the People's Republic of China. Social Transport Volume. [EB/OL].(06-30-2009). http://www.stats.gov.cn/was40/gjtjj_outline.jsp

Column 4-4: China's Sales Volume of Automobile: Infinite Potentiality

China's sales volume of automobile has stepped into a rapid development stage since the automobile industry adjustment and reinvigorating plan was implemented. The sales volume of January of 2009 held the line with that of January of 2007 and2008, nevertheless, the sales volume of February of 2009 increased sharply compared to that of the same period of last year and it has maintained high records since March. The statistics of 2007 and 2008 displays that there will be a sales plateau in three months in automobile industry, whereas no sales plateau appears in the sales volume of 2009 but always high records, which illuminates that owing to the stimulus of industry reinvigorating plan, automobile industry has overcome the shadow of the international financial crisis, and has achieved wider development space, as illustrated in Figure 4-1.

Source: Analysis on the Operation of the Automobile Industry in China from January to July, 2009. 2009-08-05, http://www.miit.gov.cn/n11293472/n11293832/n11293907/n11368223/12497380.html.

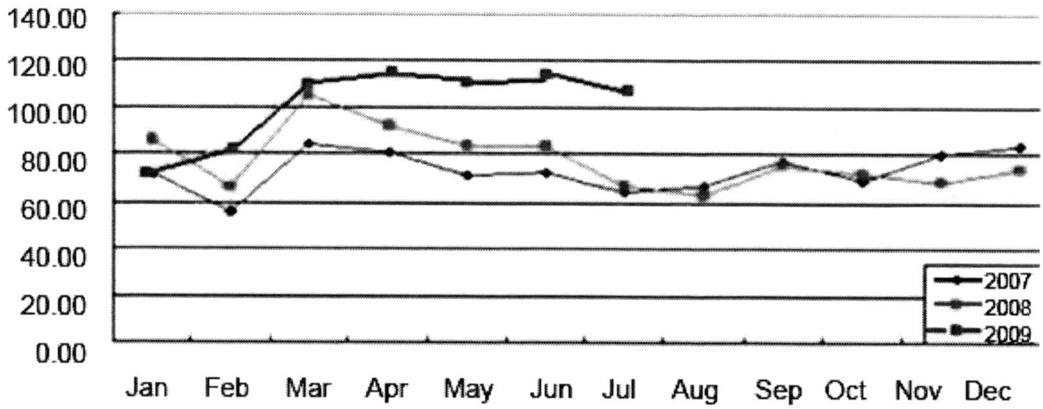

Figure 4-1. China's Monthly Automobile Sales volume from January to July of 2009 (Unit: ten thousand).

SECTION 4: IMPROVEMENT IN THE DIRECTION OF THE STIMULUS PLAN

In the medium and long term, top ten industries adjustment and reinvigorating plan is endowed with strong perceptiveness which has essential function in guiding independent innovations, industries upgrading, and adjustments of product structure and industrial organizational structure. Nevertheless, the promulgated rules for implementation request some amendments.

I. More Attention for Economic Growth Mode Transformation in Implementation of the Stimulus Plan

In recent years, Chinese economic growth depends on investment and export heavily, which is an unsustainable economic growth mode. Presently China's super-strengthening fiscal stimulus plan and investment-driven policy cannot be sustained, which denotes that real economic reinvigoration relies on economic growth mode transformation and economic structure adjustments. The economic growth mode transformation and economic structure adjustments have long-term and deep influence on industrial branches' development, and to some extent, they will change the existing pattern of industry trades. At present, there exists comparative deficiency in the study of economic growth mode transformation and economic structure adjustments, which desire for detailed and further analysis to support adjustment and improvement of specifications in the process of implementation of the reinvigorating plan.

II. More Exertion of Market Mechanism Functions in the Implementation of the Stimulus Plan

4000 billion Yuan(RMB) investment and planning have aroused local government's initiative, but in the meanwhile it can also give rise to corruption and low efficiency and crowd out private investments. As the economy recovers, it needs to consider changing short-term approaches for counteracting the crisis to long-term strategies emphasizing improvements of the market mechanism. The policy of disseminating and promoting the knowledge of technology, culture and sanitation to people in the rural areas has positive effect on the expansion of the rural market, while this kind of indirect subsidies may also cause negative influence. For this problem, it can be shifted to direct subsidies which mitigate government's control, and extend the scope of consumers' choices.

III. Avoiding Overcorrection in the Implementation of the Stimulus Plan for Promoting Industry Upgrading and Product Structure Adjustments

Insufficient consideration of structural adjustments in industry reinvigorating plan may suffocate structural adjustments that has been already taken. Considering the example of some industries in top ten industries like the steel and the textile industry, since there already exists increasing overcapacities caused by the export slump, the industry reinvigoration cannot be realized just by investment, which is likely to induce more serious overcapacity. Besides, promoting reshuffling of the enterprises is included in many plans as a significant form of industry upgrading. For instance, it is emphasized that great enterprises should play a leading role in accelerating industry reorganization in the automobiles, steel, petrifaction, and equipment manufacturing industry reinvigorating plans, which indicates that large state-owned enterprises will benefit more from those plans, while medium and small enterprises will be squeezed out and suppressed, leading to a new round of "public sectors advancing and private sectors retreating". Besides, when distributing relevant funds for supporting independent research and development and technology remolding, it should avoid the phenomenon that major state-owned enterprises and local governments use bribes to lobby the central government, and make sure to effectively use the limited funds on what should be used. All of these illuminate that China should take consideration of the above details in the reinvigorating plan in order to exert its positive effect on industry upgrading and structural adjustments.

IV. Close Watch on Environment Protection in the Implementation of the Stimulus Plan

Most of top ten global industries referred in the reinvigorating plan belong to non-environmental industry which might bring great damage to and influence on the environment. However, there are some environmental protection contents in the industry reinvigorating plans: technology reform and upgrading and elimination of backward production facilities in steel industry; plan of inputting 10 billion Yuan(RMB) special fund for technical innovation, technology remolding, and new energy vehicles development in three years; proposition about the establishment of exit mechanism, promotion of energy saving and consumption reduction, and environment protection in light industry. It is crucial to understand how to put environmental protection measures into effect. Furthermore, the main standard of elimination of backward production facilities in petrifaction, steel, shipbuilding industry is based on the equipment scale, which will force small enterprises to invest in comparatively large scale of equipments to avoid elimination, and further aggravates overcapacity. Therefore it is suggested that the standard of elimination of backward production facilities should be based on technical and economic indexes such as environmental protection and energy expenditure, but not based on the scale of enterprises and equipments, which avoids single standard of elimination.

Translator: Zeng Junxiu (SUFE)
English Version Editors: Xiao Rong (SUFE), Xu Lilei (MIT)

REFERENCES

[1] China's Government Affairs Prosperity Monitoring Center, March 2009. Ten Major Industries Stimulus Plan Will Bring Blowout to Industrial Policy (Chinese) [J]. *Information for Leadership and Decision*, 9.

[2] Zou Deping, 2009. Unscramble China's Revitalization Plan of Ten Key Industries (Chinese) [J]. *Shan Xi Statistics and Society*, 2.

[3] Zhu Guodong, January 2009. Battle of Maintaining the Growth Rate of 8% Fired, Stimulus Plan of the Ten Key industries Discharged[N]. *China Securities Journal*.

[4] Wang Yiming, 2009. Macro-control: Integration Between Short-term Policy and Mid-term Adjustment Strategies (Chinese) [J]. *Macroeconomic Management*, 3.

[5] Cao Jianhai, March 11, 2009. A Roadmap for China's industry revitalization: Short-term Growth and Long-term Structure Adjustment (Chinese). Economic Information Daily.

[6] Yuan Yuan, March 2009. The Revitalization of the Ten Key Industries Brightens Chine's Economy (Chinese)[N]. 21st Century Business Herald.

[7] Ministry of Industry and Information Technology Operation Monitoring Coordination Bureau, Institute of Industrial Economics of Chinese Academy of Social Sciences, August 2009. 2009 Summer Report of Chinese Industry Economic Operation [R].

[8] Zeng Kanhua, 2009. Tax Policy Promotes Ten Key Industries' Revitalization (Chinese) [J]. *China Taxation*, 4.

[9] Critical Count of Ten Key Global Industry Revitalization Plan [J]. *China Audit*, No.9, 2009.

[10] Xu Xiaoqing, 2009. Logistics as a Bridge Connecting Nine Key Industries: the Tenth Industry Revitalization Plan Belongs to Logistics Transportation (Chinese) [J]. *New Logistics*, 3.

In: China's New Deal
Editors: Li Xiaoxi and Hu Biliang

ISBN: 978-1-61470-815-5

Comments on Chapter 4

TEN INDUSTRIES STIMULUS PLAN CALLS FOR SMART IMPLEMENTATION: A VIEW BASED ON LESSONS OF INDUSTRIAL POLICY

Zhang Jin *

Born at a time when financial crisis hit China's economy hard, the 10 major industries stimulus plan (the stimulus plan henceforth) seems to have done its part on ensuring economic growth as China, along with the rest of the world, climbs out of the crisis. Yet, economic stimulus is just part of its mission, the long term objective is to improve the 10 industries' long-run competitiveness, and for that purpose, the mission may have just started.

In fact, from the beginning, the stimulus plan is not so much a short-term stimulus package to the economy as an opportunity of revitalization to the 10 industries themselves. The reason is simple, the crisis China faced was an abrupt drop of overseas demand when the financial crisis hit the rest of the world, and that is the reason the Chinese government has acted in a timely manner to initiate the 4 trillion stimulus package. Nevertheless, the 10 industries stimulus plan fundamentally deals with ensuring and improving supply, and does not address the issues of boosting demand directly (except for automobile industry and light industry).

At a closer look, measures aimed at structural optimization assume the most important role in this massive industries stimulus plan. Such measures fall into the following categories: First, support the development of indigenous brands, and help them achieve leadership in the respective industries; second, expand the investment of advanced technological equipment, eliminate backwards capacity; third, push for merger and acquisition among firms, improve leading firms' international competitiveness (Cao, 2009). These measures have been motivated by such issues as very low concentration, lack of competitiveness, and lack of technological innovations. Undoutedly the solutions of these problems are of great importance to the long term development of these industries.

* Zhang Jin, Ph.D. in Agricultural Economics, Texas A&M University, Associate Professor, Research Institute of Economics and Management, Southwestern University of Finance and Economics.

While the stimulus plan has been driven by very promising objectives, its implementation may not be so straightforward as it appears to be. A successful implementation would take a delicate balance between the ends and the means. In the stimulus plan, an important policy is to push for merger and acquisition, so as to produce bigger and stronger firms. This is a move that could have significant consequences and how to do it is worth thorough investigation. Focus on the major automobile industry, according to the automobile stimulus plan, major consolidation are to be made, "through merger and acquisition, form 2 or 3 automobile groups with a production and sales scale over 2 millions, 4 or 5 automobile groups with scale over 1 million, and the numbers of auto firms occupying over 90 percent of the market should drop from the current 14 to less than 10." The goal of this consolidation policy may not be hard to achieve given a rapidly growing Chinese automobile market and the power of Chinese government, however, whether we should implement it in particular cases and if so, how, may actually be another different issue. Three questions may help us better capture the relationship between the means, consolidation, and the ends, the interest of industry and the public welfare. First, does "bigger" really mean "better"? It is true that many of top auto companies such as Toyota and Volkswagen claim top production and sales scale, but it is also true that big firms such as General Motors can fail. At the same time, many smaller firms are actually quite competitive and survive well when the crisis hits. Second, like any industry, auto industry in China has its particular path to follow. Currently, China's auto market and industry are both in a stage of fast expansion and growth, if the pace of industry concentration is sped up artificially, how can we be sure the industry as a whole is not harmed? At a given stage, for a firm, too big a scale may stretch a firm's management capability and actually harms its growth; For an industry, too high a concentration will harm fair competition, product and technology innvoation, and in the end, the industry's health can be hurt; third, a more realistic question is, even if consolidation is necessary for China's auto industry, how do we ensure merger and acquistions will be carried out with reasonable chance of success? Merger and acquistions are nothing new in China including auto industry, and those among sate-owned enterprises, but not many have been successful so far. In this matter, we really should be cautious enough.

History of China's industrial policy offers some valuable lessons and we may strike a better balance between our policy and its intended purpose. Again, auto industry turns out to a good example. By the end of 2008, China's auto market reached the record high of 9.38 millions, 6.76 millions of them being cars, 26% of them by indigenous brands. In a firm perspective, the top 10 sedan firms are: FAW-VW, Shanghai-Ford, Shanghai-GM, FAW-Toyota, Dongfeng-Nissan, Chery, Guangzhou-Honda, Beijing-Hyundai, Geely, and ChangAn-Ford. Obviously, the market is dominated by joint ventures between a few state-own firms and overseas makers, except that Chery and Geely are the only two indigenous brands. Ironically, these two precious indigenous brands may not be on this top 10 list due to the industrial policy of China. Back 10 years ago, when these firms were first born, auto industrial policy of China actually forbid entry of these new firms into auto industry. Chery and Geely, and a few others, mostly private, entered this state-owne-enterprise dominated sector only after breaking serious barriers and luckily developed into today's status. These are the firms that broke the policy barrier successfully, what about those firms who did not make it? Not hard to imagine that if our auto industrial policy back then was carried out as it was designed to be, Chery, Geely, and other BYD, Great Wall, Lifan, and even more, would never have been born, then today China's auto industry would be totoally occupied by joint ventures, and there is no exsitance of any significant indigenous brands. Of course, that would never be the purpose of the 1994 Auto Industrial Policy of China.

International experience also provides a good lesson to industry stimulus plan. Compared to other industralized countries, the development of major industries of Japan in the postwar period, especially 1950s and 60s, are closer to the case of China and therefore may provide a better reference. In the 1950s and 60s, the major industries of Japan were still in their early stage of development, the problems such as small scale, price competition, over-investement, excessive capacity, and lack of technology innovations, were widely observed in Japanese industrial organizations. In order to promote a fast elevation of Japanese industrial structure, the Ministry of International Trade and Industry once pushed for the "special stimulus law" with an aim at advancing consolidation as well as "more reasonable competition". The law actually aborted as it contracted the antitrust law in Japan and failed to gain the support from industry and the opposition parties. Also during this period, out of an intention to elevate Japanese auto industry, the Ministry of Internatioal Trade and Industry in the early 1960s had intended to reorganize existing auto makers into a few groups and forbid the entry of other firms into auto industry. This plan, however, never materialized due to lack of support from industry. Soichiro Honda, the founder of Honda, even thought that Honda could be more successful were not it for the barrier from the Ministry of International Trade and Industry (Jia, 2004).

The above lessons simply tell that the making and implementation of industrial policy by government should use more caution. If a policy intervention is improper, either it aborts due to lack of support from industry, or even worse, the overall development of the particular industry and the public welfare, are harmed. In the case of China, more caution should be applied becasue state-owned enterprises make up a significant portion of the 10 key industries, and consequently the government can be more powerful when it comes to the implementation of an industrial policy. In conclusion, our industry stimulus plan, in terms of detail and implementation, should be flexible and smart. Flexibility here means that while the birth of the stimulus plan is essentially about doing something to help, in certain cases, doing less actually means doing more and being more helpful. Doing less has little to do with the government's ability to push a policy, instead, it means that too much intervention may lead to less ideal effect. Especially, when it comes to merger and acquisition policy, the government needs to decide how much to involve. The government can suggest or even encourage, where appropriate, consolidatino among firms, nevertheless, the firms themselves should be the ones to decide whether to merge and with whom. Once firms involved are willing to carry out an merger & acquisition, the government may step in to remove potential barriers and provide supports if needed. This may be a more suitable practice in an market economy environment.

On the other hand, government industrial policy does have an important role to play in the development of major industries. Policy measures of this kind have been incorporated in the stimulus plan, and they should be given priority in the following implementation process. Regarding the role of government industrial policy, Komiya and others (1988) studied the industrial policy of Japan, and their comment on the Japanese government's role is two fold. On the one side, they believe some direct intervention by government, especially those directed at advancing consolidation or correcting the so called "improper competition" were neither popular nor effective; on the other side, some supportive industrial policies, including financial investment, low interest rate loan, special tax measures, contributed positively to the development of industries. In fact, the 10 industries stimulus plan have incorporated many such measures, including the export tax rebate policies for the steel and texitile industries, the reduction of purchase tax for auto industry, increased credit support to ship buyers and extension of existing financial support policies for oceangoing vessels for the shipbuilding industry, export tax rebate and national reserve system for nonferrous metal industry, etc.

In addition, the issue of new energy is gaining increasing attention in today's China as well as the world. In the stimulus plan, RMB 10 billions are allocated to support the development of new energy vehicles. To China's auto industry, new energy vehicle means not a new type of product that can support its continued existence in the future, it also means an opportunity to catch up with the rest of world in international competition. The development of new energy vehicles depends more than new vehicle technology, the development of road infrastructure is also indispensable. For advancement on both frontiers, much can be done by the government, or precisely, some of them can be done only by the government. Auto industry aside, the elimination of bcakwards capacities in steel and shipbuilding industries, the application of new technologies in electronics and information industries, all are important measures to the improvement of industrial technical standards. All these policy measures should be given priorities in our 10 major industries stimulus plan and be done well.

REFERENCES

Cao, J., March 11, 2009, A roadmap for China's industry revitalization: Short-term growth and long-term structure adjustment, Economic Information Daily.

Jia, X. 2004, An objective view on Japan's automobile industry policy, China Industry Information Net.

Komiya, R., M. Okuno, and K. Suzumura, 1988, Industrial Policy of Japan, Academic Press.

Shirouzu, N., March 23, 2009, China Uses Green Cars To Bolster Auto Sector, The Wall Street Journal.

Wu, Y. , 2009-2-27, 10 major industries to become China's economic engines, People's Daily.

In: China's New Deal
Editors: Li Xiaoxi and Hu Biliang
ISBN: 978-1-61470-815-5

Chapter 5

REAL ESTATE: NEW CHANGES IN THE INTERNATIONAL FINANCIAL CRISIS

Zhang Qi[*], *Wang Hao, Li Hongyi and Zhang Ai*

Real estate, which was related closely with the economic and financial development, was not only the fuse and ignition point of the Sub-prime crisis in the United States, but also reinforced the financial crisis expansion. It could be described as that this financial crisis was caused by real estate, which was conversely expanded and enlarged by the financial derivatives of the sub-crime crisis, thus further deepening the depth and breadth of this crisis. Hence, it is essential to analyze the changes in real estate industry when interpreting the influence of the international financial crisis.

Since 2008, China's real estate has been fluctuating up and down influenced by the international financial crisis. In the first half of 2008, following the economic downturn in 2007, the volumes of real estate development and investment were slowly declined by months. In the second half year, the real estate market was deteriorating deeply and there was a negative market growth rate compared with the first half of 2008. However in 2009, the real estate market was gradually warmed up. Since then, the trading volume and transaction prices have been gradually restored to the level of that before the financial crisis.

SECTION 1: THE FINANCIAL CRISIS AND CHINESE REAL ESTATE INDUSTRY

Before the global financial crisis generated by the U.S. sub-prime crisis, Chinese real estate industry develops very fast, the real estate market is extremely active, and real estate finance credit grows rapidly. Moreover, the real estate industry has become a pillar industry of the national economy. Chinese real estate industry had played a very positive role to recover national economy from 1997 Asian Financial Crisis, as a driving force for economic

* Dr. Zhang Qi is professor at the Institute of Economics and Resource Management at Beijing Normal University, who earned his PhD degree from the Graduate School of the Chinese Academy of Social Sciences. Email: sxzmq588@163.com.

growth, expanding domestic demand and stimulating social investment. Rather than the causes of the US financial crisis, real estate has become an important tool for Chinese government to control its economy responding to the Asian Financial Crisis and the current world financial crisis. From the beginning of 2008, the goal of macro-control was to prevent the overheating and excessive growth of economy confronting the international financial crisis. At the same period, China underwent the rare snow disaster and the earthquake hazard. Accordingly, Chinese real estate industry was undergoing a tough test in the twists and turns.

I. Investment in the Real Estate Industry Fluctuates

Since 2008, Chinese real estate industry has come into the trough of its development circle upon the influences of the Chinese government constrains and the financial crisis. As for the real estate investment, Chinese real estate investment reached 3058 billion Yuan(RMB) in 2008, raising 530 billion Yuan(RMB) compared with the year of 2007; which is 20.9% growth rate on year-by-year basis. Compared with the 30.2% increasing rate in 2007, the growth rate has dropped by nearly 10 percentage points[1]. Also, it shows that Chinese real estate investment was in a contraction state in 2008. But the growth rate is much higher than that of fixed assets investment at the same period in the first half year. From the third quarter, total investment growth in real estate and urban aggregate fixed-asset investment moved in the same direction again (See table 5-1).

Table 5-1. the Comparison of the Growth Rate of Investment between Real Estate and Fixed Assets over the Same Period from Jan. to Dec. of 2008 in China

Year	Total Urban Investment in Fixed Assets		Total Investment in Real Estate	
2008	Billion Yuan	Growth Rate on a Year-on-year Basis	Billion Yuan	Growth Rate on a Year-on-year basis
The 1st Quarter	1831.7	25.9%	468.8	32.27%
The 2nd Quarter	4011.9	27.5%	850.8	34.14%
The 3rd Quarter	4143.6	28.8%	807.4	16.55%
The 4th Quarter	4824.1	32.3%	931	9.9%
The Whole Year	14816.7	26.1%	3058	20.9%

Source: Yearly and quarterly data, the web site of the National Statistics Bureau, http://www.stats.gov.cn/tjsj/jidusj/

The changes of the data in the table above indicate that the fluctuation on urban fixed assets investment was mainly influenced by the fluctuation on real estate investment. The

[1] Source: The People's Republic of China 2008 National Economic and Social Development Statistical Communiqué, "People's Daily", Feb. 26th 2009.

investment enthusiasm has been restrained by significant downturn in real estate development and investment, the gradually shrinking trading volume and the lower bank restrictions on credit, making the real estate markets come into recession. While in the second half year, in order to deal with the financial crisis and market depression, the government expanded the central bank's investment scale, which the increasing loan were mainly focused on accelerating the people's livelihood projects, infrastructure, ecological environment construction and post-disaster reconstruction, etc. The real estate market is largely dependent on market forces to achieve its self-adjustment and self-improvement. Although with the steady growth of China's fixed asset investment and sales-growth accelerated in the domestic market, the economic growth was still slowing down in 2008, real estate investment dropped sharply, inflation worsened and foreign trade continued to decline, hence China's economy has shrunk evidently. This phenomenon was both related with a series of austerity policies to curb rising prices and the tremendous influence on Chinese export-oriented economy caused by the financial crisis. The substantial long-term decline in international trade has seriously affected the economic growth and added additional enormous burden to China's employment, as well as the relative chain-industries (such as real estate, automobile, etc.).

II. The Growth Rate of Real Estate Prices Started to Slow Down

In 2008, China's consumer prices continued to increase, while the growth of commodity residence prices are slowing down. During this period, the consumption price index raised 5.9% from the year of 2007; residence consumption price index was 111.0 which had been increased 6.2 percent than that in 2007, which made inflation worsen. Of those, the newly-constructed house price rose up by 7.1%, the second-hand residence price rose up by 6.2% and the residence-rent price increased by 1.4%[2]. Moreover, the residence housing price was still roaring up from the real estate price point of view in 2008, 70 cities showed 6.5% up but it appeared to be a downward trend. From Jan. to Nov., the house price was still rising but the growth rate kept declining. The absolute price showed evident falling in December. The increasing scale of the selling price in that 70 cities had been the negative growth from Jan. to Dec. compared to that from Jan. to Feb (see Figure 5-1). Therefore, we can see that in 2008 China's real estate market declining prices are partly due to the buyers' pessimistic expectations on the real estate market which were caused by the financial crisis, and they became more emotionally "watching". On the other hand, since September 2007, the central bank of China (CBC) began to strengthen the management of individuals' "second apartment" home loans, thereof inhibiting the real estate market speculation and investment behavior to some extent.

[2] Source: The People's Republic of China 2008 National Economic and Social Development Statistics Bulletin, People's Daily, February 26, 2009

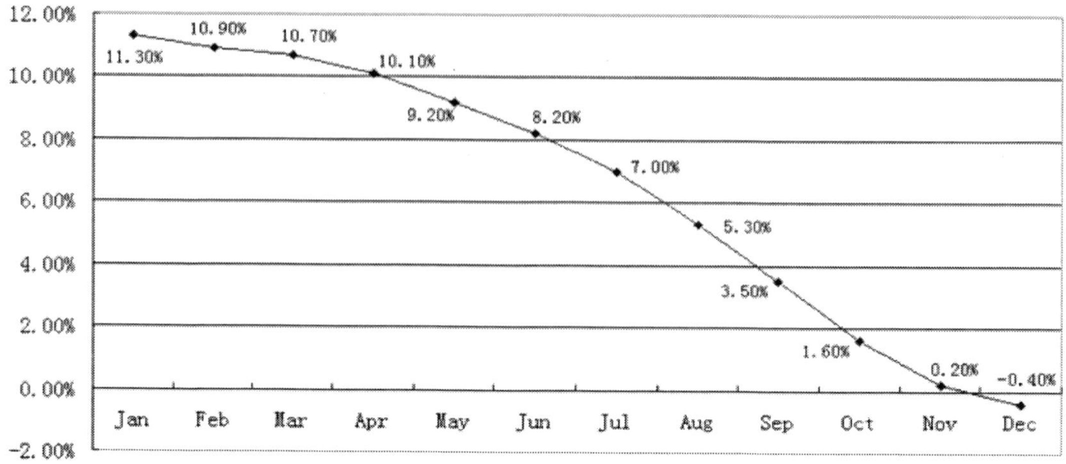

Figure 5-1. The Growth Rate of Housing Price of Year-on-year Rise in China's 70 Large and Medium Cities[3].

III. Newly-started Area and Completion Area Decline Monthly, and Land-purchase and Development Area Has Gradually Declined

The financial crisis has affected China's real estate development and investment scale, and further reduced the indicators, such as the scale of newly-started and completion area. In 2008, the growth of newly-started housing area and the completion area also has dropped significantly in china. Till the end of December 2008, the national newly-started housing area were totaled 798.89 million square meters, with an increasing rate of 1.4% compared with that of 2007. The growth rate has been reducing monthly from 24.9% in January to 1.4% in December (see Table 5-2), indicating that growth rate of the supply of real estate has reduced in a short-term.

Also it should be noted that the real estate market is closely related to the land market and is fluctuated up and down with land price change. In 2008, the national land purchase area is 370 million square meters, the year-on-year basis growth has decreased to 7.6% from Jan. to Jun.. The growth rate of Jan. and Feb was 35.8%, and from Jan. to Dec. was -8.6%. While the land purchase expenditure has increased by 10.9%, with a land exploitation area of 260 million square meters. The year-on-year basis growth rate of land development area was -5.6% of Jan.-Dec, and this rate was 21.3% from Jan. to Feb. Meanwhile, the purchased-but-undeveloped area has been up to 110 million square meters for the whole year, and the would-be exploitation area has nearly been up to 440 million square meters with the year-on-year basis growth rate of 6.5%. Compared to the land-purchasing rush and on-going emerging

[3] Source: Website of the China National Bureau of Statistics, http://www.stats.gov.cn/tjsj/jidusj/

"land king" in 2007, the national real estate land-purchasing and exploitation has been overall cooling down in 2008.

Table 5-2. Newly-started Housing in 2008 (Unit: million square meters, %)

Month	2008			
	House	On Year-on-year basis growth (%)	Newly-started area	On Year-on-year basis growth (%)
1-2	126.79	27.2	104.23	24.9
1-3	239.71	25.9	198.62	24.8
1-4	329.70	20.4	274.22	20.2
1-5	425.26	19.8	352.66	18.7
1-6	534.73	19.1	443.09	18.0
1-7	605.10	15.1	501.71	14.6
1-8	670.49	13.1	555.67	12.6
1-9	735.21	10.2	608.40	9.6
1-10	787.72	7.3	651.13	6.9
1-11	842.90	5.4	695.38	5.0
1-12	975.74	2.3	798.89	1.4

Source: 2008 China·Statistical Annual, China Statistics Press, January 2008.

IV. The Sales of Commercial Property Reduced Significantly, and Vacant Area Increased

In 2008 the real estate transaction was not satisfying apart from the large decline in the real estate investment by the influence of the international and domestic macro-economic situation. The selling price and selling area of commercial property decreased sharply. The National Real Estate Prosperity Index[4] posted by the National Bureau of Statistics fell to 98.46 from 106.59 of the beginning of the year (see fig 5-2). The area of sales declined steeply in 2008 with a figure of only 620 million square meters, from the figure of -4.2% of January-February downward to -19.7% from January to December, showing a substantial downward trend of 16.7% decrease throughout the whole year compared with 2007 which was the 1st decline in housing market. Among them, there are 560 million square meters of commercial housing, going down 20.3%, the first decline of the housing market. In addition, the gap between the completed apartment and the forward delivery house is not evident though the sales of both declined considerably.

[4] National Real Estate Prosperity Index is consisted of 6 indices of the investment of real estate development, funding source, land development area, area under construction, vacant area of commodity housing and the average price of commodity housing. And the critical point of the index is 100, above which is the prosperity space and below which the recession space.

Column 5-1: The 'Real Estate Tycoon' into 'Pig Magnate'

In the past decades, the real estate became increasingly dependent on banks support because of its huge demanding of funds, credit loans and other financial products. It can be said that affluent cash flow is the key point of the survival and development of real estate companies. However, under the influence of the international financial crisis of 2008, many banks have tightened money supply and credit. Under this background, many companies have transformed in order to develop, and the former "real estate tycoon" has transformed into "pig magnate" In Nanjing.

From early of 2008, the property market has appeared a turning point. In Nanjing, a developer has invested more than 0.2 billion Yuan(RMB) in the pig breeding industry therefore became the first "pig magnate". Since then a group of developers have followed up making dozens of developers in the pig breeding. From the "real estate tycoon" into "pig magnate", the "self-help" behavior of the developers has been the wonders in the property market.

On the 24th July, the relevant reporter called the Nanjing Agriculture and Forestry Bureau of Animal Husbandry. A staff member told the reporter that the developers developed in the pig-raising industry as a career, which has never happened before. "It appeared in recent years." The pig-raising developers have at least ten millions of funds, having more efficiency than the agricultural enterprises. A real estate company got profits of more than 30 million Yuan through five years development in real estate. July 28, the reporter found Cao Weifeng, the supervisor of the first developer Fortum Real Estate now engaging in pig farming in Nanjing. He said that the company cooperated with the Southern Agricultural University at the end of 2007 and invested 20 million Yuan(RMB) in Fengxian county in Xuzhou city to build a standard pig farms which are 200 Mu(Chinese unit) land. In this year the company invested 200 million Yuan(RMB) in building a pig farm of 1000 Mu and will invest more than 20 million to construct an ecological farming training based in Gaochun. "The pig farming land is a special approval of the local government and is leased for a period time, usually of 20 years. The goal of this year is to raise 150,000 pigs", he said. And talking about the reasons to engage in the pig farming, this line has a very good development potential and expectation.

With the same situation as Fortum, Jiangsu Jindongcheng Real Estate Group which was engaged solely in the real estate development now thinking to build a pig farm cooperated with the communities of Hengxi in Jiangsu. "Now we are making the planning. Since there are enough funds, why not look for other investment opportunities?" said a staff member.

The information gathered by the reporter showed that the "pig magnates" are mostly enterprises of sole project. Its strong speculation can make a share from the profitable real estate. But in the current market environment, they have to face many difficulties in getting lands, getting loans, developing and selling. Comparatively, pig faming investment does better than real estate. (Source: *Xinhua Daily*, August 6, 2008).

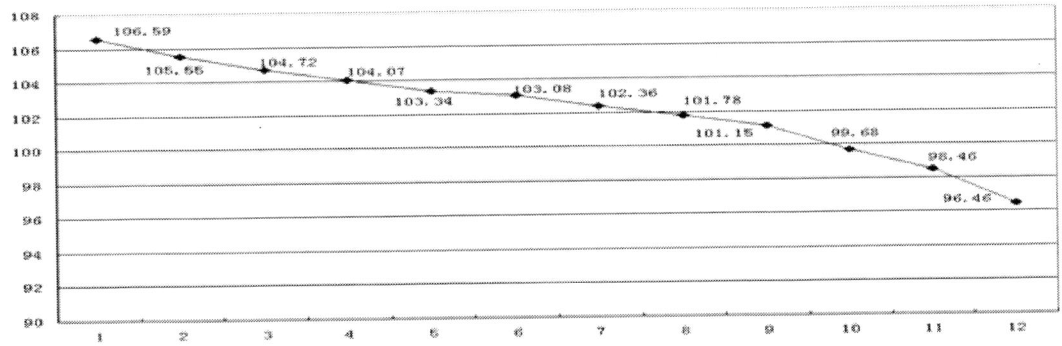

Figure 5-2. National Real Estate Prosperity Index (from January to December of 2008)[5].

And, accompanying with the decline of sales price and sales area of the commercial property, the vacant area has been increased significantly. According to the National Bureau of Statistics data, the national commercial vacant area was up to 164 million square meters since the end of the December of 2008, up to 21.8% of January—December from -3.9% of January—March, with an increase of 25.7 percentage points. Among them, the growth of vacant area of commercial property growed rapidly up to 91 million square meters with an increase of 42.8%, hitting the highest historical record since 2003. But the related indicators of the vacant area of commercial property has declined monthly[6], downward to 100.4 on December 2008, with 10.8 percentage points lower compared with the same period of last year (see Figure 5-3).

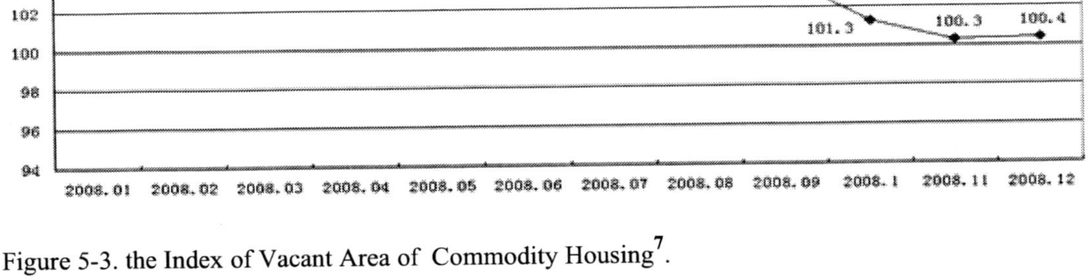

Figure 5-3. the Index of Vacant Area of Commodity Housing[7].

[5] Source: the web site of the National Bureau of Statistics throughout the yearly and quarterly data tables, http://www.stats.gov.cn/tjsj/jidusj/

[6] The higher index of commodity housing vacant area indicates the less vacant area.

[7] Source: the web site of the National Bureau of Statistics, the yearly and quarterly data tables, http://www.stats.gov.cn/tjsj/jidusj/

V. The Downward of Real Estate Brings a Chain Reaction and the related industries show a sign of recession

The financial crisis has a considerable influence on China's real estate development, which has therefore brought the knock-on effect to other industries. One impact is on the financial industry. The real estate market is closely related to the financial industry, and the real estate's slowing down is bound to affect the credit and real estate mortgage loans which are the main channels of cash flow. According to the real estate industry statistics, the domestic loans on real estate development were 725.7 billion Yuan (RMB) in the year of 2008, which declined to 3.4% from January to December and 36.9% from January to February. The growth rate was only 0.5% higher than that of 2004. From the second quarter, the downward trend was accelerated. The growth rate of January-April declined 9.1 percentage points than that of January—March[8]. At the same time, the greatest threat to the financial industry was credit loans risk. Because real estate prices declined, it will definitely bring greater risk in bank loans. And the real estate contraction in 2008 did pose a risk to the bank financial and credit loans. The decline also reflects that the banks will pay more attention to risk control.

The other impact was on the related industries and investors' confidence and their expectations. Different with other industries, the downturn of real estate will affect the related industries such as iron and steel, metallurgy, chemical industry, machinery manufacturing and so on. More importantly, the recession of real estate will motivate the pessimistic expectation. Since 2007, the U.S. sub-prime mortgage crisis and structural adjustment of China's macro-economy affected people's daily lives and expectations, making clear trend on the market "watching" and sharp decline in product transactions. The whole market was in a situation of depression.

It should be said that the most important influence on the market caused by the world financial crisis was the impact of the confidence of the investor and consumer, making them generally had a pessimistic expectation on the economic development. According to the Western point of view of economics, market expectation is the key factor in affecting the future market direction. The U.S. sub-prime crisis was a serious impact on the real estate credit system, inducing such behavior that people dare not apply for loans to purchase the housing because of their concerns of the difficulties in the loan repayment. More importantly, it prompted the Chinese government to develop more prudent credit policies of housing from the point of social and economic stability in order to ensure that macro-economy will not be affected by the fluctuations of real estate. Therefore, both the consumers and the investors and the real estate enterprises were reluctant to move on in the complex effect of these factors because of the "watching" atmosphere.

[8] Source: Chen Liuqin, 03-13-2009. The Development of China Real Estate in the Financial Crisis". Guangming Website, Guangming Observation.

SECTION 2: THE REAL ESTATE POLICY ADJUSTMENTS RESPONDING TO THE FINANCIAL CRISIS

The Chinese government continued to control the macro-economic situation from expanding investment and increasing consumption, stimulating demand and encouraging exports. The government recognized the tremendous correlation effects on investment and consumption as well as the related industries due to the introduction of the economic stimulus policies. Therefore, a prominent feature of this regulation and control was to restore the economy, starting from the recovery of the real estate and the expansion of trading volume. By lowering the interest rates, reducing taxes and lowering the threshold of the real estate development, to increase land supply and other measures can stimulate demand and supply thereby to stimulate the real estate industry so as to achieve the entire macro-economic recovery.

I. Reduce Mortgage Interest Rates

It was mentioned above that the real estate industry is related to a number of industries and plays an important role in the national economy. It is important to maintain a healthy development of the real estate industry in order to stabilize the overall economy situation. If the real estate market kept down turning, the related industries could be affected. Therefore, the first move to restore the vitality of the real estate market was to stimulate the housing needs of residents by lowering down the mortgage interest rate.

In October 2008, the Ministry of Public Finance and the People's Bank of China jointly issued measures to support the real estate industry, requiring financial institutions to reduce the mortgage lending rate to 0.7 time of the central bank benchmark interest rate. It was the first time that the proportion of down payment to purchase ordinary housing has come down to 20 %; and the local government was endowed the right to enact policies to encourage real estate transaction and tax relief. In December of the same year, the State Council introduced further measures to promote real estate development requiring that if the first apartment was lower than the local average level then the minimum interest rate of the second apartment be adjusted to 0.7 times the benchmark rate and the proportion of down-payment has dropped to 20% from 30% in the past; and to encourage financial institutions to step up credit support on housing construction for medium and small dwelling size and support the reputable developers in the process of merger and reorganization of re-financing needs. The central bank immediately lowered the five-year lending rate from 7.83% down to 5.94% with the introduction of this policy and the statutory reserve ratio from 17.5% to 15%. In addition, personal public accumulation housing fund[9] loan interest rate was reduced for the consumers of housing funds, with the five-year (or below-five-year) loan rate adjusted from the current

[9] Public accumulation fund for housing is the long-term housing savings deposit contributed by the government institutions, the state-owned enterprises, urban collective enterprises, foreign-invested enterprises, urban private enterprises and other urban enterprises and institutions as well as their employees. This is a reflection of the welfare of Chinese housing and a policy measure to solve the problem of housing for urban workers.

4.32% to 4.05%, and above-five-year loan rate from adjusted from the current 4.86% to 4.59 %, decreasing 0.27 percentage point respectively.

The changing of interest rate was commonly used by the government to regulate the supply and trading volume and regulate the demand. This measure sheds the burdens of the consumers to buy a house, stimulates the needs for housing and effectively reduces those burden who have long term loan.

II. A Variety of Tax Relief in Real Estate

In China, the current taxes which need be paid in real estate transaction are the business tax, land value increment tax, individual income tax, deed tax and stamp duties. Among them, business tax aims at the sales party, which could be exempted from the business tax if only purchasing ordinary residential and living there more than two years, otherwise it should be levied 3%—5% of the balance between the sales and purchase price. The land value increment tax is also aimed at the sales party by 30%, 40%, 50%, 60% respectively according to different conditions. The individual income tax still targets at housing sellers, the sum is 20% of the balance according to housing sales price minus the original price & the other fees occurred in the process. Deed tax and stamp duty aim at home buyers, in which the rate is between 3%—5% and stamp duty is 0.05% of the total contract value. In addition, the other related administrative fees need to be paid in the real estate transaction process. In general, this is not conducive to the real estate transaction that there are a number of tax types in the entire transaction process.

Therefore, since the second half of 2008, the government reduced the real estate mortgage rates and gradually adopted a tax rate reduction, tax exemption periodically and other measures in order to stimulate residential housing demand and promote real estate transactions. First of all, the Chinese government issued documents of stamp duty and land value-added tax free to stimulate the real estate companies to purchase land, make further development and strengthen market supply. Secondly, in order to reverse the pessimism of consumers and increase market demand, in October 2008, the Ministry of Public Finance and the People's Bank of China jointly issued measures to support the real estate industry stipulated that since November 1, 2008, the deed tax rate should be adjusted from 3%—5% to 1% for the first purchase of apartment under 90 square meters. Meanwhile, the Bank of China also lowered mortgage rates of first apartment-purchasing, the real interest rate was about 4.16% and was about 250 basis points than the average compared to that of August 2008. In addition, in order to promote the development of second-hand housing market, the State Council introduced measures that stipulated that the limit of business tax exempted has shorten to 2 years from the original 5 years with the collection of business tax to replace the sales income tax. Those tax reduction and exemption policies will facilitate to shed the burdens of the consumers to afford the housing, simplify the real estate transaction process and reduce the cost of sales and purchases of the transaction from both sides.

III. Reduce the Capital Fund of the Real Estate Development

In order to regulate and control the real estate market, both demand and supply sides are worth to be regulated. It is approached mainly by the investment of the development enterprises, construction and sales then to regulate the supply. Because in 2008 the "watching" situation in the real estate market has greatly reduced the turnover, resulting in a sharp repayment pressure on lots of real estate development enterprises, which fell into the plight of a shortage of funds and many of which even postponed or suspend the construction. In addition, in 2009 there experienced a rapid "inventory clearing", after that there appeared a shortage of supply in the real estate market of the completed housing and forward housing, which will not only relate directly to the domestic demand on the real estate but also influence the entire stability of real estate market. Therefore, it was the main content to regulate the supply and demand of the threshold of real estate investment and construction as well as to improve the enthusiasm of the enterprises to run the real estate projects.

The focus of this supply regulation was placed on the project capital of the real estate investment. The project capital system was a capital guarantee system implemented in China since 1996 on a variety of business investment projects, including the infrastructure of state-owned units, technological innovation, real estate development projects and collective investment projects in order to guarantee the construction funds before the start of the project construction. In the past few years, in order to regulate the real estate market at that time due to the real estate rush and too many projects under construction, in September 2004 the Chinese government required that the equity of the development enterprises who applied for the loans should not be lower than 35% of the total investment, which is the threshold requirement of the real estate enterprises to invest projects. Under this regulation and control measurement, in order to strengthen the development capacity of the real estate enterprises and enhance the follow-up power to supply, the Chinese government decreased the minimal equity ratio to 20% in April 2009, which was the first deregulation since September 2004, suggesting that the pressure on capital has been reduced significantly for the real estate enterprises to invest in operation project and the new policy helped to improve the initiative of the real estate business investment, construction and sales of real estate and expanded the scale of market supply.

IV. The Introduction of the Matching Measures of Local Government

The real estate market is regional. Its development status abounds a lot to various policy implementations at the project site. After launching a series macro-control policies by central government, the local governments introduced correspondingly a policies and measures to ensure the implementation of macroeconomic regulation and control from localities' own long-term interests.

In Shanghai, the government stipulated that the limit of public accumulated fund loans should be raised up to 200,000 Yuan(RMB) for those who purchased their firs apartments with more than two people (including two people) involved in the loans, and the ceiling could be up to 600.000 Yuan(RMB) for two-bread-winner family. In Chongqing, firstly, the ordinary commodities apartments that were equal to or less than 90 square meters were exempted from deed tax. Secondly, second-hand housing transaction was levied consolidated tax at the rate of 1.1%. For the purchasing of second-hand ordinary housing, the relative tax could be levied with full fiscal subsidy. In Henan, those who used the public accumulated fund to make loans need only pay 20% for the down payment and the time limit was adjusted to a maximum of 30 years. For those who purchased in Henan province can apply for the public accumulated fund loans in other provinces. For those who purchased the economically affordable housing were exempted from deed tax of real estate transaction; and for ordinary commodity housing deed tax could be reduced according to a certain percentage. In Xi'an, the government will give the buyers the fiscal subsidies directly, ranking 1.5%, 1%, 0.5%. In addition, the urban infrastructure supporting fees which were charged from real estate enterprises were reduced by 35 Yuan(RMB) on every square meter. In Nanjing, any purchase of commodity housing less than 90 square meters were given the subsidies 1% of the total purchase value; and those less than 90—144 square meters were given the subsidies 10.5% of the total value. In addition, when the real estate enterprises were in a difficulty such as shortage of funds, the construction fees such as infrastructure, civil air defense construction could be held over. In Hangzhou, the public accumulated fund loans were further increased with a ceiling of 600,000 Yuan(RMB) from 500,000 Yuan(RMB) and a maximum age of 65 years for the male or 60 years for the female. Also, the related fees were suspended such as the registration.

V. Adjusting Scale of Land supply and Demand by Land Reserving

The land transaction fee has always been the main source of revenue for the local government in China, making a net profit of more than 40% even 60% in some places apart from the costs of land acquisition, demolition, compensation and taxes. The downturn of China's real estate market in 2008 immensely restrained the desire of the developers to acquire land, then the prosperity of the primary land market fell with a rapid decline in the local revenue. In order to promote the real estate market recovery, to prosper the primary land market, to increase the revenue and ease the fiscal pressure, many local governments made efforts on developing the land bank and enhancing the land market supply. Since the half of 2008, the major cities in China started a new round to develop the land bank and to purchase, require and re-purchase all kinds of land on large scale to support the need of infrastructure construction. Among them, Beijing was very typical. According to the unified deployment to address the financial crisis in 2008, the government of Beijing intensified efforts in the land reserve and development in 2009, at the beginning of the year the carry-over was 8800 hectares with a new annual addition of 4700 hectares. Basically, the development of 3600 hectares were completed (of which the annual supply is1600 hectares), the year-end balance was 9900 hectares and land bank development funds were arranged at 100 billion. On one hand, this is a very important means to implement the land bank program and ensure the GDP growth up to 9% of economy of Beijing in 2009 against the background of the international financial crisis. On the other hand, it is the main way to carry out the major state projects and to promote the healthy development of Beijing real estate market as well as to provide the land for the livelihood projects such as price-limited housing, economically affordable housing, low-rented housing and urban infrastructure. In 2008 Shanghai also further increased the amount of the land development on the basis of land reserve of the recent years with the emphasis of government-subsidized houses. And in the same year, Shanghai (10 Municipal districts, excluding Zhabei, Putuo) planned to supply the land for residential use at a maximum of 4.9 square kilometers, and expected this would completely solve the housing problem of 60,000 urban low-income groups and would supply the land for commodity housing of 2.5—4 square kilometers at the same time.

SECTION 3: NEW CHANGES IN CHINA'S REAL ESTATE DEVELOPMENT

Since 2008, the Chinese government has made a major policy adjustment in a timely manner, and passed a 4-trillion-Yuan (RMB) economic revitalization plan, both of which were closely related with the real estate industry. Therefore, it can be observed that the Chinese government has increased the intensity of economic stimulus. The 4-trillion-Yuan (RMB) economic revitalization plan has altogether ten items, which all interrelate in numerous ways with the real estate industry. To illustrate, the acceleration of the construction of housing project for low-income urban residents, the building of rural infrastructural facilities, the railway, highway, airport and other major infrastructure constructions. Some other items, like the improving of the ecological environment and the accelerating of independent innovation and structural adjustment, boost the domestic demands by increasing investment in fixed assets. Thus, the investment established intimate links with the real estate and land supply. In addition, the economical housing, the price-limit housing, and the low-rent housing needed for the housing project for low-income urban residents were integral parts of real estate development. From these, it can be foreseen easily that the 4-trillion-Yuan (RMB) economic revitalization plan will directly be beneficial to the real estate industry. As a result, in the beginning of 2009, not only the economy heated up, but also the real estate market varied a lot, which was mainly reflected in the rapid growth rate of supply recovery, actual delivery of development and investment, prosperity of land purchasing, area under development, sources of development funds and so on.

I. China's Real Estate Investment and Supply Begin to Recover and Rebound

In 2009, along with the unveiling of the government's response policy to the financial crisis and determining of the confidence for bailout, the consumers' expectations were gradually improved. Driven by the massive governmental investment, the enterprises regained their vigor and vitality bit by bit. The market dealing and circulation re-flourished, and the quantity demand re-expanded. China's real estate industry came to a period of recovery. However, it must be pointed out that the recession and depression of the real estate industry this time lasted comparatively long, spreaded in a relatively wide range, and affected rather profoundly. For this reason, the recovery process of investment and supply will be difficult and slow. Moreover, the real estate products required a long construction period from the perception of the changes in the market to an increase in supply. Therefore, it took time to revive.

First, the growth rate of China's real estate market development and investment accomplishment underwent a quick increase. During the first half of 2009, China's real estate development investment increased month by month, and appeared a continued upward trend. The statistics from the National Bureau of Statistics shows that from January to July in 2009, the development investment accounted to 1.772 trillion Yuan(RMB), increasing by 11.6% over the same period of last year, and the growth rate rised by 1.7%, compared with the period from January to June. Among them, investment on commercial residential buildings took up 1.2427 trillion Yuan (RMB), increasing by 8.2% over the same period of last year. And the growth rate was 0.9% up compared with the period from January to June, which accounted for 70.1% of the investment in the real estate development. This was not only connected with the macroeconomic control of the government, but also on the account of the rebuilding of the consumers' confidence, the re-expansion of the demands, the rebounding of the turnover and the bright prospects of the real estate developers.

Second, the land purchasing activities began to boom. Along with the renaissance of the real estate market since February 2009, the land transaction market which has been in its downturn for many months started to heat up again. One reason is due to the market thrives; another is that the market inventory expired. The developers shifted their attention onto increasing the amount of land purchasing and hoisting the supply capacity during later period. (Table 5-3).

Table 5-3. Nationwide Real Estate Enterprises' Land Purchasing Area from January to July in 2009 (Unit: Hectare)

Time	Land purchasing area	Growth rate(year on year)
Jan—Feb2009	2288	-30%
Jan—Mar2009	4742	-40.1%
Jan—Apr2009	7266	-28.6%
Jan—May2009	9875	-28.6%
Jan—Jun2009	13644	-26.5%
Jan—Jul2009	16309	-25.8%

Sources: Yearly and quarterly data sheet, the web site of the National Bureau of Statistics, http://www.stats.gov.cn/tjsj/jidusj/

From table 3, since January 2009, the nationwide monthly land purchasing area has been undergoing a modest increase, demonstrating a monthly growth rate of more than 10%, among which, the percentage points were respectively 7.2%, 2.9% and 3.3% in March, April, and May. Moreover, the declining rate was lower over the same period of last year which indicated a constant monthly increase of land purchasing. In addition, as for the area of land developed, the absolute amount was higher compared with that of the latter half of Year 2008. Starting from December 2008 to March 2009, the area of land developed has been bigger than the land purchasing area all the while until April and May 2009, and the figure began to reverse. The total amount of land owned by development enterprises continuously keeping a modest increase that revealed a comparatively positive expectation to the market.

Third, the real estate development funds increased at a quicker pace. After the positive measures have been taken by the government since the second half of 2008, the situation of China's real estate market showed a reversed trend. Under the purchasing demands and bank credit's supports, new sources of development funds increased month by month. According to statistics, in 2009 from January to July, the real estate development fund source has increased by 28.7% over the same period of last year, and it was 6.8% higher than the development investment in the corresponding period. Among them, the growth rate of newly-added fund sources was 16.1%, and the year-end balance increased by 6%. (Table 5-4).

Table 5-4. Fund Sources for the Real Estate Development Enterprises from January to July in 2009 (Unit: a hundred million RMB)

Time	Fund sources	Growth rate(year on year)
Jan—Feb2009	6046	6.9%
Jan—Mar2009	10070	9.2%
Jan—Apr2009	13512	12.4%
Jan—May2009	17523	16.1%
Jan—Jun2009	23703	23.6%
Jan—Jul2009	28639	28.7%

Sources: Yearly and quarterly data, the web site of the National Bureau of Statistics, http://www.stats.gov.cn/tjsj/jidusj/

A substantial increase in funds for development, on one hand, accounted for the moderately loose monetary policy, the domestic loans' proportion in the real estate development funds sources gradually increases; on the other hand, subject to the increasing demands for house purchasing, the increase rates of deposit, advance payment and personal mortgage loan grow month by month. The statistics of the central bank shows that, by the end of June 2009, the balance of national commercial real estate loan was 6.21 trillion Yuan(RMB), increasing by 18.8% over the same period of last year. Among them, the balance of development loan was 2.35 trillion Yuan(RMB), increased by 20.5%; that of the purchasing loan was 3.86 trillion Yuan(RMB), up 17.8%.[10]

II. The Demands of China's Real Estate Industry Begins to Recover

Although in the second half of 2008, the demands and consumption of China's real estate industry have undergone an economic contraction and market setback, which was an inevitable response to the worldwide financial crisis. People's sensitivity to market changes has been gradually increased. However, along with the continuous introduction of real estate policies, especially China's achievements that first picked up from the worldwide economic contraction, consumers began to change their future expectation. After a short term of convergence, the long-standing rigid demand started to re-concentrate and promptly release. The accumulation of land purchasing area is a solid example.

---

[10] Sources: the People's Bank of China, August 5, 2009, Executive Report of China's Monetary Policy (the 2nd quarter of 2009).

In addition, the trading volume should be the main criterion for judging the prosperity of a market. The trading volume is not only influenced by the relation between supply and demand, but also closely related to market players' market expectations. The sharp contraction of the trading volume in the real estate market, which emerged in late 2008 and early 2009, was precisely the result of the worldwide financial crisis, coupled with the government's macroeconomic control that lead to an under-confidence of the real estate market. Since the first half of 2009, the market expectation began to improve and the market confidence has been rebuilt. With the affect of inflation expectation, the wish to preserve and increase assets value has made some investors to enter the market, so the volume keeps some stable growth. Therefore, under the stimulus of house purchasing policy and the comparatively low house price, the rigid demand which has been backlogged for nearly two years breaks out and brings a rapid increase in the market transaction volume. By the end of 2008 and early 2009, the contraction of real estate transaction volume began in Shenzhen and spread to the whole country. The pessimistic air is a strong blow to the confidence of both buyers and developers which directly leads the real estate market towards its floor. But now, affected by the government's positive measures, some consumers have rebuilt their market expectation and at the same time, their consuming desires have been aroused. The consuming demand has been increased and that is why the pre-suppressed rigid demands could be quickly released.

III. China's Real Estate Price Begins to Rise Rapidly

The changing price in the real estate market is an important indicator that shows the effects of Chinese government's financial and real estate policies. Namely, the real estate price fell and fluctuated at the bottom during the second half of 2008, and since April 2009, it began a new round of rising up. According to the statistics made by the National Development and Reform Commission and the National Bureau of Statistics, since March 2009, the sales price in 70 medium-sized cities has been running a four-month increase and the growth rate has continuously been enlarged. The average price of the whole country has hit its bottom and mildly risen up. Until June 2009, it finally skyrockets like the oil well's blowout. The statistics shows that the growth rate of the 70 medium-sized cities has reached 0.2% in June over the same period of last time, while 0.8% in July. During this current round of price-increasing, it appears a unique phenomenon, namely, the price of the second-hand houses has far surpassed the newly-built commercial houses, and the regional differences are significant. It is estimated that, ever since March 2009, the price growth rate of the second-hand houses was much higher than the newly-built houses, except for July, which made price parity. The scale of the two prices has been continuously widened. In June, the rate was even over 0.6%. (Table 5-5).

It could be observed that since the first half of 2009, the total transaction volume of commodity housing throughout the country has increased by a large rate over the same period of last year, and so is the same with cycle increment ratio. This is mainly induced by the reverse of the relation between supply and demand. The central government and local authorities have made some policies and measures to promote the healthy development of the real estate industry. Along with the introduction of these methods, residents' consuming confidence has been greatly enhanced, and the demands have been released. However, due to the market downturn of the previous period, most real estate enterprises determine to lower the market supply and reduce the numbers of the projects. Thus, the interaction between the rapid expanding demand and lagging supply leads to a nationwide increase in market volume and housing price. It should be noted that, beside rigid demands entering the market in a great quantity, the investment demands, even the speculative demands have entered the popular real estate market influenced by the expectation change of the future inflation, and in the end, push up the prices. Bank credit treats personal housing mortgage loans relatively moderate, especially in some cities, and the accessing standard has been lowered for people who own more than one apartment. This act will effectively back up the entering of the investment and speculative demands into the market in a large quantity. The statistics show that during the first half of 2009, the personal housing mortgage loans have amounted to 292.9 billion Yuan (RMB), increasing by 63.1% over the same period last year. All these factors mentioned above continuously promote the demands of the property market, and thus, form an increase of transaction volume and housing prices.

Table 5-5. Real Estate Sales Price Index of 70 Medium-sized Cities

Month	Real estate sales price index		Newly-built houses price index		Second-hand houses price index	
	Year on year	Link relative ratio	Year on year	Link relative ratio	Year on year	Link relative ratio
Mar2009	98.7	100.2	98.1	100.1	99.6	100.3
Apr2009	98.9	100.4	98.3	100.3	100.0	100.8
May2009	99.4	100.6	98.7	100.7	100.9	100.7
Jun2009	100.2	100.8	99.4	100.8	102.2	101.1
Jul2009	101.0	100.9	100.3	101.1	103.0	100.9

Sources: Yearly and quarterly data, the web site of the National Bureau of Statistics, http://www.stats.gov.cn/tjsj/jidusj/

SECTOR 4: PROSPECTS AND RECOMMENDATIONS FOR CHINA'S REAL ESTATE INDUSTRY

China's real estate industry is now in its picking-up phase. The future development depends upon not only the Chinese government's macro-economic growth and various reactions to the global financial crisis taken by other countries in the world, but also the degree of the fulfillment of China's real estate policies and the speed and direction of changes in price and market demands. In any case, the fact that China steps into the phase of economic recovery and revive ahead of other countries in the world is undeniable, and the future development of China's real estate industry entering another round of picking-up is quite optimistic. But the crucial point of promoting and affecting this process is to properly regulate the strength and opportunity. Thus, higher requirements are put forward for the state policy makers and managers.

Column 5-2: From 'Failed Estate Auction' to 'Recurrence of Land King'

Influenced by the global financial crisis in 2008, China's real estate market and land market gradually step into a phase of recession and depression. While the trading volume of real estate goes down, the phenomenon of "failed estate auction" continuously occurs.

According to the statistics, the country as a whole, in the first five months in 2008, there are 40 cases of failed estate auctions in major cities of China. In Xiamen, two occasions of concentrated transfer happened in the first half year. One was on February 9th, with all estate auctions failed; and the other was on April 10th, with four auctions failed. In Guangzhou, Nansha block failed in the auction in February; and among the three cases of land transfer in June, one of them failed, and one concluded a bargain after only one offer tender. In Fuzhou, two cases of land transfer failed in auction on May 12th due to the insufficient enrollment; on May 19th, two cases failed for the same reason even under the readjustment of the governmental land policy from 90/70 to 90/50. Besides, along with the continuous occurrence of the "failed estate auction", the central regions in major cities were out of market. In Beijing, No.15 Guangqu Road in Chaoyang District, which was known as the nearest golden region from CBD, failed in its auction. In Nanjing, No.4 land in the southern part of west-river district, which has formed a high-price area, reaching 1.77billion, failed in its auction in May.

However, one year later, affected by the governmental macroeconomic policy in response to the financial crisis, investment in fix assets and land supply were increased greatly in all parts of China. In this circumstance, the "failed estate auction" in 2008 never happened; in contrast, new land king came into stage. On June 30, 2009, No.15 land of Guangqumen in Beijing was won by Sinochem Franshion Properties Investment Management (Beijing) Ltd. at the price of 4.06 billion Yuan (RMB), becoming the newest "Land King" this year in Beijing. As it has been called "the last golden region near CBD", it attracted Vanke, Poly, SOHO China and other important developers in China, and presented us a fierce battle for land. In less than two minutes, driven up by the estate tycoons, the bidding for No.15 land of Guangqumen started from 2.112 billion Yuan (RMB) all the way to 3 billion Yuan (RMB). Therefore, the auctioneer had to reduce the bidding unit from 20 million Yuan(RMB) to 10 million Yuan(RMB). The competition at the auction scene was becoming increasingly fierce. Through a tense Muller, the price rose at the speed of one hundred million in every three minutes. No.18 Poly, No.10 SOHO China, No.15 Sino-Ocean Land, and No.28 Hui-Ming Estate refused to give in to each other. At 16:46, driven by the Sino-Ocean Land, the bidding increased from 3.27 billion directly to 3.3 billion Yuan (RMB). The audience burst into a uproar. Ten minutes later, when the price touched upon 3.67 billion, No.28 Hui-Ming estate quitted. At 17:00, Sino-Chem Franshion Properties, a dark horse, suddenly bid 3.69 billion. The competition was among No.18 Poly Estate, No.10 SOHO China and No3.Sinochem Franshion Properties. Until 17:10, SOHO China withdrew at 3.9 billion Yuan (RMB). Poly Estate and Franshion properties was all determined to win. And finally, Sinochem Franshion properties bid it at 4.06 billion. Beijing's new "Land King" was born. After 35 minutes of fierce bidding, the No.15 land of Guangqumen was sold by 4.06 billion Yuan (RMB) which broke the highest price records for "Land King" since the new auction policy implemented in Beijing.

(Source: *Beijing Times*, July 10, 2009).

I. The Real Estate Industry Will Step Into a New Round of Picking-up

Like the macro economy, the real estate industry takes on cyclic variation. In the beginning of 2009, China's real estate industry gradually steps out of recession and depression and enters a new round of recovery and prosperity. The reasons are as follows: first, the role of policy control. Since the fourth quarter of 2008, the Chinese government has reversed the direction of the former regulation and has introduced a large number of positive policies which would favorably stabilize the real estate developers' confidence and strengthen the consumers' expectations. Currently, the market rebound is only a preliminary effect of policy's guidance. As the 4 trillion's investment being put into its place, the investment effect will get an intensive and centralizing release during the third quarter and the last quarter of 2009, or even the first quarter of 2010 onwards. Second, it is the role of industrial linkage. During the period from 2008 to 2009, the introduction of China's economic stimulus plan, especially the unveiling of the stimulus package for ten sectors, all start from the global financial crisis and focus on the long-term industrial development strategy. Therefore, each sector should apply the linkage effect which would set the real estate industry in a continuously active motion. Third, the economy is in a recovery period. With the revival of real estate industry shouldering the burden of expanding domestic demands, driving investment demands and revitalizing the economy, the government would not introduce any suppressive measures that are too harsh before the sign of economic stabilization shows up, that insured a rapid pace of sustainable development of the real estate industry in the next two years. Fourth, China's urbanization lies in an accelerating phase. And this is the lasting impetus for the development of the real estate industry. China is still in the large-scale urbanization process. By the end of 2008, the urbanization rate in China is only 45%, which still lies far behind the developed countries. The rapid development of urbanization process will continuously generate strong demands for the real estate market. These demands are mainly embodied in the following aspects: first, a large number of rural residents move to the urban areas that require new houses to live; second, the urban redevelopment projects make a great many relocating households to become the consumers of the real estate market; third, as residents' consuming and investing quality has been risen up, higher requirements for improving the housing needs are put forward. Therefore, in the long-term run, China's real estate market is in a favorable developing tendency. Thus, compared with those countries which have sluggish real estate markets influenced by the credit crisis, china's performance shows its distinguishing features that depression lags behind and economy recovers fast. From recession to revive, it took less than one year, which can be hardly seen in any other countries. To sum up, China's real estate market has already accomplished its adjustment and begun a new era of picking-up.

II. Future Prices Will Go for a Steady Picking-up

The significant indicator of the picking-up phase in the real estate industry is the rise in both volume and price. And the reasons why the price is getting higher are as follows: first, the expectations for value-preserved purchasing and risk-preventing push the price-accelerating process. Since 2008, many countries have implemented expansionary monetary policies and greatly increased the money supply in response to the global financial crisis, but only result in excess liquidity and visible inflation. Although currently, the inflation hasn't occurred in China, it should be point out that the best way of value-preserving is property. Under the circumstance of the expansion of inflation expectation, for the sake of mitigating risks, consumers who hold off will upgrade their purchasing desire and infuse new vigor to the real estate market. The consequence of these actions leads to a more rapid increase of the real estate price ever since May 2009. For instance, the rising speed of real estate price in the major cities in China, like Beijing, Shanghai and Shenzhen, even exceeds the fastest rising period between 2007 and 2008. New record of price has continuously been set out. Second, the stability of real estate policies will contribute to the steadiness of the rising of real estate price. From the beginning of 2009, China's economy experiences a quick recovery: investment and consumption revive; the rising speed even exceeds that of 2007 and 2008. However, the phenomenon does not indicate that China's economy has stepped into a rapid and stable growth channel. It's our crucial time of economic recovery and the major task at present is still eliminating the negative impacts that brought by the financial crisis. Therefore, for some time in the future, the Chinese government would still implement positive fiscal policy and monetary policy. Thus, the real estate policy would undergo minute changes. A moderate rise of price would be an irresistible trend.

III. The Proportion of Policy-oriented Housing Continues to Be Enlarged

The financial crisis in 2008 not only brings about a slump of China's economy and real estate market, but also provides an opportunity for the government to adjust the market supply and demand to ease the social contradictions. In the ten measures of "the revitalization plan to expand domestic demand and promote economic growth", which was set down during the economic conference of the State Council in November 2008, to accelerate the construction of housing project for low-income urban residents was put in the first place: supporting the construction of low-rent housing more effectively; accelerating the reconstruction of the shanty town; building permanent settlements for roving herdsmen; and expanding the reform of buildings in poor condition in rural areas. First, strengthen the implementation of policy-oriented houses. It is not only the major purpose and task of housing system reform in our country, but also an effective way for a healthy development of China's real estate industry. At the same time, it is the best channel to stabilize the people's livelihood, adjust the industry structure and promote the standardized development of the real estate market. In the 4-trillion-Yuan (RMB) economic revitalization plan, the policy-oriented houses account for nearly 10%. In addition to funds-support, the government also speeds up the adjustment for residential land supply structure, cuts the proportion of common commercial land supply and increases policy-related house's land supply. Second, policy-oriented houses would contribute to stabilize the housing prices. The Chinese government has determined to speed up economic growth, stimulated consumption and promoted economic development by the means of increasing investment. So the policy-oriented houses, as fixed assets, would be one of the major targets that are striving to develop. Along with the policy-oriented houses accounting more in China's housing structure, it can be foreseen that, the effects of stabilizing the housing prices would be more impressive.

IV. The Regulation Rhythm and Intensity Should Be Paid More Attention to the Regulation for the Real Estate Market

The successful experience of managing with the real estate market in response to the financial crisis shows that the stable development of real estate industry would be beneficial for expanding domestic demands, encountering impacts from overseas financial crisis and enhancing economic development potential. It is worth to mention that, after a rational callback of housing price in 2008, the price immediately turns up a violent rebounding in 2009. In avoidance of the 2007-style irrational fluctuation and real estate speculation, and limiting the growth rate of price to a reasonable range, the government's attitude towards the real estate industry is transforming from the "comprehensive stimulus" to "partially concerned risk". Therefore, in spite of the fast starting speed and the significant impetus effect for economic recovery, the regulating rhythm and intensity should be controlled properly. First, a rational development of real estate market is a must for the healthy development of macro economy. The current real estate market could be a driven power for the recovery of iron and steel industry, cement industry and other infrastructural industries. But, the over heat of real estate would also bring about new rounds of investment which in turns would upgrade energy shortages, transportation shortages and inflation pressure. In addition, there would be an overwhelming of the cost-saving and short-term profit developing mode which consumes much, pollutes much, and outputs little. Second, the emphasis of preventing financial risks lies in the control of real estate investment growth. Currently, China's banking system takes the real estate investment risks caused by excessive dependence on bank loans, over large scale of investment, and over fast investment growth. Therefore, to control the scale of real estate investment, to alleviate the stress for the financial system, and to maintain a proper speed of growth and moderate scale of investment have become crucial aspects among the approaches of preventing financial risks in the national economy's revitalizing phases. Third, the sustainable development of real estate industry requires rhythm. Maintaining an appropriate scale of investment and regulating a proper market and developing rhythm relate to China's economic operation, the reserved rational developing space for the real estate market as well as the sustainable development of real estate market. Finally, the social stability requires a moderate growth of real estate prices. The real estate market has emerged many principal contradictions at the present stage in China, and the continuous rapid growth in housing prices is bound to break through the psychological defending line of the general public, leading to a great number of social injustice and social problems. So, while approving the positive role that the real estate industry has played, various measures to limit the excessive growth of the housing prices are bound to be taken at the aim of calming the public and stabilizing the society.

Translators: Wang Tianxue(SUFE) , Wang Cuiyun (SUFE)
English Version Editors: Jiang Jiahui(SUFE), Chang Zheng (Harvard)

References

[1] Lin Lin, 2009. Four Major Properties of Real Estate Market Operation Manifested in the First Five Month (Chinese) [J]. China Investment, 7.

[2] Zhao Guohua, Zhang Xiangjun, 2009. Price Movement in China's Real Estate Market under Financial Crisis (Chinese) [J]. Management and Administration, 7.

[3] Zhang Hong, Ma Jinjun, Kong Pei, 2007. *Real estate cycles based on dynamic econometrics model* (Chinese) [J]. Journal of Tsinghua University (Science and Technology) 12.

[4] Kaiser R, 1997. The Long Cycle in Real Estate[J]. Journal of Real Estate Research, 14, 233- 257.

[5] Grenadier S.R., 1995. The Persistence of Real Estate Cycles. Journal of Real Estate Finance and Economics, 10, 95-119.

[6] Xiao Yuanzhen, Wei Yikan, Chen Min, 2009. On Structure and political environments of China's Real Estate Market in the period of adjustment (Chinese) [J]. China Housing Facilities, 8.

[7] John M. Craig, Chen Haiqui, 2009. Research on Real Estate Market, Bubble Economy and Asian Financial Crisis (Chinese) [J]. Management Observer, 2.

[8] Wang Lina, 2009. Increasingly Evident Signs of Adjustment in Real Estate Market (Chinese) [J]. Review of Economic Research, 6.

In: China's New Deal
Editors: Li Xiaoxi and Hu Biliang

ISBN: 978-1-61470-815-5

Comments on Chapter 5

THE SUBPRIME CRISIS IN THE U.S. REAL ESTATE MARKET AND ITS IMPLICATIONS

Yuan Yan *

The chapter 5 described in detail and in depth the influence of financial crisis on China's economy. Provided that the policy and the measures taken by the Chinese government have been elaborated, it might be interesting to take a close look, from a different perspective, at the formation and development of the subprime crisis. We would like to explore its impact on the Asian and global economy in the hope that it might provide us with some insights.

Lehman Brothers, a company with a history of 158 years, filed for bankruptcy protection on September 15, 2008, indicating the worsening of subprime markets and break of the financial crisis. It was the largest bankruptcy case at that time in terms of value involved. Why could the subprime crisis in the real estate market extended to the financial institutions? Why did it escalate to the financial crisis with such a speed and spread to the global market in a short time? How did the subprime crisis happen? What has changed in China and the world ever since? What we could learn from it? The following sections will try to answer these questions.

SUBPRIME MORTGAGE IN THE US

The subprime mortgage refers to the mortgage loans to the borrowers who have lower income and whose credit ratings are imperfect. In strong contrast to traditional mortgage loans, the subprime mortgage has lower requirements on the borrowers' credit ratings and the ability to repay; therefore, it has much higher interest rates compared to the mortgage loan in general. In the United States, there are altogether three types of mortgage loans, namely, prime mortgage, alternative-A mortgage, and subprime mortgage. The standards of differentiating between prime mortgage and subprime mortgage are the credit ratings, debt to

* Yuan Yan, Ph.D. in Agricultural Economics, Texas A&M University, Associate Professor, Research Institute of Economics and Management, Southwestern University of Finance and Economics.

income (DTI) and loan to value (LTV). Generally speaking, a mortgage with DTI lower than 55% and LTV lower than 85% is a prime mortgage whereas a mortgage with DTI lower than 55% and LTV greater than 85% is categorized as a subprime loan. An alternative-A is in between prime and subprime mortgages, in which borrowers meet the requirements for prime mortgage but could not provide proof of income. The subprime mortgage is in a riskier range of loans. Once defaulted, delinquent or foreclosed, the issuing financial institutions may suffer a loss. Therefore, the subprime mortgage is an industry of higher risk and higher return.

Ever since early 2007, the US subprime market has experienced a series of events listed as follows. The subprime crisis became apparent in February 2007. The financial institutions have seen loss due to increased number of defaults. In April, the New Century Finance filed for bankruptcy. The following month, the funds of Dillon Read bankrupted because of its failure in the investment of the subprime mortgages. In June, Bear Stearns & Co informed investors in two of its funds, the High-Grade Structured Credit Strategies Enhanced Leverage Fund and the High-Grade Structured Credit Fund that it was halting redemptions. Problems in subprime mortgage impeded the operation of private equity and the investors worried about the credit crunch and the loan situation for most investment banks around the globe. The crisis spread then to the stock market. The investors' worries made the stock and bond markets to become volatile. The central banks in various countries were compelled to intervene in the market in September. From August 9 till the end of the month, the central banks injected 544.6 trillion dollars into the market, 343.4 trillion dollars by the European Central Bank and 147.3 trillion from the US Fed and 40 trillion from the Central Bank of Japan. In mid September, the Northern Rock Bank in the Great Britain experienced the first bank run in the past 140 years in British history. The Department of Treasury announced a plan to voluntarily and temporarily freeze the mortgages of a limited number of mortgage debtors holding adjustable rate mortgages (ARM) in December. In the meantime, the central banks of U.S., Europe, Britain, Canada and Swiss announced their injection in the market to the short-term lending to ameliorate the credit crunch.

ROOTS OF THE SUBPRIME CRISIS

The commentators attributed the break of the crisis to the low interest rate policy taken by the US Fed, the derailment of the financial derivatives from their real counterparts, the development of new financial instruments and the globalization.

When the real estate market is booming, the financial institutions tend to lower down payments, and to relax the restrictions on qualifications and the proof of income and property of borrowers, thus provided buyers with an easy and convenient way of obtaining mortgage loans. The financial institutions sometimes even ignore fraud by borrowers; when borrowers have difficulties in down payment, banks encourage them to apply for a second loan to pay for it. Thus, a large number of low-income families with imperfect credit ratings entered the real estate market. As estimated, the mortgage loan skyrocketed to 110 trillion dollars in 1st half of 2007 from 10 trillion dollars in 2005, over tenfold in two years. Critics in Wall Street Journal has addressed the problem and argued it reflected the loss of creditability of the US banks and rating agencies.

Financial innovations have brought an array of new financial derivatives, especially the securitization of mortgage loans. According to the Bank for International Settling, the species of derivatives in the world have risen to more than 1200 compared to only a few thirty years ago. By the end of June 2007, the total value of derivatives increased sharply to 51.6 trillion dollars. According to International Monetary Fund (IMF) Global Financial Stability Report, the value of the credit derivatives totaled 4000 trillion in 2003 and up to 26000 trillion in 2006. In the subprime market, banks used to keep the loans in the balance sheets but transferred to Collateralized Debt Obligations (CDO) in recent years and then sell them to investors. These investors create new derivatives secured with the CDOs, repack them and then sell them to other investors, so on and so forth. In recent years, 80% of the mortgage was sold through securitization and the value surged from 157.4 trillion dollars in 2004 to 549.3 trillion in 2006. This long debt link created during the process of derivatives leads the market risk to the risks of credit and liquidity and has brought instability in the financial market.

The speeding-up of financial globalization makes it natural for countries to be more interdependent on each other. The development of the financial derivatives lowers the risks of one single financial institution; however, it makes a chain interlocking a number of financial institutions all over the world. Any link that breaks in the chain will trigger the Domino effect in the global market. When the subprime market is characterized with prosperity and high return, it attracted many financial institutions in Europe and Japan, especially hedge funds to participate. Since CDO not only earns a profit and diversifies risks but also is backed by the US government thus it is perceived to have a higher creditability; foreign investors could not wait to participate in the market. This also lays potential risks for the later spread of the subprime crisis to the whole world. Besides, the difference of interest rates between U.S. and Japan makes it profitable for investors to arbitrage. They may finance a loan in Japan with low interest rates and invest in the securities in the subprime markets in the US to obtain substantial gains. Under the circumstances, banks in Germany, France, Canada, Australia and Japan and other banks and funds in Asia were immersed in trouble in varying degrees due to the break of the subprime crisis.

The collapse in the real estate market in the US and the related policy is thought to be an important factor for the crisis. After "9-11", the Fed decreased the interest rate twelve times in a row, the lowest rate ever since 1960. On the other hand, borrows are allowed to used fixed and flexible interest rates to repay the loan in the subprime market. Buyers repay the loan with low, fixed interest rates in the first several years of the purchase and with flexible interest rates in the later years. In the previous five years before 2006, the subprime mortgage market developed at an amazing speed with low interest rate and has gained sustained prosperity. The low interest rates and high housing price promoted the purchase heat in the United States, which gave birth to the securitization of mortgage loans. These tools accommodated financial convenience and short-term inexpensive loans to buyers, leading to the less restrictive credit ratings and qualifications.

From June 2004 to June 2006, the Fed started to move in the other direction. They raised interest rates 17 times consecutively, increasing it to 5.25% from 1%. In the meantime, the annual average growth rate of housing prices was 3.5%. These two factors deteriorated the situation rapidly in the real estate market. The buyers used to have net assets and got into deep debt then. Neither could they afford the interests and principle when due nor they could finance and obtain more loans from financial institutions. As a result, the default rate since 2006 has increased drastically in the subprime market. The borrowers could not repay the loan at the specified time, so banks took the houses but only to find the funds they could get were smaller than the sum of principle and interests and even the principle per se. Therefore, banks suffered huge losses, leading to higher volume of bad debts, which then caused the subprime crisis in turn.

IMPACTS ON CHINA AND WORLD ECONOMY

Under the intervention of western governments, especially central banks, the subprime crisis was once softened. In addition, the financial industry has also taken measures to avoid the deepening of the crisis. It has influenced greatly and will continue to influence the financial markets and economy in the globe. The global financial markets were then characterized with higher volatility under the influence of the crisis. A number of real estate financial institutions and funds that are closely associated with subprime market got into trouble or even broke. Major financial institutions in the western countries suffered serious losses and their debts rose by large degrees. The financial industry lost 300 to 400 trillion dollars due to the crisis. Every aspects of the international financial market experienced upheaval, and none of the stock markets, bonds markets, foreign exchange markets and commodity markets escaped. This caused the tightening of the credit markets in western countries and made the market from bad to worse, which prolongs the recovery of the real estate market in the US.

The subprime crisis slows the steps of the economic growth of the US and the world and it also has a great impact on China's economy, mainly exhibited in imports and exports sectors. China's exports in the 4th quarter of 2008 only increased by 4.3% compared to that of 2007 and imports decreased by 8.8%, the worst ever since 2002. The growth rate of China's trade with her major partners was all down, i.e., the US, the European Union, Japan and the Association of Southeast Asian Nations (ASEAN). The decline in exports of China slowed the economic growth and at the same time created high pressure in the labor markets.

Further, the crisis increases risks in China's exchange rate markets and capital markets as well. The US implements loose monetary policy and soft dollar to reduce the adverse effects of the crisis; however, it exposes China to huge exchange rate risks due to the devaluation of US dollars. The economic growth in China contrary to the economic depression in western countries in addition to the depreciation of US dollar and the appreciation expectation of Chinese yuan, international funds speeded their steps to get into China, which makes the prohibitive risks in the capital market.

MEASURES TAKEN BY CHINA

The subprime crisis has challenged the existing financial system in the United States; measures taken by the US government shed lights on the overseeing of the real estate and financial markets. How to manage risks, we think, is one of major issues we might learn from this overwhelming financial crisis. The financial industry was hit first and hard in the crisis; thus, paying attention to the potential risks behind mortgage loans should be a priority by commercial banks in China. Mortgage loans are the primary property for most commercial banks when the real estate market booms with higher yields and lower default rates. Even defaulted, banks can still auction the houses to compensate their costs. Mortgage loans occupy relatively high percentage in the assets of commercial banks in China and it is one of the major sources of income. However, once the prices of housing decrease and the interest rate increases, the default rate will increase by a large scale. Even the auctioned funds may not cover both interests and principle, which in turn causes sharp increase in bad debts. In the long run, the risks associated with mortgage loans should not be neglected and restrictive borrowing qualifications and audition procedures should be established. The financial crisis triggered by the subprime crisis initiates self-reflection of the financial regulation in the world. As described in Chapter 5 of the book, China governments have taken a series of macroeconomic policies to reduce the negative impacts of the financial crisis on the economy. These measures, including lowing interest rates, reducing property taxes and strengthening the regulation of lands, have been shown to be effective. Beware of the lessons of the crisis may help us avoid similar crisis happening in China.

In: China's New Deal
Editors: Li Xiaoxi and Hu Biliang

ISBN: 978-1-61470-815-5

Chapter 6

NEW ENERGY: NEW EXPLORATION ON ENERGY DEVELOPMENT IN CHINA

Lin Weibin**, *Li Xiaozhong, Rong Tingting and Song Tao

In recent years, as China's economy continues to grow rapidly, it also faces an increasingly severe problem of resource bottlenecks and environmental constraints. Due to this reason, the Chinese government has made a major strategic plan to build a "resource-saving" and "environment-friendly" society. Compared with traditional fossil energy, the new energy such as wind energy, solar energy and bio-energy have the advantages of abundant reserves, recyclability, cleanness and so on, which can fundamentally overcome the shortcomings of the traditional fossil energy. Traditional energy is limited in reserves and creates severe pollution; in comparison, new energy resources are more conducive to sustainable economic development. In response to imbalances between energy supply and demand and environmental pressures from social and economic development, China is stepping up to develop new energy and renewable energy, and actively exploring paths suitable for China's circumstances.

This chapter analyzes the new trend of the world's new energy development under the financial crisis and introduces the history of China's new energy development and its current situation. It also introduces public policies of promoting new energy development and the prospects of China's new energy development.

SECTION 1: RECENT DEVELOPMENT OF THE WORLD'S NEW ENERGY UNDER THE FINANCIAL CRISIS

Nowadays, the world's social and economic development is faced with severe energy constraints and environmental pressures. On one hand, the under-supply of the traditional energy such as coal, oil, and natural gas is increasingly severe. According to current energy consumption trends, the world's oil supply is enough for only about 40 years, natural gas 60

* Dr. Lin Weibin is lecturer at Institute of Economics and Resource Management of Beijing Normal University, who earned his PhD degree from School of Economics, Peking University. Email: lyland_lin@163.com

years or so, and coal resources about 120 years.[1] On the other hand, in the process of the traditional energy consumption, the large amount of greenhouse gas emission from carbon, sulfur compounds, nitrogen oxides, dust and so on greatly harms human health. New energy and renewable energy such as wind, solar, biomass, geothermal power and hydropower are considered to be the strategic choice to solve the problem of energy bottleneck and environmental constraints because of their abundant reserves, recyclability, cleanness and efficiency. Many countries boost investment in new energy and renewable energy in order to reduce their dependence on traditional energy and to enhance the sustainability of their economic development

In the context of the international financial crisis, the development of new energy will become many countries' crucial choice to turn adversity into opportunity. From a historical view, every major financial crisis will be accompanied by a new revolution of science and technology and brings about global industrial restructuring and profound changes in the global division of labor. Facing the current world financial crisis, all major countries around the world have taken action by introducing relevant policies and boosting investment in new energy to seize advantage in the scientific and technological progress and economic development.

I. Obama's New Policy and the New Energy Development in U.S.

Taking office during the financial crisis, U.S. President Barack Obama chose the development of new energy as one of his major policy measures to turn adversity into opportunity and to restore the U.S.'s economy. The short-term goal of his energy policy is to promote employment and stimulate economic recovery while the long-term goal is to reduce the dependence on foreign oil supply, get ahead in the field of new energy, and keep United States' leadership role in the world's economy. For this reason, the U.S. government issued a series of energy development plans, which includes the development of new energy and the construction of infrastructures that can be used to support alternative energy.[2]

Obama's administration accelerated the new energy decision-making and planning. Obama's core energy policy is to explore new energy and reduce the dependence on oil, which is mainly reflected in the following five aspects. First, great efforts are made to promote the development of clean energy. In the next 10 years, the government will make an annual investment of 15 billion U.S. dollars to promote the commercial development of alternative energy and energy-saving buildings and automobiles, while giving tax credits to companies engaged in the research of this field. Second, they emphasize environmental protection to cope with climate change. Through the implementation of the "Total Amount Control and Carbon Emission Trading" scheme, by 2020 greenhouse gas emission will be reduced to 1990's level and by 2050 emission will be reduced by another 80%, making the United States a leader in the field of climate change. Third, they try to achieve the diversification of energy resources. By 2010, 10% of the electricity consumption will come

---

[1] Source: BP World Energy Statistics Yearbook, June 2009

[2] Zhou Tao, November 19, 2008. New Energy of the United States under Economic Crises. China's Power Network.

from clean and new energy. By 2025 this proportion will reach to 25%. At the same time, they will develop nuclear energy under the premise of security. The fourth is to improve vehicle fuel efficiency. Economic standards of vehicle fuel will be improved by 4% annually. By 2011, they will begin to adopt new energy efficiency standard, promote the commercialization of plug-in hybrid vehicles and accelerate the introduction of low-carbon and non-fossil-fueled cars. The fifth is to achieve energy-saving, cost effectiveness and energy-efficiency. By 2020 the U.S.'s forecasted electricity demand by the Department of Energy should be reduced by 15%. The energy-saving programs include constructing energy-saving federal buildings, adjusting the current federal energy efficiency standards and implementing new stimulus packages to give public utility companies incentives to increase energy efficiency.[3]

U.S.'s new energy development is both an opportunity and a challenge for China. The current international environment will further promote international cooperation in new energy technologies and stimulate industrial restructuring and the development of new energy technology. In addition, the overflow of new energy technologies will gradually emerge, which will benefit China's new energy technology development. Barack Obama's energy policy for China is also a huge challenge. The implementation of Obama's new energy policy will increase China's pressure on emission reduction. Obama indicated that on one hand, the United States would actively join the international community's efforts to address challenges of climate change; on the other hand, he reiterated on many occasions that developing countries had to bear their corresponding responsibilities to reduce emission. China, as a developing country with the largest amount of greenhouse emission, will face the biggest pressure.

II. The EU's "Smart Investment" and Its New Energy Development

Facing the grim situation of the financial crisis, the European Union adopted an economic recovery plan named "Smart Investment" of 200 billion euro. This plan has three items directly related to environment protection and energy conservation: improvement on energy efficiency of buildings, promotion of rapid development of green products and development of clean energy for cars and buildings. This plan effectively promoted the development of low-carbon economy with emphasis on wind and solar energy. According European Wind Energy Association's report, the EU's wind energy development ranked first among all new energy source, accounting for 4% of the EU's electricity supply. At the same time, Europe alone accounted for more than 80% of the total solar power generation in the world. In April 2009, the EU announced its "Climate Action and Renewable Energy Package", which promised that by 2020 the EU would cut the greenhouse gas emissions by 20% or more from 1990's level and achieve the goal that the EU's renewable energy accounted for 20% of total energy cost.

The EU and its member states have played a dominant role in the international climate and environmental issues. The core of the EU's new actions on climate change is to comprehensively take climate change, energy and environment into account set specific new

[3] Qin Zhilai, 2009. the "Forum of Chinese Party and Government", Issue 4.

energy goals that mitigate climate change and link the objectives of domestic environment quality with climate change mitigation. After the outbreak of the financial crisis, the EU countries further increased government support for the new energy sectors in order to strengthen its already achieved comparative advantage. For example, Germany has adopted a new bill of greenhouse gas reduction, which would increase the utilization of renewable energy such as wind energy and solar energy from the current 14% to 20% by 2020. French Ministry of Environment published a scheme on November 17, 2008 aiming at renewable energy development. The French government hoped that the proportion of renewable energy consumption would substantially increase through a series of initiatives. The European Parliament approved the EU Energy and Climate Package on December 17, 2008 to ensure that the European Union's utilization of the new energy and renewable energy would reach 20% of the overall energy consumption by 2020. Almost simultaneous with Obama's new energy strategy, Germany, Spain and Denmark together launched the International Renewable Energy Agency (IRENA) in Bonn, Germany on January 26, 2009. This agency officially becomes the "New Spokesperson" of new energy whose purpose is to expand the utilization of new energy in industrial and developing countries all over the world. The organization aims to promote structural transformation of energy globally, increase new energy usage, and help developing countries acquire technology and build their own new energy industries.[4]

III. Japan's New Energy Development Plan

Amid the economic recession led by the financial crisis, the Japanese government, based on the relief methods of the U.S., is taking active measures to save Japan's key industries, and significantly reducing dependence on traditional energy by developing new energy under the consideration of long-term strategic needs.

The Japan Energy White Paper (2008 edition) points out the necessity to change Japan's energy supply structure from a petroleum-based energy system to non-fossil fuels such as solar energy, nuclear, etc. The White Paper also recommends the promotion of investment in resource-abundant countries to ensure the stability of access to energy and make full preparations for potential inflation of oil prices after global economy's recovery. The White Paper points out that although the economic recession causes declination of global oil demand in the short term, the supply shortage in emerging countries such as China and India may cause the oil prices to rise once again. For this reason, an economic structure that is less vulnerable to changes in oil prices and other existing energy must be established. From a long-term point of view, Japan should also actively promote development of energy conservation, new energy and nuclear power.

Japan's Ministry of International Trade and Industry (MITI) noted that in order to promote the development of solar power generation and other new energy, the related fiscal budget in 2009 increased by about 1,300 billion yen compared with that of 2008, which also grew by 50% compared to 2007. In addition, the budget increase also included government's support for enterprises' investment in energy-saving equipments. Japan also introduced subsidy measures for new energy. Japanese government announced the new objectives of

---

[4] Source: China Securities News , July 3, 2009.

renewable energy development on July 20, 2009 and introduced the new subsidy measures. Meanwhile financial subsidies were also increased to promote household solar power and geothermal power generation.

IV. New Energy Planning of Other Countries

Facing the severe financial crisis, many countries in the world are carrying out new energy development planning and developing new strategies energetically besides the three major economic entities: the United States, Japan and the European Union.

Renewable energy generation in Australia is expected to account for more than 20% of total energy generation by 2020. By mid-2009, the amendments of this bill is expected to be passed and within one year and half of the approval of the bill, 500 million Australian dollars of renewable energy fund will be available to encourage home installation of solar systems and investment in wind energy and solar energy. South Korea will invest 103 billion U.S. dollars for the development of new energy by 2030. India adopted a new energy security policy in December 26, 2008, one of its goals being the advocacy of clean and renewable energy.

SECTION 2: CHINA'S NEW ENERGY DEVELOPMENT STATUS

China is the second-largest countries in energy consumption in the world, but its energy efficiency is 10 percent lower than the world's average level and its energy use per unit of GDP is more than three times of the world's average level. China has also become one of the three largest carbon dioxide emitter sin the world. Can China's economy achieve sustainable development? Developing new energy will be the key measure to answer this problem. China's new energy development mainly experienced three phases. The first is the initial phase (1958-1980) which is characterized by the promotion of biogas technology. The second is the exploration and development phase (1981-2004) which aims to do follow-up research and exploration on hydropower, wind energy and solar energy. The third is a stage of scientific development from 2005 to the present. This section focuses on the current development status and technology of wind energy, solar energy, biomass, nuclear energy, hydropower, geothermal and ocean energy. In a strict sense, nuclear energy doesn't belong to new energy. However, nuclear power will be introduced as China's "new energy" in this section in view of its importance in China's energy structural adjustment and energy strategy in the future.

I. Wind Energy

The utilization of wind energy is mainly in wind power generation. The total reserves of wind energy above 10-meter altitude in China are 3.226 billion kilowatts, of which 253 million kilowatts can be actually exploited. The wind energy in China is mainly distributed in two major areas: the North and offshore areas of south-eastern coast. Specific distribution is shown in Figure 6-1.

Column 6-1: Wonders of the World in the Utilization of New Energy

1. Solar Power

The first solar power plant is Barstow No. 1. Solar Energy Power Generation Station built in Mojave Desert of California, the United States. Solar thermal power and high-temperature solar thermal power is also known as tower type generation of solar power. Several such power stations had been built in the United States, Japan, Europe, etc.

The world's largest solar power station is Barstow No. 1. Solar Energy Power Station built in Mojave Desert of California by the U.S. Department of Energy with a power of 10 MW, height 100 meters and heliostats 2 × 1818 pieces of 39.9 meters. Now the No. 2 Thermal Power Plant is being built near the No.1 Solar Energy Power Station whose power is also 10 megawatts. It is expected to be completed in 1996 with an investment of 48.5 million U.S. dollars.

2. Nuclear Power

The world's first nuclear power station was built in Russia in June 1954. It honorably retired without any serious problem in almost half a century's operation in April 30, 2002.

3. Wind Power

Denmark is the earliest country to use wind energy to generate electricity. The wind power is now commonly used. Although Denmark just has a population of about 5 million, it is a big country in wind power generation and windmill production round the world. There are 5 manufacturers of wind wheel in Denmark among 10, and more than 60% of the wind wheel manufacturers around the world's are using Danish technology. Denmark is a real "windmill power".

4. Marine Energy

The world's first pumped storage power station was Lurton Pumped Storage Power Station built in Switzerland in 1879.

The world's largest pumped-storage power station in terms of installation capacity was Bath County Pumped Storage Power Station built by the United States and on operation in 1985.

The world's first tide power station was built on the coast of the North Sea by Germany in 1913.

The world's largest tide power station is Lens Tide Power Station in France with an installation capacity of 240,000 kilowatts.

The French physicist Arsenal first proposed to use the differences of sea water to generate power. In 1926, Claude, a student of Arsenal, successfully approved this theory. In 1930, Claude built the world's first power station of sea water differences on the beach of Cuba.

Japan's Hamming testing vessel for wave power generation, which is built in 1978, is the world's first large-scale wave energy power station.

5. Hydropower

France built the world's first hydropower station in 1878.

The world's largest hydropower station in the 20th century is the Itapúa Hydropower Station built by Brazil and Paraguay jointly with an installation capacity of 12.6 million kilowatts.

At present the world's largest hydropower station is the Three Gorges Project started its construction in 1994 with an installation capacity of 17.86 million kilowatts. This project is expected to be accomplished in 2009.

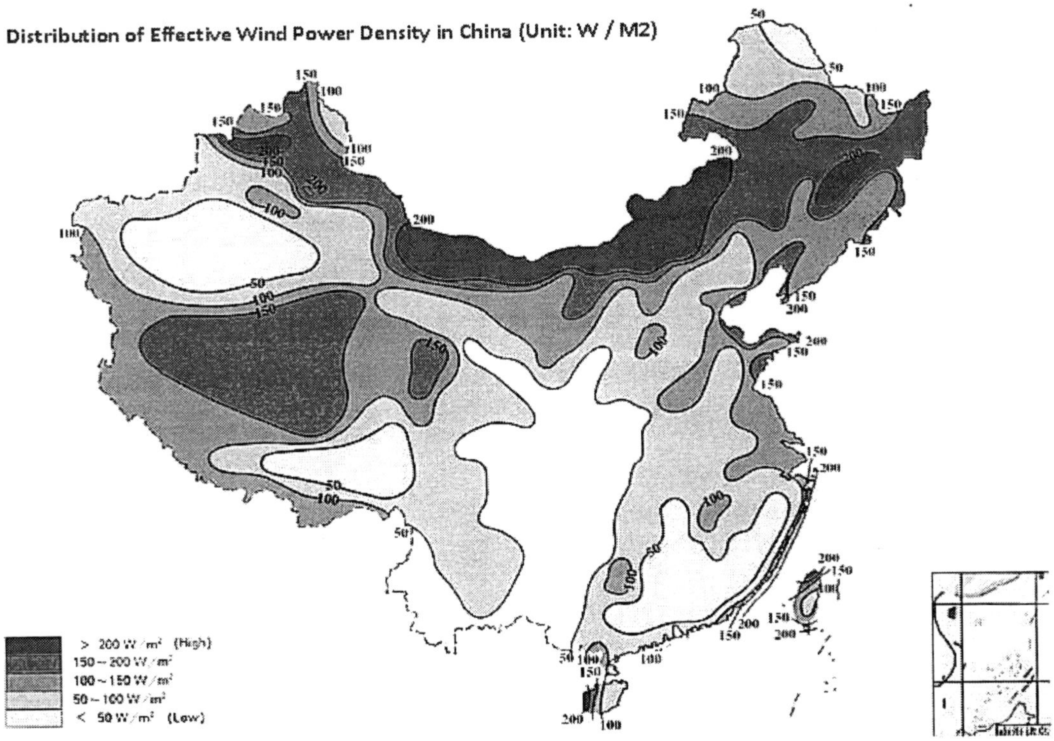

Source: http://www.xnyfd.com/kepu/html/?12367.html

Figure 6-1. Distribution of Effective Wind Power Density in China (Unit: W / M2).

China's wind power industry started in 1986 and it grew into full force in ten years. Especially since 2005, the international ranking of China's wind utilization rises year by year. The "Report on China's Wind Power Development – 2008" issued by National Energy Bureau shows that by the end of 2007 there were 20 provinces (autonomous regions and municipalities) that had developed their wind power fields (the total number of built wind power fields is 91). Installation of wind turbine units has reached 331 and the cumulative installation capacity has reached 6.05 million kilowatts, ranking fifth in the world.[5] Inner Mongolia ranked first in China with the total installation capacity of 1.59 million kilowatts.[6] By the beginning of 2009, there were 25 provinces, municipalities and autonomous regions in total having wind power installation in China. According to the latest study, by the end of June 2009 China's grid-connected wind power installation capacity will reach 11.81 million kilowatts, a growth rate of 101% compared with the same period in 2008. China's wind power installation capacity achieved "100% growth" for three consecutive years with its total installation capacity ranked fourth in the world (Table 6-1).[7]

Table 6-1. The Status of Wind Power Development in China since 2005

Year	2005	2006	2007	2008	2009
Total installation capacity (million kilowatts)	1.26	2.60	6.05	12.153	11.81[8]
World ranking	8	6	5	5	4

Source: Based on the sorting out of China's Development and Reformation Committee, China's Wind Energy Association website.

China's wind power relies on high-voltage and large-scale long-distance transmission. The instability of wind energy brings many problems in the system of peak load frequency modulation, power grid adaptability, voltage control, security and stability. China must solve these issues in its wind power development. China's wind power equipment can be basically self-produced except some core technology is still imported. Subsidies for cost of wind power is gradually resolved. By the end of July 2009, "Notice on Improving the Grid-Connected Price Policy on Wind Power" issued by China's Development and Reformation Committee divided the country into four categories of wind energy areas and nailed down the grid-connected price benchmark of wind power correspondingly. China also continued to implement wind power cost-sharing system. That is, the grid-connected price of wind power within the part of the local desulfurization coal-electricity price will be subsidized by provincial power grids, the higher parts will be covered through the national sharing of renewable energy surcharge on the electricity prices.

II. Solar Energy

China, located in the eastern part of northern Eurasia, belongs to the temperate and subtropical zone with abundant solar energy. In China, two-thirds of the land area has

[5] China's Wind Energy Association: http://www.cwea.org.cn/hynews/display_info.asp?cid=196

[6] Li Junfeng, Gao Hu, Wang Zhongying, Ma Lingjuan , Dong Luying, October, 2008. "Report on China's Wind Power Development 2008". Chinese Environment Press, 16-37

[7] Investment Consultants: http://www.ocn.com.cn/reports/2006005fenglifadian.htm

[8] The data refers to grid-connected installation capacity in the first half of 2009, "Shanghai Securities News", July 31, 2009.

sunshine of more than 2200 hours annually. The annual solar radiation in China is more than 5000 mega joules per square meter and the total solar radiation in a year is about 170 billion tons of standard coal equivalent, thus China's favorable conditions qualify for solar energy utilization. The distribution of China's solar energy is mainly concentrated in the west and the northern part, as shown in Figure 6-2. It is estimated that if the average area of solar water heater panel reaches 30 square meters per 100 people, the total usage of solar water heaters nationwide can reach 500 million square meters, which can save 65 million tons of standard coal.[9]

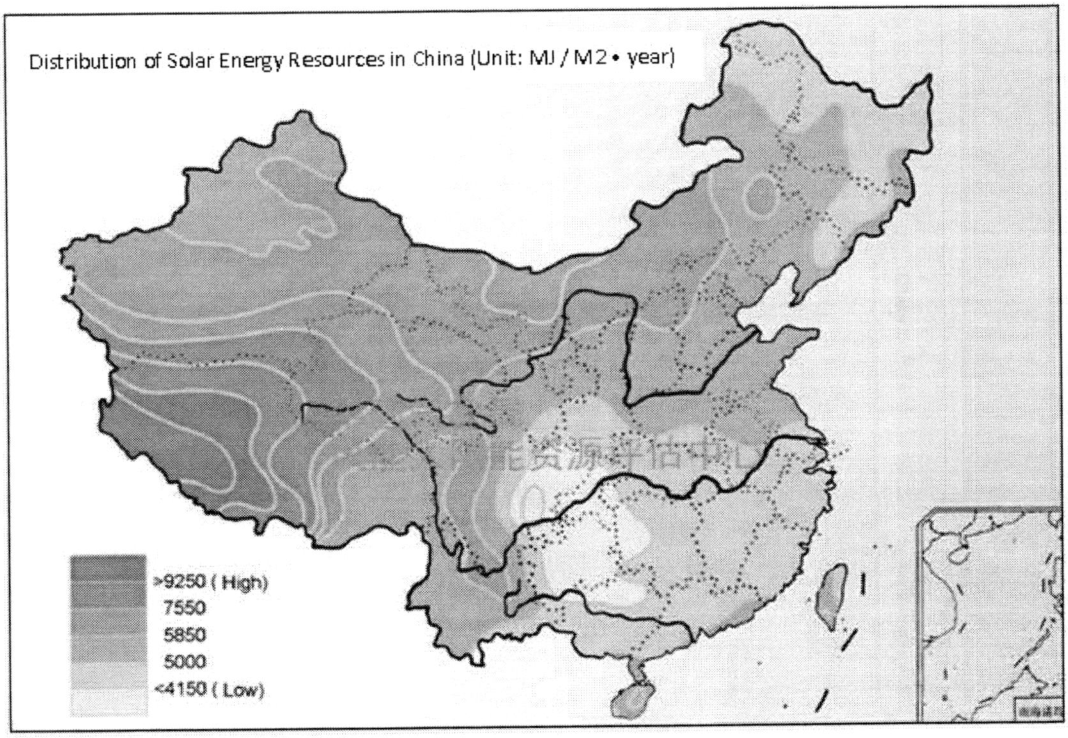

Source: China Meteorological Center for Wind and Solar Energy Resource Assessment.

Figure 6-2. Distribution of Solar Energy Resources in China (Unit: MJ / M2 • year).

China's utilization of solar energy is mainly by solar water heater. The technology of water heaters is relatively mature, which can be highly market-oriented and very suitable for China's conditions. China's solar water heater industry has developed rapidly with an annual growth rate of 20% or more since 2000. According to statistics, China's total production of solar energy was 154.8 million square meters in 2008, up 20% from 129 million square meters in 2007. Water heater products are exported to many countries in Asia, Europe and Africa.

[9] National Development and Reformation Committee, Energy Bureau, Renewable Energy and Countryside Power Office, April, 2004. Renewable Energy: The present situation and the prospect, "Venture Investment and High tech in China", 54-56.

In the aspect of manufacturing technology, China's solar water heater has formed its industrial system in which the raw material processing, product development and manufacturing, engineering design and marketing services can be accomplished independently. China's solar water heater as an emerging industry expands rapidly. China is leading the world in the aspect of its independent innovation on the heat pipe technology of vacuum tube, which is widely used in China with an annual output of more than 16 million square meters, accounting for over 90% of the world market of vacuum tube water heater. It is a global trend to integrate solar water heaters with buildings in the long run. China has just started in this regard and a number of real estate projects have been started with the construction of solar water heaters simultaneously.[10]

Table 6-2. the Utilization of Solar Water Heaters in China since 2003 (Unit: million square meters)

Year	2003	2004	2005	2006	2007	2008
Cumulative production	45	60	78	100	129	154.8
Growth rate (%)	---	33	30	28	29	20

Source: Compiled from "China's Energy Yearbook 2005/2006", "Report on China's Renewable Energy Industry Development" from China's Development and Reformation Committee, China's Renewable Energy website.

Photovoltaic (PV) power generation is mainly used in home systems and small-scale photovoltaic power stations in China, which mainly aim to solve power supply problems in remote villages without electricity. In addition, PV power generation is also utilized in public buildings and facilities such as roads, parks, and stations in many cities. In 2001, China's output of photovoltaic battery was only 3 megawatts. However, it had reached 2,000 megawatts by the end of 2007 with an increase of 600 times in six years, which made China the biggest producer in the world.[11] It is estimated that China's total applications of PV power system will grow to 40 million kilowatts by 2010 and 1.8 million kilowatts by 2020. The utilization of photovoltaic power generation in the rural and remote areas accounts for 46% of total usage, telecommunications and industry accounts for 37%, photovoltaic consumer products accounts for 12% and PV grid-connected systems accounts for 5%.[12]

The rapid development of photovoltaic power generation industry relies on the fast development of crystal silicon materials and the production capacity of solar battery. According to statistics, China's production capacity of crystal silicon was 71,500 kilowatts and China's mono-silicon production capacity was 63,500 kilowatts in 2004. The production capacity of crystal silicon and silicon chip soared to 200,000 kilowatts in 2005, a growth of 180% compared with the same period in 2004, in which single crystal silicon production capacity was 150,000 kilowatts.[13]

---

[10] Wang Zhongying, Li Junfeng, July 2009. "Report on Development of China's Renewable Energy Industry 2008", Chemical Industry Press, 33-67

[11] Nanfang Daily, August 28, 2009.

[12] Hong Kong Venture Capital Research Institute, 2008. "China Venture Investment Yearbook" Democracy and Nation-found Press, 539-541.

[13] China's Energy Yearbook Editorial Committee, August 2007. "China's Energy Yearbook 2005/2006". Science Press, 255-256.

Statistics also show that there are more than 10 poly-silicon enterprises, 60 silicon chip enterprises, 60 battery businesses and more than 330 component companies in operation in China's current PV industry. The production of solar battery in China accounts for 30% of the total production in the world .[14] Although China has already been capable of supplying equipment for the development of photovoltaic power industry, monitoring experiments and training personnel independently, it still faces some technical problems. The biggest problem is the purification technology of poly-silicon and mono-silicon, which has been relying heavily on Japanese and German technologies over a long period of time. Therefore, it is urgent to overcome the so-called bottleneck of "Technology Lease". [15] China's study on grid-connected photovoltaic power generation system is at the demonstration phase of research and all the grid-connected photovoltaic power station being built in China are low-voltage grid-client models. Therefore, China will focus on developing large and super large grid-connected photovoltaic power stations in the future.

III. Bio-energy

In theory, China's bio-energy production reaches 65 billion tons per year. If the average calorific value per kilogram is 15,000 kilojoules, China's theoretical bio-energy resources is 3.3 billion tons of standard coal equivalent, three times of the total energy consumption in China. However, the total amount of developable bio-energy resources is about 700 million tons of standard coal equivalent. With the commencement of "Conversion of Cropland to Forest and Pasture" project, afforested area is expanding and the potentially exploitable volume of bio-energy can reach 1000 million tons of standard coal equivalent. China's bio-energy mainly includes straw, firewood, agricultural production and livestock waste, municipal solid waste and industrial organic waste water and residues (Table 6-3).

Table 6-3. Volume and Potentially Exploitable Amount of China's Bio-energy Resources

Types of Bio-energy	Amount (million tons)	Potentially exploitable volume
Straw	700	350 million tons
Firewood	220	220 million tons
Industrial organic wastewater	2,500	11,000 million cubic meters
Agricultural and aquaculture waste	1,800	20,000 million cubic meters
Municipal solid waste	149	9,000 million cubic meters
Municipal wastewater	24,700	1,000 million cubic meters

Source: Lin Boqiang, September 2008. "Report on China's Energy Development 2008". China Finance and Economics Press, 331.

[14] "Economic Information Daily", August 24, 2009.
[15] "Economic Information Daily", August 26, 2009.

As shown in the table above, at present the amount of straw resources in China is more than 720 million tons, which is about 360 million tons of standard coal equivalent, of which 350 million tons can be used as energy. The annual average production of firewood from forestry, salve-cultural pruning and firewood forest is about 220 million tons, equaling to 128 million tons of standard coal. Industrial organic wastewater, agricultural and aqua cultural waste, and municipal solid waste and water waste can be developed each year amounting to 41 billion cubic meters.

Research and utilization of bio-energy technology are key projects of China's energy utilization. The central government arranged 8 billion Treasury bond accumulatively to support the construction of 8.23 million bio-gas in 73,000 villages from 2003 to 2007. By the end of the first half of 2009, the rural biogas utilization has reached 30.5 million households, with all kinds of agricultural waste biogas projects in place at 39500 locations.[16] China's bio-energy power generation got its start in 2006 with a total installation capacity of 1.284 million kilowatts, which is expected to reach 5.5 million kilowatts in 2010.[17]

IV. Nuclear Energy

Nuclear power does not belong to either renewable energy, or new energy. The reason that we explore nuclear energy in detail in this section is because of its importance in China's energy structure adjustment and China's energy strategy in the future. Nuclear power has the merits of being clean and economical; therefore, it is favored by industrial countries. Since the completion of China's first nuclear power station — Qinshan Nuclear Power Station in 1991, China's nuclear power has developed rapidly and now China already has a comprehensive system of nuclear industry. At present six nuclear power stations are in operation commercially with 11 generating units and a total capacity of 9.068 million kilowatts. Meanwhile eight generating units with 7.9 million kilowatts capacity are under construction.[18] China's nuclear power development status is show in Table 6-4.

Table 6-4. China's Nuclear Power Development Status since 2001

Year	2001	2002	2003	2004	2005	2006	2007	2008	2009
Installation capacity (million kilowatts)	2.10	4.47	6.19	6.84	6.85	6.85	8.85	8.85	9.07 [19]
Generated energy (billion kwh)	17.5	26.5	43.9	50.1	53.1	54.3	62.6	68.4	---

[16] "People's Daily", September 3, 2009, 2.

[17] "Economic Information Daily", August 26, 2009

[18] By the end of July 2009, Source: State Power Information Network, The latest results of install ion capacity of nuclear power approved.

[19] ditto

Source: Based on the sorting out of National Power Information Network, "Report on China's Energy Development 2008" (Lin Boqiang, September 2008. China Finance and Economics Press, 284-292).

According to China's "Medium and Long-term Nuclear Power Development Plan", in order to protect the security of energy supply and to optimize the power structure, it is planed that the installation capacity of nuclear power should reach 40 million kilowatts and nuclear power generating capacity will reach 260-280 billion kwh per year by 2010. These figures are calculated after considering China's overall technical strength, the construction cycle, independence of equipment manufacturing and supply conditions of nuclear fuel. The current nuclear power capacities both in operation and under construction are 16.968 million kilowatts. On this basis, the new installation capacity of nuclear power is about 23 million kilowatts. At the same time, considering the follow-up development of nuclear power, the capacity of nuclear power under construction will be about 18 million kilowatts in late 2020 (Table 6-5).

Table 6-5 .Schedule of China's Nuclear Power Construction (Unit: million kilowatts)

Planning period	New operation scale	Production scale	Carry-over next five years' scales	The overall scale of the end of the period
Before 2000	---	---	---	2.268
2001-2005	3.46	4.68	5.58	6.948
2006-2010	12.44	5.58	12.44	12.528
2011-2015	20.00	12.44	20.00	24.968
2016-2020	18.00	20.00	18.00	44.968

Note: Since the unit capacity may change, the nuclear power capacity may change actually when it is finished.

Source: Based on the sorting out of "Planning of China's Medium and Long-term Nuclear Development".

In the aspect of engineering design, China already has the capacity to independently design and construct pressurized water reactor nuclear power stations of 300,000 kW and 600,000 kW . In the course of nuclear power development in the future, China will achieve the ability of designing, constructing and operating pressurized water reactor nuclear power station of million kW. It will also establish operation and management model geared to international standards and form a relatively comprehensive industrial system of nuclear power. As for the performance of operation and nuclear safety, the nuclear power stations in China are performing well and the operation is continuously improved. The main performance indicators has matched the advanced level of World Association of Nuclear Operation (WANO). Technical route of China's nuclear power development has been clear. The present goal is to develop thermal neutron reactor nuclear power stations and the long-term goal is to develop fusion reactor nuclear power stations.

V. Hydropower

China is a country rich in hydropower resources. According to the general survey result of hydropower resources of China in 2003, the technical developable installation capacity of national hydropower resources was 540 million kilowatts, and the annually generated energy was 2.47 trillion kilowatt-hours; the economical potentially developable installation capacity of national hydropower resources was 400 million kilowatts, and the annually generated energy was 1.75 trillion kilowatt-hours. About 70% of hydropower resources are distributed in southwest China, mainly in the Yangtze River valley, the Yellow River valley, and the Pearl River valley, with total installation capacity accounting for about 60% of the economic potential installation capacity of national hydropower resources. The utilization of China's hydropower is divided into two parts: one is the traditional medium-sized hydropower stations,[20] mainly referring to China's 12 hydropower bases (see Figure 6-3); the other is a real new energy source — small hydropower, which has potential developable amount about 87 million kilowatts, accounting for 23% of the potential development amount of national hydropower resources and ranking the first in the world.[21]

Hydropower development has always been emphasized by the Chinese government. Since the 1990s, China's hydropower output has kept increasing. The generated energy in 2006 and 2007 amounted to 435.79 billion kWh and 485.26 billion kWh, but its proportion in total electricity production in that year has been falling, from 20.4% in 1990 to less than 15% in 2007, as shown in Table 6-6. It is expected that by 2020 China's installation capacity of hydropower will reach 300 million kilowatts, with small hydropower reaching 70-80 million kilowatts.[22]

Table 6-6. China's Hydropower Production Capacity and Its Proportion of Total Output of Electric Power (unit: billion kilowatt hours ;%)

Year	1990	1995	2000	2005	2006	2007
Hydropower production capacity	126.72	190.58	222.41	397.02	435.79	485.26
Hydropower's proportion of total output of electric power in that year	20.4	18.9	16.4	15.9	15.2	14.8

Source: Based on the calculation through National Bureau of Statistics Web site.

At present, the proportion of developed hydropower in China has reached about 27%, among which the traditional medium-sized hydropower resources account for more than 60%

[20] The classified criteria of China's hydropower stations: large hydropower stations' installed capacity is more than 250,000 kilowatts; medium-sized hydropower station's installed capacity is 25,000 to 250,000 kilowatts; small hydropower station's installed capacity is less than 25,000 kilowatts.

[21] Cui Minxuan, 2008. China's Energy Development Report 2008 [R]. Social Sciences Academic Press, 288.

[22] Speech of Han Wenke, Director of the National Development and Reform Commission Energy Research Institute, at the Second International Symposium on China's Distributed Energy on April 20th, 2006.

of the total hydropower resources. There is still much room for hydropower development. Currently there are 12 hydropower projects under construction with installation capacity of more than 15 million kilowatts (Table 6-7).

Photo Source: http://www.xnyfd.com

Figure 6-3. Distribution of China's Hydropower Bases.

The Chinese government has always valued the development and utilization of small hydropower. At present, there are more than 1600 countries developing the small hydropower all over the country, and nearly half of these countries' power supply is primarily from small hydropower. In 2006, hydropower installation capacity in China's rural area reached 6 million kilowatts, and the total installation capacity amounted to 50 million kilowatts, approximately accounting for 37% of the aggregate national hydropower installation capacity. The annually generated energy amounted to 150 billion kWh.[23] China's hydroelectric equipment manufacturing develops rapidly, and dam construction and hydroelectric power station construction are leading the world. The Three Gorges crew successfully achieved "the introduction, the digestion, the absorption, and the re-innovation" of technologies. Now they could almost undertake any large and middle scale hydroelectric power station's design,

[23] Speech of Han Wenke, Director of the National Development and Reform Commission Energy Research Institute, at the Second International Symposium on China's Distributed Energy on April 20th, 2006.

SECTION 3: CHINA'S NEW ENERGY DEVELOPMENT POLICIES

In order to promote the development of new energy and the renewable energy in China and guarantee the sustainable development of economy and society, in 2005, the Chinese government passed "The People's Republic of China Renewable Energy Law" which is a milestone in China's energy history. Since then, the Chinese government formulated a series of policies on new energy and renewable energy, including "Medium and Long-term Planning of Renewable Energy Development" in 2007 and policies aimed at the major fields of new energy. In order to cope with the global financial crisis of 2008, the Chinese government has made corresponding adjustments in the new energy policy.

I. "The People's Republic of China Renewable Energy Law"

China is facing prominent problems such as energy security, environment protection, and the traditional coal dependent energy structure. Along with the economic development, China's energy demand keeps increasing. Developing renewable energy can increase energy supply, optimize the energy structure, and promote the country's sustainable development. Specifically speaking, before The Reform and Opening Up Policy, there was no systematic policy to develop new energy. In order to adapt to the establishment of the socialist economic market system as well as the requirement of implementing the sustainable development, the government has made development plans of renewable energy, through fiscal allocation and project subsidy to support the research and development of renewable energy; and through low interest loans and tax credit to promote the renewable energy's industrialization. Aiming to urge the development of the renewable energy industrialization, "The People's Republic of China Renewable Energy Law" was passed officially in February 2005.

"Renewable Energy Law" primarily includes five aspects: set the overall goals and medium-long term goals of renewable energy; prioritize science and technology research of renewable energy development and utilization in the technology development and the high-tech industry development; arrange financial support for the renewable energy research, demonstration and industrialization to promote the technological advancement of renewable energy; encourages and supports the renewable energy development and utilization in rural area - the energy department of local government above county-level, joint with other department concerned, should make plans of renewable energy development in rural area, promote the transformation and utilization of bio-energy such as methane, solar energy, wind energy and hydropower energy in small scale and provides technology and financial support according to local economical and social development, zoology protection and the need of controlling the sanitation comprehensively; China's renewable energy price administration and cost allocation system; set up special fund and support policies on tax and credit to promote the science and technology research of renewable energy development and utilization, establish criterion and demonstration project.

"Renewable Energy Law" has played a significant role in China's new energy and renewable energy industrialization development. But because the accompanying policy of "Renewable Energy Law" is not perfect, legal execution and effective implementation also need to be further coordinated and optimized. There is no doubt that "Renewable Energy

Law" has provided legal and institutional safeguard for China's new energy and renewable energy development and utilization, and has brought new opportunities for China's new energy industry.

II. "Medium and Long-term Planning of Renewable Energy Development"

The Chinese government proposed explicitly in "Summary of National Economic and Social Development of the 11th Five Year Plan": "Implement the preferential finance, taxation, investment and compulsory market share policy, encourage people to produce and consume renewable energy, and enhance its proportion in energy consumption." In order to achieve this goal, based on the summary of China's renewable energy's industrial situation and drawing on the international experience of new energy and renewable energy development, "Medium and Long-term Planning of Renewable Energy Development" was published in September 2007, proposing the goals, key fields and safeguard mechanisms of China's renewable energy development from 2007 to 2020.

"Medium and Long-term Planning of Renewable Energy Development" involves many key fields of renewable energy such as hydropower, bio-energy, wind power, solar energy, and so on. (1) The hydropower: by 2010, the national hydropower installation capacity will amount to 190 million kilowatts, among which large and middle-scaled hydropower will be 140 million kilowatts, and small hydropower will be 50 million kilowatts; by 2020, the national hydropower installation capacity will amount to 300 million kilowatts, among which large and middle-scaled hydropower will be 225 million kilowatts, and small hydropower will be 75 million kilowatts. (2) The bio-energy: by 2010, the total installation capacity of bio-energy generating electricity will amount to 5.5 million kilowatts, among which the annual utilization amount of solid bio-energy fuel will be 1 million tons, the annual utilization amount of methane will be 19 billion cubic meters, adding the annual utilization amount of the non-grain raw material fuel ethanol 2 million tons, and the annual utilization amount of biological diesel oil will be 200 thousand tons; by 2020, the total installation capacity of bio-energy generating electricity will amount to 30 million kilowatts, among which the annual amount of solid bio-energy fuel utilization will be 50 million tons, the annual utilization amount of methane will be 44 billion cubic meters, the annual amount of biological fuel utilization ethanol 10 million tons, and the annual utilization amount of biological diesel oil will be 2 million tons. (3) The wind power: by 2010, the national wind power installation capacity will amount to 5 million kilowatts. The key point is to construct about 30 large-scale wind power projects with the level of 100 thousand kilowatts in the eastern coast and "Three North" areas, forming 3 wind power bases with the level of one million kilowatts in Jiangsu, Hebei, and Inner Mongolian and completing 1 or 2 offshore wind power pilot programs with the level of 100 thousand kilowatts. (4) The solar energy: by 2010, the solar energy's aggregate generation capacity will amount to 300 thousand kilowatts, and by 2020, it will amount to 1.8 million kilowatts. (5) The geothermal energy aspect: by 2010, the annual amount of geothermal energy utilization will achieve 4,000,000 tons standard coal; by 2020, the annual amount of geothermal energy utilization will achieve 12 million tons of standard coals and the tide power plant with 100 thousand kilowatts will be completed.

The formulation of "Medium and Long-term Planning of Renewable Energy Development" sets up the goal for China to carry out the Scientific Outlook on Development, to construct economical society and to realize the sustainable development. It is an important action of China to protect ecological environment, to optimize energy structure, and to cope with the climate change; it is also important for China to build a new socialist countryside, to solve the problems of electricity and energy supply for countryside residents according to local situations, to raise agricultural benefit, to increase farmers' income, to improve rural environment, and to promote economic and social sustainable development of the countryside. However, it has not yet achieved the anticipated effect, so the implementation needs to be strengthened in the future.

III. The Policies in New Energy's Key Fields

Besides the above-mentioned “The People's Republic of China Renewable Energy Law” and “Medium and Long-term Planning of Renewable Energy Development”, the Chinese government has implemented the corresponding promotion policies of new energy's key fields such as the wind energy, the solar energy, the nuclear power, etc. The main contents of these policies are shown in the table 6-8.

Table 6-8. The Policies' Data Sheet of New Energy's Key Fields

Field	Policy name	Policy content
wind energy	“Notice on Perfecting the Policy of Wind Power Generation's Online Price”	Regulate the administration of wind power price and continue to carry out the wind power price cost allocation system. Implement the bench-mark grid-connected price policy.
solar energy	“The Implementation Opinion on Speeding up the Photoelectric Construction Application of Solar Energy ”	Promote the photoelectric construction application; adopt the way of demonstration project in the initial development period; the nation brings the financial funds policy into play; implement “the solar energy roof plan”; speed up the photoelectric popularities and application.
nuclear energy	“Medium and Long-term Planning of Nuclear Energy Development (2005-2020)”	The development target is set explicitly that by 2020 China's nuclear power installed capacity in operation will amount to 10,000 kilowatts and nuclear power installed capacity under construction will amount to 18 million kilowatts. The proportion of the nuclear power accounting for the total electric power installed capacity will increase from less than 2% to 4%. Further strengthen the development of coastal nuclear power, and scientifically plan the nuclear power construction in inland area. According to the estimation based on the nuclear power installed equipment's recently proportion change, in 2020 the nuclear power's total installed capacity will possibly increase to 70 million kilowatts from 40 million kilowatts proposed in the past.

Source: Based on the sorting out of State Development and Reformation Commission website.

As the table shows, Chinese government will keep strengthening support on the new energy development policies. In the future, in the field of wind energy, China's main focus is still wind power generation and price reform; in the field of solar energy, the main focus is enhancing the heating capacity; in the field of nuclear power, the main focus is large-scale construction and application of the nuclear power.

IV. China's New Energy Policy Trend under the Global Financial Crisis

Confronted with the financial crisis, the government has made the corresponding adjustments to the new energy development policy.[30] It is mainly manifested in the following aspects:

1. Develop low-carbon economy. The industries that involve the low-carbon economy are very broad, mainly including the development and utilization of the low-carbon products, the low-carbon technology, and the low-carbon energy. Technically, the low-carbon economy involves many industries such as electricity, transportation, building, metallurgy, chemical industry, petrochemical and so on, as well as the new technologies of controlling greenhouse gas emission effectively in renewable energy and new energy. In August 2009, Wen Jiabao held the State Council Routine Conference in which "Planning the Environmental Effect Appraisal Rule (Draft)" was discussed and passed in principle, with the aim to comprehensively implement national scheme to deal with the climatic change. There are 6 key fields in the next stage of work, and the low-carbon economy will become a new point of economic growth. Since 2008, the global financial crisis greatly influenced China's economy, especially in export. The development of low-carbon economy has played a crucial role in adjusting the internal structure China's economy. China's traditional export products are high-energy consumption products, but now the main attention must be paid to restructuring the export products; otherwise, China's economy structure problem cannot be solved.
2. Strengthen the development of the nuclear power. In the nuclear power aspect, along with China's promotion of ten measures aiming to expand domestic demand and promote the economic growth, the nuclear power projects become one of the first approved projects to stimulate domestic demand. The peak of the nuclear power construction is coming. For instance, in 2009, China Guangdong Nuclear Power Group, the only energy enterprise focusing mainly on the nuclear power in China, undertakes 70% of nuclear power unit construction task of China's approved nuclear power unit construction task, ranking the first in the world right now. In sum, it will start 6 construction tasks in the 2nd phase of Ling'ao nuclear power station, Liaoning Hongyan River nuclear power station, Fujian Ningde nuclear power station, Yangjiang nuclear power station, Taishan nuclear power station, Fangchenggang nuclear power station and Hubei Xianning nuclear power project.

[30] Chenwei&Jiangwei, August 11, 2009. New Energy Industry: The Strategic Commanding Point of New Round International Competition. "Economical Reference Newspaper".

3. Deal with the surplus energy problem of new energy industry positively. China's new energy industry has begun to take shape and has great development momentum. But some new energy industries have already shown surplus.[31] In August, 2009, it was pointed out in the State Council Routine Conference that the emerging industries such as polycrystalline silicon industry and wind power equipment industry are listed in the "key strengthening instruction" industries due to their surplus energy and duplication problem. At present, China has already made the adjustment plan for industrial development. Generally speaking, China is rich in new energy resources and the new energy development is very promising. Developing new energy such as wind power, solar energy, nuclear power is not only a way to deal with the global financial crisis, but also a way for China to seize a leading position in the new round of global economic development.

SECTION 4: THE PROSPECT FOR NEW ENERGY IN CHINA

I. The Existing Problem of China's New Energy Development

Along with the perfection of related laws and regulations and the continual reinforcement of policy support, China's new energy industry layout will be gradually clear, and there will be more new energy development opportunities. However, at present, China's new energy industrialization development is also facing many restrictions such as lack of technological innovation, the lag of energy system reformation, and difficulties in financing.

1. New Energy Enterprise Lacks Independent Innovation Technology

The technology constraint is an essential factor for the slow development of China's new energy. Compared with developed countries, China's new energy utilization started later, and the average level of new energy utilization technology is lower. At present, the majority of core technologies and equipment in China's new energy depends on import and the proportion of domestic technology and equipment is still low, leading to high cost in the new energy industry. Take the wind power for example. At present, the overall equipment manufacturing capacity in the wind power industry is not high, and China's wind power manufacturing enterprises have not completely mastered the core technologies of core parts such as turbine blades, shift gear cases and generators, etc. [32] The technological inferiority becomes a great factor to restrict the development and expansion of China's new energy industry. If China wants to keep developing its new energy industry, independent innovation in technologies must be carried on continuously.

[31] Liu Qi, August 28, 2009. The new energy development just started and particular industry presented overheating. The Economic Observer Net.

[32] Song Qinghua, June 1st, 2009. New Energy Crisis. "Elite".

2. The Management System of New Energy Development and Utilization is Imperfect

First, the existing energy management system is imperfect. For instance, China has two kinds of electricity pricing mechanisms now, which cause confusion in grid- connected new energy prices, twist the price signals, and have affected the entire plan and inspection order. At the same time, under the existing management system, there are too many proposed new energy projects and the examination and approval procedure is too lengthy and complex. The responsibilities, rights and benefits of the electricity grid companies to develop new energy are also unclear, lacking incentives for electricity grid companies to develop new energy.[33]

Next, the new energy's subsidiary service mechanism is imperfect. In many places, solar cells are abandoned because local people do not know how to switch the storage battery fluid, and it's a common phenomenon that the small wind power generation equipment stops working because nobody knows how to repair the turbine.[34] Therefore the new energy enterprises should improve subsidiary service work. Only in this way could new energy enterprises become truly substantial and strong.

Finally, due to the lack of national overall arrangement, some industries have already presented the overheated phenomenon. At present, there is a serious phenomenon in China's new energy industry called "following up". The industrial development still has much blindness. Moreover, the new energy stock is overheated: massive funds enter the new energy domain continuously, and the market's non-rational ingredient inflates day by day, increasing the investment risk in new energy industry. Therefore, it is quite necessary to watch out and prevent overheating in China's new energy development.[35]

3. The Financing Channel of New Energy Development is Insufficient

The development of new energy industry primarily depends on small and medium-sized enterprises, and the risk is relatively higher for these companies than others. Therefore it is a quite common phenomenon that the enterprises have financing difficulties. The banks focus on the loan risk management and they do not want to lend money to small and medium-sized enterprises with small scale and gloomy prospect. They are cautious toward medium and small scaled new energy enterprises. China's venture capital or private equity investment pay much attention in new energy field. However, the exit channel is insufficient because it restricts the incentives of venture capital and private equity investment and affects the stable development of new energy industry.

4. The Market Is Restricted by Traditional Energy Price

The increasing international crude oil price results in a craze of new energy development and utilization. The quick development of China's new energy was closely related to the increasing oil price. However, affected by financial crisis, the oil price fell down sharply and brought challenges to the new energy development.

---

[33] Zhang Yongwei, August 31st, 2009. Straightening out Management System while Developing New Energy. "China Economic Times".

[34] Zhanglei, August 21, 2009. The Opportunity and Challenge of China's New Energy Industry. China's Energy Resources net.

[35] Zhou Xuesong, July 30, 2009. The Necessary Prevention from New Energy Foam. "China Economic Times".

Due to technological constraints, the price advantage of China's new energy product does not exist or is not remarkable. For the ideal consumer, they will not opt out from the traditional energy to choose new energy with higher price. Although the financial crisis brought a temporary drop in petroleum prices, they will remain to be high in the long run. At present the majority of energy economists believes that only when the international oil price continues to be more than 200 dollars each barrel, can this new energy technology have widespread market space to develop.

After this financial crisis, all fields of new energy industry face big problems. On one hand, the economic crisis caused global economy's decline which reduced the total demand of energy sources. On the other hand, due to the sharp drop of petroleum price caused by financial crisis, traditional energy's price advantage emerged, which had substitution effect for new energy. Around the world, demand for new energy is declining rapidly. The market is in favor of traditional energy, not the new energy with higher price. Generally speaking, this financial crisis is a great challenge for the demand of new energy.

II. The Development Prospect of China's New Energy

China's new energy industrialization development started later than most of developed countries, its technology is relatively behind and the overall industrialization level is not high. The outbreak of the financial crisis affected the energy industry seriously, but it actually provided a turning point for the development of new energy. Along with the expansion of industrial scale, the cost of new energy is decreasing Cost reduction, good policy and market environment cause the new energy to show the scale effect, which makes it more competitive than traditional energy. Moreover, in the long run, along with continuous innovation and improvement of technology, the cost advantage of China's new energy will emerge gradually compared with the conventional energy source. In this way, China's new energy product will have better competitiveness in the market.

China has rich resources to develop the renewable energy and owns certain industrial foundation. From the resources, technology and the industry angle, it has great development potential in future. The government has paid much attention to the renewable energy sources development. According to China's development targets, in 2020 electricity generation from renewable energy will be over 15% (as shown in the table 6-9), and after 2040, the proportion can reach 30% or more. The renewable energy are expected to become an important alternative energy. After recent years' development, the role of renewable energy has already emerged in China's energy supply. The following 5 to10 years will be the key phase for the development of wind power, photovoltaic generation and bio-energy in China. In order to promote large-scale application of renewable energy, it is crucial to seize the opportunity, build a firm foundation and rapidly form a renewable energy market and industry. To sum up, China's renewable energy has huge development potential and broad prospect, but in order to break through the barriers, the division and cooperation among government, related enterprises and research institutes are essential.

Table 6-9. the Development Prospect of China's New Energy

	In 2010			In 2020		
	Development capacity	**Physical energy**	**Standard coal**	**Development capacity**	**Physical energy**	**Standard coal**
Power generation	**Megawatt**	**Billion kilowatt-hour**	**Thousand tons of standard coal**	**Megawatt**	**Billion kilowatt-hour**	**Thousand tons of standard coal**
Solar PV	500	0.75	248	1,000	1.5	531
Grid wind power generation	4,000	8	2,640	30,000	60	21,240
Off-grid wind power generation	70	0.21	69.3	100	0.3	106
Small hydropower	50,000	175	57,750	75,000	262.5	92,925
Micro-hydropower	300	0.43	142	500	0.72	255
Biomass Power	5,800	31.9	10,527	20,000	110	38,940
Ocean power generation	25	0.07	23	50	0.143	51
Geothermal power generation	50	0.36	119	100	10.72	2550
Sub-total	60,745	216.72	71,518.3	126,750	435.88	151,303
2 .Gas/heat supply		Billion cubic meters			Billion cubic meters	
Rural household methane	30,000,000 (biogas digesters)	9		50,000,000 (biogas digesters)	15	
Breeding farm methane	10,000(biogas digesters)	2		2,000(biogas digesters)	4	

Table 6-9. (Continued)

	In 2010			In 2020		
	Development capacity	**Physical energy**	**Standard coal**	**Development capacity**	**Physical energy**	**Standard coal**
Power generation	**Megawatt**	**Billion kilowatt-hour**	**Thousand tons of standard coal**	**Megawatt**	**Billion kilowatt-hour**	**Thousand tons of standard coal**
Industrial organic wastewater	2,000 (plants)	4	12,576	5,000(plants)	6	1,650
City life sewage	1,000 (plans)	1		2,000 (plans)	2	
Solar-powered water heater	150 miliion square meters		18,000	300 million square meters		36,000
Geothermic heating supplying hot water	25 million square meters		700	50 million square meters		1,400
Sub-total			31,276			57,050
3. Liquid fuel			Thousand tons of standard coals	Thousand tons		Thousand tons of standard coals
Biology ethyl alcohol	500		4,290	19,000		16,300
Biology diesel oil	20		290	1,000		1,430
Sub-total	520		4,580	20,000		17,730
Sum total			107,374			226,083

Source: Lin Boqiang. "China Energy Development Reports 2008". China Financial and Economic Press, 341.

Translators: Liao Guiling(SUFE), Wang Fuming(SUFE)
English Version Editors: Jiang Jiahui(SUFE), Li Zhan (Harvard)

REFERENCES

[1] Liu Ke, 2009. A discussion on the development environment and prospects of China's new energy industry(Chinese) [J]. *Enterprise Reporter*, 5.

[2] Yang Jiejun, 2008. The new way of thinking on China's new energy and renewable energy legislation(Chinese) [J]. *Law and Business Research*, 1.

[3] Yang Jiejun & Lai Chaochao, et al. 2008. The development and utilization policy and legislation and regulations of China's new energy and renewable energy(Chinese) [J]. *Rule of law and Essays*, 23 (3).

[4] Wang Mingyuan, 2007. "Visible hand" props up a bright sky for China's renewable energy industry — based on the analysis of "The People's Republic of China Renewable Energy Law"(Chinese) [J]. *Modern law*, 29 (6), 156-165.

[5] Wang Bingchen & Wu Yuanwei, et al. 2009-4-18. Ground-source heat pump project need standardization urgently(Chinese) [N]. *Geological Survey Reporter*, 4.

[6] Wang Qingyi, 2007(4). Current situation and prospects of Renewable energy （Volume I) (Chinese) [J]. *Electric Technology and Economy*, 19 (2), 19-25.

[7] Wang Qingyi, 2007 (6) Current situation and prospects of Renewable energy （Volume II) (Chinese) [J]. *Electric Technology and Economy*, 19 (3), 23-25.

[8] Shi Lishan, 2004. An analysis on the current situation of China's energy and development plan of renewable energy(Chinese) [J]. *Renewable Energy*, 5, 1-4.

[9] Hong Kong Venture Investment Research Institute, 2008. Yearbook of China's Venture Investment [M]. Beijing: Democracy and Nation-building Press, 539-541.

[10] China's Energy Yearbook Editorial Committee, 2007, China's Energy Yearbook, 2005/2006 [M]. Beijing: Science Press, 255-256.

[11] Lin Boqiang, 2008. China Energy Development Report 2008 [R]. Beijing: China Financial and Economic Publishing House.

[12] Cui Minxuan, 2008. China's Energy Development Report 2008 [R]. Beijing: Social Sciences Academic Press.

[13] Wang Zhongying, Ren Dongming, &Gao Hu, et al. 2009. China's Renewable Energy Industry Development Report, 2008 [R]. Beijing: Chemical Industry Press.

In: China's New Deal
Editors: Li Xiaoxi and Hu Biliang

ISBN: 978-1-61470-815-5

Comments on Chapter 6

GETTING READY FOR THE COMING OF THE NEW ENERGY ERA

Liu Shiyong*

Chapter 6 initially debriefs the new energy policy in the three major economic entities i.e. European countries, United States, and Japan as well as other countries. It then systematically addresses the status quo for the development of new energy in China and existing issues left unsolved. And then, it introduces the policies related to new energy in China. Lastly, this chapter proposes the perspectives on the development of new energy in China. This comment mainly discusses, China as a major country in energy consumption in the world, how to handle the unresolved issues related to new energy by referring to the experiences and strengths in the new energy policies of the developed countries. Those issues include but are not limited to ideology, political system, policy, technology, innovation, and operation management with regard to new energy. This comment provides some constructive suggestions and advices on exploring diversified new energy sources and consuming energy in an efficient way in order to achieve the goal of sustainable development of energy in China.

Rapid economic development in China, like other industrialized and developing countries, is demanding continuous supply of cheap and clean energy. It has already become the second largest countries in energy consumption. The trend is still increasing. Therefore, the rising import dependency on energy renders China vulnerable to political and economic threats in the years to come. It is a simple logic that a country's economy is doomed to collapse with the exhaustion of traditional fossil energy if it solely or to great extend relies on the import of traditional energy. Most importantly, the problem between rising demand and limited supply is unboundedly magnified by political factor sometimes. For a county without energy, the leading producers in oil and gas might be able to control this country's economy

* Liu Shiyong, PhD in Industrial and Systems Engineering, Virginia Polytechnic Institute and State University, Associate Professor in Research Institute of Economics and Management, Southwestern University of Finance and Economics.

by suppressing the energy supply. For example, the case of gas debate between Russia and Ukraine apparently gives many countries a warning that a country's government might use energy supply as a measure to assert the sovereignty and political privilege over another country. China, with ever-increasing energy consumption for more than 1.3 billion population pursuing better quality of life, is facing even bigger challenge in security of energy than ever. China's national security is partially determined by energy security. Thereby, continued exploration and exploitation for new energy should enter the agenda of national major policies. The new energy policy should be able to consider diversifying the sources for new energy and using them in an efficient way.

1. Policy Support for Developing New Energy

By introducing the technology and operation experiences in exploration and exploitation of new energy from developed countries, China central government should draw up policies supporting diversified energies and implement them based on China' real situation and the exploitable potentials of those energies. First of all, central and state governments should provide an environment that facilitates the industrialization for full-fledged technology in new energy and associated equipment. Secondly, for energy that has potentially extensive application value, requests for investment, technology, and professionals should be given the highest priority. Last but not the least, the local governments in different locations should take advantage of their proprietary geographic strengths to develop and make use of potential new energy that characterizes local socioeconomic and natural environment.

Three facets should be given ad hoc attention in drafting and implementing new energy policy in China. First of all, it is how the governments in different hierarchies can offer support, in the policy and investment, for the research and development on the new energy which already has full-grown theoretic foundations. For example, the generation and utilization for the wind energy, solar energy, and methane have been experienced for years by many countries including China. What the government and investor should do is put focus on the Research and Development (R&D) of core technology, manufacturing of relevant equipment, and construction of infrastructure needed for producing these kinds of new energies. Core technology and equipment can be obtained through independent R&D and importing technology from developed countries. The independent R&D is to begin with the intensive investment in high-end laboratory and associated facilities. It is imperative to construct a platform by which professionals and experts can exert their knowledge and skills in the field of new energy, which requires sufficient funding to support the operation. It is the government's responsibility to build a general environment which involves building more supplementary laboratories in universities and research institutes. In order to provide a continuous supply of professionals in new energy R&D, universities should open some courses and dedicated funding for new energy research. Different governments and institutions should have incentive and reward mechanisms in place to encourage more researchers to contribute to the field of new energy. At the same time, the government should design various grant plans in order to invite foreign experts and oversea scholars in new energy field to engage in R&D in China. The financial and policy support should keep consistent and continuous because the research in energy cannot be done overnight. Central government should frame an environment to encourage the long-term risk investment, especially private capital, in new energy research by exercising preferential tax policy. In terms of introducing foreign technology and equipment, a multilateral cooperation strategy should be adopted in order to realize the win-win situation. This, consequently, can reduce the development cycle for critical technology in energy. More importantly, by sharing the new technology in energy, different countries can work together to realize the sustainable development in energy and reduction in poverty, which can also ease the regional or national conflicts caused by energy issues. In addition, the development and use of new energy, especially the clean energy such as solar energy, wind energy, and hydrogen can reduce the emission of greenhouse gases.

Secondly, national and state policy in new energy should also encourage and guide the investment in manufacturing of relevant equipment and constructing of associated infrastructure. The source of investment may come from dedicated fund from central and local fiscal income, private capital, or foreign risk investment. Central and local governments should actively provide channels to attract different investment schemes and build up a platform on which investors have confidence to play game in this field. Another important supporting source for new energy R&D is traditional energy conglomerates namely, PetroChina Company Limited[1], China Petroleum & Chemical Corporation (Sinopec Corp)[2], and China National Offshore Oil Corporation (CNOOC)[3]. These enterprises should increase their investment in R&D and expedite the industrialization on technology of new energy by using their abundant capital accumulated from operation of fossil fuel. It is going to be crucial for the energy enterprises to survive and prosper whether they have breakthrough in new energy R&D and can master the critical the technology for using new energy or not. Presently, many international oil groups are making huge investment in R&D and use of new energy. By coordinating with UC Berkeley, University of Illinois, Urbana-Champaign and the Lawrence Berkeley National Laboratory, British Petroleum (BP) put up a plan in 2007 to invest $500 million over the next ten years to establish a dedicated biosciences energy research laboratory[4]. By the same token, ExxonMobil announced plans in 2009 to invest $600 million to develop next-generation bio-fuels from algae[5].

Thirdly, the energy strategy should also support the fundamental research on the potential energies which are still in the conceptual stage. In order to diversify the new energy supply, China should explore different new energies according to the distinctions in natural environment, geologic, geographic, industrial, agricultural, forestry, hydraulic, climate factors in different regions. For example, it has been known that the deposits of Natural Gas Hydrate

[1] New energy exploration of PetroChina, http://news.cnpc.com.cn/system/2008/01/09/001150679.shtml, retrieve on Oct 5th,2009.

[2] Rivalry in new energy PetroChina vs. Sinopec, http://www.bioon.com/biology/bioengery/315151.shtml, retrieved on Oct 10th,
2009.

[3] Exploration on Natural Gas Hydrate (NGH) by China National Offshore Oil Corp and The Ministry of Land and Resource
http://www.in-en.com/newenergy/html/newenergy-200720070724112006.html, retrieved on Oct 10th, 2009.

[4] BP pledges $500 million for Energy Biosciences Institute, http://www.energybiosciencesinstitute.org, retrieved on Oct 13th, 2009.

[5] Exxon Mobil sinks $600 million into algae-based bio-fuels in major strategy shift,
http://www.guardian.co.uk/environment/2009/jul/14/green-algae-exxon-mobil, retrieved on Oct 5th, 2009.

(NGH)[6], at the north continental slop of South China Sea, are equivalent to 18.5 billion tons of oil which can provide China with 50 years' supply based on the consumption in 2008[7]. Based on the assumption that NGH can be found at tundra regions, scientists in China found another giant NGH reserve with estimated deposits being approximately equivalent to 35 billion tons of oil at place located at Qilian Mountain of Qinghai Province. Taking another example, in some regions of China, producing bio-fuel such as biodiesel through algae can not only generate alternative fuel but also cleanse the polluted water caused by eutrophication[8]. Another example of alternative bio-fuel is to use nitrogen-fixing bacteria to decompose grass or woodchips to produce hydrogen. The key benefit of burning hydrogen is that the only emission is water vapor, which significantly reduces emission of the greenhouse gas[9]. The R&D of alternative fuel needs governmental policy to encourage and support innovation. Authorities can set up a channel to expedite the industrialization of patents related to new energy. Given that the great potentials in alternative fuel, China's 12th and 13th five-year plans should give higher priority in R&D and industrialization of new energy. Most optimistically, with mastering key technology in diversified new energy and generating enough electricity power, China can consider directly exporting electricity power to adjacent countries after sufficing local needs in decades to come.

2. Policy Support for Using New Energy

China has been exploring and exploiting new energies including methane, hydraulic electricity, wind energy, and solar energy since the foundation of The People's Republic of China. The financial crisis became a turning point for China to adjust the structure of different industries. In the new act, part of the ￥4,000 billion investment should be used to replace the old equipment and upgrade infrastructure associated with these energies. It is also very necessary to adjust the way they were operated. Taking wind energy as example, in some areas, due to the instability of wind, it is hard to be combined to trunk grid. Under this circumstance, family-size and/or community-size wind energy generators are more efficient than that of centrally connected power supply. Especially for remote village or minority areas, family-size solar photovoltaic generation board and wind energy generators can be complementary to provide uninterrupted power. In addition, methane can be used for cooking and/or heating by using family-size methane tank in which organic waste and livestock dropping are used. Local government should provide allowances or other financial support to promote the participation of using new energy because this is a fundamental requirement for building harmonious society especially for the countryside. Once the family-size or community-size generators enter into different families, it is indispensable to have enough technicians to do maintenance for relevant facilities and equipment. Therefore, it is necessary to train required technicians in some vocational schools or colleges. Let's make an

---

[6] Natural Gas Hydrate http://www.newenergy.org.cn/b_ice/index.htm, retrieved on Oct 10th 2009.

[7] Chine Energy Yearbook (2007-2008)

[8] Economical biodiesel fuel from Algae, http://www.oilgae.com/, retrieved on Oct 10th, 2009.

[9] Alternative energy, http://www.alternative-energy-news.info/technology/biofuels/, retrieved on Oct 10th, 2009.

assumption that each administrative village needs 5 to 8 maintenance technicians. Optimistically, this scenario can create 3.2 million to 5.1 million employment opportunities if calculated based on the existing 640,000 administrative villages in China. Furthermore, by adding the communities in different cities and towns, the number is going to be 5 million to 7 million or even more than that. This is a double-gain game in which the policy not only realizes the continuous supply but also creates millions of employment opportunities. Except for training maintenance technicians, for enterprises which do business in wind and solar energy, they need many professionals who master skills and knowledge in both technology and operation management. Universities and vocational schools should design relevant courses and diploma program to provide appropriate talents for the coming the new energy times.

Due to the vast territory of China, the exploitation of new energy should vary by taking the regional differences and characteristics into account, so do the relevant polices. For instance, in Xinjiang Autonomous Region, windy and sunny weather can provide enough sources for generating energy; for coastal regions, energies generated by tide, wave, sea current, wind, and differences in seawater are major sources.

Given the diversity in energy sources in different regions, different local governments may not be able to implement appropriate policies due to bureaucracy or some unanticipated difficulties. There are possibly some conflicts when allocating limited resources to promote new energy or other activities. Thus, achievements and efficiency in employing new energy in different regions should become one of the key performance indicators for government officials in order to monitor and stimulate the scale of support of local governments. Central government should promote the widespread application of new energies such as bio-methanol, bio-ethanol, biodiesel, waste vegetable oil, bio-butanol, hydrogen in vehicles and machines which traditionally use fossil fuel. Therefore, enterprises have to develop new type of engines for car, ship, and airplane as well as capital machines used in plants. Consequently, the running of transportation vehicles and machines used in companies can no longer rely on traditional fossil fuel.

3. Establishing Linkage between New Energy Industry and Associated Sectors

The R&D and exploitation is, in many countries, becoming a new growth point in economy because the equipment manufacturing and construction of associated infrastructure can drive the growth of upstream and downstream industries or sectors. As a result, it can create more employment opportunities. Furthermore, the widespread use of new energy is going to bring significant changes in socioeconomic systems. For example, it will change the traditional way in using agricultural produces because crops harvested might be used for producing new energy e.g. corn is used for producing alcohol; crop stalks mixed with livestock waste are used for producing methane. By this token, the products generated from refinery towers are not gas and diesel but bio-fuel. Additionally, the exploitation of new energy put a request for constructing associated infrastructure or upgrading existing infrastructure ever used for producing fossil fuel. During the transition from using fossil fuel to new energy, there might be many issues or conflicts of interests coming from different stakeholders. As mentioned previously for rebuilding or obsolescing existing facilities and equipment, this undoubtedly adds financial burden to enterprises. Besides, the technicians and engineers also need technical training in order to operate the new equipment efficiently. In consideration of the different characteristics of new energy, traditional supply chain may be broken and old operation management skills may become obsolete. In a current supply chain of energy, a refinery plant's raw materials come from crude oil produced locally or imported from foreign countries. However, in the new energy case, the upstream for a fuel-producing might be many farms. Changes and challenges brought by the process of exploring and exploiting new energy are not limited to the above-mentioned aspects. Central government, local government, and energy enterprise should prepare a package of policies, strategies, operating tactics to handle potential issues.

4. Avoiding Under- or Over-Supply in Equipment Capacity or Infrastructure

Central government should, as necessary, employ macro-control mechanisms to avoid under- or over-supply in equipment capacity or infrastructure associated with new energy. The potential reserve for different new energies should be systematically studied in order to rationally allocate funding, investment, and fiscal subsidies. Moreover, central and local governments should deliver accurate and timely information to the public in order to get private risk capitals involved in the macro-control process on the capacities of associated equipment and facilities. The enterprises in manufacturing energy associated equipment can also look for opportunities outside China to expand their market.

5. Energy Conservation

With so large population base in China, energy consumption cannot follow the ways the developed countries have been using. According to statistics of the United State Energy Information Administration[10], the energy consumption per capita in US in 2008 is 94765.42 KW[11]. While during the same year, the number in China is 17556.32 KW[12], which means that an equivalence of 2.85 billion tons of standard coal needs to be burned in order to suffice the need. Based on this assumption, a total of 15.39 billon tons of standard coal need to be burned if all people in China were to achieve the same quality of life as in USA. This definitely will cause disastrous effects on China in terms of both energy consumption and emission. Therefore, as a country with ever-increasing economy, China should not only continuously explore various new energy sources but also increase the efficiency of using energy. Even though there are going to be breakthrough in new energy R&D, the resources are still scarce.

---

[10] United State Energy Information Administration, http://www.eia.doe.gov/, retrieved on Oct 10th, 2009.

[11] Average energy consumption in USA 2008, http://www.eia.doe.gov/emeu/mer/pdf/pages/sec1_3.pdf, retrieved on Sept 25th, 2009.

[12] Energy Consumption of China in 2008 http://www.stats.gov.cn/, retrieved on Oct 10th, 2009.

Meanwhile, in order to achieve the goal of green China, it is pressing, for major energy-consuming enterprises, to improve their overall efficiencies in energy consumption through technology upgrading in production, process reengineering and improvement of operation management.

Besides, another major demand for energy comes from transportation. Given the situation of population, China, in the long run, should consider using mass transit as transportation mode for meeting travel need instead of promoting private car driving. Even though promoting private car driving can drive domestic demand in short term, the long-run consequence is going to be the rapid increase in energy consumption and toxic emission. Moreover, too many vehicles running in cities will cause traffic congestion which makes idling vehicles consume more energy and emit more toxic gas (carbon monoxide) due to insufficient burning of fuel. Metropolitan areas such as Beijing, Shanghai, and Guangzhou can consider taking measures such as levying higher oil tax or traffic congestion pricing to mitigate traffic congestion. The revenue accumulated can be used to improve mass transit capacity and facilities accordingly. As a result, people are better off by switching to take mass transit or ride high occupancy vehicle (HOV) with improved mass transit services.

Central and local governments might consider enacting regulations and laws to regulate operation of energy-consuming enterprises and enforce environment protection. For example, it is possible to push real estate companies to install solar water heater and sufficiently utilize solar energy[13] when they construct new buildings. In addition, more strict industrial standards for energy conservation should be set for auto, electric appliances, and industrial capital equipment. Last but not the least, it is necessary to promote the energy conservation throughout the whole society through public participation and supervision. Rewards can be given to those enterprises, communities and families which have outstanding performance in energy conservation.

In a word, China government needs to employ all kinds of feasible policies and measures to develop and use new energy to improve people's quality of life and to guarantee the sustainable development of economy.

---

[13] Solar house at Virginia Tech, http://www.lumenhaus.com/, retrieved on Sept 20th, 2009.

In: China's New Deal
Editors: Li Xiaoxi and Hu Biliang

ISBN: 978-1-61470-815-5

Chapter 7

EMISSION TRADING: NEW MECHANISM FOR RESPONDING TO CLIMATE CHANGE

He Jinyu**, Xiao Boqiang and Zhang Mingming

The current global triple crisis of energy, environment and finance has made a great impact on China's economic development. Since the financial crisis has not completely faded away, it will have a profound long-term influence on the entity economy. While countries all over the world are looking for the new engine for economic growth, developing the low-carbon economy has become the common choice of most of developed countries and the impetus of China for future's development.

In the production and management activities of humans, it is evitable to discharge greenhouse gas and all kinds of pollutants, which act directly on our natural environment, change human's living environment and furthermore affect people's daily life. Emission trading is the derived trading of right and interest based on environmental factors and is designed to reduce emission through economic means. Having entered the 21st century, some developed countries, including European countries and USA, have started developing all kinds of emission trading and achieved good results. Since China is a fastest-developing country in the world, choosing the right opportunity to develop an appropriate amount of emission trading is of necessity for China to respond to climate change and provides China with the new mechanism and exploration.

SECTION 1: EMISSION TRADING IS OF NECESSITY FOR CHINA TO RESPOND TO CLIMATE CHANGE

The National Energy Program predicts that the total demand for global energy will have risend by 50% by 2030, while China's demand for primary energy will have increased from 1.2 billion tons in 2002 to 2.5 billion tons. China will face serious energy safety problems in the future. With regards to environmental crisis, the global climate change has already become an undisputable fact, and the development of high energy-consuming and emission

* He Jinyu obtained his PhD degree from the Institute of Economics and Resource Management of Beijing Normal University, and he is now working with PetroChina (CNPC). Email: hejinyu@cnpc.com.cn

industries will be seriously restricted. Developing emission trading is an effective model to realize low-carbon development. It also satisfies the need of China to realize the sustainable development of economic society, actively respond to the global climate change and achieve the goal of energy saving and emission reduction.

I. The Need for Actively Dealing with the Global Climate Change

The Intergovernmental Panel on Climate Change (IPCC) established by World Meteorological Organization (WMO) and United Nations Environment Program[1], pointed out in its forth evaluation report of 2007 on the climate change that global warming has already become an undisputable fact. The major cause is the emission of greenhouse gases, including carbon dioxide caused by human's activities, which results from the consumption of energy (mainly fossil energy) and other resources caused by the population growth. It is estimated that by 2020, the consumption of global energy will have increased by more than 70%, and the emission of carbon dioxide will have increased by more than 50% if effective measures are taken. The harsh reality of global warming has attracted attention of countries all over the world. The United Nations promulgated *Kyoto Protocol* in December, 1997, which quantitatively specifies countries' obligation of emission reduction and introduces three major market-based emission reduction mechanisms: Emission Trading, the Clean Development Mechanism and Joint Implementation. In November, 2005, the Montreal meeting secured the operation of *Kyoto Protocol* in legal terms and launched the "negotiation in the post-Kyoto times". In December, 2007, the Bali meeting promulgated the "Bali road map", which stateses that a new emission reduction agreement must have been issued by 2009 and stresses the importance of respecting the convention and principles of the protocol.

As a positive and effective method to deal with the global climate change, emission trading mechanism has been fast developed in the United States, the European Union, etc. As the world's largest developing country with the fastest economic growth and a large quantity of greenhouse gas emission, China proposes to incorporate the plan of responding to climate change into national economic and social development plan, and actively develop low-carbon economy and emission trading to a new development model, which is of great significance to dealing with global climate change.

II. The Need for Effectively Integrating Resources and Achieving China's Goal of Energy-saving and Emission Reduction

National "Eleventh Five-Year Plan" [2]clearly states the binding target that GDP energy consumption per unit in 2010 should have reduced by 20% from 2005. Energy-saving and emission reduction has already become one of the major tasks of China's reform and development. In the first three years of "Eleventh Five-Year", the central government will allocate 33.6 billion RMB from the budgetary investment and 50.5 billion RMB from the

[1] IPCC forth evaluation report.

[2] *Outline of China's the Eleventh Five-year Plan*, State Council of People's Republic of China.

central fiscal fund to support the construction of key energy-saving and emission reduction projects. Now, the energy-saving and emission reduction work in China during the first period has already taken effect. In the first half of the year 2009, the national GDP energy consumption per unit declined by 3.35%, 0.47 percent higher than the decline rate of last year; while the added-value energy consumption of above-scale industrial units declined by 11.35% compared with that of last year. It is estimated that the emission of sulfur dioxide and chemical oxygen demand (COD) will decline by 5% and 2%, respectively, among which, coal industry will decline by 3.83%; iron and steel, building materials, electricity and petroleum and petrochemical industries about 8% -9%; non-ferrous metal industry 19.59%; chemical industry 15.16%; and textile industry 11.45%. In the first half of the year, among the key enterprises whose comprehensive energy consumption of standard coal over 10,000 tons, 80% of the 25 key energy-consuming products and 108 unit consumption indicators showed downward trends.

Although the energy-saving and emission reduction work is in progressive advance, the mid-term evaluation report of “Eleventh Five-Year Plan" released at the end of 2008 shows that the decrease of two indicators, GDP energy consumption per unit of energy-saving and emission reduction and the major pollutants, is not optimistic. If we want to meet the requirement proposed in "*outline*" by 2010, there is still a long way to go. China is currently accelerating urbanization and infrastructure construction, and of depends strongly on export. Undoubtedly, the industrialization and great amount of long-term export of high-energy vector make it more difficult to save energy and reduce emission.

To achieve our goal of energy-saving and emissions reduction, traditionally, we could assign the binding quotas of energy-saving and emission reduction to the all the provinces, autonomous regions, municipalities and related businesses through administrative means. However, this single means is hard to realize lack long-term mechanism. Compared with high-cost administrative means, low-cost greenhouse emission reduction through market mechanism is much more effective. The emission trading platform encourages the capable regions and enterprises to save as much energy and reduce emission as possible, and sell their surplus quotas in the market. Regions and enterprises that have limited ability to save energy and reduce emission would also finish their producing tasks through purchasing the quotas from the market. Compared with sending the traditional administrative instructions, it is most flexible and efficient to introduce the market mechanism to save energy and reduce emission at the lowest cost and at the same time maintain the economic development.

III. The Need for Sustainable Development of Economy and Society

Since the reform and opening-up policy, China's economic growth has made universally recognized great achievement. While China enjoys high GDP growth, it exerts great pressure on the environment. In 2005, the total emission of sulfur dioxide in China was 25.494 million tons, and the acid-rain stricken area caused by sulfur dioxide counted for 6.8% of the national land. In areas stricken by comparatively serious acid-rain, such as Guiyang, Chongqing and

Changsha, the average pH value of precipitation acidity is approximately 4.53. Due to the rapid development of industrialization, the national annual emission of carbon dioxide is always at a high level throughout the year. Along with the process of industrialization, high energy-consumption and environmental problems have become increasingly prominent.

In order to achieve sustainable development of economic society, it is particularly important to solve environmental problems and actively respond to global climate change. The externality of environmental pollution and global climate change is the major cause of the spread of environmental pollution. The countermeasure is to internalize the externality. At present, countries all over the world are actively promoting emission trading, making it an important means to solve the problems of environmental pollution and global climate change. The nature of emission trading is to trade the emission trading as a commodity. Under the premise of controlling total emission, the government encourages enterprises to minimize the total pollution emission through technological progress and pollution abatement. The enterprises will be granted the legal right of pollutants emission and allowed to transfer or change the surplus quota after pollution abatement. Under this framework, our environment would be improved based on the market mechanism.

In 1999, China National Environmental Protection Administration and U.S. Environmental Protection Agency signed a cooperation agreement to unfold "the study of SO2 emission reduction through market mechanism" in China. Nantong in Jiangsu Province and Benxi in Liaoning Province became the first two pilot cities, which marks the beginning of China's solving the problems environment pollution through market mechanism. Since then, China has successively conducted pilot emission trading in seven other provinces and municipalities, such as Shandong, Shanxi, Jiangsu, Henan, Shanghai, Tianjin, and Liuzhou. Ten years of exploration and practice has shown that, as a new environmental and economic policy with the characteristics of market mechanism, emission trading system, could effectively control the environment pollution and promote sustainable development of the economy.

SECTION 2: EMISSION TRADING HAS ITS FAVORABLE FOUNDATION FOR DEVELOPMENT IN CHINA

China's research and development of emission trading has quite a long history. Since the 1990s, national environmental protection departments have been studying the emission trading of sulfur dioxide in U.S. electric power industry. With the further development in the developing countries of the Clean Development Mechanism (CDM) specified in "*Kyoto Protocol*", China has become the country that reduces the greatest amount of CO_2 emission under the mechanism of CDM. Beginning in 2008, China has also established a number of emission right exchanges, laying a favorable foundation for developing emission trading.

[3] The Eleventh Five-year Outline of China's Environmental Protection, 2006-2010,State Environmental Protection Administration, China.

I. Practice under the Framework of UN Climate Change Convention and Kyoto Protocol

The homogeneous distribution of greenhouses, such as carbon dioxide, in the Earth's atmosphere leads to the globalization of climate change. On May 9, 1992, "United Nations Framework Convention on Climate Change" ("Convention ") was passed by Intergovernmental Negotiating Committee (INC) in New York, and was signed by the participating countries during the Rio Conference on Environment. At the end of 2008, 191 countries and regional organizations had been ratified to join the Convention. TheConvention specified five major principles, including principle of differentiated responsibilities based on the common interest, precautionary principle, sustainable development principle, open-economy principle, principle of giving full consideration to the specific needs and special circumstances in developing countries. However, details about how to operate these principles have not been decided.

In 1997, "*Kyoto Protocol*" specified the law-binding quantitative quotas of emission reduction for developed countries, but not for developing countries. The core of the protocol is that from 2008 to 2012, the total emission of greenhouse in industrialized countries must be reduced by 5.2% from 1990. Due to the higher abatement cost in developed countries, "*Kyoto Protocol*" specially included three kinds of flexible implementation mechanisms: Joint Implementation (JI), Emission Trading (ET), and Clean Development Mechanism (CDM). Protocol officially came into effect on February 16th, 2005. At present, there are already 142 countries and regions that signed the Protocol. The performance of "Kyoto Protocol", however, is not so optimistic as expected. The statistics in August, 2007 shows that emission in some developed countries increased rapidly rather than declined. The emission reduction goal of both Japan and Canada is 6%, but their amount of emission increased by 7% and 27%, respectively. Though Australia ratified the Protocol in 2007, specifying that it aims to reduce emission by 8% during the first commitment period, but in contrast, its emission increased by 25%. USA, who has not ratified the Protocol, increased its emission by 16%, although its intended goal under protocol is to reduce emission by 7%.

Under the framework of "*Convention*" and "*Kyoto Protocol*", a large number of emission reduction trading systems have been established and developed worldwide. *Convention* and *Kyoto Protocol* have really become the systematic basis for emission trading. The world's major emission trading systems are summarized in Table 7-1.

The first three trading systems are derived directly from *Kyoto Protocol*, while the latter two are either directly or indirectly come into being due to the pressure of implementing *Kyoto Protocol*. Among these systems, both Clean Development Mechanism and voluntary emission-reducing system of the Chicago Climate Exchange involve the participation of Chinese enterprises. The system infrastructure introduces emission trading system into China.

II. Large Potential for China to Reduce Emission

China's potential for emission reduction is mainly shown in the following two aspects. On one side, energy demand and industrialization lead to a higher emission of greenhouse gas and major pollutants; on the other side, China has a large amount of renewable energy sources that could be used to replace traditional energy, so that the emission of greenhouse gases and major pollutants could be reduced.

According to the " *Medium and long-term planning of Renewable Energy*" of China's National Development and Reform Commission, China plans to make full use of the proven-technical, economic renewable energy resources, such as hydro-power, bio-gas, solar thermal and geothermal to accelerate its industrialized development of wind power, biomass power, solar power, and gradually improve the proportion of high-quality, clean and renewable energy in its energy mix. China should make the effort to increase percentage of the renewable energy consumption in total energy consumption to 10% by 2010, and to 15% by 2020.

Table 7-1. Major Emission Trading Systems in the World

Trading System	Trading Commodity	Buyer	Seller
Clean Development Mechanism (CDM)	Certified Emission Reduction (CER)	developed countries	developing countries
Joint Implementation (JI)	Emission Reducing Unit (ERU)	developed countries	developed countries
International Emission Trading (IET)	Assigned Amount Unit （AAU）	developed countries	developed countries
Regional Emission Trading （RET）	EU Admission (EUA)	EU entities	EU entities
Chicago Climate Exchange (CCX)	Voluntary Emission-reducing (CFI)	voluntary emission-reducing business, Charities	voluntary emission-reducing business, Charities
Regional Greenhouse Gas Initiative (RGGI)	Regional Greenhouse Gas Initiative (RGGI)	states, provinces in U.S. and Canada	states, provinces in U.S. and Canada

Source: Tianjin Climate Exchange.

According to the development program, the development goal includes that in 2020, the annual consumption of renewable resources will be equivalent to annual emission reduction of sulfur dioxide of about 8 million tons, nitrogen oxide of about 3 million tons, soot of about 4 million tons, carbon dioxide of about 1.2 billion tons, water of 20 billion cubic meters and deforestation of about 3 million acres. Therefore, market mechanism oriented emission trading can contribute to the mercerization of emission reduction products and the further of its effect. In order to achieve this development goal in 2020, the expected investment totals 2 trillions RMB.

Table 7-2. Development Scope of the Renewable Resources in China

Types of renewable resources	2005	2010	2020
Hydro-energy			
water and electricity (100 MW)	1.17	1.9	2.25
small hydropower (100 MW)	0.38	0.5	0.75
Biomass energy			
Biomass power generation(10 MW)	200	550	3000
agriculture, forestry, biomass power generation (10 MW)	170	400	2400
Biogas Power Generation (10 MW)	5(2008)	100	140
Garbage Power (KW)	20	50	300
Bio-solid fuel annual consumption(ton)	-	100	5000
Biogas utilization (100 million cubic metres)	80	190	440
Non-grain feedstock fuel ethanol (10 thousand tons)	102	200	1000
Bio-diesel (10 thousand tons)	5	20	200
Wind power			
Wind electricity (10 MW)	131	500	3000
Solar power			
Solar electrical energy generation(10 MW)	1.9	30	180
small-scale photovoltaic power stations	1.9	15	30
Solar grid-connected photovoltaic power generation		5	100
large-scale solar power stations		7	40
Commercial Photovoltaic Applications		3	10
Solar thermal use of alternative energy consumption (standard coal, 10 thousand tons)		3000	6000
other alternative renewable resources			
geothermal energy annual consumption (standard coal, 10 thousand tons)		400	1200
Tidal power stations (10 MW)			10

Source: *Medium and long-term planning of Renewable Energy*", China's National Development and Reform Commission.

III. The Government's Promotion for Emission Trading Pilots

In December 2008, after Premier Wen Jiabao proposed in his government work report "to actively carry out the emission trading pilots", the construction of provincial and municipal emission trading pilots has been speeding up. In 2003, Zhejiang province was the first to establish the Jiaxing reserve and exchange center of emission rights, followed by the release of documents on the emission trading in seven cities. In particular, the establishment of online monitoring network of key pollution sources and progressive advancement of "three quantity" account construction and pollution permits issuance laid the foundation for purchase and trading of emission rights. Encouraged by the government, Zhejiang Province has drafted the "*Zhejiang Province, sulfur dioxide emission trading management approach*"

Jiangsu Province was the first to adopt the emission trading of thermal power sulfur dioxide. Under the guidance of" *sulfur dioxide emission trading management approach in the Electric Power Industry* " of Ministry of Environmental Protection, Jiangsu Province formulated a pilot program in Taihu Lake Basin, and utilized sewage expenditure about 1.2 billion to support the establishment of the platform of emission trading in this area in 2007. On November 20, 2008, Jiangsu launched the "*Program and Rules for the Paid Use of Major Pollutants' Emission Right in Taihu Lake Basin of Jiangsu Province and its Pilot*". On February 25th, 2009, Jiangsu released "*Management Approach for the Paid Use of Major Pollutants' Emission Right in Taihu Lake basin of Jiangsu Province and Purchase of Trading Pilots for the Emission Quota*".

Hubei provincial government has been actively promoting the construction of emission trading market, and recently Wuhan Guanggu Property Exchange in Hubei Province has started the quota trading of sulfur dioxide and COD. On October 27, 2008, Hubei Province promulgated the "*Interim Measures for Emissions Trading of Major Pollutants*". In November,19,2008, Environmental Protection Agency in Hubei Province published the "*Interim Measures of Implementation of Emissions Permits in Hubei Province*", " *E-auction Trading Rules for the Emission Rights of Major Pollutants in Hubei Province (Trial)*",and " *Emission Trading Rules for Major Pollutants in Hubei Province*"(Trial). As of now, the emission trading of major pollutants in Hubei Province has already exceeded 50 million RMB. Hubei Province was officially established as the national emission trading pilot in August 2009.

In 2002, Tianjin Municipal Government established Financial Innovation Group to study the sulfur dioxide emission trading, and started to allocate the emission quotas in the same year. In 2008, it jointly set up Tianjin Climate Exchange with China National Petroleum Corporation and Chicago Stock Exchange, reported to Ministry of Finance and Ministry of Environmental Protection and obtained their joint approval to explore emission trading in Tianjin Economic and Technological Development Zone. In December 2008, Tianjin Climate Exchange conducted the first sulfur dioxide emission trading on the electronic bidding system of Chicago Climate Exchange, and then launched a voluntary emission trading system in 2009.

IV. National Policies to Support the Energy-saving and New Energy Industry

In 2007, China allocated a special fund for energy-saving and emission reduction, totaling 23.5 billion RMB, and offered subsidies to environmentally friendly energy-saving projects. That means the tax on those projects will be exempted for the first three years and halved for the latter three years. Subsidizes have been provided for energy-saving equipments of enterprises in operations. For those enterprises that purchase and physically use the required energy-saving and water-saving equipment, 10% of their investment in the equipment could be deducted from the tax of that year, or the tax of the following five years if the equipment expenditure exceeds the tax expenses.

With regards to the supporting the new energy industry, China offered a special subsidy for wind power. For the first 50 turbine groups of wind farm construction, 600 RMB / kW will be subsided, including machine manufacturers and key components manufacturers respectively 50%[4]. With regards to subsidizing renewable energy power,"*Notice on subsidies for renewable energy subsidy and quota trading program from January to June (or July-December) in the 200X-year*" jointly released by National Development and Reform Commission and National Electricity Regulatory Commission provides subsidies biannually.

China have also issued a special policy for the subsidy to energy-saving products, including subsidies ranging from 4,000 to 60,000 RMB for energy-clean cars based on the latest published "*Interim Measures for demonstration and promotion of financial subsidies to the energy-saving and new energy vehicle*". In terms of energy-saving lamps, the Development and Reform Commission and Ministry of Finance plans to allocate up to 600 million RMB, to subsidize 120 million efficient lighting products in 2010. Policy banks may provide support to energy-saving projects in business, and most of the major commercial banks are also starting to offer green credit businesses.

In addition to offering subsidies, the government also uses various tax policies to control the consumption of traditional energy and environmental resources, in order to support energy saving, emission reduction and the new energy industry. At the same time, new taxes that supports the energy-saving and new energy industry are being designed.

V. Optimistism of all Kinds of Market Players in Trading Market

The confidence of China's all kinds of market players in the emission trading are evidenced in the following two aspects: the increasing enhancement of energy-saving and emission reduction activities and full participation in CDM market of large enterprises.

1. Increasing Enhancement of Energy-saving and Emission Reduction in Large Enterprises

In 2008, China National Petroleum Corporation published the amended "*Energy-saving and water management measures in China National Petroleum Corporation*" and carried out oil associated gas' recovery projects mostly in western Tarim Oilfield and Changqing oilfields. It also developed CDM projects, and increased its capability of recycling gas to 810 million cubic meters each year through the implementation comprehensive recycle project of venting natural gas. It is estimated that 12.0 billion will be invested into 10 key energy and water-saving projects, which would save standard coal 4.6 million tons each year, substitute fuel oil 1 million tons and save water 280 million cubic meters; At the same time, with the improvement of producing and storage technologies, the consumption of crude oil and refined oil will be reduced by one million tons. At the same time, the Group planned to invest 12.15 billion RMB to establish 184 pollution abatement projects by 2008, including 66 sulfur dioxide emission reduction projects and 118 COD emission reduction projects[5].

[4] Ministry of Finance, People's Republic of China,2008.

[5] "Report of Enterprises' Social Responsibility: 2006, 2007, 2008", China National Petroleum Corporation.

In 2007, China National Offshore Oil Corporation became the first Chinese enterprise to be invited to join 3C, an international organization dealing with global climate change. In 2008, China National Offshore Oil became a member of United Nations "Global Compact" and collaborated with the international community in response to the global climate change. It proposed the "zero emission" concept and standard, which states that" Any new project must be designed and constructed according to the zero emission standard." In addition, it convened seminars to discuss the zero emission of pollutants and greenhouse gas emission control and clarified what zero emission means and how to carry it out.

By the end of 2008, new energy generation capacity of power generation groups, such as Huadian, State Power, Datang, Shenhua and etc, was approximately 500MW. By 2010, it will have reached 850MW, and by 2020, 3000MW. According to professional estimate, this figure is considered very conservative. With the ongoing implementation of industrial policies of new energy, and of solar energy in particular, the figure may double. Liu Zhenya, General Manager of the State Grid Corporation of China, said in May 2009 that the development of strong and smart grid would be carried out period after period and, a strong smart grid will have been completed by 2020.

2. China's Actively Involvement in the CDM Project

Since the development of CDM projects in China, all the provincial and municipal enterprises are actively involved in. As of September 7th, 2009, 2174 projects have been approved by National Development and Reform Commission, among which 131 have reveived the issued CER. Due to the differences in the natural environment, the number of CDM projects and the amount of emission reduction vary greatly among different provinces and cities, as shown in Table Table 7-3.

Table 7-3. Distribution of CDM Projects Approved in Different Provinces and Municipalities

Area	Quantity of projects	Area	Quantity of projects	Area	Quantity of projects	Area	Quantity of projects
Yunnan	258	Sichuan	220	Inner Mongolia	143	Hunan	128
Shandong	104	Gansu	103	Shanxi	90	Hebei	84
Zhejiang	84	Hubei	78	Henan	73	Jiangsu	72
Guangxi	71	Guizhou	70	Fujian	64	Heilongjiang	58
Liaoning	54	Guangdong	53	Jilin	50	Anhui	49
Shanxi	46	Chongqing	46	Jiangxi	45	Xinjiang	44
Ningxia	22	Qinghai	17	Hainan	15	Shanghai	13
Beijing	13	Tianjin	7	Tibet	0	Total	2174

Source : CDM database of China's National Development and Reform Commission , http://cdm.ccchina.gov.cn/web/item_data.asp?ColumnId=63.

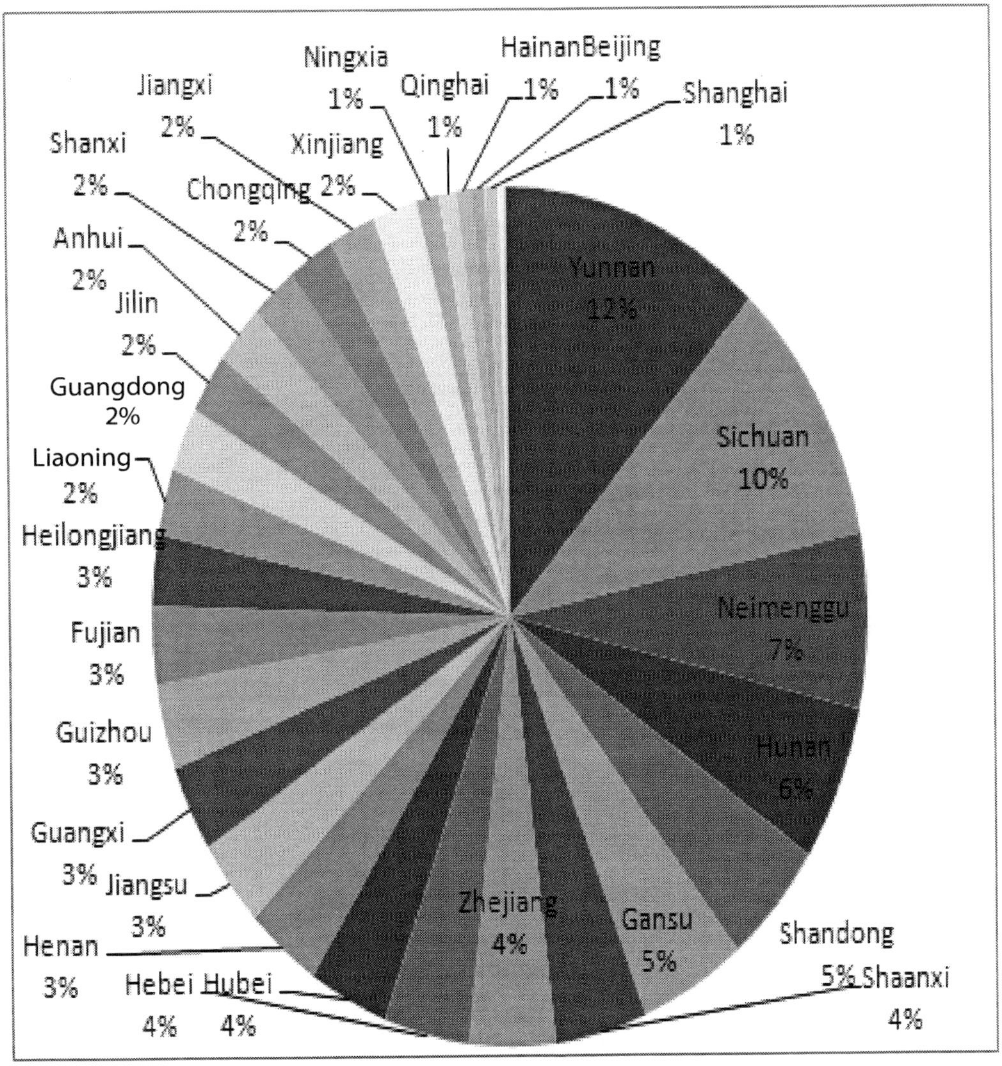

Source CDM database of China's National Development and Reform Commission, http://cdm.ccchina.gov.cn/web/item_data.asp?ColumnId=63.

Figure 7-1. Pie Chart of CDM Projects Approved in Different Provinces and Municipalities.

As shown in the figure above, Yunnan and Sichuan are abundant in water resources, and Inner Mongolia has the largest number of wind power projects. Thanks to the unique resources available for development, enterprises in these three areas are actively involved in developing CDM projects. Take Yingjiang basin in Yunnan Province as an example. There are as many as 47 CDM small-scaled hydropower projects. In contrast, the three municipalities, excluding Chongqing, rank last since they started late (no CDM projects have started yet in Tibetan). The emission reduction of CDM projects vary greatly among provinces and municipalities, as shown in Table 7-4.

Table 7-4. Estimated Annual Emission Reduction in Different Provinces and Municipalities

Area	Estimated annual emission reduction	Area	Estimated annual emission reduction
Sichuan	39,114,980.5	Jiangsu	37,393,282
Shandong	30,408,334	Yunnan	26,006,032.4
Henan	14,418,617.43	Gansu	13,062,610
Guangdong	11,602,348	Guangxi	10,515,604
Hubei	8,668,001.977	Anhui	8,681,180
Jilin	7,044,579	Guizhou	6,620,618
Beijing	4,344,351	Jiangxi	3,666,293
Tianjin	1,119,094	Hainan	816,986
Zhejiang	36,432,693.37	Shanxi	34,026,141
Inner Mongolia	25,559,287	Liaoning	21,321,627
Hunan	12,988,104.5	Hebei	12,595,576.4
Heilongjiang	10,261,251.8	Fujian	9,450,804.586
Chongqing	8,601,677	Xinjiang	8,202,206
Shanghai	6,076,237.68	Shanxi	4,756,504
Ningxia	3,224,341	Qinghai	1,703,669
Tibet	0	Total	418,683,031.643

Source : CDM database of China's National Development and Reform Commission , http://cdm.ccchina.gov.cn/web/item_data.asp?ColumnId=63.

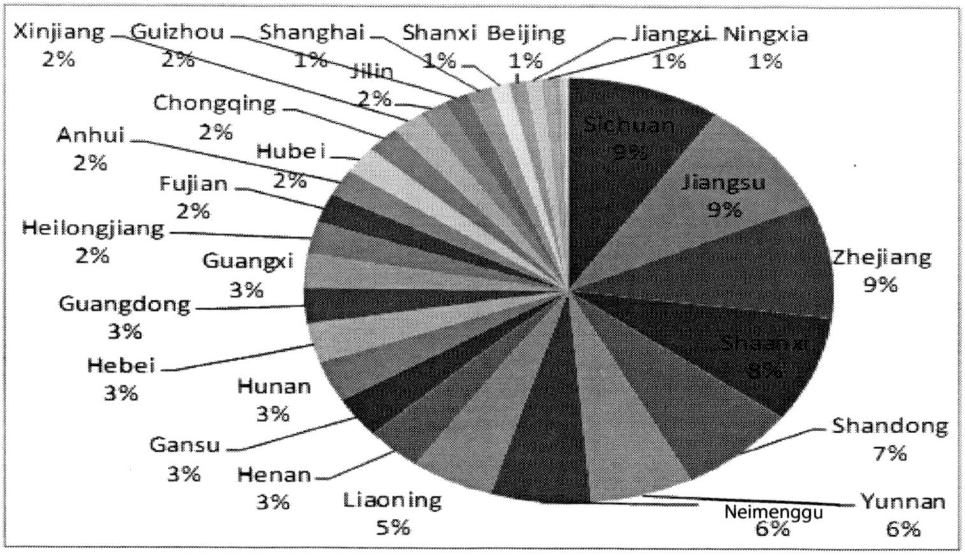

Source : CDM database of China's National Development and Reform Commission , http://cdm.ccchina.gov.cn/web/item_data.asp?ColumnId=63.

Figure 7-2. Pie Chart of Estimated Annul Emission Reduction of Approved Projects in Different Provinces and Municipalities.

Due to the small scale of hydropower projects in Yunnan province, itse annual emission reduction is estimated to be much smaller than that of Zhejiang province, where there are large-scaled key projects, such as Zhejiang Juhua. Enterprises, such as the Conch Cement Company and Landscapes Cement Group in central China, have also developed many of waste heat power generation projects, and had greater amount of estimated annual emission reduction. Except Tibet, the amounts of emission reduction in Qinghai, Tianjin, and Hainan are estimated to be the smallest.

VI. Orderly Development of International Cooperation to Respond to Climate Change

The Chinese government and the United Nations jointly held the *High-level seminar on the development and transfer of technology to deal with climate change* in December 2008. In the seminar, Premier Wen Jiabao pointed out that the Chinese government would continue to play an active and constructive role and make efforts to deal with climate change with the international community. He proposed that the international community must continue to collaborate in response to the global climate change. The Convention and Kyoto Protocol laid the legal foundation for international cooperation to deal with the climate change. And we should take full consideration of the factors specific to countries, including different countries' current situations, periods of development, historical responsibility, emission per capita, etc, focusing on our current situation and looking forward into the future. We should communicate with each other in open dialogues and cooperate with each other in the long term. Under the guidance of this principle, China's cooperation with the international community to deal with the climate change will be developed orderly.

1. Wide Development of International Cooperation in CDM Projects

As of July 14th, 2009, 2128 projects with 135 international partners have successfully registered with the National Development and Reform Commission, with an annual emission reduction of carbon dioxide totaling 345 million tons. Among the Participants in China's CDM project, World Bank ranks first in the number of registered projects with the National Development and Reform Commission and EB issuances. The Eco-securities Group jointly run by the United Kingdom and the Netherlands has the largest number of CDM projects in China, which is 139 projects. Although Goldman Sachs has been involved in nine Chinese CDM projects, its registered amount of emission reduction ranks top ten. Its oil Liaohua N2O emission reduction project is the largest CDM project in China, with an annual emission reduction of 10 million tons of carbon dioxide (Table 7-5).

Different international partners have their own technological advantages and preferred methodologies in different CDM projects. At present, most of the CDM projects approved by the National Development and Reform Commission are new energy and renewable energy projects. and Sweden Carbon Asset Management Ltd has the largest number of such projects in China (Table 7-6).

Table 7-5. Major Partners of China CDM Projects

No	Organizations	Nationality	CO2 emission reduction registered in DRC	Amount of EB issuing (t/CO2)	Quantity of projects
1	World Bank	-	26984582	33789049	17
2	Mitsubishi	Japan	26274517.6	16760738	70
3	ENEL	Italy	25800792	9299792	66
4	Eco-securities Group	UK/Holland	23627856.5	214322	139
5	4C funds	Luxemburg	18091315	8144888	39
6	UK Emissions Trading Co., Ltd.	UK	14150815	338761	52
7	Sweden Carbon Asset Management Ltd	Sweden	13241893.3	837866	120
8	Electricite de France Trading Co., Ltd.	France	12574117.14	88082	71
9	Camco International Ltd(UK)	UK	11767985	-	43
10	Goldman International	UK	11370389	4903988	9

Source: Tianjin Climate Exchange.

Table 7-6. Major Types of CDM Projects in China

Types of CDM projects in China	Quantity	1st in quantity	2nd in quantity
New and renewable resources	1250	Sweden Carbon Asset Management Ltd	Eco-securities Group
Save energy	306	Camco International Ltd(UK)	Victorinox Multi-oil group
Methane recycle	111	Netherlands International Energy Systems, Inc.	UK Emissions Trading Co., Ltd
N2O Emission reduction	23	Eco-securities Group	Mitsubishi
Fuel substitute	21	UK Emissions Trading Co., Ltd	Netherlands and China Carbon Fund
Greenhouse gas disposition HFC-23	13	ENEL	4C funds
Others	25	-	-

Source: Tianjin Climate Exchange.

2. Strategic Cooperation with Other Countries to Respond to Climate Change

China and USA and both the world's carbon dioxide emissions countries, and big producers and consumers of energy. They are complementary in terms of energy technologies. In August 2009, the two governments established a joint Sino-US Clean Energy Research Center, dedicated to promote joint research of Sino-US scientists and engineers in the field of clean energy technologies. The priority areas of the research center include: energy-efficient buildings, clean coal, clean energy vehicles, carbon capture and storage, smart grid, shale gas, the second and third generation bio-fuels, and advanced nuclear power. The research center will provide platforms and support for the relevant units in two countries to engage in bilateral cooperation of energy science and technology, and play an active role in enhancing Sino-US technological cooperation.

The European Union always takes a hard-line stance in dealing with global climate change and reducing greenhouse gas emission. EU expects that by 2020, the emission in developing countries will have reduced by 15% -20% compared with 1990, but, while China wants EU to raise their own emission reduction commitment from 20% to 30%. EU hopes that China will announce a medium-term emission target, similar to what developed countries do, in order to persuade other developed countries into making more comprises. Currently, China and the EU are experts to study the problems with transferring technologies of relevant fields. In addition, EU hopes that China will make full use of the market mechanism to solve China's funding problems in implementing emission reduction. China-EU Clean Energy Center currently being established in Beijing will be of great help to China.

India and China are both developing countries, face similar pressure in dealing with climate change and hold similar attitude on this issue. Even in the case of economic high growth, the carbon emission of developing countries is much lower than the average amount of carbon emission of developed countries. China and India both refused to accept the binding carbon emission cap. Faced with the similar pressure, these two countries may establish good cooperation relationship in the international negotiations on climate change issues.

SECTION 3: PRACTICE AND DIFFICULTIES OF CHINA'S EMISSION TRADING PILOTS

Currently, China's emission trading pilots are being carried out in some provinces and cities. Cities, including Tianjin, Hunan and Hubei, have established regional environmental exchanges. Some areas have made remarkable progress in reducing emission while others are still facing difficulties.

I. China's Practice in Emission Trading Pilots

The emission right market in China has a great potential and attracts many strategic investors. At the same time, large enterprises and government departments are actively exploring approaches to optimize the allocation of environmental resources through market mechanisms. The number of emission trading exchanges established by the environmental regulatory is gradually increasing. The current and planned emission trading pilots are shown in Table 7-7.

Table 7-7. Established or Planed Emission Trading Pilots in China

Established	Beijing Environment Exchange
	Shanghai Environment and Energy Exchange
	Tianjin Climate Exchange
	Equity Rights Exchange Center
	Emission Trading Center in Zhejiang Jiaxing
	Emission Trading System in Zhejiang Huzhou
	Environmental Resource Exchange in Changsha, Hunan
	Energy-saving and emission-reducing projects in Luliang,Shanxi
	Guanggu Property Exchange in Hubei Wuhan
	Environmental Exchange in Kunming, Yunnan
Planned	Trading System of Chemical Oxygen Demand in Jiangsu Taihu Lake Basin
	Emission Trading System for Water Pollution in Nantong ,Jiangxi
	Sulfur Dioxide Trading Center in Heilongjiang
	Emission Trading System in Shaanxi Weihe
	Emission Trading Pilot Project in Chengdu, Sichuan
	Emission Trading Pilot in Tangshan, Hebei
	Emission Trading Pilot in Shenzhen and Guangdong
	Greenhouse Gas Trading System in Guangdong and Hong Kong
	Emission Trading Pilot in Henan
	Gansu and Ningxia, Zhejiang, Wenzhou, Quanzhou, Fujian and etc
	Hong Kong Exchanges and Clearing Limited

Source: Tianjin Climate Exchange.

1. Local Support for Regional Emission Trading Pilots

The exchanges in Changsha, Wuhan, Luliang, etc are established by the local governments. Changsha, Hunan Environmental Resource Exchange was officially established and held its first auction on November 28th, 2009. It was authorized by Hunan Provincial Environmental Protection Agency to execute emission trading of sulfur dioxide and chemical oxygen demand in the entire province. On March 19th, 2009, Wuhan-Guanggu United Assets and Equity Exchange delivered chemical oxygen demand emission permits of 100 tons and sulfur dioxide emission permits of 1,000 tons. The total transaction amounted to 956,408 RMB. Shanxi Lvliang Energy Saving and Emissions Reduction Trading Center was established in Beijing in July 2008, signed "Commission Agreement of CDM Project Development" with 10 units, and selected a number of key projects.

2. Environmental Financial Market-Oriented Emission Trading Exchanges

At present, the three major exchanges designed to build environmental financial market are Tianjin Climate Exchange, Beijing Environmental Exchange and Shanghai Environmental Energy Exchange.

Beijing Environmental Exchange was funded by Beijing Equity Exchange, and the major trading modes are over-the-counter (OTC) trading and open selling. In 2008, the amount of carbon dioxide reduction listed for open selling in Beijing Environment Exchange was 8895 tons, which was verified by Tsinghua University and has been successfully sold. The source of verification is the registration of enterprises and institutions during the traffic ban of Olympic Games (July 20 to September 20), using an even-odd system based on license plate numbers that keep vehicles off the road on alternative days. Since then, Beijing Environment Exchange established the strategic cooperative relationship with BlueNext of France, the world's largest Spot EUA Stock Exchange and all the projects listed in Beijing Environmental Exchange will be simultaneously listed in BlueNext, opening direct sales channels for CER produced by the CDM projects in China.

Shanghai Environmental Energy Exchange was approved by Shanghai Municipal Government. With the support of METI in the energy-saving technologies, its transactions have exceeded 1 billion RMB, during the period beginning in 2008 and ending in September, 2008. The traded items are mainly the common stocks of the energy equipment manufacturers, provided by the Shanghai Property Rights Exchange. Soon afterwards, commissioned by the United Nations Development Program, Shanghai Environmental Energy Exchange will establish a training center for strategies to deal with global climate change in developing countries". On April 29th, 2009, Shanghai Environmental Energy Exchange signed a cooperation contract with Hangzhou Property Exchange, in which they agreed on the construction of the emission trading platform in Taihu Lake Basin. On August 5th, 2009, Voluntary Emission Reductions (VER), a voluntary emission trading mechanism and trading platform, was launched. During the 2010 Shanghai World Expo, visitors from all over the world can purchase their own carbon emission credits through this platform and reduce their reduction voluntarily.

Tianjin Climate Exchange, co-funded by China National Petroleum Corporation, Chicago Climate Exchange and Tianjin Municipality in August 2008, is China's first comprehensive emission trading institution. Since its establishment, Tianjin Climate Exchange has been actively exploring emission trading mechanism of major pollutants, and conducted its first electronic auction of sulfur dioxide emission trading in December 2008. On February 14th, 2009, Tianjin Climate Exchange became the strategic cooperator of "Enterprises switching in China - investigation on the low-capacity enterprises", dedicated to enhance the switching of low-capacity enterprises and promoting emission trading. In addition, Tianjin Climate Exchange will also shift its focus to the shareholders' resources, utilize the energy-saving and emission reduction potential of China's oil groups and offer a wide range of services, including the development of CDM projects and the construction of energy-saving and emission reduction platform.

3. Emission Trading Planning to Break the Administrative Regional Restrictions

The aggregate credits of sulfur dioxide emission in China's power industry have been allocated using a unified performance approach. Based on the aggregate credits set by the state, the administrative departments of environmental protection in provinces, autonomous regions and municipalities will allocate sulfur dioxide emission credits to power supply enterprises according to performance credits of sulfur dioxide emission and local environment situation where these enterprises are located. Affected by the acid rain, Guangdong and Hong Kong started the cross-regional negotiations for the sulfur dioxide emission permits since 2003. And in 2006 they signed a framework agreement. The specific trading model is as follows. Following the pre-set price, Hong Kong will allocate special funds for the construction of desulfurization in Guangdong Province, and the amount of desulfurization completed in Guangdong Province will be counted as Hong Kong's. According to their agreement, by 2010, both Guangdong and Hong Kong will have cut sulfur dioxide emission by 30%. Due to the high cost that Hong Kong enterprises face in FGD abatement, Hong Kong decided that it is difficult to complete its desulfurization task before deadline; hence Hong Kong applied to Hong Kong and Macro Affairs Office of the State Council for cross-regional trading. In contrast, Guangdong province has a huge potential for desulfurization and Guangdong provincial government hopes that enterprises will benefit from pollution discharge and become actively engaged in emission reduction through sulfur dioxide emission trading. Consequently, after the approval of the National Environmental Protection Administration, both of the two provinces are still in a specific stage of negotiation.

II. Difficulties of China's Emission Trading Practice

Generally speaking, the scale and degree of China's emission trading are far behind what is need to protect the environment. The obstacles to further developing emission trading are summarized as follows:

1. Lack of Legal Foundation

In recent years, Chinese government established a series of emission trading pilots and aimed to optimize the effect of environmental control through introducing market mechanism and establishing the relative incentive mechanism. China ratified United Nations *Framework Convention of the Climate Change* in January, 1993 and Kyoto Protocol on August 30, 2002. In order to carry out the regulations in Kyoto Protocol and fully utilize the CDM to promote the energy saving and emission reduction in China, on October 12, 2005, four ministries, including the National Development and Reform Commission, jointly issued *Measures for Operation and Management of Clean Development Mechanism*, which specified the operation and management of CDM projects in China and established the legal framework for the greenhouse trading based on CDM. However, it did not address whether CDM should be allowed to trade CER in China and there are no legal regulations for the emission trading of other pollutants, such as SO_2 and COD. It is difficult to decide its ownership and tradability.

2. The Immaturity of the Domestic Market

The immature nature of the domestic market for greenhouse emission is shown as follows: First, the number of unilateral projects is small. By September 7, 2009, 43 unilateral CDM projects have been approved by the National Development and Reform Commission, accounting only for 21.5% of the approved projects, while in India, this figure reach as high as 40%. In order to avoid and control the risk and ensure that at least 2% of the emission income of CDM projects could be classified into the national CDM mechanism fund, China's policies do not support unilateral projects. Consequently, most of the carbon trading in China's domestic market is open selling, with the traditional property trading as the trading target other than the real CER price competition. However, since the unilateral project started to appear in 2005, India has adopted the unilateral carbon strategy and stored CER from registered CDM projectss so that they could be available for use or sell in the future. This would help India both with controlling the market fluctuation and reducing the cost of emission reduction. Second, China's financial industry is well monitored. According to Administration of Futures Trading Tentative Regulations, the establishment of options exchanges must be approved by China Securities Regulatory. Without the approval of China Securities Regulatory, any individual or entity must not establish or establish in disguised forms futures exchanges. The trading of CER traded in CDM is similar to futures in nature. In most of the exchanges overseas, almost all the CER are traded as futures and not many of them are traded as existing goods. Consequently, China's monitoring of the financial market has limited the local trading of carbon futures to some extent. In addition, CDM projects are highly costly and difficult to register for. The certificate verification institution, DOE, charges as high as 100,000 RMB for one project, making the CDM projects in the current market with a carbon reduction amount of less than 100,000 not desirable to be further developed.

Other obstacles to the emission trading of major pollutants are listed as follows. First, local protectionism still exists. Some local governments consider reducing emission as restricting production and hence, allow enterprises to increase their emission for the interest of local economic profits. In addition, in some cross-city and cross-province emission trades, the administrative departments where the sellers are located often forbid the sellers from selling the emission credits to other regions and only allow then to trade locally. This protectionism also restricts emission trading and the effective operations of the emission market. Second, it is difficult to achieve the balance between the target amount of total emission and the pursuit for economic growth. The total emission in a region must be first verified using scientific methods and once verified, should not be adjusted with a period of time, usually 12 months; otherwise, the target total emission may not achieved. The verification of emission right in China is still in the initial period. The growth model of extensive economy and the pursuit for high rate of development has not been fundamentally changed in some regions, leading to the conflict between economic growth and control of target total emission. How to decide on the total amount of emission has become the most difficult step in the emission trading and the emission trading system is made weaker since some regions often break the bottom line of their target total emission.

3. Uncertainty under the Changing Framework of Global Market

First, there may be great changes to the existing CDM. According to the Act promulgated by EU commission in December 2008, the number of CDM projects in large developing countries, such as Brazil, China and India will be greatly reduced after 2012 Since EU enterprises are the largest purchasers of carbon credits, EU Commission has the major right of speach on the operations of carbon market in future's climate agreement. For example, EU's ban on forest carbon credits leads the projects in this industry into difficult situations----to some extent, this is because if EU enterprises do not purchase the carbon credit, the incentives from the market will be greatly reduced.

Second, emission reduction obligation can also be adjusted. The Copenhagen Conference to be held in December, 2009 aims to reach a new agreement on dealing with climate change. But in the current global financial crisis, some developed countries are unwilling to take an excess emission reduction obligation. Some developed countries, such as Britain and German, who have achieved their emission goal in the previous phase (2008-2012) are also unwilling to undertake the unilateral emission reduction obligation. This will have a direct impact on the scale and trading structure of the global emission trading market.

Column 7-1. Chinese Airlines Have Been Incorporated Into EU Emission Trading System

On January 13th, 2009, the European Commission (EC) officially passed the Act "Directive 2008/101/EC", and incorporated all airlines operating European routes into the EU Emissions Trading System (EU Emission Trading System, EUETS). On 2009 2 11, European Commission officially released the list of the major airlines and their regulator countries. 33 Chinese airlines, including Air China, Sichuan Airlines, Hainan Airlines, China Southern and other are in the list.

The Act could be summarized as follows:

1. Considering that global warming should be controlled within 2 degrees Celsius over the next 50 years, EU decided to incorporate all the airline companies that operate business and land flights in Europe into the scope of greenhouse gas emission control;
2. Having passed 12 months' verification by the European Union, 1593 airline companies will be incorporated into the management scope of the European Union Emission Trading System (EUETS);
3. In order to soften its emission reduction target, the EU will distribute 97% free EU emission Allowance (EUA) to these airline companies with 2004, 2005 and 2006 as their baselines, decrease to 95% in 2013, and after the third year remain at 95% and move the baselines onwards. But the EUA allocated to the airline companies are not tradable.
4. 15% of the EUA allocated that the European Union allocated to the airline industry will be auctioned, and the profits from the auctions will be used to deal with climate change;
5. These airline companies will be assigned to EU member countries responsible for monitoring their emission credits;
6. The airline companies incorporated into the EUETS should use certified emission- reduction (CER) and joint implementation of emission reduction (ERU) to offset the EUA that should be purchased based on that year's income from operations. But the amount of reduction must be consistent with EUETS' requirement on CER and should not exceed the offset ratio that the monitoring member countries are allowed to adopt. Currently, the average offset ratio for EU membercountries is 15%.
7. The airline companies incorporated into the EUETS should submit the energy usage data and the monitoring report of greenhouse gas emission based on airline mileage before August 31, 2009; otherwise, they will not be granted the free EUA in 2012.
8. The airline companies incorporated into the EUETS that over-emits and do not pay for EUA will be fined 100 euro / tCO_2e;
9. If the airline companies incorporated into the EUETS do not comply with this legislation and EUETS regulations issued in 2008, monitoring member countries could request the European Union to suspend those airline companies' operation in Europe.

The date for submitting monitoring report proposed in the EU Act has passed, but the National Development and Reform Commission is still in negotiating with the EU and major airlines are waiting for the outcome of the negotiations. It is estimated that a consensus will be reached a short time before or after the Copenhagen meeting. However, during this meeting, the EU is making efforts to reduce a greater amount of emission and will treat all airline companies equally. Despite the Government's' negotiations, it is still difficult to suspend the implementation of the act. Sooner or later, China's airline companies operating European routes will be incorporated into the EUETS, and became China's first batch of enterprises demanding emission reduction.

Source: Wei Yiming et.al, *The Enlightenment of EU Emission Trading System to China, Science Times*, A3, August 20, 2009; Wei Dunren, *The Enlightenment of EU Emission Trading System to China on Dealing with the Climate change*, *Wen Wei Po*, September 9, 2009, http://whb.news365.com.cn/kjwz/200909/t20090909_2457900.htm.

Third, the domestic emission markets of countries should be connected to the international emission trading market. Under the framework of the global emission trading, some countries have also started constructing domestic emission trading market, such as the voluntary emission reduction plan in Chicago Climate Exchange and the voluntary greenhouse gas emission reduction plans in some regions of China. Under the agreement made in Copenhagen Conference, the domestic emission trading market should be connected with the international trading market, in order to establish the global emission trading market, liquidate the validated emission quotas And lay a foundation for the development of environmental financial market.

SECTION 4. OUTLOOK OF CHINA'S EMISSION TRADING MARKET

According to the National Assessment Report on Climate Change, China's general guideline to mitigate climate change is as follows: under the premise of realizing the social and economic development and the goal of building a moderately prosperous society in all-round way by 2020 with basic realization of industrialization and modernization, we should update the economic growth and social consumption models, develop and promote advanced energy-saving technology, improve the efficiency of energy-use, advance renewable energy and nuclear energy technology, consume efficient, clean, low-carbon coal, utilize hydrogen energy technology to optimize the energy structure, protect the ecological environment and so on. We should take the route of developing low-carbon economy, and establish a system and mechanism to mitigate climate change step by step and reduce the emission of carbon dioxide and other greenhouse gases. Therefore, we may have some breakthroughs in the following aspects of emission trading.

I. Expectations to Establish a Market Structure on Legal and Policy Basis

To establish a reasonable and orderly emission trading market, there must be relevant laws to follow and incorporate emission trading into the scope of legal management, and policy guidance. Emission trading system is concerned with market-economy behaviors and based on traders' ownership of environmental property. This requires the authority of laws, the clear definition of property rights, and the entity of judicial interpretation right in executing the laws and regulations. An effective property right system must clearly specify the owner of the property, the right to dispose the property, the right to gain benefits from the property and undertake the uncertain all the relavant risks and costs.

Meanwhile, the emissions trading system emphasizes the role of government regulation. Especially in the economy transition, due to the lack of complete regulations, regulators should take a more active role in executing the regulations. But what the regulators do is contrary to what is expected from them .Regulators usually are not inspired to control the pollution for various profits and interests. The emission reduction system does not guarantee that these problems will be avoided. Thus, it is necessary to evaluate the responsibility of regulators in policies and impose aggravated penalties on law-breakers.

In addition, policies and regulations of emission trading must be completed. As seen from U.S. experience, Clean Air Act plays an important role in the sulfur dioxide emission trading. At present, China does not have relevant laws. Therefore, Premier Wen Jiabao pointed out in August 2009 at National Climate Change Group meeting that we should improve the legal system to deal with climate change, speed up the establishment of supplementary regulations and policy system, formulate, monitor and evaluate corresponding standards, the necessary management system and supervising mechanism. Currently in China, there are already some local rules and regulations available for emission trading and the legal conditions for emission trading in China are being built.

II. Establishment of Market Model Based on the Medium-term Goal of Global Climate Change

Setting the emission trading model based on the domestic and global medium-term goal of climate change is one of the viable models to explore. If the State targets carbon dioxide emission intensity, we should calculate the upper limit using the total GDP and increment, and conduct the difference trading. At present, many countries have already set their medium-term goal to deal with global climate change, specifically as shown in Table 7-8.

In order to meet the demand of international market for carbon offsets abroad, the exploration for China's international emission trading models can be made through establishing exchanges targeted at overseas traders and developed the same products as those in Chicago Climate Exchange(CCE) and ECE (European Climate Exchange).

III. Expand of Market through Improving the Emission Reduction Technology and Financial Support

At present, most of CDM projects in China are supported by the funds transferred from international cooperators, and little are supported through the transfer of emission reduction technology. In the global climate negotiations, developing countries have repeatedly called on developed countries to provide technical support. But these calls had not been effective, due to a higher degree of privatization in developed countries, where the emission reduction technologies are mostly held by private enterprises. On the other hand, some large enterprises in China lack advanced technology other than sufficient fund. If we could make more effort to promote the advancement of emission reduction technology, it will be beneficial for those enterprises to engage in the emission trading market and for the market to expand.

At the same time, if we want to take full advantage of the emission trading system as an economic incentive mechanism, we should keep improving the market economic system. Economic incentives in the financial markets should also play an important role. For example, only when the emission trading could be stored and its price fluctuates with the market, the emission trading can be set up. This requires coordination with the financial system. In initial phase of emission trading, credits are not required; but the long term, it would be favorable to the development of the emission trading system if banks offer business to emission enterprises to use their future emission targets, the enterprises will benefit a lot. , the enterprises that emit pollutants also need the financial support from the market. If the enterprises short of funds cannot manage to finance their projects in the markets, they will not have capital to spend on controling the emission of greenhouse gases and other major pollutants. The large-scale facilities for carbon capture, carbon separation, and pollution abatement also need the support from both money and capital markets. Without these economic incentives these facilities will not operate due to the lack of the necessary funding.

Translator: Yu Yan (SUFE)
English Version Editors: Jiang Jiahui (SUFE), Ding Mengjie (MIT)

Table 7-8. List of the Emission Situations and Emission Reduction Goals in Different Countries

Nationality	Emission situation in 2006	*Goals of the countries in Kyoto Protocol* 2008-2012	Mid-term goal since 2012 (2020)	Long-term goal since 2012 (2050)
EU	-4.6%	-8	Reduce emission 20% or 30% according to the national agreement	Reduce emission 60%-80% according to the national agreement
France	-9.4%	0	-	Reduce emission 75%
German	-19.3%	-21%	Reduce emission 40%	-
UK	-15.6%	-12.5%[20%]	Reduce emission 26%-30%	Reduce emission 80%
Norway	-28.7%	1%[-10%]	Reduce emission 30% by 2020	achieve carbon neutral by 2030
Iceland	9.8%	10%	-	Reduce emission 50%-70% and achieve carbon neutral
Canada	54.8%	-6%	Reduce emission 20% compared with year 2006	60%-70% Reduce 60%-70% compared with year 2006
Japan	5.8%	-6%	Lower 15% than in 2005	Lower 60%-80% than in 2006
News land	33.0%	0%	-	Achieve carbon neutral in energy department
Austria	6.6%	8%	Lower 5%-15% than in 2000	Lower 60% than in 2000
America	14.0%	-7%	Lower 17% than in 2005	Lower 80% than in 2005
Mexico	-	-	-	Lower 50% than in 2020
Korea	-	-	Set the emission cap in 2012	-
Costa Rica	-	-	Achieve the carbon neutral by 2021	-
South Africa	-	-	peak at 2020-2025	-

Source: Tianjin Climate Exchange.

REFERENCES

Following CDM and JI, (2009,1-9)fortnightly Carbon Point CO.,LTD.

Stern N., 2008. Key Elements of a Global Deal on Climate Change [R]. London: The London School of Economics and Political Science.

UNFCCC. Report on the Workshop on Cooperative Sectoral Approaches and Sector-specific Actions, in order to Enhance Implementation of Article 4, paragraph 1 (c), of the Convention. Summary by the chair of the workshop [R/OL]. (2008-08-25) [2008-09-19].

Aldy J, Stavins R., 2007Architectures for Agreement: Dealing with Global Climate Change in the Post-Kyoto World [M]. Cambridge, UK: Cambridge University Press: 1J27

IPCC, 2007. Climate Change 2007: Synthesis Report. Contribution of Working Groups I, II and III to the Fourth Assessment Report of the Intergovernmental Panel on Climate Change [R]. Geneva, Switzerland: IPCC.

Su Wei, Lu Xuedu, Sun guoshun, 2008.The Core Elements of Future UN Negotiation on Climate Change and Its Perspective——The interpretation and analysis on Bali Roadmap[J], *Advances in Climate Change Research*, 4 1 57-60.

Deng Liangchun, Wang Yi, Wu Changhua 2008.To Explore the Road of Low-carbon Development: Important Orientation of China’s Sustainable Development[J], *Outlook in Climate Change*,, （1） : 1-16.

Zou Ji, 2008.International Mechanism Innovation of Technology Development and Transfer in Climate Change [J], *Environment Protection*, 5A 16-17.

Editorial Committee of China s National Assessment Report on Climate Change China s National Assessment Report on Climate Change[M], Beijing : China Science Press 2007.

In: China's New Deal
Editors: Li Xiaoxi and Hu Biliang
ISBN: 978-1-61470-815-5

Comment on Chapter 7

MARKET MECHANISM AND THE LOW-CARBON ECONOMIC GROWTH IN CHINA

Huang Jeff *

Since the industrial revolution, the greenhouse gas, produced by the fossil fuel- based economy, is one of the culprits of global warming in the recent 100 years. And no country was spared the negative impacts of the more frequent extreme weather events caused by global warming. The 30 years' reform and opening-up endeavor enables China to seize the opportunities brought by the global economic integration to promote quick development and make remarkable achievements. But the high energy consumption and high emission growth pattern is widely questioned at home and abroad. As a developing country, China should actively seize the opportunities to balance, through innovation, climate change mitigation efforts and the need to grow its economy. Most importantly, market mechanism, financial innovation, and government policy all should play a role in this process.

This Chapter made comprehensive analysis of China's policies and actions in dealing with climate change, energy conservation and emission reduction, with the focus on the emission trading in China, involving both the traditional pollutants as well as the development of CDM projects which is closely related to climate change. The statistics of the CDM Project Management Center of NDRC, cited in the Chapter, has shown that China's CDM mechanism has laid a solid foundation for the project-level carbon emission reduction. Currently, China ranks the first around the world in terms of the number of the CDM projects approved and the potential projects in the pipeline. Besides, At 208, China also ranks the first in terms of the number of CDM project developers active in the market.

However, CDM, which is symbiotic with the international climate politics, is facing well-publicized bottlenecks in credit registration and issuance. The low success rate means that the CDM mechanism can not be widely applied in domestic low-carbon transformation. In addition, the global negotiation over post-2012 Kyoto protocol has great uncertainties. Thus,

* Huang Jeff is Vice president of Chicago Climate Exchange.

CDM is facing reform to solve the broader and more pronounced issues concerning greenhouse gas emission in China.

Although there is no conclusive direction yet over the next-step development of CDM mechanism, the discussion over the sectoral mechanism of carbon emission reduction appears to be gaining attention, including the mechanisms of sectoral crediting and sectoral trading. Under a sector-by-sector approach, common baselines would be established for a particular industry - such as steel or cement - and plants could be awarded with credits for cutting emissions below that baseline. Specifically, sectoral crediting features a no-lose mechanism whereas sectoral trading represents a cap and trade mechanism at the sector level. Just as the case in the aviation industry cited in the Chapter, indicates, after 2012, EU will start to implement a cap-and-trade scheme for carbon emissions targeting all airlines that operate flights in and out of Europe. This seems to imply that airlines from China will have to comply in order that planes from China will continue to be able to take off and land in European airports. The sectoral approaches in other industries in the developing economies, such as steel, power and cement, are also being studied.

Broad consensus is emerging about the need for low-carbon economic growth for both developed and developing countries. In China, Premier Wen Jiabao has made a statement recently: in the next few years, China will apply low carbon development strategy to a broader part of its economy. That means China would expand the scale of clean energy development, industrial energy efficiency projects, high-efficiency power generation and methane gas power generation, among others. But it is not easy to transition from the current fossil fuel- centered economy to low carbon economy. For instance, in terms of funding, China needs an annual additional investment of 1 trillion yuan or more[1] to finance low carbon development. Climate change and low carbon development should not be viewed as difficult challenges ONLY in technological terms. Perhaps unlike some other, developed countries, the key issue in China is how market can be allowed to play a role in achieving emission reduction. Command and control may be effective short term. But in the long run, private sector engagement, stimulated by market mechanism, is indispensible.

Benefitting from CDM's development in China, the phrase "carbon trading" has become widely used in government circles as well as in the private sector. In addition, the term "carbon neutral" appears to become increasingly popular with corporations and private individuals in the area of corporate (and individual) social responsibility. But, what a real carbon market should look like and how it may work remain unclear in China.

The fundamental goal of carbon trading is to reduce overall carbon emission at the least cost. This would require a "cap" on CO2 emission. Under total emission amount capped, carbon emission reduction credits become a scarce resource. Then enterprises or individuals are allowed, under established rules, to trade permits or credits in order to meet the cap. The cap could be both mandatory, as in EU Emission Trading Scheme and voluntary (but legally binding once agreed upon) as in the "Chicago Accord" developed by and used at the Chicago Climate Exchange in the United States.

However voluntary carbon trading is a very vague term. The carbon emission credits, which is similar to CDM credits, but is not related to a mandatory emission capping scheme, is generally called VER in China. This article seeks to establish voluntary carbon trading as a

[1] NDRC Energy Research Institute, 2050 China Energy and CO_2 Emissions Report

cap and trade scheme where the only "voluntary" part involves the capping, especially in the absence of a government-mandated target on emission. Under such a voluntary scheme, a participant (an emitting enterprise) would voluntary sign on to the emission cap. Once inside the scheme, the emitting company must comply with the pre-set emission targets and if the company emits more than the target within the required time period, the company must buy carbon credits to meet its compliance targets. A typical example of the voluntary carbon trading market is the Chicago Climate Exchange (CCX). Established at the beginning of the 21st century, CCX recruited 13 emitting companies and started the Phase I of a voluntary cap and trade market in 2003 with 14 original members. Today it has over 460 members. The aggregate emission baseline of these enterprises has reached 600 million tons, exceeding that of Germany. For Phase I (2003 to 2006), CCX emitting members are required to cut their emissions by 4%. For Phase II (2007 to 2010), this target is 6%. The accumulative emission reduction by CCX members now exceeds 460 million tons CO2, of which 400 million tons come from direct emission reduction, 51 million tons come from "carbon offsets" projects, 11 million tons come from forest projects. All these cuts are third-party audited.

CDM is fundamentally an offset mechanism. It is related to the cap under EU ETS. Contrary to the popular understanding in China, CDM represents only a part of the carbon market. The cap drives the market.

Generally speaking, what needs to be emphasized is that both mandatory and voluntary carbon markets, a cap must be established in order for the trading to work. CDM and "carbon neutralization" projects can bring forth overall social, ecological and economic benefits. In the future more, these projects may continue to play a role in the Chinese carbon markets, but a real carbon market, like anywhere else, must involve a cap.

When talking about carbon cap in China, it is inevitable that the attention is on the eternal topic— the principle of "common, but differentiated responsibilities". For developing countries in the midst of industrialization, it may not be feasible to adopt an absolute Cap on CO2 output, as differentiated from the developed countries in this regard. A responsible developing country, China has taken active mitigation actions. China must reach the peak of aggregate CO2 emission in around 2030, according to a publicly released research report by influential scholars from a government think-tank in China. And during this time period, the emission intensity (carbon emission per economic output) can drop continuously, an example being the energy intensity target per unit of GDP, as specified in China's 11th five-year plan. In a similar vein, China could conceivably set a target over carbon intensity or cap the growth of carbon emission in the future. On a pilot basis, it is conceivable that Chinese emitting enterprises from different sectors could join together to design a voluntary cap and trade scheme. China is famous for crossing the river by feeling the stones and once a large scale demonstration pilot is operational, it could inform policy makers in the future as to how market mechanism functions in achieving environmental goals.

To finance the transition to a low carbon economy, putting a price on carbon is essential. The carbon price, as reflected in the price of CERs under the CDM mechanism, is widely known and cited in China. However, the CER price is influenced by too many factors beyond China's borders. What is needed is a domestic carbon market that sets the price on carbon domestically. Take bio methane development projects as an example. As the mid to long term objective for renewable energy development, China's sets up a 4.4 million m3 target to promote bio methane in rural area with heavy governmental subsidy. But the reality is not quite encouraging: in certain area in the northeastern part of China, within one year, 90% of the rural households abandoned their government subsidized methane pits 2. This case indicates that government funds may be needed to get the projects started, but they can not seem to sustain these projects due to various reasons including poor policy design and ineffective implementation. Fundamentally, the presence of a carbon price could make things easier. Compared with this, in the US, a farmer in the rural household could sell carbon credits generated from recovering methane from the digester because there is a carbon price on CCX. Of course, there was not US government subsidy whatsoever involved. A price on carbon could drive private investments into these areas, instead of government (i.e., taxpayers) subsidy, in a sustained manner. China has 30 million methane-generating pits, increasing by 5 million annually. This space could be very well served by a well designed, properly functioning domestic carbon market.

According to the International Energy Agency (IEA), emissions worldwide in 2020 would need to be cut by 3.8 gigatonnes (Gt) worldwide to achieve the 450 ppm (parts per million) scenario, with 1.6 Gt of this reduction needed in OECD (Organization for Economic Cooperation and Development) industrialized countries. Meanwhile, policies and measures in China – already being considered by the country's government – could account for 1 Gt of emissions reductions, underlining the leading role China will play in the global combat against climate change, the IEA indicated. In light of this, the real question that needs to be asked in China is no longer "have you cut your carbon lately". The question that gets asked should be "did you manage those cuts at the least cost".

REFERENCES

[1] The Stern Review on the Economics of Climate Change by Nicolas Stern, 2006

[2] Sectoral Approaches in Electricity – Building Bridges to a Safe Climate by International Energy Agency, 2009.

---

[2] China Economic Weekly, the scrap rate of the methane-generating pit in the rural area of Heilongjiang province, worth of hundreds of millions yuan, is over 90%.
http://news.ifeng.com/society/5/200908/0831_2579_1327527.shtml

In: China's New Deal
Editors: Li Xiaoxi and Hu Biliang
ISBN: 978-1-61470-815-5

Chapter 8

New Growth Engine: The Rise of Green Economy in China

Shao Hui**, *Lin Yongsheng, Qin Jian and Xiao Huaiyang

The 2008 Financial crisis suddenly led world economy into the slow traffic lane, but a new engine for the recovery of world economy-green economy also emerged. The expressions such as "Green New Deal", "Green Recovery" and "Low Carbon Economy" have been widely discussed by many national governments. Such Emergence means that the sustainable economic growth has become the mainstream of global development in today's society, where the idea of environmental protection has been deeply rooted in the minds of the people. For all countries, green economy is not only a moral choice, but also a development opportunity. Those who have seized the development opportunity will become first movers in the future development and take the lead— This incentive is also the driving force for all countries to develop green economy.

In the present chapter, the emphasis is mainly put on the development status of "Green Economy" in China, including measures taken and results achieved by Chinese government regarding the development of green economy and its system construction, especially with regard to some related economic policies, which have formed long-term mechanism to promote and safeguard the development of China's green economy. We have also enumerated several Chinese cases about the enterprise and urban construction, which have carried out through China's green economy policy. "Green actions" taken by enterprises have showed hope for the future, and these active actions taken in green urban constructions have shown that the idea of green economy has been understood and carried out in a rather broad sense.

Section 1: The Background of The Rise of Green Economy

Green economy is not a new concept which was a hot topic discussed when financial crisis in 2008 swept all over the world. In fact, we should be aware that green economy is an

*Dr. Shao Hui is lecturer at the Institute of Economics and Resource Management of Beijing Normal University, who earned her PhD degree from the Institute of Regional & Urban Economics, Renmin University of China. Email: shaohui1250@sina.com

eternal subject along with the development of human society. Today, people all over the world advocate the implementation of green economy, which is not a whim or a fashionable saying, but an inevitable choice to meet the needs of the times.

I. What is Green Economy?

The term "Green Economy" was first put forward in Blueprint for A Green Economy[1] written by economist David Pearce, which was published in 1989. Green economy stands for establishing "An Affordable Economy" and advocates the abolishment of the traditional development model which emphasize the pursuit of economic development without consideration of environmental and social costs. Such old models has caused ecological crisis, resource exhaustion and social disruptions. By correctly handling the relationship between "men" and "nature", "man" and "man" so as to efficiently realize the sustainable use of natural resources in a civilized manner, the continuous improvement of ecological environment and continuous enhancement of quality of life have become a mode of production or an economic development pattern. The core of such pattern is the sustainable development of the economy and society, so as to ensure the continuity and development of human society.

At present, the concept of green economy which is cited widely becomes more concrete. Green economy mainly refers to the mode of economic growth which has adopted environmental friendly technology to realize the harmonious development between men and nature, and coordinated economic and social development. Green economy focuses on environmental protection and clean production, and emphasizes the sustainability of economic development.

The philosophical root of "Green Economy" is not new to China. The philosophical ideas such as "Unity of Nature and Men" and "Tao Follows Nature" in ancient China have put emphasis on the harmonious coexistence between men and nature, but not "nature acts against men. Such philosophical ideas showed that Chinese had "Green Feelings" since ancient times. Since the 16th Party Congress Meeting, the Communist Party of China (CPC) has put forward the strategic thought of "Scientific Outlook on Development", which aims to correct certain "ungreen" elements of traditional economic growth model. The traditional economic model simply emphasizes economic growth, which might intensify the inharmony between men and nature. Especially hence the economy expands excessively, it might result in the loss of social value. As an alternative, the basic connotation and characteristics of the "scientific outlook on development" is "To put people first and to aim at comprehensive, coordinated and sustainable development so as to promote the overall development of economy, society and people." The core idea is to strike the balance between men and nature on the one hand and to establish harmonious relationship among individuals on the other hand. The essence of the globalized idea of "sustainable development"also requires to embody harmonious balance between men and nature, as well as among individuals.

[1]Pearce and others.Blueprint for A Green Economy , Earthscan Publications Ltd. , London, 1989

II. The Rise of Green Economy as an international phenomenon

In October and December 2008, at the backdrop of global financial crisis, economic recession, food crisis and energy crisis, the UN Environmental Programme(UNEP) initiated multiple proposals to implement "Green Economy" and "Green New Deal" around the world and proposed that green economy as a new development strategy to maintain economic growth in the current global crises.

Global financial and economic crises have offered opportunities for the development of green economy. Economic reconstructions and recovery after crisis require huge amount of investment. Green economy advocates changes of previous investment direction and mode so as to promote a sustainable development of economy and environment. In the past, environmental concerns were on the margin during decision-making process and a large amount of investment went into real estate and financial areas, which resulted in real estate bubble and financial bubble. A huge amount of subsidies were allocated to fossil fuel heavy industry that generate a large quantity of carbon emissions, but investment in renewable energy was not sufficient. Such policy intensified ecological environment problems such as forest deterioration, water loss and soil erosion, grassland degradation, resource exhaustion worsened. The UN Environmental Programme has realized that a new development strategy must be developed as the current status needs change. This strategy not only promotes economic development and social progress by investing in environment, but also drives industry adjustment and advancemeng of the "Greening" of national economy. Such strategy can also create new green job opportunities, which helps to recover and upgrade world economy. Therefore, green economy itself is not a new concept, but it was put forward systematically for the first time by the UNEP under the background of multiple crises. The United Nation, with regard to global energy, food, finance and the like. This systematical proposal is characterized by appropriateness of timing, relevance of political implications and high degree of influence.

Green economy starts to become a new trend in the field of environment and global development. Some countries in the world have started to develop its green economy by relying on economic stimulus plan. The European Union will invest 105 billion Euros to support "Green Economy" in EU region before 2013, and all the investments will be used into environmental protection projects and its relevant employment projects in the hope of maintaining the European Union's the leading position in the field of "Green Technology" the world. The new US Administration has put forward in the US \$827 billion economic-stimulus plan that US \$100 billion (which is roughly equal to 0.7% of America's entire GDP) will be used in green economic recovery in the next two years, which will create 5 million green job opportunities, and to speed up the transformation of its economy into to a low-carbon economy, one which regards new energy as its new characteristics. South Korea has put forward in "Green New Deals" that US \$36 billion will be invested in a series of eco-friendly projects in the next four years. In Japan's green economic plan, it has put forward to expand green economic market and provide a loan policy of zero interest rates for environment-friendly enterprises that create one million new green job opportunities. Brazil and India start their economic stimulus plans and new policies including the optimization of energy structure and emission reduction.[2]

III. Challenges for China

In 2008, China experienced many natural disasters. In January 2008, an unprecedented snowstorm and frost swept more than half of China, which affected tens of millions of people and caused hundreds of billions of RMB of losses. On May 12, the 7.8-magnitude earthquake hit Wenchuan County in Sichuan Province which caused tens of thousands of deaths. Are there man-made elements behind these natural disasters? This question is worth asking. Snowstorm and earthquake are caused by natural changes. More rapidly the human society is developing, more intensely the earth change might become, which makes human beings panic. China is experiencing an unprecedented development and the conflicts between human beings and nature is becoming more obvious.

As a large developing country, China is undergoing rapid industrialization and urbanization. From 1978 to the end of 2008, the total number of China's cities increased from 193 to 655, including the increase of megacities from 27 to 118 as well as the increase of metropolises from 13 to 39 respectively. The urbanization level is 45.7%, which is expected to be increased to about 75% by 2050.[3] That is to say, there will be at least 400 million people living in cities in the upcoming urbanization process, which requires a lot of construction materials like cement and steel to be consumed. It also means that a lot of raw materials and energies will be consumed and huge quantities of carbon dioxide and various pollutants will be released. At present, China's urban building area is 17 billion square meters, increasing at an annual rate of 1 billion square meters[4]. Since the reform and opening-up policy was carried out, Chinese economy has experienced rapid development. For nearly three decades, the annual growth rate of China's GDP has reached 9.5%. The development of large-scale industrialization and extensive growth mode have made the conflicts between economic development and environmental resources more obvious. Heavy industry has always been the dominance of Chinese economy. Driven by massive infrastructure investment in the process of economic development and urbanization, heavy industry, especially high energy industry has experienced the most rapid growth in recent years. The ratio of the output value in heavy industry to China's total industrial output value was 55% in 1985 and reached 69% in 2008. This developmental mode has put China in severe shortage of energy resources: 50% of its oil are imported; the reserves of oil and natural gas only account for 7.7% and 4.1% of the world average level.[5]

---

[2] Global Financial Crisis and Green Economy Review [OL]. Reference materials for 2009 roundtable conference organized by China Council for International Cooperation on Environment and Development, chief consultant experts technical support group in Taiwan ICDF. June 25, 2009. The website of China Council for International Cooperation on Environment and Development (CCICED) is http://www.cciced.org/.

[3] Chinese Academy of Social Sciences. China Urban Development Report[R]. Social Sciences Documentation Publishing House, 2009.

[4] WWF（World Wildlife Fund）Website information [OL]. http://www.wwfchina.org/index.shtm.

[5] Si Lin. Low Carbon Economy-New Green Label for Energy-Saving and Environmental Protection. China Construction News, Version 002, May 8, 2008.

Energy-intensive economic growth and industrial structure have resulted in severe environmental pollution, especially with respect of the use and development of energies which have brought about problems such as air pollution, water pollution, solid waste, the destruction of ecological balance, "Greenhouse Effect" caused by massive carbon dioxide emissions and so forth. The influence on environmental pollution is hard to quantify, but by rough estimation, the present environmental pollution condition is quite surprising. According to a research report of green national economic accounting from State Environmental Protection Administration, the economic loss caused by environmental pollution in 2004 was RMB 511.8 billion, 3% of 2004 GDP.[6] At present, the air pollution in 20% of Chinese cities is very serious, 70% of river water system is polluted and 40% is severely polluted. The rivers running through cities are commonly polluted, 1/3 of the land is affected by the acid rain, nearly 1/5 of land area suffers from some degree of desertification, nearly 1/3 of its land faces water loss and soil erosion, and over 90% of natural grassland is degenerating. Among the world's ten most polluted cities, China has six.

China has also realized that environment and resources put strings on its economic development.。 Although China has started its development of green economy for a while, the globalization of green economy pushed China to act further. Chinese people began to gain a deeper understanding of the concept of green economy, and they take more actions to practice such concept.

For example, with respect to energy-saving, the annual output of energy-saving bulbs in China has reached 1.3 billion and its utilization rate is now higher than many developed countries. In terms of solar energy technology, China has provided hot water by solar energy for 35 million buildings.[7] By the end of 2008, more than RMB 100 billion of China's total investment is directed to coal-fired desulfurizing machine with installed capacity of 363 million kilowatts, accounting for over 60% of thermal power-generating installed capacity. China has invested more than RMB 200 billion to build more than 1,550 sewage treatment plants and the urban sewage treatment rate has reached 66%[8]. All these achievements have shown that China is making positive efforts to tackle environmental problems.

SECTION 2: THE DEVELOPMENT OF GREEN ECONOMY IN CHINA

In recent years, China has made great efforts to develop its green economy and achieved many systemic and obvious results, including energy-saving and emission-reduction, circular economy, the promotion of green consumption and so on. In 2008 world financial crisis, Chinese government promised RMB 4,000 billion investment and started a economic stimulus plan with at least RMB 210 billion Yuan to be used directly to ecological and environmental protection. From the fourth quarter in 2008 to April 2009, China distributed RMB 23 billion Yuan to be used in energy-saving and emission-reducing, ecological construction and environmental protection projects, which accounted for 10% of newly-

---

[6] Pan Yue. Green Economy : To realize the sustainable development of the society, [N]. China Labor Security News, Version 003, April 9, 2008

[7] Cui Dapeng. Promising future in developing low carbon economy [N]. People's Daily Overseas Edition, Version 003, July 5,2008.

[8] Green economy: Open up new doors for the enterprises [N]. China Enterprise Newspaper, July 6, 2009.

increased state investment.[9] This shows that China places a very high value on green economic development to guarantee its growth and expand its domestic demand.

I. Energy-Saving and Emission-Reduction

Targeting intensive energy consumption and high discharge of pollutants, Chinese government has implemented energy-saving and emission-reducing policy for several consecutive years. The government has planned that the 2010 GDP energy consumption level with respect to GDP level will reduce about 20% compared with that of 2005 and the total discharge of major pollutants reduce 10%. These two goals are compulsory for governments at all levels. It is not easy to achieve this goal either for the more developed eastern coastal region or the less-developed western region. Most of eastern provinces at present are carrying out industrial restructuring from mature light industry to heavy industry, while western provinces with advantages of resource and energy hope to develop resource-based and energy intensive industry as the main impetus to start jump-start their economy. Therefore, it is difficult to realize the goal of energy-saving and emission –reducing under the current structure.

Chinese government has taken a series of measures[10] to promote energy-saving and emission-reducing since 2006, which is mainly reflected in the following aspects:

First to promote structural adjustment and industrial upgrading, including the control of the excessive growth of high energy consumption industries with heavy pollution, and the elimination of backward production sectors. For instance, small thermal power generators with a total capacity of 34.21 million kilowatts had been shut down from 2006 to the end of 2008. Funds used in autonomous innovation and structural adjustment arranged by the state can be used to reform and promote traditional industries, to improve the level of technology and equipment, and to develop high-added value products so as to promote energy-saving and emission-reducing.

Second, to increase input and implement a full range of key projects. The central government had planned RMB 33.6 billion budgetary investment and RMB 50.5 billion central treasury funds to support the construction of the energy-saving and emission-reducing key projects.

Third, to strengthen the management of energy-saving and emission-reducing, and to establish government accountability system for the energy-saving and emission-reducing process . Relevant governmental institutions will make assessment and evaluation of the annual energy-saving goals accomplished by provincial governments and the results will be announced to the general public. For regions and enterprises which are not up to the standard, they will be asked to make rectification within a time limit and might be punished.

---

[9] Development and Reform Commission: 10% of the newly-increased central investment is used in energy-saving & emission-reducing and eco-environmental protection project [OL]. Chinese environmental protection website, August 27, 2009, http://www.chinaenvironment.com/view/ViewNews.aspx?k=20090827105951085.

[10] Achievements and major measures made in Energy-saving and emission-reducing work since this year [OL]. August 2, 2009, National Development and Reform Commission Website, http://www.sdpc.gov.cn/hjbh/hjjsjyxsh/t20090802_294569.htm

Fourth, to improve the policy, to establish a stimulating and restriction system, to advance the price reform of resource products and adjust the price which is favorable to energy-saving and emission-reducing, and the fiscal & tax policy. For instance, at the beginning of 2009, the government lowered the price of vehicle purchase tax on small passenger cars; implemented "energy-saving products Waste Management Project" so as to increase the promotion investment in high-efficient and energy-saving products such as air-conditioners, refrigerators and the like through financial subsidies; supported to carry out energy saving and new energy vehicles demonstration pilot projects in 13 provinces such as Beijing, Shanghai, Chongqing and etc. The central government spent RMB 7 billion to encourage automobile and household electrical appliances industry and to encourage people "to replace an old one by a new one ".

Fifth, to strengthen the publicity, to improve environmental consciousness of the people and to build an economical type of society. For instance, to implement the "plastic limit order" and trash sorting, to encourage the residents to use energy-saving lamps, to spread the concepts of resource-saving and environmental protection among people by widely publicizing through the media.

Table 8-1. The Reduction of Energy Consumption Per Unit GDP Since the "11th Five-Year Plan" Period

	The Change of Unit GDP Energy Consumption
2006	-1.79%
2007	-4.04%
2008	-4.59%
The First Quarter of 2009	-2.89%
The First Half of 2009	-3.35%

Data resource retrieved from the National Development and Reform Commission Website http://www.sdpc.gov.cn/hjbh/hjjsjyxsh/t20090802_294569.htm.

Chinese government has taken painstaking efforts and as a result, energy-saving and emission-reducing work has achieved tremendously. Energy consumption per-unit GDP across China decreases every year. (Table 8-1) In 2006, the number fell 1.79% compared with the previous year. In 2007, it fell 4.04% compared with the same period and fell 4.59% in 2008. For three years combined, it dropped 10.1% cumulatively, and the total energy-saving was about 290 million tons of standard coal. The total discharge of the sulphur dioxide (SO_2) and chemical oxygen demand (COD) across the country also decreased, which fell 4.66% and 3.14% respectively in 2007, and fell 5.95% and 4.42% respectively in 2008. The previous three years before "The Eleventh Five-year Plan," this number fell 8.95% and 6.61% respectively.[11] Energy consumption per-unit GDP across China in the first half of 2009 fell 2.89% in the first quarter and decreased 3.35% cumulatively. The amount of decrease over the same period increased 0.47%, the industrial added value of unit energy consumption fell 11.35% compared with a year earlier. Estimated by State Environmental Protection Administration, COD and SO_2 discharge across the country in the first half of 2009 will

[11] Work arrangement for energy-saving and emission-reducing work in 2009 [OL]. August 3, 2009. Xinhuanet, http://news.xinhuanet.com/environment/2009-08/03/content_11819408.htm.

continue to maintain the momentum of "co-drop ", the number of which are expected to fall by 2% and 5%, respectively.[12] See Table 8-1.

The above table shows that energy consumption per-unit GDP in 2006 fell 1.79% compared with the same period last year.It decreased 4.04% in 2007 and 4.59% in 2008. For the past three years combined, it dropped 10.1% cumulatively and totally energy-saving was equivalent to 290 million tons of standard coal.

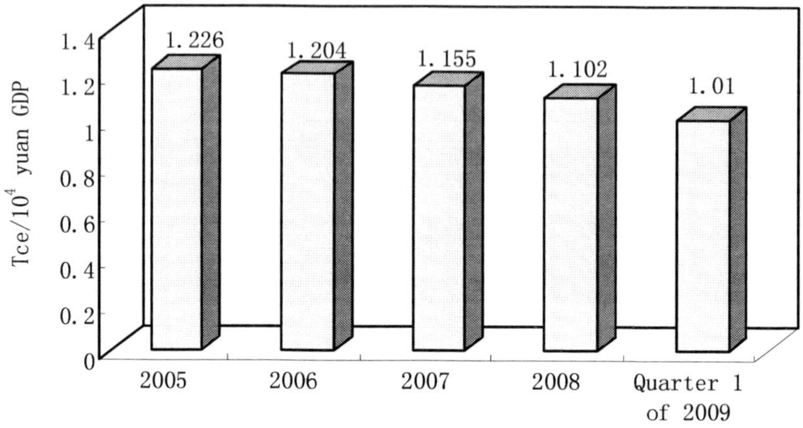

Data sources were collected and calculated according to the *Targets Announcement of Energy Consumption Per Unit GDP in each Province or Autonomous Region or Municipality in 2007 and 2008* jointly released by the National Statistics Bureau, National Development and Reform Committee(NDRC) and National Energy Board, and the quarterly statistical data is from the website of National Statistics Bureau.

Figure 8-1. The changes of energy consumption per unit GDP in China since the implementation of National 11th Five-Year Plan.

As shown in figure 8-1, RMB 10,000 GDP energy consumption in China in 2005 was about 1.226 tons of standard coal and reduced to 1.01 tons of standard coal in the first quarter of 2009.

II. Circular Economy

The difference between circular economy and traditional economy is as follows: Traditional economy is a linear economy with unidirectional flow of "resource-product-pollutants discharge", characterized by high exploitation (of resources), low utilization and high emission. Circular economy requires economic activities to organize economic production of "resource-product-renewable resource" characterized by low exploitation, high utilization and low emission, and to use materials and energies get sustainably so as to reduce the influence on natural environment caused by economic activities to a minimal level.

12 Achievements and major measures made in Energy-saving and emission-reducing work since this year [OL]. August 2, 2009, National Development and Reform Commission Website, http://www.sdpc.gov.cn/hjbh/hjjsjyxsh/t20090802_294569.htm.

Column 8-1. Plastic (Bags) Limit Order in China

Since June 1st, 2008,on a national scale all sales outlets such as shops, supermarkets and market places are forbidden to offer free plastic bags and all plastic bags must be clearly marked with prices. A nationwide ban on the production, sales and use of less than 0.025 mm thick plastic bags was implemented.

The positive influences to Chinese society caused by this implementation of the "plastic limit order" was great . First, it can save resources and serves the purpose to protect the environment. The cost of plastic bags is very low and recycling rate is extremely low. The number of plastic bags consumed in the world each year is an astronomical figure, a lot of which are contributed by Chinese consumers. Therefore, huge amounts of valuable oil resources are consumed. In addition, discarded plastic bags , which are refered to as "white trash" have resulted in serious damage to the environment,". To limit the use of plastic bags, to a large degree, can save oil resources and protect environment to a large extent.What is more, it can cultivate people's consciousness of environmental protection. "Plastic limit order" is directly related to people's daily lives, which can play a good role in spreading concepts of environmental protection and cultivating citizen's consciousness and behavior of environmental protection.

The "plastic limit order" has been implemented for one year as of June 1, 2009. The data released by China Chain Store and Franchise Association on June 20 showed that: at present, the usage rate of plastic bags has fallen by over 80% in foreign supermarkets and over 60% in domestic supermarkets. The usage rate of plastic bags in the supermarket and retail industries across the country fell 66% on average and the consumption of plastic bags reduced 40 billion. Compared with the figure before the "plastic limit order" was implemented, it has reduced 0.4~0.5 million tons of plastic consumption, which can save 2.4~3.0 million tons of oils every year and reduce 7.6~9.6 million tons of CO_2 emission. Many consumers have already formed into the habits of shopping with their own shopping bags and re-use plastic bags, and the consciousness of environmental protection is enhanced. (Data source: Lei Min, "*To use fewer than 40 billion plastic bags does not influence on the sales of the supermarket*" [N]. *Xinhua Daily Telegraph*, May 21, 2009: 7)

Aiming at correcting inefficient use of resources resulted from simple mode of resource utilization and high emission in economic activities, Chinese government started to advocate circular economy since the beginning of this century and focus on enterprises, industrial parks and regions and etc. with instant access to experiment and demonstration phase. There are 27 provinces and municipalities, 29 reproducible industrial zones, 79 enterprises, 4 agricultural towns or enterprises and 19 industrial parks, all of which have are representitives of circular economy in China.[13] In August 2008, China enacted the law of *Circular Economy Promotion Law of the People's Republic of China* . This law has the name of circular economy explicitly , and it was implemented officially in January 1 2009. This implementation means that the circular economy in China has entered to a stage of all-round developments and will play an important role in transforming the production and consumption patterns in China.

[13] Global Financial Crisis and Green Economy Review [OL]. Reference materials for 2009 roundtable conference organized by China Council for International Cooperation on Environment and Development, chief consultant experts technical support group in Taiwan ICDF. June 25, 2009. The website of China Council for International Cooperation on Environment and Development (CCICED) is http://www.cciced.org/.

China has carried out practical experiments for the circular economy at three levels and achieved remarkable results.

First, it is to actively promote clean production at the enterprise level so as to realize small circulation of circular economy and to build recycling enterprises. China is internationally acknowledged as one of the best developing countries in terms of clean production. In 2002, China enacted *Law of the Peoples Republic of China on Promotion of Clean Production.* At present, policy & law and regulations on local clean production have been stipulated in provinces such as Shanxi, Liaoning, Jiangsu, etc. and areas (cities) such as Shenyang, Taiyuan etc. According to statistics, China has started clean production audit in over 20 industries and over 400 enterprises in more than 20 provinces (regions and municipalities), and established Clean Production Center in more than 20 industries or regions with more than 10,000 people participating in different types of clean production training courses. More than 5,000 enterprises have passed the authentication of ISO14001 environmental management system and hundreds of products have environment labels.[14]

Second, ecological industrial parks should be built according to the principles of circular economy. The goal of ecological industrial park is to minimize waste materials as much as possible, to plan for scientific planning and to select scientifically and determine projects which can be entered into the park in accordance with the principles of industrial ecology, to regard by-products produced by a factory in the park as the input or raw materials of another factory. By means of waste exchange, recycling, clean production, among others, "zero-discharge" of pollution should be realized. At the same time, logistics, energy flow, technical integration, information and infrastructure sharing should be realized so as to maximize benefits. China has ratified 27 national eco-industrial demonstration parks to be built, including comprehensive industry zones and professional parks, covering industries such as sugar manufacturing, electrolytic aluminum, salt chemical engineering, coal mining, phosphorus and coal chemical, marine chemical, steel & coal chemical, etc.[15]

Third, it is to carry out the demonstration of the circular economy and to realize big circulation at social level to build resource recycling society. Approved by the National Development and Reform Commission, 27 provinces and municipalities such as Liaoning province, Guiyang city, etc have become circular economy pilot projects in two time periods. The pilot projects include the establishment of municipal solid waste, special waste and used materials, and water recycling system in the city so as to improve the utilization rate of renewable resource. The establishment of circular economy information platform can regularly issue the information on the supply and demand of the enterprise products, by-products, social waste and used materials to the society and release relevant technical catalogue and investment guidance of environment so as to promote healthy growth of circular economy.

---

[14] Wang Qiuyan (Chief Editor).China Green Development Report, [R], Page 118. Beijing: China Modern Economics Publishing House, 2009.

[15] The construction of the eco-industrial park steps into fast traffic lane [OL].September 11, 2008. The website of the Environment Protection Department of China, http://www.zhb.gov.cn/tech/stgyyq/sp/200809/t20080911_128562.htm.

III. The Start of Low Carbon Economy

At present, the viewpoints which received global recognition are: to develop low carbon economy is the basic way to solve the problems of world climate change and environment deterioration. With the increasing demand for the resource and energy consumption, China, as a developing country, faces increased international pressures from emission reduction result from greenhouse effect in the process of rapid industrialization, urbanization and modernization. However, rapid economic growth is still needed to improve the people's income and enhance their welfare thus more greenhouse gas emission is also inevitable. Under such pressures, the most urgent thing is to explore a new path to develop the low carbon economy which is in conformity with China's status quo. At present, China, taking international experience as its developing reference, is positively exploring ways to develope low carbon economy and researching into its relevant strategy, policy and technology. These are expected to enter into the pilot phase and overall propulsion phase of low carbon economy.

The Chinese government attaches great importance to climate change and considers the reduction of energy consumption per unit GDP to be measured by proxies such as obligatory targets to develop low carbon economy. China will continue to pursue the mode of production and the pattern of consumption which is favorable to preserve energy and protect the environment, and to establish recyclable and sustainable national economic system with low input and high output, low energy consumption & low emission so as to provide necessary policy foundation and condition for the development of low carbon economy. China will regard resource-saving and environmental protection as its basic national policy and the realization of sustainable development as national strategy. It plans to continue to actively respond to climate change and develop low carbon economy from three aspects including the adjustment of industrial structure, improvement of energy efficiency and development of clean and renewable energy resources.[16]

Further, China will actively adjust the economic structure and make great efforts to push forward the change in the pattern of economic growth: that is to speed up the development of the service industry, particularly modern service industries, to reduce the excessive dependence of national economic development on industrial growth and actively promote the development of high-tech industry to curb high energy, high pollution industries; to facilitate the transformation of foreign trade and to limit the exports of resource-based products simultaneously so as to accelerate the speed of the industrial structure adjustment. Also, it is to improve the standard of market access, to eliminate backward production capacity gradually, to speed up the annexation and reorganization of enterprises, to improve the intensive production level and effectively reduce the intensity of carbon emission per unit GDP. It is to continue to intensify efforts to save energy and improve energy efficiency, to reduce carbon emission intensity in high-energy-consuming industries such as steel, cement and chemicals, etc. by preserving energy and improving energy efficiency. In the future, China will also vigorously develop and use renewable and clean energy resources to develop

[16] Development and Reform Commission: Continue to develop low carbon economy from three aspects [OL]. June 19, 2009. Xinhuanet, http://news.xinhuanet.com/society/2009-06/19/content_11569119.htm.

and promote related techniques, establish necessary stimulating mechanism, carry out demonstration pilot and explore the low carbon development of key industries.

The measures to develop low carbon economy in China will mainly include five aspects: First, it is to develop and implement China's national plan to respond to climate change in terms of clear and specific goals, fundamental principles, major fields and policy measures in 2010. Second, it is to promote the transformation of the economic development pattern and the adjustment of the economic structure, to take policies and actions to eliminate backward production capacity, to encourage and promote the production mode and consumption pattern that preserves energy and resources. Third, it is to incorporate energy consumption per unit GDP as obligatory target into the "Eleventh Five-year Planning" and to establish the local enterprise responsibility system for energy saving and emission reduction which will be evaluated and assessed by level. Fourth, it is to vigorously develop low carbon energy such as water energy, nuclear energy, solar energy and rural marsh gas by enhancing policy guidance, enterprises' participation and capital input. Fifth, it is to further carry out the price reform in the fields of energy and resource as well as reforms in the fiscal and taxation system.[17]

SECTION 3: INSTITUTIONALIZED CONSTRUCTION OF GREEN ECONOMY

Entering into the new century, Chinese government has put great emphasis on the relationship between environmental protection and economic development, and has since put forward a series of new concepts such as scientific view on development. It has also made a series of reform explorations to promote the development of green economy.

I. The Construction of Legal System of Green Economy

Since the beginning of the 21st century, the goal of economic growth set by Chinese government is "good and fast". The state has put forward the scientific outlook on people-oriented, overall, harmonious and sustainable development and established a resource-saving and environment-friendly society. In order to coordinate the relationship between environmental protection and economic development, Chinese government has put forward three "historic transformations", namely the change from valuing economic growth highly and taking environmental protection lightly to attaching equal importance to both environmental protection and economic growth; the change from environmental protection which has lagged behind the economic development to environmental protection and economic development developed simutaneously; and the change from environmental protection carried out by using administrative means to the environmental problems solved by employing legal, economic and technical and necessary administrative means.

---

[17] Su Wei said: Our country will take 5 steps to develop low carbon economy [OL]. June 23, 2009.Chinaview（quoted from Economic Information Daily）, http://news.xinhuanet.com/fortune/2009-06/23/content_1158 55 68.htm

Since a long time ago, Chinese environmental protections were mainly carried out by State Environmental Protection Administration and other agencies with hige degree of administrative regulation. However, along with China's economic growth, environmental deterioration has become more severe. Government's attempt to solve the conflict between environment and development solved has become impractical and insufficient. The deficient environmental legislation obviously failed to match China's rapid economic growth. Such deficiency include lack of correspondingly assorted regulations and implementation rules, which value entity stipulations, lack of program specifications and value executive-led,. lack of public participation with weak sectors' cooperation resulted in weak operability of law and arbitrary law enforcement.

After the outbreak of the international financial crisis, Chinese government has paid more attention to the development of green economy. At first, it begins with system design and research on the cooperation between environmental protection administrations and other relevant institutions. The government restructuring reform was released in March 2008, the environmental protection agency was upgraded to the environmental protection administration, an upgraded from an institution directly under the State Council to a component of the State Council so as to better perform its environmental protection function, and the relevant laws were established and modified gradually. In 2002, China enacted *Law of the People's Republic of China on Promoting Clean Production* which facilitated enterprises to preserve resources and energies so as to realize a circular economy. In February 2008, the revised draft *Water Pollution Prevention and Control Law* passed by National People's Congress (NPC) stipulated the responsibility of local government for environment quality, which was regarded as one of its assessment criteria and imposed a heavier punishment on any illegal act regarding environment. In January 2009, *Circular Economy Promotion Law of the People's Republic of China* was officially enacted, which stimulated the development of circular economy from the entire social various professions by formulating a series of fiscal, taxation and financial policies. On April 1, 2009, the new revised *Energy Conservation Law of the People's Republic of China* went into effect, which provides a legal guarantee for the energy-saving and emission-reducing work.

II. The Tentative Exploration for the System of Green Economy Policy

In September, 2007, the framework of new environmental economy policy and roadmap were put forward for the first time in China in the "Green China Forum" organized by the State Environmental Protection Agency to construct seven environmental economic policies such as green tax, environment charge, green capital market (including direct financing--green securities and indirect financing—green credit), ecological compensation, pollution discharge right trade, green trade and green insurance within four years.[18] Subsequently, the State Environmental Protection Agency started to collaborate with the relevant *departments* such as China Banking Regulatory Commission (CBRC), China Insurance Regulatory Commission (CIRC) and China Securities Regulatory Commission (CSRC), etc. to introduce

[18] Pan Yue, Vice minister of the State Environmental Protection Administration: "Green China" nutured through environmental policy [J]. Study Monthly, 2007(19):19-21

a number of policies and started to make a tentative exploration for a system of green economy policy.

1. Green Credit Policy

China's green credit policy is that commercial bank should regard enterprises' environmental information as an important reference, and to take some measures such as to limit, stop or retreat bank loans for enterprises that have harmful impact on environment in the audit process so as to impel enterprises to control their pollution and protect environment. The practice will be helpful to curb capital expansion of the enterprises that are environmental unfriendly and to achieve the goal of pollution control using the economic means.

In July 2007, *Suggestions on the Implementation of Environmental Protection Policies & Regulations and Credit Risks Prevention* jointly released by the State Environmental Protection Agency, the People's Bank of China (PBC) and CBRC marked that the economic means of green credit had entered into China's green economic system and also marked the beginning of cooperation between the State Environmental Protection Agency and other government institutions. In November 2007, *the Guiding Opinions on the Credit Work for Energy Conservation and Emission* Reduction *formulated by* CBRC requires financial institutions such as the banking industry to combine the adjustment and optimization of the credit structure closely with national economic structure adjustment, in accordance with project classification of relevant institutions in charge of energy conservation and emission r*eduction so as to effectively prevent credit risks and adhere to* "differential treatment" *credit principle* for different enterprises.

At present, China's green credit policy has achieved certain results. Taking Industrial and Commercial Bank of China (ICBC) as an example, new policy implemented now will veto all projects which have a significant adverse effect on environment. In 2008, ICBC took the initiative to withdraw RMB 16.2 billion loan from environmental unfriendly enterprises, and cleared up 152 enterprise loans which came from "double high" industries (i.e. double high refers to high energy consumption and high pollution).[19]

However, there are many problems in green credit. For instance, information communication and sharing mechanism between environmental protection institutions and the financial departments is far from mature, which might fail to provide supports for banks to review and examine credits in a timely manner. In addition, fierce competition among banks will offer spaces for environmental unfriendly enterprises to get private loans from some banks while some environment-friendly enterprises with limited profitability are not favored by some banks. Green credit is far from adequate only by relying on the consciousness of social responsibility of banks.

The challenge now is to design an incentive mechanism for banks to remain profitable while being "green."

[19] Jiang Jianqing. RMB 16.2 billion yuan was actively returned and environmental protected was supported by "one-vote veto" [OL]. April 22, 2009. Sina Finance, http://finance.sina.com.cn/g/20090422/11026134 656. shtml.

2. Green Insurance

The introduction of China's green insurance started with *About Environmental Pollution Liability Insurance Instruction Opinion* released by the State Environmental Protection Agency (SEPA) and China Insurance Regulatory Commission (CIRC) in February 2008 This marks the second policy of green economy after the initiation of "green credit".

Environmental Pollution Liability Insurance refers to the insurance principle that insurance institutions should make timely economic compensation for the victims from accidents resulting from pollution caused by enterprises activities. This insurance is to stabilize socioeconomic order and to reduce the burden on enterprises. At the same time, insurance institutions will change the enterprise's insurance premium rate to push stronger environmental management from the enterprises' side. Proved by many countries, environmental Pollution Liability Insurance is an effective market mechanism for environmental risk management. It is explicitly stipulated in *About Environmental Pollution Liability Insurance Instruction Opinion* that: At the initial stage, the pilot work will be launched in enterprises which can produce, operate, store, transport and use hazardous chemicals, petrochemical enterprises which are more prone to cause pollution related accidents, and hazardous waste disposal enterprises, especially in those enterprises and industries which have caused serious pollutant related accidents.

At present, national pilot projects are developed and have made some interim achievements. In 2008, the pilot work in provinces and regions such as Hunan, Jiangsu, Hubei, Ningbo, Shenyang, etc., achieved remarkable results and successfully settled the first claim for compensation. In 2009, the pilot work in provinces and regions such as Shanghai, Chongqing, Ningbo, Shenzhen, etc., will start successively.[20] Chinese government hopes that, in 2015, various mechanisms such as risk assessment, loss evaluation and determination of relevant liabilities to accident handling and monetary compensation, etc. can be established by improving environmental pollution liability insurance system.

3. Green Securities

Green securities can exercise control and supervision over direct financing channel, and introduce environmental assessment and environmental information disclosure mechanism for listed companies to promote listed companies to improve their environmental performance and to guide the inflow of funds into environment-friendly enterprises, which can effectively protect the interests of investors against losses.

In February 2008, *About the Enhancement of Environment Protection Supervision of Listed Companies Instruction Opinion* officially released by the State Environmental Protection Agency focused on environmental protection verification system and environmental information disclosure system of the listed companies. Subsequently, China Securities Regulatory Commission (CSRC) issued *Notice On IPO Application Documents For Production Management Company in Heavily Polluted Industry.*

According to the requirements of the above-mentioned documents, China's green securities are mainly embodied in issuance application and post-marketing verification. On

[20] China Insurance Regulatory Commission: The pilot work for liability insurance of environmental pollution has been launched in Chongqing, Ningbo, Shenzhen and etc. [OL]. June 18, 2009. Sina Finance, http://finance.sina.com.cn/money/insurance/bxfg/20090618/15022902446.shtml.

one hand, audit advices given by environmental protection institutions will be considered as one of the required document that CSRC reviews. Such requirement is particularly needed for industries engaging in thermal power, steel, thermal power, steel and electrolytic aluminum production and enterprises operated via a cross-province network in “double high” industries (there are 13 types of heavily-polluted industries). On the other hand, after the companies are listed, environmental protection institutions will regularly report to the CSRC about environmental information of listed companies and a name list of listed companies which do not disclose their environmental information as required. In the mean time, their relevant information will also be announced to the public. Environmental Protection Agency will also release annual environmental performance indexes and rankings so that the majority of investors can discriminate and supervise the environmental performance of listed companies effectively.[21]

The policy of green security needs to be further improved, and effective implementation of green security policy will encourage funds raised from the public to go into green enterprises. Such incentive will promote listed companies to carry out their responsibilities for environmental protection, and economic rights and interests of green choice made by the majority of investors can be guaranteed and capital market can become an economic lever to promote energy-saving and emission-reducing industries and the development of green economy.

4. Green Tax

Green tax policies can be roughly classified into two groups: one is to impose an environment tax on enterprises that cause damage to the environment directly; the other one refers to tax preferences with regard to value-added tax, consumption tax and other taxes which are more preferable to environment-friendly enterprises. The export rebate rate and other tax preferences will be eliminated for environmentally unfriendly enterprises. At present, the study on China’s environment tax has already started. In addition, there are increased voices on the idea that resources such as water resource, forest resource and grassland resource should also be included into the scope of resource tax collection.

Although the time is not yet mature for China to make specific environment taxation regulation, yet it has started to introduce green tax system. In August 2008, *Preferential Catalogue of Enterprise Income Tax on Special Equipment for Energy Saving and Water Saving (2008 Edition), Preferential Catalogue of Enterprise Income Tax on Special Equipment for Environmental Protection (2008 Edition)* and *Preferential Catalogue of Enterprise Income Tax on Special Equipment for Comprehensive Utilization of Resources (2008 Edition)* were issued jointly by the Ministry of Finance, the State Administration of Taxation (SAT), National Development and Reform Commission(NDRC). In December 2008, the Ministry of Finance and SAT jointly issued *Notice on Comprehensive Utilization of Resources and Value-added Tax Policy of Other Products* and *Notice on Value-added Tax Policy of Renewable Resources.* All the above policies have introduced measures such as to encourage the comprehensive utilization of resources, to preserve energy and to reduce emission, to promote the environmental protection, etc.

[21] Journalist Chen Xiangjing. Green security policy launched by Environmental Protection Administration [N]. China Environment News, February 26, 2008.

The extensive exploration and practice of green economy policy are deepened in China, nonetheless, more interests start to be involved that complicated the exploration process. The establishment of the green economic system still has a lot of work and needs to secure supports from the government and various aspects of the society. Other green economic policies such as green trade, region and watershed environment compensation mechanism, pollution-discharge right trade need to be studied, to be improved or be formulated. The hope is that China can formulate a complicated system of environmental policy frame to ensure long-term development of the green economy.

SECTION 4: CASES OF CHINESE GREEN ECONOMIC DEVELOPMENT

I. Green City Construction-Green Beijing

Since we entered into the industrial era, the urban development became extremely quick. At present, China is in its accelerating period of urbanization. The rapid urban development will cause a series of problems especially in respect of urban environment in various aspects such as water, air, noise, waste, etc. The urban ecological system is very complex and can easily cause greenhouse effect and heat island effect, which have exerted influence on the sustainable development of any city. On the basis of more profound knowledge about the relationship between human beings and nature, green city uses environmental resources efficiently to develop a new mode of production and way of life which can realize the harmonious development between human and nature as well as economy and society. Such city also uses the principle of ecological economy and system engineering methodology to design a city so as to establish an economically developed, ecologically effective, healthy, harmonious and sustainable environment for human inhabitation.

Green city has become an irreversible green trend for the development of human beings. To establish a green city that keeps harmonious development between economy and society as well as environment has become a common subject in all societies. China has done substantive work on this and achieved great progress. A large number of demonstration projects have already been presented to the people. For example, the new eco-city of Tianjin, the first city to be different from tradition has started to be constructed in September 2008. According to the plan, Eco-city construction will be strictly carried out within a series of ecological index system. Taking “land-saving, water-saving, energy-saving emission-reducing, ecological easy living” as its target, the city will design all the buildings in the ecological zone as green buildings[22]. Caofeidian international ecological city located in Hebei Province will follow the principle of "ecological priority” and reasonably organize multiple urban ecological systems to achieve internal ecological recycling by using existing factors and adopting urban ecological design ideas to build an ideal city. The new city will make a new environment coexist with natural environment harmoniously and achieve sustainable development. The future “Caofeidian international ecological city” will promote the use of renewable energies such as wind power, solar power, ocean tidal power, geothermal energy,

[22] Jia Yihua. New eco-city planning during a reading: Green city of 30 square kilometers [EB/OL]. July 19, 2009. http://news.enorth.com.cn/system/2009/07/19/004130444.shtml.

etc. to generate electricity, while coal is regarded as an auxiliary energy. New building materials will be used instead of traditional ones, for instance, green building materials which can be permeated by water will be used in the pavement and there is a recovery system under the pavement surface that makes water backflow into the landscape lake. The transportation adopted in this future new town advocate buses, trolley buses, modem tramway, light rail, etc., with low pollution instead of private transportation and small vehicles.[23]

Beijing, as the capital of China and an international metropolis, has carried out the idea of green development into its urban construction. Especially in 2008, the Olympic Games has brought a good opportunity for Beijing to construct a green city. "Green Olympics" is one of the three principles adopted by Beijing Olympics Committee. When Beijing won its bid to host the Olympic Games, Beijing made seven promises to the world, including that the city green coverage rate would be close to 50%; mountainous area green coverage rate would achieve 70%; city green rate would be above 40%; the city would form three green ecological barriers; 23,000 hectares of greenbelts would be built in both sides of "five rivers and ten roads"; 12,000 hectares of isolated greenbelts would be built in urban area; the city natural protection area would be no less than 8% of city territory area. Through unremitting efforts, these seven promised green targets have all been achieved. Beijing has formed three green ecological barriers of mountains, plains and urban green isolation areas. Above 95% of its mountainous area has realized greening.

Beijing's green city construction is not only reflected in the "greening" of environment, but also in the seek of its sustainable development. Beijing is short of energy and resources and they are usually transported from the surrounding area. Thus, its external dependence is very high. To build a green city is also the requirement for Beijing to build an "Easy Living City", which is another local development orientation. In sum, Beijing's green city construction can be concluded by the following three aspects:

First, Beijing experienced the adjustment of industrial structure. Beijing city has accelerated the speed of high-pollution and high-energy consumption industry restructuring, and relocated Shoudu Iron & Steel Company and some other enterprises with high pollution. The orientation of future urban development and industry development has become more and more clear, and such orientation has further been transferred to high-end and high-efficiency industry. As China's political and cultural center, Beijing with its long history and profound cultural background also highlights the strength of its scientific and technical innovation with the unique advantage to develop its cultural and creative industry. Since 2005, Beijing has put forward clearly to regard cultural and creative industry as one of the most important pillars for its future development of its capital's economy and formed a set of unique cultural and creative industrial clusters. Modern service industry has also become an important economic pillar of Beijing. Beijing's service industry in 2008 amounted to 73% of GDP. In the first quarter of 2009, Beijing's aggregated economic output was ranked 10th nationally, its economic growth was the second highest, the proportion of its service industry to over all industries was ranked first. The percentage exceeded 76%, which is 32% higher than the national average, Since the service industry is a low-energy consumption industry, Beijing's

[23] Dai Chao. Tangshan scientific planning and development of the coastal areas and the establishment of international ecological city in Caofeidian [EB/OL]. June 2, 2009. Binhai news website, http://www.bh.gov.cn/bhsh/system/2009/06/02/010028053.shtml.

energy consumption per unit GDP in 2008 was 0.662 ton of standard coal, which decreased 7.36% compared to a year earlier.

Second, Beijing has highly efficient use of resource and energy which can be recycled. The external dependence for energy in Beijing remains high and the corresponding environmental requirements become higher. Therefore, in respect of hardware facilities, Beijing has paid more attention to enhance its ability to carry out environmental protection and accelerate the construction of sewage treatment facilities. During the "11th Five-Year Planning" period, 11 new sewage treatment plants will be built in the centre of the city. Beijing is also speeding up the construction of solid waste incineration plants, an important means to reuse resources and improve environmental quality. Beijing has made great efforts to use new energy. Yanqing has taken the lead in building 50,000 kilowatts of wind-power plants and filled the gap of Beijing wind-powered electricity generation. Thermal and photovoltaic use of solar energy has been greatly improved in Beijing. By the end of 2008, 110,000 solar lamps are used in Beijing.

Third, policy guidance of Beijing pays more attention to ecological environment construction. The measures for the implementation of enterprise clean production introduced in Beijing give certain amounts of subsidies and guidance for the expenses of enterprise spend to clean production. In order to further reduce traffic flow, Beijing residents are offered more options and lower prices in terms of public transportations such as subways or buses. Beijing has put forward to construct ecological conservation zone, and isolated greenbelts will be built from the outskirts of the city in accordance with the overall pattern of city planning Suburban parks around the city and ecological vegetation will be built at both sides of the 5th ring road.

II. Actions Taken By Enterprises

Enterprise is the base of market economy. All strategies and policies adopted in green economy can obtain their anticipated objectives only when they are truly implemented in the operation of enterprises and are accepted by the enterprises. In recent years, a number of excellent enterprises start to focus on sustainable development and have a strong sense of social responsibility. They actively respond to government's green economic development strategy, and explore new ways for their green development, which has achieved remarkable achievements. There are two reasons to explain the green transformation of these enterprises. On one hand, their action is caused by their responsibilities to the society, to the environment and to the public. These enterprises hope to establish a good public image and to achieve a long lasting trust from the market and consumers. On the other hand, the bottlenecks of resource and environment the whole society have to face are also the bottleneck of the enterprises' development. By making an attempt of green production, enterprises hope that they receive more opportunities in the future and to stay competitiven in the market.

In the following chapter, we have chosen three green enterprises from three different industries to illustrate the rise of green economy in the enterprise level.

The first one is the traditional energy industry, represented by Shenhua Group. The second one is advanced manufacturing industry, represented by Geely Automobile; the third one is producer service industry represented by Neusoft Group.

1. Shenhua Group- Green Exploration of Black Energy Central Enterprise

Coal, as the traditional fossil energy, is still the primary source of China's energy supply and consumption. When the world advocates “Green New Deal” and vigorously develops green economy, China’s coal industry and enterprises also have started to actively make green exploration. China’s Shenhua Group is a typical and successful enterprise among them.

China’s Shenhua Group Co., Ltd., as China's biggest coal producer and exporter, has the country's largest high-quality coal reserves. In 2006, the coal reserve of Shenhua Group was ranked the second place in the world's listed coal companies and its coal sale was also the second. Now, the Group’s business is expanded fields such as railway, harbor, power generation Shenhua Group has become a world leading coals-based integrated energy company.

Shenhua Group is the first state enterprise to publish its corporate social responsibility (CSR) reports (in 2007) and demonstrates to the community that state enterprises can fulfill their social responsibilities and take actions to undertake their social responsibilities. Green economic exploration made by China Shenhua Group is mainly manifested in two aspects[24]: One is that its coal is cleaner. For Shenhua Group, traditional energy cleaning mainly means how to use coals in a clean manner. Now, Shenhua Group has found another way: One is to undergo further transformation of coals, such as to change them into oil products and chemical products, and to develop coal oils and coal chemicals. In 2008, China Shenhua realized coal liquefaction on millions of its installations. During this process, sulphur, nitrogen and carbon dioxide can be well controlled. After carbon dioxide is extracted from coal-to-liquid plants in Ordos, Shenhua produces it into more crude oil additives along with oil companies in Ordos Basin. In such carbon dioxide capture and sequestration are realized. The other one is to vigorously develop wind energy. Shenhua group is aspired to become a leading wind power company in China and its wind energy plate has developed in large scale. In China’s coastal area and in the great western area, Shenhua Group has constructed its wind power plant on a large scale. Its future goal is that traditional energy should account for 80% in its energy supply structure. That is to say, it can supply not only coals and electricity, but also new energy such as wind power as much as 20% of its total production.

2. Green Standards Carried By “Four Wheels”— Geely Automobile Group

It's hard to imagine no cars (“four wheels”) in modem civilization, Nevertheless, automobile industry has also severe the conflicts among resource and environment and economic and social development. How to seek a balance among them needs one to implement clever strategies. After years of trials and mistakes, China Geely Automobile Group insists on its independent innovation and tries its best to seek for a solution. To be more specific, it tries to develop green technology that seeks green standards carried by four wheels.

Green concepts conceived by Geely Automobile are mainly reflected in four aspects[25]: First is the green materials. Geely promotes the usage of automotive materials that protects

---

[24] Huangqing: Shenhua has always stressed that how the natural cost is reflected in the coal price [OL]。 http://finance.sina.com.cn/hy/20090423/14146141688.shtml。 April 23, 2009, Sina finance.

[25] Li Shufu: The nation should improve automotive technical standard. [OL]. http://finance.sina.com.cn/hy/20090422/14276135569.shtml。 On April 22,2009, Sina Finance.

the environment and eliminates pollution of the environment. Briefly speaking, the standard is that when people sit in the car, no damage to their health can be made. This requirement set a high requirement for materials. Second is green spare parts. All the assorted automotive parts adopted by Geely Automobile must be green with environmental standards. When spare parts are used, the whole process must be green with environmental standard. The third is their green-manufacturing process. Geely Automobile requires that the manufacturing process be green with environmental standard, and the production process cannot harm the local environment and workers' health. At present, what Geely Automobile has done on the surface of their car is to besmear a coating. The coating Geely Automobile adopted is green with high environmental standard, unlike solvents with methanol that cause pollution to the environment.

Geely Automobile has recently launched a new green car model. The body of this new model is very small, with the length of only 2.9 meters, nonetheless clever design allows a big tall man to still sit inside the car and feels comfortable. The new model can be powered by using solar energy in sunny days, and solar energy can be supplemented by charging when it is rainy or cloudy. The air conditioner in the car has adopted the principle of electronic oscillation. Thus, the surface of car makes it look like a wagon box, but actually it functions like a refrigerator. The difference between this car and a real fridge is that this car does not have compressor and it can reduce power consumption through electron oscillation. In summer, people will feel very cool in the car, while in winter, the car can be heated by using geothermal heating using batteries on the floor. The batteries can generate heat when they are discharging. This heat can be conducted and cycled from bottom of the car all the way to the top, and people will feel warm when they sit inside the car. This car is a complete innovation without an engine. Of course, it is a green vehicle since it has zero emission.

3. Green Concepts Reflected In System Optimization-- Neusoft Group

Information management consulting industry belongs to producer service industry. In China, some enterprises engaging in producer service industry will not fall behind green economic development and try hard to infiltrate their green concepts when serving for their clients. Neusoft Group, China's largest offshore software outsource service provider has embodied the green concepts by designing software system for their clients.

Neusoft Group is a company which focuses on software technology and its main business includes industry solutions, products and engineering solutions, related software products, platform and service, etc.

This enterprise has designed an IT optimized project of medical system for its clients incorporating green concepts[26]. And this project attempts to solve the problem of inefficient use of medical resources in China. In almost all cities in China, if one goes to a big hospital, s/he will observe that big hospitals are much busier than Wal-Mart in China. Every early morning, many people stand in line waiting, which makes big hospital extremely crowded, while there are not nearly many people in the medium-small hospitals. what might possibly be the reason? One possible explanation is that most people do not trust medium-small hospitals, although they are closer to where they live. Such mentality of thinking results in

[26] Liu Jiren: The pursuit of green is the need for the enterprise to survive. [OL].http://finance.sina. com.cn/hy/20090422/14306135590.shtml. On April 22, 2009, Sina Finance.

inefficient resource allocation. According to the statistics, the best doctors in China's best hospitals spend 80% of their time with patients when professional knowledge is not required, (i.e. a disease can be cured in medium-small hospitals.) After full investigation, Neusoft Group discovers that an IT service system can be designed to form unified standard resource by connecting home computers to computer networks and collecting unified cases and resources.(check Chinese) The records and quality provided in the hospital can be guaranteed as the same as in big hospitals if a patient goes to see a doctor in a small hospital. In this system, each individual will create a health file and personal information can be obtained by researching on the internet. Some simple devices are needed at home, which can detect diabetes, chronic disease, heart disease, and pregnancy. All these data will be put into standard files in their expected hospitals. The wasted resources problem in the past have been greatly improved by introducing innovative IT models.

We believe, more and more local enterprises will become "Green Companies" which will help build a benign ecological system and gain sustainable competitiveness on the road to develop green economy. These enterprises will not only be the leaders in the business world, but also the leading force to make the society and the world a "greener" place.

Translator: Yu Xuxiang (SUFE)
English Version Editors: Jiang Jiahui (SUFE), Hou Yue (MIT)

REFERENCES

Huang Yonglv, March 31, 2009. How does green economy resolve the financial crisis? [N]. China Environment News, Version 004.

Liu Xiliang, June 2, 2009. Low Carbon Economy and the Development of Human Society, [N]. Guangming Daily, Theory Weekly Version 010.

Liu Zheng (Chief Editor), 2009. Research Review on the Sustainable Development of Chinese Economy. Shanghai: Shanghai University Press.

Pan Yue, April 9, 2008. Green Economy : To realize the sustainable development of the society, [N]. China Labor Security News, Version 003.

Wang Qiuyan (Chief Editor), 2009.China Green Development Report, [R]. Beijing: China Modern Economics Publishing House.

Yan Xingfang, 2008. Green Economy,[M].Beijing : China Industrial & Commercial United Press.

Yang Dongping (Chief Editor), 2009.China Environmental Development Report(2009) [R].Beijing : Social Sciences Documentation Publishing House.

Zhang Bingsheng , 2006. Explorations into Green Economics , [M].Beijing : China Environmental Science Press.

Zhang Chunxia , 2002. Green Economic Development Research , [M].Beijing : China Forestry Publishing House.

Edited by School of Economics in Chinese Academy of Social Sciences, 2008. Ecological Environment and Economic Development [M]. Beijing : Economic Management Press.

In: China's New Deal
Editors: Li Xiaoxi and Hu Biliang
ISBN: 978-1-61470-815-5

Comments on Chapter 8

GREEN MARKET AND GREEN ECONOMY

Mou Pu *

Rachel Carson's 《Silent Spring》, published in 1962, brought the ideology of environmental protection to the people from the academic ivory towers. Enlightened people realized the unsustainability of economic development if it is at the cost of deteriorating the environments. In the past forty+ years, environmental protection gradually came out of the laboratories of the scientists, started with people's action, to government action, and to market regulation. The process has never gone smoothly, after all, money and interests, especially the short-term interests, are too powerful drivers in the market economy that people are hardly to pay attention to the environment and the consequences of environmental destruction in their ways to pursuit interests. The terrible records of environmental destruction resulting from the 19th century and early 20th century industrialization, and the following-suit behaviors of the newly industrialized countries today demonstrated that people will do anything for the interests without check. This reminds me what the Lois XIV said: after me, the flood (Après moi, le deluge). People have tried what they would to check these roaming beasts to save the only world. Ecologists and economists, the 'reveries' with the same eco-root finally start to talk, to collaborate for computing the prices of environment for enclosing environmental protection into the market economy (Costanza et al. 1991, Costanza et al. 1997). Then, so called 'green economy' appeared, environmental protection using economic ways has been proposed and experimented. Just like an old Chinese say, untie the bell should be the ones who tied it on, once environment has its prices and can be precisely measured, the ruthless

* Prof. Mou Pu received his BS from the Northeastern Forestry University, Harbin, China in 1982, MS in forestry from University of Idaho in 1985, PhD in natural resource from Cornell University in 1991. He taught several ecology and ecology related courses in the Virginia Polytechnic Institute and State University and the University of North Carolina at Greensboro for 12 years after a 3-year period postdoctoral research associate work at Auburn University. He has been working at Beijing Normal University as a professor since 2007.
E-mail: ppmou@bnu.edu.cn; phone: 0086-01-5880-2936 (China).

economical development may be checked by the 'GREEN'. This is probably the right way of win-win situation in environmental protection and economic development.

Ms. Hui Shao's chapter, entitled "The New Engine: the Rise of Green Economy in China" provided a general description of serious environmental situation of China during the past two decades resulting from the rough managements and high resource, energy consumption, high discharge, and the heavy pressure on the environment from the high-speed economic development. The author indicated that China's economic development has already been seriously checked by the deteriorated environment and the shortage of the resources, and the green economy is the only right way to sustain the country's high-speed economic development.

The second section of the article describes the fruitful development of green economy in China in the recent years from three aspects: energy saving and discharge reduction, cyclic economy, and low carbon economy. To establish a sustainable economic system characterized low input, high yield, low energy consumption, low discharge and being able cycling, the Chinese government has conducted a series of effective measures, issued numerous government guidelines and achieved a certain degree of success.

The third section of the article discusses the establishment of a law system and an economic policy system promoting green economy in China. I feel that this section should be the main portion of the article, and should explicitly analyze the situations described. For example, in the second paragraph of the first part of the section 'A law system establishment for green economy', the description could be more vivid, more explicit and useful if some specific examples are given and analyzed to illustrate some environmental laws that have been accused as too broadly general, lacking specific regulations and implementing details. It should be recognized that many laws, particularly the environmental laws, in China are not of high quality. It is understandable to have the situation for countries like China that are undergoing the transition to rule of law. However, it is the professional articles with explicit and in-depth analyses would provide the best help for the law makers to improve the situation.

The second part of the section, 'A primary exploration of the policy system for green economy' briefly describes the implementation of the government policies on some of the green measures during the past few years, such as green loan, green insurance, green security, green taxation etc.. This part also needs some explicit and in-depth analyses, especially on the necessities, feasibilities, and the predictable promotion impacts of the policies on the economic development and environmental protections.

The last section of the article provides some cases of green economy in China: The Green Beijing as an example of a green city planning, the Shenhua Industrial Group, the Geely Auto Inc. and the Neusoft Group as the outstanding examples of the industrial companies that carrying out the green development. These models demonstrate that the green ideology of environmental protection, low consumption and high efficiency in resource and energy uses, low discharge and recycle have been implemented, in China and becoming a new growth of economic development in the country.

It should be recognized that so called green economy (Wagner et al. 2002) in China and anywhere in the world is far from mature, and still at explosive and experimental stage. The developments of the green economy are quite different and of various degrees of governmental guidance/governance among the countries. In the large government system of

China, governmental governance is particularly powerful. The purpose of green economy, however, is to achieve self-maintained, sustainable economic development under the 'green regulation' by utilizing economic rules. This needs to reduce the government directive guidance and to strengthen its roles in policy-making and rule establishment. In China, it seems more feasible to establish and improve so called 'green market' at the mean time of establishing environmental legislation, because if 'green' does not have a price, there will be no exchange, and no market, and consequently no green economy. I still remember my conversation a few years ago with Prof. Robert Costanza, the leading environmental economist who first computed the price of the global eco-services. By answering my question of why he did the computation as the eco-services are by no means perfect market commodities, he said that the politicians and policy makers evaluate everything with monetary standards, if we do not calculate the prices of eco-services, environments will no doubt become the costless sacrifice of economic development. I asked that how these eco-services could be priced accurately since no market trading as the reference standards. He answered that we'll try our best. Of course, green economy can not be established without market, without trading. If the price calculation of eco-services is mainly to provide a reference for government in policy-making, not for trading, and the best estimation is sufficient, then establishment of green market does need far more accurate ways of calculation for all so called green products.

The calculation of green products can be grouped into two types. One is pricing. The prices of some green products can be determined by the market, such as green foods, green tourism, high energy efficiency consumer products etc.(Schlegelmilch et al. 1996). While some others, relating to the public interests, having users but no specific buyers, are hardly market tradable goods, such as clean environments (i.e., water, air, scenery etc.), wildlife, plants etc. (Costanza et al. 1997), the government should involve. If the services, or green products, have the properties of public welfare and long-term interests, the government should provide subsidies (Reinhardt 1999, Wagner et al. 2002). Providing subsidies may have many ways, I feel the ways used should have the promoting impacts on green economy.

The second type of calculation is to quantify the green products. Like pricing, some green products can be easily quantified, such as the energy efficient consumer products by using new technology. Others, such as how many carbon be assimilated by one hectare of pine (oak, or maple, etc) forest at high (medium, or low) productive stand; how much pollutant (nitrogen, phosphate, lead, or other heavy metals, pesticides etc.) can be buffered by one kilometer forested (or shrub protected) riparian zone, are hard to be accurately quantified at this time. If we could not accurately quantify the products, trading them in the markets is just some hollow words. All in all, generally acknowledged metrological standards are the foundation of market society. As the progress of technology, of course, many of these are doable at endurable costs. However, this will be an enormous and complex system and is by no means an easy job. In China, the work in this field seems to be processed at a fairly low levels.

Green economy under the government guidance could greatly stimulate sustainable implementation of new technology and measures of environmental protection. The old fashioned industry, technology, and management of high pollution, high consumption, low efficiency of energy, resources uses, can be easily eliminated or replaced in the market competition. For instance, government can raise the competitiveness of the companies of low pollution, high energy/resource-using efficiency and low discharge, and promote forestation by establishing trading markets for discharge quotas of carbon and pollutants. Regulated monopoly (Raymond 2003) that has been used by some developed countries and turned out to be effective could be applied to the sectors that needs massive infrastructures, such as electricity, petroleum, natural gas and etc. to promote green technology. Of course, these enterprises should not be of state monopolized ones.

Comparing to the developed countries, China's technology is generally less advanced, and its environmental standards are relatively lower. Therefore, the large scale export-orientated industry in China is easily subjected to the limitations of ever restricting environmental standards of technology-advanced countries, or country groups. Lagging behind in technology could not only just follow, obey the rules of the games established by the developed world, but also, by no means, to occupy the high environmental moral stands. Many developed countries come out extremely restrict environmental standards with beautiful moral justifications and set up the high barriers for the input products. This is doubtless a protection measure of their own economic interests by ignoring the current state of the developing countries. Some developed countries are unwillingly to sign the 'Kyoto Protocol ', do not like to transfer the environment technology even it will benefit the whole world, because they are not willing to give up the rights in establishing rules of games with high moral position.

Markets of competition are always the drivers in promoting new technology, and green economy will be a new engine of the next round of technology revolution for China. It will lead China to be a major player in making the rule of games, and a leader of the world economy. The world-wide financial crisis of 2008 provided China a good opportunity to update its industry with the newest technology, and to promote green economy. A green economic system with green market as its solid foundation will guide China's economy to be better than ever, and sustainable.

China is the most populous large country in the world. Promoting green economy will lead to a win-win situation in economic development and environmental protection, and fit China's best long-term interests. China is large, pollution eventually will pollute its own lands and waters, it is China and its people suffer the most, while the largest beneficiary of green economy, sustainable economic development, and environmental protection is also China itself. In the recent UN environmental summit meeting and G20 meeting, The President of PR China, Hu Jintao delivered a presentations revealed the in-depth comprehension of China's leadership on green economy and environmental protection. The current round of green economy development in China indicated a good start, and the engine function of green economy on China's economic development and environmental protection has shown.

REFERENCES

Costanza, R., Daly H. E. and Bartholomew J.A. 1991 Goals, agenda, and policy recommendations for ecological economics. In: Costanza R. (ed.) Ecological Economics The Science and Management of Sustainability. Columbia Univ. Press. New York pp. 1-20.

Costanza, R., d'Arge, R., de Groot, R., Farber, S., Grasso, M., Hannon, B., Limburg, K., Naeem, S., O'Neill, R., Paruelo, J., Raskin, R.G., Sutton, P. & van den Belt, M. 1997. The value of the world's ecosystem services and natural capital. *Nature*, 387: 253-260。

Raymond L. G., 2003. "Regulatory Impressionism: What Regulators Can and Cannot Do," Review of Network Economics, *Concept Economics*, vol. 2(4): 466-479

Reinhardt, F. 1999. Market failure and the environmental policies of firms: economic rationales for 'beyond compliance' behavior. *Journal of Industrial Ecology*, 3(1), 9-21.

Wagner, M. Schaltegger, S. & Wehrmeyer, W. 2002. The relationship between environmental and economic performance of firms: what does the theory propose and what does the empirical evidence tell us? *Greener Management International*, 34: 95-108.

Schlegelmilch , B. , Bohlen, G.. & Diamantopoulos, A. 1996The link between green purchasing decisions and measures of environmental consciousness. *European Journal of Marketing*, 30(5): 35-55.

In: China's New Deal
Editors: Li Xiaoxi and Hu Biliang

ISBN: 978-1-61470-815-5

Chapter 9

THE NEW HEALTHCARE REFORM: THE NEW STARTING POINT OF HEALTHCARE DEVELOPMENT

Wang Nuo*, Yang Weibin, Ren Ran, Cong Yajing and Hou Rui

Rapid economic development accompanied by urbanization and aging of the population in China have introduced new health problems and requires transformations of the healthcare service model. In order to meet the challenges, we need to consummate the institutional arrangements including public health service system, the medical service system, the health insurance system and the pharmaceuticals provision system, etc. The health reform plan newly published in 2009 is a new starting point of China's health system. In this article, the authors present a thorough analysis on the new health reform based on the background, the key contents, the initial implementation and the existing problems, to bring forth its expected goals for the readers.

SECTION 1: THE FUTURE OF CHINA'S HEALTH SYSTEM

I. The Tortuous Exploration for China's Health System

In China's Planned Economy era, facing the fact of underdeveloped health service, the government attached great importance to basic public health inputs such as disease prevention and treatment of infectious diseases and epidemics. In urban areas, two major healthcare plans that depended totally on government funding, the "labor insurance scheme" (LIS) and "government employees insurance scheme" (GIS), were facilitated; in the countryside, the "rural cooperative medical system" using production brigade as the unit had been promoted, as well as "barefoot doctors". In this way, China achieved universal basic health insurance, and overall population health has been greatly improved. China was a poor country at that

* Wang Nuo is lecturer at the Institute of Economics and Resource Management at Beijing Normal University, who earned her PhD degree from the Graduate School of the Chinese Academy of Social Sciences. Email: jodywang@263.net

time, but its experience on improving health outcomes had been recognized by the World Health Organization (WHO).

Since the Reform and Opening up in 1978, China's health system has experienced tortuous reforms. From 1970s to 1990s, owing to the reforms of the state-owned enterprises and the promotion of the rural production contracting system, the universal basic health insurance system had been weakened or even dissolved. Moreover, fiscal decentralization, termed "power delegation and profit concession", had shifted the majority of health expenditure to local governments, resulting in not only decreased national health budget, but also money shortage caused by the local fiscal strain. The declining inputs led to the insufficient supply of health care, and problems of difficult access to quality health care, hospitalization and operations. For instance, there were 1 million malignant tumor cases each year, but only 11 thousand hospital beds to treat these patients. To solve this problem, the government decided to encourage private investment in health service, buoyed up the commercialized operation of the public hospitals and payment reforms. To some extent the efficiency of the health system was improved, but the system has also seen new problems such as "free" or unregulated market competition, profit-seeking public hospitals, private village doctors, privatized pharmaceuticals provision, and price distortion on health services.

The Reform and Opening up had produced China's economic development miracle and much higher income level, but at the same time heavier financial burden for individuals. From 1980 to 2005, the average outpatient visit and hospitalization cost for residents had increased 77 times and 116 times respectively, while the disposable income in the corresponding period increased only 16 times. 200 million Chinese did not have any health insurance in the year 2008, among which 28.1% were urban population and 7.5% were rural. The gap in access to healthcare services between the urban and rural areas has been gradually widening. In 2007, the average beds per 1000 people in the urban area are 2.83, while in rural area the number is 0.85. In Beijing, the city with the richest healthcare resources, the number is 6.90, while in Guizhou Province, the number is only 1.83. The over concentration of the limited health resources in urban or developed area deprived most of the rural residents of their basic health services. In addition to the problems of expensive health bills and difficult access to quality health services that remains unsolved, the new problem of health impoverishment has become more prevalent. Therefore, questions and doubts have been raised against the fairness of China's health system.

At the turning of the new century, China began to realize that depending wholly on "free" market competition cannot guarantee fair and effective access to basic healthcare services. The government should shoulder the indispensable responsibility in health care. Ignoring the welfare significance of health care and excessively pursuing profit directly raised the ratio of individual out-of-pocket payment on healthcare while reducing the proportion from government and society. In 1997, the nation began to undertake a healthcare financing reform and replaced the labor insurance and civil servants insurance with "the Basic Medical Insurance for Urban Employees"(BMIUE) to cover the employed population in the urban area; in 2003, the "New Rural Cooperative Medical Scheme" (CMS) improved the old rural cooperative medical system to cover the rural population; in 2007, the "New Basic Medical Insurance for Urban Residents" (BMIUR) was experimented in order to cover the unemployed population in urban areas. Thus China almost achieved universal coverage of basic health insurance; moreover, the increased proportion of government and social insurance expenditure, together with the reduction of out-of-pocket payment, are expected to improve the equity of healthcare in China.

II. The Issues for the New Scheme of Health Reform

The conflicts such as unfair access to health service, doubts on the non-profit motivation of public hospitals, the underdeveloped health insurance system and high out-of-pocket expenses indicate that it is a must for China's health system to reform. Especially in the year of 2002, the outbreak of SARS completely disclosed the crisis of public health system. In order to turn the tables, Chinese government implemented the scheme for public health system construction of 2003-2006 so as to build up a sound emergency health system that covers both urban and rural areas and can fully function in the prevention and control of diseases. In 2006, government began to improve the rural three-tiered healthcare delivery system at the county, township and village levels, and at the same time initiated the urban community healthcare reform. The above-mentioned measures reflect the central motif of "strengthen the responsibilities of the government" for the health reform in this phase.

With fast and stably growing social economy, the current government highly concentrates on people's livelihood and welfare, health care included, which provides a benign political atmosphere for the reform and development of health system. Meanwhile, China's health reform, after its 30-year tortuous exploration, has accumulated rich experience by finding and solving problems. Based on the macro-situation and the health system's own demand, the Communist Party and State Council issued the "Opinions on Deepening Reforms of the Health System" and the "Priorities in Implementing the Reforms of Health Care System (2009-2011)" in April, 2009, which unveiled a new round of reform on health system and set the new starting point for China's health system. This reform is named as the "New Health Reform" of China.

The new health reform is based on the experience of China's 30-year reforms of health system and strategic planning of future development. It defines the direction and pivot of the development framework for health system from the new starting point, and puts forward the ultimate goals of "to let everyone enjoy basic health service" and "to improve the health level of the whole nation". The scheme of investing 850 billion RMB within 3 years in health system fully reflects not only the philosophy of people-oriented development, but also the determination of the Chinese government to actively face the financial crisis.

SECTION 2: THE 2009 HEALTH REFORM SCHEME

The new health reform is a comprehensive action plan proposed for reforming the old health system, and will define a framework for future health system development. In this section, the author will focus on the analysis of "Opinions on Deepening Reform of the Health System". We start with five key aspects presented by both international and Chinese scholars, namely, health care financing, payment, the organization of health care delivery, regulation and behavior of health workers, to discuss whether the reform scheme achieved its goals.

The new health reform scheme has presented short-term, mid-term and ultimate goals respectively: 1) the short-term goal: by 2011, establish the basic public health service, improve the basic-level health service system in both urban and rural areas, implement the essential drug list, and pilot public hospital reform. The central government will input 850 billion RMB to fulfill the short-term goal; 2) the mid-term goal: by 2020, the public health system will be greatly enhanced and a universal basic health insurance system is established. The government should regulate the pharmaceuticals supply and establish the essential drug system; promote pluralistic patterns of health service provision and a sound health insurance system; 3) the ultimate goal: achieve the goal of universal coverage for the basic public health and health services, improve access and relieve the burden of out-of-pocket expenditure.

The highlight of the new health reform is its emphasis on the welfare consequence of public health care; it provides basic health service as the public product to the society. The government reiterates its indispensable responsibilities and obligations in health care.

I. Health Care Financing: the Source of Public and Private Funds

Public funding consists of government tax financing and the basic social health insurance. Public funding is more progressive in nature, especially the progressive tax financing, thus will contribute more to fairness of health care. The scheme points out that we should strengthen government's responsibility and increases its input, and insists that the public funds, especially tax financing, should play the leading role in maintaining the fairness of health care.

From 1978 to 2007, the proportion of health expenditure in GDP has risen from 3.02% in 1978 to 4.52% in 2007; the average growth per year is 0.52%. However, the growth of private funding has been its main source, among which the individual out-of-pocket payment account for the majority, with the growth rate over 2% nearly in all years whereas the growth of public funding has been no more than 2% in most of the years. Social insurance fund has been the major part in public funding, which is 1.72 times the general tax（see table 9-1）.

In response to this situation, the new health reform proposes that the growth of government input to the health system should be faster than that of frequent fiscal expenditure to raise the proportion of government health input in the frequent fiscal expenditure; this is the first time that a clear and operational public funding plan for health is proposed. Both the supply side and demand side have been fully taken into consideration and assigned respective investment plans, namely, the input on the supply side will be distributed to public health, rural health, urban community health services; the input on the demand side will be distributed to government subsidies on the basic health insurance systems. Besides, the principle that central and local governments would share health costs according to their administration level but local governments undertake the main responsibility has been clearly clarified. The central government is financially responsible for the national immunization plan, prevention and controlling for the critical trans-regional epidemic diseases, the basic health insurance for urban and rural residents, and the construction of public hospitals directly under Ministry of Health; local governments are responsible for all the rest. The earmarked payment transfer should be adopted to minimize the gap among the regions in their fiscal conditions.

Table 9-1. Health Care Expenditure in China by Source of Funding (%GDP), 1996-2007

	1996	1997	1998	1999	2000	2001	2002	2003	2004	2005	2006	2007
Health expenditure as a percentage of GDP	3.81	4.05	4.36	4.51	4.62	4.58	4.81	4.85	4.75	4.73	4.64	4.52
The proportion of public funding	1.78	1.79	1.82	1.84	1.77	1.63	1.72	1.76	1.80	1.83	1.89	2.05
1) The proportion of general tax funds	0.54	0.54	0.60	0.61	0.61	0.59	0.63	0.66	0.66	0.69	0.66	0.54
2) The proportion of social insurance funds	1.10	1.08	1.02	1.00	1.01	0.90	0.94	0.94	1.00	0.99	1.08	1.36
The proportion of private funds	2.03	2.26	2.54	2.67	2.85	2.95	3.09	3.09	2.95	2.90	2.76	2.47
1) The payment by private out-of-pocket	1.93	2.14	2.39	2.52	2.72	2.75	2.78	2.71	2.55	2.47	2.30	2.04
2) The payment by private health insurance	NA	0.02	0.03	0.05	0.03	0.06	0.10	0.18	0.16	0.17	0.18	0.15

Source：Calculated according to The Report on China's Health Expenditure 2008, P245.

NA, not available.

With respect to private funding, we should bring private sectors into supplementing the basic health service, such as exploring ways of "purchasing service" and select some service items to try out in certain regions. The new health reform proposes that the private funds (including foreign funds) should be encouraged to invest in non-profit health service, participate in the reform of public hospitals and qualified doctors are encouraged to open private clinics. In terms of health insurance (the health demand side), it proposes that the proportion of tax financing in basic health care should be raised while the share of private out-of-pocket payment should be lowered; but specialty care should be paid mainly by private cash or private health insurance.

II. Health Care Payment: the Inputs for Supply and Demand

The healthcare payment refers to the use of health fund raised from public and private channels. It is mainly concerned with payment to the supply and demand sides. In this health reform, for the supply side, the government tax funds mainly invest in the basic health service supply and help maintain the non-profit characteristics of public hospitals; for the demand side, it mainly subsidizes the basic health insurance.

Since 1978, in spite of the escalating trend of the absolute value of government health budget, China's budgeting was still influenced by the tradition of "emphasizing the supply rather than the demand", which is chiefly reflected by the heavy input into public hospitals and free healthcare to government employees, but neglected subsidizing other healthcare customers until 2003. The change of government tax funding as a percentage of total health expenditure basically keeps in line with the changes of public hospital expenditures, whereas the financial subsidy to the demand side only account for a small proportion without changing over the past 20 years. Since 2002, the growth rate of government budget on consumption is 12.2%, slower than the 20% growth of public hospital expenditure, moreover, in 2006, the government budget spent on health supply increased 18.2% from the previous year, but the subsidy to the demand side declined 0.3%.

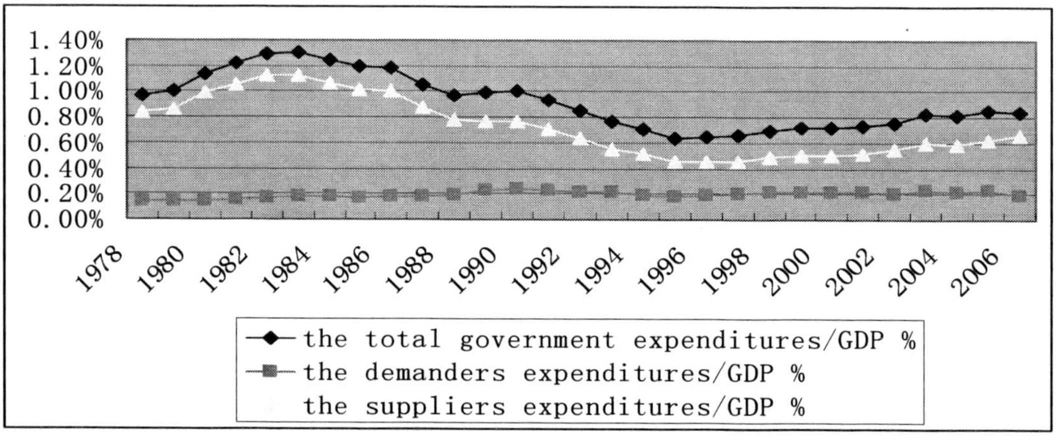

Source : Ministry of Health, *The Statistical Year Book of Health 2008*, Peking Union Medical College Press, July, 2008.

Figure 9-1.The Ratio of Government Expenditure on the Supply Side and the Demand Side.

The new health reform seeks to change this pattern of health expenditure. It proposes that at least 560 billion out of 850 billion RMB should be invested in the demand side. The government funds should be chiefly used in the subsidies for residents' health insurance and the compensation of treating serious illness. The scheme proposes that we should construct a health insurance framework with the basic health system as main body and other various health insurance and commercial (private) health insurance as the supplement. The urban worker health insurance, the urban resident health insurance and the urban-rural health aid compose the basic health security system that completely covers the urban employed and unemployed population, the rural population and the urban and rural vulnerable population. The funding of the basic health insurance should chiefly depend on the social health insurance which receives money from enterprises and individuals, and meanwhile be subsidized by government funds. In this system, the urban worker health insurance belongs to obligatory social insurances stipulated by nation and should be paid by employers and workers themselves as stipulated; the New Rural Cooperative and Urban Resident Health Insurance are voluntary under the government's instruction and financial subsidy.

To the supply side, namely, the input in the health service institutions, we should stick to the principle that government playing a dominant role with financial supports from multiple channels. In practice, the government funds should be chiefly used for subsidizing the institutions which provide the public health service and primary health care. 1) The input in the public service should be led by governments through governments' tax financing and its nature of public goods should be confirmed. 2) The input on the basic health service should be dominantly led by government and partaken by society, individual and government at a reasonable ratio. For those township health centers and urban community health centers set up directly by the central government, the government should pay for basic construction, purchase of equipment, salary of stuff and operation funds. To those urban and rural basic-level health providers that operated privately, the method of using government purchase to offer government input could be adopted. For instance, the central government should increase its input on the construction of village clinics and the public health service undertaken by village health practitioners. 3) The input on public hospitals should return to governments, and the public welfare of them should be insisted on. The government input will be used to infrastructures construction and equipment purchase, supporting the development of key research projects, health expenditure of retired staff, the subsidies on the policy-induced deficit and the special subsidies on the public health service task it undertakes. 4) The cost of specialty health care services should mainly be paid by private out-of-pocket and commercial insurance and basically without government funding.

III. The Healthcare Organizations: Co-development that Puts the Public Hospital as Primary and the Private as Secondary

The development of China's healthcare delivery system should stick to the principle of setting the non-profit health institutions as main body while allowing the for-profit ones as their supplement; putting the public hospitals in the dominant place while co-developing the private ones to a balanced health service system that covers both urban and rural areas. In 2007, the number of China's public hospital is 9832, 2.75 times as that of private hospital, and most of the health services are provided by the non-profit hospitals (see the pie Figure 9-2 & 9-3) .

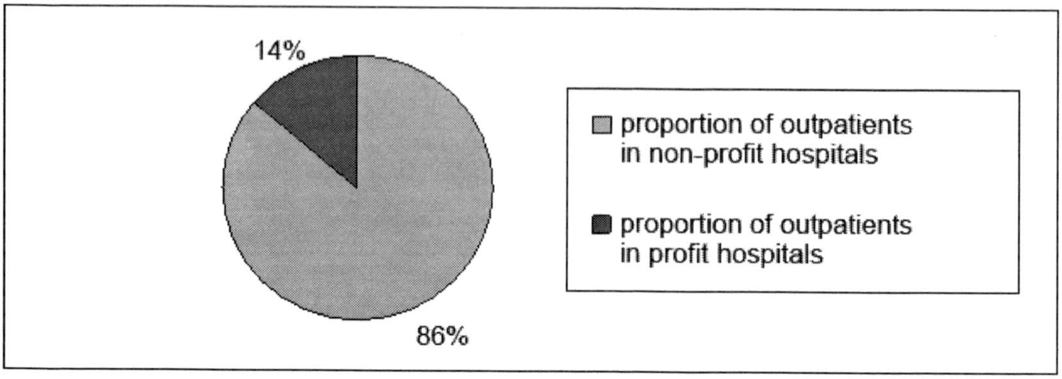

Source: Ministry of Health, *The Statistical Year Book of Health 2008,* Peking Union Medical College Press, July, 2008.

Figure 9-2. Proportion of Outpatients in Profit and Non-profit Hospitals in 2007.

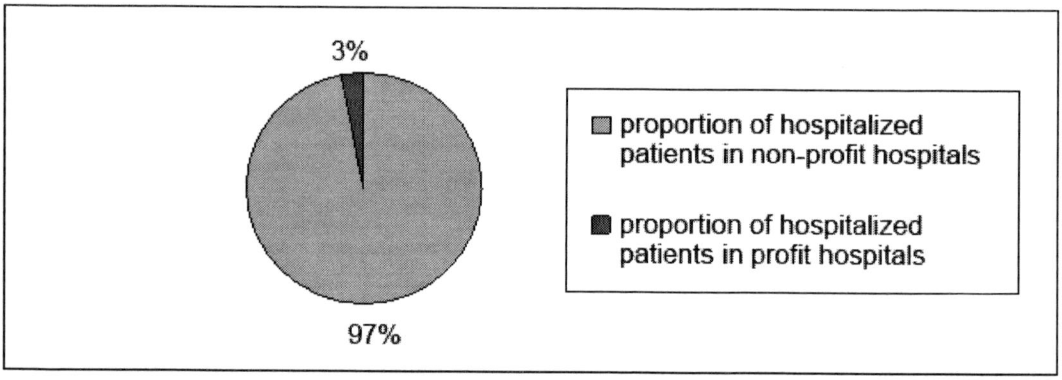

Source : Ministry of Health, *The Statistical Year Book of Health 2008*, Peking Union Medical College Press, July, 2008.

Figure 9-3. Proportion of Hospitalized Patients in Profit and Non-profit Hospitals in 2007.

The new health reform insists on the idea that health service should be mainly provided by non-profit and public hospitals but also encourages the development of for-profit hospitals and private health care providers. Therefore the healthcare delivery system will contain multiple sources and types of investment. In the public hospital reform pilots, the proportion of public health institutions has been declined and social fund has been introduced to participate in the system reform of public health institutions to facilitate the mutual stimulation and the co-development of the public and private hospitals. The qualified doctors are supported to open their private practice to benefit patients who need to receive health service.

The new health reform perfects the original tiered health service delivery system in urban and rural areas in hope of carrying out the referral mechanism and allocating health resources reasonably in the future. In cities, we should set up the new urban health service system based community health service. We should construct community health service centers to provide public health service including disease prevention and control, primary diagnosis and treatment of common diseases and chronic diseases. The city hospitals should concentrate themselves on the diagnosis and treatment of the critical emergencies and difficult and baffling diseases, medical education and research and the instruction and training on the basic-level health personnel. The community health service centers should cooperate with city hospitals, city hospitals support the community health service centers with technology and staff. Community health service centers should improve the quality of their service, lower their charge and raise the reimbursement ratio to reduce healthcare demand. In this way, initial diagnosis in community health service centers, tiered medical treatment and the two-way referral will be gradually realized in the future.

In the three-tiered rural health service system that contains village clinics, township health centers and county hospitals, village clinics should provide basic public health service and treatment of general health problems in the village; the township health centers offer comprehensive public health service and treatment of common and frequent diseases; county hospitals, as the health center in county regions, are mainly responsible for providing basic health service and emergency rescues. Meanwhile, higher-level health institutions have obligations to offer technology support and staff training the lower-level ones. The government devotes itself to operating county hospitals and assures there is at least one health center in each town, and encourages the investment from social fund to get involved in the construction of other health centers.

IV. The Management of Medical Institution

We should implement localized and industry-wide management for healthcare organizations. Whatever their ownership, forms of investment, affiliation and nature of business are, the medical institutions should all be supervised and managed by local health bureau. The central and provincial authorities could establish medical centers with the functions of research and teaching and specialty hospitals that can undertake the diagnosis and cure of the national or regional difficult and complicated diseases; the county (city) authorities are responsible to operate county hospitals, the village clinics and urban community health centers; the city authorities are in charge of operating other kinds of public hospitals.

To the operation mechanism for basic-level health service institutions, we should strictly define its service function and clearly regulate the technologies, the equipment and the basic medicine that can be used; adopt the financial management that ratifies tasks and accounting for income and expense as well as subsidizes based on performance assessment, and at the same time experiment innovative management methods like the separation of income and expense and global budget for t public health and health insurance; reform the drug price markup to sell medicine without extra profits; set up the evaluation mechanism with the quality of service as its core and the post responsibility and performance as it basis.

The central authorities should figure out the framework policy of the basic health insurance management system, and local governments should undertake the task of organizing and implementing these policies. Therefore gradually increase the overall planning level and in the end realize the united administration of the basic health insurance both in urban and rural areas.

V. The Behavior of Health Personnel

We should set up sustainable mechanism of retaining talents, and strengthening education and training of health professionals in public health, rural health and community health. We should issue benefit policy to encourage capable health professionals to work in the rural areas, the urban community and the mid-west region. We should initiate the concept of general practitioner and perfect the qualification system of it to guarantee that basic-level health institutions all have possessed qualified general practitioners.

For the personnel in public health institution, the government should ascertain the staffing, wage and outlay standard and meanwhile set up effective incentive mechanisms; for those in basic-level health institution, the government should adopt competitive employment within strict staffing limits; for those in public hospitals, the government should adopt competitive employment as well as position management to strictly manage the gross salary and performance assessment based on the quality of service and the workload of the position.

SECTION 3: THE EXECUTION AND EFFECT OF THE NEW HEALTH REFORM

In "the basic health service that everybody can enjoy", the "everybody can enjoy" means that everyone enjoys fair opportunity of healthcare and equal health outcome; the "basic health service" refers to the health service that is compatible with the development of economy and can be afforded by the country, the society and the individuals. The public health, the rural basic-level health, the urban community health and traditional medicines constitute the important contents of this scheme. To achieve the above-mentioned goals, China's government plans to input 850 billion RMB from 2009 to 2011 to implement the new health reform scheme

First, we need to understand the concepts that: 1) the 850 billion RMB will be the incremental funds that central and all levels of local governments plan to input in the health reform in 3 years, among which 39% or 331.8 billion will be input by central government; 2) the input will mainly focus on five aspects including the basic health insurance, the national essential drug system, the basic-level health service system, the equalization of basic public health service and reform pilots of public hospitals; 3) most of the funds will be used for basic-level health service. Second, we need to clarify the allocation of the funds: 1) it is the key point that 2/3 of the funds should be put into the demand side while 1/3 to the supply side; 2) the money put into the supply side should be used for improving the construction of basic-level medical service and public health; 3) priority should be given to the western region. The estimates of input is described in Figure 9-4.

Column 9-1. The Emergency Medical Response to The Wenchuan Earthquake

At 14:28, May 12, 2008, an earthquake of 8.0 on the Richter scale shook the Wenchuan County in Sichuan Province, China and hit an area of 280,000 square kilometers in Sichuan, among which the harder-hit area had reached 125,000 square kilometers (25.8% of the size of Sichuan Province), and the hardest-hit area had reached 11 thousand square kilometers (2.3% of the size of Sichuan Province). The number of disaster-affected population had reached 29.61 million (33.6% of the population in Sichuan). Moreover, the mountain and valley in the harder-hit areas, the following suspension on transportation and communication, the choking rivers, the abominable weather and the failure for rescuers, materials, vehicles and large equipment to enter part of harder-hit areas in time all add to the difficulties of disaster relief. The health rescue for calamities and accidents is mainly proceeded in following three phases.

The first phase is the emergency response. The Health Bureau of Sichuan province initiated the first-level emergency plan immediately after the earthquake. Within the first hour after the earthquake the first emergency mobilization order had been issued, and within the second, the health aid headquarter for disaster relief, which consisted of working groups for health rescue, the disease control, the supply of health devices, publicizing and reporting, the comprehensive coordination and the logistics support as well as the later-added transporting the injured and reconstruction, had been set up.

The second phase is quick response. The best timing for rescuing and remedying the injured is within 72 hours after the earthquake. Most of the injured that were rescued and had received effective treatment in this period could survive and recover. China's government and health institutions took immediate actions. On May 12, the number of health workers who participated in the medical aid in the disaster-hit areas was 540; On May 13 it increased to 1676 and on May 14 to 4265. Until May 15th, the number of the health workers in the disaster-hit area had reached 35880, among which 31000 (86.4% of the total) health workers were from Sichuan province. On the tenth day after the earthquake (May 22), as twenty health workers from Shanghai carrying first-aid medicine, equipment and satellite phones being parachuted by helicopters to the four villages in Wenchuan county, including Gengda, Sanjiang, Yinxin, Caopo, where health workers could not enter because of the traffic interruption, Sichuan's 11 cities,67 counties(districts) and 950 villages that suffered from the "5.12"earthquake had been fully covered by medical care (see Column Table 9.1.1) .

Column Table 9.1.1. The Statistics of Health Workers until May 30, 2008

Source of rescue team	Total number	Proportion
From Sichuan province	41240	81.7%
from other provinces	5969	11.8%
From military troop	3048	6.0%
From Hong Kong, Macao and outside of China	199	0.4%
Total	50456	100%

Source : Dai xiaozhou et, al. *The Record of Actual Event of Sichuan's Rescue Work in Wenchuan Earthquake*. Chinese Journal of Evidence-Based Medicine 2008, 8(10): 797-802.

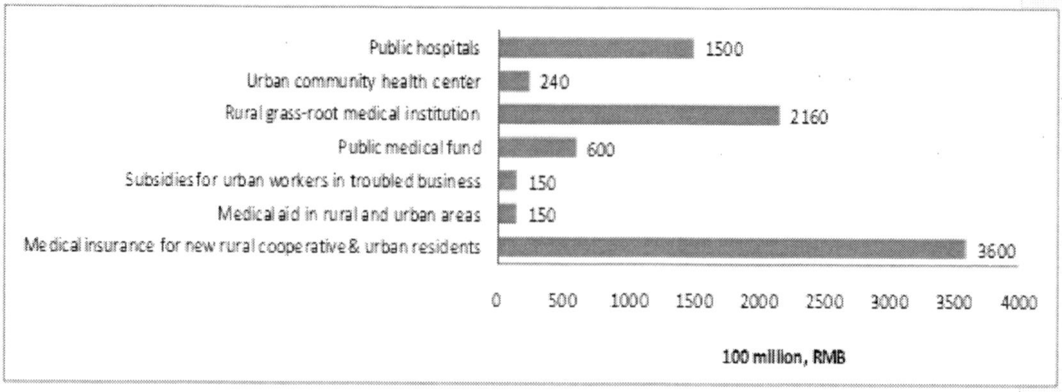

Source : http://finance.sina.com.cn/roll/20090213/02445850205.shtml

Figure 9-4. The Prediction of the Input of 850 Billion RMB in the New Health Reform.

I. The Construction of Basic Health Insurance System

From the later 1970s to the middle 1990s, the reforms of state-owned enterprises, the taxation reform and the fiscal decentralization all put the original labor insurance and government employee insurance system into difficulty. Besides, the rural reform of economic system had basically dissolved the cooperative medical system that based on collective economy. In 1997, the "Health Insurance for Urban Employees" had replaced the former labor insurance and government employee insurance system that all the employees in cities and towns began to enjoy health insurance. In 2003, the "New Rural Cooperative Medical Insurance" was implemented as a pilot program to solve the problem of health burdens for rural peasants. In 2007, the "Health Insurance for Urban Residents" and the "Urban and Rural Health Aids" was also piloted. Before the plans of health reform were promulgated in April 2009, the framework of the basic health insurance system had been basically established.

During the period of 2009-2011, the authorities plan to realize the target that the health insurance covers all residents in the whole country with an enrollmentrate over 90%. In 2010, the amount of subsidyfrom all levels of government to the health insurance for urban residents and new rural cooperative medical insurance will be raised to 120 RMB per person each year. Thus, it can be calculated that the participant number of urban and rural basic health security system will reach one billion within 3 years, and the lowest subsidizing level of the government should be 120 billion per year.

The government plans to invest 5 billion RMB to help minimum-wage employees to participate in the urban residents health insurance.RMB In 2007, 22.72 million urban residents and 41.73 million farmers benefited from the Livelihood Security System, and the sum of subsidy was 3.6 billion RMB, but 5 billion RMB will be needed if the health reform iincreases the benefit levels. In the 850 billion official funds, nearly 390 billion (46% of the total) will be used to construct the basic health insurance system; it is the focus of this scheme. (See Table 9-2)

Table 9-2. The Evolution of China's Basic Health Insurance System

Period	Phase	Urban areas	Sub-urban areas
1978-middle1990	The phase of planned economy	labor insurance and government employee insurancecontinued, but were impacted.	Rural cooperative health care gradually collapsed
1997-2000	The phase of transitional economy	In 1997, the health insurance for urban employees was experimented and gradually put into practice.	In 2003, the new rural cooperative health care started as a pilot program °
2007-2009	The phase of market economy	The basic health insurance for urban employees had been carried out, in 2007; the basic health insurance for urban residents was on the process of pilot program.	In 2008, the new rural cooperative health care had completely covered the rural areas.
April, 2009	New health reform	The health insurance system had been basically set up: the basic health insurance for urban employees, the basic health security for urban residents, the urban medical assistance program.	The new rural cooperative health care, the rural medical assistance program

Source: Gu Xin, *Analysis on the Investment Direction of New Health Reform, 21st. Century Business Herald,* February 13, 2009.

In this plan, the highest benefit of health insurance for urban employees and residents has been increased to six times as many as the local employees' average salary and the residents' disposable income; the benefit limit on new rural cooperative health care has also been raised to over six times as many as the local farmers' net income, four times higher than the present level. What's more, the plan has indicated that the surplus rate of the funds for new rural cooperative health care should not exceed 15% and the accumulated surplus not exceeding 25% of total insurance fund in that year. At the same time it plans to improve the level of fund administration. By 2011, the health insurance for urban employees and residents will realize municipal unified administration. The portable basic health insurance scheme was piloted in 2009. Take the 16 cities in the Yangtze River Delta for example; they plan to build an intra-region joint committee system of urban health insurance cooperation and launch substantial cooperative actions. At present, the healthcare cost settlement among Shanghai, Hangzhou, Jiaxin, Anji; and the cities of Hangzhou, Deqing, and Anji has been basically realized. The provinces of Hainan, Guangdong and Guizhou, together with the Guangxi Zhuang Autonomous Region have issued the *Cooperative Agreement of Off-site Health Care Cost Settlement*, striving to realize it in 2009[1].

[1]Data sources:
http://health.zjol.com.cn/05zjhealth/system/2009/03/29/015385069.shtml
http://news.pharmnet.com.cn/news/2009/06/13/257098.html

1. The Participating Situations of Health Insurance for Urban Employees and Residents in some Provinces and Municipalities

Since the new health reform, the participating rates of urban employees and residents insurance have both reached an ideal level; especially the insurance for urban residents has been speedily expanded. Like Liaoning province, its participating rate boosted 84.8% than 2008 (see Table 9-3). Shaanxi province plans to make its participating rate for urban employees and residents reach 95% and 80% respectively, and Guangdong province's plan is to exceed 95% in 3 years[2].

Table 9-3. The Summary of the Latest Participating of Health Insurance for Urban Employees and Residents in Some Provinces and Municipalities in 2009

Province (municipality)	Health insurance for urban employees			Health insurance for urban residents		
	Participating number	Increase from the previous year	Participating rate	Participating number	Increase from previous year	Participating rate
Beijing	8.72million	11.4%	93%	1.47million	0.7%	90%
Liaoning province	12.68m	4.8%	90%	5.60m	84.8%	80%
Chengdu (the capital of Sichuan province)	3.02m	15.3%	NA	6.68m	NA	NA

Data source:
http://beijing.qianlong.com/3825/2009/04/14/3042@4946802.htm
http://health.newssc.org/system/2009/04/08/011769637.shtml
http://health.syd.com.cn/html/xinwen/shenyangjianwen/2009/0812/31416.html

http://hainan.pprd.org.cn/hainannews/200906/t20090611_60629.htm
2 See 11.

2. The Participating Situations of the New Rural Cooperative Health Care in some Provinces and Municipalities

To the end of September, 2008, 2729 counties (districts) have launched new rural cooperative health care. The number stood at 814 millions, and the participating rate reached 91.5%, approaching the target of "universal coverage". After the new health reform, even the limited statistics indicates that no matter the remote provinces like Yunnan or Heilongjiang, or the large cities like Chongqing and Nanjing, the participating rate of new rural health cooperative in their territorial scope all reached 90%, demonstrating that the new system has covered China swiftly (See Table 9-4) .Jiangxi province has increase the highest benefit level of this system to that of 8 times as many as farmers' net income; Shaanxi province plans to reach the level of 90% of the participating rate; Shandong province and Guangdong province plan to reach 95% and 98% respectively[3].

Table 9-4. The Participating Situations of New Rural Cooperative Health Care in Some Provinces and Municipalities

Province (municipality)	Participating rate
Hei Longjiang	94%
Yunnan	93%
Chongqing municipality	95%
Nanjing	99%

Data sources:
http://www.cnss.cn/xwzx/yiliaobx/zcss/200902/t20090203_205020.html
http://city.cctv.com/html/chengshiyaowen/a68cf9e44e22586074ad46bfe6bb8147.html
http://www.fdinfo.org/CL0054/61513.html

The subsidy level for participating farmers from the central government and local authorities has been increased from 40 RMB to 80 RMB, and individual premium payment will increase correspondingly. In the previous three quarters in 2009, the national total expenditure of new rural cooperative medical insurance was 42.91 billion RMB, the accumulated number of beneficiaries was 370 millions[4] (14). It presents that the economic burdens of inpatient care have been relieved and the healthcare seeking rate, especially the inpatient care rate has risen.

3 See 11

4 http://www.gmw.cn/01gmrb/2009-08/24/content_968648.htm

Column 9-2. A Case of New Rural Cooperative Medical Insurance in Lufeng County, Yunnan Province [5]

The cooperative health care is a cooperative system that rural residents participate voluntarily under the reasonable guidance of the government to enroll in health insurance covering mainly the serious illness of the enrollees. In this case, the situations in a pilot area of cooperative medical insurance, Lufeng county, is going to be introduced.

Lufeng County is located in the southeast of Central Yunnan Plateau, with an area of 3536 square kilometers, and a population 415 thousand, 82.5%of them (93500) are farmers and ethinic minorities like Yi, Miao.. In 2003, the population newly fell below the poverty line is 28,434, and 12,476 people are impoverished by illness and medical cost, account for 43.9% of the total. As a pilot region of the new rural cooperative medical insurance-, 280 thousand people have participated in the program voluntarily by August, 2003. The participating rate was 87.1%; meanwhile, the participation rate of populations in extreme poverty was 100%.

The central government subsidizes 10 RMB per person each year. The province, prefecture and county governments subsidize 10 RMB at the ratio of 4:3:3. Thus, there are 20 RMB from government subsidies at all levels. As for the collective input, it subsidizes 1 RMB per person each year, and each person should pay 10 RMB premium out-of-pocket. Therefore, subsidy from the government is the main source of new rural cooperative medical insurance, accounting for 64.5% of the total fund (See pie Figure 9.2.1).

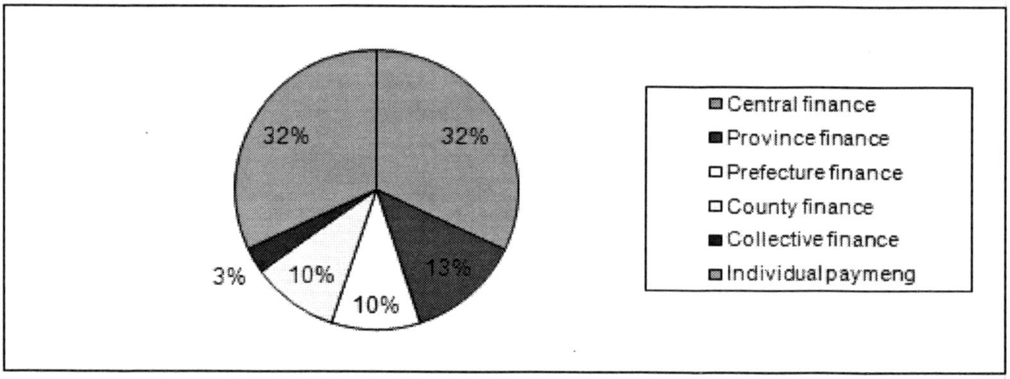

Figure 9.2.1: Funding Sources of New Rural Cooperative Medical Insurance in Lufeng, Yunnan.

The reimbursement rate was categorized by different medical institutions and service types. (See the following Figure). The highest accumulated benefit quota (including inpatient and outpatient care) was 3000 RMB , the sum of for the new cooperative medical insurance fund in 2003 was 7.95 million RMB. By March 2004, 3.406 million RMB had been disbursed. (36.9% was for outpatient care, 63.1% was for inpatient care) (See Figure 9.2.2).

[5] References: Asian Funds/Pfizer Pharmaceutical Company sponsored *The Baseline Survey Report of New Rural Cooperative Health Care in Lufeng, Yunnan Province,* Oct. 2004

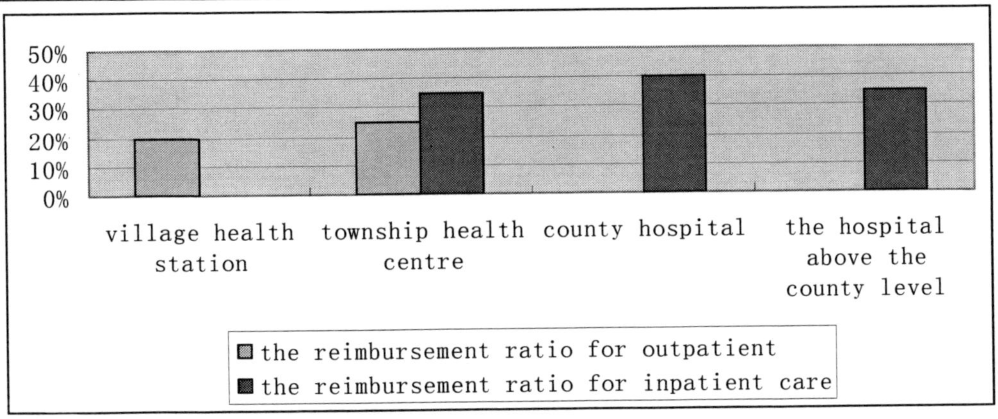

Figure 9.2.2. Regulations of Reimbursement Rate for New Rural Cooperative Medical Insurance in Lufeng.

After the new rural cooperative medical insurance has been carried out, the average prescription expenses in township health centers have risen 14.4% while the village health clinic increased 32.3%. Meanwhile, the outpatient expenses of participants are higher than that of who do not participate, and the expenses in village health clinics are higher than in township health centers. Besides, the expenses of inpatient care, medicine and surgery have increased to different extent. Therefore, it is important to control the medical expenditure, especially the surgery expenses and medicine expense in county hospitals. (See Figure 9.2.3).

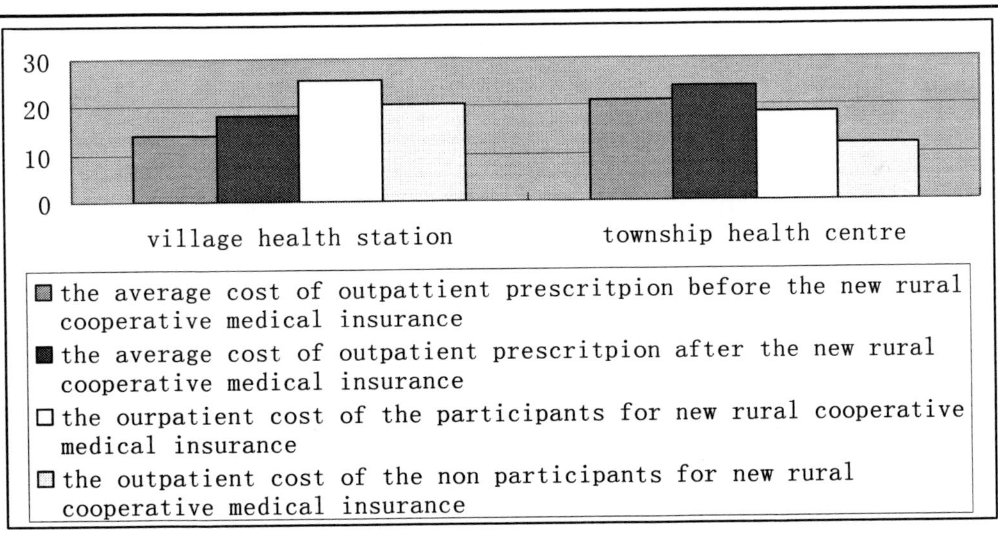

Figure 9.2.3. The Comparison of Outpatient Prescription Expenses and Patients' Prescription Expenses Before and After the New Rural Cooperative Medical Insurance.

Source: The Asia Foundation/Pfizer Pharmaceuticals supported, *A Survey Report on the Basic Implementation of New Rural Cooperative Medical Insurance in Lufeng county, Yunnan Province*, October, 2004.

II. Fairness in Access to Basic Public Health Services

In the new health reform, every Chinese will receive 15 RMB of public health subsidy each year, and in 2011, it will not be less than 20 RMB. Calculating by the number of 1.32 billion people in 2007, 19.8 billion RMB is needed. If by the standard of 20 RMB, 26 billion RMB is needed. In that 60-80 billion RMB should be put into the public health funds. The central finance, on the other hand, subsidizes the poor-stricken areas through transferring payments, and by the end of August, 2009, 15.8 billion RMB had been provided to launch major items of public health services[6].

From the beginning of 2009, the medical records system of residents have been gradually established in nation-wide scope and been under standardized management. Elderly people over 65 will receive health check and children under 3 years old will receive development check at fixed intervals; the pregnant women will receive at least 5 times of prenatal examinations and 2 post partum visits; all the patients of hypertension, diabetes, psychological diseases, AIDS or tuberculosis will be provided appropriate treatments, and thus, the major items of public health such as prevention and control of serious illness, immunization programs and rural women's reproductive health programs will be implemented. According to the scheme, from 2009, the government will offer fundamental public health services to urban and rural residents for free. For instance, replanting HBV vaccine to the groups under 15 years old; eliminating the harm of coal-burning fluorosis; providing folic acid to the rural pregnant women to prevent birth defects; curing the poor cataract patients; and construction of sewer and toilet in countryside.

III. The Completing of Basic Health Service System

In the 850 billion funds, at least 280 billion will be put into the supply side--the institutions providing health services. The sum of financial subsidy that all the public health institutions had received in 2007 was 111,855 million RMB, plus the subsidy from the higher authorities, the total number was 117,997million RMB[7]. Besides, the extra subsidies to health institutions in the new health reform are mainly used to improve the health services at the basic level. Moreover, the reform has increased the subsidy to Chinese traditional medicine in order to fully take advantage of its simple, safe, low cost characteristic to improve fundamental health and health services.

1. Urban Community Health Centers

From 2009 to 2011, 3700 community health centers and 11000 health service stations will be built or remolded. Meanwhile, the authorities will give financial support to poverty-stricken areas to build 2400 community health centers and train 160,000 health working staff, and invest 24 billion RMB within 3 years.

[6] See 11

[7] The Ministry of Health: *The Annals of Statistics on Chinese Health*, Chinese Peking Union Health College Press, July, 2009

By the end of 2008, all the cities above prefecture level and 98% districts under city administration had already launched community health services, 7232 community health centers and 21,895 community health service stations had been built; the emergency visits in some community centers had reached 30% of the outpatient care, releasing the pressure of emergency care for large hospitals to some extent[8]. A survey made by the Ministry of Health shows that the community health centers has become the top choice of people who has ailments, accounting for 71.4%[9]. The main reasons why residents choose community health centers is its convenience, good service, and cheap price.

With the implementation of new health reform, the construction of social health service continues developing. For instance, Xi'an (the capital of Shanxi province) raised 950 million RMB on the basis of sub-district administrative office and planned to build 73 community health centers and 78 health service stations. Up to August 2009, 70 community health centers and 67 health service stations had been completed respectively, 90% urban areas had been covered by the community health service system[10]. Zhengzhou, the capital of Henan province, formulated that every community health service centers should be equipped at least 6 general practitioners and 9 registered nurses[11].

2. Basic Health service in the Countryside

During the period of 2009-2011, the central finance will attach great importance on building 2000 hospitals at county level to reach the criteria that each county has at least one standard hospital. In 2009, the government supports to build 29 thousand township health centers and expand 5000 township health centers at the same time, to achieve the final goal that each county has one to three health centers. By 2011, every village would have a health station, especially in the remote villages and territories; and 360 thousand health workers for township health centers and 1.37 million for village health stations would be trained. In the next 3 years, it is estimated that the average investment into rural health institutions by the government each year may be 72 billion RMB.

To complete the system that urban hospitals aids rural health institutions, every tertiary-level urban hospital must build long-term cooperative relations with about 3 county hospitals (including some well-equipped township health centers). The project of "Ten thousand doctors supporting rural health" will be continued in the next 3 years. From 2009, the government will pay for the tuition fee and educational loans of those medical graduates who voluntarily work in the township health centers in the Middle and West areas for 3 years.

IV. The Progress of Pilot Reform of Public Hospitals

The pilot reform of public hospitals is the key and most difficult point. The plan encourages different forms of capital to take part in the reform of public hospitals, but to a reasonable extent. Although the government's input to common public hospitals may not

[8] http://chab.org.cn/NewsDetail.asp?id=5024
[9] http://www.cnr.cn/gundong/200908/t20090814_505430443.html
[10] http://www.cnwest.com
[11] http://hn.house.sina.com.cn/news/2007-03-02/08107079.html

exceed 50 billion RMB, the public hospital still need to keep its principles of public welfare and social efficiency, and the ratio of special services offered should not exceed 10% of its whole services. The plan also emphasized that the three compensating channels of the public (service charge, sale of drugs and financial subsidy) should be gradually changed to two channels: service charge and financial subsidy. Moreover, private hospital is also allowed to enjoy the same status as public hospitals like contracting to be designated healthcare providers with basic health insurances, applying for medical research funds, assessing professional titles, and continuing education. The reform of public hospitals is going to be experimented in 2009 and gradually launched in 2011.

V. The Establishment of the Essential Drug System

As the *Implemental Opinions on the Establishment of National Essential Medicine List* was issued in August 2009, the construction of essential medicine systemwas formally launched. The essential medicines fit the need of basic healthcare, the price are set at a reasonable level so that the public can obtain it fairly. The public health institutions at the basic level would completely use the basic medicines and other medical institutions would use it as required. The government would manage the choice, production, circulation, usage, pricing, reimbursement and supervision of the essential medicines. The *Items of Essential Medicines* revises once three years and all the basic medicines are included to the scope of reimbursement of health insurance with higher reimburse ratio. The National Committee of Development and Reform determines the retail prices of the medicines and the public hospitals must sale these medicines at the same prices. The 15% medicine markup held by the hospitals will be eliminated and will have significant effect on the prices of these essential drugs after the implementation. According to the relevant official estimation, the prices of these medicines have declined about 10% and the patients can save 25% of their money when they purchase medicine at basic-level health institutions[12].

The *Items of Essential Medicines (stock and use at basic-level health institutions)*, which includes 307 categories of chemical and Chinese traditional medicines was publicized on August, 2009, and was formally in effect after 34 days. In 2009, the essential medicines will be completely stocked and sold at the prices regulated by the "Items" in 30% public hospitals and health centers of every province (municipality, autonomous region); in 2011, the essential medicine system will be basically established; in 2020, the essential medicine system will completely cover the whole country.

[12] http://www.beijing.net/html/89/n-3989.html

SECTION 4: THE PROBLEMS AND TARGETS OF THE NEW HEALTH REFORM SCHEME

I. Problems Need to Make Further Definition in the New Health Reform Scheme

Even the most carefully designed systems or action plans has some non-optimal aspects which need further discussion, but the merits of the plan itself cannot be denied. In the plans of new health reform, some points are still incomplete and need further practice and prove.

1. The Coordination between Government and Market's Roles

The responsibility of the government is emphasized in this new health reform, and the purposes of realizing the fairness of health system and supplying public health and medical services as public goods are also put forward to strengthen health administration's right of dominating resources. But meanwhile, the plans implied the expectations of the combining the functions of the government and the market, which is also a perennial topic for other countries.

The government is clear about its dominant status in providing public health and basic medical services, and will ensure the public welfare of the public hospitals; but it has also announced that it will actively support the development of private health institutions and encourage different investment types and methods to participate in health services. It also encourages qualified private capitals to establish and operate non-profit hospitals. It indicates that the authorities expect to keep the market mechanism in the health system to improve or optimize the efficient allocation of health resources. However, how to distinguish concepts like public hospitals and private hospitals, for-profit- hospitals and non-profit hospitals? How many public hospitals should remain and how should they distribute? How to retain the competitiveness of the public hospitals and keep their public welfare at the same time? How should the basic health insurance pay for the health services produced by private hospitals? All these questions need to be considered in detail in the future.

It is a critical issue of how to define the role of public hospitals in the coordination between government's dominance and the adjustment of the market; this issue is also a hot topic for the public. In the reform plans, the government will put more into this field; about 150 billion RMB will be injected to public hospitals for basic construction, equipment purchasing, key subjects developing, fees for retired staff, and the compensating of policy-related losses. These are basically the same as before. In 2007, the gross income of state-owned hospitals was 375.4 billion RMB, and 28.5 billion came from the financial subsidy, which only took up 7.6% of the total[13]. Therefore, the additional 50 billion RMB, though helpful, cannot have substantive influence to the income structure and change the incentives of profit-generating of public hospitals. As the costs of staff and operation have not been dealt with in the reform plans, the main payer for public hospitals is the basic health insurance system.

[13] The Ministry of Health: *The Annals of Statistics on Chinese Health*, Chinese Peking Union Health College Press, July, 2009

The aim of establishing essential medicine system and cancelling 15% markup of pharmaceuticals is to radically cure the problem of "profiting through markup money on medicines". In the past, the income of medicines was the major source for public hospitals. In 2007, for example, the income from pharmaceuticals in public hospitals was as high as 200 billion RMB[14]. If the markup on medicine is cancelled, how should public hospitals be compensated? In improving the pilot reform of public hospitals and reducing the ratio of public healthcare providers, what principle should be the criteria, the geographic distribution or the economy of scale? What is the right proportion of public hospitals in the health system?

2. Private Village Clinics

The biggest bright spot in this reform is that the government has determined to offer basic health services to the society as public goods. As the village health stations and its health staff undertake the rural public health services, they are the main providers of rural basic health services. The plans put forward that health reform in countryside can be realized in different ways, and every administrative village should have at least one health stations; the public health services it performs can be purchased by government.

As for the commonly existed problem-the service quality in village health stations, the plans have not paid enough attention to it, and for the problem of private village doctors, the government would only deal part of it. Although the village health services are supplied by the government as public goods, whether the government will purchase them by category and number, or in other ways still need further practice.

3. The Financial Burden of Governments at Different Levels

The government has realized that financing to basic health care from government tax is the guarantee of fairness of healthcare; and the financial support to public hospitals is the guarantee of public welfare. Moreover, the government's subsidy either to the supply side (health institutions) or to the demand side (the demand for health insurance) can reduce the health burden for individuals, thus, the effect of tax in maintaining the fairness of fundamental health care should be insisted.

In the input from central and local government, the local government takes the main responsibility, but the central government will still take charge in coordination because the input from local finance varies in different regions. Though the input from central finance was emphasized in the reform plans, the input obligations of the central government were not announced specifically. As the regional disparity is obvious in China, the effect of central finance should be enforced in particular and it should undertake the obligations of paying for public health like immunization planning, severe infectious diseases preventing and controlling, fundamental health care for civilians and constructing public hospitals and so on, other responsibilities will be taken by the local governments.

In the additional 850 billion inputs, the ratio of the central finance and local finance is 4:6, but in the period of 2006-2008, the ratio was 27:73, hence the input proportion from central government in the next 3 years is increasing.

[14] See 11

4. The Proposal on General Practitioner and Referral System

When referring to standardization of rural health practice, the plans put forward the idea that basic health institutions should have general practitioners; thus bring up the concept of general practitioner for the first time. What is clear is that general practitioners are the working staff in basic health institutions, but the criteria, regulation and rights have yet to be clarified.

The referral system is the target of setting health institutions at different levels, and the breakthrough point in solving the unreasonable structure of health resource distribution. After the plans have been carried out, a system that keeps first diagnose in the basic level, then treatment in hospitals of different levels and bi-directional referral can be realized in cities from community service centre to general or specialty hospitals, in countryside from village health stations to township health centers and county hospitals. The plans also advocated guiding patients to seek healthcare at lower-level health institutions, and gradually realize the system mentioned above; but there were no specific regulations for countryside.

There are 5 intermediate targets are expected in the health reform: quality, availability and accessibility of care; fairness of healthcare financing, efficiency and cost. Through the 5 targets, the aim is to realize the ultimate goals of improving people's health,preventing risks of diseases and enhance public satisfaction . In April, 2009, the new health reform plans publicized with large amount of affliated projects, though only half a year passed and it is too early to evaluate the effects, the direction of developmental of China's health system has basically been clearly designated.

5. Re-defining the role of Chinese Traditional Medicine

In this health reform, the Chinese traditional medicine has been attached great importance; especially the prepared Chinese medicinal slices were completely brought into health insurance in the items of essential medicines. The policy purpose is to make use of the price advantage of Chinese traditional medicine to reduce people's health cost burdens. The advantages of "safe, convenience, effective, and cheap" are much suitable for the need of community and rural health services. The characteristic of general practice of Chinese traditional medicine enable it to have unique advantages in changing the situation of lack of medicines and health care in rural areas. Its advantages are also reflected in the prevention and cure of severe infectious or chronic diseases, the health protection of older people, women and children, the establishment of the rehabilitation system of critical illness, and the health economics. Therefore, the health reform plans redefined the role of the Chinese traditional medicine, but the concrete plans still need specific research.

II. The Expected Goals of the New Health Reform

Basic health services are universally accessible: the plans of the new health reform publicized in April, 2009 have clearly stated that everyone can enjoy basic health services by 2012. Universal accessibility and fairness of health care all underline fairness as a goal for the reform.

The fairness of health financing: in the new health reform, 2/3 of the 850 billion RMB input is planned to invest in the demand side, namely the official subsidy for the basic health insurance. It increases the public funds, especially the level of tax subsidy, and the insurance payment for health inpatient care and serious outpatient cases. Moreover, it accelerates expansion of health insurance coverage and improves the payment capability; thus, the individual out-of-pocket payment will decrease constantly, and the fairness of health financing will be guaranteed.

The quality of health services has further improvement: the quality of health services is the major benchmark to measure whether the health reform is successful. To deepen the reform we should not only enlarge the scope of health insurance and increase the input from the government, but also attach importance on internal management and improve service quality. It is the essential aim of health reform. The new plan guarantees the fairness of health care system and the responsibility of offering healthcare as public goods, but it also encourages market competition to improve service quality and efficiency.

The cost and efficiency are both considered: to balance the cost and efficiency is a common challenge for all the countries that undergo reform of health system. When the competition system is been brought in, the cost should be noticed too, because there are quite a few lessons from other countries.

The generalized concept of national health has been established: the generalized concept of national health should leap out of the scope of medical care, and the health problems should not be treated in isolation; because health system is only one of the decisive factors, other factors including education, housing, water and sanitation, nutrition and transportation. the level and distribution of these factors all influence the conditions of national health , and this is the problem of balanced development of macro economy and the health system.

Translators: Cai Yazhou (SUFE), Wu Shulin (SUFE)
English Version Editors: Lv Mei (SUFE), Xu Ye (Harvard)

REFERENCES

The statistic information centre of the Ministry of Health, December, 2004. the Special Study and Investigation on Health Reform : The Third Sociological Assessment Report of National Health Investigation. Chinese Peking Union Medical College Press.

The CCP and central government, April, 2009. The Views on Deepening Health System Reform.

The CCP and central government, April, 2009. The Implemental Plans on Deepening Health System Reform, 2009-2011.

The Ministry of Health, July, 2009. *The Annals of Statistics on Chinese Health*, Chinese Peking Union Medical College Press.

Cheng Tsung. M, 2008. China's Latest Health Reforms: A Conversation with Chinese Health Minister Chen Zhu, *Health Affairs*, Vol.27, No.4 (2008):1103-1110.

Hsiao William.C and Maynard A., 2009.Foreword, *Health Economics,* Vol. 18, No S2, July 2009: 1-2.

Wagstaff A., YW., Lindelow M. and Hsiao William.C, 2009. China's Health System and Its Reform: A Review of Recent Studies, *Health Economics,* Vol.18, No S2, July 2009: 7-23.

In: China's New Deal
Editors: Li Xiaoxi and Hu Biliang

ISBN: 978-1-61470-815-5

Comments on Chapter 9

CHINA'S HEALTHCARE REFORM: NEW BEGINNING AND NEW CHALLENGES

Liu Yuanli*

The paper by Wang et al on China's new round of healthcare reforms gives a comprehensive review of China's healthcare reform initiatives in the past, and provides an overview of the major contents of the State Council Decisions on Deepening Healthcare Reform as well as the Implementation Plan of Deepening Healthcare Reform for 2009-2011, both of which were announced by the Chinese central government in April 2009. In addition, Wang's paper also discusses some major unresolved issues including the proper roles of the government and market in healthcare, while offering an optimistic outlook of the potential impact of the reform measures. Although I share many of the views of the authors as well as their sense of optimism about the positive impact of the reform plan when it is successfully implemented, I like to point out that our sense of optimism is based on the assumption that the major decision makers will continue their efforts in sorting out major unsolved conceptual and practical issues, and face new challenges head on throughout the whole implementation and policy adjustment process.

I usually use three criteria to evaluate "quality" of a new public policy: a. measurability of the policy goals, b. operational feasibility, c. effectiveness in terms of the actual impact. Applying the framework to analyzing the five major policy provisions for the next three years of the reform plan, we can develop a set of different realistic expectations about the policies because five policies would score different points on the 3 accounts. For example, some policies (e.g. Establishing Essential Drug System) do not have quantifiable goals, thus it would be hard to judge the success of the policies. Nonetheless, I believe that the reform would stand a good chance of success, if the policy makers pay adequate attention to strengthening three aspects of China's healthcare system: capacity-building, incentive structure (making major-stake-holders motivated to do the job they are supposed to do), and accountability (holding major-stake-holders accountable for their actions). In the remainder of

* Dr. Liu Yuanli teaches at Harvard School of Public Health as well as at Tsinghua School of Public Policy and Management.

this short essay, I will use the above frameworks to analyzing 5 major questions facing China's next steps of healthcare reform.

First, how to reform China's provider payment system? Fee-for-service is the predominant payment method currently used in China to pay providers. Under this system, doctors are paid more for doing more. In addition, the prices of medical services are set by the Chinese government rather low, relative to the costs. Meanwhile, doctors in China both prescribe and dispense drugs. Drug sales, where the mark-up are as high as 15-20%, account for about 45% of the total healthcare expenditure in China. Therefore, unless the payment system is reformed, China's healthcare cost escalation problem will continue to exist, in large party due to the strong financial incentive for the providers to over-prescribe. Of course, there are pros and cons with different alternative payment methods such as DRGs, capitation payment and so on. China needs to conduct some well designed and evaluated pilot projects to test out new payment systems that would work in China.

Second, how to regulate uses of the "essential drugs"? If cost-effectiveness is used as the sole principle to select essential drugs, some drugs, even though they may be necessary for certain patients such as patients suffering stroke, may be excluded from the formulary because they are expensive. Since China has remarkable inter-regional disparities, implementing the "Essential Drug System" uniformly across China may turn out to be difficult.

Third, how to increase effective demand for the community health se4rvices? The Chinese planed to use about one third of the increased government investment in health sector (about US$125 billion over the next three years) to strengthen China's primary healthcare facilities. This would include building or refurbishing many community health centers. The idea is that once the "quality" of community health centers is improved, people will start using them more. However, since community health centers currently do not have the kind of trust people have in big hospitals, the effective demand for community health services may not increase as significantly as the policy makers have imagined. So, the challenge is how to increase people's perceived quality of and trust in the reformed community health centers, so that people will choose them as their trusted source of care, rather than going directly to big hospitals. One way to do it is to develop vertically integrated healthcare systems. Policy makers can encourage such pilot initiatives to test their effectiveness in improving healthcare efficiency and quality.

Fourth, how to strengthen the prevention and public health function of medical organizations. China's current healthcare reform plan includes a provision to strengthen China's CDC system. However, since hospitals and other healthcare organizations are places, where people interact with the system most frequently, they represent a particularly challenge as well as opportunity to carry out preventive strategies such as health education. Meanwhile, most of the clinical doctors have minimal training in public health and hospitals do not pay adequate attention to prevention and public health.

Fifth, how to strengthen the social mission of China's public hospitals? Currently, China's public hospitals behave like a profit-maximizer, because government budget allocation only accounts for about 7% of their income and they have to generate revenue to pay doctors' salaries and bonuses etc. Therefore, public hospitals often welcome rich patients while turning away patients who cannot pay the bills. How to develop a social mechanism to finance the "uncompensated care" remains a critical issue to be worked out.

In answering the above questions, government intervention and more regulation will be of course a major approach, especially when it comes to improving equity in healthcare. But in light of existing regulatory failures and government corruption etc, mechanisms of market competition can and should also be used to help increase efficiency, especially in the area of healthcare delivery. A better healthcare system is the one, where both government and market interventions play roles in areas, in which they must and can play an important role.

REFERENCES

Wilensky GR. Healthcare reform: a work in progress. *Healthc Financ Manage*. 2009 Aug; 63(8):28, 30.

Chen Z. Launch of the health-care reform plan in China. *Lancet*. 2009 Apr 18; 373(9672):1322-4.

Yip W, Wagstaff A, Hsiao WC. Economic analysis of China's health care system: turning a new page. *Health Econ*. 2009 Jul;18 Suppl 2:S3-6.

Wagstaff A, Yip W, Lindelow M, Hsiao WC. China's health system and its reform: a review of recent studies. *Health Econ*. 2009 Jul;18 Suppl 2:S7-23.

Liu Y. The anatomy of China's public health system. In Freeman CW. Eds. China's capacity to manage infectious diseases. Center for Strategic and International Studies, Washington, DC: 2009.

Liu Y. Reforming China's healthcare: for the people, by the people? *Lancet 2009*; Vol. 373; 281-283.

Liu Y, Rao K, Wu J, Gakidou E. China's health system performance. *Lancet, 2008*; Vol. 372 No. 9653 pp 1914-1923.

Hu S, Tang S, Liu Y, Zhao Y, Escobar M, De Ferenti D. Reforming how healthcare is financed and paid for in China: challenges and opportunities. *Lancet, 2008*; Vol. 372 No. 9652 pp 1846-1853.

In: China's New Deal
Editors: Li Xiaoxi and Hu Biliang
ISBN: 978-1-61470-815-5

Chapter 10

PILOT ZONES FOR SUPPORTING COMPREHENSIVE REFORMS: A NEW PLATFORM FOR FURTHER INSTITUTIONAL REFORMS *

Zhao Shaoqin*, Fan Lina and Wang Xiwei

In the process of reform and opening up during the past 30 years, explorations by "experimental units" have played a key role in China. For major reforms, the Chinese government usually first experiment with projects and gradually expand to other areas only after gaining useful experience. This is the most valuable piece of experience that leads to the success of China's reform and opening up.

As early as in 1980, the Chinese central government decided to develop Special Economic Zones (SEZs) on a trial basis, in specific areas in Shenzhen, Zhuhai, and Shantou in Guangdong province and Xiamen in Fujian province. Hereafter, from coastal areas to areas along the Yangtze River and boarder areas, from eastern areas to central and western areas, the level of China's reform and opening up has been ever deepening and the scope been ever widening. In 1980s and 1990s, China has launched two sessions of comprehensive supporting and coordinated reform trial programs in urban areas which speeded up the reform of economic structure in their areas and the whole country. Since the beginning of 21st century, China's reform of economic structure has entered a stage where critical problems need to be addressed. Carrying out comprehensive supporting reforms in areas which are typical, representative and with good economic foundation becomes a must for deepening the reform and opening up in the period.

In June 2005, Chinese government approved the Shanghai Pudong New Area to carry out comprehensive supporting reform on a trial basis. In accordance with the principle of careful selection and strict control, the central government has selected certain regions in eastern,

---

* This paper was completed with the guidance of Professor Li Xiaoxi--Dean of Research Institute of Economics and Resource Management, Beijing Normal University. In the writing process, Researcher Wang Jianing from Chongqing Academy of Social Science has offered ardent instruction and deliberation. Gratitude to them all.

* Dr. Zhao Shaoqin earned his PhD degree from the Institute of Economics and Resource Management of Beijing Normal University. Email: zhaoshaoqin@gmail.com

central and western areas to launch pilot reforms of different types. In April 2006, the State Council released *Proposals on Issues for Further Developing and Opening the Binhai New Area in Tianjin* to prove the Tianjin Binhai New Area to be a National Comprehensive Supporting Reform Pilot Zone (CSRPZ). In June 2007, Chengdu and Chongqing were approved to build national CSRPZs for Balanced Urban and Rural Development have been approved. In December 2007, Wuhan city circle and Changsha, Zhuzhou and Xiangtan city cluster were approved to build national CSRPZs for Building a Resource-efficient and Environment-friendly Society have been approved. Shenzhen City, as a special economic zone, has been playing the role of a "demonstrative window" for reform and opening up. In December 2008, the State Council officially confirmed Shenzhen as a aational CSRPZ by authorizing *the Development Plan of reform in the Pearl River Delta (from 2008 to 2020).* By now, the echelon distribution layout of 7 national CSRPZs from the eastern areas to the central and western areas has basically taken shape.

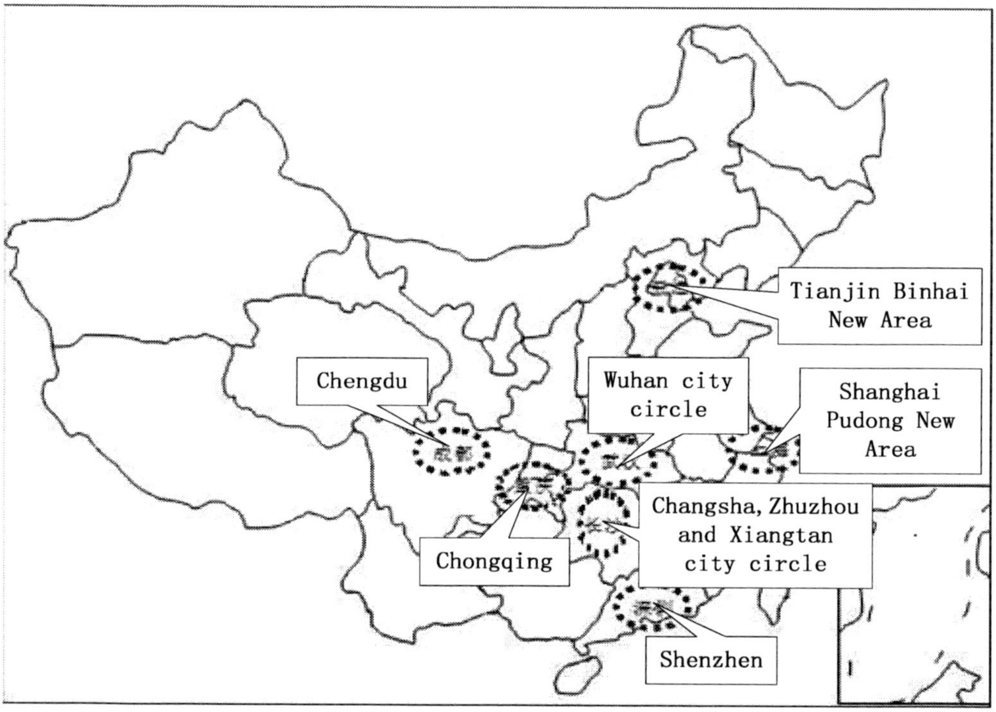

Figure 10-1. Distribution Sketch Map of the Seven National CSRPZs .

Representative domestic researches suggest that understandings of national CSRPZs can be from three levels: firstly, the "national level" indicates that the trial zones should play an exemplary and leading role in the regional economic development of the country. Secondly, the "comprehensive supporting reforms" suggests that the reforms are comprehensive and coordinated rather than some scattered single ones in order to realize multi-level and in-depth coordinated development. Thirdly, the "pilot zone level" implies the head start of comprehensive coordinated reforms, which specific economic zones carry out trial reforms in all respects concerned with social economics and lives. With a view to systematic innovation, reforms will be promoted through overall systematic development.[1]

Chinese scholars usually divide the seven pilot zones into two types. One type is to assume responsibilities of comprehensive and all-round reforms, this type of zones include Shanghai Pudong New Area, Tianjin Binhai New Area and Shenzhen[2]. The other type is to shoulder reform responsibilities that are thematic and in key areas, such as Chongqing and Chengdu that aim at balanced urban and rural development, also Wuhan city circle and Changsha, Zhuzhou and Xiangtan city cluster that aim at systematic reform innovation related to building of a resource-efficient and environment-friendly society.[3]

SECTION 1: SHANGHAI PUDONG NEW AREA, TIANJIN NEW SEASHORE AREA AND SHENZHEN CITY--LEAD THE WAY IN BUILDING A SOCIALIST MARKET ECONOMIC SYSTEM

China has a vast territory, so that it is hard for different regions's level of economic development to develop in line with each other, and the degree of systematic perfection are highly varied. Under such circumstances, it is necessary to choose some areas with good economic foundation to lead the way in overall reform and building a socialist market economic system. Shanghai Pudong New Area and Shenzhen as vanguards in China's economic development during the past 30 years, and Tianjin Binhai New Area which is greatly expected to be the "future polar of economic increase" are appointed as national CSRPZ.

I. Shanghai Pudong New Area: Building a Core Area of International Financial and Shipping Center

Shanghai Pudong New Area is located in the eastern part of Shanghai, near the mouth of the Yangtze River, and face the Shanghai downtown in distance. In April 1990, the Chinese government announced the reform and opening up of Shanghai Pudong New Area. In the short span of the last 19 years, earthshaking changes have taken place in this area. The State Council ratified Nanhui District is assimilated into Pudong New Area. The enlarged Pudong

1 Hao Shouyi, Gao Jintian: An Analysis of CSRPZ in issue 2, 2006 of China Opening Herald. See also Wang Jianing, Hu Xinhua: Management System Investigation of CSRPZ : Shanghai Pudong and Tianjin Binhai in issue 8, 2009 of Reform

2 Some scholars suggest that the reforms in Shenzhen are concentrated in administrative management thus they can't be listed in the first category. This paper considers that Shenzhen as a Special Economic Zone shoulders tasks of leading the way in completing socialist market economic system. The plan of Shenzhen comprehensive supporting reform involves the entire major reforms which elaborates efforts to make breakthroughs in public institution and foreign-related system reform besides administrative system reform. Thus this paper lists Shenzhen into the first category.

3 Some scholars suggest that the status and function of CSRPZ varies as the approval authorities are different. For example, the status and function of Shanghai Pudong New Area and Tianjin Binhai New Area which are approved by the State Council are higher than other Pilot Zones that are approved by the National Development and Reform Commission. This paper doesn't accept this kind of division.

New Area covers an area of 1210 square meters with a resident population of 4.12 million, amounting to roughly a fifth that of Shanghai, and its GDP totaled RMB367.6 billion accounts for 26.8% of Shanghai. Since 2005, the Shanghai Pudong New Area have launched dozens of reforms in government administration system, financial system, technology system, social service, foreign economic relations, the urban-rural dichotomy and so on, significant progresses and breakthroughs have been achieved in some key areas and links. Especially after the annexation with Nanhui District, Pudong New Area "starts a new undertaking" further intensifies reform, in a bid to build a core area of international financial and shipping center.

1. Positioning Objectives as "Small Government, Big Society, Big Service", Service-oriented Government Function is Highlighted.

Shanghai Pudong New Area explores a new governance model with institutional framework establishment as the basis and ground floor as the reform unit. Pudong New Area takes systematic reforms in strengthening the administration and performance management, establishment of grass-roots accountability mechanisms and public oversight mechanisms, exploring new governance models in industry organizations and new type community-based organizations, optimizing the work of administrative examination and approval, e-administration Promotion. It is in a bid to realize the so-called "two highs and one low"—high administrative efficiency and transparency, with low charge.

By establishing a "1 plus 23"public service system, which radiates across the whole district, consisting of a district level civic center and 23 community level service centers for accepting affairs, Pudong New Area promotes the functions of the street towards the provision of public services, monitoring the activities of government power.

2. Based on Financial Innovation, Making Efforts to Optimize the Financial Ecology

The reform of the financial industry is a breakthrough in the overall comprehensive reform of Pudong New Area. With Lujiazui Financial and Trade Zone as the carrier, Pudong makes frequent "bright spots" in its financial reform. Shanghai Stock Exchange launched the Shanghai 50ETF fund products and warrants product; Shanghai Futures Exchange has also launched metallic zinc and gold futures products. The inter bank foreign exchange market maker system and the Shanghai Inter Bank Offered Rate (SHIBOR) have also been implemented.

Podong's reforms put on stage a number of "Firsts". Mainland China's first financial derivatives exchange was formally established in Pudong; Pudong New Area People's Court established the nation's first financial division. Besides, Hua'an Fund Management Corporation was approved as the first domestic company to conduct pilot QDII fund.

The financial ecological environment has been significantly optimized. People's Bank of China Shanghai headquarters and many other financial institutions, including the only two domestic currency brokerage company--Shanghai Guoli Currency Brokerage Company and Shanghai International Monetary brokerage Company--have settled in Pudong. The enactment of *Measures for the Implementation of Pudong New Area Financial Professionals Gather* facilitated the gathering of talents.

Column 10-1. Shanghai Pilot of 'Governance, Handling, Assessment' Reform in Educational Sector [4]

In June 2005, Shanghai Pudong New Area government signed *Agreement of State-run Donggou High School's Management be Commissioned* with Shanghai Victory Education Management Service Center (hereinafter referred to as Management Center)—a private–run education management institution locates in Puxin, raised the curtain of "Governance, Handling, Assessment" gear reform in social affairs.

According to the Agreement, during commissioned management period, the nature of the school, its subordinate relationship with the Bureau of New Area Social development, the grant from government and its other duties remain unchanged; students' education expense will not increase. The New Area government pays annual management fee to the Management Center for its government service.

The Management Center practices self-management in appointing head teacher, adopting educational philosophy, innovating management model, training teachers, organizing teaching after its taking charge of Donggou High School. The Pudong New Area also establishes the Shanghai Pudong Development Education Assessment Center. Ultimately, the quality of the school is assessed by professional intermediaries assessment bodies rather than by the government or the Management Center.

One year later, the Assessment Center took the first assessment over Donggou High School which shows that great changes have been taken place in the school's teaching achievement compared with that before the mandatory administration. Anonymous survey indicates that teachers', students' and parents' degree of satisfaction towards the school's victorious education have reached 92.2%, 94.7%, 97.7% respectively.

Donggou High School's taking lead of the "Governance, Handling, and Assessment" comprehensive reform had a rather big impact in the education sector. The Ministry of Education sent staffs to appraise this model and gave it a high evaluation

The core of the "Governance, Handling, and Assessment" comprehensive reform is that it clarfies the relationships among government, school and society, crystallizes all sectors' functional positioning with the premise of transformation of government functions.

Governance is no longer that the government takes on all things no matter big or small. Instead, the government implements macro-management through planning and guidance, policy design, supervision and inspection. Some detailed professional service and affair management are left to social professional bodies by buying their services.

Handling means the schoolmaster is fully empowered in resources of human resource, capital, and material assets to run a school more freely which will fully motivate the initiative and creativity of a school.

Assessment is to establish specialized non-governmental assessment organizations which remark impartially on educational decisions, teaching achievements and so on.

Through the establishment of a "Governance, Handling, Assessment" mechanism in which sectors are separated but coordinated, working potentials of government, school and society have completely been unleashed, and the enthusiasms from all parts have been motivated.

Source: Website of China Education Pioneer Net,
http://www.ep-china.net/content/news/c/20070823104846.htm

---

4 Quote from related reports from PCPChina,
http://www.ep-china.net/content/news/c/20070823104846.htm

3. Trinity of "Financing, Services, Rights Protection", the Park Formed a Delegation to Promote Scientific and Technological Innovation

The Pudong New Area played its unique advantage and have built up a science and technology investment and financing mechanism. The government funds played a leading role in driving various types of social capital, entrepreneurial venture capital and bank credit to blend, thus created a venture investment guide fund, special funds for the pilot of intellectual property rights pledge financing, and the SME credit guarantee system for the building of special funds to support enterprises technological innovation.

It has built a multi-level technology platform for public service consists of nonprofit, both nonprofit and operational and corporate-owned shared facilities. Currently, the Pudong New Area has owned the most intensive public service platform of technology, the highest level of technical services, and the greatest extent of shared public services.

It has basically built up an integration of intellectual property rights protection mechanism for dispute resolution that is composed of judicial protection, administrative law enforcement and industry self-regulation.

It has set up an intellectual property rights dispute settlement procedure, and established a trial model of "Three Traditional Procedures be consolidated to One System"; established an emergency mediation mechanism for disputes of intellectual property rights protection. The Area has set up an Intellectual Property Rights Protection Center and Association to continuously improve the intellectual property rights protection mechanism that is business-oriented and private self-regulated.

"Focus on Zhangjiang" is an act of co-partnership between industry, teaching and research. Six science and technology innovation Pilot Parks, including the National Torch innovative experiments in urban areas, National Hi-tech Standardization Pilot Park, and National Intellectual Property Pilot Park in Zhangjiang Hi-Tech Park, and 11 state-level industrial bases including National Base of Integrated Circuit Industry have entered Zhangjiang. In the past six months, 159 national key scientific and technological special projects with a total investment of over 7.7 billion, including Chinese-made large aircraft R&D Center, the National Broadband Center, the National Antibody Engineering Center, and Shanghai Synchrotron Radiation Facility took root in Pudong.

4. A Good Wind Relies on Outside Force; the Comprehensive Reforms Kindled a New Economic Impetus to Develop Open Economy.

Pudong first carried out the pilot project of "Nine Measures" in reform of multinational companies' foreign exchange management, and gave an effective solution to the knotty problem of multinational corporations' management in foreign exchange capital.

Positively building a public service platform of outsourcing, Pudong founded the county's first national service outsourcing research center-- China Service Outsourcing Research Center. It has formed an industrial layout with information technology service as the lead, financial background service as the highlight, R&D design service as the speciality, and consultancy, logistics, human resource, finance, exhibition and other professional services as complementarities.[5]

---

[5] Pudong Service and Innovation: Impetus from Service Outsourcing, Pudong Service Outsourcing Net, 2008.11

Moreover, Pudong actively promote reform of the port management system, and constantly enhance the service functions and competitiveness of Pudong port. Take the combination of service outsourcing and customs reform as example. In 2009, the Pudong Area chose some biomedical R&D outsourcing companies in Zhangjiang Hi-Tech Park to carry out pilot measures for convenient customs clearance such as "early customs and cargo clearance as soon as it arrives", "24-hour reservation clearance". These measures have greatly speeded up the clearance of raw materials for pharmaceutical R&D.

Table 10-1. Pilot Reforms List in Pudong Port Management Mode

Sequence number	Content of Pilot Reforms
1	Exploring the Pudong International Airport as a pilot to carry out "three forms amalgamated into one", that is amalgamate the customs, border control, inspection and quarantine declaration form
2	Promoting the "one document for two customs declarations" in import and export of goods, and network monitoring, expanding the green channel of foreign trade
3	Carrying out the pilot project of integrity enterprises in exempting 3C
4	Exploring inspection and quarantine supervision model innovation in service center and logistics center
5	Innovating customs supervision model in exhibition industry. Based on Shanghai Pudong International Exhibits Monitoring Service Center, Pudong implement "unified customs, unified storage, and unified supervision" to provide one-stop service for the international exhibition in international exhibition record, customs clearance, inspection, leaving purchase, cancellation and other services. Carrying out reforms in popularizing paperless customs clearance and online filing system of international exhibitions
6	Carrying reforms of inspection and quarantine of imported biological materials in Zhangjiang Hi-Tech Park

Source: Shanghai Customs of the People's Republic of China:
http://shanghai.customs.gov.cn/publish/portal27/tab20382/module51836/info130022.htm

5. Making Efforts in Reforming Rural Management System, Coordinating Urban-rural Integration.

In planning the layout, Pudong broke the situation of separated management between urban and rural areas, coordinated regional development planning and new town planning and construction; promoted the common development of urban and rural areas; actively promoted the "village-owned separation", "one branch of two committees", "Union Village management" as the main content of reforming the rural management system, deepened the village collective asset management system, enhanced the management level in rural areas; promoted the merger of management system of urban and rural social undertakings and infrastructure investment integration. At present, elementary education has achieved standardization in four areas of allocation criteria, hardware equipment, information platform and teachers training.

II. Tianjin Binhai New Area: Building into the Growth Pole in Northern China

Binhai New Area is located in Tianjin's eastern coastal part, stands at the heart area of Bohai Rim Area, including 3 administrative districts of the Tanggu District, Hangu District, Dagang District and the Development Zone, the Bonded Area, Tianjin Port, and part of the Dongli District, Jinnan District, with a planned area of 2270 square kilometers and the resident population of 2.03 million.

In accordance with the positioning by the State Council' s *Proposals on Issues for Further Developing and Opening the Binhai New Area in Tianjin*, the Binhai New Area will be based on Beijing, Tianjin and Hebei, serves Bohai Rim Area, radiates the "Three North Region", and faces the Northeast Asia. It strives to build itself into a portal in opening up in northern China, a high level base of modern manufacturing and R&D transformation, the international shipping center and international logistics center in northern China, and gradually become a livable ecological new city with economic prosperity, social harmony and environmental beauty.

In recent years, Tianjin Binhai New Area launched a series of important reform measures in finance, foreign affairs, science and technology, and land. It has made important progress in areas of expanding direct financing channels, promoting foreign exchange reform, improving the financial environment, management innovation in Dongjiang Bonded Port, service outsourcing base construction, promoting regional economic cooperation and technology platform construction, linkage of increase in urban construction land and reduction in rural construction land.

1. Financial Reform Docks Corporate Finance to Build a Platform for Financial Transactions

Binhai New Area has organized "China International Private Equity Forum" twice, and actively promoted the abutment of fund's equity investment and capital of corporate equity financing. In addition, Tianjin Equity Investment Fund Association and the Bohai Industrial Investment Fund were established. Moreover, Binhai New Area carried out Small and Medium Equity Fund pilot projects and preparation of the Shipbuilding Industry Investment Fund. Registered small and medium equity investment funds in Tianjin totaled 149 with the registered capital (contractual capital) of nearly RMB 40 billion.

Binhai International Equity Exchange is conducting trial operation with 427 identified founding members, 183 enterprises applicants to lists for financing and 52 strategic cooperation agencies. Based on this platform, Binhai New Area has basically established a corporate equity market and private equity transaction market.

Binhai New Area has preliminarily formed a comprehensive financial service system with banking and insurance as the body, including trust, leasing, funds, securities, futures, factoring.

2. With Port Construction as the Carrier, Innovating Port Management System

In 2007, the world's largest artificial deep-water port--Tianjin Port built a 250,000-ton deep-water channel. Since then, Tianjin Binhai New Area has set up 10 inland "non-water ports", opened an e-port and logistics information platform, and practiced 24-hour customs clearance system. Binhai New Area was the first in China launched the "non-water port" regulatory approach. The companies only have to handle once the customs procedures in the "non-water port"; goods can be directly cleared through customs after their arrival. So, processing time of integrated customs clearance operations can be shortened from 60 hours to 2 hours.

Dongjiang Bonded Port Area develops fast. The Port Area introduced accredited agencies like Customs, Inspection and Quarantine, Maritime, Border, Foreign Exchange, Taxation to form a Co-supervision Coordination Committee. With the former Dongjiang Bonded Area Administrative Committee, the port area is building an international logistic operation center that radiates and drives the "Three North" region, opens up to the domestic and overseas, and fully connects the Northeast Asia Economic Circle.

3. Actively Expanding Multilateral Cooperation, Promoting Contacts between Scientific and Technological Research and Industrial Investment

Six science and technology innovation platform including National Biomedical Joint Research Institute, National Engineering Research Center of Cell Products has basically been completed and 20 provincial-level R&D Center for Conversion were put into operation.

The Binhai New Area cooperated with the Ministry of Science and Technology and China National Offshore Oil Corporation to develop Binhai Hi-tech Zone and the National Bio-pharmaceutical Industry Demonstration Zone. It also carried out scientific and technological R&D cooperation with the Chinese Academy of Sciences, Tsinghua University and other well-known universities and national research institutes in Italy, United States, Sweden and other countries. Moreover, Binhai Area keeps close relations with big state-owned corporations to promote scientific and technological industrialization.

Binhai New Area are exploring new business venture capital model, and have set up two venture capital guide funds. It has already attracted 10-plus famous venture capital organizations including National Science Ruihua and Redpoint Ventures and possesses a total volume of RMB17 billion in venture capital fund.

4. Implementing "Separation of Expropriation and Conversion" in Land Acquisition, Protecting the Rights of Farmers Whose Lands have been Expropriated.

In 2008, 14 batches, 510 hectares of land have been expropriated under the model of "Separation of Expropriation and Conversion". By implementing experiments of linkage of increase and reduction in urban and rural construction lands and carrying out two groups of pilot projects of homestead exchanged for commodity house, 100 thousand of farmers have moved into small towns.

5. Developing a Recycling Economy and Creating an Eco-city

Binhai established six national recycling economy pilot projects including the Economic and Technological Development Zone, Lingang Industrial Zone and Beijiang Power Plants. Based on the conclusion of five kinds of recycling economy model--the Taida mode, Ziya mode, Lingang mode, Beijiang model, Huaming mode, the Binhai New Area promulgated *Tianjin Energy Conservation Ordinance* and issued *Tianjin Energy Supervision and Testing Regulations* and other supporting administrative regulations. The Binhai New Area set up the Eco-city Administrative Committee, promulgated the *"China-Singapore Tianjin Eco-City regulations"*, and formulated the eco-city index system, innovated operating mode of eco-city construction and management and established environment access mechanism and the environmental monitoring warning system. The Binhai New Area also supported business innovation of the Tianjin Emissions Exchange and carried out pilot projects of total amount control of pollutants and emissions trading.

III. Shenzhen: Creating New Advantages, Attaining a Higher Goal

Shenzhen is located in the south of Guangdong, on the east bank of the Pearl River and is adjacent to Hong Kong. In 1979, the Chinese Central Government made the decision to establish the city of Shenzhen. In August 1980, the Standing Committee of the National People's Congress approved to set up Shenzhen Special Economic Zone. Shenzhen's total land area is 1952.84 square kilometers of which 395.81 square kilometers is the Shenzhen Special Economic Zone. By the end of 2007, Shenzhen has a permanent population of 8.6155 million of which the registered population is 2.1238 million and the non-permanent population is 6.4917 million.

From a small South China Sea coastal town to co-building a world-class metropolis with Hong Kong, Shenzhen is the wonderful epitome of China's Reform and Opening up and modernization drive. As the best developed and the most influential Special Economic Zone, Shenzhen fully played the demonstrative role of "window" and "experimental field" of reform and opening up over the years, and made great contributions to the establishment and improvement of the socialist market economic system.

In recent years, Shenzhen actively promoted comprehensive supporting reforms and continuously achieved new breakthroughs in key areas and key links. On December 31st, 2008, the State Council officially confirmed Shenzhen as a National Comprehensive Supporting Reform Pilot Zone by authorizing *the Development Plan of reform in the Pearl River Delta (from 2008 to 2020)*. This indicates a new era of Shenzhen's comprehensive supporting reforms.

1. With Transforming the Government Functions as the Starting Point to Promote Administrative Reform

Actively Explore Super-Ministry System Reforms

In recent years, in accordance with the principle of functional unity, Shenzhen actively adjusted the functions of relevant departments and reasonably settled over overlappging duties among authorities, formed Urban Management Bureau, Communications Bureau, and Bureau of Agriculture, Forestry and Fisheries and other functional bureaus under the Super-Ministry System.

Establish a Reasonable Division of the Urban Powers

Through the clear allocation and division of powers in urban areas, basic-level government's social management functions have been strengthened, and center of social management have been downward shifted. In 2007, Shenzhen began to comprehensively implement the new system of streets law enforcement, 21 specific law enforcement tasks are handed to the street comprehensive enforcement unit to carry out. This has effectively solved the problems of weak enforcement and law enforcement involving authorities at various departments at the same level in the basic-level government, and greatly enhanced the efficiency of administrative law enforcement and urban management level.

Actively Carry Out Pilot Projects to Streamline the Administrative Hierarchy

In 2006, Buji and Longhua Street were separated into 6 streets of Buji, Bantian, Nanwan, Longhua, Minzhi, and Dalang after which the street community's economic management function has gradually been weakened but its social management function has been strengthened. On June 1st, 2007, on the basis of Guangming Industrial Park, the Guangmin New District has been established and pilot projects to streamline the administrative hierarchy have been carried out.

Actively Promote the Construction of a Society Under the Rule of Law

In 2006, Shenzhen began the fourth round of the administrative examination and the approval of system reform. 697 non-administrative licensing examination and approval projects of 37 municipal government departments were thoroughly examined of which 348 projects were reserved, 98 projects were cancelled, and 251 other projects were pegged as non-administrative licensing examination and approval ones. Shenzhen took the lead in establishing the administrative examination and approval of electronic monitoring system to monitor the whole process of administrative approved items' acceptance, undertake, approval, completion, notification and other links. In the end of 2008 the State Council Legislative Affairs Office and Shenzhen Municipal Government signed a cooperation agreement to accelerate the building of a government under the rule of law, developed the construction index system and introduced a set of policy documents to speed up the building of a government the rule of law.

Investigate to Further the Reform of the Civil Service Classification Management

In August 2008, National Civil Service Agency formally approved and agreed to carry out the civil service classification management reforms in Shenzhen. At present, the relevant departments are studying programs of classified reforms, to work out coordinated management systems according to different characteristics and requirements of the three categories of civil servants for administration, administrative law enforcement and professional technology.

Comprehensively Push Forward the Reform of Public Institutions

In July 2006, Shenzhen began its classified reforms in public institutions. 124 enterprises and institutions are transformed to business units and 270 companies operated by government organs and institutions have been complete stripping and transference which involving 19000 staffs.

2. With the Deepening of the Economic Reform as a Focal Point to Promote Transformation of Economic Growth Mode

Deepening the Investment and Financing System Reform

Shenzhen stipulated the "government Investment Project Management Regulations and Implementation Details" to act as the Construction Agent for the Government's investment projects. The progress of major investment project approval is significantly shortened with the processing time compressed from 386 days to the current 100 days.

Deepening State-Owned Enterprise Reform and Innovating the Classified Supervision System of State-Owned Assets

Through deepening systematic reforms on labor, personnel and distribution of the municipal government-owned enterprises, standardization of payment distribution system, and the establishment of the performance evaluation system that suits to legal person management system, the healthy development of state-owned economy has been promoted. Through regulatory regime innovation, clear thinking of official performance, improvement of official performance system, adjustment of official performance institutions, improvement of official performance way and timely change in official performance focus, a complete set of new type of state-owned assets supervision and administration system has initially been formed.

Strengthening Hong Kong- Shenzhen Financial Cooperation, Increasing the Opening of the Financial Sector

In 2008, the Shenzhen Municipal People's Congress passed *Regulations to Promote Financial Development in Shenzhen Special Economic Zone*. Hong Kong and Shenzhen in-depth cooperation in the monetary field, as well as the exploration of the possibility to establish oil and other commodities futures exchanges and other cooperatives were included in the Ordinance. By accelerating financial market integration between Shenzhen and Hong Kong, and expanding business cooperation scope between Shenzhen and Hong Kong financial institutions, the common prosperity and development in Hong Kong and Shenzhen's financial industries have been greatly promoted.

3. With Building a Modern Social Welfare and New Social Management System as its Focus, Shenzhen Continuously Deepened Reforms in the Field of Social Livelihood

Shenzhen actively promoted and regulated the development of social organizations. In September 2009, Shenzhen issued Shenzhen City issued *proposals on furthering and standardizing development of social organization of the city*, explored direct registration of industrial and commercial, public welfare and social charity organizations, and tried community social organizations to implement a two-track system of registration and record management.

Shenzhen innovated population and social organization management system. Since 2008, the full implementation of local residence permit laid solid foundation for strengthening dynamic management of a liquid population, promoting residential welfare of non-registered population, achieving coordinated development among population, society, economy, environment and resources, leading the way in realizing equal basic public services. By the end of 2008, accreditation of residence permit has passed 7 million.

The city's comprehensive security system has constantly been improved. Through the improvement of city, district, street, community public employment service system, Shenzhen has established a multi-level network of public employment service system. Five major social insurance enrollment rates of residential workers is 98%, the number of children's health insured and co-ordinate medical insured people up to 517,000, and the number of labor workers health insurance enrollment reached 6.102 million. Shenzhen's Labor workers pension, the number of work-related injuries and health insured ranked first among large and medium cities.

4. With Promoting Implementation and Innovation of Functional Opening Up Areas as the Focal Point, Shenzhen Furthers Expansion of Opening-up and Domestic Co-operation

Implementing and innovating functional opening up areas. Actively promote development and supervision mode transformation of the present bonded area, sea port and airport. By implementing development strategy of the "One District with Two Wings", Shenzhen carried out "zone-port interaction" with the eastern Yantian Port and western airport. Shenzhen straightened out the large industrial area management system to fully play the leading role of Export Processing Zone.

Promote close cooperation between Hong Kong and Shenzhen. Under the framework of the nine Cooperation Programs with the *memorandum of strengthening cooperation between Shenzhen and Hong Kong* (referred to as "1 +8" Cooperation Agreement) as the general provisions, Shenzhen further push the Shenzhen-Hong Kong co-operation with the concept of "learn from Hong Kong and serve Hong Kong". Shenzhen continuously refined the implementation of CEPA, through stipulating coordinated policies for infrastructure construction, port clearance, industrial cooperation, the Hetao area development, public investment and other fields.

SECTION 2: THE CITY OF CHONGQING AND CHENGDU - THE EXPLORATION OF THE NEW WAYS OF URBAN AND RURAL DEVELOPMENTAL COORDINATION

Urban-rural dual structure has become a prominent conflict of China's comprehensive and coordinated development of economy and society. Although China has largely held the foundations and conditions of dismantling its urban-rural dual structure after years of development, systematical reform and innovation is still needed in order to fundamentally solve this problem and substantially change the institutional arrangements of emphasizing urban areas and ignoring rural ones when offering public resources and services. It is in this context that the state chooses the city of Chongqing and Chengdu due to their typical features in western regions to carry out the comprehensive reform experiment and to realize the national urban-rural harmonious development.

I. The City of Chongqing: to Drive Large Rural Areas through Development of Large Urban One

The city of Chongqing, located in the southwestern part of China's interior and the upper reaches of Changjiang River, is a multi-centered and delegation-styled city. In 1997, it became China's fourth directly governed city region. The whole city of Chongqing covers an area of 8.2 million square kilometers, 2.39 times that of Beijing, Shanghai and Tianjing in total, and has the largest jurisdiction cover in China's four directly governed cities. By the end of 2007, Chongqing has held a registrated population of 32.35 million. Chongqing is greatly featured by its large city, large country, large reservoir, big mountainous areas and minority- congregated regions, and features intense conflict because of its urban-rural dual structure. It is a microcosm of China's national conditions. Around the urban-rural harmonious development, Chongqing city has selected the region of Jiulongpo, Dianjiang county and Liangping county as experimental points due to the better conditions and the quality of representative, model and leading. Meanwhile, other districts and counties also practise positively, boldly and from their own reality. They start from fields they can afford and have made breakthroughs in the reforms of such domains as land systems, identity transformations of peasants, urban-rural integrations of basic public services and new rural constructions. Through these practices, they have explored some new ideas and initiatives to overcome institutional obstacles of coordinating urban and rural development.

1. Conducting Urban-rural Multi-level Programming and Building Comprehensive Systems of Programming and Management

They take the "Master Plans of Chongqing Urban and Rural Areas" approved by the State Council as the guide, and make prominent efforts to complete the three-level programming systems of district and county, township and village. The planning departments collectively managed the planning of urban and rural areas, changed the status quo of separation, and accelerated the pace of extending planning and management functions to villages. They make amendment in guidelines of village planning, and initiate 105 municipal key towns to carry out authorizing plan work. Priority has been given to "boosting villages of new rural construction" to conduct village authorizing plan experiment. In 6 districts and counties such as Jiulongpo and Jiangbei, "four rules superimposed" experiment has been expanded in economic and social development, urban and rural construction, land use and environmental planning.

2. "Ring Wing" Interaction and Promoting the Economic and Social Integration of Urban and Rural Areas

They built an economic circle taking the main city as the core and an hour's drive as radius and form the Chongqing North-East Region taking Wanzhou as the center and Chongqing Three Gorges reservoir area as the main zone. They set up a "one ring, two wings" starting point taking Qianjiang as the center, including the minority segregated south-east poor zones. They established a helping relationship between districts and counties. Such measures as direct financial support, cadres sending, teachers exchange, medical training, science and research consultation, standard workshop building support have been taken to promote the countryside extension of city's quality public service resources. Mutual assistance mechanism has been established from the following 8 aspects: industrial linkage, employment transfer, mutual assistance on education, scientific and technological cooperation, health sharing, personnel exchanges, financing support and poverty alleviation and development .In 2008 the first batch of 10 demonstration projects to extend the city resources into countryside was confirmed.

3. Multi-field Coordination in Administration, Employment, Social Insurance, Targeting the Implementation of Integrated Strategies

They have reasonably divided the administrative rights of the three level governments of city, county, and township, and integrated the agriculture-related sectors of municipal governments to establish Agriculture Commission. According to the principle of match of financial and power, they established a public financial institutional framework integrating the urban and rural areas, and ensured that 75 percent of financials is used upon the development of districts, counties and villages. They launch the rule of "village financials is managed by county but used for villages" and gradually resolve the historical debts of grass roots governments.

The original policy to help "three kinds of persons", that is, emigrants of Three Gorges reservoir area, retired soldiers and children of poor families in rural areas to enter secondary vocational technical schools, has been extended cover "five kinds of persons" adding two other types: school-age children of the state's welfare institutions and minimal insurance persons of cities and towns. The average years of education rises from 7.7 in 2002 to the current 8.4, and they have firstly paid off the debt of "universal nine-year compulsory education" for1.96 billion RMB.

They have opened up a "green passage" through which peasant workers can transfer their household registration into cities and towns. Qualified migrant peasants and workers are encouraged to settle their whole families into cities and towns. The National Labor E-commerce Platform, the country's largest labor information platform for peasants and workers, has also been built up. By the end of 2008, it has signed employment service contracts with 7000 enterprises in total, through which 450,000 peasant workers have registered with information, and 55,000 peasant workers have found employment.

As for urban and rural cooperative medical insurance, the rule of "a platform, two sets of standards" has been reached, that is, the contribution menu has just two levels of 100 and 200 RMB in all, which people can choose voluntarily which one to pay. Two kinds of people, that is, 587,000 urban- transferred peasants whose lands have been acquired and 109,000 residents who have worked under urban employers but exceeded statutory retirement age, have been brought into the system of urban old-age insurance. Urban-rural minimal insurance regulations has been promulgated, through which all people in urban and rural areas are covered, the urban-rural linkage adjustment mechanism has been set up, and the ratio of minimal insurance between urban and rural areas has been narrowed to 2:1.

4. Combining Farmland Transfer and Scale Management, and Standardizing the Management of Urban and Rural Construction Lands

They have established three-level land transfer service agencies including county, township and village to guide and standardize the farmer and rural collective economic organizations to transfer lands legally, voluntarily and with compensations. Some complementary measures to support scale operation have also been implemented, for example, to support major producers of grain by 15% of direct subsidy funds for grains, to subsidy major and households specializing in standardized breeding with 1 to 8 million RMB. By the end of 2008, scale operation rate upon farming lands in Chongqing city has amounted to 17%, while scale breeding rate reached 40%.

While maintaining the basic amount of arable lands and ensuring the self-sufficiency of grains, they have developed reform experimental point such as a steady increase or decrease in linking urban and rural construction land and rural land exchange. By August 2009, rural land exchange point has held five trading land tickets with a total of 5800 acres in lands and 400 million RMB in turnover.

Column 10-2. 'Land Ticket' Trade in Chongqing [6]

Since the establishment of Rural Land Exchange in Chongqing, much public attention has been attracted. It has been commented as "China's rural land reform initiative" and "as important as the once-established Stock Exchange." Chongqing Rural Land Exchange mainly includes barter exchange and land ticket exchange. Barter mainly refers to the exchange of the use right and contractual operational right of arable land, forest land and other agricultural land. Land ticket transaction is a major trading type of land exchange and an important innovation.

The so-called "land ticket" mainly deals with such types of lands: the land that people re-cultivate on the spared rural homestead and the ancillary facility, the land of township and village enterprises, the land occupied by rural public facilities and public welfare undertakings, and other rural collective construction lands. Those lands have been turned into lands available for planting crops. After strict examinations of land management departments, they are transferred as construction land index, certificates are issued by homeland and house management sector of the city. This certificate is called "land ticket".

There are 4 steps for the usage of a land ticket:

Re-cultivation. People re-cultivate the spared rural homestead and the ancillary facility land, the land of township and village enterprises, the land occupied by rural public facilities and public welfare undertakings, and other rural collective construction lands.

Examination. After strict examinations of land management departments, they are transferred as construction land index. The "land ticket" is issued by the municipal land administrative departments to persons who have land use rights according to the corresponding areas.

Exchange. Chongqing Rural Land Exchange is established to carry out "land ticket" transactions. All legal persons and natural persons with independent civil capacity can buy land tickets through a public bid. Land ticket transaction volume is regulated and controlled by the principle of not exceeding 10% of the state-assigned increased construction land plan of that year.

Use. While it is used in urban areas, the land ticket can be brought into newly increased construction land plan. They can increase the same amount of urban construction land and offset the added construction land use-compensated fees and farmland reclamation fees. However, it has to comply with the overall layout of land use and urban-rural overall plans. Expropriation and transfer procedures have to be handled and compensation and resettlement of farmers has to be completed. While lands are levied as state-owned, people have to obtain land use right via legal procedures of "tender, auction and publicity"

6 Source: Dengli, *Chongqing Starts 'land ticket' exchange towards rural lands,* online website about "rural area, agriculture and farmers". http://www.farmer.com.cn/news/nyxw/200904/t20090423_441133.htm

Legally ensuring peasants` rights of possessing, using and proceeding towards lands is the focusing point of Chongqing Rural Land Exchange while they are designing schemes. Comprehensive designs have been made in this aspect:

The process of applying for land reclamation: farmers who apply for homestead land reclamation must have other stable housing. Besides, they must have stable jobs and sources of livelihood. Other rules are as follows: newly added arable lands through re-cultivation and settlement continue to be contracted and operated by original homestead peasants, and those who give up running can acquire corresponding incomes through the lands` next transfer.

The process of price assessment: after a comprehensive consideration of reclamation fee, added construction land use-compensated fees and other factors, Chongqing Municipal Government has constituted a unified price norm upon the urban-rural construction land-linked index exchange for traders to consult.

The process of income distribution: in addition to paying a small amount of taxes, the vast majority of "land ticket" transaction proceedings are distributed to peasants. All the transaction income of the right of contraction and operation towards arable lands and forest lands belongs to peasant families. The land income obtained by rural collective economic organizations is mainly used as peasants` social insurance and new rural construction.

The implementation of "land ticket" promotes arable land protection. The land using model of "occupying first" was substituted for "compensating first". The long-distance and large-scale transfer between rural collective construction land and urban construction land has largely raised the rural land values of remote areas. It has realized the objective of city nurturing rural areas and developed areas supporting backward ones, contributing to the establishment of uniform land markets between urban and rural areas. The reach of change linkage of urban and rural construction land, the establishment of urban-rural unified land markets and the drive of rural factor market nurturing have significantly promoted the construction of rural capital, technology and other factor markets. The peasants` income increase through land ticket exchange has enhanced their social insurance level and development potential after their entering into urban areas.

II.The City of Chengdu: The Pioneer of Urban-rural Coordination

Chengdu is the capital of Sichuan province, with a total area of 12,390 square kilometers, the downtown area of which is 283.86 square kilometers. It is located in southwest of China, the heart area of Chengdu Plain. It has a history of more than 2300 years, and has been playing a significant role since ancient times. In 2007, Chengdu city's household registration population was 11.123 million, while the resident population 12.579 million. In recent years, Chengdu city has taken the following measures as the main contents of urban and rural coordination reform and path designing, encompassing the reform of rural land property right system, the innovation of rural grass-roots government structure and the establishment of the urban-rural integrations of public service system. They strive to boost Chengdu experimental area as the pilot pattern of reform deepening and urban-rural developmental coordination of the whole country, the model window of harmonious society construction and the successful example of post-disaster reconstruction.

1. The Breakthrough of Rural Property Right Reform

In order to establish a modern rural property rights system, featured by clear ownership, clear rights and responsibilities, strict protection and smooth transfer, and to form dynamic rural factor markets, Chengdu city officially launched the pilot reform of rural property right system in early 2008.

The Innovation of Arable Land Protection Mechanism

Chengdu has established the farmland protection fund by using new construction land use-compensated fees and part of land transfer fees and issued the "Use and Management Methods of Chengdu Farmland Protection Funds (for Trial)". Those peasants who take responsibility for farmland protection are provided with old-age insurance subsidies in accordance with following standards: 400yuan for every Mu of basic farmland per year while 300RMB for common arable land. As such, the peasants` enthusiasm of protecting farmland has been mobilized, the protection quality has been enhanced and the comprehensive production capability of farmland has been raised.

The Completion of the Rural System of Property Right-Related Policies

Chengdu municipal committee and government have issued the "Advices on Strengthening the Farmland Protection to Further Reform and Perfect the System of Rural Land and Housing Property Rights (for Trial)". a series of other supplementary files have also been set, including "Checks and Registrations of Chengdu Collective Land Ownership Interim Provisions", " Registration and Management Methods of Contraction and Operation Rights of Chengdu Rural Lands (Trial) "," Checks and Registrations of Chengdu Collective Construction Land Access Interim Provisions(Trial)". Reform of property rights in rural areas has effectively boosted the tenor of post-earthquake reconstruction. In 2008, 10315 households of those peasants who has selected self-construction in Dujiangyan City, Pengzhou City, Chongzhou City, Dayi County, Qionglai county and other damaged counties and cities has finished check of rights and award of certificates. 5225 rural quake-damaged households have acquired 204million loan via pledge and assurance of the homestead right certificates while they were self-building houses after the calamity. 1.714 billion RMB of social capital has been introduced through the collective focus on the use of construction land.

The Establishment of the Service Systems of the Rural Property Right Transfer

Through the right check and certificate award of rural property right, the platform of rural property right transfer has been formed, while the rules constituted. Besides, a series of files have been issued such as the "Management Methods of Chengdu Rural Land Contraction and Operation Right Transfer (for Trial)". The rural property right exchange has initially been built up in Chengdu, while the rural property rights exchange branches have been set up in district(city) and county and the service stations of the rural property right exchange has also been established in village and township. As a result, the city, county and village three-level property right transfer service system has been preliminarily formed. It has created conditions for the rural collective construction land use rights, the land contracting and management rights and the smooth transfer of the forest rights and house ownerships.

2. The New Achievements of "Three Concentration" Boost

Since 2003, Chengdu has achieved remarkable results in promoting urban and rural integration and boosting "Three Concentration" (industry concentrate towards parks, peasants concentrate towards cities and towns, and farmlands concentrate towards scale-operation). In 2008, Chengdu issued another two files: "advices on Further improving the quality of farmers` living together" and "advices on promoting the transfer of the migrant rural workers to urban residents ", to further improve the systematic mechanism of the "Three Concentration" boost.

Vigorously Promote the Shift of Industry to Focusing Development Zones

Adhering to the new industrialization road, Chengdu has been improving the industrial layout and development planning, and continuously optimizing the industrial structure. 116 development zones of the whole city have been integrated as 21 industry-centered development zones. 9 industry-centered points of key towns with preferable industrial foundations and resource conditions have also been layed out. It has nailed down the spatial scale and industry positioning of the industry-centered development zones and the industry-centered points of key towns. It is required that the newly added projects, technical transformation and land expansion projects have to be brought into industry-centered development zones in principle, and fundamentally change the situation of "every village ignites (in a factory), and every household emits smokes", which has laid foundations upon the centralized and intensive development of Chengdu industry. By the end of 2008, the settled scaled industrial enterprises amount to 1775, with industrial concentration rate around 68.2% and added value 87.164 billion RMB. The heavily earthquake-hit areas positively carry on the industrial transfers of their aiding counterparts. Chuan- min Industrial Garden" has been programmed and established in Pengzhou industry-centered development zone to carry on the industrial transfer from the aiding province Fujian; "Chuan-su Science and Technology Industrial Garden" in Dujiangyan industry-centered development zone has brought a batch of enterprises to settle down such as Pushi Ning Jiang, Hua Han and Huimin Building Material.

The Encouragement and Guide Towards Farmers to Gather into Cities and Towns

Chengdu sticks to the road of new urbanization and follows the principle of "local adaptation, farmer self-determination, paid in accordance with law and steadily push forward". Peasants are guided into cities and towns orderly and sequentially, and adapted to live together in accordance with local conditions. A city and town system including 1 megacity,8 medium-sized cities, 30 key towns, 60 new cities and towns and 2000 rural new types of communities has been programmed and established. In the downtown of the city, they carry out the complete combination of rural and urban community, set up new types of communities according to the criteria of urban ones and boost the shift of farmers to urban residents; They take county seat and key towns as focusing point to build up new types of township communities according to the criteria of urban ones. They crack the housing problems of the rural migrant workers and those whose lands have been levied and boost the peasants` transformation to urban residents. In rural areas, they set up new houses according to the principle of "concentrate where it is suitable, scatter where it is suitable". In 2008, Chengdu farmers` per capita net income reached 6178 RMB, with an increase of 9.5%. The area of the "new housing projects" and "new community" that has been started reaches 4.12 million square meters, and the urbanization rate has increased to 63.5%.

Positive Promotion of Lands' Scale Operation

Adhere to the path of agricultural modernization with Chinese characteristics, and the basis of stabilizing rural household contraction and management, according to the principle of "being lawful, voluntary and compensatory", Chengdu steadily boost the lands to center upon the agricultural leading enterprises, rural collective economic organizations and farmers specialized and cooperative economic organizations and great growers.

Following the requirement of "Three Concentration" boost and modern agricultural development, Chengdu has constituted and completed the industrial development layout of agriculture. Around such confirmed competitive industries as grain and oil, livestock and poultry, fruit and vegetable, flower and nursery stock, tea and mulberry, bamboo and forest, aquatic products, they have established large-scale base of competitive agricultural products. In the outskirt of the city, they strive to promote the land-based scale business mainly containing biology, leisure and tourism. In the outer suburbs of Plain areas, efforts have been made to promote land-based scale business mainly containing high-quality grain and oil, vegetables, flowers and nursery stock .In the outer suburbs of uplands and Basin-surrounded mountainous areas, efforts have been made to promote high-quality fruits, vegetables, teas, mulberry, upland crops, authentic Chinese medicinal herbs, forests, bamboos, and aquatic products. At the same time, they focus upon the improvement of the processing level of advantageous agricultural products, actively develop modern agricultural logistics and convert the model of agricultural development. In 2008, the scaled agricultural leading enterprises have reached 657, and the dragging area of agricultural industrialization has amounted to 66.7%.

3. Balanced Development of Urban-rural Public Service Construction

In 2008, Chengdu issued the "Advices on Deepening the Urban and Rural Integration and Further Enhancing the Level of Public Services and Social Management at Village Level (Trial)." They set up the target of supply classification, financial security, facility construction integration, democratic management and personnel improvement towards village-level public services and social management. They systematically put forward the contents embraced in village-level public services and social management, and clearly defined the responsibilities of government, village-level autonomy organizations in public services and social management.; The outlay of village-level basic public services and social management has been brought under financial budget at all levels. There is a clear stipulation saying: "based in 2008, the annual increase in governmental investment at all levels upon the construction of public utilities and infrastructure should be mainly used in rural areas, until the urban and rural public service basically reached parity", which has comprehensively improved the financial security level of village-level public services and social management. Peasants can practice democratic appraisal, democratic decision-making and democratic supervision as for the implementation of village-level public services and social management. The government should fully respect the will of farmers and safeguard the rights of them, and enhance the efficiency of public services and social management.

At present, experimental reforms of village-level public service and social management are held in 10 districts (city) and counties such as Qingyang district, Wenjiang district, Pixian,

Qionglai city. After it achieves effects and experience, the reform will be pushed through in the city-wide.

4. Further In-depth Boost of Standardized Service-type Government Construction

Since 2003, Chengdu launched the deepening of administrative reform, and promote the construction of standardized and service-oriented government comprehensively, which aims to speed up the process of administrative reform, explore innovations of government management and promote the coordinated development of urban and rural areas. After the approval of pilot zone, the construction of standardized and service-type government goes in-depth.

The Establishment of Major-Sector Management System to Coordinate Urban and Rural Development

With an eye to the coverage and spread of public management and services to rural areas, the integration of departmental functions and the establishment of major-sector management system to coordinate urban and rural development, the administrative systems of more than 30 departments such as planning, agriculture, water supply, finance, transportation, gardening and forestry has successively been reformed and adjusted, which has formed the preliminary establishment of administrative system to coordinate the urban and rural development.

The Reform of Administrative Approval System

They carefully clean up and simplify the projects and procedures of administrative approval. 845 approval items have been adjusted and cancelled and 74 business links and53 declaring materials have been eliminated, 6665 working days of processing time has been compressed. They have explored and established a parallel approving mode containing "consent foretelling, service forward shift, one window acceptance, internal operation, parallel processing, time limited, monitoring evaluation" and a centralized processing mode principled by "one window processing, one room auditing, one leader approving and one seal ending".

The Establishment and Completion of Decision-Making Mechanism and the Implementation of Scientific and Democratic Decision-Making

"Publicizing and Hearing Methods of Chengdu Major Administrative Decision-making Matters" and "Expert Consultation and Reasoning Methods of Chengdu Major Administrative Decision-making Matters" has been successively constituted, which has formed the preliminary establishment of "three combination" scientific and democratic decision-making mechanism linking public participation, expert demonstration, and legal determination of government. The municipal government has established a policy advisory committee .900 experts advisory reserves have been set up. Those departments with policy-making functions and strong quality of specialization have set up their own expert advisory reserves. Significant decisions made by the municipal committee and government and major affairs of departments all need consulting and debate of related experts. Local laws and draft regulations which are closely related to public interests are all processed by inviting the public to participate in the hearing, publicity, discussion, online counseling and other means to extensively solicit views of the community. At present, over 100 price adjusting programs, more than 10 drafts of local laws and regulations as well as IC card of bus have been publicly heard.

SECTION 3: WUHAN CITY CIRCLE AND CHANGSHA, ZHUZHOU AND XIANGTAN CITY CLUSTER

The establishment of a resource-saving and environment-friendly society is an inevitable choice to fundamentally convert the mode of economic development, take a new industrialization road, radically ease the resource constraints, alleviate the environmental pressure, strengthen the overall quality and competitiveness of the national economy and realize the goal of comprehensive well-off society construction. Especially for the central and western regions which are undertaking the eastern industrial transfer accelerating the industrialization process, the old path of "pollution first, treatment later" can no longer be followed. With a view to this, Chinese Government choose the Wuhan city circle and the CZX city group to carry out comprehensive and supplementary experimental reform in allusion to the resource-saving and environment-friendly society construction.

I. Wuhan City Circle: 1 Plus 8 > 9

The so-called Wuhan City Circle refers to the urban community with Wuhan as the center and 100 km as the radius, including Wuhan and Huangshi, Ezhou, Xiaogan, Huanggang, Xianning, Xiantao, Qianjiang, Tianmen, etc. 8 surrounding cities. The circle accounts for 33% of the land in Hubei province and 51.6% of the population. It has a high urban density level, a favorable economic foundation and a preferable environmental and natural condition. It is the biggest city domain of not only the Hubei province, but also the mid-Yangtze River region. The entry point of the experimental reform of Wuhan city circle lies in energy resources saving and ecological environmental protection. It aims to take a new way to develop industrialization and urbanization which distinguish itself with traditional mode via reform and innovation of systematical mechanism and transformation of developmental model.

1. Simultaneously Push “9 Innovative Mechanisms” and Jointly Build the "Two-oriented Society" System Platform

According to the overall reform program, the circle will prosecute the following 6 mechanism innovations, including resource conservation, environmental protection, science and technology, optimization and upgrading of industrial structure ,urban and rural developmental integration and conservative and intensive land use. Another 3 systematical innovation including financials and taxation, inward and outward opening-up and administrative management will also be supplementarily boosted.

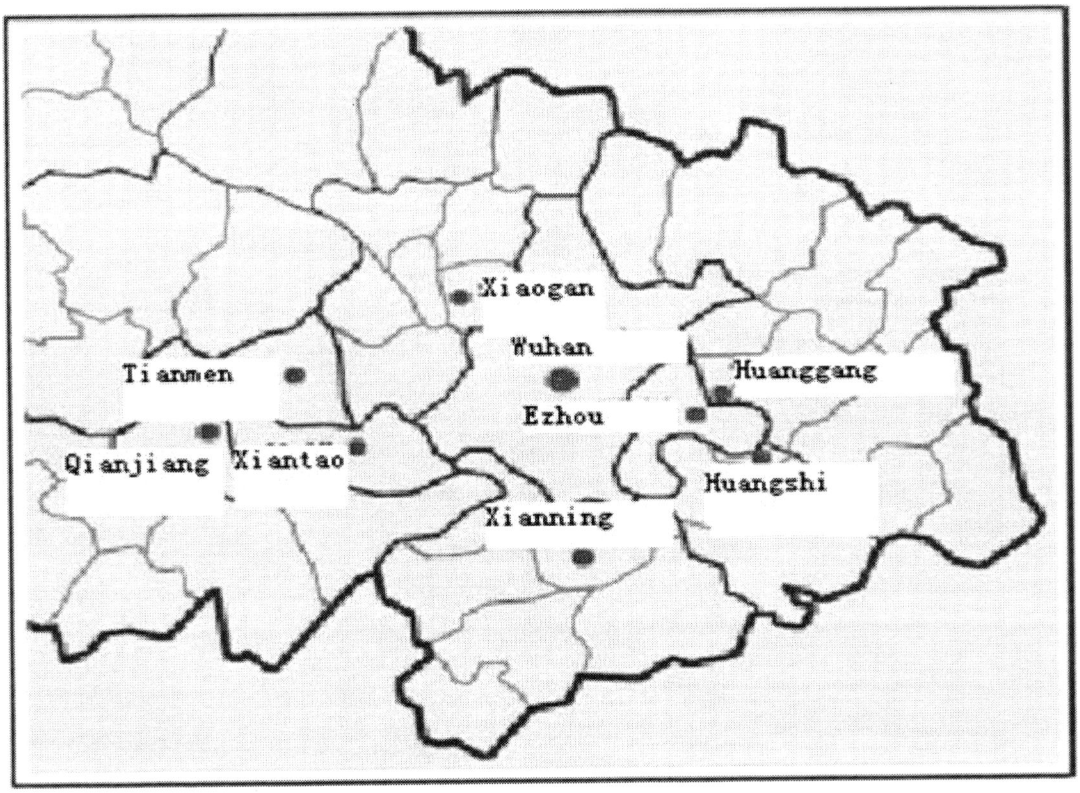

Figure 10-2. The Sketch Map of Wuhan City Circle.

2. First Push of Investment, Financing and Public Infrastructure Construction to Lay a Solid Foundation of Reform

In September 2008, Wuhan city circle set up the Hubei Province Joint Development and Investment Corporation, the registered capital of the company is 3.2 billion RMB. It is positioned as the Wuhan city circle cross-regional investment and financing platform and also China's first regional government-enterprise co-investment and financing platform. It is composed of 16 shareholders, and all the 9 city members of the circle infuse capital and form shareholders, the other four are Wuhan-located central government's enterprises. The largest shareholder is the SASAC, Hubei Province, investing 1.3 billion RMB and accounting for 40.63% shares. In order to integrate the financial resources of the circle, Wuhan city circle embarks upon the built-up of regional banks to form a unified financial market. In 2007, Wuhan Commercial Bank changed its name to Air Bank and became a regional bank, attracting attentions of the background centers of the country's 9 major banks. It lays a solid foundation for the raising of financing function.

On December 26, 2007, the 5th Yangtze River Bridge was completed and brought into use. The 7 urban freeways connecting the circle members have also basically finished. 9 cities in the circle can be accessible within 1 hour, achieving the geographical city integration and barriers eliminated.

3. To Start Compiling Five Special Plans, Focusing on the Promotion of "Five Integration"

In 2009, Wuhan City Circle is prepared for the compiling and implementation of five special plans including space, integrated transport, industrial development, social undertakings and ecological environment. In addition, the "five integration" including the optimization and reasonable transfer of industries, linkage sharing of social resources, fast-speed access of circles and domains, industrialization of modern agriculture, chain management of commence is the focus of this year's work.

4. To Promote Co-operation between Ministries and the Province and to Set up Policy Support Platform for the City Circle Construction.

Joint establishment between Ministries and province is an important strategic initiative to promote "Resource-efficient and environment -friendly society" construction of Wuhan City Circle. Hubei province has taken a number of forms to push the establishment of ministry-province cooperation mechanism in order to let national ministries and units carry out pilot projects and coordination on some important reforms, items and policies in Wuhan city circle. By now, up to 47 national ministries and units have signed cooperative agreements and memorandums with Hubei provincial government.

II. Changsha, Zhuzhou and Xiangtan City Cluster: Building a New Platform of Regional Integration

The Changsha, Zhuzhou and Xiangtan City Cluster is situated in the northeast of Hunan Province. The spatial distribution of the three cities Changsha, Zhuzhou, Xiangtan takes on the form of the Chinese character"品"(the form of an equilateral triangle) . The distance between every two cities is less than 40 kilometers with a compact structure. From history's perspective, the urban agglomeration, as a national old industrial base and economic center of Hunan Province, has long been influenced by the concept of "superior development but inferior governance" and the structural pollution has been comparatively conspicuous. The City Cluster, as the country's comprehensive and supplementary reform pilot area of "Resource-efficient and environment -friendly society" establishment, focuses upon the exploration of the new roads of industrialization and urbanization.

1. Striving to Reform the Feasibility of Policy, and Pursue the Trend of "High End" upon Programming

Through the combination of administrative and professional level, the comprehensive, thorough and long-term planning towards the reform and construction of pilot areas has been performed. They sum up and draw the international experience, and provincial and municipal departments perform their own duties respectively: the provincial departments have conducted 12 specific reform programs and the authorizing promotion work of 14 regional specific planning. Changsha, Zhuzhou and Xiangtan are responsible for overall reform of authorization work, implementing programs of 12 specific reforms, and 42 city-level specific programming. The surrounding 5 cities, Yueyang, Changde, Yiyang, Hengyang and Loudi respond positively and perform active docking from aspects of planning, infrastructure and industries.

The program will be divided into three phases of reform :the first phase is scheduled during the years of 2008-2010 with the task of preliminary establishment of "two-type" social framework and form of regional economic integration inducting the harmonious development of the formation of Changzhutan harmonious development of Changsha, Zhuzhou, Xiangtan and the surrounding cities. The second phase is scheduled during the years of 2011-2015 with the task of the preliminary set-up of "two-type" society's industrial structure, increasing mode and consumption patterns. The third phase is schemed during the years of 2016-2020 aiming to finish the reform task and form the mechanism of "two type" society and the development pattern of new-type industrialization and urbanization.

2. Taking Environmental Simultaneous Management as the Guide and Unveiling the Purdah of Urban Fusion

2008 is the starting year of the City Cluster's "Resource-efficient and environment -friendly society" construction. The regional coordination mechanism construction first unveiled the curtain in the implementation of all measures of environmental simultaneous management.

In the reform of environmental protection, resource conservation ,the Ministry of Land Resources carry out intensive and economical use of land in the Changsha, Zhuzhou and Xiangtan Urban Agglomeration and has preliminarily formed four patterns of intensive economical use of land. "Collaboration Agreement upon Joint Push of the 'Resource-efficient and environment -friendly society' establishment in Changsha, Zhuzhou and Xiangtan City Cluster" has been signed between the Ministry of Environmental Protection and Hunan province. The Xiangjiang River and Dongting Lake has been brought under the system of water pollution prevention and planning of Yangtze medium-east water basin. The Changsha, Zhuzhou and Xiangtan City Cluster has been wholly brought in the national water-saving-city pilot project.

Hunan Provincial Government has established the system of duty assessment and supervision towards energy-saving and emission-reduction targets and started scientific and technological support of energy-saving and emission-reduction. In June 2008, the "Long Xiangjiang Clean Water Action" has started the tough battle of comprehensive environmental renovation upon Xiangjiang River. It is budgeted that 17.4 billion capitals will be injected within 3years to improve production quality. Those industries will be banned-off, washed-out, closed, deadline-managed, or moved away. New projects with heavy pollutions are strictly prohibited to build within 10 km from two sides of Xiangjiang River. 18 heavily polluting enterprises or production lines in Qingshuitang area of Zhuzhou city has been shut down within one year. 73 enterprises of Xiangtan city has been renovated and 17 heavily polluted enterprises has been closed down via the "Blue Sky Green Water" project. Environmental protection departments of Changsha, Zhuzhou and Xiangtan will carry out joint efforts to implement linkage mechanism of pollution handling in the emergency state of Xiangjiang dry season. Those key pollution sources would be managed by measures of being shut down, limiting production or emission-limited in the state of emergency.

3. The Implementation of Multi-dimensional Push and the Joint Establishment of Regional Integration

The coming period will be a new phase of city cluster construction. Regional integration reform will be centered upon five mechanism innovation of resource conservation, environmental protection, industrial optimization, technological innovation and land management. Investing and financing, external development, finance and taxation, urban and rural co-ordination and administrative management will be supplementarily promoted.

Hunan Provincial Government has specially funded the establishment of the "two-type" social construction, concentrating upon the cross-regional infrastructure construction, the "two-type" industrial development, ecological environmental management and other tasks. Provincial Land Capital Operation Company has been reorganized as the Hunan Development and Investment Group Co., Ltd. to build the main platform for provincial investing and financing.

As for infrastructure construction, telephone area-code of the three cities has been incorporated and upgraded. China Mobile, China Telecom and the provincial government have signed a strategic cooperation agreement. Among them, China Telecom and the provincial government jointly push the "Ten Information Engineering"; The core area of the urban agglomeration will focus on building the "seven vertical and seven horizontal" inter-city main roads to form the inter-city traffic framework. “2 rings and 6 radials” highway network established as keys in the “3 plus 5” urban agglomerations will set up the urban agglomeration traffic framework.

As for the construction of the "two-type" industry, in the coming 2 or 3 years, the City Cluster will initially form a "two-type" industrial system taking high-tech and modern service industry as the leader, advanced manufacturing as the base, circular economy as the characteristic. According to the plan, by 2010, the urban agglomeration will form two leading industries of equipment manufacturing and light industry and a batch of pillar industry groups. It will produce 10 high-tech leading enterprises with the sales income of each over 10 billion RMB and 20 enterprises over 5 billion RMB. They will strive to set up the “one city six bases” including Changsha engineering machinery city, rail transportation, wind power equipment and other bases. The proportion of the added value of high-tech industries over GDP is expected to exceed 20%.

Regional port crossing reforms will also be pushed step by step, form a clearing model of interior and inter-port "declaration at origin and inspection in port" and establish the fast-transit access of "Hunan, Guangdong and Hong Kong Express" for custom clearance.

CONCLUSION

Through experimental units to further reforms is not only a piece of experience for success in China’s 30 years’ reform and opening up but also a natural choice for accomplishing key tasks in reforms. By initially experimenting locally and gaining useful experience and then gradually expand and further reforms can effectively control risks and prevent deviations. What is more, it is conducive to mobilize the grass root and all sources in the society to reform and innovate with initiative and motivation. At present, besides the seven National CSRPZs, many provinces have carried out provincial experiments of comprehensive supporting reforms.

Here, one point should be mentioned. In recent years, the Chinese government has established and implemented several economic zones' regional plans which involve Guanzhong-Tianshui Economic Belt, Guangxi Beibu Gulf Economic Zone, Fujian West Taiwan Strait Economic Zone, Jiangsu Coastal Economic Belt, Liaoning Coastal Economic Belt and so on. These economic zones' regional plans mostly aim at solving problems in development direction, space planning, priority industry, supporting policies and the like. Although furthering reforms is needed in the implement process of these plans, these economic zones are not National CSRPZs.

It can be predicted that, through the implementation of comprehensive supporting reforms experiments which combines solving regional problems and exploring overall common challenges, realizing key breakthroughs and overall innovation, economic systematic reforms and other aspects, we can carry out reforms safely, smoothly and orderly, and can conclude and popularize the continuous new achievements and experiences. This will greatly advance the process of improving socialist market-oriented economic systems nationwide.

Translators: zhu Jun (SUFE), Ma Mingchao (SUFE)
English Version Editors: Lv Mei (SUFE), Chen Mo (Harvard)

REFERENCES

Hao Shouyi, 2008(9). Researches on Comprehensive Supporting Reform Pilot Zones [M], Science Press.

Kong Jingyuan, 2009(9). Report on China's Economic System Reform in 2008 [M], People's Press.

Wang Jianing, Hu Xinhua, 2009(8). Study on Administrative System of Comprehensive Supporting Reform Pilot Zones: Shanghai Pudong New Area and Tianjin Binhai New Area [J]. *Reform*.

Li Jiaxiang, Dai Chao, 2008(6). On Hot Issues in Comprehensive Supporting Reform Pilot Zones [J]. China Opening Herald.

Li Jiaixiang, 2007(1). Theoretical Value and Stage Characteristic on China's Comprehensive Supporting Reform Pilot Zones [J]. Economic Perspectives.

Yang Jianwen, Hu Xiaopeng, 2007(3). Comprehensive Supporting Reform: On Comparative Researches of Tianjin, Shanghai and Shenzhen [J]. Shanghai Economic Review.

Shi Hongxing, 2007(2). Systematic Design of Reform in Comprehensive Supporting Reform Pilot Zones [J]. *China Opening Herald*.

Hao Shouyi, Gao Jintian, 2006(2). On State Pilot Zone for Overall Reform [J]. *China Opening Herald*.

Chen Zhenming, Li Deguo, 2008(11). Practical Exploration and Developmental Trends in the State Comprehensive Supporting Reforms Pilot Area [J]. *Chinese Public Administration*.

Zhang Huanzhao, Hao Shouyi, 2007(1). National Pilot Zone for Overall Reform and Spatial Evolution Analysis of Institution [J]. *Journal of Finance and Economics.*

Niu Lichao, 2008(6). Reform and Innovation: Adaptive Evolution of Government Administration System—Comparative Analysis of Binhai New Area, Shenzhen Special Economic Zone and Pudong New Area [J]. *Journal of Tianjin Administration Institute.*

Zhao Xiuchun, 2007(4). State Comprehensive Supporting Reforms Pilot Area is a Major Step to Follow a Scientific Approach of Development [J]. *China Reform.*

Website of National Development and Reform Commission: tgs.ndrc.gov.cn.

Website of Shanghai Pudong New Area: www.pudong.gov.cn.

Website of Tianjin Binhai New Area: www.bh.gov.cn.

Shenzhen Government Online: www.sz.gov.cn.

Website of Chongqing Development and Reform Commission: www.cqdpc.gov.cn

Website of Chengdu Municipal Government: www.chengdu.gov.cn.

Website of Wuhan City Cluster: www.whcsq.gov.cn.

Dedicated Homepage of CSRPZ for Building a Resource-efficient and Environment-friendly Society in Changsha-Zhuzhou-Xiangtan City Cluster in the Website of Hunan Provincial Government: http://www.hunan.gov.cn/zhuanti/cztlxsh/bjjs/gk/

In: China's New Deal
Editors: Li Xiaoxi and Hu Biliang

ISBN: 978-1-61470-815-5

Comments on Chapter 10

ON THE SETTING-UP AND CONSTRUCTION OF THE PILOT ZONES FOR SUPPORTING COMPREHENSIVE REFORMS

Huang Lin *

Promotion after experiment is the usual practice for Chinese government to conduct social and economic reform. Several Special Economic Zones (SEZs) established in the early period of the country's reform and opening up have not only achieved substantial economic and social progress, but also accumulated significant experience for the whole country's reform. Although the success of SEZs gives Chinese government more ambitions to enlarge the reform based on the trial and experimental approach, many other side effects start to show, which includes: the backward administration system restricting further economic and social reform, the unbalanced economic development introduced by the limited marketization reform between experimental and other areas, the long-standing dual economic structure between the rural and urban economy, over-extensive use of natural resources during economy development and the problem of environmental damages, and so on. In order to continue gaining rational identification from the society for the current institutional arrangement, the preferred way is to expand the economic experiment scale and to further economic and social reform. The State Council has set up various comprehensive supporting reform pilot zones (CSRPZs), which is another important attempt in this regard. These new experiment platforms, based on their own distinct characteristics and region-specific purpose of development, shall in the first place carry on some overall significantly, in-depth economic and social policies, accumulate new experiences in reform pattern, and explore new development methodology for the country in the new era.

Since 2005, China has set up three national multiple-targeted CSRPZs (m-CSRPZs) in Shanghai Pudong new district, Tianjin Binhai new area, Shenzhen city respectively, and established four special–targeted CSRPZs (s-CSRPZs), which are "Comprehensive

* Huang Lin, PHD, University of Essex, Associate Professor, Research Institute of Economics and Management, Southwestern University of Finance and Economics. Email:lhuang@swufe.edu.cn.

Supporting Reform Pilot Zone for Balanced Urban and Rural Development" in Chengdu and Chongqing, and "Comprehensive Supporting Reform Pilot Zone for Building a Resource-efficient and Environment-friendly Society" in Wuhan and Changsha-Zhuzhou-Xiangtan city cluster. The basic distinction between the first three m-CSRPZs and the rest four s-CSRPZs is on the range of reform. Those m-CSRPZs enjoy the capability with full horizon of reforms, while s-CSRPZs are subject-specified reform area. From the purpose of those CSRPZs, Shanghai Pudong, Tianjin Binhai and Shenzhen aim at radiating the reform to the Yangtze River Delta, Pan-Bohai and Pearl River Delta respectively. All m-CSRPZs focus on how to break the institutional defects impeding the full realization of market function. They are designed to perfect the market economic system. Their functions include many aspects such as the construction of modern market institution, market administration system, financial system, legal regulations on the emerging markets, etc. All those aspects directly relate to the question of how to deepen the reform to construct a free market system that closely interconnected with the global market, to fit the international rules and to become the important part of the world economy system. Furthermore, through the experiment of the three m-CSRPZs, the experiences and knowledge should be accumulated not only to assist to build an ideal market system with coordinated institutional reform, but also to probe new development pattern for other areas. The s-CSRPZs in Chengdu and Chongqing city mainly focus on unifying the development of urban and rural area to fundamentally solve the development problems of "rural area, agriculture and farmers" and to promote the urban – rural integration, especially on the construction of a modern rural market to narrow the economic and social gap between the urban and the rural. Setting-up s-CSRPZs in Wuhan and Changsha-Zhuzhou-Xiangtan city clusters is an important measure in implementing "the Rise of the Central Region" strategy of the Central Government. They are designed to build the capacity for a long term development path mainly through the innovation of new-type industrialization, new-type urbanization, endogenous and scientific development with features of energy-and-resource-saving and environmental-and-ecological friendly. By optimizing industrial structure and transformation of developmental mode, these two s-CSRPZs are expected to build an efficient and sustainable development mechanism and to become the growth engines for the Central Rise and new growth pole of the state. Above all, these seven comprehensive supporting reform pilot zones are set-up so that China will establish comprehensive reform experimental fields from economy to society, to administration, and to environment, covering different typical area across the east, the west and the central area. These CSRPZs share two common features which are comprehensive and complementary, that is, on the one hand the reforms require coordination and inter-conditioning in various fields; on the other hand, the reforms require mutual supporting between different reform measures.

The setting structure of CSRPZs reflects the macro layout of the Chinese government's effects to deepen reform, and if those experiments succeed, it will directly encourage the administrative authority to efficiently make relevant policies and complete necessary legislation. However, since the setting-up of CSRPZs is a large scale economic and social experiment, some problems are needed to pay attentions to when planning CSRPZs' reform.

First, the border issue of the "Trial Right" or "Pilot Power" of CSRPZs. Pilot power refers to a special right authorized by the state, based on the needs of reform, to set regulation by their own standards in certain areas before the unified provisions made by the national level legislation. It is worth notifying that the pilot power is not a statutory right, it is just a practical method to carry on the breakthrough and the innovation to the national routine policy making process in the area that lack of the superior legislation. Exercising the trial right must comply with the law priority doctrine. Due to the non-legislation nature of the trial right, it is unavoidable that some local piloting polices may conflict with the law. For example, Chongqing Municipality in 2007 introduced the measure "with regard to comprehensively implement the spirit of the third party congress in Chongqing, the implementation opinion on serving the comprehensive development balancing the urban and rural (Chongqing Gong Sang Fa (2007) No.17) ", which allowed the farmers to contribute shares to a limited company using land contractual rights. However, the land shares of a limited company may be transferred to a person or entity other than the member of the rural collective. Moreover, if the limited company was in bankruptcy, the land in shares may be liquidated, and the farmers who hold the contractual right could be forced to lose their land. This potential risk will endanger the basic living condition of the farmers, and possibly undermine the stability of rural economic system. In review of the aforesaid, the central authority suspended the implementation of this practice.

Although the pilot power is not an explicitly statutory right, the State Council in fact transfer some policy-making privileges to CSRPZs, which eventually expand CSRPZs autonomous power in certain areas. The CSRPZs may misuse or overuse the increased power of policy-making to design the new policy which conflicts with the statutory law, and may lead to unexpected consequence that violate the policy orientation of the whole country. To avoid such conflicts or consequence, firstly, the CSRPZs should be familiar with the relative laws and keep a close relationship with the central administrative authorities from which to gain the supports. Secondly, while the expansion of regulatory rights facilitates CSRPZs to access in-depth reform, the improper use of such rights may stimulate greater risks. The CSRPZs should be very cautious in implementing new polices beyond the scope of existing regulations. Before the state introduce formal legislation on the content and border of such pilot power, the CSRPZs need pay special attention to legal risks that may arise during the reforming process.

Second, the issue of the unbalanced development introduced by the setting of various CSRPZs. The trial rights of CSRPZs not only bring a greater and more flexible policy space and policy supporting given by the central government, but also facilitate themselves to attract more types of external resource inflows such as capital, high-profile intelligent, information, international attention, etc. It is the very reason that local governments are enthusiastic about applying to be CSRPZs. Those factors helpfully upgrade CSRPZs into at least a regional development center. One obvious result is that the granted trial right and relatively preferential reform policies render an advantage for CSRPZs at start point. While the setting-up of CSRPZs give these areas new development opportunities, but from the country's overall view, the inequality in policy-making sphere among different regions will inevitably induce gaps in regional development, which is contrary to the principle of equitable development opportunity between different regions in the state.

Therefore, the next reasonable question raised is why to set up so many CSRPZs with very large geographic coverage. This is mainly the result of interest-seeking games between regions. There are many officials and scholars have discussed the benefits from setting up comprehensive pilot zones, less people think of the adverse effects of this trial approach. In fact, if the number of CSRPZs exceeds a reasonable scope, different CSRPZs may undertake the same kind of polices and reform approach, the experimental results may converge and they will lose their experimental value. To some extent, the setting up of CSRPZs will discourage these regions that failed in the game of chasing the member of CSRPZs.

Thridly, the cost issue of CSRPZs. One feature of China's reform is characterized by its gradual nature. The gradual reform is likely to cause unequal opportunity in the possession of capital, resources, and unequal access to markets among different groups or individuals. This inequality will virtually form a variety of interest groups which gradually become the barrier to the reform. Comprehensive supporting reform is undoubtedly a systematic project covered a wide range of social levels. It needs to break the long-established pattern of interest groups and will inevitably encounter resistance from them. For example, one reform strategy in Chengdu to foster rural growth and break the urban-rural gap is to identify and certificate the property rights of rural lands and real estates to secure the interests of farmers and rural collectives. Those active measures will help to promote the establishment of clear land ownership in rural areas and capitalize the rural assets. For years China's local fiscal revenue heavily depends on the land transfer income, which means that to increase the local government fiscal position, it has to enlarge the scope of land acquisition. If the farmers are given the clear identification of the property right of land and homestead, it will raise the cost of local government to requisition land from rural collectives. It will also increase the cost of leaders of rural collectives that usually benefiting from the illegal deprivation of the property right of farmer's land to collude with the government in land acquisition. Therefore the local officials and leaders of rural collectives are reluctant to kindly implement such reform policies. Till now, many reform measures in CSRPZs start out with a bang and end with a whimper. The government enthusiastically advocates the reform, while the grass-roots who are in desperate need of support encounter various resistances. If reform program could not gain enough support from grass-roots and establish a corresponding institutional foundation, the reform will be hardly to promote and face the risk of reinventing the wheel. The specific reform program, its foundation, promoting schedule, all are needed to be carefully, scientifically evaluated and planned.

If we say the special economic zone is a kind of Pareto-marginal increasing reform, the comprehensive supporting reform is a more difficult undertaking and may bring more interest bargaining. It needs reasonable adjustment to interests of all vested groups. This challenge will undoubtedly be a new test to the reform planners and promoters.

REFERENCES

[1] Annual Report on Chengdu Balanced Rural and Urban Development Comprehensive (2007-2008), Research Institute on Scientific Development, Sichuan University, 2009 (in Chinese).

[2] Brandt, Loren, Scott Rozelle, and Matthew Turner, Local Government Behavior and Property Rights Formation in Rural China, UC Davis Working Paper No. 02-004, 2002.

[3] Bromley, Daniel and Yang Yao (2006), Understanding China's Economic Transformation: Are There Lessons for the Developing World? *World Economics* Volume 7, Issue 2, Pages 73-96.

[4] Kong, Jingyuan (2008), Thirty years in China's Comprehensive Reform Experiment, *Liao Wang*, Issue 10 (in Chinese).

[5] Li, Guo, Scott Rozelle and Loren Brandt (1998), Tenure, Land Rights, and Farmer Investment Incentives in China, *Agricultural Economics*, Volume 19, Issues 1-2, Pages 63-71.

[6] Lin, Yifu and Yang Yao eds. (2005), China's Miracle: Retrospect and Prospect, Peking University Press (in Chinese).

[7] Niu, Wenyuan eds. (2009), The Report on China's New Urbanization, Science Publishing House (in Chinese).

In: China's New Deal
Editors: Li Xiaoxi and Hu Biliang

ISBN: 978-1-61470-815-5

Chapter 11

NEW STAGE: CURRENT ECONOMIC DEVELOPMENTS AND PROSPECTS IN ETHNIC REGIONS

Zhang Wenguang**, Zhang Qingan, Dong Xiaoyu and Wang Ying

In this chapter, we will first introduce the economic development status quo in Chinese ethnic regions and the economic development since the reform and opening-up policy thirty years ago. Especially the western development in the last ten years will be comprehensively described. On this basis, the characteristics of the economic development in ethnic regions under current financial crisis will be analyzed. In the second part, great emphases will be put on analyzing and summarizing the systems and effects of ethnic economic policy. In the third part, an outlook will be made on the economic development and policy of ethnic regions.

SECTION 1: STATUS QUO OF ECONOMIC DEVELOPMENT IN REGIONS INHABITED BY ETHNIC GROUPS

Since reform and opening-up policy in China thirty years ago, people in Chinese ethnic regions have overcome the influence of geographic disadvantages and macroeconomic policy of unbalanced development, given full play to the initiative and enthusiasm in ethnic regions, steadily carried out and deepened each system changes, resolutely promoted the opening to both the inside and outside world, held on to regional system of ethnic minorities, coped with the relationship between help provided by the country and self-help, constantly adjusted and upgraded industrial structure. In this way, they have gained significant achievements in socialist modernization drive and realized the historic leap from starvation to an overall well-off life.

Under the influence of global financial crisis, China's economic growth has been slowed down since 2008. But in the second quarter in 2009, the economy started to rebound, because

*Dr. Zhang Wenguang is associate professor of School of Management, Beijing Normal University. Email: zhwg318@263.net

the industrial structure in ethnic regions is less export-oriented, the impact of financial crisis is limited and GDP keeps increasing significantly.

I. The Overall Characteristics of Economic Growth in Ethnic Regions

The overall characteristics of economic growth in ethnic regions are the bases for us to understand the economic status quo here. Since advancement in the development of the western region ten years ago, the economic power in ethnic regions is obviously strengthened. The economic structure is more optimized. The economic benefit is greatly increased and the status in national economy has raised a lot to catch up with the more developed regions. New progress has been made in building a well-off society.

Economy in ethnic regions refers to economy in certain regions inhabited by ethnic groups in multinational country. In China, the economy in ethnic regions is the same with economy in regions inhabited by ethnic groups, which refers to economy in minority autonomous areas and in areas inhabited by relatively more minorities. They are special economic areas with homogeneous characteristics. China's ethnic regions have clear administrative divisions. Most of them are in the western, southwestern and northern border areas, comprised of five autonomous regions which are Xizang, Xinjiang, Inner Mongolia, Guangxi and Ningxia and three provinces which are Qinghai, Guizhou and Yunnan.

1. The Economy in Ethnic Regions Grows Continuously, Steadily and Quickly with Surpassing Tendency.

Since the reform and opening-up policy, the economy in ethnic regions have rapidly improved. GDP has increased from 32.4 billion in 1978 to 2476.8 billion in 2007 with a 15.2 times of comparable price increase and an average annual increase of speed of 10.1%. Compared with the 841.1 billion in 2000 when began to carry out western development, GDP in ethnic regions has increased 1.4 times in comparable price with an average annual increase of speed of 11.7%. After the GDP breakthrough of 100 billion in ethnic regions in 1994, people in ethnic regions took ten years to achieve the breakthrough of 200 billion in GDP increase in 2004, two years to 300 billion in 2006 and one year to 400 billion in 2007. As one of the greatest growth poles in ethnic regions, the development of Inner Mongolia autonomous region explains the process of great-leap-forward development in ethnic regions in the 30 years of reform and opening-up. Now the economic aggregate of Inner Mongolia has ranked among the middle level all over the country and the GDP per capita has been in the very front of the country. In 1978, the total output value of Inner Mongolia is 5.804 billion, ranking the 25th in the country. In 2007, the total output value is 609.1 billion, ranking the 16th. The rise in speed of region total output value in Inner Mongolia has been the top all over the country for six years. Until the end of 2007, the GDP per capita calculated by permanent population has firstly exceeded 3000 dollars, which marks the golden period of economic development of Inner Mongolia and the strength of the economy has increased stability.

Table 11-1. GDP in Ethnic Regions 1998-2008
Unit: 100 million Yuan

Region	1998	2000	2002	2004	2006	2008
Guizhou	841.9	993.5	1185	1677.8	2267.43	3333.40
Yunnan	1793.9	1955.1	2232.3	3081.9	4001.87	5700.1
Tibet	91.2	117.5	161.4	220.3	290.05	395.91
Qinghai	220.2	263.6	341.1	466.1	641.05	961.53
Ningxia	227.5	265.6	329.3	537.1	706.98	1098.51
Xinjiang	1116.7	1364.4	1598.3	2248.8	3018.98	4203.41
Inner Mongolia	1192.3	1401.0	1734.3	3020.0	4790.00	7761.80
Guangxi	1903.0	2050.1	2437.2	3433.5	4801.98	7171.58
Ethnic regions	7386.7	8410.8	10018.9	14685.5	20518.34	30626.24
Proportion in the country	8.75%	8.48%	8.33%	9.19%	9.68%	10.19%

Source: data of up to 2006 are from *China Statistic Yearbook* in each year, data of 2008 are from the statistical bulletin of each province in 2009.

The above table shows that the proportion of total output value in ethnic regions accounting for domestic GDP increases from 8.75% in 1998 to 10.19% in 2008. Thus, the fact of constant disparity between the eastern part since the reform and opening-up policy has been reversed and reduced. The economy in ethnic regions grows with a tendency to surpass the other regions.

2. The Revenue in Ethnic Regions Greatly Increases, the Social Undertakings Continuously Develops and the Ecological Construction Achieves First-stage Success.

Since the reform and opening-up 30 years ago, the revenue in ethnic regions has been greatly increased. Local finance general budget revenue has been increased from 5.2 billion in 1978 to 212.4 billion in 2007 with an increase of 39.6 times and an average annual growth rate of 13.6%. The revenue per capita in ethnic regions has been increased from 40 in 1978 to 1110 in 2007 with an increase of 26.8 folds. In Guangxi, for example, the fiscal income had been broken through 10 billion in 1994, 20 billion in 1999 and from then on got a new level in each year. In 2007, the fiscal income in Guangxi reached 70.4 billion with an increase of 49 times comparing with that of the early reform and opening-up. The increasing amount exceeded 10 billion for the first time. The local financial capability in ethnic regions continuously grows so as to improve the ability of local government in regulating economy and lay a solid foundation for improving people's livelihood and constructing infrastructure.

The social undertakings in ethnic regions have been continuously developed. The "two basics" program[1] has been completed on schedule. The population coverage rate of "two

[1]"Two basics" refers to popularizing nine-year compulsory education basically and eliminating illiteracy among young and middle-aged people basically. It is proposed by the Ministry of Education of the state. To implement *Decisions of the State Council on Further Strengthening the Work of Rural Education* (Promulgated by the State Council [2003], No.13),promote the Western Development, realize the objectives of popularizing nine-year compulsory education basically and eliminating illiteracy among young and middle-

basics" has been raised from 77% in 2003 to 99%. The population rate of illiterate and half-illiterate over 15 years old has been decreased from 9% to below 5%. The "two exemptions and one allowance" policy helps 0.15 billion students and poor boarding school students from 780 families. The new system of rural cooperative medical and health care services has been improved constantly.

The ecological construction and environmental conservation have achieved gratifying results. Since the western development, the total control of water loss and soil erosion is 16 million hectares, the area of ecological and natural restoration is 28 million hectares, the total area of returning land for farming to forestry is 5.26 million hectares, the area of forestation in waste mountains and lands is 7.65 million hectares, the area of returning husbandry to grassland is 19.33 million hectares and the population that has been moved is 1.2 million. With the implementation of returning land for farming to forestry, forestation in waste mountains and lands, returning husbandry to grassland, natural forest protection, Beijing and Tianjin sandstorm-source control, ecological protection at river source and so on, gratifying achievements have been gained in ecological protecting and constructing in ethnic regions.

Table 11-2. The proportions of major industrial and agricultural products in ethnic regions accounting for the country's output in 2006

Product Name	Unit	Output	The Proportion in the Country(%)
Cotton	(10 thousand ton)	219	32.5
Sugar Materials	(10 thousand ton)	8399	76.1
Cured Tobacco	(10 thousand ton)	111	44.7
Turpentine	(10 thousand ton)	50	55.6
Tungoiltree Seed	(10 thousand ton)	16	43.0
Mutton	(10 thousand ton)	203	43.2
Milk	(10 thousand ton)	1206	37.8
Wool	(10 thousand ton)	22	57.3
Cashmere	(ton)	9576	58.4
Sugar	(10 thousand ton)	776	81.7
Cigarette	(100 million)	5132	25.4
Run Coal	(100 million ton)	5.8	24.4
Crude Oil	(100 million ton)	2701	14.6
Natural Gas	(100 million m^3)	189	32.3
Output of Plant	(100 million kwh)	4718	16.5
Electrolytic Aluminium	(10 thousand ton)	338	36.4
Alumina	(10 thousand ton)	205	26.4

Source of data : sorted out from the materials provided by the State Ethnic Affairs Commission.

aged people basically in western area, *"two basics" Plan in Western Area (2004-2007)* is formulated especially.

3. The Industrial Structure in Ethnic Regions has been Constantly improved, and Distinctive Agricultural and Industrial Systems have seen initial achievement.

The added value from the industries of energy, metallurgy, chemical engineering, equipment manufacturing, agricultural and animal products processing, high and new technology in Inner Mongolia accounts for over 90% of the total added value produced in the whole autonomous region. The eight pillar industries of machinery, sugar manufacturing, metallurgy, food, petrifaction, nonferrous metal, electrical power, and building materials of Guangxi produce 70% of the total added value of the autonomous region. In Yunnan a multi-pillar industry has been formed, breaking down the previous industrial structure which excessively depends on resources. The contribution rate to the industrial growth of the four major industries -- electrical power, metallurgy, nonferrous metal, wine, and tobacco-- of Guizhou is 57.7%, among which the electricity industry has contributed 24.3%. The industries of mine, green drink processing, Tibetan medicine, national handicrafts, and building and materials have improved steadily in Tibet. The saline chemical engineering, electrical power, petroleum, gas and nonferrous metal industries in Qinghai have been further strengthened. A distinctive industry group has been formed in Xinjiang which is dominated by petroleum and petrochemical industry and supported by spinning, steel, nonferrous metal, building materials and food industries.

4. The Poverty Reduction is Remarkable, the Population Below the Poverty Line Declines Steadily, and the Standard of Living Has Been Improved Significantly

Since the economic reform took in place 30 years ago, the population below the poverty line has steadily declined. The country has made more efforts on poverty reduction, strengthened poverty related policies and measures. In this way, the extremely poor population has reduced from 100 million in 1978 to 7.74 million now. During 1994 to 2000, the "87" anti-poverty project resolved the problem of food and clothing of 36 million people. The poverty reduction has entered into a new phase in 2000. The State Council issued and implemented *China Rural Poverty Reduction and Development Program (2001-2010)*, in which it is decided to intensify the poverty reduction. It has set the poor villages as primary targets, and paid close attention to those extremely poor. At the end of the year 2007, 9.13 million people in these autonomous regions have been able to meet their basic need.

The standard of living has been improved remarkably. A historic leap from starvation to overall well-off society has been achieved. The disposable income per capita of urban residents has increased from 414 in 1980 to 11490 in 2007 with an increase of 26.8 times. The net income per capita of rural residents has increased from 168 in 1980 to 2937 in 2007 with an increase of 16.5 times. The GDP per capita in ethnic regions has increased from 248 in 1978 to 12954 in 2007 with an average annual increase of 14.6%. It exceeded 1000 dollars in 2005 for the first time. The consumption of foodstuffs per capita in ethnic regions increased from 288kg in 1980 to 407kg in 2006.

5. The Fixed Assets Investment Grows Rapidly, and the Basic Industry and Infrastructure have been Further Improved

Since the opening-up reform, the fixed assets investment in ethnic regions have grown rapidly. The fixed assets of the whole society in ethnic regions increased from 7.7 billion in 1978 to 1470.6 billion in 2007 with an increase of 192 times and an average annual increase rate of 19.9%. Especially since the western development plan implemented, the ethnic regions have become the new priority of investment destination gradually. Only from 2000 to 2005 has the country implemented 70 major projects which include Qinghai-Tibet railway, west-east natural gas transmission, west-east electricity transmission and so on with a total amount over 970 billion. In 2005, projects of western main and branch line airport, the second phase of Inner Mongolia Yimin electric power plant, western major coal mine, rebuilding and adding an extension of Xinjiang Dushanzi petrifaction, the safety of drinking water in rural areas and so on begun to be implemented. The infrastructure projects in traffic, water conservancy, energy, communication and so on have achieved substantive progress in ethnic regions. The infrastructure in farming and pastoral area has been significantly strengthened. The production and living conditions of people in farming and pastoral area have been greatly improved. In Tibet, for example, the comprehensive system of transport has shown an initial shape marked by the opening of Qinghai-Tibet railway. This comprehensive system is composed of three airports in Gongga, Bangda and Linzhi and two airports in Rikaze and Elikunsha which are under construction. The highway distance has reached 49,000km with an increase of 33,000km compared with that in 1978. The total postal and telecommunications service has reached 3.1 billion customers with an increase of over 140 times compared with that in 1978. The number of landline and mobile phone users has reached 1.44 million with an increase of over 300 times compared with that in 1978. The objective of connecting every village with optical cable and connecting every county with telephones had been achieved in 2004. The cultural infrastructure constructions such as "Tibet-Xinjiang Project", "Every Village Connection Project", rural movie "2131 Project" and so on have been fullfilled. The population coverage of broadcast and TV has reached 88% and 89% respectively.

6. A Big Gap Still Remains between the Average Level of the Country and the Ethnic Regions

The overall level of development in ethnic regions is still backward. The proportion of GDP gross in ethnic regions accounted for 10.19% of the whole country. The sum of the gross GDP of the eight provinces is even smaller than that of Guangdong province. The GDP per capita of ethnic regions is 66.73% of the average level of the whole country. The absolute gap has expanded from 3204 RMB in 2000 to 8577 RMB in 2008. The income level of urban and rural residents in ethnic regions is relatively low. In 2008, the net income per capita of rural residents in ethnic regions was 89% of the average level of the whole country with an absolute gap of 512 RMB.

The financial system of the ethnic regions is still weak and the problems of livelihood are relatively more frequent than other places. In 1993, the fiscal revenue per capita in ethnic regions was 97.8% of the whole country and the local fiscal revenue covered the 80.0% of the expense. In 2007, the fiscal revenue per capita in ethnic regions was only 63.5% of the whole country and the local fiscal revenue only met the 38.1% of the expense. The ratio of fiscal revenue in ethnic regions to the whole country was declined from 14% in 1933 to 9.4% in 2007. The other problems of livelihood consists mostly in education, health, culture and so on. There are still 42 towns which haven't fulfilled the "Two Basics" objective. The area of dilapidated school building accounts for over 80% of that of the whole country. Many administrative villages are lack of medical center, doctors and medicine. About 60% of the health clinics in towns and townships need to change conventional equipment. About 25 thousand towns and townships are lack of cultural stations all over the country, drinking water of 2500 billion peasants is lack of safety, nearly one hundred towns and townships don't have highway, and about 2 million household of rural population don't have access to electrical power. These problems concentrate in ethnic regions.

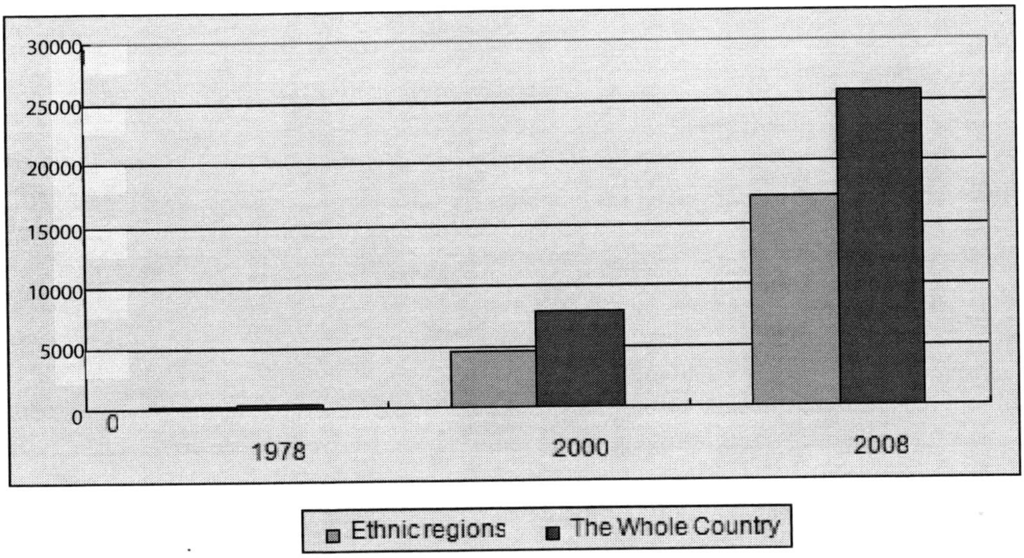

Figure 11-1. Comparison of per capita total output value between ethnic regions and the whole country.

Data source: *China Statistical Yearbook 2000, China Statistical Communiqué 2008.*

II. The Economy in Ethnic Regions under the International Financial Crisis

Since early 2008, the financial crisis originated in America has constantly spread and made important influence on China's economy. The growth rate of China's GDP has declined from 10.6% of the first quarter in 2008 to 6.1% of the first quarter in 2009. In order to deal properly with the negative influence, Chinese government has released several measures comprising 4000 billion rescue package of investment, ten programs of industrial invigoration and so on. According to the macro economic data of the second quarter in 2009 published by Chinese government, Chinese economy begins to stop dropping and go upward. The growth of GDP in the second quarter reached 7.9% and until then the GDP of first half of had increased 7.1% compared with the previous year. From January to July, the industrial added value above nationwide scale grew 7.5% compared with the previous yeas, accelerating by 0.5% than the first half. The momentum of recovering and stabilizing the economy was very clear. In this macro context, the economic development of ethnic regions showed not exactly the same trend as the overall economic situation, mainly in the following areas.

1. GDP Grows on a High Level in Ethnic Regions, Which is Higher than the National Average, but will slow down.

From the fundamental perspective, the impact of financial crisis on the economic development of ethnic regions in western area is limited. From 2008, the national GDP grows by 9% of the previous year, while the average rate of economic growth in ethnic regions is 12.15%. The rate in some regions such as Inner Mongolia is as high as 17.2% and in Tibet is 10%. However, most research institutions forecast that the growth rate of China's economy in 2009 will further decrease to about 8%, which is a severe test for the economic growth in ethnic regions.

2. The Investment in Fixed Assets in Ethnic Regions Maintains Rapid Growth.

The structural adjustment of China's economic stimulus program of 4 trillion shows the following main features: first, increase the investment in fundamental industry, infrastructure and social works in order to lay the foundation for future economic development. Second, increase scientific and technological innovation and investment to enhance the competitiveness of enterprises. Third, accelerate the development of recycling economy and promote energy-saving and emission-reduction action. Fourth, increase investment in central and western areas and promote coordinated development among regions. It is clear that the western area belonging to ethnic regions is one important composition in this economic stimulus program. According to the latest statistics, the gross investment in fixed asset of Tibet in 2008 was 30.333 billion with an increase of 12.5% over the previous year, the gross investment in fixed asset of Inner Mongolia in 2008 was 559.645 billion with an increase of 27.1% over the previous year,that of Guangxi in 2008 was 377.81 billion with an increase of 27.2% over the previous year, that of Xinjiang in 2008 was 231.4 billion with an increase of 25.0% over the previous year, and that of Guizhou in 2008 was 185.832 billion with an increase of 24.8% over the previous year.

3. The Import and Export Maintain Steady Growth in Ethnic Regions

In the financial crisis, the foreign import and export maintain steady growth in ethnic regions. In 2008, the total trade volume in ethnic regions achieved 60.75 billion dollars with an increase of nearly 37% compared with that in 2007, and it was higher than the national growth rate. Along the 22,000 kilometers of China's boundary, 135 boundary counties, banners, cities are located, among which 107 belong to ethnic autonomous regions. Since the opening–up reform, the State Council opened 13 border trade centers including Manzhouli City in the Inner Mongolia Autonomous Region, Yining City in Xinjiang Uygur Autonomous Region, Ruili City in Yunnan, Pingxiang City in Guangxi Zhuang Autonomous Region and so on. Preferential policies are given there and trade in the border areas rapidly developed.

4. The Consumer Demand in Ethnic Regions Maintains Growing

From 2008, the consumer demand in ethnic regions continuously grows. The consumption structure shows positive changes. The consumption of vehicle, housing, interior design and so on has been expanded constantly, and the upgrades of household electrical appliances and communication goods have been accelerated. The total retail sales of consumer goods in the Inner Mongolia in 2008 was 236.333 billion with an increase of 24.1% over the previous year with 4.7 percentage points higher than the previous year. The value of retail sales of vehicle increased 28.8% and that of household electrical appliances and sound appliances increased 19.8%. Inner Mongolia's consumer market in rural and pastoral areas has been effectively boosted due to the pilot program of "home appliances to the countryside". Throughout the year, the value of retail sales of towns and townships achieved growth of 22.3% with 5.1 percentage increase over the previous year.

SECTION 2: THE ECONOMIC POLICY IN ETHNIC REGIONS

Thanks to the economic policy implemented in ethnic regions by Chinese government, balanced development is taking place among regions,. The Chinese government has always set the ethnic issue as an important component in the national development strategy. In addition to consistent adherence to the policy of regional autonomy of ethnic minorities, the Chinese government formulates various corresponding and specific ethnic policies according to the reality and characteristics of ethnic regions with the objective to sincerely help minority develop economy. Economic policies in ethnic minority areas highlight economic development as the main task of current ethnic affairs and implementing government supportive policy of regional autonomy of ethnic minorities as the core. With the increasing national comprehensive power, the Chinese government will strengthen its efforst to supporting economic and social development of minorities in ethnic regions.

Economic policies in ethnic regions at this stage can be divided into several aspects. First, with the system of regional autonomy of ethnic minorities as the core, conscientiously implement specific plans will promote fast development of economy and society of ethnic regions. Second, uphold development as the fundamental way to resolve ethnic problems. Persist in handling the relationship between national aid and self-reliance and handle the relationship among reform development and stability. Third, keep insisting on the principle of opening to the domestic market and the outside world and steadily promote the action of "vitalizing border areas and enriching the people's life". Fourth, adhere to the principles of "taking actions that fit local circumstances", "taking actions that meet ethnic groups' demand" and "adapting different guidance to different areas". Implement a variety of preferential and supportive policies and strengthen the support on minority nationalities with relatively less population. Fifth, implement preferential policies for commerce and goods in ethnic regions and cultivate market competitiveness of manufacturing enterprises in ethnic regions.

I. Set the Autonomous Administration of Ethinic Regions as the Core

Since the opening-up 30 years ago, the Chinese government has always insisted on upholding and improving the autonomous system of ethnic regions to promote economic and social development in ethnic regions and enhance national unity. Deng Xiaoping pointed out that, "China solves the ethnic problem with the autonomous administration of ethnic regions instead of federal system. We believe that this system is better suited to the situation in China."

Countries in the world vary in geography, history and ethnic composition. Ethnic relations and ethnic issues are different, so there is a great variety of policies toward ethnic issue with specific features. But they can be summed up in four types, namely, restricting development, laissez-faire, passive protection and positive promotion. The positive promotion policy toward ethnicity applied by China is to help ethnic minorities develop society, economy and culture in an all-round way, make all the ethnicities harmonious, realize the real equality of ethnic minorities in the politics, economy and culture; eliminate the development gap among various ethnic groups; promote modernization process in areas inhabited by the minorities; respect the cultural traditions religious beliefs and customs of various ethnic minorities in order to effectively safeguard national integrity and unity; ensure common prosperity, progress and development of all nationalities. In addition to consistent adherence to the autonomous administration of ethnic regions, China formulates several specific national policies according to the reality and characteristics of each ethnicity in order to help them develop economy and culture. The formation and evolution of these national policies have gone through four phases with the influences of international environment of different periods and specific domestic politics, economy and culture in different historical stages. These four phases are initial formation phase (1949-1957), tortuous development phase (1958-1965), stagnating and wandering phase (1966-1976) and restoring and optimizing phase (1977-present). [2]

Formulating and implementing specific policies and measures from the reality of ethnic regions are basic experiences of Chinese policy on ethnic issues. Compared with the whole country, China's ethnic regions have differences in natural conditions and economic development stages, and characteristics in cultural traditions and spiritual life. Chinese government constantly studies and formulates preferential policies, specific policies and specific plans according to these realities in order to accelerate development in ethnic regions. Chinese government implements specific fiscal systems which is different from other provinces and cities in the five autonomous regions of Inner Mongolia, Xinjiang, Tibet, Guangxi and Ningxia and three multi-national provinces of Yunnan, Guizhou and Qinghai. Chinese government also establishes special funds as "subsidy to ethnic regions", "minorities education subsidies", "frontier construction subsidies", "infrastructure subsidies in ethnic regions", "development fund for supporting the economically underdeveloped areas", "food and clothing fund for impoverished ethnic minority area", "fixed fiscal subsidy" and so on.

II. Development as the Fundamental Way to Solve Ethnic Problems

The core issue of upholding and improving the autonomous system of ethnic regions is to promote the comprehensive, coordinated and sustainable development of ethnic autonomy areas. Hu Jintao stressed that, "the difficulties and problems of ethnic areas should be solved by development"; "We must fully understand the extreme importance of accelerating development in ethnic areas from the global and strategic perspective". Development is the fundamental need and way to solve the current ethnic problems. Only by accelerating the development of economy and society of minorities groups can national unity and social harmony achieve a solid foundation. Since the opening up reform 30 years ago, ethnic regions have always focused on the major task of economic construction. On the one hand, persist in

[2] National Situation Analysis and Research Group in Chinese Academy of Sciences, The Seventh Report on National Conditions Research: Ethnic Groups and Development—Research on Accelerating the Social and Economic Development in the Central and Western Areas, Liaoning People's Press, June, 2000.

formulating various policies and measures with the standard of productivity development. On the other hand, give full consideration to the economic development of the real situation, choose priorities, find the direction and depth of development based on different locations, and make efforts to promote the comprehensive, coordinated and sustainable development of economy and society.

Column 11-1. China's Important Policies for Supporting Economic Development of Ethnic Regions in Recent Years

For the past few years, Chinese government has issued a number of policy documents to support development of ethnic regions, such as *Chinese Communist Party Central Committee and the State Council's Decision on Further Strengthening the National Work, Accelerating Economic and Social Development of Ethnic Minorities and Ethnic regions*, and *Provisions for the Implementation of the Law on Ethnic Regional Autonomy by the State Council*. The State Council has also issued and implemented three special plans, which are *Development Plan on Supporting Ethnic Minorities with Less Population (2005-2010), the Eleventh-Five Plan on Ethnic Affaurss*, and *the Eleventh-Five Plan on the Action of "Vitalizing Border Areas and Enrich the People Living there"*. The State Council has studied and formulated regional economic policies for accelerating development in ethnic autonomous areas. Since the opening reform, the Party has held four Tibet conferences, formulated policies for Tibetan development, and promulgated *Opinions on Resent Support for Development of Tibetan Economy and Society*. In recent years, the State Council and related departments have also formulated A Number of *Opinions on Further Promotion of Development of Xinjiang Economy and Society ((Promulgated by the State Council [2008]), Opinions on Support for Yunnan's Project of Vitalizing Border Areas and Enriching the People Living there (Development and Reform Commission [2008] No. 2327)*, and are now formulating opinions on support for development of Tibetan areas in Qinghai Province, etc. In the early 2008, the State formally approved the implementation of *Development Planning on the Northern Gulf Economic Zone in Guangxi*, which marked that the development of the northern gulf economic zone had been formally included in the national development strategy. It laid a foundation for the development with historic leap in Guangxi. The northern gulf economic zone in Guangxi has gradually become an essential strategic highland to driven and support the western development and a new pole of China's coastal development.

Accelerating the development of ethnic regions can not live without the strong support and help from the country. Since the reform and opening-up 30 years ago, China has continuously increased the strength of transferring payment from the exchequer and special financial support, and the distribution of each fund has been tilted to the ethnic regions. In the medium-term and long-term plan and program, China has always attached importance and support to accelerate development in ethnic regions, given full consideration to the special difficulties of ethnic regions, set the ethnic regions as priority, and continuously strengthened support. But more important things are enthusiasm, creativity, self-reliance, hard work in ethnic regions which will constantly enhance their self-development ability. Since the reform and opening-up 30 years ago, people in ethnic regions persist in combining national aid with giving full play to the enthusiasm and creativity of ethnic people, combining national aid with strengthening the internal vitality and self-development ability of ethnic people, combining government's major policies and principles with specific realities of ethnic regions, opening mind, breaking with the conventional ideas, innovating ideas, and constantly promoting the comprehensive, coordinated and sustainable development of economy and society in ethnic regions.

In the 30 years of reform and opening up, people in minority areas have always persisted in taking economic construction as the core, and unswervingly pushing forward the each reform course. Reform is to liberate and develop labor force. Reform can not only stimulate the enthusiasm and creativity of the masses to provide strong power for development, but also provide a favorable system and institutional guarantee for the fast economic development. Reform is a powerful driving force for economy development and the only way to build up a powerful nation and making the people rich. To speed up the economic and social development in ethnic areas, we must adhere to the policy of promoting development through reform, constantly deepen reform, and insist on expanding opening up policy.

In the 30 years of reform and opening up, the reform and opening-up in ethnic regions has always been active and steady, step by step, and combined with local circumstances. Stability is the basic prerequisite for reform and development. People in ethnic regions persist in steady development. In the developing process, people in ethnic regions actively and safely handle the relationship among "reform, development and stability". Not only development should be accelerated and each reform be promoted properly, but also should balance the strength and deepness of the reform with maintaining the stability as a priority.

III. Adherence to the Basic Policy of Opening to the Internal and External World

In the 30 years of reform and opening up, people in minority areas have always persisted in opening to the domestic market and outside world, giving play to the geographical advantages of abutting on border, constantly deepening open areas and promoting economic development. They have formulated flexible open-door policies which created more favorable investment environment, constantly expanded foreign trade contracts and economic and technical cooperation, introduced foreign capital and advanced technology and management experiences. It promoted the opening to outside world to a higher level and wider areas. Ethnic regions are horizontally joined with coastal and inland developed cities in order to realize complementary advantages, mutual benefit and common prosperity.

With the implementation of opening-up strategy which is promoted in trapezoid from the coastal to the border, along the Yangtze River, along the railway lines and inland provincial capital cities, the opening-up level of ethnic regions is constantly advanced, the opening-up areas continuously expande and the opening-up level deepened. Since the 1980s, there having been established one coastal open city, 14 river-shore open cities, 7 testing places of reform and opening-up in ethnic regions, 14 border economic cooperation zones and 71 national first-grade ports in China's ethnic regions and border areas.

China is bounded by 14 countries and most of the border areas are regions inhabited by ethnic groups. Of 22,000 kilometers of boundary line, 19,000 kilometers are in the ethnic autonomous area. The border area is 2.12 million square kilometers and the ethnic area account for 92%. There are 135 border countries and 107 of them are minority autonomous areas. The total population of border area is 22.65 million, in which the ethnic minority population accounts of 10.67 million. To promote the comprehensive development of economy and society in border areas, China makes great efforts to implement the project of “vitalizing border areas and enriching the people’s life”, aiming at accelerating development in border areas and promoting the well-off living standards. To support this project, “vitalizing border areas and enriching the people’s life” project fund is set up in the ethnic minority development fund by the central government; help and support are provided to the pilot and key towns and townships; a set of “vitalizing border areas and enriching the people’s life” programs are implemented; and social funds are secured for the project. The funding has gradually increased from 15 million in 2000 to 180 million in 2007 and 360 million in 2008. From 2000 to 2008, the total action fund of “vitalizing border areas and enriching the people’s live” is 1.025 billion. At the same time, a large number of social funds have been driven and attracted to the border areas. In this way, a great many of programs for resolving special difficulties of people’s work and life have been set up, including in the field of economic and social development, that of infrastructure, agricultural industry, ecological construction, cultural education and so on. Favorable economic benefit and social benefit have been taken place. This project has become “work to the people’s heart” and “benevolent governance project” in border areas.

After the implementation of “vitalizing border areas and enriching the people’s life” project, people in border areas seize the opportunity to accelerate development. At the end of 2006, the total output value of 135 border countries had reached 229.3 billion with an increase of 145% of 93.5 billion in 2000. The total output value per capita had reached 10126 with an increase of 122% of 4563 in 2000. The local fiscal revenue has been 13 billion with an increase of 65% of 7.9 billion in 2000. The large-scale industrial output value had been 104.1 billion with an increase of 213% of 33.3 billion in 2000.

IV. Adherence to the Principles of “Taking Actions that Fit Local Circumstances, Taking Actions that Fit Ethnic Groups and Apply Different Guidance to Different Types of Areas”

The actual situations in ethnic regions vary widely and the mode and pattern of development are different, so the policy formulation and implementation shouldn’t be uniform in all cases. People should act properly based on the reality, considering the advantages, avoid weakness, and develop agriculture and industry in proper areas at proper rate, so as to promote the healthy and stable development of minority regions. Since the reform and opening up 30 years ago, the supporting policy for ethnic regions has always adhered to the principles of “taking actions that fit local circumstance, taking actions that fit ethnic groups and apply different guidance to different types of areas”

For ethnic regions with more difficulties in production and daily life, policy regarding tax relief is implemented which is appropriate, for recuperating and reconstruction. The procurement prices of agricultural, husbandry, sideline and local products in ethnic regions are adjusted and raised, the areas implementing “three preferences” policy are expanded, and village fair trade and border trade are opened for peasants and herdsmen to sell their own products. Supplies are provided and price fixing policy is carried out for the capital goods and living goods which are closely related to the production and life of ethnic minorities. The policies of low-interest fund, tax relief and special investment are carried out for enterprises which produce ethnic goods. Preferential policy is also given to industrial enterprises both domestic and foreign joint-venture. Preferential treatment is applied in foreign exchange retention ratio of exporting product of Inner Mongolia, Guangxi, Xinjiang, Ningxia, Tibet, Qinghai, Yunnan and Guizhou, as well as retention ratio of earning from foreign exchange by export product in each autonomous region. Economically developed provinces and cities are organized to give counterpart support to ethnic regions, and experienced managers and technical experts are sent to ethnic regions to carry out the program of "aiding the border areas with intellectual forces". In arranging production layout and formulating long-term economic development plans, the reality in ethnic regions is taken into consideration from the national strategic objective. In order to help ethnic areas formulate economic and social development plans. A number of development experimental zones are established in ethnic regions as examples of construction and open of ethnic regions.

Of China's 55 ethnic minorities, 22 are with population below 100,000. The total population of ethnic minorities is 630,000 (according to the fourth census in 1990) and all of them are called minority groups. After the founding of People's Republic of China, although the politics, economy and society of these minoritis have developed sustainably, the development level is still relatively low and the poverty problems are still outstanding. The country has implemented *development plan for supporting ethnic groups with relatively small population (2005-2010)*, adopted special policies and measures in order to concentrate strength on helping these ethnic groups with relatively small population to accelerate the pace of development, change the backwardness and achieve common prosperity.

From 2002-2008, a total of 2.133 billion RMB funding has been put into supporting ethnic groups with smaller population, including 0.775 billion RMB of special construction funding for development of minority groups, 0.805 billion RMB of special development fund for ethnic groups with smaller population arranged by ethnic minority development fund, 0.329 billion RMB of special or matching fund at provincial level, and 0.224 billion RMB of central government poverty reduction fund, counterpart helping fund, international cooperation fund and so on. At present, of 640 villages inhabited by ethnic groups with smaller population, 260 have reached inspecting criterion and achieved the objective of "four getting through, five possessing, three reaching" in advance. A favorable situation of production developing, living standard improving and national unity is presented in areas of ethnic groups with smaller population.

V. Implementation of Preferential Policy on Trade and Product in Ethnic Regions

The special support policy implemented by the country to the ethnic trade and enterprises of producing ethnic commodities advances them in management reform and technological transformation. It can enhance their market competitiveness, increase the service ability for ethnic minorities, meet the special needs and daily supply of all ethnic groups, promote the commodity circulation and market boom in ethnic minority areas, and achieve favorable economic and social benefits.

The policies enjoyed by ethnic trade and ethnic commodity production enterprises have been perfected constantly. According to the survey, in the "Tenth-Five" period, the three preferential policies for ethnic trade were implemented properly. The ethnic trade enterprises above the township level enjoyed value-added tax refunded 50% which amounts to 1.2 billion, the ethnic trade enterprises below the township level exempted from value-added tax of about 1 billion, and the dark tea enterprises exempted from value-added tax of about 50 million nationwide. The ethnic trade and specially designated enterprises of ethnic commodity production enjoyed a total of about 2.2 billion of preferential lending rate. The discount loan fund of 14.45 million was arranged by the central government for constructing ethnic trade network and technological transforming of ethnic commodity production enterprises.

The development of ethnic commodity production and the commercial distribution activeness with ethnic trade enterprises as the main channel ensure a basic balance commodity supply in ethnic regions, and the problem of "hard buying and selling" which is caused by poor commodity circulation channel is resolved in most areas. Particularly in recent years, related department of the government has strengthened the support in important goods of Moslem food, dark tea, ethnic galosh and so on. Therefore, ethnic people can have "quality-assured meat", "quality-assured tea" and "quality-assured shoes". Their rights and interests can be protected. And the stable supply of production and daily life necessities in ethnic regions can be secured.

The implementation of preferential policy on ethnic trade greatly eases the difficulty of capital turnover of ethnic trade and production enterprises, reduces the production and operation costs of enterprises, and enhances enterprise competitiveness. The development of ethnic trade enterprises shows three distinct trends. First, a set of medium and large sized commercial enterprises with comprehensive competitiveness are gradually formed, such as Guizhou Hualian Group, Tianchi Department Store in Yanbian, Jilin, and so on. Second, a set of regional specialized markets are gradually developed which radiate more and more widely and of greater and greater driving force for ethnic people. It helps casting off poverty, such as Sanjiaji Tea and Livestock Market in Guanghe, Gansu, Ningxia Laoheqiao Market, and so on. Third, the operating procedure of ethnic trade becomes closer to the production and livelihood of peasants and herdsmen. Chain-store operations and uniform distribution extend to remote rural and pastoral areas, such as Xingxing Ethnic Trade Company in Tongliao, Inner Mongolia, Jilin Yanbian International Trade Company and so on. Meanwhile, the enterprises producing ethnic goods with special needs continuously grow. With the improvement of productive force in ethnic regions and living standard of ethnic groups, new types of ethnic products with special needs which suit people's consuming fashion and create healthy lifestyle are constantly emerged. Technology content and ethnic culture taste are improved step by step, such as ethnic language software, solar electrical energy generation equipment used in pasturing area, sprinkler irrigation equipment for home using and so on. A group of enterprises producing ethnic goods with special needs, such as Changchun Haoyue, Grassland Xingfa and so on becomes to develop multi-operational management and specialize in production with industrial chains constantly spreading.

The development of ethnic trade and production of ethnic commodity with special needs activate the circulation, expand domestic demand and accelerate the development pace of individual and private economy in ethnic regions. It also becomes the essential channel for accumulating private capital, the major industry of individual and private economic development and new economic growth opportunities in ethnic regions. It helps more and more greatly in promoting national economic development in ethnic regions, driving the growth of other related industries, arranging for employment of ethnic people and leading ethnic people to get rich. For example, in Hefeng Country, Enshi in Hubei Province, there are 9 ethnic trade and commodity enterprises. They got 280 million of loan with preferential interest rates, 8.19 million discount loans, accounting for 28% of the total credit of the country in 2004. During the "Tenth-Five" period, the nine ethnic trade and commodity enterprises achieved a net increase of 240 billion in total output value, accounting for 63% of the country's net increase. The annual average growth was as high as 14% and the contribution rate of the country's tax revenue growth was 38%. 38,000 peasant households benefited from them, accounting for 66% of the country's peasant households.

SECTION 3: AN OUTLOOK OF ECONOMIC DEVELOPMENT IN ETHNIC REGIONS

At present, we should pay much attention on the continuing implementation of ethnic cause in the 11th "Five-Year Plan", strongly promoting the equalization of public services, focusing on solving major problems and difficulties among ethnic minorities and in ethnic regions, an effort to achieve an overall, coordinated and sustainable development in ethnic regions.

I. Improved System of Ethnic Policy

In the early founding days of the New China, Chinese government established an ethnic policy with ethnic equality, unity, regional autonomy and common prosperity as the core purpose. The reform and opening-up put forward new demands for China's ethnic policy and created social conditions for development of ethnic policy. After the 60-years' continuous development and improvement, especially the development of 30-years' reform and openness, China has formed an ethnic policy system with Chinese characteristics which is well tailored for the country's realities. China's ethnic policy includes politics, economy, culture, society and other aspects in terms of content, including policies formulated by the central government, related departments and local government down the hierarchy. The cornerstone of China's ethnic policy is ethnic equality. The basic contents are maintaining national unity and country unification, carrying out ethnic regional autonomy, developing economy and preserving cultures, training various talented persons of minority, and respecting the religious beliefs and customs of ethnic minorities. And the staring point and goal are achieving common prosperity of all ethnic groups.

China's ethnic policies are based on the actual conditions. China takes full account of both the basic fact that ethnic groups take the socialist route and that differences of development levels, culture and customs among 56 ethnic groups; not only summaries the historical experience and lessons dealing with the ethnic issues deeply, but also draws lessons from other counties; not only maintains the stability, consistency and continuity of basic contents, but also enriches and improves policies constantly with social development and plenty practice. In this way, the policies are endowed with historical and realistic scientific basis.

For a long period into the future, the Chinese government will continue developing and perfecting ethnic economic policy based upon the present ethnic policy system. The related departments of central government will continue policies fostering investment, finance, taxation, price, land, mineral resources, foreign trade and so on in ethnic regions. The government will formulate the special planning and development ideas of corresponding areas, in order to promote the development of economy in ethnic regions. To act in concert with the western development strategy and provide further support on the development of ethnic areas, special funding was implemented by central government in ethnic regions since 2000 to resolve special difficulties there. Considering the special expense in ethnic regions, the central government also implements policy of subsidy, such as increasing the subsidy magnitude and amount in ethnic regions and so on. The main objective of general transfer payment is to realize equalization of basic public services without setting specific purpose. It is arranged by local government and is a typical balanced fund. From 2000 to 2007, the central Ministry of Finance totally arranged 209.683 billion of general transfer payment in eight ethnic provinces. The increasing strength on transfer payment shows that the central government pays great attention on the development of ethnic regions. According to the assumption of reforming transfer payment system, the financial transfer payment will be more inclined to the central and western areas, especially the western and ethnic areas, in order to improve the capacity of public service in central and western areas and make all people share the reform and development outcomes.

The plan of "vitalizing border areas and enriching the people living there" will be promoted continuously. Various departments will be coordinated according to the *Action of Vitalizing Border Areas and Enriching the People Living There in the Eleventh Five-Year-Plan*. The infrastructure and ecological construction in border areas should be strengthened. We should improve producing and living conditions of people living on the border, especially solvng the poverty problem, widening channels for increasing income, strongly promoting the development of border trade and regional economic cooperation, greatly developing the education, health and culture structures in border areas, strengthen national unity, and maintain stability in border areas.

II. Comprehensive and Coordinated Development of Ethnic regions and Other Parts of the Country

Since the implementation of western development, to promote the interaction between the east and the west, Chinese government enhances various economic subjects in various areas to achieve production factors flowing across administrative regions in accordance with rules of market economy. With the support of various departments of the State Council, the eastern and western regions work together to build platforms of inter-provincial support, Western China International Economy and Trade Fair and so on, find a scientific new way of "government guided, free market, corporate oriented, social participation" for promoting interaction between the east and west. The interaction between the east and west shows a favorable trend of constantly expanding scale, widening areas, innovating mechanism and growing benefit.

As the important area of the Western Development, ethnic regions are ushered a strategic opportunity period for overtaking and developing in the strategy of coordinating regional economic development in east-west interaction. On the basis of current achievement in east-west interaction, people in ethnic regions will see a lot of industry transfers from the east, introducing advanced technology, ideas and management experience, so as to effectively expand the scale of economic growth and accelerate the quality and efficiency of development. People in ethnic regions should develop themselves, root out the potential value, and strengthen cooperation in the fields of infrastructure, competitive industry with distinguishing features, economic and trade contact, etc. The initiative and enthusiasm of east-west interaction will also be improved. People in ethnic regions will continue to optimize the environmental development to improve the level of technological innovation, optimize industrial layout, obey demand of the main functional areas, start from the reality, take actions that suit local circumstances, maximize favorable factors and minimize unfavorable ones, guide the industries to transfer to important economic zones, central cities, resource-rich areas and important port cities in ethnic regions, promote industrial concentration distribution, intensive land use, resource conservation use and comprehensive environmental improvement, and enhance industrial cluster development.

In the new historical period of east-west interaction, the guiding function of the government must be strengthened. The government will also improve the development framework and policies of east-west interaction, strengthening the driving force in examination and approval of project and land, fiscal and financial support and so on, guiding enterprises for higher competitiveness through cross-region reorganization, and promote the mature development of east-west interaction mechanism. The scale of interaction should be expanded, the areas should be broadened, and the level should be improved.

III. Promoting Industrialization in Ethnic Regions by Specialty Industries

In recent years, the proportion of primary industry in ethnic regions decreases greatly, the proportion of secondary and tertiary industries increases stably, and the industrialization develops rapidly. In the whole western area, the proportion of primary industry accounting for GDP decreased from 25.45% in 1998 to 15.97% with a decline of nearly 10 percentage points. The output values of secondary and tertiary industries grew by 5.3 and 4.16 percentage points. The paces of industrial restructuring and urbanization have been accelerated, positive changes have appeared in economic restructuring, and the coordination among regions, between city and countryside, and among industrial development have been further strengthened. Economic development has pulled tens million of ethnic people out of poverty. The net income of peasants and herdsmen in the Tibet Autonomous Regions has maintained a double-digit growth for three years. The secondary industry in ethnic regions has a rising trend except for Guangxi and Yunnan. The proportion of tertiary industry also rises in general.

With the preference and guidance of national related policies, the resource advantages of ethnic regions have gradually transformed into economic advantages. A number of distinctive industries grow with a strong momentum. First, characteristic agricultural bases are formed. The duration of sunshine in ethnic regions is long, the temperature difference is big, and the grassland is large. In this way, ethnic regions are ideal areas for development of green industry, animal husbandry and dry land agriculture. At present, the dairy product, beef, mutton and other industries in Xinjiang, Qinghai, Inner Mongolia and Ningxia have begun to take shape. A large number of well-known enterprises and high quality products have entered domestic and international markets. Second, characteristic industry has been formed. Inner Mongolia and Xinjiang which are rich in coal, oil and natural gas resources have become national bases of energy and heavy chemical industry. The industries of tobacco, bio-products and hydroelectric power in Yunnan and Guizhou, the industries of nonferrous metal and sugar in Guangxi, and the national carpet industry in Qinghai, etc. have become industrial clusters with leading enterprises. The industrial division level has been improving and the industrial value chain has been extending constantly. It is learned that in the next five years, China will adopt preferential policies to accelerate the development of infrastructure and basic industry in ethnic regions, an effort to optimize industrial structure, and narrow the gap between ethnic regions and the relatively more developed regions.

IV. Improving the Enterprises Competitiveness in Ethnic Regions

People in ethnic regions will continue to earnestly implement the national trade preferences, improve the industrialization level and market competitiveness of enterprises, and complete the ethnic trade system and commodity producing system of national products, and the service level for the minority people.

First, the regulation and guidance roles of ethnic trade preferential policies should be given full play of. People should strengthen the publicity of ethnic trade preferential policies, to combine the policies with the local realities creatively, to formulate and detail management practices, standardize procedures, to make good, sufficient and flexible use of policies and to strengthen the supervision and inspection of policies to ensure the their strict implementation.

Second, the market order of commodities with national special needs should be safeguarded. People should persist on "stress on both hands"—one hand is on policy supporting of ethnic trade and ethnic commodities enterprises, and the other one hand is on supervision of respect paid by the enterprises for the ethnic life and customs. People should also standardize the market order by law, strike counterfeit and shoddy products, guide the enterprises to combine economic benefit with better service and better products and effectively protect the consumer rights of ethnic minorities. And the government will implement dynamic and hierarchical management on ethnic trade and ethnic commodities enterprises, support the superior and eliminate the inferior, manage strictly, actively support the guild to regulate business competition by law, and maintain a normal market economic order.

Third, enterprises should be promoted to transfer growth mode. People should improve the quality, science and technology content and national culture grade of commodities with national special needs through supporting technological transformation of enterprises, attracting more private capital into ethnic trade and producing of commodities with national special needs through the guiding role of preferential policies toward capital movements.

Fourth, a group of commodity firms matched with national special needs should be made stronger and larger. Moslem food, ethnic medicine, dark tea and ethnic handicrafts are the most important industries with good industrial base, large market capacity and broad development prospect. People should carefully review the development effect and experience of the four industries, encouraging enterprises to move toward the market, supporting a number of key enterprises towards upper scale, creating brands by technological improvement and introduction of advanced management experience. People should also drive the development of four industries by relying on the brand-name enterprises, so as to revitalize the national economy.

Translator: Yu Xuxiang (SUFE)
English Version Editors: Lv Mei (SUFE), Zhang Jingyi (Harvard)

REFERENCES

[1] Provisions for the Implementation of *the Law on Ethnic Regional Autonomy* by the State Council.

[2] *The Eleventh-Five Plan on Ethnic Undertakings.*

[3] *The Eleventh-Five Plan on Action of Vitalizing Border Areas and Enriching the People Living there.*

[4] *Development Plan on Supporting Ethnic Minorities with Less Population.*

[5] *China Statistic Yearbook (2000-2008).* The Ethnic Publishing House.

[6] Zhao Xianren, Ge Zhongxing. *Action of Vitalizing Border Areas and Enriching the People Living there (from the first to the fourth volume)*, The Ethnic Publishing House.

[7] The Ethnic Problem Researching Center of the State Nationalities Affairs Commission, *Development Assessment Report on China's Ethnic Autonomous Areas*, The Ethnic Publishing House.

[8] Tie Muer, Zhao Xianren. *Statistical Analysis and Strategy Research on China's Ethnic Villages*, The Ethnic Publishing House.

[9] National Situation Analysis and Research Group in Chinese Academy of Science, June, 2000*The Seventh Report on National Conditions Research: Ethnic Groups and Development—Research on Accelerating the Social and Economic Development in the Central and Western Areas*, Liaoning People's Press.

[10] *China's Statistical Summary in 2009*. China Statistical Publishing House.

In: China's New Deal
Editors: Li Xiaoxi and Hu Biliang
ISBN: 978-1-61470-815-5

Comments on Chapter 11

LET ONE HUNDRED FLOWERS BLOOM: THE ECONOMIC DEVELOPMENT IN THE ETHNIC REGIONS OF CHINA

Yuan Jingan*

The topic of ethnic regions in China is still a strange concept to most Westerners; therefore, it is necessary to give a brief introduction here first.

Contrary to most Westerners' belief, China is not an ethnically homogeneous nation. In China, there are as many as 55 officially recognized ethnic minority groups. Minority refers to all ethnic groups other than *Han*. There are significant ethnic and cultural differences among these minorities. For example, millions of Koreans live in the northeast of China, which shares the border with North Korea. They are Chinese citizens, but usually they can speak both Chinese and Korean fluently. In the northwest of China, which shares border with Central Asian countries, nine million *Hui* people live in the regions such as Xinjiang, Gansu, and Ningxia. They are Muslims, and visit Mosques regularly. The minority group that has the largest population is *Zhuang*: they live in various places in China, and have a total population of over 16 million. In total, the population of all minorities in China is 132 million, which accounts for 9.4% of the total population in China.

Different from many other countries, in China most minorities have lived with *Han* people for more than two thousand years and have long-term mutual influences on each other. Although there have been conflicts now and then in the past thousands of years, in general, during most of the time the minorities accepted the administration of the central government of China (the emperor and royal court), and the communications among all ethnic groups has never stopped. During a few dynasties, minorities even took control of the central government and were the administrators of the whole China. The most influential one was the Qing dynasty, which was founded by *Manchu*, one of the minorities. It governed the entire China,

* Yuan Jingan, PHD, University of Alabama, Associate Professor Southwestern University of Economics and Finance.

including the current Mongolia, Taiwan, and part of Siberia, for 268 years from 1644 to 1912. Due to the characteristics mentioned above, the minorities in China have played an important role in Chinese history. Also because of the long-term co-existence of all the ethnic groups, both the dominant *Han* people and the minority people identify themselves as "Chinese".

After the founding of the People's Republic of China in 1949, the Chinese government followed the Soviet Union government's doctrine and set up five ethnic minority autonomous regions in the areas where large number of minority people live, and named the regions after the dominant minority groups. Those five ethnic minority autonomous regions are, Xinjiang Uyghur Autonomous Region, Tibet Autonomous Region, Inner Mongolia Autonomous Region, Ningxia *Hui* Autonomous Region and Guangxi *Zhuang* Autonomous Region. The political status of these autonomous regions is equal to that of provinces or municipalities, or states in the U.S. The total land area of the five autonomous regions accounts for three fifth of the total land area of China. In these autonomous regions, at least the Vice Governors must be from the minority group. But actually, the population of *Han* people accounts for more than half of the total population in four out of those five autonomous regions.

For a long time, the minorities in China have received preferential treatment by national policies and enjoyed more privileges than *Han* people. For example, when minorities compete for government positions or seats at People's Congress, they have priorities; the descendants of minorities' can be admitted into universities with lower scores than *Han* students'. It is also worth mentioning that even though the Constitution of China sets strict "One Child Policy", this policy does not apply to minority people. Therefore, the proportion of minorities in the total population of China has an increasing tendency.

In the regions where minority and *Han* people live together, like most areas in Yunnan province, the existence of minorities provides a variety of thriving cultures. The *Lijiang* Ancient City, which has a prestigious position among American and European tourists, is located in this province. *Lijiang* has a history of over 800 years and is well known for its abundant architectural styles and music of *Naxi* ethnic group. It is on the list of UNESCO World Heritage. For backpackers, this kind of place is even more attractive than those sparkling big cities like Beijing and Shanghai. In fact, there are many other ancient cities like *Lijiang* in Yunnan province and they make Yunnan a major tourism destination in China. These activities have greatly pushed Yunnan's economic development up. In provinces like this, the minorities and *Han* people get along with each other very well.

In some other regions like Xinjiang and Tibet, the available natural resources are very limited. The area of Xinjiang Autonomous Region accounts for one sixth of the land area of China, but most of this region is disserts and only a small part is arable land. For a long time, people's living standard was very low. After the founding of the People's Republic of China, the central government has invested large amount of fund to build highways and railroads, and to develop agriculture and oil industry in this autonomous region.

Since the reform-and-open policy was adopted in China in 1979, as mentioned in Chapter 11, the millions of Muslims living in Xinjiang Autonomous Region took advantage of their natural connections with the Muslims in Central Asia, and created tight business and trade relations. In fact, because Xinjiang has better infrastructure than most Central Asian countries, like Pakistan, Afghanistan and Turkmenistan, it has become an important business center in this area, and the old Silk Road is revived.

Certainly, the rapid development of Xinjiang took place only in the past decades and its accumulation of wealth still lags behind compared to other areas which have developed prosperously for hundreds or even thousands of years. However, this area has its unique strategic importance. As mentioned above, Xinjiang shares thousands of kilometers' borders with Central Asian countries like Afghanistan and Pakistan. The influences of the extremists of Muslim (fundamentalists) enter Xinjiang unavoidably. In today's world, when most countries are pursuing peace and development, and when any form of terrorism is resolutely opposed, developing the economy in areas like Xinjiang is especially important. The Chinese government has realized long ago that cutting the connections with the rest of the world is not the way to solve problems, and that the minority areas especially need to strengthen and extend their communications with other areas in China and the world. Only in this way can the different ethnic groups understand each other, learn from each other and develop together.

In Chapter 11, the authors, from the perspectives of economists and with the support of a large number of statistic data, conduct a detailed review of the economic development of ethnic regions in China for the past thirty years after the open policy was established. This chapter analyzes in details the income per capita, government's financial income and expenditure, the structure of industries, infrastructure, and the efforts that are made to eliminate poverty, and summarizes the great achievements and progress in the economic development in these ethnic regions.

Meanwhile, this chapter points out that these ethnic regions are still lagging behind. Furthermore, this chapter analyzes the economic development in these ethnic regions during the Financial Crisis of 2008-2009 and describes the relatively fast developments in economic growth rate, investment in fixed assets, import and export, and consumption. This chapter also proposes constructive comments and suggestions for the future.

Then from the policy perspective, this chapter illustrates in details the reason behind the rapid economic development in the ethnic regions in China in the past decades. That includes: the political structure which takes "autonomous regions" system as the core, and the fundamental economic approach which focuses on comprehensive, coordinated and sustainable development. The economic policy in ethnic regions should follow the basic principle of insisting open policy internally and externally and the principles of "Adapt to local conditions, use different measures according to different ethnic groups' needs and provide guidance in accordance with different needs". Also, preferential and supportive policies should be implemented to the trade and commodity-producing industry in ethnic regions to accelerate their development.

In the end, this chapter proposes specific suggestions on how to perfect the policy system of the ethnic regions, how to enhance the connections between ethnic regions and the rest of the country, how to improve the competitiveness of businesses in ethnic regions, and how to develop featured industries of each region. This chapter also expresses confidence in the future of ethnic regions' economic development.

REFERENCES

[1] Transnational Linkages and Development Initiatives in Ethnic Korean Yanbian, Northeast China: "Sweet and Sour" Capital Transfers. (2009). *Pacific Affairs V. 82 No. 3 (Fall 2009) P. 427-46*, *82*(3), 427-446.

[2] Li, L. (2009). A Garden of Growing Imams. *Beijing Review V. 52 No. 20 (May 21 2009) P. 42*, *52*(20), 42.

[3] Hartley, L. (2008). [On the Margins of Tibet]. *History of Religions V. 48 No. 1 (August 2008) P. 77-9*, *48*(1), 77-79.

[4] Market Watch. (2007). *Beijing Review V. 50 No. 45 (November 8 2007) P. 42-3*, *50*(45), 42-43.

[5] Winkler, D. (2008). Yartsa Gunbu (Cordyceps Sinensis) and the Fungal Commodification of Tibet's Rural Economy. *Economic Botany* V. 62 No. 3 (November 2008) P. 291-305, 62(3), 291-305.

[6] Davis, E. (2008). Uyghur Muslim Ethnic Separatism in Xinjiang, China. *Asian Affairs (New York) V. 35 No. 1 (Spring 2008) P. 15-29*, *35*(1), 15-29.

[7] Torode, G., & Dubai, R. (2009). Al-Qaeda Leader Targets China Terrorist Group Calls for Jihad in Xinjiang. South China Morning Post; Hong Kong, *(October 08 2009) P. 1.*

In: China's New Deal
Editors: Li Xiaoxi and Hu Biliang

ISBN: 978-1-61470-815-5

Chapter 12

THE NEW MEASURES: HONG KONG AND MACAO TACKLING THE FINANCIAL CRISIS

Zheng Yanting*, Zhao Zheng, Wu Wenshi and Du Yamin

In September 2008, the global financial crisis worsened, which made Hong Kong and Macao's sustained strong economic growth of many years screeched to a halt. In the first quarter of 2009, various economic indicators were in overall negative growth. Subsequently, Hong Kong and Macao Special Administrative Region Governments have issued a number of measures ranging from financial support to policies concerning people's basic needs in order to weather the financial crisis. Meanwhile, the central government also supported Hong Kong and Macao's economy in many ways. Many economic indicators have signified that these countermeasures have achieved positive results. It is believed that with the mainland's economy rebounding to stabilization and the global economy gradually showing signs of recovery, the resumption of Hong Kong and Macao's economic growth is just around the corner. This chapter begins with the analysis of the impact of financial crisis on Hong Kong and Macao's economy from two aspects: the overall economy and representative industries, and then summarizes the measures that Hong Kong and Macao adopted to deal with the financial crisis, as well as the central government's supporting policies to Hong Kong and Macao.

SECTION 1: THE IMPACT OF FINANCIAL CRISIS ON HONG KONG AND MACAO'S ECONOMY

Since the beginning of the new century, relying on the mainland's preferential policies, Hong Kong and Macao are in rapid economic growth. Since CEPA ("Closer Economic Partnership Arrangement") was signed six years ago in 2003, Hong Kong's GDP has an average annual growth of 6.3%. Since the opening up of betting rights in 2002, driven by the gaming industry, Macao's economy realized double digit annual GDP growth except the 6.9% growth in 2005.

* Zheng Yanting is lecturer of the Institute of Economics and Resource Management, Beijing Normal University, who earned her PhD degree from the Department of Geography, the University of Hong Kong. Email: zhengyt@bnu.edu.cn

However, in September 2008, with the bankruptcy of the fourth-largest U.S. investment bank Lehman Brothers, there has been a serious turmoil in global financial markets. Nowadays, with rapid international capital flows and transmission of information, the rapid spread of the financial crisis in the United States triggered the global financial crisis. Hong Kong and Macao, which are highly export-oriented, are inevitably influenced by the financial crisis.

I. The Impact of Financial Crisis on Hong Kong and Macao's Overall Economy

Due to the global financial crisis, Hong Kong and Macao's economy shows a downward trend. Imports and exports suffer a heavy blow, consumption turned to be weak, investment sentiment went down, and unemployment rate increases. This section analyzes how the GDP of Hong Kong and Macao and their major expenditure components changed from 2006 to the first half of 2009, seeking to reveal the impact of global financial crisis on Hong Kong and Macao's economy.

Data show that there were already downward signs of Hong Kong's economy in the second quarter of 2008 (see Table 12-1). In the second quarter, while gross domestic product increased by 4.1% compared with the same period of 2007, the seasonally adjusted gross domestic product decreased by 1% compared with the previous quarter. In the third quarter, GDP continued to decline, only increased by 1.5% over the same period of the previous year, and fell by 0.8% sequentially compared with the previous quarter. The fourth quarter, when compared with same period of the previous year, showed the first negative growth (-2.6%), and fell by 9% compared with the previous quarter. In the first quarter of 2009, recent economic downturn reached the maximum amplitude, decreased by 7.8% compared with the same period of the previous year, and fell by 4.3% compared with the previous quarter. As the international economic situation stabilized, in the second quarter of 2009, the drop range was narrower (-3.8%). In view of seasonal comparisons, the economy of the second quarter is better than that in the first quarter, with an increase of 3.3%.

Through the comparison of gross domestic product by the change of major expenditure components, we can make a more in-depth analysis of economic changes. In the first quarter of 2009, the reasons of the accelerated decline of Hong Kong's economy mainly involved: 1) Exports of goods fell sharply. Exports decreased 145.7 billion Hong Kong dollars compared with the same quarter of the previous year, with a decrease of 22.7%. Among these, more than 90% are re-exports, while the vast majority of re-exports were from the Pearl River Delta. Thus, to a large extent, the impact of weak consumption caused by financial crisis on the manufacturing sector of the Pearl River Delta region was also a blow to Hong Kong's economy. 2) The apparent decline in private consumption expenditure. Since the third quarter of 2008, Hong Kong's economy showed a gradual decline in private consumption expenditure. In the first quarter of 2009, the decline even reached as much as -6% with a decrease of 14.74 billion Hong Kong dollars in comparison with the same period of the previous year. Its contribution to the GDP drop is more than 50%, reflecting general public's lack of confidence in the economy in the first quarter of 2009. 3) Fixed capital formation continued to fall. Since the fourth quarter of 2008, fixed capital formation continued to fall at lower than -10% and the decrease lasted until the second quarter of 2009, when the local business climate remained poor.

Table 12-1. Hong Kong's GDP and its major expenditure components, 2006-2009Q2 (100 million Hong Kong dollars)

Year	2006	2007	2008	2008			2009	
Quarter				2	3	4	1	2
GDP*	15185.41	16154.31	16536.36	3978.47	4228.09	4245.26	3765.93	3826.09
Annual change rate (%)	7.0	6.4	2.4	4.1(-1.0)	1.5(-0.8)	-2.6(-1.9)	-7.8(-4.3)	-3.8(3.3)
Private consumption expenditure	8955.54	9720.27	9863.17	2500.10	2430.17	2476.41	2309.07	2474.29
Annual change rate (%)	5.9	8.5	1.5	3.0	-0.2	-4.1	-6.0	-1.0
Government consumption expenditure	1266.33	1303.98	1326.69	313.95	324.60	335.65	357.59	319.09
Annual change rate (%)	0.3	3.0	1.7	3.1	1.6	1.8	1.4	1.6
Gross domestic fixed capital formation	3145.34	3253.49	3238.20	859.53	832.76	716.28	716.29	739.50
Annual change rate (%)	7.1	3.4	-0.5	5.1	2.9	-17.8	-13.7	-14.0
Exports of goods	2,5216.81	26988.50	27511.40	6889.19	7303.66	6897.08	4964.77	6032.41
Annual change rate (%)	9.3	7.0	1.9	4.4	1.3	-4.9	-22.7	-12.4
Exports of services	5792.55	6607.28	6981.41	1618.79	1849.52	1805.36	1599.47	1527.02
Annual change rate (%)	10.1	14.1	5.7	8.4	4.8	0.4	-6.3	-5.7

Table 12-1. (Continued)

Year	2006	2007	2008	2008			2009	
Quarter				2	3	4	1	2
Imports of goods	2,6213.51	28525.22	29049.64	7406.92	7641.96	7192.65	5352.71	6463.98
Annual change rate (%)	9.2	8.8	1.8	4.8	2.0	-6.4	-21.4	-12.7
Imports of services	2963.83	3322.40	3430.60	825.60	899.58	837.81	784.29	779.63
Annual change rate (%)	8.1	12.1	3.3	3.8	2.2	-3.4	-9.6	-5.6

* Measured in chain volume of 2007.

\# The figures in brackets are the percentage changes of seasonally adjusted GDP compared with the previous quarter. The seasonally adjusted GDP series have been removed in terms of statistical methods due to seasonal effects. Therefore the figures connecting two quarters could make a more meaningful comparison. The annual change figures of original series of GDP and the connected change figures of quarterly GDP which are adjusted seasonally are analyzed together, so we can see more clearly the trend of GDP, especially at the economic turning points.

Source: Hong Kong SAR Census and Statistics Department - Hong Kong statistics.

Data show rebound of Hong Kong's economy in the second quarter of 2009, mainly due to: 1) Decrease in exports of goods narrowed down. Export of goods in the second quarter of 2009 decreased 85.7 billion Hong Kong dollars compared with the same period of the previous year, which is greatly reduced compared with the previous quarter's drop of 145.7 billion Hong Kong dollars, and the rate of decline narrowed from -22.7% to -12.4%. 2) Recovery in private consumption expenditure. Compared with the first quarter which had a decline of 14.7 billion Hong Kong dollars, the decline narrowed down to only 2.58 billion Hong Kong dollars, which shows that Hong Kong people gradually restored confidence in the economy. 3) Government investment was in sustained growth. In the first and second quarter of 2009, government consumption expenditure each increased by 5 billion compared with the same quarter of the previous year. Therefore, the local government's increasing investment, and implementation of measures which stimulated consumption in the Mainland, as well as the global economy which picked up, all gave impetus to Hong Kong's economic recovery while enhancing Hong Kong people's confidence in the economy. However, it is worth noting that despite the rebound of economy in the second quarter, the prospects for Hong Kong's real economy is still worrisome. The drop of gross fixed capital formation over the same period of the previous year expanded from 11.3 billion Hong Kong dollars in the first quarter to 12.0 billion Hong Kong dollars. This shows that it might take a long time for Hong Kong's economy to move out of the financial crisis completely.

Since the second half of 2008, Macao's economy has taken a downward trend (see Table 12-2). Compared with the same period of the previous year, in 2008, GDP of the first quarter grew 32.5%; this rate dropped to 22.4% in the second quarter; in the third quarter it dropped to 10.4%; and in the fourth quarter, GDP growth started to fall by 7.6%. In the first quarter of 2009, the rate of decline even extends to -11.9%. Unlike Hong Kong, in the second quarter of 2009, Macao's GDP growth continued to decline and the rate further expanded to reach -13.7% which is different from Hong Kong's. It can be shown that Macao's economy has not recovered due to benefits from the current external economic upturn.

As for Macao, about 90% of GDP comes from export of services, while the majority of exports of services are gaming revenues. Therefore, exports of services can be broadly equivalent to gaming revenues. In the fourth quarter of 2008, exports of services had a 5.5% negative growth for the first time in the past five years. The drop rate in the first quarter of 2009 expanded to -14.6% and in the second quarter continued to drop by 14.6%. In the second quarter, exports of services decreased by 5 billion compared with the first quarter, which narrowed compared with the decrease of 5.24 billion in the first quarter. In comparison, exports of goods fell even more rapidly. Since the fourth quarter of 2008, the rate of decline has been close to 50%. In the second quarter of 2009 it decreased by 2.57 billion Macao patacas, which is 6 billion more compared with the first quarter of 1.94 billion. Similar to the previously described situation in Hong Kong, re-exports of goods from Guangdong province accounted for the major part of exports of goods in Macao. Therefore, in addition to a decline in gaming revenues, another main reason for Macao economy's continued slide in the second quarter is that the export had not been restored. Furthermore, rate of decline in gross fixed capital formation has remained at about 20% since the second quarter of 2008, which even exceeded 30% in the first quarter of 2009 and continued to decline in the second quarter. In response to the financial crisis, Macao's government spending in the first and second quarter of 2009 both continued to increase over the previous year by 15.8% and 6.2%, respectively. To summarize, since the second quarter of 2009, Macao's economy is still not

Table 12-2. Macao's GDP and its main expenditure components, 2006-2009Q2 (100 million Macao patacas)

Year	2006	2007	2008	2008			2009	
Quarter				2	3	4	1	2
GDP*	992.45	1243.21	1407.49	376.53	338.55	324.09	324.41	324.76
Annual change rate (%)	16.5	25.3	13.2	22.4	10.4	-7.6	-11.9	-13.7
Private consumption expenditure	264.35	295.07	317.30	78.76	80.94	81.01	81.00	79.28
Annual change rate (%)	8.2	11.6	7.5	7.0	5.1	9.4	5.8	0.7
Government consumption expenditure	81.16	91.65	91.45	23.74	19.42	32.31	18.50	25.20
Annual change rate (%)	3.8	12.9	-0.2	6.1	-7.8	-4.4	15.8	6.2
Gross domestic fixed capital formation	267.29	333.58	271.12	72.46	69.69	67.67	42.17	52.63
Annual change rate (%)	44.5	24.8	-18.7	-19.4	-25.0	-15.1	-31.2	-27.4
Exports of goods	204.03	204.99	152.91	44.14	39.35	30.41	19.56	18.43
Annual change rate (%)	2.8	0.5	-25.4	-16.0	-26.6	-44.6	-49.9	-58.3
Exports of services	782.23	1057.90	1304.34	343.34	313.45	290.08	305.11	293.21
Annual change rate (%)	18.7	35.2	23.3	38.5	20.0	-5.5	-14.6	-14.6

Year	2006	2007	2008	2008			2009	
Quarter				2	3	4	1	2
Imports of goods	458.01	522.12	468.96	116.25	118.26	117.34	90.91	90.24
Annual change rate (%)	18.3	14.0	-10.2	-8.1	-16.5	-14.3	-22.4	-22.4
Imports of services	157.34	225.54	267.10	71.31	67.67	61.59	52.22	54.96
Annual change rate (%)	19.2	43.3	18.4	31.6	14.0	-3.1	-21.6	-22.9

* Estimated based on constant (2002) prices.
Source: Macau Statistics and Census Service

optimistic. Goods and services exports are still declining, consumer confidence has been affected, and further deterioration is taking place in the business environment.

In addition, the financial crisis also affected employment in Hong Kong and Macao to different extents. In the second quarter of 2008, unemployment rate in Hong Kong and Macao continued to rise (see Table 12-3). In Hong Kong, in the second quarter of 2009, the number of unemployed exceeded 200,000, which reached 214,000 during the period from May to July. In the fourth quarter of 2008, the unemployment rate exceeded 4%, 0.8% up compared with the third quarter. In the first quarter of 2009, the unemployment rate was more than 5%, with an increase of as much as 1.1%. In the second quarter of 2009, the latest data showed the unemployment rate continued to rise to 5.4%, but the upward trend slowed down. Similarly, in the third quarter of 2008, Macao's unemployment rate increased from 2.8% of the previous quarter to 3.1%, and then rose to 3.8% in the first quarter of 2009. Different from which in Hong Kong, Macao's unemployment rate remained at 4% or less, and in the second quarter of 2009, the unemployment rate began to decline.

Table 12-3. Unemployment rate in Hong Kong and Macao, 2008-2009 (%)

Year	Quarter	Hong Kong *	Macao
2008	1	3.3	2.9
	2	3.3	2.8
	3	3.4	3.1
	4	4.1	3.3
2009	1	5.2	3.8
	2	5.4	3.6

* The seasonally adjusted unemployment rate

Source: Census and Statistics Department of Hong Kong, Macau Statistics and Census Service.

To sum up, since the second half of 2008, with the extension of the global financial crisis, Hong Kong and Macao's overall economy has been greatly affected. This effect in the first quarter of 2009 is the most serious. Since the second quarter of 2009, with the improvement of external economic situation as well as implementation of various local response measures, downward trend in Hong Kong's economy has slowed down, supported by goods exports, private consumption recovery, and sustained growth in investments from Hong Kong Government. However, as for Macao, due to the impact of the economic decline of the goods, service exports and fixed investment in 2009, the downward trend of economy in the second quarter continues.

II. The Impact of Financial Crisis on Major Sectors of Hong Kong and Macao

While the financial crises affected the overall economy of Hong Kong and Macao, it also made certain impact on local industries. This section makes an analysis of the impact of financial crisis on major sectors in Hong Kong and Macao since the second half of 2008.

1. The Impact of Financial Crisis on Hong Kong's Major Businesses

The impact of financial crisis on Hong Kong's main industry, which began from the banking sector, also spread to other industries. As Table 12-4 shows, in the second quarter of 2008, the financial crisis first hit the banking sector and led to its decline of 6.8%. In the third quarter, the banking sector decline rate expanded to 13.1%; in the fourth quarter, decline furthered to 45.5%. Sharp drop occurred in the performance of the financial sectors, and the employment situation of the industry was severe. HSBC, Citigroup, Standard Chartered, DBS and other major Hong Kong banks announced lay-offs. Among them, HSBC has cut more than 500 employees[1]. In the third quarter of 2008, in addition to banks, other financial industries have been affected with amplitude (-27.2%) exceeding that of the banking sector. Decline in other financial sectors reached -57.6%, and the financial markets and asset management industry had a decline of as high as 60.9%.

The influence of the financial crisis also spread to other industries. The real estate sector was particularly obvious. In the fourth quarter of 2008, the real estate revenue index fell by 26.9%. As one of the pillars of economy, real estate accounted for more than 20% of GDP, which is the economy's "barometer." The financial crisis makes the volume of Hong Kong's real estate industry decline rapidly, and the real estate price declines have hit record highs. The statistics released by Hong Kong Land Registry shows that in October 2008 sale and purchase agreements for all types of building accounted for a total of 6,054, down by 17.8% compared with September, and the annual decrease was 54.2%. In October, sale and purchase contract was worth 18.9 billion, reduced by 19.6% on a monthly basis, and down by 62.6% on a yearly basis. November's volume was only 5,000 or so, close to May 2003 which was the SARS period. At the same time, the drop of property market turnover brought down the selling prices and rental prices. Hong Kong residential rents were also lower in late October and the number of negative equity in late November showed an upward trend.[2]

In the aspect of transportation, the long distance water and air transportations are affected much more severely than land transportation, showing that water and air transportations are mainly affected and burdened by the external economic environment. Logistics industry has also suffered a rare cold snap. Statistics indicates that compared to the same period of 2008, in the first quarter of 2009, the turnover of Hong Kong logistics companies declined by nearly 40%. More than 300 small logistics companies have been shut down, and the operation rate of the survived logistics company was very low Drivers of some companies only worked 10 days in one month and many vehicles were out of operation. At the same time, according to the statistics released by Hong Kong Container Trucks Transportation Employees General Union, the number of Guangdong-Hong Kong cross-boundary freight traffic reduced from 16,500 in July 2008 to 12,500 in March of 2009, with 4,000 truck outages during the 8-month period, which are 24.2% of the total number.[3] Huangpu Port also lost its usual busy scenes and fell into a business depression.

---

1 Source: Xinhua - "by the global financial crisis began to affect Hong Kong's financial industry layoffs" http://www.sc.xinhuanet.com/content/2008-11/21/content_14979529.htm

2 Source: Hexun - "the financial crisis hit Hong Kong property market, the Pearl River Delta has been selling real estate" http://money.hexun.com/2008-12-01/111834402.html

3 China News Net - "cross-border logistics industry has been severe influenced by the financial crisis with outage of 4000 trucks"http://www.chinanews.com.cn/ga/fzsj/news/2009/03-23/1613608.shtml

In the first quarter of 2009, with the continous support of Hong Kong government on the local banks, the decline rate of banking sector narrowed to -15.5%; while the other financial sectors continued to decline (-46.9%). Among these, the financial markets and asset management industry, which achieved rapid growth of 49% and 71%in 2006 and 2007 respectively, declined in the first quarter by as much as -51.8%. Insurance (-10.3%) and real estate industry (-13.2%), which are closely related to the financial sector, had slightly reduced their rates of decline. But this time, the economic decline began to spread to other non-financial sectors. For example, imports and exports recorded a 23.1% decline, -25.7% for wholesale trade, and -22.8% for transportation. With the exception of catering services, telecommunications and other industry related to basic needs of people's lives, all other sectors fell. So far, the impact of this financial crisis has involved all aspects of the Hong Kong society.

In the second quarter of 2009, Hong Kong business income indices for service industries show that the performance of various sectors in Hong Kong has fully recovered. The most impressive is the financial industry (except banking) with an increase of 39.2%; the financial markets and asset management industry rebounds by 45.9%. Driven by the warming external environment, all the export-oriented industries have resumed growth, including import and export trade appreciation rate of 20.1%; wholesale increased by 22.1%; goods and warehouses grew by 9.6% compared with the same period of the previous year; long distance transportation (water transport increased by 4.4% and air transportation increased by 1.4%) and express industry (15.1%) have also increased. The above analysis shows that in the second quarter of 2009, against the background of a gradual warming external environment, the shadow that the financial crisis casted on Hong Kong is being wiped out.

In addition, the financial crisis also created a huge influence on the development of Hong Kong tourist industry and business and trade sectors. According to Hong Kong Tourism Board statistics, visitor arrivals in 2008 reached 29.5 million, increased by 4.7 compared with last year's 28.16 million, which is lower than forecasted tourist's visits level of 30 million at the beginning of the year. In February 2009, mainland tourists visiting Hong Kong in the form of personal tours reached a total of 773,145 passengers, and the number declined by 9.6% from a year earlier. Visitor arrivals from most major regional markets declined compared with the same period of last year; Taiwanese tourist arrivals declined most severely, by more than 17%. Arrivals from North Asia decreased by more than 14%, while arrivals from Europe, Africa and the Middle East fell by more than 11%. Affected by the financial crisis, along with the overall weakness in the consumer market, Hong Kong's commerce industry also fell into a depression. Some old large chain stores had operating difficulties, and companies including the listed company "Youhui" operating "U-right" clothing chain management, "Tai-Lin" which is a 62-year-old electronics chain stores, and "Hejun Group " which is one of the world's largest toy factory, closed down one after another. Many commercial enterprises have begun advancing the holiday sales and introduced a number of activities to promote the price reduction for instant cash.

Table 12-4. Rate of change of profitability indices of Hong Kong service industries / different business domains (%)

Year	2006	2007	2008	2008			2009	
Quarter				2	3	4	1	2
Import and export trade	10.3	8.1	7.4	14.4	9.6	-5.8	-23.1	
Wholesale	7.9	10.4	6.3	14	10	-6.9	-25.7	
Retail	7.2	12.8	10.6	14.3	10.5	0.9	-3.9	
Transportation	10	10.5	4.4	12	6.1	-10.3	-22.8	
Of which: Land transport	4.3	3.7	2.2	5.1	1.2	-2.2	-2.9	
Water transport	16.4	16.8	6.1	14.8	4.5	-10.3	-24.7	
Air Transport	6.8	7.3	3.5	11.4	8.8	-12.3	-26.7	
Goods and warehouse	10.2	15.9	6.6	3.8	7.1	5.7	-10.1	
Express	6.9	5.5	2.1	8.8	7.1	-10.8	-20.1	
Accommodation	13.8	15.2	3.8	9.6	5.9	-6.5	-18.1	
Catering services	9.6	13.4	13.1	15.7	13.2	8.2	1.5	
Information and Communication	6.1	8.4	6.6	7.9	5.4	2.5	-4.6	
Of which: Telecommunications	-0.7	11	9.8	14.5	9.5	3.6	6.4	
Movies	1.7	6.1	-0.7	5.9	1.2	-10.7	-18.4	
Banks	19.5	38.3	-16.9	-6.8	-13.1	-45.5	-15.8	
Finance (except banking)	47.9	68.8	-19.4	4.7	-27.2	-57.6	-46.9	
Of which: Financial markets and asset management	49.3	71.7	-20	7.4	-29.2	-60.9	-51.8	
Of which: Asset Management	51.8	56.8	-5.2	21.5	-15.2	-45.9	-48.1	
Insurance	21.3	28.8	—	12.1	-0.4	-22.4	-10.3	
Properties	-0.3	39.5	-3.7	10.1	-1.8	-26.9	-13.2	
Professional, scientific and technical services	13.2	12.3	6.8	7.7	5.3	1.1	-9	
Administrative and voluntary service	20.9	11.5	9.4	9.4	13.5	2.8	-15.9	
Tourism, conference and exhibition services	14.1	18.9	10.1	12.6	12.8	1.7	-5.5	
Computer and information technology services	9.8	6.8	5.3	16.2	2.4	-12.2	-27.5	

Source: Hong Kong SAR Government Financial Analysis and Business Facilitation and Analysis Division, "Half-yearly Economic Report 2009", 2009.8.

2. The Impact of Financial Crisis on Major Industries of Macao

For Macao, the impact of the financial crisis on the industries of Macao is mainly embodied in the gaming industry and tourism.

The financial crisis had its earliest and most severe impact on the booming gaming industry, which not only directly affected the Macao SAR Government's main revenue sources – gaming tax revenue, but also caused concerns in betting industry about the employment situation. From Table 12-5 we can see that due to the impact of the financial tsunami, Macao's gross gaming revenues in the third and fourth quarter of 2008 continued to fall, by 26.2 billion patacas and 24.3 billion patacas respectively. However, because in the first three quarters of 2008, gross gaming revenue had already reached 85.468 billion patacas, even more than 2007 annual total of 83.8 billion patacas, this shows that if it did not reflect the impact of the financial tsunami, Macau gaming gross income in 2008 would have had a greater increase. In addition, in the first and second quarter of 2009, the gross gaming revenue fell by 12.7% and 12.2% respectively compared with the same period of previous year. The continued decline of gross gaming revenues of Macao's gaming industry in the first two quarters of 2009 has shown that the sector still continues to slow down with the impact of the financial tsunami.

Table 12.5. Macao's gross gaming revenue for the period of 2006-2009

Year	Quarter	Gaming gross income (millions Macao Pataca)
2006		57 521.3
2007		83 846.8
2008		109 826.3
2008	1	30 084.9
	2	29 179.2
	3	26 204.3
	4	24 357.7
2009	1	26 252.2
	2	25 618.8

Source: Macao Statistics and Census Service.

After Macao's gaming industry encountered its first winter since liberalization, the overall employment situation of Macao has also deteriorated immediately. A number of gaming companies have announced plans to suspend or slow down investment expansions. Cutting costs and reducing manpower is the most common practice for corporations to respond to the financial crisis, and thus triggering a series of layoffs and payment cuts.

In the first quarter of 2009, gross fixed capital formation experienced a sharp decline of 32.1%, largely due to the suspension of international investments following decline in external demand. For example, the U.S. corporation Sands, was once very famous in Macau gaming industry. Due to the problem of indebtedness and plummeted stock prices caused by the impact of the financial crisis, in November 2008, the corporation put on an emergency stop on the fifth and sixth large-scale projects of Macao Bright Avenue. The number of construction workers who were immediately affected reached 10,000, among whom about 2,000 local workers in Macao were deployed one by one to other jobs, and the remaining more than 9,000 employees from outside were laid off, in addition to internal suggestions to cut payments for employees or ask them to go on leaves without payment. Due to the impact from the parent company Sands, operation and investment plans of a subsidiary of the Venetian Macao gaming were also thrown into a halt.

The financial crisis is sweeping the globe, and the tourism as a pillar of Macau's service industries is inevitably affected to a greater extent. The number of Macao's visitor arrivals and Macau's gaming revenue are in a high positive correlation, with the correlation coefficient as high as 0.809.[4] This shows that changes in the number of visitors and consumer preferences will have a direct impact on Macao's gaming revenue. As Table 12-6 shows, in 2008, the total number of Macao's visitor arrivals was 12.07 million, down by 55.30% compared with the same period in 2007. Among the main sources of tourists, mainland tourist visits to Macao fell by 38.89%. Hong Kong and Taiwan have fallen even more than half, reaching 77.50% and 60.77% respectively. Up until 2009, when the world is facing the financial turmoil and the new flu outbreak is also spreading rapidly around the world, in Macao, during the first 6 months of 2009, the total number of visitors compared with last year's first 6 months has reduced by 11.39%, down to 10.37 million. Based on travelers' original residence, visitors from mainland China reached 5.13 million (accounting for 49.45% of the total number of visitors), declined by 17.31% compared with the same period of last year; among these, mainland tourists on personal tours to Macao significantly reduced by 51% from 0.47 million of last June to a current 0.23 million. In addition, in June 2009, the number of group visitors to Macao has also further reduced to 0.18 million , which, compared with the same month last year, fell significantly by 46%.[5] In the meantime, in June 2009, the number of Macao residents traveling abroad in groups compared with the same month in 2008 fell significantly by 40%. In the first half of 2009, the total number of group travelers of Macao residents traveling outside was 90,000, decreased by 14.0% compared with the same period in 2008.[6]

It is noteworthy that in addition to the sudden drop of the number of visitors to Macao, the consumption by visitors also decreased. In 2008, the total spending by tourists fell by a big margin of 16.7%. In the first quarter of 2009, the average per capita spending of visitors was 1,638 Macao patacas with a reduction of 5% over the same quarter of 2008. The per capita consumption by visitors that stay overnight and visitors that do not decreased 1% and 15% respectively, compared with the same quarter of 2008.[7]

The decrease in visitors and tourist consumers to Macao greatly affected the overall economic development of Macao, which is one of the main reasons of significant drop of GDP in the first quarter. In addition, the industry chain as extension of gaming tourism industry, such as hotel and catering services, wholesale, retail and exhibition industry, also suffered in different degrees because of turmoil in the gaming industry.

4 Zeng Zhonglu, Dong-Mei Zhang. Macau visitors: Analysis and Forecast [M]. Macao Polytechnic Institute publication. June 2007, P5-6.
5 Source: Macau Statistics and Census Service: the number of visitors from entry.
6 Source: Macau Statistics and Census Service: the tour and hotel occupancy rates.
7 Source: Macau Statistics and Census Service: Passenger consumption surveys.

Table 12-6. Total number of inward tourists to Macao based on original place of residence, 2007-June 2009

Origin		2007	2008	Rate of change (%)	January to June, 2008	January to June, 2009	Rate of change (%)
Total (Person time)		27 003 370	12 069 291	-55.30	11 704 139	10 370 979	-11.39
Eastern Asia	The Republic of Korea	225 417	28 079	-87.54	158 500	95 217	-39.93
	China Mainland	14 873 490	9 089 578	-38.89	6 201 538	5 127 933	-17.31
	Hong Kong	8 176 964	1 839 848	-77.50	3 337 162	3 326 335	-0.32
	China Taiwan	1 444 342	566 618	-60.77	672 075	617 545	-8.11
	Japan	299 406	38 965	-86.99	166 779	169 127	1.41
	Others	4 803	2 494	-48.07	2 671	2 370	-11.27

Source: Macau Statistics and Census Bureau, the number of visitors by entry.

SECTION 2: THE MAJOR MEASURES OF HONG KONG AND MACAO IN DEALING WITH THE FINANCIAL CRISIS

Since the outbreak of the financial crisis, Hong Kong and Macao governments have introduced a number of policies and measures, including supporting the banks, improving people's livelihood as well as focusing on solving the employment problems to deal with the impact of financial crisis on the local economy.

I. The Main Measures of Hong Kong Government in Dealing with the Financial Crisis

Facing the tremendous impact of the global financial crisis, in early 2009, Hong Kong Government's overall strategic plan was changed from "Market Leads, Government Facilitates" in 2008 to "stick to 'stabilizing financial system, supporting businesses and preserving employment.'" Hong Kong Government took a number of measures to maintain financial system's stability, ease employment pressure and promote economic development, which shows the SAR government's determination to combat the financial crisis.

1. To Support Banks and to Ensure Financial Stability

The financial sector is one of major pillars of Hong Kong. As a world-renowned international financial center, Hong Kong's financial industry was seriously affected by the rapid spread of financial crisis. In the period from September to October in 2008, Hong Kong interbank interest rates were soaring, which was somewhat eased only after the authorities' provision of liquidity to the banking sector. Subsequently, Hong Kong SAR Government has strengthened the prudential supervision of the financial industry. The HKMA has introduced measures to reduce banks' liquidity pressures, stabilize banks' customer deposit base and enhance confidence in the banking system. These measures made Hong Kong's financial system relatively stable, which reduced the negative impact of the financial crisis. In the first half of 2009, Hong Kong financial industry's order is gradually restored and the rebound in the stock market is more significantly evident.

Hong Kong Monetary Authority took som measures including: foreign exchange operations within the Convertibility Zone and a total of 24.8 billion injections of funds to the banking system; announcing the launch of five temporary measures and providing liquidity support according to demands of licensed banks in Hong Kong; providing term financial support to banks through the discount window or outside the discount window in the case that the licensed banks provide a wide range of collateral with credit quality approved by the board; announcing an adjustment to the Base Rate formula, valid for 2008.10.9-2009.3.31; the Financial Secretary announcing the availability of two preventive measures to strengthen confidence in the banking system of Hong Kong. These two measures are: to provide a temporary full deposit protection, and to establish a contingent bank capital facility. Two measures will remain in force until the end of 2010; implement expansion of the Exchange Fund Bills and Notes Program; publish the measures of providing term loans to licensed banks in the case of individual licensed banks providing credit quality of collaterals which are acceptable to the HKMA, and introduce optimization of the arrangements. Bank of China and the HKMA signed the 200 billion Yuan (227 billion Hong Kong dollars) currency swap agreement, which supports short-term liquidity for the two branches of banks located in one and another when necessary to ensure Hong Kong financial stability. [8]

2. Multiple Simultaneous Measures to Protect the People's Livelihood

Employment is the core of people's livelihood. Since the financial crisis, Hong Kong's overall employment situation has been deteriorating and the unemployment rate broke sustained lowest point in the past decade. To this end, the SAR Government set increasing employment as its main task, and has taken many positive measures to promote employment and protect the people's livelihood.

(A) Financial Support Directly Eases the Predicament

Hong Kong SAR government has increased direct financial support, and announced a number of fiscal measures to ease the predicament in February 2008, July 2008, and May 26, 2009 respectively, , with a total aggregate of approximately 87.6 billion Yuan to allocate resources, which is equivalent to 5.2% of Hong Kong's GDP.[9] At the same time, in Hong Kong Special Administrative Region Government's budget, the section about "short-term

8 Source: Hong Kong SAR Government Financial Analysis and Business Facilitation and Analysis Division, "Half-yearly Economic Report 2009", 2009.8.

9 2009-2010 Annual Budget - Budget Speech http://www.budget.gov.hk/2009/chi/speech.html

strategy - jobs security" was increased, including setting aside 400 million Yuan of non-recurrent funding to strengthen the Labor Department's implemented employment programs; an additional 13 million Yuan to help the Labor Department to actively assist the corporate layoffs and closure-affected employees due to the financial crisis.

(B) Projects Create Employment Opportunities

Hong Kong SAR Government made efforts to promote the construction of infrastructure projects to accelerate and promote 10 major infrastructure construction projects, such as the Hong Kong-Zhuhai-Macao Bridge, Guangzhou-Shenzhen-Hong Kong Express Rail Link, and the MTR South Island Line. These projects involve funding of 250 billion Hong Kong dollars, which can create 250,000jobs, accounting for 7% of total employment. At the same time, Hong Kong SAR Government has actively carried out the constructions of commercial buildings, residential buildings and hotels and building renovation, maintenance and other minor areas of residential construction projects. In the 2009 legislative year, 100 projects have been funded by the Legislative Council with the total cost of 70 billion Hong Kong dollars, which is expected to create 40,000 jobs.

(C) To Make Efforts to Solve Graduates Employment Problems

In February 2009, Hong Kong Special Administrative Region Government announced the first trainee program for university students. The funds of 140 million Yuan for the Training Scheme are mainly provided by the SAR government. The Plan provides for this year's graduating university students in Hong Kong with preferential policies. If a student graduates from college and stays to work in Hong Kong, HKSAR Government will give 2,000 Yuan subsidy per person per month; if the graduate works in the mainland, HKSAR Government will provide 3,000 Yuan subsidy per person per month. After the implementation of the scheme, employers pay a monthly salary to the graduates, including government subsidy and the amount of subsidy paid by businesses, and then the corporations submit rosters to the government to obtain the appropriate funding. It is estimated that the money could be allocated to more than 4,000 students.

3. Vigorously Support the Development of SMEs

SMEs account for 98% of the total number of enterprises in Hong Kong and absorb more than half of the local working population. Under the influence of the financial crisis, banks tightened credit for the sake of self-protection, which leads to many SMEs facing the difficulties of survival. To alleviate the plight of small and medium enterprises, Hong Kong Special Administrative Region Government has taken two major steps to ease the credit problems faced by small and medium enterprises with an aim to strengthen support for SMEs. The first is "Special Finance Scheme for SMEs". Until July 2009, the total loans granted by the plan were about 36.5 billion, benefiting 11,000 companies and more than 290,000 employees. The second is that Hong Kong ECIC deals with export credit insurance applications by SMEs with flexibility, and increases the sum insured; improving six emerging markets' underwriting limits and reducing the premium; shortening the application procedures for small credit limit to 4 days; the ECIC undertakes the supreme law of responsibility from the present 15 billion Yuan to 30 billion Yuan.

4. Structural Change, Cultivate New Industries

Through the promotion of industrial restructuring, Hong Kong SAR Government also found a new economic growth point and developed new industries to deal with financial crises. On the basis of supporting the traditional industries, Hong Kong Special Administrative Region Government makes efforts to cultivate six new competitive industries, including testing and certification, medical services, technology innovation, culture and creative industries, environmental protection industries and educational services. To take cultural and creative industries for example, there are 32,000 institutions engaging in creative industries, including film, television, design, architecture, urban planning, comics, animation, digital entertainment and many other areas. The annual output value is 60 billion Hong Kong dollars, which accounts for about 4% of the Gross Domestic Product. Creative industries have shown unlimited business opportunities and good prospects for development. In this case, Hong Kong Special Administrative Region Government based on the reality and made the creative industry as an important starting point to achieve industrial restructuring and address the financial crisis. In order to establish a creative capital of the Asia-Pacific region , the government set up a "Creative Hong Kong" office to help Hong Kong's creative industries to develop the mainland and international markets, and to display the achievements in order to promote the development of creative industries in Hong Kong. It provides a new impetus for Hong Kong's response to the financial crisis and attempts to seek for continued prosperity and stability in the region.

II. Macao Government's Main Measures in Responding to the Financial Crisis

Macao SAR Government actively takes effective measures to combat the financial crisis. The specific points are as follows:

1. To Expand Public Investment and Introduce a Number of Policies on People's Livelihood and Welfare

In response to the financial crisis, Macao SAR Government has introduced a series of economic measures and policies on people's livelihood, including a modest increase in public investment. In April 2009, SAR Government started to increase input in public construction projects. The "public investment program" soared from 376 million in the first 5 months to 648 million in June of the same year with 154.7 percent surge. In addition, in order to ease the impact of the financial crisis on the real estate market, the SAT government made an introduction of incentives to support the home ownerships in a timely manner. First-time home buyers who purchase housing units with value less than 3 million patacas will receive four per cent interest subsidies, so that Macao's real estate market has not been plunged into recession because of financial crisis. In the aspect of people's livelihood and welfare policy, it will continue to reduce 25% of the occupational tax of the overall employed population; to continue to waive all sales taxes, hawker license fees, market stall rentals, life and non-life insurance stamp duty, bank charges and stamp duty, etc; The SAR government will continue to provide 150 Yuan electricity subsidies for residential units per month per unit; starting from 2009/2010 school year, SAR government has launched a new book allowance system to pay 1,500 Macau patacas allowance per school year to all students of Macau identity. Among a number of policies on people's livelihood and welfare, the more remarkable is that the Macao SAR Government has introduced cash-sharing and voucher scheme. In 2008, the Macao SAR Government disbursed 6,000 patacas and 3,600 Macau dollars in cash grants to permanent and non- permanent local residents respectively, as well as health care vouchers of 500 patacas per person. Although due to the impact from external economic environment, external demand reduction is the main reason of GDP decline in first quarter of 2009, there is actually an increase in domestic demand in Macao. For example, private consumption expenditure raised by 3.4%,[10] reflecting that the released cash-sharing plan and a number of welfare policies have had a positive effect.

2. To Actively Create Employment Opportunities and Protect Local Employment

In order to create more employment opportunities, SAR Government announced the launch of 77 small and medium sized projects to drive 5,500 construction workers jobs. And in 2008, there were 7,000 public housing units that government has been completed or building under construction, which will bring life to employment for construction workers in the near future. In addition, in order to ease the employment pressure, Macao government reduced the number of outside employees to protect the employment of local workers.

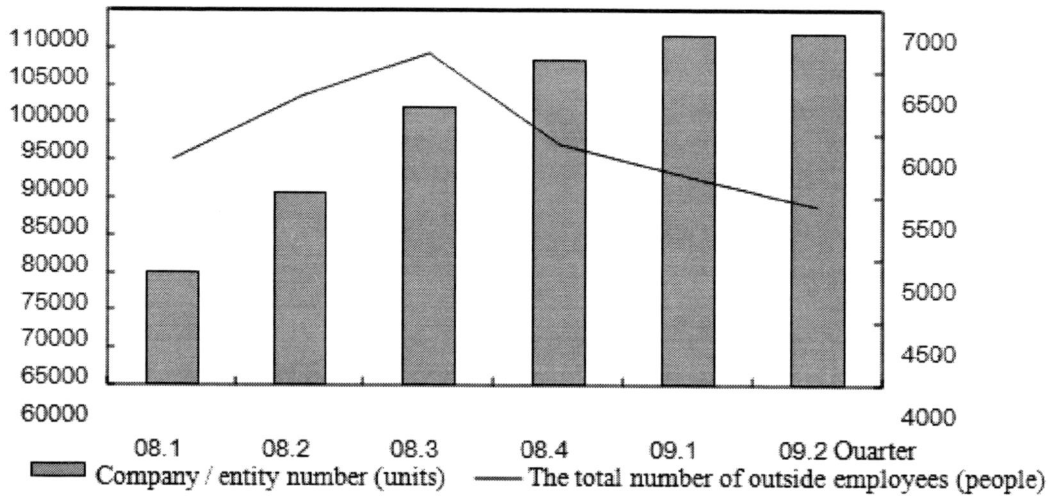

Source: Labor Affairs Bureau of Macao.

Figure 12-1. 2008-2009Q2 number of Macao enterprises / entities and foreign employees on a quarterly basis.

As shown in Figure 12-1, in the fourth quarter of 2008, total number of employees from the outside was 92,000, with a decrease of 11.62% compared with 104,000 people in the previous quarter. Although from 2008 to the first and second quarters of 2009, the number of Macao's business / entities are constantly increasing, and in the second quarter of 2009, the enterprise / entity number rose to 6,804, the number of outside employees declined further to

10 Source: Macau Statistics and Census Service: Gross Domestic Product

84,000 during the same period, mainly because a substantial part of outside employees have been laid off in the financial crisis or have not renewed the contracts after they expired.

3. Strongly Support the Non-Gaming SMEs to Overcome Difficulties

SMEs in Macao have been occupying a pivotal position, but in previous years due to the rapid development of Macao's gaming economy, SMEs are faced with the severe challenges of restructuring and operating difficulties. Coupled with the financial crisis, the situation of SMEs worsened. In this situation, Macao SAR Government emphasized the support for SMEs through difficult times and assisting their restructurings to explore overseas markets. The governmental policies of 2009 announced that SAR Government will allocate 1.5 billion patacas budget to the commercial and industrial development fund to assist small and medium enterprises' development; SAR government will also deliver government proposals to the Legislative Council with an aim to increase the limit for governmental guaranteed loans for small and medium enterprises from banking institutions, from 300 million patacas rise to 3.5 billion patacas.[11] The "14th Macao International Trade and Investment Fair (MIF)" which will be held in October 2009 encourages the participation of Macao SMEs to open up more business opportunities. The government will launch a special enterprise exhibitors' concession for SMEs in Macao, in which each 9 square meters booth costs only 3,600 patacas, which received exhibitors' warm response.

III. The Central Government's Support Measures for Hong Kong and Macao

Hong Kong and Macau were influenced by the serious impact of current financial crisis and the slowdown of global economy. As China's two Special Administrative Regions, the central government attaches great importance to Hong Kong and Macao economic stability and supports Hong Kong and Macao Special Administrative Region Governments to face the challenges of the global financial crisis.

1. Measures to Support Hong Kong

As a strong backing for Hong Kong, the central government attaches great importance to Hong Kong's economic stability and provides Hong Kong SAR full support to properly deal with the challenges of the global financial crisis. The central government has launched financial cooperation, economic cooperation, infrastructure and many other effective measures to promote the comprehensive development of Mainland-Hong Kong economic cooperation and exchanges. In conjunction with Hong Kong, the central government deals with the financial crisis and revitalizes the economy.

11Source: The People's Republic of China Macao Special Administrative Region Government address in 2009 fiscal year.

(A) To Expand RMB Business and Consolidate the Role of Hong Kong as an International Financial Center

In the financial aspect, allowing qualified corporations to conduct RMB business in Hong Kong and make trade payments in RMB will strengthen Hong Kong's status as an international financial center. In the current financial crisis, RMB is gradually growing into a strong currency and demonstrating an international potential. It is foreseeable that, RMB will gradually be internationalized, but at the same time, historical experience shows that any currency changing from a closure currency to a freely convertible currency needs a transition process over time. Hong Kong's prominent position and important commercial value in international economic development mainly result from its proximity to Mainland China which is home to a large population and rich resources. The basic driving forces for Hong Kong's economic development are its roles as a bridge to help China communicate with the world economy and as an intermediary between Chinese and foreign countries. As an intermediary link, Hong Kong has the opportunity to become the center of outside RMB clearing and circulation business. Thus, RMB business in Hong Kong has become the key foreign investment's expectation for its functionality as a financial center. During the financial crisis, allowing qualified corporations to conduct RMB business in Hong Kong will undoubtedly inject new vitality to financial development of Hong Kong trade payments. By May 2009, there were 41 banks offering RMB business in Hong Kong. Hong Kong's total RMB deposits were up to 53.449 billion; Hong Kong banks have transferred a cumulative total of 29.996 billion Yuan to the mainland; the total amount of aggregate liquidation of mainland cards used in Hong Kong was equivalent to 92.85 billion Yuan; the total amount of aggregate liquidation of Hong Kong's RMB cards used in the mainland was equivalent to 5.344 billion Yuan. Five mainland financial institutions including State Development Bank have issued RMB bonds for 7 times in Hong Kong, and the total value amounts to 22 billion Yuan. Hong Kong and Mainland's cross-boundary trade settlements are going forward, which will play a huge role in the consolidation of Hong Kong's status as an international financial center.

(B) The Building of the Hong Kong-Zhuhai-Macao Bridge Enhances the Space Value of Hong Kong

In the aspect of infrastructure, Hong Kong-Zhuhai-Macao Bridge under construction is the biggest highlight. The construction period of Hong Kong-Zhuhai-Macau Bridge project is six years, and the estimated total investment is more than 70 billion Yuan, among which the co-constructed main project has been invested with about 38 billion Yuan. Mainland (including the financial support from the central government), Hong Kong and Macau invested a combined total amount of 15.73 billion Yuan. Hong Kong-Zhuhai-Macao Bridge upon completion is a great bridge connecting Hong Kong, Zhuhai and Macao. The design driving speed is of 100 km per hour and the driving duration from Hong Kong to Zhuhai will be reduced from more than three hours to half an hour. Hong Kong-Zhuhai-Macao Bridge upon completion will not only expand space for Hong Kong's further development, but also improve Hong Kong's own transportation situation, connecting social, cultural and commercial activities with more efficient and convenient transportation system. More importantly, it can strengthen the rail and road transportation links between Hong Kong and neighboring Shenzhen in the Pearl River Delta and enhance the integration between Hong

Kong and surrounding areas to consolidate Hong Kong's position as a international city, thus comprehensively enhancing Hong Kong's space value.

(C) To Promote Service Industry Liberalization and Highlight the Features of Hong Kong's Industries

China's central government continued to promote the openness of services in the mainland to Hong Kong in making economic cooperation between the mainland and Hong Kong closer. According to CEPA Supplement 6, from October 1 of 2009, based on the mainland's commitment stated in the original CEPA and its supplementary agreement on promising liberalization in service trade to Hong Kong, China's central government will further relax market entry criteria in rail transportation, research and development and other fields. The implementation of CEPA further escalated the industrial integration between Hong Kong and the Pearl River Delta region. The main feature is to accelerate the services integration with each other. The entry of Hong Kong's manufacturing service enterprises provides Pearl River Delta's manufacturing industry with an international service system, which accelerated the upgrade of Pearl River Delta region's manufacturing structure and the extension of the service industry chain. Meanwhile, as Pearl River Delta's manufacturing industry aggregates, the demands for financial services, scientific and technological research and development, design management, storage and transportation, logistics and distribution, accounting, legal and other production-related services are increased. The production-related services in Hong Kong will also play an increasingly important leading role, which provides more options to further promote the leader and influential role for Hong Kong's service industries and help consolidate Hong Kong's leadership in finance, trade, logistics and professional services.

(D) Expansion of the "Free Travel" Propelled Hong Kong's Tourism Development

The growth of Mainland residents' "free travel" visits to Hong Kong has provided a strong support for tourism development in Hong Kong. Central Government states that residents of Shenzhen whose official residence in identification book is not Guangdong can apply for the Scheme of Individual Visit to Hong Kong in Shenzhen. Shenzhen residents, whose official residence in identification book is Shenzhen and meet certain criteria, can apply for one-year multiple-entry tourist visa. At the same time, the Central Government decides to increase the number of cities in the Mainland with Individual Visit Scheme. At present, the number of cities whose residents can enjoy "free travel" to Hong Kong and Macao has reached 49; the total number of people has increased to about 250 million to 300 million. The largest-ever "free travel" expansion has brought over one hundred million mainland new tourists resources for Hong Kong.

2. Measures to Support Macao

Policy measures that the Central Government used to support Macau are divided into six specific areas, mainly including: 1. promote and strengthen financial cooperation between mainland and Macao (i.e. expand pilots that offer RMB business in Macao); 2. Speed up construction of related infrastructure projects in Macao; 3. Promote and deepen economic cooperation between Guangdong and Macao; 4. Help ease the operational difficulties of Macao-owned small and medium enterprises; 5. Further expand the opening up of service industries to Macau; 6. Make every effort to ensure food supply security for Macao.[12] In the six major policy measures, the focuses of social concern have been encouraging Macao's participation in the development and construction of Heng Qin Island, the Hong Kong-Zhuhai-Macao Bridge, and deepening economic cooperation between Macao and Guangdong.

First of all, the development of Heng Qin promotes moderate economic diversification. In June 2009, the State Council passed "HENG QIN overall development plan" in principle, and proposed to attach great importance to Macao's industry docking, including the focus on the development of tourism and leisure industry which docks and sometimes dislocate with the Macau gaming industry. The plan also includes emphasis on the development of the modern service industry to match with tourism services; supporting the relocation of the University of Macau in Heng Qin; emphasis on the development of science, research and development industry adapting to the plan; exploring new cooperation mechanisms for technology innovation within Macao-Zhu Hai based on the platform of Heng Qin. At the same time, the central government has also agreed on the relocation of University of Macau to Heng Qin on the request of Macau SAR Government. Heng Qin Island, which covers an area of 1.09 square kilometers, will be leased to the Macao SAR Government as a new campus of Macau University in the way of leasing. NPC Standing Committee made a breakthrough in the management of authorization. The project becomes the first landmark project for Heng Qin development.[13] The important strategic value of Heng Qin development lies in its ability to cover up the shortage of Macao's land resources and address the issue of the overly-single industrial structure, and provide new opportunities to promote the diversity and sustainable development of Macao's economy. With Heng Qin, Macao can overcome the limitation of regional space and integrate into the greater region, so that the whole industry chain can be extended. The hotels, wholesale and retail industries as well as the new convention and exhibition industry associated with the gaming industry are the main direction to Macao's economic diversification in recent years. Based on the existing industrial resources, to extend or expand the industrial chain is the root of resolving the problem of Macao's economic diversity.

Second, Hong Kong-Zhuhai-Macao Bridge project will effectively strengthen the tripartite inter-regional cooperation. The initiation of Hong Kong-Zhuhai-Macao Bridge marked a new orientation for Pearl River Delta regional cooperation, from Macau-Zhuhai or Hong Kong-Macao cooperation in the past to the entire Pearl River Delta regional cooperation, which really links the three together for better development. Take tourism for example, after the opening of Hong Kong-Zhuhai-Macao Bridge, tourist industry of the three places gradually shifted from the local diversified development of tourist products to diversified development of regional cooperative tourist products. As for Macao, it can take its comparative advantage and focus on gaming and entertainment, history and culture, special exhibition and feature cuisines to expand the advantages of related projects. Macao's tourism positioning will not be limited to the land of 28 square kilometers, but as part of the overall regional tourism. Guangdong, Hong Kong and Macao will develop middle- and long-term

12 .Local events recalled in 2008, the Macao Radio and Television Co., Ltd. http://www .tdm.com.mo/c_news/news.php?p=recalling2008

13. Regional cooperation group of Macau Development Strategy Research Center, Heng Qin development and new opportunities for Macau - "HENG QIN overall development plan,", P.2-9 .

target markets for inbound tourism in the future. Visitors will not be limited to multi-day visit to Hong Kong and Macao; their itinerary can include Hong Kong and Guangdong Province. The completion of Hong Kong-Zhuhai-Macao Bridge is bound to further promote the overall economic activity in regional tourism development. [14]

Column 12-1. Hong Kong-Zhuhai-Macao Bridge

The construction concept of Hong Kong-Zhuhai-Macao Bridge was originally proposed in 1983.

The Hong Kong-Zhuhai-Macao Bridge in planning is a large bridge connecting Hong Kong, Zhuhai and Macao in the form of highway bridges. The starting point is Hong Kong's Dayu hill. Passing Tai O, across the Pearl River estuary, and divided into Y-shaped, one end of the bridge connects Zhuhai, and the other connects Macau.

The bridge is designed to be about 29 km throughout with the standard of six-lane expressway and the vehicle speed of 100 km per hour. Upon completion, it will closely follow Lake Pontchartrain Bridge and Ningbo Hangzhou Bay Bridge, and become the world's third longest bridge.

Hong Kong-Zhuhai-Macao Bridge is estimated to cost 2 billion U.S. dollars, among which ports and connection cables are invested by the governments of Guangdong, Hong Kong and Macao with a total investment of about 35 billion Yuan; the total investment of the main bridge is 37.6 billion Yuan, among which the Central Government and Guangdong, Hong Kong and Macao government jointly invested 15.73 billion Yuan, and bank loans consist of approximately 22.0 billion Yuan.

The start of Hong Kong-Zhuhai-Macao Bridge construction is set on December 20, 2009.

Third, the interactive collaboration between Guangdong and Macao is the basic approach to maintain sustainable economic development of Macao. First, Guangdong-Macao cooperation ensures the basic livelihood of the residents of Macao. For example, the "Water Transfer from the Western Region to the Eastern Region" project in 2006, to a certain extent, improved the water supply situation in two places during the salt tide. In addition, Macao government and southern electricity companies are scheduled for electricity purchase project in 2015. By then Macau can purchase 10% power from Guangdong Province. Furthermore, Guangdong and Macao's trade and economic cooperation is mainly embodied in the "9+2" cooperation and Zhuhai-Macao cooperation. The proposal of "9+2" concept can provide better protection to the sustainable development of the Pearl River Delta in Guangdong Province and a lot of space for Macao's economic restructuring and development diversification. Macao Government positions Macau as a service center in the western region of the Pearl River Delta and develops Macau into an attractive gaming center in the world and a platform for regional business service cooperation. Because of this, it is necessary for Macao to take full advantage of the opportunities of the Pan-Pearl River Delta economic integration to expand its space. Especially under the influence of the financial tsunami, it

14 Tang Jizong, the discipline of the fourth meeting of Macao and Zhuhai Development Forum, Macau Research No. 50, P.185-186.

should leverage its advantage as free port and actively play the role of a platform for foreign countries such as the European Union or Portuguese-speaking areas to improve its overall competitiveness and industrial diversification. On the basis of mutual benefit, Macao will expand the scale of its economy and ensure genuine sustainable development.

To sum up, Heng Qin development, Hong Kong-Zhuhai-Macao Bridge project, and Guangdong-Macao cooperation are all proposals on regional cooperation. The ultimate goal of regional economic cooperation is to build economic growth based on a number of economic development points, and explore new growth opportunities in a timely manner, so that the region's economic development can be continued and diversified. As for Macao, actively seeking and participating in regional cooperation is not just in response to the financial crisis. The consideration of how to make the appropriate extension of the gaming tourism industry chain through regional cooperation, and the realization of industry diversification are both the inevitable choices for Macao's sustained economic development.

CONCLUSION

The global financial crisis brought Hong Kong's economy a huge trauma. The overall economy, employment, import and export are all facing the impact. As an open economic system, Hong Kong is highly integrated with the world economy. Therefore, the future performance of Hong Kong's economy is greatly influenced by external environment. However, from the above analysis we can see that under effective financial supervision, Hong Kong is in a more stable financial order. With the resumption of economic growth in the mainland as well as slightly stabilized external environment, Hong Kong's economy has improved in the second quarter of 2009. In the future, Hong Kong's economic recovery has not been solid yet, the adverse effects of global financial crisis on Hong Kong has not weakened, the situation of sharp decline in external demand continues, and the production and management of a number of industries and enterprise are still relatively difficult. However, as the government has introduced measures for financial markets stabilization, enterprise support, job creation, in addition to the further growth of the mainland economy, the impact of the measures that the central government has introduced to support Hong Kong's economic development is gradually showing up. We have reasons to believe that in the current financial crisis, Hong Kong's long-term economic prospects remain optimistic.

Macao's economy has also been tremendously influenced by the global financial crisis. As we mentioned earlier, from the fourth quarter in 2008 to the first two quarters in 2009, a number of Macao's major economic indicators continued to drift downwards, including GDP, gross fixed capital formation, exports of goods and gaming gross income, etc., which reflects that Macao's economy has not yet come out of the haze of the financial tsunami. As a highly externally-oriented micro-economic system, in addition to the impact of the reduction of external demand on Macao's economic slowdown, other factors include Macao's own structural problems, the continued spread of H1N1 influenza and the tightening central policy for Macao's "free travel" (the abolition of simultaneous access to the Mainland visitors with Hong Kong visas to visit Macau without another sign), and so on. The opening-up of Macao's gaming industry has brought rapid economic growth and many challenges, such as pressure from excessive visitors and infrastructure bottlenecks. The impact of financial crisis on

Macao's economy further demonstrates the high risk of depending on a single industry. Therefore, the financial crisis provided a opportunity for Macao's heating economy to take a breath and adjust the industrial structure, improve infrastructure and actively leverage the surrounding resources and markets to make up for its own shortage. In the long run, this is undoubtedly an opportunity to expand markets and achieve moderate diversification. In general, Macao should make development strategies for relevant industries based on its specific situation, and strive to minimize the negative impact of the current financial crisis. We believe that Macao has the ability and confidence to achieve this.

Translator: Shi Jing (SUFE)

English Version Editors: Lv Mei (SUFE), He Yiming (Harvard)

REFERENCES

Since CEPA has been signed for six years, the average annual growth of the total import and export in Hong Kong and mainland is 18.4%.

http://finance.sina.com.cn/roll/20090625/02086395193.shtml

Hong Kong SAR Government Census and Statistics Department - Hong Kong Statistics

http://sc.info.gov.hk/gb/www.censtatd.gov.hk/home/index_tc.jsp

China Financial Planning net - " financial business income of Hong Kong's banking plummeted nearly 30% in the crisis "

http://www.51cfp.net/CFP/skill/200812/32729.html

Tencent Finance - "the crisis affect Hong Kong's financial industry, business incomes significantly shrunk in size in the first quarter "

http://finance.qq.com/a/20090612/017536.htm

China Travel Monthly net - Hong Kong tourism review in 2008 and outlook in 2009 http://www.hkcts.com/yuekan/200904/ywsx1.htm

China News net - "cross-border logistics industry has been severe influenced by the financial crisis with outage of 4000 trucks" http://www.chinanews.com.cn/ga/fzsj/news/2009/03-23/1613608.shtml

Market research report net - "Hong Kong real estate was hit by financial crisis,"http://www.164.com.cn/industrynews/200811/xianggangfangdichan301315.htm

2009-2010 Annual Budget - Budget Speech

http://www.budget.gov.hk/2009/chi/speech.html

People's net – "Central Government introduced 14-point package to support Hong Kong" http://politics.people.com.cn/GB/1026/8550889.html

Pan-PRD co-operation Information net – "the central government and Hong Kong join hands to take positive measures to cope with the global financial crisis" http://www.pprd.org.cn/zhuanti/twelve/dkjrwj/200906/t20090630_61676.htm

He Zhong, Xie Ya-xuan, Zhao Wen-li, 2008. Overseas macroeconomic Comments - the Central Government put forward 14-point package in seven aspects to support Hong Kong to cope with the financial crisis [R] *China Merchants Securities (HK) Macro report*,12:5-6

Baidu Wikipedia – "Hong Kong-Zhuhai-Macao Bridge"
http://baike.baidu.com/view/157066.htm

China net – "Tung Chee-hwa: Pan-PRD regional cooperation is conducive to play our strengths"
http://www.china.com.cn/chinese/2004/Jun/576363.htm

Hong Kong International Investment Promotion Council's official website - "the central government added 1.6 billion to help small and medium enterprises crack financing difficulties"
http://www.hkiipa.com/shownews.asp?id=246

The People's Republic of China National Development and Reform Commission's official website - "CEPA Supplementary Agreement 6 was signed in Hong Kong and 9 measures are on a pilot in Guangdong "
http://www.sdpc.gov.cn/dffgwdt/t20090512_278135.htm

Interactive Wikipedia - "CEPA"
http://www.hudong.com/wiki/CEPA

Joint Morning paper – "signing CEPA Supplement 6, the mainland of China opens up services trade and financial and securities industry to Hong Kong"
http://www.zaobao.com/special/china/hk/pages2/hk090510.shtml

Xinhua net – "motherland backing is as strong as a mountain- the documentary of Central Government supporting prosperity development of Hong Kong's economy"
http://news.xinhuanet.com/gangao/2009-06/30/content_11627944_3.htm

Li Xiaoxi, in April 2009. International Symposium on Macao's return for a decade: My understanding of moderate diversified development of Macau's economy [R]. Beijing Normal University, Institute of Economic and Resources Management.

Wu Wenshi, January 2009. Macao Study Series: The study of the scale and development speed of Macau gaming industry and tourism [R]. Beijing Normal University, Institute of Economic and Resources Management.

Economic News, July 31, 2009. SMEs participate in exhibitions [N]. Macao Daily News, A10 economic.

Ryu Ji-Yi. The Problems and Solutions of Human Resources of Macao Special Administrative Region in financial tsunami [R]. Macau Research No. 51, P.102-107.

Tang Jizong, the minutes of the fourth meeting of Macau and Zhuhai Development Forum [R]. Macau Research No. 50, P.185-186.

Guo Yongzhong. The challenges and opportunities faced by the third term Chief Executive [R]. Macau Research No. 50, P.155-156.

Liang Angel. Local events Recalled in 2008 [Z]. Macau, Radio and Television Limited.
http://www.tdm.com.mo/c_news/news.php?p=recalling2008

Zeng Zhong-Lu, Zhang Dong-Mei, 2007. Macau visitors: Analysis and Forecast [M]. Macao Polytechnic Institute Publishing.

Regional cooperation group of Macau Development Strategy Research Center, Heng Qin development and new opportunities for Macau - "HENG QIN overall development plan", P.2-9.

Macau Research Editorial Board. "Macau and Zhuhai Cooperation and Development under 'One Country Two Systems'" Round Table Minutes [R]. Macau Research No. 51, P.155-162.

ZHANG Yang& KWAN Fung. MACAO'S ECONOMY IN THE FINANCIAL CRISIS[J], EAI Background Brief No.456, P.1-18.

In: China's New Deal
Editors: Li Xiaoxi and Hu Biliang

ISBN: 978-1-61470-815-5

Comments on Chapter 12

HOW HONG KONG AND MACAO TACKLE THE 2008 GLOBAL FINANCIAL CRISIS

Sun Yong*

The Chapter 12 of the book describes the impact on the economy of Hong Kong and Macao by the 2008 global financial crisis, and how the Hong Kong and Macao local governments deal with the global financial crisis. Finally, the policies issued by the Chinese central government to support the economy of Hong Kong and Macao are presented. There are three parts in this chapter: the first part describes how the 2008 global financial crisis had effect on the economy of Hong Kong and Macao. When describing the impact, the authors made the analysis from two aspects: one is the whole economy of Hong Kong and Macao; another is the main industries in Hong Kong and Macao. The second part introduced the detailed measures adopted by Hong Kong and Macao local governments to deal with the 2008 global financial crisis, including supporting the banking industry, reducing the unemployment rate, increasing the public finance, and improving the welfare of the residents. The third part is the measures adopted by the Chinese central government to support Hong Kong and Macao, including strengthening the status of Hong Kong as an international finance center by increasing RMB business, building a bridge across the sea to connect Hong Kong, Zhuhai and Macao, and accumulating the development of tourism in Hong Kong and Macao by permitting more tourists from mainland China to travel freely in Hong Kong and Macao. In general, there are three outstanding features in this chapter:

First, the authors use detailed data, tables and figures to describe the significant impact on Hong Kong and Macao's economy brought by the global financial crisis. By using those detailed data, tables and figures, the authors present a picture of Hong Kong and Macao's economy. Based on the detailed data, tables and figures, we find that not only the whole economy of Hong Kong and Macao, but also the major industries in Hong Kong and Macao decline significantly. For example, the table 12-4 in this chapter reflects the percentage change of revenue in the service industry in Hong Kong. From the table, we can easily note

* Sun Yong, PhD, Old Dominion University, USA, Associate Professor of Finance, Research Institute of Economics and Management, Southwestern University of Finance and Economics, China.

that almost all industries decline significantly since the fourth quarter of 2008, especially the finance industry, declined by about 60 percent. These data strongly explain that the negative impact of the global financial crisis on the economy of Hong Kong and Macao. As an international financial center, Hong Kong's finance industry received the largest loss. Another example is the table 12-6. The table presents the change of the number of tourists in Macao. According to the data of the table, the number of tourists visiting Macao declined significantly since 2008. Based on a lot of data, tables and figures, the authors strongly support the conclusion of this chapter, that is, as the open economy of Hong Kong and Macao, almost all indexes the economy declined significantly, not only including the whole economy of Hong Kong and Macao, but also including the main industries. The global financial crisis has a huge negative effect on the economy of Hong Kong and Macao. Thus, for both Hong Kong and Macao local government and the Chinese central government, it is necessary to apply powerful measures to deal with the negative impact brought by the global financial crisis.

Second, in this chapter, the authors introduce the new deals adopted by the Hong Kong and Macao local governments and the Chinese central government step by step. It provides a very clear clue for the authors to understand the reasons behind those new deals. It helps the readers to better understand it is very necessary for the governments to take very quick and powerful actions to deal with the negative effect of the global financial crisis. At the beginning of this chapter, the authors made analyses of the negative impact brought by the global financial crisis. In this aspect, the authors focus on not only the whole economy of Hong Kong and Macao, but also the representative industries, such as the banking industry. When describing the economy of Hong Kong and Macao, the authors cited a lot of data, tables, and figures, just like a picture showing us how the global financial crisis has effect on the economy of Hong Kong and Macao. Next, the authors respectively introduce the measures adopted by the Hong Kong and Macao local governments for dealing with the global financial crisis. Due to the different economic structure between Hong Kong and Macao, the measures used by the Hong Kong and Macao local governments are also different. As an international finance center, Hong Kong government focuses more on the stability of financial markets, the protection of residents' benefits, the development of small and medium-sized enterprises (SMEs); while the Macao government focuses more on expanding the public finance, improve the welfare of residents, and the development of non-gaming SMEs. Although the measures adopted by Hong Kong and Macao local governments are different, both the Hong Kong government and Macao government focus on the protection of residents' benefits and the development of SMEs. The last part of this chapter mentioned the policies issued by the Chinese central government, including expanding business involving RMB to strengthen the status of Hong Kong as the international finance center, developing the tourism of Hong Kong by permitting more people from mainland China to travel in Hong Kong freely, constructing the Hong Kong-ZhuHai-Macao bridge, and developing Heng Qin Island. In particular, the Chinese central government emphasizes the cooperation between Hong Kong and Macao and Pearl River Delta (PRD) region. Furthermore, the Chinese central government develops the transportation to reduce the distance between Hong Kong Macao and mainland China to become the basis of further link and cooperation in economy between Hong Kong Macao and mainland China. The structure of this chapter is as follows: at the beginning of this chapter the authors introduce how the global financial crisis affects the economy of Hong Kong and Macao. Next, the powerful measures adopted by the Hong Kong and Macao governments are presented. Finally, the policies issued by the Chinese central government to support the economy of Hong Kong and Macao are introduced. The step by step structure of this chapter like this makes readers better understand and accept the reasons behind the policies issued by the Hong Kong and Macao local governments and Chinese central government.

Third, through this chapter the authors focus on the major clue, that is, the policies issued by both the Hong Kong and Macao local government and the Chinese central government, which is the most important part among the measures to deal with the global financial crisis. One of the most significant features of this chapter is that all topics are related to the measures and policies issued by the government. This is because the government is the most important entity when dealing with the global financial crisis. Not just because the government has the most resources and is more powerful than other entities like firms and individuals, but also because the government usually thinks about the problem from the standpoint of the whole society., not from the standpoint of any firm or individual. Therefore, following the policies and measures adopted by the government is equal to following the most significant part among all measures dealing with the global financial crisis. This chapter first describes the impact on the economy of Hong Kong and Macao brought by the global financial crisis. Based on the description of the impact, the authors introduce all kinds of specific measures and policies issued by the Hong Kong and Macao local governments and the Chinese central government. From some specific industries to the whole economy, from the Hong Kong and Macao local governments to the Chinese central government, the authors deliberate the efforts made by the Hong Kong and Macao local governments and the Chinese central government to deal with the global financial crisis. When presenting the specific policies and measures adopted by the Hong Kong and Macao local governments to the Chinese central government, the authors describe the difference in the measures and policies issued by the Hong Kong and Macao local governments to the Chinese central government, and also show us the common points in those measures and policies. For example, as the international finance center, Hong Kong's local government focus more on the stability of the finance industry, the protection of residents' benefits, and the development of SMEs; while the Macao local government focuses more on expanding the public finance, the improvement of the welfare of residents, and the development of non-gaming SMEs. Although the measures adopted by Hong Kong and Macao local governments are different, both the Hong Kong government and Macao government focus on the protection of residents' benefits and the development of SMEs. The last part of this chapter mentioned the policies issued by the Chinese central government, including expanding business involving RMB to strengthen the status of Hong Kong as the international finance center, developing the tourism of Hong Kong by permitting more people from mainland China to travel in Hong Kong freely, constructing the Hong Kong-ZhuHai-Macao bridge, and developing Heng Qin Island. In particular, the Chinese central government emphasizes the cooperation between Hong Kong and Macao and Pearl River Delta. Furthermore, the Chinese central government develops the transportation to reduce the distance between Hong Kong Macao and mainland China to become the basis of further cooperation in economy between Hong Kong Macao and mainland China. The Chinese central government focuses more on the whole economy of China, including mainland China, Hong Kong and Macao. The policies issued by the Chinese central government focus more on the further cooperation between mainland China and Hong Kong and Macao, which helps Hong Kong and Macao to deal with the negative impact brought by the global financial crisis. At the same time, the opening economy of Hong Kong and Macao helps the development of mainland China. Hence, in this chapter the authors focus on the main clue, that is, the measures and policies issued by the Hong Kong and Macao local governments and the Chinese central government. In this way, the readers can feel the significant and key role played by the government in dealing with the global financial crisis.

Although the features of this chapter are outstanding, we also find this chapter ignored something important. This chapter focuses only on the measures and policies issued by the Hong Kong and Macao local governments and the Chinese central government, however, it ignored any measures adopted by other entities, such as banks, firms, or even individuals. From the viewpoint of microeconomics, banks, firms, and individuals also adopt some measures to deal with the global financial crisis in order to protect their own benefits. For example, the banking industry may strengthen the risk management to reduce the possible loan losses. Individuals may save more money to deal with possible risks in the future. In essential, all the measures and policies issued by the Hong Kong and Macao local governments and the Chinese central government will benefit the banks, the firms, and individuals finally. The way includes improving the liquidity of the banking industry, promoting individuals' consumption, and finally developing the economy. Furthermore, the effect of the measures and policies issued by the Hong Kong and Macao local governments and the Chinese central government also depends on the benefits of banks, firms, and individuals. The most important thing is the expectation and confidence of banks, firms, consumers about the future of the economy. Only after consumers are more optimistic and would like to buy more products or services, the market can recover and the economy can go upward.

As a free and open economy, Hong Kong and Macao are inevitably affected by the 2008 global financial crisis stemming from the US sub-prime mortgage meltdown, resulting in moderation in exports and hence dragging the overall economic performance. In the long run, to combat possible financial crisis in the future, the government, the industries or firms, and even individuals need to make further efforts. In particular, the government is the most important institute in tackling the global financial crisis.

First, Hong Kong and Macao local governments need to maintain a healthy and stable financial system, help design better systems in risk management, and improve the ability to defend any possible global financial crisis in the future. The economy of Hong Kong and Macao is very free and open, which means the government seldom intervenes the market. Nevertheless, economies need to take extraordinary measures at exceptional times like this and the Government should be prepared to intervene in a timely and decisive manner. To support small and medium-sized enterprises (SMEs), the credit crunch has caused great difficulties among SMEs. In early November 2008, enhancement measures to SME Loan Guarantee Scheme and SME Export Marketing Fund were implemented; A time-limited \$100 billion Special Loan Guarantee Scheme (SpGS) was launched on December 15, 2008 to provide up to 70% guarantee to participating lending institutions granting loans to Hong Kong companies; Under the SpGS, the maximum amount of loan that each enterprise may obtain is \$6 million, within which \$3 million can be used as a revolving credit line such as commercial overdraft and letter of credit. All companies except listed companies may apply. To continue public spending and stimulate domestic demand, despite the expected economic downturn, the Hong Kong and Macao Government do not cut public expenditure, especially its commitments in healthcare, social welfare, education and improving the environment. To boost employment, 60,000 jobs would be created in 2009 through infrastructure projects, advanced recruitment of civil servants, creation of temporary positions and expediting minor works in the 18 districts. Since Hong Kong and Macao are very open to the world, the economy of Hong Kong and Macao is also easily affected by any possible financial crisis from other countries or areas. Thus, a healthy and reasonable regulatory framework set by the government is the most important basis of defending any possible global financial crisis in the future.

Second, a strong backing from the Chinese central government is very necessary for Hong Kong and Macao to combat the global financial crisis. In fact, the Chinese central government has been always the strongest backup for Hong Kong and Macao. The Chinese central government issues a series of policies to support the economy of Hong Kong and Macao. For example, On December 20, 2008, the Chinese Central Government announced a 14-point package to bolster Hong Kong's economy. The package is made in the interest of Hong Kong and is also an advancement of the development of the nation as whole. Key measures include: (1) Strengthen cooperation between the Mainland China and Hong Kong in financial services sector. (2) Allow qualified enterprises to settle their trade obligation in Hong Kong in RMB, making the city one of the pilot areas for transforming RMB into an international currency. (3) Allow currency swap arrangements between the People's Bank of China and the HKMA, a measure that will enable each side to help stabilize the other's currency if necessary. (4) Encourage Mainland enterprises to use Hong Kong as a platform to develop international businesses. (5) Support the listing of Mainland firms in Hong Kong. (6) To expedite infrastructure projects related to Hong Kong, such as the Guangzhou-Shenzhen-Hong Kong Express Rail Link, the Hong Kong-Zhuhai-Macao Bridge and the rail link between airports in Hong Kong and Shenzhen.

Third, as ignored by the chapter 12, further cooperation among banks, corporations, and even individuals between mainland China and Hong Kong Macao are also very important. There are several favorable policies issued by the Chinese central government to enhance economic cooperation between Hong Kong Macao and the Pearl River Delta and even the whole mainland China. For example, to alleviate difficulties faced by Hong Kong SMEs on the Mainland, the Chinese Central Government will further adjust the tax rebate rates for exports, improve the labor system to reduce burden on enterprises, facilitate the development of domestic market for processing trade enterprises and develop a guarantee mechanism for SME financing. Other policies include expanding the Individual Visit Scheme, further the opening up of the Mainland services sector to Hong Kong service providers through CEPA, and enhancing economic co-operation between Hong Kong and the Pearl River Delta. In fact, more and more companies from Hong Kong and Macao invest in mainland China, and people in mainland China know Hong Kong and Macao better since 1997. The closer link and tighter cooperation in all kinds of aspects between Hong Kong Macao and mainland China help Hong Kong and Macao to tackle the global financial crisis and promote the development of Hong Kong and Macao. A very good example is that during the 1997-98 Asian financial crisis, the China's central government issued a policy to permit individuals from mainland China to visit Hong Kong and Macao freely. The policy brought Hong Kong and Macao a lot of money and helped Hong Kong and Macao recover from the 1997-98 Asian financial crisis rapidly, comparing with other countries or areas suffering by the 1997-98 Asian financial crisis, such as South Korea, Thailand, and Indonesia.

Finally, Hong Kong is known for its resilient, hardworking spirit and will endeavor to turn crisis into opportunity by vigorously pursuing economic development in areas such as financial services, cross-boundary integration, infrastructure projects, creative industries and scientific research. As long as Hong Kong and Macao continue to uphold the principle of "Big Market, Small Government" in promoting economic development, with sound fundamentals, a strong regulatory framework, a robust Exchange Fund, a production base and market on the Mainland and a strong backing from the Central Government, Hong Kong and Macao's economy is well equipped to withstand the downturn. In a word, the chapter 12 of this book provides us with a picture of describing how the Hong Kong and Macao local governments and China's central government tackle the 2008 global financial crisis.

REFERENCES

"Asia's Las Vegas, Macau, Fights The Economic Crisis", *The Independent*, January 2009

"HK Stock Market Capitalization Halved in 2008", *Xinhua net*, 2009-03-03

Hong Kong Census & Statistics Department

Hong Kong's Economy in the Financial Crisis by Zhang Yang and Sarah Y. Tong, April 2009, *East Asian Institute Background Brief No. 448*

"How Hong Kong Tackles Financial Crisis", Global Financial Crisis IPA December 2008

Macao's Economy in the Financial Crisis by Zhang Yang and Kwan Fung, June 2009, *East Asian Institute Background Brief No. 456*

"Macau economy: Unemployment creeps up", *THE ECONOMIST INTELLIGENCE UNIT*, April 2009

Macao Special Administrative Region Economic Services

Statistics and Census Service (DSEC), Macao SAR Government

POSTSCRIPT

It has taken three months to finish this book. From the beginning of July to early October, I was just like leading a troop fighting in the battlefield to the extent that I almost lost my breath. I can still remember that we put together the preliminary outline of this report on July 9th. Then on July 12th, a meeting was held among people in charge of sub-projects to amend the outline. Another meeting was held among leaders responsible for each chapter on the 22nd of the same month, to pin down the division of work and requirements on writing style. On August 28th, having finished the first draft of all chapters, all members on this project got together to discuss about it. Based on the proposed changes, all groups finalized their first draft by September 9th, which were handed over to the scholars of Southwestern University of Finance and Economics to be commented. Then the comments and the revised draft of each chapter were handed in on September 22nd for a general review. During the National Day holidays, I finished the General Remarks of this report, along with the final reviewed draft of the third volume of The Handbook of Natural Resources and Energy Economics. I went over the draft for a last time with Editor Qiu Yuefang from Encyclopedia of China Publishing House on October 12th, and supplemented it with the preface and the postscript.

As I submit this copy, I would like to first extend my thanks to New York based Nova Science Publishers, Inc. in the U.S. and Encyclopedia of China Publishing House in China. They signed the publishing contracts with us on July 6th and 8th respectively, with favorable conditions, and agreed on delivery of the Chinese and the English drafts to the publishing houses in October and November respectively. That started the engine for us to begin this project. In particular, I am deeply indebted to Frank Columbus, Chief Editor of Nova Science Publishiers, Inc. and Guo Yinxing, Director of the Academic Works Department of Encyclopedia of China Publishing House. They have given us their utmost support and trust by allowing us to determine and amend the outline, as well as to specify the content all by ourselves. We owe them our success if this report is recognized by the society.

I also wish to sincerely thank the leaders of both Beijing Normal University and Southwestern University of Finance and Economics, for they have extended great to us on researches and writings of this project, and provided us enough funds to carrying this work out smoothly. Professor Shi Peijun and Professor Bian Huimin, both vice presidents, were delegated by the management of the two universities to supervise this project, thereby promoting coordination and speeding it up.

Here I would also like to express my thanks and congratulations to all experts, teachers and students on this project. Thank you for your efforts and cooperation in good faith to ensure that the Chinese and English drafts were finished in high quality. Scholars in charge of each chapter are all very serious about the work. They tried to, introduce the current situation of China's economic development comprehensively and profoundly in the context of the financial crisis, marking the highlights of China's economic development, analyzing the most recent and influential economic incidents and cases, pointing out the policy orientation and trend of development, and using tables, figures and columns to make the articles easier to understand. I really appreciate the hard work and dedicated teamwork by each and every scholar, and would like to congratulate you all for climbing up the ladder of academic research. Here I would like to make a special introduction of Professor Hu Biliang, who has just completed his joint research project at Harvard University. His performance at organizing this project really impresses me. He is quick thinking, passionate, diligent and extremely efficient. He has painstakingly and efficiently advised on the title, the outline, the writing style, and the revision on the comments. Revised drafts from him at Harvard University were often received at midnight in U.S. time. To make a joke of his name, "the moon from ancient times hangs over Beijing, while one light is sure to be on at Harvard." (In Chinese, the family name "Hu" can be divided into two parts, "gu" and "yue", which means "ancient" and "moon", while the first name "Bi liang" can be interpreted as "sure to be" "turned on.")

Coordinated by Dean Gan Li and Dean Liu Fangjian, and organized by Dr. Fan Gangzhi, a group of overseas educated scholars at Southwestern University of Finance and Economics took an active role in the commentary work of this report. They associated the current situation at home with that abroad to make comments from an international perspective. Their work was finished in a timely and efficient way. Furthermore, I would especially like to express my thanks to Dr. Liu Yuanli from Harvard School of Public Health, and Mr. Huang Jiefu who is the vice president of the Chicago Climate Exchange, and Professor Mou Pu from the School of Life Sciences, Beijing Normal University. They have accorded great importance to this joint program and written some fantastic critical comments on parts of this report, which significantly added to the academic value of our book.

There are too many people to be appreciated and praised, and I will list some of them: Editor Qiu Yuefang, who is extremely responsible; Madam Nadya Gotsiridze and Stephanie Gonzalez from Nova Science Publishiers, Inc. who are highly efficient with their jobs; the teachers and students from the Translation Center of Southwestern University of Finance and Economics who are engaged in the Chinese-English translation, including Jiang Jiahui, Lv Mei, Xiao Rong et al.; Associate Professor Zeng Xuewen, who has been working so hard in organizing and coordinating this project; Mr. Wang Jianing, Chief Editor of the magazine Reform, who put forward some valuable amendment suggestions; Dr. Zhao Shaoqin, who gave us inspiring suggestions on the improvement of the outline; Dr. Wang Nuo, who put together the invitation for comments on behalf of me; Dr. Zheng Yanting, Dr. Ren Ran and other teachers and students, who translated the paper on Emerging Market Economies with great efficiency; Liu Tao, Tan Song, Wang Xuelei and Zhang Kun, who provided important and timely material for this report; Song Tao and Zhang Ai, who assisted in translating NOVA's publishing contract with us and Nova Science manuscript Form and Style Guideline; Jia Limin, Wang Ying, Yan Ling et al., who worked toward allocation of the project fund and reimbursement.

I would also like to attribute special credit to my assistants Lin Yongsheng and Liu Yimeng, who worked as liaison for the project. Lin Yongsheng is in charge of communication between all authors and commentators, and worked with a great sense of responsibility and high efficiency. Liu Yimeng is in charge of communication with the Nova Publishers, including filing applications, translating the contract, submitting our manuscript, etc. with care and patience. Both of them also assisted me with the completion of another article about Emerging market: its orientation, new developments, and new investment at the request of Nova Publishers, which is the prior cooperative work with Nova before this report.

I would particularly express my sincere gratitude to a group of doctoral and master students from Harvard University and MIT include Chang Zheng, Chen Jinzhu, Chen Mo, Ding Mengjie, Du Wenxin, He Yiming, Hou Yue, Jin Tao, Jin Xiaoshi, Li Zhan, Liu Xuzhi, Xu Lilei, Xu Ye, Zhang Jingyi who helped us finalizing the English version of this book within very limited time in November. They contributed tremendously to the improvements of the quality of the translation of the Chinese version into this English version of the book.

So many people and things have impressed me that it is hard to list them all here, so let me conclude in anticipation of our future cooperation with success.

Li Xiaoxi

December 9th, 2009

Translator: Zhang Li (SUFE)

English Version Editors: Xiao Rong (SUFE), Chen Jinzhu (Harvard)

INDEX

A

B

C

D

E

F

P

Q

R

S

T

U

V

W

Z